ENVIRONMENTAL SCIENCE, ENGINEERING AND TECHNOLOGY

ENCYCLOPEDIA OF ENVIRONMENTAL RESEARCH
(2 VOLUME SET)

ENVIRONMENTAL SCIENCE, ENGINEERING AND TECHNOLOGY

Additional books in this series can be found on Nova's website under the Series tab.

Additional E-books in this series can be found on Nova's website under the E-books tab.

ENVIRONMENTAL REMEDIATION TECHNOLOGIES, REGULATIONS AND SAFETY

Additional books in this series can be found on Nova's website under the Series tab.

Additional E-books in this series can be found on Nova's website under the E-books tab.

Environmental Science, Engineering and Technology

Encyclopedia of Environmental Research

Alisa N. Souter
Editor

Nova Science Publishers, Inc.
New York

Copyright © 2011 by Nova Science Publishers, Inc.

All rights reserved. No part of this book may be reproduced, stored in a retrieval system or transmitted in any form or by any means: electronic, electrostatic, magnetic, tape, mechanical photocopying, recording or otherwise without the written permission of the Publisher.

For permission to use material from this book please contact us:
Telephone 631-231-7269; Fax 631-231-8175
Web Site: http://www.novapublishers.com

NOTICE TO THE READER

The Publisher has taken reasonable care in the preparation of this book, but makes no expressed or implied warranty of any kind and assumes no responsibility for any errors or omissions. No liability is assumed for incidental or consequential damages in connection with or arising out of information contained in this book. The Publisher shall not be liable for any special, consequential, or exemplary damages resulting, in whole or in part, from the readers' use of, or reliance upon, this material. Any parts of this book based on government reports are so indicated and copyright is claimed for those parts to the extent applicable to compilations of such works.

Independent verification should be sought for any data, advice or recommendations contained in this book. In addition, no responsibility is assumed by the publisher for any injury and/or damage to persons or property arising from any methods, products, instructions, ideas or otherwise contained in this publication.

This publication is designed to provide accurate and authoritative information with regard to the subject matter covered herein. It is sold with the clear understanding that the Publisher is not engaged in rendering legal or any other professional services. If legal or any other expert assistance is required, the services of a competent person should be sought. FROM A DECLARATION OF PARTICIPANTS JOINTLY ADOPTED BY A COMMITTEE OF THE AMERICAN BAR ASSOCIATION AND A COMMITTEE OF PUBLISHERS.

LIBRARY OF CONGRESS CATALOGING-IN-PUBLICATION DATA
Encyclopedia of environmental research / [edited by] Alisa N. Souter.
p. cm.
Includes bibliographical references and index.
ISBN 978-1-61761-927-4 (hardcover : alk. paper)
1. Environmentalism--Research--Encyclopedias. I. Souter, Alisa N.
GE195.E53 2010
333.72--dc22
2010037348

Published by Nova Science Publishers, Inc. † New York

CONTENTS

Preface — xiii

Volume 1

Short Communications

A	Seasonal Patterns of Nitrogen and Phosphorus Losses in Agricultural Drainage Ditches in Northern Mississippi *Robert Kröger, Marjorie M. Holland, Matt T. Moore and Charlie M. Cooper*	1
B	Global Change-Induced Agricultural Runoff and Flood Frequency Increase in Mediterraenan Areas: An Italian Perspective *Marco Piccarreta and Domenico Capolongo*	13

Research and Review Articles — 25

Chapter 1	Soil Organic Matter in an Altitudinal Gradient *Bente Foereid and Kim Harding*	27
Chapter 2	The Biodiversity Potential of Riparian Field Margins in Intensively Managed Grasslands *L. J. Cole, D. I. McCracken, D. Robertson and W. Harrison*	37
Chapter 3	Chemotaxonomical Analyses of Herbaceous Plants Based on Phenolic Patterns: A Flexible Tool to Survey Biodiversity in Grasslands *Nabil Semmar, Yassine Mrabet and Muhammad Farman*	43
Chapter 4	Framework for Integrating Indigenous Knowledge and Ecological Methods for Implementation of Desertification Convention *Hassan G. Roba and Gufu Oba*	177

Chapter 5	Montane Grasslands of the Udzungwa Plateau, Tanzania: A Study Case about Its Herpetological Importance within the Eastern Afromontane Hotspot *Roberta Rossi, Raffaele Barocco, Sebastiano Salvidio and Michele Menegon*	221
Chapter 6	Evaluation of Grazing Pressure on Steppe Vegetation by Spectral Measurement *Tsuyoshi Akiyama, Kensuke Kawamura, Ayumi Fukuo, Toru Sakai, Zuozhong Chen and Genya Saito*	243
Chapter 7	South Brazilian Campos Grasslands: Biodiversity, Conservation and the Role of Disturbance *Alessandra Fidelis*	265
Chapter 8	Relationship of Management Practises to the Species Diversity of Plants and Butterflies in a Semi-Natural Grassland, Central Japan *Masako Kubo, Takato Kobayashi, Masahiko Kitahara and Atsuko Hayashi*	283
Chapter 9	Towards the Influence of Plant-Soil-Microbes Feedbacks on Plant Biodiversity, Grassland Variability and Productivity *A. Sanon, T. Beguiristain, A. Cébron, J. Berthelin, I. Ndoye, C. Leyval, Y. Prin, A. Galiana, E. Baudoin and R. Duponnois*	309
Chapter 10	Arbuscular Mycorrhizal Fungi: A Belowground Regulator of Plant Diversity in Grasslands and the Hidden Mechanisms *Qing Yao and Hong-Hui Zhu*	345
Chapter 11	Carnivorous Mammals in a Mosaic Landscape in Southeastern Brazil: Is It possible to Keep Them in an Agro-Silvicultural Landscape? *Maria Carolina Lyra-Jorge, Giordano Ciocheti, Leandro Tambosi, Milton César Ribeiro and Vânia Regina Pivello*	359
Chapter 12	Genetic Diversity of Festuca Pratensis Huds. and Lolium Multiflorum Lam. Ecotype Populations in Relation to Species Diversity and Grassland Type *Madlaina Peter-Schmid, Roland Kölliker and Beat Boller*	375
Chapter 13	Sustainable Waste Management and Climatic Change *Terry Tudor*	389
Chapter 14	Environmental Uncertainity and Quantum Knowledge? *Christine Henon*	401
Chapter 15	Soil: A Precious Natural Resource *C. Bini*	413
Chapter 16	Carbon Sequestration in Soil: The Role of the Crop Plant Residues *Silvia Salati, Manuela Spagnol and Fabrizio Adani*	461

Chapter 17	Collagen Waste as Secondary Industrial Raw Material *F. Langmaier, P. Mokrejs*	483
Chapter 18	Biodiversity Conservation and Management in the Brazilian Atlantic Forest: Every Fragment Must Be Considered *Roseli Pellens, Irene Garay and Philippe Grandcolas*	513
Chapter 19	Conservation of Araucaria Forest in Brazil: The Role of Genetics and Biotechnology *Valdir Marcos Stefenon, Leocir José Welter*	549
Chapter 20	Mexican Fish: A Fauna in Threat *Marina Y. De la Vega-Salazar*	565
Chapter 21	Application of LCA to a Comparison of the Global Warming Potential of Industrial and Artisanal Fishing in the State of Rio De Janeiro (Brazil) *D. P. Souza, K. R. A. Nunes, R. Valle, A. M. Carneiro and F. M. Mendonça*	581
Chapter 22	The Conservation Conundrum of Introduced Wildlife in Hawaii: How to Move from Paradise Lost to Paradise Regained *Christopher A. Lepczyk and Lasha-Lynn H. Salbosa*	593
Chapter 23	Utilization of Bovids in Traditional Folk Medicine and their Implications for Conservation *Rômulo Romeu da Nóbrega Alves, Raynner Rilke Duarte Barboza, Wedson de Medeiros Silva Souto and José da Silva Mourão*	603
Chapter 24	Theoretical Perspectives on the Effects of Habitat Destruction on Populations and Communities *Matthew R. Falcy*	619
Chapter 25	Freshwater Ecosystem Conservation and Management: A Controlled Theory Approach *Y. Shastri and U. Diwekar*	637
Chapter 26	Science and Non-Science in the Biomonitoring and Conservation of Fresh Waters *Guy Woodward, Nikolai Friberg and Alan G. Hildrew*	683
Chapter 27	Natural Renewable Water Resources and Ecosystems in Sudan *Abdeen Mustafa Omer*	697
Chapter 28	Strategy for Sustainable Management of the Upper Delaware River Basin *Piotr Parasiewicz, Nathaniel Gillespie, Douglas Sheppard and Todd Walter*	741
Chapter 29	Outcomes of Invasive Plant-Native Plant Interactions in North American Freshwater Wetlands: A Foregone Conclusion? *Catherine A. McGlynn*	757

Chapter 30	Morphology and Biomass Allocation of Perennial Emergent Plants in Different Environmental Conditions- A Review *Takashi Asaeda, P.I.A. Gomes, H. Rashid and M. Bahar*	769
Chapter 31	Life-History Strategies: A Fresh Approach to Causally Link Species and Their Habitat *Wilco C.E.P. Verberk*	799
Chapter 32	Nutritional Approaches for the Reduction of Phosphorus from Yellowtail (Seriola Quinqueradiata) Aquaculture Effluents *Pallab Kumer Sarker and Toshiro Masumoto*	819
Chapter 33	Influence of Vitamin E Supplementation on Dermal Wound Healing in Tilapia, *Oreochromis niloticus* *Julieta Rodini Engrácia de Moraes, Marina Keiko Pieroni Iwashita, Rodrigo Otávio de Almeida Ozório, Paulo Rema and Flávio Ruas de Moraes*	839
Chapter 34	Fisheries Economics Impacts on Cooperative Management for Aquaculture Development in the Coastal Zone of Tabasco, Mexico *Eunice Pérez-Sánchez, James F. Muir, Lindsay G. Ross, José M. Piña-Gutiérrez and Carolina Zequiera-Larios*	855

Volume 2

Chapter 35	Aquarium Trade as a Pathway for the Introduction of Invasive Species into Mexico *Roberto Mendoza Alfaro, Carlos Ramírez Martínez, Salvador Contreras Balderas, Patricia Koleff Osorio and Porfirio Álvarez Torres*	871
Chapter 36	Biodiversity, Succession and Seasonality of Tropical Freshwater Plankton Communities Under Semi-Field Conditions in Thailand *Michiel A. Daam and Paul J. Van den Brink*	887
Chapter 37	Prospects and Development in Fish Sperm and Embryo Cryopreservation *Vanesa Robles, Elsa Cabrita, Vikram Kohli and M. Paz Herráez*	913
Chapter 38	Ecosystem Functioning of Temporarily Open/Closed Estuaries in South Africa *R. Perissinotto, D. D. Stretch, A. K. Whitfield, J. B. Adams A. T. Forbes and N. T. Demetriades*	925
Chapter 39	Paleoecological Significance of Diatoms in Argentinean Estuaries: What Do They Tell Us about the Environment? *Gabriela S. Hassan*	995

Chapter 40	Anthropogenic Impacts in a Protected Estuary (Amvrakikos Gulf, Greece): Biological Effects and Contaminant Levels in Sentinel Species *C. Tsangaris, E. Cotou, I. Hatzianestis and V. A. Catsiki*	1073
Chapter 41	Santos and São Vicente Estuarine System: Contamination by Chlorophenols, Anaerobic Diversity of Degraders under Methanogenic Condition and Potential Application for Ex-Situ Bioremediation in Anaerobic Reactor *Flavia Talarico Saia*	1101
Chapter 42	Physico-Chemical Characteristics of Negative Estuaries in the Northern Gulf of California, Mexico *Hem Nalini Morzaria-Luna, Abigail Iris-Maldonado and Paloma Valdivia-Jiménez*	1125
Chapter 43	Implications for the Evolution of Southwestern Coast of India: A Multi-Proxy Analysis Using Palaeodeposits *B. Ajaykumar, Shijo Joseph, Mahesh Mohan, P. K. K. Nair, K. S. Unni and A.P. Thomas*	1149
Chapter 44	Human Impacts on the Environment of the Changjiang (Yangtze) River Estuary *Baodong WANG, Linping XIE and Xia SUN*	1171
Chapter 45	Integrated Approaches to Estuarine Use and Protection: Tampa Bay Ecosystem Services Case Study *James Harvey, Marc Russell, Darrin Dantin and Janet Nestlerode*	1187
Chapter 46	Integrative Ecotoxicological Assessment of Contaminated Sediments in a Complex Tropical Estuarine System *D.M.S. Abessa, R.S. Carr, E.C.P.M. Sousa, B.R.F. Rachid, L.P. Zaroni, M.R. Gasparro, Y.A. Pinto, M.C. Bícego, M.A. Hortellani, J.E.S Sarkis and P. Muniz*	1203
Chapter 47	Production Fresh Water Fish with Unconventional Ingredients in Egypt *Magdy M.A. Gaber*	1239
Chapter 48	The Poultry Litter Land Application Rate Study – Assessing the Impacts of Broiler Litter Applications on Surface Water Quality *Matthew W. McBroom and J. Leon Young*	1301
Chapter 49	Agricultural Runoff: New Research Trends *Víctor Hugo Durán Zuazo, Carmen Rocío Rodríguez Pleguezuelo, Dennis C. Flanagan, José Ramón Francia Martínez and Armando Martínez Raya*	1327

Chapter 50	Effects of Agricultural Runoff versus Point Sources on the Biogeochemical Processes of Receiving Stream Ecosystems *Gora Merseburger Eugènia Martí, Francesc Sabater and Jesús D. Ortiz*	1351
Chapter 51	Processes for the Treatment of Dirty Dairy Water: A Comparison of Intensive Aeration, Reed Beds and Soil-Based Treatment Technologies *Joseph Wood and Trevor Cumby*	1373
Chapter 52	Evaluation of Intertwined Relations between Water Stress and Crop Productivity in Grain-Cropping Plain Area by Using Process-Based Model *Tadanobu Nakayama*	1409
Chapter 53	Runoff and Ground Moisture in Alternative Vineyard Cultivation Methods in the Center of Spain *M.J. Marques, M. Ruiz-Colmenero and R. Bienes*	1439
Chapter 54	Multi-Functional Artificial Reefs for Coastal Protection *Mechteld Ten Voorde, José S. Antunes Do Carmo and Maria Da Graça Neves*	1455
Chapter 55	Massive Sedimentations at Coastal and Estuarine Harbors: Causes and Mitigating Measures *Yu-Hai Wang*	1513
Chapter 56	Relative Sea Level Changes in the Lagoon of Venice, Italy. Past and Present Evidence *Rossana Serandrei-Barbero, Laura Carbognin and Sandra Donnici*	1551
Chapter 57	Sediment Dynamics and Coastal Morphology Evolution *François Marin*	1577
Chapter 58	Biodiversity of Sourdough Lactic Acid Bacteria *Luca Settanni and Giancarlo Moschetti*	1581
Chapter 59	An Overview of the Biodiversity and Biogeography of Terrestrial Green Algae *Fabio Rindi, Haj A. Allali, Daryl W. Lam and Juan M. López-Bautista*	1625
Chapter 60	Conservation and Management of the Biodiversity in a Hotspot Characterized by Short Range Endemism and Rarity: The Challenge of New Caledonia *Roseli Pellens and Philippe Grandcolas*	1643
Chapter 61	Macroinvertebrate Distribution on Erosional and Depositional Areas Including A Former Gravel-Pit. Biodiversity and Ecological Functioning *A. Beauger, N. Lair and J. L Peiry*	1657

Chapter 62	Plant Biodiversity Hotspots and Biogeographic Methods *Isolda Luna-Vega and Raúl Contreras-Medina*	1673
Chapter 63	Identifying Priority Areas for Biodiversity Conservation in Northern Thailand: Land use Change and Species Modeling Approaches *Yongyut Trisurat, Naris Bhumpakphan, Utai Dachyosdee, Boosabong Kachanasakha and Somying Tanhikorn*	1685
Chapter 64	China: A Hot Spot of Relict Plant Taxa *Jordi López-Pujol and Ming-Xun Ren*	1709
Chapter 65	An Assessment of Biodiversity Status in Semen Mountains National Park, Ethiopia *Behailu Tadesse and S.C. Rai*	1725

Chapter Sources   1735

Index   1737

Chapter 35

AQUARIUM TRADE AS A PATHWAY FOR THE INTRODUCTION OF INVASIVE SPECIES INTO MEXICO

Roberto Mendoza Alfaro, Carlos Ramírez Martínez, Salvador Contreras Balderas, Patricia Koleff Osorio and Porfirio Álvarez Torres
Facultad de Ciencias Biológicas de la UANL, Monterrey, Nuevo León, México

ABSTRACT

The total ornamental fish industry worldwide (including dry goods) is valued at approximately US $15 billion dollars and it has been estimated that approximately one billion ornamental fish are commercialized every year with a value in the order of US $6 billion dollars. Freshwater species constitute the bulk of the trade; 90 percent of these are obtained from aquaculture and only 10 percent are wild captured. Around 800 to 1000 species and varieties are traded worldwide. Ornamental fish trade has been recognized as an important pathway for the introduction of non native species in several countries and present trends indicate that this pathway may turn into the main source of exotic invasive species in North America. Invasive species are characterized by posing different threats to the environment, the economy and human health. Among the main impacts provoked in the aquatic environment by invasive species are: competition with native species, hybridization, predation, introduction of diseases, habitat disruption and trophic webs modification. The introduction of exotic species has been related to the extinction of 54 percent of aquatic native species worldwide. Overall, 70 percent of the extinctions of North American fishes and 60 percent of those from Mexico are related to non native species, totally or partially. Aquarium trade has shown an accelerated increase during the last decade with a trade value of US $160 million dollars. This increase parallels the boost of exotic species in the country. In fact, in the 80's only 55 non native fish species were registered in Mexico and by 2004 the number raised to 118, of which 67 (58.26 percent) have turned invasive. Several facts contribute to explain this: i) The low amount of varieties cultured in Mexico (61 varieties pertaining to 19 species), compared to the huge number of varieties imported (more than 700 from 117 families), ii) The number of fish imported in Mexico; 40 million ornamental fish are traded annually, of which 45 percent are imported (nearly 18 million fish were imported in 2006) while 55 percent are captive bred iii) there is a lack of official regulations for the establishment and operation of farms producing ornamental fish and for the translocation of ornamental fish within the country. As a result, ornamental fish species have been established in 9 out of 10 continental aquatic regions of Mexico. Some of these species have already severely impacted the environment and the economy in most regions of the country.

INTRODUCTION

The Ornamental Fish Industry Worldwide

The global ornamental fish industry (including dry goods) is valued at approximately US $15 billion dollars (Bartley, 2000) and currently about one billion ornamental fish are traded anually (Whittington and Chong, 2007), with a value of US $6 billion dollars (Holthus and Gamain, 2007) of which Asia provides 78 percent of such exportations worldwide (UNEP, 2008).

The main trading countries obtain most of their product from developing countries such as Philippines, Colombia, Peru, Brazil and Sri Lanka to later be re-exported to the rest of the world. However, the ornamental fish trade is not just important in terms of its share in international trade. The sector is frequently a welcome provider of employment opportunities (Wijkström, 2002) and an important source of income for these economically depressed rural, coastal and insular communities where this industry manage to account under the proper developmental scenario, up to 60 percent of the local economy as in the case of the Amazon (Chao and Prang, 1997; Dowd and Tlusty, 2000; Huanqui-Canto, 2002).

Most of the production of cultured ornamental fish focuses on freshwater species. Approximately 90 percent of freshwater ornamental fish are captively bred, compared to only 10 percent collected in the wild (Dawes, 1998, Chao and Prang, 2002; Wabnitz et al., 2003). This contrasts with marine ornamental species which are mostly (99 percent) collected in the wild and just 1 percent are reared (Hershberger, W.K., 2003).

The variety of fish involved in this industry is very significant. The estimated number of species artificially reared in the freshwater ornamental trade range from 800 to more than 1,000 species, despite the scarce studies existing concerning their culture (Hernández, 2002; Tlusty, 2002). This is an important figure when it is compared to the total number of finfish species used for food or stock enhancement in commercial aquaculture which is estimated to be only about 180 (Williams 1997). In contrast, it is estimated that a total of about 2,600 species are harvested in industrial and artisanal fisheries. (Hershberger, 2003)

The Ornamental Fish Industry as a Pathway for Aquatic Invasive Species

There is a clear and direct relationship between international trade and the spread of invasive alien species (Burgiel., et al., 2006) and the most important pathway for non-indigenous introductions into North America has been intentional or unintentional importations of organisms associated with international trade (Rixon et al., 2005). In this regard, at the international level, aquarium trade has been widely recognized as an important pathway for the introduction of alien aquatic species into new environments (Taylor et al., 1984; Welcomme, 1992) and the current trends indicate that it may become soon the main source of introductions of invasive exotic species in North America (Courtenay, 1995; Courtenay and Williams, 1992). Following the criteria Burgiel and co-workers (Burgiel., et al. 2006), several facts support this contention:

- The rate of introductions of invasive alien species likely correlates with the volume of trade. More than one billion ornamental fish are traded every year (Whittington and Chong, 2007)
- More introductions lead to a greater probability that an invasive alien species will become established. The rate of movement of species between countries has accelerated since 1900 (Welcomme, 1992). From approximately 100 species of ornamental fish that have been recorded as introduced into USA and Canada natural waters via the aquarium trade, 40 have established populations (Rixon et al., 2004), while in the case of Mexico from 115 introduced exotic species 67 have become established (Contreras, In Press).
- An increasing variety of goods and means of transport increases both the potential array of species that may be moved and their pathways for transfer. Only the state of Florida in the US produces 800 varieties of freshwater fish (Tlusty, 2002) and it has been stated the high potential risk posed by the tremendous taxonomic diversity of the ornamental fish industry (Weigle et al., 2005). On the other hand, ornamental fish are transported by planes, ships and trucks all over the world.
- More frequent delivery of goods from and to a wider range of countries and habitats increases the rate and variety of potential introductions. Data from the last decade show that 133 countries imported ornamental fish, while 146 countries exported them (Huanqui-Canto, G, 2002).
- Faster modes of transport may improve an organism's chance of survival while in transit. Fuller (In press) points out that aquarium trade has become the most important pathway in the introduction of invasive species in the Gulf of Mexico and South Atlantic region and that potential invasive species are more likely to survive their journey now than in the past because the increased speed of travel and better packaging techniques.

Also, internationally, ornamental fish farms constitute one of the main pathways for the introduction of exotic species into the natural environment. This is in part due to the high number of continuous changing species and varieties of fish produced and the closeness to natural freshwater bodies (Copp et al., 2005). Fish are known to have escaped or were purposefully released from aquarium fish culture facilities, and others were introduced by well-meaning but misguided aquarists. The escape or release of aquarium fishes into open waters has never resulted in beneficial introductions. In many such instances, there have been negative impacts on native fishes and habitats (Contreras-Balderas and Escalante, 1984; Courtenay and Stauffer, 1990).

The Ornamental Fish Industry in Mexico: History, Size and Scope

The ornamental fish industry in Mexico initiated in the 50's, when ornamental fish were reared for the first time and important exhibitions in public aquaria were mounted. It is also at this time, that the first ornamental fish production and commercialization association was established (Ortiz, 1997).

However, the development of the industry through the time was incipient, as stated by the fact that during the 70's there were only 5 producers and 100 retailers. This industry started culturing cyprinids, brought from Asia by the federal government, as an alternative to agriculture. The value of the industry at the time was US $500,000 dollars, while the importations, represented only US $21,000 dollars (INP, 1974). It is worthy to point out that during the 70's there were severe restrictions to import exotic species from Asia and Africa and the availability of equipment for tanks was minimal.

The effort to culture ornamental fish continued and during the 80's several fish farms, originally created to culture tilapias and freshwater prawns, were reconverted to culture ornamental fish due to former cultured species low yields. However, the growth of the industry kept on being low not only because the government stopped subsidizing the production, but also due to the lack of experience and technical skills, together with the poor existing sanitary control measures at that time. (SEPESCA, 1984).

During the early 90's the demand increased and the federal government began providing proper financial support to establish farms. However, the most important contribution to the industry development and expansion was the nation economic crisis during the mid 90's, which limited the importers and pushed local producers to sustain the market with fish accessible to consumers.

In the same sense, from the official records of imported fish and those of registered farmers in the states of Morelos, Yucatán, Hidalgo and Puebla, it can be assessed that ornamental fish commerce grew more than 100 percent in the past decade (Fig. 1) responding to the continuous growing demand, particularly in the three main large cities (México City, Guadalajara and Monterrey), where owing to urbanization and lack of space fish rapidly became the preferred pets.

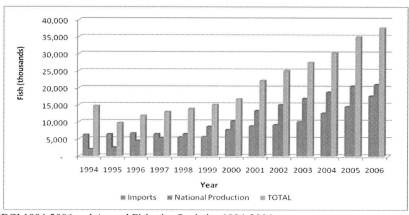

Source: INEGI 1994-2006 and Annual Fisheries Statistics 1994-2004.

Figure 1. Sales of ornamental fish in México in the period of 1994-2006.

According to information on the Annual Statistical Reports from Foreign Commerce of Mexico (1994-2007), and the Annual Fisheries Statistics (1994-2003) the fresh water aquarium fish trade in México, increased more than 100 percent, going from approximately 13 million fish to nearly 37.5 million from 1994-2006, representing thus an annual growth rate of 8 percent (Fig. 1).

However, accordingly to a spoken statement from five of the main aquarium fish traders from México City, it is a common practice to declare less than the amount of fish specimens than are actually received in shipments, leading to the inaccuracy of official reports. In this way, importations may be underestimated by almost 25 percent. This situation may indicate that the number of freshwater fish species traded in México may actually be a bit above 40 million individuals a year (Fig. 2).

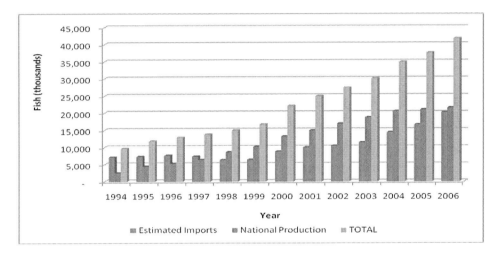

Figure 2. Estimated sales of fresh water aquarium fishes in México, during 1994-2006.

Approximately 45 percent of these fish are imported, while the remaining 55 percent are raised in ornamental fish farms located around several states of the country. These farms are the source of employment to more than 1,250 persons (Ramírez and Mendoza, 2005). According to the last economical census there are 5,126 aquarium stores (INEGI, 2005), however unofficial figures point to the existence of 20,000 aquarium stores, employing around 41,000 persons.

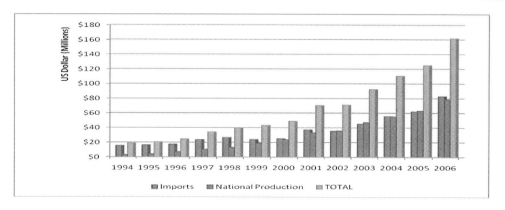

Figure 3. Retail amount of sales from ornamental freshwater fish in México during the period 1994-2006.

Ornamental freshwater fishes generate sales accounting to US $161,500,000 dollars at retail prices, of which 51 percent comes from importations and 49 percent were provided by national production, with an average annual increase of 18 percent in the last 12 years (Fig. 3)

As a consequence of freshwater ornamental fish industry rapid growth, diverse risks have also increased. Unfortunately, a large amount of those exotic organisms accidentally and/or intentionally have been released to natural aquatic environments and many of them became invasive.

National Production

The bulk of the national production is concentrated in the state of Morelos, which would be the equivalent of the State of Florida in the USA.

The number of ornamental fish farms at the national level is uncertain, but it is estimated to be around 250. To find official figures a request to the National Institute of Access to Information (IFAI) was done, and the National Commission of Aquaculture and Fisheries answered that "…according to the Law of Transparency there is not a systematized register of ornamental fish production or producers, however there would be a gross estimate of 128 farms in the state of Morelos…" Nevertheless according to the SAGARPA Coordinator of Fisheries Promotion at Morelos State there would be at least 200 farms. Moreover, from a survey made at the national level there are farms located in other 20 states (Fig.4), but these would not surpass 50. Most of these farms are rustic and are operated at the family level.

Aquarium Trade as a Pathway for the Introduction of Invasive Species into Mexico   877

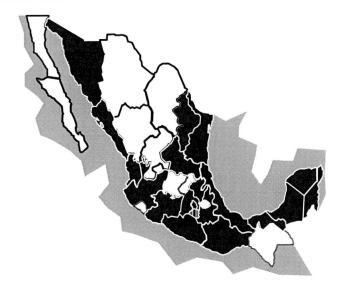

Figure 4. Mexican states where facilities for the culture of ornamental freshwater fish are located (marked in black).

The state of Morelos produces around 17 million individuals/year (CONAPESCA, 2005). These farms are concentrated in 17 of the 33 municipalities of the State and their size varies from 2,000 m^2 to 1.5 has. Most of them (85 percent) are considered small (between 2,000 to 40,000 m^2). The majority of the ponds are concrete made, but the larger ones are rustic earthen ponds (Fig. 5). The bulk of these fish production is sold mainly in Mexico City, Jalisco, Nuevo León, Baja California and Guanajuato States. Fish are sold directly in the farms or sent to the wholesalers market in Mexico City (SAGARPA, 2005).

Figure 5. Typical ornamental fish farm in the state of Morelos, Mexico.

Nearly all the producers are organized in two main associations. Of the 128 farms officially registered, 45 belong to the APPOEM Ornamental Fish Farmers Association of Morelos (Asociación de Productores de Peces de Ornato del Estado de Morelos) and 54 to "La Perla" (Productores de Peces de Ornato "La Perla" de Morelos A.C) however there are 29 independent farms. The first of these associations was created under an initiative of ornamental farmers, while the second was created as a request from the state government, so the producers could benefit from the different government support programs and economic subsidies. Ornamental fish farmers have benefited from government official programs not only by obtaining significant tax discounts, but also by being eligible to financial support and subsidies of up to US $7,000 dollars, only if their farms are officially registered. Nowadays, besides representing the producers they serve as a discussion forum to deal with different problems such as the adaptation of their facilities to new regulations, and homogeneity of prices to face intermediaries. The greatest part of the production takes place from March to November, as the winter temperatures are rather low for the production of goldfish which have the highest demand. One of the producers' associations (La Perla) has been looking for technical assistance of the Morelos State University to increase their production levels. The main problem faced by producers is the supply of good quality broodstock and the control of genetic quality of the different varieties of fishes produced.

The recent success of the ornamental fish industry has attracted the interest of several entrepeneurs from other sectors, willing to invest on this activity, which will be reflected in its growth, in the short term.

The variety of ornamental fish produced in Mexico can be considered low when compared to the international standard, as there are only 61 varieties from 19 species currently reared. In contrast, it has been estimated that nearly 700 varieties pertaining to 117 families are imported each year into the country (Alvarez-Jasso, 2004). However, the economic accessibility to these varieties is restricted and contributes to explain the drop in consumption during the times of economical crisis.

The varieties currently reared are: Livebearers (mollies, guppies, platies and swordtails) constituting the bulk of the production, however many usual oviparous fish such as cichlids (angels and different African cichlids), anabantids (gouramies and betas), characins (tetras), silurids (catfish, corys and plecos) and chiefly a vast array of goldfish and other cyprinids (barbs, rasboras, sharks) are also cultured.

Importations

The ornamental fish industry in Mexico has been operating for more than 50 years and although at the beginning, this industry depended mostly on species captured from the natural environment and some imported from Central and South America, imports, for year 2006 have been estimated in nearly 18 million fish.

The main reasons to import fish into Mexico are:

- The high retailers' price of several alien species
- The quality of many imported species, that can't be attained by national producers
- The variety of organisms, which is a fundamental aspect, since the demand in the ornamental fish industry is fashion-based

- The availability of many species during the whole year, while in Mexico some species are just available during a part of the year (e.g. goldfish)
- The support of the academic sector from exporting countries

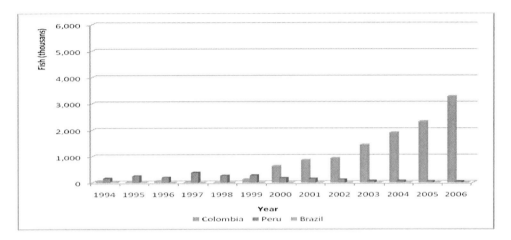

Figure 6. Importations of ornamental freshwater fish from South America to México during 1994-2006.

In the case of importations coming from South America, especially from the Amazons region pertaining to Brazil, Colombia and Peru, a significant change in the volume of fish exported has been noticed during the last five years. Not only has the total number of individuals increased in an important way, but also the participation of Colombia over those of Brazil and Peru has increased over time. This is in great part due to the Treaty on Free Trade Between the Republic of Colombia, the Republic of Venezuela and Mexico also known as G3. In particular, this treaty not only eases the importations of goods coming from Colombia, but also gives this country an important advantage over the rest due to the tariff elimination (Fig. 6)

On the other hand, Almenara (2001) points out a disadvantage between importers in relation to the NAFTA (North American Free Trade Agreement), signed between Mexico, Canada and the United States of America. According to this agreement, aquatic ornamental organisms whose origin is either Canada or the USA are duty free, in comparison with 23 percent import duty from other countries; this also applies to the duty on freight and packing paid to transport them. As all fishes coming from Asia pass through the USA, this allows some cheating importers to turn China goldfish into 'American bred' goldfish and, thus end up paying less than US $15.00 dollars in customs fee per shipment, compared to the more than US $2,500.00 dollars per shipment with legal papers that honest importers pay. The lack of professional staff at the Customs Department cannot control this tax and duty evasion that is harmful to the country. Since it is virtually impossible to determine the precise origin of tank-bred species of fish, the Customs authorities rely only on the exporter's paperwork.

Within this context, in spite of the substantial increase in commercial exchange between Mexico and the USA after the NAFTA treaty establishment, more than ten years ago, it is interesting to note that ornamental fish importations follow a cyclic pattern in which every six years when the federal government changes (1994, 2000, 2006) there is an economic constriction, which coincides with the drop of importations. Notwithstanding, importations remain to be the second source of ornamental fish in Mexico

The more frequent imported freshwater fish families are:

Table 1. Imported families of ornamental fish according to the number of species and number of individuals

Families	Species	Families	Number of individuals
Cichlidae	107	Characidae	1,187,788
Characiidae	64	Cyprinidae	1,092,506
Cyprinidae	27	Osphoronemidae	689,776
Callichthyidae	24	Poeciliidae	456,563
Loricariidae	20	Cichlidae	161,276
Conaitidae	15	Ambassidae	133,003
Aplocheilidae	13	Callicthyidae	130,385
Anabantidae	12	Pangasiidae	118,739
Pimelodidae	11	Loricariidae	102,743

The importance of these families changes when the number of individuals per family is taken into account. This shows that in some cases, such as the cichlids, a relatively high number of species, but a small number of individuals is imported, which contrasts with characids in which a high number of individuals and a low number of species is imported, obeying to the market demand.

It has been suggested that the risk of establishment largely depends on the biogeographic origin of the organisms (Ramírez y Mendoza, 2005). Thus, many fish imported from Asia, Europe and the USA are actually captively bred and consequently are less prone to survive in the wild; however the threat associated to the introduction of these organisms are diseases caused by exotic pathogens and parasites. In contrast, most of those organisms imported from South America, particularly from the Amazons, are captured in the wild and consequently represent a high risk of establishment in a similar environment, as has been the case with some cichlids and fish of the Loricariidae family. In relation to this, it should be noticed that ornamental fish importations from South America have been increasing, while those from Florida are decreasing, due to import substitutions made by domestic breeders (Almenara, 2001). However, a large amount of fish has been historically imported from the USA.

Problems of the Industry

During the last decade the culture and commercialization sectors of the ornamental fish industry have been experiencing some difficulties due to the following aspects:

- Lack of promotion
- No research support
- Lack of internal organization
- Lack of an efficient regulation

The industry lacks a real and reliable organization due to fierce competition between and among producers, importers, wholesalers and retailers ("changarreros"). To this should be added the scarce research on different aspects of ornamental fish species (particularly on reproduction, nutrition, pathology and genetics), the lack of communication (only a minority gathers in an annual congress), lack of training or information within the context of sustainability and responsibility towards the environment (there are very few national aquarium magazines). The official figures and statistics concerning the sector (number of producers, areas of production, volume and origin of imports or countries where fish are exported) are inaccurate and often non trustworthy.

As a consequence of the above mentioned problems, a continuous negative impact to the environment has been registered, particularly due to the absence of a specific and efficient legal framework. Indeed, the absence of regulatory measures requesting farmer owners to have an adequate infrastructure has originated the continuous escapes of native or exotic species into the wild. An example is the state of Morelos, where most of ornamental fish farms are concentrated. In the rivers of this state there are a total of 22 species registered, 8 of which are native while 14 are exotic (Contreras-MacBeath, 1998).

In addition, several voids in the environmental laws together with the lack of necessary enforcement have given place to illegal trade (importation and exportation) and translocation within the country of these species. In fact, despite the existence of a seven day quarantine period prior to selling the product after importation, in many cases the law is not observed by non-ethical people whose businesses are based on just moving their fishe rapidly without caring for quality. The results of these improper practices are disastrous to those importers who have been investing their money, effort and time in order to offer healthy organisms according to the prevailing rules. They are at a total disadvantage against those people who keep on cheating the law and doing whatever possible to keep giving away their products without any consideration. An undesirable, yet expected consequence is that all importers are pushed to ignore the quarantine period as well. Even though the authorities know that the law is not being properly observed (or not observed at all), they cannot act, due to the lack of resources to enforce the law. In consequence, this leaves the legal importers sector abandoned at the mercy of unscrupulous merchants (Almenara, 2001). To this should be added the difficulty to obtain the permits, the frequency to get them (every three months), the strict specificity of permits (which increase the price with each new species) and the geographical distance from Mexico City to Mazatlan city (1,085 Km) to obtain the permit where fisheries authorities are based.

Ecological Risks

In Mexico, fish are commonly released by farmers when prices drop or when a disease is detected, because due to the cost of antibiotics they will not obtain the desired commercial value. This situation implies a high risk of escapes generating important problems, such as competition for food resources and space, predation, native species displacement and alteration of communities' trophic structure, hybridization, incidence of diseases and parasites and habitat alteration, (Contreras-Balderas, 1969, Contreras-Balderas and Escalante, 1984, Lassuy, 2000; MIT Sea Grant, 2002; Goldburg, 2001; Hopkins, 2001)

At the national level, most of the ornamental fish farms lack of adequate biosecurity systems, implying the risk of disease propagation and escapes of native and non-native species. In general, operating farms use extensive culture systems, with low or non-existent technology and are operated by untrained personnel (Ramírez, 1999).

Furthermore, most of the farms are not constructed under the official standards and regulations established by the environmental authority, such as conducting environmental impact assessments to demonstrating that the farm will not pose any threats to the environment, as stated by the environmental law (DOF, 2000) and often farms do not adopt current sanitary regulations, (NOM-010-PESC-1993, NOM-011-PESC-1994). As a result, many farms are located in places where they represent a high ecological risk and are operating with production systems far from being adequate.

This situation is critical as Mexico is one of the five most diverse countries in the world and among the 12 considered as megadiverse (CONABIO, 2006), that hosts 60 and 70 percent of the known biological diversity in the planet. This high biodiversity is due in part to the biogeographic conditions of the country, in a transitional zone between the two large neotropical and neartic regions. As a result, Mexico has a rich source of flora and fauna and particularly a high species diversity of freshwater fish. There are more than 47 families, composed by approximately 506 species, 375 of which are exclusively restricted to freshwater environments (Contreras-Balderas, et al., 2008; Torres-Orozco, y Kobelkowsky, 1991). Additionally to this high diversity, the important variety and isolation of the different basins has given place to several endemic species which ecological importance is even higher.

One of the reasons why the impact of alien invasive species has passed inadvertently in Mexico is the high biological diversity of the country. In this regard, it has been stated that because some ecosystems are richer in number of species than are others, it is likely that such ecosystems would be less hospitable to establishment and invasion by nonnative species through species packing over time (Courtenay, 1990). In the same sense Moyle and Light (1996) mentioned that major community effects of invasions are most often observed where the number of species is low and that invaders are most likely to extirpate native species in aquatic systems with extremely low variability.

Unfortunately, it should be pointed out that according to the National Fishery Chart (DOF 2004) 9 of the 10 aquatic provinces of the country host several species that have been introduced due to aquarium trade.

Some of these exotic species have reached critical areas such as Mexico's Natural Protected Areas, characterized by a high endemism, e.g. African jewel cichlid in Cuatro Cienegas Basin (Contreras-Balderas and Ludlow, 2003). Unfortunately, according to Contreras-Balderas, (1971) and Hamilton (2001) native species, especially indigenous species that evolved in isolated ecosystems are often very poorly equipped to survive once invasive alien species are introduced. Additionally, most of the endemic fishes are small and therefore, subject to more adverse impacts through introductions of small bait or ornamental fishes than with larger fishes (Courtenay and Deacon, 1982). The prediction of an invasion by non-native fishes in Cuatro Cienegas and its consequences was reported since 1983 (Courtenay, 1983).

There are two reasons to expect invasion threats from the aquarium industry to increase with time. Firstly, the pool of potential invaders is ever expanding as the industry searches for new, potentially popular species to market. Secondly, because most aquarium species are of tropical and subtropical origin, the probability of their establishment in North America will increase with climatic warming (Rixon et al., 2004).

Finally, it should be noted the tight relationship existing between a highly deteriorated aquatic ecosystems, as those of the state of Morelos, and the establishment of invasive species as the plecos or the convict cichlid. With this regard, it is difficult to estimate to which extent the severe pollution was the direct cause of extinction of native species and what role was played by the invasive species. Anyhow, it has been reported that most impacts, to date, have been in areas where habitats modification has already stressed fish populations (Minckley and Deacon 1968; Deacon 1979; Courtenay, et al, 1985; Welcomme, 1988; Courtenay and Robins, 1989). Moreover, the environmental tolerance of non-indigenous fish combined with increasing habitat disruption in streams and lakes assures their continued dispersal into formerly unoccupied areas (Boydstun, et al, 1995). This is also true regarding their establishment, as in aquatic systems with intermediate levels of human disturbance, any species with the right physiological and morphological characteristics can become established and long-term success (integration) of an invading species is much more likely in an aquatic system permanently altered by human activity than in a lightly disturbed system (Moyle and Light, 1996).

CONCLUSION

Several aspects in the aquarium trade industry in Mexico need urgently to be reviewed and modified from current trade and farming practices to those related to norms, standards and regulations applicable to the industry. Among these the need to adequately regulate importations and adopt preventive measures such as risk analysis and the elaboration of white and black lists. In the case of production there is a strong need of biosecurity standards and preventive measures such as HACCP (Hazard Analysis to Control Critical Points) plans for each farm. While in the case of commercialization international measures such as those proposed by the Ornamental Aquatic Trade Association should be adopted.

REFERENCES

Almenara Roldán,S. 2001.The Current Status of the Aquarium Industry in Mexico. *OFI Journal Issue* 35. 3pp

Alvarez-Jasso, M. 2004. "*La introducción de peces ornamentales en México a través de las importaciones durante el año 2001 y su ordenamiento.* Tesis. De Licenciatura. Facultad de Ciencias. UNAM. México. 75 p.

Bartley, D.M. 2000. Responsible ornamental fisheries. *FAO Aquat. Newsl.* 24, 10–14.

Boydstun, Ch., Fuller, P. and Williams, J.D. 1995. *Nonindigenous Fish. In: Our living resources.* A report to the nation on distribution, abundance, and health of U.S. Plants, Animals, and Ecosystems. U.S. Department of the Interior. National Biological Service. USGS, Florida. USA.

Burgiel, S., G.Foote, M. Orellana and A. Perrault. 2006. *Invasive Alien Species and Trade: Integrating Prevention Measures and International Trade Rules.* Center for International Environmental Law and Defenders of Wildlife. 54pp

Chao, N.L., Prang, G., 1997. Project Piaba—towards a sustainable ornamental fishery in the Amazon. *Aquarium Sci. Conserv.* 1, 105–111.

Chao, N.L., Prang, G., 2002. Decade of project Piaba: reflections and prospects. *OFI Journal* (39):1-9.

Conabio 2006. *Capital natural y bienestar social.* Comisión Nacional para el Conocimiento y Uso de la Biodiversidad, México. 70pp

Conapesca (Comisión Nacional de Acuacultura y Pesca). 2005. *Respuesta a la consulta número 00008000050305 del Instituto Federal de Acceso a la Información Pública*. Documento Oficial.

Contreras-Balderas, S. and M.A. Escalante. 1984. *Distribution and Known Impacts of Exotic Fishes in México*. Chapter 6.102=130. In: Distribution, and Management of Exotic Fishes. Eds. W.R. Courtenay y J.R. Stauffer. John Hopkins University Press

Contreras-Balderas, S. 1971. El hombre, el desierto y la conservación de los recursos naturales. *Revista Fauna Silvestre y Cacería*, 3:20-24./ Armas y Letras,(Universidad de Nuevo León), 1974(1):43-50.

Contreras-Balderas, S. 1969. Perspectivas de la ictiofauna en las zonas áridas del Norte de México. Mem. Primer Simp. Internacional de Aumento de Producción de Alimentos en Zonas Áridas. *ICASALS, Texas Tech. Publ.* 3:294-304

Contreras-Balderas, S. En Edición. *Peces Exóticos Invasivos de agua dulce en México*. En trámite de impresión.

Contreras-Balderas, S. y Ludlow, A. 2003. *Hemichromis guttatus*, Nueva Introducción en Cuatro Ciénegas, Coahuila, Primer Reporte para México (Pisces: Cichlidae). *Vertebrata Mexicana*, 12:1-5.

Contreras-Balderas, S.y Medina Gándara, J. 1971. *Ecología de Zonas Áridas de México, Para: Inventario de Recursos Humanos, Materiales y Financieros de Ciencia y Tecnología para Zonas Áridas*. Informe Consejo Nacional de Ciencia y Tecnología, México.

Contreras-Balderas, S., G. Ruiz-Campos, J. J. Schmitter-Soto, E. Diaz-Pardo, T. Contreras-McBeath, M. Medina-Soto, L. Zambrano-Gonzalez, A. Varela-Romero, R. Mendoza-Alfaro, C. Ramirez-Martinez, M. A. Leija-Tristan, P. Almada-Villela, D. A. Hendrickson, & J. Lyons, 2008. Freshwater fishes and water status in Mexico: a country-wide appraisal. *Aquatic Ecosist. Health & Managmnt.*, 11(3):246-256.

Contreras-MacBeath, T. H. Mejia Mojica and R. Carrillo Wilson. 1998. Negative impact on the aquatic ecosystems of the state of Morelos, Mexico from introduced aquarium and other commercial *Fish. Aquarium Sciences and Conservation*, 2: 67-78.

Copp, G.H., K. Wesley and L. Vilizzi. 2005. Pathways of ornamental and aquarium fish introductions into urban ponds of Epping Forest (London, England): the human vector. *J. Appl. Ichtyol.* 21:263-274

Courtenay, W.R.Jr. 1983. Fish introductions in the American Southwest: a case history of Roger Springs, Nevada. *The Southwestern Naturalist* 28:221-224.

Courtenay, W.R., Jr,. 1990. Fish conservation and the enigma of introduced species. *Proc. Aust. Soc. Fish Biol* 8: 11-20

Courtenay, W.R., Jr,. 1995. The case for caution with fish introductions. *American Fisheries Society Symposium* 15: 413-424.

Courtenay, W.R., Jr, J.E. Deacon, D.W. Sada, R. C. Allan and G.L. Vinyard. 1985. Comparative status of fishes in along the course of the pluvial White River, Nevada. *Southwestern Naturalist* 28:221-224.

Courtenay, W.R., Jr. and J.E. Deacon .1982. Status of introduced fishes in certain spring systems in Southern Nevada. *Great Basin Naturalist* 42:361-366.

Courtenay, W.R., Jr, and C.R. Robins. 1989. Fish introductions: good management? *Critical Reviews in Aquatic Science* 1: 159-172.

Courtenay, W.R., Jr, and J.R. Stauffer, Jr. 1990. The introduced fish problem and the aquarium fish industry. *Journal of the World Aquaculture Society* 21: 149-159.

Courtenay, W.R., Jr. And J.D. Williams. 1992. Dispersal of exotic species from aquaculture sources, with emphasis on freshwaters fishes. In: A. Rosenfield and R. Mann, eds. *Dispersal of living organisms into aquatic ecosystems.* Maryland Sea Grant College, College Park. pp 49-81

Dawes, J., 1998. International experience in ornamental marine species management. Part 1: perspectives. *Ornamental Fish Int. J.,* 26, February 1999. http://www.ornamental-fish-int.org/marinespecies1.htm.

Deacon, J.E. 1979. Endangered and threatened fishes of the West. In: The endangered species: a symposium. *Great Basin Naturalist Memoirs* 3: 41-64.

DOF (Diario Oficial de la Federación). 2000. *Ley General del Equilibrio Ecológico y Protección al Ambiente.* Ultima Reforma. 7 de Enero, 2000. México.

DOF (Diario Oficial de la Federación). 2004. *Carta Nacional Pesquera.* México

Dowd, S., Tlusty, M.F. 2000. *Project Piaba—working toward a sustainable natural resource in Amazon freshwater fisheries.* Endangered Species Update 17, Univ. MI. School Natural Resources. 88– 90.

Fuller, P. In Press. *Nonindigenous aquatic pathways analysis for the Gulf and South Atlantic region.* Report produced for the Aquatic Nuisance Species Taskforce. 55 pp (In press)

Goldburg, R.J., M.S. Elliott, R.L. Taylor. 2001. *Marine Aquaculture in the United Status: Enviromental IMpacts and Policy Options.* Pew Oceans Commission, Arlington, Virginia

Hamilton, L.A. 2001. Invasive alien species: a major continuing threat to the health and integrity of freshwater ecosystems. *Resource Futures International, Ottawa.* (9): 1-9.

Hernández, O. D. 2002. *Producción de Peces de Ornato. Hypatia.* (4). http://hypatia.morelos.gob.mx/no4/peces ornato.htm

Hershberger, W.K. 2003. *Aquaculture: Its Role in the Spread of Aquatic Nuisance Species.* National Center for Cool and Cold Water Aquaculture. USA. 13 pp

Hopkins, C.C.E. 2001. *A review of introductions and transfers of alien marine species in the North Sea area*

Holthus, P. y Gamain, N. 2007. *Del arrecife al minorista.* Conservación Mundial. http://www.iucn.org/publications/worldconservation/docs/2007_01/12_world_conservation_2007_01_es.pdf

Huanqui Canto, G, 2002. *El comercio mundial de peces ornamentales.* Prompex. Perú. 10 p.

INEGI (Instituto Nacional de Estadística, Geografía e Informática). 2001.. *Censos Económicos* 1999. Resultados Definitivos. México.

INEGI (Instituto Nacional de Estadística, Geografía e Informática). 2005_a.. *Censos Económicos* 2004. Resultados Generales. México.

INEGI (Instituto Nacional de Estadística, Geografía e Informática). 2005_b.. *Anuario Estadístico del Comercio Exterior de los Estados Unidos Mexicanos.* 2004.

INEGI (Instituto Nacional de Estadística, Geografía e Informática). 2005_c. *Anuario Estadístico del Comercio Exterior de los Estados Unidos Mexicanos* 2004.

INP 1978. Estado de la acuicultura en México. En: la acuicultura en América Latina. Instituto Nacional de la Pesca Mexico y FAO, *Informes de Pesca*, 159, Vol. 3: S/C 13

Lassuy, D.R., 2002, I*ntroduced Species as a factor in extinction and endangerment of native fish species.* Workshop: Management, implications and co-occurring native and introduced fishes proceedings, Portland Oregon: 27-28,

Minckley, W.L. and J.E. Deacon. 1968. Southwestern fishes and the enigma of "endangered species" *Science* 159: 1424-1432.

MIT Sea Grant,. 2002. *Exotic species an ecological roulette with nature.* MIT Sea Grant College Program, Costal Resources. Fact sheet.

Moyle, P.B. and T. Light. 1996. Biological invasions of fresh water: empirical rules and assembly theory. *Biological Conservation* 78: 149-161.

NOM-010-PESC-1993. *Norma Oficial Mexicana que establece los requisitos sanitarios para la importación de organismos acuáticos vivos en cualesquiera de sus fases de desarrollo, destinados a la acuacultura u ornato, en el territorio nacional.* Diario Oficial de la Federación. 20 de Julio, 1994.

NOM-011-PESC-1993. *Norma Oficial Mexicana para regular la aplicación de cuarentenas, a efecto de prevenir la introducción y dispersión de enfermedades certificables y notificables, en la importación de organismos acuáticos vivos en cualesquiera de sus fases de desarrollo, destinados a la acuacultura y ornato en los estados unidos mexicanos.* Diario Oficial de la Federación. 20 de Julio, 1994.

Ortiz, G.C. 1997. *Esfuerzos que se desarrollan para la difusión de la acuariofilia. En: Primer Congreso Nacional Acuariofilia. Dirección General de Acuacultura.* SEPESCA. México.

Ramírez-Martínez, C 1999. *"La perspectiva de la acuacultura social en México"* En: Memorias del IV encuentro de empresas sociales de pesca y acuacultura. Secretaría de Desarrollo Social. México.

Ramírez-Martínez, C. y Mendoza, R.2005. *" La producción y comercialización de peces de ornato de agua, como vector de introducción de especies acuáticas invasivas en México"* En: *Memorías del 37° Symposium del Desert Fishes Council.* Cuatrociénegas, Coah.

Rixon, C., I. Duggan, N. Bergeron, A. Ricciardi and H. Macisaac. 2005. Invasion risks posed by the aquarium trade and live fish markets on the Laurentian Great Lakes. *Biodiversity and Conservation* 14: 1365-1381.

SAGARPA (Secretaría de Agricultura, Ganadería, Pesca y Alimentación). 2005. Delegación *Estatal en el estado de Morelos* http://www.sagarpa.gob.mx/dlg/morelos/pesca/informacion.htm

SEPESCA (Secretaria de Pesca). 1984. Reporte sobre el desarrollo de la acuacultura de subsistencia en Morelos. Reporte Técnico. *Dirección General de Acuacultura.* México, D.F. 45 p.

Taylor,J.N., W.R. Courtenay,Jr.and J.A. McCann. 1984. *Knows impacts of exotic fishes in the continental United States.* In: W.R. Courtenay, Jr. and J.R. Stauffer, Jr., eds. *Distribution, biology, and management of exotics fishes.* The Johns Hopkins University Press, Baltimore, MD. USA.

Tlusty, M., 2002. *The benefits and risks of aquacultural production for the aquarium trade Aquaculture* 205: 203– 219

Torres-Orozco, B.R y Kobelkowsky, A. 1991. *Los Peces de México.* AGT Editor, S.A. México, D.F. 235 p.

UNEP (United Nations Environment Programme). 2008. *Monitoring of International Trade in Ornamental Fish.* European Commission. Cambridge, UK 43 p.

Wabnitz, C., Taylor, M., Green, E., Razak, T. 2003. *From Ocean to Aquarium.* UNEP-WCMC, Cambridge, UK. 64 p

Welcomme, R. 1992. *Pesca Fluvial.* FAO Documento Técnico de Pesca, Roma.303 p.

Weigle, S. L.D. Smith, J.T. Carlton and J. Pederson. 2005. *Assesing the risk of introducing exotic species via the live marine species trade. Conservation Biology.* 19(1): 213-223

Williams, M. J. 1997. *Aquaculture and sustainable food security in the developing world.* Pages 15-51. *In* John E. Bardach, editor. *Sustainable Aquaculture*. John Wiley and Sons, Inc., New York, NY.

Wijkström, U., A. Gumy and R. Grainger. 2002. *The State of World Fisheries and Aquaculture* (Sophia) FAO, 150pp.

Whittington, R.J. and R. Chong. 2007.Global trade in ornamental fish from an Australian perspective: The case for revised import risk analysis and management strategies *Preventive Veterinary Medicine*. (81)1-3: 92-116.

In: Encyclopedia of Environmental Research
Editor: Alisa N. Souter

ISBN: 978-1-61761-927-4
© 2011 Nova Science Publishers, Inc.

Chapter 36

BIODIVERSITY, SUCCESSION AND SEASONALITY OF TROPICAL FRESHWATER PLANKTON COMMUNITIES UNDER SEMI-FIELD CONDITIONS IN THAILAND

Michiel A. Daam[1] and Paul J. Van den Brink[2]

[1] Instituto Superior de Agronomia / Technical University of Lisbon, Lisbon, Portugal
[2] Alterra, Wageningen University and Research centre, Wageningen, The Netherlands

ABSTRACT

Although biodiversity has often been indicated to increase towards the equator, taxonomic expertise and research efforts have focused mainly on temperate regions. Hence, biodiversity and limnology of tropical freshwaters are presently poorly understood. Also in Thailand, literature on limnology is scarce, and many of the studies carried out are published in the grey literature or in local (university) journals with extremely limited circulation.

Two outdoor microcosm studies were previously carried out in Thailand to evaluate the fate and effects of the insecticide chlorpyrifos (CPF) and the herbicide linuron (LIN) on tropical freshwater communities under semi-field conditions. The present chapter lists and discusses the periphyton, phytoplankton and zooplankton species communities encountered in the controls of these experiments. The periphyton community was studied in the LIN experiment using microscopic slides that were placed approximately 10 cm under the water surface of the microcosms. After an incubation period of four and six weeks, the periphyton community existed solely of *Chamaesiphon* sp. whereas in later stages (eight and ten weeks incubation), contributions of chlorophytes and diatoms increased. The LIN phytoplankton community was dominated by Chlorophyta with a radiation of *Scenedesmus* species, whereas a bloom of the cyanobacterium *Microcystis aeruginosa* was noted in the CPF experiment. The large cladoceran *Diaphanosoma* sp., one of the dominating cladocerans in the LIN experiment, was absent in the CPF experiment and replaced by smaller cladoceran species. Possible underlying influences of tropical seasonality and climatic conditions on the observed community structures, primary producers-zooplankton interactions and implications for conservation of tropical freshwaters are discussed.

INTRODUCTION

The diversity of species has often been discussed to increase towards the tropics, indicating that the number of species that could potentially be affected by pollutants is also greater (Gaston et al., 1995; Mares, 1997; Lacher and Goldstein, 1997; Kwok et al., 2007). On the other hand, a greater diversity of species also implies that the potential for functional redundancy is greater. Thus, if the protection goal is to assure ecosystem functioning and possible effects on ecosystem structure are considered acceptable, tropical ecosystems may have an advantage over their temperate counterparts in maintaining their functionality after chemical stress. Although species protection is mostly the aim in environmental risk assessments, this so called functional redundancy principle has been considered suitable to evaluate the acceptability of the impact of pesticides in areas with as main function the production of crops and food, like rice paddies and fish breeding ponds (Van der Linde et al., 2006; Brock et al., 2006).

Species diversity, however, appears not to be greater in the tropics for all environmental communities. For example, although freshwater fish species richness increases towards the equator (e.g., Lévêque et al., 2008; Table 1), freshwater plankton communities do not show a marked latitudinal trend in species diversity, and there may even be a minor trend towards a lower diversity at low latitudes (Kalff and Watson, 1986; Lewis, 1987; Fernando, 2002). In line with this, the number of cladocerans and especially rotifers species in the Afrotropical, Neotropical and Oriental zoogeographic regions appear to be lower than in the Palaearctic and Nearctic zoogeographic regions (Table 1). The insect classes *Chironomidae* and *Culicidae*, both belonging to the insect order Diptera, have a respectively greater and lesser species diversity in the Palearctic and Nearctic regions compared to the Neotropical, Afrotropical and Oriental regions (Balian et al., 2008a; Table 1).

Table 1. Total species diversity of selected groups of freshwater animals by zoogeographic region. Data for rotifers were obtained from Segers (2008) and other data from Balian et al. (2008a)

Animal group/region*	Palaearctic	Nearctic	Afrotropical	Neotropical	Oriental
Crustaceans	4499	1755	1536	1925	1968
Cladocera	245	189	134	186	107
Insects	15190	9410	8594	14428	13912
Chironomidae	1231	1092	618	406	359
Culicidae	492	178	1069	795	1061
Vertebrates	2193	1831	3995	6041	3674
Fish	1844	1411	2938	4035	2345
Other phyla	3675	1672	1188	1337	1205
Rotifera	978	804	563	453	486

* Biogeographic regions according to Balian et al. (2008b): Palearctic: Europe and Russia, North Africa (not including the Sahara) and Northern and Central Arabian Peninsula, Asia to south edge of Himalayas; Nearctic: North America, Greenland and the high-altitude regions of Mexico; Afrotropical: Africa south of the Sahara, the Southern Arabian Peninsula and Madagascar; Neotropical: Southern and coastal parts of Mexico, Central America, and the Caribbean islands together with South America; Oriental: India and Southeast Asia south of Himalayas (including lowland southern China) to Indonesia down to the Wallace's Line.

A comparison of species diversity between the temperate and tropical zone has often been reported to be biased by the fact that taxonomic expertise and research efforts have centered on temperate regions (Mares, 1997; Dudgeon, 2000; Sarma et al., 2005; Balian et al., 2008a). Of particular concern is the extent to which freshwater biodiversity has been neglected in tropical Asia (the Oriental Region), and this region does not seem to invoke the same concern over biodiversity conservation compared to the Neotropics and Africa (Dudgeon, 2000). Consequently, the limnology and biodiversity of tropical (Asian) freshwaters are poorly understood (Foran, 1986; Mares, 1997; Dudgeon, 2003; Gopal, 2005; Sarma et al., 2005; Balian et al., 2008a). This lacune in knowledge is due, in part, to a lack of basic research, but is also a reflection of a scattered, highly fragmentated literature, some of which is inaccessible (Dudgeon, 2000). For example, most of the little literature on limnology in Thailand is published in the grey literature or local (university) journals with extremely limited circulation (Peerapornpisal et al., 2000; Campbell and Parnrong, 2001). Furthermore, analysis of the available data is impeded by fuzzy taxonomy and the questionable reliability of many records (Segers, 2001).

Two microcosm experiments were carried out at the hatchery of the Asian Institute of Technology (AIT), located approximately 42 km north of Bangkok (Thailand). These experiments were initiated to evaluate the fate and effects of a single application of the herbicide linuron (LIN; Daam et al., 2009a,b) and single versus multiple applications of the insecticide chlorpyrifos (CPF; Daam et al., 2008). The aim of the present chapter was to contribute to the knowledge of tropical freshwaters by i) listing the plankton communities encountered in the control microcosms of these experiments; ii) evaluating possible influences of season and climatic conditions on the observed community structures; iii) analyzing primary producers-zooplankton interactions and iv) discussing implications of study findings for the conservation of tropical freshwaters.

MATERIALS AND METHODS

Experimental Design

Some information on the experimental design of the microcosm experiments is summarized in Table 2. Circular (diameter 0.75 m, height 0.65 m) concrete microcosms were allocated to the experiments and set-up outdoors. Before each experiment, tanks were newly coated with watertight non-toxic epoxy paint to avoid any influence from previous experiments. Subsequently, water was taken from the canal surrounding AIT and added to the microcosms. The canal water was passed though a net (mesh size, 0.1 mm) to prevent fish and prawns from entering the experimental systems.

Over an acclimatization period of one (CPF) to five (LIN) weeks, a biocoenosis was allowed to develop in the microcosms. During this period, water was circulated twice a week to achieve similarity between the communities in the systems. Nutrient additions were made twice a week during the entire experimental period to stimulate plankton growth and to compensate for possible losses (Table 2). In the LIN experiment, the herbicide was applied to eight microcosms in four duplicate treatments. In the CPF experiment, six microcosms were treated with 1 µg chlorpyrifos/L of which three microcosms received a second application two weeks after the first application. In both experiments, four microcosms were not treated with pesticide to serve as controls.

Table 2. Information on the experimental design and some physicochemical characteristics of the water (mean values ± SD measured in the morning over the experimental period) in the chlorpyrifos (CPF) and linuron (LIN) experiments. For a detailed explanation of the experimental set-up, the reader is referred to the materials and methods section of this chapter and/or the references indicated in the table

	CPF	LIN
Pre-treatment period (wks)	1	5
Date of application	26 September 2003	1 March 2005
Monitoring period post (first) application (wks)	6	8
Water layer: height (cm)/volume (L)	56/250	56/250
Nutrients added		
Nitrogen (as urea; mg/L biweekly)	1.4	1.4
Phosphorous (as TSP; mg/L biweekly)	0.35	0.18
Water chemistry		
Dissolved oxygen (mg/L)	5.8 ± 2.8	5.1 ± 1.9
pH	9.3 ± 0.6	8.9 ± 0.4
Electrical conductivity ($\mu S\ cm^{-1}$)	441 ± 49	701 ± 93
Temperature (°C)	27.6 ± 1.0	27.3 ± 2.0
Reference	Daam et al., 2008	Daam et al., 2009a, b

Plankton Sampling

At several moments during the course of the experiments, a 10-L bulk water sample was collected in a bucket by taking several depth-integrated water samples using a perspex tube. From this bulk sample, a subsample of 1-L was stained with lugol and concentrated after sedimentation for 6 days to study the phytoplankton community. Additional lugol was added when considered necessary to assure conservation of the samples. Another 5 L was transferred through a zooplankton net (mesh size 60 μm). The concentrated zooplankton sample was subsequently fixed with formol in a final concentration of 4%.

The periphyton communities were studied in the LIN experiment using glass slides that served as artificial substratum. The slides were positioned in a glass frame that was suspended 2 weeks before application at approximately 10 cm below the water surface. At 2-week intervals, the periphyton biomass of five slides was collected by brushing the slides visually clean in tap water. Preservation of the periphyton samples was done as described above for phytoplankton.

Identification of the Plankton Samples

Subsamples of the plankton samples were counted with an inverted microscope (magnification 100–400) and numbers were recalculated to numbers per litre microcosm water (phytoplankton and zooplankton) or numbers per cm^2 (periphyton). Colony forming algae except *Microcystis aeruginosa* and *Microcystis incerta* were quantified by counting the number of colonies. *M. aeruginosa* and *M. incerta* form large 3-dimensional colonies that, especially when occurring in high abundances, are difficult to quantify with high precision. Therefore, these two species were quantified as single cells in subsamples of the phytoplankton samples after disintegration of the colonies by ultrasonication as described by Kurmayer et al. (2003).

CLIMATIC CONDITIONS AND SEASON DURING THE EXPERIMENTS

The tropical climate in Thailand is regulated primarily by the monsoon winds that produce three seasons: the cool, hot and rainy season. From May through mid-October the Southwest Monsoon brings warm moist air across Southeast Asia. The surface temperature of the land is higher than the arriving air mass, resulting in thunderstorm formations and the start of the rainy season (June – October). Rainfall is intense but of short duration, accompanied by much lightning and high winds. The rainfall has a moderating effect on the air temperature and direct sunlight is often blocked by cloud cover. Except immediately before and during thunderstorms, surface winds are very light during the wet monsoon. Beginning in July, rainfall of longer duration and greater regularity replaces the less dependable thunderstorm precipitation. In mid-October the Southwest winds are replaced by winds from the Northeast, which brings cool, dry air from Central Asia across Thailand. The rains abruptly cease, and the cool season (November – February) follows. The air becomes clear and the direct sunlight warms the earth rapidly. The cool season is followed by the hot season (March – May). A gradual decrease in wind velocity begins in February, leading eventually to a period of air stagnation and high daytime temperatures. The hot season ends with the arrival of the Southwest Monsoon in May. This annual seasonal cycle occurs with great regularity (Heckman, 1979).

The CPF experiment was carried out from mid July till the end of October 2003. This experiment was thus carried out during the rainy season, with average daily temperatures of 29 ± 0.6 °C and a cumulative rainfall of 116 mm. Sunlight was often blocked by the relatively high cloud cover values (4.8 ± 1.7 octas), resulting in a radiation levels of 18.8 ± 2.5 mJ/m^2 (Figure 1). Linuron was evaluated between mid-January and the end of April 2005, so the experimental period covered a part of the cool season (January – February) and a part of the hot season (March - April). Average daily temperatures increased accordingly in the course of the experiments from 24 °C at the start to 33 °C by the end of the experiments (Figure 1). To compensate for the preceding unusual dry period, the cloud seeding technique described in European Patent Office (2004) was applied in the first semester of 2005 by the Thai government to artificially produce rain. Consequently, cloud cover showed a rather high variation and ranged from 1.4 to 6.7 octas, accompanied with unusual low radiation levels for the time of the year (15.1 ± 2.7 mJ/m^2).

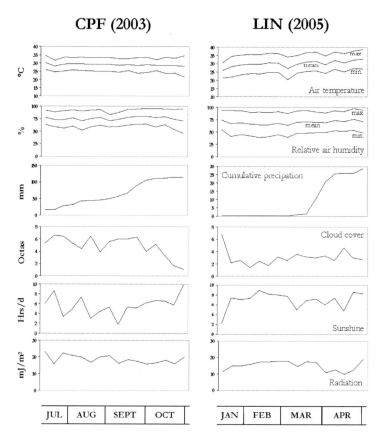

Figure 1. Climatic conditions during the experiments Meteorological conditions and physical/chemical water characteristics during the course of the chlorpyrifos (CPF) experiment in 2003 and the linuron (LIN) experiment 2005. Data were obtained from the meteorological station at AIT.

PLANKTON COMMUNITY STRUCTURES

Primary Producers

A total number of 93 taxa were identified from the phytoplankton and periphyton samples (Table 3). Chlorophyta was the most diverse algae division with 55 taxa, followed by Cyanophyta (19), Bacillariophyta (9), Cryptophyta (5), Euglenophyta (3) and Crysophyta (2). Remarkably, approximately 40% (author's calculation) of the species recorded in a checklist of freshwater algae in Thailand belong to the division Chlorophyta (Wongrat, 1995). Several identified species belonging to the genera *Scenedesmus* were not recorded in this checklist (Wongrat, 1995) nor in several field studies carried out in different parts of Thailand (Peerapornpisal et al, 2000; Ariyadej et al., 2004; Pongswat et al., 2004). These include *S. bicaudatus* and *S. dispar*, which combined numbers made up approximately two-third of total phytoplankton numbers in the controls of the LIN study. Based on a comparison between the number of phytoplankton species known in the world and in Thailand, Baimai (1995) concluded that the discovery of new species for Thailand may indeed be expected. This is supported by the fact that in studies of the Mae Sa stream in Chiang Mai (Northern Thailand), a total of 68 new species were recorded in field samples between 1997 and 1998 and another 51 between 1998 and 1999 (Peerapornpisal et al., 2000; Pekthong and Peerapornpisal, 2001). In addition, *S. bicaudatus* has previously been recorded in Asia (Ling and Tyler, 2000) and several varieties of *S. dispar* were even first described for Vietnam and India (Hegewald and Silva, 1988).

Table 3. List of all taxa identified from the phytoplankton and periphyton samples and a semi-quantative indication of their abundance/dominance

	Phytoplankton		Periphyton
	CPF (39)	LIN (74)	LIN (37)
Total number of taxa (93)			
Chlorophyta (55)	26	44	20
Ankistrodesmus falcatus	*	*	
Ankistrodesmus nannoselene	*	*	
Botryococcus braunnii		**	
Coelastrum astroideum	**	**	*
Coelastrum cambricum	*	*	*
Coelastrum microporum	**	**	*
Coelastrum reticulatum		*	*
Coelastrum sphaericum	*	*	
Crucigenia apiculata	*		
Crucigenia rectangularis	*		
Crucigenia tetrapedia	*		
Elakatothrix gelatinosa		*	
Gloeocapsa sp			*
Golenkinia radiata	*	*	
Hydrodictyon sp			*
Kirchneriella obesa		*	
Micractinium pusillum	**		
Monoraphidium sp	*	*	
Oocystis borgei	***	**	
Oocystis elliptica		*	
Oocystis lacustris	*	**	
Oocystis pusilla	*	**	
Oocystis rupestris		*	
Pediastrum duplex		***	
Pediastrum simplex		*	*
Pediastrum tetras		***	***
Scenedesmus aristatus		**	
Scenedesmus bernardii		*	
Scenedesmus bicaudatus	***	***	**
Scenedesmus bijuga	*	**	*
Scenedesmus bijuga var alternans	*		
Scenedesmus denticulatus		**	
Scenedesmus denticulatus var linearis	*	*	*
Scenedesmus dimorphus	**	*	*
Scenedesmus dispar	*	***	**

Table 3. Continued

	Phytoplankton		Periphyton
Total number of taxa (93)	**CPF (39)**	**LIN (74)**	**LIN (37)**
Scenedesmus javanensis			*
Scenedesmus longispina		*	
Scenedesmus maximus		**	*
Scenedesmus obliquus		*	
Scenedesmus opoliensis		**	
Scenedesmus perforatus		*	
Scenedesmus quadricauda	**	**	**
Scenedesmus quadrispina		*	
Scenedesmus tropicus		**	
Schizochlamys sp		*	
Schroederia sp	*	*	*
Sphaerocystis schoeteri		*	
Staurastrum sexangulare		*	
Staurastrum sp		*	
Stigeoclonium sp			***
Tetraedron caudatum		*	*
Tetraedron minimum		**	**
Tetraedron trigonium	*	*	
Tetrastrum staurogeniaeforme	*		
Trebouxia	*		
Cyanophyta (19)	5	13	9
Anabaena sp			*
Aphanocapsa sp			
Chamaesiphon sp			***
Chroococcus dispersus		*	
Chroococcus dispersus var minor		*	
Chroococcus limneticus			*
Coelosphaerium sp		*	
Gloeotrichia sp			*
Gomphosphaeria sp		*	
Merismopedia minima		*	
Merismopedia tenuissima	*	**	*
Microcystis aeruginosa	***	*	
Microcystis incerta	**	*	
Oscillatoria limnetica	*	*	**
Oscillatoria tenius	**	*	*
Phormidium mucicola		*	*
Pseudoanabaena limnetica		*	
Spirulina laxissima		*	
Spirulina subsalsa			*

Table 3. Continued

	Phytoplankton		Periphyton
Total number of taxa (93)	CPF (39)	LIN (74)	LIN (37)
Bacillariophyta (9)	3	8	6
Amphora sp		*	
Cocconeis sp		**	***
Cyclotella sp	*	*	**
Frustulia sp		*	
Gomphonema parvulum		**	***
Gomphonema sp			***
Nitzschia amphibia	*	*	*
Nitzschia palea	*	**	**
Surirella tenera		*	
Cryptophyta (5)	3	5	0
Campilomonas spp		*	
Chilomonas paramecium	*	*	
Chroomonas acuta		*	
Cryptomonas ovata	**	*	
Cryptomonas pyrenoidifera	*	*	
Crysophyta (2)	1	2	0
Mallomonas caudata		*	
Mallomonas spp	*	*	
Euglenophyta (3)	1	1	2
Euglena pisciformis			*
Phacus longispina		*	*
Trachelomonas sp	*		

* = rare: occurs in less than half of sampling occasions in low numbers; ** = common: occurs in half of sampling occasions or more in low numbers; *** = dominant: occurs in more than half of sampling occasions in high numbers.

After comparing the phytoplankton biomass and community structure of tropical and temperate lakes, Kalff and Watson (1986) concluded that the proportion of phytoplankton in the major taxonomic and size divisions is not so much a function of latitude as of nutrient levels and mixing regime. For example, these authors discussed that the relative importance of diatoms, with their heavy frustules and rapid sinking rate, is more related to levels of silica, phosphorous and mixing regime than to latitude (Kalff and Watson, 1986). The fact that lentic systems were used without mixing (besides water sampling moments) may thus be related with the low abundances of diatoms. In line with this, diatoms were slightly more abundant on the periphyton substrates in the LIN experiment, even though diversity was poor (Table 3). Relatively large species belonging to Dinophyceae (e.g., *Ceratium*, *Peridinium*, *Peridiniopsis*) and Zygnemaphyceae (especially *Staurastrum*), frequently reported in Thai field studies (e.g., Ariyadej et al., 2004; Wongrat, 1995) may have been respectively absent and rare in the phytoplankton for the same reason. The absence of these species in the periphyton community may have been due to the fact that they appear to be mostly found in phytoplankton samples rather than being part of the periphyton community (Peerepornpisal et al., 2000a). The nutrient additions made in the experiments may also have played a role since, for example, *Staurastrum* is known to be dominant in phosphorous-poor environments (Huszar et al., 1998). No silica (Si) additions were made although phosphorous (P) and nitrogen (N) were applied twice per week (Table 2). Consequently, Si:P and Si:N ratios were probably low, indicating a decreasing competitive advantage of diatoms (Calijuri et al., 2002). The periphyton community structure was similar to that of the phytoplankton community with the exception of the filamentous periphytonic species *Stigeoclonium* sp., *Chamaesiphon* sp., and *Gloeotrichia* sp. (Table 3).

Zooplankton

Due to year-round predation by the great diversity of fish and invertebrate predators and increased metabolic costs with increasing temperatures, large cladocerans like *Daphnia* have been reported to be practically absent in the tropics (e.g., Duncan, 1984; Fernando, 1994; Dumont, 1994). Tropical water bodies typically contain cladocerans from the genera *Moina*, *Ceriodaphnia*, *Macrothrix* and *Diaphanosoma* (Sarma et al., 2005) and the limnetic species *Ceriodaphnia cornuta*, *Moina micrura* and *Diaphanosoma excisum* are eurytopic in tropical Asian freshwaters.

Table 4. Zooplankton taxa identified in the samples taken over the course of the chlorpyrifos (CPF) and linuron microcosm experiments. * = rare: occurs in less than half of sampling occasions in low numbers; ** = common: occurs in half of sampling occasions or more in low numbers; *** = dominant: occurs in more than half of sampling occasions in high numbers

	CPF	LIN		CPF	LIN
Total number of taxa (21)	19	14			
Cladocera (7)	5	3	**Rotifera (20)**	14	11
Alona sp	*		*Brachionus angularis*	*	*
Ceriodaphnia cornuta	**	***	*Brachionus calyciflorus*	***	***
Diaphanosoma sp		***	*Brachionus falcatus*	*	*
Dunhevedia crassa	**		*Brachionus quadridentatus*	*	
Moina micrura	***	***	*Brachionus urceolaris*	**	**
Streblocerus pygmaues	**		*Colurella sp*	*	
			Filinia longiseta	*	*
Copepoda			*Filinia opoliensis*	*	
Nauplii	*	***	*Hexarthra mira*	**	
Cyclopoida	*	***	*Keratella tropica*	*	***
Calanoida	*	***	*Lecane bulla*	*	**
			Lecane closterocerca		*
Ostracoda	**	***	*Lecane luna*	*	*
			Lepadella patella	*	*
			Trichocerca sp	*	**

The high tropical temperature determines the distribution in tropical water bodies of mainly warmwater adapted rotifer species of which *Brachionus caudatus*, *B. plicatilis*, *B.*

falcatus, B. calyciflorus, Filinia opoliensis, Hexarthra mira, Keratella tropica, Anuraeopsis fissa, Asplanchna brightwelli are the most dominant (Kutikova, 2002). The Asian rotifer community has been reported to show in general a typical tropical species composition with many *Brachionus* and *Lecane* species (Dussart et al., 1984; Segers, 2001). *Keratella quadrata* is very common in the temperate climatic zones and is replaced by *K. tropica* in tropical climates (De Ridder, 1981). Interestingly, *K. tropica* was indeed the only *Keratella* species present in the microcosms (Table 4)

Ostracoda are rare in temperate freshwaters but quite common in tropical waters (Dussart et al., 1984; Victor, 2002). In line with this, numbers counted in control samples taken over the course of the experiment averaged approximately 170 per liter. Based on the information outlined above and the species listed in Table 4, it can be concluded that the microcosms contained a zooplankton community typical for that described for tropical Asia.

PERIPHYTON SUCCESSION

Over the entire incubation period, a total number of 20 chlorophyte, 9 cyanophyte, 6 diatom and 2 euglenophyte taxa were identified from the periphyton substrated (Table 3). The succession of the control periphyton community is visualized in Figure 2A. After an incubation of 2 weeks, the periphyton community in controls consisted of comparable numbers of Chlorophyta, Bacillariophyta and Cyanophyta. Euglenophyta were absent and were also only found in very low numbers as compared to the other periphyton divisions during the rest of the experimental period. Microscopic slides that were incubated for 4 to 8 weeks were completely (99.5%) dominated by the cyanobacterium *Chamaesiphon sp.* This species, however, had completely disappeared after an incubation period of 8 weeks and was replaced by chlorophytes and diatoms (Figure 2A).

Studies into the community composition and succession of periphyton in tropical freshwaters has been minimal, although periphyton serves as food for many invertebrates and fishes (Talling and Lemoalle, 1998; Pekthong and Peerepornpisal, 2001). In temperate freshwaters, the succession of periphytic algae on a pristine surface is from small, flat cells with a large surface area attached to the substrate like diatoms to standing or stalked forms and eventually filamentous forms (Brönmark and Hansson, 2005). In line with this, dominance of the filamentous colony-forming cyanobacterium *Chamaesiphon* sp. was only reached after four weeks. Two possible causes for the complete disappearance of this species eight weeks after incubation of the substrates may be i) an increase in the number of snails grazing on the substrates; and ii) the occurrence of two tropical rainshowers.

Firstly, snails were observed to be scarce to absent in periods that *Chamaesiphon* sp. did not dominate the periphyton community. Four and eight weeks after incubation, however, an average number of respectively two and five snails were counted in the periphyton samples (snails not identified; data not shown). This grazing on *Chamaesiphon* sp. may have given the opportunity for green algae and diatom species colonizing the substrates. Indeed, the succession stages as described in the previous paragraph have been reported to be reversed at high grazing pressure (Brönmark and Hansson, 2005). In an indoor study with experimental channels evaluating the effects of snails on periphyton communities, Munõz et al. (2001) also concluded that grazing simplified the algal taxonomic composition and physiognomic structure. For example, crustose, stalked and prostrate forms like *Achnanthes* and *Cocconeis* decreased with grazing by snails (Munõz et al., 2001). Remarkably, *Cocconeis* sp. was the species dominating the diatom community after an incubation of eight and ten weeks, which raises the question why snails were observed to be scarce to absent in that period. Low periphyton biomass as quantified by chlorophyll-a content and hence a low amount of available food may be related with this. Furthermore, alkalinity levels dropped to a large extent over the experimental period which may have hampered shell development (Daam et al., 2009a). Interestingly, total numbers of rotifers classified as periphytic (i.e., taxa belonging to the *Lecane*, *Lepadella* and *Trichocerca* genera: Pourriot, 1977; Martinéz et al., 2000; Green, 2003) had greater abundances in those periods that snails were scarce or absent (Figure 2B). Besides the reduction in competition for food due to the absence of snails, colonies of *Chamaesiphon* sp. may have been too large to be consumed by these rotifers.

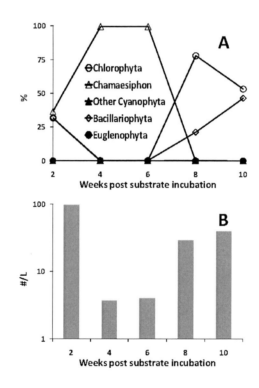

Figure 2. Dynamics of relative contributions (% of total counts) of Chlorophyta, Bacillariophyta, the dominant cyanobacterium *Chamaesiphon* sp., other Cyanophyta and Euglenophyta to the periphyton community in the linuron experiment (A). Corresponding dynamics in total numbers of periphytic rotifers are also presented (B).

Secondly, three tropical rainshowers occurred after six weeks of incubation (65, 58 and 32 mm on the days corresponding to respectively 25, 33 and 45 after incubation). Scouring of substratum by spates have been indicated to cause reduced standing stocks of periphyton in tropical freshwaters (Talling and Lemoalle, 1998; Dudgeon, 1999). Alterations in physicochemical (e.g., pH and nutrient levels; Daam et al., 2009a), either related with the spates or not, and climatic (e.g., cloud cover and radiation values; Figure 1) conditions may evidently also have played a role.

The above may illustrate the complexity of the periphyton community structure and possible interactions with grazers as well as physicochemical and climatic conditions. Especially considering the importance of periphyton for tropical freshwaters mentioned earlier, studies into tropical periphyton communities and their interrelationships are needed to come to a better understanding of this biological component.

Seasonality

Contrary to the common believe of many temperate limnologists, regular seasonal fluctuations in production and life histories are characteristic of tropical freshwaters (Nilssen, 1984). In fact, the seasonal or annual biomass oscillations in the tropics are not systematically lower than in the temperate zone (Kalff and Watson, 1986; Khan, 1996). Because the variation of phytoplankton succession is strongly linked to meteorological and water stratification mixing processes, patterns in temperate ecosystems differ considerably from those in tropical waters (Wetzel, 2001). Investigation of seasonal changes of phytoplankton in tropical areas with more or less regular patterns of dry and rainy seasons as this is the case in the monsoonal areas of Southeast Asia deserves special interest, since the situation differs clearly from both temperate areas with the four clearly separated seasons and from areas of the central tropical belt in which no or little seasonal changes of precipitations and wind circulations occur (Rott et al., 2002).

Seasonal patterns in tropical plankton communities are related with the monsoon (Peerapornpisal, 1996; Khan, 1996; Dudgeon, 2000). For example, Dudgeon (2000) reported that phytoplankton and zooplankton abundances in Asian streams are lower in the wet season compared to the dry season due to washout, dilution and greater turbidity. Peereporpsisal (1996) demonstrated that whereas the cyanobacterium *Cylindrospermopsis raciborskii* dominated reservoirs in Chiang Mai (Thailand) during the rainy season and the cold part of the dry season, other groups of algae dominated in the warm part of the dry season. The author attributed this to a decrease in water volume and a subsequent increase in nutrient levels in the warm part of the dry season (Peerepornpisal, 1996).

In the present study, dominant algal groups differed between the LIN and CPF experiments. In the LIN experiment, which was carried out in the cool season, the phytoplankton community was largely dominated by chlorophytes (Figure 3). The CPF experiment was conducted during the rainy season and showed a complete dominance by the cyanobacterium *Microcystis aeruginosa* (Figure 3). This dominance was explained by the fact that *Microcystis* colonies can regulate their buoyancy, implying that during periods of water stability they have an advantage over other phytoplankton for utilizing nutrients and especially light (Dokulil and Teubner 2000; Bonnet and Poulin 2002). In the rainy season, direct sunlight is often blocked by cloud cover (Heckman 1979; Figure 1), indicating that light may indeed be a limiting factor during this time of the year (Daam et al., 2008). In line with this, Vijanakorn et al. (2004) reported *M. aeruginosa* blooms in a reservoir in Thailand during the rainy season. *M. aeruginosa* has also been reported to be characteristic for stable hot-rainy periods with more prolonged stratification in Brazilian tropical reservoirs (Tundisi, 1990). Greater phosphorous additions in the CPF experiment as compared to the LIN experiment (Table 2) may have enhanced the *Microcystis* bloom, because the optimum N:P ratio for *M. aeruginosa* (5.0-8.5) is much lower than for other *Microcystis* species (Dokulil and Teubner, 2000).

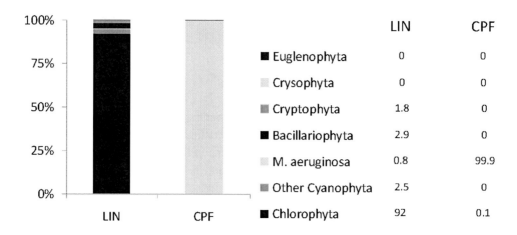

Figure 3. Averaged relative contributions (% of total counts) of phytoplankton groups to the phytoplankton community in the chlorpyrifos (CPF) and linuron (LIN) experiments.

Long-term studies into changes of tropical zooplankton community structure over different seasons are scarce (Pinto-Coelho, 1998). Duncan (1984) reported that the density of tropical rotifer populations can change with season and in different types of waterbodies and also with depth and time of the day. Duncan and Gulati (1981) noted a decrease in rotifer density in the rainy season related to either high flushing rates or dilution with water containing few rotifers. Other studies, however, reported an increased species richness during the wet season as compared to the dry season (e.g., Koste and Robertson, 1983; Green, 2003). In an abandoned meander lake (Lago Amapá, Brazil), Keppeler and Hardy (2004) noted a greater diversity and abundance of rotifers at high-water and low-water, respectively. Pinto-Coelho (1998) concluded that some tropical lakes and reservoirs undergo large but nonrecurring temporal variations in their abundance, while other tropical lakes are permanently dominated by a simple community of small organisms such as small cladocerans and rotifers. After studying the zooplankton community in seven tropical reservoirs, Sampaio et al. (2002) also concluded that different rotifer species occurred in succession, being abundant in different periods, with no defined pattern. Pinto-Coelho (1998) further stipulated that some zooplankters, such as *Diaphanosoma*, exhibited temporal variations that seem to reflect the long-term changes in the trophic status of the reservoir. Opportunistic zooplankters, including *Moina* and *Ceriodaphnia*, show neither a recurrent seasonal trend nor long-term trend; these populations peaks seem to occur over short periods of time under specific (and unstable) conditions (Pinto-Coelho, 1998). The latter seems applicable to the present study; regardless of the season in which the experiments were carried out, *Moina micrura* and *Ceriodaphnia cornuta* were the dominant cladocerans (Table 4). Interestingly, *Diaphanosoma* sp. was one of the dominant cladocerans in the experiment carried out in the cool season (LIN), whereas this species was not encountered in the wet season experiment (CPF). Evidently, another significant factor influencing zooplankton species richness is the variety of available food (Green, 2003), which will be discussed in the next section.

RELATION BETWEEN ZOOPLANKTON AND PHYTOPLANKTON

Basis ecological processes are similar in tropical and temperate lakes including grazing, competition, predation and abiotic adaptation; the major difference being the greater speed of the processes in the warm tropics (Nilssen, 1984; Lewis, 1987; Kutikova, 2002). However, the temperate limnology is not immediately applicable to tropical countries because the nature of the biota and characteristic pathways and biological processes differ to a large extent (Schiemer, 1995). According to Fernando (1994), the major difference between fish and zooplankton in tropical and temperate lakes is the predominance of rotifera and herbivorous fish in tropical lakes versus crustacea and non-herbivorous fish in temperate lakes (Figure 4). The relative biomass of zooplankton/phytoplankton is low in the tropics and therefore, unlike in temperate lakes, zooplankton does not control phytoplankton biomass in the tropics (Piyasiri and Perera, 2001). In this respect, Talling (2003) reported that the typical inverse relationship of chlorophyll-a and cladocerans is lost when relatively inedible large phytoplankters (e.g., filamentous cyanobacteria) are abundant and chlorophyll-a minima are less pronounced. In line with this, no correlation could be demonstrated between total number of cladocerans and chlorophyll-a levels in the CPF experiment, when *Microcystis aeruginosa* completely dominated the phytoplankton community (Pearson correlation test, $r = 0.03$; $p > 0.05$; Figure 5A). In the by chlorophytes dominated LIN control microcosms, however, chlorophyll-a levels and total number of cladocerans were positively correlated (Pearson correlation test, $r = 0.67$; $p < 0.01$; Figure 5B).

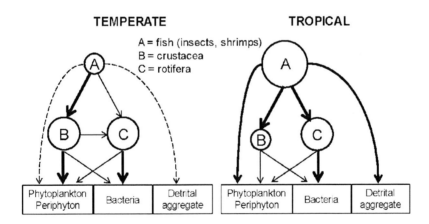

Figure 4. Main freshwater ecosystem components and their food interrelationships in temperate versus tropical freshwaters. Modified from Fernando, 1994.

Bergquist et al. (1985) studied the phytoplankton community structure during grazing by contrasting zooplankton assemblages in polyethylene enclosures. These authors concluded that grazing by a mixture of small zooplankters led to increased growth of phytoplankters with greatest axial linear dimensions < 25 µm and ratios of surface area to volume < 2.6. Larger phytoplankton taxa declined in the presence of small zooplankters. In contrast, a mixture of large zooplankters caused declines in phytoplankters with greatest axial dimensions < 60 µm and ratios of surface area to volume < 2.75, while larger algae increased (Bergquist et al., 1985). The greater abundance of rotifers and smaller-sized cladocerans in the tropics than in temperate areas implies a greater exploitation of smaller food particles in tropical than temperate waters (Sarma et al., 2005). The dominance of the LIN phytoplankton community by the small green algae *Scenedesmus bicaudatus*, *S. dispar* and *Pediastrum tetras*, whose combined numbers made up approximately two-third of phytoplankton counts, may thus be related with this.

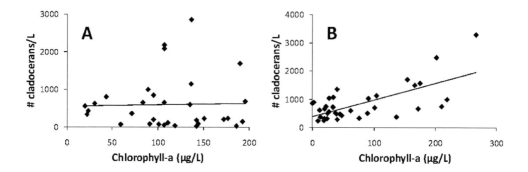

Figure 5. Correlation between chlorophyll-a levels and total number of cladocerans in the chlorpyrifos (A) and linuron (B) microcosm studies.

In the CPF experiment, a complete dominance of *Microcystis aeruginosa* was observed over the experimental period. Various studies have indicated that cladocerans are especially vulnerable to *Microcystis* blooms, whereas rotifers are less vulnerable (Lampert, 1987; Ferrão-Filho, 2002). For example, the growth and reproduction of *Moina micrura* have been reported to be severely reduced when reared with *Microcystis*, even when mixed with *Chlorella* (Hanazato and Yasuno, 1987). The (misconception of a) lower species diversity of cladocerans in the tropics has even been associated with a permanent infestation of many tropical water bodies with toxic cyanobacteria like *Microcystis* spp. (Zafar, 1986). Although the large cladoceran *Diaphanosoma* sp., which occurred in high numbers in the LIN experiment, was indeed absent in the CPF experiment, abundances of *M. micrura* over the course of both experiments averaged approximately 600 per liter. In addition, three small cladoceran taxa not recorded in the LIN experiment (*Alona* sp., *Dunhevedia crassa* and *Streblocerus pygmaues*) were fairly common in the CPF experiment (Table 4). These findings show a remarkable resemblance with those from an enclosure experiment in China by Liu et al. (2002). These authors reported that *M. aeruginosa* suppressed copepods and the larger cladoceran *Diaphanosoma brachyurum* but enhanced the development of smaller cladocerans and rotifers. Filaments or colonies may be too large for rotifers and small cladocerans to ingest. Consequently, they would be less likely than large cladocerans to both collect and ingest these, and hence, the inhibitory effect of *M. aeruginosa* endotoxins and/or mechanical interference with food collection could be lessened (Liu et al., 2002). In line with this, various other studies have demonstrated that the existence of long-lasting blooms of algae, such as cyanobacteria or filamentous diatoms, creates favorable conditions for some smaller cladocerans such as *Bosmina longirostris*, *Bosmina hagmanii* (Branco and Senna, 1996) and *Ceriodaphnia cornuta* (Ferrão-Filho et al., 2000). Liu et al. (2002) further stipulated that the actively growing *M. aeruginosa* probably promoted bacterial production by releasing extracellular products, which can decompose *M. aeruginosa* into detritus. These bacteria and organic matter are probably utilized by rotifers and small cladocerans (Liu et al., 2002). Other studies also indicated that bacteria associated with high cyanobacterial biomass may be a good food source for cladocerans (Hanazato and Yasuno, 1987; Hanazato, 1991; Ferrão Filho 2009). Although bacteria and organic matter were not quantified, composition of the rotifer community in terms of food-preferences (according to Pourriot, 1977) indeed implies that significance of bacteria and detritus as food source increased with increasing abundance of *M. aeruginosa* (Figure 6).

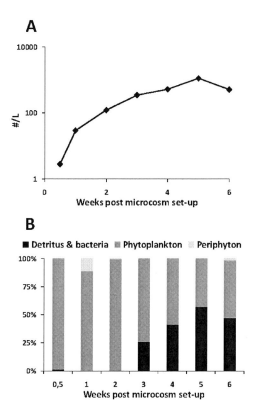

Figure 6. Dynamics of the cyanobacterium *Microcystis aeruginosa* (A) and relative contributions of detritus & bacteria, phytoplankton and periphyton-feeding rotifers (B) in the CPF experiment.

LIMNOLOGY OF TROPICAL FRESHWATERS AND THEIR CONSERVATION

Habitat degradation and loss of biodiversity in the tropics has increased substantially over the past decades. Major underlying factors responsible for this include increasing population densities and developmental pressures, socio-cultural factors, economic considerations and lack and/or weak enforcement of policies concerning land and water resources (Dudgeon, 2003; Gopal, 2005; Sarma et al., 2005). For example, phosphate inputs to inland waters in many European countries and in North America have been reduced by stringent nutrient reduction measures. Such control measures have, however, so far not been initiated in many tropical countries. Therefore, eutrophication is not only rampant but still on the rise in tropical waters (Sarma et al., 2005).

Information on the prevailing situation with respect to habitat integrity and biodiversity is rarely available (Dudgeon, 2000). In a survey by Dudgeon (2003), scientists from tropical Asia authored fewer than 2% of papers dealing with freshwater biology, and less than 0.1% of freshwater biology papers dealt with the conservation of biodiversity in tropical Asia. Research on tropical freshwater bodies has mainly been directed towards describing species; only a few studies are available where the structure and trophic relationships of freshwater communities have been described in detail (Nilssen, 1984; Fernando, 2002). Hence, the present study was initiated to contribute to this lacune in knowledge.

Our ability to ameliorate or mitigate the effects of human activities is predicated upon an adequate understanding of freshwater ecology (Dudgeon, 2000). Differences in the global distribution patterns of species will result in spatial differences in community composition and hence potentially to spatial differences in sensitivity to environmental contaminants (Brock et al. 2008a). For example, the most sensitive taxa to the fungicide carbendazim in a microcosm conducted in Thailand (water boatmen) were different from those recorded to be most sensitive in temperate studies ("worm-like" taxa; Daam et al., 2009c). Therefore, biodiversity studies on community-level are needed to increase our understanding of tropical freshwater community structures and their sensitivity to potential threats outlined in the first paragraph.

As discussed in the present chapter, periphyton and detritus/bacteria are important food sources in the tropics but have hardly been studied in tropical freshwaters. Evidently, alterations in the availability of these components may have severe consequences for higher trophic levels. In this context, Fernando (2002) reported that our knowledge of trophic relationships under natural and experimental conditions in the tropics remains meagre though some recent work has been done in this field. Given the discussed importance of fish and shrimps in structuring tropical freshwaters (Figure 4), it would be interesting to carry out model ecosystem studies that include indigenous representatives of these organisms to study trophic interrelationships on a semi-field level . Chemicals like pesticides could be used to knock down a defined (i.e., depending on the mode of action of the chemical) part of the foodweb so ecosystem interactions can be studied.

Natural freshwater communities are characterized as being temporally dynamic systems. Therefore, an important issue is whether ecosystem level experiments performed in certain periods of the year and with certain exposure regimes can be extrapolated in time (Brock et al., 2008b). In the microcosm experiment evaluating repeated chlorpyrifos treatments, effects on zooplankton indeed differed between the first and second application (Daam et al., 2008). Furthermore, the phytoplankton community in the microcosms was also discussed to differ depending on the season the experiment was carried out: the LIN experiment was largely dominated by a variety of green algae whereas the CPF phytoplankton community was completely dominated by *M. aeruginosa*. Thus, to ensure an adequate representation of the most sensitive endpoint, future outdoor model ecosystems studies evaluating herbicides under a tropical monsoon climate (e.g., Thailand) should be carried out in the end of the cold season or the beginning of the hot season (Daam et al., 2009a). For the same reason, outdoor microcosm and mesocosm studies in temperate countries are recommended to be carried out in spring to mid-summer (Giddings et al., 2002).

The importance of phycological knowledge on tropical freshwaters may further be illustrated by various applications of such knowledge. For example, top-down control of eutrophication by the use of filter-feeding fish (e.g. by hybrids of Oreochromis niloticus and O. mossambicus; Piyasiri and Perera, 2001) was found to have a good potential in tropical countries. Phycological knowledge was found to be highly relevant for the monitoring of toxic phytoplankton taxa in several reservoirs used for drinking water supply in Thailand. In a multi-purpose canyon-shaped reservoir in N-Thailand studied for a longer period, health risks by a bloom of toxic Cyanobacteria/ Cyanophytes could be minimized by optimization of the water intake procedure (Peerapornpisal et al., 2002). Phycological information was also very relevant to optimize capture fisheries in many reservoirs in Sri Lanka based on optimization of hydraulic conditions to avoid phytoplankton blooms (Silva, 1999; Rott et al., 2002).

The looming global climate change may only worsen the situation for tropical freshwater conservation unless remedial measures are taken on a large scale and urgency (Gopal, 2005). Possible scenarios of global climate change include wetter wet seasons and drier dry seasons, with an increased frequency of extreme flow events. Lower flows during the dry season will concentrate pollution loads. The combination of extreme flow events and the effects of deforestation of drainage basins on run-off can be anticipated to spur construction of even more dams and flood-control projects to the detriment of tropical aquatic biodiversity (Dudgeon, 2000). Another consequence of climate change may be the introduction of tropical species across continents (Fernando 2002). For example, Havens et al (2000) found the exotic cladoceran *Daphnia lumholzii* dominating a subtropical lake in Florida during summer months. Knowledge on tropical freshwater structure and functioning may thus become relevant for detecting and understanding possible effects of climate change on temperate freshwaters.

REFERENCES

Ariyadej C, Tansakul R, Tansakul P, Angsupanich S (2004). Phytoplankton diversity and its relationship to the physico-chemical environment in the Banglang reservoir, Yala province. *Songklanakarin Journal of Science and Technology* 26: 595-607.

Baimai V (1995). *Status of Biological Diversity in Thailand* (in Thai). TRF Publishing, Bangkok, Thailand. 254 pp.

Balian EV, Segers H, Lévêque C, Martens K (2008a) The freshwater animal diversity assessment: an overview of the results. *Hydrobiologia* 595: 627-637.

Balian EV, Segers H, Lévêque C, Martens K (2008b). An introduction to the Freshwater Animal Diversity Assessment (FADA) project. *Hydrobiologia* 595: 3-8.

Bergquist AM, Carpenter SR, Latino JC (1985). Shifts in phytoplankton size structure and community composition during grazing by contrasting zooplankton assemblages. *Limnol Oceanogr* 30: 1037-1045.

Bonnet MP, Poulin M (2002). Numerical modelling of the planktonic succession in a nutrient-rich reservoir: environmental and physiological factors leading to *Microcystis aeruginosa* dominance. *Ecol Model* 156: 93-112.

Branco CWC, Senna PAC (1996). Relations among heterotrophic bacteria, chlorophyll-a, total phytoplankton, total zooplankton and physical and chemical features in the Paranoá reservoir, Brasília, Brazil. *Hydrobiologia* 337: 171–181.

Brock TCM, Arts GHP, Maltby L, Van den Brink PJ (2006). Aquatic risks of pesticides, ecological protection goals, and common aims in European Union legislation. *Integrated Environ Assessment Manage* 2: e20-e46.

Brock TCM, Maltby L, Hickey CH, Chapman J, Solomon KR (2008a). Spatial extrapolation in ecological effect management of chemicals. In: Solomon KR, Brock TCM, De Zwart D, Dyer SD, Posthuma L, Richards SM, Sanderson H, Sibley PK, Van den Brink PJ (eds). *Extrapolation Practice for Ecotoxicological Effect Characterization of Chemicals*, SETAC Europe Press, Brussels, Belgium, pp 223-256.

Brock TCM, Solomon K, van Wijngaarden R, Maltby L (2008b). Temporal extrapolation in ecological effect assessment of chemicals. In: Solomon KR, Brock TCM, De Zwart D, Dyer SD, Posthuma L, Richards SM, Sanderson H, Sibley PK, Van den Brink PJ (eds). Extrapolation Practice for Ecotoxicological *Effect Characterization of Chemicals*, SETAC Europe Press, Brussels, Belgium, pp 187-222.

Brönmark C, Hansson LA (2005). *The biology of lakes and ponds*, 2nd edition. Oxford University Press, Great Britain.

Calijuri MC, Dos Santos ACA, Jati S (2002). Temporal changes in the phytoplankton community structure in a tropical and eutrophic reservoir (Barra Bonita, S.P.-Brazil). *J Plankton Res* 24: 617-634.

Campbell IC, Parnrong S (2001). Limnology in Thailand: present status and future needs. *Verh. Int. Ver. Limnol.* 27: 2135-2141.

Daam MA, Van den Brink PJ, Nogueira AJA (2008). Impact of single and repeated applications of the insecticide chlorpyrifos on tropical freshwater plankton communities. *Ecotoxicology* 17: 756–771.

Daam MA, Rodrigues A, Van den Brink PJ, Nogueira AJA (2009a). Ecological effects of the herbicide linuron in tropical freshwater microcosms. *Ecotox Environ Saf* 72: 410-423.

Daam MA, Van den Brink PJ, Nogueira AJA (2009b). Comparison of fate and ecological effects of the herbicide linuron in freshwater model ecosystems between tropical and temperate regions. *Ecotox Environ Saf* 72: 424-433.

Daam MA, Satapornvanit K, Van den Brink PJ, Nogueira AJA (2009c). Sensitivity of macroinvertebrates to carbendazim under semi-field conditions in Thailand: Implications for the use of temperate toxicity data in a tropical risk assessment of fungicides. *Chemosphere* 74: 1187-1194.

De Ridder M (1981). Some considerations on the geographical distribution of rotifers. *Hydrobiologia* 85: 209-225

Dudgeon D (1999). *Tropical Asian Streams: Zoobenthos, Ecology and Conservation*. Hong Kong University Press, Hong Kong.

Dudgeon D (2000). The ecology of tropical Asian rivers and streams in relation to biodiversity conservation. *Annu Rev Ecol Syst* 31: 239-263.

Dudgeon D (2003). The contribution of scientific information to the conservation and management of freshwater biodiversity in tropical Asia. *Hydrobiologia* 500: 295-314.

Dokulil MT, Teubner K (2000). Cyanobacterial dominance in lakes. *Hydrobiologia* 438: 1-12.

Duncan A (1984). Assessment of factors influencing the composition, body size and turnover rate of zooplankton in Parakrama Samudra, an irrigation reservoir in Sri Lanka. *Hydrobiologia* 113: 201-215.

Duncan A, Gulati RD (1981). Parakrama Samudra (Sri Lanka) Project – a study of a tropical lake ecosystem, 3. Composition, density and distribution of the zooplankton in 1979. *Verh. Int. Ver. Limnol.* 21: 1001-1006.

Dumont HJ (1994). On the diversity of the Cladocera in the tropics. *Hydrobiologia* 272: 27-38.

Dussart BH, Fernando CH, Matsumura-Tundisi T, Shiel RJ (1984). A review of systematics, distribution and ecology of tropical freshwater zooplankton. *Hydrobiologia* 113: 77-91.

European Patent Office (2004). *European Patent Application nr. EP1491088. European Patent Bulletin* 53. 18 pp.

Fernando CH (1994). Zooplankton, fish and fisheries in tropical freshwaters. *Hydrobiologia* 272: 105-123.

Fernando CH (2002). *Zooplankton and tropical freshwater fisheries*. In: Fernando CH (ed). A guide to tropical freshwater zooplankton, Backhuys Publishers, Leiden, The Netherlands, pp 255-280.

Ferrão-Filho A, Azevedo SMFO, DeMott W (2000). Effects of toxic and non-toxic cyanobacteria on the life history of tropical and temperate cladocerans. *Freshwater Biol* 45: 1–20.

Ferrão-Filho A, Domingos P, Azevedo SMFO (2002). Influences of a *Microcystis aeruginosa* Kützing bloom on zooplankton populations in Jacarepaguá Lagoon (Rio de Janeiro, Brazil). *Limnologica – Ecology and Management of Inland Waters* 32: 295-308.

Ferrão-Filho AS, Soares MCS, Magalhães VF, Azevedo SMFO (2009). Biomonitoring of cyanotoxins in two tropical reservoirs by cladoceran toxicity bioassays. *Ecotox Environ Saf* 72: 479–489.

Foran JA (1986). The relationship between temperature, competition and the potential for colonization of a subtropical pond by *Daphnia magna*. *Hydrobiologia* 134: 103-112.

Gaston KJ, Williams PH, Eggleton P, Humphries CJ (1995). Large scale patterns of biodiversity: spatial variation in family richness. *P Roy Soc Lond B Biol* 260: 149-154.

Giddings JM, Brock TCM, Heger W, Heimbach F, Maund SJ, Norman SM, Ratte HT, Schäfers C, Streloke M, eds (2002). Community-level aquatic systems studies - Interpretation criteria. Proceedings from the CLASSIC Workshop, held 30 May - 2 June 1999 at Fraunhofer Institute-Schmallenberg, Germany. SETAC publication. Brussels, Belgium, *Society of Environmental Toxicology and Chemistry* (SETAC) (2002). 60 pp.

Gopal B (2005). Does inland aquatic biodiversity have a future in Asian developing countries? *Hydrobiologia* 542: 69-75.

Green J (2003). Associations of planktonic and periphytic rotifers in a tropical swamp, the Okavango Delta, Southern Africa. *Hydrobiologia* 490: 197-209.

Hanazato T, Yasuno M (1987). Evaluation of *Microcystis* as food for zooplankton in a eutrophic lake. *Hydrobiologia* 144: 251-259.

Hanazato T (1991). Interrelations between *Microcystis* and Cladocera in the highly eutrophic Lake Kasumigaura, Japan. *Int. Rev. Hydrobiol.* 76: 21–36.

Havens KE, East TL, Marcus J, Essex P, Bolan B, Raymond S, Beaver JR (2000). Dynamics of the exotic Daphnia lumholtzii ad native macro-zooplankton in a subtropical chain-of-lakes in Florida, USA. *Freshwater Biol* 45: 21-32.

Heckman CW (1979). *Rice field ecology in Northeastern Thailand*. The effect of wet and dry seasons on a cultivated aquatic ecosystem. Dr. W. Junk bv Publishers, The Netherlands.

Hegewald E, Silva PC (1988). Annotated catalogue of Scenedesmus and nomenclaturally related genera, including original descriptions and figures. *Bibliotheca Phycologica* 80. 254 pp.

Husnar VLM, Silva LHS, Domingos P, Marinho M, Melo S (1998). Phytoplankton species composition is more sensitive than OECD criteria to the throphic status of three Brazilian tropical lakes. *Hydrobiologia* 369/370: 59-71.

Kalff J, Watson S (1986) Phytoplankton and its dynamics in two tropical lakes: a tropical and temperate zone comparison. *Hydrobiologia* 138: 161-176.

Keppeler EC, Hardy ER (2004). Abundance and competition of Rotifera in an abandoned meander lake (Lago Amapá) in Rio Branco, Acre, Brazil. *Revista Brasileira de Zoologia* 21: 233-241.

Khan MA (1996). The phytoplankton periodicities of two warm-climate lakes subject to marked seasonal variability. *J Trop Ecol* 12: 461-474.

Koste W, Robertson B (1983). Taxonomic studies of the Rotifera 8Phylum aschelminthes) from a central Amazonian varzea lake, Lago Camaleao (Ilha de marchataria, Rio Solimoes, Amazonas, Brazil). *Amazoniana* 8: 225-254.

Kurmayer R, Christiansen G, Chorus I (2003). The abundance of microcystin-producing genotypes correlates positively with colony size in *Microcystis* sp. and determines its microcystin net production in Lake Wannsee. *Appl Environ Microb* 69: 787–795.

Kutikova LA (2002). Rotifera. In: Fernando CH, ed. *A guide to tropical freshwater zooplankton*. Backhuys Publishers, Leiden, the Netherlands. pp: 23-68.

Lacher TE, Jr, Goldstein MI (1997). Tropical Ecotoxicology: status and needs. *Environ Toxicol Chem* 16: 100-111.

Lampert W (1987). Laboratory studies on zooplankton-cyanobacteria interactions. *New Zeal. J. Mar. Fresh.* 21: 483-490.

Lévêque C, Oberdorff T, Paugy D, Stiassny MLJ, Tedesco PA (2008). Global diversity of fish (Pisces) in freshwater. *Hydrobiologia* 595: 545-567.

Lewis WM Jr (1987). Tropical limnology. *Annu Rev Ecol Syst* 18: 159-184.

Ling HU, Tyler PA (2000). Australian freshwater algae (exclusive of diatoms). *Bibliotheca Phycologica* 105: 1-643.

Liu H, Xie P, Chen F, Tang H, Xie L (2002). Enhancement of planktonic rotifers by *Microcystis aeruginosa* blooms: an enclosure experiment in a shallow eutrophic lake. *J Freshwater Ecol* 17: 239-247.

Kwok KWH, Leung KMY, Chu VKH, Lam PKS, Morritt D, Maltby L, Brock TCM, Van den Brink PJ, Warne MStJ, Crane M (2007). Comparison of tropical and temperate freshwater species sensitivities to chemicals: implications for deriving safe extrapolation factors. *Integrated Environ Assessment Manage* 3: 49-67.

Mares MA (1997). Tropical ecotoxicology: explorers for a new era. *Environ Toxicol Chem* 16: 1-2.

Martinéz JCC, Canesin A, Bonecker CC (2000). Species composition of rotifers in different habitats of an artificial lake, Mato Grosso do Sul State, Brazil. *Acta Scientiarum* 22: 343-346.

Muñoz I, Real M, Guasch H, navarro E, Sabater S (2001). Effects of atrazine on periphyton under grazing pressure. *Aquat Toxicol* 55: 239-249.

Nilssen JP (1984). Tropical lakes – functional ecology and future development: The need for a process-orientated approach. *Hydrobiologia* 113: 231-242.

Peerapornpisal Y (1996). *Phytoplankton seasonality and limnology of the three reservoirs in the Huai Hong Khrai Royal Development Study Centre*, Chiang Mai, Thailand. Doctoral thesis Innsbruck University, Austria.

Peerapornpisal Y, Pekthong T, Waiyaka P, Promkutkaew S (2000). Diversity of phytoplankton and benthic algae in Mae Sa stream, Doi Suthep-Pui National Park, Chiang Mai. *Natural History Bulletin of the Siam Society* 48: 193-211.

Peerapornpisal Y, Sonthichai W, Sukotiratana M, Lipigorngoson S, Ruangyuttikam W, Ruangrit K, Pekkoh J, Prommana R, Panuvanitchakorn N, Ngearnpat N, Kiatpradab S, Promkutkaew S (2002). Survey and monitoring of toxic Cyanobacteria in water resources for water supplies and fisheries in Thailand. *Chiang Mai Journal of Science* 29: 71-79.

Pekthong T, Peerapornpisal Y (2001). Fifty one new record species of freshwater diatoms in Thailand. *Chiang Mai Journal of Science* 28: 97-112.

Pinto-Coelho RM (1998). Effects of eutrophication on seasonal patterns of mesozooplankton in a tropical reservoir: a 4-year study in Pampulha Lake, Brazil. *Freshwater Biol* 40: 159-173.

Piyasiri S, Perera N (2001). *Role of Oreochromis hybrids in controlling Microcystis aeruginosa blooms in the Kotmale Reservoir*. In: S. S. de Silva (ed.), Reservoir and Culture-Based Fisheries: Biology and Management. Proceeding of an International Workshop. 15-18 February, 2000, 2001, Bangkok, Thailand. pp. 137-148.

Pourriot R *(1977)*. Food and feeding habits of Rotifera. *Arch. Hydrobiol. Beih. Ergebn. Limnol.* 8: 243–260.

Pongswat S, Thammathawom S, Peerapornpisal Y, Thanee N, Somsiri C (2004). Diversity of phytoplankton in the Rama IX lake, a man-made lake, Pathumthani province, Thailand. *ScienceAsia* 30: 261-267.

Rott E, Peerapornpisal Y, Ingthamjitr S, Silva EIL (2002). Phytoplankton seasonality in reservoirs under monsoon climate in Sri Lanka and Thailand. Abstracts 4th Internat. Reservoir Limnol. *Conf. Ceske Budejovice*: 290-291.

Sampaio EV, Rocha O, Matsumura-Tundisi T, Tundisi JG (2002). Composition and abundance of zooplankton in the limnetic zone of seven reservoirs of the Paranapanema river, Brazil. *Braz. J. Biol.* 62: 525-545.

Sarma SSS, Nandini S, Gulati RD (2005). Life history strategy of cladocerans: comparisons of tropical and temperate taxa. *Hydrobiologia* 542: 315-333.

Segers H (2001). Zoogeography of the Southeast Asian Rotifera. *Hydrobiologia* 446/447: 233-246.

Segers H (2008). Global diversity of rotifers (Rotifera) in freshwater. *Hydrobiologia* 595: 49-59.

Schiemer F (1995). *Bottom-up vs. top-down control in tropical reservoir management*. In: Timotius, K.H. and Golthenboth, F. (eds). Tropical Limnology 1, Satya Wacanan Christian University, Indonesia, pp. 57–67.

Silva EIL (1999). Status of surface water quality in Sri Lanka. *Scope* 82: 11-124.

Talling JF (2003). Phytoplankton-zooplankton seasonal timing and the "clear-water phase" in some English lakes. *Freshwater Biol* 48: 39-52.

Talling JF, Lemoalle J (1998). *Ecological dynamics of tropical inland waters*. Cambridge University Press, UK.

Tundisi JG (1990). Distribuição espacial, sequência temporal e ciclo sazonal do fitoplâncton em represas: factores limitantes e controladores. *Revta Bras Biol* 50: 937-955.

Van der Linde AMA, Boesten JJTI, Brock TCM, Van Eekelen GMA, De Jong FMW, Leistra M, Montforts MHMM, Pol JW (2006). *Persistence of plant protection products in soil; a proposal for risk assessment*. RIVM Report 601506008, Bilthoven, The Netherlands.

Vijaranakorn T, Nutniyom P, Chantara S (2004). Distribution of *Microcystis aeruginosa* Kutz., microcystin concentrations and water quality of Mae Kuang Udomtara Reservoir, Chiang Mai Province. *Chiang Mai Journal of Science* 31: 69-84.

Victor R (2002). Ostracoda. In: Fernando CH, ed. *A Guide to Tropical Freshwater Zooplankton. Identification, Ecology and Impact on Fisheries*. Backhuys Publishers, Leiden, The Netherlands. pp: 189-233.

Wetzel RG (2001). *Limnology* (3rd edition). Academic Press, California, USA.

Wongrat L (1995). Freshwater algae in Thailand. In: Lewmanomont K, Wongrat L, Supanwanid C (eds). *Algae in Thailand.* Office of Environmental Policy and Planning, Ministry of Science Technology and Environment, Bangkok, Thailand. pp: 96-292.

Zafar AR (1986). Seasonality of phytoplankton in some South Indian lakes. *Hydrobiologia* 138: 177-187.

In: Encyclopedia of Environmental Research
Editor: Alisa N. Souter

ISBN: 978-1-61761-927-4
© 2011 Nova Science Publishers, Inc.

Chapter 37

PROSPECTS AND DEVELOPMENT IN FISH SPERM AND EMBRYO CRYOPRESERVATION

Vanesa Robles[1], Elsa Cabrita[2], Vikram Kohli[3] and M. Paz Herráez[4]

[1] Instituto de Sanidad Animal y Desarrollo Ganadero, University of León, Spain
[2] Center for Marine Science-CCMAR, University of Algarve, Portugal
[3] Department of Electrical and Computer Engineering, University of Alberta, Edmonton, Alberta
[4] Department of Cell Biology, University of León, Spain

ABSTRACT

The Industry of Aquaculture has been profiting from fish sperm cryopreservation for many years. Sperm from more than two hundred fish species, most of them with commercial value, has been cryopreserved, allowing year round availability of sperm and the creation of sperm banks from those individuals with special interest. However, to date, fish embryo cryopreservation has not been successfully achieved in any teleost species. The benefits of fish embryo cryopreservation are numerous, not only for aquaculture but also for conservation purposes. This technique will allow for the maintenance of a constant supply of animals, reduce the facilities required on the fish farm, facilitate animal transportation between different farms and enable the preservation of valuable lines. These practical advantages will also report a cost reduction and minimize the impact of epidemics on productivity. This chapter presents an overview of the recent advances in this field and their impact on Aquaculture.

SPERM CRYOPRESERVATION

Cryopreservation is a technique with undoubted interest, not only for fish farming but also for the conservation and genetic improvement of broodstocks. This technique has been established in some freshwater fish species mainly, salmonids, sturgeons and carps, however, only in the last decade intensive work was done in marine species. The benefits of sperm cryopreservation include: 1- synchronization of gamete availability of both sexes: sperm can be stored and used when the eggs are available; 2- sperm economy: it allows the use of total volume of available sperm. This is useful when semen is difficult to obtain and also when low volume of semen can be stripped in captivity; 3- simplifies broodstock maintenance: off season spawning can be induced only in females and cryopreserved sperm can be used to fertilize the eggs, and 4- allows the transport of gametes from different fish farms, avoiding animal transportation and their requirements and costs. Moreover, due to the high domestication of cultured species, several inbreeding problems have been identified in broodstocks, being necessary more concerned about broodstocks genetic improvement and selection of specific strain lines.

Thus, the use of sperm cryobanks would be of undoubtedly interest not only from a productive point of view, but also can benefit the reposition of stocks from a genetic point a view. Sperm cryobanking has been used in conservation programs, genetic improving of wild and captive populations and in the selection of specific fish characteristics based on their reproductive performance. This last issue would certainly benefit aquaculture industry.

Successful attempt has been done to cryopreserve sperm from most commercialized fish species, such as turbot (*Scophthalmus maximus*) [1,2], gilthead seabream (*Sparus aurata*) [3,4], European seabass (*Dicentrarchus labrax*) [5,6], European and Japanese eel (*Anguilla anguilla, Anguilla japonica*) [7,8], striped bass (*Morone saxatilis*) [9], sharpsnout seabream (*Diplodus puntazzo*) [10], halibut (*Hippoglossus hippoglossus*) [11], Winter flounder (*Pleuronectes americanus*) [12], dusky and malaba grouper (*Epinephelus marginatus*, *Epinephelus malabaricus* [13, 14] cod and haddock (*Gadus morhua, Melanogrammus aeglefinus*) [15]. Indeed, regardless marine species, some good results have been achieved, but there is not a concern about the standardization of protocols including an exhaustive description of procedures and evaluation of sperm quality before and after freezing, giving special attention to initial sperm quality and to specific damage occurring during cryopreservation. From a production, conservation and specially, research point of view this is a crucial mater since sperm quality must be guarantee.

Standardized methods have been proposed in human spermatozoa, however, no such procedures have been developed to standardize sperm analysis of domestically species, especially fish. The development of standardized protocols for the analysis of fish sperm quality before and after cryopreservation would be crucial to guarantee reproducible results within species worldwide. Nonetheless, this is far from being achieved in fish. Although sperm from a wide range of fish species have been cryopreserved, few protocols were adapted and standardize for commercial scale [16]. This could be one of the principal bottlenecks in the cryopreservation of fish sperm and in the spreadness and commercialization of samples.

Semen characteristics vary among species, stocks and even within samples from the same animal, depending on the collection period during the reproductive season. Therefore, the assay of seminal quality and the development of tools to characterize this quality should be used extensively for quality control in research on seminal physiology and in commercial artificial breeding programs. Most of these quality parameters should allow to predict fertilizing potential as well as to establish stocks of good quality. It is well known that sperm samples have different liability for freezing, according to their quality. Thus, it is crucial an evaluation of sperm in order to select the best quality samples for freezing.

Cryopreservation produce several damage in cells, affecting plasma membrane, organelles (in particular the mitochondrion) and it can also produce damage at chromatin level. Cell damage can be produced by mechanical forces due to ice crystals formation, both within the cells and in the external medium or to osmotic stress [17]. The conservation of each cell structure as well as its functionality will depend on the viability of the cryopreservation protocol. Plasma membrane is the principal target of cell damage. In order to fertilize the eggs, sperm must conserve their integrity and cells must maintain the control of the osmotic regulation during the fertilization process. Thus, membrane integrity and functionality can not be compromised. Moreover, spermatozoa need to maintain its moving capacity when released into water in order to achieve the oocyte. Spermatozoa motility is crucial and depends on different aspects of the cell, such as mitochondria status, ATP production, plasma membrane channels and flagella structure. Any change in these aspects will compromise motility. Also, DNA integrity must be guaranteed to accomplish successful embryo development after fertilization.

The assessment of sperm quality using these tools will be decisive for the selection of good sperm samples and for the standardization of the designed cryopreservation protocols.

EMBRYO CRYOPRESERVATION

Fish embryo cryopreservation has yet to be achieved. Successful preservation of teleost embryos would undoubtedly impact the aquaculture industry. The ability to cryopreserve embryos would be important to genetic resource banking for the maintenance of transgenic fish lines and the re-population of exotic species for wildlife conservation. A resource bank of embryos would simplify the transfer of animals between fish farms and improve hatchery operations by reducing the spread of epidemics while maintaining storage of valuable genetic lineages. Despite these benefits, fish embryo preservation has been tedious, with active research being pursued in the cryopreservation of numerous teleost species including the winterflounder (*Pleuronectes americanus*), the Japanese flounder (*Paralichthys olivaceus*), the zebrafish (*Danio rerio*), the turbot (*Scophthalmus maximus*), the gilthead seabream (*Sparus aurata*) and the medaka (*Orizias latipes*). Zebrafish and medaka have received particular interest among vertebrate embryologists, zoologists and developmental biologists as an important model system in medical and biological research, while others like turbot and gilthead seabream have significant commercial value in the fisheries industry. Specific to the zebrafish, this model system has been used in the study of genetics, drug monitoring, human disease, cardiac function and blood disorders [18-25]. While the race for cryopreserving fish embryos continues, the field of cryobiology has expanded eliciting the help of embryologists, developmental biologists, biophysicists, physicists and engineers, with the common intent of preserving teleost embryos. It is through this multidisciplinary pursuit that we hope to achieve successful fish embryo cryopreservation in the near future. Challenges hindering embryo cryopreservation include the embryo's low surface-to-volume ratio, low membrane permeability, differing osmotic properties of the cells and yolk, the large size of the yolk and cells, the high chilling sensitivity and the relatively large osmotically inactive water fraction volumes [26-34]. Preservation requires a proper balance between cooling rates, thawing rates and cryoprotectant concentration [35-37]. Intracellular cryoprotectants function to lower the intracellular ice formation temperature allowing cells to be cooled to lower temperatures with the avoidance of ice formation [38, 39]. Several membrane permeable and impermeable cryoprotectants have been used in the attempted cryopreservation of fish embryos including: methanol (MeOH), dimethyl sulfoxide (DMSO), glycerol, propane-1,2-diol, propylene glycol (PG), ethylene glycol (EG), trehalose and sucrose [27, 32-34, 40-46]. Using the above cryoprotectants several studies have examined the toxicity, permeability and chilling sensitivity in various species of teleost embryos.

The chilling sensitivity and survivability of embryos exposed to various cryoprotectants has been actively studied in order to determine the optimal parameters for embryo cryopreservation. For instance, in the zebrafish, Zhang et al. [27] examined the survival of blastula (3 hrs), epiboly (7 and 10 hrs), optic vesicle (15 hrs), auditory placode (20 hrs) and hartbeat stage (27 hrs) embryos in the absence of cryoprotectants. These embryo stages were held in a low temperature bath at 0°C for 24 hrs. Zhang and colleagues [27] found that early developmental stages were extremely sensitive to chilling. However, the chilling sensitivity was found to decrease as the embryo aged. To determine if cryoprotectants can improve the chilling sensitivity, Zhang and colleagues [27] examined the survival (% hatch) of heartbeat stage embryos stored at 0, -5 and -15°C in a cocktail of cryoprotectants. Among the different cryoprotectants tested 1.0 M MeOH provided the best protection with a maximum survival of 79 and 69% at 0°C for 18h and 24h of storage . A combination of MeOH and sucrose was also used, with embryo survival reaching a maximum of 84%. Using this combination of cryoprotectants had minor effects on survival when the storage temperature was decreased to –15°C. In another study by Zhang et al. [32], the effect of developmental arrest on chilling injury in zebrafish embryos was investigated. Embryos were exposed to methanol or anoxic conditions before they were cooled with different cooling rates: 1, 30 and 300°C/min. Zhang found that anoxia and developmental arrest had no effect on the chilling sensitivity of these embryos.. Hagedorn et al. [33] also preformed chilling resistance studies on zebrafish embryos Similar to Zhang et al. [27] results, early stage embryos (1.25, 1.5 and 1.75 and 2 hrs), were found to be the most sensitive to chilling, while 50%, 75%, 100% epiboly and 3-somite staged embryos had higher survival rates. They also demonstrated that enzymatic removal of the chorion did not alter the pattern of sensitivity to chilling [33].

The above chilling resistance studies indicate that later embryo developmental stages are more resistant to chilling injury. It is well known that chilling sensitivity observed at slow cooling rates might be avoided using vitrification. Therefore, cryobiologists have attempted to use this method for cryopreserving teleost embryos. However, the problem with vitrification is that rapid cooling could induces intracellular ice formation since it is not allowing water removal by cell dehydration. Formation of intraembryonic ice causes irreparable cellular damage leading to embryo death. The addition of high concentration cryoprotectants in vitifrication is meant to prevent ice formation. However, the high cryoprotectant concentrations required in vitrification make embryo toxicity an issue. Hagedorn et al. [33] showed that exposure of 3-somite embryos to 1.5 M glycerol or EG for up to 30min at RT was deleterious to zebrafish embryos. Glycerol caused 100% embryo mortality, and EG resulted in disassociation of the blastoderm from the yolk [33]. Suzuki et al. [47] measured the toxicity of DMSO in several fish embryos including the rainbow trout, medaka, pejerrey and carp. Suzuki and colleagues did not observe any deleterious effects on embryo survival using 2.0 M DMSO. However, prolonged exposure to 2.0 M and higher DMSO concentrations resulted in increased embryo mortality. Liu et al. [41] attempted to vitrify zebrafish embryos using MeOH in a two-step process. Early and late cleavage (1-cell and 64-cell), 50 % epiboly, 6-somite and prim-6 embryos were first exposed to 2.0 M MeOH at room temperature for 10 (1-cell), 20 (64-cell) and 30 min (50 %, 6-somite and prim-6). After, embryos were re-exposed to MeOH at a concentration of 10.0 M for 1 (1-cell), 3 (64-cell), 10

(50% epiboly), 5 (6-somite) and 10 min (prim-6), respectively. Embryos were vitrified in liquid nitrogen or nitrogen slush for 30 sec. Following vitrification, Liu and colleagues [41] report that 80% of the later stage embryos (50% epiboly, 6-somite and prim-6) and 50% of the early stage embryos (1-cell and 64-cells) were morphologically intact immediately post-thaw. However, within 20 min the plasma membrane collapsed and the yolk ruptured for all embryo stages [41]. Intraembryonic ice formation could not be avoided despite the high concentration of MeOH used in the study [41]. In similar work, Zhang et al. [48] attempted to vitrify 6-somite and heartbeat stage (27 hrs) embryos. Following thawing and dilution of the cryoprotectants, the vitelline membrane collapsed and the yolk ruptured with the observance of intraembryonic ice [48].

As previously mentioned, intraembryonic ice formation causes cellular damage and embryo death. Ice formation has been observed in both vitrification and slow cooling studies. Hagedorn et al. [49] recently reported that when zebrafish are cooled at a rate of 2 °C/min, intraembryonic ice forms within seconds after the appearance of external ice in the extracellular solution. This ice front propagates and causes the embryo to freeze. However, intraembryonic ice should be avoided with the presence of cryoprotectants, and the lack of conferred protection indicates a permeability barrier. Several studies have attempted to elucidate the permeability barrier hindering the intraembryonic accumulation of cryoprotectants and the movement of water. Harvey et al. [40] measured the permeability of chorionated and dechorionated 50% epiboly zebrafish embryos to labelled DMSO ([^{14}C]DMSO) and glycerol ([^{3}H]Glycerol). Using a scintillation counter, Harvey and colleagues [40] were able to theoretically estimate permeation or percentage equilibration of both labelled cryoprotectants. Harvey et al. [40] found that after 2 hrs of incubation, glycerol permeation reached a maximum of 8% while DMSO was slightly greater than 2%. Dechorionated 50% epiboly embryos showed increased permeability (~ 9%) indicating that the chorion could represent a barrier to the permeation of cryoprotectants [40]. Hagedorn et al. [28] determined the permeability and distribution of water in 40 % epiboly to 6-somite stage zebrafish embryos, using transport equations and volumetric measurements of the embryos suspended in nonpermeating hypertonic solutions. Hydraulic conductivitiy (L_p) measurements revealed that the permeability of water changed as function of developmental stage, with L_p ranging from 0.022 to 0.040 µm•min^{-1}•atm^{-1} for 75% epiboly to 3-somite stage L_p increased by two folds for embryos at 6-somite (0.049 to 0.100 µm•min^{-1}•atm^{-1}). However, Hagedorn and colleagues [28] are unaware if this twofold change in L_p was a result of an increase in water permeability or an artefact due to morphological complexity of the embryo stage. Water was also found to be unequally distributed within the embryo, for instance, in 6-somite embryos, the yolk contained 61% of the total volume and 42% water while the blastoderm contained a total volume of 39% and 82% water [28]. Further complicating permeability was the measurement of the osmotically inactive volume (V_b) fraction (volume not participating in osmosis), which was found to be 55% for the entire 6-somite stage embryos [28]. The small L_p values indicate that zebrafish have low membrane permeabilities (compare to Drosophila L_p = 0.722 µm•min^{-1}•atm^{-1} [50]) at later developmental stages. The permeability of water and cryoprotectants of one-cell and 6-somite stage zebrafish embryos were also investigated by Zhang et al [34]. Using video microscopy and transport equations, the solute permeability (P_s) of 2.0 M MeOH to both developmental stages was evaluated. P_s was found to be significantly higher for one-cell stage embryos, 0.45

± 0.19 µm/s, than 6-somite embryos, 0.04 ± 0.018 µm/s [34]. Zhang measured higher L_p and V_b values than Hagedorn for 6-somite embryos, 0.35 µm•min^{-1}•atm^{-1} and 82.6% respectively. These discrepancies likely arise from differences in the volume and weight measurements. In a study by Susuki et al. [47], the authors used high-performance liquid chromatography (HPLC) for determining intraembryonic DMSO cryoprotectant concentrations in rainbow trout, pejerrey, medaka and carp. HPLC measurements in rainbow trout and medaka suspended in 4.0 M MeOH for 2 hrs reached a maximum permeating concentration of 1.25 M for the former and 3.0 M for the latter [47]. Only 1.0 and 1.5% permeation was measured in pejerrey and carp exposed to 3.0 M MeOH for 2 hrs [47]. These relatively low permeating concentrations, especially for rainbow trout, indicate that these embryos have similar cryoprotectant permeabilities to that of the zebrafish. Cabrita et al. [45] also used HPLC to estimate the DMSO concentration in the yolk, blastoderm and perivitelline space in turbot embryos. Cabrita and colleagues [45] found that DMSO had a low permeability to the yolk and blastoderm, but could be readily permeate into the perivitelline space. Hagedorn and Zhang hypothesized on the specific embryonic barriers in the zebrafish that might hinder cryoprotectant permeability. Hagedorn et al. [51] used magnetic resonance (MR) microscopy and spectroscopy to determine the kinetics of cryoprotectant permeation in multiple compartments of 3 to 6-somite stage zebrafish embryos. Using chemical-shift (CS) imaging, these authors were able to generate a map of cryoprotectant distribution within the embryos [51]. The permeability of three cryoprotectants were studied, 2.0 M DMSO, 2.0 M PG and 2.25 M MeOH. From CS imaging of the embryos in each solution, DMSO and PG images appeared dark within the yolk, indicating an absence of cryoprotectants, for over 2 hrs, whereas MeOH maps appeared relatively bright within 15 min, indicating permeation of this cryoprotectant throughout the embryo [51]. To determine the compartmental barrier hindering cryoprotectant permeation, Hagedorn and colleagues [29, 51] microinjected DMSO and PG into the yolk of 6-somite stage zebrafish embryos. It was found that the cryoprotectants could not diffuse outside the yolk, but could diffuse freely within the yolk, indicating that this compartment was not a cryoprotectant barrier [29]. Through MR spectroscopy the blastoderm was found not to be a cryoprotectant barrier as well. This lead Hagedorn et al. [29] to suggest that the yolk syncytial layer (YSL) (compartment between the yolk and blastoderm) was the primary barrier blocking cryoprotectant entry into the yolk cell. In addition to the YSL, Zhang and colleagues [34] suggested that the yolk mass may also represent a barrier that could prevent cryoprotectant penetration. Both hypotheses are currently being actively pursued.

The combination of chilling sensitivity and permeability barrier has been the main difficulties hindering successful fish embryo cryopreservation. While the chilling sensitivity can be reduced using later stage embryos, the development of the YSL during development (in the case of the zebrafish beginning at mid blastula) has been implicated as a permeability barrier to cryoprotectants. Liu et al. [52] reported that yolk-reduced embryos could improve embryo chilling sensitivity in zebrafish. Partial removal of the yolk in prim-6 and high-pec were less sensitive to chilling injury at 0 °C than control embryos [35]. Since the yolk provides nutrients to the developing blastomeres, partial removal of the yolk may be detrimental to embryo development. Lower embryo survival was observed in yolk-reduced embryos for all developmental stages studied [52]. As described by Harvey et al. [40], dechorionating embryos does improve the permeability of cryoprotectants, however, protease digestion of the chorion may not be suitable in vitrification and slow-cooling studies, as dechorionated embryos are very sensitive making handling tedious. Several reports have addressed methods for overcoming the membrane permeability problems in fish embryos. In an attempt to increase both the permeability of water and cryoprotectants, Hagedorn et al. [30] microinjected early cleavage (1 to 4-cell) stage zebrafish embryos with mRNA encoding the aquaporin-3 (AQP3) water channel protein. These embryos were subsequently cultured to 50% epiboly. L_p and P_s measurements revealed that the AQP3 treated embryos had a hydraulic conductivity and solute permeability that was 6 and 2.5 times greater than those values for control embryos (cryoprotectant used was 2.0 M ^{14}C PG) [30]. However, while this approach to solving the permeability barrier was interesting, the addition of AQP3 in the zebrafish membrane did not improve the embryo freezing response, with control embryos behaving similar to modified embryos [49]. As an alternative approach, Janik et al. [53] microinjected PG and DMSO/amide into the yolk of 100% epiboly zebrafish embryos. To determine if direct injection into the yolk confers protection during vitrification and dehydration, Janik and colleagues [53] microinjected 30 nl of PG (final concentration 5.9 M) into the yolk of 100% epiboly embryos. Post-vitrification and fixing, electron micrographs revealed damaged mitochondria, the disorganization of ribosomes and the disruption of the YSL [53]. Similar results were observed in control embryos. These findings indicate that bypassing the permeability barrier by direct injection into the yolk was not sufficient enough to obtain normal embryo morphology and high survival post-vitrification. Kohli et al. [54] recently demonstrated the application of femtosecond (fs) laser pulses as method for introducing cryprotective carbohydrates (i.e. sucrose) into mammalian cells. When fs laser pulses were focused onto the plasma membrane, transient pores (reversible permeabilization) were formed exposing the extracellular solution to the intracellular environment. The kinetics of the pores, as measured through volumetric responses, was found to seal within less than 270 ms [54]. Theoretical estimates of the delivered intracellular concentration were estimated to be 72.3 and 66% for 0.2, 0.3 and 0.5 M sucrose to a diffusion length of 10 μm [54]. For the highest delivered concentration (0.2 M; 72.3% loading efficiency) cell survival was found to be 91.5 ± 8% (as determined through membrane integrity assays), indicating that the application of fs laser pulses was non-invasive [54]. Kohli and colleagues [55] further demonstrated the application of fs laser pulses by introducing a fluorescence reporter molecule, streptavidin-conjugated quantum dots or DNA into the blastomere cells of developing zebrafish embryos. Transient pores were formed in the embryonic cells of chorionated and dechorionated embryos; where in the former, pore formation in the blastomeres occurred without damage to the chorion [55]. Using this method of delivery, perivitelline captured reporter molecules were introduced into the blastomere cells of chorionated embryos without disrupting this layer. As reported by the authors [38], fs laser pulses may benefit the field of cryobiology, using this technique as a method for introducing cryoprotectants. Perivitelline captured cryoprotectants could be introduced directly into the blastomeres, or the solute permeability of blastomere-permeable cryoprotectants could be increased by transient pore formation in the cells [55].

In addition to the above techniques, cryobiologists have explored the use of antifreeze proteins (AFP) as a cryprotectant to reduce intraembryonic ice formation and improve chilling sensitivity. AFPs function by binding to ice crystals preventing ice growth and recrystallization. The proteins are known to lower the freezing temperature and potentially protect cellular membranes from damage due chilling [56-58]. Several types of AFP exist

[59] (AFP I, II, III; there is also a sub-class of antifreeze proteins called antifreeze glycoproteins (AFGP)), however, AFP I from the winter flounder has received particular attention in cryopreservation studies [60-64]. Robles et al. [42] found that when 11-day-old winter flounder embryos were vitrified and thawed, a small percentage continued to survive with intact morphology and continued pigmentation for 6 days [42]. Robles and colleagues hypothesized that survival likely arose from the presence of AFPs in the winter flounder [42]. Encouraged by these results, Robles et al. [65-66] microinjected AFP I and III in different embryonic compartments of turbot and seabream embryos. These authors found that the presence of AFP I improved the chilling sensitivity of seabream embryos microinjected at 2-cell stage with a hatching rate approaching 100% after chilling exposure.

Despite over 50 years of research, the cryopreservation of teleost embryos has still been unsuccessful. As an alternative approach to embryo cryopreservation, cryobiologists have considered other options for fish genome cryobanking, such as the storage of somatic fish cells suggested by Mauger et al. [67]. Cryopreservation of primordial germ cells (PGCs) and blastomeres have already been achieved [68, 69]. Recently, Kobayashi et al. [68] reported that cryopreserved/tawed PGCs could be transplanted into the peritoneal cavities of allogenic trout hatchlings and differentiate into mature spermatozoa and eggs in the recipient gonads. These authors have also reported that the fertilization of eggs derived from cryopreserved PGCs by cryopreserved spermatozoa resulted in the development of fertile F1 fish. [68]. This PGC cryopreservation technique represents an important tool for Aquaculture Industry but also for re-population of fish species and wildlife conservation.

REFERENCES

[1] Suquet, M., Dreanno, C., Petton, B., Normant, Y., Omnes, M.H., and Billard, R. (1998). Long-term effects of the cryopreservation of turbot (*Psetta maxima*) spermatozoa, *Aquat. Living Resour.* 11, 45- 48.

[2] Chereguini, O., Garcia de la Banda, I., Herrera, M., Martinez, C., and De la Hera, M. (2003). Cryopreservation of turbot *Scophthalmus maximus* (L.) sperm: fertilization and hatching rates, *Aquac. Res.* 34, 739-747.

[3] Cabrita, E., Robles, V., Cuñado, S., Wallace, J.C., Sarasquete, C., Herráez, M.P. (2006). Evaluation of gilthead seabream, S*parus aurata*, sperm quality after cryopreservation in 5 ml macrotubes. *Cryobiology* 50, 273-284.

[4] Fabbrocini, A., Lavadera, L., Rispoli, S., and Sansone, G., (2000). Cryopreservation of sea bream (*Sparus aurata*) spermatozoa. *Cryobiology* 40, 46-53.

[5] Fauvel, C., Suquet, M., Dreanno, C., Zonno, V., and Menu, B. (1998). Cryopreservation of sea bass (*Dicentrarchus labrax*) spermatozoa in experimental and production simulating conditions. *Aquatic Living Resources* 11, 387-394.

[6] Sansone, G., Fabbrocini, A., Ieropoli, S., Langellotti, A., Occidente, M., and Matassino, D. (2002). Effects of extender composition, cooling rate, and freezing on the motility of sea bass (*Dicentrarchus labrax*, L.) spermatozoa after thawing, *Cryobiology* 44, 229-239.

[7] Asturiano, J.F., Pérez, L., Marco-Jiménez, F., Olivares, L., Vicente, J.S., and Jover, M. (2003). Media and methods for the cryopreservation of European eel (*Anguilla anguilla*) sperm. *Fish Physiol. Biochem.* 28, 501-502.

[8] Tanaka, S., Zhang, H., Horie, N., Yamada, Y., Okamura, A., Utoh, T., Mikawa, N., Oka, H.P., and Kurokura, H. (2002). Long-term cryopreservation of sperm of Japanese eel. *J. Fish Biol.* 60, 139-146.

[9] He, S. and Woods III, L.C. (2003). Effects of glycine and alanine on short-term storage and cryopreservation of striped bass (*Morone saxatilis*) spermatozoa, *Cryobiology* 46, 17-25.

[10] Taddei, A.R., Barbato, F., Abelli, L., Canese, S., Moretti, F., Rana, K.J., Fausto, A.M., and Mazzini, M. (2001). Is cryopreservation a homogeneous process? Ultrastructure and motility of untreated, prefreezing and postthawed spermatozoa of *Diplodus puntazzo* (Cetti), *Cryobiology* 42, 244-255.

[11] Babiak, I., Ottesen, O., Rudolfsen, G.,and Johnsen, S.(2006). Chilled storage of semen from Halibut, Hippoglossus hippoglossus: Optimizing the protocol. *Theriogenology* 66, 2025-2035.

[12] Rideout, R.M., Litvak, M.K., and Trippel, E.A. (2003). The development of a sperm cryopreservation protocol for winter flounder *Pseudopleuronectes americanus* (Walbaum): evaluation of cryoprotectants and diluents, *Aquac. Res.* 34, 653-659.

[13] Cabrita, E., Engrola, S., Conceição, L.E.C., Pousão-Ferreira, P., Dinis, M.T. Successful cryopreservation of sex-reversed sperm from *Epinephelus marginatus*. *Aquaculture (in press)*

[14] Gwo, J-C. (1993). Cryopreservation of black grouper (*Epinephelus malabaricus*). *Theriogenology* 39, 1331-1342.

[15] Rideout, R.M., Trippel, E.A., and Litvak, M.K. (2004). The development of haddock and Atlantic cod sperm cryopreservation techniques and the effect of sperm age on cryopreservation success, *J. Fish Biol.* 65, 299-311.

[16] Tiersch, T.R. (2000). Introduction. In: *Cryopreservation in Aquatic Species*. Tiersch, T.R. and Mazik, P.M.(Eds). World aquaculture Society, Baton Rouge, Louisiana, XIX-XXVI.

[17] Watson, P.F. and Morris, G.J. (1987). Cold shock injury in animal cells. *Symp. Soc. Exp. Biol.* 41, 311-40.

[18] Sar, A.M.v.d., Appelmelk, B.J., Vandenbroucke-Grauls, C.M.J.E., and Bitter, W. (2004). A star with stripes: zebrafish as an infection model. *Trends Microbiol* 12, 451-457.

[19] Barut, B.A. and Zon, L.I. (2000). Realizing the potential of zebrafish as a model for human disease. Physiol. *Genomics* 2, 49-51.

[20] Warren, K.S., Wu, J.C., Pinet, F., and Fishman, M.C. (2000). The genetic basis of cardiac function: dissection by zebrafish (*Danio rerio*) screens. *Phil. Trans. R. Soc. Lond. B.* 355, 939-944.

[21] Dooley, K. and Zon, L.I. (2000). Zebrafish: a model system for the study of human disease. *COGD* 10, 252-256.

[22] Hill, A.J., Teraoka, H., Heideman, W., and Peterson, R.E. (2005). REVIEW: Zebrafish as a Model Vetebrate for Investigating Chemical Toxicity. *Toxicol. Sci.* 86, 6-19.

[23] Jagadeeswaran, P., and Sheenhan, J.P. (1999). Analysis of Blood Coagulation in the Zebrafish. *Blood Cells Mol. Dis.* 25, 239-249.

[24] Nasevicius, A., and Ekker, S.C. (2000). Effective targeted gene 'knockdown' in zebrafish. *Nat. Genet.* 26, 216-220.

[25] Thisse, C., and Zon, L.I. (2002). Organogenesis-Heart and Blood Formation from the Zebrafish Point of View. *Science* 295, 457-462.

[26] Kasai, M. (1996). Simple and efficient methods for vitrification of mammalian embryos. *Anim.Reprod. Sci.* 42, 67-75.

[27] Zhang, T. and Rawson, D.M. (1995). Studies on Chilling Sensitivity of Zebrafish (*Brachydanio rerio*) Embryos. *Cryobiology* 32, 239-246.

[28] Hagedorn, M., Kleinhans, F.W., Freitas, R., Liu, J., Hsu, E.W., Wildt, D.E., and Rall, W.F. (1997). Water Distribution and Permeability of Zebrafish Embryos, *Brachydanio rerio. J. Exp. Zool.* 278, 356-371.

[29] Hagedorn, M., Hsu, E., Kleinhans, F.W., and Wildt, D.E. (1997). New Approaches for Studying the Permeability of Fish Embryos: Toward Successful Cryopreservation. *Cryobiology* 34, 335-347.

[30] Hagedorn, M., Lance, S.L., Fonseca, D.M., Kleinhans, F.W., Artimov, D., Fleischer, R., Hoque, A.T.M.S., Hamilton, M.B., and Pukazhenthi, B.S. (2002). Altering Fish Embryos with Aquaporin-3: An Essential Step Toward Successful Cryopreservation. *Biol. Reprod.* 67, 961-966.

[31] Harvey, B., and Chamberlain, J.B. (1982). Water Permeability In The Developing Embryo Of the Zebrafish, *Brachydanion Reiro. Can. J. Zoolog.* 60, 268-270.

[32] Zhang, T., Liu, X.-H., and Rawson, D.M. (2003). Effects of methanol and developmental arrest on chilling injury in zebrafish (*Danio rerio*) embryos. *Theriogenology* 59, 1545-1556.

[33] Hagedorn, M., Kleinhans, F.W., Wildt, D.E., and Rall, W.F. (1997). Chill Sensitivity and Cryoprotectant Permeability of Dechorionated Zebrafish Embryos, *Brachydanio rerio. Cryobiology* 34, 251-263.

[34] Zhang, T., and Rawson, D.M. (1998). Permeability of Dechorionated One-Cell and Six-Somite Zebrafish (*Brachydanio rerio*) Embryos to Water and Methanol. *Cryobiology* 37, 13-21.

[35] Diller, K.R. (1975). Intracellular Freezing: Effect of Extracellular Supercooling. *Cryobiology* 12, 480-485.

[36] Fahy, G.M., MacFarlane, D.R., Angell, C.A., and Meryman, H.T. (1984). Vitrification as an Approach to Cryopreservation. *Cryobiology* 21, 407-426.

[37] Mazur, P. (1984). Freezing of living cells: mechanisms and implications. *Am J Physiol Cell Physiol* 247, C125-C141.

[38] Hubalek, Z. (2003). Protectants used in the cryopreservation of microorganisms. *Cryobiology* 46, 205-229.

[39] Rall, W.F., Mazur, P., and McGrath, J.J. (1983). Depression of the ice-nucleation temperature of rapidly cooled mouse embryos by glycerol and dimethyl sulfoxide. *Biophys. J.* 41, 1-12.

[40] Harvey, B., Kelley, R.N., and Ashwood-Smith, M.J. (1983). Permeability of Intact and Dechorionated Zebra Fish Embryos to Glycerol and Dimethyl Sulfoxide. *Cryobiology* 20, 432-439.

[41] Liu, X.-H., Zhang, T., and Rawson, D.M. (1998). Feasibility of vitrification of zebrafish (*Danio rerio*) embryos using methanol. *Cryo Letters* 19, 309-318.

[42] Robles, V., Cabrita, E., Fletcher, G.L., Shears, M.A., King, M.J., and Herráez, M.P. (2005). Vitrification assays with embryos from a cold tolerant sub-arctic fish species

The first report of fish embryo survival following vitrification. *Theriogenology* 64, 1633-1646.

[43] Robles, V., Cabrita, E., Paz, P.d., Cunado, S., Anel, L., and Herráez, M.P. (2004). Effect of a vitrification protocol on the lactate dehydrogenase and glucose-6-phosphate dehydrogenase activities and the hatching rates of Zebrafish (*Danio rerio*) and Turbot (*Scophthalmus maximus*) embryos. *Theriogenology* 61, 1367-1379.

[44] Robles, V., Cabrita, E., Real, M., Alvarez, R., and Herráez, M.P. (2003). Vitrification of turbot embryos: preliminary assays. *Cryobiology* 47, 30-39.

[45] Cabrita, E., Robles, V., Chereguini, O., Paz, P.d., Anel, L., and Herráez, M.P. (2003). *Dimethly sulfoxide influx in turbot embryos exposed to a vitrification protocol. Theriogenology* 60, 463-473.

[46] Cabrita, E., Robles, V., Chereguini, O., Wallace, J.C., and Herráez, M.P. (2003). Effect of different cryoprotectants and vitrificant solution on the hatching rate of turbot embryos (*Scophthalmus maximus*). *Cryobiology* 47, 204-213.

[47] Suzuki, T., Komada, H., Takai, R., Arii, K., and Kozima, T.T. (1995). Relation between Toxicity of Cryoprotectant DMSO and Its Concentration in Several Fish Embryos. *Fish. Sci.* 61, 193-197.

[48] Zhang, T., and Rawson, D.M. (1996). Feasibility Studies on Vitrification of Intact Zebrafish (*Brachydanio rerio*) Embryos. *Cryobiology* 33, 1-13.

[49] Hagedorn, M., Perterson, A., Mazur, P., and Kleinhans, F.W. (2004). High ice nucleation temperature of zebrafish embryos: slow-freezing is not an option. *Cryobiology* 49, 181-189.

[50] Lin, T.-T., Pitt, R.E., and Steponkus, P.L. (1989). Osmometric Behavior of *Drosophila melanogaster* Embryos. *Cryobiology* 26, 453-471.

[51] Hagedorn, M., Hsu, E.W., Pilatus, U., Wildt, D.E., Rali, W.F., and Blackband, S.J. (1996). Magnetic resonance microscopy and spectroscopy reveal kinetics of cryoprotectant permeation in a multicompartmental biological system. *Proc. Natl. Acad. Sci USA* 93, 7454-7459.

[52] Liu, X.-H., Zhang, T., and Rawson, D.M. (1999). The Effect of Partial Removal of Yolk on the Chilling Sensitivity of Zebrafish (*Danio Rerio)* Embryos. *Cryobiology* 39, 236-242.

[53] Janik, M., Kleinhans, F.W., and Hagedorn, M. (2000). Overcoming a Permeability Barrier by Microinjecting Cryoprotectants into Zebrafish Embryos (*Brachydanio rerio*). *Cryobiology* 41, 25-34.

[54] Kohli, V., Acker, J.P., and Elezzabi, A.Y. (2005). Reversible permeabilization using high-intensity femtosecond laser pulses: Applications to biopreservation. *Biotechnol. Bioeng.* 92, 889-899.

[55] Kohli, V., Robles, V., Cancela, M.L., Acker, J.P., Waskiewicz, A.J., and Elezzabi, A.Y. (2007). An alternative method for delivering exogenous material into developing zebrafish embryos. *Biotechnology and Bioengineering* 98, 1230-1241.

[56] Fletcher, G.L., Hew, C.L., and Davies, P.L. (2001). Antifreeze proteins of teleost fishes. *Annu. Rev. Physiol.* 63, 359-390.

[57] Tomczak, M.M., Vigh, L., Meyer, J.D., Manning, M.C., Hincha, D.K., and Crowe, J.H. (2002). Lipid unsaturation determines the interaction of AFP type I with model membranes during thermotropic phase transitions. *Cryobiology* 45, 135-142.

[58] Wang, J.H. (2000). A comprehensive evaluation of the effects and mechanisms of antifreeze proteins during low-temperature preservation. *Cryobiology* 41, 1-9.

[59] Yeh, Y., and Feeney, R.E. (1996). Antifreeze Proteins: Structures and Mechanism of Function. *Chemical Reviews* 96.

[60] Shaw, J.M., C.Ward, and Trounson, A.O. (1995). Evaluation of propanediol, ethylene glycol, sucrose and antifreeze proteins on the survival of slow-cooled mouse pronuclear and 4-cell embryos. *Human Reproduction* 10, 396-402.

[61] Carpenter, J.F. and Hansen, T.N. (1992). Antifreeze protein modulates cell survival during cryopreservation: Mediation through influence on ice crystal growth. *Proc. Natl. Acad. Sci USA* 89, 8953-8957.

[62] Chao, H., Davies, P.L., and Carpenter, J.F. (1996). Effects of antifreeze proteins on red blood cell survival during cryopreservation. *The Journal of Experimental Biology* 199, 2071-2076.

[63] Tomczak, M.M., Hincha, D.K., Estrada, S.D., Wolkers, W.F., Crowe, L.M., Feeney, R.E., Tablin, F., and Crowe, J.H. (2002). A mechanism for stabilization of membranes at low temperatures by an antifreeze protein. *Biophysical J.* 82, 874-881.

[64] Arav, A., Rubinsky, B., Fletcher, G., and Seren, E. (1993). Cryogenic protection of oocytes with antifreeze proteins. *Molecular reproduction and development* 36, 488-493.

[65] Robles, V., Barbosa, V., Herraez, M.P., Martinez-Paramo, S., and Cancela, M.L. (2007). The antifreeze protein type I (AFPI) increases seabream (*Sparus aurata*) embryos tolerance to low temperatures. *Theriogenology* 68, 284-289.

[66] Robles, V., Cabrita, E., Anel, L., and Herraez, M.P. (2006). Microinjection of the antifreeze protein type III (AFPIII) in turbot embryos: toxicity and protein distriubtion. *Aquaculture* 261, 1299-1306.

[67] Mauger, P.-E., Le, P.-Y.B., and Labbe, C. (2006). Cryobanking of fish somatic cells: Optimizations of fin explant culture and fin cell cryopreservation. *Comp. Biochem. Physiol B Biochem Mol. Biol.* 144, 29-37.

[68] Kobayashi, T., Takeuchi, Y., and Takeuchi, T. Yoshizaki, G., (2007). Generation of viable fish from cryopreserved primordial germ cclls. *Mol. Reprod. Dev.* 74, 207-213.

[69] Strüssmann, C.A., Nakatsugawa, H., Takashima, F., Hasobe, M., Suzuki, T. and Takai, R. (1999). Cryopreservation of isolated fish blastomeres: Effects of Cell Stage, Cryoprotectant Concentration, and cooling rate on postthawing survival. *Cryobiology* 39, 252-261.

Chapter 38

ECOSYSTEM FUNCTIONING OF TEMPORARILY OPEN/CLOSED ESTUARIES IN SOUTH AFRICA

R. Perissinotto[1], D. D. Stretch[2], A. K. Whitfield[3], J. B. Adams[4], A. T. Forbes[1,5], and N. T. Demetriades[5]

[1] School of Biological & Conservation Sciences, University of KwaZulu-Natal, Westville Campus, P Bag X54001, Durban 4000, South Africa
[2] Centre for Research on Environmental, Coastal & Hydrological Engineering (CRECHE), University of KwaZulu-Natal, Howard College Campus, Durban 4041, South Africa
[3] South African Institute for Aquatic Biodiversity (SAIAB), P Bag 1015, Grahamstown 6140, South Africa
[4] Department of Botany, Nelson Mandela Metropolitan University, PO Box 77000, Port Elizabeth 6031, South Africa
[5] Marine & Estuarine Research, PO Box 417, Hyper by the Sea 4053, Durban, South Africa

ABSTRACT

Apart from representing the vast majority (71%) of South Africa's 258 functional estuaries, temporarily open/closed estuaries (TOCEs) are common in Australia, on the southeastern coasts of Brazil and Uruguay, the southwestern coasts of India and Sri Lanka, but are poorly represented in North America, Europe and much of Asia. The regular change between open and closed mouth phases makes their physico-chemical dynamics more variable and complicated than that of permanently open estuaries. Mouth states are driven mainly by interplay between wave or tide driven sediment transport and river inflow. Mouth closure cuts off tidal exchanges with the ocean, resulting in prolonged periods of lagoonal conditions during which salinity and temperature stratification may develop, along with oxygen and nutrient depletion. Mouth breaching occurs when water levels overtop the frontal berm, usually during high river flow, and may be accompanied by scouring of estuarine sediment and an increased silt load and turbidity during the outflow phase. Microalgae are key primary producers in TOCEs, and while phytoplankton biomass in these systems is usually lower than in permanently open estuaries, microphytobenthic biomass is often much higher in TOCEs than in permanently open systems. During the closed phase, the absence of tidal currents, clearer water and greater light penetration can result in the proliferation of submerged macrophytes. Loss of tidal action and high water levels, however, also result in the absence or disappearance of mangroves and have adverse effects on salt marsh vegetation. Zooplankton are primary consumers both in the water-column and within the upper sediment, due to diel migrations. A prolonged period of TOCE mouth closure leads to poor levels of zooplankton diversity, but also to the biomass build-up of a few dominant species. Benthic meiofaunal abundance is usually greater during closed phases and is generally dominated by nematodes. Macrobenthic densities, and occasionally even biomass, in TOCEs are higher than in permanently open systems. The dominance of estuarine and estuarine-dependent marine fish species in TOCEs is an indication of the important nursery function of these systems. Marine juvenile fish recruit into TOCEs not only when the mouth opens, but also during marine overwash events when waves from the sea wash over the sand bar at the mouth. The birds that occur in TOCEs are mostly piscivorous, able to catch a variety of fish species either from the surface or by diving underwater. Waders are absent or uncommon because of the infrequent availability of intertidal feeding areas when the mouth is closed. Addressing the challenges facing the sustainable management of TOCEs is critical, as in some cases their ecological integrity, biodiversity and nursery function have already been compromised.

INTRODUCTION

The South African coast has 258 functional estuaries and 182 of these (71%) are currently classified as temporarily open/closed systems, or TOCEs (Whitfield 2000). These are essentially the same as the "wet and dry tropical/subtropical"Australian estuaries, as defined by Eyre (1998) and more recently renamed "Intermittently Closed and Open Lakes and Lagoons (ICOLLs)" (Roy et al. 2001). Similar estuaries are found on the southeastern coasts of Brazil and Uruguay (Bonilla et al. 2005) and the southwestern coasts of India and Sri Lanka (Ranasinghe & Pattiaratchi 2003), where they are often referred to as "seasonally open tidal inlets/coastal lagoons". There are also a few seasonally open systems along the USA south and west coast, particularly in California and Texas, and occasionally as far north as Long Island, New York (Gobler et al. 2005, Kraus et al., 2008). These estuaries are very different from most northern hemisphere systems, particularly those on the European and North American coasts, where a steady and often strong freshwater inflow ensures a virtual permanent connection with the sea.

The vast majority of South African TOCEs (178) are situated along the eastern seaboard (Indian Ocean) (Figure 1). Conversely, the west or Atlantic coast exhibits only 10 functional estuarine systems, 5 of which are TOCEs, although Heydorn & Tinley (1980) identified a number of other small non-estuarine systems on this coast. Many of these systems comprise dry riverbeds that only carry water at times of exceptional rainfall (Heydorn 1991).

In terms of biogeography, South African estuaries can be subdivided into three main climatic regions: (1) cool temperate from the Orange (Gariep) River to the Cape Peninsula; (2) warm temperate from False Bay to the Mbashe River; and (3) subtropical from the Mbashe to Kosi Bay. There are thus only 5 TOCEs in the cool temperate region, while the warm temperate and the subtropical regions exhibit 84 and 93 such systems, respectively (Whitfield 2000) (Figure 1).

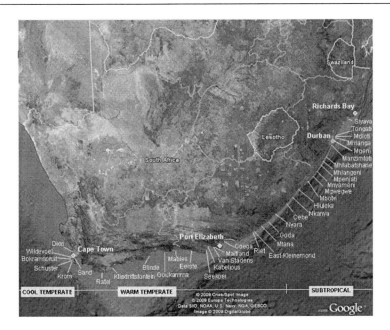

Figure 1. Map of the South African coastline, highlighting the 3 different ocean climatic regions and indicating the position of some of the most prominent among its 182 temporarily open/closed estuaries (background imagery: Google Earth™).

A key feature of most TOCEs is their small river catchments (Whitfield 1992). For example, in the subtropical region about two-thirds of TOCEs have catchments smaller than 100 km^2. As a result, during the dry season and under low river flow conditions they are closed off from the sea by a sandbar (beach berm), which forms at the mouth. Flow rates during this period are generally low, well below the mean annual runoff (MAR) (Huizinga & van Niekerk 2002). Following periods of high rainfall and freshwater runoff, the water level inside the estuary may rise until it equals or exceeds the height of the sandbar at the mouth (Whitfield 1992, Wooldridge & McGwynne 1996). Once an outflow channel has formed, a rapid drop in the water level will occur, often exposing large areas of substratum that may have been submerged for long periods and colonised by a rich community of plants and animals. River conditions may briefly dominate the estuary during and immediately after breaching events, when floodwaters leave the system. However, when the river freshwater inflow decreases, a typical estuarine open phase can occur, with regular tidal exchange and seawater penetration into the lower, middle and sometimes upper reaches. The tidal prism often remains small relative to the estuary's closed phase volume (Whitfield 1992). The open phase ends when the sandbar at the mouth is regenerated by along-shore, cross-shore and on-shore sand movement in the surf-zone, usually within days to weeks. This leads to another closed phase during which the only seawater inflow is provided by berm wash-over at the peak of the spring tide or during storm surges. Depending on the climatic conditions and rainfall patterns in the catchment area, closure periods may vary naturally from days to months or even years. This extremely dynamic situation has important implications for the full range of physical and chemical parameters of TOCEs and ultimately for their biological structure and ecological functioning.

1. Hydrodynamics & Sedimentology

The term "hydrodynamics" is used here in a broad sense to refer to all processes concerning the availability and movement of water through these systems – for example catchment hydrology is included since it is central to characterizing the inflow to estuaries. Hydrodynamics, in this broad sense, is the primary driver of all estuarine functions.

A key distinguishing feature of TOCEs is their intermittent link to the sea. Mouth state has a profound influence on the physico-chemical state of these systems, i.e. water level (and depth), flow velocity, turbidity, salinity, temperature, residence time, etc., that can have major impacts on their biological structure and function. The mouth dynamics of TOCEs and how it may be affected by changes, such as climate variability, land-use development, and other human or natural factors, is therefore key to understanding them.

In South Africa there are many TOCEs, particularly in the sub-tropical biogeographic region, that are "perched". The term "perched" is used here to refer to estuaries where the minimum water level (which usually ocurrs in open mouth conditions) is elevated above mean sea level (MSL). This also implies that average bed levels in the mouth region are at or above MSL. This is not a precise concept, however, since both bed levels and minimum water levels can vary considerably, e.g. due to scouring by episodic large floods.

There is a paucity of published quantitative hydrodynamic information on South African TOCEs. This is in sharp contrast to the much more extensive information available on the biological structure and functioning of these systems. Even basic information concerning mouth status (open/closed) is not available for the majority of the TOCEs in the country. Furthermore, streamflow data for small catchments is also mostly unavailable or of low confidence, since they must be inferred by segregating the information available for larger catchments.

1.1. Catchment Processes

South Africa is a semi-arid country - the mean annual precipitation (MAP) averaged over the whole country is about 500 mm, which is significantly below the global average of about 800 mm. The distribution of rainfall is uneven and ranges from 700-1000 mm along the eastern subtropical and warm temperate southern coastal areas, to less than 200 mm along the northern parts of the cool temperate western coast (Figures 1 & 2). The seasonality of rainfall also varies across the country: the southern coast has a mediterranean climate with winter rainfall, while the north-eastern coast is sub-tropical with summer rainfall. The central coastline has a warm temperate climate with rainfall all year around.

An elevation map for South Africa (Figure 3) provides an overview of the general catchment features on the sub-continent. Note the narrow coastal plain stretching up the east coast. The estuaries on the eastern coast are situated on a narrow coastal plain and therefore typically have relatively small steep catchments. For example, in the sub-tropical region about 70% have catchments less than 100 km^2. There is a broadening of the coastal plain in the sub-tropical north-east and temperate south-west, where catchments can be much larger. For instance, the Mfolozi River in the north-east has a catchment area of about 10000 km^2, the Sundays River in the south-west has a catchment area of 22000 km^2, while the Goukou catchment is 1550 km^2.

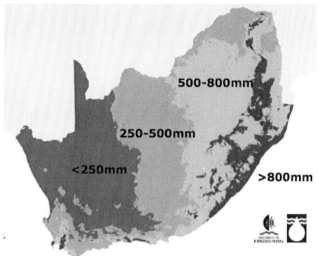

Figure 2. Distribution of mean annual precipitation (MAP) in South Africa (source: Schulze *et al.* 1996).

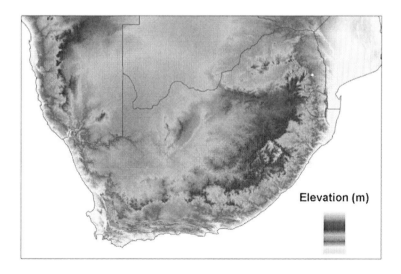

Figure 3. Elevation map of South Africa illustrating the overall features of catchments (derived from SRTM satellite data, e.g. Jarvis *et al.* 2008).

The physical characteristics of catchments, including area, average slope, and shape can have a significant effect on river runoff, both in terms of quantity and in terms of temporal distribution. These features impact the hydrodynamics of the estuaries that they supply. For example, a small, steep catchment responds quickly to rainfall events such that the streamflow increases rapidly, but is not sustained for long after the rainfall ends, i.e. runoff hydrographs tend to have high peaks but short durations, and sustained baseflows may be small. TOCEs with small catchments usually breach frequently following rainfall events, but close again within a short period (days to weeks) because sustained baseflows are important for maintaining open mouth conditions.

Land cover and/or vegetation in a catchment also has an effect on run-off characteristics. Densely vegetated catchments trap water and release it more slowly than those covered by relatively impervious material. Midgely *et al.* (1994) applied rainfall-runoff modelling to estimating the effects of land-use changes on river flows, which allows simulated natural streamflow to be used as a reference condition for management decisions.

Flow duration curves (FDC) are useful statistical tools for analyzing the characteristics of streamflow into TOCEs. An FDC is a plot of flow magnitudes versus their exceedance probabilities. Simulated monthly flows are available for all "quaternary" level catchments in South Africa (Midgely *et al.* 1994), which typically have areas of at least 500 km^2. Area scaling is often used to infer the FDCs for smaller sub-catchments. However monthly flows do not have a fine enough temporal resolution for application to small TOCEs, where hydrodynamic changes often occur on time-scales of 1-day or less. For example, the time of run-off concentration for a small catchment of 100 km^2 is typically a few hours. This issue has been addressed by Smakhtin (2000) who devised an empirical method to derive daily FDCs from monthly FDCs. Samples of monthly and daily FDCs for catchments on the east coast of South Africa are shown in Figure 4 and were derived using the above-mentioned methods. Note how the daily flow FDC has a steeper slope, particularly for high flows, as is expected when the averaging time reduces.

Smakhtin and Masse (2000) developed a method to derive continuous daily flow series using FDCs in combination with rainfall data. The latter are much more readily obtainable for small catchments than are streamflow data. These daily flow sequences are useful for modelling the water budget for small TOCEs, which can in turn be used for detailed analysis of the hydrodynamic functioning of these systems including the mouth dynamics.

Catchment sediment yields can have important effects on estuaries. TOCEs are typically closed for significant durations and during these periods they act as repositories for the accumulation of sediments carried down by inflowing rivers. Sediment yields are related to climatic factors (e.g. rainfall amount, intensity and duration) and catchment characteristics such as geology, steepness and land cover. Rooseboom (1992) estimated the sediment yields for South African catchments, but there are few detailed and reliable measurements available and actual values may vary both seasonally and/or annually. For example, recent work by Grenfell *et al.* (2009a, 2009b) for the Mfolozi River in the north-east, has highlighted this variability including hysteresis effects due to seasonal changes in land cover.

Human activities can have major impacts on catchment systems. Changes in land use due to agriculture, water abstractions (e.g. for irrigation or other human consumption), river impoundments for water storage and use, and effluent discharges from factories or wastewater treatment facilities, are some of the factors that can change the quantity and quality of water reaching estuaries. These issues and their impacts are discussed further in section 5.

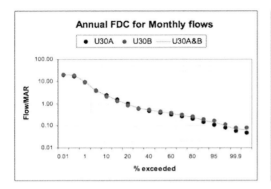 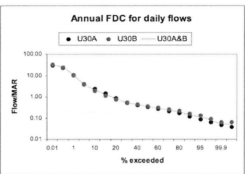

Figure 4. Samples of normalized flow duration curves for monthly flows (left) and daily flows (right). Simulated flows for quaternary catchments U30A & U30B were used to derive the monthly FDC and the method of Smakhtin (2000) was used to derive the daily FDC.

1.2. Coastal Processes

The wave climate of the South African coast is energetic and has been reviewed in detail by Rossouw (1984). Wave characteristics do not vary dramatically along the southern African coastline, but the angle of the coastline in relation to the dominant ocean swell directions varies considerably. This has an effect on long-shore sediment transport. The ocean climate off southern Africa is dominated by eastward tracking, mid-latitude cyclones. These systems generate southerly storm waves and swells that propagate to the South African coastline and can cause major shoreline sediment re-arrangement. South-easterly tracking tropical cyclones from the Madagascar region have also been associated with episodic generation of significant storm waves and long-period swells (10-20 seconds) that have affected the north-eastern coastline. These storm systems and the wave energy they generate can affect TOCEs by causing major reworking of beach sediments that could trigger the closure of an open estuary inlet, or could cause overtopping of the beach berms of already closed systems. On the South African coast, prevailing winds are roughly aligned with the coastline. Wind-driven short-period waves contribute to a persistently energetic wave regime for the whole coast.

Tides affect the functioning of TOCEs through tide-driven exchange flows at open inlets and by their influence on beach characteristics. The South African coastline is microtidal (spring tide range ~ 2 m) and is semi-diurnal (M2 dominant forcing), with a period of 12.4 hrs. There are now over 20 years of data available from some of the major ports around the coast and experience has shown that calibrated harmonic tide models can provide reliable predictions, except in the presence of extreme weather conditions. The height datum for reporting tide data is the Lowest Astronomical Tide (sometimes referred to as Chart Datum). Published tide tables provide the vertical offset between Lowest Astronomical Tide and Mean Sea Level. There are only small variations in tide characteristics along the South African coastline.

Nearshore currents are important to the functioning of TOCEs because of the role they play in sediment dynamics, which in turn has a key role in influencing mouth states.

Longshore currents are generated by the interaction between breaking waves and the coastline. When waves propagate obliquely to a beach there is a mean flux of long-shore momentum (the radiation stress) that provides the driving force for longshore currents. These currents are major factors in the littoral transport of sand along the coastline. Depending on the magnitude and direction of incident wave energy, they can transport large volumes of sand. In South Africa, these currents are generally upcoast (i.e. in a general northerly direction).

Rip currents are part of wave-driven cellular circulation patterns formed within the breaker zone, usually involving interactions with offshore bars. They occur during shore normal wave conditions and can be intensified by interaction with man-made or natural shoreline protrusions. Rip currents typically have high velocities and can contribute to a significant re-arrangement of beach sediments, particularly off-shore transport of sediments. Near-shore currents may also be wind-driven. Along the South African coastline, the prevailing winds typically have strong along-shore components and can therefore generate significant long-shore currents that can (in combination with wave action) cause significant sediment transport.

Tidal currents on the continental shelf are generally small along the South African coastline because of its micro-tidal range. However, tidal currents become very important in tidal inlets where they are strongly concentrated and are responsible for major sediment transport effects. In effect, it is these tide-driven flows that are key to the mouth dynamics of TOCEs.

Sediment dynamics in the littoral zone is a major factor in the mouth dynamics of TOCEs and therefore plays a key role in the overall functioning of these systems. Longshore and cross-shore sediment transport in the littoral zone are driven by wave action and are, therefore, related to both prevailing average wave conditions and to episodic extreme storm conditions. Longshore sediment transport rates along the open South African coastline vary from about 400 000 to 1 200 000 m^3 per annum, increasing from the southern end of the coastline northwards, which is also the direction of the net transport. There are, however, large inter-annual variations (Schoonees 2000). Along rocky shorelines or sheltered areas the net average longshore transport rate may be much lower, mostly between about 10 000 m^3 and 400 000 m^3 per annum (Theron 2004). Cross-shore sediment transport rates are also expected to be highly variable and depend strongly on wave conditions. Stretch & Zietsman (2004) estimated that cross-shore transport rates in average wave conditions were about 15 – 20 m^3 day^{-1} per metre of coastline. These are rough estimates based on the observed rebuilding of a breached sandbar at a small TOCE on the east coast of South Africa. Extreme cross-shore transport rates have been estimated to be as high as 150 m^3 day^{-1} per metre of coastline (Theron *et al.* 2003).

Wave energy, sediment transport rates and beach morphology are interlinked. High energy storm waves tend to erode and flatten beach slopes. In these conditions, there may be a net offshore transport of sediment in the littoral zone, with formation of offshore bars. Lower energy waves reverse this process rebuilding beaches and increasing beach slopes. Beach slopes are also linked to sediment characteristics: coarse-grained sediments are associated with steeper beach slopes and fine-grained sediments with shallower beach slopes. Along the coast of South Africa, beach sediments in the south are generally finer and beach slopes shallower than those in the north. Shallow slope beaches are more effective at dissipating wave energy than steep beaches, which reflect more of the incoming wave energy. Wave run-up is therefore higher for steeper beaches and this in turn leads to higher beach berms. This general trend is evident along the South African coastline and explains the prevalence of perched TOCEs in the north-eastern sector of the coastline. In section 1.1 above, the term "perched" was defined as the superelevation of the minimum water level in the estuary, relative to mean sea level (MSL).

1.3. Estuarine Processes

The size (area, depth, volume), shape and hypsometric (surface area versus elevation) characteristics of estuaries can have significant effects on hydrodynamics. For example, long narrow systems (relative to the tidal wavelength $L = T.(gh)^{1/2}$ where T is the tidal period, h is an average depth, and g is the acceleration due to gravity) respond differently to tidal forcing from those that are short compared to the tidal wavelength. Similarly, shallower systems have greater frictional resistance to flow and tend to be more mixed than deeper systems. Whether an estuary is perched above mean sea level or not is another important factor that influences the magnitude of tide-driven exchange flows when the mouth is open, and also the hydrodynamic response to breaching events. Perched estuaries tend to drain most of their water when they breach (Figure 6), which greatly reduces the available pelagic habitat in these systems. Perched TOCEs are common on the north-eastern coastline of South Africa.

Water, salinity and nutrient budgets can provide a means to interpret observations and gain insight into the overall functioning of an estuarine system. In the case of small TOCEs along the South African coastline, the estimation of these budgets currently relies on sparse data, which makes them subject to large uncertainties. There are several timescales associated with the dynamics of the water budget that can be helpful in understanding and comparing the functioning of different TOCEs. For example:

- residence time, T_R, is an indication of the average time that water spends in an estuary before being flushed out to sea and, in the case of a TOCE, this is largely determined by the duration of the closed mouth periods;
- inflow timescales, T_Q, represent the time for a given river flow to fill the estuary, i.e. $T_Q = V/Q$, where V is the volume of the estuary and Q is the flow rate;
- mouth closure time-scales, T_C, represent the time required for the mouth to re-close after a breaching event.

Estuaries with similar residence times can be expected to have similar functioning, in terms of hydrodynamics and perhaps biological structure. Also, when the flow timescale T_Q are smaller than the mouth closure timescale T_C, then the mouth may be expected to remain open due to the dominance of river flow.

The salinity budget for a TOCE depends critically on the mouth dynamics and the tide driven exchange flows when the mouth is open. The tidal prism (or volume of water exchanged with the ocean during a tidal half-cycle) depends on the tidal range in the sea, the size and shape of the estuary basin, the degree of inlet constriction and other factors. Marine overwash events can also influence the salt budget of TOCEs, as can seepage through the sandbar into the sea.

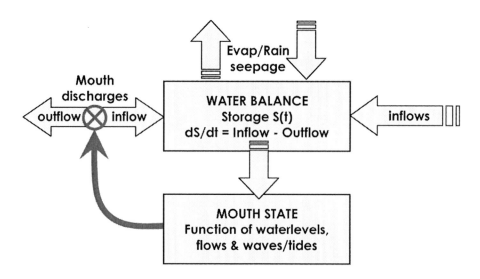

Figure 5. Schematic representation of the water budget for a TOCE.

During open mouth conditions, a longitudinal salinity gradient normally exists, from seawater at the mouth (35 ‰) to freshwater at the head of the estuary. The distribution of the haloclines depends on the tidal regime and the amount of freshwater inflow. In some cases, vertical stratification can develop, usually in systems with deeper water (> 2 m). If freshwater inflow rates are small, they may flow over saline intrusions without mixing, thus maintaining strong vertical density/salinity stratification. Entrainment of freshwater into deeper layers, e.g. due to wind mixing, causes the water column to gradually change into a homogenous brackish water body. Salinity intrusions can indirectly affect sedimentation patterns in TOCEs by promoting flocculation of fine silt and clay particles (Santiago 1991).

Nutrient budgets in estuaries are complex because of the geochemical and biochemical processes involved. Nutrients are transported and can accumulate in TOCEs during closed phases and can be internally recycled, e.g. by sequestration in sediments and subsequent release after bacterial breakdown. Mouth breaching can scour and release large quantities of these accumulated nutrients into the nearshore zone, which may be an important nutrient source for the oligotrophic waters of the South African east coast (van Ballegoyen *et al.* 2008).

Mouth state is perhaps the most important factor influencing the hydrodynamic functioning of TOCEs. When the mouth of a TOCE closes, exchange with the ocean essentially ceases (apart from some occasional wave overtopping events). Residence time in the estuary becomes extended and water level and depth typically increase substantially from the open mouth state. These factors significantly change the physico-chemical environment and, therefore, drive specific biological responses. Once the river flow is sufficient to exceed losses from the system due to seepage and evaporation, the water level will rise and lead to a breaching of the sand berm and restoration of the link with the sea. To understand the overall functioning of TOCEs it is fundamental to understand the factors that control the mouth state of these systems.

In terms of the proportion of time that TOCEs in South Africa are open or closed, it has been observed (Perissinotto *et al.* 2004) that there is a bi-modal distribution – there are groups of estuaries that are either open for less than $1/3$ of the time or for more than $2/3$ of the time. Only a few systems have balanced open/closed regimes. This distribution seems to indicate that some systems only open during episodic extreme events such as river floods, while others only close due to extreme events such as high waves associated with storms.

Two important mechanisms of mouth closure have been suggested by Ranasinghe & Pattiaratchi (1998, 2003) and Ranasinghe *et al.* (1999), which are applicable to high wave energy microtidal coastlines with seasonal runoff discharges. The first mechanism concerns interaction between the inlet currents and longshore sediment transport, which leads to the formation of an updrift shoal and the eventual development of a spit across the inlet entrance, if the inlet current is not strong enough to erode it. The second mechanism involves interaction between the inlet currents and onshore sediment transport during wave conditions that favour beach accretion. Provided the inlet current is weak (i.e. the tidal prism and river flow are small), this process can lead directly to the accumulation of sediment in the mouth and closure of the inlet. In practice, a combination of these two mechanisms may typically apply, with longshore transport providing a supply of sediment and cross-shore transport contributing to depositing this sediment in the inlet. Most TOCEs in South Africa are small and have small tidal prisms. Furthermore, the longshore transport rates are high and given the small size of most TOCE inlets, the limiting process in terms of time to closure is usually cross-shore transport (Stretch & Zietsman 2004).

The breaching of the sand-barriers of TOCEs, when they open, also has important implications for the overall functioning of these systems. Firstly, breaching leads to a sudden change in water level as stored water is flushed from the system. Secondly, significant scouring of sediment can accompany such events. The study by Stretch & Parkinson (2006) and Parkinson & Stretch (2007) has provided important scaling relationships for the breach size and the outflow magnitudes and hydrograph. For example, they have shown that the size of the breach is proportional to the cube-root of the volume that drains out of the breach. They have also provided simple scalings for the peak discharge and duration of the outflow hydrograph that can be used to estimate sediment scouring. For example, the peak discharges for a documented breaching of a small perched estuary on the east coast (Figure 6) typically involves a peak outflow that is similar to that of a 25 year return period flood event. It is clear that these events can contribute significantly to the sediment budget of TOCEs. Case studies of sediment transport during breaching events have also been reported by Beck *et al.* (2004, 2008).

Changes to the flows into estuaries can have a significant impact on mouth dynamics. Increasing pressure on scarce water resources has generally reduced the amount of river flow into estuaries, thereby increasing the proportion of time that TOCEs are closed. However, in some cases, discharges from waste-water treatment works may significantly increase flows (particularly the low flow regime), which can also impact natural breaching patterns. The implications for water quality and the general ecological health of these systems can be severe, although they are not yet fully understood. Case studies have been reported by De Decker (1987), Reddering (1988), Whitfield & Bruton (1989), Whitfield (1993), Whitfield & Wooldridge (1994), Schumann & Pearce (1997), Grange *et al.* (2000), Perissinotto *et al.* (2004) and Lawrie *et al.* (2009). General relationships between flow and estuarine morphology are discussed in reviews by Cooper *et al.* (1999), Schumann *et al.* (1999), Harrison *et al.* (2000) and Cooper (2001).

Figure 6. The breaching of a small perched TOCE, Mhlanga Estuary, on the east coast of South Africa. The estuary essentially empties when it breaches (photo: DD Stretch).

Figure 7. A perched outlet channel of a small TOCE during a partially open state (photo: DD Stretch).

The relationship between flows and the mouth states of TOCEs is particularly relevant to the management of these ecosystems. Recent studies by Perissinotto *et al.* (2004) and Smakhtin (2004) have started to address this issue in detail within the context of South African systems, using more detailed analysis and modelling of the water balance of TOCEs. Gillander *et al.* (2002) reviewed the issue with a focus on Australian systems.

Case studies, such as those noted above, have provided information on the flow rates associated with specific mouth conditions. While low flows are generally associated with closed mouth conditions and high flows (in particular episodic flood events) are associated with open mouth conditions, the actual flow magnitudes are specific to each case study and have little relevance to others. However, the general principles that govern the functioning of the mouth are encapsulated in the water budget for each system, which determines (for example) the time scale required for water levels to overtop the berm and cause the breaching and subsequent re-opening of the mouth.

The wave climate and tidal regime at an estuary's location, and associated cross-shore and long-shore sediment transport, are key factors in determining the detailed functioning of the mouth. On the sub-tropical north-east coast, northward long-shore sediment movement has been estimated to exceed 1000 m^{3}day^{-1} (Schoonees 2000). Allowing several tidal cycles for cross-shore redistribution of the sediment, the time scale for inlet closure is short, perhaps 2 - 5 days. This is consistent with the observations made by Perissinotto *et al.* (2004) in their field studies. Of course, high river flow rates (i.e. short flow time scales T_Q) will contribute to erosion of these sediments and can prolong the open period. Other factors that prolong an open mouth include features that dissipate or refract wave energy and reduce their ability to carry sediment into an inlet. For example, an inlet may be partially protected by a reef or rocky promontory.

When an estuary is closed, some marine 'overwash' may occur during high wave or storm events. There may also be a semi-closed state, with only a shallow 'trickle' of water flowing out to sea. An example is shown in Figure 7. This intermediate state may occur for a few days in the transition from open to closed state, but may also persist for much longer periods (weeks or months), provided the river flow rates remain in a narrow range.

Once the mouth of a TOCE has opened, tide-driven exchange flows can occur that play an important role in the salinity budget of the system. Furthermore, these flows also play a key role in sediment transport within and near the inlet and, therefore, influence the dynamics of the mouth re-closure process. All TOCEs essentially have an unstable inlet (by definition) that will tend to re-close when it is open. However, as a general principle, the larger the tidal prism, the longer the mouth will stay open due to the flushing effects of the tidal flows that scour sediment from the inlet. Most TOCEs in South Africa are small and have constricted inlets, which limits the tidal prism and gives rise to asymmetric tides in the estuary with flood dominance. Many TOCEs in South Africa are also perched, particularly on the north-east coast where beaches are steeper and beach berms higher (up to about 3 m above mean sea level), which also severely restricts the tidal prism. These features of South African TOCEs give rise to some surprising characteristics for the tidal exchange flow.

Figure 8 shows a time history of water levels in a typical perched TOCE in an open state. The tide levels in the sea are also shown on the same vertical datum. The water levels in the estuary are significantly super-elevated above those in the sea, and yet still show a significant tidal signal. The tide range in the estuary is about 50% of that in the sea, but reduces over several tidal cycles as the mouth becomes more constricted and eventually re-closes. The explanation for these observations is that wave action generates a significant set-up of the sea water levels due to the steep, reflective beach slopes. Essentially, the estuary "sees" an elevated sea level due to wave action, which is depicted by the upward shifted sea-tide levels shown in Figure 8. These shifted tide levels provide the required hydraulic gradient to drive the exchange flows associated with the water level changes in the estuary. There is also an asymmetry in the estuary tide, with the flood tide about half as long as the ebb tide (i.e. flood dominant). The times marked A and C in Figure 8 correspond to changes in the direction of the exchange flows and are the low and high tide times for the estuary. These times are significantly offset from the high and low tide times in the sea (points B and D in Figure 8), especially the latter. It is also evident that the maximum hydraulic gradient occurs at low tide in the sea.

A physical illustration of these features is provided in Figure 9, which shows photographs of the inlet flows during the flood and ebb tide, respectively. The flood tide fills the entire width of the inlet channel, with the inflow being enhanced by wave action. Ebb flows are mainly constrained to a narrower ebb channel with higher friction (smaller hydraulic radius).

Within estuarine water bodies, depending on their size and/or orientation, wind can play an important role in generating short period waves and associated mixing. The amplitude of these waves depends on the available fetch – for example a fetch of 1 km can produce waves capable of mixing the water column to depths of about 2 m. In large systems, winds also produce set-up effects with associated surface and bottom currents, but in small TOCEs these processes may not be dominant. At St Lucia (an important estuarine lake in the north-east with an area of 300 km^2) they play a very significant role in mixing and circulation. Here, wind setups of about 0.5 m have been observed, with wind waves rapidly mixing the shallow (~1 m) lake and increasing the turbidity of the water column (Taylor 2006).

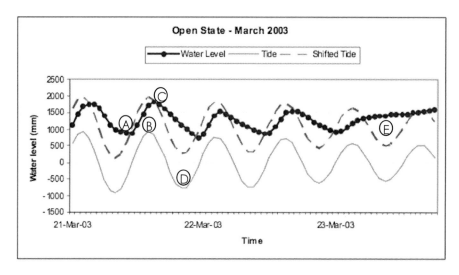

Figure 8. Sample time-history of water levels and tide heights for a small perched TOCE on the east coast of South Africa. Water levels are relative to mean sea level.

Figure 9. Inflow and outflow during the flood (left) and ebb (right) tide respectively. Note the wider channel during the flood tide and the action of waves in pushing water through the open mouth of a perched estuary. This contributes to the asymmetry of the estuary tide (photo: DD Stretch).

Sedimentation of estuaries is an important issue that is intrinsically related to hydrodynamics. Specific case studies of South African TOCEs are discussed by Cooper

(1989, 1990, 1993a, b, c, 1994a, b, 2001), Cooper *et al.* (1990, 1999), Reddering (1988) and Reddering *et al.* (1990). A general review and morphological classification of South African estuaries is given by Harrison *et al.* (2000). Roy *et al.* (2001) have provided a similar overview of these issues for estuaries in south-eastern Australia. The breaching of TOCEs is thought to play an important role in maintaining an approximate equilibrium with respect to sedimentation (Reddering & Rust 1990, Harrison *et al.* 2000). High flow rates can occur during sudden breaching events and lead to significant scouring of sediments (Stretch & Parkinson 2006, Parkinson & Stretch 2007, Beck *et al.* 2004, 2008). Changes to breaching patterns due to flow variations could therefore have a significant impact on the morphology of the estuaries concerned.

Estuarine sediments include organic and mineral particles of a wide range of sizes and composition. The organic content of sediment tends to increase with the fineness of the deposit (Gray 1981). Marine sediment input is almost exclusively inorganic and occurs through three main processes: tidal inflow (e.g. formation of flood-tidal deltas); barrier overwash; and wind action (Cooper *et al.* 1999). Fluvial inputs, on the other hand, include a large organic component, mainly in the form of plant detritus and an inorganic component of sand, silt and clay that is generally proportional to the rate of precipitation, the degree of soil erosion in the area, the geology of the catchment and its agricultural management. Warm temperate TOCEs often exhibit flood-tidal deltas, with marine sand constituting the bulk of estuarine sediments in the lower reaches (Reddering & Rust 1990). In subtropical east-coast TOCEs, sediments are dominated by river-borne deposits as a result of the high relief of their drainage basins, the high rainfall within the region and the perched nature of many estuaries.

During the open phase, the sediment distribution in TOCEs is very similar to the pattern observed in permanently open systems. This means an abundance of coarse/medium sand (0.25-2 mm) towards the mouth and a predominance of silt (4-63 µm) and mud/clay (< 4 µm) in the middle and upper reaches (Day 1981). This pattern is reflected in the results obtained in the Mhlanga Estuary in 1978 (Whitfield 1980b). At that time, the lower reaches of the Mhlanga were dominated by medium/coarse sand, the middle reaches by medium/fine sand and the upper reaches by very fine sand/silt (Whitfield 1980b). A significantly different situation was, however, observed in the nearby Tongati and Mdloti estuaries during the 1989/81 study of Blaber *et al.* (1984). While the lower and middle reaches of these two systems exhibited the same particle size composition of their Mhlanga counterparts, there was a virtual absence of silt in their upper reaches. The substratum was also relatively uniform throughout their length. This was attributed to an increase in the flow rate and frequency of mouth breaching in these two systems, compared to the Mhlanga (Blaber *et al.* 1984). Grains of different sizes are found in any sediment type, however well-sorted sediments tending towards an homogeneous type are typical of high current velocity situations. Conversely, poorly sorted or highly heterogeneous sediments are generally encountered in areas, or during periods, of low current activity (Gray 1981).

Wave overtopiing events, especially during high (spring) tides, can cause substantial re-arrangement of the beach sediments on the berm that separates TOCEs from the sea. An example of this process is shown in Figure 10, where it is evident that significant changes to the berm morphology can result from these events. Mouth breachings can scour and release large quantities of accumulated sediments and nutrients into the nearshore zone. An example of a turbidity plume emanating from a small TOCE during a flood event is shown in Figure

11. Such mouth discharges may be important nutrient sources for the oligotrophic waters of the South African east coast.

Figure 10. Effects of wave overtopping on the re-distribution of sediment on the beach berm of the Mhlanga TOCE (photo: DD Stretch).

Figure 11. A coastal turbidity plume from a small TOCE on the east coast of South Africa during a large flood event in September 1987 (source: CSIR 1989).

2. PHYSICO-CHEMICAL ENVIRONMENT

2.1 Temperature

Seasonality and regional climate are dominant factors affecting water temperatures in TOCEs. Seasonal ranges are largest in the cool temperate region on the Atlantic coast and smallest in the subtropical part of the eastern seaboard. Typical annual temperature ranges are from approximately 11-24 °C in cool temperate TOCEs (e.g. Rietvlei/Diep, Day 1981), however the minimum can drop to 9 °C if the estuary is open when strong upwelling events occur on the west coast (Monteiro & Largier 1999, Snow & Taljaard 2007). In the warm temperate region, temperature ranges of 13-27 °C (East Kleinemonde, Whitfield *et al.* 2008) have been reported. As expected, the range is slightly higher in subtropical TOCEs, from about 15-28 °C (Mpenjati, Perissinotto *et al.* 2002) and 15-30 °C (Mdloti, Thomas *et al.* 2005). These TOCE ranges are similar to those observed in permanently open estuaries, although they tend to be generally smaller than the latter owing to reduced riverine inflow and isolation from the sea for much of the year (Day 1981). Coastal upwelling events may be locally important in affecting the temperature of west and south coast TOCEs during their open phase (Schumann *et al.* 1999).

The above factors also affect the formation of horizontal and vertical thermal gradients. The surface temperature gradients established along an estuary when the mouth is open do not persist after its closure, but the vertical gradients often do, because density depends on salinity (Day 1981). For this reason, during their closed phase, TOCEs tend to lack much temperature gradient along the length of the estuary, with differences of < 2 °C recorded from mouth to head both in winter and summer (Oliff 1976, Perissinotto *et al.* 2000, Perissinotto *et al.* 2002). On the other hand, vertical temperature differences can be very substantial, and more marked than in nearby permanently open systems. This is because of poor vertical mixing, leading to the establishment of vertical density gradients as a result of the denser high-salinity waters entering the estuary during the closing phase and settling at the bottom of the water-column. In the subtropical estuaries of KwaZulu-Natal, this can result in a winter/spring temperature gradient increasing with depth. On the west coast, in the winter rainfall area, this pattern is likely to be reversed as warmer waters develop during the summer closed phase over the denser and colder ocean water that penetrated during the winter/spring open phase.

2.2 Salinity

Mouth state is a key factor controlling salinity in TOCEs. During their open, tidal dominated phase, TOCEs experience similar horizontal and vertical salinity gradients to those typically observed in permanently open estuaries. At times these may be very marked, with a vertical salinity stratification prevalent in deep TOCEs. For instance, in the Mtamvuna Estuary, an isohaline of 15 normally lies at a depth of 1-2 m, with a halocline between 2-3 m and the salinity gradually increasing to 30-35 below this layer to the bottom, which may reach a maximum depth of approximately 10 m (Oliff 1976, Day 1981). In broad shallow TOCEs, however, (e.g. Mdloti, Mpenjati), the wind fetch is generally sufficient to ensure complete mixing from surface to bottom and no salinity stratification is noted during either the closed or open phases (Blaber *et al.*1984, Grobbler *et al.* 1987, Thomas *et al.* 2005). When TOCEs are breached, riverine conditions normally prevail for some time, with freshwater replacing all previous water types. Whitfield *et al.* (2008) recently defined this as the "outflow phase", in contrast to the "tidal phase", which is typically dominated by seawater moving in and out of the estuary, with a longitudinal salinity gradient shaped by the balance between tidal penetration and the magnitude of river inflow (Snow & Taljaard 2007).

Salinity in a closed TOCE will vary as in open systems, but typically on a time scale ranging from days to weeks, rather than hours as in open systems where tidal effects are significant. Horizontal salinity gradients tend to break down once the marine influence is cut off, although some contrast between the mouth and head areas may remain, depending on seepage and marine overwash at the bar and the magnitude of freshwater input at the head. Mesohaline (5-18) and oligohaline (0.5-5) conditions often prevail during the closed phase in shallow systems and in the upper layers of deeper systems. In areas of high seasonal rainfall, such as along the Western Cape and KwaZulu-Natal coasts, the sustained freshwater inflow into TOCEs results in the salinity dropping to near freshwater conditions. On the KwaZulu-Natal coast, this effect is compounded by the perched nature of TOCEs and the seepage of saline water out of the estuary being replaced by inflowing river water. Limnetic conditions (0.1-0.5) may then prevail through most of the closed phase, with salinity seldom exceeding 1 even in the lower reaches (e.g. Mdloti, Nozais *et al.* 2001, Thomas *et al.* 2005). At the opposite end, in areas of low rainfall and high evaporation rate, or during periods of severe droughts, hypersaline conditions (> 40) may occur. An example of this was recorded in the Seekoei Estuary in the Eastern Cape, where hypersaline levels of up to 98 resulted in the mortality of many of its fish species (Whitfield & Bruton 1989, Whitfield 1995).

2.3 Irradiance

Photosynthetic available radiation (PAR, 400-700 nm) in TOCEs may vary widely, in response to weather conditions, time of day and suspended solids in the water. Maximum levels of surface irradiance recorded in South African estuaries are in the range of 1638-2300 μmol m^{-2} s^{-1} Perissinotto *et al.* 2000, Nozais *et al.* 2001, Perissinotto *et al.* 2002). During the open tidal phase, as a general rule, the suspended solids concentration in the water-column decreases from high to low tide and from the mouth to the upper reaches (Day 1981). As flocculation occurs at salinity values between zero and 5 ‰, there is a turbidity maximum at the river-estuary interface (Kemp 1989). Flocculated material settles out of the water-column quite rapidly, unless there is powerful wind-mixing or other turbulence in the area (Snow & Taljaard 2007).

Light attenuation (K_d) through the water-column shows a highly significant and positive relationship with rainfall, obviously reflecting the increase in turbidity as a result of runoff from the catchment area (Nozais *et al.* 2001, Thomas *et al.* 2005). During the closed phase of a TOCE, K_d may vary between 0.5 and 4 m^{-1} (Perissinotto *et al.* 2002, Thomas *et al.* 2005), with > 1% of the surface light intensity reaching the bottom of the water-column at any time. Conversely, when the opening of an estuary leads to heavy silt loading and a considerable increase in turbidity levels, K_d may attain maximum values of 4-29 m^{-1} (Perissinotto *et al.* 2000, Nozais *et al.* 2001, Thomas *et al.* 2005). During this period, the percentage of surface light intensity reaching the bottom is often < 0.1%, thus resulting in aphotic conditions at the sediment surface as well as parts of the water-column. In other words, the depth of the euphotic zone drops to values much shallower than the estuary total depth. This has important negative implications for the photosynthetic efficiency of microphytobenthos.

Available Secchi disc measurements from past surveys indicate that values may range from minima of 5-10 cm during the outflow phase to maxima in excess of 3 m during the driest part of the closed phase (Day 1981, Begg 1984). In the East Kleinmonde Estuary, Secchi depths ranging from 1.1 to 1.75 m during the closed phase and from 0.1 to 0.65 m during the open phase were recorded (van Niekerk *et al.* 2008). Measurements of turbidity, in units of nephelometric turbidity (NTU), show a wide range of values for TOCEs, ranging from 0.2-7 NTU during the closed phase to 75-90 NTU at the onset of mouth breaching (Cooper et al. 1993, Harrison unpubl. data). During the outflow phase, when river flooding occurs, turbidity levels often increase to > 100 NTU for brief periods, while in the subsequent tidal phase these may drop to < 10 NTU, especially in the cool and warm temperate regions where the incoming seawater is usually clear (Snow & Taljaard 2007, Whitfield et al. 2008). Recent investigations in the subtropical Mhlanga and Mdloti estuaries and the warm temperate East Kleinemonde Estuary have shown that average turbidity differences between closed and open phases are not as marked as earlier hypothesized (Thomas et al. 2005, Whitfield et al. 2008). Turbidity is controlled by factors such as sediment composition (e.g. fine mud versus coarse sand), turbulence (e.g. wind-mixing, current speed) and water depth. These in turn will depend on the geology of the catchment, the bathymetry of the broader basin, the prevailing climate, agricultural practices and other regional characteristics.

2.4 Dissolved Oxygen

In the absence of tidal currents and strong freshwater inflow, oxygen levels in closed systems may decline in bottom water, particularly if stratification is present and there is benthic organic accumulation. During the closed phase, the oxygen concentrations in the deeper layers of TOCEs depend on the ratio of area to depth and the circulation due to the wind (Day 1981). Systems like the Umgababa in KwaZulu-Natal are relatively broad and shallow and show little evidence of substantial oxygen depletion, except in the deepest areas. This is so because wind action is sufficient to prevent the formation of thermoclines and the water is generally well oxygenated throughout the water column (Oliff 1976, Day 1981, Begg 1984). However, recent investigations on the East Kleinemonde and the Maitland TOCEs of the Eastern Cape (Gama et al. 2005, Whitfield et al. 2008), as well as the Mhlanga and Mdloti of KwaZulu-Natal (Perissinotto et al. 2004), have shown that hypoxic to anoxic conditions can occur in these shallow systems, particularly after prolonged periods of mouth closure and calm weather. In the Mhlanga and the Mdloti, dissolved oxygen concentrations ranged from minima of 0.08-1.8 mg L^{-1} to maxima of 12.2-13.2 mg L^{-1}, respectively (Perissinotto et al. 2004).

Narrower and more sheltered TOCEs, on the other hand, regularly experience anoxic or hypoxic conditions, particularly towards the end of a prolonged closed phase. This situation is well illustrated by the results obtained from the Nyara Estuary in the Eastern Cape (Walker et al. 2001). In March, at the onset of the closed phase, the upper 2 m of the water column of the estuary was well oxygenated (5-8 mg L^{-1}), while lower oxygen conditions characterised the deeper water near the sediments (3-4 mg L^{-1}). These values declined 6 months later to less than 1 mg L^{-1} at a depth of 3.5 m (Walker et al. 2001). Similar situations have been reported for subtropical TOCEs in KwaZulu-Natal, e.g. during 1980/81 the Tongati Estuary experienced semi-anoxic waters (< 25% oxygen saturation) throughout its length during much of the closed phase (Blaber et al. 1984). Saturation levels of < 20% have also been recorded below the 2 m halocline during periods of mouth closure in the Mtamvuna Estuary (Oliff 1976, Begg 1984). Semi-anoxic conditions, with dissolved oxygen concentrations of ≤ 1 mg L^{-1}, were documented at the bottom of several KwaZulu-Natal TOCEs during the Estuarine Health Index project (Harrison et al. 2000).

Detailed studies on dissolved oxygen responses have recently been undertaken in the Wamberal Lagoon and Smiths Lake, two southeastern Australian ICOLLs (Gale *et al.* 2006). Results show that oxygen concentrations in bottom water decrease when buoyancy frequency in the water-column exceeds 0.1 s^{-1}. A higher vertical stability was consistently associated with persistent stratification in the larger and deeper (max 5 m depth) Smiths Lake, while the shallower Wamberal Lagoon (max 2 m depth) experienced only short and irregular stratification events. Stratification events, with consequent deterioration in bottom water oxygen levels, appeared to be controlled by a combination of rainfall, solar insolation, wind stress and tidal mixing (Gale *et al.* 2006).

2.5 Inorganic Nutrients

The dynamics of macronutrient cycling in TOCEs are, in general, poorly understood since for only a few estuaries are there more than snapshot data available. Given the predominantly oligotrophic nature of the Agulhas Current, nutrient concentrations in TOCEs situated along the warm temperate and subtropical east coastal regions of South Africa are influenced mainly by the inflow of freshwater from their catchment, rather than seawater exchanged through tidal action. On the cool temperate west coast, on the other hand, prevailing marine upwelling conditions provide an adequate supply of nutrients to coastal waters and this is expected to significantly affect the trophic dynamics of the local estuaries (van Ballegooyen *et al.* 2007). Also, it is only in this region that silicate (SiO_2) concentrations may be of importance to estuarine productivity when large Si-dependent diatom blooms occur. Elsewhere along the coast, the availability of SiO_2 is generally far in excess of phytoplankton demand (van Ballegooyen *et al.* 2007), and it is mainly low dissolved inorganic nitrogen (DIN: sum of nitrate, NO_3 and ammonia, NH_4) and phosphorous (DIP: orthophosphate, PO_4) levels that control productivity.

Prior to the recent high-frequency sampling studies of the Mhlanga, Mdloti, Mpenjati and East Kleinmonde TOCEs, a snapshot survey of the nutrient status of the Mdloti and Tongati conducted in November 1981 had concluded that the lowest soluble reactive phosphate values of 0.65-0.97 µM recorded in the Mdloti may have been sufficiently low to inhibit primary production (Blaber *et al.* 1984). Most of the nitrogen in Tongati was in the form of ammonia, possibly derived from sewage discharges or agricultural activities encroaching on its banks. High nitrate values were recorded in the Mdloti (up to 25 µM). These were attributed to the use of agricultural fertilizer on adjacent sugarcane fields. In the Western Cape, the Groot Brak Estuary exhibited mean nitrate values of 0.8 µM in August 1992, increasing to 1.7 µM in November (late spring) as a result of an increase in freshwater inflow during this period (Slinger *et al.* 1995). Further snapshot measurements of nutrient concentrations in TOCEs were made by Harrison *et al.* (2000) in virtually all South African systems during the period 1992-1999. DIN values obtained during those surveys ranged from < 1 to 396 µM, while DIP ranged from < 0.1 to 206 µM.

Recently, annual surveys with monthly sampling frequency have been carried out at the Mdloti, the Mhlanga and Mpenjati estuaries in KwaZulu-Natal, and at the East Kleinomende Estuary in the Eastern Cape. While both Mdloti and Mhlanga are located in the peri-urban region of Durban and are, therefore, largely impacted with significant input of treated sewage waters (Nozais *et al.* 2001, Perissinotto *et al.* 2004), the Mpenjati and East Kleinemende are more peripheral and in a relatively pristine state (Perissinotto *et al.* 2002, Kibirige 2002, Whitfield *et al.* 2008). Results indicate that water-column macronutrient concentrations are now much higher in the Mdloti/Mhlanga than in the Mpenjati/East Keinemonde estuaries. Also, in all systems there is generally a marked increase in macronutrient concentrations during the open phase, compared to the closed phase. Indeed, after prolonged periods of mouth closure the water-column may even become depleted in macronutrients, probably as a result of sustained algal uptake. DIN concentrations ranged from 0.14 to 204 µM in the Mdloti and from 14 to 418 µM in the Mhlanga. DIP concentrations varied between 0.1 and 13.5 µM in the Mdloti and between 4.7 and 81.7 µM in the Mhlanga (Nozais *et al.*, 2001, Thomas *et al.* 2005). Both systems have, therefore, become heavily impacted by anthropogenic/cultural eutrophication, with the former receiving a volume of about 8 ML of treated sewage water per day and the latter about 20 ML (Perisssinotto *et al.* 2004). By comparison, in the Mpenjati Estuary DIN concentrations ranged from 1.6 to 17.7 µM and DIP from 0.2 to 1.2 µM (Perissinotto *et al.* 2002). In the East Kleinemonde, values were approximately between < 1 and 200 µM for DIN and < 0.3 - 1 µM for DIP (van Niekerk *et al.* 2008).

The few estimates of porewater nutrient concentration that are available, mainly for the Mdloti and the Mhlanga estuaries, show a range between 5.3 and 310 µM for DIN and 0.1-45.1 µM for DIP (Perissinotto *et al.* 2006). This indicates that groundwater input within TOCE sediments may contribute substantially to the total nutrient budget of these estuaries, possibly through steady remineralization and diffusion across the sediment-water interface. Phosphate is also absorbed onto resuspended clay particles whenever their concentration in the water-column exceeds the equilibrium point with the sediment. On the other hand, nitrogen-fixing cyanobacteria may occasionally enhance the nitrogen load of estuarine waters, provided that salinity remains below 8-10 ‰ (Conley *et al.* 2009).

When DIN:DIP molar ratios have been estimated in South African TOCEs, these have shown values well above the Redfield ratio of 16:1 during the open phase, but often below this critical value during the closed phase. This has led to the suggestion of a potential limiting effect of nitrogen during the closed phase and phosphate during the open phase (Nozais *et al.* 2001).

3. PRIMARY PRODUCERS

3.1 Phytoplankton

Snapshot measurements of phytoplankton chlorophyll-a biomass were made by Harrison *et al.* (2000) in virtually all South African TOCEs during the period 1992-1999 and ranged from < 1 to 10.92 µg L^{-1}. Detailed studies on water-column microalgae (phytoplankton) have recently been completed only for five warm temperate TOCEs in the Eastern Cape (Van Stadens, Maitlands, Nyara, Kasouga and East Kleinemonde) and four subtropical systems on the KwaZulu-Natal coast (Mdloti, Mhlanga, Mpenjati and Tongati).

Data on the taxonomic composition and distribution of phytoplankton communities in South African TOCEs is very limited. Flagellates and cyanobacteria (blue-green algae) have been reported as dominant components of the phytoplankton of the Groot Brak Estuary, with smaller numbers of diatoms, dinoflagellates and euglenoids also present (Adams *et al.* 1999). Diatom species were found to occur over a wide salinity range from freshwater to virtual marine conditions (salinity 35 ‰). In the Nyara study of 1997, diatoms were dominated by the genera *Navicula, Amphiprora, Gyrosigma* and *Nitzschia* (Walker *et al.* 2001). Their concentrations ranged from 10^3 to 10^5 cells L^{-1} and were consistently lower than picoplankton and dinoflagellate numbers (Walker *et al.* 2001). The phytoplankton community of the East Kleinemonde, on the other hand, was dominated by diatoms (62 % of the total) during the closed phase, but by dinoflagellates and cryptophytes during the subsequent open phase (Whitfield *et al.* 2008).

In terms of size classes, nanophytoplankton (2-20 μm) is generally the dominant fraction in South African TOCEs (e.g. Perissinotto *et al.* 2000, Foneman 2002, Thomas *et al.* 2005). At the Mdloti and Mhlanga estuaries this fraction accounted for 81 % and 80 % of the total phytoplankton biomass, respectively (Thomas *et al.* 2005). This was followed by picophytoplankton (< 2 μm) and microphytoplankton (> 20 μm), with respective values of 13 % and 6 % respectively in the Mdloti, and 12 % and 8 % at the Mhlanga. In the Eastern Cape warm temperate region, however, Froneman (2002) found that picophytoplankton was actually dominant throughout the closed phase of the Kasouga Estuary, with nano- and microphytoplankton becoming more important only after mouth breaching and the inflow of river water. A similar pattern was observed in the East Kleinemonde, where after a 4-week period following mouth breaching, nanophytoplankton had attained dominance with 67% of total biomass, followed by picoplankton with 27% (Whitfield *et al.* 2008).

Recent studies in relatively pristine South African TOCEs have shown that phytoplankton biomass ranges from 0.09-15.4 mg(chl-a) m^{-3} (Oliff 1976, Perissinotto *et al.* 2000, Nozais *et al.* 2001, Perissinotto *et al.* 2002, Whitfield *et al.* 2008). These values are lower than those measured in permanently open estuaries, where maximum values often exceed 20 mg(chl-a) m^{-3} (Adams *et al.* Bate 1999).

However, TOCEs located in peri-urban regions are showing progressive signs of eutrophication, with regular phytoplankton blooms observed in the eThekwini metropolitan area, around the city of Durban. For instance, during the study of March 2002 - March 2003, phytoplankton biomass ranged from 0.87-111 mg(chl-a) m^{-3} in the Mdloti and from 0.73-303 mg(chl-a) m^{-3} in the Mhlanga Estuary (Thomas *et al.* 2005) (Figure 12). This last value is actually the highest so far recorded in any southern African estuary and an indication of the degradation that the Mhlanga TOCE has undergone in recent years as a result of the large and steady volume of treated sewage effluents that it receives (Perissinotto *et al.* 2004). During closed phases, when the residence time of water inside the estuary is sufficiently long for nutrient utilization, dense algal blooms form and are often followed by anoxic or hypoxic events.

The first phytoplankton productivity study carried out in a TOCE was by Oliff (1976) at the Fafa Estuary, with values obtained ranging from 0.3-7.5 mg(C) $m^{-3} h^{-1}$ (^{14}C method) and from 1.6-29.9 mg(C) $m^{-3} h^{-1}$ (Oxygen method) (Oliff 1976, Grindley 1981). More recently, the comparative productivity of phytoplankton has been investigated in the Mdloti and Mpenjati estuaries (Perissinotto *et al.* 2003, Anandraj *et al.* 2007). Results have shown a phytoplankton production range of 0.02-409 mg(C) $m^{-2} h^{-1}$ at the Mdloti and 0.3-206 mg(C) $m^{-2} h^{-1}$ at the Mpenjati, with no clear evidence of substantial differences between open and closed phases. In the oligotrophic Van Stadens Estuary, the highest phytoplankton productivity occurred both in the spring (14 mg(C) $m^{-2} h^{-1}$) and summer (13.4 mg(C) $m^{-2} h^{-1}$)

seasons, during an open mouth phase and two weeks following mouth closure, respectively (Gama et al. 2005).

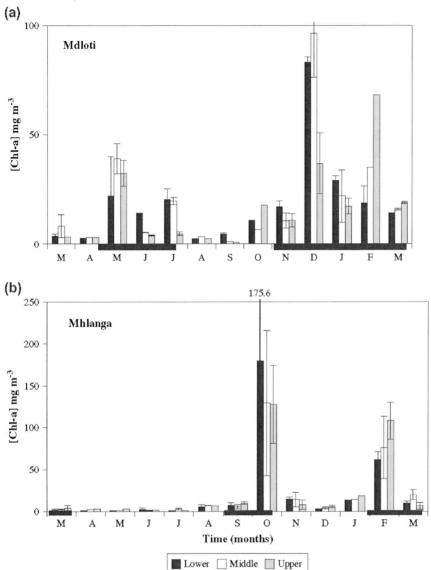

Figure 12. Spatio-temporal variations in phytoplankton chl-*a* biomass (mean ± SD) in the Mdloti (a) and the Mhlanga (b) estuaries during the period March 2002 - March 2003. Black horizontal bars indicate periods of mouth closure (source: Thomas et al. 2005).

An interesting issue that has been addressed recently deals with the recovery of phytoplankton productivity and biomass following mouth breaching and flushing of a TOCE. It was observed that while 94-99% of phytoplankton biomass can be washed out to sea through flushing, pre-breaching levels were re-established within 35-40 days following a breaching event (Anandraj et al. 2008, Whitfield et al. 2008). Remarkably, pelagic primary production reached its maximum rates during the open phase, however this was not followed immediately by biomass accumulation inside the estuary because of the continual flushing and tidal export (Anandraj et al. 2008).

Phytoplankton not only constitutes an important carbon source for local benthic and pelagic food webs, but also provides an essential link between inorganic compounds and organic matter available to the higher trophic levels and top predators (Miller et al. 1996; Mortazavi et al. 2000). Studies at the Mpenjati and Nyara estuaries have indicated that the grazing impact of zooplankton on phytoplankton is substantial, at times even in excess of phytoplankton production (Perissinotto et al. 2003). Current research is now estimating the degree of phytoplankton utilization as an energy source by other components of the food web, such as meroplanktonic larvae and macro- and meiobenthic organisms. The ultimate fate of these microalgae in the ecosystem and their degree of direct consumption by primary and higher level consumers is of fundamental importance to the energy balance and, therefore, the ecological functioning of TOCEs.

3.2 Microphytobenthos

An extensive taxonomic study of benthic diatoms has been carried out at a number of South African estuaries as part of a dedicated Water Research Commission project. Among these are two TOCEs: the Mpekweni in the Eastern Cape and the Manzimtoti in KwaZulu-Natal (Bate et al. 2002). This study was largely aimed at evaluating the usefulness of these microalgae in the assessment of ecosystem health, by showing associations between estuarine water quality and dominant microphytobenthic species. In the sediments of the Tongati and Mhlanga, Blaber et al. (1984) reported the common presence of unidentified species of the genera *Navicula, Nitschia, Synedra, Spirulina, Ocillatoria,* as wells as euglenoid flagellates. A whole year, spatio-temporal study of the benthic diatom community of the Mdloti Estuary has shown a clear dominance of four genera and 26 species: 13 *Navicula* spp., 6 *Nitschia* spp., 3 *Amphora* spp. and 2 *Achnanthes* spp. (Mundree 2001). Species composition and dominance have been related to water quality in a comparative study of the Mhlabatshane and the Manzimtoti (Watt 1998). The naturally oligotrophic Mhlabatshane sediments were found to be dominated by the diatom genera *Acanthidium* and *Diploneis*, while the nutrient enriched and polluted Manzimtoti sediments contained mainly small species of *Navicula* and *Nitzschia*. A total of 44 diatom species were identified in the Manzimtoti and 21 in the Mhlabatshane (Watt 1998). In the warm temperate Van Stadens TOCE, a total of 29 microphytobenthic taxa were identified, including 24 diatoms, 3 cyanophytes, one flagellate and an euglenoid (Skinner et al. 2006). A similar trend was found in the East Kleinemonde, where bacillariophytes accounted for 75%, chlorophytes 15% and cyanobacteria for 10% of the total microphytrobenthic assemblage (Whitfield et al. 2008).

Distribution data have indicated that dominance of a diatom taxon is largely dependent on sediment type and the open/closed state of an estuary. In the Mdloti, microphytobenthic cells were found in substantial numbers at depths of up to 5 cm below the sediment surface, well below the photic zone (Mundree 2001). Their distribution at such depths may represent an important stock of potential primary producers in TOCEs and is likely to be related to active migration, physical hydrodynamic conditions and/or bioturbation by deposit feeders. In terms of relationships with sediment type, vertically homogenous microphytobenthic chl-a concentrations were recorded in the upper 5 cm of sediment in the lower reaches of the

Mdloti Estuary, where coarse sand dominated the sediment structure. Conversely, chl-a was concentrated in the top 2 cm in the upper reaches of the estuary, where fine silt made up the bulk of the sediment (Mundree 2001).

Contrary to what has been observed for the microalgae inhabiting the water-column, microphytobenthic biomass has been found to be generally higher in TOCEs than in permanently open estuaries, with values ranging from 1.4-616 mg(chl-a) m^{-2} (Adams & Bate 1994, Nozais et al. 2001). The maximum concentrations obtained are the highest values ever recorded for any South African estuary and among the highest reported in the literature (Cahoon et al. 1999, Adams et al. 1999, Perissinotto et al. 2002). In the southwestern Australian Peel-Harvey estuarine system, Lukatelich & McComb (1986) obtained values of 107-202 mg(chl-a) m^{-2} during summer and 151-163 mg(chl-a) m^{-2} in winter. Skinner et al. (2006) reported that in the South African warm temperate Van Stadens TOCE microphytobenthic concentration was higher during the open than during the closed phase, with the respective values ranging from 1.4 to 3.6 µg(chl-a) g^{-1} sediment and from 0.3 to 1.9 µg(chl-a) g^{-1} sediment. The same pattern was observed in the East Kleinemonde, where the highest biomass of about 200 mg(chl-a) m^{-2} occurred in the lower reaches during an open mouth state (Whitfield et al. 2008). Conversely, in subtropical TOCEs, microphytobenthic biomass appears to be higher under closed mouth conditions than during open phases. In the Mpenjati, for instance, Perissinotto et al. (2002) recorded concentrations in the range 19.6-616 mg(chl-a) m^{-2}, with significantly higher values during the closed phase. In the Mdloti, values ranged from 1.3-131 mg(chl-a) m^{-2} during the open phase and from 18-391 mg(chl-a) m^{-2} during the closed phase. In the Mhlanga Estuary the highest value of 313 mg(chl-a) m^{-2} was obtained during an open phase (Perissinotto et al. 2006) (Figure 13), but this system is heavily impacted by eutrophication and flow alterations as a result of large-scale treated sewage water disposal in its upper reaches (Thomas et al. 2005).Microphytobenthic biomass has been shown to contribute a significant fraction of the total primary biomass in TOCEs, equaling or even exceeding that of phytoplankton in the overlying water. Studies in the Mdloti, Mpenjati and Nyara estuaries have reported microalgal biomass in the sediment to be 1 to 3 orders of magnitude higher than in the water-column (Nozais et al. 2001; Perissinotto et al. 2003). However, despite the much larger biomass, microphytobenthos in TOCEs actually exhibits lower production rates than water-column microalgae. Productivity measurements of benthic microalgae have been taken in the Mpenjati and Mdloti estuaries and these show rates of 0-7 mg(C) $m^{-2} h^{-1}$ at Mpenjati and 0-16 mg(C) $m^{-2} h^{-1}$ at Mdloti, i.e. about an order of magnitude lower than rates measured for phytoplankton in the water-column at the same time (Anandraj et al. 2007). Such results have been attributed to the optimal light conditions prevailing in the water-column when compared to the sediment surface, as well as the higher grazing impact by zooplankton in the water-column and settling of phytoplankton cells out of the water column, both of which will result in a reduced phytoplankton biomass.

General spatial and temporal patterns of microphytobenthic biomass and production have been found to be largely controlled by factors such as salinity, exposure (desiccation), water currents, sediment hydrodynamcs, granulometric composition (Adams et al. 1999, Mundree 2001), and by interactions between light, nutrient availability and grazing pressure (Mundree 2001, Perissinotto et al. 2000, Nozais et al. 2001, Perissinotto et al. 2003).

3.3 Macrophytes

Macrophytes in TOCEs respond to changes in water level, salinity, sediment input, turbidity and nutrients and are represented by submerged plants, reeds and sedges, salt marsh, mangroves and swamp forest. Along the east coast of South Africa, Colloty *et al.* (2002) found that salinity influences community composition, e.g. between the Great Fish and Great Kei rivers salt marsh is present in TOCEs because of high water column salinity (13-23 ‰) and wide intertidal and supratidal areas (15-186 ha). Reeds and sedges are dominant north of the Great Kei River because in that area TOCEs are small (< 50 ha) and the salinity is generally low (< 18 ‰) due to the higher rainfall along this section of the coast. Swamp forest, represented by the freshwater swamp tree, *Barringtonia racemosa,* and lagoon hibiscus, *Hibiscus tiliaceus*, occur north of the Mngazana Estuary because of the sub-tropical climate. Walker (2004) also showed that salinity and the state of the estuary mouth influenced macrophyte species composition (Figure 14). The permanently open estuaries had approximately 11 dominant macrophyte species, whereas the TOCEs had less than six dominant species. Generally, TOCEs have lower diversity and cover of emergent macrophytes compared to permanently open estuaries. A comparative study of four Eastern Cape estuaries (Adams *et al.* 1992) showed that this can be attributed to the lack of suitable habitat, fluctuating water levels and periodic hypersaline conditions occurring in TOCEs.

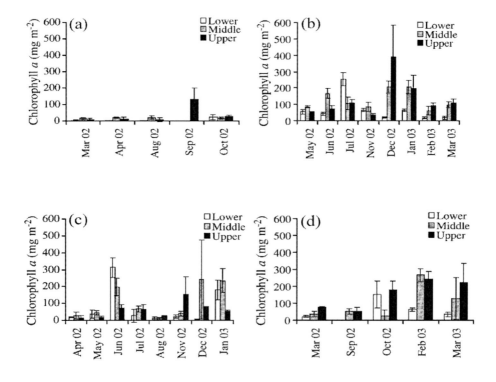

Figure 13. Temporal and spatial distribution of microphytobenthic biomass (mean ± SD) in the Mhlanga (a: open phase; b: closed phase) and Mdloti (c: open phase; d: closed phase) estuaries during the period March 2002 – March 2003 (source: Perissinotto *et al.* 2006).

Large, permanently open, saline estuaries (e.g. Kwelera, Nahoon)	Saline, often open TOCEs (e.g. Bulura, Cefane, Cintsa, Ncera)	Small, freshwater dominated TOCEs (Blind, Hlaze, Igoda, Mtendwe)
Avicennia marina --- *Spartina maritima* --- --------*Zostera capensis* ------- -----------*Halophila ovalis* ------- ----------------------*Sarcocornia spp.*------------------------------------ -----------------------*Salicornia spp.*--------------------------------- ---*Sporobolus virginicus*--- ------------------------------------*Stenotaphrum secundatum*----------------------------------- --*Phragmites australis*------------- --*Juncus kraussii*------------------- ----------*Potamogeton sp.*-----------		

Figure 14. The distribution of dominant macrophyte species in different estuary types in the Eastern Cape, South Africa. The position of the species in the chart indicates the estuarine type in which it is most dominant, while the dashed line indicates the full extent of its range (adapted from Walker 2004).

During the closed phase, the absence of tidal currents, clear water and good light penetration can result in the proliferation of submerged macrophytes and macroalgae (Adams et al. 1992, 1999). *Ruppia cirrhosa* is the dominant submerged macrophyte, particularly in warm-temperate TOCEs. These plants grow and expand in cover when the mouth is closed and the water level is greater than 1.5 m amsl for more than two consecutive months (Riddin & Adams 2008). Laboratory studies have shown that *Ruppia cirrhosa* can survive salinity ranging from freshwater to 75 ‰, can tolerate fluctuating salinity and can recover from high salinity treatments. It is therefore well adapted to the fluctuating environment characteristic of TOCEs (Adams & Bate 1994). The lack of submerged macrophytes in KwaZulu-Natal estuaries has often been attributed to an increase in catchment and bank erosion, sedimentation and lack of available light (Howard-Williams & Allanson 1979, Begg 1984). Submerged macrophytes play an essential role in nutrient trapping and recycling in estuaries. In the Wilderness Lakes, *Potamogeton pectinatus* contributed more than 70 % of the total submerged biomass of the lake (Weisser & Howard-Williams 1982). Similarly, in the Bot River submerged macrophytes contributed 72 % of the total production of 13 000 tons dry mass per year (Branch et al. 1985).

Prolonged mouth closure can result in a loss of tidal action and a high water level, which can have adverse effects on the macrophytes. Where there is a reduction of tidal and saline influence, salt marsh areas are often colonised by other species, such as reeds and sedges which grow under fresher and more inundated conditions. O'Callaghan (1990) identified restriction in tidal input and reduced saline input as a serious threat to salt marshes in the estuaries of False Bay, Western Cape. Loss of salt marsh as a result of increased flow due to urbanization is common in most developed countries (Greer & Stow 2003).

The scouring of estuarine sediment that follows mouth breaching events, often results in removal of macrophytes, particularly the submerged community which is associated with the water column. Floodwater can also remove reeds and sedges that have encroached into the main channel. A rapid drop in water level associated with mouth breaching also results in the exposure of submerged macrophytes and the partial or complete loss of biomass within hours due to desiccation (Adams & Bate 1994, Tyler-Walters 2001). They can, however, show rapid growth from surviving rootstock and the seed bank once an optimum water level returns, e.g. *Ruppia cirrhosa* can complete its life-cycle within 8 to 12 weeks (Gesti *et al.* 2005, Calado & Duarte 2000).

Macrophytes in TOCEs are thus adapted to extreme environmental conditions. In a recent study, habitats exposed following mouth breaching in the East Kleinemonde Estuary were rapidly colonized by the salt marsh succulent, *Sarcocornia perennis*. The plants germinated from a large seed bank (mean = 7929 seeds m^{-2}) three days after the mouth breached and germinated intermittently thereafter. These plants are, however, sensitive to prolonged inundation (> 3 months) following mouth closure and once dead the habitat is again colonized by submerged macrophytes (Riddin & Adams 2008). An assessment of the seed banks of two Eastern Cape TOCEs showed that they are characterised by low species diversity but high seed density (Riddin & Adams 2009). Large sediment seed reserves ensure macrophyte survival and persistence, even after prolonged periods of unfavourable conditions. However, in disturbed estuaries water level would need to be managed to ensure that the plants flower and set seed, as inundation slows flowering and subsequent seed production (O'Callaghan 1990).

Long periods of exposure during open mouth conditions can result in desiccation of reed, sedge and swamp forest habitat. The swamp forest tree, *Hibiscus tiliaceus,* seems to be more adversely affected by periods of prolonged exposure during open mouth conditions than by extended periods of root immersion (Perissinotto *et al.* 2004). Mangroves are, however, sensitive to flooding and major die-backs in response to inundation have been reported in the St Lucia (Steinke & Ward 1989), Mgobezeleni (Bruton & Appleton 1975), Mlalazi (Hill 1966) and Mgeni (Steinke & Charles 1986) estuaries.

In some estuaries, increased nutrient input coupled with prolonged mouth closure has resulted in excessive macroalgal growth. Eutrophication occurs in certain TOCEs, with increased epiphyte, phytoplankton and macroalgal growth reducing light available to submerged macrophytes. Decaying mats of filamentous algae and other macrophytes impact the social acceptability of water and is often the reason for the manipulated opening of the estuary mouth. Increasing availability of inorganic nutrients (especially N and P) is known to stimulate the abundance of ephemeral and epiphytic macroalgae in shallow coastal waters (Sand-Jensen & Borum 1991, Duarte 1995, Valiela *et al.* 1997, Karez *et al.* 2004). *Ulva, Enteromorpha* and *Cladophora* often form mass accumulations due to their filamentous nature and higher nutrient uptake rates than thicker algae (Raffaelli *et al.* 1998 cited in Karez *et al.* 2004). These mass accumulations can reduce estuarine water quality, not only by depleting the oxygen within the water column upon decomposition, but also by causing anoxic sediment conditions when large mats rest on the sediment at low flow conditions.

Although TOCEs are dynamic and resilient systems, changes in macrophyte species richness, abundance and community composition occur in response to changes in the controlling environmental factors. A future threat is the loss of plant diversity due to reed encroachment, which has already occurred in certain TOCEs, e.g. Siyaya Estuary in KwaZulu-Natal.

The common reed, *Phragmites australis,* is dominant in most TOCEs and its abundance is increasing due to sedimentation, estuarine shallowing and an increase in freshwater and nutrient run-off from surrounding developments.

4. CONSUMERS & TOP PREDATORS

4.1 Zooplankton

The prolonged periods of mouth closure that TOCEs experience, generally leads to low levels of zooplankton diversity. This is in contrast to the situation observed in their permanently open counterparts, where neritic marine species penetrating with flood tides often increase the species richness, particularly in the mouth region (Grindley 1981). The few studies conducted in South African TOCEs show that during the closed phase (usually coinciding with the winter/spring seasons on the eastern coast) the mesozooplankton community is composed of few taxa (10-20), with 2-5 species normally accounting for ≥ 80% of total numerical abundance (Perissinotto *et al.* 2000, Kibirige 2002). These include the calanoid copepods *Pseudodiaptomus hessei* and *Acartia natalensis* as well as a few unidentified species of harpacticoids, cyclopoids such as *Halicyclops* spp. and *Thermocyclops emini*, the amphipod *Grandidierella lignorum* and isopods such as *Cirolana fluviatilis* (Whitfield 1980a, Connell *et al.* 1981, Grindley 1981, Blaber *et al.* 1984, Schlacher & Wooldridge 1995, Perissinotto *et al.* 2000, Kibirige 2002). Their larval stages, and particularly copepod nauplii, are always the most numerically abundant group during this phase. Where conditions at the mouth are favourable for the maintenance of salinity levels above 10 ‰, the mysids *Gastrosaccus brevifissura* (e.g. Mpenjati Estuary, Kibirige & Perissinotto 2003a), *Mesopodopsis africana* (e.g. Mdloti and Mhlanga estuaries, Kibirige *et al.* 2006) and *Rhopalophthalmus terranatalis* (e.g. Kabeljous Estuary, Schlacher & Wooldridge 1995) are also among the dominant components of the mesozooplankton community during the closed phase. Elsewhere, where limnetic or oligohaline conditions prevail throughout the estuarine closed phase, such as in the perched TOCEs on the KwaZulu-Natal north coast, mesozooplankton may include large proportions of *Prionospio* spp. polychaetes (Blaber *et al.* 1984), the cladocerans *Moina micrura* and *Ceriodaphnia reticulata* (Connell *et al.* 1981), as well as ostracods and veligers of the bivalve *Musculus virgiliae* (Blaber *et al.* 1984).

A study on the diel vertical migration of the mesozooplankton during the closed phase was undertaken in the Kabeljous Estuary in February 1992 (Schlacher & Wooldridge 1995). This has shown that in the absence of any horizontal-displacing movement (tidal or freshwater flows) the pattern of diel vertical ascent/descent for all species is substantially simpler than under open mouth conditions. Predictably though, there are still some marked differences between species. For example, the calanoid copepod *P. hessei* and the mysid *R. terranatalis* exhibit only one bout of upward migration after sunset, while harpacticoid copepods have a more bimodal behaviour, with one peak after sunset and a second one around midnight. Other species show a short peak after sunset, followed by either a progressive disappearance from the water-column during the rest of the night (e.g. *G. lignorum*) or by marked fluctuations throughout the night (e.g. *C. fluviatilis*) (Schlacher & Wooldridge 1995).

A totally different situation is observed during the outflow phase, normally in the spring and summer seasons on the eastern and southern coasts of South Africa. The initial period of river dominance is typified by the sudden, temporary removal of the dominant zooplankton components of the closed phase and the appearance of freshwater species, all the way to the mouth. These may include chironomid and other insect larvae (Blaber *et al.* 1984), several species of cladocerans, ostracods and rotifers (Connell *et al.* 1981, van der Elst *et al.* 1999, Kibirige *et al.* 2006). Following the onset of the tidal phase, with substantial penetration of seawater during flood tides, there is a major change in zooplankton composition with the appearance of numerous neritic taxa. These include many species of copepods (mainly calanoids), gastropod and bivalve veligers, barnacle cypris, chaetognaths and other taxa. Larval forms, including those of fish, anomuran prawns, caridean shrimps, penaeid prawns and brachyurans (Whitfield 1980a, Wooldridge 1991, Forbes *et al.* 1994, Kibirige & Perissinotto 2003a) are prominent and contribute the bulk of the recruitment process into these estuaries (Figure 15).

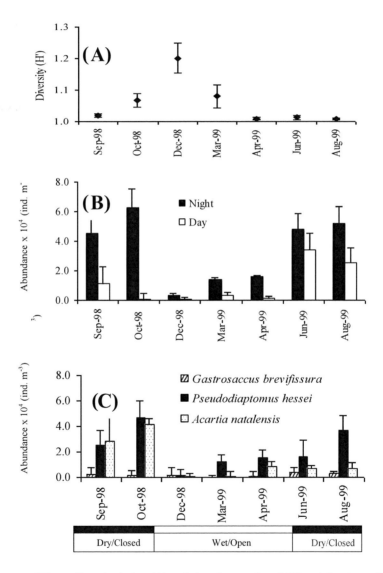

Figure 15. Shannon-Wiener diversity index (A) and abundance of total (B) and dominant (C) components of the zooplankton community of the Mpenjati Estuary during the period September 1998 – August 1999 (source: Kibirige & Perissinotto 2003).

Table 1. Standing stock of zooplankton, mg(dry mass) m^{-3}, in some South African estuaries. *: no data available; POE: permanently open estuary; TOCE: temporarily open-closed estuary; EB: estuarine bay (adapted from Deale 2009)

Zooplankton standing stock, mg(dry mass) m^{-3}				
Estuary	Type	Mean	Max	Reference
East Kleinemonde	TOCE	21	41	Whitfield et al. (2008)
Great Fish	POE	1597	11681	Grange (1992)
Keiskamma	POE	1627	7497	Allanson & Read (1995)
Kariega	POE	38	108	Grange (1992)
Mbotyi	POE	87	109	Wooldridge (1974)
Mdloti	TOCE	127	705	Kibirige et al. (2006)
Mhlanga	TOCE	52	431	Kibirige et al. (2006)
Mpenjati	TOCE	280	2180	Kibirige & Perissinotto (2003a)
Msikaba	POE	15	35	Wooldridge (1976)
Nyara	TOCE	*	2030	Perissinotto et al. (2000)
Richards Bay	EB	174	344	Grindley & Wooldrige (1974)
Swartkops	POE	17	95	Grindley (1981)

The microzooplankton survey carried out in the warm temperate Nyara Estuary in 1997 (Walker et al. 2001) showed relatively large concentrations of ciliates, tintinnids and dinoflagellates during the closed estuarine phase, with numerical abundances of these taxa ranging between 10^4 and 10^5 cells L^{-1}. In terms of carbon content, the most important group was the dinoflagellates, particularly an unidentified species of *Protoperidinium* (Walker et al. 2001). By comparison, total microheterotroph and bacterial densities estimated in the Kasouga Estuary during its closed phase ranged between 5.8-7.0 x 10^3 cells L^{-1} and 1.3-1.7 x 10^5 cells L^{-1}, respectively (Froneman 2006). This appears to be a condition typical of an ecosystem strongly affected by the microbial loop and driven to a large extent by *in-situ* regenerated, rather than new, nutrients. Such a situation will, of course, be reversed during the open phase, when allochthonous nutrients enter the system once again via fresh water inflow and tidal influence.

The high zooplankton diversity observed in permanently open estuaries, compared to TOCEs, is in sharp contrast to the relatively low biomass that individual taxa normally exhibit within their zooplankton communities. On the other hand, the low average diversity of TOCEs is more conducive to the biomass build-up of a few dominant species (Grindley 1981, Wooldridge 1999). Thus, it may be expected that TOCEs achieve higher gross zooplankton productivity rates when compared to permanently open estuaries, especially given their long residence time during an average annual cycle. Prolonged closed mouth phases would minimize competition with neritic species and also reduce the losses due to exchange with the ocean. This would enhance their role as nursery areas for marine and estuarine invertebrate species when compared with permanently open estuaries (Perissinotto et al. 2002, Kibirige & Perissinotto 2003a). However, the recruitment capacity for species with a marine planktonic phase is probably hampered by the short-term and small-scale nature of their communication with the open ocean. The environmental stability of TOCEs and the ability of certain estuarine invertebrates (e.g. specialist estuarine copepods) to conduct their entire life cycle under closed mouth conditions may explain the very high zooplankton biomass levels that have recently been observed in some of these systems, particularly the Nyara in the Eastern Cape (Perissinotto et al. 2000, Walker et al. 2001) and the Mpenjati on the KwaZulu-Natal south coast (Kibirige & Perissinotto 2003a). In the Nyara, night-time abundance estimates ranged from 3.8×10^4 to 6.5×10^6 individuals m^{-3} and dry biomass from 0.002-2.03 g m^{-3}. The Mpenjati exhibited night-time abundance levels of 1.8×10^4 - 2.2×10^5 individuals m^{-3} and a dry biomass of 0.39-2.18 g m^{-3}. In all cases, values were much higher during the closed phase than during the open phase of the estuary. The above biomass levels show a large variation, with values ranging from the highest ever recorded in South African estuaries to almost the lowest (Table 1). Very high zooplankton abundances have also been recorded in Australian ICOLLs, with peak values of about 4.5×10^5 individuals m^{-3} observed in the temperate Wilson Inlet (Gaughan & Potter 1995) and close to 10^6 individuals m^{-3} in the Hopkins River Estuary, Victoria (Newton 1996). These peaks have been attributed to high nutrient loading in response to flooding events and/or to the stability of the water column as a result of the lack of tidal exchanges with the sea. This suggests that there are zooplankton productivity bursts during the closed phase followed by periods of restricted productivity during the open phase (Whitfield 1980a, Perissinotto et al. 2000). The above view is supported by the fact that, on average, the biomass values obtained in TOCEs are 2-5 times higher than the average values reported for the most and the least productive South African permanently open estuaries (Table 1) (Wooldridge 1999, Perissinotto et al. 2000). Combined with the relatively poor phytoplankton stocks observed in the water-column of TOCEs (see earlier section on microalgae), this is a situation that may result in prolonged periods of imbalance between autotrophic food availability and herbivore food demands in the pelagic subsystem.

Mesozooplankton grazing studies were carried out at the Mpenjati between 1998 and 2001 (Kibirige 2002, Kibirige & Perissinotto 2003b). During the winter closed phase of 1999, autotrophic food consumption by the three dominant species of zooplankton at the Mpenjati ranged between 34 and 70 % of water-column biomass, thus suggesting that at times these rates may exceed the total daily phytoplankton production of the estuary (Perissinotto et al. 2003). These rates are very high when compared with those reported from similar studies undertaken in South African permanently open estuaries, where grazing rates are normally within the range of 4-40 % of available phytoplankton biomass (Froneman 2000, Grange 1992). The influence of phytoplankton size-class availability on mesozooplankton grazing in a warm temperate TOCE has recently been investigated by Froneman (2006). Results show that mesozooplankton is able to feed directly on phytoplankton cells only when the assemblage is dominated by the nano size-class. Under other circumstances, and particularly when picophytoplankton dominates the autotrophic stock, mesozooplankton meets its energy budget through a link with the microbial loop. In the process, microheterotrophs act as the main consumers of picophytoplankton and bacteria and, in turn, are ingested by mesozooplankton (Froneman 2006).

Thus, in TOCEs phytoplankton alone may not be able to sustain the entire energetic demands of the consumers throughout the year. Measurements of $\delta^{13}C$ and $\delta^{15}N$ ratios in the three dominant zooplankton species of the Mpenjati and in their possible food sources (particulate organic matter, detritus and microphytobenthos) show that each grazer derives most of its energetic requirements from a specific and unique food source within the same trophic level (Kibirige et al. 2002). This strategy may minimize inter-specific competition and hence improve the utilization of the food sources available in these estuaries.

4.2 Meiofauna

Meiobenthos includes the fraction of benthic invertebrates that falls between macro- and microbenthos, and conventionally refers to organisms that pass through a 1.0 or 0.5 mm sieve but are retained by a 0.045 mm mesh (Mare 1942). Separation on this basis results in the inclusion of larval forms of certain macrobenthic species, which eventually outgrow the meiofauna size range. The permanent meiofauna are usually dominated by nematodes and harpacticoid copepods, followed by mystacocarids, ostracods, halacarid mites, kinorhynchs, gastrotrichs, archiannellids, tardigrads, turbellarians and rotifers. There are also some specialized forms of typically larger-sized organisms such as hydrozoans, nemerteans, bryozoans, gastropods, solenogasters, holothurians, tunicates, priapulids and sipunculids (Giere 1993).

Estuarine meiofaunal invertebrates are almost exclusively marine in origin and both their diversity and abundance tend to decline sharply as physico-chemical conditions depart from those of typical seawater (Coull 1999). The nature of the sediment is also of critical importance to meiofauna; fine silt/mud tend to trap large amounts of clay and organic matter, which in turn will promote the exclusion of specific taxa and also limit the depth at which meiofaunal organisms are able to penetrate (Dye & Barros 2005). As a result of these factors, estuarine meiofauna generally decreases progressively from the mouth (euhaline salinity and coarse sediment) to the upper reaches (oligohaline and fine sediment). TOCEs, in particular, are expected to exhibit the lowest meiofaunal abundance and diversity during the outflow phase when sediment scouring occurs on a large scale.

Very little information is available on meiofaunal diversity and ecology within South African estuaries. An early review by Dye & Furstenburg (1981) was based largely on their own work in the Swartkops Estuary but they also refer to the Kromme and Berg estuaries, all of which are permanently open systems. A later review by de Villiers et al. (1999) added further information from the permanently open Mngazana Estuary (Dye 1983a, b) and the Botriviervlei estuarine lake (de Decker & Bally 1985). Dye & Furstenburg (1981) calculated an average meiofaunal density in open estuaries of ca. 1000 individuals 10 cm^{-2}.

The only published data from a South African TOCE were derived from the Mdloti Estuary in KwaZulu-Natal (Nozais et al. 2005). Meiofauna were composed mainly of nematodes, harpacticoid copepods, crustacean nauplii, mites, turbellarians, polychaetes, oligochaetes, ostracods and chironomid larvae. Abundance was higher during closed mouth periods, peaking at 1785 individuals 10 cm^{-2} but declining to less than 500 individuals 10 cm^{-2} in all months under open mouth conditions (Figure 16). On average, nematodes contributed 61 % numerically to the total meiofauna during the year. These TOCE values are generally lower than meiofaunal abundances recorded in South African permanently open estuaries

(Nozais et al. 2005), as well as those reported for lagoonal systems in the northern hemisphere (Castel 1992, McArthur et al. 2000). Densities of the Mdloti Estuary meiobenthos during closed mouth phases appear to be comparable with figures obtained from permanently open systems such as the Swartkops (Dye & Furstenburg 1981), but are much lower during open mouth conditions.

Potential ingestion rates, estimated from allometric equations, indicated an average impact on the local microphytobenthic biomass of about 11 %, with nematodes, mites and harpacticoid copepods contributing the highest rates. The total range of estimated microphytobenthic biomass consumption by the meiofaunal assemblages of the Mdloti Estuary was from a low of 0.1 % to a maximum of 254 % of the total microphytobenthic stock available (Nozais et al. 2005). The unexpected appearance of the astigmatid mite *Tyrophagus putrescientiae* in the meiobenthic samples of the Mdloti Estuary was reported by Marshall et al. (2001). This species occurred in most months and constituted 98 % of the samples in November 1999, a period when the mouth was open.

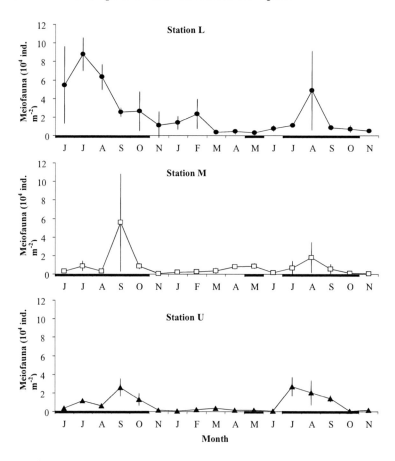

Figure 16. Temporal variations in total meiofaunal abundance (average ± SD) during the period June 1999 – November 2000 in the upper (Stations U), middle (Station M) and lower (Station L) reaches of the Mdloti Estuary. Thickening on the horizontal axis indicates periods of mouth closure (source: Nozais et al. 2005).

In Australian ICOLLs, Dye & Barros (2005) found that meiofaunal abundance generally decreased with increasing distance from the sea. Average total density ranged from 467 ± 28 individuals 10 cm^{-2} at the mouth, to 314 ± 35 individuals 10 cm^{-2} in the middle reaches and 369 ± 30 individuals 10 cm^{-2} in the upper reaches. Also, community structure changed from one dominated by nematodes, copepods and turbellarians at the mouth of the estuary, to one dominated by polychaetes and oligochaetes towards the upper reaches. These authors also found considerable differences in the meiobenthos of natural ICOLLs compared to similar systems that are breached artificially on a regular basis. In particular, mouth manipulation resulted in reduced diversity and abundance, as well increased spatial variability in the meiofaunal assemblages of the mouth region and, surprisingly, also in the upper estuarine reaches (Dye & Barros 2005).

4.3 Macrobenthos

The macrobenthic fauna (defined here as invertebrates larger than 500 μm) of South African estuaries is dominated by errant and sedentary polychaetes; cirripedes, amphipods and isopods amongst the lower Crustacea; penaeid, caridean and anomuran prawns and brachyurans amongst the higher Crustacea; and finally bivalve and gastropod molluscs (Perissinotto *et al.* 2004). While other groups occur, they are far less common. Subtropical species such as the penaeid prawns and some of the crabs tend to become more common further north but many other species are remarkably widespread (Day 1981).

Although there is no simple relationship between species richness and the frequency or duration of the open mouth phase, the macrobenthic fauna of closed systems represents a subset of the much larger number of species that are recorded in permanently open systems (Perissinotto *et al.* 2004). Species that disappear from closed estuaries are not replaced by other more specialized species, and the estuary-associated invertebrates that are found in TOCEs are also present in permanently open estuaries (Teske & Wooldridge 2001).

The most obvious species absence in TOCES are the fiddler crabs *Uca* spp., the soldier crab *Dotilla fenestrata*, *Macrophthalmus* spp. and the marsh and mangrove crabs of the genus *Sesarma*. All of these species, with the possible exception of some of the sesarmids, have a strong dependence on intertidal areas for feeding and the loss of this habitat through the absence of tidal action will be a strong excluding factor (Perissinotto *et al.* 2004). The occurrence of migratory species, which have an obligatory marine phase in their life cycles, such as penaeid prawns, the anomuran mudprawn *Upogebia africana* and the mangrove crab *Scylla serrata* will be dependent on the coincidence of migration periods with an open mouth state. If this condition occurs, substantial recruitment may follow as in the case of *S. serrata* in the West Kleinemonde Estuary (Hill 1975).

The life cycle of *U. africana* has been elucidated by Wooldridge & Loubser (1996), in terms of the migration of the larval phases from the spawning estuarine population to the inshore marine environment and then back into the estuary. If the duration of the TOCE open phase is too short, the *U. africana* late larvae are unable to return to the estuary and the cohort will be lost to the system. Conversely, the sand prawn *Callianassa kraussi* which, like *U. africana*, is a burrower in estuarine sediments, does not have a planktonic larval phase and has no marine phase in its life cycle (Forbes 1973). It is therefore capable of survival in both open and closed estuarine systems where it is usually a major component of the sandy benthos (Perissinotto *et al.* 2004).

On the basis of the above arguments, it is clear that the number of invertebrate taxa surviving in TOCEs is progressively reduced by both the loss of tidal action and the intermittent link with the marine environment. Assuming that the surviving species are independent of the above factors, the next parameter of significance is salinity (de Villiers *et al.* 1999). Low salinity, or even freshwater conditions, do not appear to be a problem for some of the more common TOCE associated species. The benthic fauna of Lake Sibaya, formerly an estuary but now a completely isolated freshwater lake, includes the typical estuarine polychaete *Ceratonereis keiskama* and several crustaceans, such as the tanaid *Apseudes digitalis,* the amphipods *Corophium triaenonyx, Grandidierella lignorum* and *Orchestia ancheidos* and the crab *Hymenosoma orbiculare* (Allanson *et al.* 1966). Therefore low salinity in closed estuaries may be a contributing factor to a reduction in zoobenthic richness when compared to marine dominated permanently open estuaries, but some species are pre-adapted to survive prolonged low salinity or even fresh water conditions in TOCEs. It is perhaps no coincidence that these same species tend to be the most abundant zoobenthic taxa in TOCEs.

The number of species recorded in any one estuary actually depends on a combination of biogeographic region, the predominant mouth state and the degree of riverine influence. Greater invertebrate diversity is typically associated with large, tropical permanently open tidal systems with a strong marine influence. Conversely, permanently open estuaries and river mouths that are dominated by freshwater tend to have a much lower zoobenthic diversity (de Villiers *et al.* 1999), probably due to lower sediment stability and exclusion of some of the more marine taxa. Similarly, TOCEs also have a species diversity about 50 % of that of nearby permanently open estuaries that are marine dominated (Teske & Wooldridge 2001), thus indicating the importance of both mouth phase and salinity regime on macrobenthic diversity.

Although fewer species are usually present in TOCEs, overall invertebrate densities can be higher when compared to permanently open systems (Teske & Wooldridge 2001). In the subtropical Mhlanga Estuary, Whitfield (1980c) recorded nine zoobenthic taxa, with the polychaetes *Ceratonereis erythraensis* and *Dendronereis arborifera* and the amphipod *Corophium triaenonyx* being the most numerous. Blaber *et al.* (1984) recorded nine and 16 taxa respectively in the subtropical Tongati and Mdloti estuaries, with oligochaetes being common in both systems. The combination of the three polychaetes *C. erythraensis, D. arborifera* and *Desdemona ornata* was significant in the Mdloti, while a fourth polychaete *Prionospio* sp. was dominant in the benthos of the Tongati. In all cases the only significant group of freshwater origin was the Chironomidae.

The seven most common zoobenthic taxa in the warm temperate East Kleinemonde Estuary were the amphipods *Corophium triaenonyx* and three species of *Grandidierella*, the tanaid *Apseudes digitalis*, cumacean *Iphinoe truncata* and bivalve *Macoma litoralis* (Wooldridge & Bezuidenhout 2008). The nature of the substratum is the single most important factor influencing the composition of euryhaline subtidal benthic invertebrate assemblages (Teske & Wooldridge 2003), with an estuarine sand and mud fauna being clearly distinguished where other horizontal physico-chemical parameters are more uniform. In the East Kleinemonde Estuary, which does not have a marked vertical or horizontal salinity gradient when closed, the sandy zoobenthic invertebrate community of the mouth region differed from the mud-prefering zoobenthic community found in the rest of the estuary (Wooldridge & Bezuidenhout 2008). Even when the mouth was open and a strong horizontal salinity gradient was created, the composition of the estuarine and euryhaline zoobenthic components of the sandy and muddy areas remained relatively constant, thus indicating the dominant role of sediment type in structuring these TOCE zoobenthos assemblages.

Teske & Wooldridge (2001) investigated 13 Eastern Cape estuaries of which 7 were TOCEs. They recorded 72 taxa, of which a total of 38 (range 14 – 22) occurred in TOCEs, where total individual benthic densities were often two to four times higher than in nearby permanently open systems. Forbes & Demetriades (2008) investigated 12 TOCEs over the 80 km of the sub-tropical Durban coast and recorded a total of 45 taxa (range 10 – 26) in grab samples (Table 2). As in the case of the Eastern Cape estuaries the dominant groups were the polychaete annelids and amphipods, followed by a smaller variety of isopods and other crustaceans as well as bivalve and gastropod molluscs. The low numbers, *viz.* 9, 10 and 11, in three of the Durban systems arguably reflect some anthropogenic degradation of these estuaries. The total of 45 also included two species of penaeid prawns, which would not be strictly part of the burrowing or smaller epibenthic fauna as well as the invasive alien snail *Tarebia granifera* (Appleton *et al.* 2009), which is spreading rapidly through the estuaries and coastal freshwater bodies of the region. However, the benthic grab used in both these studies does not effectively sample deep burrowing species such as the thalassinid prawns *C. kraussi* and *U. africana* or large bivalves such as *Solen capensis*. An overall assessment of differences in zoobenthic biomass and productivity between TOCEs and permanently open estuaries can only be conducted once these deep burrowing taxa have also been assessed in terms of their influence on benthic communities. Recently published work (Pillay *et al.* 2007a, b, c, 2008) has shown that *C. kraussi* can have a major effect on associated benthic communities via impacts on feeding and larval settlement with cascading effects on higher trophic levels.

The loss of epibenthic invertebrates to the sea during the TOCE outflow phase following river flooding may reduce zoobenthic density and biomass during this phase (Whitfield 1980c). In addition, large areas of zoobenthic habitat are exposed in most TOCEs during the outflow and tidal phases, thus temporarily reducing the available area for colonization by benthic invertebrate species, e.g. the surface area of the East Kleinemonde Estuary ranges from only 35 000 m^2 after mouth breaching to 477 000 m^2 when the water level is high during the closed phase (van Niekerk *et al.* 2008). Surface areas of subtropical TOCEs are likely to vary even more widely due to the perched nature of many of these systems, thus making benthic habitat fluctuations a greater factor in this region.

Although TOCE zoobenthic studies in other parts of the world are very limited (Hutchings 1999), indications are that river flooding can be sufficient to flush all salt water from this type of estuary in Australia, thus creating a temporary freshwater system (Sherwood & Rouse 1997). These conditions are sufficient to cause an annual mass mortality of macrobenthos such as the bivalve *Soletellina alba* in the seasonally open Hopkins Estuary (Matthews & Fairweather 2004). Similar macrobenthic mortality in South African TOCEs during the river outflow phase has not been documented, although mass mortality of the bivalve *Solen cylindraceus* under hypersaline conditions and lowered water levels during the closed mouth phase have been observed in the Kasouga Estuary in the Eastern Cape (AKW pers. observ.).

Table 2. Macrobenthic invertebrate taxa of the temporarily open/closed estuaries within the Durban Metropolitan area

Higher taxon	Nearest identifiable taxon	TO	MD	MH	MB	MA	LM	LO	MS	UM	NG	ML	MO
CNIDARIA			■	■		■			■	■		■	
NEMATODA			■	■									
ANNELIDA													
Hirudinea			■							■			■
Oligochaeta			■										
Polychaeta	*Ceratonereis keiskama*	■	■										
	Dendronereis arborifera		■										
	Desdemona ornata		■	■									
	Ficopomatus enigmaticus			■									
	Prionospio malmgreni (?)			■									
	Prionospio sp.		■			■	■						
	Scololepis squamata									■			
	Capitellidae		■	■		■	■	■	■	■	■		
	Cirratulidae				■								
	Glyceridae								■				
	Spionidae												
	Tubificidae	■											
CRUSTACEA													
Amphipoda	*Bolttsia minuta*				■								
	Corophium triaenonyx		■			■							
	Grandidierella lignorum								■				
	Grandidierella lutosa	■								■			
	Grandidierella sp.		■										
	Melita zeylanica									■			
Cumacea					■					■		■	■
Isopoda	*Cirolana* sp.				■	■							
	Cyathura carinata (aestuaria?)				■					■			
	Eurydice longicornis									■			
	Leptanthura laevigata									■			
	Munna sp.									■			
	Pontogeloides latipes								■	■	■		
	Unid. Isopods				■								
Tanaidacea	*Apseudes digitalis*												■
Anomura	*Callianassa kraussi*							■	■	■			
	Upogebia africana							■		■			
Penaeidae	*Metapenaeus monoceros*									■			
	Penaeus indicus			■				■					
	Penaeus monodon	■			■								
Brachyura	*Hymenosoma orbiculare*											■	
	Paratylodiplax blephariskios												
	Rhynchoplax bovis									■			
	Thaumistoplax spiralis				■				■	■			
MOLLUSCA													

Table 2 (Continued)

Higher taxon	Nearest identifiable taxon	TO	MD	MH	MB	MA	LM	LO	MS	UM	NG	ML	MO
Bivalvia	Brachiodontes virgiliae		■						■		■	■	
	Dosinia hepatica												
	Hiatula lunulata								■				■
	Solen cylindraceus								■				
	Tellina prismatica		■						■		■		
Gastropoda	Assiminea ovata												
	Lymnaea sp.												
	Melanoides tuberculata	■	■						■				
	Tarebia granifera					■		■					
	Thiara amarula											■	
INSECTA													
Diptera	Chironomidae	■	■	■	■		■	■		■		■	■
	Total taxa	10	17	18	17	11	9	23	26	21	17	26	22

Source: Forbes & Demetriades 2008

TO: Tongati; MD: Mdloti; MH: Mhlanga; MB: Mbokodweni; MA: Manzimtoti; LM: Little Manzimtoti; LO: Lovu; MS: Msimbazi; UM: Umgababa; NG: Ngane; ML: Mahlongwana; MO: Mahlongwa.

4.4 Fish

Published scientific information from at least 100 South African TOCEs exists on fish studies. These studies have covered a variety of topics, including basic ichthyofaunal surveys (e.g. Millard & Scott 1954, Harrison 1997a, 1997b, 1998, 1999a, 1999b, James & Harrison 2008) and comparative studies including seasonal variations in fish community structure (e.g. Begg 1984a, Bennett 1989, Dundas 1994, Harrison & Whitfield 1995, Cowley & Whitfield 2001a, Vorwerk *et al.* 2001). Other research has examined trophic interactions (e.g. Whitfield 1980b, 1980c), larval fish composition (e.g. Strydom *et al.* 2003, Montoya-Maya & Strydom 2009) and recruitment mechanisms (e.g. Whitfield 1980a, Harrison & Cooper 1991, Cowley *et al.* 2001, Kemp & Froneman 2004, James *et al.* 2007a). In addition to reviews by Perissinotto *et al.* (2004) and James *et al.* (2007b), several studies have focused on the degradation and rehabilitation of estuaries (e.g. Blaber *et al.* 1984, Ramm *et al.* 1987, van der Elst *et al.* 1999). The following is a summary of the most relevant findings from these studies.

From the results of basic fish surveys of west coast estuaries, Harrison (1997a) concluded that the majority of the smaller outlets along the Atlantic Ocean coast are of little value as a habitat for estuary-associated fishes. Due to the arid conditions of the area many of these systems comprise dry riverbeds and only carry water at times of exceptional rainfall (Heydorn 1991). On the southwest Cape coast, Millard & Scott (1954) described the physico-chemical characteristics of the Diep Estuary; fish communities were also included in that study. Heavy rainfall during winter and spring results in the Diep Estuary opening to the sea, with salinity being reduced and turbidity increasing; in summer the estuary mouth closes and the system may become hypersaline (Millard & Scott 1954). In spite of these highly variable environmental conditions, the fishes in the Diep are dominated by estuarine and estuarine-dependent marine species, thus indicating a viable nursery function (Millard & Scott 1954, Harrison 1997b). Indeed, TOCE fish surveys indicate a dominance of estuarine and estuarine-associated marine species, with these systems playing an important nursery role (e.g. Harrison 1998, 1999a, 1999b).

Clark *et al.* (1994) compared the ichthyofauna of the Sand and Eerste estuaries in the Western Cape Province and that of the adjacent surf-zones. These two estuaries were dominated by a few species, with the abundance of marine migrant species being higher in the Eerste and this was attributed to differences in the duration of connection with the sea. Although estuarine and adjacent surf-zone ichthyofauna assemblages were similar, fish densities in the estuaries were considerably higher than those recorded in the adjacent surf-zones and this was ascribed to the higher spring and summer water temperatures in the estuaries, together with the greater protection offered from marine piscivores.

Bennett (1989) compared the fish communities of a permanently open, a seasonally open (TOCE) and a normally closed estuary in the southwestern Cape and reported marked seasonal changes in the fish fauna of the Kleinmond TOCE. In winter, freshwater input breaches the estuary, river currents are strong and temperatures are low; few fish species and individuals are present during this period. With the onset of the summer dry season, freshwater discharge declines, temperature rises and species enter the estuary in increasing numbers. When the system closes, these fishes remain in the estuary until it opens again the following winter (Bennett 1989). A similar cycling of marine larval fish recruitment during the spring and summer open phases in KwaZulu-Natal TOCEs has been recorded, followed by an overall increase in fish biomass during the winter closed phase and then emigration when the mouth breaches, usually the following spring (Whitfield 1980a, Harrison & Whitfield 1995).

River flooding usually leads to TOCE mouth breaching, but mouth opening events can also be caused by high waves lowering the sand bar at the mouth to a level that enables an outflow channel to form. During the outflow phase of the Seekoei, Kabeljous and Van Stadens estuaries in St Francis Bay, juvenile and adult fishes generally decrease in abundance, primarily because of the emigration of large marine migrant species from these systems (Dundas 1994). Conversely, the abundance of larval and early juvenile fishes migrating into these same estuaries increases during mouth opening events (Strydom 2003). Evidence from the subtropical Mhlanga (Whitfield 1980a) and warm-temperate East Kleinemonde (James *et al.* 2008a) estuaries suggest that open mouth phases of approximately one week are sufficient to allow high levels of fish recruitment to occur, provided the estuary opening coincides with seasonal larval and early juvenile abundance in the adjacent coastal zone. Similarly, Griffiths (2001a) has also shown that short open mouth phases (< 7 days) are sufficient to promote fish community stability in a southwestern Australian ICOLL.

Cowley & Whitfield (2001a) described the ichthyofauna characteristics of the East Kleinemonde TOCE. Estuarine resident and estuarine-dependent marine fishes were well represented in the system, with the former group dominating the fish community numerically and the latter group in terms of biomass. The East Kleinemonde was found to have lower species diversity relative to similar subtropical KwaZulu-Natal TOCEs and nearby permanently open estuaries. The reasons given for the lower species diversity were the biogeographical position of the system (warm-temperate), as well as the estuary being closed for much of the year. Indeed, Young *et al.* (1997) concluded that the duration and size of mouth opening, together with the estuary salinity regime, are the single most important factors governing ichthyofaunal composition in seasonally and intermittently closed estuaries.

The dominance of certain estuarine-dependent marine species, such as *Rhabdosargus holubi* in TOCEs has been attributed to their extended spawning season and the ability of postflexion larvae to recruit into the estuary, not only when the mouth opens but also during marine overwash events (Cowley & Whitfield 2001a). Cowley *et al.* (2001) noted that the larval fish assemblage in the surf-zone adjacent to the East Kleinemonde Estuary was mainly one of estuarine-associated marine fishes and that species such as *R. holubi* use overwash events as a mechanism to recruit into estuarine nursery areas (Bell *et al.* 2001, Kemp & Froneman 2004, Vivier & Cyrus 2001). In contrast, overwash events have also been recorded in Australian ICOLLs but no evidence of estuary-associated marine fishes entering these estuaries during those events has been recorded (Griffiths 2001a). The ability of postflexion larvae and early juveniles of estuarine-dependent species to locate TOCEs (Strydom 2003) has been attributed to olfactory cues (James *et al.* 2008b). These olfactory cues may be riverine and/or estuarine in origin and can either enter the adjacent coastal environment following mouth breaching or seepage through the TOCE sandbar at the mouth.

Larval fish assemblages in TOCEs are strongly influenced by mouth state. Whitfield (1980a) highlighted the increased abundance of marine postflexion larvae and early juveniles entering the subtropical Mhlanga Estuary when it opened and Cowley *et al.* (2008) recorded a similar situation in the warm-temperate East Kleinemonde Estuary. Densities of estuarine resident larvae were significantly higher in both estuaries during the closed phase, as breeding by species such as *Gilchristella aestuaria* is favoured by the stable environment prevailing during this phase (Whitfield 1980c, Strydom *et al.* 2003).

Cowley & Whitfield (2001b) found that overall populations of marine migrant species associated with the East Kleinemonde Estuary are characterized by a high degree of inter-annual variability. For example, the total population size increased almost eightfold between the spring of 1994 and the spring of 1995 and this variability was attributed to differences in both abiotic (e.g. estuary mouth state) and biotic conditions (e.g. differential bird predation) between the two periods. The importance of a spring mouth opening for species such as *Lithognathus lithognathus* was highlighted by the study of James *et al.* (2007a), which demonstrated that some fish taxa appear unable to recruit into the estuary during closed phase overwash events.

Vorwerk *et al.* (2001) compared the ichthyofauna of ten estuaries on the former Ciskei coast of the Eastern Cape. Eight of these estuaries were TOCEs and dominated by resident estuarine and estuarine-dependent marine fishes. They noted little seasonal change in the ichthyofaunal structure and attributed this to the predominantly closed nature of these systems preventing large immigrations or emigrations of species. Vorwerk *et al.* (2003) also suggested that the timing, duration and frequency of mouth opening events play an important role in determining fish species composition, diversity and seasonality within TOCEs. Prolonged closed phases in small TOCEs result in low recruitment potential for juvenile marine fish and effectively prevent the emigration of adults back to sea (James *et al.* 2007b). However, these closed phases appear to promote the growth of marine fish species that recruit into the TOCE during the open phase (Griffiths 2001b).

Large TOCEs usually have higher nutrient input, a greater range in water depth, positive salinity gradients and are invariably open for longer periods than small TOCEs. The longer open mouth phase, together with the greater habitat variation in large TOCEs, promotes fish diversity within these systems (Vorwerk *et al.* 2003, Jones & West 2005). Despite the higher fish diversity in permanently open Australian lagoons compared to intermittently open lagoons, the latter systems support a higher ichthyofaunal biomass than the former (Pollard 1994).

Begg (1984a, 1984b) conducted a comparative study of 62 KwaZulu-Natal estuaries using gear which selected small fishes and concluded that predominantly open estuaries were more important nursery areas for estuarine-dependent marine species, while predominantly closed estuaries contained a higher proportion of resident estuarine and freshwater species. From a study of three TOCEs in KwaZulu-Natal, Harrison & Whitfield (1995) found that estuarine-dependent marine species were also an important component of the ichthyofauna of these latter systems. They also noted seasonal changes, with different fish assemblages dominating the ichthyofauna at different periods which was linked to the spawning and migration patterns of the various species, as well as the hydrological regime of each estuary. During winter, these estuaries are normally closed and water level and food and habitat availability are high; at this time, estuarine-dependent marine fishes dominate the fish community. With the onset of the summer rains, these systems breach and adult and sub-adult estuarine-dependent marine species migrate to sea, while juveniles begin recruiting into these systems. Spring and summer are also peak breeding periods for resident estuarine and freshwater species, resulting in an increase in their contribution to the overall ichthyofauna. The reduction in water levels (and water volume) following breaching also results in the concentration of fishes in the lower reaches, further contributing to the increase or dominance in estuarine and freshwater species. The rise in water level, following mouth closure in autumn allows the redistribution of estuarine and freshwater fishes into the upper reaches, leaving estuarine-dependent marine species to dominate the middle and lower reaches (Harrison & Whitfield 1995).

Whitfield (1980a) examined factors influencing the recruitment of juvenile fishes into the Mhlanga Estuary, KwaZulu-Natal. He found that the extended spawning strategy of many estuarine-dependent marine species enabled juveniles to enter the system, which was closed in winter and open in spring. High numbers of species were also reported in November, coinciding with the peak immigration period for this group of fishes (Whitfield 1980a). Harrison & Cooper (1991) described a mechanism whereby juvenile estuarine-dependent marine fishes actively recruit into TOCEs during short open phases, even under strong outflow current conditions. Resident estuarine species also have an extended breeding season, which serves as a buffer against sudden breaching events that lead to a loss of eggs and larvae during the outflow phase (Whitfield 1999). Many estuarine resident species breed during the stable closed phase, thus enabling the larvae to utilise the peak in zooplankton that coincides with the prolonged closed phase which may occur in winter (Whitfield 1980c). A similar versatile spawning strategy by estuarine resident fish species in South Australian ICOLLs has also been recorded, which coincides with stable hydrological conditions and elevated zooplankton food availability (Newton 1996).

The distribution of many fish species in the Mhlanga Estuary was also governed largely by the abundance of the preferred food items (Whitfield 1980b). The most important food source in the Mhlanga was benthic floc (detritus and associated micro-organisms), which supported more than 90 % of the fish biomass in the system (Whitfield 1980c). The highest standing crops of fish food resources were also found to occur during the closed phase and this was attributed to the stability of the physical environment. Following breaching, food resources declined considerably due to the prolonged exposure of large areas of the estuary bed and the flushing effect of floodwaters (Whitfield 1980c).

Whitfield *et al.* (2008) have highlighted the changes in fish community structure of the East Kleinemonde Estuary between the open and closed mouth phases (Figure 17). They have shown that the open phase is vital to the successful recruitment of the full spectrum of estuary-associated marine species and that prolonged mouth closure will lead to an eventual decline in this fish component within the estuary. However, despite the decreasing abundance of these species during the closed phase (Blaber 1973), fish biomass and productivity actually increases under closed mouth conditions (Cowley & Whitfield 2002).

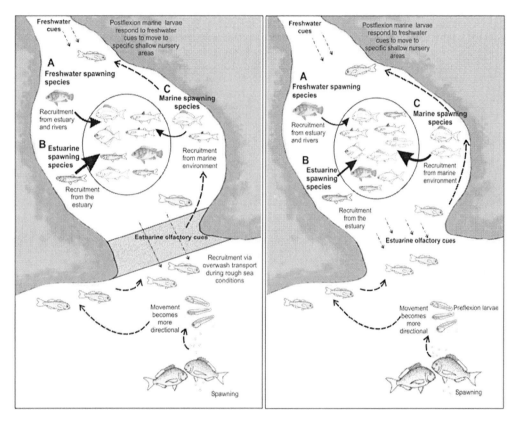

Figure 17. Major contributors to fish community structure in the East Kleinemonde Estuary during a prolonged closed phase (A) and during an open mouth phase (B) (source: Whitfield *et al.* 2008).

Recent research has shown that the supratidal floodplain of Australian ICOLLs can be an important habitat (Becker & Laurenson 2008) and feeding ground (Becker & Laurenson 2007) for certain estuary-associated fish species once the floodplain becomes inundated. Similar studies have not been conducted in South African TOCEs, but Whitfield (1980c) highlighted the connectivity of the Mhlanga Estuary floodplain to the fish productivity of this system during the closed phase. Trophic flexibility by fishes feeding in TOCEs has been emphasized both in South Africa and Australia (Whitfield 1980b, Hadwen *et al.* 2007).

4.5 Birds

Avifaunal counts have been conducted on both permanently open and temporarily open/closed South African estuaries (e.g. Ryan *et al.* 1988, Turpie 2004) yet no attempt has been made to contrast bird composition and utilization in these two estuarine types. Similarly, reviews of the estuarine avifauna by Siegfried (1981) and Hockey & Turpie (1999) have also not covered the topic.

Most bird studies have been concentrated on permanently open estuaries (Kalejta & Hockey 1994) or estuarine lakes (Berruti 1980). More recently, a dedicated study by Terörde (2008) compared avian use of four TOCEs in the Eastern Cape, thus providing a basis to answering many questions around the value of these systems to different bird components. Perissinotto *et al.* (2004) suggested that waders would be poorly represented in TOCEs compared to permanently open estuaries due to the limited intertidal areas in the former estuarine type. The majority of waders use intertidal areas for foraging on surface or shallow burrowing invertebrates and these prey organisms are generally unavailable in closed, non-tidal estuaries. Despite the limited shallow foraging areas for waders, this group comprised approximately 25 % of the bird community during both the open and closed mouth phases in the East Kleinemonde Estuary (Whitfield *et al.* 2008). Piscivorous birds dominated the community, both in terms of absolute numbers and frequency of occurrence (Figure 18).

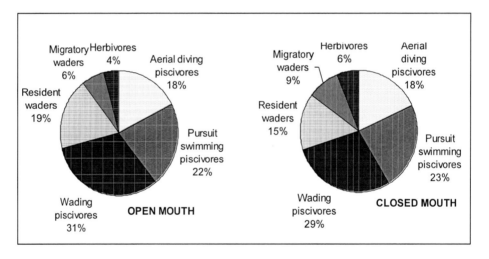

Figure 18. Community composition of the avifauna of the temporarily open/closed East Kleinemonde Estuary during the period February 2005- October 2006 (source: Whitfield *et al.* 2008).

Perissinotto *et al.* (2004) also hypothesized that the greater water clarity in TOCEs would favour piscivorous birds and that this component would tend to dominate the avifauna in TOCEs. Piscivorous birds use a variety of foraging techniques to access fish prey, ranging from wading (e.g. grey heron *Ardea cinerea*), aerial diving (e.g. pied kingfisher *Ceryle rudis*) to pursuit swimming species (e.g. reed cormorant *Phalacrocorax africanus*). In the East Kleinemonde Estuary study, a slight decrease in the abundance of wading piscivores occurred as water levels rose during the closed phase but this was offset by an increase in the abundance of pursuit swimming species as the water became deeper (Terörde 2008). Overall, piscivorous birds were the dominant avifaunal group in the four TOCEs studied by Terörde (2008).

The ability of piscivorous birds to reduce the abundance of favoured fish prey species has been noted by Blaber (1973) in the West Kleinemonde Estuary, where white-breasted cormorant *Phalacrocorax carbo* significantly reduced the population of *Rhabdosargus holubi*. Similarly Cowley (1994) recorded a major decline in the abundance of marine migrant fish species in the East Kleinemonde Estuary following heavy predation by Cape cormorants *Phalacrocorax capensis*.

5. THE FUTURE OF TOCES

A detailed review of the management aspects of South African estuaries is given by Morant & Quinn (1999). Apart from some of the more general and typical management problems, TOCEs are also exposed to some unique issues related to their periodical isolation from the oceanic environment. A preliminary assessment of the conditions of all South African estuaries was made by Whitfield between 1995 and 2000 (Whitfield 2000). This indicates that 112 (61.5 %) out of a total of 182 TOCEs are in good to excellent conditions, with the majority of them situated in the former homeland regions (i.e. Transkei & Ciskei), where little industrial, agricultural and residential development would have occurred compared to the rest of the coastline. The remaining 70 TOCEs (38.5 %) fall within the fair to poor category and are mainly situated in KwaZulu-Natal (44). A more thorough assessment was carried out by Harrison *et al.* (2000), on the basis of estuarine geomorphology, ichthyofauna, water quality and aesthetics. In this case, only 23 TOCEs scored good to excellent on all accounts, while 14 TOCEs scored moderate to very poorly. All the others exhibited either a mixture of low and high scores (92) or had not been sampled for all the components required (53). Under the assumption that the sustainable management of TOCEs is aimed at protecting their ecological integrity, biodiversity and nursery function, the following can be regarded as key issues to be addressed in future plans, in order to ensure that they continue to provide a wide range of goods and services.

5.1 Catchment and Flow Management

Deforestation, wetland destruction and overgrazing in catchment areas are major causes of soil erosion in South Africa. The consequences of this are particularly important for TOCEs, as increased sedimentation and silt loading can have considerable effects on benthic communities and can also reduce the availability of radiation for photosynthetic microalgae within the water-column (Morant & Quinn 1999, Nozais *et al.* 2001). Land use practices can also impact on the runoff characteristics (volumes and time scales) of TOCEs. These issues are particularly important for the management of TOCEs because of their sensitivity to factors that influence mouth dynamics. Most TOCEs in South Africa have small catchments, which enhances their sensitivity to land use changes. It is therefore particularly important for management decision-making to adopt a holistic approach that includes the entire catchment system.

Estuarine water quality can also be affected by leaching of fertilizers and pesticides from the numerous types of plantations that have replaced the natural vegetation in most catchments. Citrus, pineapple and sugar cane are probably the cultures most directly responsible for the agrochemical pollution of South African TOCEs (Begg 1978, Morant & Quinn 1999).

Combinations of abstractions, impoundments and discharges from sewage treatment works have the capacity to change flow regimes, water quality and mouth behaviour, thereby affecting seasonality of habitat availability and fish migration opportunities. Upstream impoundments can have major impacts on runoff inputs into estuaries. Total flow volumes are reduced and the temporal characteristics of the flow are altered. Normally high peak, short duration flows have reduced peaks and longer time scales after routing through a storage reservoir. This can change the flow-duration characteristics of the inflows and thereby the mouth dynamics of TOCEs. Impoundments also change the sediment loadings in rivers by trapping sediment. This may reduce sediment loads into a downstream estuary. Abstractions (e.g. for irrigation) also reduce inflows to estuaries, although in a somewhat different way. Whereas impoundments tend to have their most profound effects on high flows (floods), abstractions tend to occur mostly in low rainfall times and therefore have their most significant impact on low flows. Abstraction can disproportionally influence the base flow, resulting in these systems closing for longer and longer periods and only being opened by episodic extremes. In recent years, even permanently open systems in South Africa (e.g. Mfolozi and Mlalazi estuaries in KwaZulu-Natal) have experienced significant periods of closed mouth conditions as a result of abstraction and impoundments. This could lead to an increased number of such systems in the future being regarded as true TOCEs.

Discharges from wastewater treatment works can significantly increase flows into small estuaries, usually by providing a near constant discharge volume. By contributing significantly to the flow, especially in low-flow periods, these discharges can reduce residence times and play a key role in the mouth dynamics during those periods. Discharge volumes are generally small relative to flood (high) flows and therefore do not significantly affect the influence of these flows on estuary functioning. In assessing the affects of flow changes on the functioning of TOCEs for management purposes, it can therefore be important to distinguish the exact nature of the flow changes, as indicated in the above examples. In the case of changes in high flows, the effect on mouth dynamics can be predicted with high confidence because the water balance of the estuary is inflow-dominated during high flow events, which usually lead to open mouth conditions. The effect of changes to the lower flow regimes can be more uncertain, since in those situations the water balance may no longer be dominated by inflows. Outflows and losses (e.g. due to seepage and evaporation) must also be accounted for in order to predict the affects on water levels and mouth dynamics. Thomas *et al.* (2005) and Lawrie *et al.* (2010) have done detailed investigations of the effects of wastewater discharges on a typical small TOCE, and have shown how the functioning of the estuary can be completely altered by these discharges.

5.2 Artificial Breaching

Until recently, it used to be common practice in South Africa to artificially breach closed estuaries when the water level became what was considered to be too high. This was generally undertaken with the purpose of safeguarding properties, farmland and other manmade structures built below the normal breaching level of the particular system. Thus, the entire functioning of an ecosystem could be jeopardised by township planning undertaken in ignorance of the natural envelope of variability of an estuary (Morant & Quinn 1999). In other cases, artificial breaching was used to flush the system from a build-up of contaminants or sediments (Stretch & Parkinson 2006).

The interference with the mouths of estuaries by artificial breaching and the construction of sandbars is known to have adverse effects on the natural functioning of these systems (Begg 1984b, de Decker 1987, Morant & Quinn 1999, Harrison & Connell 2002). Available evidence strongly indicates that artificial breaching may have long term impacts upon the sediment dynamics, biota and basic functioning of an estuary. Artificial opening of an estuary mouth, when the water level is below that at which breaching naturally occurs, results in a reduced scour potential. In the long term, this may lead to accumulation of sediments in the estuary mouth, thereby compounding the original problem. The broad ecological implications of natural and artificial breaching were highlighted for KwaZulu-Natal estuaries by the 1980s (Begg 1984b). In particular, the community structure of the benthos of the Mdloti Estuary was heavily impacted by the frequent artificial breaching of the mouth, observed 16 times over a two year period (Begg 1984b). This is supported by more recent studies on the meiofauna of the Mdloti Estuary, which have shown that the physico-chemical changes associated with breaching impact heavily on meiofaunal abundance and biomass (Nozais *et al.* 2005). A similar situation has been reported from New South Wales ICOLLs, where meiobenthos is significantly less diverse and more spatially variable in the mouths of artificially breached lakes than in the more natural systems (Dye & Barros 2005). In the Western Cape, artificial breaching of the Bot River Estuary was found to have an immediate impact on the resident benthic community, as well as long-term detrimental effects on species richness and diversity (de Decker 1987). This is indicated by an average occurrence of only about 30 macrobenthic species in this estuary, with only six species consistently abundant and comprising 95 % of the biomass (Bally 1987). This is in contrast to similar but less manipulated Cape estuaries, such as the Klein River, where maximum macrobenthic species diversity can peak at over 100 species (Perissinotto *et al.* 2004).

Many TOCEs along the eastern seaboard of South Africa are utilized by estuarine-dependent marine fishes and, therefore, provide an important nursery function for this group (Harrison & Whitfield 1995, Harrison & Connell 2002). The spawning and migration patterns of these fishes are linked to the natural seasonal breaching pattern of these estuaries (Whitfield 1999). Spawning takes place mainly in late autumn, winter and spring and recruitment into estuaries takes place when increased rainfall and river flow breaches the mouths of these systems (Harrison & Connell 2002). Full breaching is in fact not required, as juvenile fish are able to use wave-overtopping events at high tide as a means of entry into estuaries (Cowley *et al.* 2001). Once the estuaries close, habitat, nutrients and food availability increase thereby providing ideal conditions for the growth and survival of these fishes. Unseasonable flushing of these systems reduces the nursery function by removal of food resources and premature flushing of juveniles. Artificial breaching can have similar impacts on the bird community that utilizes estuarine resources (Quinn *et al.* 1998) and huge changes in avian biomass have been reported following breaches. For instance, in the Bot River Estuary bird biomass fell from a peak of 40000 kg to only 700 kg after two consecutive breaches (Heÿl & Currie 1985).

Figure 19. Controlled artificial breaching undertaken at the Mdloti Estuary on 11 February 2004, in response to a prolonged period of closure, development of anoxic conditions and a fish kill (photo: Nicolette Demetriades).

Currently, artificial breaching is allowed only under specific conditions and authorized by permit and is seen as a last resort management tool under exceptional circumstances, in order to address the effects of reduced inflow due to dams upstream, or to prevent the development of anoxic conditions. In the first instance, this may be required because reduced inflow can greatly increase the closure periods of TOCEs (Stretch & Parkinson 2006). This in turn can lead to the slow deterioration of the ecosystem, and in particular to the deprivation of juvenile fish and invertebrate recruits from the ocean, as well as the prevention of adult migrations back to the ocean. Similarly, the controlled breaching of TOCEs is implemented when oxygen levels become too low and fish and invertebrates start to die, normally as a result of disposal of sewage outlets and other pollutants into the estuarine basin (Stretch & Parkinson 2006) (Figures 19).

The exact timing and detailed consequences of artificial breaching require an understanding of the processes and its impact on the structure and functioning of estuarine ecosystems (Stretch & Parkinson 2006). Recommendations on the methods to be used for artificially breaching an estuary have been layed out. Ideally, the water level in the estuary should be as high as possible prior to the breach. This will result in maximum amounts of sediment being flushed out, thus preventing future sediment build-up in the estuary. The estuary should be breached as late in winter and/or spring as possible and three to four days before spring tide. The position where the mouth should be breached depends on local conditions. If possible, a deeper and wider channel is better than a small and narrow trench. The actual period of breaching during the tidal cycle is at high tide, or as soon after high tide as possible, waves permitting (Van Niekerk *et al.* 2005).

5.4 Flood Plain Encroachment

Due to the shelter they provide against coastal winds and waves, as well as the scenic views over the water, TOCEs have become favourite sites for residential development. In the western half of the Eastern Cape and on the KwaZulu-Natal south coast, the process has escalated to the point that most TOCEs in these areas now exhibit urban/resort settlements, often with dense human populations (Morant & Quinn 1999).

Encroachments on the flood plains of estuaries for agriculture are common in the South African context, e.g. in KwaZulu-Natal sugar cane has often been grown right down to the banks of the normal estuary channel. Such developments generate pressure for artificial mouth breaching during times of high water levels (i.e. closed mouth conditions). Other encroachments onto the estuarine floodplain, such as the construction of road/rail bridges and their associated approach embankments, can influence local scour and sedimentation patterns resulting in mouth-breaching patterns deviating from natural. Since mouth dynamics is a critical driver of the physico-chemical environment of TOCEs, such developments can have a severe impact on the overall functioning of the estuaries concerned and should be included in future management/planning decisions. Unfortunately, this was generally not the case in the past and a large number of road railway and pipeline bridges were constructed over TOCEs, often very close to the sea, thus compromising the ecological functioning of entire estuarine sections (e.g. Heydorn & Bickerton 1982).

Recently, a protocol has been developed to define the core estuarine area of KwaZul-Natal estuaries and the subsequent application of this protocol has shown that major habitat loss has resulted from the floodplain encroachments described (Demetriades & Forbes 2008).

5.5 Sand Mining Operations

Sand mining operations are widespread in South African rivers, particularly in KwaZulu-Natal. Sand is extracted from river beds usually for concrete and fill used in construction and building (Hay *et al.* 2005). Demetriades (2007) found that 18 of the 64 rivers surveyed in KwaZulu-Natal had sand mining activity within the coastal zone. There were 16 individual mining operations in TOCEs. Sand mining causes disturbance and destruction of instream and riparian habitat. The excavation of large amounts of sand removes the ecologically important sediment associated microorganisms. Thus sand mining has a direct impact on the structure and biodiversity of important groups, such as microphytobenthos, meio- and macrofauna, and a cascading negative impact on riverine and estuarine ecosystem functioning. The higher trophic levels, like fishes and birds, are generally driven out of the system due to the noise and the sediment disturbance caused by the sand mining operation. The release of fine sediment into the water column reduces water clarity and smothers benthic plants and animals. There are also the associated oil and fuel spillages which would impact the estuarine biota (Demetriades 2007).

In the 1980's, about 80 000 tons of building sand was removed per annum from the Mgeni Estuary. Apart from the high impact of the extraction, this can fundamentally change the sediment budget and the nature of the sediment available to the estuary (Hay *et al.* 2005).

Rivers and estuaries may deepen due to the depletion of sand which would change the channel hydraulics and influence vertical and longitudinal physico-chemical gradients.

With the growing demand for sand in the building industry, sand mining in estuaries is likely to become more widespread. The national Department of Minerals and Energy (Pretoria) has guidelines in place to ensure minimum negative impact on the environment, as well as rehabilitation upon closure of the sand mining operations. However, the destruction of riparian and associated habitat is extensive and although rehabilitation can take place, the loss of key components of the biotic ecosystem may be irreversible.

Management of sand mining operations should include a thorough assessment of the ecological sensitivity of the proposed site(s) prior to a permit being issued. Currently, impact assessments only consider marginal and adjacent vegetation, with no consideration for biotic components of the river and estuary. There is also a need to understand the total sand yield of an estuary or river, so that the amount of sand that can be removed from a system is determined before it becomes irreversibly degraded. Another important recommendation is to implement regular monitoring of the river/estuary in which sand mining is occurring. This would ensure that the ecosystem is protected and also help address any problems that may arise, either through direct or indirect impacts.

5.6 Exploitation of Bioresources and Recreational Activities

South Africa does not have an abundance of large estuaries and in the global context, with the exception of systems such as St Lucia and Kosi in northern KwaZulu-Natal, is poorly endowed. In addition, its high energy coastline, while often spectacular, cannot be described as user-friendly by comparison with the calm coastal waterways found in many other parts of the world. South Africa's natural inland water bodies are limited and those which do exist are largely artificial impoundments. However, even the smallest of these water bodies present attractions to the boating and/or fishing community, a fact which immediately presents management problems arising out of overexploitation of living resources and user conflict. Estuaries, including TOCEs, many of which constitute very small bodies of water, have not escaped this impact and where access and depth of water permit one can expect that boating, fishing, water ski-ing and the use of jetskis will very often be found, sometimes all on the same systems.

Fishing or the indications of fishing activity in the form of litter, car tracks or campfire sites are generally apparent in and around estuaries. Bait, typically the thalassinid sandprawn *Callianassa kraussi*, is extensively collected from TOCEs and the evidence of this activity is in the form of disturbed sediments in the lower reaches. Most fishing is conducted using a rod and line but a developing problem with far-reaching implications in many TOCEs is the use of gill nets. Lamberth and Turpie (2003) reported that illegal gillnetting is currently responsible for the largest fish catches in South African estuaries, accounting for about 47 % of the total yield. The small size and shallow nature of most TOCEs, which makes virtually the entire habitat accessible and vulnerable, presents a particular problem that requires urgent management attention.

5.7 Cultural Eutrophication

In the last two decades, a rapid population increase in South Africa has coincided with an improvement in the provision of water and hygiene to previously unserviced or poorly serviced communities, particularly those located in peri-urban areas. This has led to unprecedented amounts of treated sewage effluents being disposed of into estuaries, most of them TOCEs. Inorganic nutrient concentrations in the water-column of the recipient estuaries, as well as the sediments, have consequently increased several fold, particularly in terms of dissolved inorganic nitrogen (DIN) and phosphorous (DIP). Agricultural land use in the catchment of these estuaries has also been regarded as responsible for periodical and/or regular peaks in nutrient loading (Begg 1978, Morant & Quinn 1999).

Cases that have been investigated recently in the eThekwini metropolitan area, near Durban, include the Mhlanga and the Mdloti TOCEs (Thomas *et al.* 2005, Perissinotto *et al.* 2006, Kibirige *et al.* 2006). Both systems are currently receiving large volumes of treated sewage water, particularly the Mhlanga where approximately 20 ML of effluent are discharged every day in the upper reaches from two different waste water treatment plants. DIN concentrations in the estuaries have now attained peaks of about 200 µM at the Mdloti and 400 µM at the Mhlanga (Thomas *et al.* 2005). Similarly, DIP concentrations have escalated to maxima of about 15 µM in the Mdloti and 80 µM in the Mhlanga. Prior to the proliferation of sewage treatment works in the area, nutrient loading was seldom in excess of 100 µM for DIN and 5 µM for DIP (Nozais *et al.* 2001), leading to maximum chl-a concentrations in the water-column of 10 µg L^{-1}. Currently, chl-a levels in these estuaries can attain peaks of 100-300 µg L^{-1}, especially during closed mouth conditions, when the residence time of water is sufficiently long to promote the uptake of nutrients by microalgae (Thomas *et al.* 2005).

Because of this, and the more stable water-column that TOCEs experience during closed phases, anoxic or hypoxic conditions develop with regular frequency in such systems. As eutrophication-induced phytoplankton blooms deteriorate and sink towards the bottom, the biological oxygen demand in the water-colum escalates. This leads to the formation of pockets of anaerobic sludge in the deeper parts of the estuary, often associated with a rapid deterioration of oxygen through much of the water-column and the accumulation of hydrogen sulfide within and above the sediment. Fish and invertebrate kills are then inevitable (Figure 20), increasing even further the biological demand for oxygen, hence compounding and amplifying the initial problem.

Shifts in phytoplankton and zooplankton community structure in response to cultural eutrophication have been observed in many coastal areas around the World (e.g. Livingston 2001). Cyanobacteria blooms have already been observed on a few occasions in South African TOCEs, involving mainly a toxic *Microcystis* species (e.g. Sipingo Estuary, Begg

1984b). Winter blooms of filamentous algae, *Cladophora* spp., have also been associated with eutrophication events at the Sipingo and the Mdloti (Begg 1984b, McLean 2008). In the Mhlanga Estuary, it has recently been shown that hypereutrophic conditions may impact negatively on the microzooplankton community, while promoting the dominance of larger mesozooplankton groups (Kibirige *et al.* 2006). These shifts have fundamental implications for the functioning of TOCEs, as an entire food-web may become dysfunctional or significantly impaired in a relatively short period of time.

Figure 20. Large scale fish kill documented at the Mdloti Estuary in February 2004, prior to the controlled artificial breaching of the mouth directed at flushing its anoxic waters out of the estuary (photo: Nicolette Demetriades).

5.8 Climate Change Impacts

Climate changes over long timescales, whether of anthropogenic origin or natural, are expected to have major impacts on the functioning of TOCEs. In particular, changes to rainfall and run-off patterns, together with associated changes in vegetation cover will alter the water budget and mouth dynamics of estuaries. For example, a previously permanently open system may evolve into a temporarily open system or vice versa. Regional climate impact studies for South Africa by Fauchereau *et al.* (2003) and Schulze (2000, 2005, 2006) have started to elucidate the types of effects that may be expected. These effects vary across the country with decreased rainfall predicted in some places and increases in others. They also include predictions of an increase in the frequency of extreme events. These factors and their specific impacts on TOCEs remain an important area for future research.

Sea level rise associated with climate change is also expected to have a significant impact on estuaries. Recent studies by Mather (2007) and Mather *et al.* (2009) estimated the trends in sea levels along the southern African coastline using historical tide gauge records. The records suggest a rising sea level trend of $2 - 3$ mm yr^{-1}. If the trend continues, it will be associated with episodic coastal erosion that can affect the status of TOCEs by shifting the sediment transport balances that determine mouth dynamics and tide-driven exchange flows.

CONCLUSION

South African temporarily open/closed estuaries (TOCEs) and similarly-named systems along the coastline of other regions of the world, especially Australia, are among the most productive aquatic ecosystems. In South Africa, under natural conditions these systems shift seasonally from an open state during the rainy season to a closed phase during the dry part of the year. This allows a whole range of juvenile forms of estuarine-dependent marine species (e.g. 35 species of fish alone) to be recruited into these sheltered and productive systems, where they complete their growth to maturity. They then return back to the ocean once they have reached sexual maturity and are ready to spawn. TOCEs, therefore, play a fundamental nursery function for this biota by fluctuating between an ecological state characterized by high biodiversity but low biomass (open phase) and another with high biomass but low diversity (closed phase). It is the low diversity/competition that allows a few euryhaline species to dominate the ecosystem during the closed phase, thus resulting in the optimal biomass conversion of the resources available. While many of these systems are small, their collective value as a regional resource is highly significant.

Unfortunately, under the impact of global change TOCEs are now undergoing rapid and critical changes that, in some cases, have already led to the loss or the significant deterioration of their natural condition. Land use and erosion in their catchments, fresh water impoundments and abstractions in their river tributaries, mining operations, eutrophication and other types of chemical pollution, overexploitation of their living and non-living resources and invasions by introduced alien species are possibly the main threats that TOCEs are currently facing in South Africa. National and regional legislation has recently been put in place with the specific aim of protecting these invaluable ecosystems. However, it is now essential that enforcement is implemented with the vigour necessary to avoid the loss of the irreplaceable ecosystem goods and services that TOCEs provide.

ACKNOWLEDGMENTS

We thank the Water Research Commission (WRC), the Consortium on Estuarine Research & Management (CERM), the WWF, the National Research Foundation (NRF) and Marine & Coastal Management (MCM) for their financial and logistical support for our estuarine research. S Mitchell, G Bate, T Wooldridge, D Cyrus, PW Froneman, L van Niekerk, S Taljaard and other members of WRC steering committees for Programme K5/1247 and K5/1581 are thanked for their advice. Professors Guy Bate (Nelson Mandela Metropolitan University, Port Elizabeth, South Africa) and Christian Nozais (Université du Québec à Rimouski, Canada) are especially thanked for their critical reviews of this manuscript.

REFERENCES

Abril, G., Commarieu, M. V., Sottolichio, A., Bretel, P. & Guérin, F. (2009). Turbidity limits gas exchange in a large macrotidal estuary. *Estuarine, Coastal and Shelf Science*, *83*, 342-348.

Adams, J. B. & Bate, G. C. (1994). The freshwater requirements of estuarine plants incorporating the development of an estuarine decision support system. *Water Research Commission Report*, No. 292/1/94, 1-151.

Adams, J. B., Knoop, W. T. & Bate, G. C. (1992). The distribution of estuarine macrophytes in relation to freshwater. *Botanica Marina*, *35*, 215-226.

Adams, J. B., Bate, G. C. & O'Callaghan M. (1999). Primary producers. In B.R. Allanson, & D. Baird, (Eds.), *Estuaries of South Africa*, (91-117). Cambridge, UK: Cambridge University Press.

Allanson, B. R., Hill, B. J., Boltt, R. E. & Schulz, V. (1966). An estuarine fauna in a freshwater lake in South Africa. *Nature*, *299*, 532-533.

Allanson, B. R. & Read, G. H. L. (1995). Further comment on the response of Eastern Cape Province estuaries to variable freshwater flows. *Southern African Journal of Aquatic Sciences*, *21*, 56-70.

Anandraj, A., Perissinotto, R. & Nozais, C. (2007). A comparative study of microalgal production in a marine versus a river-dominated temporarily open/closed estuary, South Africa. *Estuarine, Coastal and Shelf Science*, *73*, 768-780.

Anandraj, A., Perissinotto, R., Nozais, C. & Stretch, D. (2008). The recovery of microalgal production and biomass in a South African temporarily open/closed estuary, following mouth breaching. *Estuarine, Coastal and Shelf Science*, *79*, 599-606.

Appleton, C. C., Forbes, A. T. & Demetriades, N. T. (2009). The occurrence, bionomics and potential impacts of the invasive snail *Tarebia granifera* (Lamarck, 1822) (Gastropoda: Thiaridae) in South Africa. *Zoologische Mededelingen Leiden*, *83*, 525-536.

Bate, G. C., Smailes, P. A. & Adams, J. B. (2002). Diatoms as indicators of water quality in South African river and estuarine systems. *Water Research Commission Report*, No. K5/1107, 1-245.

Bally, R. (1987). Conservation problems and management options in estuaries: the Bot River Estuary, South Africa, as a case-history for management of closed estuaries. *Environmental Conservation*, *14*, 45-51.

Beck, J. S. & Basson, G. R. (2008). Klein River Estuary (South Africa): 2D numerical modelling of estuary breaching. *Water SA*, *24*, 33-38.

Beck, J. S., Kemp, A., Theron, A. K., Huizinga, P. & Basson, G. R. (2004). Hydraulics of estuarine sediment dynamics in South Africa: Implications for Estuarine Reserve determination and the development of management guidelines. *Water Research Commission Report*, No. 1257/1/04, 1-188.

Becker, A. & Laurenson, L. J. B. (2007). Seasonal and diel comparisons of the diets of four dominant fish species within the main channel and flood-zone of a small intermittently open estuary in south-eastern Australia. *Marine and Freshwater Research*, *58*, 1086-195.

Becker, A. & Laurenson, L. J. B. (2008). Presence of fish on the shallow flooded margins of a small intermittently open estuary in south eastern Australia under variable flooding regimes. *Estuaries and Coasts*, *31*, 43-52.

Begg, G. W. (1978). The Estuaries of Natal. *Natal Town and Regional Planning Report, 41*, 1-657.

Begg, G. W. (1984a). The comparative ecology of Natal's smaller estuaries. *Natal Town and Regional Planning Report, 62*, 1-182.

Begg, G. W. (1984b). The estuaries of Natal, Part 2. *Natal Town and Regional Planning Report, 55*, 1-631.

Bell, K. N. I., Cowley, P. D. & Whitfield, A. K. (2001). Seasonality in frequency of marine access to an intermittently open estuary: implications for recruitment strategies. *Estuarine, Coastal and Shelf Science, 52*, 305-325.

Bennett, B. A. (1989). A comparison of the fish communities in nearby permanently open, seasonally open and normally closed estuaries in the south-western Cape, South Africa. *South African Journal of Marine Science, 8*, 43-55.

Berruti, A. (1980). Birds of Lake St Lucia. *Southern Birds, 8*, 1-60.

Blaber, S. J. M. (1973). Population size and mortality of the marine teleost *Rhabdosargus holubi* (Pisces: Sparidae) in a closed estuary. *Marine Biology, 21*, 219-225.

Blaber, S. J. M., Hay, D. G., Cyrus, D. P. & Martin T. J. (1984). The ecology of two degraded estuaries on the north coast of Natal, South Africa. *South African Journal of Zoology, 19*, 224-240.

Bonilla, S., Conde, D., Aubriot, L. & Perez, M. D. (2005). Influence of hydrology on phytoplankton species composition and life strategies in a subtropical coastal lagoon periodically connected with the Atlantic Ocean. *Estuaries, 28*, 884-895.

Branch, G. M., Bally, R., Bennett, B., De Decker, H. P., Fromme, G. A. W., Hëyl, C. W. & Willis, J. P. (1985). Synopsis of the impact of artificially opening the mouth of the Bot River estuary: Implications for management. *Transactions of the Royal Society of South Africa, 45*, 465-483.

Breen, C. M. & Hill, B. J. (1969). A mass mortality of mangroves in the Kosi Estuary. *Transactions of the Royal Society of South Africa, 38*, 285-303.

Bruton, M. N. (1980). An outline of the ecology of the Mgobezeleni lake system at Sodwana, with emphasis on the mangrove community. In M.N. Bruton, & K.H. Cooper (Eds.), *Studies on the Ecology of Maputaland*, (408-426). Grahamstown, South Africa: Rhodes University.

Bruton, M. N. & Appleton C. C. (1975). Survey of Mgobezeleni lake-system in Zululand, with a note on the effect of a bridge on the mangrove swamp. *Transactions of the Royal Society of South Africa, 41*, 283-294.

Cahoon, L. B., Nearhoof J. E. & Tilton, C. L. (1999). Sediment grain size effect on benthic microalgal biomass in shallow aquatic ecosystems. *Estuaries, 22*, 735-741.

Calado, G. & Duarte, P. (2000). Modeling growth of *Ruppia cirrhosa*. *Aquatic Botany, 68*, 29-44.

Castel, J. (1992). The meiofauna of coastal lagoon ecosystems and their importance in the food web. *Vie et Milieu, 42*, 125-135.

Clark, B. M., Bennett, B. A. & Lamberth, S. J. (1994). A comparison of the ichthyofauna of two estuaries and their adjacent surf zones, with an assessment of the effects of beach-seining on the nursey function of estuaries for fish. *South African Journal of Marine Science, 14*, 121-131.

Colloty, B. M., Adams, J. B. & Bate, G. C. (2002). Classification of estuaries in the Ciskei and Transkei regions based on physical and botanical characteristics. *South African Journal of Botany, 68*, 312-321.

Conley, D. J., Paerl, H. W., Howarth, R. W., Boesch, D. F., Seitzinger, S. P., Havens, K. E., Lancelot, C. & Likens, G. E. (2009). Controlling Eutrophication: Nitrogen and Phosphorous. *Science, 323*, 1014-1015.

Connell, A. D., McClurg, T. P., Stanton, R. C., Engelbrecht, E. M., Stone, V. C. & Pearce, Z. N. (1981). *The Siyaya River study (nutrient balance in an estuarine system): Progress Report for the period up to July 1980. Cooperative Scientific Programmes, Inland Water Ecosystems Annual Report 1981*. Pretoria, South Africa: Council for Scientific and Industrial Research.

Cooper, J. A. G. (1989). Fairweather versus flood sedimentation in Mhlanga lagoon, Natal: implications for environmental management. *South African Journal of Geology, 92*, 279-294.

Cooper, J. A. G. (1990). Ephemeral stream-mouth bars at flood-breach river-mouths: comparison with ebb-tidal deltas at barrier inlets. *Marine Geology, 95*, 57-70.

Cooper, J. A. G. (1993a). Sedimentation in a river-dominated estuary. *Sedimentology, 40*, 979-1017.

Cooper, J. A. G. (1993b). Sedimentation in the cliff-bound, microtidal Mtamvuna estuary, South Africa. *Marine Geology, 112*, 237-256.

Cooper, J. A. G. (1993c). Sedimentary processes in the river-dominated Mvoti estuary, South Africa. *Geomorphology, 9*, 271-300.

Cooper, J. A. G. (1994a). Sedimentation in a river-dominated estuary. *Sedimentology, 40*, 979-1017.

Cooper, J. A. G. (1994b). Lagoons and microtidal coasts. In R.W.G. Carter, & C. Woodroffe (Eds.), *Coastal Evolution*, (219-265). Cambridge, UK: Cambridge University Press.

Cooper, J. A. G. (2001). Geomorphological variability among microtidal estuaries from the wave dominated South African coast. *Geomorphology, 40*, 99-122.

Cooper, J. A. G., Harrison, T. D., Ramm, A. E. L. & Singh, R. A. (1993). *Refinement, enhancement and application of the estuarine health index to Natal's estuaries, Tugela-Umtamvuna. Technical Report Ref. No. EFP 05.930401*. Pretoria, South Africa: Department of Environmental Affairs.

Cooper, J. A. G., Mason, T. R., Reddering, J. S. V. & Illenberger, W. K. (1990). Geomorphological effects of catastrophic fluvial flooding on a small subtropical estuary. *Earth Surface Processes and Landforms, 15*, 25-41.

Cooper, A., Wright, I. & Mason, T. (1999). Geomorphology and sedimentology. In B.R. Allanson, & D. Baird (Eds.), *Estuaries of South Africa*, (5-25). Cambridge, UK: Cambridge University Press.

Coull, B. C. (1999). Role of meiobenthos in estuarine soft-bottom habitats. *Australian Journal of Ecology, 24*, 327-343.

Cowley, P. D. (1994). *Fish population dynamics in a temporarily open/closed South African estuary*. PhD thesis. Grahamstown, South Africa: Rhodes University.

Cowley, P. D. & Whitfield, A. K. (2001a). Ichthyofaunal characteristics of a typical temporarily open/closed estuary on the southeast coast of South Africa. *Ichthyological Bulletin of the J.L.B. Smith Institute of Ichthyology, 71*, 1-19.

Cowley, P. D. & Whitfield, A. K. (2001b). Fish population size estimates from a small intermittently open estuary in South Africa, based on mark-recapture techniques. *Marine and Freshwater Research, 52*, 283-290.

Cowley, P. D. & Whitfield, A. K. (2002). Biomass and production estimates of a fish community in a small South African estuary. *Journal of Fish Biology, 61*, 74-89.

Cowley, P. D., Whitfield, A. K. & Bell, K. N. I. (2001). The surf zone ichthyoplankton adjacent to an intermittently open estuary, with evidence of recruitment during marine overwash events. *Estuarine, Coastal and Shelf Science, 52*, 339-348.

Cowley, P. D., Muller, C. M., James, N. C., Strydom, N. A. & Whitfield, A. K. (2008). An Intermediate Ecological Reserve Determination Study of the East Kleinemonde Estuary, Appendix J, Specialist Report: Fish. *Water Research Commission Report, No. 1581/2/08*, 180-191.

C. S. I. R. (1989). The impact of the September 1987 floods on the estuaries of Natal/KwaZulu: a hydro-photgraphic perspective. *Coucil for Scientific and Industrial Research Report, 640*, 1-22.

Day, J. H. (1981). *Estuarine Ecology with particular reference to Southern Africa*. Cape Town, South Africa: A. A. Balkema.

Deale, M. (2009). *Recovery dynamics of zooplankton following a mouth-breaching event in the temporarily open Mdloti Estuary*. MSc Thesis. Durban, South Africa: University of KwaZulu-Natal.

de Decker, H. P. (1987). Breaching the mouth of the Bot River estuary, South Africa: Impact on its benthic macrofaunal communities. *Transactions of the Royal Society of South Africa, 46*, 231-250.

de Decker, H. P. & Bally, R. (1985). The benthic macrofauna of the Bot River estuary, South Africa, with a note on its meiofauna. *Transactions of the Royal Society of South Africa, 45*, 379-396.

Demetriades, N. T. (2007). *An inventory of sandmining operations in Kwazulu Natal estuaries: Mtamvuna to Thukela. Investigational report for Coastwatch*. Durban, South Africa: Wildlife and Environment Society of Southern Africa.

Demetriades, N. T. & Forbes, A. T. (2008). *Biodiversity Management Guidelines for Estuarine Ecosystems in KwaZulu-Natal. Specialist Report for Land Resources International, December 2008*. Durban, South Africa: Land Resources International.

de Villiers, C. J., Hodgson, A. N. & Forbes, A. T. (1999). Studies on estuarine macroinvertebrates. In B.R. Allanson, & D. Baird (Eds.), *Estuaries of South Africa*, (167-207). Cambridge, UK: Cambridge University Press.

Duarte, C. M. (1995). Submerged aquatic vegetation in relation to different nutrient regimes. *Ophelia, 41*, 87-112.

Dundas, A. (1994). *A comparative analysis of fish abundance and diversity in three semi-closed estuaries in the Eastern Cape*. MSc Thesis. Port Elizabeth, South Africa: University of Port Elizabeth.

Dye, A. H. (1983a). Composition and seasonal fluctuations of meiofauna in a southern African mangrove estuary. *Marine Biology, 73*, 165-170.

Dye, A. H. (1983b). Vertical and horizontal distribution of meiofauna in mangrove sediments in Transkei, southern Africa. *Estuarine, Coastal and Shelf Science, 16*, 91-598.

Dye, A. H. & Barros, F. (2005). Spatial patterns in meiobenthic assemblages in intermittently open/closed coastal lakes in New South Wales, Australia. *Estuarine, Coastal and Shelf Science*, *62*, 575-593.

Dye, A. H. & Furstenburg, J. P. (1981). Estuarine meiofauna. In J.H. Day (Ed.), *Estuarine Ecology with particular reference to southern Africa*, (179-186). Cape Town, South Africa: A. A. Balkema.

Eyre. B. (1998). Transport, retention and transformation of material in Australian estuaries. *Estuaries*, *21*, 540-551.

Fauchereau, N., Trzaska, S., Rouault, M. & Richard Y. (2003). Rainfall variability and changes in southern Africa during the 20th century in the global warming context. *Natural Hazards*, *29*, 139-154.

Forbes, A. T. (1973). An unusual abbreviated larval life in the estuarine burrowing prawn *Callianassa kraussi* (Crustacea: Decapoda: Thalassinidea). *Marine Biology*, *22*, 361-365.

Forbes, A. T. & Demetriades, N. T. (2008). *Estuaries of Durban, KwaZulu-Natal. Report for the Environmental Management Department, Ethekwini Municipality.* Durban, South Africa: Ethekwini Municipality.

Forbes, A. T., Niedinger, S. & Demetriades, N. T. (1994). Recruitment and utilisation of nursery grounds by penaeid prawn postlarvae in Natal, South Africa. In K. R. Dyer, & R. J. Orth, (Eds), *Changes in fluxes in estuaries: implications from science to management*, (379-384). Fredensborg, Denmark: Olsen & Olsen.

Froneman, P. W. (2000). Feeding studies on selected zooplankton in a temperate estuary, South Africa. *Estuarine, Coastal and Shelf Science*, *51*, 543-552.

Froneman, P. W. (2002). Response of the plankton to three different hydrological phases of the temporarily open/closed Kasouga Estuary, South Africa. *Estuarine, Coastal and Shelf Science*, *55*, 535-546.

Froneman, P. W. (2006). The importance of phytoplankton size in mediating trophic interactions within the plankton of a southern African estuary. *Estuarine, Coastal and Shelf Science*, *70*, 693-700.

Gale, E., Pattiaratchi, C. & Ranasinghe, R. (2006). Vertical mixing processes in Intermittently Closed and Open Lakes and Lagoons, and the dissolved oxygen response. *Estuarine, Coastal and Shelf Science*, *69*, 205-216.

Gama, P. T., Adams, J. B., Schael, D. M. & Skinner T. (2005). Phytoplankton chlorophyll *a* concentration and community structure of two temporarily open/closed estuaries. *Water Research Commission Report, No. 1255/1/05*, 1- 91.

Gaughan, D. J. & Potter, I. C. (1995). Composition, distribution and seasonal abundance of zooplankton in a shallow seasonally closed estuary in temperate Australia. *Estuarine, Coastal and Shelf Science*, *41*, 117-135.

Gesti, J., Badosa, A. & Qunitana, X. D. (2005). Reproductive potential in *Ruppia cirrhosa* (Pentagna) Grande in response to water permanence. *Aquatic Botany*, *81*, 191-198.

Giere, O. (1993). *Meiobenthology – the Microscopic Fauna in Aquatic Sediments*. Berlin, Germany: Springer.

Gillanders, B. & Kingsford, M. J. (2002). Impact of changes in flow of freshwater on estuarine and open coastal habitats and associated organisms. *Oceanography and Marine Biology, an Annual Review*, *40*, 233-309.

Gobler, C. J., Cullison, L. A., Koch, F., Harder, T. M. & Krause, J. W. (2005). Influence of freshwater flow, ocean exchange and seasonal cycles on phytoplankton-nutrient dynamics in a temporarily open estuary. *Estuarine, Coastal and Shelf Science, 65*, 275-288.

Grange, N. (1992). The influence of contrasting freshwater inflows on the feeding ecology and food resources of zooplankton in two Cape estuaries, South Africa. *PhD Thesis.* Grahamstown, South Africa: Rhodes University.

Grange, N., Whitfield, A. K., De Villiers, C. J. & Allanson, B. R. (2000). The response of two South African east coast estuaries to altered river flow regimes. *Aquatic Conservation: Marine and Freshwater Ecosystems, 10*, 155-177.

Gray, J. S. (1981). The ecology of marine sediments, an introduction to the structure and function of benthic communities. *Cambridge Studies in Modern Biology, 2*. Cambridge, UK: Cambridge University Press.

Greer, K. & Stow, D. (2003). Vegetation type conversion in Los Peñasquitos Lagoon, California: An examination of the role of watershed urbanization. *Environmental Management, 31*, 489-500.

Grenfell, S. E. & Ellery, W. N. (2009). Hydrology, sediment transport dynamics and geomorphology of a variable flow river: Mfolozi River, South Africa. *Water SA, 35*, 271-282.

Grenfell, S. E., Ellery, W. N. & Grenfell, M. C. (2009). Geomorphology and dynamics of the Mfolozi River floodplain, KwaZulu-Natal, South Africa. *Geomorphology, 107*, 226-240.

Griffiths, S. P. (2001a). Factors influencing fish composition in an Australian intermittently open estuary. Is stability salinity-dependent? *Estuarine, Coastal and Shelf Science, 52*, 739-751.

Griffiths, S. P. (2001b). Recruitment and growth of juvenile yellowfin bream, *Acanthopagrus australis* Günther (Sparidae), in an Australian intermittently open estuary. *Journal of Applied Ichthyology, 17*, 240-243.

Grindley, J. R. (1981). Estuarine plankton. In J.H. Day (Ed.), *Estuarine Ecology, with particular reference to southern Africa*, (117-146). Cape Town, South Africa: A. A. Balkema.

Grindley J. R. & Wooldridge, T. H. (1974). The Plankton of Richards Bay. *Hydrobiological Bulletin, 8*, 201-212.

Grobbler, N. G., Mason, T. R. & Cooper, J. A. G. (1987). *Sedimentology of Mtati Lagoon. S.E.A.L. Report 3*. Durban, South Africa: University of Natal.

Hadwen, W. L. & Arthington, A. H. (2008). Food webs of two intermittently open estuaries receiving ^{15}N-enriched sewage effluent. *Estuarine, Coastal and Shelf Science, 71*, 347-358.

Hadwen, W. L., Russell, G. L. & Arthington, A. H. (2007). Gut content- and stable isotope-derived diets of four commercially and recreationally important fish species in two intermittently open estuaries. *Marine and Freshwater Research, 58*, 363-375.

Harrison, T. D. (1997a). A preliminary survey of coastal river systems on the South African west coast, Orange River – Groot Berg, with particular reference to the fish fauna. *Transactions of the Royal Society of South Africa, 52*, 277-321.

Harrison, T. D. (1997b). A preliminary survey of coastal river systems on the south-west coast of South Africa, Cape Columbine – Cape Point, with particular reference to the fish fauna. *Transactions of the Royal Society of South Africa, 52*, 323-344.

Harrison, T. D. (1998). A preliminary survey of coastal river systems of False Bay, south-west coast of South Africa, with particular reference to the fish fauna. *Transactions of the Royal Society of South Africa, 53*, 1-31.

Harrison, T. D. (1999a). A preliminary survey of estuaries on the south-west coast of South Africa, Cape Hangklip – Cape Agulhas, with particular reference to the fish fauna. *Transactions of the Royal Society of South Africa, 54*, 257-283.

Harrison, T. D. (1999b). A preliminary survey of estuaries on the south coast of South Africa, Cape Agulhas – Cape St Blaize, Mossel Bay, with particular reference to the fish fauna. *Transactions of the Royal Society of South Africa, 54*, 285-310.

Harrison, T. D. & Connell, A. (2002). Biotic response to closed estuary mouth breaching. *Final Report to the Department of Environmental Affairs and Tourism, Project Number ERD14*. Pretoria, South Africa: Department of Environmental Affairs and Tourism.

Harrison, T. D. & Cooper, J. A. G. (1991). Active migration of juvenile grey mullet (Teleostei: Mugilidae) into a small lagoonal system on the Natal coast. *South African Journal of Science, 87*, 395-396.

Harrison, T. D., Cooper, J. A. G. & Ramm, A. E. L. (2000). *State of South African Estuaries: Geomorphology, Ichthyofauna, Water Quality and Aesthetics*. Pretoria, South Africa: Department of Environmental Affairs and Tourism.

Harrison, T. D. & Whitfield, A. K .(1995). Fish community structure in three temporarily open/closed estuaries on the Natal coast. *Ichthyological Bulletin of the J.L.B. Smith Institute of Ichthyology, 64*, 1-80.

Hay, D., Huizinga, P. & Mitchell, S. (2005). Managing Sedimentary Processes in South African Estuaries: A Guide. Prepared for the Water Research Commission by the Institute of Natural Resources, *Water Research Commission Report, No. TT 241/04*, 1-68.

Heydorn, A. E. F. (1991). The conservation status of southern African estuaries. In B.J. Huntley, (Ed.), *Biotic Diversity in Southern Africa. Concept and Conservation*, (290-297). Cape Town, South Africa: Oxford University Press.

Heydorn, A. E. F. & Bickerton, I. B. (1982). Report No. 9: Uilkraals (CSW 17). In: Heydorn AEF, Grindley JR (eds.), Estuaries of the Cape, Part 2. Synopses of available information on individual systems. *Council for Scientific and Industrial Research Report, 408*, 1-37.

Heydorn, A. E. F. & Tinley, K. L. (1980). Estuaries of the Cape. Part I. Synopsis of the Cape Coast – Natural Features, Dynamics and Utilization. *Council for Scientific and Industrial Research Report, 380*, 1-97.

Heÿl, C. W. & Currie, M. H. (1985). Variations in the use of the Bot River estuary by waterbirds. *Transactions of the Royal Society of South Africa, 45*, 397-418.

Hill, B. J. (1975). Abundance, breeding and growth of the crab *Scylla serrata* in two South African estuaries. *Marine Biology, 32*, 119-126.

Hill, B. J. (1966). A contribution to the ecology of the Umlalazi Estuary. *Zoologica Africana, 2*, 1-24.

Hockey, P. & Turpie, J. (1999). Estuarine birds in South Africa. In B. R. Allanson & D. Baird, (Eds.), *Estuaries of South Africa*, (235-268). Cambridge, UK: Cambridge University Press.

Howard-Williams, C. & Allanson, B. R. (1979). *The ecology of Swartvlei: research for planning and future management*. Pretoria, South Africa: Water Research Commission.

Huizinga, P. (2000). Introduction and Discussion on Water Flow and Mouth Management of Estuaries. *MCM/DEAT National Estuaries Workshop, UPE, May 2000.* Pretoria, South Africa: Department of Environmental Affairs and Tourism.

Huizinga, P. & van Niekerk, L. (2002). *Mouth Dynamics and Hydrodynamics of Estuaries.* Coastal Engineering Course, University of Stellenbosch, Sept. 2002. Stellenbosch, South Africa: University of Stellenbosch.

Hutchings, P. (1999). Taxonomy of estuarine invertebrates in Australia. *Australian Journal of Ecology, 24,* 381-394.

James, N. C., Cowley, P. D. & Whitfield, A. K. (2007a). Abundance, recruitment and residency of two sparid fishes in an intermittently open South African estuary. *African Journal of Marine Science, 29,* 527-538.

James, N. C., Cowley, P. D., Whitfield, A. K. & Lamberth, S. (2007b). Fish communities in temporarily open/closed estuaries from the warm- and cool-temperate regions of South Africa – A review. *Reviews in Fish Biology and Fisheries, 17,* 565-580.

James, N. C., Whitfield, A. K. & Cowley, P. D. (2008a). Long-term stability of the fish assemblages in a warm-temperate South African estuary. *Estuarine, Coastal and Shelf Science, 76,* 723-738.

James, N. C., Cowley, P. D., Whitfield, A. K. & Kaiser H. (2008b). Choice chamber experiments to test the attraction of postflexion *Rhabdosargus holubi* larvae to water of estuarine and riverine origin. *Estuarine, Coastal and Shelf Science, 77,* 143-149.

James, N. C. & Harrison, T. D. (2008). A preliminary survey of the estuaries on the south coast of South Africa, Cape St Blaize, Mossel Bay – Robberg Peninsula, Plettenberg Bay, with particular reference to the fish fauna. *Transactions of the Royal Society of South Africa, 63,* 111-127.

Jarvis, A., Reuter, H. I., Nelson, A. & Guevara, E. (2008). *Hole-filled seamless SRTM data V4,* International Centre for Tropical Agriculture (CIAT), available from http://srtm.csi.cgiar.org.

Jones, M. V. & West, R. J. (2005). Spatial and temporal variability of seagrass fishes in intermittently closed and open coastal lakes in southeastern Australia. *Estuarine, Coastal and Shelf Science, 64,* 277-288.

Kalejta, B. & Hockey, P. A. R. (1994). Distribution of shorebirds at the Berg River estuary, South Africa, in relation to foraging mode, food supply and environmental features. *Ibis, 136,* 33-239.

Karez, R., Engelbert, S., Kraufvelin, P., Pedersen, M. F. & Sommer, U. (2004). Biomass response and changes in composition of ephemeral macroalgal assemblages along an experimental gradient of nutrient enrichment. *Aquatic Botany, 78,* 103-117.

Kemp, W. M. (1989). Estuarine Chemistry. In J.H. Day, C.A.S. Hall, W.M. Kemp (Eds.), *Estuarine Ecology,* (79-143). London, UK: John Wiley & Sons.

Kemp, J. O. G. & Froneman P. W. (2004). Recruitment of ichthyoplankton and macrozooplankton during overtopping events into a temporarily open/closed southern African estuary. *Estuarine, Coastal and Shelf Science, 61,* 529-537.

Kibirige, I. (2002). *The structure and trophic role of the zooplankton community of the Mpenjati Estuary, a subtropical and temporarily-open system on the Kwazulu-Natal coast.* PhD Thesis. Durban, South Africa: University of Durban-Westville.

Kibirige, I. & Perissinotto, R. (2003a). The zooplankton community of the Mpenjati Estuary, a South African temporarily open/closed system. *Estuarine, Coastal and Shelf Science*, *58*, 727-741.

Kibirige, I. & Perissinotto, R. (2003b). In situ feeding rates and grazing impact of zooplankton in a South African temporarily open estuary. *Marine Biology*, *142*, 357-367.

Kibirige, I., Perissinotto, R. & Nozais, C. (2002). Alternative food sources of zooplankton in a temporarily-open estuary: evidence from $\delta^{13}C$ and $\delta^{15}N$. *Journal of Plankton Research*, *24*, 1089-1095.

Kibirige, I., Perissinotto, R. & Thwala, X. (2006). A comparative study of zooplankton dynamics in two subtropical temporarily open/closed estuaries, South Africa. *Marine Biology*, *148*, 1307-1324.

Kraus, N. C., Patsch, K. & Munger, S. (2008). Barrier beach breaching from the lagoon side, with reference to Northern California. *Shore & Beach*, *76*, 33-43.

Lamberth, S. J. & Turpie, J. K. (2003). The role of estuaries in South African fisheries: economic importance and management implications. *African Journal of Marine Science*, *25*, 1-27.

Lawrie, R. A., Stretch, D. D. & Perissinotto, R. (2010). The effects of wastewater discharges on the functioning of a small temporarily open/closed estuary. *Estuarine, Coastal and Shelf Science* (in press).

Livinston, R. J. (2001). *Eutrophication Processes in Coastal Systems*. Boca Raton, FA: CRC Press.

Lukatelich, R. J. & McComb, A. J. (1986). Distribution and abundance of benthic microalgae in a shallow southwestern Australian estuarine system. *Marine Ecology Progress Series*, *27*, 287-297.

Mare, M. F. (1942). A study of a marine benthic community with special reference to the micro-organisms. *Journal of the Marine Biological Association UK*, *55*, 517-554

Marshall, D. J., Perissinotto, R., Nozais, C., Haines, C. J. & Proches, S. (2001). Occurrence of the astigmatid mite *Tyrophagus* in estuarine benthic sediments. *Journal of the Marine Biological Association UK*, *81*, 889-890.

Mather, A. A. (2007). Linear and nonlinear sea level changes at Durban, South Africa. *South African Journal of Science*, *103*, 509-512.

Mather, A. A., Garland, G. G. & Stretch, D. D. (In press). Southern African sea levels: corrections, influences and trends. *African Journal of Marine Science*, *31*, XX-XX.

Matthews, T. G. & Fairweather, P. G. (2004). Effect of lowered salinity on the survival, condition and reburial of *Soletellina alba* (Lamarck, 1818) (Bivalvia: Psammobiidae). *Austral Ecology*, *29*, 250-257.

McArthur, V. E., Koutsoubas, D., Lampadariou, N. & Dounas, C. (2000). The meiofaunal community structure of a Mediterranean lagoon (Gialova Lagoon, Ionian Sea). *Helgoland Marine Research*, *54*, 7-17.

McLean, C. T. (2008). *Seven estuaries within the Ethekwini Municipal Area: Aspects of their ecology and implications for management.* MSc Thesis. Durban, South Africa: University of KwaZulu-Natal.

Midgley, D. C., Pitman, W. V. & Middleton, B. J. (1994). Surface Water Resources of South Africa 1990. *Water Research Commission Report, No. 298/5.1/94*, 1-263.

Millard, N. A. H. & Scott, K. M. F. (1954). The ecology of South African estuaries, Part VI: Milnerton estuary and the Diep River, Cape. *Transactions of the Royal Society of South Africa*, *34*, 279-324.

Miller, D. C., Geider, R. J. & MacIntyre, H. L. (1996). Microphytobenthos: The ecological role of the "secret garden" of unvegetated, shallow-water marine habitats. II: Role in sediment stability and shallow water food-webs. *Estuaries*, *19*, 202-212.

Monteiro, P. M. S. & Largier, J. L. (1999). Thermal stratification in Saldanha Bay (South Africa) and subtidal, density-driven exchange with the coastal waters of the Benguela upwelling system. *Estuarine, Coastal and Shelf Science*, *49*, 877-890.

Montoya-Maya, P. H. & Strydom, N. A. (2009). Description of larval fish composition, abundance and distribution in nine south and west coast estuaries of South Africa. *African Zoology*, *44*, 75-92.

Morant, P. & Quinn, N. (1999). Influence of Man and management of South African estuaries. In B. R. Allanson, & D. Baird, (Eds.), *Estuaries of South Africa*, (289-320). Cambridge, UK: Cambridge University Press.

Mortazavi, B., Iverson, R. L., Landing, W. M., Lewis, F. G. & Huang, W. (2000). Control of phytoplankton production and biomass in a river-dominated estuary: Apalachicola Bay, Florida, USA. *Marine Ecology Progress Series*, *198*, 19-31.

Mundree, S. (2001). *Dynamics of the microphytobenthic community of a temporarily-open estuary: Mdloti, KwaZulu-Natal*. MSc Thesis. Durban, South Africa: University of Durban Westville.

Newton, G. M. (1996). Estuarine ichthyoplankton ecology in relation to hydrology and zooplankton dynamics in a salt-wedge estuary. *Marine and Freshwater Research*, *47*, 99-111.

Nozais, C., Perissinotto, R. & Mundree, S. (2001). Annual cycle of microalgal biomass in a South African temporarily-open estuary: nutrient versus light limitation. *Marine Ecology Progress Series*, *223*, 39-48.

Nozais, C., Perissinotto, R. & Tita, G. (2005). Seasonal dynamics of meiofauna in a South African temporarily open/closed estuary (Mdloti Estuary, Indian Ocean). *Estuarine, Coastal and Shelf Science*, *62*, 325-338.

O'Callaghan, M. (1990). The ecology of the False bay estuarine environments, Cape, South Africa, 1. The coastal vegetation. *Bothalia*, *20*, 105-112.

Oliff, W. D. (1976). *National Marine Pollution Monitoring Program. First Annual Report (509 pp) & Second Annual Report, (172)*. Durban, South Africa: National Institute of Water Research.

Parkinson, M. G. & Stretch, D. D. (2007). Breaching timescales and peak outflows for perched temporary open estuaries. *Coastal Engineering Journal*, *49*, 267-290.

Perissinotto, R., Iyer, K. & Nozais, C. (2006). Response of microphytobenthos to flow and trophic variation in two South African temporarily open/closed estuaries. *Botanica Marina*, *49*, 10-22.

Perissinotto, R., Nozais, C. & Kibirige, I. (2002). Spatio-temporal dynamics of phytoplankton and microphytobenthos in a South African temporarily-open estuary. *Estuarine, Coastal and Shelf Science*, *55*, 47-58.

Perissinotto, R., Nozais, C., Kibirige, I. & Anandraj, A. (2003). Planktonic food webs and benthic-pelagic coupling in three South African temporarily-open estuaries. *Acta Oecologica*, *24*, S307-S316.

Perissinotto, R., Walker, D. R., Webb, P., Wooldridge, T. H. & Bally, R. (2000). Relationships between zoo- and phytoplankton in a warm-temperate, semi-permanently closed estuary, South Africa. *Estuarine, Coastal and Shelf Science*, *51*, 1-11.

Perissinotto, R., Blair, A., Connell, A., Demetriades, N. T., Forbes, A. T., Harrison, T. D., Iyer, K., Joubert, M., Kibirige, I., Mundree, S., Simpson, H., Stretch, D., Thomas, C., Thwala. X. & Zietsman, I. (2004). Contributions to information requirements for the implementation of Resource Directed Measures for estuaries. Volume 2. Responses of the biological communities to flow variation and mouth state in two KwaZulu-Natal temporarily open/closed estuaries. *Water Research Commission Report*, *No. 1247*, 1-166.

Pillay, D., Branch, G. M. & Forbes, A. T. (2007a). Effects of *Callianassa kraussi* on microbial biofilms and recruitment of macrofauna: a novel hypothesis for adult–juvenile interactions. *Marine Ecology Progress Series*, *347*, 1-14.

Pillay, D., Branch, G. M. & Forbes, A. T. (2007b). The influence of bioturbation by the sandprawn *Callianassa kraussi* on feeding and survival of the bivalve *Eumarcia paupercula* and the gastropod *Nassarius kraussianus*. *Journal of Experimental Marine Biology and Ecology*, *344*, 1-9.

Pillay, D., Branch, G. M. & Forbes, A. T. (2007c). Experimental evidence for the effects of the thalassinidean sandprawn *Callianassa kraussi* on macrobenthic communities. *Marine Biology*, *152*, 611-618.

Pillay, D., Branch, G. M. & Forbes A. T. (2008). Habitat change in an estuarine embayment: Anthropogenic influences and a regime shift in biotic interactions. *Marine Ecology Progress Series*, *370*, 19-31.

Pollard, D. A. (1994). A comparison of fish assemblages and fisheries in intermittently open and permanently open coastal lagoons on the south coast of New South Wales, south-eastern Australia. *Estuaries*, *17*, 631-646.

Quinn, N. W., Breen, C. M., Hearne, J. W. & Whitfield, A. K. (1998). Decision support systems for environmental management: A case study on estuary management. *Orion*, *14*, 17-35.

Raffaelli, D., Limia, J., Hull, S. & Pont, S. (1991). Interactions between the amphipod *Corophium volutator* and macroalgal mats on estuarine mudflats. *Journal of the Marine Biological Association UK*, *71*, 899-908.

Ramm, A. E. L., Cerff, E. C. & Harrison, T. D. (1987). Documenting the recovery of a severely degraded coastal lagoon. *Journal of Shoreline Management*, *3*, 159-167.

Ranasinghe, R. & Pattiaratchi, C. (1998). Flushing characteristics of a seasonally-open tidal inlet: A numerical study. *Journal of Coastal Research*, *14*, 1405-1421.

Ranasinghe, R. & Pattiaratchi, C. (2003). The seasonal closure of tidal inlets: causes and effects. *Coastal Engineering Journal*, *45*, 601-627.

Ranasinghe, R., Pattiaratchi, C. & Masselink, G. (1999). A morphodynamic model to simulate the seasonal closure of tidal inlets. *Coastal Engineering Journal*, *34*, 1-36.

Reddering, J. S. V. (1988). Prediction of the effects of reduced river discharge of the estuaries of the south-eastern Cape Province, South Africa. *South African Journal of Science*, *84*, 726-730.

Reddering, J. S. V. & Rust, I. C. (1990). Historical changes and sedimentary characteristics of southern African estuaries. *South African Journal of Science*, *86*, 425-428.

Riddin, T. & Adams, J. B. (2008). Influence of mouth status and water level on the macrophytes in a small temporarily open estuary. *Estuarine, Coastal Shelf Science, 79,* 86-92.

Riddin, T. & Adams, J. B. (2009). The seed banks of two temporarily open / closed estuaries in South Africa. *Aquatic Botany, 90,* 328-332.

Rooseboom, A. (1992) Sediment Transport in Rivers and Reservoirs – A Southern African Perspective. *Water Research Commission Report, No.297/1/92,* 1-92.

Roussouw, J. (1984). A review of existing wave data, wave climate & design waves for SA and SWA. *Counsil for Scientific and Industrial Research Report, T/SEA 8401,* 1-65.

Roy, P. S., Williams, R. J., Jones, A. R., Yassini, I., Gibbs, P. J., Coates, B., West, R. J., Scanes, P. R., Husdon, J. P. & Nichol, S. (2001). Structure and function of south-east Australian estuaries. *Estuarine, Coastal and Shelf Science, 53,* 351-384.

Ryan, P. G., Underhill, L. G., Cooper, J. & Waltner, M. (1988). Waders (Charadrii) and other waterbirds on the coast, adjacent wetlands and offshore islands of southwestern Cape Province, South Africa. *Bontebok, 6,* 10-19.

Sand-Jensen, K. & Borum, J. (1991). Interactions among phytoplankton, periphyton and macrophytes in temperate freshwaters and estuaries. *Aquatic Botany, 41,* 137-175.

Santiago, A. E. (1991) Turbidity and seawater intrusion in Laguna de Bay. *Environmental Monitoring & Assessment, 16,* 185-195.

Schlacher, T. A. & Wooldridge, T. H. (1995). Small-scale distribution and variability of demersal zooplankton in a shallow, temperate estuary: tidal and depth effects on species-specific heterogeneity. *Cahiers de Biologie Marine, 36,* 211-227.

Schulze, R. E., Maharaj, M., Lynch, S. D., Howe, B. J. & Melvill-Thomson, B. (1996). South African Atlas of Agrohydrology and Climatology. *Water Research Commission Report, No. TT 82/96,* 1-58.

Schulze, R. E. (2006). Climate change and water resources in southern Africa. *Water Research Commission Report, No. 1430/1/05,* 1-49.

Schulze, R. E. (2005). "Adapting to climate change in the water resources sector in South Africa". Climate Change and Water Resources in Southern Africa: Studies on Scenarios, Impacts, Vulnerabilities and Adaptation, *Water Research Commission Report, No. 1430/1/05,* 423-449.

Schulze, R. E. (2000) Modelling hydrological responses to land use and climate change: A southern African perspective. *AMBIO: A Journal of the Human Environment, 29,* 12-22.

Schoonees, K. (2000). Annual variation in the net long-shore sediment transport rate, *Coastal Engineering Journal, 40,* 141-160.

Schumann, E. H. & Pearce, M. W. (1997). Freshwater inflow and estuarine variability in the Gamtoos Estuary, South Africa. *Estuaries, 20,* 24-133

Schumann, E., Largier, J. & Slinger, J. (1999). Estuarine Hydrodynamics. In B.R. Allanson, & D. Baird (Eds.), *Estuaries of South Africa,* (27-52). Cambridge, UK: Cambridge University Press.

Sherwood, J. E. & Rouse, A. (1997). Estuarine chemistry – primordial soup down under. *Chemistry in Australia, 64,* 19-22.

Siegfried, W. R. (1981). The estuarine avifauna of southern Africa. In J.H. Day (Ed.), *Estuarine Ecology with particular reference to southern Africa,* (223-250). Cape Town, South Africa: A. A. Balkema.

Skinner, T., Adams, J. B. & Gama, P. T. (2006). The effect of mouth opening on the biomass and community structure of microphytobenthos in a small ologotrophic estuary. *Estuarine, Coastal and Shelf Science, 70,* 161-168.

Slinger, J. H., Taljaard, S. & Largier, J. (1995). Changes in estuarine water quality in response to a freshwater flow event. In K. R. Dyer, & R. J. Orth, (Eds.), *Changes in fluxes in estuaries,* (51-56). Fredensborg, Denmark: Olsen & Olsen.

Snow, G. C. & Taljaard, S. (2007). Water quality in South African temporarily open/closed estuaries: a conceptual model. *African Journal of Aquatic Science, 32,* 99-111.

Smakhtin, V. U. (2000). Estimating daily flow duration curves from monthly streamflow data. *Water SA, 26,* 13-18.

Smakhtin, V. U. (2004). Simulating the hydrology and mouth conditions of small, temporarily closed/open estuaries. *Wetlands, 24,* 123-132.

Smakhtin, V. U. & Masse, B. (2000). Continuous daily hydrograph simulation using duration curves of a precipitation index. *Hydrological Processes, 14,* 1083-1100.

Steinke, T. D. & Charles, L. M. (1986). Litter production by mangroves. I: Mgeni Estuary. *South African Journal of Botany, 52,* 552-558.

Steinke, T. D. & Ward, C. J. (1989). Some effects of the cyclones Domoina and Imboa on mangrove communities in St Lucia Estuary (South Africa). *South African Journal of Botany, 55,* 340-348.

Stretch, D. D. & Zietsman, I. (2004). The Hydrodynamics of Mhlanga and Mdloti Estuaries: Flows, Residence Times, Water Levels and Mouth Dynamics. *Water Research Commission Report, No. K5/1247,* 71-111.

Stretch, D. D. & Parkinson, M. G. (2006).The breaching of sand barriers at perched temporarily open/closed estuaries a model study. *Coastal Engineering Journal, 48,* 13-30.

Strydom, N. A. (2003). Occurrence of larval and early juvenile fishes in the surf zone adjacent to two intermittently open estuaries, South Africa. *Environmental Biology of Fishes, 66,* 349-359.

Strydom, N. A., Whitfield, A. K. & Wooldridge, T. H. (2003). The role of estuarine type in characterizing early stage fish assemblages in warm temperate estuaries, South Africa. *African Zoology, 38,* 29-43.

Taylor, R. H. (2006). *Ecological Responses to Change in the Physical Environment of the St Lucia Estuary.* Norwegian University of Life Sciences. Department of Plant and Environmental Sciences. 2006: 7, Doctor Scientarium Thesis. Aas, Norway: Norwegian University of Life Sciences.

Terörde, A. I. (2008). *Variation in the use of intermittently-open estuaries by birds: a study of four estuaries in the Eastern Cape, South Africa.* MSc Thesis. Cape Town, South Africa: University of Cape Town.

Teske, P. R. & Wooldridge T. H. (2001). A comparison of the macrobenthic faunas of permanently open and temporarily open/closed South African estuaries. *Hydrobiologia, 464,* 227-243.

Teske, P. R. & Wooldridge, T. H. (2003). What limits the distribution of subtidal macrobenthos in permanently open and temporarily open/closed South African estuaries? Salinity vs sediment particle size. *Estuarine, Coastal and Shelf Science, 57,* 225-238.

Theron, A. K. (2004). *Sediment transport regime at East London.* MEng thesis. Stellenbosch, South Africa: University of Stellenbosch.

Theron, A. K., Schoonees, J. S., Huizinga, P. & Phelp, D. T. (2003). *Beach diamond mining design at the rocky Namaqualand coast*. Proceedings, 4th Coastal Structures Conference, , Portland, Oregon: ASCE.

Thomas, C. M., Perissinotto, R. & Kibirige, I. (2005). Phytoplankton biomass and size structure in two South African eutrophic, temporarily open/closed estuaries. *Estuarine, Coastal and Shelf Science*, *65*, 223-238.

Turpie, J. K. (2004). Existing data on area, plants, invertebrates, fish and birds of South African estuaries. *Water Research Commission Report, No. 1247/1/04*, 7-36.

Tyler-Walters, H. (2001). *Ruppia maritima*. Beaked tasselweed. *Marine Life Information Network: Biology and Sensitivity Key Information Sub-programme* [on-line]. Plymouth: Marine Biological Association of the United Kingdom. [cited 06/06/2005]. Available from: http://www.marlin.ac.uk/species.

Valiela, I., McClelland, J., Hauxwell, J., Behr, P. J., Hersh, D. & Foreman, K. (1997). Macroalgal blooms in shallow estuaries: controls and ecophysiological and ecosystem consequences. *Limnology and Oceanography*, *42*, 1105-1118.

van Ballegooyen, R. C., Taljaard, S., van Niekerk, L., Lamberth, S. J., Theron, A. K. & Weerts, S. P. (2007). Freshwater flow dependency in South African marine ecosystems: a proposed assessment framework and initial assessment of South African marine ecosystems. *Water Research Commission Report, No. KV191/07*, 1-88.

van der Elst, R. P., Birnie, S. L. & Everett, B. I. (1999). Siyaya catchment demonstration project: An experiment in estuarine rehabilitation. *Oceanographic Research Institute Special Publications*, *6*, 1-99.

van Niekerk, L., Bate, G. C. & Whitfield, A. K. (2008). An Intermediate Ecological Reserve Determination study of the East Kleinemonde Estuary. *Water Research Commission Report, No. 1581/2/08*, 1-204.

Van Niekerk, L., van der Merwe, J. H. & Huizinga, P. (2005). The hydrodynamics of the Bot River revisited. *Water SA*, *31*, 73-85.

Vivier, L. & Cyrus, D. P. (2001). Juvenile fish recruitment through wave-overtopping into a closed subtropical estuary. *African Journal of Aquatic Science*, *26*, 109-113.

Vorwerk. P. D., Whitfield, A. K., Cowley, P. D. & Paterson, A. W. (2001). A survey of selected Eastern Cape estuaries with particular reference to the ichthyofauna. *Ichthyological Bulletin of the J.L.B. Smith Institute of Ichthyology*, *72*, 1-52.

Vorwerk, P. D., Whitfield, A. K., Cowley, P. D. & Paterson A. W. (2003). The influence of selected environmental variables on fish assemblage structure in a range of southeast African estuaries. *Environmental Biology of Fishes*, *66*, 237-247.

Walker, D. R., Perissinotto, R. & Bally, R. P. A. (2001). Phytoplankton/protozoan dynamics in the Nyara Estuary, a small temporarily open system in the Eastern Cape (South Africa). *African Journal of Aquatic Sciences*, *26*, 31-38.

Walker, D. R. (2004). *Plant and algal distribution in response to environmental variables in selected Eastern Cape estuaries*. PhD Thesis. Port Elizabeth, South Africa: University of Port Elizabeth.

Ward, C. J. & Steinke, T. D. (1982). A note on the distribution and approximate areas of mangroves in South Africa. *South African Journal of Botany*, *1*, 51-53.

Watt, D. A. (1998). Estuaries of contrasting trophic status in KwaZulu-Natal, South Africa. *Estuarine, Coastal and Shelf Science*, *47*, 209-216.

Weisser, P. J. & Howard-Williams, C. (1982). The vegetation of the Wilderness lakes system and the macrophyte encroachment problem. *Bontebok, 2,* 19-40.

Whitfield, A. K. (1980a). Factors influencing the recruitment of juvenile fishes into the Mhlanga estuary. *South African Journal of Zoology, 15,* 166-169.

Whitfield, A. K. (1980b). Distribution of fishes in the Mhlanga estuary in relation to food resources. *South African Journal of Zoology, 15,* 159-165.

Whitfield, A. K. (1980c). A quantitative study of the trophic relationships within the fish community of the Mhlanga estuary, South Africa. *Estuarine and Coastal Marine Science, 10,* 417-435.

Whitfield, A. K. (1992). A characterization of southern African estuarine systems. *Southern African Journal of Aquatic Sciences, 18,* 89-103.

Whitfield, A. K. (1993). *Preliminary evaluation of the impact of the proposed dam on the ecology of the Haga Haga River and Estuary.* Impact assessment: evaluation of proposed dam site on the Haga Haga River. Report No. 2/1993. East London, South Africa: Pollution Control Technologies.

Whitfield, A. K. (1995). Mass mortalities of fish in South African estuaries. *Southern African Journal of Aquatic Sciences, 21,* 29-34.

Whitfield, A. K. (1999). Ichthyoplankton diversity, recruitment and dynamics. In B. R. Allanson, & D. Baird, (Eds.), *Estuaries of South Africa* (pp. 209-218). Cambridge, UK: Cambridge University Press.

Whitfield, A. K. (2000). Available scientific information on individual southern African estuarine systems. *Water Research Commission Report, 577/3/00,* 1-217.

Whitfield, A. K. & Bruton, M. N. (1989). Some biological implications of reduced fresh water inflow into eastern Cape estuaries: a preliminary assessment. *South African Journal of Science, 85,* 691-695.

Whitfield, A. K. & Wooldridge, T. H. (1994). Changes in freshwater supplies to southern African estuaries: some theoretical and practical considerations. In K. R. Dyer, & R. J. Orth, (Eds.), *Changes in fluxes in estuaries: implications from science to management* (pp. 41-50). Fredensborg, Denmark: Olsen & Olsen.

Whitfield, A. K., Adams, J. B., Bate, G. C., Bezuidenhout, K., Bornman, T. G., Cowley, P. D., Froneman, P. W., Gama, P. T., James, N. C., Mackenzie, B., Riddin, T., Snow, G. C., Strydom, N. A., Taljaard, S., Teröfde, A. I., Theron, A. K., Turpie, J. K., van Niekerk. L., Vorwerk, P. D. & Wooldridge, T. H. (2008). A multidisciplinary study of a small, temporarily open/closed South African estuary, with particular emphasis on the influence of mouth state on the ecology of the system. *African Journal of Marine Science, 30,* 453-473.

Wooldridge, T. H. (1974). *A study of the zooplankton of two Pondoland estuaries.* MSc Thesis. Port Elizabeth, South Africa: University of Port Elizabeth.

Wooldridge, T. H. (1976). The zooplankton of the Msikaba Estuary. *Zoologica Africana, 11,* 23-44.

Wooldridge, T. H. (1991). Exchange of two species of decapod larvae across an estuarine mouth inlet and implications of anthropogenic changes in the frequency and duration of mouth closure. *South African Journal of Science, 87,* 519-525.

Wooldridge, T. H. (1999). Estuarine zooplankton community structure and dynamics. In B. R. Allanson, & D. Baird, (Eds.), *Estuaries of South Africa,* (141-166). Cambridge, UK: Cambridge University Press.

Wooldridge, T. H. & Bezuidenhout, K. (2008). An Intermediate Ecological Reserve Determination Study of the East Kleinemonde Estuary. Appendix H, Specialist report: zoobenthos. *Water Research Commission Report, No. 1581/2/08*, 154-179.

Wooldridge, T. H. & Loubser, H. (1996). Larval release rhythms and tidal exchange in the estuarine mudprawn *Upogebia africana*. *Hydrobiologia, 337*, 113-121.

Wooldridge, T. H. & McGwynne, L. (1996). *The Estuarine Environment.* Report prepared for the Western Region District Council, Report No. C 31. Port Elizabeth, South Africa: SAB Institute for Coastal Research.

Young, G. C., Potter, I. C., Hyndes, G. A. & de Lestang, S. (1997). The ichthyofauna of an intermittently open estuary: Implications of bar breaching and low salinities on faunal composition. *Estuarine, Coastal and Shelf Science, 45*, 53-68.

In: Encyclopedia of Environmental Research
Editor: Alisa N. Souter
ISBN: 978-1-61761-927-4
© 2011 Nova Science Publishers, Inc.

Chapter 39

PALEOECOLOGICAL SIGNIFICANCE OF DIATOMS IN ARGENTINEAN ESTUARIES: WHAT DO THEY TELL US ABOUT THE ENVIRONMENT?

Gabriela S. Hassan[*]

Conicet - Instituto de Geología de Costas y del Cuaternario, Universidad Nacional de Mar del Plata. Mar del Plata, Buenos Aires, CC722 (7600), Argentina.

ABSTRACT

Diatoms are an important and often dominant component of the microalgal assemblages in estuarine and shallow coastal environments. Given their ubiquity and strong relationship with the physical and chemical characteristics of their environment, they have been used to reconstruct paleoenvironmental changes in coastal settings worldwide. The quality of the inferences relies upon a deep knowledge on the relationship of modern diatom species and their ecological requirements, as well as on the taphonomic constrains that can be affecting their preservation in sediments. In Argentina, information on estuarine diatom ecology is scattered and fragmentary. Studies on estuarine diatoms from the 20th century have been mostly restricted to taxonomic descriptions of discrete assemblages. Given the lack of detailed studies on the distribution of modern diatoms in local estuarine environments and their relationship with the prevailing environmental conditions, most paleoenvironmental reconstructions were based on the ecological requirements of European diatoms. However, studies on diatom distribution along estuarine gradients from Argentina have increased in recent years, constituting a potential source of data for paleoecologists. In this chapter, the literature on modern estuarine diatoms from Argentina is revised in order to synthesize the available ecological information and to detect possible modern analogues for Quaternary diatom assemblages. The main objective is to build bridges between ecology and paleoecology, and to discuss the reaches and limitations of the different approaches to diatom-based paleoenvironmental reconstructions. Further studies exploring the relationship between

---

[*] Corresponding author: E-mail: ghassan@mdp.edu.ar

estuarine diatom distribution and environmental characteristics are necessary in order to increase the precision of paleoenvironmental inferences in the region and to generate new hypothesis for further study.

INTRODUCTION

Estuaries are transitional environments located between rivers and the sea, characterized by widely variable and often unpredictable hydrological, morphological and chemical conditions (Day et al., 1989). Given these particular environmental characteristics, estuarine organisms are often restricted to limited sections of estuarine gradients, resulting in well-developed distribution patterns (Moore & McIntire, 1977; Ysebaert et al., 2003; De Francesco & Isla, 2003).

Diatoms are the main source of primary production in shallow estuarine systems (Admiraal, 1984; Colijn et al., 1987; Wolfstein et al., 2000; Rybarcyk & Elkaïn, 2003), serving as an essential supply of food for numerous species of zooplankters and deposit feeders (Bianchi & Rice, 1988; Bennett et al., 2000; Rzeznik-Orignac et al., 2003) and forming biofilms that increase the resistance of sediment surface to erosion (Paterson, 1989; Underwood & Paterson, 1993; Underwood, 1997; Austen et al., 1999; Bergamasco et al., 2003). Laboratory experiments showed that different diatom species have different levels of tolerance to salinity, nutrient concentrations, temperature and light availability (Admiraal, 1977a,b,c,d; Admiraal & Peletier, 1980; Admiraal et al., 1982). Distribution patterns observed in the field usually respond to a combination of these variables (Moore & McIntire, 1977; Amspoker & McIntire, 1978; Oppenheim, 1991; Underwood, 1994; Gómez et al., 2004). Moreover, the distribution of diatoms in estuarine environments is the result of a complex set of interactions between environmental variables and interspecific competitive interactions (Underwood, 1994).

Given their sensitivity to environmental variables and abundance in sediments, diatoms constitute useful indicators for the study of paleoenvironmental changes (Cooper, 1999). This has been well known since the late 1890s, when the pioneering studies of Cleve (1894/1895) demonstrated that benthic diatom assemblages from surface sediments reflect the physical and chemical characteristics of the overlying water masses (Maynard, 1976). However, only after the 1920s the value of diatom analysis in paleoenvironmental reconstructions was recognized (Denys & De Wolf, 1999). Since the identification of salinity as a major determinant of diatom distribution, the remains of these organisms have become widely used as paleoenvironmental indicators in coastal deposits. Furthermore, a variety of problems in coastal geology were tackled by applying diatom-based methods, covering fields such as stratigraphy, coastal processes, paleogeography, sea-level and climate changes (Denys & De Wolf, 1999). In estuarine systems, they have also been used to define the naturally occurring state of the ecosystem, in order to infer historical changes due to human influences (Cooper, 1999).

The methods used in paleoenvironmental reconstructions rely on the general assumption that the environmental requirements of the fossils used as bioindicators have remained constant during the period considered and, consequently, are similar to those of their closest living representatives. In this way, the environmental information obtained from living organisms can be used as modern analogous and extrapolated to the fossil record, particularly in Quaternary research. This approach is based on a strict substantive application of the principle of Taxonomic Uniformitarianism (*the ecology of modern organisms is the key to that of past organisms*; Dodd & Stanton, 1990). Estuarine diatom-based paleoenvironmental reconstructions have been based in autoecological or synecological techniques. In autoecological studies, the composition of modern diatom assemblages is analyzed, and relevant environmental requirements of each species or group of species are considered (De Wolf, 1982; Vos & De Wolf, 1988, 1993; Denys & De Wolf, 1993). In the last decades, the great volume of autoecological data available for European diatoms has been summarized as a series of ecological codes (De Wolf, 1982; Vos & De Wolf, 1988, 1993; Denys, 1991/1992; Van Dam et al., 1994). The most commonly used diatom classifications in coastal areas were based on salinity tolerances (polihalobous, mesohalobous, oligohalobous halophilous, oligohalobous indifferent and halophobous; Hustedt, 1953) and life forms (plankton, epiphytes, benthos, and aerophilous; De Wolf, 1982). Later, Vos & De Wolf (1988) combined both classifications in order to define autoecological groups (i.e. marine/brackish epiphytes, brackish/freshwater tychoplankton) characteristic of different coastal habitats. Specific sedimentary environments in coastal wetlands were characterized on the basis of the relative frequencies of the 16 ecological groups defined (Vos & De Wolf, 1988).

Besides its usefulness, the application of autoecological techniques to the interpretation of past environmental changes has limitations and needs to be interpreted with caution. This methodology is based on the classification of single taxa in autoecological categories delimited by general ecological borderlines (Vos & De Wolf, 1993). Although to some extent such borders can be drawn, there are many cases of gradual species turnover along environmental gradients in nature, and many taxa have large adaptability to changing environmental conditions (Denys & De Wolf, 1999). This is particularly true for estuarine environments, where most taxa usually show wide salinity tolerances, making it difficult their placement into discrete categories (Licursi et al, 2006; Hassan et al., 2009). In fact, this difficulty of assigning a taxon unambiguously to an individual class constitutes one of the main problems of the autoecological classification (Battarbee et al., 1999).

In contrast to the use of generalized autoecological concepts, synecological techniques are based on the application of statistical inference models derived from modern contemporaneous species-environment relations, allowing quantitative inference of important parameters. A set of regional observations seems imperative in this, since hydrographic and ecological conditions differ between study areas (Denys & De Wolf, 1999). The statistical calibration of selected environmental variables and dead diatom assemblage composition (*transfer functions*) constitutes the most precise method, since it is based on the study of the entire diatom assemblage rather than on individual taxa (Juggins, 1992; Ng & Sin, 2003). In the last decades the need for quantification in Quaternary research has increased and a great number of diatom-based transfer functions have been developed in coastal and estuarine environments of the Northern Hemisphere (Juggins, 1992; Campeau et al., 1999; Sherrod, 1999; Zong & Horton, 1999; Gehrels et al., 2001; Ng & Sin, 2003; Sawai et al., 2004; Horton et al., 2006).

In Argentina, information on estuarine diatom ecology is scattered and fragmentary, and there is a lack of detailed distributional studies. As most diatom taxa are cosmopolitan, the autoecological information necessary to carry out local paleoenvironmental reconstructions has been historically gathered from European datasets (e.g. Espinosa, 1998, 2001; Espinosa et al., 2003). Studies on modern estuarine diatoms from Argentina during the 20th century have been mostly restricted to taxonomic descriptions of discrete assemblages (see Vouilloud, 2003). Works on diatom distribution along estuarine gradients have increased during the 21th century, constituting a potential source of data for paleoecologists. However, the information provided by these ecological studies has not always been applied to infer paleoenvironmental conditions from fossil diatoms. This points to the question if the lack of contact between paleoecological and ecological studies may be responding to methodological barriers between both disciplines rather than to a real scarcity of information.

In this chapter, the literature on modern estuarine diatoms from Argentina is reviewed in order to summarize the available ecological information and to evaluate its usefulness as modern analogues for Quaternary diatom assemblages. The main objective is to build bridges between ecology and paleoecology, and to discuss the reaches and limitations of the different approaches to diatom-based paleoenvironmental reconstructions. Although the discussion will focus on estuarine settings from Argentina, it could be useful for guiding the debate in other regions or environmental settings with similar research histories. The main questions to be addressed are: 1) Do estuarine diatoms reliably reflect estuarine environmental conditions? 2) How much information about ecological requirements of estuarine diatoms do we have? 3) How can researchers improve the quality of diatom-based paleoenvironmental inferences in coastal settings?

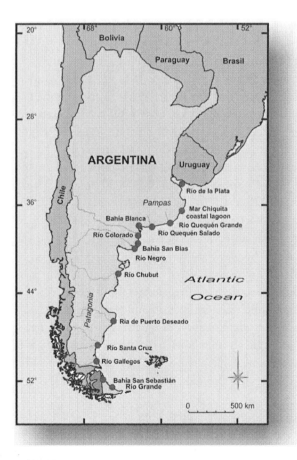

Figure 1. Location map showing the main Argentinean estuaries.

DO ESTUARINE DIATOMS RELIABLY REFLECT ESTUARINE ENVIRONMENTAL CONDITIONS?

The first issue to take into account in order to transfer ecological information to the past is to understand how accurately fossil organisms reflect their living environment and how much environmental information become lost in their transition from live to dead assemblages. This subject is particularly essential in the study of sedimentary diatom assemblages, since they are the result not only of ecological processes that drive the distribution of living diatoms along the environmental gradients, but also of taphonomic processes (i.e., the postmortem history of dead remains) that alter dead frustules after their deposition. Therefore, their distribution within a locality may not necessary constitute an accurate representation of their living habitat (Juggins, 1992; Vos & De Wolf, 1993; Sherrod, 1999). In highly variable and energetic environments, such as coastal and estuarine areas, taphonomic processes can so drastically alter the species composition of a diatom assemblage that the original ecological signals reflected by the *in situ* assemblage may be either obscured or obliterated (Sherrod, 1999). Thus, the assessment of how accurately dead diatom assemblages preserve the original environmental information becomes a main requisite in order to evaluate the applicability of modern data sets.

When looking for modern analogues of paleoenvironments, most researchers turn to the surface sediment diatom thanatocoenoses (dead diatoms, both autochthonous and allochthonous remains, present at a particular place in the sediment; Sherrod, 1999), which are assumed to integrate small-scale temporal and spatial perturbations into more defined assemblages; consequently, they are assumed to be more accurate indicators of general environmental conditions than biocoenoses (living communities). The use of diatom thanatocoenoses as modern analogous is based on the general assumption that dead diatom assemblages faithfully reflect the environmental conditions prevailing at the sampling point. Hence, they are considered reliable indicators of environmental parameters, without requiring time consuming seasonal studies (Juggins, 1992).

The most common approach to the evaluation of the ecological fidelity of fossil assemblages has been the testing of agreement between living communities and the locally accumulating dead assemblages in modern environments. This method has led to powerful guidelines for paleoecological reconstruction in foraminifers (e.g., Goldstein & Watkins, 1999; Horton, 1999; Murray & Pudsey, 2004), ostracodes (Alin & Cohen, 2004), mollusks and brachiopods (e.g., Kidwell, 2001, 2002; Kowalewski et al., 2003). However, there is a general lack of detailed quantitative works attempting to evaluate the fidelity of coastal diatom assemblages.

In Argentina, this approach has been recently applied by Hassan et al. (2008), who analyzed the environmental fidelity of dead diatom assemblages along two microtidal estuaries (Mar Chiquita coastal lagoon and Río Quequén Grande; Figure 1) and discussed their potential use as modern analogues in paleoenvironmental reconstructions. A good agreement between live benthic communities and total surface assemblages was found in both estuaries. The comparison between live cells and empty frustules did not allow the recognition of a significant allochthonous component. Although relatively high percentages of empty frustules were found in the tidal inlet zone from Mar Chiquita coastal lagoon, they originated mainly from taxa found alive in the same site. Similar results were obtained in tidal flats from salt marshes of Japan, where only 3% of the empty frustules present in surface sediments of the littoral zone were found to be allochthonous (Sawai, 2001). The investigation about possible and net effects of transport on population composition in other groups, led to the general conclusion that out-of habitat postmortem transport does not constitute an overwhelming taphonomic problem in ordinary depositional settings (Kidwell & Flessa, 1995; Horton, 1999; Behrensmeyer et al., 2000; Alin & Cohen, 2004). These results, together with the good preservation shown by diatom valves, suggest that benthic diatom assemblages are not under significant alteration by biostratinomic and early-diagenetic processes along the estuarine foreshore: although mixing of autochthonous and allochthonous diatoms does occur, estuarine dead diatom assemblages still reflect the environmental gradient with high fidelity. As a consequence, they constitute useful modern analogues for paleoenvironmental reconstructions and provided advantages over the use of live communities. Moreover, since paleoecologists have only total sedimentary assemblages available to examine and interpret (Scott & Medioli, 1980), the understanding of taphonomic alterations suffered by them leads to an increase in the precision of paleoenvironmental interpretations.

In contrast to the use of benthic diatoms, the application of modern ecological data gathered from phytoplanktonic assemblages becomes a more problematic issue. According to a strict definition (Birks & Birks, 1980) the term allochthonous refers to those individuals transported away from their life position before burial. It has been proposed that only benthic taxa should be considered autochthonous and used in palaeoecological reconstruction, since plankton forms are by definition allochthonous and, thus, more subject to lateral transport by tides and currents (Simonsen, 1969). Vos & De Wolf (1993) also emphasized life form as an important variable to interpret paleoenvironments, pointing out that marine plankton and tychoplankton diatoms are basically allochthonous components, whilst epiphytic and epipsammic diatoms are probably autochthonous. Accordingly, a wide distribution of empty valves and frustules of the tychoplanktonic *Paralia* sp. was observed throughout the entire tidal zone in marshes from Japan as a consequence of their transport by currents action (Sawai, 2001).

The representation of phytoplanktonic diatom species in surface sediments of Argentinean estuaries has not been systematically assessed. Frenguelli (1935, 1941) remarked the large differences in the salinity tolerances of diatom assemblages of sedimentary and plankton net samples from Río de la Plata and Mar Chiquita estuaries (Figure 1), which were attributed to taphonomic biases (see Río de la Plata and Mar Chiquita sections below). Licursi et al. (2006) recorded up to 70% of empty frustules in plankton net samples from Río de la Plata, which were closely related to bathymetry. These high percentages of empty frustules belonged mainly to freshwater diatoms, which were probably allochthonous riverine elements transported from the headwaters (Gómez et al., 2004; Licursi et al., 2006). High percentages of tychoplanktonic taxa were found in sediment samples from Mar Chiquita and Río Quequén Grande estuaries, but as their distribution along the estuarine gradient was consistent with their salinity tolerances, they were not ecologically out of place (Hassan et al., 2008). Moreover, as tychoplanktonic diatoms are closely associated to the sediment, they are less prone to lateral transport than true plankton. To sum up, systematic studies comparing the diatom assemblage composition in surface sediments and the overlying water column are needed in order to estimate their grade of preservation and environmental fidelity.

Meanwhile, caution is needed when paleoenvironmental inferences in estuaries are derived from phytoplanktonic diatom assemblages.

HOW MUCH INFORMATION ABOUT ECOLOGICAL REQUIREMENTS OF ESTUARINE DIATOMS DO WE HAVE?

The Argentina coastline has a wide variety of estuaries ranging from the widest in the world (Río de la Plata) to very small ones located in areas of very difficult access (Figure 1). Due to the different climates that characterize the Argentinean territory, the estuaries show different discharges, being the Río de la Plata the largest one. Tidal amplitudes also vary significantly, being microtidal between the Río de la Plata and the Río Quequén Salado, mesotidal in the coast between Bahía Blanca estuary to Río Chubut, and macrotidal along the rest of the Patagonian estuaries (Piccolo & Perillo, 1999).

Vouilloud (2003) published a review listing of the publications about Argentinean diatoms from the 19th century to the '90 decade. Of the revised literature, only a small proportion of the articles (see Figure 2 in Vouilloud, 2003) dealt with modern estuarine diatoms. Among them, taxonomic studies were the most numerous, although some ecological articles (mainly focused on the whole phytoplanktonic assemblage) were also published. Only recently, some distributional studies on estuarine diatoms were published, particularly for Río de la Plata (Licursi et al., 2006; Gómez et al., 2009); Mar Chiquita coastal lagoon (Hassan et al., 2006; 2009), Río Quequén Grande and Río Quequén Salado (Hassan et al., 2007; 2009).

In the following sections, the state of the knowledge on each of the main estuaries from Argentina is reviewed, focusing mainly on the ecological requirements of the dominant diatom taxa, and stressing the value of the information presented for paleoecological purposes. In order to summarize the available information, tables listing all the reviewed works of recent publication (Appendix I) and the diatom taxa cited in them (Appendix II) were constructed. Comprehensive lists of the diatom taxa cited in older works can be found in Ferrando et al. (1962), Ferrario and Galván (1989), Vouilloud (2003) and Sar et al. (2009). In order to allow the comparison of data among the different reviewed works, all diatom names and their authorities were updated to their currently accepted name following Algaebase (Guiry & Guiry, 2009) and WoRMS (SMEBD, 2009) taxonomic databases.

Rio De La Plata Estuary

The Argentina coast starts in the Río de la Plata estuary (Figure 1), located at about 35ºS on the Atlantic coast of South America. The river drains the second largest basin of this continent, following that of the Amazon (Piccolo & Perillo, 1999). Its drainage area covers ca. 3.1×10^6 km^2, which represents about 20% of the South American continental area (Acha et al., 2008). It forms one of the most important estuarine environments in South America, being a highly productive area that sustains fisheries in Uruguay and Argentina. The estuary is characterized by a salt-wedge regime, low seasonality in the river discharge, low tidal amplitude (<1m), a broad and permanent connection to the sea, and high susceptibility to atmospheric forcing, due to its large extension and shallow water depth (Acha et al., 2008 and references therein).

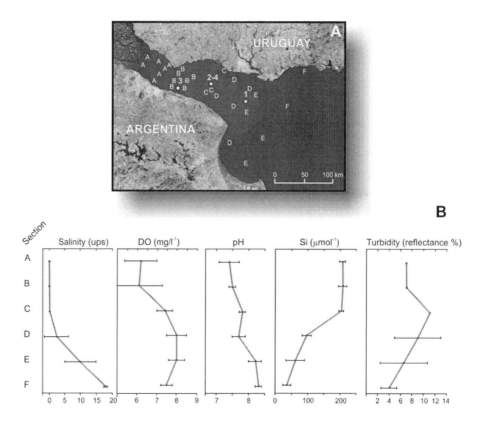

Figure 2. A) Location of sampling sites from Frenguelli (1941, numbers) and Licursi et al. (2006, letters) at the Río the la Plata estuary, and B) Summary of environmental information provided by Licusi et al (2006).

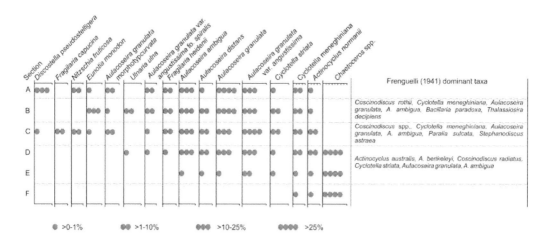

Figure 3. Relative frequencies of diatom assemblage composition for Río de la Plata estuary (based on data from Licursi et al., 2006). Dominant taxa found by Frenguelli (1941) in each section are listed.

The Río de la Plata estuary and its oceanic front has been the most extensively studied of Argentina. The first phytoplanktonic diatom from the Río de la Plata, *Caloneis bivittata* var. *rostrata*, was mentioned by Heiden (Schmidt et al., 1874-1959). Tempère and Peragallo (1907-1915) mentioned 8 new forms. The list increased to 68 forms during the 1920s and the 1930s, with a series of taxonomic studies which focused on plankton samples of the estuary (Carbonell & Pascual, 1924; Hentschel, 1932; Thiemann, 1934; Carbonell, 1935; Cordini, 1939).

Frenguelli (1941), studied 3 plankton and 1 bottom sediment samples collected in three different points of the estuarine gradient (inner estuary, middle estuary and mouth, Figure 2A). A total of 309 taxa, present at very low abundances, were observed. The dominance of these taxa, mostly benthic, epiphytic and aerophilic forms, was related to the transport of littoral diatoms from the headwaters and the adjacent coast. The assemblage composition of the plankton samples was homogeneous and characteristic of estuarine environments. They were dominated by *Aulacoseira granulata* and *A. ambigua*, accompanied by some freshwater and marine taxa (Figure 3). The bottom sediment sample, on the other hand, showed very scarce diatom frustules, mostly marine neritic forms, with only one species (*Paralia sulcata*) classified as frequent. Detailed taxonomical descriptions of the dominant taxa were provided, together with information on their ecological preferences. The later data, however, were taken from European floras (particularly Hustedt 1937/1938), and no *in situ* measurements of the main environmental parameters from the sampling site were provided. Guarrera (1950) analyzed the composition of the phytoplanktonic assemblage in two sampling stations located near Buenos Aires city, identifying 16 genera. Müeller Melchers (1945, 1952, 1953, 1959) worked on plankton samples from the Río de la Plata maritime front, listing and providing taxonomic descriptions for 69 diatom taxa. Although the number of studies on phytoplankton increased significantly since the 1970s, most studies focused on the coastal areas and maritime front (Balech, 1976, 1978; Martínez Macchiavelo, 1979; Lange, 1985; Baysee et al., 1986; Elgue et al., 1990; EcoPlata Team, 1996; Gayoso, 1996), being scarce studies on the estuarine zone of the river (Roggiero, 1988, CARP-SIHN-SOHMA, 1989).

Many works on the composition and dynamics of the phytoplankton have been carried out in the estuarine zone of the river in the last decades (Gómez & Bauer, 1998a, 1998b, 2000; Cervetto et al., 2002; Gómez et al., 2002, 2004; Carreto et al., 2003, 2008; Calliari et al., 2005, 2009). In most of these studies diatoms represent one of the dominant groups. The centric taxa *Aulacoseira granulata* var. *angustissima*, *A. granulata*, *A. distans*, *A. ambigua*, *Actinocyclus normanii*, *Thalassionema nitzschioides*, *Stephanodiscus hantzschii* and *Skeletonema costatum* were mentioned among the dominant diatoms in most of these studies. The dominance of these taxa was explained as a consequence of their capability for exploiting this low light environment owing to their efficient light-harvesting mechanisms (Gómez & Bauer, 1998). Unfortunately, although lists of the dominant taxa and environmental information are provided, none of these works presents information on the patterns of distribution of each taxon along the estuarine gradient or their ecological preferences. Hence, the way in which the data are presented limits their usefulness for paleoenvironmental applications.

From an autoecological point of view, the most valuable information on phytoplanktonic diatom ecology and distribution in the Río de la Plata estuary was provided by Licursi et al. (2006), who studied the factors affecting the composition and structure of diatom phytoplankton across the estuarine gradient. Samples were collected with plankton nets from

29 sites distributed along a gradient of estuarine conditions from the headwaters to the estuary mouth. The estuarine gradient was divided into 6 zones of 50 km long and sites grouped according to them (Figure 2A). For each zone, data on environmental variables were also provided (Figure 2B). As reported in previous works, the assemblages were dominated by chains of centric diatoms (Figure 3). Canonical Correspondence Analysis (CCA) was performed in order to relate diatom assemblages to environmental variables, allowing recognizing two groups of taxa: the first group was related to low values of salinity, pH and concentrations of dissolved oxygen and higher amounts of suspended solids and nutrients (sections A to C, Figure 3). The second group clustered taxa that tolerate higher salinity and alkalinity (sections D to F, Figure 3). This assemblage was characteristic of marine environments, and had a lower limit of salinity tolerance of 7-8. No taxa exclusive of brackish waters were identified, but some freshwater and marine taxa presented wide salinity tolerances. Despite the taphonomic limitations to the use of phytoplanktonic taxa as modern analogues in estuarine environments, the information on diatom distribution and environmental preferences provided in this work is of great utility for coastal paleoenvironmental reconstructions.

Microphytobenthic diatom communities from Río de la Plata, on the other hand, have received little attention. Metzeltin and García-Rodríguez (2003) published a book on the taxonomy of the Uruguayan diatoms based on the analysis of samples of periphyton collected along the Uruguayan coast of the estuary, listing and illustrating 295 species. Bauer et al. (2007) assessed the usefulness of biofilms covering *Schoenoplectus californicus* (a bulrush widely distributed along the shore of the Río de la Plata) as indicators of water quality. They selected three sampling sites in the freshwater tidal zone of the estuary (salinity <0.5) subjected to different grades of human impact and analyzed the taxonomic composition and tolerances of the taxa present over *S. californicus* stems. Diatoms constituted one of the dominant organisms in the biofilms, and their distribution was mainly conditioned by turbidity, pH, salinity and water-quality variables. Two assemblages were defined: one related to the highest turbidity values (average 50±22 NTU), dissolved oxygen (average 7.5±1 mg l^{-1}) and pH (average 7.4±0.5), and included pollution sensitive species such as *Encyonema silesiacum, Navicula erifuga, N. rynchocephala, Neidium dubium, Nitzschia fonticola, N. nana, Placoneis clementis* and *Pleurosira laevis,* and less tolerant species such as *Gomphonema augur, Luticola ventricosa* and *Nitzschia brevissima*. The second assemblage was related to high conductivities (average 740±200 $\mu S\ cm^{-1}$), ammonia (average 1.7±1.1 mg l^{-1}), nitrates (average 0.08±0.04 mg l^{-1}) and phosphates (average 0.87±0.42 mg l^{-1}) concentrations. This group included mainly taxa characteristic of polluted sites such as *Nitzschia palea*.

In a recent contribution, Gómez et al. (2009) analyzed the seasonal and spatial distribution of microbenthic communities in 10 sites located along 155 km of the estuarine shoreline. Diatoms were abundant, particularly during autumn. *Navicula novaesiberica, N. erifuga, Fallacia pygmaea, Nitzschia palea, Amphora lybica* and *Sellaphora pupula,* were the most abundant taxa (>60%). According to their relationship with environmental variables, the whole assemblage was separated into two groups by CCA: the first group was composed by *L. ventricosa, Stauroneis brasiliensis* and *Fallacia omissa*, and related with the highest nitrite (0.14±0.10 mg l^{-1}) and ammonia (0.30±0.23 mg l^{-1}) values. The second group of species included *Amphora acutiuscula, A. lybica, Pleurosira laevis, Actinocyclus normanii, Staurosirella pinnata, Hantzschia amphioxys, Hippodonta hungarica,* and *Navicula tenelloides*, associated with high conductivity (1657 ± 1597 µS cm^{-1}), and *Nitzschia lacunarum* linked to high concentrations of nitrates (0.94 ± 0.17 mg l^{-1}). Although this study covered a large portion of the estuarine gradient and provided detailed environmental data for each sampling point, the information on the distribution of single taxa in each sampling station was not presented. Hence, it is not possible to extract information on single taxa environmental preferences, which would be very useful in autoecological paleoenvironmental reconstructions.

Mar Chiquita Coastal Lagoon

Mar Chiquita is the only coastal lagoon of Argentina that is chocked with a long inlet (Piccolo & Perillo, 1999). It is a brackish-water body with a surface area of 46 km^2 and a mean depth of 0.6 m, extending along the microtidal Argentinean coast (Figures 1 and 4). From a hydrological point of view, the coastal lagoon can be divided into an innermost shallow zone, where the tidal effect is not significant, and an estuarine channel subjected to tidal action (Reta et al., 2001). Sediments are mainly composed of sand and silt with high proportions of mollusk shells. The shallow depth and particular dynamics of the coastal lagoon induces sediment reworking and prevents the development of a stable salinity gradient (Fasano et al., 1982). Nutrients and suspended sediment concentration are higher in the inner areas of the lagoon than in the tidal channel, whereas salinity, current speed and depth show the opposite pattern (Schwindt et al., 2004).

The study of diatoms from Mar Chiquita began with Frenguelli (1935) who described the assemblages present in two samples collected from the inlet of the coastal lagoon (Table 1). The first was a sediment sample taken from the bottom of the inlet, which contained relatively scarce diatom frustules of marine-neritic origin. In the second sample, which was taken with plankton net, diatom frustules were conspicuous, and consisted in a mixed assemblage of fluvial, lacustrine and estuarine taxa, characteristic of both oligohaline and mesohaline conditions. The significant differences between both assemblages were taphonomicaly explained: whereas than in the plankton sample diatom assemblages reflected an average of the living communities that succeeded in the very changing ecological environment, the diatom composition of the sediment sample was interpreted as a reworked fossil assemblage which indicates that in the past the zone was a marine bay (Frenguelli, 1935).

No new studies on Mar Chiquita diatoms were conducted until the 21st century. Recently, the temporal and spatial dynamics of the phytoplankton and its relation to nutrient concentrations were studied (De Marco, 2002; De Marco et al., 2005). Although diatoms constituted the dominant assemblage, taxa were identified only at the genus level.

Espinosa et al. (2006) analyzed the distribution of surface diatom assemblages across the marsh in a sampling station located in the Mar Chiquita tidal inlet (site 6, Figure 4A). The marsh was divided into five subenvironments: floodplain, distant and closer high marshes, levee/chenier, and mudflat. In the flood plain, where tidal submersion is infrequent and of short duration, the assemblage was dominated by brackish/epiphytic and aerophilous taxa (Figure 5). Brackish/freshwater epiphytic and tychoplanktonic diatoms dominated the distant high marsh, whereas the closer high marsh was dominated by the brackish aerophilous *Diploneis interrupta,* a taxon typical of supratidal environments. The levee and chenier zone, where the tidal flooding is frequent, was dominated by marine planktonic, benthic and epiphytic taxa. The diatom assemblage of the mudflat was dominated by a mixture of marine (epiphytic and benthic) and freshwater (planktonic and tychoplanktonic) taxa. Overall, the composition of diatom assemblages in this microtidal marsh was related to morphology, duration and frequency of tidal exposure, and the consequent salinity fluctuations.

Table 1. Diatom assemblage composition and environmental significance of the two samples collected by Frenguelli (1935) at Mar Chiquita coastal lagoon

	Bottom sample	Plankton net sample
Abundant species	*Paralia sulcata*	------------
Frequent species	*Actinocyclus vulgaris*	*Aulacoseira granulate, Bacillaria paradoxa, Cyclotella meneghiniana, Navicula peregrina, Nitzschia circumsuta, Tropidoneis lepidoptera* var. *proboscidea*
Ecological Conditions	Marine/neritic assemblage. Fossil and reworked, probably indicating the presence of a marine bay in the past.	Mixed assemblage of mesohalobous and oligohalobous taxa, of lacustrine, fluvial and estuarine origin. Planktonic and benthic. Assemblage composition reflects the mean environmental conditions of the basin.

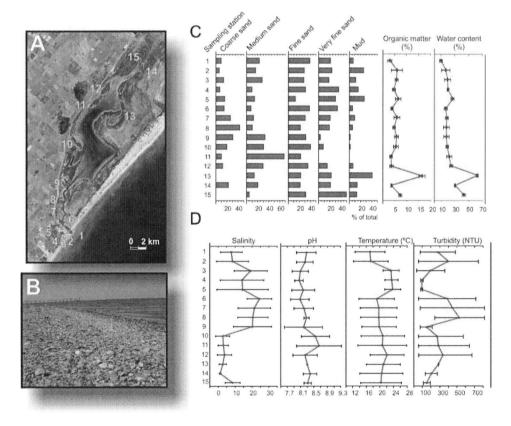

Figure 4. Location of sampling sites (A), view of the estuarine zone (B), and the corresponding sedimentary (C) and water quality (D) parameters at Mar Chiquita coastal lagoon (modified after Hassan, 2008).

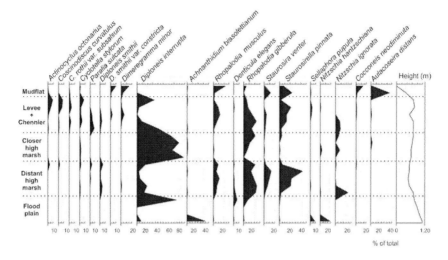

Figure 5. Distribution of the dominant diatom taxa across the Mar Chiquita lagoon marsh (modified after Espinosa et al., 2006).

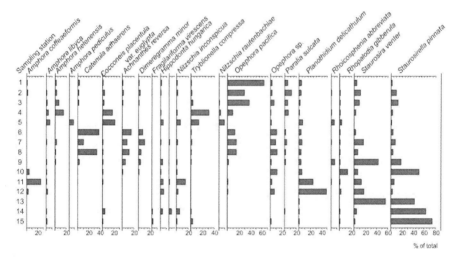

Figure 6. Diatom assemblage composition at Mar Chiquita coastal lagoon (modified after Hassan et al., 2009).

Hassan et al. (2006) studied the diatom assemblages dominating in surface sediments along a transect from the inlet to the inner reaches of the coastal lagoon in relation to the main environmental parameters (Figure 4). Most diatom species found were highly euryhaline taxa, adapted to the great salinity and tidal ranges that characterize the lagoon. Besides salinity, other environmental factors such as turbidity, temperature and sediment properties were important in explaining diatom assemblage composition. The marine/brackish diatoms *Catenula adhaerens* and *Opephora pacifica* dominated in the tidal channel, whereas the inner lagoon was dominated by the brackish/freshwater tychoplanktonic diatoms *Staurosira venter* and *Staurosirella pinnata* (Figure 6). Similar distributional patterns, characteristic of environments with fluctuating salinity regimes, have been observed in other coastal lagoons from the Atlantic Ocean coasts (e.g., Sylvestre et al. 2001; Bao et al., 2007; Witkowski et al.,

2009). In these environments, taxa are selected according to their ability to adapt to changing salinity rather than to their salinity optima (Snoeijs, 1999). The diatom assemblages from Mar Chiquita coastal lagoon are of particular importance for the paleoenvironmental reconstruction of the many estuarine lagoons developed in the microtidal Argentinean coast during the Holocene marine transgression (Espinosa et al., 2003; Hassan et al., 2009).

Quequén Grande Estuary

The Río Quequén Grande is a partially mixed estuary discharging at the microtidal coastline of northern Argentina (Figures 1 and 7). Mean depth is 2–3 m and width is 150–200 m. Most of the river runs on Pleistocene partly cemented loessic sediments. Due to the sediment characteristics – silty loess with caliche levels – large portions of the river flow within a canyon whose walls reach up to 12 m high (Perillo et al., 2005). There is no significant accumulation of sediment on the bottom, and the river is well known by its rapids, composed of indurate levels of caliche. However, the river carries large amounts of silt during floods. Salinity decreases significantly along the estuarine gradient, the highest salinities (20–25) are found within the first 2–3 km of the inlet; approximately 10 km upstream, salinity decreases to 0–1 (Figure 7C). Given its economic and strategic importance, the estuary has been the focus of many man-made modifications (i.e., dredging, jetty and harbour construction, etc.) that have reduced water circulation producing strong reductive and even anoxic conditions (Perillo et al., 2005).

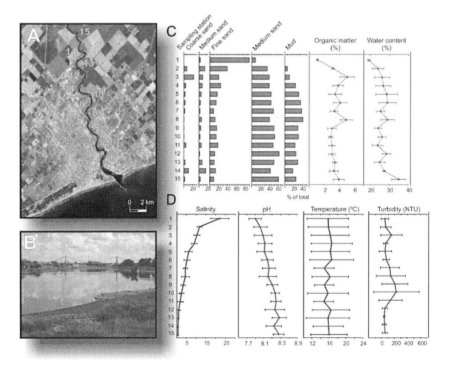

Figure 7. Location of sampling sites (A), view of the estuarine zone (B), and the corresponding sedimentary (C) and water quality (D) parameters at Quequén Grande river (modified after Hassan, 2008).

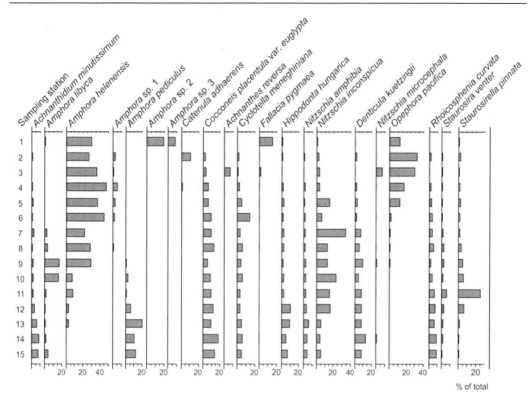

Figure 8. Diatom assemblage composition at Quequén Grande river (modified after Hassan et al., 2009).

The composition of the diatom assemblages present in surface sediments from the estuary have been recently studied (Hassan et al., 2006, 2007). Diatom composition was significantly related to salinity, and the assemblages showed gradual turnovers along the stable salinity gradient that characterizes the estuary. The marine/brackish diatoms *Amphora helenensis* and *Opephora pacifica* dominated in the inlet, while the brackish/freshwater diatoms *Cocconeis placentula* var. *euglypta* and *Nitzschia inconspicua* increased their relative frequencies towards the middle estuary. A diverse freshwater assemblage, characterized by *Achnanthidium minutissimum*, *Amphora pediculus*, *Hippodonta hungarica*, *Denticula kuetzingii* and *Rhoicosphenia abbreviata*, dominated the upper estuary (Figure 8). Similar diatom zonations were recorded in estuaries characterized by stable salinity gradients (Moore & McIntire, 1977; Ampsoker & McIntire, 1978; Juggins, 1992; Debenay et al., 2003; Resende et al., 2005). As salinity is one of the main environmental factors controlling diatom distribution in estuaries (Cooper, 1999), the diatom zonation observed in the Quequén Grande estuary was explained by the existence of a stable salinity gradient. Hence, the strong relationship between diatoms and salinity in the estuary makes them useful analogues for inferring past salinity changes in the region.

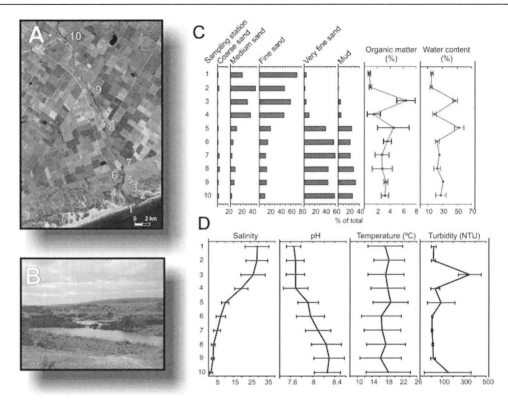

Figure 9. Location of sampling sites (A), view of the estuarine zone (B), and the corresponding sedimentary (C) and water quality (D) parameters at Quequén Salado river (modified after Hassan et al., 2007).

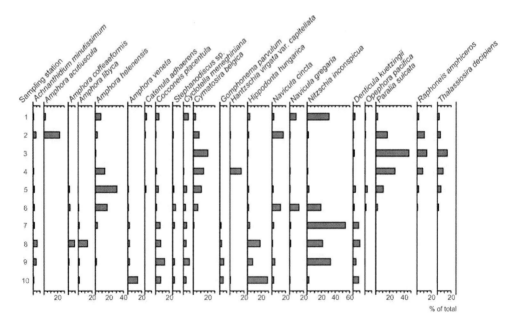

Figure 10. Diatom assemblage composition at Quequén Salado river (modified after Hassan et al., 2009).

Quequén Salado Estuary

The Río Quequén Salado is located 100 km westwards of the Río Quequén Grande, constituting the northernmost estuary subjected to a mesotidal regime in the Argentinean coast (Figures 1 and 9). The estuary has been minimally impacted by human activity because of the absence of large urban settlements, bridges, jetties or harbors. Moreover, it has been suggested that, although a bit smaller, the Quequén Salado estuary presently represents similar conditions to those of the Quequén Grande estuary prior to the anthropogenic influence (Perillo et al., 2005). The lower valley is oriented to the SSE, with steep walls of 8–15 m high. This portion of the river is also characterized by rapids caused by resistant caliche levels. In the last 5 km the river runs across a sandy barrier composed of vegetated dunes (Marini & Piccolo, 2000).

The study of surface sediment diatom assemblages from Río Quequén Salado estuary, which started very recently, yielded distributional patterns very similar to those found in Quequén Grande estuary, as both present stable salinity gradients (Hassan et al., 2007). Marine and marine/brackish diatoms, such as *Paralia sulcata*, *Cymatosira belgica* and *Amphora helenensis*, dominated the lower and middle estuary, and were gradually replaced by the brackish/freshwater and freshwater taxa *Nitzschia inconspicua* and *Hippodonta hungarica* towards the headwaters (Figure 9). However, the marine/brackish diatom assemblage was more widely distributed in Río Quequén Salado and had no analogues when compared to the assemblages represented in Quequén Grande. This difference between both estuaries may be related to differences in salinity and grain size distribution. In fact, the range of salinities and sediment grain sizes in the first kilometers of the Quequén Salado estuary were higher than those recorded at Quequén Grande estuary, where polyhaline conditions and sandy sediments were recorded only in the first meters of the inlet. The differences between both estuaries were attributed to the tidal range and the grade of human impact on each estuary: whereas many modifications have produced major consequences altering the original geomorphology and circulation in the Quequén Grande estuary in the last 100 years, particularly the obstruction of the incoming tidal wave (Perillo et al., 2005), the Quequén Salado mouth dynamics has remained almost unaltered. Since diatom distribution is mainly influenced by the salinity range and sediment type in these estuaries, their morphological differences originated by human modification constitute a key factor in explaining the observed differences in diatom distribution. Hence, diatom assemblages from Río Quequén Salado constituted useful analogues of salinity in low impacted estuaries. Moreover, the data sets from Mar Chiquita, Quequén Grande and Quequén Salado estuaries have been recently used by Hassan et al. (2009) to develop a regional diatom-based salinity transfer function to quantitatively infer past salinity values from fossil diatoms, which will be described below.

Bahía Blanca Estuary

Bahía Blanca estuary is a geomorphologicaly complex environment derived from a Late Pleistocene-early Holocene delta complex (Piccolo & Perillo, 1999). It is formed by a series of NW-SE tidal channels separated by extensive intertidal flats, low marshes and islands (Popovich & Marcovecchio, 2008). The northern area is geomorphologicaly dominated by the Main Channel (main navigation channel), while the southern area is dominated by the channels named Bahía Falsa and Bahía Verde, which are the largest within the estuary (Figure 11). The dominant sedimentology is based on silty clays on the flats and sand in most of the deeper parts of the channels (Piccolo & Perillo, 1999). Mean annual (13°C), summer (21.6°C), and winter (8.5°C) surface water temperatures in the Main Channel are always slightly higher at the head of the estuary (Piccolo et al., 1987), while mean surface salinity increases exponentially from the head to the mid-reaches of the estuary. The water column is vertically homogeneous all throughout the estuary although it may be partially mixed in the inner zone, depending on freshwater runoff conditions. Bahía Blanca estuary includes the largest deepwater harbor system in Argentina, a fact that makes it economically important. This area gathers important urban centers as well as large industrial companies such as a petrochemical industrial park, a thermoelectric plant, fertilizer plants and a commercial duty-free zone on its northern coast (Popovich & Marcovecchio, 2008).

The phytoplankton of Bahía Blanca has been intensively studied during the past decades (Gayoso 1981, 1988, 1998, 1999; Popovich, 2004; Popovich et al., 2008). These studies were mainly focused on the seasonal succession patterns in a fixed station located at the inner part of the Main Channel (Puerto Cuatreros, Figure 11). The site was characterized by its shallowness and extremely high turbidity (secchi depth <0.5 m), and seasonally changing salinity (22.8 to 41). In these long-term studies, the genus *Thalassiosira* was found to be the most conspicuous component of the phytoplankton in the area. *T. curviseriata* was the most abundant species, followed by *T. anguste-lineata*, *T. pacifica*, *T. rotula* and *T. hibernalis*. *Chaetoceros* (*Chaetoceros* sp., *C. diadema*, *C. ceratosporus* var. *brachysetus* and *C. subtilis* var. *abnormis*) was the second most abundant genus. Other important taxa mentioned were *Skeletonema costatum*, *Ditylum brightwellii*, *Guinardia delicatula*, *Asterionellopsis glacialis*, *Thalassiosira eccentrica*, *Cyclotella striata*, *Cerataulina pelagica*, *Thalassiosira hendeyi*, *Paralia sulcata*, and *Gyrosigma attenuata*.

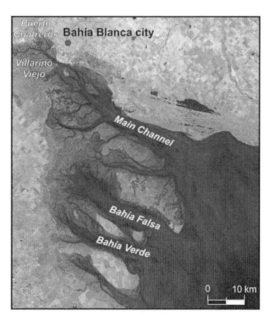

Figure 11. Location map of Bahía Blanca estuary, showing the main channels and the two points studied for diatoms: Puerto Cuatreros and Villarino Viejo.

Table 2. Compositon of the microphytobenthic diatom assemblage and environmental variables in the two sampling sites studied by Parodi and Barria de Cao (2003) at the inner part of Bahía Blanca estuary

	PUERTO CUATREROS	VILLARINO VIEJO
N total (%)	0.21	0.19
P extractable (ppm)	8.82	8.71
pH	8.60	8.50
Salinity	35.6	34.5
Temperature (°C)	9.20	9.40
Dominant to abundant diatoms	*Nitzschia* sp., *Pleurosigma fasciola*, *Navicula* spp,. *Surirella gemma*, *Amphripora alata*, *Stauroneis* sp., *Scoliopleura* sp., *Cocconeis* sp.	*Nitzschia sigma*, *Scoliopleura* sp., *Cocconeis* sp.
Rare diatoms	*Paralia sulcata*?	*Nitzschia* sp,. *Gyrosigma attenuata*, *Entomonoeis amphyprora*, *Pleurosigma fasciola*, *Navicula* spp., *Petrodictyon gemma*, *Cylindrotheca closterium*, *Amphiprora alata*, *Stauroneis* sp., *Paralia sulcata*?

Literature on the ecology and dynamics of the phytoplankton along the Bahía Blanca estuarine gradient is rather poor, especially towards the outer part of the estuary. In a recent attempt, Popovich & Marcovecchio (2008) studied the spatial and seasonal variation in physical and chemical characteristics and phytoplankton biomass in 9 sites located from the inner to the outer reaches of the estuary. Phytoplankton abundance and nutrient levels (N, P and Si) showed a marked decreasing trend from the head to the mouth of the Bahía Blanca estuary. Mean salinity was relatively constant, from 31.6 in the innermost part to 32.9 in the estuary mouth. Distributional tendencies were exposed in a general qualitative way: the inner and middle zones exhibited a seasonal pattern in diatom assemblages composition: whereas than *Thalassiosira curviseriata, T. anguste-lineata, T. pacifica, T. rotula, T. hibernalis, T. eccentrica, Chaetoceros ceratosporus, C. diadema, C. debilis* and *Skeletonema costatum* dominated these regions in winter, summer and autumn assemblages were dominated by *Cerataulina pelagica, Guinardia delicatula* and *Cylindroteca closterium*. On the other hand, the occurrence of several marine species such as *Corethron criophilum, Odontella mobiliensis, Coscinodiscus* spp. and *Actinoptychus* spp. at the outer region indicated a higher influence of euhaline offshore waters on this zone of the estuary. Unfortunately, the abundances of individual diatom taxa along the nine sampling sites were not detailed in this contribution, preventing a more precise inference of their autoecological characteristics.

Sedimentary and microphytobenthic diatoms from Bahía Blanca received much less attention than their phytoplanktonic counterparts. Only one preliminary work (Parodi & Barría de Cao, 2003), which focused on the taxonomic composition of the microalgal mats from the inner part of the estuary (Puerto Cuatreros and Villarino Viejo stations, Figure 11), was published. Puerto Cuatreros site was closer to the harbor, and hence, more influenced by the suspended sediments and the impact of dredging than Villarino Viejo site. Although both sites exhibited similar values for the physical and chemical parameters measured, the species assemblage of the superficial sediment layers showed important differences (Table 2). Whereas than in Puerto Cuatreros diatoms were the dominant microalgae, in Villarino Viejo mats were dominated by blue-green algae. These differences were attributed to the major disturbance of the former due to the deposition of particles of the fluid mud layer produced by the nearby dredging.

Overall, the analysis of the relatively numerous publications on algae from Bahía Blanca leads to the general conclusion that, although information on single species distribution and environmental preferences does exists, this is presented in a very qualitative and descriptive way that prevents its application in diatom-based paleoenvironmental studies.

Estuaries from Patagonia and Tierra Del Fuego

Rivers in the Patagonian region are fed by water originated from the precipitation and/or snow melting on the Andes. They flow across the arid and desert Patagonia region, where practically no tributaries are received. Some of the rivers are considered to be the largest in the country both in valley size and river discharge, such as the Río Colorado, Río Negro and Río Santa Cruz. The climate is semiarid to arid, characterized by strong westerly winds throughout the year (Piccolo & Perillo, 1999). Unfortunately, little is known about the diatoms (and the biota in general) of these estuaries. Only a few contributions on microalgal assemblage composition are available for Río Negro, Bahía San Blas, Río Chubut, Ría de Puerto Deseado and Bahía San Sebastián (see Figure 1), which are described in the following sections.

Río Negro estuary: The Río Negro drains a large basin of 115,800 km^2, and its valley is of great importance both for economical and hydrological reasons (Figures 1 and 12A). River width varies between 500 and 800 m but close to the mouth it has a width of 1 km and flows along a valley of approximately 12 km. Depth ranges from 5 to 10 m. Two banks are found in its mouth, forming an open ebb delta (Piccolo & Perillo, 1999). The river receives the domestic and industrial effluents of the several cities located along their margins, and is regulated by a number of damps and hydroelectric plants located in their tributaries (Pucci et al., 1996). Only one published study is available for the microalgae of this estuary, which is focused on phytoplanktonic communities (Pucci et al., 1996). In that work, samples were collected from three sampling stations located along the last 30 km of the river, in two seasons (spring and autumn; Figure 12A). The composition of the assemblages was homogeneous between sites in spring, being *Aulacoseira granulata* and *Asterionella formosa* the dominant diatoms. Sampling in spring was conducted during low tide. Hence, salinity values were low in the three sampling stations (between 0.052 and 0.32). Autumn samples were taken during high tide, and consequently salinity rose up to 26 in the station closer to the mouth, whereas it decreased to values under 0.19 in the other two stations. Accordingly, diatom assemblages were more diverse, and dominated by brackish-freshwater forms in the two inner stations; and by coastal-marine taxa in the outer station (Figure 12B). Information on nutrients, pH and temperature were also provided (Table 3). Although scarce, the information provided in this work is the only information on modern diatoms from Río

Negro. Detailed studies on diatom distribution and variability across the estuarine gradient should be conducted in order to provide useful analogues for paleoenvironmental reconstructions in this estuary.

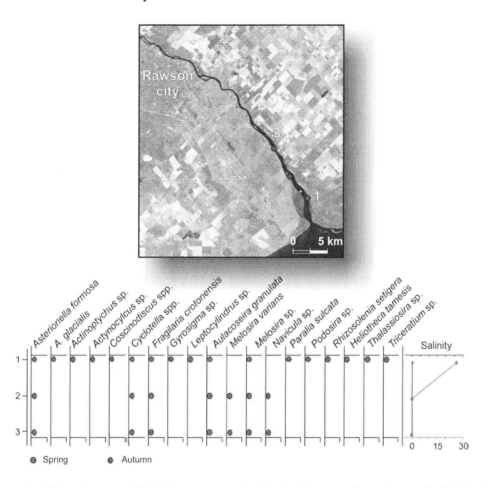

Figure 12. A) Location map of Rio Negro estuary, showing the three sampling points studied by Pucci et al. (1993); B) Distribution of diatom taxa in the sampling sites and the corresponding salinity values in autumn (red) and spring (green), based on data from Pucci et al. (1993).

Table 3. Measurements of environmental variables at Río Negro estuary (modified after Pucci et al., 1993). Numbers correspond to sampling sites signaled in Figure 12. A: autumn; and S: spring, measurements

Station	1		2		3	
Season	A	S	A	S	A	S
Nitrates (µatgN/L)	1.83	0.22-0.92	0.45	6.49	33.8	0.53
Nitrites (µatgN/L)	0.43	0.05-0.09	0.06	0.08	0.09	0.07
Phosphates (µatgP/L)	0.8	0.13-0.14	0.18	0.26	0.53	0.33
Silicates (µatgSi/L)	47.9	165-181	165	140	147	162
Salinity (ppm)	26.09-26.04	0.052-0.31	0.02-0.19	0.206-0.237	0.03	0.261
pH	8.55	8.25-8.3	7.95-8.1	7.8-7.85	7.2-7.8	8.2-8.25

Bahía San Blas estuarine complex: Only two works on diatoms were performed at Bahía San Blas estuarine complex. The first was conducted by Frenguelli (1938), who analyzed the diatom content of plankton samples and surface sediments in the Jabalí creek (which outflows into the southernmost part of the complex) and in Bahía San Blas harbor (Figure 13). At Jabalí creek, 6 samples from diverse origins were collected and analyzed (3 of surface sediments, 1 of macroalgae, 1 of ascidians and 1 of bryozoans, Table 4). Samples were dominated by a mixture of brackish and coastal-marine taxa, constituting an estuarine assemblage under a strong tidal influence. In San Blas harbor, one sediment and three plankton samples were analyzed. The assemblage was dominated by marine taxa, although low proportions of brackish diatoms were also recorded (Table 5). Overall, the analyzed samples always showed relatively high proportions of freshwater diatoms (47 taxa were listed), probably transported from the headwaters by the strong winds blowing from the west. Recently, Isla & Espinosa (2005) described the diatom assemblages from a core taken at the Jabalí creek. The top sample of the core (that represents the modern assemblage) was dominated by *Cymatosira belgica* (~20%), *Paralia sulcata* (~10%), *Achnanthes lacus-vulcani* (~10%) and *Planothidium delicatulum* (~20%). Although measurements of salinity (38), pH (7.72) and turbidity (20 NTU) were provided, the application of this punctual datum in paleoenvironmental analyses is limited, since it does not represents the spatial or temporal variability in the composition of diatom assemblages.

Río Chubut estuary: The river has a meandering channel, which varies from 70 to 200 m in width, and averages 2 m in depth (Figure 14A). The river bed shows several sigmoidal bars constituted by medium to coarse sand that divide the river into channels. Waters are rich in silica, and high gradients of Cl^-, Na^+, SO_4^{-2}, K^+ and Mg^{2+} are found from the mouth to about 2 km upstream. The river has been dammed at about 120 km from the mouth. It passes through several cities being impacted by agricultural and industrial activities and receiving urban sewages with no or little treatment (Piccolo & Perillo, 1999). A series of works attempted to define the composition of the phytoplankton community in the estuary of the Chubut river (see Appendix I). Sastre et al. (1990) and Villafañe et al. (1991) described the taxa present in the last 9 km of the estuary as a function of salinity. In the inner estuary, salinity was under 3, and the phytoplankton was dominated by *Aulacoseira granulata*, which accounted for more than 80% of the total cells. Although this zone was characterized by a low light penetration (secchi depth: 0.4 m), this did not affect the growth of *A. granulata*, which was able to produce large numbers of individuals under these conditions because it is adapted to low levels of light. Sastre et al. (1994) reported the dominance of this species up to 150 km away from the estuary mouth, where it produced blooms and constituted up to 96% of the total cells. Phytoplankton abundance decreased towards the middle estuary, where salinity ranged between 3 and 30. *A. granulata* was also the most abundant taxon in this zone, although other species of planktonic and benthic diatoms, such as *Biddulphia alternans, B. antediluviana, Gramatophora marina, Triceratium favus, Odontella aurita, Actinoptychus* spp., and *Surirella* spp., were also present. In the outer estuary salinity was higher than 30, and the phytoplanktonic assemblage was dominated by *Odontella aurita*, which accounted for more than 80% of the total cells. Santinelli et al. (1990) analyzed the composition of the community in the mouth of the estuary during two years. Salinity values varied significantly during the tidal cycle from fluvial (0-10) to marine (25-35) conditions. Diatoms were the dominant phytoplanktonic group, being identified 39 taxa, which were grouped by cluster analysis and related to their salinity tolerances (Table 6). One of the groups showed a significant association to low salinity values (group 1, Table 6). The second group (group 2, Table 6) comprised euryhaline taxa, which were distributed all over the estuary. The third group, on the other hand, showed a marked association to the higher salinity values prevailing at the estuary mouth (group 3, Table 6). The defined groups constitute potential analogues useful for paleosalinity reconstructions in Patagonian estuaries.

Puerto Deseado estuary: It has a general WSW-ENE orientation, and has an elongated 40 km funnel form (Figure 14B). Freshwater input comes from the Río Deseado, which used to carry much water during the Pleistocene, but is now reduced to a temporary river. The estuary width varies from 2.5 to 0.4 km, while depth ranges from 5 m in the inner part to 20 m in its mouth. Mean tidal amplitudes range between 4.2 and 2.9 m, and salinity variation is small (<2; Ferrario, 1972; Piccolo & Perillo, 1999). Diatoms from Puerto Deseado estuary were studied by Müller Melcher (1959), who mentioned 12 taxa. In a series of recent contributions (Ferrario, 1972, 1981; Ferrario & Sar, 1984; Ferrario, 1984a,b,c) the list of taxa was expanded to 88 species. For each taxon, a series of taxonomical, ecological and distributional observations were provided. The sampled area was typically marine; salinity ranged between 32 and 34 and pH between 7.5 and 8.4. Nitrates and phosphastes concentrations were of 0.5 mg/l and 0.1 mg/l, respectively. The complete list of diatom taxa mentioned in these works is presented in Appendix II.

Figure 13. Location map of Bahía San Blas estuary.

Table 4. Diatom assemblage composition of the Jabalí creek samples analyzed by Frenguelli (1938)

Substrate	Jabalí Creek Samples					
	Estuarine sediment 1 (mud)	Estuarine sediment 2 (mud)	Beach sediment (sandy mud)	Inside ascidia coenobium (Julinia sp.)	Epiphytes under macroalgae (Stipocaulon sp. and Cladophora sp.)	Epibiotic under bryozoans (Gemellaria sp.)
Freshwater diatoms	Cocconeis placentula C. placentula var. lineata Coscinodiscus lacustris Epithemia adnata Pinnularia borealis	Epithemia adnata Aulacoseira granulata Luticola mutica Nitzschia frustulum Opephora martyi Pinnularia borealis	Planothidium lanceolatum, Discostella stelligera Epithemia adnata Gomphonema gracile Hantzschia amphioxys var. xerophila, Aulacoseira granulata, Luticola mutica, Navicula peregrina, Nitzschia frustulum var. Perpusilla, Martyana martyi, Rhopalodia gibba, R. gibberula	Amphora perpusilla Epitemia adnata Staurosira construens Aulacoseira italica Nitzschia frustulum Rhopalodia gibba	-	Encyonema turgidum Epithemia adnata
Brackish diatoms	Achnanthes brevipes var. intermedia Planothidium delicatulum Caloneis permagna Cyclotella striata	Nitzschia clausii Rhopalodia musculus	Achnanthes brevipes var. intermedia Planothidium delicathulum Gyrosigma balticum Nitzschia N. sigma var. rigida	Achnanthes brevipes var. intermedia Planothidium delicatulum Gyrosigma spenceri var. exilis Nitzschia clausii Nitzschia sigma var. rigida	-	Cyclotella baltica Gyrosigma balticum Bacillaria paradoxa B. paradoxa var. tropica

Table 4. (Continued)

	Jabali Creek Samples					
Substrate	Estuarine sediment 1 (mud)	Estuarine sediment 2 (mud)	Beach sediment (sandy mud)	Inside ascidia coenobium (Julinia sp.)	Epiphytes under macroalgae (Stipocaulon sp. and Cladophora sp.)	Epibiotic under bryozoans (Gemellaria sp.)
Marine diatoms	*Amphora granulata* *Auliscus sculptus* *Cocconeis scutellum* *C. scutellum var. Parva* *Paralia sulcata*	*Paralia sulcata*	*Amphora angusta* *Campilosira cymbelliformis* *Cocconeis scutellum var. ornata* *C. scutellum var. parva* *Coscinodiscus excentricus var. minor*	*Amphora granulata* *Cocconeis scutellum* *C. scutellum var. parva* *Rhoicosphaenia marina* *Navicula gourdoni* *N. oceanica* *N. platyventris*	*Cocconeis scutellum var. ornata* *C. scutellum var. minor*	*Cocconeis scutellum var. ornata* *C. scutellum var. parva* *Corethron criophilum*

Bahía San Sebastián: It is a wide bay located in northern Tierra del Fuego, having a semicircular shape partly closed by a long and narrow gravel spit (Figure 14C). The bay is 55 km long and 40 km wide. The spit has a length of 17 km, and the open mouth is about 20 km wide. Freshwater input into the system is provided by the Río San Martín, which discharges at the southwestern part of the bay. Tidal range is 10 m and wind influence is from the west (Piccolo & Perillo, 1999). Frenguelli (1923, 1924) described the diatom taxa found in a sediment sample collected in San Sebastián bay. Diatoms frustules were scarce. The assemblage was dominated by *Paralia sulcata*, whereas than other ten less frequent taxa were also mentioned (Table 7). No environmental characterization of the sampling point was provided in these studies.

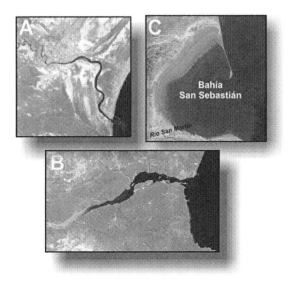

Figure 14. Location maps of A) Río Chubut, B) Puerto Deseado, and C) Bahía San Sebastián estuaries.

Table 5. Diatom assemblage composition of Bahía San Blas samples analyzed by Frenguelli (1938)

	Bahía San Blas Samples	
	Sediment (mud)	Plankton net
Brackish diatoms	*Achnanthes brevipes, A. brevipes* var. *intermedia, Planothidium delicatulum, Gyrosigma balticum, Nitzschia clausii, N. sigma, N. sigma* var. *sigmatella, Rhopalodia musculus*	-
Marine diatoms	*Cocconeis scutellum, C. scutellum* var. *parva, Paralia sulcata*	*Biddulphia chilensis, Odontella mobiliensis, Lithodesmium undulatum, Thalassiosira decipiens, Rhizosolenia imbricata, Thalassiosira javanica, Rhaphoneis amphiceros, Thalassiosira eccentrica*

Table 6. Compositon of the phytoplanctonic diatom assemblages along the salinity gradient in the Río Chubut estuary (modified from Santinelli et al., 1990)

Group	Salinity range	
	0-10 ppm	10-25 ppm
Freshwater/ Brackish	*Navicula radiosa, Navicula* spp., *Cymbella cystula, Cymbella* spp., *Epithemia sorex, Rhopalodia gibba, Cocconeis placentula, Cocconeis* sp., *Cymatopleura solea, Surirella* spp., *Asterionella formosa*	
Eurihaline	*Paralia sulcata, Odontella aurita, Gomphoneis herculeana, Biddulphia alternans, Aulacoseira granulata Nitzschia* spp., *Thalassiosira* spp., *Gramatophora marina, Rhabdonema adriaticum, Biddulphia antediluviana, Ulnaria ulna, Synedra* spp., *Melosira varians*	
Marine		*Triceratium favus* *Actinoptychus vulgaris*

Table 7. Diatom assemblage composition of the Bahía San Sebastián sediment sample analyzed by Frenguelli (1923, 1924)

Diatom taxa	
Actinoptychus senarius	*Psammodictyon panduriforme* var. *parva*
Thalassiosira eccentric	*Raphoneis amphiceros*
Hyalodiscus radiatus	*Surirella striatula*
Paralia sulcata	*Surirella tuberosa* var. *costata*
P. sulcata var. *biseriata*	*Triceratium scitulum*
P. sulcata var *crenulata*	

Table 8. Diatom assemblage composition of the samples from Río Grande estuary, according to Cleve (1900)

Marine and brackish taxa	Freshwater taxa
Actinoptychus undulatus	*Amphora pediculus*
Amphora lineolata	*Cymbella aspera*
Biddulphia aurita	*Frustulia rhomboides*
Biddulphia rhombus	*Hantzschia elongata*
Cocconeis scutellum var. *genuina*	*Melosira* sp.
Coscinodiscus decipiens	*Neidium oblique striatum* var. *magellanicum*
Coscinodiscus excentricus	*Pinnularia borealis*
Coscinodiscus oliverianus	*Pinnularia commutata*
Entyopyla incurvata	*Pinnularia elliptica*
Epithemia musculus	*Pinnularia gibba*
Hantzschia virgata	*Pinnularia lata*
Hyalodiscus radiates	*Pinnularia latevittata*
Hyalodiscus scoticus	*Pinnularia gibba* var. *luculenta*
Melosira nummuloides	*Pinnularia major* var. *linearis*
Navicula anglica var. *subsalsa*	*Pinnularia nodosa*
Navicula arenacea	*Pinnularia stauroptera*

Table 8. (Continued)

Marine and brackish taxa	Freshwater taxa
Navicula cincta	*Pinnularia viridis*
Navicula gregaria	*Rhoicosphenia curvata*
Navicula pygmaea, Navicula salinarum, Navicula sub-inflata, Navicula tumida, Nitzschia apiculata, Ntzschia constricta var. *subconstricta, Nitzschia panduriformis, Nitzschia sigma, Paralia sulcata* var. *radiata, Pleurosigma normanii, Pleurosigma nubecula* var. *intermedia, Pleurosigma rigidum, Podosira maxima, Rhabdonema arcuatum, Rhabdonema minutum, Rhaphoneis amphiceros, Stauroneis salina, Surirella gemma, Surirella striatula, Triceratium affine*	*Rhopalodia gibba* *Stauroneis phoenicenteron* var. *amphilepta* *Surirella guatemaliensis* *Surirella splendida* var. *tenera*

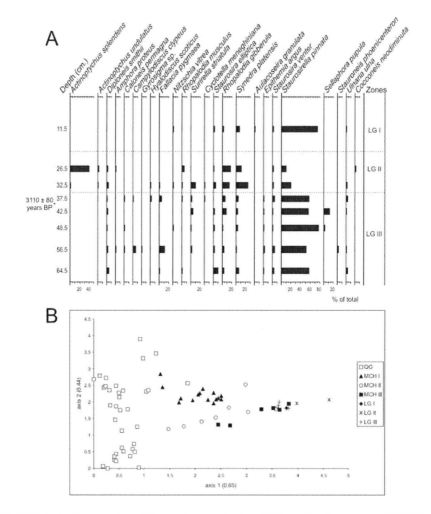

Figure 15. A) Relative frequencies of diatom taxa in the Las Gallinas Creek sequence; B) DCA of combined surface (QG: Quequén Grande, MCHI, MCHII, MCHIII: Mar Chiquita tidal inlet, inner lagoon and headwaters, respectively), and fossil diatom samples (LG: Las Gallinas Creek). Diatom zones were delimited through cluster analyses (reproduced from Hassan et al., 2006; with permission).

Río Grande estuary: The Río Grande flows from west to east, receiving tributaries from the south and the north. Before discharging into the Atlantic Ocean, the river makes a long bend to the south around gravel beach barriers on which the Río Grande is built (Figure 1). The inlet is therefore constrained by gravel spits that have a significant morphologic variability. The mean tidal range in Río Grande outer estuary is 4.16 m (Isla & Bujalesky, 2004). The only work on diatoms from the Río Grande estuary was carried out by Cleve (1900), who analyzed a series of samples of the estuarine area and provided lists of marine-brackish (38 taxa) and freshwater (22 taxa, Table 8) forms. This work is taxonomic and do not includes environmental information.

HOW CAN RESEARCHERS IMPROVE THE QUALITY OF DIATOM-BASED PALEOENVIRONMENTAL INFERENCES IN COASTAL SETTINGS OF ARGENTINA?

The application of diatom autoecology to paleoenvironmental reconstructions has a long history in the Argentinean coast. Pioneer studies were conducted by Frenguelli (1924, 1925, 1945), who described diatom assemblages present in Holocene successions outcropping in estuaries along the Pampean coast. Besides some of the major estuaries described in the previous section (Río Quequén Grande, Río Quequén Salado and Bahía Blanca), many small streams that flow into the Río de la Plata or the Atlantic coast were included. A total of 276 diatom taxa were listed, from which only 11 species were present in high proportions and formed the dominant assemblage. These were: *Campylodiscus clypeus, Cocconeis placentula, Denticula valida, Diploneis argentina, Hyalodiscus subtilis, Nitzschia vitrea, Rhopalodia gibberula, R. argentina, Surirella striatula* and *Synedra platensis*. Overall, diatom assemblages indicated the presence of environments under marine influence that evolved to brackish/freshwater continental conditions, and ended in swamps which finally got dry as a consequence of the climatic aridization.

During the last 20 years, the paleoenvironmental evolution of the southern Pampas coast and its relationship to the Holocene sea-level fluctuations have been inferred from the detailed study of sedimentary successions originated by the infilling of estuarine sediments. The analyses were based on the diatoms autoecological classifications of salinity and life form taken from De Wolf (1982), Vos and De Wolf (1988, 1993), and Denys (1991/1992), and allowed to infer the presence of sedimentary environments characterized by different salinities and depths. Between ca. 6700 and 3900 ^{14}C yr BP, the marine influence related to the sea-level high stand was the dominant forcing on paleosalinity trends, occurring at different times and magnitudes according to the characteristics of each basin (Isla et al., 1986; Espinosa, 1998). In the area of Arroyo La Ballenera (see Figure 1) an estuarine lagoon with small or no tidal range was inferred for the interval between ca. 6200 and 4800 ^{14}C yr BP, whereas in Arroyo Las Brusquitas (Figure 1) estuarine conditions lasted up to ca. 3900 ^{14}C yr BP (Espinosa et al., 2003). In Punta Hermengo area (Figure 1), a tidal channel infilling was inferred at ca. 6700 ^{14}C yr BP (Espinosa, 2001). In Río Quequén Grande, the maximum saline influence was detected between ca. 7100 and 5350 ^{14}C yr BP at 2 km from the river mouth in relation to the development of an estuarine lagoon (Espinosa, 1988, 1998). This marine influence was not recorded in synchronic deposits outcropping 32 km upstream from the previous site (Zárate et al., 1998). In Río Quequén Salado, the analysis of diatom assemblages from two sequences outcropping at 20 and 30 km from the estuary mouth revealed the presence of fluvial-lacustrine environments during the late Pleistocene, followed by alluvial plains with a pulse of marine influence, which finally evolved to lacustrine environments that became brackish and shallower towards the early Holocene (Schillizzi et al., 2006). In the sector of the Pampas coast located between Río Quequén Salado and Bahía Blanca, diatom analyses allowed to infer the development of estuarine lagoon environments during the middle Holocene (between *ca.* 6500 and 6900 years BP). These estuarine lagoons were transgressed by the sea towards the late Holocene (*ca.* 5300-4800 ^{14}C years BP; Gutiérrez Téllez & Schillizzi, 2002; Aramayo *et al.*, 2005).

In contrast to the abundant information available on coastal Holocene diatoms from the Pampean region, data from Patagonia are scarce and studies initiated only recently (see Espinosa, 2008). Isla & Espinosa (2005) analyzed the evolution of southern Bahía San Blas during the late Holocene. The dominance of marine and marine/brackish diatom assemblages in a sediment core obtained in the Jabalí Creek suggested that the zone maintained a hypersaline regime during the last 4700 years. Escandell et al. (in press) analyzed the diatom assemblages from a core obtained 9 km upstream from the Río Negro mouth, in order to reconstruct the late Holocene paleoenvironmental evolution of the estuary. In this contribution, both European ecological codes as well as modern information provided by Hassan (2008) for pampean estuaries were applied. The sequence, which comprised the interval between 2027±34 ^{14}C years BP and the present, recorded the evolution of a shallow vegetated brackish-freshwater environment at the bottom, which evolved towards a tidal channel that declined gradually in depth and salinity to the middle, to finally end in a marsh influenced by tides and floods towards the top of the sequence.

The first attempt to apply data on local modern diatom distribution to paleoenvironmental reconstruction in estuarine settings of Argentina was carried out by Hassan et al. (2006). In this work, modern data from Mar Chiquita and Quequén Grande estuaries were compared with fossil data obtained from a late Holocene sequence outcropping at the headwaters of the Mar Chiquita coastal lagoon (Arroyo Las Gallinas) through the application of semi-quantitative techniques (DCA ordination). The sequence had been previously studied through autoecological techniques (Espinosa, 1994). All diatom assemblages were dominated by oligohalobous indifferent taxa (*Staurosirella pinnata* and *Staurosira venter*), accompanied by some oligohalobous halophilous and mesohalobous taxa (such as *Staurosira elliptica*, *Fallacia pygmaea* and *Campylodiscus clypeus*), except for a level located near the middle of the sequence that was dominated by the polyhalobous *Actinoptychus splendens*, the mesohalobous *Rhopalodia musculus* and the oligohalobous halophilous *R. gibberula* (Figure 15). DCA ordination of modern and fossil samples showed that, except for this level, fossil diatoms from Arroyo Las Gallinas were analogue to modern diatom assemblages living today in the inner lagoon of Mar Chiquita (sites 14 and 15 in Figure 4A), representing a shallow brackish/freshwater environment, with low salinity fluctuations (1-9) and no tidal influence. Espinosa (1994) proposed tidal channel conditions for the basal levels of Las Gallinas sequence, based on the presence of silty clays and the dominance of tychoplankton. Espinosa (1998) reinterpreted Las Gallinas paleoenvironments as shallow brackish environments with low tidal influence and significant freshwater inflow. On the basis of modern data analysis, Hassan et al. (2006) discarded the tidal influence, since there was no similarity between fossil levels and modern assemblages from Mar Chiquita tidal zone.

In an attempt to increase the accuracy of coastal paleoenvironmental reconstructions in southern Pampas, Hassan et al. (2009) conducted the first quantitative reconstruction of past environmental parameters in estuarine environments of Argentina. In this contribution, the modern data sets provided by Hassan et al. (2006, 2007) for Mar Chiquita coastal lagoon, Río Quequén Grande and Río Quequén Salado were integrated to construct a diatom-based salinity calibration model, based on Weighted Averaging Partial Least Squares techniques (WA-PLS, ter Braak & Juggins, 1993). WA-PLS, together with its simpler version Weighted Averaging (WA), constitute the most robust and simple regression techniques available for quantitative reconstructions based on unimodal distributions (ter Braak et al., 1993; Birks, 1995). In a first step, the relationship between the 48 dominant diatom taxa and salinity was evaluated, and optima and tolerances for each taxon were calculated (Figure 16). According to their salinity optima, diatom taxa were divided into three groups: a freshwater group, with salinity optima in oligohaline waters (up to 5); a brackish group, distributed in mesohaline waters (5–18) and a polyhalobous group, restricted to polihaline waters (18–30; Day, 1981). According to their salinity tolerances, most taxa can be regarded as markedly euryhaline (Denys, 1991/1992), since they tolerated salinity changes between 2.3 and 11.6. Taxa located at both ends of the diagram (freshwater and polihalobous taxa) showed the narrowest tolerance ranges, whereas mesohaline taxa showed the widest ones (Figure 16). The salinity transfer function constructed on the basis of this data set showed a good performance, with an error of 4.42, comparable to the obtained in salt marshes from North America (Sherrod, 1999).

The modern data set was applied to the paleoenvironmental reconstruction of a sedimentary sequence outcropping at the left margin of the Río Quequén Grande (Puente Taraborelli section, site 13 in Figure 7A), 12 km upstream from the estuary mouth. Diatom assemblages of the basal and medium sections of the sequence (0.8–1.8 m in depth) were dominated by *Fragilariforma virescens*, *Staurosira venter*, *Cocconeis placentula* var. *euglypta*, *Denticula kuetzingii*, *Nitzschia inconspicua* and *Planothidium delicatulum*. Samples from the top of the sequence (0–0.8 m in depth) were dominated by *Staurosirella pinnata*, accompanied by *Staurosira venter*, *Catenula adhaerens* and *Paralia sulcata* (Figure 17). In a semi-quantitative approach, modern and fossil samples were ordered in a two dimensional space through DCA (Figure 18). Results of DCA ordination showed that Holocene diatom assemblages were more similar to the modern diatom assemblages from Mar Chiquita than those living today at Quequén Grande river, suggesting the presence of an estuarine lagoon rather than an estuary of lotic characteristics. The application of the transfer function to the fossil diatom assemblages allowed the quantitative reconstruction of Holocene salinity fluctuations (Figure 17). Maximum salinity values, estimated at about 13, were detected between ca. 7500±90 and 6040±90 ^{14}C yr BP. Therefore, the integration of these results to those obtained in previous works (Espinosa, 1998) suggested that the marine influence in Quequén Grande occurred since ca. 7500 ^{14}C yr BP, extending up to 12 km from the present coastline through ca. 7000 ^{14}C yr BP, in relation to the development of an estuarine lagoon of large dimensions. In contrast to these quantitative results, the application of the autoecological classifications (*sensu* Vos & De Wolf, 1993) only allowed the recognition of two main sedimentary environments within broad salinity compartments: diatom assemblages indicated a brackish/freshwater environment of continental characteristics in the basal and medium sections of the sequence, and a marine/brackish environment subjected to small tidal range towards the top of the sequence (Figure 19).

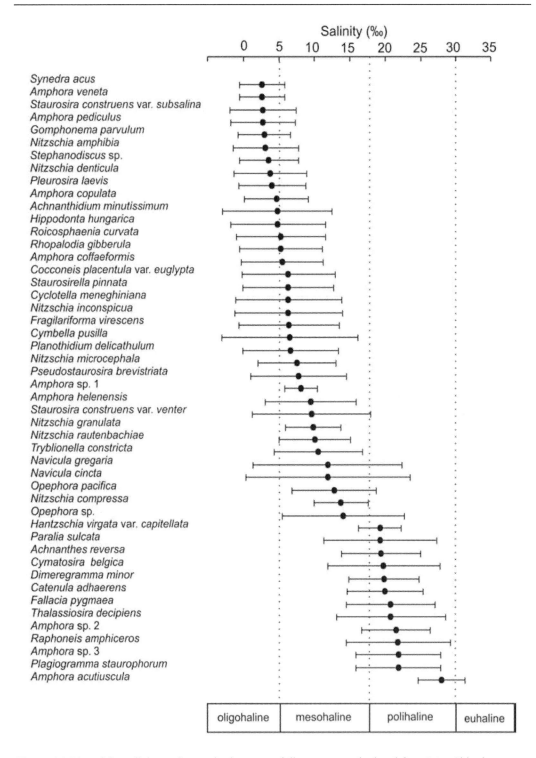

Figure 16. Plot of the salinity optima and tolerances of diatom taxa calculated from Mar Chiquita, Quequén Salado and Quequén Grande datasets. Salinity classification follows Day, 1981 (reproduced from Hassan et al., 2009; with permission).

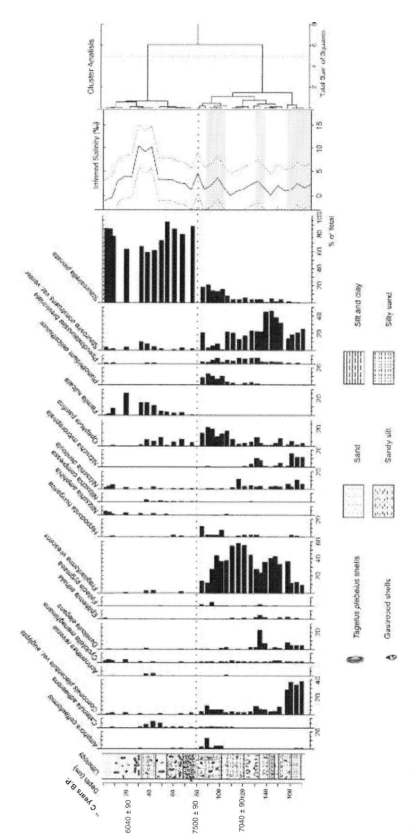

Figure 17. Lithology, relative frequency diagram of diatom composition and inferred salinity values at Puente Taraborelli profile. Grey shadows indicate salinity values inferred from samples that lack good analogues in the training set (reproduced from Hassan et al., 2009; with permission).

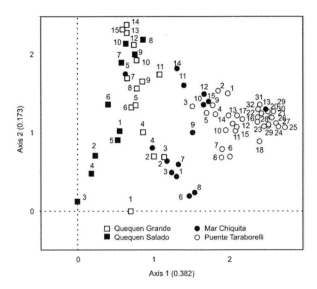

Figure 18. Results of DCA ordination of modern diatom samples from Mar Chiquita, Quequén Grande and Quequén Salado estuaries, and fossil diatom samples from Puente Taraborelli sequence. MCH, QG and QS numbers correspond to sampling sites showed in Figs. 4, 7 and 9 (reproduced from Hassan et al., 2009; with permission).

It becomes clear that to the general characterization of sedimentary environments provided by the autoecological techniques widely applied in the region, the transfer function approach adds a method to map both temporal and spatial variations in paleosalinity values. The wide salinity tolerances of estuarine diatoms found in the three studied estuaries restricts the accuracy of the paleoenvironmental reconstructions based on their autoecology, even when autoecological data are obtained from local environments. A clear example of this is the euryhaline species *Staurosirella pinnata*, which dominates Holocene sucessions of both estuarine (e.g. Hassan et al., 2009) and freshwater (e.g. Espinosa, 1994) origin, limiting the paleoenvironmental inferences that can be done from the assemblage. This limitation is strongly linked to the impossibility of classifying individual taxa into narrow salinity classes, problem that is saved by applying synecological techniques, since they are based on a weighted average of the optima and tolerances of all taxa present in a fossil sample. Accordingly, researchers can improve the quality of diatom-based paleoenvironmental reconstructions by incorporating quantitative approaches to their projects. Furthermore, it would be useful to generate modern data sets that allowed a semi-quantitative analysis of fossil data by detecting and identifying modern environments that could possibly be analogue to the ones that developed during the Holocene. Even when this approach does not provides quantitative estimates of past environmental variables it supplies a useful tool to assess the paleoenvironmental significance of fossil diatom assemblages dominated by taxa with broad salinity tolerances.

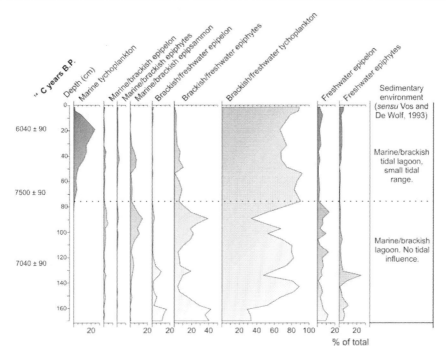

Figure 19. Relative frequency diagram of diatom ecological groups and their environmental significance according to Vos and De Wolf, 1993 (modified from Hassan, 2008).

CONCLUSIONS

The bibliographic analysis carried out in the previous sections evidences that information on modern diatoms from Argentinean estuaries is very scarce and fragmentary, a fact that clearly contrasts with the abundance, magnitude and economic importance of these environments in the region. In general, most of the reviewed works focused on diatom assemblages from the more densely inhabited Pampean coast, whereas estuaries from Patagonia, in some cases less accessible, received less attention. From the analysis of Appendix II, a general tendency of increasing diatom richness towards the south can be recognized: the highest number of taxa was mentioned for Río de la Plata (n= 356), whereas the lowest values were recorded in Bahía San Blas (n=15) and Río Negro (n=19). Although some geographical component could be invoked to explain this apparent tendency, many problems with the dataset pose serious limitations to the formulation of general biogeographical conclusions. Works contrasted significantly in sampling strategy and intensity, as well as in the ecological compartment studied. For example, while some studies only dealt with one sample (e.g. Frenguelli 1923, 1924), others included more than 50 samples (e.g. Licursi et al., 2006; Hassan et al., 2009). Moreover, studies focused either on phytoplanktonic (e.g. Licursi et al., 2006; Pucci et al., 2006), sedimentary (e.g. Parodi & Barría de Cao, 2003; Hassan et al., 2009) or epiphytic (e.g. Bauer et al., 2007) assemblages. It is evident that the more samples are analyzed, the more different and numerous taxa that can be found. It is also obvious that different habitats contain different diatom floras. Hence, the methodological inconsistency underlying the data set does not allow performing comparisons on diatom diversity and biogeography aspects.

Many of the reviewed works supplied some kind of environmental information (particularly on salinity), which is one of the main requisites to apply the information on diatom assemblage composition to reconstruct past environments. However, studies differed significantly in the quality of the datasets provided. The most complete and useful works were those in which the research was guided by autoecological or paleoecological objectives. These detailed distributional studies were carried out in the Río de la Plata (Licursi et al., 2006), Mar Chiquita, Quequén Grande and Quequén Salado (Hassan et al., 2009) estuaries. They provided information not only on single diatom taxa distribution but also on environmental parameters along the estuarine gradient, allowing extracting either autoecological or quantitative data applicable to the fossil record.

Unfortunately, a great number of the available works included only punctual samplings, restricted either in time or in space, which do not reflect the high variability of diatom assemblages. The most significant example was the Bahía Blanca estuary, where a relatively large number of detailed studies on diatom seasonality were conducted in the last decades, but mostly restricted to one single site located in the inner estuary (Puerto Cuatreros). In other cases, studies covered neither spatial nor temporal variability on diatom assemblage composition, since samples were taken only in one (e.g. Bahía San Blas; Isla and Espinosa, 2005) or two (e.g. Río Negro; Pucci et al., 1996) moments of the year. This is an essential issue when working in estuarine environments, which are subjected to significant environmental fluctuations to which diatoms must become adapted. Consequently, it is not possible to assess the environmental preferences of the diatom taxa present in a sample if the whole range of variability in estuarine conditions has not been covered by the sampling strategy. Hence, the report of the presence of a taxon at a given salinity value in a sole sampling point, as provided in many of the reviewed studies, constitutes only an anecdotal data of restricted applicability for paleoenvironmental reconstructions.

There were also some works which, although based on detailed and well-planned sampling strategies, did not present the results in an accessible way. Examples of these are found in the Río de la Plata (Gómez et al., 2009) and Bahía Blanca (Popovich & Marcoveccio, 2009) estuaries, where although diatom assemblages from a set of sampling sites distributed along the estuarine gradient were studied and environmental data presented, the frequencies or abundances of taxa in each site were not provided. This omission prevented the linking of each taxon to the values of the environmental parameters at which they were found, information that would have resulted very useful for paleoenvironmental reconstructions. In other words, there is a large amount of information but it is unavailable to the reader. This constitutes one of the most surprising findings of the present review, since evidences a lack of contact between ecologists and paleoecologists that may lead to an unnecessarily doubling of research efforts.

If the problems listed above are taken into consideration, the information summarized in Appendix II can be reliably applied to paleoenvironmental reconstructions. However, it is necessary to be aware of the fact that the salinity information listed represents the ranges at which each taxon was found in Argentinean estuaries, and not its optimal and tolerance (excepting for a few exceptions in which these parameters were statistically calculated). Likewise, the type of sample (plankton, sediment, or vegetation) at which each taxon was recorded in each estuary does not necessarily coincide with the habitat of that species. In some cases, taphonomic processes can resuspend benthic diatoms and incorporate them into the water column, while in others plankton forms can be found deposited in surface sediments (Juggins, 1992). Examples of these are presence of planktonic taxa (such as *Actinoptychus splendens* or *Actinocyclus octonarius*) in sediments of Mar Chiquita coastal lagoon, as well as the finding of non-planktonic species (such as *Cocconeis placentula* and *Gomphonema parvulum*) in plankton samples from the Río de la Plata estuary.

Finally, it should be noted that progress in Holocene estuarine diatom paleoecology in Argentina will greatly depend on further study of all aspects of modern diatom ecology and distribution, as well as of the taphonomic processes that alter dead diatom frustules before and during its deposition. In that way, there are many issues that need to be investigated, such as the nature and extent of the taphonomic biases suffered by plankton assemblages; the detailed distribution patterns of diatom assemblages along the environmental gradient of most of the Argentinean estuaries; the single taxa optima and tolerances for key environmental factors, and the biogeographical distributional patterns. The observation of modern environments would not only allow a better knowledge of the environmental significance of fossil assemblages, but also to construct new hypothesis to guide future investigations in paleoecological research.

ACKNOWLEDGMENTS

I am very grateful to Frank Columbus for inviting me to participate of this publication. This work would not have been possible without the help, patience, and critical comments of Claudio De Francesco. I would also like to acknowledge Silvia De Marco, Hugo Freije, Gerardo Perillo, and Viviana Sastre for gently providing copies of their works. José María Guerrero, Patricio Rivera, Fabricio Idoeta, and Eleonor Tietze helped in the compilation of the bibliography. Drs. Felipe García-Rodríguez and Marcela Espinosa made helpful commentaries and suggestions on the manuscript. Landsat 5 and 7 satellite images were freely supplied by the Comisión Nacional de Energía Atómica (CONAE) through its webpage (http://www.conae.gov.ar). Financial support for this work was provided by University of Mar del Plata (EXA 457/09). G.S.H. is member of the Scientific Research Career of the Consejo Nacional de Investigaciones Científicas y Técnicas (CONICET).

REFERENCES

Acha, M. E., Mianzan, H., Guerrero, R., Carreto, J., Giberto, D., Montoya, N. & Carignan, M. (2008). An overview of physical and ecological processes in the Rio de la Plata Estuary. *Palaeogeography, Palaeoclimatology, Palaeoecology*, 275, 77-91.

Admiraal, W. (1977a). Influence of light and temperature on the growth rate of estuarine benthic diatoms in culture. *Marine Biology*, 39, 1-9.

Admiraal, W. (1977b). Salinity tolerance of benthic estuarine diatoms as tested with a rapid polarographic measurement of photosynthesis. *Marine Biology*, 39, 11-19.

Admiraal, W. (1977c). Influence of various concentrations of orthophosphate on the division rate of an estuarine benthic diatom, *Navicula arenaria*, in culture. *Marine Biology, 42*, 1-8.

Admiraal, W. (1977d). Tolerance of benthic diatoms to high concentrations of ammonia, nitrite ion, nitrate ion and orthophosphate. *Marine Biology, 43*, 307-315.

Admiraal, W. (1984). The ecology of estuarine sediment inhabiting diatoms. In F. E. Round, & D. J. Chapman (Eds.), *Progress in Phycological Research, 3*, (269-322). Bristol: Biopress Limited.

Admiraal, W. & Peletier, H. (1980). Distribution of diatom species on an estuarine mud flat and experimental analysis of the selective effect of stress. *Journal of Experimental Marine Biology and Ecology, 46*, 157-175.

Admiraal, W., Peletier, H. & Zomer, H. (1982). Observations and experiments on the population dynamics of epipelic diatoms from an estuarine mudflat. *Estuarine Coastal and Shelf Science, 14*, 471-487.

Alin, S. R. & Cohen, A. S. (2004). The live, the dead, and the very dead: taphonomic calibration of the recent record of paleoecological change in Lake Tanganyika, East Africa. *Paleobiology, 30*, 44-81.

Ampsoker, M. C. & McIntire, C. D. (1978). Distribution of intertidal diatoms associated with sediments in Yanquina Estuary, Oregon. *Journal of Phycology, 14*, 387-395.

Aramayo, S. A., Gutierrez-Tellez, B. & Schillizzi, R. A. (2005). Sedimentologic and paleontologic study of the southeast coast of Buenos Aires province, Argentina: a late Pleistocene-Holocene paleoenvironmental reconstruction. *Journal of South American Earth Sciences, 20*, 65-71.

Ayestaran, M. G. & Sastre, A. V. (1995). Diatomeas del curso inferior del Río Chubut (Patagonia Argentina). Pennales I: Naviculaceae. *Boletin de la Sociedad Argentina de Botánica, 31*, 57-68.

Austen, I., Andersen, T. J. & Edelvang, K. (1999). The influence of benthic diatoms and invertebrates on the erodibility of an intertidal mudflat, the Danish Wadden Sea. *Estuarine, Coastal and Shelf Science, 49*, 99-111.

Balech, E. (1976). Fitoplancton de la campaña convergencia 1973. *Physis, 35*, 47-58.

Balech, E. (1978). Microplancton de la campaña productividad IV. *Revista del Museo Argentino de Ciencias Naturales, Hidrobiología, 5*, 137-201.

Bao, R., Alonso, A., Delgado, C. & Pagés, J. L. (2007). Identification of the main driving mechanisms in the evolution of a small coastal wetland (Traba, Galicia, NW Spain) since its origin 5700 cal yr BP. *Palaeogeography, Palaeoclimatology, Palaeoecology, 247*, 296-312.

Battarbee, R. W., Charles, D. F., Dixit, S. S. & Renberg, I. (1999). Diatoms as indicators of surface water acidity. In E. F. Stoermer, & J. P. Smol, (Eds.), *The diatoms: applications for the environmental and Earth Sciences*, (85-129). London: Cambridge University Press.

Bauer, D. E., Gómez, N. & Hualde, P. R. (2007). Biofilms coating *Schoenoplectus californicus* as indicators of water quality in the Río de la Plata Estuary (Argentina). *Environmental Monitoring Assessment, 133*, 309-320.

Bayssé, C., Elgue, J. C., Burone, F. & Parietti, M. (1986). Campaña de invierno 1983. II Fitoplancton. *Publicaciones de la Comisión Técnica Mixta del Frente Marítimo, 1*, 218-229.

Behrensmeyer, A. K., Kidwell, S. M. & Gastaldo, R. A. (2000). Taphonomy and paleobiology. *Paleobiology (supplement)*, *26*, 103-147.

Bennett, J. R., Bianchi, T. S. & Means, J. C. (2000). The effects of PAH contamination and grazing on the abundance and composition of microphytobenthos in salt marsh sediments (Pass Fourchon, LA, USA). II. The use of plant pigments as biomarkers. *Estuarine, Coastal and Shelf Science*, *50*, 425-439.

Bergamasco, A., De Nat, L., Flindt, M. R. & Amos, C. L. (2003). Interactions and feedbacks among phytobenthos, hydrodynamics, nutrient cycling and sediment transport in estuarine ecosystems. *Continental Shelf Research*, *23*, 1715-1741.

Bianchi, T. S. & Rice, D. L. (1988). Feeding ecology of *Leitoscoloplos fragilis*. II. Effects of worm density on benthic diatom production. *Marine Biology*, *99*, 123-131.

Birks, H. J. B. (1995). Quantitative palaeoenvironmental reconstructions. In D. Maddy, & J. S. Brew, (Eds.), *Statistical modelling of Quaternary science data*, (161-254). Cambridge: Quaternary Science Association.

Birks, H. J. B. & Birks, H. H. (1980). *Quaternary palaeoecology*. Baltimore: University Park Press.

Campeau, S., Pienitz, R. & Héquette, A. (1999). Diatoms as quantitative paleodepth indicators in coastal areas of the southeastern Beaufort Sea, Arctic Ocean. *Palaeogeography, Palaeoclimatology, Palaeoecology*, *146*, 67-97.

Calliari, D., Gómez, M. & Gómez, N. (2005). Biomass and composition of the phytoplankton in the Río de la Plata: large-scale distribution and relationship with environmental variables during a spring cruise. *Continental Shelf Research*, *25*, 197-210.

Calliari, D., Brugnoli, E., Ferrari, G. & Vizziano, D. (2009). Phytoplankton distribution and production along a wide environmental gradient in the South-West Atlantic off Uruguay. *Hydrobiologia*, *620*, 47-61.

Carbonell, J. J. (1935). Some micrographic observations of the waters of the River Plate. *Verhandl. Internat. Vereinig. F. theor. u . angewandle Limnologie*, *7*.

Carbonell, J. J. & Pascual, A. (1925). Una *Melosira* nueva para el Río de la Plata. *Physis*, *8*, 106-107.

Carp-Sihn-Sohma (Com. Adm. Río de la Plata –Serv. Hidrog. Naval Argentina –Serv. Oceanog. Hidrol. Meteorol. Armada Uruguay). (1989). *Estudios para la evaluación de la contaminación en el Río de la Plata. Buenos Aires-Montevideo*, 422.

Carreto, J. I., Montoya, N. G., Benavides, H. R., Guerrero, R. & Carignan, M. O. (2003). Characterization of spring phytoplankton communities in the Río de La Plata maritime front using pigment signatures and cell microscopy. *Marine Biology*, *143*, 1013-1027.

Carreto, J. I., Montoya, N., Akselman, R., Carignan, M. O., Silva, R. I. & Cucchi Colleoni, D. A. (2008). Algal pigment patterns and phytoplankton assemblages in different water masses of the Río de la Plata maritime front. *Continental Shelf Research*, *28*, 1589-1606.

Cervetto, G., Mesones, C. & Calliari, D. (1987). Phytoplankton biomass and its relationship to environmental variables in a disturbed coastal area of the Río de la Plata, Uruguay, before the new sewage collector system. *Atlântica*, *24*, 45-54.

Colijn, F., Admiraal, W., Baretta, J. W. & Ruardij, P. (1987). Primary production in a turbid estuary, the Ems-Dollard: field and model studies. *Continental Shelf Research*, *7*, 1405-1409.

Cleve, P. T. (1894/1895). Synopsis of the naviculoid diatoms. *Kgl. Sven. Vet. Akad. Handl* *26/27*, 1-219.

Cleve, P. T. (1900). Report on the diatoms of the Magellan territories. *Svenska Expeditionen till Magellansläderna, 3*, 273-283.

Cooper, S. R. (1999). Estuarine paleoenvironmental reconstructions using diatoms. In E. F. Stoermer, & J. P. Smol, (Eds.), *The diatoms: Applications for the environmental and Earth Sciences*, (352-373). London: Cambridge University Press.

Cordini, J. M. (1939). El seston del Río de la Plata y su contenido diatómico. *Revista del Centro de Estudios de Doctorado de Ciencias Naturales, 2*, 157-169.

Day, J. H. (1981). *Estuarine ecology, with particular reference to southern Africa.* Rotterdam: A.A. Balkema.

Day, J. W., Hall, C. A. S., Kemp, W. M. & Yáñez-Arancibia, A. (1989). *Estuarine Ecology.* John Wiley & Sons, Inc., (Eds). New York, Chichester, Brisbane, Toronto, Singapore, 558.

De Francesco, C. G. & Isla, F. I. (2003). Distribution and abundance of hydrobiid snails in a mixed estuary and a coastal lagoon, Argentina, *Estuaries, 26*, 790-797.

De Marco. S. G. (2002). *Características hidrológicas y bioópticas de las aguas de la laguna costera Mar Chiquita y su relación con el fitoplancton.* Doctoral Thesis, Universidad Nacional de Mar del Plata.

De Marco, S. G., Beltrame, M. O., Freije, R. H. & Marccovecchio, J. E. (2005). Phytoplankton dynamic in Mar Chiquita coastal lagoon (Argentina), and its relationship with potential nutrient sources. *Journal of Coastal Research, 21*, 818-825.

De Wolf, H. (1982). Method of coding of ecological data from diatoms for computer utilization. *Mededelingen Rijks Geologische Dienst, 36*, 95-99.

Debenay, J. P., Carbonel, P., Morzadec-Kerfourn, M. T., Cazaboun, A., Denèfle, M. & Lézine, A. M. (2003). Multi-bioindicator study of a small estuary in Vendée (France). *Estuarine, Coastal and Shelf Science, 58*, 843-860.

Denys, L. (1991/1992). A check-list of the diatoms in the Holocene deposits of the Western Belgian coastal plain with the survey of their apparent ecological requirements, *I. Introduction, ecological code and complete list.* Berchem: Service Geological of Belgium Professional Paper, 246.

Denys, L. & De Wolf, H. (1999). Diatoms as indicators of coastal paleo-environments and relative sea-level change. In E. F. Stoermer, & J. P. Smol, (Eds.), *The diatoms: Applications for the Environmental and Earth Sciences*, (277-297). London: Cambridge University Press.

Diodato, S. L. & Hoffmeyer, M. S. (2008). Contribution of planktonic and detritic fractions to the natural diet of mesozooplankton in Bahía Blanca estuary. *Hydrobiologia, 614*, 83-90.

Dodd, J. R. & Stanton, E. J. Jr. (1990). *Palaeoecology, Concepts and Applications*, Second Edition, Wiley-Interscience Publication, 553.

EcoPlata Team, (Eds.), (1996). *The Río de la Plata. An Environmental Overview.* An EcoPlata Project background report. Working draft, November 1996. Dalhouse University, Halifax, *Nova Scotia, 242.*

Elgué, J. C., Bayseé, C., Burone, F. & Parietti, M. (1990). Distribución y sucesión especial del fitoplancton de superficie de la zona común de pesca Argentino-Uruguaya (Invierno de 1983). *Frente Marítimo, 6*, 67-107.

Escandel, A., Espinosa, M. A. & Isla, F. I. (in press). Diatomeas como indicadoras de variaciones de salinidad durante el Holoceno tardío en el estuario del río Negro, Patagonia Norte, Argentina. *Ameghiniana.*

Espinosa, M. A. (1988). Paleoecología de diatomeas del estuario del Río Quequén (Prov. de Buenos Aires, Argentina). *Thalassas*, *6*, 33-44.

Espinosa, M. A. (1994). Diatom paleoecology of the Mar Chiquita lagoon delta, Argentina. *Journal of Paleolimnology*, *10*, 17-23.

Espinosa, M. A. (1998). *Paleoecología de diatomeas en sedimentos cuaternarios del sudeste bonaerense*, Doctoral Thesis, Universidad Nacional de Mar del Plata.

Espinosa, M. A. (2001). Reconstrucción de paleoambientes holocenos de la costa de Miramar (provincia de Buenos Aires, Argentina) basada en diatomeas. *Ameghiniana*, *38*, 27-34.

Espinosa, M. A., De Francesco, C. G. & Isla, F. I. (2003). Paleoenvironmental reconstruction of Holocene coastal deposits from the southeastern Buenos Aires province, Argentina. *Journal of Paleolimnology*, *29*, 49-60.

Espinosa, M. A., Hassan, G. S. & Isla, F. I. (2006). Diatom distribution across a temperate microtidal marsh, Mar Chiquita coastal lagoon, Argentina. *Thalassas*, *22*, 9-16.

Espinosa, M. A. (2008). Diatoms from Patagonia and Tierra del Fuego. In J. Rabassa (Ed.) *The Late Cenozoic of Patagonia and Tierra del Fuego*, (383-392). Ciudad: *Developments in Quaternary Sciences*, 11.

Fasano, J. L., Hernández, M. A., Isla, F. I. & Schnack, E. J. (1982). Aspectos evolutivos y ambientales de la laguna Mar Chiquita (provincia de Buenos Aires, Argentina). In P. Lasserre, & H. Postma, (Eds.), *Coastal lagoons: Proceedings of the International Symposium of Coastal Lagoons*, (285-292). Bordeaux: Oceanologica Acta 4.

Ferrando, H. J., de Castro, T. M. & Tenyn, E. (1964). Clave para las principales diatomeas planctónicas del Atlántico Sudoccidental, *Revista del Instituto de Investigaciones Pesqueras*, *1*, 185-225.

Ferrario, M. E. (1972). Diatomeas pennadas de la ría de Puerto Deseado (provincia de Santa Cruz, Argentina). I. Araphidales. *Anales de la Sociedad Científica*, *193*, 135-176.

Ferrario, M. E. (1981). Diatomeas centrales de la ría de Puerto Deseado (Santa Cruz, Argentina). IV.- S.O. Biddulphiineae, Fam. Eupodiscaceae y Fam. Lithodesmiaceae. *Darwiniana*, *23*, 475-488.

Ferrario, M. E. (1984a). Diatomeas centrales de la ría de Puerto Deseado (Santa Cruz, Argentina). I. S. O. Rhizosoleniineae Familia Rhizosoleniaceae y S.O. Biddulphiineae, Familia Chaetoceraceae. *Revista del Museo de La Plata, Nueva Serie 8, Botánica*, *83*, 247-254.

Ferrario, M. E. (1984b). Diatomeas centrales de la ría de Puerto Deseado (Santa Cruz, Argentina). II.- S.O. Coscinodisciineae Familia Hemidiscaceae y Familia Melosiraceae. *Revista del Museo de La Plata, Nueva Serie 8, Botánica*, *84*, 267-289.

Ferrario, M. E. (1984c). Diatomeas centrales de la ría de Puerto Deseado (Santa Cruz, Argentina). III.- S. O. Coscinodisciineae Familia Coscinodiscaceae, Familia Heliopeltaceae y Familia Thalassiosiraceae. *Revista del Museo de La Plata, Nueva Serie 8, Botánica*, *85*, 291-311.

Ferrario, M. E. & Galván N. M. (1989). *Catálogo de las diatomeas marinas citadas entre los 36° y los 60° S con especial referencia al Mar Argentino*. Instituto Antártico Argentino, Publicación N°20, 327, Buenos Aires.

Ferrario, M. E. & Sar, E. A. (1984). Diatomeas pennadas de la ría de Puerto Deseado (provincia de Santa Cruz). II. S.O. Raphidiineae. *Revista del Museo de La Plata, Nueva Serie, 8, Botánica*, *80*, 213-230.

Ferrario, M. E. & Sastre, V. (1990). Ultraestructura, polimorfismo y ecología de *Odontella aurita* (Lyngbye) Agardh (Bacillariophyceae) en el estuario del Río Chubut, Argentina. *Revista de la Facultad de Oceanografía, Pesca y Ciencias Alimentarias, 2*, 98-106.

Frenguelli, J. (1921). Los terrenos de la costa atlántica en los alrededores de Miramar (provincia de Buenos Aires) y sus correlaciones. *Boletín de la Academia de Ciencias de Córdoba, 24*, 325-485.

Frenguelli, J. (1923). Diatomeas de Tierra del Fuego. *Anales de la Sociedad Científica Argentina, 95*, 225-263.

Frenguelli, J. (1924). Diatomeas de Tierra del Fuego. *Anales de la Sociedad Científica Argentina, 98*, 5-89.

Frenguelli, J. (1925). Diatomeas de los arroyos del Durazno y Las Brusquitas en los alrededores de Miramar (provincia de Buenos Aires). *Physis, 8*, 129-185.

Frenguelli, J. (1935). Diatomeas de la Mar Chiquita al norte de Mar del Plata. *Revista del Museo de La Plata, Nueva serie 1, Botánica, 5*, 121-141.

Frenguelli, J. (1938). Diatomeas de la Bahía San Blas (provincia de Buenos Aires). *Revista del Museo de La Plata, Nueva serie 1, Botánica, 5*, 251-337.

Frenguelli, J. (1941). Diatomeas del Río de la Plata. *Revista del Museo de La Plata, Nueva serie 3, Paleontología, 16*, 77-251.

Frenguelli, J. (1945). Las diatomeas del Platense. *Revista del Museo de La Plata, Nueva serie 3, Botánica, 15*, 213-334.

Gayoso, A. M. (1981). *Estudio del fitoplancton del estuario de Bahía Blanca*. Bahía Blanca: Instituto Argentino de Oceanografía.

Gayoso, A. M. (1988). Variación estacional del fitoplancton de la zona más interna del estuario de Bahía Blanca (prov. Buenos Aires, Argentina). *Gayana, Botanica, 45*, 241-247.

Gayoso, A. M. (1996). Pytoplankton species composition and abundance off Río de la Plata (Uruguay). *Arch. Fish. Mar. Res., 44*, 257-265.

Gayoso, A. M. (1998). Long-term phytoplankton studies in the Bahía Blanca estuary, Argentina. *ICES Journal of Marine Science, 55*, 655-660.

Gayoso, A. M. (1999). Seasonal succession patterns of phytoplankton in the Bahía Blanca estuary (Argentina). *Botanica Marina, 42*, 367-375.

Gehrels, W. R., Roe, H. M. & Charman, D. J. (2001). Foraminifera, testate amoebae and diatoms as sea-level indicators in UK saltmarshes: a quantitative multiproxy approach. *Journal of Quaternary Science, 16*, 201-220.

Goldstein, S. T. & Watkins, G. T. (1999). Taphonomy of salt marsh foraminifera: an example of coastal Georgia. *Palaeogeography, Palaeoclimatology, Palaeoecology, 149*, 103-114.

Gómez, N. & Bauer, D. E. (1998a). Phytoplankton from the Southern Coastal Fringe of the Río de la Plata (Buenos Aires, Argentina). *Hydrobiologia, 380*, 1-8.

Gómez, N. & Bauer, D. E. (1998b). Coast phytoplankton of the Río de la Plata river and its relation to pollution. *International Association of Theoretical and Applied Limnology, 26*, 1032-1036.

Gómez, N. & Bauer, D. E. (2000). Diversidad fitoplanctónica en la franja costera sur del Río de la Plata. *Biología Acuática, 19*, 7-26.

Gómez, N., Hualde, P. R., Licursi, M. & Bauer, D. E. (2004). Spring phytoplankton of Río de la Plata: a temperate estuary of South America. *Estuarine, Coastal and Shelf Science, 61*, 301-309.

Gómez, N., Licursi, M. & Cochero, J. (2009). Seasonal and spatial distribution of the microbenthic communities of the Rio de la Plata estuary (Argentina) and possible environmental controls. *Marine Pollution Bulletin*, *58*, 878-887.

Guarrera, S.A. (1950). Estudios hidrobiológicos en el Río de la Plata. *Instituto Nacional de Investigación de las Ciencias Naturales, Ciencias Botánicas*, *2*, 62.

Guiry, M. D. & Guiry, G. M. (2009). AlgaeBase, World-wide electronic publication, National University of Ireland, Galway. http://www.algaebase.org; searched on July, 2009.

Gutiérrez Tellez, B. M. & Schillizzi, R. A. (2002). Asociaciones de diatomeas en paleoambientes cuaternarios de la costa sur de la provincia de Buenos Aires, Argentina. *Pesquisas en Geociências*, *29*, 59-70.

Hassan, G. S. (2008). *Diatomeas estuáricas del sudeste bonaerense: distribución, composición, diversidad y su aplicación en paleoecología*. Doctoral Thesis, Universidad Nacional de Mar del Plata.

Hassan, G. S., Espinosa, M. A. & Isla, F. I. (2006). Modern diatom assemblages in surface sediments from estuarine systems in the southeastern Buenos Aires Province, Argentina. *Journal of Paleolimnology*, *35*, 39-53.

Hassan, G. S., Espinosa, M. A. & Isla, F. I. (2007). Dead diatom assemblages in surface sediments from a low impacted estuary: the Quequén Salado river, Argentina. *Hydrobiologia*, *579*, 257-270.

Hassan, G. S., Espinosa, M. A. & Isla, F. I. (2008). Fidelity of dead diatom assemblages in estuarine sediments: how much environmental information is preserved? *Palaios*, *23*, 112-120.

Hassan, G. S., Espinosa, M. A. & Isla, F. I. (2009). Diatom-based inference model for paleosalinity reconstructions in estuaries along the northeastern coast of Argentina. *Palaeogeography, Palaeoclimatology, Palaeoecology*, *275*, 77-91.

Hentschel, E. (1932). Die biologischen methoden und das biologische beobachtungsmaterial der "meteor-expedition". *Wissenschaftl. Ergbn. d. deustch. atlant. Exped. a. d. Forsh. – Vermessungsschiff "Meteor" 1925-1927*, *10*, 1-174.

Horton, B. P. (1999). The distribution of contemporary intertidal foraminifera at Cowpen Marsh, Tees Estuary, UK: implications for studies of Holocene sea-level changes. *Palaeogeography, Palaeoclimatology, Palaeoecology*, *149*, 127-149.

Horton, B. P., Corbett, R., Culver, S. J., Edwards, R. J. & Hillier, C. (2006). Modern saltmarsh diatom distributions of the Outer Banks, North Carolina, and the development of a transfer function for high resolution reconstructions of sea level. *Estuarine, Coastal and Shelf Science*, *69*, 381-394.

Hustedt, F. (1937/1938). *Systematische und ökologische Untersuchungen über die Diatomeen. "Flora von Java, Bali und Sumatra nach dem Material der Deutschen Limnologischen Sunda-Expedition"*, Archive für Hydrobiologie supplement, Band, *15*, 131-506.

Hustedt, F. (1953). Die Systematik der diatomeen in ihren beziehungen zur geologie und ökologie nebst einer revision des halobien-systems. *Svensk Botaniska Tidskr*, *47*, 274-344.

Isla, F. I. & Espinosa, M. A. (2005). Holocene and historical evolution of an estuarine complex: the gravel spit of the Walker creek, Southern Buenos Aires. *XVI Congreso Geológico Argentino, Actas*, 149-154.

Isla, F. I., Fasano, J. L., Ferrero, L., Espinosa, M. A. & Schnack, E. J. (1986). Late Quaternary marine-estuarine sequences of the southeastern coast of Buenos Aires Province, Argentina. *Quaternary of South America and Antarctic Peninsula, 4*, 137-157.

Isla, F. I. & Bujalesky, G. G. (2004). Morphodynamics of a gravel-dominated macrotidal estuary: Rio Grande, Tierra del Fuego. *Revista de la Asociación Geológica Argentina, 59*, 220-228.

Juggins, S. (1992). *Diatoms in the Thames estuary, England: ecology, paleoecology, and salinity transfer function.* Berlin: J. Cramer.

Kidwell, S. M. (2001). Preservation of species abundance in marine death assemblages. *Science, 294*, 1091-1094.

Kidwell, S. M. (2002). Mesh-size effects on the ecological fidelity of death assemblages: a meta-analysis of molluscan live-dead studies. *Geobios, 24*, 107-119.

Kidwell, S. M. & Flessa, K. W. (1995). The quality of the fossil record: populations, species, and communities. *Annual Reviews of Ecology and Systematics, 26*, 269-299.

Kowalewski, M., Lazo, D. G., Carroll, M., Messina, C., Casazza, L., Puchalski, S., Gupta, N. S., Rothfus, T. A., Hannisdal, B., Sälgeback, J., Hendy, A., Stempien, J., Krause Jr., R., Terry, R. C., LaBarbera, M. & Tomašových, A. (2003). Quantitative fidelity of brachiopod-mollusk assemblages from modern subtidal environments of San Juan Islands, USA. *Journal of Taphonomy, 1*, 43-65.

Lange, K. B. (1985). Spatial and seasonal variations of diatom assemblages off the Argentinean coast (South Western Atlantic). *Oceanologica Acta, 8*, 361-369.

Licursi, M., Sierra, M. V. & Gómez, N. (2006). Diatom assemblages from a turbid coastal plain estuary: Río de la Plata (South America). *Journal of Marine Systems, 62*, 35-45.

Marini, M. F. & Piccolo, M. C. (2000). El balance hídrico en la cuenca del río Quequén Salado, Argentina. *Papeles de Geografía, 31*, 39-53.

Martínez Macchiavelo, J. C. (1979). Diatomeas del Océano Atlántico frente a la provincia de Buenos Aires, Argentina (Bacillariophyceae-Eupodiscales). *Revista del Museo Argentino de Ciencias Naturales Bernardino Rivadavia, Botánica, 5*, 229-239.

Maynard, N. G. (1976). Relationship between diatoms in surface sediments of the Atlantic Ocean and the biological and physical oceanographic of the overlying waters. *Paleobiology, 2*, 99-121.

Metzeltin, D. & García-Rodríguez, F. (2003). *Las diatomeas Uruguayas.* DIRAC: Montevideo, Uruguay.

Moore, W. W. & McIntire, C. D. (1977). Spatial and seasonal distribution of littoral diatoms in Yanquina estuary, Oregon (USA). *Botanica Marina, 20*, 99-109.

Müller-Melchers, F. E. (1945). Diatomeas procedentes de algunas muestras de turba del Uruguay. *Comunicaciones Botánicas del Museo de Historia Natural de Montevideo, 1*, 1-25.

Müller Melchers, F. E. (1952). Biddulphia chilensis Grev. as indicator of ocean currents. *Comunicaciones Botánicas del Museo de Historia Natural de Montevideo, 2*, 1-25.

Müller Melchers, F. E. (1953). New and little known diatoms from Uruguay and the South Atlantic coast. *Comunicaciones Botánicas del Museo de Historia Natural de Montevideo, 3*, 1-25.

Müeller Melchers, F. (1959). Plancton diatoms of the southern Atlantic Argentina and Uruguay coast. *Comunicaciones Botánicas del Museo de Historia Natural de Montevideo, 3*, 1-45.

Murray, J. W. & Pudsey, C. J. (2004). Living (stained) and dead foraminifera from the newly ice-free Larsen Ice Shelf, Weddell Sea, Antarctica: ecology and taphonomy. *Marine Micropaleontology*, *53*, 67-81.

Ng, S. L. & Sin, F. S. (2003). A diatom model for inferring sea level change in the coastal waters of Hong Kong. *Journal of Paleolimnology*, *30*, 427-440.

Oppenheim, D. R. (1991). Seasonal changes in epipelic diatoms along an intertidal shore, Berrow Flats, Somerset. *Journal of the Marine Biological Association of the United Kingdom*, *71*, 579-596.

Parodi, E. R. & Barría de Cao, S. (2003). Benthic microalgal communities in the inner part of the Bahía Blanca estuary (Argentina): a preliminary qualitative study. *Oceanologica Acta*, *25*, 279-284.

Paterson, D. M. (1989). Short-term changes in the erodibility of intertidal cohesive sediments related to the migratory behavior of epipelic diatoms. *Limnology and Oceanography*, *34*, 223-324.

Perillo, G. M. E., Pérez, D. E., Piccolo, M. C., Palma, E. D. & Cuadrado, D. G. (2005). Geomorphologic and physical characteristics of a human impacted estuary: Quequén Grande River Estuary, Argentina. *Estuarine, Coastal and Shelf Science*, *62*, 301-312.

Piccolo, M. C. & Perillo, G. M. (1999). The argentinean estuaries: a review. In G. M. E., Perillo, M. C. Piccolo, & M. Pino Quivira, (Eds.), *Estuaries of South America: their geomorphology and dynamics*, (101-132). Berlin: Springer-Verlag.

Piccolo, M. C., Perillo, G. M. & Arango, J. M. (1987). Hidrografía del estuario de Bahía Blanca, Argentina. *Geofísica*, *26*, 75-89.

Popovich, C. A. (2004). Fitoplancton. In M. C. Piccolo, & M. Hoffmeyer, (Eds.), *El ecosistema del estuario de Bahía Blanca* (pp. 69-78). Bahía Blanca: Instituto Argentino de Oceanografía.

Popovich, C. A. & Marcovecchio, J. E. (2008). Spatial and temporal variability of phytoplankton and environmental factors in a temperate estuary of South America (Atlantic coast, Argentina). *Continental Shelf Research*, *28*, 236-244.

Popovich, C. A., Spetter, C. V., Marcovecchio, J. E. & Freije, R. H. (2008). Dissolved nutrient availability during winter diatom bloom in a turbid and shallow estuary (Bahía Blanca, Argentina). *Journal of Coastal Research*, *24*, 95-102.

Pucci, A. E., Hoffmeyer, M. S., Freije, H. R., Barría, M. S., Popovich, C. A., Rusansky, C. & Asteasuain, R. (1996). Características de las aguas y del plancton en un sector del estuario del río Negro (República Argentina). In J. Marcovecchio, (Ed.), *Pollution Processes in Coastal Environments*, (146- 151). Mar del Plata: Universidad Nacional de Mar del Plata.

Resende, P., Azeiteiro, U. & Pereira, M. J. (2005). Diatom ecological preferences in a shallow temperate estuary (Ria de Aveiro, Western Portugal). *Hydrobiologia*, *544*, 77-88.

Reta, R., Martos, P., Perillo, G. M. E., Piccolo, M. C. & Ferrante, A. (2001). Características hidrográficas del estuario de la laguna Mar Chiquita. In O.O. Iribarne (Ed.), *Reserva de Biosfera Mar Chiquita: características físicas, biológicas y ecológicas*, (31-52). Mar del Plata: Editorial Martin.

Roggiero, M. F. (1988). Fitoplancton del Río de la Plata, I. *Lilloa*, *37*, 137-152.

Rybarcyk, H. & Elkaïm, B. (2003). An analysis of the trophic network of a macrotidal estuary: the Seine Estuary (Eastern Channel, Normandy, France). *Estuarine, Coastal and Shelf Science*, *58*, 775-791.

Rzeznik-Orignac, J., Fichet, D. & Boucher, G. (2003). Spatio-temporal structure of the nematode assemblages of the Brouage mudflat (Marennes Oléron, France). *Estuarine, Coastal and Shelf Science, 58,* 77-88.

Sar, E. A., Sala, S. E., Sunesen, I., Henninger, M. S. & Montastruc, M. (2009). Catalogue of the genera, species, and infraspecific taxa erected by J. Frenguelli. *Diatom Monographs, 10,* 419.

Santinelli, N., Sastre, A. V. & Caille, G. (1990). Fitoplancton del estuario inferior del río Chubut y su relación con la salinidad y la temperatura. *Revista de la Asociación de Ciencias Naturales del Litoral, 21,* 69-79.

Sastre, A. V., Santinelli, N. H. & Caille, G. (1990). Diatomeas y dinoflagelados del estuario del río Chubut (Patagonia, Argentina). II. Estructura de las comunidades. *Revista Fac. de Ocean., Pesq. y Cs Alimentarias, 2,* 181-192.

Sastre, A. V., Santinelli, N. H. & Sendin, M. (1994). Floración de *Aulacoseira granulata* (Ehr.) Simonsen (Bacillariophyceae) en el curso inferior del río Chubut. *Revista Brasileira de Biologia, 54,* 641-647.

Sastre, A. V., Santinelli, N. H., Otaño, S. H. & Ivanissevich, M. E. (1998). Water quality in the lower section of the Chubut River, Patagonia, Argentina. *Verhandlungen Internationale Vereiningen Limnolgie, 26,* 951-955.

Sawai, Y. (2001). Distribution of living and dead diatoms in tidal wetlands of northern Japan: relations to taphonomy. *Palaeogeography, Palaeoclimatology, Palaeoecology, 173,* 125-141.

Sawai, Y., Horton, B. P. & Nagumo, T. (2004). The development of a diatom-based transfer function along the Pacific coast of eastern Hokkaido, northern Japan – an aid in paleoseismic studies of the Kuril subduction zone. *Quaternary Science Reviews, 23,* 2467-2483.

Schillizzi, R., Gutiérrez Tellez, B. & Aramayo, S. (2006). Reconstrucción paleoambiental del Cuaternario en las barrancas del río Quequén Salado, provincia de Buenos Aires, Argentina. *III Congreso Argentino de Cuaternario y Geomorfología, Actas,* 649-658.

Schmidt, A., Schmidt, M., Fricke, F., Heiden, H., Müller, O. & Hustedt, F. (1874-1959). *Atlas der Diatomaceen Kuncle.* Serie I-X, Leipzig.

Schwindt, E., Iribarne, O. O. & Isla, F. I. (2004). Physical effects of an invading reef-building polychaete on an Argentinean estuarine environment. *Estuarine, Coastal and Shelf Science, 59,* 109-120.

Scott, D. B. & Medioli, F. S. (1980). Living vs. total foraminiferal populations: their relative usefulness in paleoecology. *Journal of Paleontology, 54,* 814-831.

Sherrod, B. L. (1999). Gradient analysis of diatom assemblages in a Puget Sound salt marsh: can such assemblages be used for quantitative paleoecological reconstructions? *Palaeogeography, Palaeoclimatology, Palaeoecology, 149,* 213-226.

Simonsen, R. (1969). Diatoms as indicators in estuarine environments. *Veröffentl. Inst. Meersforsch. Bremerhaven, 11,* 287-291.

SMEBD (2009). WoRMS: The World Register of Marine Species. Available online at http://www.marinespecies.org. Accessed on July, 2009.

Snoeijs, P. (1999). Diatoms and environmental change in brackish waters. In E. F. Stoermer, & J. P. Smol, (Eds.). *The diatoms: Applications for the environmental and Earth Sciences,* (298-333). London: Cambridge University Press.

Sylvestre, F., Beck-Eichler, B., Duleba, W. & Debenay, J. (2001). Modern benthic diatom distribution in a hypersaline coastal lagoon: the Lagoa de Arauama (R.J.), Brazil. *Hydrobiologia, 443*, 213-231.

Tempère, M. & Peragallo, H. (1907-1915). *Diatomees du Monde Entier*. Edition 2, 30 fasc., Arcachon, Graz-sur-Loing.

ter Braak, C. J. F. & Juggins, S. (1993). Weighted averaging partial least squares regression (WA-PLS): an improved method for reconstructing environmental variables from species assemblages. *Hydrobiologia, 269/270*, 485-502.

ter Braak, C. J. F., Juggins, S., Birks, H. J. B. & van der Voet, H. (1993). Weighted averaging partial least squares regression (WA-PLS): definition and comparison with other methods for species-environment calibration. In: G. P. Patil, & C. R. Rao, (Eds.), *Multivariate Environmental Statistics*, (525-560). Amsterdam: Elsevier Science Publishers.

Thiemann, K. (1934). Das Plankton der Flussmündungen. *Biologische Sonderuntersuchungen Wissensch. Ergebn. Atlant. Exped. Forsch. Vermessungsschiff "Meteor", 12*, 199-273.

Underwood, G. J. C. (1994). Seasonal and spatial variation in epipelic diatom assemblages in the Severn Estuary. *Diatom Research, 9*, 451-472.

Underwood, G. J. C. (1997). Microalgal colonization in a saltmarsh restoration scheme. *Estuarine, Coastal and Shelf Science, 44*, 471-481.

Underwood, G. J. C. & Paterson, D. M. (1993). Seasonal changes in diatom biomass, sediment stability and biogenic stabilization in the Severn Estuary. *Journal of the Marine Biological Association of the United Kingdom, 73*, 871-887.

Van Dam, H., Mertens, A. & Sinkeldam, J. (1994). A coded checklist and ecological indicator values of freshwater diatoms from The Netherlands. *Netherlands Journal of Aquatic Ecology, 28*, 117-133.

Villafañe, V., Hebling, E. W. & Santamarina, J. (1991). Phytoplankton blooms in the Chubut river estuary (Argentina): influence of stratification and salinity. *Revista de Biología Marina (Valparaíso), 26*, 1-20.

Vos, P. C. & De Wolf, H. (1988). Methodological aspects of paleo-ecological diatom research in coastal areas of the Netherlands. *Geologie en Mijnbouw, 67*, 31-40.

Vos, P. C. & De Wolf, H. (1993). Diatoms as a tool for reconstructing sedimentary environments in coastal wetlands; metodological aspects. *Hydrobiologia, 269*, 285-296.

Vouilloud, A. A. (2003). *Catálogo de diatomeas continentales y marinas de Argentina*. La Plata: Asociación Argentina de Ficología.

Witkowski, A., Cedro, B., Kierzek, A. & Baranowski, D. (2009). Diatoms as a proxy in reconstructing the Holocene environmental changes in the south-western Baltic Sea: the lower Rega River Valley sedimentary record. *Hydrobiologia, 631*, 155-172.

Wolfstein, K., Colijn, F. & Doerffer, R. (2000). Seasonal dynamics of microphytobenthos biomass and photosynthetic characteristics in the Northern German Wadden Sea, obtained by the photosynthetic light dispensation system. *Estuarine, Coastal and Shelf Science, 51*, 651-662.

Ysebaert, T., Herman, P. M. J., Meire, P., Craeymeersch, J., Verbeek, H. & Heip, C. H. R. (2003). Large-scale spatial patterns in estuaries: estuarine macrobenthic communities in the Schelde estuary, NW Europe. *Estuarine, Coastal and Shelf Science, 57*, 335-355.

Zárate, M. A., Espinosa, M. A. & Ferrero, L. (1998). Palaeoenvironmental implications of a Holocene diatomite, Pampa Interserrana, Argentina. *Quaternary of South America and Antarctic Peninsula, 5*, 135-152.

Zong, Y. & Horton, B. P. (1999). Diatom-based tidal-level transfer functions as an aid in reconstructing Quaternary history of sea-level movements in the UK. *Journal of Quaternary Science, 14*, 153-167.

APPENDIX I

List of the recent publications containing information about estuarine diatoms from Argentina used to construct Appendix II, type of sample analyzed (P: plankton; S: sediment; E: epiphytes under vegetation), and number of taxa mentioned (n/a: not available). Works providing of environmental data are marked (+).

Estuary	Author	Sample	N° of taxa	Env. Data
Rio de la Plata	Gómez and Bauer (1998)	P	32	+
	Carreto et al. (2003)	P	8	+
	Metzeltin & García-Rodríguez (2003)	S	295	-
	Gómez et al. (2004)	P	15	+
	Calliari et al. (2005)	P	10	+
	Licursi et al. (2006)	P	87	+
	Bauer et al. (2007)	E	44	+
	Carreto et al. (2008)	P	4	+
	Calliari et al. (2009)	P	11	+
	Gómez et al. (2009)	S	52	+
Mar Chiquita coastal lagoon	De Marco (2002)	P	n/a	+
	De Marco et al. (2005)	P	n/a	+
	Espinosa et al. (2006)	S	20	+
	Hassan et al. (2006)	S	31	+
	Hassan et al. (2008)	S	15	+
	Hassan et al. (2009)	S	28	+
Rio Quequén Grande	Hassan et al. (2006)	S	37	+
	Hassan et al. (2008)	S	18	+
	Hassan et al. (2009)	S	36	+
Rio Quequén Salado	Hassan et al. (2007)	S	32	+
	Hassan et al. (2009)	S	30	+
Bahía Blanca	Gayoso (1981)	P	30	+
	Gayoso (1988)	P	23	+
	Gayoso (1998)	P	n/a	+
	Gayoso (1999)	P	19	+
	Andrade et al. (2000)	P	1	+
	Parodi and Barría de Cao (2003)	S	13	+
	Parodi (2004)	S	12	+
	Popovich (2004)	P	48	+
	Diodato and Hoffmeyer (2008)	P	13	+
	Popovich et al. (2008)	P	20	+
	Popovich and Marcovechio (2009)	P	14	+
Rio Negro	Pucci et al. (1996)	P	19	+
Bahía San Blas	Isla and Espinosa (2005)	S	15	+
Río Chubut	Ferrario and Sastre (1990)	P	1	+
	Sastre et al. (1990)	P	40	+
	Santinelli et al. (1990)	P	39	+
	Villafañe et al. (1991)	P	12	+
	Sastre et al. (1994)	P	1	+
	Ayestarán and Sastre (1995)	P	28	-
	Sastre et al. (1998)	P	10	+
Ría Puerto Deseado	Ferrario (1972, 1981, 1984 a, b) Ferrario and Sar (1984)	P	88	+

APPENDIX II

List of the diatom taxa cited for estuaries of Argentina, based on the publications listed in Appendix I. 1: Río de la Plata; 2: Mar Chiquita coastal lagoon; 3: Río Quequén Grande; 4: Río Quequén Salado; 5: Bahía Blanca; 6: Río Negro; 7: Bahía San Blas; 8: Río Chubut; 9: Ría Puerto Deseado. Letters indicate the type of sample in which each taxon was found; P: plankton; S: sediment; E: epiphytes under vegetation. In the last column a summary of the salinity ranges reported in all works is provided. WA: optima ± tolerance calculated by weighted averaging; n/a: no data available.

TAXA NAME/AUTHORITY	1	2	3	4	5	6	7	8	9	SALINITY
Achnanthes brevipes Agardh		S	S	S			S			9-38‰ (marine/brackish, euryhaline)
Achnanthes brevipes var. *intermedia* (Kützing) Cleve									P	32-34‰ (marine)
Achnanthes elata (Leud.-Fort.) Gandhi	S									n/a
Achnanthes exigua Grunow	S									n/a
Achnanthes inflata (Kützing) Grunow	P/S									0-0.8‰ (freshwater)
Achnanthes inflata var. *gibba* Gandhi	S									n/a
Achnanthes inflatagrandis Metzeltin, Lange-Bertalot & Garcia-Rodriguez	S									n/a
Achnanthes lacus-vulcani Lange-Bertalot & Krammer							S			38‰
Achnanthes reversa Lange-Bertalot		S	S							WA: 19.5±5.5 (marine/brackish)
Achnanthes subelata Metzeltin, Lange-Bertalot & Garcia-Rodriguez	S									n/a
Achnanthidium biasolettianum (Grunow) Round & Bukht.		S								n/a
Achnanthidium coarctatum Brébisson ex Smith	S									n/a
Achnanthidium lanceolatum spp. *biporoma* Lange-Bertalot	S									n/a

Appendix II (Continued)

TAXA NAME/AUTHORITY	1	2	3	4	5	6	7	8	9	SALINITY
Achnanthidium lanceolatum spp. *miota* Lange-Bertalot	S									n/a
Achnanthidium lanceolatum spp. *frequentissima* Lange-Bertalot	S									n/a
Achnanthidium minutissimum (Kützing) Czarnecki	S	S	S	S						WA: 4.8±7.7 (brackish, euryhaline)
Achnanthidium parexigua (Metzeltin & Lange-Bertalot) Metzeltin	S									n/a
Actinocyclus spp.						P				-
Actinocyclus actinochilus (Ehrenberg) Simonsen		S	S					P		n/a
Actinocyclus divisus (Grunow) Hustedt		S	S							8.4‰ (brackish)
Actinocyclus kutzingii (A. Schmidt) Simonsen					P					n/a
Actinocyclus normanii (Gregory) Hustedt	P/S/E									0-18‰ (brackish/marine)
Actinocyclus normanii f. *subsalsus* (Juhlin-Dannfelt) Hust.	S									n/a
Actinocyclus octonarius Ehrenberg	P	S							P	0-34‰ (marine/brackish, euryhaline)
Actinocyclus subocellatus (Grunow) Rattray									P	32-34‰ (marine)
Actinocyclus subtilis (Gregory) Ralfs									P	32-34‰ (marine)
Actinoptychus spp.	P					P		P		-
Actinoptychus adriaticus Grunow					P					n/a
Actinoptychus campanulifer Schmidt									P	32-34‰ (marine)
Actinoptychus frenguellii Müller Melchers									P	32-34‰ (marine)
Actinoptychus senarius (Ehrenberg) Ehrenberg	P				P			P	P	0-34‰ (marine/brackish, euryhaline)
Actinoptychus splendens (Shadbolt) Ralfs		S		S	P				P	32-34‰ (marine)

Appendix II (Continued)

TAXA NAME/AUTHORITY	1	2	3	4	5	6	7	8	9	SALINITY
Actinoptychus splendens var. *glabrata* (Grunow) Pantocsek									P	32-34‰ (marine)
Actinoptychus vulgaris Schumann								P	P	25-35‰ (marine/brackish)
Amphipleura lindheimeri Grunow	S									n/a
Amphipleura pellucida (Kützing) Kützing	S									n/a
Amphipleura rutilans var. *antarctica* (Grunow) Grunow									P	32-34‰ (marine)
Amphitetras antediluviana Ehrenberg	S									n/a
Amphora spp.		S	S	S						-
Amphora acutiuscula Kützing	S	S	S	S						WA: 28±3‰ (marine/brackish)
Amphora coffeaeformis (Agardh) Kützing		S	S	S						WA: 5.5±5.75‰ (brackish/freshwater)
Amphora commutata Grunow	S									n/a
Amphora exigua Gregory									P	32-34‰ (marine)
Amphora frenguelli Forti		S	S	S						0.5-3‰ (brackish/freshwater)
Amphora helenensis Giffen		S	S	S						WA: 9.5±6.5‰ (brackish/marine)
Amphora libyca Ehrenberg	S	S	S	S						WA: 4.5±5.5‰ (brackish/freshwater)
Amphora montana Krasske	S	S	S							<1‰ (freshwater)
Amphora normanii Rabenhorst	S									n/a
Amphora ovalis (Kützing) Kützing			S	S						30-40‰ (marine)
Amphora pediculus (Kützing) Grunow		S	S							WA: 2.8±4.6‰ (brackish)
Amphora veneta Kützing		S	S	S						WA: 2.6±3‰ (brackish)
Anomoeoneis sphaerophora Pfitzer		S								0-2‰ (brackish/freshwater)
Asterionella formosa Hassall						P		P		0-26‰ (marine/brackish, euryhaline)

Appendix II (Continued)

TAXA NAME/AUTHORITY	1	2	3	4	5	6	7	8	9	SALINITY
Asterionellopsis glacialis (Castracane) Round					P	P			P	26-34‰ (marine)
Aulacoseira ambigua (Grunow) Simonsen	P									0-15‰ (brackish/freshwater)
Aulacoseira distans (Ehrenberg) Simonsen	P/S	S								0-15‰ (brackish/freshwater)
Aulacoseira granulata (Ehrenberg) Simonsen	P/S/E	S	S		P	P		P		0-15‰ (brackish/freshwater)
Aulacoseira granulata var. *angustissima* (Müller) Simonsen	P									0-15‰ (brackish/freshwater)
Aulacoseira granulata var. *valida* (Hustedt) Simonsen	S									n/a
Aulacoseira italica (Ehrenberg) Simonsen									P	32-34‰ (marine)
Aulacoseira muzzanensis (Meister) Krammer	P									0-0.2‰ (freshwater)
Auliscus sculptus (Smith) Ralfs			S						P	32-34‰ (marine)
Bacillaria paradoxa Gmelin		S	S	S						0.5-14‰ (brackish/freshwater)
Bacteriastrum furcatum Shadbolt								P		n/a
Berkeleya rutilans (Trentepohl) Grunow		S	S	S					P	32-34‰ (marine)
Biddulphia alternans (Bailey) Van Heurck								P	P	0-35‰ (marine/brackish, euryhaline)
Biddulphia antediluviana (Ehrenberg) Van Heurck								P	P	0-35‰ (marine/brackish, euryhaline)
Biddulphia biddulphiana (Smith) Boyer	S									n/a
Biddulphia rhombus (Ehrenberg) Smith		S								24‰ (marine/brackish)
Brachysira neoexilis Lange-Bertalot	S									n/a
Brebissonia lanceolata (Agardh) Mahoney & Reimer		S	S	S						0-5‰ (brackish/freshwater)
Caloneis amphisbaena (Bory) Cleve								P		<1‰ (freshwater)
Caloneis bacillum (Grunow) Cleve	P/S/E	S	S	S						0-5‰ (brackish/freshwater)

Appendix II (Continued)

TAXA NAME/AUTHORITY	1	2	3	4	5	6	7	8	9	SALINITY
Caloneis brevis	S									n/a
Caloneis hyalina Hustedt	S									n/a
Caloneis permagna (Bailey) Cleve		S		S						0-3‰ (brackish/freshwater)
Caloneis tenuis (Gregory) Krammer	S									n/a
Caloneis westii (Smith) Hendey		S	S	S				P		<1‰ (freshwater)
Campilosira spp.		S		S						-
Campylodiscus clypeus (Ehrenberg) Ehrenberg			S	S						0-7‰ (brackish/freshwater)
Capartogramma crucicula (Grunow) Ross	S									n/a
Catacombus gaillonii (Bory) Williams & Round	S								P	32-34‰ (marine)
Catenula adhaerens (Mereschkowsky) Mereschkowsky		S	S	S						WA: 20±5.4‰ (marine/brackish)
Cavinula lapidosa (Krasske) Lange-Bertalot	S									n/a
Cavinula monoculata (Hustedt) Mann	S									n/a
Cerataulina pelagica (Cleve) Hendey					P					23-30‰ (marine/brackish)
Chaetoceros spp.	P				P			P		-
Chaetoceros affinis Lauder	P									5-18‰ (brackish/marine)
Chaetoceros brevis Schütt	P									17.5-18.5 (brackish/marine)
Chaetoceros ceratosporus Ostenfeld					P					30.4-32.8‰ (marine)
Chaetoceros ceratosporus var. *brachysetus* Rines & Hargr.					P					30.4-32.8‰ (marine)
Chaetoceros convolutus Castracane									P	32-34‰ (marine)
Chaetoceros debilis Cleve					P					20-33‰ (marine/brackish)
Chaetoceros decipiens Cleve									P	32-34‰ (marine)

Appendix II (Continued)

TAXA NAME/AUTHORITY	1	2	3	4	5	6	7	8	9	SALINITY
Chaetoceros diadema (Ehrenberg) Gran										30.4-32.8‰ (marine)
Chaetoceros similis Cleve					P				P	25-34‰ (marine/brackish)
Chaetoceros socialis Lauder									P	32-34‰ (marine)
Chaetoceros	P				P				P	18-40‰ (marine/brackish)
Chaetoceros subtilis var. *abnormis* Prosckina-Lavrenko					P					n/a
Chaetoceros teres Cleve									P	32-34‰ (marine)
Cocconeis spp.					S/P			P		-
Cocconeis grunowii Pantocsek									P	32-34‰ (marine)
Cocconeis guttata Hustedt & Aleem							S			38‰ (marine)
Cocconeis neodiminuta Krammer	S	S							P	32-38‰ (marine)
Cocconeis	S									n/a
Cocconeis pellucida var. *minor* Grunow									P	32-34‰ (marine)
Cocconeis placentula Ehrenberg (+ vars.)	P	S	S	S				P	P	0-34‰ (freshwater to marine, euryhaline)
Cocconeis scutellum Ehrenberg									P	32-34‰ (marine)
Cocconeis scutellum var. *parva* (Grunow) Cleve					P		S			38‰ (marine)
Corethron criophilum Castracane					P					n/a
Coscinodiscus spp.	P					P				-
Coscinodiscus argus Ehrenberg	S									n/a
Coscinodiscus asteromphalus Ehrenberg	S									n/a
Coscinodiscus bispculptus Rattray	S									n/a
Coscinodiscus concinnus Smith									P	32-34‰ (marine)

Appendix II (Continued)

TAXA NAME/AUTHORITY	1	2	3	4	5	6	7	8	9	SALINITY
Coscinodiscus curvatulus Grunow		S						P	P	32-34‰ (marine)
Coscinodiscus granii Gough					P					n/a
Coscinodiscus janischii Schmidt									P	32-34‰ (marine)
Coscinodiscus jonesianus (Greville) Ostenfeld									P	32-34‰ (marine)
Coscinodiscus marginato-lineatus var. *antarctica* Manguin									P	32-34‰ (marine)
Coscinodiscus marginatus Ehrenberg					P				P	32-34‰ (marine)
Coscinodiscus nitidus Gregory									P	32-34‰ (marine)
Coscinodiscus obscurus Schmidt									P	32-34‰ (marine)
Coscinodiscus oculus-iridis (Ehrenberg) Ehrenberg					P				P	32-34‰ (marine)
Coscinodiscus perforatus var. *cellulosa* Grunow									P	32-34‰ (marine)
Coscinodiscus radiatus Ehrenberg	P	S							P	32-34‰ (marine)
Coscinodiscus rothii (Ehrenberg) Grunow					P					n/a
Coscinodiscus rothii var. *subsalsum* (Juhlin-Dann.) Hustedt		S								n/a
Cosmioneis pusilla var. *incognita* (Krasske) Aboal	S									n/a
Craticula accomoda (Hustedt) Mann	P									0.1-0.4‰ (freshwater)
Craticula ambigua (Ehrenberg) Mann	S									n/a
Craticula cuspidata (Kutzing) Mann	P/S/E	S		S				P		<1‰ (freshwater)
Craticula halophila (Grunow) Mann	P/S/E									<1‰ (freshwater)
Craticula pampeana (Frenguelli) Lange-Bertalot	S									n/a
Craticula submolesta (Hustedt) Lange-Bertalot	S									n/a
Ctenophora pulchella (Ralfs) Williams & Round	S									n/a

Appendix II (Continued)

TAXA NAME/AUTHORITY	1	2	3	4	5	6	7	8	9	SALINITY
Cyclotella spp.	P				P	P				-
Cyclotella atomus Hustedt	P									0-0.2‰ (freshwater)
Cyclotella meneghiniana Kützing	P/S/E	S	S	S	P			P	P	WA: 6.3±7.5‰ (brackish, euryhaline)
Cyclotella striata (Kützing) Grunow	P	S	S	S						0-15‰ (brackish/freshwater)
Cyclotella stylorum Brightwell		S					S			38‰ (marine)
Cylindrotheca closterium (Ehrenberg) Reimann & Lewin					S/P				P	30-36‰ (marine)
Cymatopleura solea (Brébisson) W. Smith		S	S	S				P		0-10‰ (freshwater/brackish)
Cymatosira belgica Grunow		S	S	S			S			WA: 19.8±8‰ (marine/brackish)
Cymbella spp.								P		-
Cymbella affinis Kützing	P		S					P		0-0.2‰ (freshwater)
Cymbella australica (Schmidt) Cleve	S									n/a
Cymbella cistula (Hemprich & Ehrenberg) Kirchner		S		S				P		0-10 (freshwater/brackish)
Cymbella cymbiformis Agardh		S	S	S						0-6‰ (freshwater/brackish)
Cymbella cymbiformis var. *nonpunctata* Fontell								P		<1‰ (freshwater)
Cymbella neocistula Krammer	S									n/a
Cymbella prostrata (Berkeley) Cleve								P		<1‰ (freshwater)
Cymbella proxima Patrick & Reimer	S									n/a
Cymbella tumida (Brébisson) Van Heurk								P		<1‰ (freshwater)
Cymbella turgidula Grunow	S									n/a
Cymbopleura naviculiformis (Auerswald) Krammer	S									n/a
Dactyliosolen fragilissimus (Bergon) Hasle									P	32-34‰ (marine)

Appendix II (Continued)

TAXA NAME/AUTHORITY	1	2	3	4	5	6	7	8	9	SALINITY
Delicata nepouiana Krammer	S									n/a
Denticula elegans Kützing		S	S	S						0-3‰ (freshwater/brackish)
Denticula kuetzingii Grunow		S	S	S						W/A: 3.8±5‰ (freshwater/brackish)
Denticula tenuis Kützing			S	S						0-2‰ (freshwater/brackish)
Denticula valida (Pedicino) Grunow	S									n/a
Diadesmis contenta (Grunow) Mann	S									n/a
Diatoma moniliformis Kützing			S							<1‰ (freshwater)
Diatoma vulgaris Bory		S	S	S						0-5‰ (freshwater/brackish)
Dickieia subinflata (Grunow) Mann		S	S	S						15-21‰ (marine/brackish)
Dimeregramma minor (Gregory) Ralfs		S	S	S						W/A: 20±5‰ (marine/brackish)
Diploneis caffra (Giffen) Witkowski	S									n/a
Diploneis chilensis (Hustedt) Lange-Bertalot	S									n/a
Diploneis interrupta (Kützing) Cleve		S	S	S						8.3-29‰ (marine/brackish)
Diploneis ovalis (Hilse) Cleve		S	S	S						0.5-4‰ (freshwater/brackish)
Diploneis puella (Schumann) Cleve		S	S	S						0.5-28‰ (marine/brackish)
Diploneis smithii (Brébisson) Cleve		S	S							0.5-2‰ (freshwater/brackish)
Diploneis smithii var. *constricta* Heiden		S								n/a
Diploneis subovalis Cleve	S									n/a
Discostella pseudostelligera (Hustedt) Houk & Klee	P				P					0-0.2‰ (freshwater)
Ditylum brighwellii (West) Grunow								P	P	30.4-34‰ (marine)
Ditylum sol (Schmidt) Cleve									P	32-34‰ (marine)

Appendix II (Continued)

TAXA NAME/AUTHORITY	1	2	3	4	5	6	7	8	9	SALINITY
Encyonema mesiana (Cholnoky) Krammer	S									n/a
Encyonema minutum (Hilse) Mann	S									n/a
Encyonema silesiacum (Bleisch) Mann	P/S/E	S	S							<1‰ (freshwater)
Encyonema sprechmannii Metzeltin, Lange-Bertalot & García-Rodriguez	S									n/a
Encyonopsis microcephala (Grunow) Krammer	S									
Entomoneis alata (Ehrenberg) Ehrenberg					S/P					34.5-35.6‰ (marine)
Entopyla australis (Ehrenberg) Ehrenberg									P	32-34‰ (marine)
Epithemia adnata (Kützing) Brébisson		S	S	S						<1‰ (freshwater)
Epithemia argus (Ehrenberg) Kützing		S	S							0.5-5‰ (freshwater/brackish)
Epithemia sorex Kützing		S						P		0-10‰ (freshwater/brackish)
Epithemia turgida var. *granulata* (Ehrenberg) Brun	S									n/a
Eunotia arcus Ehrenberg	P									0-1.2‰ (freshwater)
Eunotia bilunaris (Ehrenberg) Schaarschmidt	P									0-10.5‰ (freshwater/brackish)
Eunotia biseriata Hustedt	S									n/a
Eunotia camelus Ehrenberg	S									n/a
Eunotia formica Ehrenberg	P									0-0.2‰ (freshwater)
Eunotia hexaglyphis Ehrenberg	P									0-0.25‰ (freshwater)
Eunotia implicata Nörpel, Lange-Bertalot & Alles	S									n/a
Eunotia incisa Smith	S									n/a
Eunotia larra Frenguelli	S									n/a

Appendix II (Continued)

TAXA NAME/AUTHORITY	1	2	3	4	5	6	7	8	9	SALINITY
Eunotia luna var. *aequalis* Hustedt	S									n/a
Eunotia major var. *gigantea* Frenguelli	S									n/a
Eunotia major var. *major* (Schmith) Rabenhorst	S									n/a
Eunotia monodon Ehrenberg	P									0–0.2‰ (freshwater)
Eunotia monodon var. *bidens* (Gregory) Hustedt	S									n/a
Eunotia odebrechtiana Metzeltin & Lange-Bertalot	S									n/a
Eunotia pectinalis var. *undulata* (Ralfs) Rabenhorst	P									0–0.5‰ (freshwater)
Eunotia pyramidata var. *monodon* Krasske	S									n/a
Eunotia praerupta Ehrenberg	P									0–0.25‰ (freshwater)
Eunotia praerupta var. *excelsa* Krasske	P									0.2–0.4‰ (freshwater)
Eunotia tecta Krasske	S									n/a
Eunotia tridentula Ehrenberg	S									n/a
Eunotia veneris (Kützing) De Toni	S									n/a
Fallacia monoculata (Hustedt) Mann	S									n/a
Fallacia omissa (Hustedt) Mann	S									n/a
Fallacia pygmaea (Kützing) Stickle & Mann	S/E	S	S	S						W/A: 20.8±6.3‰ (marine/brackish)
Fistulifera saprophila (Lange-Bertalot &. Bonik) Lange-Bertalot	S									n/a
Fragilaria capucina Desmazières	S/P									0–1.2‰ (freshwater)
Fragilaria capucina subsp. *rumpens* (Kützing) Lange-Bertalot	S									n/a
Fragilaria capucina var. *vaucheriae* (Kützing) Lange-Bertalot	S									n/a

Appendix II (Continued)

TAXA NAME/AUTHORITY	1	2	3	4	5	6	7	8	9	SALINITY
Fragilaria crassa Metzeltin & Lange-Bertalot	S									n/a
Fragilaria crotonensis Kitton						P				0–26‰ (brackish/marine, euryhaline)
Fragilaria goulardii (Brébisson) Lange-Bertalot	S									n/a
Fragilaria heidenii Østrup	P/S									0–6.5‰ (freshwater/brackish)
Fragilaria tenera (Smith) Lange-Bertalot	S									n/a
Fragilariforma virescens (Ralfs) Williams & Round		S	S	S						WA: 6.4±7‰ (freshwater/brackish)
Frankophila similioides Lange-Bertalot & Rumrich	S									n/a
Frustulia neomundana Lange-Bertalot & Rumrich	S									n/a
Frustulia rhomboides (Ehrenberg) De Toni			S	S						<1‰ (freshwater)
Frustulia rhomboides var. *viridula* (Brébisson) Cleve									P	32–34‰ (marine)
Frustulia vulgaris (Twaites) De Toni								P		<1‰ (freshwater)
Geissleria decussis (Østrup) Lange-Bertalot & Metzeltin	S									n/a
Geissleria ignota (Krasske) Lange-Bertalot & Metzeltin	S									n/a
Geissleria perelegans (Hustedt) Metzeltin & Lange-Bertalot	S									n/a
Geissleria schmidiae Lange-Bertalot & Rumrich	S									n/a
Gomphoneis minuta (Stone) Kociolek & Stoermer								P		<1‰ (freshwater)
Gomphoneis herculeana (Ehrenberg) Cleve								P		0–35
Gomphonema spp.								P		-
Gomphonema abbreviatum (Agardh) Kützing		S	S							0.5–20‰ (marine/brackish)
Gomphonema acuminatum Ehrenberg								P		<1‰ (freshwater)

Appendix II (Continued)

TAXA NAME/AUTHORITY	1	2	3	4	5	6	7	8	9	SALINITY
Gomphonema affine Kützing	S									n/a
Gomphonema affine var. *rhombicum* Reichardt	S									n/a
Gomphonema anglicum Ehrenberg	S									n/a
Gomphonema angustatum (Kützing) Rabenhorst		S	S	S						8-22‰ (marine/brackish)
Gomphonema apicatum Ehrenberg	S									n/a
Gomphonema augur Ehrenberg	P/S/E									0-0.2‰ (freshwater)
Gomphonema auritum Braun	S									n/a
Gomphonema capitatum Ehrenberg	S									n/a
Gomphonema clavatum Ehrenberg	P									0.2-0.4‰ (freshwater)
Gomphonema gracile Ehrenberg	P									0-0.2‰ (freshwater)
Gomphonema lagenula Kützing	S									n/a
Gomphonema laticollum Reichardt	S									n/a
Gomphonema olivaceum (Lyngbye) Kützing		S	S	S				P		0-5‰ (freshwater/brackish)
Gomphonema parvulum (Kützing) Grunow	P/S/E	S	S	S						WA: 3±3.7‰ (freshwater/brackish)
Gomphonema pseudotenellum Lange-Bertalot								P		<1‰ (freshwater)
Gomphonema salae Lange-Bertalot & Reichardt	S									n/a
Gomphonema truncatum Ehrenberg	P	S	S					P		0-0.1 (freshwater)
Gomphonema turris Ehrenberg	S									n/a
Gomphonema turris var. *brasiliensis* (Fricke) Frenguelli	S									n/a
Grammatophora angulosa Ehrenberg									P	32-34‰ (marine)
Grammatophora hamulifera Kützing									P	32-34‰ (marine)

Appendix II (Continued)

TAXA NAME/AUTHORITY	1	2	3	4	5	6	7	8	9	SALINITY
Grammatophora marina (Lyngbye) Kützing								P	P	3-35‰ (marine/brackish)
Grammatophora oceanica Ehrenberg		S	S							32-34‰ (marine)
Grammatophora serpentina Ehrenberg									P	32-34‰ (marine)
Grammatophora undulata Ehrenberg	S									n/a
Guinardia delicatula (Cleve) Hasle					P					30-33‰ (marine)
Guinardia flaccida (Castracane) Peragallo					P					30-33‰ (marine)
Gyrosigma spp.		S	S	S		P		P		-
Gyrosigma acuminatum (Kützing) Rabenhorst								P		<1‰ (freshwater)
Gyrosigma attenuata (Kützing) Rabenhorst	P				S/P	P				0-35‰ (euryhaline)
Gyrosigma fasciola (Ehrenberg) Griffith & Henfrey					S					24-26‰ (brackish/marine)
Gyrosigma scalproides (Rabenhorst) Cleve	P									0-0.2‰ (freshwater)
Gyrosigma spencerii (Bailey) Griffith & Henfrey	P									0-15‰ (freshwater/brackish)
Hantzschia amphioxys (Ehrenberg) Grunow	P/S/E		S	S						0-0.2‰ (freshwater)
Hantzschia amphioxys var. *capitellata*	S									n/a
Hantzschia uruguayensis Metzeltin, Lange-Bertalot & García-Rodríguez	S									n/a
Hantzschia virgata var. *capitellata* Hustedt				S						WA: 19.3±3‰ (brackish/marine)
Hantzschia vivax (Smith) Tempère	S									n/a
Helicotheca tamesis (Shrubsole) Ricard						P				26‰ (marine/brackish)
Hemiaulus sinensis Greville					P					n/a
Hippodonta capitata (Ehr.) Lange-Bert., Metz. & Witk.	P/S/E									0-0.2‰ (freshwater)

Appendix II (Continued)

TAXA NAME/AUTHORITY	1	2	3	4	5	6	7	8	9	SALINITY
Hippodonta hungarica (Grun.) Lange-Bert., Metz. & Witk.	P/S	S	S	S						WA: 4.9±6.7‰ (freshwater/brackish)
Hippodonta linearis (Østrup) Lange-Bert, Metz & Witk.		S	S							7-22‰ (marine/brackish)
Hippodonta luneburgensis (Grun.) Lange-Bert., Metz. & Witk.		S	S							7-22‰ (marine/brackish)
Hippodonta subtilissima Lange-Bertalot	S									n/a
Hyalodiscus radiatus (O' Meara) Grunow									P	32-34‰ (marine)
Hyalodiscus scoticus (Kützing) Grunow									P	32-34‰ (marine)
Hyalodiscus subtilis Bailey		S	S	S					P	32-34‰ (marine)
Karayevia clevei (Grunow) Round & Bukhtiyarova	S									n/a
Lemnicola hungarica (Grunow) Round & Basson	S/E									<1‰ (freshwater)
Leptocylindrus sp.						P				-
Licmophora sp.								P		-
Licmophora abbreviata Agardh									P	32-34‰ (marine)
Licmophora flabellata Agardh									P	32-34‰ (marine)
Lithodesmium undulatum Ehrenberg					P			P		n/a
Luticola charcotii var. *magelanica* (Hustedt) Metzeltin	S									n/a
Luticola claudiae Metzeltin, Lange-Bertalot & García-Rodríguez	S									n/a
Luticola cohnii (Hilse) Mann	S/E									<1‰ (freshwater)
Luticola dapalis (Frenguelli) Mann	S									n/a
Luticola dapaloides (Frenguelli) Metzeltin & Lange-Bertalot	S									n/a

Appendix II (Continued)

TAXA NAME/AUTHORITY	1	2	3	4	5	6	7	8	9	SALINITY
Luticola frenguellii Metzeltin & Lange-Bertalot	S									n/a
Luticola goeppertiana (Bleisch) Mann	S/E									<1‰ (freshwater)
Luticola mutica (Kützing) Mann	S	S	S	S				P		1-7‰ (brackish/freshwater)
Luticola nivalis (Ehrenberg) Mann	S									n/a
Luticola ventricosa (Kützing) Mann	S/E									<1‰ (freshwater)
Luticola saxophila (Bock) Mann	S									n/a
Luticola undulata (Hilse) Mann	S									n/a
Luticola undulata var. *chilensis* (Hustedt) Metzeltin	S									n/a
Lyrella david-mannii Witkowski, Lange-Bertalot & Metzeltin	S									n/a
Lyrella lyra (Ehrenberg) Karajeva									P	32-34‰ (marine)
Mastogloia belaensis Voigt		S	S							3-9‰ (brackish)
Mastogloia elliptica (Agardh) Cleve		S	S	S						3-28‰ (marine/brackish, euryhaline)
Mayamea atomus (Kützing) Lange-Bertalot	P									0-0.25‰ (freshwater)
Melosira sp.						P				-
Melosira fausta Schmidt									P	32-34‰ (marine)
Melosira moniliformis (Müller) Agardh					P					n/a
Melosira moniliformis var. *octagona* (Grunow) Hustedt	S									n/a
Melosira nummuloides Agardh									P	32-34‰ (marine)
Melosira varians Agardh	S	S	S	S		P		P		<1‰ (freshwater)
Navicella pusilla (Grunow) Krammer	S	S	S							WA: 6.5±9.5‰ (freshwater/brackish)

Appendix II (Continued)

TAXA NAME/AUTHORITY	1	2	3	4	5	6	7	8	9	SALINITY
Navicula spp.					S/P	P		P		-
Navicula angusta Grunow	S									n/a
Navicula antonii Lange-Bertalot	S									n/a
Navicula atomus (Kützing) Grunow	S									n/a
Navicula breitenbuchii Lange-Bertalot	S									n/a
Navicula capitatoradiata Germain								P		n/a
Navicula caterva Hohn & Hellermann			S							2‰ (brackish)
Navicula cincta (Ehrenberg) Kützing		S	S	S						WA: 12±11.6‰ (brackish/marine)
Navicula constans Hustedt	P									0-0.25‰ (freshwater)
Navicula cryptocephala Kützing	P	S	S							0-6‰ (brackish/freshwater)
Navicula cryptotenella Lange-Bertalot	S									n/a
Navicula cryptotenelloides Lange-Bertalot	S									n/a
Navicula digitatoradiata (Gregory) Ralfs		S								2.5-8.5‰ (brackish)
Navicula eichhorniaephila Manguin	S									n/a
Navicula elmorei Patrick	P									0-0.3‰ (freshwater)
Navicula endophytica Hasle		S	S							<1‰ (freshwater)
Navicula erifuga Lange-Bertalot	P/S/E									0-0.4‰ (freshwater)
Navicula exigua Gregory	P									0-0.3‰ (freshwater)
Navicula forcipata var. *densestriata* Schmidt									P	32-34‰ (marine)
Navicula gregaria Donkin	S/E	S	S	S				P		WA: 12±10.5‰ (brackish/marine)
Navicula		S	S	S						8-28‰ (marine/brackish)

Appendix II (Continued)

TAXA NAME/AUTHORITY	1	2	3	4	5	6	7	8	9	SALINITY
Navicula laterostrata Hustedt	S									n/a
Navicula longicephala Hustedt	S									n/a
Navicula microcari Lange-Bertalot	S									n/a
Navicula notha Wallace	P									0-0.3‰ (freshwater)
Navicula novaesiberica Lange-Bertalot	S									n/a
Navicula peregrina (Ehrenberg) Kützing	P	S	S	S				P		0-6.5‰ (brackish/freshwater)
Navicula peregrinopsis Lange-Bertalot & Witkowski	S									n/a
Navicula pseudotenelloides Krasske	S									n/a
Navicula radiosa Kützing								P		0-10‰ (brackish/freshwater)
Navicula rhynchocephala Kützing	P/S/E									0-0.2‰ (freshwater)
Navicula rostellata Kützing	S									n/a
Navicula sanctaecrucis Østrup	S									n/a
Navicula schroeteri Meister	S									n/a
Navicula symmetrica Patrick	S									n/a
Navicula tackei f. *major* Maidana & Herbst								P		<1‰ (freshwater)
Navicula tenelloides Hustedt	S									<1‰ (freshwater)
Navicula tripunctata (Müller) Bory		S	S	S				P		0-2‰ (freshwater/brackish)
Navicula trivialis Lange-Bertalot	P/S/E	S	S	S						0-6.5‰ (brackish/freshwater)
Navicula veneta Kützing	S/E							P		<1‰ (freshwater)
Neidium affine (Ehrenberg) Pfitzer	P/S/E									0-0.4‰ (freshwater)
Neidium affine var. *longiceps* (Gregory) Cleve	S									n/a

Appendix II (Continued)

TAXA NAME/AUTHORITY	1	2	3	4	5	6	7	8	9	SALINITY
Neidium amphirhynchus (Ehrenberg) Pfitzer	S									n/a
Neidium ampliatum (Ehrenberg) Krammer	S									n/a
Neidium catarinense (Krasske) Lange-Bertalot	S									n/a
Neidium dubium (Ehenberg) Cleve	S/E									<1‰ (freshwater)
Neidium hercynicum Mayer	S									n/a
Neidium iridis (Ehrenberg) Cleve	S									n/a
Neidium iridis var. *amphigomphus* (Ehrenberg) Tempere & Peragallo	S									n/a
Neidium iridis var. *intercedens* Mayer	P									0-0.4‰ (freshwater)
Neidium magellanica var. *minor* Frenguelli	S									
Neocalyptrella robusta (Norman) Hern-Bec. & Meave									P	32-34‰ (marine)
Nitzschia spp.		S		S	S			P		-
Nitzschia acicularis (Kützing) Smith	P/S/E									0-0.4‰ (freshwater)
Nitzschia amphibia Grunow	S	S	S	S						WA: 3.2±4.6‰ (brackish/freshwater)
Nitzschia angularis Smith									P	32-34‰ (marine)
Nitzschia brevissima Grunow	P/S/E									0.2-0.4‰ (freshwater)
Nitzschia capitellata Hustedt	S									n/a
Nitzschia clausii Hantzsch	S/E	S	S	S						0-7‰ (brackish/freshwater)
Nitzschia commutata Grunow	S									n/a
Nitzschia commutatoides Lange-Bertalot	P									0-0.4‰ (freshwater)
Nitzschia constricta (Gregory) Grunow	P							P	P	0-34‰ (marine/brackish, euryhaline)
Nitzschia draveillensis Coste & Ricard	P/S/E									0-0.4‰ (freshwater)

Appendix II (Continued)

TAXA NAME/AUTHORITY	1	2	3	4	5	6	7	8	9	SALINITY
Nitzschia filiformis (Smith) Hustedt	P/S/E									0-0.5‰ (freshwater)
Nitzschia filiformis var. *conferta* (Richt) Lange-Bertalot	S									n/a
Nitzschia fonticola (Grunow) Grunow	S/E									<1‰ (freshwater)
Nitzschia frustulum (Kützing) Grunow	P/S/E									0-7‰ (brackish/freshwater)
Nitzschia fruticosa Hustedt	P									0-0.2‰ (freshwater)
Nitzschia gracilis Hantzsch	P									0-0.5‰ (freshwater)
Nitzschia habirshawii Febiger									P	32-34‰
Nitzschia hantzschiana Rabenhorst		S								n/a
Nitzschia	S									n/a
Nitzschia inconspicua Grunow	S	S	S	S						WA: 6.4±7.6‰ (brackish/freshwater)
Nitzschia lacunarum Hustedt	S									n/a
Nitzschia linearis (Agardh) Smith	P/S/E									0-0.2‰ (freshwater)
Nitzschia	S									n/a
Nitzschia microcephala Grunow		S	S	S						WA: 7.6±5.5‰ (brackish/freshwater)
Nitzschia nana Grunow	P/S/E	S								0-0.2‰ (freshwater)
Nitzschia palea (Kützing) Smith	P/S/E									0-0.2‰ (freshwater)
Nitzschia paleacea Grunow	P/S									0-0.2‰ (freshwater)
Nitzschia perminutum (Grunow) Peragallo	S									n/a
Nitzschia pumila Hustedt	S									n/a
Nitzschia rautenbachiae Cholnoky		S	S	S						WA: 10±5‰ (brackish)
Nitzschia	S									n/a

Appendix II (Continued)

TAXA NAME/AUTHORITY	1	2	3	4	5	6	7	8	9	SALINITY
Nitzschia	S									n/a
Nitzschia sigma (Kützing) Smith	P/S	S		S	S					0-34.5‰ (marine/brackish, euryhaline)
Nitzschia	S									n/a
Nitzschia sinuata var. *delongei* (Grunow) Lange-Bertalot	S			S						n/a
Nitzschia socialis Gregory										29‰ (marine)
Nitzschia subconstricta Grunow	S									n/a
Nitzschia umbonata (Ehrenberg) Lange-Bertalot	S									n/a
Nitzschia vermicularis (Kützing) Hantzsch	P									0-0.4‰ (freshwater)
Nitzschia vitrea Norman		S								2-20‰ (brackish/marine)
Nupela lesothensis (Schoeman) Lange-Bertalot	S									n/a
Odontella sp.				S						-
Odontella aurita (Lyngbye) Agardh								P	P	25-35‰ (marine/brackish)
Odontella mobiliensis (Bailey) Grunow					P			P		25-35‰ (marine/brackish)
Odontella obtusa Kütz.									P	32-34‰ (marine)
Odontella sinensis (Greville) Grunow					P					n/a
Opephora sp.		S	S							-
Opephora marina (Gregory) Petit							S			38‰ (marine)
Opephora pacifica (Grunow) Petit		S	S	S						WA: 13±6‰ (brackish/marine)
Orthoseira roeseana (Rabenhorst) O'Meara	S									n/a
Paralia sulcata (Ehrenberg) Cleve		S	S	S	S/P	P		P	P	WA: 26±3‰ /0-35‰ (marine/brackish, euryhaline)

Appendix II (Continued)

TAXA NAME/AUTHORITY	1	2	3	4	5	6	7	8	9	SALINITY
Petrodictyon gemma (Ehrenberg) Mann					S					30-36‰ (marine)
Petroneis monilifera (Cleve) Stickle & Mann		S								20-25‰ (marine/brackish)
Pinnularia acrosphaeria (Brébisson) Smith	S									n/a
Pinnularia acrosphaeria f. *maxima* Cleve	S									n/a
Pinnularia borealis Ehrenberg	S	S	S	S				P	P	3-20‰ (brackish/marine, euryhaline)
Pinnularia borealis var. *islandica* Krammer	S									n/a
Pinnularia borealis var. *scalaris* (Ehrenberg) Rabenhorst	S									n/a
Pinnularia borealis var. *sublinearis* Krammer	S									n/a
Pinnularia brevicostata Cleve								P		<1‰ (freshwater)
Pinnularia carambolae Frenguelli	S									n/a
Pinnularia divergens var. *elliptica* Grunow	S									n/a
Pinnularia divergens var. *malayensis* Hustedt	S									n/a
Pinnularia divergens var. *sublinearis* Cleve	S									n/a
Pinnularia divergens var. *undulata* Peragallo & Héribaud	S									n/a
Pinnularia divergens var. *protracta* Krammer, & Metzeltin	S									n/a
Pinnularia doehringii Frenguelli	S									n/a
Pinnularia dubitabilis Hustedt	S									n/a
Pinnularia ehrlichiana Metzeltin, Lange-Bertalot & García-Rodríguez	S									n/a
Pinnularia fistuciformis Metzeltin, Lange-Bertalot & García-Rodríguez	S									n/a
Pinnularia gibba Ehrenberg	S/E									<1‰ (freshwater)

Appendix II (Continued)

TAXA NAME/AUTHORITY	1	2	3	4	5	6	7	8	9	SALINITY
Pinnularia hyalina Hustedt	S									n/a
Pinnularia aff. *joculata* (Manguin) Krammer	S									n/a
Pinnularia latevittata Cleve	S									n/a
Pinnularia maior (Kützing) Cleve	P									0-0.4‰ (freshwater)
Pinnularia marchica Ilka Schönfelder	S									n/a
Pinnularia mesolepta (Ehrenberg) Smith	P									0-0.5‰ (freshwater)
Pinnularia microstauron (Ehrenberg) Cleve	P							P		0-0.5‰ (freshwater)
Pinnularia neomajor Krammer	S									n/a
Pinnularia neuquina Frenguelli	S									n/a
Pinnularia nitzschiophila Rumrich	S									n/a
Pinnularia rabenhorstii var. *franconia* Krammer	S									n/a
Pinnularia schweinfurthii (Schmidt) Patrick	S									n/a
Pinnularia subacoricola Metzeltin, Lange-Bertalot & Garcia-Rodriguez	S									n/a
Pinnularia subanglica Krammer	S									n/a
Pinnularia cf. *subcapitata* Gregory	S/E									<1‰ (freshwater)
Pinnularia spec. cf. *stomatophora* var. *salina* Krammer	S									n/a
Pinnularia tabellaria Ehrenberg	S									n/a
Pinnularia viridiformis Krammer	S									n/a
Pinnularia viridis (Nitzsch) Ehrenberg		S								20‰ (brackish)
Placoneis clementis (Grunow) Cox	S/E									<1‰ (freshwater)

Appendix II (Continued)

TAXA NAME/AUTHORITY	1	2	3	4	5	6	7	8	9	SALINITY
Placoneis disparilis (Hustedt) Metzeltin & Lange-Bertalot	S									n/a
Placoneis gastrum (Ehrenberg) Mereschkovsky	S	S								14‰(brackish)
Placoneis placentula (Ehrenberg) Mereschkowsky	P/S									0-0.3‰ (freshwater)
Placoneis parelginensis (Gregory) Cox	S									n/a
Placoneis serena (Frenguelli) Metzeltin	S									n/a
Plagiogramma staurophorum (Gregory) Heiberg			S						P	WA: 22±6‰ (marine/brackish)
Planothidium delicatulum (Kützing) Round & Bukht.	S/E	S	S	S			S			WA: 6.7±6.7 (brackish)
Planothidium lanceolatum (Brébisson) Lange-Bert.		S	S	S			S			0-38‰ (marine/brackish, euryhaline)
Pleurosigma spp.		S		S				P		-
Pleurosigma angulatum Smith						P				25-35‰ (marine/brackish)
Pleurosigma elongatum Smith	P									18‰ (marine/brackish)
Pleurosigma normanii Ralfs	P								P	32-34‰ (marine)
Pleurosigma strigosum Smith									P	32-34‰ (marine)
Pleurosira laevis (Ehrenberg) Compère	S/E	S	S	S				P		WA: 4±4.7‰ (brackish/freshwater)
Podosira sp.						P				-
Podosira maxima (Kützing) Grunow									P	32-34‰ (marine)
Podosira montagnei Kützing									P	32-34‰ (marine)
Podosira stelligera (Bailey) Mann		S	S	S	P		S			33-38‰ (marine)
Psammodictyon constrictum (Gregory) Mann		S	S	S						WA: 10.5±6‰ (brackish/marine)
Psammodictyon panduriforme (Gregory) Mann		S	S	S						20-28‰ (marine/brackish)
Pseudo-nitzschia spp.	P									-

Appendix II (Continued)

TAXA NAME/AUTHORITY	1	2	3	4	5	6	7	8	9	SALINITY
Pseudo-nitzschia seriata (Cleve) Peragallo					P					n/a
Pseudostaurosira brevistriata (Grunow) Williams & Round	S	S	S	S						WA: 7.8±6.75‰ (brackish/freshwater)
Reimeria sinuata (Gregory) Kociolek & Stoermer	S	S	S	S				P		0-5‰ (freshwater/brackish)
Reimeria uniseriata Sala, Guerrero & Ferrario	S									n/a
Rhabdonema adriaticum Kützing								P	P	30-35‰ (marine)
Rhabdonema arcuatum (Lyngbye) Kützing	S								P	32-34‰ (marine)
Rhabdonema minutum Kützing									P	32-34‰ (marine)
Rhaphoneis amphiceros (Ehrenberg) Ehrenberg	S	S	S	S	P		S	P	P	15-38‰ (marine/brackish)
Rhizosolenia sp.	P					P				-
Rhizosolenia hebetata Bailey									P	32-34‰ (marine)
Rhizosolenia setigera Brightwell	P					P			P	25-35‰ (marine/brackish)
Rhizosolenia styliformis Brightwell									P	32-34‰ (marine)
Rhoicosphenia abbreviata (Agardh) Lange-Bertalot	S	S	S	S				P		WA: 5.25±6.3‰ (brackish/freshwater)
Rhopalodia brebissonii Krammer	S/P	S	S	S						0-2‰ (freshwater/brackish)
Rhopalodia gibba (Ehrenberg) Müller	S		S	S				P		0-0.5‰ (freshwater)
Rhopalodia gibberula (Ehrenberg) Müller	S	S	S	S			S			WA: 5.3±5.8‰ (brackish/freshwater)
Rhopalodia musculus (Kützing) Müller		S	S	S						0-10‰ (brackish/freshwater)
Rhopalodia operculata (Agardh) Hák.	S									n/a
Scoliopleura sp.					S					-
Sellaphora laevissima (Kützing) Mann	S									n/a

Appendix II (Continued)

TAXA NAME/AUTHORITY	1	2	3	4	5	6	7	8	9	SALINITY
Sellaphora nyassensis (Müller) Mann	P/S									0-0.3‰ (freshwater)
Sellaphora pupula (Kützing) Mereschkovsky	P/S/E	S		S						0-0.4‰ (freshwater)
Sellaphora rectangularis (Gregory) Lange-Bertalot & Metzeltin	S									n/a
Sellaphora seminulum (Grunov) Mann		S	S							<1‰ (freshwater)
Skeletonema costatum (Greville) Cleve	P				P				P	2-34‰ (marine/brackish, euryhaline)
Skeletonema subsalsum (Cleve) Bethge	S/E									<1‰ (freshwater)
Stauroneis spp.		S			S					-
Stauroneis anceps Ehrenberg	S									n/a
Stauroneis brasiliensis (Zimmermann) Compère	S									n/a
Stauroneis cf. *javanica* (Grunov) Cleve	S									n/a
Stauroneis obtusa Lagerstedt	S									n/a
Stauroneis phoenicenteron (Nitzsch) Ehrenberg	S									n/a
Stauroneis producta Grunow		S	S	S						<1‰ (freshwater)
Stauroneis schinzii var. *maxima* Frenguelli	S									n/a
Stauroneis cf. *schroederi* Hustedt	S									n/a
Stauroneis subgracilis Lange-Bertalot & Krammer	S									n/a
Staurosira tackei (Hustedt) Krammer & Lange-Bertalot		S	S	S						0-2.5‰ (freshwater/brackish)
Staurosira altiplanensis Lange-Bertalot & Rumrich	S									n/a
Staurosira construens Ehrenberg	P	S	S							0-0.4‰ (freshwater)
Staurosira elliptica (Schumann) Williams & Round		S	S							WA: 2.7±4.5‰ (brackish/freshwater)

Appendix II (Continued)

TAXA NAME/AUTHORITY	1	2	3	4	5	6	7	8	9	SALINITY
Staurosira fernandae Garcia-Rodriguez, Lange-Bertalot & Metzeltin	S									n/a
Staurosira cf. *leptostauron* (Ehrenberg) Hustedt	S									n/a
Staurosira longirostris Frenguelli	S									n/a
Staurosira martyi (Hérib.) Lange-Bertalot	S									WA: 9.7±8.3‰ (brackish/freshwater)
Staurosira venter (Ehrenberg) Kobayasi		S	S	S						WA: 9.7±8.3‰ (brackish/freshwater)
Staurosirella pinnata (Ehrenberg) Williams & Round	S	S	S	S						WA: 6.3±6.4‰ (brackish/freshwater)
Stellarima stellaris (Roper) Hasle & Sims					P					n/a
Stephanodiscus spp.	P	S	S	S				P		-
Stephanodiscus hantzschii Grunow	P/S									0–0.4‰ (freshwater)
Stephanodiscus parvus Stoermer & Håkansson	P		S					P		0–6‰ (freshwater/brackish)
Surirella spp.		S	S					P		-
Surirella angusta Kützing	S									n/a
Surirella biseriata Brébisson	S									n/a
Surirella brebissonii Krammer & Lange-Bertalot	S									n/a
Surirella guatimalensis Ehrenberg	S									n/a
Surirella inducta Schmidt		S	S	S						2–10‰ (brackish)
Surirella minuta Brébisson	S	S	S	S						0–2.5‰ (freshwater/brackish)
Surirella minuta var. *peduliformis* Frenguelli	S									n/a
Surirella ovalis Brébisson	S/P	S	S	S						0–6‰ (freshwater/brackish)
Surirella ovalis var. *apiculata* Müller		S	S							<1‰ (freshwater)

Appendix II (Continued)

TAXA NAME/AUTHORITY	1	2	3	4	5	6	7	8	9	SALINITY
Surirella splendida (Ehrenberg) Kützing	S							P		n/a
Surirella striatula Turpin	S	S	S	S						0.5-10‰ (brackish)
Synedra sp.								P		-
Synedra fulgens (Greville) Smith									P	32-34‰ (marine)
Synedra platensis Frenguelli		S	S	S						0-3‰ (freshwater/brackish)
Synedra tortuosa Williams & Metzeltin	S									n/a
Synedra ulna var. *claviceps* Hustedt	S									n/a
Tabularia investiens (Smith) Williams & Round	S									n/a
Tabularia tabulata (Agardh) Snoeijs		S	S	S					P	5-34‰ (marine/brackish, euryhaline)
Terpsinoe americana (Bailey) Ralfs	S									n/a
Terpsinoe musica Ehrenberg	S									n/a
Thalassionema nitzschioides (Grunow) Mereschkowsky	P				P					25-35‰ (marine/brackish)
Thalassiosira spp.	P				P	P		P		-
Thalassiosira anguste-lineata (Schmidt) Fryxell & Hasle	P				P			P		20-33‰ (marine/brackish)
Thalassiosira curviseriata Takano					P					25-35‰ (marine/brackish)
Thalassiosira decipiens (Grunow) Jørgensen		S	S	S						WA: 21±7‰ (marine/brackish)
Thalassiosira eccentrica (Ehrenberg) Cleve		S		S	P			P		28-33‰ (marine)
Thalassiosira hendeyi Hasle & Fryxell					P					25-35‰ (marine/brackish)
Thalassiosira hibernalis Gayoso					P					30-35‰ (marine)
Thalassiosira leptopus (Grunow) Hasle & Fryxell					P					n/a
Thalassiosira minima Gaarder					P					n/a

Appendix II (Continued)

TAXA NAME/AUTHORITY	1	2	3	4	5	6	7	8	9	SALINITY
Thalassiosira pacifica Gran & Angst					P					n/a
Thalassiosira rotula Meunier	P				P					18-33‰ (marine/brackish)
Thalassiosira simonensii Hasle & Fryxell								P		n/a
Trachyneis aspera (Ehrenberg) Cleve									P	32-34‰ (marine)
Trachyneis aspera var. *perobliqua* Cleve									P	32-34‰ (marine)
Triceratium sp.						P				-
Triceratium favus Ehrenberg		S		S				P	P	25-35‰ (marine/brackish)
Tryblionella acuminata Smith	S/E								P	32-34‰ (marine)
Tryblionella angustata Smith	P/S/E									0-0.2‰ (freshwater)
Tryblionella apiculata Gregory	S									n/a
Tryblionella coarctata (Grunow) Mann	S									n/a
Tryblionella compressa (Bailey) Poulin	S/E	S	S	S			S			WA: 14±4‰ (brackish)
Tryblionella debilis Arnott	S									n/a
Tryblionella gracilis Smith		S	S	S						0-0.5‰ (freshwater)
Tryblionella granulata (Grunow) Mann		S								WA: 10±4‰ (brackish)
Tryblionella hungarica (Grunow) Frenguelli	P/S/E									0-0.2‰ (freshwater)
Tryblionella levidensis Smith	P/S	S	S				S			0-0.4‰ (freshwater)
Tryblionella perversa (Grunow) Mann	S									n/a
Ulnaria delicatissima var. *angustissima* (Grunow) Aboal & Silva	S									n/a
Ulnaria acus (Kützing) Aboal	S									n/a
Ulnaria ulna (Nitzsch) Compère	S/P	S	S	S				P		0-6.5‰ (brackish/freshwater)
RICHNESS	356	140	122	106	62	19	15	74	88	
Nº REVISED WORKS	10	6	3	2	11	1	1	7	6	

In: Encyclopedia of Environmental Research
Editor: Alisa N. Souter

ISBN: 978-1-61761-927-4
© 2011 Nova Science Publishers, Inc.

Chapter 40

ANTHROPOGENIC IMPACTS IN A PROTECTED ESTUARY (AMVRAKIKOS GULF, GREECE): BIOLOGICAL EFFECTS AND CONTAMINANT LEVELS IN SENTINEL SPECIES

C. Tsangaris[1], E. Cotou[2], I. Hatzianestis[1] and V. A. Catsiki[1]

[1] Institute of Oceanography, Hellenic Center for Marine Research (HCMR), 46.7 km, Athinon-Souniou Ave, P.O. Box 712, 190 13 Anavyssos, Greece.
[2] Institute of Aquaculture, Hellenic Center for Marine Research (HCMR), Agios Kosmas, Elliniko 16610, Greece.

ABSTRACT

This chapter aims to highlight possible pollution impacts in a protected estuarine ecosystem, Amvrakikos Gulf, considered one of the most important wetlands in Greece. Although land-based discharges in the area are low, Amvrakikos Gulf receives inputs by riverine transport, mainly contaminants related to agricultural practices. In order to assess pollution impacts, biological effects (determined with biomarkers and bioassays) and contaminant levels were measured in sentinel species (mussels *Mytilus galloprovincialis*) over a two-year period. Biomarkers at the sub organism level (Scope for Growth [SFG]) revealed stress conditions and provided early warning signals of possible consequences at higher levels of biological organization. Biochemical markers (acetylcholinesterase, glutathione peroxidase and metallothionein) suggested the presence of organic contaminants but absence of elevated metal levels and indicated a risk for pesticide contamination. Likewise, the Microtox bioassay applied in fluid extracts from mussels demonstrated stress conditions and the presence of organic contaminants based on the production of distinctly different light level-time response curves of the luminescent bacteria (*Vibrio fidceri*). Chemical analysis in the mussel tissues confirmed contamination of agricultural origin showing moderately elevated ΣDTT concentrations, low ΣPCB and low heavy metal concentrations.

INTRODUCTION

Estuaries and lagoons are productive ecosystems of high ecological value. They are important habitats for various species, and those of highest value are protected by national and international conventions such as the Ramsar Convention. However, even if protective actions are taken to avoid environmental deterioration of ecologically important areas, anthropogenic activities, often from distant sources, may put such ecosystems at risk. Estuaries worldwide are subjected to environmental pressure due to riverine transport of contaminants from various land-based sources by industrial, urban and agricultural activities in the rivers' drainage areas. Thus, even when there are no known pollution sources within the vicinity of estuarine areas, assessment of contaminant impacts is of major importance for the evaluation of possible risks incurred by inputs from distant sources.

Biological effects at different levels of biological organization, from the molecular to the community level, can be used as tools for the evaluation of risk by environmental contamination (Peakal, 1994). It is generally believed that sub organism and organism responses occur prior to alterations at the population and community levels; thus, biological effects of pollution in sentinel organisms can be used as early warning signals of ecosystem-level damage (Walker et al., 2006). Biomarkers and bioassays are common methods of biological effect measurements at the sub organism and organism level. Biomarkers represent molecular, biochemical, physiological or behavioural changes in organisms that can be related to exposure and/or effects of chemical contaminants. Whole organism bioassays are measures of lethal and sub lethal toxicity of environmental samples that can be used as tools to assess environmental quality. On the other hand, measurement of contaminant levels in the tissues of sentinel organisms is indicative of contaminant bioavailability. Combined measurements of bioaccumulation of contaminants and biological effects in sentinel species can provide information on causes of possible impacts (Widdows et al., 2002; Galloway et al., 2004; Damiens et al., 2007).

This chapter aims to highlight possible pollution impacts in a protected estuarine ecosystem, Amvrakikos Gulf, Greece, by means of biological effect and contaminant level measurements in sentinel species.

Mytilus galloprovincialis mussels were chosen as the sentinel species because of their economical importance in the study area and their extensive use in pollution biomonitoring programmes worldwide (Viarengo et al., 2007).

To detect possible stress effects, a biomarker indicative of health status of the organism, i.e., Scope for Growth (SFG), and the Microtox bioassay were used. SFG integrates estimates of major physiological responses (feeding, food absorption, respiration and excretion) converted to energy equivalents into an index that represents the energy available for growth and reproduction (Widdows & Donkin, 1992). SFG provides a general response to environmental stressors and is considered an ecologically relevant biomarker since adverse effects on the energy budget of an organism can be predictive of long-term consequences to the growth and survival of individuals and populations (Widdows & Donkin, 1992). The Microtox bioassay can be used for the determination of potential toxicity of biological fluids extracted from bivalves (Lau-Wong, 1990; Cotou et al., 2002; Bihari et al., 2007). The Microtox is a luminescent bacteria bioassay that measures the effect of potential inhibitors on luminescence under defined conditions. Reduction in luminescence indicates a stress response that reflects the presence of chemical contaminants.

To identify exposure to certain types of contaminants, biochemical biomarkers of exposure, i.e., acetylcholinesterase (ACHE), metallothionein (MT) and glutathione peroxidase (GPX), and the light level-time response curves of the Microtox bioassay were employed. The biochemical markers measured reflect exposure to different types of contaminants. ACHE is a biomarker of exposure to organophosphate and carbamate pesticides (Bocquené and Galgani, 1991), MT is a biomarker of exposure to heavy metals (Viarengo et al., 1999), and GPX is an antioxidant enzyme that provides an indication of the presence of contaminants capable of reactive oxygen species production (Doyotte et al., 1997). The light level-time response curves of the Microtox bioassay distinguish between metal and organic contaminants and were thus used to identify the type of contaminants in the mussel fluids responsible for potential toxicity.

To verify the type and degree of contamination and relate contaminant levels to biological effects, organochlorine compounds and heavy metals were determined in the tissues of the mussels.

AREA DESCRIPTION AND BACKGROUND

Amvrakikos Gulf is a shallow (max. depth 60m), semi-enclosed embayment of 405 km^2 in western Greece (Figure 1). It is connected to the open Ionian Sea through a narrow channel (width 800m, depth 12 m) of 5 km length. A complex system of lagoons and an extensive delta originating from the rivers Louros and Arachthos that outflow into the Gulf are located in its northern part (Guelorget et al., 1986). Three major lagoons situated between the delta of Louros and Arachthos rivers (Rodia, Tsoukalio and Logarou), smaller lagoons along the coastline and a variety of wetlands (salt marshes, reedbeds, mudflats and shallow bays) around the lagoons comprise this complex ecosystem, one of the largest wetlands in Greece. It is considered eutrophic, as relatively low salinity combined with nutrient enrichment from the rivers favour phytoplankton growth and result in relatively high chlorophyll *a* biomass (Panayotidis et al., 1994). The area is of great ecological importance as the variety of habitat types sustains diverse flora and fauna and significant number of endangered species, especially birds (Crivelli et al., 1998; Diapoulis et al., 2001; Sarika et al., 2005). Dolphins and sea turtles are also found in the Gulf (Rees & Margaritoulis, 2006; Bearzi et al., 2008). The Amvrakikos Gulf was designated a Ramsar site in 1975, a Special Protection Area of the NATURA 2000 network under EC Directive 79/409 on the conservation of wild birds, and a Specially Protected Area (SPA) under Protocol 4 of the Barcelona Convention. In 2008 the Amvrakikos Gulf was declared a National Park.

However, there is growing concern about pollution in the Amvrakikos Gulf, mostly by contaminants transferred by the rivers. The drainage areas of Louros and Arachthos rivers are 785 km^2 and 1894 km^2, respectively, and consist predominantly of agricultural land. Agriculture in these areas is intense and includes citrus fruits, olives, corn, alpha-alpha and cotton. Land-based pollution discharges in the Gulf are considered low and include municipal wastes from coastal cities and waste waters from small industries. In the greater area there are cities with populations between 10,000 and 100,000, the largest of which are served by municipal wastewater treatment plants. Industrial activities involve processing of agricultural and live-stock products located mostly in the vicinity of Preveza, the largest city in the area. In the broader area there are pig breeding units, dairy farms, oil mills and cheese factories.

Chemical analysis of contaminants in water and sediments imply moderate pesticide contamination (Readman et al., 1993; Albanis et al., 1995a). Herbicides (atrazine, simazine, alachlor, metolachlor, trifluralin, diuron) and organochlorine insecticides (a-BHC, b-BHC, lindane, DDT and metabolites) have been found in water and sediments of river estuaries and wetlands of the Amvrakikos Gulf (Albanis et al., 1995a). The annual flux of all pesticides through Louros Estuary into the marine environment of the Amvrakikos Gulf has been estimated at 95.5 kg for 1995 and 149.3 kg for 1996 (Albanis & Hela, 1998). Peak concentrations of herbicides were observed during their application from May to August (Albanis & Hela, 1998) whereas organochlorine insecticides were found in a stable level in sediments representing a reservoir acting as a source for seawater contamination for years after their usage ceased (Albanis et al., 1995a). Organochlorines have also been detected in eggs of pelicans although concentrations were considered inadequate to cause a risk of harmful effects (Albanis et al., 1995b).

Earlier studies on heavy metal concentrations in sediments of the Amvrakikos Gulf and the adjacent lagoons showed levels of Fe, Cr, Zn, Co, Ni, Cu and Pb comparable to those seen in unpolluted areas of Greece (Voutsinou-Taliadouri & Balopoulos, 1991). More recent studies indicate that Louros River is not heavily polluted by heavy metals although it is clear that the metal burden of the river has increased considerably over the past 20 years (Giokas et al., 2005). The dissolved metal profile (Cd, Pb, Cr, Cu, Zn, Fe) in the wider area of Louros discharge basin shows elevated concentrations near the harbor of Preveza dropping significantly at remote locations (Giokas et al., 2005). Concentrations of V, Cr, Ni and Zn in Amvrakikos lagoon sediments are relatively high but this is attributed to the natural weathering of metal-bearing ultra-basic rocks, rather than anthropogenic activities (Karageorgis, 2007). On the other hand, a study of Cu, Cd and Fe in marine organisms from the estuarine area of Amvrakikos Gulf showed that in some species concentrations ranged into the upper limits of those mentioned in the literature for the Mediterranean (Panayotidis & Florou, 1994).

Studies on community parameters indicate undisturbed conditions in some of the lagoons (Reizopoulou et al., 1996; Nicolaidou et al., 2006), however information on potential effects of chemical contaminants in individual organisms inhabiting Amvrakikos Gulf is lacking.

MATERIALS AND METHODS

Sampling

Wild mussels of similar shell length (46.5±0.8 mm) were collected from two sites in Amvrakikos Gulf, Salaora and Mazoma (Figure 1). The Salaora site is located in the northern part of the Gulf in proximity to the lagoons formed in the delta of Louros and Arachthos rivers, whereas the Mazoma site is located in the western part of the Gulf. Six samplings were carried out during 1996–1998 (Autumn 96, Winter 97, Spring 97, Summer 97, Winter 98 and Spring 98).

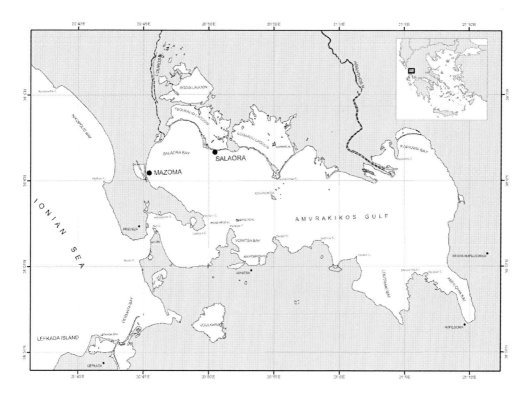

Figure 1. Study area and location of the two sampling sites, Salaora and Mazoma, in Amvrakikos Gulf.

Supplementary mussels that were previously used as reference by our laboratory were collected from a site in East Saronikos Gulf (Agios Kosmas), in Winter 98 and Spring 98 samplings. These were farmed mussels immersed in plastic cages for 1 month, sufficient time for their physiological adjustment to natural environmental conditions. Previous biomarker studies reflect low stress levels in caged mussels at this site (Cotou & Papathanassiou, 1996; Cotou et al., 2002; Tsangaris et al., 2007).

All mussels were transported to the laboratory in moist/cool conditions within 24 hours. Contaminant analyses in the mussel tissues were performed during all samplings. Scope for growth was also measured in all samplings except Spring 98. Biochemical biomarkers and the Microtox bioassay were applied in Winter 98 and Spring 98 samplings.

ANALYTICAL METHODS

Condition Index

The condition index was used as a measure of the nutritional and reproductive status of the mussels. The condition index was calculated according to Bayne et al. (1985) as dry whole tissue weight (g) versus shell cavity volume (ml) X 1,000 per individual.

Chemical Analyses

Chlorinated compounds

The mussel samples for the chlorinated compounds determination were stored at -20°C prior analysis. The soft tissues of mussels were removed and lyophilized. For each station, a composite sample was prepared, consisting from 10 to 40 individuals. A quantity of 5 g from the homogenized dry samples was Soxhlet extracted with a mixture of hexane–dichloromethane 1:1 for 9h. The lipids content determination (as the extractable organic material) was performed in an aliquot of this extract and the rest of it was cleaned up and fractionated by column chromatography with 5% deactivated alumina (Satsmadjis et al., 1988). The eluant was n-hexane. The fractions were concentrated under a stream of pure nitrogen to a final volume of 0.5 ml and the organochlorines were determined in a Varian Star 3400 Cx gas chromatograph equipped with an ECD detector. An optima-δ3 capillary column of 30 m X 0.25 mm i.d. 0.25 μ film thickness was used and the oven temperature program was from 120°C (1 min) to 220°C at 15°C/min and then to 270°C at 4°C/min (10 min). The following compounds were quantified by using external standards: p,p'-DDT, its metabolites p,p'-DDD and p,p'-DDE, lindane, and seven PCB congeners (No 101, 118, 153, 105, 138, 156 and 180). The detection limits were 0.01 ng/g dry wt for the organochlorine insecticides and 0.02 ng/g dw for the PCBs.

Heavy metals

Totally 380 mussel specimens were examined. During each sampling occasion five to six replicates of composite samples consisting of the soft parts of 6 specimens of similar size were prepared. After freeze-drying with a CHRIST GAMMA 1-20 lyophilisator, about 0.5 g of dried tissue was digested with 5.5 ml of HNO3 in a microwave-device (CEM MDS 2100). The drying factor varied as 22–25% of the fresh weight. After digestion the samples were diluted with distilled water to 10 ml. A Varian Spectr AA 20 Plus Atomic Absorption Spectrophotometer with flame was used for the determination of Cu, Cr, Ni, Zn, Fe and Mn concentrations, while Cd was measured with a graphite furnace PERKIN ELMER 4100 Atomic Absorption Spectrophotometer. The accuracy and precision of the analytical methodology was tested with the reference material BCR (No 279, *Ulva lactuca*).

Biomarkers

Scope for growth

The procedure described by Widdows & Salked (1992) with the modification of Cotou et al. (2002) was used to measure SFG. Clearance rate, respiration rate and food absorption efficiency were measured for each mussel individually under laboratory standardized conditions. Prior to all measurements mussels were left undisturbed in the chambers for 12 hours to recover from transportation and resume feeding. For clearance rate and absorption efficiency measurements, mussels were placed in individual chambers in synthetic seawater filtered through 1 μm and 3 μm activated carbon cartridge filters in a closed circulated system at a flow rate of 160–180 mL/min. The unicellular algae *Dunaniella tertiolecta* was added in the chambers at a concentration of 6500–7000 cells/mL. Subsequently algal cell concentrations were measured from the outfall of all chambers including the controls (two chambers without mussels), four times at 45 min intervals using a Z1 Coulter Counter (Coulter Electronics, Luton, UK) calibrated to count particles above 3 μm.

Clearance rate (Cl) was calculated as follows:

$$C_l = Fl \times (C_0 - C_1)/C_0 \text{ where}$$

Fl=flow rate, C_0=inflow concentration (control chamber), C_1=outflow concentration (experimental chamber).

Faeces were collected from the chambers after mussels remained in the system for 12 hours. Feaces produced during the initial 12 hour recovery period were discarded. Absorption efficiency (AE) was measured by the ratio of Conover (1966):

$$AE = (F-E)/((1-E) \times F) \text{ where}$$

F=ash free dry mass/dry mass of food, E=ash free dry mass/dry mass of faeces.

Respiration rate was measured for each mussel in a transparent plastic respirometer (i.e., modified Quickfit chamber) containing 750 mL of air saturated synthetic seawater circulated with a magnetic stirrer. Each mussel was settled in the respirometer for 15 min and subsequently the decline in oxygen tension was monitored over a period of 30–45 min using a Strathkelvin Model 781 oxygen meter (Strathkelvin Instruments, Glasgow, UK).

Physiological rates were converted to energy equivalents and were used to calculate SFG, the energy available for growth and reproduction, according to the equation:

$$\text{Scope for growth (SFG)} = A - R = (C \times \text{absorption efficiency}) - R \text{ where}$$

A = energy absorbed from food, C = energy consumed from food, R = energy lost via respiration.

Energy lost via excretion was not included in the above equation because it consisted less than 5% of the total energy loss. Calculation of C, A and R were as follows:

$$C (J/g/h) = \text{clearance rate (L/g/h)} \times \text{algae cells concentration (mg ash-free dry weight/L)} \times 23.5 \text{ J/mg ash free dry weight}^1$$

$$A (J/g/h) = C \times \text{absorption efficiency}$$

$$R (J/g/h) = \text{respiration rate } (\mu mol\ O_2/g/h) \times 0.456 \text{ J}/\mu mol\ O_2{}^2$$

[1] 23.5 J/mg ash free dry weight is the energy content of phytoplankton cells (Widdows et al., 1979)

[2] 0.456 J/μmol O_2 is the energetic equivalent of respiratory oxygen consumption (Gnaiger, 1983).

Acetylcholinesterase

ACHE activity was measured in the adductor muscle as ACHE has an essential role in nerve transmission processes at neuromuscular junctions and this tissue has been previously used for ACHE measurements in mussels (Bocquené et al., 1993). Samples were prepared as described by Bocquené et al. (1993). Tissues from 4–6 individuals (4–5 replicates per site) were pooled and homogenised using a Potter-Elvehjem homogeniser (Heidolph Electro GmbH, Kelheim, Germany) in 1:5 (w:v) 0.1 M Tris-HCl buffer, pH 7. Homogenates were centrifuged at 15 000g for 20 min. All preparation procedures were carried out at 4°C. ACHE activity was assayed at 412 nm on a Beckman DU-64 spectrophotometer by the colorimetric method of Ellman (Ellman et al., 1961). ACHE activity was expressed as nmol of acetylthiocholine hydrolysed/min/mg protein. Total protein content in the homogenate supernatants was measured by the Bradford method (1976).

Metallothionein

Metallothionein concentration was determined in the digestive gland, which is the proposed target tissue in mussels for MT analysis, according to Viarengo et al. (1997). The method is based on the estimation of the sulphydryl content of MT proteins by spectrophotometric determination of the -SH groups using Ellman's reagent. Samples of 1 g (pooled tissues of 4–6 individuals, 4–5 replicates per site) were homogenized in a Potter-Teflon homogenizer and centrifuged as previously described (Viarengo et al., 1997; Cotou et al., 2001). The MT content was evaluated spectrophotometrically at 412 nm according to Ellman's reaction (Ellman, 1958) utilizing reduced glutathione (GSH) as the reference standard. The amount of metallothionein was calculated assuming an arbitrary SH content of 21 SH/mole with a molecular weight of 7 kDa (Mackay et al. 1993). Metallothionein concentration was expressed as μg MT/g wet weight.

Glutathione peroxidase

GPX was measured as described by Livingstone et al. (1992) in the digestive gland which is the major detoxication tissue in mussels, and where the highest GPX activities are found (Gamble et al., 1995). Digestive glands from 4–6 individuals were pooled (4–6 replicates per site) and homogenised using a Potter-Elvehjem homogeniser in 1:4 (w:v) 20 mM Tris-HCl buffer, pH 7.6, containing 1 mM EDTA, 1 mM dithiothietrol, 0.5 M sucrose and 0.15 M KCl. Homogenates were centrifuged at 500g for 15 min and subsequently at 10 000g for 30 min. All preparation procedures were carried out at 4°C. GPX activity was assayed in the 10 000g supernatant by the rate of NADPH oxidation monitored at 340 nm by the coupled reaction with glutathione reductase using H_2O_2 as a substrate (GSSG + NADPH +H^+ → 2GSH + $NADP^+$, NADPH extinction coefficient = 6.2 mM^{-1}/cm). GPX activity was expressed as nmol NADPH consumed/min/mg protein.

Microtox Bioassay

The Microtox bioassay is based on re-hydrated freeze-dried luminescent bacteria (*Vibrio fischeri*) and measures the effect of potential inhibitors on luminescence under defined conditions. Different chemicals affect the organisms in the Microtox Acute Test Reagent (*Vibrio fischeri*) at different rates, producing distinctly different Light Level-Time Response curves. Multiple light readings (It) at 5, 15 and 30 minutes are recommended for testing samples whose average characteristics are not well documented. As indicated below three types of Light Level-Time Response curves for the Microtox Acute Test Reagent typically occur (Microbics Corporation Manual, 1995). The phenol type, where light output drops sharply, then levels off or rise slightly over time (Figure 2a), the organic compound type where the light output curve drops more gradually than the phenol, then levels off (Figure 2b) and the heavy metal type where the decay rate of the light is essentially constant for an extended period of time and depends on concentration (Figure 2c).

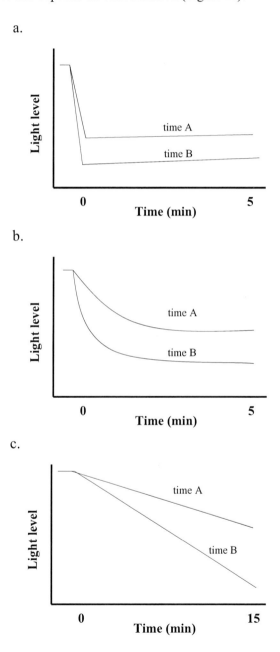

Figure 2. Types of Light Level-Time Response curves for the Microtox Acute Test Reagent (Microbics Corporation Manual, 1995) a. the phenol type, b. the organic compound type, c. the heavy metal type.

Biological fluid was extracted from two to three mussels. Whole bodies were dissected, pooled (1 g tissue per sample) and homogenized in a Potter-Teflon homogenizer with a specially prepared nontoxic solution of 2% NaCl in a ratio 1:3 (w/v). This solution was used to dilute the sample and the reagent (bacteria) and was provided by the supplier. The homogenate was centrifuged at 5000 rpm for 15 min. The liquid phase (supernatant) was tested with the Microtox System. As representative models for the light level-time response curves types for organic compounds and heavy metals we used diazinon and zinc solutions respectively as well as a solution of a mixture of diazinon and zinc. Nontoxic solution of 2% NaCl was used as control. Light outputs were measured every minute for 30 minutes with the Model 500 Analyzer. The bacteria (*Vibrio fischery*) were obtained from AZUR Environmental as freeze-dried lyophilized cells. The presented data are the average of three independent samples determination. The duplicate basic test procedure was used to demonstrate the level of toxicity between the two sites (Salaora, Mazoma). Five concentrations of the supernatant were used in a 1:2 serial dilutions. The Initial Concentration (IC) of the extracted fluid was calculated based on the equation:

$$\text{Sample Concentration (\% SC)} = \text{gram of the wet tissue} \times 100$$
$$IC\% = (\% SC \times \text{ml of sample})/(\text{ml of sample} + \text{ml diluent} + \text{ml reagent})$$

The bacterial light outputs were detected with the Microtox Analyzer (model M 500) at 15 C after 5, 15 and 30 min of exposure. The results were expressed as %EC50 (Effect Concentration 50) which refers to the concentration of the extracted fluid that could cause a 50% decrease in the bacterial light output. EC50 was calculated based on the regression of log-transformed percent effect values using the MicrotoxOmni™ Software. The presented data are the average of three independent determinations of the %EC50 values.

Statistical Analysis

Table 1. Condition index of *M. galloprovincialis* mussels collected from Salaora and Mazoma sites in Amvrakikos Gulf

Sampling period	Site	N	Shell length (mm)	Dry weight (g)	Condition index
Autumn 96	Salaora	7	54.9±1.2	0.476±0.05	48.8±3.7 [a]
	Mazoma	7	54.3±0.8	0.405±0.05	44.7±6.1 [a]
Winter 97	Salaora	6	46.0±0.5	0.656±0.04	107.2±4.4 [b]
	Mazoma	6	49.5±0.8	0.675±0.05	84.9±6.2 [b]
Spring 97	Salaora	7	45.6±0.4	0.580±0.09	97.8±13.9 [b]
	Mazoma	7	46.1±0.5	0.324±0.05	54.3±10.6
Summer 97	Salaora	6	46.0±0.6	0.464±0.11	67.6±14.1
	Mazoma	6	46.5±0.3	0.435±0.06	67.3±6.0 [b]
Winter 98	Salaora	6	36.1±0.7	0.410±0.04	105.8±10.0 [b]
	Mazoma	5	37.3±0.8	0.301±0.03	70.6±8.2

Mean±SE
[a, b]: significant differences (Mann-Witney U test, $P < 0.05$) between sampling periods within each site

Data are presented as mean ± standard error of the mean. The Kolmogorov-Smirnoff test and Levene's test were applied to test normal distribution and homogeneity of variance respectively. Two-way analysis of variance (ANOVA) was applied to determine differences between stations and sampling periods. Pairwise comparisons (t-test) were made to determine which values different significantly when a significant overall ANOVA was found. When there was no homogeneity of variance the Mann-Witney U test was applied. Correlations between SFG and contaminant concentrations were examined by Pearson's correlation coefficient. Statistical analysis was performed using the SPSS statistical package. Significance level was set at P<0.05.

RESULTS

Condition Index

Condition index was similar in mussels from the Salaora and Mazoma sites in all sampling periods (Mann-Witney U test, P>0.05) (Table 1) while seasonal differences were evident with higher values in winter and spring samplings (Mann-Withey U test, P< 0.05).

Contaminant Levels in the Mussel Tissues

Chlorinated compounds

The organochlorine concentrations in the whole soft tissues of mussels from Amvrakikos gulf are presented in Figures 3, 4 and 5. The major organochlorine compounds identified were p,p'-DDT and its metabolites p,p'-DDE and p,p'-DDD. The sum of the concentrations of these compounds (ΣDDTs) ranged between 19.7 and 287.7 ng/g dry wt in Salaora (mean value: 95.7 ng/g) and between 27.9 and 139.5 ng/g in Mazoma (mean value: 66.2 ng/g) (Figure 3). Comparing the two sampling sites, Salaora seems to present higher levels of DDTs with an exception in winter 1997, when very low DDTs concentrations were measured in the mussels from Salaora. In all the samples analyzed, p,p'-DDE was the dominant compound, accounting for 70–90% of the ΣDDTs, while the parent p,p'-DDT accounted for only 3–22% of the ΣDDTs. The highest values of DDTs in both areas were recorded during autumn 1996, while relatively high values were also measured during winter 1998.

The sum of concentrations of the individual polychlorinated biphenyls (PCBs) ranged from 5.2 to 31.5 ng/g dry wt in Salaora (mean value: 12.8 ng/g) and from 3.0 to 16.3 ng/g in Mazoma (mean value: 8.1 ng/g) (Figure 4). It should be mentioned here that the seven CB congeners (IUPAC Nos 101, 118, 153, 105, 138, 156, 180) determined in this study were selected in accordance with the recommendation of ICES and taking into consideration that they occur in relatively high concentrations in the technical PCB mixtures. As in the case of ΣDDTs, the mussels collected from Salaora were more contaminated than those from Mazoma in most sampling periods, whereas the highest values were recorded during autumn 1996. The hexachlorobiphenyls 138 and 153 were always the dominant congeners (average 37% and 27% of the total PCBs) followed by the pentachloro- 118 and 101.

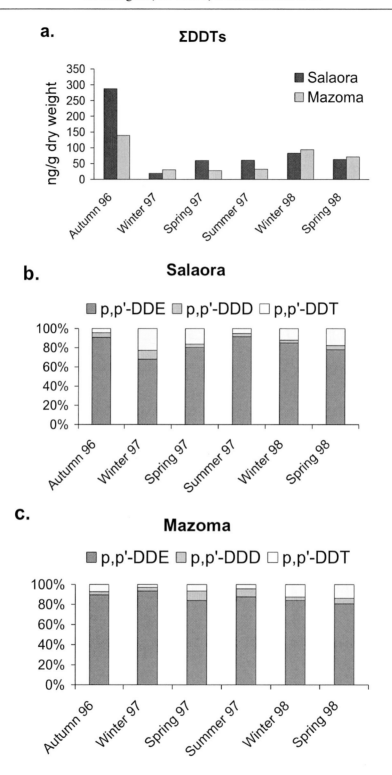

Figure 3. DDT and metabolites concentrations (ng/g dry wt) in the tissues of *M. galloprovincialis* mussels collected from Salaora and Mazoma sites in Amvrakikos Gulf. a. ΣDDTs, b. and c. % of p,p'-DDT p,p'-DDE and p,p'-DDD.

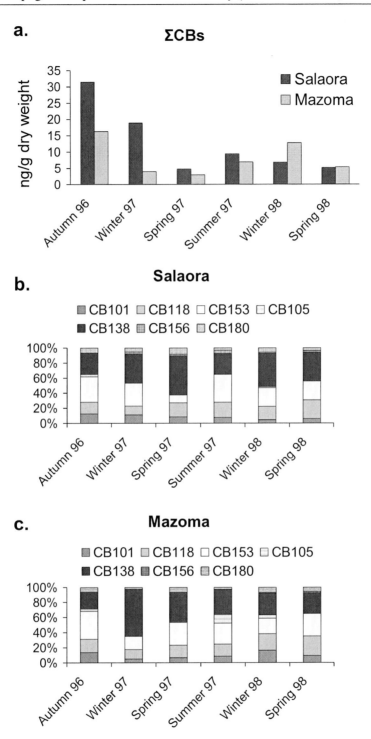

Figure 4. Concentrations of polychlorinated biphenyls (ng/g dry wt) in the tissues of *M. galloprovincialis* mussels collected from Salaora and Mazoma sites in Amvrakikos Gulf. a. ΣCBs, b. and c. % of individual CB congeners.

Lindane was measured in quite low concentrations in all samples (maximum value: 2.3 ng/g dw) (Figure 5).

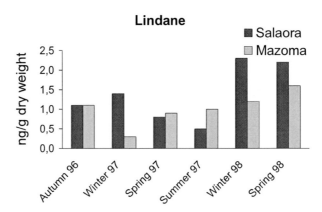

Figure 5. Concentrations of lindane (ng/g dry wt) in the tissues of *M. galloprovincialis* mussels collected from Salaora and Mazoma sites in Amvrakikos Gulf.

Heavy metals

Heavy metal concentrations in the whole soft bodies of mussels from the two sites in Amvrakikos Gulf are shown in Figure 6. Mean concentrations ranged from 0.47 to 8.61 μg/g for Cu, from 0.61 to 5.04 μg/g for Cr, from 1.66 to 8.61 μg/g for Ni, from 43 to 264 μg/g for Zn, from 72 to 626 μg/g for Fe, from 5.2 to 35.5 μg/g for Mn, from 0.37 to 8.57 μg/g for Cd and did not show significant differences between the two sites. However mussels from Salaora tended to present higher maximum concentrations than those from Mazoma, as well as larger concentration ranges. Metal concentrations varied between sampling periods. Cd, Ni and Fe concentrations were highest in Autumn 1996 sampling, whereas Cr and Zn showed highest concentrations in Spring 1998. Cu levels were highest in Winter 98.

Biomarkers

SFG

SFG was lower in mussels from Salaora site compared to mussels from Mazoma site (t-test, P<0.05) (Figure 7) and showed negative values in three sampling periods (-5.2 to -2.3 J/h/g). SFG varied seasonally in all mussels (Two-way ANOVA, P<0.05) with higher values in Spring 97 and this increase was significant in mussels from the Salaora site (t-test, P<0.05). An increase in SFG was also noted in mussels from the Salaora site in the Winter 98 sampling (t-test, P<0.05). SFG showed no significant correlations (P>0.05) with contaminant concentrations in the mussel tissues.

The physiological parameters (i.e., clearance rate, respiration rate, and absorption efficiency) used to calculate the integrated SFG responses are shown in Table 2. Clearance rates were lower in mussels from Salaora site compared to mussels from Mazoma site in three sampling periods (Mann-Witney U test, P<0.05), respiration rates were higher in Salaora site than Mazoma site in four sampling periods (Mann-Witney U test, P<0.05) whereas there were no significant differences in absorption efficiencies among the sites (Mann-Witney U test, P>0.05). Seasonal variations in clearance rates, respiration rates, and absorption efficiencies were also evident as shown in Table 2.

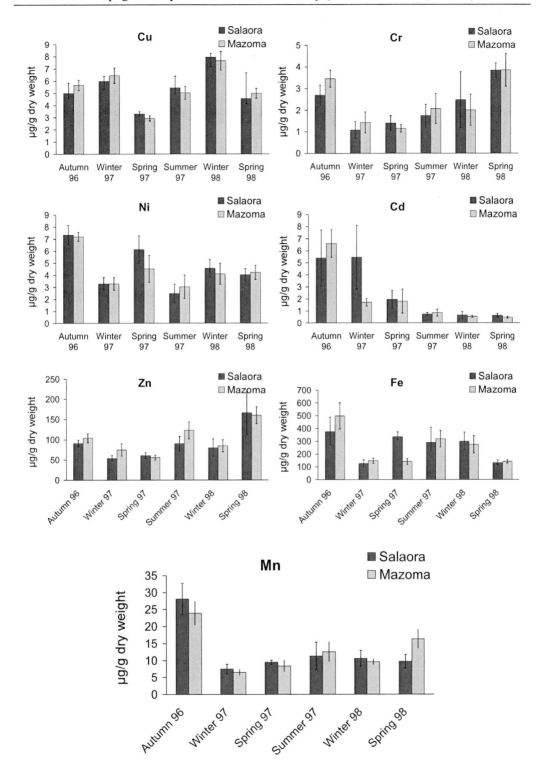

Figure 6. Heavy metal concentrations (μg/g dry weight) in the tissues of *M. galloprovincialis* mussels collected from Salaora and Mazoma sites in Amvrakikos Gulf.

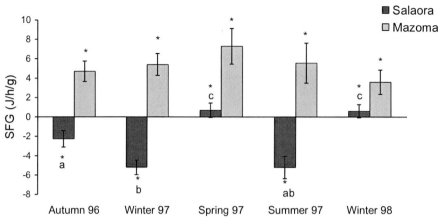

Mean±SE, N=5-7.
* indicate significant difference between the two sites within one sampling period
a b c indicate significant differences between sampling periods for each site

Figure 7. SFG (J/h/g) in *M. galloprovincialis* mussels collected from Salaora and Mazoma sites in Amvrakikos Gulf.

Table 2. Clearance rate (l/g/h), respiration rate (μmol O_2/g/h) and absorption efficiency of *M. galloprovincialis* mussels collected from Salaora and Mazoma sites in Amvrakikos Gulf

Sampling period	Site	N	Clearance rate (l/g/h)	Respiration rate (μmol O_2/g/h)	Absorption efficiency
Autumn 96	Salaora	7	2.96±0.36 [a]	22.19±1.13 [*ac]	0.32±0.30
	Mazoma	7	3.91±0.29 [a]	16.85±2.07 [*]	0.38±0.11
Winter 97	Salaora	6	2.48±0.36 [*]	18.19±0.98 [*b]	0.26±0.04
	Mazoma	6	4.29±0.31 [*a]	12.11±1.29 [*a]	0.52±0.01 [a]
Spring 97	Salaora	7	1.92±0.20 [b]	17.04±1.62 [b]	0.57±0.07
	Mazoma	7	3.71±0.65	23.81±5.58	0.63±0.05 [b]
Summer 97	Salaora	6	1.42±0.30 [*bc]	29.11±3.54 [*c]	0.70±0.01 [a]
	Mazoma	6	2.58±0.41 [*b]	16.13±1.42 [*]	0.62±0.03
Winter 98	Salaora	6	1.23±0.09 [*c]	12.33±0.97 [*d]	0.40±0.04 [b]
	Mazoma	5	2.02±0.24 [*b]	16.63±1.01 [*b]	0.44±0.03 [c]

Mean±SE
*: significant differences (Mann-Witney U test, $P < 0.05$) between the two sites in each sampling period
a, b, c, d: significant differences (Mann-Witney U test, $P < 0.05$) between sampling periods within each site

ACHE activity

ACHE activities were lower in mussels from Salaora site compared to mussels from Mazoma site though differences were not statistically significant (t-test, P>0.05) (Figure 8). ACHE activities in mussels from Amvrakikos Gulf were lower than ACHE activities of reference mussels and these differences were significant for Salaora mussels (t-test, P<0.05) (Figure 8). No significant seasonal differences in ACHE activities were found (Two-way

ANOVA, P>0.05) although in reference mussels ACHE activities tended to increase in the spring (Figure 8).

MT content

MT levels were similar in mussels from Salaora and Mazoma sites and there were no differences in MT levels between these mussels and reference mussels (Two-way ANOVA, P>0.05) (Figure 9). MT levels were higher in the winter sampling (Figure 9).

GPX activity

GPX activities were similar in mussels from Salaora and Mazoma sites but were different from those in reference mussels (Figure 10). In the winter sampling, GPX activities in mussels from the two sites in Amvrakikos gulf were significantly lower than in the reference mussels (t-test, P<0.05) while in the spring sampling they were higher than in the reference mussels and these differences were significant for the Mazoma site (t-test, P<0.05). GPX activities varied with respect to sampling season (Figure 10) and in mussels from Amvrakikos gulf they were higher in the spring sampling (t-test, P<0.05).

Microtox Bioassay

The %EC50 values representing the influence of the biological fluids extracted from the mussels on the luminescence bacteria are shown in Table 3. The %EC50 values of the mussel extracts show significantly higher toxicity for the Salaora samples compared to Mazoma at 5, 15 and 30 min of exposure (t-test, P<0.05). All samples showed higher toxicity in the winter sampling.

The light level response curves produced by the biological fluid extracts of the mussels evidenced organic contamination for both sites as indicated in Figure 11. The light level response curves produced by the mussel extracts were similar to the diazinon type light level response curve (organic type).

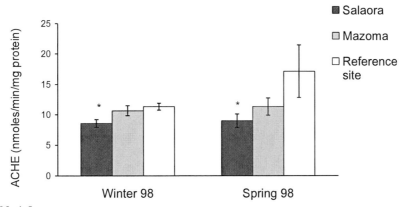

Mean±SE, N=4-5
*indicate significant differences from the reference site within one sampling period

Figure 8. ACHE activities (nmoles/min/mg protein) in *M. galloprovincialis* mussels collected from Salaora and Mazoma sites in Amvrakikos Gulf and a reference site.

Table 3. Toxicity levels expressed as %EC50 of biological fluids extracts of *M. galloprovincialis* mussels collected from Salaora and Mazoma sites in Amvrakikos Gulf

Sampling period	Site	Exposure Time (min)	%EC50	95% Confidence Limits
Winter 98	Salaora	5	1.89	1.72 – 2.07
		15	1.21	1.11 – 1.32
		30	0.54	0.38 – 0.77
	Mazoma	5	3.86	3.54 – 4.22
		15	1.70	1.47 – 1.96
		30	0.96	0.80 – 1.14
Spring 98	Salaora	5	6.32	5.41 – 7.37
		15	3.66	3.23 – 4.14
		30	2.55	2.16 – 3.02
	Mazoma	5	6.95	5.90 – 8.19
		15	7.32	6.12 – 8.76
		30	5.59	5.21 – 6.0

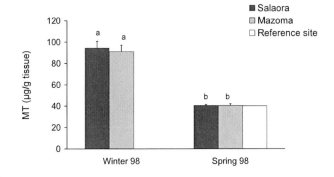

Mean±SE, N=4-5
a b indicate significant differences between sampling periods for each site

Figure 9. MT (μg/ g tissue) in *M.* galloprovincialis mussels collected from Salaora and Mazoma sites in Amvrakikos Gulf and a reference site.

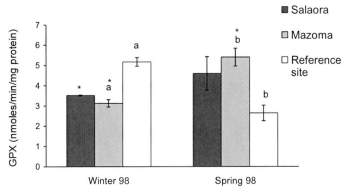

Mean±SE, N=4-6
*indicate significant differences from the reference site within one sampling period
a b indicate significant differences between sampling periods for each site

Figure 10. GPX activities (nmoles/min/mg protein) in *M. galloprovincialis* mussels collected from Salaora and Mazoma sites in Amvrakikos Gulf and a reference site.

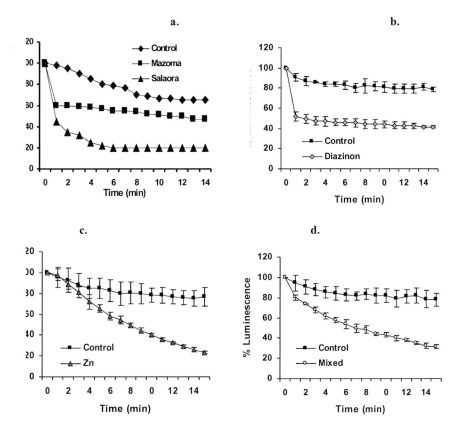

Figure 11. Light level-time response curves produced by the biological fluid extracts of
M. galloprovincialis mussels collected from Salaora and Mazoma sites in Amvrakikos Gulf (a);
and representative light level-time response curves types for organic compounds and heavy metals
produced by diazinon (b); zinc (c); and diazinon-zinc solutions (d).

DISCUSSION

The use of combined measurements of biological effects and contaminant bio-accumulation is recognised as an important approach for the assessment of pollution, as chemical analysis of environmental samples alone does not provide evidence of contaminant impacts in biota. This approach was applied to assess risk from possible pollution impacts in the protected area of Amvrakikos Gulf using mussels as sentinel species. Chemical analyses in the tissues of mussels from Amvrakikos Gulf reveal contamination by organochlorine pesticides (DDTs), while biomarker responses and bioassay results indicate stress conditions.

The DDTs concentrations in Amvrakikos mussels which are clearly higher than those measured in mussels from other Greek coastal regions (Hatzianestis et al., 2000, 2001) indicate moderate pollution. In marine organisms p,p'-DDT easily undergoes a degradative dehydrochlorination, resulting in a less toxic but more persistent compound, the p,p'-DDE which is the main DDT metabolite (Kumblad et al., 2001). Thus the clear predominance of p,p'-DDE in the mussels suggests the absence of "fresh" DDT inputs in the Gulf. Although the use of organochlorine pesticides has been ceased, these compounds are still found in the marine environment of Amvrakikos Gulf (Albanis et al., 1995) and their increased load can be considered as an indication of agricultural related pollution.

The total PCBs concentration levels in Amvrakikos mussels were much lower than DDTs levels and were also lower than those reported in other coastal areas (Picer & Picer, 1994; Hatzianestis et al., 2000). This is in accordance with the absence of major industries in the greater area. The pattern of dominant congeners (hexachlorobiphenyls 138 and 153 followed by the pentachloro- 118 and 101) is commonly found in marine organisms as it is related to enzymatic metabolic processes (Boon & Eijgenraam; 1988).

Mean metal levels in mussels of Amvrakikos Gulf are not considered high since they are similar to those in mussels from other Greek (NCMR, 1998, Tsangaris et al., 2004) and Mediterranean coastal areas (Giordano et al., 1991 & UNEP, 1996). Compared to a previous study in Amvrakikos Gulf (Panayotidis & Florou, 1994), Cu and Cd concentrations in the mussels are lower, while those of Fe are similar to those of the earlier study.

Organochlorine compounds indicate a higher degree of pollution in Salaora compared to Mazoma site and similarly SFG and the Microtox bioassay revealed higher stress conditions in Salaora. SFG represents the energy budget of an organism and provides an integrated measure of stress caused by environmental conditions (Bayne et al., 1985). Negative SFG values indicate high level of stress when the organism utilizes body reserves for maintenance, while maximum positive values indicate optimum environmental conditions. SFG values of mussels from the Salaora site (-5 to 1 J.h-1.g-1) are comparable to values reported in the literature for high levels of pollution stress (<5 J.h-1.g-1: Widdows et al., 1995; Widdows et al., 1997; Widdows et al., 2002) while at the Mazoma site SFG values (4 to 7 J.h-1.g-1) correspond to literature values reflecting moderate stress levels (5-15 J.h-1.g-1: Widdows et al., 1995; Widdows et al., 1997; Widdows et al., 2002). The Microtox bioassay measures luminescence inhibition of the bacteria *Vibrio fischeri* by chemical substances. Reduction in luminescence indicates a stress response that reflects the presence of contaminants. When applied on the biological extracts of bivalves, luminescence reduction is related to the degree of their toxic content (Lau-Wang, 1990). Thus low EC50 values in the Microtox bioassay are indicative of stressed areas while high EC50 values denote from low to none pressure conditions (Cotou et al., 2002). The lower SFG and EC50 values found in mussels from Salaora site compared to Mazoma site are overall in agreement with generally higher levels of organochlorine contaminants in the tissues of these mussels. On the other hand, the lack of correlations between SFG or EC50 values and the concentrations of contaminants in the mussel tissues indicate that additional stressors may have affected both parameters. Luminescence reduction is mainly influenced by chemical contaminants in the mussel extracts whereas SFG is affected both by chemical contaminants, and biotic and abiotic factors such as reproductive condition, food availability, temperature (Bayne et al., 1985). Therefore the condition index was used to examine the nutritional and reproductive state of mussel populations from the two sites. The similar condition index values in mussels from Salaora amd Mazoma sites suggest that the differences in stress levels shown by SFG are related to variations in contaminant levels and not to different nutritional or reproductive condition of the mussels. It has to be noted that in addition to organochlorine pesticides that were moderately elevated in the mussel tissues, various types of pesticides including triazines, organophosphates, carbamates and substituted ureas are found in significant amounts in sediments of Amvrakikos Gulf (Albanis et al., 1995; Hela et al., 2000). Such pesticides also negatively affect SFG (El Shenawy et al., 2006) and bacteria luminescence (De Lorenzo et al., 2001). Consequently the lack of correlations between organochlorine compound levels and SFG or Microtox EC50 values is likely due to the presence of other contaminants such as currently used pesticides taken up by the mussels that additionally influenced SFG and bacteria luminence.

Biochemical biomarkers showed no differences between the two sites in Amvrakikos Gulf nevertheless ACHE and GPX revealed differences in comparison to reference mussels. Due to the lack of a reference native population in Amvrakikos Gulf, mussels of aquaculture origin that were caged at a reference site in Saronikos Gulf served as reference mussels for ACHE, GPX and MTs, as previous biomarker measurements of our laboratory reflect low stress levels in these mussels (Cotou & Papathanassiou, 1996; Cotou et al., 2002; Tsangaris et al., 2007). The decreased ACHE activities in mussels from Amvrakikos Gulf compared to the reference mussels indicate ACHE inhibition by organophosphate and carbamate pesticides that are found in the area (Albanis & Hela, 1998; Hela et al., 2000). GPX responses in mussels from Amvrakikos Gulf can be related to elevated concentrations of DDT and relevant metabolites in the mussel tissues since DDTs are compounds able to stimulate ROS production (Lemaire & Livingstone, 1993; Livinstone, 2001). GPX has been previously correlated with organochlorinated compound body concentrations including DDTs, in mussels and other bivalve species (Solé et al., 1994). GPX is a component of a complex antioxidant defense system and its response is possibly accompanied by responses of other antioxidant enzymes and scavenger molecules, nevertheless its activity variations may provide an indication of presence of contaminants capable of ROS production (Cossu et al., 1997; Doyotte et al., 1997; Reid & MacFarlane, 2003). GPX activities in mussels from Amvrakikos Gulf were lower than in the reference mussels in the winter and higher in the spring suggesting inhibition in the winter and induction in the spring. Similar results have been reported for the antioxidant enzyme catalase in mussels from polluted sites in the Venice Lagoon (Nesto et al., 2004). Antioxidant enzymes can be induced by enhanced production of ROS as a protection mechanism against oxidative stress or inhibited when deficiency of the system occurs predicting toxicity (Winston & Di Giulio 1991; Cossu et al., 1997). During winter decrease in activity of the antioxidant defence system corresponds to an enhanced susceptibility of mussels to oxidative stress in this period (Sheehan & Power, 1999). GPX responses in mussels from Amvrakikos Gulf suggest susceptibility to oxidative stress in the winter and compensation in the spring. This is in agreement with SFG results showing poor overall health condition in the winter and improved condition in the spring. The high GPX activity observed in the spring may be related to intense overall metabolism induced by the increased availability of food. On the contrary, MTs in mussels from Amvrakikos Gulf showed similar low levels to those of the reference mussels in agreement with the low metal tissue concentrations also suggesting low metal contamination in the area. MT values of mussels from Amvrakikos Gulf were similar to values obtained from a previous study in mussels from coastal areas of the Saronikos Gulf that showed metal concentrations comparable to those found in non-polluted coastal areas (Tsangaris et al., 2004).

The light level-time response curves of the Microtox bioassay on the biological fluids of the mussels provided additional evidence of organic contamination for both sites in Amvrakikos Gulf in accordance with contaminant tissue concentrations and biochemical biomarker results.

The seasonal trends observed in Amvrakikos Gulf for SFG and GPX with higher values in the spring sampling (early June) and for MT with higher values in the winter sampling

(February) are in accordance with previous studies (Bayne & Widdows, 1978; Sole et al., 1995; Raspor et al., 2004; Ivanković et al., 2005; Bocchetti & Regoli, 2006). Seasonal trends in contaminants levels measured in the mussels tissues were not related to those of biomarkers indicating influences of other factors on biomarker seasonality. Seasonal variation in biomarkers can be attributed to changes in a variety of abiotic and biotic factors such as temperature, food supply and reproductive status including inputs of contaminants (Sheehan & Power, 1999). The fact that ACHE levels were not increased in the spring in Amvrakikos gulf as would be expected in a non contaminated environment (Escartin & Porte, 1997) can be attributed to increased concentrations of pesticides in the area during this season just after their application in the agricultural fields (Albanis et al., 1995; Albanis & Hela, 1998). Similar results on seasonal ACHE activities were reported by Moreira & Guilhermino (2005) at a site in the Portuguese coast influenced by pesticide contamination.

CONCLUSION

The combined use of Scope for Growth, biochemical markers of exposure, the Microtox bioassay and bioaccumulation levels in *M. galloprovincialis* was a useful approach to assess contaminant impacts in the estuarine environment of Amvrakikos Gulf. Our results show stress effects that may provide an early warning signal of potential consequences at higher levels of biological organization. Chemical analyses in the mussel tissues imply increased pesticide concentrations and low metal levels. The negative energy budgets in mussels from the Salaora site indicate that the population is under severe stress for most of the year and thus might be at risk, while at the Mazoma site the population is still capable of normal growth and reproduction as demonstrated by the positive, although low, energy budgets of these mussels throughout the year. The EC50 values of the Microtox bioassay indicate higher stress caused by chemical contaminants in the biological extracts of these mussels at the Salaora site. Biochemical biomarker responses (ACHE, MT and GPX) and the light level-time response curves of the Microtox bioassay are consistent with chemical data indicating exposure to organic contaminants, possibly pesticides. These findings point out the need for biomonitoring in Amvrakikos Gulf with particular focus on pesticide contamination, since although the area was declared a National Park in 2008, contamination via riverine transport of agricultural contaminants still occurs.

ACKNOWLEDGMENTS

This research was performed in the framework of the programme "Development and Implementation of New Techniques for Pollution Control, Protection and Management of Amvrakikos Gulf" financed by the General Secretariat for Research and Technology, Ministry of Development, Greece.

REFERENCES

Albanis, T. A., Danis, T. G. & Hela, G. H. (1995a). Transportation of pesticides in estuaries of Louros and Arachthos rivers (Amvrakikos gulf, N.W. Greece). *The Science of the Total Environment*, *171*, 85-93.

Albanis, T. A., Hela, D. G. & Hatzilakos, D. (1995b). Organochlorine residues in eggs of *Pelicanus crispus* and its prey in wetlands of Amvrakikos Gulf, North-Western Greece. *Chemosphere*, *31*, 4341-4349.

Albanis, T. A. & Hela, D. G. (1998). Pesticide concentrations in Louros river and their fluxes into the marine environment. *International Journal of Environmental and Analytical Chemistry*, *70*, 105-120.

Bayne, B. L. & Widdows, J. (1978). The physiological ecology of two populations of *Mytilus edulis*. *Oecologia*, *37*, 137-162.

Bayne, B. L., Brown, D. A., Burns, K., Dixon, D. R., Ivanovici, A., Livingstone, D. R., Lowe, D. M., Moore, M. N., Stebbing, A. R. D. & Widdows J. (1985). *The effects of stress and pollution on marine animals.* New York: Praeger Pres.

Bearzi, G., Agazzi, S., Bonizzoni S., Costa, M. & Azzelino, A. (2008). Dolphins in a bottle: abundance, residency patterns and conservation of bottlenose dolphins *Tursiops truncatus* in the semi-closed eutrophic Amvrakikos Gulf, Greece. *Aquatic Conservation: Marine and Freshwater Ecosystems*, *18*, 130-146.

Bihari, N., Fafanđel, M. & Piškur V. (2007). Polycyclic aromatic hydrocarbons and ecotoxicological characterization of seawater, sediment, and mussel *Mytilus galloprovincialis* from the Gulf of Rijeka, the Adriatic Sea, Croatia. *Archives of Environmental Contamination and Toxicology*, *52*, 379-387.

Bocchetti, R. & Regoli, F. (2006). Seasonal variability of oxidative biomarkers, lysosomal parameters, metallothioneins and peroxisomal enzymes in the Mediterranean mussel *Mytilus galloprovincialis* from Adriatic Sea. *Chemosphere*, *65*, 913-921.

Bocquené, G. & Galgani. F. (1991). L'acétylcholinésterase chez les organismes marins, outil de surveillance des effets des pesticides organophosphorés et carbamates. *Océanis*, *17*, 439-448.

Bocquené, G., Galgani, F., Burgeot, T., Le Dean, L. & Truquet, P. (1993). Acetylcholinesterase levels in marine organisms along French coasts. *Marine Pollution Bulletin*, *26*, 101-106.

Boon, J. P. & Eijgenraam, F. (1988). The possible role of metabolism in determining patterns of PCB congeners in species from the Dutch Wadden Sea. *Marine Environmental Research*, *24*, 3-8.

Bradforb, M. (1976). A rapid and sensitive method for the quantitation of microgram quantities of protein utilizing the principle of protein-dye binding. *Analytical Biochemistry*, *772*, 248-64.

Conover, R. J. (1966). Assimilation of organic matter by zooplankton. *Limnology and Oceanography*, *11*, 338–354.

Cossu, C., Doyotte, A., Jacquin, M. C., Babut, M., Exinger, A. & Vasseur, P. (1997). Glutathione reductase, selenium-dependent glutathione peroxidase, glutathione levels and lipid peroxidation in freshwater bivalves, *Unio tumidus*, as biomarkers of aquatic contamination in field studies. *Ecotoxicology and Environmental Safety*, *38*, 122-131.

Cotou, E. & Papathanassiou, E. (1996). Physiological responses of marine indicator organisms to global pollution. In *MAP Technical Report Series*, *103* (43-55). Athens, UNEP.

Cotou, E., Papathanassiou, E. & Tsangaris, C. (2002). Assessing the quality of marine coastal environments: comparison of scope for growth and Microtox ® bioassay results of pollution gradient areas in eastern Mediterranean (Greece). *Environmental Pollution*, *119*, 141-149.

Cotou, E., Vagias, C., Rapti, T. & Roussis, V. (2001). Metallothionein levels in the bivalves *Callista chione* and *Venus verrucosa* from two Mediterranean sites. *Zeitschrift fuer Naturforschung, C. Biosciences*, *56*, 848-852.

Crivelli, A. J., Hatzilacou, D. & Catsadorakis, G. (1998). The breeding biology of the Dalmatian Pelican *Pelecanus crispus*. *Ibis*, *140*, 472-481.

Damiens, G., Gnassia-Barelli, M., Loquès, F., Roméo, M. & Salbert, V. (2007). Integrated biomarker response index as a useful tool for environmental assessment evaluated using transplanted mussels. *Chemosphere*, *66*, 574-583.

DeLorenzo, M. E., Scott, G. I. & Ross P. E. (2001). Toxicity of pesticides to aquatic microorganisms: A review. *Environmental Toxicology and Chemistry*, *20*, 84-98.

Diapoulis, A., Koussouris, T. H, Bertahas, I. & Photis, G. (1991). Ecological stresses on a delta area in western Greece. *Toxicological & Environmental Chemistry*, *31*, 285-290.

Dimosthenis L. Giokas, D. L., Evangelos K., Paleologos, E. K. & Miltiades I. Karayannis, M. I. (2005). Optimization of a multi-elemental preconcentration procedure for the monitoring survey of dissolved metal species in natural waters. *Analytica Chimica Acta*, *537*, 249-257.

Doyotte, A., Cossu, C., Jacquin, M. C., Babut, M. &Vasseur, P. (1997). Antioxidant enzymes, glutathione and lipid peroxidation as relevant biomarkers of experimental or field exposure in the gills and the digestive gland of the freshwater bivalves *Unio tumidus*. *Aquatic Toxicology*, *39*, 93-110.

Ellman, G. L. (1958). A colorimetric method for determining low concentrations of mercaptans. *Archives of Biochemistry and Biophysics*, *74*, 443-450.

Ellman, G. L., Courtney, K. D., Andres, V. Jr. & Featherstone, R. M. (1961). A new and rapid colorimeric determination of acetylcholinesterase activity. *Biochemical Pharmacology*, *7*, 88-95.

El-Shenawy, N. S., Greenwood, R., Ismail, M., Abdel-Nabi, I. M. & Nabil, Z. I. (2006) Effect of atrazine and lindane on the scope for growth of marine mussel *Mytilus edulis*. *Acta Zoologica Sinica*, *52*, 712-723.

Escartin, E. & Porte, C. (1997). The use of cholinesterase and carboxylesterase activities from *Mytilus galloprovincialis* in pollution monitoring. *Environmental Toxicology and Chemistry*, *16*, 2090-2095.

Galloway, T., Brown, R. J., Browne, M., Dissanayake, A., Lowe, D., Jones, M. B. & Depledge, M. (2004). A multibiomarker approach to environmental assessment. *Environmental Science and Technology*, *38*, 1723-1731.

Gamble, S. C., Goldfarb, P. S., Porte, C. & Livingstone, D. R. (1995). Glutathione peroxidase and other antioxidant enzyme function in marine invertebrates (*Mytilus edulis*, *Pecten maximus*, *Carcinus maenas* and *Asterias rubens*). *Marine Environmental Research*, *39*, 191-195.

Giordano, R., Arata P., Rinalda S., Giani M., Cicero A. M. & Costantini S. (1991). Heavy metals in mussels and fish from Italian coastal waters. *Marine Pollution Bulletin*, *22*, 10-14.

Gnaiger, E. (1983). Heat dissipation and energetic efficiency in animal anoxibiosis: economy contra power. *Journal of Experimental Zoology*, *228*, 471-490.

Guelorget, O., Frisoni, G. F., Monti, D. & Perthuisot, J. P. (1986). Contribution a l'etude ecologique des lagunes septentrionales de la baie d'Amvrakia (Grece). *Oceanologica Acta*, *9*, 9-17.

Hatzianestis, I., Sklivagou, E. & Georgakopoulou, E. (2000) Organic contaminants in sediments and mussels from Thermaikos gulf, Hellas, *Toxicological and Environmental Chemistry*, *74*, 203-216

Hatzianestis, I., Sklivagou, E. & Georgakopoulou, E. (2001). Levels of chlorinated compounds in marine organisms from Greek waters. In *Proceedings of the 7th international conference on environmental science and technology*, Ermoupolis, Syros I, Greece, 3-6 September (*Vol. 1*, 310-317).

Hela, D. G., Albanis, T. & Anagnostou, C. (2000). Concentration levels of pesticide residues in coastal sediments of Amvrakikos Gulf. In *Proceedings of the 6th Hellenic symposium on oceanography and fisheries*, Chios, Greece, 23-26 May (*Vol. 1*, 517-522).

Karageorgis, A. (2007). Geochemical study of sediments from the Amvrakikos Gulf lagoon complex, Greece. *Transitional Waters Bulletin*, *3*, 3-8.

Kumblad, L., Olsson, A., Koutny, V. & Berg, H. (2001). Distribution of DDT residues in fish from the Songkhla Lake, Thailand. *Environmental Pollution*, *112*, 193-200.

Lau-Wong M. M. (1990). Assessing the effectiveness of depuration of polluted clams and mussels using the Microtox bioassay. *Bulletin of Environmental Contamination and Toxicology*, *44*, 876-833.

Lemaire, P. & Livingstone, D. R. (1993). Pro-oxidant/antioxidant processes and organic xenobiotic interactions in marine organisms, in particular the flounder *Platicthys flesus* and the mussel *Mytilus edulis*. Trends in *Comparative Biochemistry and Physiology*, *1*, 119-1150.

Livingstone, D. R. (2001). Contaminant-stimulated reactive oxygen species production and oxidative damage in aquatic organisms. *Marine Pollution Bulletin*, *42*, 656-666.

Livingstone, D. R., Lips, F., Garcia Martinez, P. & Pipe, R. K. (1992). Antioxidant enzymes in the digestive gland of the common mussel *Mytilus edulis*. *Marine Biology*, *112*, 265-276.

Mackay, E. A., Overnell, J., Dumbar, B., Davidson, I., Hunziker, P. E. & Kägi, J. H. (1993). Complete amino acid sequences of five dimeric and four monomeric forms of metallothionein from the edible *Mytilus edulis*. *European Journal of Biochemistry*, *218*, 183-194.

Microbics Corporation, 1995. Microtox® Acute Toxicity Basic Test Procedures, Entire Manual© 6-19-95, 1-63

Moreira, S. M. & Guilhermino, L. (2004). The use of *Mytilus galloprovincialis* acetylcholinesterase and glutathione S-transferases activities as biomarkers of environmental contamination along the Northwest Portuguese coast. *Environmental Monitoring and Assessment*, *105*, 309-325.

NCMR (1998). Pollution research and monitoring programme in Saronikos Gulf. MED-POL programme. *Technical Report*, Athens, November 1998, 10-94.

Nesto, N., Bertoldo, M., Nasci, C., & Da Ros, L. (2004). Spatial and temporal variation of biomarkers in mussels (*Mytilus galloprovincialis*) from the Lagoon of Venice, Italy. *Marine Environmental Research, 58,* 287-291.

Nicolaidou, A., Petrou, K., Kormas, K. A. & Reizopoulou, S. (2006). Inter-Annual Variability of Soft Bottom Macrofaunal Communities in Two Ionian Sea Lagoons. *Hydrobiologia, 555,* 89-98.

Panayotidis, P. & Florou, H. (1994). Copper, cadmium and iron in marine organisms in a eutrophic estuarine area (Amvrakikos Gulf, Ionian Sea, Greece). *Toxicological and Environmental Chemistry, 45,* 211-219.

Panayotidis, P., Pancucci, A., Balopoulos, E. & Gotsis-Skretta, O. (1994). Plankton Distribution Patterns in a Mediterranean Dilution Basin: Amvrakikos Gulf (Ionian Sea, Greece). *Marine Ecology, 15,* 93-104.

Peakall, D. B. (1994).The role of biomarkers in environmental assessment (1). Introduction. *Ecotoxicology, 3,* 157-160.

Picer, M. & Picer, N. (1994). Levels and long-term trends of polychlorinated biphenyls and DDTs in mussels collected from the middle Adriatic coastal waters. *Chemosphere, 29,* 465-475.

Raspor, B., Dragun, Z., Erk, M., Ivankovič, D. & Pavičič, J. (2004). Is the digestive gland of *Mytilus galloprovincialis* a tissue of choice for estimating cadmium exposure by means of metallothioneins? *The Science of the Total Environment, 333,* 99-108.

Readman, J. W., Albanis, T. A., Barcelo, D., Galassi, S., Tronczynski, J. & Gabrielides, G. E. (1993). Herbicide Contamination of Mediterranean Estuarine Waters: Results from a MED POL Pilot Survey. *Marine Pollution Bulletin, 26,* 613-619.

Rees A. F. & Margaritoulis D. (2006). Telemetry of loggerhead turtles (*Caretta caretta*) in Amvrakikos Bay, Greece. In Pilcher N. J., (Compiler), *Proceedings of the twenty-third annual symposium on sea turtle biology and conservation.* (235-238), NOAA Technical Memorandum NMFS-SEFSC-536.

Reid, D. J. & Mac Farlane, G. R. (2003). Potential biomarkers of crude oil exposure in the gastropod mollusc *Austrocochlea porcata*: Laboratory and manipulative field studies. *Environmental Pollution, 123,* 147-155.

Reizopoulou, S., Thessalou-Legaki, M. & Nicolaidou, A. (1996). Assessment of disturbance in Mediterranean lagoons: an evaluation of methods. *Marine Biology, 125,* 189-197.

Sarika, M., Dimopoulos, P. & Yannitsaros, A. (2005). Contribution to the knowledge of the wetland flora and vegetation of Amvrakikos Gulf, W Greece. *Willdenowia, 35,* 69-85.

Satsmadjis, J., Georgakopoulos-Gregoriades, E. & Voutsinou-Taliadouri, F. (1988). Separation of organochlorines on alumina, *J. Chromatogr., 437,* 254-259.

Sheehan, D. & Power, A. (1999). Effects of seasonality on xenobiotic and antioxidant defence mechanisms of bivalve molluscs. *Comparative Biochemistry and Physiology, 123C,* 193-199.

Solé, M., Porte, C. & Albaiges, J. (1995). Seasonal variation in the mixed-function oxygenase system and antioxidant enzymes of the mussel *Mytilus galoprovonciallis*. *Environmental Toxicology and Chemistry, 14,* 157-164.

Solé, M., Porte, C. & Albaiges, J. (1994). Mixed function oxygenase components and antioxidant enzymes in different marine bivalves: Its relation with contaminant body burdens. *Aquatic Toxicology, 30,* 271-283.

Tsangaris, C., Papathanassiou & Cotou, E. (2007). Assessment of the impact of heavy metal pollution from a ferro-nickel smelting plant using biomarkers. *Ecotoxicology and Environmental Safety, 66*, 232-243.

Tsangaris, C., Strogyloudi, E. & Papathanassiou, E. (2004). Measurement of biochemical markers of pollution in mussels *Mytilus galloprovincialis* from coastal areas of the Saronicos Gulf (Greece). *Mediterranean Marine Science, 5*, 175-186.

UNEP (1996). Assesement of the state of pollution in the Mediterranean Sea by zinc, copper and other compounds. MAP Technical Report Series, 105, (65-70), Athens, UNEP.

Viarengo, A., Burlando, B., Dondero, F., Marro, A. & Fabbri, R. (1999). Metallothionein as a tool in biomonitoring programmes. *Biomarkers, 4*, 455-466.

Viarengo, A., Lowe, D., Bolognesi, C., Fabbri, E. & Koehler, A. (2007). The use of biomarkers in biomonitoring: A 2-tier approach assessing the level of pollutant induced stress syndrome in sentinel organisms. *Comparative Biochemistry and Physiology Part C, 146*, 281-300.

Viarengo, A., Ponzano, E., Dondero, F. & Fabbri, R. (1997). A simple spectrophotometric method for metallothionein evaluation in marine organisms: Application to mediterranean and antartic molluscs. *Marine Environmental Research, 44*, 69-84.

Voutsinou-Taliadouri, F. & Balopoulos, E. T. (1991). Geochemical and Physical Oceanographic Aspects of the Amvrakikos Gulf (Ionian Sea, Greece). *Toxicological and Environmental Chemistry, 31/32*, 177-185.

Walker, C. H., Hopkin, S. P., Sibly, R. M. & Peakal, D. B. (2006). *Principles of Ecotoxicology*, Third Edition. Boca Raton, FL: CRC Press,Taylor & Francis Group.

Widdows, J. & Donkin, P. (1992). The mussel *Mytilus*: Ecology, Physiology, Genetics and Culture. In E. M. Gosling (Ed.), *Mussels and environmental contaminants: bioaccumulation and physiological aspects.* (1-589), Amsterdam: Elsevier Press.

Widdows, J. & Salked, P. (1993). Practical procedures for measurement of Scope for Growth. In UNEP/FAO/IOC, *Selected techniques for monitoring biological effects of pollutants in marine organisms. MAP technical report series 71*, (147-172). Athens, UNEP.

Widdows, J., Donkin, P., Brinsley, M. D., Evans, S. V., Salked, P. N., Franklin, A., Law, R. J. & Waldock, M. J. (1995). Scope for growth and contaminant levels in North Sea mussels *Mytilus edulis*. *Marine Ecology Progress Serie, 127*, 131-148.

Widdows, J., Donkin, P., Staff, F. J., Matthiessen, P., Law, R. J., Allen, Y. T., Thain, J. E., Allchin, C. R. & Jones, B. R. (2002). Measurement of stress effects (scope for growth) and contaminant levels in mussels (*Mytilus edulis*) collected from the Irish Sea. *Marine Environmental Research, 53*, 327-356.

Widdows, J., Fieth, P. & Worrall, C. M. (1979). Relationships between seston, available food and feeding activity in the common mussel *Mytilus edulis*. *Marine Biology, 50*, 195-207.

Widdows, J., Nasci, C. & Fossato, V. U. (1997). Effects of pollution on the scope for growth of mussels (*Mytilus galloprovincialis*) from the Venice Lagoon. Italy. *Marine Environmental Research, 43*, 69-79.

Winston, G. W. & Di Giulio, R. T. (1991). Prooxidant and antioxidant mechanisms in aquatic organisms. *Aquatic Toxicology, 19*, 137-161.

In: Encyclopedia of Environmental Research
Editor: Alisa N. Souter

ISBN: 978-1-61761-927-4
© 2011 Nova Science Publishers, Inc.

Chapter 41

SANTOS AND SÃO VICENTE ESTUARINE SYSTEM: CONTAMINATION BY CHLOROPHENOLS, ANAEROBIC DIVERSITY OF DEGRADERS UNDER METHANOGENIC CONDITION AND POTENTIAL APPLICATION FOR EX-SITU BIOREMEDIATION IN ANAEROBIC REACTOR

Flavia Talarico Saia

Laboratory of Biological Processes, Department of Hydraulic and Sanitation,
São Carlos School of Engineering, University of São Paulo, Brazil

1. INTRODUCTION

1.1. Area Characterization

Estuarine System: Geographical and Morphological Characteristics; History of the Estuarine Contamination and its Environmental Restoration Programme

The Santos-São Vicente Estuarine System is located in the central coast of São Paulo State, Brazil, in a metropolitan region known as Baixada Santista. This region has a permanent population of over 1,200,000 and an estimated fluctuating population of over 780,000 (Hortellani et al., 2005), and comprehends the municipalities of Santos, São Vicente, Praia Grande, Cubatão and Guarujá (Figure 1). The area bears the largest commercial harbour in Latin America, the Port of Santos, as well as one of the most important petrochemical, chemical and metallurgical industrial complexes of Brazil, the region of Cubatão city. Baixada Santista region has become famous as one of the most relevant examples of degradation of coastal environments as a consequence of anthropogenic activities (Cetesb, 2001).

Climate in the region is hot and wet, with a yearly average temperature of 22°C and relative humidity rates that reach more than 80% throughout the year. Rainfall rates range between 2,000 to 2,500mm, with a more intense regime in summer months in the south hemisphere - January to March (Cantagalo et al., 2008).

The Santos-São Vicente estuarine system is confined between the Serra do Mar cliff and the Atlantic Ocean. Rivers coming down from the Serra do Mar flow fast and intensely, losing energy when they reach the plains, which have little or no declivity. As a result, a complex pattern of streams and creeks are formed, transforming vast regions in wetlands covered by mangroves. The São Vicente Estuary is located at the Western outlet of the system and the Santos Estuary, where the Santos Port is situated, lies at the eastern outlet. Both estuaries are interconnected at the upper part of the system (Hortellani et al., 2005; Rodrigues et al., 2009), where another port can be found, the Cubatão Maritime Terminal, a private port to the *Companhia Siderurgica Paulista* – Steelworks Company of São Paulo (COSIPA).

The hydrodynamic in the region is influenced by the tides and water currents, as well as by the winds and fluvial discharge. At the upper part of the estuarine system, the water and solid fluxes from the mountain and the rivers Cubatão, Perequê, Jurubatuba and Quilombo predominate. However, the mangrove belt that surrounds the estuary retains most of the transported material, and only the suspended matter reaches the estuarine channels (Fulfaro & Ponçano, 1976). At the lower part of the system, the movement of the tides and its currents are the factors that most interfere in the sedimentation process. High sedimentation rates occur only in isolated sites, as Bertioga Channel, Caneus Lake and south edges of São Vicente and Port of Santos channels (Fulfaro & Ponçano, 1976). Navigation to and from the Port of Santos is guaranteed by periodic dredging along the channel main axis, in order to maintain a depth of more than 10m (Hortellani et al., 2005).

The environmental characteristics of the Baixada Santista make this region poorly capable of absorbing impacts caused by anthropogenic activities. Nevertheless, the region has undergone cycles of intense urban and industrial development since the end of the 19th Century, which resulted in a critical state of environmental degradation.

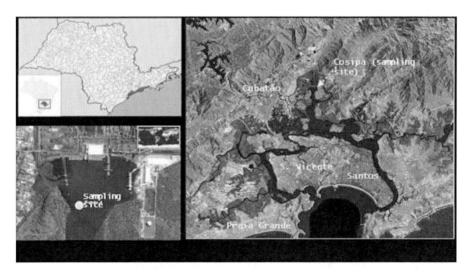

Figure 1. (a) Map of Brasil (small white figure) showing São Paulo State (highlighted in red) and the location of Santos São Vicente Estuary in São Paulo State (highlighted in red), (b) Satellite image of Santos and São Vicente Estuary (c) Sediment sampling site at Port of Cosipa (Miranda & Coutinho, 2004 *apud* Domingues 2007).

The intense process of urbanization and industrialization on the banks of the estuary, mainly over the last 50 years, has been responsible for the degradation of the mangrove vegetation, the emission of industrial effluents and the discharge of domestic sewage and solid residues. In 1999, the total population of the cities of Santos and São Vicente was about 800,000 (IBGE, 2000) of which 90% benefited from sewage systems. The other 10% released untreated wastewater into drains or channels flowing out into the estuary. Domestic sewage with a daily discharge rate of 369,038 m^3 and an organic load of 110,712 kg (MMA 2001 apud Bicego et al. 2006) also represents a substantial contribution to the degradation of the Santos -São Vicente estuarine system. Furthermore, tourism activities in the region bring twice the number of visitors than that of residents during the summer months and holidays, with a consequent increase in the number of vehicles (Martins et al., 2007).

The activities of the port of Santos and Cubatão industrial complex also represent an important contamination source to the region. In 1995, the port handled over 35 million tons of freight and around 50% of the total container shipping volume of the country (Zanardi et al., 2000). The harbour effluents have been dumped directly into the estuary for several decades, thus leading to the contamination of water, sediments and biota. Because of the intense harbour activities, more than 5,000 tons of sediment must be dredged every year, and the resulting sludge is disposed of in the ocean. As a consequence, persistent organic pollutants (POPs) such as organochlorine compounds (OCs), petroleum hydrocarbon and polycyclic aromatic hydrocarbons (PAHs) are present all over the Santos - São Vicente estuarine environment (Bícego et al. 2006; Piza et al., 2004; Rodrigues et al., 2009). The Cubatão industrial complex, in the Cubatão river basin, extends towards the estuary and comprises 23 medium and large-sized factories including a steel mill, an oil refinery, and fertilizer, cement and chemical/petrochemical plants summing up to 260 pollutant emission sources (CETESB, 1999). Monthly discharges of pollutants such as zinc, phenol and mercury reach values of 100,000kg in the waters of the estuarine system (Zanardi et al., 2000).

The impacts originated from the industrial development in the region include the removal of more than 80% of the vegetation in Serra do Mar cliff, the degradation of mangrove areas in the estuarine system, accidental spills of oil and other chemical substances, and the release of industrial non-treated wastewater, as well as residues and sewage from the Port of Santos and cities in Baixada Santista. These impacts caused a serious environmental problem, with major reflexes on social and public health areas (Cetesb, 1990, 2001). In view of this critical situation, the São Paulo State Environmental Company – Companhia Ambiental do Estado de São Paulo - CETESB initiated in 1984 a programme of intense control of air, water and soil pollution in the Cubatão industrial complex. Industrial wastewater treatment systems were installed in all plants in the region, drastically reducing the input of contaminants in the water bodies (CETESB, 2001). However, the first signs of environmental recovery could only be detected after further control measures were taken during the 1990s, which included the recirculation of liquid effluents, increase in the reutilization of served water, and implementation of groundwater collection and treatment systems to recover hydrocarbon and organochlorines contaminated sediment.

In 1999, CETESB carried out a new assessment of water, sediment and fauna contamination in 26 sites including the Santos-São Vicente estuarine system and adjacent sea. Although the reduction of contaminant input in the estuarine system resulted in 93 and 97% decrease of organic compounds and heavy metal, respectively, the results of the assessment still showed the presence of organochlorinated pesticides and aromatic compounds, polycyclic aromatic hydrocarbons (PAHs), aromatic and halogenated solvents, dioxins, furans, and heavy metals in concentrations above the limits determined by related regulations (Monteiro Filho, 2002). Sediment was the compartment presenting the highest variety and frequency of contaminants. The area around the private port to COSIPA (Figure 1b,c) was the most critically contaminated site, showing the highest concentrations of PAHs and polychlorinated biphenyls (PCBs), as well as heavy metals and phenolic compounds in concentrations above Canadian and American toxicity limits (Cetesb, 2001; Piza et al., 2004; Rodrigues et al., 2009). In the water, the main contaminant detected was cadmium, followed by benzene, chloroform, endosulfan B, phenol, 2,4-dimethylphenol, hexachlorobenzene (HCB), hexachlorocyclohexane (BHC) and copper. Analysis in fish, crabs, mussels, oysters, shrimps, and stout razor clams also detected the bioaccumulation of PCBs, PAHs, dioxins, and furans, reaching, in some cases, levels above the limits for human consumption, according to Brazilian and American regulation (Cetesb, 2001).

1.2. Chlorophenols: Relevance as Environmental Contaminant, Usages, Health Impacts, Biodegradation and Fate in the Environment; Microbial Metabolic Pathways to Degrade Chlorophenols; Anaerobic Biotechnological Application to CPs Bioremediation Purposes

Chlorophenols can be of natural or anthropogenic sources. For example, strains of the fungus *Penicillium* sp. can naturally produce 2,4-DCP, and 2,6-DCP is a pheromone for some tick species. However, the amounts produced in nature are not relevant if compared to anthropogenic sources, especially for chlorophenols with higher numbers of chlorines (WHO, 1989).

Man can produce chlorophenols for direct use as biocides, antiseptics, herbicides, dyes and drugs, or they can be subproducts in processes such as paper pulp bleaching, disinfection of industrial effluent and water in treatment plants, incomplete combustion of chlorophenols containing wastes, and bioconversion of chlorobenzenes and related compounds (WHO, 1989; British Columbia, 2001). Pentachlorophenol was first produced during the 1930s as a biocide for wood-preserving industry. Since then, it has also been used as general-purpose biocide, fungicide and insecticide. The widespread use of PCP and PCP treated products has led to its broad distribution in the environment (Wild et al., 1992; British Columbia, 2001).

Chlorophenols in the aquatic environment can be degradaded by photolysis, especially if water bodies are shallow and subject to high radiation indexes. This is also one of the main processes of PCP removal in estuarine environments, although transformation rates can be influenced by radiation intensity and water turbidity (British Columbia, 2001). However, most chlorophenols in estuaries adsorb to particulate suspended matter in water and are incorporated into the sediment, changing the compounds availability to other degradation processes (Wild et al., 1992; Mc Allister et al., 1996).

Even though physical and chemical processes play an important role in chlorophenols' transformation and transport through ecosystems, the aerobic and anaerobic microbial degradation is the major and most complete removal process in the environment. Despite chlorophenols' biocidal properties, several microbial species are able to degrade and use these compounds as substrate for growth, raising the interest in their use for bioremediation of contaminated sites or residues (Mc Allister *et al*., 1996; Magar *et al*., 2000; Zou *et al*., 2000).

As many PCP-contaminated sites are oxygen depleted, an understanding of anaerobic PCP biodegradation is important. Under anaerobic conditions, the main pathways for the transformation of chlorophenols are: (1) co-metabolism, in which enzymes or cofactors from the regular metabolism of the microorganism fortuitously degrade the chlorophenol, at the expenses of substrate. (2) biosynthesis, in which the compound is used as source of carbon and energy, usually C1 and C2 chlorinated compounds; and (3) halorespiration, in which the compounds are used as electron acceptors and energy for cellular growth is generated from dehalogenating exergonic reactions (Figure 2) (El Fantroussi *et al*., 1998). Hydrogen and glucose, for example, are good substrates for anaerobic biodegradation of PCP, trichlorophenol (TCP) and dichlorophenol (DCP), as well as other halogenated organic compounds (Magar *et al*., 2000; Hendriksen *et al*. 1992; Tartakovsky *et al*. 2001; Saia *et al*., 2007).

Anaerobic microbial degradation of chlorophenols has been reviewed and there are many results in regards to PCP, as for example research on microbial consortia and co-cultures that showed complete removal of the compound through reductive dehalogenation (Mikesell and Boyd, 1986; Tartakovsky *et al*., 2001; Montenegro *et al*., 2002). Chlorinated compounds degrading strains have also been isolated, such as the chlorobenzoate dehalogenating and sulfate reducing bacteria *Desulfomonile tiedjei* (Mohn and Kennedy, 1992), *Desulfitobacterium dehalogenans* (Utkin *et al*, 1994), *Desulfitobacterium hafniense* (Madsen and Licht, 1992) and *Desulfitobacterium frappieri* (Bouchard *et al*., 1996).

The use of biological reactors is one of the main strategies for bioremediation of contaminated sites. Such bioreactors can be used in pump-and-treat strategies with the unique unit or combined with chemical or physical processes (Droste *et al*, 1998; Cattony *et al*., 2005; Robles-González *et al*., 2008). Foresti *et al*. (1995) developed the horizontal-flow anaerobic immobilized biomass (HAIB) reactor, which contains polyurethane foam as biomass immobilization support. It is a tubular reactor, whose configuration, in compact units, enables increase of useful volume as a consequence to volume reduction meant for gas separation. Studies on bench scale have been showed the potential of application of HAIB reactor for treatment of hazardous compounds (de Nardi *et al*., 2004), such as pentachlorophenol (Saia *et al*., 2007; Baraldi *et al*., 2008; Damianovic *et al*., 2009), phenol (Bolaños *et al*., 2001) and BTEX compounds (Cattony *et al*., 2005; de Nardi *et al*., 2007).

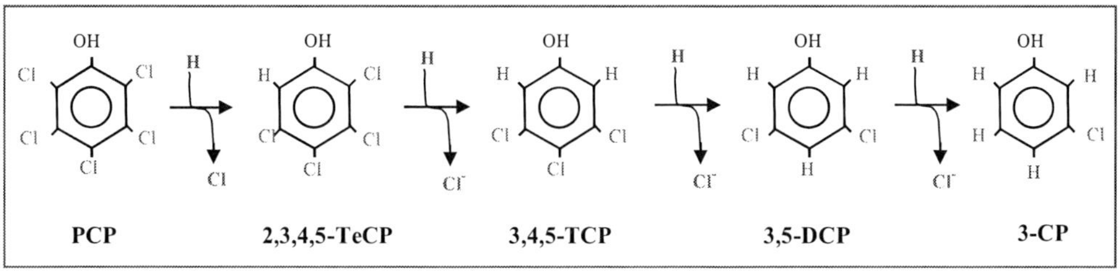

Figure 2. PCP reductive dehalogenation showing the replacement of Chlorine (Cl) by Hydrogen (H) (adapted from Madsen & Aamand, 1990).

1.3. Synthesis of the Study

The peculiar characteristics of the Santos-São Vicente Estuarine System make it interesting as a case study for the research of the microbial diversity associated to contaminated sites, biodegradative processes and potential for development of bioremediation technologies. In the year 2000, the thematic project "Molecular Ecology and Polyphasic Taxonomy of Bacteria of Industrial and Agroindustrial Interest" was approved as part of the Biota Program, an initiative of the São Paulo State Research Foundation (Fapesp) to collect, organize and disseminate information about the biodiversity in São Paulo State. This project allowed for the integration of several research lines with the objective of studying the diversity of autochthonous microbial communities involved directly or indirectly in the aerobic and anaerobic degradation of hydrocarbons and organochlorinated compounds, and its potential for technological application.

In this chapter, we present a synthesis of the research involving chlorophenols degradation by anaerobic microbial communities in sediment from the area around the private port to the siderurgical company COSIPA, the most contaminated site in the Santos-São Vicente Estuarine System. The study involved the following stages: (1) enrichment and characterization of chlorophenols degradating anaerobic communities through controlled experiments under methanogenic conditions, and molecular techniques; (2) determination of optimum N:P ratios for PCP degradation in batch cultures (3) technological application of the autochthonous microbiota to treat PCP and other organic substrates in bench scale horizontal-flow anaerobic immobilized biomass (HAIB) operated under halophylic and methanogenic conditions. Methane production *in situ*, MPN counts of methanogens and its preferably substrates were also performed to obtain information on the methanogenic activity in the site studied.

2. EXPERIMENTAL SETUP AND METHODS

2.1. Sediment Sampling and Determination of Methane and Carbon Dioxide Fluxes

Sediment samples were collected at the COSIPA port area on October 2002 (S' 23°52'65.3''; W' 46°22'67.5'') and on July 2004 (S' 23°52'401''; W' 46°22'727''), at 2m and 1m depth, respectively, using a gravity corer device (Pedersen *et al*. 1985). Sediment samples collected in 2002 were used in enrichment 1 (Nakayama, 2005; Araujo 2002) and HAIB reactor 1 (Saia, 2005) in order to obtain methanogenic cells and chlorophenol degrading cultures under methanogenic conditions. Sediment samples collected in 2004 were used as inoculum of enrichment 2 and HAIB reactor 2 (Brucha, 2007).

Air in the interface with the sea was sampled according to Rosa *et al.* (2002) to determine local methane and carbon dioxide fluxes on July 2004.

2.2. Enrichment Cultures Strategies and Enumeration of Methanogens

Due to lack of preterit data on anaerobic chlorophenols degradation and methanogenic microorganisms in Santos - São Vicente Estuarine System, different conditions were tested to guarantee that chlorophenol degrading and methanogenic cultures were obtained. Two kinds of experiments were performed: (1) enrichment cultivation to obtain chlorophenol degrading cultures, with a broad range of chlorophenols and electron acceptors and donors, and (2) enrichment cultivation with different Nitrogen:Phosphorus (N:P) ratios, in order to optimize nutritional conditions according to the Resource Ratio Theory (Smith et al., 1998). In both enrichments, anaerobic mineral medium adapted for cultivation of estuarine microorganisms was prepared according to Saia et al. (2007). PCP and other chlorophenols, as well as electron donors and acceptors, were added from sterile anaerobic sodium salts stock solutions.

Enrichment 1 was done in replicate of 100-mL serum bottles, inoculated with 20% (v/v) sediment collected on October 2002. Glucose ($1g.L^{-1}$), pyruvate (20mM) and formate ($2g.L^{-1}$) were added as electron donors, sodium sulphite (5mM) as electron acceptor, L-cystein as reducing agent and yeast extract (0.05%) as a nutrient supplement to stimulate growth. After 16 hours, 2,6-DCP, 2,3,4-TCP and PCP were added to batches in concentrations of 27.6μM; 21.6μM and 18.77μM, respectively and were re-added when they were depleted. Controls for culture medium (without chlorophenols addition) and chlorophenol degradation (with chlorophenols and autoclaved inoculum) were also prepared.

Enrichment 2 was done in replicate of 2000-mL bottles, inoculated with 10% (v/v) sediment collected on July 2004. Carbon was expressed as COD (chemical oxygen demand). Glucose ($0.5g.L^{-1}$) and formate ($0.5g.L^{-1}$) were added as electron donors, resulting in a COD of 1 $g.L^{-1}$. PCP was added to batches in concentration of $1mg.L^{-1}$. To verify nutritional condition for PCP degradation, 13 different combinations of nitrogen and phosphorus amounts were used, resulting in COD:N:P ratios of: (1) 1000:0:0 – control without nitrogen and phosphorus; (2) 1000:5:1; (3) 1000:5:2; (4) 1000:10:0; (5) 1000:10:1; (6) 1000:10:2; (7) 1000:15:2; (8) 1000:20:5; (9) 1000:50:10; (10) 1000:130:45; (11) 1000:130:45 with inoculum autoclaved; (12) 0:130:45 with $1mg.L^{-1}$ PCP and (13)) 0:130:45 with 2 $mg.L^{-1}$ PCP. The ratios relation studied as indicated in 2 to 6 was around the ratio (1000:10:2) established by Speece (1996) as limit for growth of anaerobic microorganisms. The ratios studied as indicated in 7 and 8 represent an intermediary condition for aerobic and anaerobic growth. The ratio in 9 was used in order to stimulate growth of aerobic microorganisms and the ratio in 10 corresponds to the original condition of the mineral medium described in Saia et al. (2007)

Methanogenic *Archaea* in the sediment was quantified by the most probable number technique – MPN (Alexander, 1982) using tenfold serial dilutions from 10^{-1} to 10^{-7} in anaerobic basal medium (Zinder et al., 1984) amended separately with sodium acetate (20mM), methanol (20mM) or maintained under atmosphere of $H_2:CO_2$ (80:20). Positive results were determined by detection of methane in the atmosphere of the cultures and the presence of fluorescent cells as observed under UV light in a fluorescence microscope.

2.3. Horizontal-Flow Anaerobic Immobilized Biomass (HAIB) Reactors

Technological application of the autochthonous microbiota to degrade PCP and other organic substrates with different amounts of nitrogen and phosphorus was carried out in two identical compact HAIB reactor units filled with 20.8 g of polyurethane foam cubes (3 mm in size, apparent density of 23 kg m^{-3} and bed porosity of 40%) for biomass attachment. The borosilicate glass reactors were 1 meter long, with a 5-cm diameter (D), a length/diameter ratio (L/D) of 20, final volume of 1991 ml and liquid volume of 800 ml. Each reactor was equipped with a gas collector and five screw-cap sampling ports along its length (L), at L/D of 4, 8, 12, 16 and 20, (HAIB 1) and at L/D 3.3; 6.6, 9.9, 13.2, 16.5 (HAIB 2) allowing for sampling of the packing material. Installed in a temperature-controlled chamber (30±2°C), each reactor was fed with a peristaltic pump and operated with a retention time of 18 hours. (Figure 3). A N_2 atmosphere was supplied to the influent feeding flask, using a gas balloon device.

Both reactors were inoculated with sediments previously enriched for methanogenesis, as described in Saia *et al.* (2007). COD:N:P ratios used for enrichments as well as throughout operation of HAIB reactors were 1000:130:45 for HAIB 1 and 1000:10:1 for HAIB 2. COD was due to 1g.L^{-1} of glucose (HAIB 1) and 0.5 g.L^{-1} of glucose plus 0.5 g.L^{-1} of formate (HAIB 2).

The inoculation and immobilization of the biomass was performed using closed feed-circulation. Ten percent of enriched sediment in a halophilic medium (Saia *et al.*, 2007) with glucose (0.5 g.L^{-1}) or glucose (0.5 g.L^{-1}) plus formate (1.0 g. L^{-1}), was inoculated into the reactors for 6 days. After the inoculation, the systems operated continuously. For a period of 26 days the reactors were fed with influent without PCP. Thereafter, in HAIB reactor 1, PCP was added in increasing concentrations of 5, 13, 15 and 21 mg.L^{-1}, at intervals of 36, 36, 16 and 12 days, corresponding to phases 1, 2, 3 and 4,respectively. In HAIB reactor 2, PCP was added in increasing concentrations of 0.5, 5, 10 and 12.4 mg.L^{-1}, at intervals of 16, 08, 12 and 16 days, corresponding to phases 1, 2, 3 and 4, respectively.

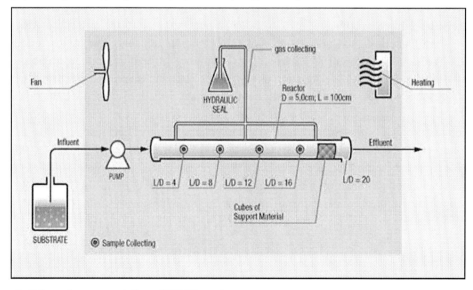

Figure 3. Schematic representation of HAIB reactor.

2.4. Microbial Community Characterization

Morphology of microbial cells was observed by phase-contrast microscopy and methanogenic *Archaea* was detected by fluorescence microscopy in all enrichment, MPN cultures, and in HAIB reactors. Scanning electron microscopy was also used to observe morphology and cell arrangement in the HAIB reactor 1, using the procedures described in Araújo *et al.* (2003). Characterization of archaeal and bacterial communities was also carried out by fluorescence *in situ* hybridization (FISH), denaturing gradient gel electrophoresis (DGGE) and 16S rDNA libraries, as described below.

FISH analyses were performed to identify methanogenic archaeal cells in MPN cultures The probes used were ARC915 (*Archaea* Domain; Stahl & Amann, 1991), MSMX860 (*Methanosarcinales*; Raskin *et al.*, 1994) and MB310 (*Methanobacteriaceae*, Raskin *et al.*, 1994). The NON 338 probe was used as a negative control (Manz *et al.*, 1992). The fixation protocol and hybridization conditions have been described previously (Araújo *et al.*, 2000). Samples were examined in a Zeiss Axiovert S100 fluorescence microscope.

DNA for DGGE analysis and 16S rDNA libraries was extracted from the biofilms and enrichment cultures using a phenol-chloroform protocol adapted from Griffiths *et al.* (2000). PCR amplifications for DGGE analysis were carried out with the bacterial primers 758R-341FGC (Tartakovsky *et al.* 2001) and 968F-GC/1392R (Nielsen *et al.*, 1999) for samples from HAIB reactors and chlorophenols enrichments, respectively, and with the archaeal primers 1400R-1100FGC (Kudo *et al.* 1997). For 16S rDNA libraries, the primers 20F/958R for *Archaea* Domain (Orphan *et al.*, 2001) and the primers 27F/907R for *Bacteria* Domain (So & Young, 1999) were used. PCR amplifications were performed with a PCR thermal cycler (Perkin Elmer, Gene Amp PCR system 2400), according procedures described in Saia et al. (2007) and Domingues (2007).

DGGE was conducted as described in Muyzer *et al.* (1993) using a DCodeTM Universal Mutation Detection System (Bio-Rad Laboratories). PCR products were applied directly onto 8% (wt/v) polyacrylamide gel in 0.5 x TAE with gradients containing 20-70% and 30-60% denaturant (urea and formamide), for samples from HAIB reactors and chlorophenols enrichments, respectively. Band patterns were photographed under UV light (254nm) and analyzed in a Eye TMIII gel documentation system (Stratagene). The obtained band patterns were used to analyze the structure of the archaeal and bacterial communities. DGGE bands of interest were excised, re-amplified by PCR with the archaeal and bacterial primers 1100F/1400R and the 968F/1392R, respectively, and sequenced as described in Domingues (2007).

PCR products for the 16S rDNA libraries were cloned and sequenced according to Domingues (2007), using the primers T7 and SP6 (Gibco BRL). Sequence quality was analyzed with the SeqMan software (Lasergene Sequence Analysis Software – DNAStar) and high quality sequences checked for chimeras with the Bellerophon software (Huber *et al.*, 2004). Consensus sequences were compared to databases of the Ribosomal Database Project (RDP-II) (Cole *et al.*, 2003) e the National Center for Biotechnology Information, using the BLAST software (Basic Local Alignment Search Tool) (Altschul *et al.*, 1997). Phylogenetic trees were constructed as described by Domingues (2007), using the maximum likelihood (ML) method and bootstrap of 1735 and 1000 for *Archaea* and *Bacteria* Domains, respectively. *Thermotoga maritima* (*Bacteria*) and *Sulfolobus acidocaldarius* (*Archaea*) were used as external groups.

2.5. Organic Compounds Analyses

Organic matter concentration was analyzed as COD by the closed-reflux method (Standard Methods for Examination of Water and Wastewater, 1998) with adaptation for estuarine samples as described in Saia et al. (2007). Volatile fatty acids were determined using a HP 6869 gas chromatograph equipped with a HP Innovax column of 30 m x 0.25 mm x 0.25 µm and a flame ionization detector, according to Moraes et al. (2000). Methane was measured by gas chromatography (Gow-Mac chromatograph with a thermal conductivity detector and Porapak Q column-2 m x ¼ in 80 to 100 mesh) as described in Cattony et al. (2005) or in a HP6850A chromatograph equipped with a HP-Plot Al_2O_3 "S" column of 50m x 0.53mm x 0.15µm and a flame ionization detector, according to Nakayama (2005). PCP, 2,3,6-trichlorophenol (TCP); 2,4,6-TCP and 2,3,4-TCP; 2,3-dichlorophenol (DCP); 2,6-DCP and 2,4-DCP concentrations were determined with a HP 5890 series II gas chromatograph equipped with an electron capture detector and a HP-5 (5% diphenyl, 95% dimethylpolysiloxane) column of 30 m x 0.32 mm x 0.25 µm, according to the acetylation method described by Damianovic et al. (2007).

3. CHARACTERIZATION OF CHLOROPHENOLS DEGRADERS AND METHANOGENIC COMMUNITIES FROM THE ESTUARINE SEDIMENT SAMPLES

The enrichment of sediment samples in the presence of 2,6-DCP, 2,3,4-TCP or PCP, and a broad range of organic substrates (enrichment 1) carried out by Nakayama (2005) resulted in complex methanogenic microbial communities able to degrade up to 10mg.L^{-1} chlorophenols in periods of up to 15 days (Figure 4). Formate was the best electron donor to PCP degradation, while pyruvate was the best substrate to 2,6-DCP and 2,3,4-TCP degradation. It was no possible to quantify all possible PCP metabolites and organic acids were not measured in the chlorophenols enrichments, but degradation curves evidenced the accumulation of metabolites in PCP degrading cultures, especially after the time period between PCP amendments were reduced (Figure 4). This result indicates the presence of subproducts more slowly metabolized than PCP. Madsen & Aamand (1992) observed similar response from cultures studied, which comprised a microbial consortium able to degrade 2,4,6-TCP in 4 days, and plus 37days to metabolize the products from the trichlorophenol degradation, 2,4-DCP and 4-CP. Accumulation of lower chlorinated phenols and/or organic acids produced during anaerobic digestion eventually led to inhibition of PCP degradation in some enrichment cultures, probably due to the high concentrations reached or the production of intermediates toxic to the microbial community. Similar results were also shown by Juteau et al. (1995), who observed that semi-continuous PCP feeding regime resulted in progressively larger PCP consumption times.

The 2,6-DCP degrading culture was confirmed as a methanogenic consortium which growth under the chlorophenol breakdown with the predominance of gram-negative vibrios form, while the PCP degradation showed the predominance of gram-negative rods (Nakayama, 2005). DGGE analysis of PCP and 2,6-DCP degrading cultures indicated the presence of bacterial and methanogenic groups. DGGE bands were cloned and sequenced, revealing the presence of bacteria mostly related to the order *Clostridiales*, including sequences related to *Sedimentibacter* sp. and *Clostridium* sp (Nakayama, 2005; Domingues, 2007). Several authors reported the chlorophenol degrading activity of *Clostridium* sp. and *Sedimentibacter* sp., either in consortia with other bacteria (Tartakovsky et al., 2001; van Doesburg et al., 2005), in mixed cultures (Bedard et al., 2006) or as pure strains (Zhang et al., 1994; Breintenstein et al., 2002), with pyruvate being commonly used as electron donors or fermenting substrate. Other groups of bacteria, as *Alkalibacter* sp., which have fermentative metabolism and require yeast extract to grow were also identified (Domingues, 2007).

Archaeal sequences analysed by Domingues (2007) were related to hydrogenotrophic and acetoclastic methanogens belonging to the order *Methanobacteriales* (7.1%), such as *Methanobacterium* sp.; *Methanosarcinales* (14.3%), such as *Methanosaeta* sp.; *Methanomicrobiales* (57.1%), such as *Methanocalculus* sp. e non-cultived members of *Euryarchaeota* (21.4%). These organisms are able to utilize bacterial fermentation sub-products such as acetate, $H_2:CO_2$ and formate as substrates for methanogenesis, contributing for the maintenance of the balance of other reactions that occurred in the system. Previous characterization of the Santos-São Vicente estuarine system sediment by 16S rDNA libraries (Piza, 2004) revealed the presence of the orders *Methanobacteriales*, *Methanomicrococcales*, *Methanomicrobiales* and *Methanosarcinales*, including the clone DI_A01, identified as *Methanosaeta* sp., whose sequence was also found in the PCP enrichment culture. These results indicate that the culture conditions were adequate for the cultivation of autochthonous methanogenic *Archaea*. Sequences of both *Bacteria* and *Archaea* Domains bearing little similarity with the identified groups present in the databases were also found, which indicated that new groups may be present in the enrichment cultures.

MPN counts for methanogenic *Archaea* using sediment as inoculum and $H_2:CO_2$, acetate or methanol amended separately as organic sources showed that hydrogenotrophs and acetotrophs belonging to the families *Methanosarcinaceae* and *Methanobacteriaceae* (Figures 5 and 6) were the most abundant methanogenic groups ($3.0.10^4$ and $2.3.10^4$ cells.mL^{-1}, respectively), corroborating the results obtained with chlorophenol degrading enrichments (Domingues, 2007) and previous sediment molecular characterization (Piza, 2004). Methylotroph methanogens belong to the *Methanosarcinaceae* family (Figure 5) were also present in the sediment, but in lower numbers ($3,0.10^3$ cells.mL^{-1}) (Araújo, 2002).

Methane fluxes determinations in the area (S' 23°52'401''; W' 46°22'727'') showed that the site is a source of the gas to the atmosphere, at a rate of 23.90 ±13.38 mgCH$_4$.m^{-2}.d^{-1}. (Brucha, 2007). Methane fluxes in estuaries with the same salinity, 2% as NaCl, as those found in Santos - São Vicente estuarine system were smaller, showing rates of 1.6 (Scheldt Estuary, Belgium/Holland) and 2.08 mgCH$_4$.m^{-2}.d^{-1} (Dipper Harbour, Canada) (Middelburg et al., 1996). However, data from Brazilian rivers and freshwater reservoirs show values reaching 57.4 mgCH$_4$.m^{-2}.d^{-1} (Rosa et al., 2002). These data suggest that salinity as well as temperature, water circulation and winds may be factors influencing local methanogenesis. On the other hand, the severe pollution observed in the area may be contributing to increase methane emission to the atmosphere, as higher organic input leads to faster oxygen depletion and to the establishment of anaerobic processes. The quantitative numbers of methanogenic *Archaea* obtained and the methane fluxes determinations confirmed that methanogenesis has been an important process in the studied area, and for that, it is possible to suggest that chlorophenol anaerobic degradation under methanogenesis could occurred in this site, as it was detected in the enrichment cultures metabolism.

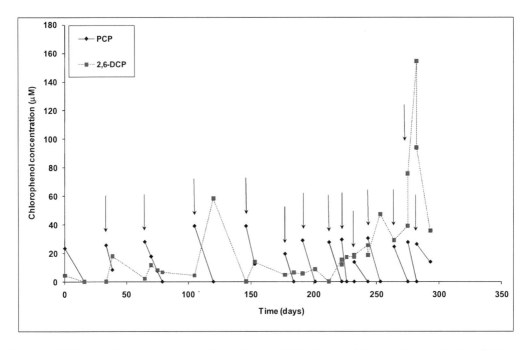

Figure 4. PCP degrading enrichment 1 culture showing PCP biodegradation and accumulation of 2,6-DCP after reduction time between PCP amendments. Arrows indicate addition of 10mg.L^{-1} (18.77μM) PCP to the culture medium.

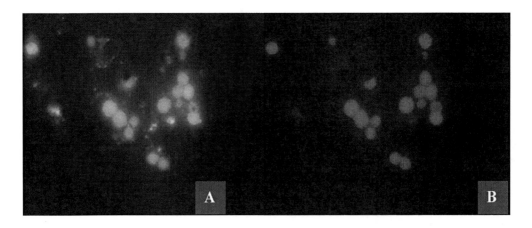

Figure 5. Morphologies resembling *Methanosarcina* sp., stained with DAPI (A) and hybridized with probe MS860 (*Methanosarcinaceae*) (B) from methanogenic cultures amended with acetate (20mM), methanol (20mM) or $H_2:CO_2$ (80:20%) Magnification, x1,000.

4. BIOTECHNOLOGICAL STUDIES FOR PCP BIOREMEDIATION PURPOSE

4.1. Determination of N:P Ratios for Optimal PCP Degradation

The enrichment cultures experiments to evaluate the effects of N:P ratios in the structure of microbial community under PCP (1mg.L^{-1}) and other organics supplies (0.5g.L^{-1} of glucose plus 0.5 mg.L^{-1} of formate) showed that at 5:1 ratio neither PCP nor other organics removal was achieved during the incubation period of 90 days. At the end of the experimental time, acetic acid was detected up to 482 mg.L^{-1}. The PCP degradation was observed under 5:2 ratio after 26 days of incubation time (Brucha, 2007). The DGGE characterization of the enriched sediment showed a microbial community composed of two or even fewer specific bands (Figure 7).

On the other hand, the enrichments carried out under 10:1, 10:2, 15:2, 20:5, and 50:10 ratios under the same amounts of PCP and organics, such as glucose and formate, showed biodegradation activities measured by PCP consumption and methane production up to 9.3 mMol/L at 13th day of incubation time at 30^{0}C (Brucha, 2007). DGGE analyses of these enrichments showed a microbial community composed of higher number of bands than observed at N:P ratios of 5:1, 5:2 and 10:0 (Figure 7). These results showed that N:P ratios exerted an influence on the Domain *Bacteria* and PCP degradation under methanogenic conditions. Based on these results, the HAIB reactor (2) was operated with N:P ratio of 10:1.

4.2. HAIB Reactors

Technological studies with the autochthonous microbiota to treat PCP under N:P ratio selected by the previous studies, was carried out in bench scale horizontal-flow anaerobic immobilized biomass (HAIB) reactors .The operational conditions were running under 30^{0}C, with halophylic synthetic culture medium plus glucose and glucose plus formate, respectively amended to HAIB reactors R1 (C:N:P=1000:130:45) and R2 (C:N:P = 1000:10:1), and PCP concentration up to 21 mg.L^{-1} (R1) and 15mg.L^{-1}(R2). Both reactors were filled with polyurethane foam inoculated with culture previously enriched for methanogenesis activity with glucose (0.5 g.L^{-1}) or glucose (0.5g.L^{-1}) plus formate (0.5g.L^{-1}). The HAIB reactor R1 was operated under strict conditions present in the halophylic medium and HAIB reactor R2 was also fed with this medium but without addition of yeast extract.

The HAIB reactor R1 fed with halophylic medium (Saia *et al*. 2007) with COD:N:P = 100:130:45, gradual increasing amount of PCP up to 21 mg.L^{-1} did not inhibit methanogens nor PCP dehalogenation. The performance of the HAIB reactor in respect to PCP removal was 100% for all PCP additions (Figure 8a) and the average percentage of methane in the biogas was maintained at 64.6±6.6%. The average of organic matter removal, as COD, was of 92±9.2%. Until PCP 13 mg.L^{-1} COD removal was nearly 100%. At 15 and 21 mg.L^{-1} COD removal ranged from 67 to 80% (Figure 8b), indicating a possible alteration in the electron flux in the anaerobic metabolism for degrading organic matter. Propionate was detected in the effluent (40 mg.L^{-1}) indicating that propionate degraders were the most sensitive to PCP (Saia *et al.*, 2007).

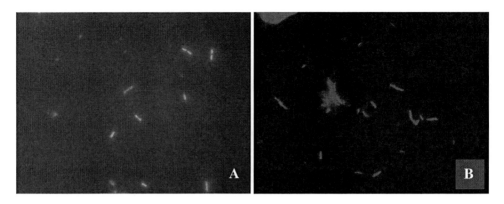

Figure 6. Bacili stained with DAPI (A) and hybridized with probe MB310 (*Methanobacteriaceae*) (B) from methanogenic cultures amended with acetate (20mM) or $H_2:CO_2$ (80:20%)Magnification, x1,000.

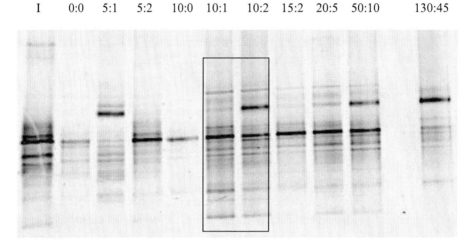

Figure 7. DGGE-profiling of *Bacteria* at the end of enrichment 2 in a semi-continuous system amended with COD of 1000 and different N:P ratios of 0:0. 5:1; 10:0; 10:1; 10:2; 15:2; 20:5; 50:10 and 130:45, respectively. I- inoculum. Square indicates the ratios were highest number of bands were observed.

The high chlorine removal rate in the first portion of the reactor in the presence of acetate in a range of 240 to 280 mg.L^{-1} (Saia *et al*, 2007) as well as the selection of specific bands of Domain *Bacteria* in the presence of PCP (Figure 9a) suggest that PCP dehalogenation occurred preferring acetogenic conditions, and members of the Domain *Bacteria* were directly involved with this metabolism. The approach using Archaeal primers showed a DGGE profile uniform throughout the reactor during all experimental phases (Figure 9b), suggesting that methanogenic archaeas were not directly implicated in PCP dechlorination. Microscopic characterization of the biofilms revealed several distinct microbial morphologies including cocci, rods and filament cells (Figure 10). The *Bacteria* and *Archaea* Domain profiles were similar to those previously reported in the literature, which found that groups of bacteria are responsible for the initial PCP dehalogenation, while the methanogenic archaeas are associated with the metabolism of the products formed, showing the importance of the syntrophic associations in PCP degradation (Juteau *et al*. 1995; Tartakovisky *et al*. 2001; Lanthier *et al*. 2005).

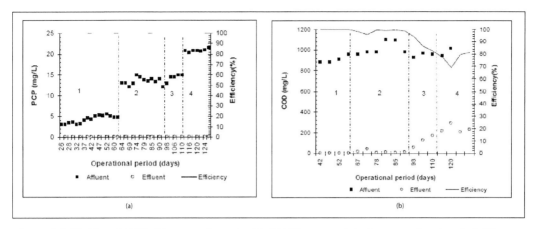

Figure 8. PCP (a) and COD (b) removal acquired in HAIB reactor 1 during operational period. R1 was amended with COD:N:P of 1000:130:45 and PCP up to 21 mg.L^{-1}.

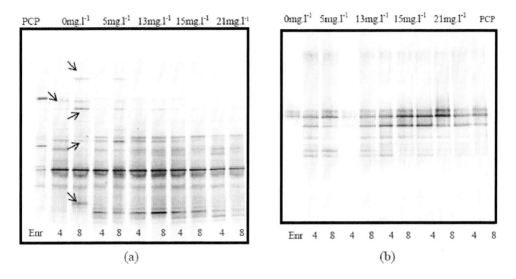

Figure 9. DGGE-profiling of *Bacteria* (a) and *Archaea* (b) rDNA samples for enriched sediment (Enr) and for all PCP amendments (0, 5, 13, 15 and 21 mg l^{-1}) at the sampling points L/D=4 and L/D=8 of HAIB reactor R1. The first column in graphs a; b shows the DGGE-profile of rDNA of the inoculum. Arrows and squares indicate, respectively, populations that disappeared and predominated during PCP additions (Saia et al., 2007).

The HAIB reactor R2 fed with halophylic medium with COD:N:P=1000:10:1 and without addition of yeast extract, PCP up to 12 mg.L^{-1} was removed in the range of 98 to 100% (Figure 11a) and COD removal efficiency was around 80% for COD (Figure 11b). High PCP removal was also observed in the first portion of the reactor (data not shown). Different from the results obtained in HAIB reactor 1, DGGE-profiling of *Archaea* showed a slight variation during the different operational phases of HAIB reactor 2 in the presence of PCP (Figure 12). Selection of specific members of *Archaea*, most probably was related not only with PCP loads but also with nutritional condition used. Sequencing of DGGE bands revealed the presence of hydrogenotrophic methanogenic *Archaea* 100% related to uncultured *Methanocalcullus* sp. and of acetoclastic methanogenic *Archaea* 99% related to *Methanosaeta* sp. Piza (2004), during studies on molecular diversity of *Archaea* in Santos-

São Vicente estuary at COSIPA site discovered that 26.5% of the obtained clones belong to *Methanosaeta* genus, suggesting that members of this genus are an important and predominant group in this site. Similar results were observed in sediments of different origins (Dojka *et al.*, 1998; Grosskopf *et al.* 1998, Von Wintzingerode *et al.* 1999; Purdy *et al.*, 2002), indicating that this group is an active member of the microbial community of sediments. The detection of these microorganisms as an important component of the microbial community in a site contaminated with hydrocarbons, highlight the idea that the acetoclastic methanogenesis can be the main methanogenic process associated with the degradation of hydrocarbons (Dojka *et al.*, 1998; Ficker *et al.*, 1999; Watanabe *et al.*, 2002). Members of *Methanocalcullus* genus, such as *Methanocalcullus halotolerans* (Ollivier *et al.*, 1998), *Methanocalcullus pumilus* (Mori *et al.*, 2000) and *Methanocalcullus Taiwanensis* (Lai *et al.*, 2002) are capable to use carbon dioxide and formate as substrate to methanogenesis in saline environments. The high identity with the lineages characterized by Piza (2004) showed the importance of these autochthonous microorganisms in the microbial syntrophism involved in the dehalogenation of toxic compounds present in the estuary.

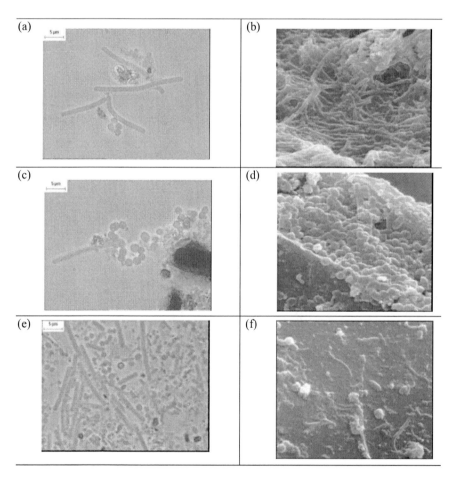

Figure 10. Contrast-phase microscopy and SEM image of the biofilm of HAIB reactor 1 exposed to 21 $mg.L^{-1}$ of PCP: (a,b) Morphology of *Methanosaeta*-like cells; (c,d) Cocci cells; (e,f) rods, cocci and morphology of *Methanosaeta*-like cells. Magnification, x1,500 (Contrast-phase) and x5,000 (SEM) (Saia, 2005).

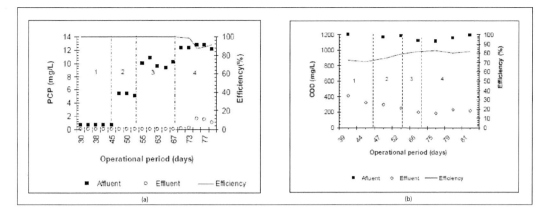

Figure 11. PCP (a) and COD (b) removal acquired in HAIB reactor R2 during operational period. R2 was amended with COD:N:P of 1000:10:1 and PCP up to 13 mg.L^{-1}.

These results also show that N:P ratios tested in continuous feed system were suitable to select microorganisms able to degrade PCP under methanogenic and halophylic conditions. This is relevant under both ecological and economical points of view, since it shows that the microbial community studied was adapted to the PCP and nutritional conditions

5. CONCLUDING REMARKS

The present research is a relevant source of information to understand the potential of anaerobic microbial processes in the sediments of Santos - São Vicente Estuarine System, as well as to contribute with data that can allow for the implementation of remediation technologies to the restoration of sites in that area. Also, the results highlight information on the anaerobic microbial diversity of a tropical ecosystem severely polluted with toxic compounds and confirm the occurrence of methanogenesis in this environment. The practical management of the sediments considering controlled engineering anaerobic processes is, therefore, a possible solution.

The combined use of cultivation dependent methods and molecular biology techniques either in batch and continuing systems (HAIB reactors) confirmed the reductive dehalogenation activity associated to methane production in the estuarine sediment from the most contaminated site in the region. The results showed that dehalogenation of the compounds is possibly carried out by the bacterial communities, and groups related to the *Clostridiales* family, such as *Clostridium* sp. and *Sedimentibacter* sp., may be involved in this process. The archaeal community is most probably involved in the mineralization of the organic acids resulting from chlorophenol degradation, playing an important role in keeping the balance of the system. Predominant archaeal groups in batch enrichments and reactors included the genera *Methanocalcullus* sp. and *Methanosaeta* sp. These genera were also present in previous molecular characterization of sediment samples from the area of study (Piza, 2004), showing that the cultivation parameters were adequate to reproduce *in situ* occurring processes. Finally, the studies on N:P ratios for optimizing PCP degradation showed that the local microbial community is versatile and able to degrade the compound under diverse nutritional combinations.

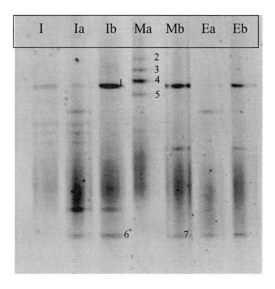

Figure 12. DGGE-profiling of *Archaea* rDNA samples at the sampling points along the HAIB reactor 2. Subtitle: I-Inoculum; I, M, E – samples of initial (L/D = 3.3), medium (L/D = 9.9) and end (L/D = 16.5) of HAIB reactor points, respectively. (a) before addition of PCP, (b) after 54 days of PCP addition, at PCP concentration of 12 mg.L^{-1}. Bands 1 to 6 were sequenced (Brucha, 2007).

ACKNOWLEDGMENTS

The authors acknowledge the "Fundação de Amparo à Pesquisa do Estado de São Paulo (FAPESP)" for their financial support to Biota Program (grant 98/05068-0) and scholarships (grants 00/08323-3; 00/11900-2; 02/03486-7; 03/07946-5). We are also thankful to researchers and technicians from CETESB for providing infra-structure for the sample collections.

REFERENCES

Alexander, M. (1982). *Most Probable Number Method for Microbial Populations in Methods of Soil Analysis.*, Part 2. Chemical and Microbiological Properties-Agronomy no. 9 (2a. ed), 815-820.

Araújo, J. C., Brucha, G., Campos, J. R. & Vazoller, R. F. (2000). Monitoring the development of anaerobic biofilms using fluorescent *in situ* hybridization and confocal laser scanning microscopy., *Wat SciTechnol.*, *14*, 69-77.

Araújo, A. C. V. (2002). Avaliação quantitativa de arqueas metanogênicas presentes em sedimento do Estuário de Cubatão, São Paulo, Brasil. Contribuição aos estudos de diversidade funcional de microrganismos realizados no Programa BIOTA FAPESP. Bacharel thesis, Instituto de Ciências Biomédicas, Universidade de São Paulo.

Araújo, J. C., Téran, F. C., Oliveira, R., Nour, E. A. A., Montenegro, M. A. P., Campos, J. R. & Vazoller, R. F. (2003). Comparison of hexamethyldisilazane and critical-point drying treatments for scanning electron microscopy analysis of anaerobic biofilms and granular sludge. *J Electron Microsc.*, *52*, 429-433.

Altschul, S. F., Madden, T. L., Schaffer, A. A., Zhang, J., Zhang, Z., Miller, W. & Lipman, D. J. (1997). Gapped BLAST and PSI-BLAST: a new generation of protein database search programs. *Nucleic Acids Research.*, *25*, 3389-3402.

Baraldi, E. A,. Damianovic, M. H. R. Z., Manfio, G. P., Foresti, E. & Vazoller, R. F. (2008). Performance of a horizontal-flow anaerobic immobilized biomass (HAIB) reactor and dynamics of the microbial community during degradation of pentachlorophenol (PCP). *Anaerobe.*, 14, 268-274.

Bícego, M. C., Taniguchi, S., Yogui, G. T., Montone, R. C., da Silva, D. A. M., Lourenço, R. A., Martins, C. C., Sasaki, S. T., Pellizari, V. H. & Weber, R. R. (2006). Assessment of contamination by polychlorinated biphenyls and aliphatic and aromatic hydrocarbons in sediments of the Santos and São Vicente Estuary System, São Paulo, Brazil. *Mar Pollut Bullet.*, 52, 1784-1832.

Bolaños, R. M. L., Varesche, M. B. A., Zaiat, M. & Foresti, E. (2001). Phenol degradation in horizontal-flow anaerobic immobilized biomass (HAIB) reactor under mesophilic conditions. *Water Sci Technol.*, *44*, 167-174.

Bouchard, B., Beaudet, R., Villemur, R., Mcsween, G., Lépine, F. & Bisaillon, J. G. (1996). Isolation and characterization of *Desulfitobacterium frappieri* sp. nov., an anaerobic bacterium which reductively dechlorinates pentachlorophenol to 3-chlorophenol. *International Journal of Systematic Bacteriology.*, *46*, 1010-1015.

British Columbia. *Water, air and climate change branch* (internet web site: <http://wlapwww.gov.bc.ca/wat/wq/Bcguidelines/chlorophenols/>)

Brucha, G. (2007). Influência dos nutrients nitrogênio e fósforo na degradação anaeróbia do pentaclorofenol e na diversidade microbiana dos sedimentos enriquecidos do Estuário de Santos e-São Vicente, Estado de São Paulo. PhD thesis, Escola de Engenharia de São Carlos, *Universidade de São Paulo*.

Cattony, E. B. M., Chinalia, F. A., Ribeiro, R., Zaiat, M., Foresti, E. & Varesche, M. B. A. (2005). Ethanol and toluene removal in a horizontal-flow anaerobic immobilized biomass reactor in the presence of sulphate. *Biotechnol Bioeng.*, *91*, 244-253.

Cantagallo, C., Garcia, G. J. & MIilanelli, J. C. C. (2008). Mapeamento de sensibilidade ambiental a derramamentos de óleo do sistema estuarino de Santos, Estado de São Paulo. *Braz J Aquat Sci Technol.*, 12, 33-47.

Cetesb (1990). Contaminantes na Bacia do Rio Cubatão e seus reflexos na biota aquática. Government of São Paulo Sate, *Technology Company for Environmental Sanitation (CETESB)*, 81p (www.cetesb.sp.gov.br-site in portuguese).

Cetesb (2001). *Estuarine System of Santos and São Vicente – Final Report*. Government of São Paulo Sate, Technology Company for Environmental Sanitation (CETESB), 178p (www.cetesb.sp.gov.br-site in portuguese).

Cole, J. R., Chai, B., Marsh, T. L., Farris, R. J., Wang, Q., Chandra, S., McGarrel, D. M., Schmidt, T. M., Garrity, G. M. & Tiedje, J. M. (2003). The ribosomal Database project (RDP-II): previewing a new autoaligner that allows regular updates and the new prokaryotic taxonomy. *Nucleic Acids Research.*, *31*, 442-443.

Damianovic, M. H. R. Z., Saia, F. T., Moraes, E. M., Landgraf, D., Rezende, M. O. O., Vazoller, R. F. & Foresti, E. (2007). Gas chromatographic methods for monitoring of wastewater chlorophenol degradation in anaerobic reactors. *J Environ Sci Health B.*, 42, 45-52.

Damianovic, M. H. R. Z., Moraes, E. M., Zaiat, M. & Foresti, E. (2009). Pentachlorophenol (PCP) dechlorination in horizontal-flow anaerobic immobilized biomass (HAIB) reactors. *Bioresource Technology.*, 100, 4361-4367.

de Nardi, I. R., Damianovic, M. H. R. Z., Ribeiro, R., Bolanos, M. L., Oliveira, S. V. W. B., Zaiat, M. & Foresti, E. (2004). Potential use of an anaerobic fixed-bed reactor for hazardous organic compounds degradation. *Proceedings of the Anaerobic Digestion, Montreal.*, 4, 2350-2353.

de Nardi, I. R., Zaiat, M. & Foresti, E. (2007). Kinetics of BTEX degradation in a packed-bed anaerobic reactor. *Biodegradation.*, 18, 83-90.

Dojka, M. A., Hugenholtz, P., Haack, S. K. & Pace, N. R. (1998). Microbial diversity in a hydrocarbon and chlorinated-solvent-contaminated aquifer undergoing intrinsic bioremediation. *Appl Environ Microbiol.*, 64, 3869-3877.

Domingues, M. R. (2007). *Investigação sobre a diversidade e a filogenia de arquéias e bactérias em amostras de sedimentos estuarinos enriquecidos na presença de clorofenóis.* PhD thesis, Escola de Engenharia de São Carlos, Universidade de São Paulo.

Droste, R. L., Kennedy, K. J., Lu, J. & Lentz, M. (1998). Removal of chlorinated phenols in upflow anaerobic sludge blanket reactors. *Water Science Technology.*, 38, 359-367.

El Fantroussi, S., Naveau, H. & Agathos, S. N. (1998). Anaerobic dechlorinating bacteria. *Biotechnology Progress.*, 14, 167-188.

Ficker, M., Krastel, K., Orlicky, S. & Edwards, E. (1999). Molecular characterization of toluene-degrading methanogenic consorptium. *Appl Environ Microbiol.*, 65, 5576-5585.

Foresti, E., Zaiat, M., Cabral, A. K. A. & Nery, V. (1995). Horizontal-flow anaerobic immobilized sludge (HAIS) reactor for paper industry wastewater treatment. *Braz J Chem Eng.*, 12, 157-163.

GrossKopf, R., Janssen, P. H. & Losack, W. (1998). Diversity and structure of the methanogenic community in anoxigenic rice paddy soil microcosms as axamined by cultivation and direct 16S RNA gene sequence retrieval. *Appl Environ Microbiol.*, 64, 960-969.

Fúlfaro, V. J. & Ponçano, W. L. (1976). Sedimentação atual do estuário e Baía de Santos: um modelo geológico aplicado a projetos de expansão da zona portuária. In: *Anais of Brazilian Congress of Geology and Engineering*, Rio de Janeiro, Brasil.

Griffiths, R. I., Whiteley, A. S., O'donnell, A. G. & Bailey, M. J. (2000). Rapid method for co-extraction of DNA and RNA from natural environments for analysis of ribosomal DNA and rRNA-based microbial community compositon *Appl Environ Microbiol.*, 66, 5488-5491.

Hendriksen, H. V., Larsen, S. & Ahring, B. K. (1992). Influence of a supplemental carbon source on anaerobic dechlorination of pentachlorophenol in granular sludge. *Appl Environ Microbiol.*, 58, 365-370.

Hortellani, M. A., Sarkis, J. E. S., Bonetti, J. & Bonetti, C. (2005). Evaluation of mercury contamination in sediments from Santos - São Vicente Estuarine System, São Paulo State, Brazil. *J Braz Chem Soc.*, 16, 1140-1149.

Huber, T., Faulkner, G. & Hugenholtz, P. (2004). Bellerophon: a program to detect chimeric sequences in multiple sequence alignments. *Bioinformatics.*, *20*, 2317-2319.

IBGE (Instituto Brasileiro de Geografia e Estatı́stica), 2000. <http:// www.ibge.gov.br>.

Juteau, P., Beaudet, R., McSween, G., Lépine, F., Milot, S. & Bisaillon, J. G. (1995). Anaerobic biodegradation of pentachlorophenol by a methanogenic consortium. Applied. *Microbiology and Biotechnology.*, *44*, 218-224.

Kudo, Y., Nakajima, T., Miyaki, T. & Oyaizu, H. (1997). Methanogen flora of paddy soils in Japan. *FEMS Microbiology Ecology.*, *22*, 39-48.

Lai, M. C., Chen, S. C., Shu, C. M., Chiou, M. S., Wang, C. C., Chuang, M. J., Hong, T. Y., Liu, C. C., Lai, L. J. & Hua, J. J. (2002). *Methanocalculus taiwanensis* sp. nov., isolated from an estuarine environment. *International Journal of Systematic and Evolutionary Microbiology.*, *52*, 1799-1806.

Madsen, T. & Aamand, J. (1992). Anaerobic transformation and toxicity of trichlorophenols in a stable enrichmente culture. *Applied and Environmental Microbiology.*, *58*, 557-561.

Madsen, T. & Licht, D. (1992). Isolation and characterization of an anaerobic chlorophenols – transforming bacterium. *Applied and Environmental Microbiology.*, *58*, 2874-2878.

Magar, V. S., Stensel, H. D., Puhakka, J. & Fergunson, J. F. (2000). Characterization Studies of an Anaerobic Pentachlorophenol –Dechlorinating Enrichment Culture. *Bioremediation Journal.*, *4*, 285-293.

Martins, C. C., Mahiques, M. M., Bícego, M. C., Fukumoto, M. M. & Montone, R. C. (2007). Comparison between anthropogenic hydrocarbons and magnetic susceptibility in sediment cores from the Santos Estuary, Brazil. *Marine Pollution Bulletin.*, *54*, 240-246.

Manz, W., Amann, R., Ludwig, W., Wagner, M. & Schleifer, K. H. (1992). Phylogenetic oligodeoxynucleotide probes for the major subclasses of proteobacteria: problems and solutions. *Syst Appl Microbiol.*, *15*, 593-600.

McAllister, K. A., Lee, H. & Trevors, J. T. (1996). Microbial degradation of pentachlorophenol. *Biodegradation.*, *7*, 1-40.

Middelburg, J. J., Soetaert, K., Herman, P. M. J. (1996). Evaluation of the nitrogen isotope-pairing method for measuring benthic denitrification: a simulation analysis. *Limnol Oceanog.*, *41*, 1839-1844.

Mikesell, D. M. & Boyd, A. S. (1986). Complete reductive dechlorination and mineralization of pentachlorophenol by anaerobic microorganisms. *Applied and Environmental Microbiology.*, *52*, 861-865.

Miranda, E. E. & Couninho, A. C. (2004). Brasil visto do espaço. Embrapa monitoramento por satellite, Campinas (http://www.cdbrasil.cnpm.embrapa.br).

MMA (2001). Brazilian national programme of action for the upper Southwest Atlantic Region. <http://www.gpa.unep.org/Documents/ NPA/NPA_BRAZIL.pdf>.

Mohn, W. W. & Kennedy, K. J (1992). Reductive dehalogentaion of chlorophenols by *Desulfomonile tiedjei* DCB-1. *Applied and Environmental Microbiology.*, *58*, 1367-1370.

Montenegro, M. A. P., Moraes, E. M., Soares, H. M. & Vazoller, R. F. (2002). Hybrid reactor performance in pentachlorophenol (PCP) removal by anaerobic granules. *Water Sci Technol.*, *44*, 137-144.

Monteiro Filho, M. (2002). *Águas degradadas*. Revista Problemas Brasileiros, n. 354 (<http://www.sescsp.org.br/sesc/revistas/revistas>).

Moraes, E. M., Adorno, M. A. T., Zaiat, M. & Foresti, E. (2000). A gas chromatographic determination approach for total volatile acids in effluents of anaerobic reactors treating liquid and solid wastes (Determinação de ácidos voláteis totais por cromatografia gasosa em efluentes de reatores anaeróbios tratando resíduos líquidos e sólidos). *Proceeding of the VI Latin American Workshop and Symposium on Anaerobic Digestion, Recife.*, 2, 235-238

Mori, K., Yamamoto, H., Kamagata, Y., Hastu, M. & Takamizawa, K. (2000). *Methanocalculus pumilus* sp. nov., a heavy-metal-tolerant methanogen isolated from a waste-disposal site. *Int J Syst Evol Microbiol.*, 50, 1723-1729.

Muyzer, G., De Waal, E. C. & Uitterlinden, A. G. (1993). Profiling of complex microbial populations by denaturing gradient gel electrophoresis analysis of polymerase chain reaction-amplified genes coding for 16S rRNA. *Applied and Environmental Microbiology.*, 59, 695-700.

Nakayama, C. R. (2005). *Degradação anaeróbia de pentaclorofenol (PCP), 2, 3, 4-triclorofenol (2,3 4-TCP) e 2, 6-diclorofenol (2,6-DCP) por culturas enriquecidas a partir de sedimento contaminado do Sistema Estuarino Santos e São Vicente.* pHD thesis. Instituto de Biociências, Universidade de São Paulo, São Paulo.

Nielsen, T. A., Liu, W. T., Filipe, C., Grady, L. & Molin, S. D. A. (1999). Identification of a novel group of bacteria in sludge from a deteriored biological phosphorus removal reactor. *Applied and Environmental Microbiology.*, 65, 1251-1258.

Ollivier, B., Fardeau, M., Cayol, J., Magot, M., Patel, B. K. C., Prensier, G. & Garcia, J. (1998). *Methanocalculus halotolerans* gen. nov., sp. nov., isolated from an oil-producing well. Int J Syst. *Bacteriol.*, 48, 821-828.

Orphan, V. J., Hinrichs, K. U., Ussler, W., Paul, C. K., Taylor, L. T., Sylva, S. P., Hayes, J. M. & DeLong, E. F. (2001). Comparative analysis of methane-oxidizing archaea and sulfate-reducing bacteria in anoxic marine sediments. *Applied and Environmental Microbiology.*, 67, 1922-1934.

Pedersen, T. F., Malcolm, S. J. & Sholkovitz, E. R. (1985). A lightweight gravity corer for undisturbed sampling of soft sediments. *Can J of Earth Sciences.*, 22, 133-135.

Piza, F. F., Prado, P. A. & Manfio, G. P. (2004). Investigation of bacterial diversity in Brazilian tropical estuarine sediments reveals high actinobacterial diversity. *Antonie van Leeuwenhoek.*, 86, 317-328.

Piza, F. F. (2004). Ecologia molecular microbiana associada a sedimentos do Estuário de Santos-São Vicente (SP, Brazil). Tese de Doutorado. Universidade Estadual de Campinas, *Campinas, São Paulo.*

Purdy, K. J., Munson, M. A., Nelwell, D. B., Embley, T. M. (2002). Comparation of molecular diversity of the methanogenic community at the brackish and marine ends of the UK estuary. *FEMS Microbial Ecology.*, 39, 17-21.

Raskin, L., Stromley, J. M., Rittman, B. E. & Stahl, D. A. (1994). Group-specific 16S rRNA hybridization probes to describe natural communities of methanogens. *Appl Environ Microbiol.*, 60, 1232-1240.

Rodrigues, D. F., Sakata, S. K., Comasseto, J. V., Bícego, M. C. & Pellizari, V. H. (2009). Diversity of hydrocarbon-degrading Klebsiella strains isolated from hydrocarbon-contaminated estuaries. *J Appl Microbiol.*, 106, 1304-1314.

Robles-González, I. V., Fava, F. & Poggi-Varaldo, H. M. (2008). A review on slurry bioreactors for bioremediation of soils and sediments *Microbial Cell Factories.*, 7, 1-16.

Rosa, L. P., Sikar, B. M., Santos, M. A., Monteiro, J. L., Sikar, E. M., Silva, M. B., Santos, E. D. & Junior, A. P. (2002). Emissões de gases do efeito estufa derivados de reservatórios hidrelétricos Projeto BRA/00/029 166p, Rio de Janeiro, COPPE/UFRJ.

Saia, F. T. (2005). Contribuição à exploração tecnológica dos estudos microbianos realizados no Programa BIOTA FAPESP: Avaliação do potencial da degradação anaeróbia de pentaclorofenol em reator anaeróbio. PhD thesis, Escola de Engenharia de São Carlos, Universidade de São Paulo.

Saia, F. T., Damianovic, M. H. R. Z., Cattony, E. B. M., Brucha, G., Foresti, E. & Vazoller, R. F. (2007). Anaerobic biodegradation of pentachlorophenol in a fixed-film reactor inoculated with polluted sediment from Santos-São Vicente Estuary, Brazil. *Appl Microbiol Biotechnol.*, 75, 665-672.

So, M. C. & Young, L. Y. (1999). Isolation and characterization of a sulfate-reducing bacterium that anaerobically degrades alkanes. *Applied and Environmental Microbiology.*, 65, 2969-2976.

Standard methods for the examination of water and waste water (1998). 19th edition: American Public Health Association/ American Water Association/ Water Environment Federation, Washington, DC, USA.

Smith, V. H., Graham, D. W. & Cleland, D. D. (1998). Application of Resource-Ratio Theory to Hydrocarbon Biodegradation. *Environmental Science and Technology.*, 32, 3386-3395.

Speece, R. E. (1996). *Anaerobic Biotechnology for industrial wastewater Nechville*: Archaea Press., 394.

Stahl, D. A. & Amman, R. I. (1991). Development and application of nucleic acid probes. In: STACKEBRANDT, E.; GOODFELLOW, M. eds., *Nucleic acid techniques in bacterial systematics*, John Wiley & Sons, Ltd, London, England., 8, 207-248.

Tartakovisky, B., Manuel, M. F. J., Beaumier, D., Greer, C. W. & Guiot, S. R. (2001). Enhanced selection of an anaerobic pentachlorophenol-degrading consortium. *Biotechnol Bioeng.*, 73, 476-483.

Utkin, I. B., Woese, C. & Wiegel, J. (1994). Isolation and characterization of *Desulfitobacterium dehalogenans* gen.nov.sp.nov. an anaerobic bacterium which reductively dechlorinates chlorophenolic compounds. *International Journal of Systematic Bacteriology.*, 44, 612-619.

Watanabe, K., Kodama, Y., Hamamura, N. & Kadu, N. (2002). Diversity abundance and activity archaeal population in oil-contaminated graundwater accumulated at the botton of an undergraound crude oil storage cavity. *Appl Environ Microbiol.*, 62, 4299-4301.

Wild, S. R., Harrard, S. J. & Jones, K. C. (1992).. Pentachlorophenol in the UK environment in: A budget and source inventory. *Chemosphere.*, 24, 833-845.

World Health Organization-WHO (1989). Pentachlorophenol. *Environmental Health Criteria*, v. *71*. 236.

Zanardi, E., Bıcego, M. C., Castro Filho, B. M., Miranda, L. B. & Prosperi, V. (2000). Southern Brazil. In: Shepard, Charles. (Org.). Seas At Millenium: In *Environmental Evaluation, vol. 1*. Amsterdam., 731-747.

Zinder, S. H., Cardwell, T., Anguish, M., Lee, M. & Koch, M. (1984). Methanogenesis in a thermophilic (58^{0}C) anaerobic digestor: *Methanothrix sp.* as an important acetoclastic methanogen. *Appl Environ Microbiol.*, 47, 796-807.

Zou, S., Anders, K. M. & Fergunson, J. F. (2000). Bioestimulation and Bioaugmentation of Anaerobic Pentachlorophenol Degradation in contaminated Soils. *Bioremediation Journal.*, *4*, 19-25.

In: Encyclopedia of Environmental Research
Editor: Alisa N. Souter

ISBN: 978-1-61761-927-4
© 2011 Nova Science Publishers, Inc.

Chapter 42

PHYSICO-CHEMICAL CHARACTERISTICS OF NEGATIVE ESTUARIES IN THE NORTHERN GULF OF CALIFORNIA, MEXICO

Hem Nalini Morzaria-Luna[1], Abigail Iris-Maldonado[2] and Paloma Valdivia-Jiménez

[1]Centro Intercultural de Estudios de Desiertos y Océanos, A.C. Edif. Agustín Cortés s/n. C.P. 83550 Puerto Peñasco. Sonora, México.

[2] Current address: Posgrado en Ecología Marina. Centro de Investigación Científica y de Educación Superior de Ensenada, BC. Km. 107 Carretera Tijuana - Ensenada C.P. 22860, Ensenada, B.C. México.

ABSTRACT

We describe water quality in two hypersaline negative estuaries in the Northern Gulf of California, along the coast of Sonora, Mexico, over a two-year period. In the Northern Gulf, non-mangrove salt marshes known as esteros (negative estuaries) cover 134, 623 ha. Esteros are characterized by an extreme tidal range, higher salinity at their head than at their mouth due to high evaporation, limited freshwater input and a mixed semi-diurnal tidal regime. Between 2005 and 2007, we sampled surface temperature, dissolved oxygen, salinity, pH, depth, chlorophyll, nutrients (NH_{+4}, NO_3^-, NO_2^-, and PO_4), and total suspended solids across one wetland, Estero Morúa. We also led a participatory monitoring effort, where oyster farmers took daily measurements of surface temperature, salinity, pH, and dissolved oxygen in both Estero Morúa (31°17′09" N; 113° 26′19" W) and Estero Almejas (31°10′15" N; 113°03′53" W). These sites allow comparisons between distinct habitats and levels of oceanic influence. Estero Morúa is a high energy lagoon with a narrow mouth and a prominent permanent channel restricted by spits, while Almejas is an open bay with a large intertidal mudflat. Among the main findings are 1) Surface temperatures follow a seasonal pattern, with highest temperatures in June to September and lowest from December to February. 2) Low rainfall and runoff together with high seawater input and evaporative loss results in high salinities. 3) High dissolved oxygen concentrations and low nutrient levels are indicative of the recharge rate between the wetlands and the sea, and are characteristic of oligotrophic systems. It is likely that residence time and tidal dynamics are the main factors dictating the physicochemical dynamics of these systems.

INTRODUCTION

Negative or inverse estuaries are those where seawater is concentrated by the removal of fresh water, either by evaporation that exceeds precipitation, or by freezing and the occurrence of sea ice [1,2]. The former type is usually associated with arid climates and occurs in both hemispheres including in the Red Sea, the Gulf of Suez, the Mediterranean Sea, the Adriatic Sea, the Arabian Gulf, South Australia, and the Gulf of California [2]. Negative estuaries can be permanent or can be negative only during part of the year where rainfall is strongly seasonal [3].

According to Miller et al. [4], salinity in negative estuaries can increase as one moves horizontally from the mouth of the estuary to the head, especially in the bottom layers. Gravitational circulation can lead to the export of high density lagoonal water, at the same time that ocean water fills into the lagoon as a low-density surface layer. The water exchange between the lagoon and the ocean is enhanced in the flood and ebb cycles. The physical and chemical characteristics of water within estuaries are heavily influenced by meteorological conditions [5], the amount of freshwater runoff [6] and tidal hydrology [7]. The interaction among tidal hydraulics, open water, benthic and marsh surface processes yields unique water chemistry profiles [7].

In Mexico, negative estuaries are hypersaline salt marshes where salinities may exceed > 40, pH ~ 9, oxygen is close to saturation and total nitrogen is low [8]. These wetlands are found in both the Gulf of Mexico and Pacific coasts of the country, and are common in the states of Baja California, Sonora and Oaxaca [9]. Negative estuaries in the Northern Gulf of California, above 29 °N, are non-mangrove salt marshes locally known as *esteros* [10] that connect to the open gulf by channels or inlets. The Northern Gulf is very arid, with mean annual rainfall < 125 mm and mean maximum annual temperature of 22 - 26 °C [11]. There are no perennial rivers [12], and the lack of freshwater input and excess evaporation result in an increasing salinity gradient towards the head of wetlands [13]. Tidal range in the Northern Gulf may exceed 8 m [14], so the wetlands empty out during low tides and at high tides the water floods the marsh surface, resulting in varied areas of open water habitat, exposed mudflat, and shallow water available daily [15]. Thus, tidal hydrology plays an important role in connecting different areas of the marsh that remain isolated during most of the tidal period.

Environmental conditions can be challenging for the organisms that inhabit *esteros*. Many species use negative estuaries opportunistically, entering only during high tides to rest, feed and reproduce [3]. Other species are residents with special adaptations to withstand the harsh environmental conditions. *Esteros* represent one of the hottest zones occupied by fish [16]; eurythermal species such as the gobies *Gillichthys miriabilis*, the longjaw mudsucker and the endemic *Gillichthys seta*, the shortjaw mudsucker are some of the most abundant residents [17,18]. These fish can withstand the daily temperature fluctuations in the wetlands that can range from 5° to 7° C seasonally [19]. Other species, such as the halophyte marsh plants *Sarcocornia pacifica*, *Mönantochloe littoralis,* and the endemic *Distichlis palmeri*, use the controlled uptake of Na^+ into cell vacuoles to drive water into the plant tissues against a low external water potential.

The description of water quality in negative estuaries can be useful to understand the ecology of species that inhabit them and represents important background information on the dynamics of physical and chemical parameters. We evaluated seasonal variability of water quality in Estero Morúa and Estero Almejas in the state of Sonora, Mexico as part of a project to determine the causes of oyster mortality. We recorded monthly measurements in Estero Morúa, as well as daily measurements in Estero Morúa and Estero Almejas as part of participatory monitoring, where members of local oyster cooperatives measured key parameters in their farms.

METHODS

Survey Location

We surveyed water quality in Estero Morúa (31°17'09" N; 113° 26'19" W; Figure 1) and Estero Almejas (31°10'15" N; 113°03'53" W; Figure 1). Estero Almejas and Morúa are negative estuaries with a higher salinity at the head than at the mouth [1]. These sites are located in the Northern Gulf of California, in the state of Sonora in Northwestern Mexico. Estero Morúa and Estero Almejas are within the municipality of Puerto Peñasco; the main city in the area is Puerto Peñasco (Pop. 44,875, [22].

The Northern Gulf of California is characterized by climatic conditions of a low-altitude coastal desert (Foster 1975), with scarce rain during the summer averaging 70 – 90 mm [24,10]. The annual temperature average is 20.1°C, fluctuating between 11.3°C in January and 30°C in August [10]. High evaporation rates, coupled with the lack of freshwater input, results in the increasing salinity gradient inside the estuaries [12]. The Sonoran coast experiences prevailing northerly and northwesterly winds in the winter, and southern and southeasterly winds in the summer. During the winter, offshore winds predominate and in the summer onshore winds are dominant [24]. Maximum wind speeds vary from 43 km h-1 (maximum one-day mean) in winter to maximum 21.6 km h-1 in summer, with occasional gusts reaching 118 km h-1 [25].

Estero Morúa (Figure 2) extends over 1,097 ha [13], and is of deltaic origin; it formed as part of an estuarine system fed by the Sonoyta river, although the flow is now diverted for agriculture and no longer reaches the sea [26]. In the northeastern side of the wetland, remnant riparian vegetation remains in the dry river bed maintained by subsurface flow [13]. Estero Morúa is a high energy embayment with a narrow mouth and a central channel restricted by spits [27] that measures 11.58 km in length with a mouth 0.5 km wide. Rock outcroppings protrude through the sand at the mouth of the estero. Sediment grain at the mouth is coarse sand (0.5 mm) and particle size decreases inwards to fine sand in the side channels and silt in the upper reaches at the head [27]. The main lagoon is separated from the Gulf by barrier dunes that used to reached heights of ~100 feet [28], but have been partially been flattened during the construction of the surrounding residential developments. Estero Morúa houses four aquaculture cooperatives that farm Japanese oysters (*Cassostrea virginica*) along the main tidal channel.

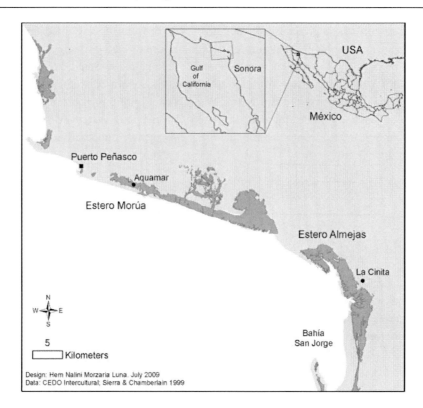

Figure 1. Location of Estero Morúa and Estero Almejas, hypersaline negative estuaries in the state of Sonora, northwestern Mexico and (●) oyster cooperatives, where the participatory water quality monitoring program was implemented.

Figure 2. Estero Morúa, Sonora, México at neap tide, July 2009. Photo: Alejandro Castillo. CEDO Intercultural.

Figure 3. Estero Almejas, Sonora, México at neap tide, July 2009. Photo: Alejandro Castillo. CEDO Intercultural.

Estero Almejas (Figure 3; also known as Bahía Salina) is located in Bahía San Jorge, and extends over 2,286.02 ha [10]. The wetland is a semi-open embayment with a sand bar to the north composed of consolidated dunes. When the tide recedes, it exposes extensive mudflats. The landward area is covered in salt flats, where salt was harvested in evaporative lagoons (Salinera del Desierto Rojo). The coastal beaches and mudflats of Estero Almejas formed during the Holocene, ~ 5,000 BP, when sea level reached its maximum. The beaches and mudflats were formed by wind-blown sediments and silt deposited by the tides and currents [13]. Estero Almejas has one oyster farm and a fishing camp in the vicinity (Pop. 41) [22] which sustains important artisanal fisheries for blue crab (*Callinectes arcuatus* and *Callinectes bellicosus*), rays, and pelagic fish [23]

In both Estero Morúa and Almejas, the wetland channels are surrounded by salt marsh vegetated with halophytes that rarely grow beyond 50-60 cm [29]. The immediate wetland margin is covered by low shrubs including Palmer's frankenia (*Frankenia palmeri*), white bur sage (*Ambrosia dumosa*), creosote bush (*Larrea tridentata*), salt bush (*Atriplex polycarpa*, *Atriplex canescens*), and desert thorn (*Lycium* spp.) [30]. Beyond, we find vegetation of the Sonoran Desert - Lower Colorado River Valley subdivision [29].

The tidal regime in the Northern Gulf is mixed semidiurnal (two high and two low tides of different elevations in 24 hours), with tidal amplitudes of 5-10 m [31]. In the Puerto Peñasco region tidal amplitude is 7.04 m, with lowest tides in April and highest in October [21]. During the highest spring tide the marshes flood completely, and at the lowest neap tide the wetlands empty, uncovering extensive mudflats where some water remains forming pools [15].

We describe sample collection and analysis for the monitoring program in Estero Morúa, where data were collected monthly during neap tide. We then describe the participatory monitoring program in Esteros Morúa and Almejas, for which data were collected daily.

Monitoring Program in Estero Morúa

Sample collection – In Estero Morúa, samples were collected between 2005 and 2007. We used a stratified random sampling design, to account for the difference in channel and creek area across the marsh. Starting in August 2005, for each month we selected ten random points across six sections. Each such set of ten points is referred to hereafter as a "survey". In total, we sampled 220 points throughout 22 months (Figure 4). Samples were collected at the lowest tide of the neap tide, to minimize oceanic influence. Thus, collection times varied monthly. Sampling points were visited in 1-2 days. The time for sampling was selected using tidal prediction software (WXTide 32 Versión 4.7; http:\\wxtide32.com).

At each point we took three replicate measurements of each parameter: water temperature, dissolved oxygen (D.O.), salinity, and pH (Figure 3). We also recorded water depth, time and location. Dissolved oxygen, water temperature, and pH were measured with an Oakton multimeter 300 (Oakton Instruments, Vernon Hills, Il. EUA). Sensors were immersed in the water column until a constant value was obtained (Figure 5). Dissolved oxygen is expressed in mg/L and temperature in °C. Oxygen measurements were compensated by temperature but not for salinity, as salinity values were routinely higher than the instrument's compensation limits. A pipette was used to collect 2-3 drops of water, which were placed on a Vista refractometer to measure salinity on a 0-100 scale. Location was recorded with an eTrex Garmin GPS (Garmin Ltd., Olathe, KS, EUA), in the WSG1984 coordinate system.

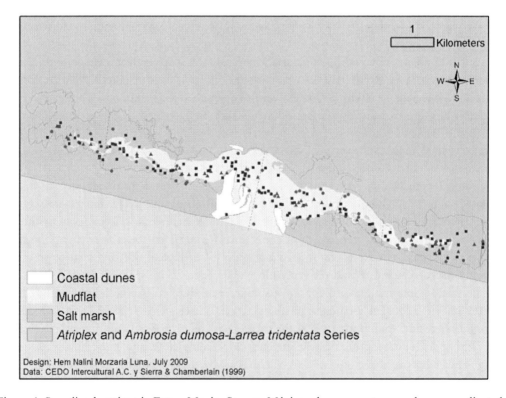

Figure 4. Sampling locations in Estero Morúa, Sonora, México where seawater samples were collected in 2005 (●), 2006 (■), and 2007 (▲). Dashed lines indicate the six sections where random points were selected monthly using a stratified random sampling design. Figure also shows habitat types.

Figure 5. Data collection in Estero Morúa, Sonora, México. Photo: Alejandro Castillo. CEDO Intercultural.

Starting in September 2005, we also collected 4 liters of water at six alternate points (usually points 1, 3, 5, 6, 8 and 10) for determination of dissolved nutrients (NO_2^-, NO_3^-, NH_{+4} and PO_4) and particulate compounds (total suspended solids, TSS, chlorophyll a, b and c). Samples for TSS collected in September and October were lost, so data are only available for this parameter from November 2005 to May 2007.

We also measured sea surface temperature on the open coast and ambient air temperature. Sea surface temperature was measured daily at approximately 9 am outside of Estero Morúa, at 31°17′21" N; 113°29′47" W. Between August 2005 and April 2006 temperature was measured using a partial immersion mercury thermometer. After this date temperature was measured using an electronic VWR waterproof remote probe thermometer with ± 0.1 °C precision (VWR International LLC, Batavia IL, USA). Ambient air temperature was recorded at CEDO's field station, ~ 0.6 km from Estero Morúa, using a Weather Monitor II station (Davis Instruments, Haywood, CA, USA). We report monthly averages based on hourly data.

Sample analysis - We analyzed TSS using the gravimetric method [32]. We filtered 700ml of seawater or less, as necessary to saturate an ash-free Whatman GF/C filter of known weight (previously burned at 400 °C for four hours). The filters were then dried at 60 °C until of constant weight. The filters were re-weighed and the total solids were reported as the weight difference between the two weights adjusted for volume filtered in mg/L.

Chlorophylls were analyzed following Arar [33]. We analyzed 700 ml of seawater or less, as necessary to saturate a Whatman GF/C filter. We saved 20 ml of the filtered water for nutrient analysis; these samples were frozen until analysis. While filtering the last 10 ml of water we added 1 ml of saturated $MgCO_3$ suspension. The filters were wrapped in aluminum foil and frozen until analysis. For chlorophyll extraction, the filters were ground with 12 ml of acetone with a glass rod and incubated for 4 h in the dark (shaking every hour), and centrifuged at 1000 g for 5 min. The supernatant was read in a spectrophotometer at 750, 664, 647 and 630 nm. Concentrations of chlorophylls a, b and c were determined based on the trichromatic equations of Jeffrey & Humphrey [34] and expressed as mg/m^3.

The nutrient analysis was carried out in the Centro de Investigaciones Biológicas del Noroeste, in Hermosillo, Sonora. Water samples were thawed and analyzed using the microplate technique described by Hernández-López & Vargas-Albores [35]. This method is based on the formation of colored complexes in 96-well plates that are then measured by spectrophotometry, reducing costs while maintaining precision. Nutrient concentrations are presented in μM.

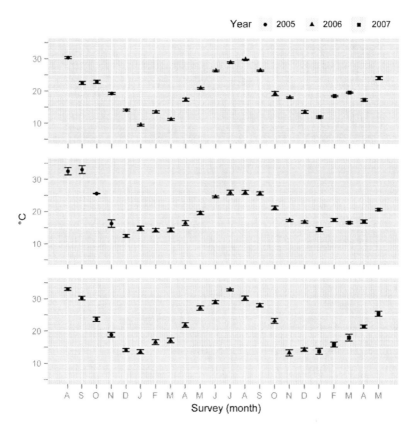

Figure 6. Water temperature (means ± SE) between August 2005 and May 2007. a. Monthly measurements in Estero Morúa (n = 30); b. Participatory monitoring in Estero Morúa, daily readings. c. Participatory monitoring in Estero Almejas, daily readings.

Participatory monitoring

Sample collection - The participatory monitoring program collected daily data but had no spatial replication, as all samples were taken in a single point across time. Measurements were recorded between August 14, 2005 and May 15, 2007. In Estero Almejas, the Sociedad Cooperativa de Producción Pesquera y Acuícola (SCPPA) La Cinita collected data throughout the study period. In Estero Morúa, the SCPPA Unica de Mujeres collected data between August and November 2005, after this time it was replaced by the SCPPA Aquamar. Members of the cooperatives received a pH meter, a D.O. meter, and a refractometer from the Aquaculture Institute of the State of Sonora. We prepared a manual and training course on water quality monitoring, that were offered to members of each cooperative prior to the start of the project. We also assisted the cooperatives with equipment maintenance and calibration and maintained the database for the duration of the program.

Each day, a member of the cooperative registered near their oyster trays sea surface temperature, D.O., salinity and pH. Oxygen measurements were compensated for temperature but not for salinity, as salinity readings were routinely higher than the instrument's compensation limits. We requested that measurements be carried out at or near the same time every day. In Estero Almejas, the measurements generally occurred between 6-8 am, and were recorded by 13 people who alternated. Readings were recorded on 95% of the days of the study period. In Estero Morúa, one person was routinely responsible for data collection, parameters were measured at variable times and compliance was only 77 %. Several problems occurred during the participatory monitoring program, including equipment breakdown and errors while measuring and recording data. Thus, the quantity of data available by parameter and site is variable.

Data Analysis

We used linear regression to study the relationship between temperature, salinity, D.O., pH, total suspended solids, chlorophylls (*a*, *b* and *c*), and nutrients (NO_{-2}, NO_{-3}, NH_{+4} and PO_4) across the months surveyed. When negative values were present for parameters that conceptually can only be ≥0, such as nutrients and chlorophylls, we corrected the data by substituting negative values for 0. The Akaike Information Criterion (AIC) was used on the linear models to select the variables that best explained the structure of the data [36]. Data summarized in tables and graphs is untransformed. Data analysis was carried out in the R Open Source system (R Development Core Team).

RESULTS AND DISCUSSION

Temperature, Salinity, pH and Dissolved Oxygen

Mean water temperature in Estero Morúa throughout the study period was 19.8 ± 0.2 °C (Table 1). Temperature followed a seasonal pattern with highest temperatures in summer (30.3 ± 0.3 °C in August 2005) and lowest in winter (9.4 ± 0.3 °C in January 2006; F = 362.27, P < 0.001; Figure 6). Previous studies in Estero Morúa found similar temperature intervals. Place & Hofmann [37] found that on average water temperature in the estuary ranged from 5° to 30°C during tidal cycles in winter and from 18° to 36°C during tidal cycles in summer. Buckley & Hofmann [19] recorded temperature in the water column of Estero Morúa between January and June 2001, and found a range from < 5°C in early January to higher than 33 °C in June. We also found significant variability in water temperature between sampling locations (F = 12.7, P < 0.001). Spatial variations in temperature in the wetland result from differences in water depth, tidal exchange and bottom sediments, which may play a role in the absorption or transfer of heat to overlaying waters [38].

Water temperatures from the participatory monitoring program were in agreement with monthly data, although the latter average measurements throughout the tidal cycle. Mean temperature was 22.2 ± 0.3 °C in Estero Morúa, and there was variation between surveys (F = 95.39, P < 0.001), with the highest mean in September 2005, 33 ± 0.4 °C, and lowest in November 2006, 13.3 ± 1 °C. In Estero Almejas, temperature varied between 33 ± 1.2 °C in August 2005, to 12.5 ± 1 °C in December 2005 (Figure 6), and there was significant variation amongst surveys (F = 61.36, P < 0.001). Water temperature in both wetlands follows the general patterns of sea surface and ambient temperature measured outside Estero Morúa (Figure 7). The main exception is in winter, when water temperatures in the estuary are lower because tidal areas are shallow and lack sufficient heat storage capacity. The pattern of variation in water temperature is the result of variations of solar heating, which is extreme in the desert climate of Northwest Mexico [1] and can also be related to seasonal changes in water masses [39].

Table 1. Physical and chemical parameters of water in two coastal wetlands of the Northern Gulf of California, measured between August 2005 and May 2007

Model	Estero Morúa			Participatory monitoring					
				Estero Morúa			Estero Almejas		
	Mean	Max	Min	Mean	Max	Min	Mean	Max	Min
Temp. (°C)	19.77 ± 0.24	33.1	6	19.53 ± 0.28	39.6	10	22.19 ± 0.31	37.7	8.05
Sal.	46.13 ± 0.38	> 100	31	43.83 ± 0.16	54	34	47.12 ± 0.16	60	38
pH	8.06 ± 0.02	10.25	5.09	7.88 ± 0.02	8.98	6.08	7.91 ± 0.19	9.62	4.73
D.O. (mg/L)	8.68 ± 0.07	16.12	3.81	8.31 ± 0.07	14.64	6.01	8.14 ± 0.06	12.14	5.25
NO_3^- (μM)	2.92 ± 0.16	6.52	0						
NO_2^- (μM)	0.001 ± 0.0002	0.01	0						
NH_4^+ (μM)	0.006 ± 0.0006	0.02	0						
PO_4 (μM)	0.002 ± 0.0004	0.04	0						
TSS (mg/L)	18 ± 0.45	40.95	7.52						
Chl a (mg/m^3)	0.13 ± 0.02	0	0.02						
Chl b (mg/m^3)	0. 03 ± 0.01	2.54	0						
Chl c (mg/m^3)	0.07 ± 0.02	2.92	0						

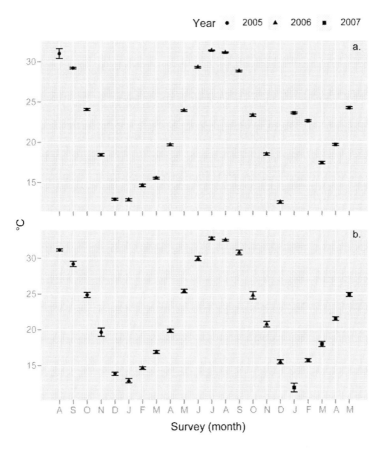

Figure 7. Ambient and sea surface temperature (b.) measured in the CEDO field station and outside of Estero Morúa during the study period (means ± SE).

Salinity values were high in Estero Morúa, overall 46.2 ± 0.4 (Table 1), and varied between 53.2 ± 3.2 in April 2006 to 41.1 ± 0.45 in May 2007 (Figure 8). These hypersaline values are likely due to excess evaporation and lack of freshwater input. The Puerto Peñasco region receives < 100 mm of rain per year [24]. Although there was variation among surveys (F = 10.15, P < 0.001) there was no seasonal pattern. We also found significant variation between sampling locations (F = 83.8, P < 0.001), a product of the spatial variability in salinity, with higher values at the head of the estuary compared to the mouth, where there is more water exchange with the Gulf. Salinity increases in the inner arms of bays and lagoons are typical in regions where evaporation exceeds river flow and precipitation [8]; in the Northern Gulf evaporation exceeds precipitation by 250 cm/yr-1 [1].

In the participatory monitoring program, daily salinity readings in Estero Morúa (43.8 ± 0.2; 45.21 ± 0.56 April 2006 – 38.9 ± 0.5 August 2005) and Estero Almejas (47 ± 0.2; 49.7 ± 0.7 May 2006 – 42.4 ± 0.5 August 2005) were equivalent to monthly data. At both sites, salinity varied among surveys (Estero Morúa: F = 6.44, P < 0.001; Estero Almejas: F = 20.24, P < 0.001; Figure 8). Salinity values recorded in both Estero Morúa and Almejas are similar to those reported for other inverse estuaries. In Bahía San Quintín, on the Pacific Coast of the Baja California Península, México, average salinities range from 34.7 in summer to 33.8 in winter [40]; while in Estero La Cruz, in the central Gulf of California, salinity averages 36.8

[41]. We found water salinity values as low as 31 in October 2005 and as high as >100 in June 2006. These extreme values are common in negative estuaries. Valenzuela-Siu et al. [42] found salinity varied between 35 – 42 in summer and 37 – 39 in winter at Estero Lobos.

pH values in Estero Morúa were on average 8.4 ± 0.2 (Table 1), with a high of 9.1 ± 0.1 in February 2006 and a low of 7.4 ± 0.2 in September 2006. The overall mean is slightly higher than pH values for seawater in equilibrium with atmospheric CO_2 (8.1-8.3) [38]. There was no significant variation in pH values among sample locations or surveys (Figure 9). We found that pH readings were very sensitive to small changes in how measurements are collected, which could explain some of the high values. It is unlikely that high pH values are related to plant growth which reduces CO_2 content, as we did not sample the marsh surface. pH data from the participatory monitoring program showed less variation than monthly data (Figure 9). pH values in Estero Morúa varied between 8.1 ± 0.02 in November 2005 to 7.6 ± 0.03 February 2007 (7.9 ± 0.02 overall). In Estero Almejas, mean pH during the study period was 7.9 ± 0.02, and varied from 8.2 ± 0.1 November 2005 to 7.4 ± 0.1 in October 2006 (Figure 9). Lower pH values, especially in the summer, could reflect the production of CO_2 by the decay of organic matter in bottom muds [8].

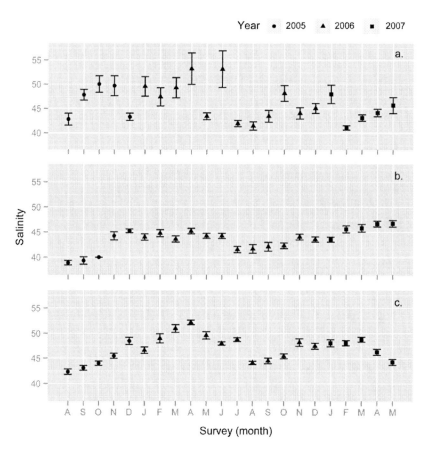

Figure 8. Water salinity (means ± SE) between August 2005 and May 2007. a. Monthly measurements in Estero Morúa (n = 30); b. Participatory monitoring in Estero Morúa, daily readings. c. Participatory monitoring in Estero Almejas, daily readings.

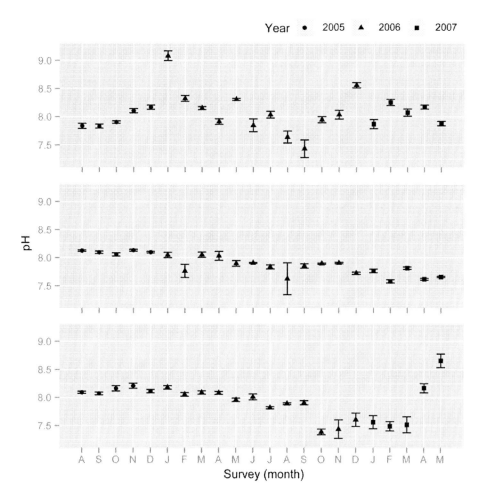

Figure 9. pH (means ± SE) between August 2005 and May 2007. a. Monthly measurements in Estero Morúa (n = 30); b. Participatory monitoring in Estero Morúa, daily readings. c. Participatory monitoring in Estero Almejas, daily readings.

Dissolved oxygen concentrations in Estero Morúa were high (8.68 ± 0.07 mg/L), and varied between 11.34 ± 0.18 mg/L in July 2006 to 6.21 ± 0.22 mg/L in August 2005 (Figure 10). We found that D.O. values varied with location (F = 2.46, P < 0.001) and survey (F = 64.5, P < 0.001), with a general pattern of increasing concentrations in warmer months, but no definite seasonal pattern (Figure 10). In other coastal lagoons, such as Estero La Cruz, Sonora, D.O. shows a clear seasonal pattern with higher values in winter (8.58 ml/L) to (2.58 ml/l) in spring and summer [41].

The D.O. values we found are within the intervals previously reported in other coastal lagoons in the Gulf of California [40]. Dissolved oxygen in Estero Almejas and Estero Morúa measured during the participatory monitoring showed a similar pattern, with low values of 6.1 ± 0.08 and 6.09 ± 0.08 mg/L respectively in August 2005, to highs of 10.91 ± 0.22 mg/L on April 2007 in Morúa and 10.61 ± 0.11 mg/L on December 2006 in Almejas (Figure 10). The high D.O. concentrations can be attributed to the high exchange rate between the wetland and the sea, and the low depth of the water column during sampling, while high values in winter are related to increased solubility at low temperatures [41].

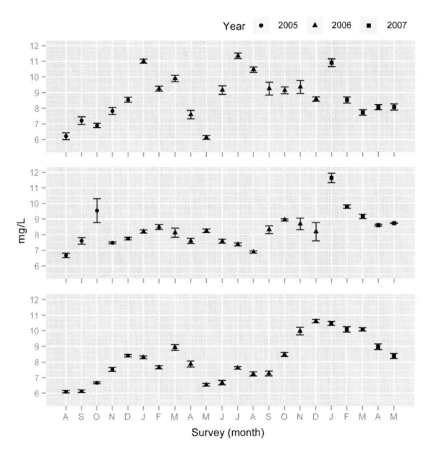

Figure 10. Dissolved oxygen (means ± SE) between August 2005 and May 2007. a. Monthly measurements in Estero Morúa (n = 30); b. Participatory monitoring in Estero Morúa, daily readings. c. Participatory monitoring in Estero Almejas, daily readings.

Dissolved Nutrients

We found that nutrient concentrations in Estero Morúa were low, likely a result of low precipitation and runoff. Figure 11 shows the fluctuations of the various forms of nitrogen and of phosphate in water throughout the study period. Nitrate was the main inorganic N form in Estero Morúa, with concentrations of 3.17 ± 0.15 µM (Table 1). The other nutrients were present in lower concentrations, NO_2^-, 0.001 ± 0.0003 µM, PO_4, 0.002 ± 0.0004 µM, and NH_{4+}, 0.006 ± 0.0006 µM. Nitrogen concentrations are lower than values reported for other hypersaline wetlands in the Northern Gulf. In Estero Lobos, $NO_2^- + NO_3^-$ ranged from 1.2 – 1.1 µM seasonally and NH_{4+} increased from 0.6 µM in summer to 1.2 µM in winter [42]. The dynamics of biophilic elements, such as nitrogen and phosphorus, in tidal estuaries are strongly related to the variation in tidal cycle [43]. These changes are strongly dependent on the spring-neap stage or amplitude [7], current velocity [44] winds [45] and precipitation rate, which affect runoff [6]. The nutrients found in Estero Morua probably stem from internal recycling and wind; desert dust may provide nutrient inputs to wetlands in arid areas [46]. Runoff during rain is most certainly an insignificant factor, as in 2006, Puerto Peñasco received only 2.27 cm of rain and between January and June 2007 only 0.13 cm (Data: CEDO Intercultural). In Northwest Sonora, there are no other nutrient sources, such as rivers or the shrimp aquaculture farms that are present in coastal lagoons in the southern part of the state [47]. The observed increase in NH_{4+} and NO_2^- during 2007 (Figure 11) could be related to upwelling frequently present on the eastern coast of the Gulf of California during winter-spring [48].

The low nutrient levels observed in Estero Morúa be an artifact from collecting samples at low tide in wetlands nutrients can be rapidly sequestered by microbial activity in benthic surfaces, in which case water column monitoring such as ours can result in an underestimation of nutrient availability [50].

NO_2^- and NH_{4+} concentrations were higher in winter and spring 2007 (Figure 11) relative to other surveys (F = 11.28, P < 0.001; F = 34.72, P < 0.001), while PO_4 showed few differences between surveys (Figure 11). Nitrate concentrations were higher, 3.17 ± 0.15 µM, than for other nutrients and showed high variability as a function of sampling location (F = 2.76, P = 0.02) and survey (F = 11.03, P < 0.0001), but there is no clear monthly or yearly pattern. Patterns in the variability of total dissolved N (NO_3^- + NO_2^- + NH_{4+}) in Estero Morúa will reflect the concentrations of NO_3^-, since levels of the other forms of nitrogen were so low.

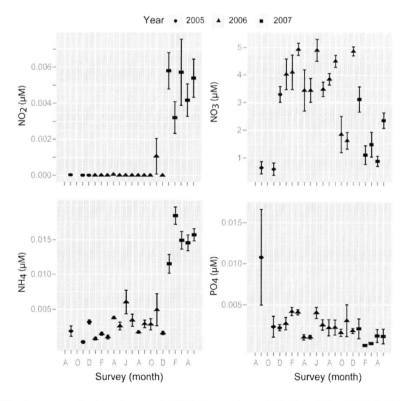

Figure 11. Nutrient concentrations (means ± SE) between August 2005 and May 2007 in Estero Morúa (n = 6).

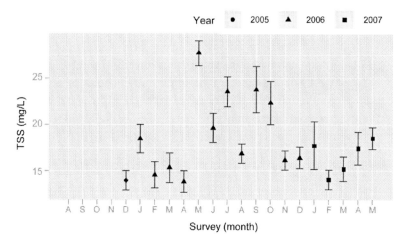

Figure 12. Total dissolved solids (means ± SE, n=12) measured monthly in Estero Morúa between October 2005 and May 2007.

Particulate Compounds

The average TSS concentration was 18.06 ± 0.45 mg/L (Table 1). We sampled in low tide where water depth during the study period was 15.26 ± 0.43 cm. In shallow areas, light can penetrate to the entire water column and hydrodynamics are affected by bottom morphology and wind, which promote the resuspension of materials, nutrients and small biota from the bottom surface into the water column [50]. TSS is also influenced by the type of vegetation, land cover and land use surrounding the site. Estero Morúa is surrounded by Sonoran Desert, with low lying, scarce vegetation [29], and nearby residential developments (Figure 2) have contributed to erosion by destroying vegetation and flattening the dunes that surrounded the site on the seaward edge [28].

We found that TSS concentration varied between survey (F = 6.77, P < 0.001; Figure 12) and sampling location (F = 4.36, P < 0.001). A high TSS concentration was present in May 2006 (27.75 ± 1.33 mg/L) and some summer months (Figure 12), but there was no clear seasonal pattern. In estuaries, the cycling of suspended particles depends on processes such as re-suspension and settling, mudflat processes and particle-particle interactions (i.e. Brownian motion, differential settling and coagulation), which are driven by river flow, tidal energy and storms and tend to be accentuated in shallow environments [51]. In Estero Morúa, the variation in TSS is likely driven by wind and the tidal regime, as river flow and runoff are not important factors in the region [10].

Chlorophylls

The chlorophyll a (Chl a) concentrations observed in Estero Morúa indicate a system with low primary productivity. Although the Gulf of California is very productive [13]. The Northern Gulf exhibits patches of high Chl (up to > 2 mg/m3) even during summer due to tidal mixing which carries cold and nutrient rich waters to the surface, throughout the year [52].

In esteros of the Northern Gulf, the oceanic influence is strong, resulting from the exchange of water during high tides [41]. A stable isotope analysis of trophic food webs in Estero Morúa found that marine derived phytoplankton contributes the most carbon to the system [53]. Mean Chl a concentrations were on average 0.13 ± 0.02 mg/m^3 for the study period (Table 1). Chl a concentrations showed a significant variability between sampling locations (F= 2.96; P = 0.009) and surveys (F= 2.26; P = 0.004; Figure 13). The highest concentration was found in June 2006 (0.50 ± 0.32 mg/m^3; Figure 13) but there was no seasonal pattern. The average Chl a concentrations in Estero Morúa are on the low range of previously reported values for other arid wetlands. Gilmartin and Relevante [49] found values between $0.2 - 19.9$ mg/m^3 in 12 coastal lagoons of the Gulf of California. Negative estuaries in the Northern Gulf are usually less productive than positive estuaries because of the scarcity of organic matter, especially at their head waters [1].

The concentrations of chlorophyll b (Chl b) and chlorophyll c (Chl c), were low (0.03 ± 0.01 and 0.07 ± 0.02 mg/m^3 respectively). These parameters were only detected in a few surveys. Both were found in January, February, and June 2006. Chlorophyll b was also found in April 2007 and Chl c was detected in August 2006, and January and March 2007.

Relationship between Parameters

Explanatory linear models for the variables measured are found in Table 2. We found a significant negative relationship between temperature and dissolved oxygen, likely due to the relationship between water temperature and gas saturation that results in less oxygen being available at higher temperatures [41]. We also found a negative relation of temperature with pH, and a positive response in relation to depth. There was a positive relationship between variability of NO_2^- and NO_3^-, and a negative relation between NO_3^- and NH_4^+, as expected since NH_4^+ is sequentially oxidized to NO_2^- and then to NO_3^- during bacterial denitrification [54]. In some cases these nitrogen forms were also related to D.O., pH, temperature, depth, salinity. Strong positive relationships were found between PO_4, salinity, D.O. and NO_3^-. Phosphate concentrations and dissolved oxygen are linked and controlled by the rate of water exchange and biological processes [55]. Lower TSS values were associated with lower depths, salinity, temperature, and Chl a. Finally, Chl a concentrations showed a negative relation with TSS and were a positive function of PO_4, water temperature, depth, and salinity, underscoring the fact that microalgal biomass is sensitive to changes in environmental conditions [43], and is regulated by light [41].

CONCLUSIONS

Estero Morúa and Almejas are hypersaline systems, where temperature, dissolved oxygen concentrations, pH, and salinity are highly dependent on depth, tidal circulation, and solar heating. The hypersaline conditions are prevalent year-round because of the low precipitation and high evaporation present in the region [1]. We found that environmental conditions resulting from physical and chemical factors can be extreme: temperatures ranged between 6°

and ~40 °C and salinity reached > 100. Overall levels of dissolved oxygen were high > 8 mg/L and nutrient concentrations were low, except for nitrate, particularly in summer and autumn, resulting in oligotrophic conditions. The complex interactions between biological, physical and chemical parameters of water in these estuaries can exert stress in the organisms found in different environments in the marsh, many of which are important commercial species in the region.

Table 2. Linear models for the selected physical and chemical variables in Estero Morúa, Sonora. Asterisks indicate explanatory variables (ANOVA, ns: not significant, . P < 0.1, * P < 0.05, ** P < 0.01, ***: P < 0.001). Models with lowest AIC values were selected. Negative relationships are underlined.

Model	Temp.	Sal.	pH	Depth	D.O.	NO_3^-	NO_2^-	NH_4^+	PO4	TSS	Chl a
Temp.		.	***	***	***						
Sal.											
pH	***										
D.O.	***										
NO_3^-		ns					***	***	***		
NO_2^-	ns		ns	**		***		***			
NH_4^+	ns		***		***	***	***				
PO4		***				***	***	*		.	
TSS	***	*		**							*
Chl a	*	***		*		ns	ns	ns	*	*	

We found that seasonal cycles clearly determine physicochemical variables such as temperature and salinity, but the seasonal effect was not as clear for other variables such as TSS (Figure 12) and Chlorophylls (Figure 13). These variables might experience higher spatial variability or be influenced by the spring-neap variations in tidal cycle [41]. On a time scale of hours, the entry of sea water could be responsible for changes in salinity, nutrient concentrations, and suspended solid content [43]. Mixing and strong currents due to the macrotidal regime likely preclude the formation of vertical gradients during high tide [42]. Current velocities in the mouth of Estero Morúa can reach 50 cm/s and decrease to 10 cm/s in the arms of the wetland and residence time is only 6 hours [21]. As a result, during high tides, physicochemical parameters likely approximate seawater. For example, water salinity during spring tides in Estero Morúa is closer to sea water, 38.2 ± 0.4 (A. Iris Maldonado, unpublished data).

Our study is a first step in describing the environmental conditions of negative estuaries in northwest Sonora. Further studies are needed that consider the spring-neap tide, depth profile variations and fluxes of nutrients and net metabolism [i.e. 42] to evaluate changes in the system and the effect of human activities. Coastal wetlands in the Gulf of California, like Esteros Morúa and Almejas, are under increasing stress from coastal zone development [56] and other human activities that cause changes in tidal hydrology, increased pollution, and changes in flora and fauna [57]. Particular attention should be placed on waterborne health risks, since estuaries in the region contribute to the local economy through fisheries production and oyster culture [23]. Our sampling did not reveal the high levels of nutrients common in other wetlands in Sonora, that have been impacted by the shrimp farms or other human activities [47], but erosion of coastal dunes surrounding Estero Morúa and other local wetlands could increase sediment deposition. In Estero Morúa and Almejas, a state water quality monitoring program continues to ensure that farmed oysters are safe for human consumption (more information available at ttp://www.cosaes.com).

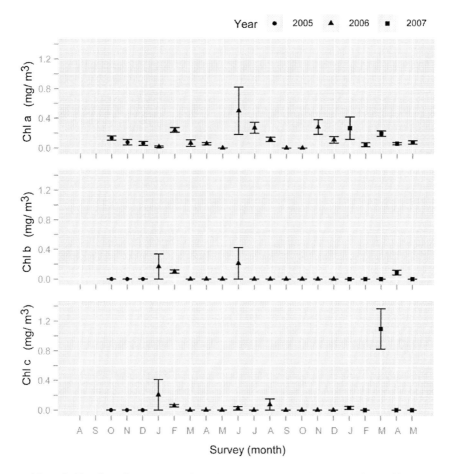

Figure 13. Chlorophyll *a, b,* and *c* concentrations (means ± SE, n=12) measured monthly in Estero Morúa between October 2005 and May 2007.

ACKNOWLEDGMENTS

This study was funded by The David and Lucile Packard Foundation grants 2004-26759 and 2006-30328 to CEDO Intercultural and by the Secretaría de Agricultura, Ganadería, Desarrollo Rural y Pesca - Comisión Nacional de Acuacultura y Pesca - Instituto de Acuacultura del Estado de Sonora, O.P.D. through the Grupo Interinstitucional de Investigación en Moluscos Bivalvos and the project "Determinación de agentes causales de alta mortalidad en los cultivos del Ostión Japonés, *Crassostrea gigas*, de las costas de Sonora". A. Castillo López, R. Fraser, A. García Sanchez, S. López Alvirde, S. Mason, S. Reyes Fiol, and A. Rosemartin participated in sample collection. J. Hernández-López at CIBNOR facilitated the analysis of nutrient samples. I.C. Kaplan provided valuable comments to the manuscript. We would specially like to thank the members of the oyster farm cooperatives for patiently collecting daily data, at La Cinita: I. Zepeda, G. Hernández P., S. Álvarez P., J. Curiel, J.Villalobos, E. Duarte O., V. Rodríguez G., F. Núñez R., E. López B., J. Bernal, R. Porra, G. Curiel, and H. Bernal. Unica de Mujeres: M. Esther, X. Tanori, M. del R. Luna, J.R. Luna, Ma. C. García S., and I. Cervantes. Aquamar: C. Castañeda and Don Carlos.

REFERENCES

[1] Thomson, DA; Mead, A; Schreiber, J. Probable environmental impact of heated brine effluents from a nuclear desalination plant on Northern Gulf of California., Tucson, AZ, University of Arizona, Marine Science Committee. U.S. Dept. of Interior. Office of Saline Water, *Research and Development.*, 1968.

[2] Ansell, A; Barnes, M. *Oceanography and marine biology*, CRC Press, 1997.

[3] Little, C. *The biology of soft shores and estuaries*, Oxford University Press, 2000.

[4] Miller, JM; Pietrafesa, LJ; Smith, NP. *Principles of hydraulic management of coastal lagoons for aquaculture and fisheries,* Food & Agriculture Organization of the United Nations, 1990.

[5] Magni, P; Montani, S. Development of benthic microalgal assemblages on an intertidal flat in the Seto Inland Sea, Japan: effects of environmental variability, *La Mer.* 35(n.d.) 137-148.

[6] Page, HM; Petty, RL; Meade, DE. Influence of Watershed Runoff on Nutrient Dynamics in a Southern California Salt Marsh, *Estuarine, Coastal and Shelf Science*, 1995, 41, 163-180.

[7] Vörösmarty, C; Loder, T. Spring-neap tidal contrasts and nutrient dynamics in a marsh-dominated estuary, *Estuaries and Coasts*, 1994, 17, 537-551.

[8] Contreras-Espinosa, F; Warner, BG. Ecosystem characteristics and management considerations for coastal wetlands in Mexico, *Hydrobiologia*, 2004, 511, 233-245.

[9] Contreras-Espinosa, F. *Ecosistemas Costeros Mexicanos,* México, D.F., Universidad Autónoma Metropolitana. Unidad Iztapalapa, 1993.

[10] Valdes-Casillas, C; Carrillo-Guerrero, Y; Zamora-Arroyo, F; Hinojosa-Huerta, O; Camacho-López, M; Delgado-García, S; et al., *Mapping and management of coastal wetlands of Puerto Peñasco, Sonora: A multinacional project,* Sonora, Center for Conservation of Natural Resources (CECARENA), Instituto Tecnológico y de Estudios Superiores de Monterrey – Campus Guaymas (ITESM-CG); Pronatura. Arizona State University, 1999.

[11] Cartron, JE; Ceballos, G; Felger, RS. *Biodiversity, ecosystems, and conservation in northern Mexico,* New York, Oxford University Press, 2004.

[12] Brusca, RC. *Common intertidal invertebrates of the Gulf of California*, Tucson, University of Arizona Press, 1980.

[13] Glenn, EP; Nagler, PL; Brusca, RC; Hinojosa-Huerta, O. Coastal wetlands of the northern Gulf of California: inventory and conservation status, *Aquat*, 2006, 16, 5-28.

[14] Brusca, RC; Bryner, GC. A case study of two Mexican biosphere reserves: The Upper Gulf of California and Colorado River Delta and the El Pinacate and Gran Desierto de Altar Biosphere Reserves, in: NE; Harrison, GC. Bryner, (Eds.), *Science and Politics in*

the International Environment, Lanham, MD, Rowman & Littlefield Publishers, 2004, 28-64.

[15] Pepe, P. Estero Morua: Tour Through a Living Estuary: Nursery Ground and Sanctuary, *CEDO News*. Vol2(1) (1999) 9-11.

[16] Huang, D; Bernardi, G. Disjunct Sea of Cortez-Pacific Ocean *Gillichthys mirabilis* populations and the evolutionary origin of their Sea of Cortez endemic relative, *Gillichthys seta, Marine Biology*, 2001, 138, 421-428.

[17] Barlow, GW. Gobies of the genus *Gillichthys,* with comments on the sensory canals as a taxonomic tool, *Copeia*, 1961, 423-437.

[18] Buckley, BA; Hofmann, GE. Thermal acclimation changes DNA-binding activity of heat shock factor 1 (HSF1) in the goby *Gillichthys mirabilis*: implications for plasticity in the heat-shock response in natural populations, *Journal of Experimental Biology*, 2002, 205, 3231-3240.

[19] Buckley, BA; Hofmann, GE. Magnitude and duration of thermal stress determine kinetics of hsp gene regulation in the goby *Gillichthys mirabilis, Physiological and Biochemical Zoology*, 2004, 77, 570-581.

[20] Flowers, TJ; Troke, PF; Yeo, AR. The mechanism of salt tolerance in halophytes, *Annu. Rev. Plant. Physiol*, 1977, 28, 89-121.

[21] Juárez Romero, L. Determinación de agentes causales de alta mortalidad en los cultivos del Ostión Japonés, *Crassostrea gigas,* de las costas de Sonora, Hermosillo, Sonora, Instituto de Acuacultura del Estado de Sonora. Q.P.D. Universidad de Sonora. Centro de Investigaciones Biológicas. Centro de Investigación en Alimentación y Desarrollo. Centro Intercultural de Estudios de Desiertos y Océanos. Centro de Estudios Superiores del Estado de Sonora. Comité de Sanidad Acuícola del Estado de Sonora., 2007.

[22] INEGI, Conteo de población y vivienda 2005, Instituto Nacional de Estadística, *Geografía e Informática*, 2005.

[23] Green, CR. Metereological conditions, in: DA; Thomson, A; Mead, J. Schreiber, (Eds.), *Probable Environmental Impact of Heated Brine Effluents from a Nuclear Desalination Plant on Northern Gulf of California.,* Tucson, AZ, University of Arizona, Marine Science Committee. U.S. Dept. of Interior. Office of Saline Water, Research and Development, 1968, 7-15.

[24] Lavín, MF; Godinez, VM; Alvarez, LG. Inverse-estuarine features of the Upper Gulf of California, *Estuarine Coastal & Shelf Science*, 1998, 47, 769-795.

[25] Gifford, EW. Archaeology in the Punta Penasco region, Sonora, *American Antiquity*, 1946, 11, 215-221.

[26] Brown, K. Sedimentologic and morphologic analyses of modern macrotidal deposits, *Puerto Peñasco*, Sonora, Mexico, M.S. thesis. Northern Arizona University, 1989.

[27] Johnson, AF. Dune vegetation along the eastern shore of the Gulf of California, *Journal of Biogeography*, 1982, 9, 317-330.

[28] Cudney-Bueno, R; Turk Boyer, PJ. *Pescando entre mareas del Alto Golfo de California: Una guia sobre pesca artesanal, su gente y sus propuestas de manejo.,* Puerto Peñasco, Sonora. Mexico, CEDO Intercultural, A.C., 1998.

[29] Felger, RS. Flora of the Gran Desierto and Rio Colorado of Northwestern Mexico, Tucson, AZ, University of Arizona Press, 2000.

[30] Morzaria-Luna, H; Polanco-Mizquez, E; López-Alvirde, S; Reyes-Fiol, S. Caracterización de la vegetación de los esteros de Bahía Adair, Estero Morúa, y La

Salina y predios circundantes. ANEXO, Reporte Final. Proyecto FMCN EFC-06-002 A. Castillo López, responsable técnico. Puerto Peñasco, Sonora, Centro Intercultural de Estudios de Desiertos y Océanos, *A.C.*, 2007.

[31] Alvarez-Borrego, S. Gulf of California, in: BH. Ketchum, (Ed.), *Ecosystems of the World 26. Estuaries and Enclosed Seas,* New York, Elsevier Scientific, 1983, 427-449.

[32] Mackie, GL. *Applied aquatic ecosystem concepts*, Dubuque, Iowa, Kendall/Hunt Publishing Company, 2001.

[33] Arar, E. Method 446.0. *In vitro determination of chlorophylls a, b, c1 + c2 and pheopigments in marine and freshwater algae by visible spectrophotometry,* National Exposure Research Laboratory. Office of Research and Development. U.S. Environmental Protection Agency, 1997.

[34] Jeffrey, S; Humphrey, G. New spectrophotometric equations for determining. Chlorophylls a, b, c + c in higher plants, algae and natural phytoplankton, *Biochemistry and Physiology Pflanzen*, 1975, 167, 191-194.

[35] Hernández-López, J; Vargas-Albores, F. A microplate technique to quantify nutrients (NO2, NO3, NH4 and PO3:4) in seawater, *Aquaculture Research*, 2003, 34, 1201-1204.

[36] Venables, WN; Ripley, BD. *Modern applied statistics with S*, New York, Springer, 2002.

[37] Place, SP. Hofmann, GE. Temperature interactions of the molecular chaperone Hsc70 from the eurythermal marine goby Gillichthys mirabilis, *Journal of Experimental Biology*, 2001, 204, 2675-2682.

[38] Nichols, M. *Composition and environment of recent transitional sediments on the Sonoran coast*, Mexico, 1969.

[39] Roden, G; Emilsson, I. *Oceanografía física del Golfo de California*, Centro de Ciencias del Mar y Limnología. UNAM, 1980.

[40] Smith, S; Ibarra-Obando, SE; Boudreau, P; Camacho-Ibar, V. *Comparison of carbon, nitrogen and phosphorus fluxes in Mexican coastal lagoons,* Texel, The Netherlands, LOICZ Core Project Office, 1997.

[41] Valdez-Holguín, J. Variaciones diarias de temperatura salinidad, oxígeno disuelto, y clorofila *a* en una laguna hipersalina del Golfo de California, *Ciencias Marinas*. 1994, 20, 129-137.

[42] Valenzuela-Siu, M; Arreola-Lizárraga, J; Sánchez-Carrillo, S; Padilla-Arredondo, G. Flujos de nutrientes y metabolismo neto de la laguna costera Lobos, México, *Hidrobiológica*, 2007, 17, 193-202.

[43] Magni, P; Montani, S; Tada, K. Semidiurnal dynamics of salinity, nutrients and suspended particulate matter in an Estuary in the Seto Inland Sea, Japan, during a spring tide cycle, *Journal of Oceanography*, 2002, 58, 389-402.

[44] Dyer, KR; Christie, MC; Feates, N; Fennessy, MJ; Pejrup, M; Lee, WVD. An investigation into processes influencing the morphodynamics of an intertidal mudflat, the dollard estuary, The Netherlands: I. hydrodynamics and suspended sediment, estuarine, *Coastal and Shelf Science*, 2000, 50, 607-625.

[45] Yin, K; Harrison, PJ; Pond, S; Beamish, RJ. Entrainment of nitrate in the fraser river estuary and its biological implications. III. Effects of Winds, *Estuarine, Coastal and Shelf Science*, 1995, 40, 545-558.

[46] Okin, G; Gillette, D; Herrick, J. Multi-scale controls on and consequences of aeolian processes in landscape change in arid and semi-arid environments, *Journal of Arid Environments*, 2006, 65, 253-275.

[47] Paez-Osuna, F; Guerrero-Galvan, SR; Ruiz-Fernandez, AC. Discharge of nutrients from shrimp farming to coastal waters of the Gulf of California, *Marine Pollution Bulletin*, 1999, 38, 585-592.

[48] Lluch-Cota, SE. Coastal upwelling in the eastern Gulf of California, *Oceanologica Acta*, 2000, 23, 731-740.

[49] Gilmartin, M; Revelante, N. The phytoplankton characteristics of the barrier island lagoons of the Gulf of California, *Estuarine and Coastal Marine Science*, 1978, 7, 29-47.

[50] Diamantopoulou, E; Dassenakis, M; Kastritis, A; Tomara, V; Paraskevopoulou, V; Poulos, S. Seasonal fluctuations of nutrients in a hypersaline Mediterranean lagoon, *Desalination*, 2008, 224, 271-279.

[51] Bianchi, TS; Pennock, JR; Twilley, RR. *Biogeochemistry of Gulf of Mexico estuaries*, John Wiley and Sons, 1998.

[52] Caffrey, J; Harrington, N; Ward, B. Biogeochemical processes in a small California estuary. 2. Nitrification activity, community structure and role in nitrogen budgets, *Marine Ecology Progress Series*, 2003, 248, 27-40.

[53] Spackeen, J. *Analysis of food web dynamics in the Northern Gulf of California using stable isotopes,* Bachelor's degree with departmental honors, University of Miami. *Rosenstiel School of Marine and Atmospheric Science*, 2009.

[54] Duxbury, AC. Orthophosphate and Dissolved Oxygen in Puget Sound, *Limnology and Oceanography*, 1975, 20, 270-274.

[55] Brusca, RC; Cudney-Bueno, R; Moreno-Báez, M. Gulf of California esteros and estuaries. Analysis, state of knowledge, and conservation and priority recommendations, *Arizona-Sonora Desert Museum*, 2006.

[56] Vitousek, PM; Mooney, HA; Lubchenco, J; Melillo, JM. Human Domination of Earth's Ecosystems, *Science*, 1997, 277, 494-499.

In: Encyclopedia of Environmental Research
Editor: Alisa N. Souter

ISBN: 978-1-61761-927-4
© 2011 Nova Science Publishers, Inc.

Chapter 43

IMPLICATIONS FOR THE EVOLUTION OF SOUTHWESTERN COAST OF INDIA: A MULTI-PROXY ANALYSIS USING PALAEODEPOSITS

B. Ajaykumar[1], Shijo Joseph[1], Mahesh Mohan[1], P. K. K. Nair [2]*, K. S. Unni[1] and A.P. Thomas[1]

[1]School of Environmental Sciences, Mahatma Gandhi University, Priyadarshini Hills P.O., Kottayam, Kerala, India – 686 560
[2]Environmental Resource Research Centre, NCC Nagar, Peoorkada, Thiruvananthapuram, Kerala, India

ABSTRACT

The southwestern coast of India was under the influence of local marine environments to a minor extent in the geological past. The sea level along this part of the coast stood around 60–100 m below the present MSL during the last glacial maxima (around 20,000 YBP) and the rivers flowing at that time incised their valley to this base level. Later a humid climate with maximum representation of mangrove vegetation around 10,000 YBP was reported in this area, which suggested strengthened Asiatic Monsoon till the first part of Atlantic period. Initial subsidence and consequent flooding due to transgression, which had occurred 8000-6000 YBP, destroyed the mangrove vegetation giving rise to peaty soil. The pattern of rivers and geomorphological set up suggested that coastline was much towards east during the geological past and the rivers were flowing to the west and debouching in to the sea. The occurrences of peat sequence in the sediments in the low-lying area around the present Vembanad Lake and their radio carbon dating studies indicated their formation was from submerged coastal forest, especially mangrove vegetation. An event of regression (5000- 3000 YBP) was occurred along Kerala coast during the late Holocene. Contemporaneous to this, one of the major backwater systems of the southwestern coast- the Vembanad Lake- was developed. The shell deposits of Kerala, which form a very rich source of carbonate, are formed by the accumulation of dead shell-bearing organisms and it is suggested that these organisms after being trapped in their ecological niche were destroyed due to marine regression. The

---

* Corresponding author: Email: jemnair@gmail.com, Telephone : 919447062674, Fax : 91 481 2732620

present day backwaters and estuaries that occur behind the sandbars in Kerala thus possibly owe their origin to the regression during the past. Present study on the formation of palaeodeposit of sand in the Meenachil River basin lying along the southwest coast of India point towards the morphometric rearrangement and a multi-proxy analysis of the palaeodeposits could portray the Holocene geomorphological modifications of the southwestern coast. Sedimentological, geochemical, palynological and palaeobotanical analysis were done to evaluate the influence of palaeoclimatic factors on the modifications of the earth's surface configuration along this part of the Indian sub-continent and it is suggested that geomorphological modifications of the southwestern coast of India shall be classified into three categories (1) Pre-Vembanad Lake formation, (2) Contemporaneous to Lake formation and (3) Post-Lake formation.

Keywords: South-western coast, Atlantic period, Asiatic Monsoon, palaeoenvironmental analysis, Vembanad Lake.

INTRODUCTION

A number of landforms along the southwestern coast of India owed their genesis from the Quaternary episodes of sea level oscillations. The synthesis of these landforms has revealed a great deal of information on the transgressive and regressive events, which in turn, helped in hypothesizing the possible evolution of the southwest coast of India. These events have also been interrupted and modified by tectonic crustal flexure (Nair, 2005) in the form of down-warping and up-arching. Hence the evolution of the coastal landscape of the study area can well be understood by analysing the three major factors (1) fluvial sedimentation, (2) tectonism, and (3) Holocene sea level variations.

According to Menon (1967), southwestern coast of Indian peninsula was under the influence of local marine environments to a minor extent in the geological past (Table 1). The sea level along this part of the coast stood around 60 – 100 m below the present MSL during the last glacial maxima (around 20,000YBP) the coastal region had witnessed a five stage transgressive and regressive episodes (Nair, 2005). Palaeopalynological analysis showed humid climate with maximum representation of mangrove vegetation (around 10,000YBP) in the Arabian Sea area (Van Campo, 1986). According to Rajendran *et al.* (1989) and Prakash *et al.* (2001) the formation of the backwater system found along the Indian southwest coast was linked to a series of transgression (8000-6000YBP) and regression (5000-3000YBP) events. Several studies shown that neotectonic/eustatic event in the southern coastal stretches of peninsular India caused the modification of the coast during the Quaternary period (Soman *et al.*, 2002; Narayana *et al.*, 2002) and the rivers were also followed these changes (Narayana *et al.*, 2001). Vaidhyanadhan (1971) reported that Kerala coast indicates filling up of a series of bays with mud during the Holocene and later covered by sand and the radiocarbon dating of one such emerged areas has given a date of 6460YBP. Along with other river systems of the coast, Meenachil River had also adjusted its morphology according to the coastal changes. These geomorphological modifications along with channel shifts might have caused for the unique and extensive distribution of palaeo-deposit of sand, identified recently from the watershed area of Meenachil River (Mohan *et al.,* 2005)- one of the five river systems debouching into the Vembanad lake. The present study aims to throw light into the palaeoenvironmental conditions of south western coast of India during the Holocene through multi-proxy analysis of the palaeo-deposits present in Meenachil River basin.

Table 1. Stratigraphic sequence of southwestern coast (Najeeb, 1999)

Age	Name	Lithological assemblage
Quaternary (0.0 to 1.6ma BP)	Vembanad formation	Sand, clays, molluscan shell beds, riverine alluvium and flood plain deposits, laterite capping of crystallines and Tertiary sediments
Tertiary (1.66 to 66.4ma BP)	Warkalli formation	Sandstone and clay with lignite seams
	Quilon formation	Limestone, marl, clays/calcareous clays with marine and lagoonal fossils
	Vaikom formation	Sandstones with pebble and gravel beds, clays and lignite and carbonaceous clay
Mesozoic to Archaean		Intrusives with veins of quartz, pegmatite, granites, granophyres, dolerite and gabbro, garnet sillimanite gneiss, hornblende biotite gneiss, garnet biotite gneiss, quartzo feldspathic gneiss, charnockites, charnockite gneiss.

STUDY AREA AND METHODOLOGY

The study area (Figure 1) falls in the lower reaches of the Meenachil River (9° 25' to 9° 55' N latitudes and 76° 20' to 76° 55' E longitudes). The palaeodeposits (sediment and carbonaceous material) present in the Meenachil river basin were subjected to sedimentological (American Standards for Testing Materials sieve method), geochemical (standard methods), carbon dating, palaeo-palynological (Faegri and Iversen, 1975) and palaeo-botanical analysis in order to portray the inter-relationship with the evolution of the southwestern coast.

RESULTS AND DISCUSSION

Sedimentological Analysis

Sedimentation is inherently a discontinuous process (Sadler, 1981) and the depositional systems are controlled by sea level changes (Amorosi *et al.,* 1999), subsidence of basins, climate, sediment production and input which vary with time. Several Studies on sediment texture have attempted to relate grain size distribution to mode of transport and to specific depositional environments (Middleton, 1976; Nemec and Kazanci, 1999). The study of sedimentation processes has gained prime importance in recording the depositional sequence in glacio-fluvial sediments (Makinen, 2003). The works of Folk and Ward (1957), Friedman (1961), Shamsuddin (1986) have established that environment of deposition can be inferred from statistical treatment of sieve data.

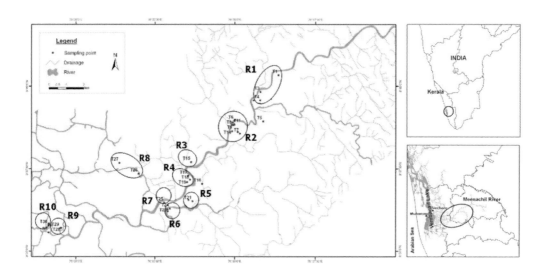

Figure 1. Study area showing Palaeodeposit regions and sampling sites.

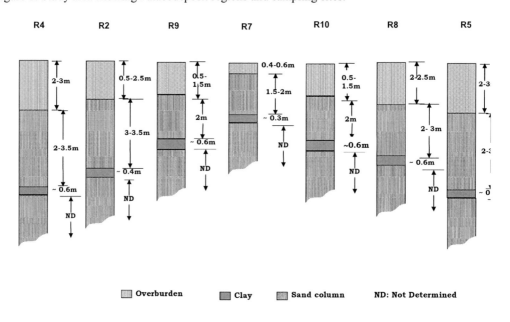

Figure 2. Succession of deposits at different regions.

Mud content in all the palaeodeposit samples was very low (<2%) indicating strong winnowing action of water in flushing away the mud fraction. Textural analysis of the palaeodeposits expressed certain unique characteristics. The samples collected from the T30, T31 and T32 sites were gravelly sand, while those from T28 and T29 were sand, even though the former sites were located further headward than the locations T30, T31 and T32. Variation in the textural characteristics might have owed to the influence of the labyrinth network of the distributaries. The textural pattern of the deposits of site T26 and T27 revealed that the former was of slightly gravelly sand and later was of gravelly sand. This was because of the change in the depositional environment. The site T26 was situated near to a small rivulet and at a distance of 150m away from one of the northerly flowing distributaries, while the site T27 situated near to that distributary. The sites T16, T20, T21, T22, T23, T24 and T25 have sediments with textural class of gravelly sand.

The textural class of sites T17 and T19 was gravelly sand and this was sandy gravel for site T18, which was located at the innermost part of a meandering loop. Three distinct horizons analysed for site T15, revealed that the bottom sample (T15c) was of sandy gravel, middle sample (T15b) with gravelly sand and top sample with slightly gravelly sand (T15a). The samples from site T8, T9, T10, and T11 were falling in the textural class of sand and other sites T6, T7, T12, T13 and T14 have shown slightly gravelly sand to sandy gravel.

The textural analysis showed that the samples of palaeodeposit have ranged from sand to sandy gravel. The palaeodeposits lying far away from the existing channel of the Meenachil River were coarser than those lying near to the course. The most conspicuous thing observed is that the coarse grained deposits were found deposited near to the small rivulets running into the adjacent main channel as seen at regions R2, R5 and R8. Another important observation is that the bed sediments collected from the bottom level (below the layer of carbonaceous clay) were finer than those collected from the top level at locations of R4 and R5. But there was a general observation of gradual decrease in grain size pattern from bottom to top in R3 and R6. The samples collected from the three distinct horizons of the T15 deposit varied from sandy gravel at the bottom to slightly gravelly sand at the top, indicating a gradual ceasing of a fluvitile realm. The sediments of R1 were of sandy nature, but in location no. T2, it was of gravelly sand suggestive of another abandonment of a stream channel. Another noteworthy thing was that the bed sediments collected from an abandoned channel (T5) was sandy gravel.

The statistical analysis (measures of skewness, standard deviation and kurtosis) of the sediment samples (37 samples, sites T15 (2), T19 (1), T21 (1), T24 (1) have extra samples from various depths) clearly indicates different sets of environment for their deposition (Figure 3, 4 and 5). The samples of the regions R2, R4, R5 and R8 were formed due to the shift in the main channel or its abandonment, while R6, R9 and R10 were formed as point bar deposits. Similarly regions R1, R2 and R3 provided depositional environment for a particular span of time and might be linked with then existed tributaries.

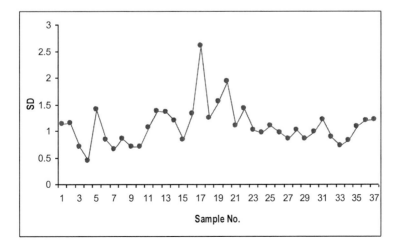

Figure 3. Standard deviation of palaeodeposit samples (along the down stream direction).

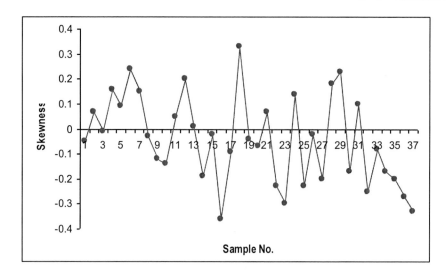

Figure 4. Skewness of palaeodeposit samples (along the downstream direction).

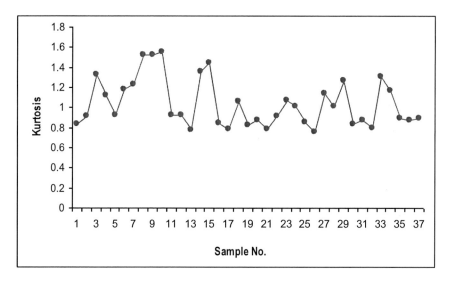

Figure 5. Kurtosis of palaeodeposit samples (along down stream direction).

The non-uniform distribution pattern of the palaeodeposit of sand along the flood plain regions of the Meenachil River is suggestive of as formed by the northward and southward shifting of the main channel according to palaeoenvironmental changes including global eustatic sea level fluctuations. The absence of repeated graded bedding in the sequence of sedimentary column rules out its formation as flood plain deposit while graded bedding, supposed to be formed under a single realm, is seen in point bar deposits. Besides, the occurrence of palaeodeposits of sand well above the present riverbed is clearly suggestive of elevated base level of erosion (Photo 1) during the Holocene.

Photo 1. Occurrence of palaeodeposits well above the present base level of erosion.

The above results strongly suggest that the south western coast of India also had experienced intensified and prolonged Asiatic Monsoon during the earlier part of the Atlantic chronozone (8500 to 6000YBP) as observed from other parts of the western coast (Borole et al., 1982; Thamban et al., 2001; Ramesh, 2001; Ansari and Vink, 2007) and the causative factors which triggered this mechanism were likely to be a function of mid-Holocene insolation regime (Morimoto et al., 2007) and shifts in the position of the Inter Tropical Convergence Zone (ITCZ) (Ansari and Vink, 2007). Besides, the sedimentological analysis revealed that the mud content in the palaeodeposit of sand was very low (<2%) indicating strong winnowing action during the phase of deposition (Mohan et al., 2005) and the presence of granules at the bottom level of the palaeodeposit marks the high velocity hydrological regime triggered by intensified monsoonal precipitation during the earlier part of the Holocene (Photo 2).

Photo 2. Presence of granules and current bedding suggest high velocity hydraulic regime during the early Holocene.

Geochemical Analysis

It has been proved that the analysis of soil organic carbon content and pollens as well as the ^{14}C dating can excellently be incorporated in depicting the palaeo-ecological reconstruction of long-term landscape and vegetation changes. It is revealed that the lowest level of organic sediments represents the earliest phase of plant growth. Brown et al. (1999 and 2000) recorded that the sedimentary organic matter content is generally low in Pleistocene deposits and high in Cretaceous deposits. According to Sifeddine and Wirrmann (2004), the organic matter present in the sedimentary deposits records several types of information depending on the abundance, provenance and preservation of the constituent components. Generally speaking, two main sources of organic matter can be recognized in carbonaceous sediments. One source is derived from the aquatic organisms, while the other corresponds to plant debris coming from within the watershed. Although the petrography observation of the organic matter enables a distinction between phytoplanktonic components and vegetal debris originating within the watershed (Patience *et al.*, 1995 and Sifeddine *et. al.*, 1996), it does not allow discrimination of the different types of terrestrial vegetal material. Similarly the C/N ratio can be used to distinguish two main types of organic matter, the first type originates from algae and phytoplankton in lacustrine environments and does not exhibit cellulose structures; it is characterised by C/N ratios in the range 4-10 and the second type has a cellulosic structure and is produced by terrestrial plants; it is characterised by C/N ratios greater than or equal to 20 (Meyers and Ishiwatari, 1993 and Meyers, 1994)

The geochemical analysis of the sandwiched carbonaceous clay (Photo 3) was done for depicting the conditions of deposition as well as the palaeo-environmental scenario of the region during the time of deposition. The sampling locations were fixed as given in Figure 6.

The results of the textural analysis (Table 2) revealed that their deposition has a regime of water stagnancy. The percentage-wise comparison of the constitutional fragments indicate that the samples collected from P7 has more sand fractions, while those collected from P2 (S3) has the dominance of clay. The conditions of deposition clearly indicated that the elevated base level of erosion (sea level rise?) was responsible for the water stagnancy by closing down the process of one fluvitile realm in the region.

Table 2. Particle proportions of sand-silt-clay in samples collected from the sandwiched clay layer

Name of the Location	Sample Number	Clay	Silt	Sand	Terminology
P1	S2	13.86	1.07	85.07	Clayey sand
P2	S3	53.95	8.99	37.06	Sandy clay
P2	S4	11.65	0.53	87.82	Clayey sand
P3	S6	29.03	1.65	69.32	Clayey sand
P6	S11	39.66	4.18	56.16	Clayey sand
P7	S14	07.59	2.17	90.24	Sand

Photo 3. Carbonaceous clay as sandwiched between the sand column.

Photo 4. Yellow ochre within the sand deposit indicated arid condition during the late Holocene.

Table 3. Interpretation of the level of Organic Carbon in sandwiched clay layer

Region	Sample No.	% Organic Carbon	% Total Nitrogen	C/N Ratio
P1	S1	0.55	0.06	9.17
	S2	5.83	0.17	34.29
P2	S3	5.11	0.19	26.89
	S4	4.38	0.20	21.90
P3	S5	2.19	0.21	10.43
	S6	4.38	0.18	24.33
P4	S7	5.83	0.21	27.76
	S8	2.92	0.19	15.37
	S9	4.74	0.17	27.88
P5	S10	8.75	0.21	41.67
P6	S11	1.02	0.09	11.33
	S12	3.65	0.09	40.56
	S13	6.20	0.14	44.29
P7	S14	4.74	0.18	26.33
P8	S15	4.38	0.12	36.50
P9	S16	3.65	0.16	22.81
	S17	4.02	0.17	23.65
Mean		4.25	0.16	26.19

The geochemical analysis of the carbonaceous clay revealed that level of organic carbon is very high in all samples (Table 3). The samples collected from P5-S10 (8.75%), P6-S13 (6.20%), P1-S2 (5.83%), P4-S7 (5.83%) and P2 – S3 (5.11%) have very high value and they lie well above the mean organic carbon percentage of all locations (4.25%). Comparing to all other samples, those collected from P6 – S11 (1.02%) and P1-S1 (0.55%) have very low organic carbon content (Table 3). The percentage of Total Nitrogen is very high in samples of P5 (0.21%), P3 – S5 (0.21%), P4 – S7 (0.21%) and P2– S4 (0.204), while the same is lower than the average value in samples of P6 (0.09%-0.14%) and P1-S1 (0.06%) when compared with the mean value of all the locations (0.16%). The values of the C/N ratio is remarkably very high in the samples collected from P6 – S13 (44.29), P5 (41.67) and P6 – S12 (40.56) and the same is lower than the mean value of all samples (26.40) in P4 – S8 (15.37), P6 – S11 (11.33), P3 – S5 (10.43) and P1 – S2 (9.17).

The average value of organic carbon for all the sampling locations is extraordinarily comparable with those recorded from some other regions of the world (Table 4). The global distribution of organic carbon, total nitrogen and C/N ratio are fluctuating depends on the climatic variability and monsoon patterns. Many have stood on a contention that the relativity of Inter Tropical Convergence Zone (ITCZ) played a major role in the Holocene climatic variability, hence the vegetational pattern and distribution. The correlation matrix of the samples in relation with sand-silt-clay fractions and the organic carbon- nitrogen proportions were analysed using SPSS software. The analysis showed that the sand fraction in a sample has significant negative correlation with clay and silt fractions, while the level of organic carbon shown a positive significant correlation with level of nitrogen.

Table 4. Comparison of present result with some other Palaeodeposits

Location	Age	Org. C (%)	Tot. N (%)	C/N	Remarks
Meenachil river basin	Holocene	4.254	0.162	26.116	Present study
Kochi off shore, Kerala	Mid Holocene	0.710	ND	ND	Reddy, N.P.C. (2003)
South-west margin, India	Late Pleistocene	1.4 – 3.0	0.02- 0.15	>15	Kessarkar and Rao (2007)
South-west margin, India	Early Holocene	1.7 – 2.5	0.02-0.15	8.5-27.2	Kessarkar and Rao (2007)
South-west margin, India	Late Holocene	2.5 – 4.4	0.05-0.43	14.1-40.5	Kessarkar and Rao (2007)
Eastern Atlantic off Ghana	Pliocene – Pleistocene	0.5 – 1.4	0.06-0.14	4.0-18.0	Wagner (1998)
Rhone Delta, France	Late Holocene	7.64	ND	ND	Stanley (2000)
Sangla Valley, Himalaya	Holocene	0.04-1.6	0.04-0.12	8 - 18	Chakraborty et al. (2006)
Lake Ossa, Cameroon	Middle-Late Holocene	3.0-6.0	0.25-0.50	12	Sifeddine and Wirrmann (2004)

ND : Not Detected.

The level of organic carbon (C) and total nitrogen (N) and the C/N ratio in almost all samples, except P1-S1, P3-S5, P4-S8 and P6-S11, have shown a remarkable comparison with other Holocene deposits. The level of organic carbon is a measure of productivity, detrital input and preservation/degradation processes and the level of total nitrogen is higher in lower plants such as aquatic phytoplankton and in bacteria, since the organic nitrogen occurs preferentially in proteins and nucleic acids. The high ratio of C/N rules out the formation of organic carbon by algae and phytoplankton in a closed environment. But, it clearly is produced by terrestrial plants. As the lignin and cellulose are the dominant components of terrestrial higher plants, allochthonous and submerged organic matter have high C/N ratio. The warm and humid climate as recorded all over the tropical regions during 10,000 – 4000YBP (related with the position of ITCZ?) with high intensity rainfall (Jayalakshmi *et al.,* 2005; Van Campo, 1986 and Thamban et al., 2001) have attributed the high productivity in the study area The aridity which was witnessed during the late Holocene has reflected in the low productivity in locations like P3-S5 and P4-S8. The presence of yellow ochre within the top sand column (Photo 4) at P2 also revealed the arid climate during the receding of sea level.

The radiocarbon dating suggested that the palaeodeposits at the southern part of the Meenachil River were formed well before the formation of the Vembanad Lake. Limeshell from the bottom of Vembanad Lake yielded ages of 3710±90 (Vechoor) and 3130±100 (Muhamma) ^{14}C YBP and it was also suggested to be interpreted that foundering of the estuaries/lake systems along the southwestern coast of India was sequential with the northern estuaries/lake systems being earlier than the southern ones. In relation with this observation, it is well established that the southern palaeodeposit regions (R1, R3 and R4) of Meenachil river basins had the origin even before the formation of the Vembanad lake system along the southwestern coast. The global eustatic sea level change had contributed a lot for its formation, especially the formation of the sandwiched carbonaceous clay within the sand column. The phenomena like river avulsion and channel abandonment were also developed inconsequent to changing morphology. The shift in the main channel of Meenachil River from south to north is attributed to the tectonism around 4000YBP as suggested by Soman (1997). The study also coincides with the observation of Joseph and Thrivikramji (2002) that the different sectors of the Kerala coastal land had experienced non-uniform tectonic activity.

The formation of the palaeo-deposits of the northern side of the Meenachil River is of later age, contemporaneous to the development of the Vembanad Lake and local base level of erosion. An elevated sea level than the present is suggested during 3000YBP. Apart from the normal block faults along the Kerala coastal belt, the possibility of a tectonic combination of horst-graben development around Cochin estuary during the Middle to Late Holocene can not be ruled out not only by the present study but also on the basis of the prominence in channel development in the case of the northern distributaries of the Meenachil and Muvattupuzha Ar (an important river system lying adjacent and north to Meenachil river), shrinking of the southern distributary of Periyar River (the most important river system along the south western coast). Narayana *et al.,* (2001) identified a palaeodelta near the mouth of Periyar River, one of the major rivers of Kerala, formed by its more youthful northern branch. A series of cymatogenic uparching and downwarping in Peninsular India, developed during the Pleistocene (Ramasamy and Balaji, 1995), might also have influenced in defining the channel morphometry during the transgressive and regressive episodes. Hence apart from the global eustatic sea level fluctuations; the tectonism also had played a decisive role in shaping the channel geometry of rivers of central Kerala, which includes the Meenachil River also.

Palaeo-Palynological Analysis

Pollen analysis is a powerful tool for reconstruction of the past vegetation patterns. Sedimentary pollen analysis (Palaeopalynology) is one of the most studied methods for reconstruction of palaeoenvironment. Pollen analysis has the distinction of being the most widely used micropaleontological technique- at least for the Quaternary (Birks and Birks, 1980 and Williams *et al.*, 1998). Duly considering the limitations and shortcomings, palynological data has now been accepted as an established proxy record for the palaeoenvironmental and palaeo-climatic reconstruction during Quaternary (Bhattacharyya *et al.*, 2005). The most common source of data used in Holocene models of continental climate and vegetation is pollen analyses from lake and wetland sediments (Bartlein *et al.*, 1998; Webb, 1988; Webb *et al.*, 1993).

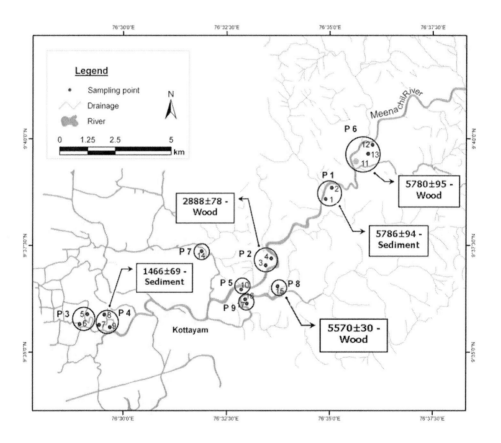

Figure 6. Sampling locations for geochemical and palynological analysis. Radiocarbon dates of the sediment and wood samples collected from the sand deposits were also given.

The representative sediment samples for palynological analysis were collected from the carbonaceous clay bed found sandwiched between the palaeodeposit of sand in four regions (Figure 5). The collected sediment samples were subjected to chemical treatment and pollen grains were analysed as per the methodology suggested by Faegri and Iversen (1975).

Table 5. Plant family and species found in the palaeodeposits of different regions

Family	Species			
	P1	P2	P8	P6
Amaranthaceae	*Amaranthus* sp.		*Amaranthus* sp	*Alternanthera* sp., *Amaranthus* sp.
Amaryllidaceae	U			
Caesalpiniaceae	*Bauhinia* sp.			
Casuarinaceae	*Casuarina equisetifolia*			*Bauhinia* sp.
Compositae (Asteraceae)	*Tribe astreae*			*Tribe astreae*
	Tribe vernoneae			*Tribe vernonea*
Cyperaceae	U			
Poaceae	Grass sp.	*Oryza sativa*	Grass sp.	U
Icacinaceae	*Ilex* sp.			
Leguminosae	*Delonix regia*		U	
Liliaceae	U	U		U
Meliaceae	*Azadirachta indica*			
Myrtaceae	*Eugenia* sp.		*Eugenia* sp.	*Eucalyptus* sp.
Papilionaceae	U			U
Plantaginaceae	*Plantago* sp.			
Acanthaceae	*Barleria* sp.	*Barleria* sp.		
Asteraceae	U	U	*Tribe Astreae*	
Nymphaeaceae	*Nymphaea* sp.	*Nymphaea* sp.	*Nymphaea* sp.	*Nymphaea* sp.
Piperaceae	*Piper* sp.	*Piper* sp.		
Pteridophyte		U		U
Trapaceae		*Trapa natans*	*Trapa natans*	
Arecaceae			*Cocos nucifera*	*Borassus flabellifer, Cocos nucifera*
Bombacaceae			*Bombax ceiba*	
Chenopodiaceae			*Chenopodium* sp.	
Combretaceae			*Terminalia* sp.	
Mimosaceae			*Albizia lebbeck*	
Plumbaginaceae			*Plumbago* sp.	
Lemnaceae				*Lemna* sp.
Lentibulariaceae				*Utricularia* sp.

U : Unidentified sp.

Implications for the Evolution of Southwestern Coast of India 1163

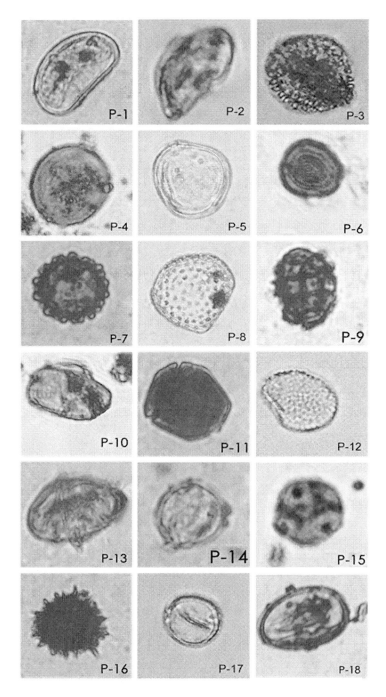

P-1: Cyperaceae, P-2: Cocos nucifera, P-3: Ilex sp, P-4: Amaranthus sp., P-5: Grass, P-6:Bombax ceiba, P-7:Bauhinia sp. P-8: Chenopodiacae, P-9: Delonix regia, P-10: Nympheae, P-11: Azadirachta sp, P-12: Lemna sp, P-13: piper sp, P-14: Casuarina equisetifolia, P-15: Tribe vernoneae, P-16-Tribe astreae, P-17: Grass, P-18: Terminalia sp.

Figure 7. Photomicrographs of pollen grains.

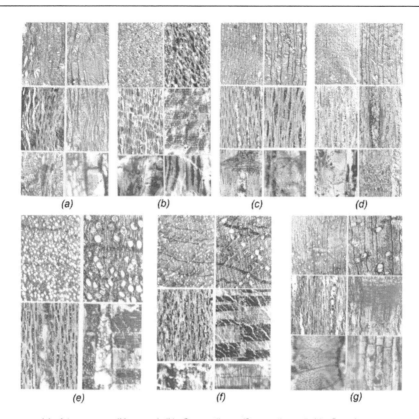

(a) - *Artocarpus sp. (Moraceae)*; (b) - *Sonneratia sp. (Sonneratiaceae)*; (c) - *Canarium sp. (Burseraceae)*; (d) - *Spondias sp. (Anacardiaceae)*; (e) - *Calophyllum sp. I (Clusiaceae)*; (f) - *Calophyllum sp. II (Clusiaceae)*; (g) - *Holigarna sp. (Anacardiaceae)*

Figure 8. Photomicrographs of fossil phytolith assemblages.

The palynological analysis showed that the floral elements composed of angiosperms (monocots and dicots), and pteridophytes. Of the 46 families identified, 32 are trees. The carbonaceous clay formation of P-1 contained a total of 49 species of pollen grains, of which 43 species were identified (Table 5). These pollens belong to 15 angiosperm families. The presence of Cyperaceae, Compositae, and Amaranthaceae, Amarylladiaceae, Liliaceae and Papilionaceae indicated the herbaceous weedy habitat of the region. The presence of *Casuarina equisetifolia* (P-14 of Figure 7), *Azadirachta* sp. and *Eugenia* sp. reveals the possibility of cultivation in this region. In P-2, twenty two species of pollen grains were found, of which 16 species were identified (Table 5) belonging to five angiosperm families. Two spores of Pteridophytes were also found in the sample. The presence of *Nymphaea* (P-10 of Figure 7) pollen and *Oryza sativa* pollen of Poaceae indicated the possible wetland habitat of the region. The presence of *Piper* sp (P-13 of Figure 7) and *Oryza sativa* pollens clearly indicated the cultivation in the region. Of the total 39 species of pollens recorded from the P-8, 38 species were identified (Table 5). The presence of Poaceae, Cyperaceae (P-1 of Figure7) and Nympheaceae indicated the existence of wetland system, and the dominance of grass species in this region denoted the possible water logging habitat. Besides, other weedy elements belonged to Amaranthaceae, Chenopodiaceae and Compositae were also observed. Pollens of *Bombax ceiba* (P-6 of Figure 7) which are characteristic to moist deciduous habitat were also recorded. The sediment samples of P-6 contain 34 species of pollen grains, out of which 28 species were identified (Table 5). The collected pollens of this region belong to 14 angiosperm families. The presence of Nympheaceae, clearly signified a shallow aquatic habitat. Possibly it was a wetland and the absence of mangrove/ mangrove associated plants indicates its freshwater nature. The presence of *Cocos nucifera* (P-2 of Figure 7) indicated cultivation in this region.

The fossil pollen and spores have been of special importance in gaining knowledge on the Quaternary vegetational history leading to the present day vegetation (Nair, 1977). Among the herbaceous component, diversity of grass pollen as seen by their size and morphological features are suggestive of the preponderance of the group in the flora. Besides, other weedy elements belonged to Amaranthaceae, Chenopodiaceae were also present. The presence of freshwater herbs like *Lemna, Nymphaea,* and marshy aquatic herbs of Cyperaceae and Poaceae showed the existence of a possible wetland system in the study area. It seems that riverbed vegetation was dominated by weeds, grasses and other ferns. There could have been pools of water formed due to the transgression of sea and such an ecological condition seems to have lead to the occurrence of aquatic species of *Nymphaea* growing along the boundaries.

The most conspicuous geomorphological modification at the south western coast of India during the late Holocene was the origin and development of the Vembanad Lake system. The palaeodeposit regions P1, P8 and P6 (Figure 6) were developed even before the formation of the Lake system. As stated earlier, the formation of the Vembanad Lake system was an outcome of a very prominent neotectonic/eustatic event, which caused the Meenachil River to shift from south to north than its position during the time of pre-Vembanad lake formation. Later the region has received another episode of maximum precipitation, which led to the formation of P2 (age is 2888 ± 78 ^{14}C YBP) followed by an event of aridification. The results of the present work going hand in hand with the previous observations that several other parts of the western coast of India enhanced by winter monsoon during 3900 to 3000 ^{14}C YBP followed by an onset of aridification at about 3000 YBP (Sarkar *et al.*, 2000; Luckge *et al.*, 2001; Ramesh, 2001;). Kumaran *et al.*, (2005) suggested that after about 5000 or 4000 ^{14}C YBP, the monsoon became gradually decreased leading to drying up of many marginal mangrove ecosystem. It has been proposed by Jayalakshmi *et al.* (2005) also that there was the presence of a dry climate during 2180 ± 70 ^{14}C yr B.P. The absence of evergreen forest elements in pollen spectra also indicated that there was a transition of a wetland and deciduous forest system. The presence of deciduous forest trees like *Terminalia* sp*, Bombax ceiba* and the occurrence of *Eugenia* species also support the observation. There seems to have been farming activities in the area with the floodplain being fed with humus rich sediments caused by flooding and the rich riverbed seems to have provided an ideal ground for growing crops.

Palaeo-Botanical Analysis

Fossil phytolith assemblages from soils and lake sediments have been used to reconstruct palaeovegetation patterns, especially forest/grassland ecotones (Alexandre *et al.*, 1997). The modern phytolith assemblages record local as well as regional vegetation (Barboni *et al.*, 1999). The study of the fossil remains can give not only the species distribution, but the influence of the past environmental conditions on that particular species also, which ultimately resulted in its bloom or extinction.

Well preserved and carbonised wood samples were observed from the palaeodeposits of sand at many regions lying within the lower reaches of the Meenachil River Basin. The P8 and P9 regions (Table 6 and Figure 6) of the Meenachil River are presently located at distance of about 15km inland of Vembanad Lake and are remarkably characterised by the presence of carbonized wood fragments at certain depth. No carbonized wood has been reported so far from the Meenachil River basin and the present study deals with the identification and significance of these woods.

Table 6. List of carbonised woods identified from the palaeodeposits

Location No	Sample No	IDENTIFICATION	
		Genus	Family
P8	MIA	*Artocarpus*	Moraceae
	MIB	*Sonneratia*	Sonneratiaceae
	MIIA	*Holigarna*	Anacardiaceae
	MIIB	*Calophyllum*	Clusiaceae (Guttiferae)
	MIIIA	*Calophyllum*	Clusiaceae (Guttiferae)
	MIIIB	*Sonneratia*	Sonneratiaceae
	MIIIC	*Spondias*	Anacardiaceae
P9	PIIA	*Canarium*	Burseraceae

The thin sections of the samples were prepared and photomicrographs were taken to observe the characteristic features (Figure8). The characteristic features like vessels, parenchyma, rays and fibres were taken into consideration and the carbonised woods were identified as given in Table 6.

Among the above genera, *Calophyllum, Spondias* and *Sonneratia* are inhabitants of coastal area and indicated near shore conditions particularly the last one. *Sonneratia* is a mangrove tree that occurs in the tidal creeks and littoral forests. *Calophyllum inophyllum*, a comparable species is found all along the coast above high water mark and in the evergreen forests of Western Ghats along the river banks. Likewise, *Artocarpus* (the jack fruit tree), *Holigarna* and *Canarium* are found in the evergreen forests of Western Ghats including Kerala. The assemblage indicated that the area was covered by dense forest with high rainfall, which might be turned into warm and humid as revealed from the later deposition of Spondias. The fungal infection in some of the woods further substantiates the existence of warm and humid conditions. So climatically there is no significant difference since the time of deposition of these woods. However, occurrence of *Sonneratia,* specially indicated the proximity of sea in the area at the time of deposition. Obviously, the sea level was much higher at that time than at present and later it receded. Thus the carbonized woods have thrown light into the palaeoenvironmental conditions and sea level fluctuations in the area. It is suggested that P 8 and P 9 region, which are located about 15km inland of present Vembanad Lake, had witnessed an episode of marine transgression, regression and contemporaneous geological modifications.

CONCLUSION

Georeferenced locations and the nature of the terrestrial sand deposits and their proximity to the existing water courses and ^{14}C dating of sediment and carbonaceous wood samples are suggestive of modification and shifting of stream channels of Meenachil River in relation to marine transgression/regression coupled with tectonism. The south-western coast of India had experienced intensified Asiatic Monsoon during the earlier part of the Holocene and the formation of the palaeo-deposit of sand had its origin even before the formation of the Vembanad Lake system ($\cong$ 4000yrBP). It is also suggested that the presence of clay deposit found sandwiched between the sand deposits is indicative of the decline of one fluvitile condition, stagnancy and the beginning of the other. The dominance of deciduous species obtained from the palaeo-palynological studies indicated that the wet climate, which nourished the evergreen species and mangroves, had declined during the late Holocene and a drier climate was experienced later along the western coast of India. The region had experienced another episode of elevated base level of erosion around 3000 YBP and the morphological modification of the region could be classified under (1) Pre-Vembanad Lake formation, (2) Contemporaneous to the Lake formation and (3) Post-Lake formation.

REFERENCES

Alexandre, A., Meunier, J. D., Lezine, A. M., Vincens, A. & Schwartz, D. (1997). Phytolith: indicators of grassland dynamics during the late Holocene in intertropical Africa. *Palaeogeogaphy, Palaeoclimatology Palaeoecology, 136*, 213-229.

Amorosi, A., Colalongo, M. L., Pasini, G. & Preti, D. (1999). Sedimentary response to Late Quaternary sea-level changes in the Romagna coastal plain (northern Italy). Sedimentology, *46*, 99-121.

Ansari, M. H. & Vink, A. (2007). Vegetation history and palaeoclimate of the past 30kyr in Pakistan as inferred from the palynology of continental margin sediments off the Indus Delta. *Review of Palaeobotany and Palynology, 145(3-4)*, 201-216.

Barboni, D., Bonnefille, R., Alexandre, A. & Meunier, J. D. (1999). Phytoliths as palaeoenvironmental indicators, west side Middle Awash valley, Ethiopia. *Palaeogeography, Palaeoclimatology, Palaeoecology., 152*, 87-100.

Bartlein, P. J., Anderson, K. H., Anderson, P. M., Edwards, M. E., Mock, C. J., Thompson, R. S., Webb, R. S., Webb, T. & Whitlock, C. (1998). Paleoclimate simulations for North America over the past 21,000 years: Features of the simulated climate and comparison with palaeoenvironmental data. *Quaternary Science Reviews, 17*, 549-585.

Bhattacharya, A., Ranhotra, P. S., Sharma, J. & Shah, S. K. (2005). Quaternary palynological scenario of India and its future perspective. *Abstract volume of diamond jubilee National Conference, BSIP, Lucknow*, 15-16.

Birks, H. J. B. & Birks, H. H. (1980). *Quaternary palaeoecology*. Arnold, London.

Borole, D. V., Rao, K. K., Krishnamurthy, R. V. & Somayajulu, B. L. K. (1982). Late Quaternary faunal change in coastal Arabian sea sediments. *Quaternary Research, 18(2)*, 236-239.

Brown, C. J., Coates, J. D. & Schoonen, M. A. A. (1999). Localized sulfate-reducing zones in the Magothy aquifer, Suffolk County, New York. *Ground Water, 37(4)*, 505-515.

Brown, C. J., Rakovan, J. & Schoonen, M. A. A. (2000). Heavy minerals and sedimentary organic matter in Pleistocene and cretaceous sediments on Long Island, New York, with emphasis on pyrite and marcasite in the Magothy Aquifer. *Water-resources investigations report 99-4216*, U. S. Geological Survey, New York.

Chakraborty, S., Bhattacharya, S. K., Ranhotra, P. S., Bhattacharyya, A. & Bhushan, R. (2006). Palaeoclimatic scenario during Holocene around Sangla valley, Kinnaur northwest Himalaya based on multi proxy records. *Current Science, 91(6)*, 777-782.

Faegri, & Iversen. (1975). *Textbook of pollen analysis* (4th edition). John Wiley & Sons, 328.

Folk, R. L. & Ward, W. C. (1957). Brazos river bar: A study in the significance of grain size parameters. *Journal of Sedimentary Petrology, 27*, 3-26.

Friedman, G. M. (1961). Distinction between dune, beach and river sands from their textural characteristics. *Journal of Sedimentary Petrology, 31*, 514-529.

Jayalakshmi, K., Kumaran, K. P. N., Nair, K. M. & Padmalal, D. 2005. Late Quaternary environmental changes in South Kerala Sedimentary Basin, Southwestern India. *Geophytology, 35(1&2)*, 25-31.

Joseph, S. & Thrivikramaji, K. P. (2002). Kayals of Kerala coastal land and implication to Quaternary sea level changes. In: A. C. Narayana, (Eds.), *Late Quaternary geology of India and sea level changes*. Geological Society of India, Banagalore: 51-64.

Kessarkar, P. M. & Rao, V. P. (2007). Organic carbon in sediments of the southwestern margin of India: Influence of productivity and monsoon variability during late Quaternary. *Journal of Geological Society of India., 69*, 42-52.

Kumaran, K. P. N., Nair, K. M., Shindikar, M., Limaye, R. B. & Padmalal, D. (2005). Stratigraphical and palynological appraisal of the Late Quaternary mangrove deposits of the west coast of India. *Quaternary Research, 64(3)*, 418-431.

Luckge, A., Doose-Rolinski, H., Khan, A. A., Schulz, H. & van Rad, U. (2001). Monsoonal variability in the northeastern Arabian Sea during the past 5000 years: geochemical evidence from laminated sediments. *Palaeogeography, Palaeoclimatology, Palaeoecology, 167(3-4)*, 273-286.

Makinen, J. (2003). Time-transgressive deposits of repeated depositional sequences within interlobate glaciofluvial (esker) sediments in Koylio, SW Finland. *Sedimentology, 50*, 327-360.

Menon, K. K. (1967). Warkalli beds at Kolathur, Trivandrum District. *Current Science, 36*, 102-103.

Meyers, P. (1994). Preservation of elemental and isotopic source identification of sedimentary organic matter. *Chemical Geology, 114*, 289-302.

Meyers, P. & Ishiwatari, R. (1993). The early diagenesis of organic matter in lacustrine sediments. In: M. H. Engel, & S. A. Macko, (Eds.) *Organic geochemistry*. Plenum Press, New York, 185-209.

Middleton, G. V. (1976). Hydraulic interpretation of sand size distribution. *Journal of Geology, 84*, 405-426.

Mohan, M., Ajaykumar, B., Satheesh, R. & Benno Joseph. (2005). Textural analysis of bed sediments of terrestrial sand mining sites of Meenachil river basin and their implication to the status of Meenadom Ar. *Indian Mineralogist, Spl. Vol. 1*, 37-44.

Morimoto, M., Kayanne, H., Abe, O. & McCulloch, M. T. (2007). Intensified mid-Holocene Asian monsoon recorded in corals from Kikai Island, subtropical northwestern Pacific. *Quaternary Research, 67(2)*, 204-214.

Nair, A. S. K. (2005). Evolutionary model of the Holocene coastlines of the South-west coast of India. *Proceedings of the Kerala Environment Congress.*, 113-121.

Nair P. K. K. (1977). Pollen morphology of Indian plants and its bearing on micropalaeontology. *Proceedings of the 4th Indian colloquium on Micropalaeontology and Stratigraphy*, 157-163.

Najeeb, K. M. (1999). *Ground water exploration of Kerala as on 31-03-1999*. Report of Ground water Board, Kerala region.

Narayana, A. C., Priju, C. P. & Chakrabarthi, A. (2001). Identification of palaeo-delta near the mouth of Periyar River in Central Kerala (short communication). *Journal of Geological Society of India, 57*, 544-547.

Narayana, A. C., Priju. C. P. & Rajagopalan, G. (2002). Late Quaternary peat deposits from Vembanad Lake (lagoon), Kerala, SW coast of India. *Current Science, 83(3)*, 318-321. Nemec, W. & Kazanci, N. (1999). Quaternary colluvium in west-central Anatolia: sedimentary facies and palaeoclimatic significance. *Sedimentology, 46*, 139-170.

Patience, A. J., Lallier-Verges, E., Sifeddine, A. & Guillet, B. (1995). Organic fluxes and early diagenesis in the lacustrine environment. In: E., Lallier-Verges, N. Tribovillard, & P. Bertrand, (Eds.) *Organic matter accumulation: the organic cyclicities of the Kimmeridge Clay Formation* (Yorkshire, GB), and the Recent Maar sediments. Springer – Verlag, Heidelberg (Lecture notes in Earth Sciences, 57), 145-156.

Prakash, T. N., Nair, M. N. M., Kurian, N. T. & Vinod, N. V. (2001). Developing a District coastal management plan and training for environmental improvement, Ashtamudi estuary, Kollam, India. Geology and Sediment characteristics, *Technical report No.9.CESS, Kerala*, 1-24.

Rajendran, C. P., Rajagopalan, G, & Narayanaswamy. (1989). Quaternary Geology of Kerala, Evidence from Radiocarbon dates. *Journal of Geological Society of India, 33(3)*, 218-222.

Ramasamy, S. M. & Balaji, S. (1995). Remote sensing and Pleistocene tectonics of Southern Indian peninsula. *International Journal of Remote Sensing, 16(13)*, 2375-2391

Ramesh, R. (2001). High resolution Holocene monsoon records from different proxies: an assessment of their consistency. *Current Science, 81(11)*, 1432-1436.

Reddy, N. P. C. (2003). Organic matter distribution in the continental shelf sediments, off Kochi, West Coast of India. *Indian Journal of Petroleum Geology, 12(2)*, 41-47.

Sadler, R. M. (1981). Sediment accumulation rates and the completeness of stratigraphic sections. *Journal Geology, 89*, 569-584.

Sarkar, A., Ramesh, R., Somayajulu, B. L. K., Agnihotri, R., Jull, A. J. T. & Burr, G. S. (2000). High resolution Holocene monsoon record from the eastern Arabian Sea. *Earth and Planetary Science Letters, 177(3-4)*, 209-218.

Shamsuddin, M. (1986). Textural differentiation of foreshore and breaker zone sediments on the Northern Kerala coast, *Indian Journal of Sedimentary Geology, 46*, 135-145.

Sifeddine, A. & Wirrmann, D. (2004). Palaeoenvironmental reconstruction based on lacustrine organic matter: examples from the tropical belt of South America and Africa. In: L. D., De Lacerda, R. E., Santelli, E. K. Duursma, & J. J. Abraao, (Eds.)

Environmental Geochemistry in tropical and subtropical environments. Springer-Verlag, New York, 7-18.

Sifeddine, A., Bertrand, P., Lallier-Verges, E. & Patience, A. J. (1996). The relationship between lacustrine organic sedimentation and palaeoclimatic variations: Massif Central, France. *Quaternary Science Reviews, 15*, 203-211.

Soman K. (1997). Geology of Kerala (1st Edition). *Geological Society of India*, Bangalore: 191

Soman, K., Chattopadhyay, M., Chattopadhyay, S. & Krishnan, Potti, G. (2002). Occurrence and water resource potential of fresh water lakes in South Kerala and their relation to the Quaternary geological evolution of the Kerala coast. Geological Society of India, Bangalore. *Memoir, 49*, 21.

Stanley, D. J. (2000). Radiocarbon dating the artificially contained surfaces of the Rhone deltaic plain, Southern France. *Journal of Coastal Research, 16(4)*, 1157-1161.

Thamban, M., Rao, V. P., Schneider, R. R. & Grootes, P. M. (2001). Glacial to Holocene fluctuation in hydrography and productivity along the western continental margin of India. *Palaeogeography, Palaeoclimatology, Palaeoecology, 165(1-2)*, 113-127.

Vaidhyanadhan, R. (1971). Evaluation of drainage of Cauvery basin in South India. *Journal of Geological Society of India, 12(1)*, 14-23.

Van Campo, E. (1986). Monsoon fluctuations in the Two 20,000 YBP oxygen isotope/ pollen records of South- West India. *Quaternary Research, 26(3)*, 376-388.

Wagner, T. (1998). Pliocene-Pleistocene deposition of carbonate and organic carbon at site 959: palaeoenvironmental implications for the eastern equatorial Atlantic off the Ivory coast/Ghana. *Proceedings of the Ocean Drilling program, Scientific results, 159*, 557-574.

Webb, T., III, Bartlein, P. J., Harrison, S. P. & Anderson, K. H. (1993). Vegetation, lake levels, and climate in eastern North America for the past 18,000 years, in H. E., Wright, Jr., J. E., Kutzbach, T., Webb, III, W. F., Ruddiman, F. A. Street-Perrott, & P. Bartlein, (Eds)., Global climates since the last glacial maximum: *Minneapolis*, University of Minnesota Press, 415-467.

Webb, T., III. (1988). Eastern North America, in B. Huntley, & T. Webb, III (Eds)., *Vegetation History: Dordrecht*, Kluwer Academic Publishers, 385-414.

Williams, M., Dunkerly, D., De Deckker, P. & Chappel, J. (1998). Quaternary *Environments* (2nd edn) Arnold, London, *329*.

In: Encyclopedia of Environmental Research
Editor: Alisa N. Souter

ISBN: 978-1-61761-927-4
© 2011 Nova Science Publishers, Inc.

Chapter 44

HUMAN IMPACTS ON THE ENVIRONMENT OF THE CHANGJIANG (YANGTZE) RIVER ESTUARY

Baodong Wang, Linping Xie and Xia Sun*
The First Institute of Oceanography, State Oceanic Administration, Qingdao, China.

ABSTRACT

The Changjiang (Yangtze) River is known to contribute significantly to the ecosystems of the Changjiang River estuary and adjacent waters. In this chapter we present some long-term data of freshwater discharge, sediment load, nutrient concentrations and compositions in the river and estuary waters, as well as the long-term response of the ecosystem in the estuary.

The freshwater discharge from the Changjiang River fluctuated but no trend of increase or reduction in the past six decades; however, sediment load decreased continuously from the 1950s to present, and a sharp decrease was observed after the closure of the Three Gorges Dam.

The concentrations of dissolved inorganic nitrogen and phosphate increased in the Changjiang River water by a factor of six from the 1960s to present, and a reduction in dissolved silicate by two thirds over the same period. Concomitantly, an increase in DIN concentration and a reduction in silicate concentration both by a factor of two were observed in the estuarine water.

As an ecological consequence to such nutrient changes, the chlorophyll *a* concentration increased by a factor of four since the 1980s, and both the number and affected area of harmful algal blooms increased rapidly since 1985 in the Changjiang River estuary and adjacent sea areas. Also, the area of hypoxic zone off the Changjiang River estuary increased in the recent decades.

We predict that, the freshwater discharge and sediment load from the Changjiang River would be reduced further due to continuous construction of dams and water division projects on the river; the symptoms of eutrophication associated with nutrients would worsen in the Changjiang River estuary in the near future due to continuously increasing of the riverine nutrient pressure.

---

[*] Corresponding author: 6 Xianxialing Road, Hi-Tech Industrial Park, Qingdao 266061, P. R. China. E-mail: wangbd@fio.org.cn

INTRODUCTION

Anthropogenic perturbations have caused a worldwide increase in river inputs of nitrogen (N) and phosphorus (P) to the coastal waters by more than a factor of four, leading to considerable eutrophication and an increase in the frequency of nuisance and toxic algae blooms (Nixon, 1995; Anderson and Garrison, 1997). The considerable changes in trophic composition and ecosystem structure (e.g., plankton community) of the coastal environments adjacent to large-size and medium-size river plumes have been attributed to the changes in Si:N:P ratios caused by excess of N and P, that is, cultural eutrophication (Admiraal et al., 1990; Humborg et al., 1997; Turner and Rabalais, 1994).

The enormous Changjiang (Yangtze) River drainage basin lies between 91°E and 122°E, and between 25°N and 35°N, with a total area of 1.81×10^6 km^2 and a population of 400 million (Chen et al., 2001). The Changjiang River is ranked the third largest river in the world in flow volume and is a major pathway of nutrients, which channels anthropogenic impacts in the catchments to the estuary and adjacent coastal waters (Chen et al., 2003). In the dry season (November through April), a combination of low discharge and prevailing northeasterly wind confines the influence of the Changjiang diluted water (CDW) to a narrow band southward along the coast. In the wet season (May to October), under the combined actions of the higher runoff and the prevailing southerly winds, the CDW flows to the northeast towards Cheju Island and may reach as far as 126°E (Beardsley *et al.*, 1985; Hu, 1994). The plume front defined as the area with salinity between 18 and 28 usually swings around 122°40'~123° in the wet season and moves back landward in the dry season (Hu et al., 2002).

In this chapter, we present some long-term data of freshwater discharge and sediment load as well as nutrient discharges from the Changjiang River and the subsequent changes in nutrient concentrations and Si:N:P ratios in the estuary. We also discuss the long-term responses of the trophic composition and ecosystem structure in the river plume to such changes.

MATERIALS AND METHODS

Downstream from Datong to the estuary, no large tributary flows in to the Changjiang River so that the additional input of water into this part of river is small. The tidal fluctuations, however, can influence the water level at Datong. Therefore, the water discharge at Datong is commonly used to represent the total discharge from the Changjiang River to the sea, although it is still 680 km from the sea [Chen et al., 2001a]. Freshwater discharges, sediment loads and nutrient concentrations in the Changjiang River water were based on the monthly or seasonally routine monitoring data at the Datong Gauging Station (Duan & Zhang, 2001; Li & Cheng, 2001; Liu et al., 2002), sporadic surveys (Liu et al, 2003; Shen, 1993; Shen, 2001) as well as the gazettes of river sedimentation and discharge of China (MWR, 2009).

Though there have been many surveys in the Changjiang River estuary, different campaigns have involved different areas and different station locations. For the sake of the comparability of the data, a limited study area of the Changjiang River plume (Figure 1) was chosen for this study and only those data with high spatial coverage and reasonable sampling frequency within this limited study area were selected for this study. Additionally, the annual means of nutrient (i.e., DIN (= nitrate + nitrite + ammonia), phosphate, and silicate) concentrations in the Changjiang plume surface water were calculated based on at least four seasonal survey data in a year or two neighboring years; the average chlorophyll *a* concentrations in the Changjiang plume surface water in the summer were calculated based on the data in August. For the macrozoobenthos, since it was difficult to find the long-term data of a fixed station with the same sampling time (e.g., month), the annual averaged total biomass of macrozoobenthos within the limited study area were calculated based on at least two seasonal survey data in a year or two neighboring years. The detailed data sources for nutrients, chlorophyll *a* and macrozoobenthos in this limited study area are listed in Table 1. During these cruises, seawater samples collected for the analysis of nutrients and chlorophyll *a* were taken with a Niskin sampler or Teflon-coated GO-FLO bottles at each station. Nutrients were measured spectrophotometically using methods previously described by Wang et al. (2003). Water samples for chlorophyll *a* analysis were immediately filtered through GF/F filter paper and stored at -20°C until analysis. The chlorophyll *a* retained on the GF/F filters was determined fluorometrically (Strickland and Parsons, 1972) using a fluorometer (Turner BioSystems Inc., USA). Duplicate or triplicate macrozoobenthic samples at each station were collected with 0.1 m^{-2} sediment grab or box corer and sieved on 1 mm screens.

Data on the incidents of harmful algal blooms (HABs) are from published reports (e.g. Guan & Zhan, 2003; Yan et al., 2004) and relevant gazettes (SOA, 2009).

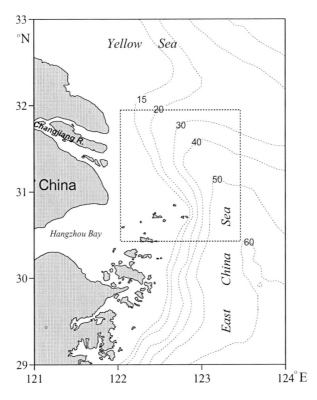

Figure 1. Map of the Changjiang estuary and adjacent sea areas showing the studied area (the dashed square).

Table 1. Data sources used for nutrients, chlorophyll *a* and macrozoobenthos in the Changjiang River estuary.

Variables	Time	No. of Stations within the Limited Area	Spatial Coverage within the Limited Area	References
Nutrients (Nitrate, nitrite, ammonia, phosphate, and silicate)	Monthly from Jan. to Dec. 1959	27	100 %	Office of Integrated Oceanographic Survey of China, 1961a
	Jun., Sept. and Nov. 1981, Mar. 1982	8	80 %	Sun et al., 1986
	Aug. 1981, Jan. 1982	17	90 %	Huang et al., 1986
	Monthly from Aug. 1985 to Jul. 1986 excluding Feb.	33	100 %	Shen et al., 1992
	Nov. 1997, May and Aug. 1998, Jan. 1999	10	80 %	Wang et al., 2003
	Sept. 1998	24	90 %	Ye et al., 2000
	May, Sept. and Nov. 2002, Feb. 2003	16	100 %	Han et al., 2003; Wang et al., 2004b
	Jul. and Dec. 2006, Apr. and Oct. 2007	72-83	100 %	Unpublished data
Chlorophyll *a*	Aug. 1984	12	80 %	Editorial Board of Marine Atlas, 1991
	Aug. 1985	33	100 %	Guo & Yang, 1992
	Jul. 1986	20	85 %	Ning & Coudé, 1991
	Aug. 1988	36	100 %	Shen & Hu, 1995
	Aug. 1991	13	80 %	Shen & Hu, 1995
	Aug. 2000	40	100 %	Zhu, 2004
	Aug. 2002	16	100 %	Zhou et al., 2004
	Jul. 2006	77	100 %	Unpublished data
Macrozoobenthos	Jan., Apr., Jul. and Oct. 1959	27	100 %	Office of Integrated Oceanographic Survey of China, 1961b
	Aug. and Dec. 1978, Apr. 1979	29	90 %	Wu et al., 1984
	Aug. and Oct. 1982, Feb. and May 1983	27	100 %	Dai, 1991
	Aug. and Nov. 1984, Feb. and May 1985	25	100%	Editorial Board of Marine Atlas, 1991
	Monthly from Aug. 1985 to Jul. 1986 excluding Feb.	33	100 %	Liu et al., 1992
	Apr. and Oct. 1988	28	100 %	Sun et al., 1992
	May 2000, May and Aug. 2001, May 2002	12	80 %	Wang et al., 2004a

Long-Term Variations in Freshwater Discharge and Sediment Load of the Changjiang River

Dam construction in the Changjiang River basin began to flourish in the late 1950s. Up to now, altogether 48,000 reservoirs has been built in the Changjiang River basin, of which 966 reservoirs are of medium or large size including the Three Gorges Reservoir, the largest reservoir in China, with a total storage capacity of 180×10^9 m^3 (Liu et al., 2003; Wang and Brockmann, 2008), which represents 20% of the annual water discharge from the Changjiang River into the estuary. However, the freshwater discharge of the Changjiang River has not been reduced by these dams. Indeed, the variations of annual runoff show periodic fluctuations but no long-term decline (Figure 2). Moreover, the decadal-averaged annual discharge during the 1990s was slightly higher than ever before (Figure 2). One of the possible explanations for this lack of effect may be that water retention in reservoirs only regulates water volume in the time domain and has no significant influence on total runoff volume, for most of the dams on the Changjiang River were for power generation rather than water consumption; another reason may be attributed to the slightly increasing trend of precipitation in the Yangtze River catchment area over the past decades (Chen and Chen, 2003; Liu et al., 2008).

The variation of the annual sediment load from the Changjiang River in to the estuary demonstrated a different pattern with that of freshwater discharge. The annual sediment load from the Changjiang River maintained at a level of $4.94 \pm 0.75 \times 10^8$ t a^{-1} in the 1950/60s, but it has shown a significant decrease since the 1970s. The annual sediment load was 3.47×10^8 t a^{-1} in the 1990s, which was one-third lower than that in the 1960s; and it decreased to 2.11×10^8 t a^{-1} in the 2000s, which was only 40% of that in the 1960s.

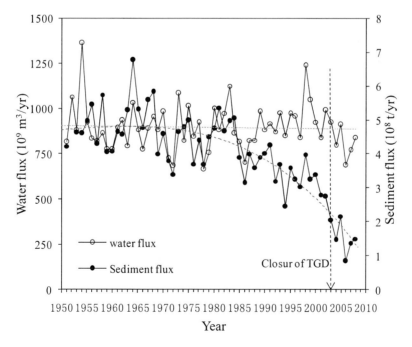

Figure 2. Long-term variations of freshwater discharge and sediment load from the Changjiang River (1951-2008).

The inter-annual fluctuations in annual sediment flux can be attributed to cyclic changes in climate, because they were in phase with the fluctuations of the annual water discharge (Figure 2) (Yang et al., 2004). Nevertheless, the decreasing trend in annual sediment flux was caused by human activities, because the water discharge did not show any decreasing trend. It has been evidenced that the total amount of sediments trapped by reservoirs was closely related with the storage capacity of the reservoirs in the Changjiang River basin (Yang et al., 2004). This could be confirmed by the case of the Three Gorges Dam. A sharp reduction of sediment load into the estuary was observed after closure of the Three Gorges Dam in 2003 (Figure 2). The sediment load decreased from 3.24×10^8 t a^{-1} of five-year average value (1998-2002) just before closure of the Three Gorges Dam to 1.58×10^8 t a^{-1} of five-year average value (2003-2007) after closure of the Three Gorges Dam, a reduction by more than a half. In addition to sediment trapping by dams, other human activities should be also responsible for the reduction of sediment load from the Changjiang River, such as the impact of water and soil conservation programs (the Project of Yangtze Upstream Water and Soil Conservation) which begun in 1988 (Xu et al., 2006) and increased sand mining (5×10^7 t a^{-1} from 1990 to 2002) within the Yangtze River watershed (Liu et al., 2008).

The decreased riverine sediment discharges would reduce the turbidity and increase the light availability, and thus most likely increase the primary productivity in the Changjiang River estuary. On the other hand, however, clear water below the dam will enhance erosion of the coasts in the lower reaches of the river and by this contribute to the retreating coastline, erosion of the delta, and modification of the entire ecosystem especially the benthic community (Chen, 2000; Nixon, 2003). Furthermore, reduction of sediment load would reduce the fluxes of organic matter as well as other pollutants absorbed on the sediment from the river into the estuary.

Changes in Nutrient Concentrations and Compositions in the Freshwater and Estuarine Water

Nutrient concentrations, i.e., DIN and phosphate, of the Changjiang River water increased exponentially and by a factor of five from the 1960s to the end of the 1990s (Figure 3a), mainly due to the increased amount of fertilizer application and effluent from cities in the river basin (Duan et al., 2001). In contrast, silicate concentrations decreased exponentially by two thirds from the 1960s to the 1990s (Figure 3a). Silicate concentration in the river was much higher than that of DIN before 1990, but it declined to approximately the same level as that of DIN by the middle of the 1990s. As a result of the increased N and P concentrations and the decreased Si concentration, the nutrient balance, i.e., N:Si:P ratios, of the Changjiang River water have been changed greatly (Figure 3b). Despite the high variability in the N/P ratios, an overall increase in N/P ratio can be assumed for the Changjiang River water from the existing data (Figure 3b). The N/P ratio was approximately 50 during the period of 1965 to 1975, but it increased sharply to 125 in 1985 and has stayed nearly constant since then. In contrast, the Si/N ratio decreased drastically from levels greater than 15 in the 1960s to close to 1.0 in the 1990s (Figure 2b). Besides, a sharp increase in DIN concentration and a sharp decrease in silicate concentration in the Changjiang freshwater can be observed from 1984 to1985 (Figure 3a). Concomitantly, a sharp increase in N/P ratio and a sharp decrease in Si/N ratio can also be observed at the same time (Figure 3b).

Following these changes in riverine nutrients, the nutrient concentrations and composition in the Changjiang plume also changed significantly (Figure 4). The annual average concentration of DIN increased by a factor of two from 1959 to 2002, slowly from 1959 to 1985 but sharply from 1985 to 2002, and it maintained at nearly constant level from 2002 to 2007. In contrast, the annual average concentration of Si decreased by a factor of two from 1959 to 2002, sharply from 1959 to 1985 but slowly from 1985 to 2002, and it also maintained at nearly constant level after 2002. The annual average concentration of phosphate was almost constant with no long-term trend of increase or decline in the past 50 years. In terms of the nutrient composition, changes of N/P and Si/N ratios in the estuarine water were found to be similar to those in the river water in the long-term trend but were less variable. The N/P ratio in the Changjiang plume water was 18.5 in 1959, which was close to the Redfield ratio of 16, but increased by two-fold to 35 in 2002, whereas the Si/N ratio decreased almost linearly from 3.8 in 1959, which was much greater than the Redfield ratio of 1 to a value of 0.85 in 2002, which was less than the Redfield ratio.

Very high N/P ratios in the freshwater and the estuarine water have resulted in an excess amount of DIN and thus P-limited phytoplankton growth in the estuarine and coastal waters of ECS (Harrison et al., 1990; Wang et al., 2003; Wong et al., 1998). Moreover, the reduction in silica discharge may have significant implications for the planktonic food-web structure in the coastal zone (Conley et al., 2000; Humborg et al., 2000). Generally, a non-limiting Si and DIN supply at a Si/DIN ratio of >1 predominantly results in the development of diatoms, which promotes the transfer of organic matter through the trophic food chain to copepods and finally to fish, while lower Si/DIN ratios lead to reduced zooplankton grazing (Officer and Ryther, 1980; Fransz et al, 1992). For instance, the resulting changes in the nutrient composition (Si:N:P ratio) of river discharges seem to be responsible for dramatic shifts in phytoplankton species composition in the Black Sea and the Baltic Sea (Humborg et al., 1997). Since the average Si/DIN ratio decreased from 3.8 to 0.85 in the Changjiang plume water, as a consequence, the chemical environment would shift from a state that allows diatoms to compete effectively with other algal classes, to a different state that gives selective advantage to the nondiatom taxa species not requiring silicate for growth.

Response of the Pelagic Ecosystem in the Changjiang River Estuary

The chlorophyll a (Chl a) concentration was used as an indicator for phytoplankton biomass or primary productivity of the Changjiang River estuary. Since the annual variation of the Chl a concentration in the Changjiang plume surface water could be well described with a normal distribution curve with the peak in the summer (Hu et al., 2002), the inter-annual Chl a concentrations in the summer were used to evaluate the response of phytoplankton to the nutrient enrichment. Since no measurement of Chl a was made before 1980, only Chl a concentrations in the recent two decades were presented in Figure 5. The Chl a concentration increased by a factor of four from 1984 to 2002 (Figure 5). A sharp increase in Chl a concentration in the Changjiang plume water can be observed from 1984 to1985 (Figure 5). High concentrations of Chl a occurred mostly in the plume front area where the Sechi depth was usually greater than 5 m (Hu et al., 2002).

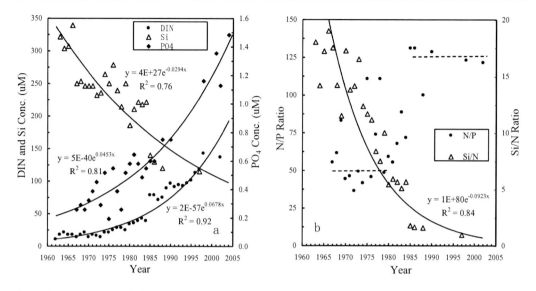

Figure 3. Long-term variations of nutrient concentrations (a) and Si/N/P ratios (b) in the Changjiang freshwater (1962-2007).

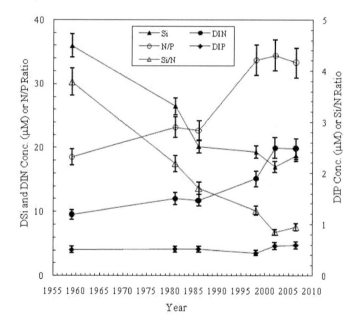

Figure 4. Long-term variations of nutrient concentrations and Si/N/P ratios in the surface water of the Changjiang River estuary (1959-2007).

Increased nutrient inputs promote a complex array of symptoms, beginning with the excessive growth of algae, which, in turn, may lead to other, more serious symptoms (e.g., nuisance and toxic algal blooms). The elevated phytoplankton biomass as indicated by the Chl *a* concentration has caused an increase of bloom events in the river plume. Incidents of harmful algal blooms (HABs) in the Changjiang plume and adjacent coastal areas were rare before 1985 but have increased rapidly since then (Figure 6). For instance, 58 HAB incidents were observed in 2003 in the Changjiang plume and adjacent coastal areas, 10 times more than the average number in the 1990s. Also, the annual total affected area of HABs was less than 1000 km² in the 20th century, but it increased rapidly from 2000 to 2005; however, it had a decreasing tendency from 2005 to present (Figure 6). Incidents of HABs occurred mostly in the spring and the summer with a maximum in May, but rarely in the winter (Guan & Zhan, 2003). The durations of the HAB incidents and sizes of the affected areas have also increased (Guan & Zhan, 2003). Toxic species of HABs such as *Alexandrium* and *Gymnodrium* were often observed in these areas (Guan & Zhan, 2003; Yan et al., 2004).

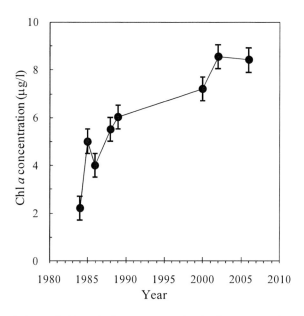

Figure 5. Long-term variations of chlorophyll *a* concentration in the summer (August) in the surface water of the Changjiang River estuary (1984-2007).

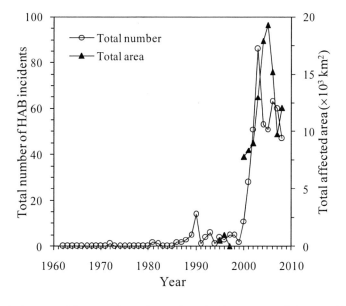

Figure 6. Historical changes of the annual number and total affected area of HAB incidents in the East China Sea (1962-2008).

Eutrophication is a natural process by which productivity of a water body increases as a result of increased nutrient inputs. Cultural eutrophication is the enhanced accumulation of organic matter, particularly algae, that is caused by human-related increases in the amount and composition of nutrients being discharged to the water body (Bricker et al., 2003). The Changjiang River, as a major pathway of nutrients, channels anthropogenic impacts in the catchments to the estuary. The remarkable positive linear correlativity of Chl a concentration in the Changjiang River estuary versus the Changjiang River freshwater DIN concentration (n = 8, r^2 = 0.89) and phosphate concentration (n = 7, r^2 = 0.76) during the recent two decades suggests that the dramatic increase in nutrient inputs was responsible for the elevated phytoplankton biomass as indicated by the Chl a concentration in the Changjiang River estuary. The significant increase of nutrient inputs from the Changjiang River as well as the sharp increase in N/P ratio and decrease in Si/N ratio both in the freshwater and plume water seem to be responsible for the drastic increase of bloom events in the estuary from 1985.

Response of Macrozoobenthos in the Changjiang River Estuary

The variation of the number of species and biomass may be more suitable for evaluating long-term eutrophication effects. However, the available data of the number of macrozoobenthic species and biomass were from different campaigns which involved different spatial coverage and frequency, as well as different sampling method; it is difficult to evaluate reliably the long-term variation of the number of macrozoobenthic species in the Changjiang plume area. Here we examine the eutophication effect mainly based on the long-term variation of total biomass of macrozoobenthos (Figure 6). The annual average total biomass of macrozoobenthos was $ca.$ 16 g m^{-2} before 1980, it increased sharply by 60% to 26 g m^{-2} in 1982, then decreased sharply until 1988, and had a slightly deceasing trend since then. Compared to the levels of 1959 and 1982, the current level of total macrozoobenthic biomass is lower by 30% and 50%, respectively.

Macrozoobenthos data may be more suitable for evaluating long-term eutrophication effects compared to those of short-lived pelagic species, due to the macrozoobenthos' more stationary way of life and their potential to integrate chronic stress and enrichment situations. However, the macrobenthic species composition, abundance and biomass, as the expressions of species and population growth rates and predatory losses, are not only the expression of temperature- and food-limited growth, but also the product of physical impacts (e.g., climate-related variations in mortality, substrate limitation, sediment displacements, wave action, exposure time, and bottom trawling) and biological controls (predation by birds, fish, and crustaceans, parasite-induced mortality, and migrations to suitable environment). However, it seems that there were no such changes of the physical impacts and biological controls in the Changjiang plume area. Nevertheless, the work of Beukema and Cadée (1997) and Madsen and Jensen (1987) pointed to the high relevance of type of location for the species' potential for growth and reproduction: only species communities not exposed to severe stress due to long exposure times, wave impact, or other stress factors do have the potential to react to enhanced feeding conditions. If this was the case in the Changjiang plume area, the response of the total macrozoobenthic biomass to the increased input of organic matter in the Changjiang plume may be well described by the model of Pearson and Rosenberg (1978): in the first stage of moderate eutrophication before 1982, macrozoobenthic biomass increased due to enhanced organic matter; with the organic load continuing to increase after 1982, the macrozoobenthic biomass decreased due to harmful effects such as oxygen deficiency and related stress factors. Oxygen deficiency in the bottom water has been widely considered to be one of the most negative effects of eutrophication on macrozoobenthic communities.

It is worth noting that all the variables, i.e., the nutrient concentrations and N/P/Si ratios both in the freshwater and the plume water, Chl a concentration, bloom events, and macrobenthic biomass, presented themselves sharp changes in the middle of the 1980s. This suggests that i) the sharp changes of the nutrient concentrations and N/P/Si ratios both in the freshwater and the plume water may be responsible for the sharp changes of the Chl a concentration, bloom events and the macrobenthic biomass in the Changjiang River estuary area in the middle of the 1980s, and ii) the middle of the 1980s might be a critical juncture in the eutrophication process in the Changjiang River estuary area with a low or moderate eutrophication stage existing before that time but a high eutrophication stage after that time, especially in the latest decade.

Variation of Hypoxic zone off the Changjiang River Estuary

A historical picture of oxygen conditions off the Changjiang River estuary indicates that hypoxia in the near-bottom water off the Changjiang River estuary was first recorded in the summer of 1958; hereafter there were episodes of hypoxia in the past 50 years but not every year. The overall probability of occurrence of summer hypoxia was 60% of the total number of summer cruises in the past 50 years, but up to nearly 90% after the 1990s, demonstrating an increasing trend (Wang, 2009). The minimum oxygen concentrations observed in the area off the Changjiang River estuary were different in different years, but there was no obvious trend of decline or increase (Figure 7). However, although the size of the hypoxic zone also differed greatly in different years., all the events of hypoxia with affected area greater than 5000 km^2 occurred after the end of the 1990s, indicating that the extent of summer hypoxia became more severe in recent years, possibly due to the increasing input of anthropogenic nutrients to the estuary (Wang, 2006).

Future Outlook

Based on China's strategic planning for development, the Changjiang drainage basin is expected to provide $ca.$ $10^7 \sim 10^8$ $t \cdot y^{-1}$ more foodstuffs in order to satisfy the demand from the population increase within the next 50 years, which will probably cause a further increase in fertilizer application in a region of dense population and intensive agriculture (Zhang et al., 1999). If the DIN concentrations and DIN loads of the Changjiang increase at the same rate as shown in Figure 3a in the near future, the predicted DIN load in 2010 would be $Ca.$ 290×10^9 mol yr^{-1}, twice as much as that in 1998, whereas the silicate concentration is expected to decrease continually with the construction of large dams on the river, such as the Three Gorges Dam (Zhang et al., 1999). This would cause a further increase of the N/P ratio and a decrease of the Si/N ratio in the estuarine water. Accordingly, the nutrient-related symptoms of eutrophication are expected to continue to worsen in the Changjiang River estuary plume in the near future.

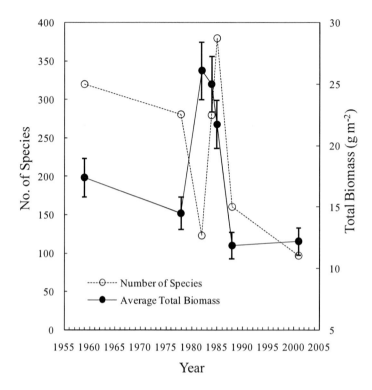

Figure 7. Long-term variation of the total biomass of macrozoobenthos in the Changjiang River estuary.

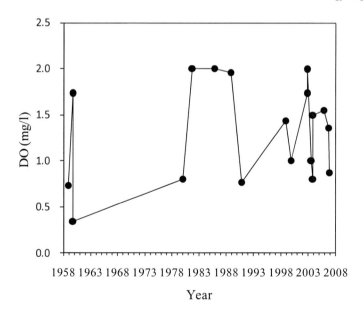

Figure 8. Long-term variation in the minimum dissolved oxygen content in the hypoxic zone off the Changjiang estuary (1958-2006).

ACKNOWLEDGMENT

This study was funded by the National Key Basic Research Program, Ministry of Science and Technology of China (Grant No. 2010CB429004) and the Program of Investigation and Assessment of Coastal Seas of China (Contract No. 908-JC-I-03).

REFERENCES

Admiraal, W., Breugem, P., Jacobs, D. M., De Ruyer, L. H. A. & Van Stevenick, E. D. (1990). Fixation of dissolved silicate and sedimentation of biogenic silicate in the lower river Rhine during diatom blooms. *Biogeochemistry*, *9*, 175-185.

Anderson, D. M. & Garrison, D. J. (Eds.). (1997). The ecology and oceanography of harmful algal blooms. *Limnol. Oceanogr.*, *42*, 1009-1305.

Beardsley, R. C., Limeburner, R., Yu, H. & Cannon, G. A. (1985). Discharge of the Yangtze River (Changjiang) into the East China Sea. *Continental Shelf Research*, *4*, 57-76.

Beukema, J. J. & Cadée, G. C. (1997). Local differences in macrozoobenthic response to enhanced food supply caused by mild eutrophication in the Wadden Sea area: Food is only locally a limiting factor. *Limnol. Oceanogr*, *42(6)*, 1424-1435.

Bricker, S. B., Ferreira, J. G. & Simas, T. (2003). An integrated methodology for assessment of estuarine trophic status. *Ecol. Modell.*, *169*, 39-60.

Chen, C. T. A. (2000). The Three Gorges Dam: reducing the upwelling and thus productivity in the East China Sea. *Geophy Res Lett.*, *27(3)*, 381-383.

Chen, J. Y. & Chen, S. L. (2003). Ecological environmental changes in the Changjiang estuary and suggestions for countermeasure. *Water Resources Hydropower Engineering*, *34(1)*, 19-25 (in Chinese).

Chen, X., Zong, Y., Zhang, E., Xu, J. & Li, S. (2001a). Human impacts on the Changjiang (Yangtze) River basin, China, with special reference to the impacts on the dry season water discharges into the sea. *Geomorphology*, *41*, 111-123.

Chen, Z., Li, J., Shen, H. & Wang, Z. (2001b). Changjiang of China: historical analysis of discharge variability and sediment flux. *Geomorphology*, *41*, 77-91.

China-HAB Network. (2001). Records of the historical harmful algal blooms in China Seas. http://www.china-hab.cn/chinese/ccht/ccht2001-2.htm.

Conley, D. J., Stalnacke, P., Pitkänen, H. & Wilander, A. (2000). The transport and retention of dissolved silicate by rivers in Sweden and Finland. *Limnol. Oceanogr.*, *45(8)*, 1850-1853.

Dai, G. L. (1991). Ecological characteristics of macrobenthics of the Changjiang River estuary and adjacent waters. *Journal of Fisheries of China*, *15(2)*, 104-116 (in Chinese with English abstract).

Duan, S. W. & Zhang, S. (2001). Material fluxes and their variations from the Changjiang River into the sea. In: *Land-Sea Interaction in Estuaries of the Changjiang River, Zhujiang River and their Adjacent Sea Areas*. Edited by Zhang, S. & Shen, H. T. Beijing: China Ocean Press, 10-26 (in Chinese).

Editorial Board for Marine Atlas of China. (Ed.). (1991). *Marine Atlas of Bohai Sea, Huanghai Sea and East China Sea: Biology*. Beijing: China Ocean Press, 254.

Fransz, H. G., Gonzalez, S. R., Cadée, G. C. & Hansen, F. C. (1992). Long-term change of Temora longicornis (Copepoda, Calanoida) abundance in a Dutch tidal inlet (Marsdiep) in relation to eutrophication. *Neth. Inst. Sea Res., 30*, 23-32.

Guan, D. M. & Zhan, X. W. (2003). Red-tide disaster in coastal water of China and its prevention suggestions. *Mar. Environ. Sci., 22(2)*, 60-63 (in Chinese with English abstract).

Guo, Y. J. & Yang, Z. Y. (1992). Variation of phytoplankton quantity and ecological analysis in the Changjiang estuary. *Studia Marina Sinica, 33*, 167-189 (in Chinese with English abstract).

Han, X. R., Wang, X. L., Sun, X., Shi, X. Y., Zhu, C. J., Zhang, C. S. & Lu, R. (2003). Nutrient distribution and its relationship with occurrence red tide in coastal area of the East China Sea. *Chinese Journal of Applied Ecology, 14(7)*, 1097-1101(in Chinese with English abstract).

Harrison, P. J., Hu, M. H., Yang, Y. P. & Lu, X. (1990). Phosphate limitation in estuarine and coastal waters of China. *J. Exp. Biol. Ecol., 140*, 79-87.

Hu, D. X. (1994). Some striking features of circulation in the Huanghai Sea and the East China Sea. In: *Oceanology of China Seas*, Vol.1 (D. Zhou, Y. B. Liang, & C. K. Zeng Eds.). Kluwer Academic Publishers, Netherlands, 27-38.

Hu, F. X., Hu, H. & Gu, G. C. (Eds.). (2002). *The Study of Fronts in the Changjiang Estuary*. Shanghai: The East China Normal University press, 220 (in Chinese).

Huang, S. G., Yang, J. D., Ji, W. D., Yang, X. L. & Chen, C. X. (1986). Spatial and temporal variation of reactive Si, N, P and their relationship in the Changjiang estuary water. *Taiwan Strait, 5(2)*, 114-123 (in Chinese with English abstract).

Humborg, C., Conley, D. J., Rahm, L., Wulff, F., Cociasu, A. & Ittekkot, V. (2000). Silicon retention in river basins: far-reaching effects on biogeochemistry and aquatic foodwebs in coastal marine environments. *Ambio, 29*, 45-50.

Humborg, C., Ittekkot, V., Cociasu, A. & Bodungen, B. (1997). Effect of Danube River dam on Black Sea biogeochemistry and ecosystem structure. *Nature, 386*, 385-388.

Li, D. & Daler, D. (2004). Ocean pollution from land-based sources: East China Sea, China. *Ambio, 33(1-2)*, 107-112.

Li, M. T. & Cheng, H. Q. (2001). Changes of dissolved silicate flux from the Changjiang River into sea and its influence over the last 50 years. *China Environ. Sci., 21(3)*, 193-197 (in Chinese with English abstract).

Liu, C., Sui, J. & Wang, Z. (2008). Sediment load reduction in Chinese rivers. *International Journal of Sediment Research, 23*, 44-55.

Liu, R. Y., Xu, F. S., Sun, D. Y., Cui, Y. H. & Wang, H. F. (1992). Benthic fauna in the Changjiang River estuary and prediction of the influence of the Three Gorges Project on it. *Studia Marina Sinica, 33*, 237-247 (in Chinese with English abstract).

Liu, S. M., Zhang, J., Chen, H. T., Wu, Y., Xiong, H. & Zhang, Z. F. (2003). Nutrients in the Changjiang and its tributaries. *Biogeochemistry, 62*, 1-18.

Liu, X. C., Shen, H. T. & Huang, Q. H. (2002). Concentration variation and flux estimation of dissolved inorganic nutrient from the Changjiang River into its estuary. *Oceanologia Limnologia Sinica, 33(5)*, 332-340.

Madsen, P. B. & Jensen, K. (1987). Population dynamics of Macoma balthica in the Danish Wadden Sea in an organically enriched area. *Ophelia, 27*, 197-208.

Ministry of Water Resources (MWR) of China (2009). China Gazette of River Sedimentation 2000-2007 http://www.mwr.gov.cn/shuiwen/gb/hlnsgb.asp (in Chinese).

Ning, X. R. & Coudé, C. (1991). Interrelationships among chlorophyll *a*, bacteria, ATP, POC and microbial respiration rate in the Changjiang estuary and plume area. *Acta Oceanologica Sinica, 13(6)*, 831-838.

Nixon, S. W. (1995). Coastal eutrophication: A definition, social cause and future concerns. *Ophelia, 41*, 199-220.

Nixon, S. W. (2003). Replacing the Nile: Are anthropogenic nutrients providing the fertility once brought to the Mediterranean by a great river? *Ambio, 32(1)*, 30-39.

Office of Integrated Oceanographic Survey of China (Ed.). (1961a). *Dataset of the National Integrated Oceanographic Survey (Vol. 1): Survey data of hydro-meteorological and chemical elements in the Bohai, Huanghai and East China Seas*. Beijing, 811.

Office of Integrated Oceanographic Survey of China (Ed.). (1961b). *Dataset of the National Integrated Oceanographic Survey (Vol. 5): Distribution records of the biomass and dominant species of the phytoplankton and benthos in the Bohai, Huanghai and East China Seas*. Beijing, 908 pp.

Officer, C. B. & Ryther, J. H. (1980). The possible importance of silicon in marine eutrophication. *Mar. Ecol. Prog. Ser., 3(1)*, 83-91.

Pearson, T. H. & Rosenberg, R. (1978). Macrobenthic succession in relation to organic enrichment and pollution of the marine environment. *Oceangr. Mar. Biol. Ann. Rev., 16*, 229-311.

Shen, H. T. (Ed.). (2001). *Material Flux of the Changjiang Estuary*. Beijing: China Ocean Press, 176.

Shen, X. Q. & Hu, F. X. (1995). Basic characteristics of distribution of chlorophyll *a* in the Changjiang estuary. *Journal of Fishery Sciences of China, 2(1)*, 71-80 (in Chinese with English abstract).

Shen, Z. L. (1993). A study of the relationships of the nutrients near the Changjiang river estuary with the flow of the Changjiang river water. *Chinese Journal Oceanologia Limnologia, 11*, 260-267.

Shen, Z. L., Lu, J. P., Liu, X. J. & Diao, H. X. (1992). Distribution characters of the nutrients in the Changjiang River estuary and the effect of the Three Gorges Project on it. *Studia Marina Sinica, 33*, 109-129 (in Chinese with English abstract).

State Oceanic Administration of China (SOA). (2009). China Gazette of Marine Disaster 1989-2008. http://www.soa.gov.cn/hyjww/hygb/A0207index_1.htm. (in Chinese).

Strickland, J. D. H. & Parsons, T. R. (Eds.). (1972). *A Practical Handbook of Seawater Analysis*. Fisheries Research Board of Canada, Ottawa, Canada.

Sun, B. Y., Yu, S. R. & Hao, E. L. (1986). Comprehensive survey and research report on the sea areas adjacent to the Changjiang estuary and Chejudo Island: Chapter 4: Marine Chemistry. *Journal of Shandong College of Oceanology, 16(1)*, 132-210 (in Chinese with English abstract).

Sun, D. Y., Xu, F. S., Cui, Y. H., Sun, B. & Wang, H. F. (1992). Seasonal variation of distribution of macrobenthos in the Changjiang River estuary. *Studia Marina Sinica, 33*, 217-235 (in Chinese with English abstract).

Turner, R. E. & Rabalais, N. N. (1994). Coastal eutrophication near the Mississippi River delta. *Nature, 368*, 619-621.

Wang, B. (2009). Hydromorphological Mechanisms Leading to Hypoxia off the Changjiang Estuary. *Marine Environmental Research*, *67*, 53-58.

Wang, B. & Brockmann, U. (2008). Potential impacts of the Three Gorges Dam on the ecosystem of the East China Sea. *Acta Oceanologica Sinica*, *27(1)*, 67-76.

Wang, B., Wang, X. & Zhan, R. (2003). Nutrient conditions in the Yellow Sea and the East China Sea. *Estu.Coast. Shelf Sci.*, *58*, 127-136.

Wang, J. H., Huang, X. Q., Liu, A. C. & Zhang, Y. F. (2004a). Tendency of the biodiversity variation nearby the Changjiang estuary. *Marine Science Bulletin*, *23(1)*, 32-39 (in Chinese with English abstract).

Wang, X. L., Sun, X., Han, X. R. Zhu, C. J., Zhang, C. S., Xin, Y. & Shi, X. Y. (2004b). Comparison in micronutrient distributions and composition for the high frequency HAB occurrence areas in the East China Sea between summer and spring 2002. *Oceanologia et Limnology Sinica*, *35(4)*, 323-331 (in Chinese with English abstract).

Wong, G. T. F, Gong, G. C., Liu, K. K. & Pai, S. C. (1998). Excess nitrate in the East China Sea. *Estu.Coast. Shelf Sci.*, *46*, 411-418.

Wu, Q. Q., Jin, Y. K., Jin, Q. Z. & Zhu, K. T. (1984). Report of pollution survey in the East China Sea: Biology. In: *Report of Pollution Survey in the East China Sea: 1978-1979*. Edited by The Pollution Survey and Monitoring Group of the East China Sea. Beijing: China Ocean press, 101-145.

Xu, K., Milliman, J. D., Yang, Z. & Wang, H. (2006).Yangtze sediment decline partly from Three Gorges Dam. *EOS*, *87(19)*, 185-190.

Yan, T., Zhou, M. J. & Zou. J. Z. (2004). A national report on harmful algal blooms in China. www.pices.int/publications/ scientific_reports/Report23/HAB_China.pdf.

Yang, S. L., Gao, A., Hotz, H. M., Zhu, J., Dai, B. S. & Li, M. (2005). Trends in annual discharge from the Yangtze River to the sea(1865-2004). *Hydrological Sciences Journal*, *50(5)*, 825-836.

Yang, S. L., Shi, Z., Zhao, H. Y., Li, P., Dai, S. B. & Gao, A. (2004). Effect of human activities on the Yangtze River suspended sediment flux into the estuary in the last century. *Hydrology and Earth System Sciences*, *8(6)*, 1210-1216.

Ye, X. S., Zhang, Y. & Xiang, Y. T. (2000). Characteristics of nutrient distributions in the Changjiang River estuary and its formation mechanism. *Marine Science Bulletin*, *19(1)*, 89-92 (in Chinese with English abstract).

Zhang, J., Zhang, Z. F., Liu, S. M., Wu, Y., Xiong, H. & Chen, H. T. (1999). Human impacts on the large world rivers: Would the Changjiang (Yangtze River) be an illustration. *Global Biogeochemical Cycles*, *13(4)*, 1099-1105.

Zhou, W. H., Yuan, X. C. & Huo, W. Y. (2004). Distributions of chlorophyll a and primary productivity in the Changjiang estuary and adjacent sea areas. *Acta Oceanologia Sinica*, *26(3)*, 143-150 (in Chinese with English abstract).

Zhu, J. R. (2004). Distribution of chlorophyll a off the Changjiang estuary and its dynamic formation mechanism. *Science in China (Ser. D)*, *34(8)*, 757-762.

In: Encyclopedia of Environmental Research
Editor: Alisa N. Souter

ISBN: 978-1-61761-927-4
© 2011 Nova Science Publishers, Inc.

Chapter 45

INTEGRATED APPROACHES TO ESTUARINE USE AND PROTECTION: TAMPA BAY ECOSYSTEM SERVICES CASE STUDY

James Harvey, Marc Russell, Darrin Dantin and Janet Nestlerode

US EPA, Office of Research and Development,
Gulf Ecology Division, Gulf Breeze FL 32561, USA

ABSTRACT

The Tampa Bay region faces projected stress from climate change, contaminants, nutrients, and of human development on a natural ecosystem that is valued (economically, aesthetically and culturally) in its present state. With fast-paced population increases, conversion and development of open land, and other stresses, redressing past damage to Bay habitats and protecting them in the future will remain the greatest challenge for managers in this region. Maintaining water quality gains of recent decades and sustaining ecological services requires careful thought and planning to compensate for these stressors. Approaches piloted during this study will be applicable to many urbanized estuaries, particularly along the U.S. Gulf of Mexico coastline, because they face similar stresses.

Regional and local planners, managers, and decision-makers, like those in the Tampa region, require new or modified ways to address questions regarding the production, delivery, and consumption of ecosystem services under various projected future scenarios of climate change and urban development. Approaches that we are developing collaboratively must address human well-being endpoints (the focus of ecosystem services) as well as more traditional measures of ecosystem integrity, sustainability, and productivity. Predictive estimates of ecosystem services value require 1) Alternative future scenarios, especially population growth and effects from climate change 2) Models translating environmental conditions set by these scenarios into ecosystem services production and their sustainability at different scales from local neighborhoods to watershed and larger and 3) An approach to visualize and link model outputs to a common currency for prioritization and valuation. Collectively, these comprise tools that can be used to make well-informed, thoughtful, and publically transparent decisions.

DISCLAIMER

The information in this document has been funded wholly (or in part) by the U.S. Environmental Protection Agency. It has been subjected to review by the National Health and Environmental Effects Research Laboratory and approved for publication. Approval does not signify that the contents reflect the views of the Agency, nor does mention of trade names or commercial products constitute endorsement or recommendation for use. This is contribution number XXXX from the Gulf Ecology Division.

ACKNOWLEDGMENTS

We thank our partners, Holly Greening of the Tampa Bay Estuary Program, and Suzanne Cooper of the Tampa Bay Regional Planning Council for assistance and their invaluable collaboration in this project.

INTRODUCTION

Measurement of ecosystem services– such as the provision of clean air, clean water, productive soils, and opportunities for recreation in vibrant natural places – is an important strategic focus for EPA's Ecosystem Services Research Program (ESRP). ESRP scientists believe that making the evaluation of these services a routine part of decision-making will more fully illuminate the ways in which our choices affect our well-being and will transform the way we understand and respond to environmental issues. Toward that end, the ESRP's mission is to conduct innovative ecological research that provides the information and methods needed by decision makers to assess the benefits of ecosystem services to human well-being, and, in turn, to shape policy and management actions at multiple spatial and temporal scales.

Therefore, the ESRP is initiating research to characterize ecosystem services and to enable their routine consideration in environmental management. Research is organized around two types of foci: ecosystem type (wetlands and coral reefs will be studied) or geographic place (four place-based studies are being initiated). Research themes cutting across these systems and places will include monitoring, modeling and mapping; future-scenario analysis and valuation of services; impacts of reactive nitrogen; relationships to human health; development of appropriate decision support systems; and education. Some ecosystem services produce tangible goods valued by humans through existing markets, such as fisheries, crop harvests, and timber (Daily, 1997). Ecosystems also provide "life support services" such as water filtration, air quality regulation, and storm water mitigation that support market-valued ecosystem services and are necessary for civilization (Costanza et. al., 1997; Daily et. al., 1997). These services do not have traditional markets to establish their value and despite their crucial role, it is difficult for society to properly consider their worth (Bingham et. al., 1995; Costanza et. al., 1997; Daily et. al., 2000; Farber et. al., 2002). The 5-year Tampa Bay pilot, initiated in 2008, requires that EPA's ecological researchers develop new partnerships across disciplines (e.g., with economists and social scientists). It will enable decisions that better account for the full value of ecosystem services, and expected alterations of those services.

Studying Ecosystem Services in the Tampa Bay Region

The Tampa Bay Ecosystem Services Demonstration Project will focus on the region defined as the Tampa Bay Estuary Watershed (Figure 1). Tampa Bay, Florida's largest open-water estuary, covers 398 square miles at high tide and comprises six major sub-watersheds. Popular for sport and recreation, the Bay also supports one of the world's most productive natural systems. Estuaries like Tampa Bay are nurseries for young fish, shrimp, and crabs. More than 70 percent of all fish, shellfish, and crustaceans spend some critical stage of their development in near-shore waters, protected from larger predators that swim the open sea. The Bay also is home to dolphins, sea turtles, and manatees as well as water bird nesting colonies. The Tampa Bay markets value certain ecosystem services, such as fishing, tourism, and shipping, which are supported by other ecosystem services that cannot be bought and sold like traditional goods (Ronnback, 1999; Barbier, 2007).

Along with this extremely productive natural ecosystem is a very large urban center of commerce, transportation, and industry. The Port of Tampa is Florida's largest port and consistently ranks among the top 10 ports nationwide in trade activity. Tampa's ports contribute an estimated $15 billion to the local economy and support 130,000 jobs. The Tampa-St. Petersburg area annually contributes billions to the region's economy and is home to more than 2.3 million people. Tampa Bay is also a focal point for the region's premier industry – tourism.

The number of people in the Tampa area is expected to grow by nearly 19 percent by the year 2015, as approximately 500 people each week move to one of the watershed's three counties. According to a study completed by the University of Florida, the west-central region of Florida will experience "explosive" growth, with continuous urban development from Ocala to Sebring, and St. Petersburg to Daytona Beach (Zwick and Carr, 2006). The I-75 and I-4 corridors are expected to be fully developed. Most of Florida's heartland will convert to urban development, resulting in a dramatic loss of agricultural character and native Florida landscape that define this region today. Virtually all the natural systems and wildlife corridors in this region will be fragmented, if not replaced, by urban development.

Along with competing population and development stressors, climate change imposes yet another stress from rising sea level and changes in precipitation patterns. The 2006 National Wildlife Federation report "An Unfavorable Tide" presents maps and discusses the projected effects of a 38 cm sea-level rise on the Tampa Bay area (NWF, 2006). Climate change may be a significant driver of change on Tampa Bay shorelines, with projections of sea level rise virtually eliminating coastal salt marshes and potentially increasing the area of mangrove wetlands.

Regional and local managers, like those in the Tampa Region, need new or modified tools to address questions on the production, delivery, and consumption of ecosystem services under various projected future scenarios of urban development. Scenarios have been developed for population growth projections to 2050 for the Tampa Bay Region (Zwick and Carr, 2006). Tools we will collaboratively develop need to address human well-being endpoints (the focus of ecosystem services) as well as ecosystem integrity, sustainability, and productivity.

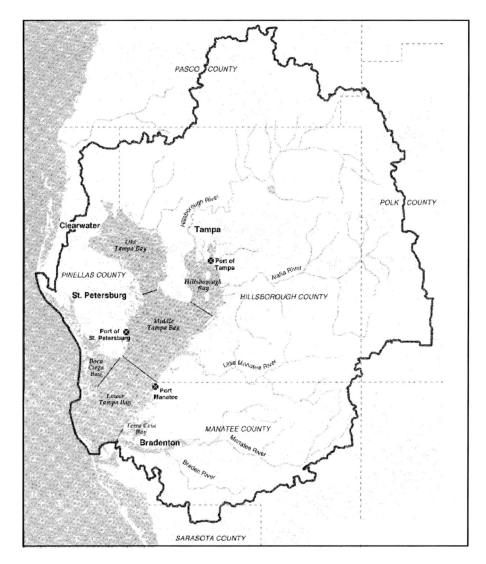

Figure 1. Tampa Bay estuary watershed (Source: Tampa Bay Estuary Program, 2006).

The Tampa Bay Region is home to a wide assortment of research entities that have been generating datasets that will be used for completion of this project. This research has resulted in vast quantities of available data. These datasets are providing the initial foundation for ecosystem services model development. We will further develop or use existing stressor response models for various defined ecosystem types to predict the changes in ecological production of services. Initially this will be completed using only existing data, but some models may require new datasets to properly calibrate them to local conditions. Our strategy includes the translation of these stressor response models into a common dynamic modeling framework so that they can be linked to inputs of the conditions predicted to exist under various alternative future scenarios and their outputs can be linked to economic and other human well-being translators. Another key part of the project strategy is to present our research outputs in a spatially explicit manner; relying on maps, connectivity networks, and web based visualization systems to present model results. Our work will provide the framework for producing a functional approach that will demonstrate how ecological stressor response models can be translated into usable endpoints for decisions affecting ecosystem services production, and also allows local partners and collaborators to contribute their extensive body of knowledge and past research findings to this ecosystem services pilot effort.

APPROACH

Uncertainties in both the ecological and economical valuation of services at these watershed and sub-watershed scales requires that we approach this effort in an iterative way so that after refinement and calibration, the information and tools we generate will become fully integrated with current management practices in Tampa Bay and other regions. Our project examines connectivity between those ecosystems that produce valued ecosystem services, through hydrologic, atmospheric, and human transportation networks, and the utilization of those services by humans at the local to regional scale. This is the key step in translating ecological functions, as described in existing datasets, into ecosystem services relevant to human well-being.

Our approach has been discussed and vetted with local and regional stakeholders and partners, so that our endpoints relate to the decision making process already well established in the Tampa Bay Region.

A major limitation of this project is the reliance on existing datasets and models. The strategy depends on the availability of existing knowledge and datasets to develop our models of the production of ecosystem services, their drivers, and their value. The Tampa Bay region was chosen as a placed-based demonstration site due to the availability of data, but past research was not designed to address the production of ecosystem services and so may not be ideally suited for model development needs. A large amount of uncertainty may exist in translating existing datasets into an ecosystem services framework.

Before any samples are taken, or any experiments are suggested, ecosystem services important to stakeholders of the Tampa area must be defined and inventoried. We are accomplishing this first step through identifying potential partners, collaborators, and stakeholders. Primary partners for the Tampa Bay Demonstration Pilot are the people who have the responsibility, obligation, and authority to manage and make decisions regarding services, use, and growth in the Tampa Bay watershed to provide a high quality of life and a sustainable future for citizens and businesses of the Tampa Bay watershed. Secondary clients are the citizens and businesses that depend on our primary clients to make decisions for the greater good of the community and Tampa, and tertiary clients include citizens and decision-makers of other coastal cities, who will benefit from our approach.

We envision education and outreach as key components in building partnerships to meet the goals and objectives of the Tampa Bay Ecosystem Services Pilot Project. It is EPA's duty to fully explain the concepts of ecosystem services and how they apply to a vision of the Tampa Bay community in the future. It is also incumbent on us to listen and become educated by the participants from Tampa Bay because they are in the best position to relate their scientific and regulatory needs. Any products or models we co-develop within the partnership must acknowledge models and decision tools already successfully in use. To maintain credibility, legitimacy, and support good working relationships, our partners must understand what the needs are and let those needs guide development of the project. The information must be available and transparent to communicate to the working groups and any interested parties. The partners are not only the developers of these models, maps and decision platforms, we are the users.

The Tampa Bay Estuary Program's Community Advisory Committee (CAC) conducted a public opinion poll indicating that citizens have a good grasp of key issues affecting the health of Tampa Bay. The poll was created to assess perceptions and attitudes about the bay, how residents use the bay, and what major challenges they foresee to continued restoration and protection. More than 164 responses were received between February 2001 and January 2002. Some findings of that poll are shown in Figure 2.

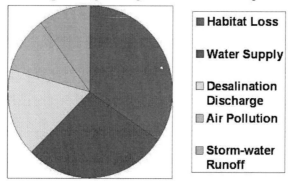

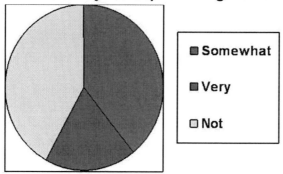

Figure 2. Results of Tampa Bay Community Advisory Committee public opinion poll.

Our approach for quantifying and valuing ecosystem services in the Tampa Bay watershed has been to identify the relevant parties that have needs to use ecosystem services in their decision making process, or a mandate to manage resources with linkages to ecosystem services. Once identified, we surveyed these potential end users of our work to inventory methods, models, and approaches currently in use to manage resources or services. During the identification of a set of ecosystem services most relevant to our partners, we have been establishing a working group that is using conceptual and systems dynamic models to build a working database of reference materials, which documents existing research on ecosystem services and their production. The working group has been identifying which services are most ecologically important, are best understood, and are most socially relevant or economically valuable. As the working group moves forward and adds more research partners we will identify, from our combined conceptual and dynamic models, which pathways and services are most ecologically critical, valuable to the Tampa Bay region, and for which we can advance knowledge over the project lifetime. We must communicate the linkages between projected future development scenarios and ecosystem services trade-offs. Consideration must also be made for the wide range of interactions that may affect the production of individual services in the context of the full suite of bundled ecosystem services.

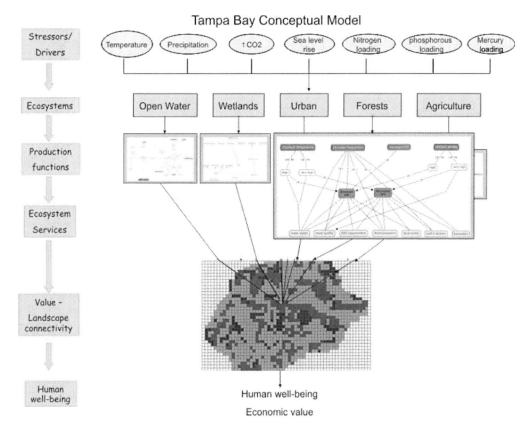

Figure 3. Conceptual model for Ecosystem Services Modeling in space and time.

Services for individual ecosystems within and around the Tampa Bay area (wetlands, open water, agriculture, forests, and urban) are assessed and weighted with respect to prioritized research needs, to develop forecasting tools that model service fidelity and sustainability under different urban development and climate change scenarios. The ecosystem services include: water supply and quality, CO_2 sequestration and storage, flood and storm surge protection, food and fiber production, habitat/refuge and biodiversity, recreation, toxicant regulation, and nitrogen regulation.

A literature search and review is being conducted to determine the extent of existing research on the impacts of human and natural stressors on each ecosystem service. Based on this review, conceptual models are being constructed and refined using the Cmap software program developed by the Institute for Machine and Human Cognition (IHMC) in Pensacola Florida (http://cmap.ihmc.us/conceptmap.html) to determine connections between land use type, eco-service, ecosystem functions, and drivers. These Cmap models are providing the backbone structure for construction of dynamic models using the Multiscale Integrated Model of Earth System's Ecological Services (MIMES) developed by the Gund Institute at the University of Vermont (http://www.uvm.edu/giee/mimes/). MIMES is programmed within Simile, a declarative visual modeling environment developed by Simulistics, LTD, (http://www.simulistics.com/) with inclusion of legacy sub-models where appropriate.

Effects of stressors on services, and ultimately human well-being, will be quantified and valued to develop forecasting tools that will allow city and regional planners the capacity to assess the impact of urban development and climate change predictions. The Tampa Bay conceptual model (Figure 3) illustrates how the research for each ecosystem will be integrated to quantify the effects of human and natural stressors on ecosystem services. For each ecosystem, production functions will describe the impact of stressors on services. Functions for each ecosystem will be connected within a spatially-explicit modeling framework to display landscape connectivity of ecosystems and synergistic effects of stressors and drivers.

For convenience, the project was broken down into five major areas, each based on an ecosystem type (Agriculture, Forest, Open-Water, Urban, and Wetlands). Extensive discussions with partners allowed us to group Agriculture, Forest, and Urban ecosystem into a terrestrial/watershed group. The research efforts associated with this task plan fall within these three major areas and focus EPA project partner research strengths to address critical research gaps or areas of greatest uncertainty.

Creation and Refinement of Conceptual Maps

Concept maps are an effective means of representing and communicating knowledge (Novak, 1977). Novak proposed that the primary elements of knowledge are concepts and that relationships between concepts are propositions (Novak, 1998). A concept map is a graphical two-dimensional display of concepts connected by directional lines that are labeled with short phrases characterizing the relationships between pairs of concepts, thus forming propositions.

We are using "Cmap Tools," a software environment developed at the Institute for Human and Machine Cognition, which enables users to represent knowledge using concept maps and to share and cooperatively develop them with peers and colleagues (Cañas et. al., 2004). The software enables the graphical display of propositional pathways and also provides for linking supporting documentation to the concepts and relationships.

The refinement of Cmaps is an ongoing effort within the Tampa Bay study; as the scenarios continue to be established more drivers and services will be identified and quantified. It will be important to continue to update documentation behind these linkages, and to periodically reevaluate pathways that may or may not prove substantial within the overall study. It is also a goal of the Cmap work to identify gaps in knowledge and modeling opportunities. As the study proceeds and different models are utilized, it will be beneficial to use these Cmaps to guide the interaction between modeling efforts, as the Cmaps provide a way to visualize interactions at larger spatial scales. The Cmaps for the Tampa Bay study are being developed as part of a knowledge gap and sensitivity analysis, and therefore include a complex web of interactions and linkages. As the study progresses, certain interactions will be seen as the most critical and the Cmaps can be simplified, but for now it is necessary to maintain complexity to avoid potential omissions.

Process for Evaluation of Pathways

The conceptual models are being used to evaluate individual pathways from scenarios to services. The Cmaps have the following advantages:

Using the Cmap software capability to include links to reference material, one can document existing research on specific links within the pathways.
Based on this documentation of links and pathways, one can identify which pathways and services are most critical to ecosystem services, which pathways and services will not be included in analysis, and highlight the rationale for these choices.
One can communicate the full pathways from scenarios to services, while also considering indirect interactions that may affect the pathway.

Another issue related to the changes in and evaluation of services is the correlation of ecosystem services. There may be "bundles" of services that can be grouped by their responses to common drivers and will increase or decrease together. Improvements in one ecosystem service may lead to additional improvements or degradation of other ecosystem services. Therefore, one of the goals in using conceptual maps is not only to evaluate changes in services, but to identify which services are tightly connected. Understanding trade-offs among services requires not only understanding the relative value of each service, but also understanding how tightly connected one service is to another, and the nature of those interactions to avoid unintentional consequences of decisions or actions.

In futures modeling, the term "baseline" is sometimes used to denote a modeling scenario in which current trends and policies are assumed to continue into the future. Such a baseline scenario is then compared with a policy scenario in which a policy change is simulated. By contrast, we use the term to denote a specific, recent year (2006) for which a landscape will be constructed to serve as a basis of comparison for future landscapes. For clarity, therefore, we will usually refer to this landscape as the foundation-year landscape since we will build any development scenarios based on the conditions existing at this time.

We have selected 2006 as the best period for constructing a foundation-year landscape for the following reasons:

In 1986, the Tampa Bay Regional Planning Council completed "Documenting the Economic Importance of Tampa Bay," quantifying the ecological service benefits of Tampa Bay. The report states that "Tampa Bay constitutes the central geographic feature most responsible for, both historically and at present, shipping, industrial development, and aesthetic and recreational values that encompass the overall attractiveness of the region to population influx." and that one of the six attributes and uses included in their economic analysis was ecosystem services.

Current projections of growth and development in the region already exist for the year 2025. Xian and Crane (2005) used the SLEUTH model to estimate how much and where impervious surfaces could increase as the population grows.

A detailed land cover map for the entire landscape of the watershed is publicly available for 2006. We will use information housed in the Florida Land Use/Cover Classification System dataset to create spatially explicit representations of ecological service production, delivery, and consumption.

Research Prioritization

First we separated our research team into sub-groups each identifying with a specific ecosystem type (Agriculture, Forest, Open water, Urban, and Wetlands) in the Tampa Bay Region. Each group then completed extensive literature searches on research involving production ecosystem services under different stressor levels. For some ecosystem types the link between existing data and resulting services was strong and information pertaining to our approach was available. This resulted in very complete approaches to those tasks (Urban storm water mitigation services, mercury deposition and recreational fisheries). In other cases, the linkages between existing data and service production is more complicated and lack of existing approaches or data resulted in less complete approach sections (Agriculture, Open water, Forest, Wetlands).

Second we generated conceptual maps with our partners for each subgroup that conceptualized the pathways from each stressor, through intermediate steps, to the generation of ecosystem services. This conceptual framework was then expanded to show the general functional relationships (ecological response functions) at each node along each pathway, the amount of reference support we found for each node, and the relative importance of each node to the production of each service after accounting for interactive effects.

The third piece of the prioritization work involved developing an initial valuation of the ecosystem services currently produced in the Tampa Bay Region. Our economist partners, using primarily benefit transfer methods, placed values on the annual production of services for those ecosystems that we could currently place a dollar value on. This per area valuation was then applied to a landscape characterization map for the Tampa Region which quantified the spatial extent of each ecosystem.

To visualize priority topics with in each sub-group we used the information in the Cmaps to create three-dimensional plots of economic value, ecological importance, and ecological support. With the information in place on each node's level of research support, its importance to generation of each ecological service, and its monetary value we were able to prioritize our research focus. The economic value and ecological importance were the first two factors to determine high priority topics. The third factor, ecological support (amount of existing data and information based on literature surveys), was then used to identify priority topics and research gaps for partners to address. Figures 4 and 5 show the Urban Forest and Urban Wetlands three-dimensional plots as examples of our approach. Land cover maps are also used to identify the spatial extent of (urban) forests and wetlands. Once mapped, services can be extrapolated and assigned to these areas.

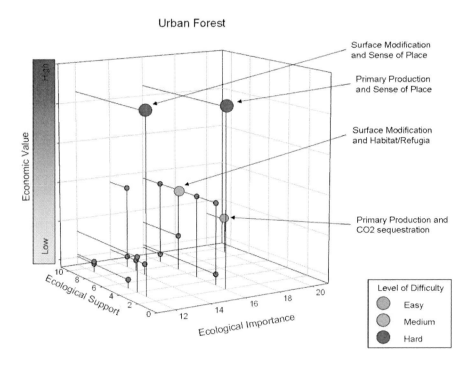

Figure 4. Urban Forest 3-D plot.

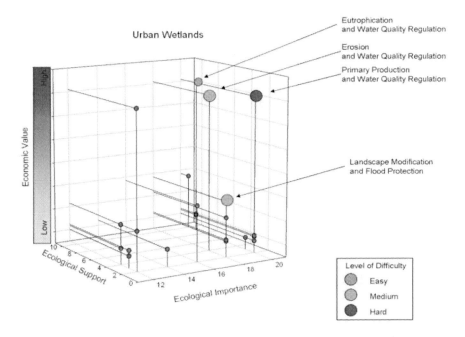

Figure 5. Urban Wetland 3-D plot.

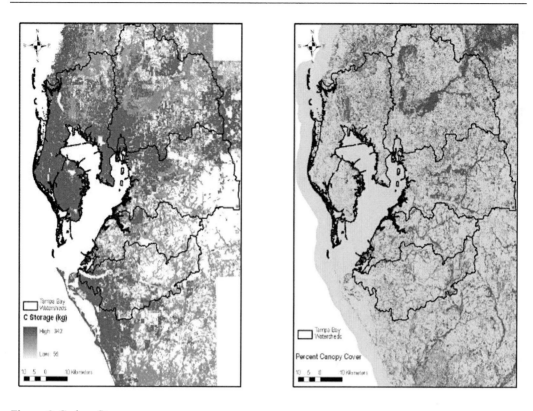

Figure 6. Carbon Storage.

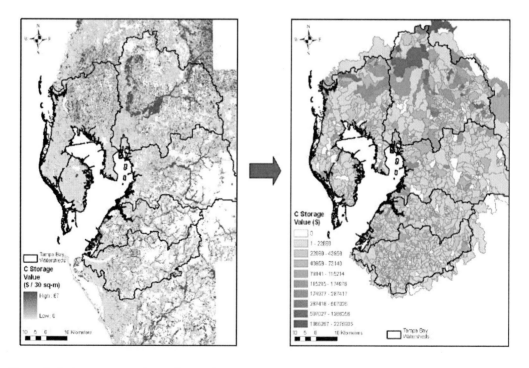

Figure 7. Carbon storage value estimates for Tampa Bay.

Figure 6 demonstrates how we can use this information to address the ecological service of carbon storage and its value. Using GIS maps of current soil carbon storage values and percent canopy cover, potential carbon storage values can be estimated.

Applying the current value for carbon storage from the Chicago Carbon Exchange ($15/ton C) and apportioning that value to the NHD + basins maps provides a carbon storage value of $1.8 billion for the Tampa area currently (Figure 7). Underlying economic values came from our partner economists' benefit transfer estimates. These are our initial estimates and will need to be better refined and calibrated to reflect local conditions. By providing this type of information to partners, including decision-makers, choices regarding competing uses, i.e. development vs. carbon storage, can be made more transparently. Ultimately, we will produce a decision framework that incorporates most, if not all, ecosystem services for the Tampa area in a similar fashion. Research partners are building maps such as these for each important ecosystem service in Tampa Bay and will be producing an atlas of the work in 2010.

CONCLUSION

The Tampa Bay region is faced with some tough decisions of how to manage the system as population continues to grow and the watershed becomes more urbanized, but these pressures will continue against the back drop of climate change. As our work has shown, approaches for predicting changes in ecosystem services must include knowledgeable local and regional experts and must consider trade-offs in services. The demand for the final service of drinking water provisioning, for example, is predicted to increase as population does and many areas may be faced with tough decisions on how to produce needed supply. The lack of available drinking water may be compounded by changing precipitation patterns due to regional climate change effects. Climate change projections for the Southeast U.S. predict that significant alterations in wetland hydrology will alter wetland nitrogen and carbon transformation rates, and wetland water recharge capabilities, resulting in drastic changes in the production of valued ecosystem services (e.g., carbon storage, nitrogen regulation, water provisioning, habitat support, flood protection, recreation, and aesthetics). Biogeochemical removal of increased nitrogen loads to watersheds associated with increased agricultural and urban development, wildlife habitat usage, and protection of human development from storm surge and flooding are three of the most critical services provided by wetlands in today's society. Specific climate change effects (e.g., sea-level rise, saltwater intrusion into aquifers, and increased drought and storm frequencies) have been implicated throughout the world in compromising key ecosystem services. Habitat, an intermediate service required for the production of many valued final services, may be threatened by both development and climate change. Millions of acres of habitat for charismatic species will be lost to new development if current population trends and building practices continue with the potentially more important loss of functions related to these habitats.

To better understand and predict the effects of climate change and sea level rise, we are conducting an assessment of water level's influence on both biogeochemical cycles and water supply recharge within the major river watershed in Tampa. Water level meters will track the connectivity between river stage height and corresponding soil saturation depths in riparian wetlands. Simultaneous measurements of biogeochemical rates will allow us to couple soil redox zones, influenced by ground water saturation depth, with measurements related to the production of valuable services. Services of focus include carbon storage and sequestration, water recharge, habitat support for biodiversity, and nitrogen and other "chemicals of concern" removal. This concerted effort will integrate scientists from disparate scientific disciplines such as ecology, microbiology, hydrologic modelers, and wildlife biologists and should lead to a greater understanding and improved models of wetland and associated habitats dynamic responses to changing environmental conditions. Our work also provides a means for estimating a value for many ecosystem services that traditionally have been very difficult to relate in terms of monetary value. As our models become more refined, decision-makers and regulators will be able to better consider the value of ecosystem services, perhaps eventually at the same level as economics have driven past decisions.

REFERENCES

Barbier, E. B. 2007. Valuing ecosystem services as productive inputs. *Economic Policy.* 22: 177-229.

Bingham, G., Bishop, R., Brody, M., Bromley, D., Clark, E., Cooper, W., Costanza, R., Hale, T., Hayden, G., Kellert, S., Norgaard, R., Norton, B., Payne, J., Russell, C., Suter, G. 1995. Issues in ecosystem valuation: improving information for decision making. *Ecological Economics.* 14: 73-90.

Cañas, A. J., (ed). 2004. CmapTools: A Knowledge Modeling and Sharing Environment, In: Concept Maps: Theory, Methodology, Technology, Proceedings of the First International Conference on Concept Mapping, Universidad Pública de Navarra: Pamplona, Spain. p. 125-133.

Costanza, R., d'Arge, R., de Groot, R., Farber, S., Grasso, M., Hannon, B., Limburg, K., Naeem, S., O'Neill, R. V., Paruelo, J., Raskin, R. G., Sutton, P., van den Belt, M. 1997. The value of the world's ecosystem services and natural capital. *Nature.* 387: 253-260.

Daily, G. (ed). 1997. Nature's Services: Societal Dependence on Natural Ecosystems. Island Press, Washington, DC.

Daily, G. C., Alexander, S., Ehrlich, P. R., Goulder, L., Lubchenco, J., Matson, P. A., Mooney, H. A., Postel, S., Schneider, S. H., Tilman, D., Woodwell, G. M. 1997. Ecosystem services: Benefits supplied to human societies by natural ecosystems. *Issues in Ecology.* 1: 1-18.

Daily, G. C., Soederqvist, T., Aniyar, S., Arrow, K., Dasgupta, P., Ehrlich, P. R., Folke, C., Jansson, A.- M., Jansson, B. – O., Kautsky, N., Levin, S., Lubchenco, J., Maeler, K. – G., Simpson, D., Starrett, D., Tilman, D., Walker, B. 2000. The value of nature and the nature of value. *Science.* 289: 395-396.

Farber, S. C., Costanza, R., Wilson, M. A. 2002. Economic and ecological concepts for valuing ecosystem services. *Ecological Economics.* 41: 372-392.

National Wildlife Federation. 2006. An Unfavorable Tide- Global Warming, Coastal Habitats and Sportfishing in Florida.

Novak, J. D. 1977. A Theory of Education. Ithaca, NY: Cornell University Press.

Novak, J. D. 1998. Learning, Creating, and Using Knowledge: Concept Maps as Facilitative Tools in Schools and Corporations. Mahwah, NJ: Lawrence Erlbaum Associates.

Ronnback, P. 1999. The ecological basis for economic value of seafood production supported by mangrove ecosystems. *Ecological Economics.* 29: 235-252.

Tampa Bay Regional Planning Council Final Report. 1986. Documenting the Economic Importance of Tampa Bay.

Xian, G. and Crane, M. 2005. Assessments of urban growth in the Tampa Bay watershed using remote sensing data. *Remote sensing of Environment.* 97:203-215.

Zwick, P. D. and Carr, M. H. 2006. Florida 2060, A population distribution scenario for the state of Florida. Report to 1000 Friends of Florida, University of Florida, Gainesville, Florida.

In: Encyclopedia of Environmental Research
Editor: Alisa N. Souter

ISBN: 978-1-61761-927-4
© 2011 Nova Science Publishers, Inc.

Chapter 46

INTEGRATIVE ECOTOXICOLOGICAL ASSESSMENT OF CONTAMINATED SEDIMENTS IN A COMPLEX TROPICAL ESTUARINE SYSTEM[*]

D.M.S. Abessa[†1,] R.S. Carr[2], E.C.P.M. Sousa[3], B.R.F. Rachid[4], L.P. Zaroni[3-5], M.R. Gasparro[3], Y.A. Pinto[6], M.C. Bícego[3], M.A. Hortellani[7], J.E.S Sarkis[7] and P. Muniz[8]

[1]UNESP Campus Experimental do Litoral Paulista. Praça Infante Dom Henrique, s/n. Parque Bitaru, São Vicente, SP, Brazil. 11330-900

[2]US Geological Survey, CERC, Marine Ecotoxicology Research Station, TAMU-CC, NRC Ste. 3200, 6300 Ocean Dr., Corpus Christi, Texas, 78412 USA

[3]Instituto Oceanográfico da USP. Praça do Oceanográfico, 191. Cidade Universitária, São Paulo, SP, Brazil. 05508-900

[4]Fundação de Estudos e Pesquisas Aquáticas, FUNDESPA. Avenida Afrânio Peixoto, 412, 05507-000, São Paulo, SP. Brazil

[5]Aplysia Tecnologia para o Meio Ambiente. Rua Júlia Lacourt Penna, nº 335, Jardim Camburi. Vitória, ES, Brazil, 29090-210

[6]Ministério do Meio Ambiente. Esplanada dos Ministérios - Bloco "B", 70.068-900, Brasilia, DF, Brazil.

[7]Instituto de Pesquisas Energéticas Nucleares, Av. Prof. Lineu Prestes, 2242, 05508-900 São Paulo - SP, Brazil

[8]Sección Oceanología, Facultad de Ciencias – UdelaR, Iguá 4225, Montevideo, 11400, Uruguay

---

[*] Chapter 9 was also published in "Marine Pollution: New Research", edited by Tobias N. Hofer, Nova Science Publishers. It was submitted for appropriate modifications in an effort to encourage wider dissemination of research.

[†] Corresponding author: Email: dmabessa@csv.unesp.br.

Abstract

The Santos Estuarine System (SES) is a complex of bays, islands, estuarine channels and rivers located on the Southeast coast of Brazil, in which multiple contaminant sources are situated in close proximity to mangroves and other protected areas. In the present study, the bottom sediment quality from the SES was assessed using the Sediment Quality Triad approach, which incorporates concurrent measures of sediment chemistry, toxicity and macrobenthic community structure. Elevated concentrations of metals were detected in the inner parts of the estuary, in the vicinity of outfalls, and in the eastern zone of Santos Bay. PAHs were found at high concentrations only in the Santos Channel. Anionic detergents were found throughout the system, with higher concentrations occurring close to the sewage outfall diffusers and in the São Vicente Channel. Sediments were considered toxic based on whole sediment tests with amphipods and porewater tests with sea urchin embryos. The observed toxicity appeared to coincide with proximity to contaminant sources. The macrobenthic community for the entire study area showed signs of stress, as indicated by low abundance, richness and diversity. The integrative approach suggested that both environmental factors and contaminants were responsible for the altered benthic community structure. The most critically disturbed area was the Santos Channel (upper portion), followed by the São Vicente and Bertioga Channels, and the immediate vicinity of the sewage diffusers.

Keywords: Sediment Quality Triad, Santos Estuarine System, pollution, toxicity, benthic communities, estuary, Brazil.

Introduction

The Santos Estuarine System (SES) is situated on the central coast of the state of São Paulo, Brazil (23°30'- 24°00'S e 46°05'- 46°30'W), and was once considered one of the world's most polluted estuaries (CETESB, 1985; Tommasi, 1979). The deterioration of SES began in the 1950s, with the concomitant installation of the Santos Organized Port and the Cubatão industrial complex, accelerated urbanization, increased domestic and industrial wastes dumping, and the expansion of tourism without proper infrastructure. In 1984, the State Environmental Agency initiated a pollution control program, which reduced the amounts of pollutants discharged into the SES water bodies by industrial effluents up to 90% (Lamparelli et al., 2001). However, this program was not nearly as effective in reducing contamination to the sediments and biota of the SES (Lamparelli et al., 2001), as demonstrated by recent studies (Abessa et al., 2005; Cesar et al., 2006; Medeiros and Bícego, 2004).

The SES sediment contamination is now publicly recognized as an environmental, socio-economic and human health problem; however its ecological effects and bioavailability to the resident organisms are not well known. Therefore, there was a need for a comprehensive ecotoxicological assessment in this region, which would allow the most degraded areas and their relationships with the contaminant sources to be identified, and also to provide data to determine the effectiveness of policies for environmental recovery and pollution control.

In the present study, the sediment quality from SES was assessed using the Sediment Quality Triad (SQT) approach, which consists of concurrent measurements of sediment chemistry, toxicity and macrobenthic community structure (Long and Chapman, 1985). This weight of evidence approach reduces the uncertainties in interpretation when these methods are used together (Zamboni and Abessa, 2002) and may be used in monitoring programs, ecological risk assessments, and environmental diagnosis, with the objectives of identifying the degree to which an area is affected by contaminants.

Secondary objectives were: to estimate the size of the area affected by the contamination and presenting benthic system degradation along the Santos and São Vicente channels, towards Santos Bay; to estimate the influence of the contaminants discharge on the quality of the sediments of Bertioga Channel; and to rank the sites based on the sediment degradation degree.

MATERIALS AND METHODS

Sediment Sampling

The sediments and biological material were collected at 25 sampling sites distributed along the Santos Estuarine System (Figure 1). The choice for sites selection considered the existing natural gradients and the presence of multiple contamination sources. The exact locations for sampling were determined using a Garmin 38 GPS, and the geographical coordinates are showed in Abessa (2002). The reference sediment was collected at the Palmas Bay, in the Anchieta Island State Park (PEIA), which is about 180 km distance from the SES. This site has previously been used as a reference and did not exhibit toxicity (Abessa *et al.*, 1998; 2001; Rachid, 2002). There is not a suitable reference site in the SES because the sediments are strongly influenced by anthropogenic activities.

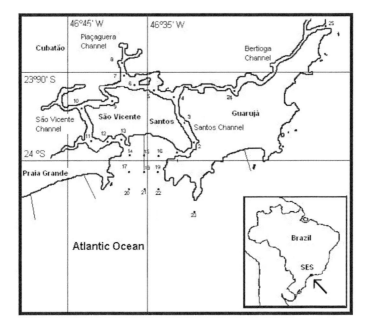

Figure 1. Map of the Santos Estuarine System - SES (SP, Brazil) showing the sampling stations.

Sediment samples were collected between March 20 and 23, 1998, by the use of a stainless steel Petersen grab sampler, with 0.026 m^2 of sampling area.

For the physical-chemical analysis and bioassays, only the surficial first 2-cm layer from the retained material was collected. The samples were composited by pooling the surficial layer from approximately 10 grabs replicates. The composited sediment was homogenized and aliquots taken for separate analyses. Sub-samples for grain-size distribution, total organic carbon (TOC) and physical-chemical analyses were frozen until they were analyzed, whereas those for toxicity tests were kept refrigerated at 4 ± 2 °C for 3 days before being tested.

The pore water was extracted by the suction method (Winger and Lasier, 1991), followed by centrifugation of the pore water for 20 minutes at 4200 x g (Carr *et al.*, 1996a; 2001). The supernatant was transferred to amber glass flasks and stored frozen.

For the macrofauna, three independent replicates were taken to about 5-10 cm sediment depth. The collected material was sieved through a 0.5 mm mesh. The biological material retained was fixed in 4% formaldehyde and then transferred to 70% alcohol. In each sampling station, bottom water (i.e., at less than 1 meter from the bottom) samples were collected using Nansen bottles, and their physical-chemical properties were measured in the field. The salinities were analyzed by an ATAGO-S/Mill refractometer, with 0.5% precision. The dissolved oxygen content was measured with a TOA oximeter, model DO-14P, with 0.01 mg/L precision. The pH values were estimated by Macherey-Nagel colorimetric indicator sticks, with 0.1 precision. Water temperature was measured with thermometers with 0.1 °C precision.

GRAIN SIZE DISTRIBUTION AND TOTAL CARBONATE CONTENT

The carbonate content was estimated according to Gross (1971). Sediments were weighed (30 g), digested in 10% nitric acid for 2 days, washed with distilled water and dried at 60 °C for 24 h, and then weighed again. Weight differences corresponded to the carbonate contents in each sample.

Grain size distribution was analyzed by the dry sieving method (Suguio, 1973). Sediment (30 g) aliquots were dried at 60 °C for 48 h, then sieved for 15 min using a set of sieves with 0.5 ∅ intervals in the Went-Worth scale. Fractions retained in each sieve were weighed, allowing estimates of sand and mud (silt and clay) content and the granulometric classification of the samples (Shepard, 1954; Folk and Ward, 1957).

TOTAL ORGANIC CARBON, NITROGEN AND SULFUR CONTENTS

Total organic carbon content, nitrogen and sulfur were measured using a LECO CNS 2000 automated analyzer, which uses the Micro Kjeldahl method (McKenzie and Wallace, 1954). According to the protocol, 100 mg of sediment were separated, and maintained for 48h in 10% HCl to eliminate carbonates. The samples were then washed with distilled water and lyophilized, and 0.5 g aliquots were separated and introduced into the analyzer.

Metals

The concentrations of the following metals were analyzed: Al, Fe, Cd, Cr, Co, Hg, Ni, Pb and Zn. Before the extraction, the sediment samples were dried at ambient temperature and sieved through a 1 mm mesh. For Al, Cr and Fe, total digestion in high pressure microwaves system was used (CEM Corporation, model MDS – 2000). The acid extraction solution consisted of a mixture containing Milli-Q water, HNO_3, HF, and HCl. For Cd, Co, Pb and Zn, Aqua Regia and $HClO_4$ were employed in the extraction. Analyses were made using a fast sequential Atomic Absorption Spectroscope. For Hg, extractions were made in volumetric flasks heated for 30 min at 90 °C (Akagi and Nishimura, 1991), using Aqua regia and $HClO_4$. The extracted solutions were then introduced into a system of Flux Injection for Cold Vapour generation (FIA-CV-AAS) (Fostier et al., 1995), and analyzed with an Atomic Absorption Spectroscope (VARIAN, AAS 220-FS) at a wavelength of 253.7 nm. As Quality Assurance and Quality Control procedure (QA/QC), the methods were verified using certified reference sediment (Buffalo River Sediment®).

Hydrocarbons

The chemical analyses followed the protocol described by UNEP (1991). The sediment samples were dried at 50 °C in an oven and then homogenized. A 25g aliquot of each sample was separated and received 1 ml mixture of internal standards of aliphatic and aromatic hydrocarbons.

The extracts were prepared in Soxhlet, with a mixture of dichlorometane /n-hexane, concentrated to 1 ml in a vacuum rotary evaporator. Then, each one was treated with activated copper strings, to eliminate the copper sulphate. The extracts were submitted to column chromatography, containing deactivated silica gel and alumina. Afterwards, the two fractions were successively eluted: F1, with 10 ml n-hexane; and F2, with 20 ml of 30% dichloromethane in n-hexane. Then, the fractions were concentrated to 1 ml, and 1 µl aliquots of each were injected into the gas chromatograph.

The F1 fraction was used to measure the n-alkanes, and was analyzed in a Hewlett Packard Gas Chromatographer, model 5890 II, equipped with the flame ionization detector maintained at 325 °C. The F2 contained the polyaromatic hydrocarbons (PAHs). This fraction was analyzed in a similar gas chromatograph, coupled to a mass spectrometer V.G. Masslab – Fisons, model TRIO 1000. The analysis was made in the IMS (Ion Monitoring System) mode.

The final calculations were made by using calibration curves, considering the response factor calculated for each hydrocarbon and its comparison to the mixture analysis. The method detection limit (MDL) was determined as 3 times the standard deviation from 7 replicates of a sample which contained all the compounds. Such MDLs were 0.07 $\mu g.g^{-1}$ for the total n-alcanes and 0.24 $ng.g^{-1}$ for the total aromatics. The detection limits for individual compounds ranged from 0.0002 to 0.0112 $\mu g.g^{-1}$ for n-alkanes and from 0.002 to 0.022 $ng.g^{-1}$ for aromatics. The QA/QC procedure consisted in the analysis of NIST reference material.

Anionic Detergents

Surfactants concentrations in sediments were estimated by the Methylene Blue Active Substance (MBAS) method (APHA/AWWA/WEF, 1998) after extraction by elutriation in distilled water (Abessa and Sousa, 2004). This method consists of transferring methylene blue, a cationic dye, from an aqueous solution into an organic liquid, through the ion pair formation by MBAS anion and the methylene blue cation. The intensity of the resulting blue color in the organic phase is a measure of MBAS. The procedure is simple and comprises three successive extractions from acid aqueous medium containing methylene blue into chloroform, followed by an aqueous backwash and measurement of the blue color in the chloroform by spectrophotometry at 652 nm.

Whole Sediment Toxicity Test

Whole sediment toxicity tests were conducted using the amphipod *Tiburonella viscana*, following the protocol described by Melo and Abessa (2002). One day before the start of the test, each sediment sample was thoroughly homogenized within its storage container by stirring, and aliquots were distributed into the test chambers (1-L polyethylene beakers). Sediments were not sieved. The test chambers were filled to 2 cm depth with the test sediments and filtered seawater up to 750 ml and then maintained overnight at 25 ± 2 °C with gentle aeration. On the next day, 10 amphipods were added to each test chamber and the test was initiated. Animals that did not bury within 1 hour were removed and replaced. Three replicates per test sediment were prepared. The test was conducted at 25 ± 2°C, under constant aeration and lighting. After ten days, the contents of the test-chambers were gently sieved through a 0.5-mm screen and the surviving amphipods were counted. Missing organisms were considered dead. The mortalities in the test and reference sediments were compared by Student t'-test (Zar, 1984), and sediment samples were considered toxic when the amphipod survival in the respective sample differed statistically from the observed for the animals exposed to the reference sediment ($p<0.05$) . The dissolved oxygen concentration, salinity and pH of the overlying water in the test chambers were measured at the beginning and end of the test. Water temperature was monitored daily.

Porewater Toxicity Test

Porewater samples were evaluated by the early life stage bioassay with embryos of the sea urchin *Lytechinus variegatus* (CETESB, 1992), which is similar to the protocol recommended by the USEPA (1988) for *Arbacia punctulata*, *Strongylocentrotus purpuratus* and other related species. Six hours before the test, samples were thawed at 25 °C. Afterwards, the following physical-chemical variables were checked: salinity, pH, dissolved oxygen and temperature. The total ammonia concentration was measured by a colorimetric method (Koroleff, 1970). Using these data, the unionized ammonia contents were estimated, using the method reported by Whitfield (1974).

Prior to the beginning of the test, adult individuals were collected at rocky reefs, in Ubatuba, a clean site, and taken to the laboratory. The spawning was induced by osmotic stimulation, i.e. through the injection of 2-3 ml KCl into the coelomic cavities of the animals (CETESB, 1992).

Gametes of 3 animals of each gender were collected. Eggs were collected by precipitation in beakers filled with filtered seawater, whereas the sperm was collected dry and transferred to beakers kept on ice. The sperm was activated by dilution in filtered seawater and the eggs were fertilized by adding 2 ml sperm solution to the eggs solution. The fertilization success was confirmed by examination under the microscope. The toxicity test was conducted in glass test tubes containing 10 ml test-solution. Four replicates were used for each concentration. Three porewater concentrations were prepared as described in Carr et al., (2001): 100, 50 and 10%. The experiment was kept in a temperature controlled room, at 25 ± 2 °C. After 24h, the test was finished by adding 0.75 mL of 10% buffered formaldehyde to each replicate.

The embryos were analyzed microscopically for morphological anomalies and retarded development (100 per replicate; 4 replicates per tested solution, totalizing 400 organisms per test-solution). All the embryos which did not reach a well developed pluteus larvae were considered affected. The results were statistically analyzed by ANOVA, followed by the Dunnett's t'test comparison, with the SAS statistical package (SAS, 1992).

Macrobenthic Community

The biological material collected was sorted by major taxonomic groups under microscope, separated and stored in 70% alcohol solution. Then, the organisms were identified to the lowest possible taxonomic level (i.e., to species when possible). Identification keys were used for polychaetes (Day, 1967; Amaral, 1980; Nonato, 1981; Amaral and Nonato, 1982; 1984; Lana, 1984; Bolívar and Lana, 1986; Amaral and Nonato, 1994; 1996), and mollusks (Rios, 1984). Voucher samples for some groups were identified by experts.

The results obtained for each sampling station were expressed as number of organisms collected in a 0.078 m^2, which represented the sum of the 3 replicates. Organisms densities were normalized to orgs/m^2, in order to allow comparisons to other studies. For each station, the following indices were estimated, by using the software package Bio Diversity Professional (Lambshead et al., 1997): Species Richness, Shannon-Weaver Diversity, Simpson Dominance, Pielou Evenness and Polychaetes Density. Such indices are recommended as part of the "BIBI" *Benthic Index of Biotic Integrity*" (Van Dolah et al., 1999; Weisberg et al., 1997).

Results Integration

Initially, the data of each SQT component were evaluated separately, and then the results each component were combined, by using different integrative approaches. The interpretation of SQT results is challenging, due to the complexity and different nature of the data. Different methods to integrate the data have been used (Chapman et al., 1991; Green, 1993; Green and

Montagna, 1996). However there is not a consensus if one method is more reliable than the others. For this reason, the use of more than one integration method may be recommended, which allows the analyses of the data by different views and a deep exploration of the results. Since the SQT is a "weight of evidence" based approach, this recommendation follows this guiding principle.

Ratio to Mean Values

To combine the chemical, toxicological and ecological data, the first selected approach was the Ratio-to-Mean Values method (RTMV), which is an adaptation of the Ratio-to-Reference (Long and Chapman, 1985; Chapman, 1990) and Ratio-to-Maximum Values methods (Del Valls and Chapman, 1998; Del Valls *et al.,* 1998), and was proposed by Abessa (2002) and Cesar *et al.* (in press).

The values obtained for each variable were converted to not dimensional values (RTMVs[1]), by dividing the value obtained in each station by the arithmetic mean obtained for all stations, including the reference. Then, for each SQT component, the RTMVs were joined, by the calculation of a mean for each of them[2], producing one new single value for the chemistry (Nic), one for the ecotoxicological data (Nit) and one for the benthos (Nib). Such values were normalized to those estimated for the reference station, and then plotted in 3-axis graphics, forming triangles. The area of each triangle represented the relative degree of degradation of a single sediment sample.

To allow an easy visualization of the sediment quality in SES, according to RTMVs, the values were digitalized, interpolated by the krigging method and then displayed graphically on a map using Surfer software (Golden Software, 1995).

Despite its limitations due to the loss of information during the data reduction to a single index, and the fact that the triangle areas increase exponentially, this method was adopted due to its simplicity, and the RTMVs were used only to rank the samples, based on their degradation indices.

Non-Statistical Comparative Method

Another integrative approach involves an independent interpretation of the data among stations for each of the SQT components, after which the independent assessments are combined. The limitations of this approach concern the subjectivity of the relative comparisons and the difficulty in making comparisons with other studies.

---

[1] RMTVs were calculated for Fe; Al; Cr; Cd; Co; Hg; Ni; Pb; Zn; Σn-alkanes; ΣPAHs; detergents (for the chemistry component - RMTVc); amphipod mortality; sea-urchin embryos abnormal development (for the toxicity component - RMTVt); species richness; diversity; dominance; polychaetes abundance (for the benthic community component - RMTVb).

[2] Example for Station 1: Nic = ΣRMTV (Fe+Al+Cr+Cd+Co+Hg+Ni+Pb+Zn+n-alkanes+PAHs+detergents)/12; Nit = ΣRMTV (amphipod mortality + abnormal embryonic development)/2; Nib = ΣRMTV (S richness + diversity + dominance + polychaetes abundance)/4.

Chemical contamination

The chemical contamination data for each sample were compared to the Canadian sediment quality guidelines (Environment Canada, 1999; Smith *et al.*, 1996), which have been used by the São Paulo State Environmental Agency (Lamparelli *et al.*, 2001). Such comparison followed the Sediment Quality Guidelines Quotients (SQGQ) (Fairey *et al.*, 2001), based on the Probable Effect Level (PEL) values. According to the mentioned authors, the sample ranking criteria was:

Minimal contamination: SQGQ value between 0 and 0.1;
Moderate contamination: SQGQ value between 0.1 and 0.25;
Strong contamination: SQGQ value greater than 0.25.

Ecotoxicological data

The analysis of the ecotoxicological data was simpler, being made as following:

No Toxicity: absence of toxicity in both assays;
Moderate toxicity: occurrence of toxicity in only one assay;
Strong toxicity: occurrence of toxicity in both assays.

Macrobenthic community data

The first step in this analysis was the definition of which ecological indices would be used. Abessa (2002) proposed the use of four parameters for the Santos Estuarine System: Species Richness, Shannon-Weaver Diversity, Simpson Dominance and Polychaete Densities; all of them commonly applied in the different Benthic Indices of Biotic Integrity (BIBI).

The second step was the classification of the values as indicators of the degree of environmental degradation. An arbitrary classification criterion was recommended by Abessa (2002) and adopted for this investigation, based on theoretical references (Odum, 1971; Magurran, 1988), on data previously obtained for Santos (Tommasi, 1979) and on particular characteristics of the data (Table 1). The status for each sample was given by a comparison of the conclusions achieved for each index. The conclusion for the majority of the indices was considered the decision for the sample.

Table 1. Criterion for classifying the sediment degradation based on benthic community descriptors (Abessa, 2002)

INDICES	Degradation Degree		
	Minimal	Moderated	Strong
Specific Richness	≥ 20	$10 \geq x > 20$	< 10
Simpson Dominance	≤ 0.333	$0.333 < x \leq 0.666$	> 0.666
Shannon Diversity (H')	> 2.36	$1.18 > x \geq 2.36$	≤ 1.18
% Polychaetes	$< 60\%$	$60 \leq x < 80$	≥ 80

After the independent data evaluation for each SQT parameter, the results were represented in pie diagrams, plotted on a map, allowing a visualization of the conditions in each sampling station.

Table 2. Physical-chemical parameters of water collected in each sampling station at Santos Estuarine System and at the reference station

Sampling date	Sampling Station	Salinity (‰)		pH*	Dissolved Oxygen (mg/l)*	Temperature (°C)	
		Surface	Bottom	Bottom	Bottom	Surface	Bottom
03/20/1998	1	27	31	7.9	6.2	25.5	24.9
	2	17	32	7.9	6.8	25.1	21.5
	3	18	32	7.8	5.2	24.3	21.4
	4	13	30	7.8	4.2	24.5	21.7
	5	9	30	7.8	3.6	25.1	22.9
	6	3	30	7.8	3.0	27.7	23.4
	7	4	30	7.8	3.2	25.4	23.3
	8	8	29	7.7	3.8	24.1	23.5
	9	10	29	7.9	6.6	24.3	22.9
	10	13	29	7.9	5.9	24.2	22.3
	11	24	29	7.9	6.2	24.8	25.7
	12	23	29	8.0	6.6	23.4	27.8
	13	31	31	8.0	6.4	27.2	27.2
	14	28	31	8.0	6.9	27.0	25.4
	15	28	32	8.0	7.0	26.7	25.2
03/21/1998	16	27	32	7.9	7.0	25.6	25.5
	17	29	31	8.0	5.2	27.3	25.2
	18	29	31	7.8	6.0	26.9	25.4
	19	29	31	8.1	6.4	25.6	25.5
	20	31	31	8.2	6.0	27.7	26.8
	21	29	32	8.1	7.0	28.0	25.3
	22	30	32	8.1	6.6	27.0	24.8
	23	32	32	8.1	6.9	26.5	23.6
	24	9	30	7.9	3.8	25.2	23.3
	25	32	32	8.0	5.1	24.7	24.7
03/23/1998	Ref.	33	33	8.2	6.8	25.3	25.3

* Dissolved Oxygen and pH measured only for bottom water.

Principal Component Analysis

The use of sequential principal component analyses (PCAs) have been recommended to integrate the SQT results (Green, 1993). Firstly, from the results obtained separately, the variables were selected, and the sheets were prepared. This procedure led to 4 data matrices:

Matrix 1) Environmental data and organic contamination;

Matrix 2) Metals contamination;
Matrix 3) Ecological indices;
Matrix 4) Toxicological data.

Each of these matrices was analyzed independently. For each result, from the axes which explained about 80% of the variance (2 or 3 axes), the corresponding PC scores were selected. A new data matrix was prepared using these scores, and then a new PCA was run.

RESULTS AND DISCUSSION

When conducting a study regarding ecological and ecotoxicological of aquatic communities, the knowledge and comprehension of the natural factors which may influence on the benthic community structure and/or on the toxicity is required, because such factors have important roles, not only to the phenomena related to the contaminants adsorption and releasing by the sediments but also on the biological responses (Abessa, 2002). Thus, these natural factors were studied together with the anthropic ones, because the combination of both influences produced the observed results, at least in a preliminary instance.

Water Column

The salinity of surface water (Table 2) ranged from 3 to 35 ‰, with lower values in the inner parts of the estuary. In the bottom water, however, salinities were always high, with values ranging from 29 to 35 ‰ (Table 2). Such result shows the existence of a saline intrusion into the estuary, by the Santos Channel, reaching the Piaçaguera Channel (Figure 1). Also, it is evidenced that the freshwater drainage from basin flows through the surface, which was already predicted (Harari et al., 2000).

Water temperatures were generally high, as expected for summer conditions. Surface water temperatures ranged from 23.4 to 28.0 °C, whereas bottom water temperatures were slightly lower, ranging between 21.4 and 27.8 °C (Table 2). At stations 11 and 12 the bottom water temperatures were higher than the surface ones, possibly due to tidal turbulence or the influence of the saline intrusion. The estuarine waters tended to be cooler than those from Santos Bay. The thermal heterogeneity may be attributed to some independent factors, as freshwater input, tidal currents, saline intrusion, and sunlight irradiation, among others.

The pH of the bottom waters ranged from 7.7 to 8.2 (Table 2), within the values expected for seawater. Typical marine values were found on the Santos Bay, whereas lower values, more typical for estuaries or areas with organic matter inputs, were observed within the estuary and close to the sewage outfall diffusers, e.g., station 18. A similar influence of sewage outfall discharges was detected by Rachid (2002).

The dissolved oxygen levels in the bottom water ranged from 3.0 to 7.2 mgO_2/L (Table 2). The lowest values occurred in the inner parts of Santos and Bertioga Channels (stations 4, 5, 6, 7, 8 and 24).

Water characteristics were similar to those already described for the SES (Bonetti, 2000; Moser, 2002; Tommasi, 1979). In the estuarine channels, there is a stronger continental influence, and the waters exhibit lower pH, salinities and DO levels. The saline intrusion in the Santos Channel has been reported previously (Bonetti, 2000; Tommasi, 1979).

Table 3. Sediment grain-size and organic characteristics in the SES and reference areas

Station	% Sand	% Mud	%CaCO$_3$	TOC (%)	S (%)	N (%)
1	14.82	85.18	11.20	1.39	0.55	0.12
2	6.40	93.60	21.20	2.53	0.43	0.27
3	0.80	99.20	21.20	2.37	0.64	0.22
4	29.70	80.73	23.90	1.03	0.34	0.09
5	2.12	97.88	20.30	2.14	0.62	0.16
6	8.82	91.18	11.90	0.79	0.13	0.07
7	11.46	88.54	11.20	1.39	0.55	0.12
8	56.41	43.59	14.17	2.76	0.41	0.27
9	60.60	39.40	11.50	2.62	0.41	0.11
10	93.33	6.77	12.10	2.03	0.12	0.09
11	91.16	8.84	11.20	2.51	0.49	0.08
12	98.33	1.77	6.38	0.31	0.01	0
13	98.34	1.76	11.33	1.22	0.02	0.01
14	97.03	2.97	5.20	0.12	0.01	0
15	95.73	4.27	6.29	0.14	0.01	0
16	72.74	27.26	11.33	0.70	0.13	0.06
17	92.17	7.83	7.43	0.23	0.03	0
18	28.84	71.16	21.66	1.39	1.43	0.12
19	33.26	66.74	17.30	1.55	0.67	0.14
20	97.43	2.57	5.29	0.14	0.01	0
21	41.97	58.03	21.27	1.17	0.55	0.09
22	88.03	11.97	8.35	0.29	0.04	0.02
23	45.53	54.47	7.70	0.21	0.03	0.05
24	9.82	90.18	11.30	0.87	0.09	0.06
25	98.74	1.21	4.73	0.77	0.05	0.07
Reference	98.33	1.77	13.17	0.86	0.13	0.05

Sand = particles > 0.0625 mm; Mud = particles ≤ 0.0625 (silts + clays).

Sediment Characteristics

The sedimentological analyses showed that the sediments presented a variable composition, in terms of grain-size distribution and levels of total organic carbon, total nitrogen and total sulphur (Table 3).

The stations situated in the Santos Channel and in the inner portion of the Bertioga Channel exhibited muddy sediments, excepting the sample from station 8, which was composed of muddy-sand. On the other hand, at the inner part of the São Vicente Channel, in the mouth of Bertioga Channel and on the west side of Santos Bay, sediments were predominantly sandy, as well as the reference sediment. In the central and east portions of

Santos Bay, sediments were composed by mixtures of mud and sands. These areas are under the influence of the Santos Sewage Outfall System (SSOS) and the Santos Channel. Station 23, located at the former disposal site of sediments dredged from the port, exhibited a mixture of mud and sand while the reference sediments were sandy.

The contents of sand around the SSOS diffusers (station 18 and 21) were slightly lower than those previously reported (Fúlfaro and Ponçano, 1976; Fúlfaro et al., 1983; Tommasi, 1979), suggesting that particles which are discharged by the SSOS deposit around the diffusers.

The carbonate content was variable, ranging from 4.73% to 23.9%. The central portion of the bay, the inner part of the Santos Channel and the reference station exhibited sediments with higher $CaCO_3$ content. Muddy sediments tended to have more carbonates, suggesting that such levels corresponded to deposition areas. The TOC contents ranged between 0.12 and 2.76 %. The higher concentrations occurred inside the estuary and close to the SSOS diffusers.

The levels of total sulphur (S) in the sediment ranged from 0.01 to 1.43 %. The sediments collected close to the SSOS diffusers presented a concentration much higher than was observed for the other samples. However, high concentrations were also present in the sediments from the estuary. Elevated hydrogen sulfide concentrations are expected in areas influenced by sewage discharges (Thompson et al., 1991), and may contribute to the sediment toxicity (Pardos et al., 1999).

The total nitrogen contents were higher in the sediments from the inner part of the estuary and close to the SSOS diffusers, and lower in the interior part of São Vicente Channel and west part of the bay.

It was possible to detect a trend, in which the distribution of TOC, sulphur and nitrogen in the SES sediments tended to follow the pattern of muds distribution. However, this pattern probably was due to the co-influences of the mangroves and the anthropogenic sources of contamination.

Sediment characteristics were similar to those observed previously (Fúlfaro and Ponçano, 1976; Fúlfaro et al., 1983; Tommasi, 1979). The sediment distribution in the SES is complex, with finer and organically enriched sediments inside the estuary. Moreover, the region can be divided in two sectors: eastern, with silty and organically rich sediments, and western, with sandy sediments. According to Fúlfaro and Ponçano (1976), suspended solids are transported from the west to the east part of the bay, especially during the winter. These authors showed that the sedimentation rates in the Santos Channel were low, with most of the solids being retained in the upper portions of the estuary. However, Fúlfaro et al. (1983) warned that human interventions in the adjacent areas could change such processes, resulting in economic losses to the port management.

Metals

The results of the analyses of metals concentrations are shown in Table 4. The highest values tended to occur in the sediments from the inner portion of the SES.

Aluminum (Al) ranged from 0.1 to 7.9%. The higher concentrations were observed in the sediments from the upper estuary, in a decreasing gradient towards the sea. A similar distribution was observed for iron (Fe), with concentrations ranging from 0.5 to 8.0%..

Similar results were found previously for these elements (Bonetti, 2000; Siqueira et al., 2001). The distribution of Al and Fe suggests that there are inputs to the system, which may have industrial or terrestrial origin. The influence of a steel plant near the most contaminated portion of the SES warrants further investigation.

Table 4. Concentrations of metals (in % and µg/g) in sediments from the Santos Estuarine System

Stations	Metals								
	Al	Fe	Zn	Ni	Pb	Cd	Cr	Co	Hg
	%		µg/g						
1	3.04	1.21	40.1	9.5	10.9	< 0.50	18.7	6.0	0.11
2	3.00	0.87	47.6	8.9	11.2	< 0.50	17.6	5.2	0.12
3	3.11	0.92	44.5	7.0	10.8	< 0.50	7.5	4.2	0.36
4	5.34	2.77	180.0	21.8	204.8	0.75	37.9	10.7	0.74
5	5.21	2.68	284.4	22.2	23.5	0.92	44.1	10.3	0.23
6	6.07	2.75	86.9	25.0	19.2	0.99	44.8	12.3	0.32
7	7.91	4.99	152.8	34.1	39.7	0.42	65.8	17.0	0.92
8	6.19	7.99	312.0	44.2	89.9	0.98	97.5	15.3	0.75
9	3.33	1.76	77.6	13.2	19.6	1.49	22.8	5.1	0.50
10	0.58	0.63	14.2	2.5	3.7	< 0.50	5.0	0.9	0.11
11	5.91	2.85	37.9	10.2	10.3	< 0.50	53.6	4.8	0.31
12	0.08	0.52	7.6	1.3	2.5	< 0.50	5.0	0.2	0.04
13	0.06	0.62	10.9	2.4	17.0	< 0.50	5.0	1.1	<0.03
14	3.42	1.27	34.0	9.1	6.5	< 0.50	12.5	4.1	0.05
15	2.71	1.77	41.4	11.3	8.3	< 0.50	18.8	5.8	0.04
16	1.64	0.83	23.8	4.9	5.3	< 0.50	10.0	2.3	0.04
17	2.96	1.25	35.9	10.3	7.8	< 0.50	18.4	5.4	0.04
18	3.56	1.60	61.7	12.5	16.8	< 0.50	28.4	7.1	0.19
19	3.75	1.70	44.7	13.4	11.8	< 0.50	29.0	6.5	0.06
20	2.70	1.11	49.6	7.9	5.25	< 0.50	9.5	3.2	0.04
21	2.60	1.23	32.2	14.7	5.75	< 0.50	19.6	4.1	0.04
22	4.53	2.07	55.5	17.9	18	< 0.50	40.9	8.5	0.076
23	2.28	1.14	29.7	8.1	5.5	< 0.50	5.0	3.8	<0.030
24	5.53	2.35	74.4	21.2	24.5	0.85	69.5	11.6	0.11
25	1.00	0.61	16.8	5.9	< 2.0	< 0.50	5.0	1.6	0.037
Reference	0.78	0.91	20.8	7.9	8.29	0.85	5.0	3.9	<0.03
TEL	-	-	124.0	15.9	30.2	0.70	52.3	-	0.130
PEL	-	-	271.0	42.8	122.0	4.21	160.0	-	0.696

Cadmium (Cd) concentrations in the sediments varied between < 0.5 and 1.49 µg/g. The higher levels occurred at Santos, São Vicente and Bertioga Channels, where concentrations exceeded the TEL. At Santos Bay, the levels were below 0.5 µg/g. The results are similar to those obtained by Lamparelli et al. (2001), and these authors suggested that possible sources of Cd to the system were the industries and the numerous sources of sewage. Unexpectedly, a

Cd concentration above the TEL was observed in the reference sediment. Since there is not an obvious source of Cd in the vicinity of this site, this observation warrants additional investigation.

The contents of Pb ranged from 2.0 to 204.8 µg/g, with higher concentrations occurring in the confluence of Bertioga and Santos channels (above PEL). Concentrations above TEL were observed at inner portions of the Santos Channel. This distribution differed from that observed by Lamparelli et al (2001), where the most contaminated sediments were collected close to the steel plant. The levels were consistent with those observed in recent studies (Bonetti, 2000; Siqueira *et al.,* 2001).

Chromium (Cr) concentrations ranged from 5 to 97.5 µg/g. The higher levels occurred in Bertioga, Santos and São Vicente channels, at the East of the Santos Bay (station 22) and close to the SSOS diffusers. Violation of TEL occurred at stations 7, 8 and 24. These results were similar to previous data reported for the same area (Lamparelli *et al.*, 2001; Prósperi *et al.*, 1998; Siqueira *et al.*, 2001).

Results showed that the contamination by mercury (Hg) is widespread in the SES sediments. The Hg concentrations in the samples ranged from 0.02 and 0.92 µg/g. Sediments from the estuary exhibited the higher levels, with concentrations exceeding the PEL in sediments from stations 4, 7 and 8, and levels above TEL at stations 3, 5, 6, 9, 11 and 18. The observed distribution was similar to that detected by Lamparelli *et al.* (2001), and may be attributed to the industrial discharges, terminal operations, sewage disposal and industrial landfills.

Nickel (Ni) contents in the studied sediments ranged between 1.3 and 44.2 µg/g, with higher values in the Santos, São Vicente and Bertioga channels. Moderated levels were recorded at the East of Santos Bay. Sediment from station 8 presented Ni concentration exceeding the PEL , whereas in the sediments from the stations 4, 5, 6, 7, 22 and 24 the levels were above TEL. This distribution was in accordance to those reported previously (Bonetti, 2000; Lamparelli *et al.,* 2001; Prósperi *et al.*, 1998).

Zinc (Zn) concentrations ranged from 7.6 to 312.0 µg/g, with the highest concentrations again in the Santos Channel, exceeding PEL and TEL at stations 5 and 8, and 4 and 7, respectively. The concentrations were lower than those observed by Lamparelli *et al.* (2001) and Prósperi *et al.* (1998), but very similar to those reported by Bonetti (2000).

Cobalt (Co) levels ranged from 0 to 16 µg/g, showing highest concentrations at the upper estuary and moderate levels close to the sewage diffusers and at the West of the Santos Bay.

The distribution of metals in the SES sediments seems to be well defined, with elevated concentrations in the Santos Channel (stations 4, 5, 6, 7 and 8). Other areas exhibited signs of degradation, as Bertioga Channel (station 24), São Vicente Channel (stations 9 and 11), the vicinity of the SSOS (station 18) and the East of Santos Bay (stations 22 and 23); all of them situated close to contaminant sources.

Regarding the temporal trends, for most parts of the SES, the levels are not increasing, however in the Santos Channel, apparently the levels of Pb, Cd and Cr are increasing, and at the same time the concentrations of Al, Co, Fe, Ni, Hg and Zn remain high. This shows that, despite the enforcements implemented by the State Environmental Agency, the situation has not improved, at least for the sediments. Only the Hg contamination exhibited an unusual pattern, in which the highest levels are similar to those observed in the past (Tommasi, 1979;

CETESB, 1985), but nowadays the contamination is more widespread throughout the system. A similar phenomenon was observed in the Trieste Gulf (Horvat et al., 1999).

Sediment contamination by metals is serious, due to their incorporation into the biota and through the trophic chain. Lamparelli et al. (2001) detected contamination by Cu, Ni and Zn in tissues of mollusks, fishes and crustaceans collected in the SES, which are consumed by the local population and commercialized to other regions of the state.

Organic Compounds

The results obtained for n-alkanes and PAHs are shown in the table 5. The presence of n-alkanes is frequently related to the contamination by aliphatic hydrocarbons, corresponding to a fraction of the total aliphatic hydrocarbons (Medeiros and Bícego, 2004). Inside the estuary, these compounds occurred systematically, reaching levels up to 9.0 µg/g. The higher concentrations were observed in São Vicente Channel sediments, followed by the Santos and Bertioga Channels and the SSOS vicinity (Table 5).

In most samples, n-alkanes were mainly from biogenic sources, as showed by the relationships between odd/even compounds and pristane/phytane. This is expected in an estuarine system surrounded by mangroves (Nishigima et al., 2001). However, in the sediments of stations 1, 2, 3, 4, 7, 8, 10, 18 and 25, the contribution of petrogenic hydrocarbons was greater. The values obtained in the present study were similar to those observed by Bonetti (2000) and Medeiros and Bícego (2004).

The distribution of PAHs was different than that observed for the n-alkanes, probably because their respective sources are not exactly the same. The PAHs were restricted to the Santos Channel (Table 5), where the maximum concentration reached 42.39 µg/g (sum of 30 compounds). Moreover, the concentrations of some individual PAHs were elevated, above PELs and/or TELs. A decreasing gradient towards the lower estuary was observed, suggesting that the sources are located in its upper portion (close to station 8). The most frequent compounds in higher levels were anthracene, fluoranthene, pyrene, chrysene, fluorene, acenapthalene, acenaphthene and phenanthrene. This result is similar to those reported previously (Medeiros and Bícego, 2004; Bonetti, 2000), but differs from the obtained by Lamparelli et al. (2001), in which benzo(a)pyrene was the major contaminant.

Although the PAH contamination is not wide in the SES, its presence is problematic, not only due to their toxicity (Driscoll et al., 1998; Fishelson et al., 1999) but also to their bioaccumulative properties (Hatch and Burton, 1999). Lamparelli et al. (2001) observed high PAHs concentrations in tissues of fishes, mollusks and crustaceans from SES.

Compared to other sites, in general the concentrations of hydrocarbons in sediments of the SES were higher than those found in São Sebastião, Brazil (Weber et al., 1998), and comparable to the levels found in Montevideo Harbour (Muniz et al., 2004) and Todos os Santos Bay (Venturini and Tommasi, 2004).

Detergents concentrations ranged from 0.997 to 13.678 µg/g (Table 6), with highest concentrations close to the SSOS diffusers, followed by the stations from São Vicente channel (stations 13, 10 and 15). The minimum and maximum concentrations were within the ranges observed by Tommasi (1979), however this author found the highest content at the

Table 5. Concentrations of poly aromatic and n-alkane hydrocarbons in sediments from the Santos Estuarine System (µg/g)

Compounds (ug/g)	Sample																									Ref
	1	2	3	4	5	6	7	8	9	10	11	12	13	14	15	16	17	18	19	20	21	22	23	24	25	
Total n-alkanes	1,05	1,83	1,48	2,12	3,06	2,87	3,71	4,09	1,77	4,29	9,31	0,02	0,03	1,4	0,49	0,53	1,03	4,09	2,08	0,31	1,29	1,58	0,88	3,21	0,4	0,2
Pristane/Phytane	-	1,01	0,23	0,5	0,57	-	0,66	0,57	-	0,4	0,25	-	-	1	-	1	1	1,05	1	3	2	1	1	0,05	0,4	-
odd/even	1,81	1,42	2,03	2,53	4,61	3,75	1,65	1,49	4,42	1,24	5,81	NC	2	3,69	4,33	3,9	3,59	1,67	4,69	3	3,54	3,61	3,72	4,37	1,46	2,8
odd/even <23	1	1,42	2,28	1,18	1,36	1,74	1,27	1,88	1,8	1,76	1,52	NC	1	1,5	1,5	1,5	1,5	1,49	1,7	1,25	1,83	1,25	2,5	1,67	1	2
odd/even >24	2,13	1,43	1,96	3,38	5,53	3,97	1,74	1,44	4,89	1,13	6,83	NC	NC	4,26	6,6	5,5	4,79	1,74	5,85	4,75	4	4,36	4,07	5,3	2,5	4
UCM	NM	NM	NM	NM	NM	NM	NM	NM	4,32	NM	54,4	-	0,59	10,2	3,2	4,52	8,82	72	21,2	2,59	7,51	15	6,21	15,3	8,18	1,49
Sum of PAHs*	0,08	0,19	2,91	1,68	39,8	28,8	11	42,4	2,12	1,38	0,03	0,00	0	0,02	0,01	0	0,01	0,51	0,01	0	0,01	0,01	0	0,03	0	0

NC = Not calculable; NM = not measured; UCM = Unresolved Complex Misture
*Sum of 30 PAHs.

Santos Channel. The difference can be attributed to the decrease in the pumping of contaminated waters from the Billings Reservoir to the estuary and the expansion of shantytowns over the mangroves at the estuary banks. The results obtained in the present study show that detergents are important contaminants to the SES.

Table 6. Detergent (MBAS) concentrations in sediments from the SES

Station	Concentration (ug/g)	Station	Concentration (ug/g)
1	0.977	14	3.908
2	2.931	15	7.816
3	1.466	16	1.467
4	Not measured	17	Not measured
5	3.419	18	13.678
6	5.373	19	Not measured
7	3.419	20	6.839
8	2.931	21	4.397
9	2.442	22	6.351
10	7.328	23	4.885
11	4.885	24	5.373
12	2.442	25	1.954
13	11.235	Reference	1.954

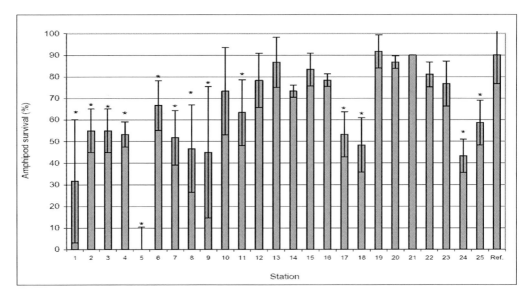

Figure 2. Whole sediment toxicity to the amphipod *T. viscana*, (* = significant difference (p<0.05).

Toxicity

In the whole sediment toxicity test, the samples from the estuarine channels were considered toxic, as well as those from the stations 17 and 18, located close at the SSOS diffusers (Figure 2). In the test chambers, the physical-chemical parameters of the water were kept within the recommended limits. The saturation of dissolved oxygen was always above 60%; the pH ranged from 8.0 to 8.4; the salinity varied between 34 and 35 ‰ and the total ammonia concentrations ranged from not detectable (< 0.01 mg/l) to 0.01 mg/L.

Table 7. Presence or absence of toxicity and estimations of the NH_3 contents in pore water samples extracted from sediment collected in SES (T = toxic; NT = not toxic)

Estação	100% Dilution		50% Dilution		25% Dilution	
	Toxicity	NH_3 (mg/l)	Toxicity	NH_3 (mg/l)	Toxicity	NH_3 (mg/l)
1	T	0,11	T	0,055	T	0,0275
2	T	0,07	T	0,035	T	0,0175
3	T	0,11	T	0,055	NT	0,0275
4	T	0,11	T	0,055	T	0,0275
5	T	0,14	T	0,070	T	0,0350
6	T	0,27	T	0,135	T	0,0675
7	T	0,22	T	0,110	T	0,0550
8	T	0,14	T	0,070	T	0,0350
9	T	0,04	T	0,020	T	0,0100
10	T	0,23	T	0,115	NT	0,0575
11	T	0,07	T	0,035	T	0,0175
12	T	0,05	T	0,025	T	0,0125
13	T	0,10	T	0,050	T	0,0250
14	T	0,07	T	0,035	T	0,0175
15	T	0,32	T	0,160	NT	0,0800
16	T	0,03	T	0,015	NT	0,0075
17	T	0,04	NT	0,020	NT	0,0100
18	T	0,18	T	0,090	T	0,0450
19	T	0,07	T	0,035	T	0,0175
21	T	0,07	T	0,035	NT	0,0175
22	T	0,13	T	0,065	T	0,0325
23	T	0,07	T	0,035	T	0,0175
24	T	0,11	T	0,055	T	0,0275
Reference	NT	0,00	NT	0,000	NT	0,0000

Yellow marked cells = unionized ammonia content above 0.05 mg/l
Orange marked cells = toxic samples when NH_3 < 0.05 mg/l

The results of the porewater toxicity test are shown in (Table 7). The porewater salinities ranged from 28 to 35‰, whereas the pH of the samples ranged between 7.6 and 8.5. Total ammonia contents ranged from 0.0 to 7.5 mg/l; corresponding to estimated concentrations of unionized ammonia from 0.0 to 0.32 mg/l. Samples from stations 20 and 25 were lost and not tested. At 100% concentration, all the porewater samples produced significant reduction in the larval development. At 50% concentration, almost all the samples, with exception of that from the station 17, caused significant effects on the exposed larvae. At 25% porewater dilution, only the samples from the stations 10, 15, 16, 17 and 21 did not affect the larval development.

The interpretation of the ecotoxicological results must be preceded by the analysis of the confounding factors. For the whole sediment toxicity test, the grain size is considered a major factor; however the compositions of the samples were not extreme, so we assumed that the granulometry did not affect the amphipod survival. According to Carr et al. (2001), the unionized ammonia can contribute to the toxicity of porewater samples. The NOEC for NH_3 for the *L. variegatus* embryonic development test in water only exposures is 0.05 mg/L (Prósperi, 2002).

Ammonia accounted for toxicity of the whole samples, and the dilutions were used to help to interpret the data. The results indicated that for the samples from the stations 1, 2, 4, 5, 8, 9, 11, 12, 13, 14, 16, 17, 18, 19, 21, 22, 23 and 24 toxicity occurred even when the NH_3 levels were below 0.05 mg/L, suggesting that other contaminants may have produced the toxicity, beyond ammonia, as further discussed in the SQT data integration section. For station 3, no toxicity was observed in the lower dilutions, thereby implicating NH_3 as being responsible for the toxicity.

Regarding the geographical positioning, stations from the estuary and also those from the SSOS vicinities were observed to be the most toxic. The amphipods showed a response that seemed to be more associated with the measured contamination, which is complex and varies along the study area. Such response seemed to be associated with high levels of metals, n-alkanes and PAHs, especially those from the inner portions of the estuary. Also, the toxicity exhibited by the sediments collected close to the SSOS diffusers possibly was due to the combination of metals, aliphatics, total S and detergents. However, some toxic sediments did not exhibit significant contamination (stations 1, 2, 17 and 25) whereas some moderately contaminated samples were not toxic (stations 22, 23, 10 and 13); results for 3 of these samples possibly were influenced by the TOC (stations 10 and 13) and the fines content (station 23), which may immobilized the contaminants. These results are also in agreement with previous studies made in the Santos region (Abessa et al., 2001; 2005; Abessa and Sousa, 2001; Sousa et al., 2007; Cesar et al., 2006; Prósperi et al., 1998).

There was relatively good agreement between the two tests in designating the most toxic areas of the SES, as well as distinguishing zones with better environmental conditions, especially when the ammonia influence is interpreted. The most contaminated and toxic sediments were collected near the main contaminant sources, as the industrial complex, the port, the dumping sites, the intermittent sewage discharges and the SSOS diffusers. A similar situation was found in San Francisco Bay (Thompson et al., 1999), where sediment toxicity was associated with the contaminant sources, but could not be related to specific contaminants. These authors suggested that several reasons could explain this result including: 1) only one contaminant in concentrations high enough to produce toxicity but this contaminant was not measured; 2) different contaminants in low concentrations, which interact to produce toxicity; 3) many contaminants in high concentrations; 4) mixtures of contaminants with confounding factors, as unionized ammonia or sulphur.

Benthic Community

Most of the samples exhibited few species, with predominance of polychaetes, followed by bivalve mollusks. About 60 *taxa* of polychaetes were identified, 27 bivalves, 13 gastropods, 18 crustaceans, 5 ophiuroids, among others. The original data for benthos is available in Abessa (2002). Polychaete dominance was usually above 60%, whereas crustaceans occurred only at the stations 1, 13, 17 and 19.

Fifty-eight species, distributed among 12 taxonomic groups were observed at the reference station (Anchieta Island State Park). Except for the reference, the most species rich station was station 16, at which 37 species were identified.

Station 22 had the highest abundance, with 10,564.1 individuals per m^2. Abundance was also relatively higher at stations 1, 16, 23, 24, 25 and the reference. No organisms were encountered at stations 5 and 12. The abundance was low at stations 4, 8, 10, 13, 14, 15, 17 and 20 (<100 indiv./m^2).

Most of the species were considered rare, occurring at only one, two or three stations. The more common *taxa* were polychaetes, especially *Capitella capitata* and *Owenia fusiformis*, which were present in 10 and 11 stations, respectively. In addition to these two species, other relatively common polychaetes were *Rhodine* sp, *Magelona posterolongata*, *Ninoe brasiliensis*, *Ophioglycera* sp, *Hemipodus* sp, *Nephtys* sp, *Lumbrineris* sp and *Diopatra cuprea*. From them, at least *C. capitata*, *O. fusiformis*, *M. posterolongata*, and *Nephtys* are pollution tolerant, suggesting that the contamination influences on the benthic species composition. Among the other taxonomic groups, the more frequent species were *Anachis obesa* (gastropod), *Chione cancelata*, *Tagellus* sp and *Ctena pectinella* (bivalves), and the ophiuroid *Microphiopholis atra*. Except for *Anachis obesa*, which is typically found in estuarine areas, the other species were typical of coastal marine regions or cosmopolitan, (e.g., *C. capitata*).

The polychaete *Cirrophorus* sp was the most abundant, however it occurred only at stations 22 and 25. Capitellid polychaetes were dominant at 5 stations (7, 9, 23, 24 and 25), whereas *Magelona posterolongata* dominated at stations 11, 18 and 21. Mollusks were dominants at stations 2, 3, 4 and 16, whereas macrofauna from station 13 was composed only of two individuals of the isopod, *Excirollana armata*.

Generally, the species richness was low, as expected in estuarine systems, subject to extreme natural environmental variations (Table 8). Species richness was high at stations located on the East side of Santos Bay (stations 16, 19, 22, 23), in the mouth of Santos Channel (stations 1 and 2) and in Bertioga Channel (station 24). Organisms were absent at stations 5 and 12, two exhibited only one species (stations 4 and 13) and two had only two species (stations 10 and 20). Stations situated to the west of the bay tended to exhibit few species (stations 14, 17, 20), as well as the neighboring area formed by the stations from the São Vicente Estuary stations (12 and 13). Inside the estuary (stations 3, 4, 5, 6, 8, 9 and 10), few species were observed, excepting for station 7, where 9 species were recorded. The estuarine channels tended to have fewer species, when compared to the stations located in Santos Bay, however, stations 1 and 7 (Santos Channel) exhibited a higher variety of species; these exceptions may be related to the influence of clean waters streams from the ocean (Station 1) and from the Quilombo River (in front of Station 7). The west portion of the Bay (stations 14, 17 and 20) tended to be poorer than those from the east side (stations 16, 19, 22 and 23). Moreover, in the latter stations, ophiuroids, amphipods and other crustaceans were collected, together with more polychaetes species. Such difference between the west and east portions of the bay have been described previously by Tommasi (1979), but the species assemblages found in each sector differed from those observed in the present study.

Shannon diversity (H') at the SES was low as well, ranging between 0 and 3.362, and the value for the reference station was 3.553. The values were generally lower than those observed previously (Tommasi, 1979).

Table 8. Ecological Indices (Specific Richness; Simpson Dominance, Shannon Diversity (H') and polychaete abundances, in each sampling station

Station	Ecological índices			
	Specific Richness	Diversity Shannon	Simpson dominance	Polychaetes abundance (%)
1	16	1.611	0.399	58.54
2	15	2.380	0.134	37.93
3	9	1.942	0.257	80.00
4	1	0	1.000	0
5	0	0	nc	0
6	5	1.560	0.211	75.00
7	10	1.033	0.610	94.91
8	4	1.330	0.268	83.33
9	3	0.656	0.641	100
10	2	0.693	0.480	100
11	8	1.774	0.204	92.86
12	0	0	nc	nc
13	1	0	1.000	0
14	3	1.099	0.316	66.67
15	4	1.332	0.269	80.00
16	37	3.362	0.047	50.77
17	3	1.040	0.363	0
18	6	0.911	0.560	91.30
19	14	2.432	0.104	88.00
20	2	0.693	0.480	50.00
21	9	1.946	0.190	85.71
22	18	0.815	0.673	99.16
23	18	2.642	0.088	85.71
24	17	2.383	0.127	83.05
25	13	2.108	0.154	97.32
Reference	59	3.553	0.045	25.92

nc = not calculable.

This pattern was similar to that exhibited for species richness. However, the Simpson dominance was low in the majority of stations. This possibly was due to the occurrence of few individuals in most of the samples, which may have biased the interpretation because they are calculated as rare species. High dominances were obtained in stations with only one species (4 and 13) and in stations 7, 9 and 18, which were dominated by the opportunistic polychaetes *Capitella capitata*, *Heteromastus filiformis* and *Magelona posterolongata*, respectively, and at station 22, where *Cirrophorus* sp was dominant (c.a. 700 ind./m^2).

Another important factor to consider is the polychaete dominance, which sometimes is an indicative of alteration on the benthic community (Van Dolah et al., 1999; Weisberg et al., 1997). However, in naturally unstable environments such as estuaries, polychaete dominance is expected (Reish, 1986). In Santos, the polychaete dominance was generally high, and inside the estuary this group represented the majority of the benthic macrofauna present. In the bay, more species were found, although most of them were opportunistic.

The SES is known to be a complex, unstable and dynamic environment (Tommasi, 1979) in which the abiotic variables are more important than the interspecific relationships in determining the benthic community structure. In this kind of environment, the opportunistic organisms are favored, as they respond rapidly to natural changes. This is supported by the dominance of opportunistic species, such as *Capitella capitata*, *Magelona posterolongata*, *Owenia fusiformis*, *Rhodine* sp, *Anachis obesa*, *Chione cancelata*, among others.

The influence of natural environmental variables makes it difficult to discern the effects of contaminants on the benthic community structure of the SES, especially when there are multiple contamination sources to consider. Moreover, as previously mentioned, the benthic community structure appears to follow the pattern described by Tommasi (1979), where the estuarine system was divided between the east and the west of the bay, and the differences in the benthos corresponded to the environmental variables which indicated the differing hydrodynamic influences (e.g., sediment grain size distribution, salinity, and levels of nutrients in sediments). Such previous study observed "gradients" in physiological stress, which increased from the bay to the estuary. The patterns observed in the present study are similar to those reported previously (Tommasi, 1979) and confirmed that the system is very heterogeneous, where there are considerable variations for numerous environmental factors, such as sediment grain size, bottom water temperature, salinity and dissolved oxygen, TOC, and carbonates as well as the presence of many kinds of contaminants.

SQT Data Integration

Each SQT component was has been evaluated independently and then all the components were combined in a weight-of-evidence approach. Other integrative approaches (Green and Montagna, 1996; Carr et al., 2000) were also applied and the results were compared. Thus, it was possible to determine more accurately the degree of degradation in each station and the relationships among toxicity, contamination and benthic community structure.

Linear correlations

Linear correlations were used to observe general trends, because the species distribution normally is influenced by several variables, especially in a complex system like the SES. Significant correlation coefficients ($p \leq 0.05$) are presented in Table 9.

Table 9. Correlation coefficients between variables, when p = 0.05

Variable		r
Specific richness	Bottom water salinity	0.6519
Specific richness	Pore water toxicity 50%	0.5207
Specific richness	Pore water toxicity 100%	0.5272
Diversity (H')	Bottom water salinity	0.5971
Simpson dominance	Tensoactives contents	0.5601
Simpson dominance	Pb concentrations	0.5004
Whole sediment toxicity	Mud content	0.5302
Whole sediment toxicity	Hg concentration	0.53728
Pore water toxicity (25% dilution)	Mud content	0.5617
Pore water toxicity (25% dilution)	Hg concentration	0.5180
Pore water toxicity (25% dilution)	Total PAHs concentration	0.5431
Pore water toxicity (25% dilution)	Zn concentration	0.505

The coefficients tended to be low with "*r*" values in the 0.5 to 0.6 range. Heterogeneous and unstable environments tend to exhibit this pattern, due to interactions among variables. The species richness and diversity correlated positively with the bottom water salinity, suggesting that areas with higher salinities (more influenced by marine waters) exhibited more species, which is in agreement with the hypothesis that marine areas are more stable and allow the colonization of a greater diversity of benthic species (Tommasi, 1979). Simpson dominance correlated positively to Pb and surfactants concentrations, suggesting that benthic communities are affected by such contaminants (possibly by elimination of sensitive species and selection of the resistant ones).

The mortality of *T. viscana* in the toxicity tests correlated positively to the mud content and Hg concentration. The rate of abnormal development in *L. variegatus* at the 25% dilution correlated positively to mud contents, Hg, PAH and Zn concentrations. Such correlations suggest that these contaminants were in part responsible for the observed toxicity. Also, fine-grained sediments tend to accumulate higher concentrations of contaminants, and may transfer them back to the pore water, producing toxicity to benthic organisms. These results suggest that bottom water salinity is an important factor influencing the benthic community structure in the SES, but contaminants play an important role as well, particularly Hg, PAHs, Zn, Pb and surfactants.

Principal component analyses

The sequential application of PCA has been recommended to integrate SQT data (Green, 1993). This method allows the observation of the interaction among various combinations of variables (Table 10). The first three principal components (PCs) explained 66.09% of the data variance. The first PC explained 30.86%, and correlated with the following eigenvectors: *Tox1PC* (explained by the porewater toxicity at 25 and 50% dilutions), *Chem1PC* (explained by Fe, Ni, Cr and Co concentrations), and *Amb1PC* (primarily explained by the bottom water DO contents). The second PC explained 20.2% of the variance and correlated with *Tox2PC* (explained by the toxicities of whole sediment and 100% porewater), *Eco1PC* (explained by diversity), and *Amb2PC* (explained by the bottom water salinity, MBAS and n-alkane concentrations). The third PC explained 15.03% of the variance and correlated with

Chem2PC (explained by the Pb and Cd concentrations) and to *Amb3PC* (explained by the sediment TOC contents, bottom water DO, and PAHs concentrations).

Table 10. Autovectors and *Eigenvectors* obtained by the PCA

Var	EIGENVECTORS					
	1	2	3	4	5	6
Tox1PC	0.5134	-0.0503	0.1138	0.0162	0.2706	-0.4793
Tox2PC	0.1722	0.5594	0.2201	0.1442	-0.4716	-0.2029
Chem1PC	-0.4529	-0.2538	0.3407	0.0950	0.1609	-0.1812
Chem2PC	-0.0083	-0.1206	-0.5075	-0.6369	-0.1185	-0.4666
Eco1PC	-0.2870	0.5017	0.0146	-0.4018	-0.2041	0.1893
Eco2PC	-0.3347	0.2194	-0.2247	0.4848	-0.0656	-0.5910
Amb1PC	-0.5206	-0.2234	-0.0468	-0.0064	-0.1313	-0.0547
Amb2PC	0.1747	-0.5044	0.0676	0.1061	-0.7753	0.0033
Amb3PCA	0.0749	0.0372	-0.7124	0.3956	0.0104	0.3007

Grey Marked values indicate significant correlations ($p<0.05$).

Table 11. RTM values estimated for sediments from Santos

Station	RTM value
1	14,48
2	8,44
3	12,87
4	50,44
5	nc
6	19,99
7	31,10
8	40,04
9	33,73
10	17,18
11	16,67
12	nc
13	17,62
14	12,10
15	9,93
16	3,63
17	11,29
18	26,39
19	6,71
20	8,41
21	7,37
22	14,43
23	5,99
24	15,13
25	8,98
Reference	1,31

Each PC explained a low percentage of the total variance, which This may be due to the heterogeneity of the environment, where the interactions of the variables are complex. A similar result was obtained by Long et al., (2002), when applying a PCA for a SQT study in Biscayne Bay, Florida. However, this method allows the identification of the main interactions, which are evident when the variables related to the PCs are identified. This method also shows that environmental (DO, salinity and TOC) and anthropogenic (Fe, Ni, Cr, Co, MBAS, n-alkanes, Pb, Cd, PAHs, whole sediment and porewater toxicity) factors explain the variability in the benthic community structure in the SES. The PCA method, which considered the interaction among the different variables, allowed a better comprehension of the complexity of the interactions that influence the benthos at the study area, providing evidence that natural factors as well as anthropogenic ones influence the benthic community health at the SES.

RTMV Method

The data obtained for chemistry, toxicity and macrobenthos were also integrated by the RTMV method (Table 11), in order to rank the stations. There was a trend for the biological factors to have a greater influence on the RTMV than the other variables. Macrobenthos was most important for stations 4, 5, 9, 10, 12, 13, 14, 15 and 20, and was the major factor in combination with toxicity for stations 17, 18, 19, 21, 22 and 23. The three factors had similar weight for stations 6, 7, 8, 11 and 27, whereas toxicity was more important for stations 1, 2, 3, 16, 24 and 25. The influence of the biological factors is explained by the fact that the SES is a physically controlled estuary, favoring the occurrence of opportunistic species, indicating that the alterations in the benthos could be due not only to the pollution but also to natural factors.

The RTMV values were interpolated and displayed in a geo-referenced map, allowing the degraded and non-degraded areas to be visualized (Figure 3). Inner portions of the estuary were considered most degraded, with some areas exhibiting moderate degradation, such as the area close to the SSOS diffusers. Non-degraded areas were also observed.

When interpreting the RTMVs from areas such as Santos, where many stations exhibited degradation to some degree, care must be taken, because impacted sites may appear to be unaltered using this method (Type II error – false negative). Abessa (2002) compared different approaches to calculate sediment quality values (as Ratio to Reference; Ratio to Maximum Value, Ratio to Mean Vales, among others) and considered RTMVs as appropriate for SES, and stated that they could be used to rank the sediments based on their relative quality. It must be highlighted that RMTVs should not be used with purposes of addressing the difference of sediment quality among the samples, once such values are based on the triangle areas, which increases exponentially when any component increases linearly.

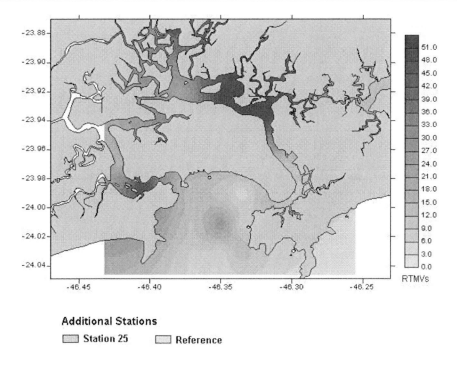

Figure 3. Distribution of the sediment quality in the SES, according to RTM values.

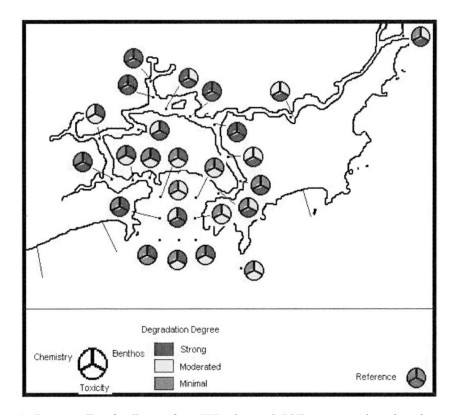

Figure 4. Sediment quality of sediments from SES, when each SQT component is evaluated separately (see Non-statistical comparative method in the Material and Methods Section).

Pie Diagrams

The data were also integrated using pie diagrams, which require a previous interpretation of the separate SQT components. The chemistry data were analyzed based on the Sediment Quality Guideline Quotients (SQGQ) calculated to PEL (Fairey et al., 2001), and some estuarine stations (3 to10 and 24) and station 18 (SSOS outfall) were considered altered. Almost all the stations exhibited some toxicity, except station 20. The most toxic sediments were collected inside the SES. However, since the ammonia influenced the results of the sea urchin bioassay, an overall interpretation was adopted, taking into account the conclusion for each sample when the 3 dilutions are considered.

As previously mentioned, the analysis of the macrobenthic data consisted of using some ecological indices, and their interpretation was based on previous data (Abessa, 2002; Tommasi, 1979). However, alterations due to natural factors were not considered in this interpretation, thus the alteration in the benthos may be overestimated.

The conclusions for the three components were combined in a pie diagram (Figure 4). The combination of the data showed that most of the samples exhibited alterations in at least one SQT component. Stations 3, 4, 5, 6, 7, 8, 9, 10, 24 (estuary) and 18 (SSOS diffusers) presented more obvious signs of degradation.

Data Integration and Conclusions

The results obtained by each integrative method were compared. The isolated analysis of each SQT component, the physical-chemical, sedimentological data, and the multivariate results indicated that, like the environmental variables, contamination can also influence the benthic community structure. This information was used when determining the degree of degradation for each area studied.

There was good agreement among the different techniques used to classify the stations. Sediments from stations 3, 4, 5, 6, 7, 8, 9, 10, 18 and 24 showed strong evidence of degradation. They correspond to the inner portions of the estuary and to the region close to the SSOS diffusers. For these stations, the three SQT components indicated negative effects. Moreover, the contaminants present in the sediment from station 24 (Bertioga Channel) were similar to those observed in the Santos Channel (stations 4 and 5, in particular). This suggests that the discharge of contaminants into Santos Channel influences the Bertioga Channel, which contradicts the conclusion from circulation studies which proposed a low influence of Santos Channel waters on Bertioga Channel (Harari et al., 2000).

Sediments considered low or not degraded were those from stations 14, 15, 16, 19, 20, 21, 23 and 25. These stations correspond to the mouth of São Vicente Channel, Santos Bay and mouth of Bertioga Channel.

Six stations exhibited moderately degraded sediments (stations 1, 2, 11, 13, 22 and 17). For stations 1 and 2, the macrobenthic community was rich and diverse, and the contamination was low, but toxicity was present. Further studies (e.g., toxicity identification evaluation studies) would be required to understand the causes of this observed toxicity. Sediments from the stations 11 and 22 exhibited toxicity and moderate levels of contaminants. Station 13 exhibited high levels of MBAS (detergents), porewater toxicity and the presence of only one macrofaunal species (an isopod). Direct discharge of untreated sewage was observed at this station when the sampling was carried out. Further studies are also required to define if the toxicity was due to contaminants associated with the discharge or by natural factors. Station 17 did not exhibit high levels of contaminants, but showed toxicity and a degraded macrobenthic community. However, organisms considered pollution sensitive were found there, the sea pansy *Renilla* sp and the mysid *Promysis atlantica* (Tommasi, 1979), suggesting that the low density and richness were due to natural factors, as the grain size, the hydrodynamics (Harari *et al.*, 2000), and sediment remobilization.

The inconclusive data for stations 1, 2, 13 and 17 could be expected. Nipper *et al.* (1998) reported that for moderately degraded sediments, the SQT results may be too variable to allow alterations to be detected. For the stations which were considered degraded, the SQT data provided indications of the priority areas, and also gave information that can be used in the establishment of management strategies, decision making and development of remediation and control plans.

The present research showed the existence of contamination and toxicity gradients along the Santos and São Vicente Channels, with association between contaminants and toxicity. Most degraded sediments occurred close to the contamination sources listed by the State Environmental Agency – CETESB (Lamparelli *et al.*, 2001). The most critical area was the Santos Channel, especially its inner portion (Piaçaguera Channel), which is close to the industrial complex, (e.g., the steel and fertilizers industries), and the region close to the oil terminal (Barnabé Island); these regions need remediation actions. In the São Vicente Channel, moderated to high sediment degradation was observed, and the pollution sources are the industrial and domestic landfills, the shantytowns and the discharges of sewage; due to the nature of contamination and the existing levels of degradation, the needed actions are more related to the control of contamination sources. In the Santos Bay, degradation was observed close to the SSOS and in the disposal area of dredged sediments from the Santos Channel; in the first case an abatement plan would be desirable, whereas the dredging operations require more appropriate management actions.

ACKNOWLEDGMENTS

We would like to thank FAPESP (98/00808-6 Process) for the financial support; the IOUSP North Base staff for the logistic and general assistance; and the help of Dr. Luiz Roberto Tommasi, Biol. Ylara Almeida Pinto, Dr. Kátia Christol dos Santos, Dr. Maria Teresa Valério Belardo, Dr. Maria Fernanda Lopes dos Santos, Dr. Ana Maria Setúbal Pires Vanin and Dr. Thaís Navajas Corbisier, who kindly identified some taxonomic groups (respectively, echinoderms, mysids and decapods, tanaids, amphipods, cumaceans, isopods and anthozoans). We greatly appreciate the constructive technical reviews provided by Dr. Marion Nipper and Ed Long.

REFERENCES

Abessa, D. M. S. (2002). *Avaliação Da Qualidade De Sedimentos Do Sistema Estuarino De Santos*. Doctorate Thesys. Universidade De São Paulo, Instituto Oceanográfico, 290.

Abessa, D. M. S. & Sousa, E. C. P. M. (2001). Preliminary Studies On The Acute Toxicity Of Marine Sediments Collected Close To The Sewage Outfalls From Baixada Santista, Sp, Brazil. In: *Anais Do 1º Congresso Brasileiro De Pesquisas Ambientais. 03* De Setembro De 2001. Santos, Sp. Cd-Rom, 59-61.

Abessa, D. M. S. & Sousa, E. C. P. M. (2004). Adaptação De Método De Extração De Tensoativos Aniônicos De Sedimentos Hídricos Marinhos Para Análise Por Mbas: Aplicação Em Avaliações Ambientais Preliminares. *O Mundo Da Saúde, 28(4)*, 431-435.

Abessa, D. M. S., Sousa, E. C. P. M., Rachid, B. R. F. & Mastroti, R. R. (1998). Use Of The Burrowing Amphipod *Tiburonella Viscana* As Tool In Marine Sediments Contaminantion Assessment. *Brazilian Archives Of Biology And Technology, 41(2)*, 225-230.

Abessa, D. M. S., Sousa, E. C. P. M., Rachid, B. R. F. & Mastroti, R. R. (2001) Sediment Toxicity In Santos Estuary, Sp-Brazil: Preliminary Results. *Ecotoxicology And Environmental Restoration, 4(1)*, 6-9.

Abessa, D. M. S., Carr, R. S., Rachid, B. R. F., Sousa, E. C. P. M., Hortelani, M. A. & Sarkis, J. E. (2005). Influence Of A Brazilian Sewage Outfall On The Toxicity And Contamination Of Adjacent Sediments. *Marine Pollution Bulletin, 50*, 875-885.

Akagi, H. & Nishimura, H. (1991). *Advances In Mercury Toxicology.* Ed. Plenum, New York, 53-63.

Amaral, A. C. Z. (1980). Breve Caracterização Dos Gêneros Da Família Capitellidae Grube (Annelida Polychaeta) E Descrição De *Nonatus Longilineus* Ge. Sp. N. *Boletim Do Instituto Oceanográfico, S Paulo, 29(1)*, 99-106.

Amaral, A. C. Z. & Nonato, E. F. (1982). *Anelídeos Poliquetos Da Costa Brasileira. 3 – Aphroditidae E Polynoidae.* Brasília, Cnpq/Coordenação Editorial, *46*.

Amaral, A. C. Z. & Nonato, E. F. (1984). *Anelídeos Poliquetos Da Costa Brasileira. 4 – Polyodontidae, Pholoidae, Sigalionidae E Eulepethidae.* Brasília, Cnpq/Coordenação Editorial, *54*.

Amaral, A. C. Z. & Nonato, E. F. (1994). *Anelídeos Poliquetos Da Costa Brasileira. 5 – Pisionidade, Chrysopitellidae, Amphinomidae E Euphrosinidae.* Revista Brasileira De Zoologia, Curitiba, *11(2)*, 361-390.

Amaral, A. C. Z. & Nonato, E. F. (1996). *Annelida Polychaeta: Características, Glossário E Chaves Para Famílias E Gêneros Da Costa Brasileira.* Editora Da Unicamp. Campinas, S. 1240.

Apha/Awwa/Wef, (1998). 5540 - Surfactants. In: *American Public Health Association, Water Environment Federation.* Standard Methods For Examination Of Water And Wastewater. 20th Ed., USA. *1269*.

Bolívar, G. A. & Lana, P. C. (1986). Padrões De Distribuição De *Spionidae* E *Magelonidae* (Annelida: Polychaeta) Do Litoral Do Estado Do Paraná. Dissertação De Mestrado. Universidade Federal Do Paraná, *Setor De Ciências Bioógicas, 116*.

Bonetti, C. (2000). Foraminíferos Como Bioindicadores Do Gradiente De Estresse Ecológico Em Ambientes Costeiros Poluídos. Estudo Aplicado Ao Sistema Estuarino De Santos - São Vicente (Sp, Brasil). Tese De Doutorado. Universidade De São Paulo, Instituto Oceanográfico. *São Paulo*, 229 + Anexos.

Carr, R. S., Chapman, D. C., Howard, C. L. & Biedenbach, J. M. (1996a). Sediment Quality Triad Assessment Survey Of The Galveston Bay, Texas System. *Ecotoxicology, 5*, 341-364.

Carr, R. S., Montagna, P. A., Biedenbach, J. M., Kalke, R., Kennicutt, M. C., Hooten, R. & Cripe, G. (2000). Impact Of Storm Water Outfalls On Sediment Quality In Corpus Christi Bay, Texas. *Environ. Toxicol. Chem.*, *19*, 561-574.

Carr, R. S., Nipper, M. G., Adams, W. J., Berry, W. J., Burton Jr., A. G., Ho, K., Macdonald, D. D., Scroggins, R. & Winger, P. V. (2001). Summary Of A Setac Technical Workshop: Porewater Toxicity Testing: Biological, Chemical, And Ecological Considerations With A Review Of Methods And Applications, And Recommendations For Future Areas Of Research. 18-22 March 2000. Pensacola, Fl. *Society Of Environmental Toxicology And Chemistry (Setac)*, 38.

Cesar, A., Pereira, C. D. S., Santos, A. R., Abessa, D. M. S., Fernández, N., Choueri, R. B. & Delvalls, T. A. (2006). Ecotoxicology Assessment Of Sediments From Santos And São Vicente Estuarine System – Brazil. *Brazilian Journal Of Oceanography*, *54(1)*, 55-63.

Cesar, A., Abessa, D. M. S., Pereira, C. D. S., Santos, A. R., Fernández, N., Choueri, R. B. & Delvalls, T. A. (2007). A Simple Approach To Integrate Ecotoxicological And Chemical Data For The Establishment Of Environmental Risk Levels. *Brazilian Archives Of Biology And Technology*.

Cetesb. (1985). Baixada Santista - Memorial Descritivo. Carta Do Meio Ambiente E De Sua Dinâmica. Relatório Técnico Cetesb. *São Paulo*, *Sp. 33*.

Cetesb. (1992). Água Do Mar - Teste De Toxicidade Crônica De Curta Duração Com *Lytechinus Variegatus* Lamarck, 1816 (Echinodermata: Echinoidea). Norma Técnica L5.250. *São Paulo, Cetesb, 16*.

Chapman, P. M. (1990). The Sediment Quality Triad Approach To Determining Pollution-Induced Degradation. *The Science Of The Total Environment*, 97/98, 815-825.

Chapman, P. M., Dexter, R. N., Anderson, H. B. & Power, E. A. (1991). Evaluation Of Effects Associated With An Oil Platform, Using The Sediment Quality Triad. *Environmental Toxicology And Chemistry*, *10(3)*, 407-424.

Day, J. H. (1963). *Polychaeta Of Southern Africa, Part 1. Errantia And Part 2. Sedentaria*. London, British Museum (Nat. Hist.) Publ. 656.

Del Valls, T. A. & Chapman, P. M. (1998). Site-Specific Quality Values For The Gulf Of Cádiz (Spain) And San Francisco Bay (Usa), Using The Sediment Quality Triad And Multivariate Analysis. *Ciencias Marinas*, *24(3)*, 313-336.

Del Valls, T. A., Forja, J. M. & Gómez-Parra, A. (1998). The Use Of Multivariate Analysis To Link Sediment Contamination And Toxicity Data To Establish Sediment Quality Guileines: An Example In The Gulf Of Cádiz (Spain). *Ciencias Marinas*, *24(2)*, 127-154.

Driscoll, S. B. K., Schaffner, L. C. & Dickhut, R. M. (1998). Toxicokinetics Of Fluoranthene To The Amphipod, *Leptocheirus Plumulosus*, In Water-Only And Sediment Exposures. *Marine Environmental Research*, *45(3)*, 269-284.

Environment Canada, (1999). *Canadian Sediment Quality Guidelines For The Protection Of Aquatic Life. Summary Tables*. Http://Www.Ec.Gc.Ca.

Fairey, R., Long, E. R., Roberts, C. A., Anderson, B. S., Phillips, B. M., Hunt, J. W., Puckett, H. R. & Wilson, (2001). An Evaluation Of Methods For Calculating Mean Sediment Quality Guideline Quotients As Indicators Of Contamination And Acute Toxicity To Amphipods By Chemical Mixtures. *Environmental Toxicology And Chemistry*, *20(10)*, 2276-2286.

Fishelson, L., Bresler, V., Manelis, R., Zuk-Rimon, Z., Dotan, A., Hornung, H. & Yawetz, A. (1999). Toxicological Aspects Associated With The Ecology Of *Donax Trunculus*

(Bivalvia, Mollusca) In A Polluted Environment. *The Science Of The Total Environment*, *226*, 121-131.

Folk, R. L. & Ward, W. C. (1957). Brazon River Bar: A Study In The Significance Of Grain Size Parameters. *Journal Of Sediment And Petrology*, *27(1)*, 3-27.

Fostier, A. H., Ferreira, J. R. & De Andrade, M. O. (1995). Digestion For Mercury Determination In Fish-Tissues And Botttom Sediments By Automated Cold Vapour Atomic Absorption Spectrometry. *Química Nova*, *18(5)*, 425-430.

Fúlfaro, V. J. & Ponçano, W. L. (1976). Sedimentação Atual Do Estuário E Baía De Santos. Um Modelo Geológico Aplicado A Projetos De Expansão Da Zona Portuária. In: *Congresso Brasileiro De Geologia E Engenharia*, 1. 1976. Anais. Belo Horizonte/Mg, *Sociedade Brasileira De Geologia*, 67-90.

Fúlfaro, V. J., Requejo, C. S., Landim, P. M. B. & Fúlfaro, R. (1983). Distribuição De Elementos Metálicos Nos Sedimentos Da Baía De Santos, Sp. In: *Simpósio Regional De Geologia*, 4. 1983. Atas. São Paulo/Sp, *Sociedade Brasileira De Geologia*, 275-289.

Golden Software, (1995). Surfer (Win 32) Version 6.01. Surface Mapping System. Golden Software Inc, Golden, Colorado.

Green, R. (1993). Application Of Repeated Measures Designs In Environmental Impact And Monitoring Studies. *Australian Journal Of Ecoology*, *18*, 81-98.

Green, R. & Montagna, P. (1996). Implications For Monitoring: Study Designs And Interpretation Of Results. *Canadian Journal Fisheries And Aquatic Science*, *53*, 2629-2636.

Gross, M. G. (1971). Carbon Determination. In: Carver, R.E. (Ed) *Procedures In Sedimentary Petrology*. New York, *Wiley-Interscience*, 573-596.

Hatch, A. C. & Burton Jr., A. G. (1999). Photo-Induced Toxicity Of Pahs To *Hyalella Azteca* And *Chironomus Tentans*: Effects Os Mixtures And Behavior. *Environmental Pollution*, *106*, 157-167.

Harari, J., Camargo, R. & Cacciari, P. L. (2000). Resultados Da Modelagem Numérica Hidrodinâmica Em Simulações Tridimensionais Das Correntes De Maré Na Baixada Santista. *Revista Brasileira De Recursos Hídricos*, *5(2)*, 71-87.

Horvat, M., Covelli, S., Faganelli, J., Logar, M., Mandic, V., Rajar, R., Sirca, A. & Zagar, D. (1999). Mercury In Contaminated Coastal Environments; A Case Study: The Gulf Of Trieste. *The Science Of The Total Environment*, 237/238, 43-56.

Koroleff, F. (1970). Direct Determination Of Ammonia In Natural Waters As Indophenol Blue. *Cons. Int. Explor. Mer*, Information On Techniques And Methods For Sea Water Analysis, N.3.

Lambshead, P. J. D., Paterson, G. L. J. & Cage, J. D. (1997). Biodiversity Professional. Software Package. The National Hystory Museum And The Scottish Association For Marine Science. In: Burton, G. A. (Ed.). *Sediment Toxicity Assessment*. Chelsea, Lewis Publishers, *Inc.*, *P.*183-211.

Lamparelli, M. L., Costa, M. P., Prósperi, V. A., Bevilácqua, J. E., Araújo, R. P. A., Eysink, G. G. L. & Pompéia, S. (2001). Sistema Estuarino De Santos E São Vicente. Relatório Técnico Cetesb. *São Paulo. 178.*

Lana, P. C. (1984). Anelídeos Poliquetos Errantes Do Litoral Do Estado Do Paraná. Tese De Doutorado. Universidade De São Paulo, *Instituto Oceanográfico, 275.*

Long, E. R. & Chapman, P. M. (1985). A Sediment Quality Triad: Measures Of Sediment Contamination, Toxicity And Infaunal Community Composition In Puget Sound. *Marine Pollution Bulletin, 16(10)*, 405-415.

Long, E. R., Macdonald, D. D., Severn, C. G. & Hong, C. B. (2000). Classifying Probabilities Of Acute Toxicity In Marine Sediments With Empiricall Derived Sediment Quality Guidelines. *Environmental Toxicology And Chemistry, 19(10)*, 2598-2601.

Long, E. R., Hameedi, M. J., Sloane, G. M. & Read, L. B. (2002). Chemical Contamination, Toxicity, And Benthic Community Indices In Sediments Of The Lower Miami River And Adjoining Portions Of Biscayne Bay, Florida. *Estuaries, 25(4a)*, 622-637.

Magurran, A. E. (1988). *Ecological Diversity And Its Measurements*. Chapman And Hall, Princeton. *179*.

Mckenzie, H. A. & Wallace, H. S. (1954). The Kjeldahl Determination Of Nitrogen: A Critical Study Of Digestion Conditions. *Australian Journal Of Chemistry, 7*, 55.

Melo, S. L. R & Abessa, D. M. S. (2002). Testes De Toxicidade Com Sedimentos Marinhos Utilizando Anfípodos Como Organismo-Teste. In: I., Nascimento, E. C. P. M. Sousa, M. G. Nipper, (Eds.). *Ecotoxicologia Marinha: Aplicações No Brasil*. Editora Artes Gráficas, Salvador/Ba. Cap.Xiv, 163-178.

Medeiros, P. M. & Bícego, M. C. (2004). Investigation Of Natural And Anthropogenic Hydrocarbon Inputs In Sediments Using Geochemical Markers. I. Santos, Sp—Brazil. *Marine Pollution Bulletin, 49*, 761-769.

Moser, G. A. O. (2002). Aspectos Da Eutrofização No Sistema Estuarino De São Vicente-Santos: Distribuição Espaço Temporal Da Biomassa E Produtividade Primária Fitoplânctonica E Transporte Instantâneo De Sal, Clorofila-A, Material Em Suspensão E Nutrientes. Tese De Doutorado. Universidade De São Paulo, *Instituto Oceanográfico. São Paulo, Sp. 426* + Anexos.

Muniz, P., Danulat, E., Yannicell, B., García-Alonso, J., Medina, G. & Bícego, M. C. (2004). Assessment Of Contamination By Heavy Metals And Petroleum Hydrocarbons In Sediments Of Montevideo Harbour (Uruguay). *Environment International, 29*, 1019-1028.

Nipper, M. G., Roper, D. S., Williams, E. K., Martin, M. L., Van Dam, L. F. & Mills, G. N. (1998). Sediment Toxicity And Benthic Communities In Mildly Contaminated Sediments. *Environmental Toxicology And Chemistry, 17(3)*, 2-38.

Nishigima, F. N., Weber, R. R. & Bícego, M. C. (2001). Aliphatic And Aromatic Hydrocarbons In Sediments Of Santos And Cananéia, Sp, Brazil. *Marine Pollution Bulletin, 42(11)*, 1064-1072.

Nonato, E. F. (1981). Contribuição Ao Conhecimento Dos Anelídeos Poliquetas Bentônicos Da Plataforma Continental Brasileira, Entre Cabo Frio E O Arroio Chuí. Tese De Livre-Docência. Universidade De São Paulo, *Instituto Oceanográfico, 246*.

Odum, E. P. (1971). *Fundamentos De Ecologia*. 4th Ed. Lisboa. *Fundação Calouste Gulbenkian, 927*.

Pardos, M., Benninghoff, C., Thoms, R. L. & Khim-Heang, S. (1999). Confirmation Of Elemental Sulfur Toxicity In The Microtox® Assay During Organic Extracts Assessment Of Freshwater Sediments. *Environmental Toxicology And Chemistry, 18(2)*, 188-193.

Prósperi, V. A. (2002). *Comparação De Métodos Ecotoxicológicos Na Avaliação De Sedimentos Marinhos E Estuarinos*. Tese De Doutorado. Universidade De São Paulo, Escola De Engenharia De São Carlos. São Carlos. 118p. + Anexos.

Prósperi, V. A., Eysink, G. G. J. & Saito, L. M. (1998). Avaliação Do Grau De Contaminação Do Sedimento Ao Longo Do Canal De Navegação Do Porto De Santos. *Relatório Técnico Cetesb*. São Paulo, Sp. 33p + Anexos.

Rachid, B. R. F. (2002). Avaliação Ecotoxicológica Dos Efluentes Domésticos Lançados Pelos Sistemas De Disposição Oceânica Da Baixada Santista. Tese De Doutorado. Universidade De São Paulo, *Instituto Oceanográfico*. São Paulo. *286*.

Reish, D. J. (1986). Benthic Invertebrates As Indicators Of Marine Pollution: 35 Years Of Study. Iee *Oceans' 86 Conference Proceedings*. Washington, Dc. 885-888.

Rios, E. (1984). *Sea Shells Of Brazil*. 2nd Edition. Editora Da Furg. Rio Grande, Rs. *368*.

Sas Institute Inc. (1992). *Sas/Lab® Software: User's Guide, Version 6, First Edition*, Sas Institute Inc., Cary, Nc, *291*.

Shepard. F. P. (1954). Nomenclature Based On Sand Silt-Clay Ratios. *Journal Sediment Petroleum, (24)*, 151-158.

Siqueira, G. W., Ducatti, G. M. F. & Braga, E. S. (2001). Avaliação De Metais Pesados (Pb, Cr, Cu, Fe) No Sistema Estuarino De Santos/São Vicente E Baía De Santos (São Paulo, Brasil).). In: *Ix Colacmar – Congreso Latinoamericano Sobre Ciencias Del Mar*. San Andrés Isla, Colombia, 16-20 Octubre. Resumenes Ampliados.

Smith, S. L., Macdonald, D. D., Keenleyside, K. A. & Gaudet, C. L. (1996). The Development And Implementation Of Canada Sediment Quality Guidelines. In: M. Munawar And G. Dave (Eds.). *Development And Progress In Sediment Quality Assessment: Rationale, Challenges, Techniques And Strategies,* Spb Academic Publishing, Amsterdam, The Netherlands. 233-249.

Sousa, E. C. P. M., Abessa, D. M. S., Gasparro, M. R., Zaroni, L. P. & Rachid, B. R. F. (2007). Ecotoxicological Assessment Of Sediments From The Port Of Santos And The Disposal Sites Of Dredged Material. *Brazilian Journal Of Oceanography*, *55(2)*, 75-81.

Suguio, K. (1973). *Introdução À Sedimentologia*. Edgar Blücher, Edusp, São Paulo. *317*.

Thompson, B., Bay, S., Greenstein, D. & Laughlin, J. (1991). Sublethal Effects Of Hydrogen Sulfide In Sediments On The Urchin *Lytechinus Pictus*. *Marine Environment Research*, *31*, 309-321.

Thompson, B., Anderson, B., Hunt, J., Taberski, K. & Phillips. (1999). Relationships Between Sediment Contamination And Toxicity In San Francisco Bay. *Marine Environmental Research*, *48*, 285-309.

Tommasi, L. R. (1979. Considerações Ecológicas Sobre O Sistema Estuarino De Santos, São Paulo. Tese De Livre Docência. Universidade De São Paulo, Instituto Oceanográfico. *São Paulo. 2vols*.

Unep. (1991). *Determinations Of Petroleum Hydrocarbons In Sediments*. United Nations Environment Programme. *Reference Methods For Marine Pollution Studies*, *97*.

Usepa. (1988). Short-Term Methods For Estimating The Chronic Toxicity Of Effluents And Receiving Waters To Marine And Estuarine Organisms. Epa-600/4-87-028. U.S. *Environmental Protection Agency*, Cincinnati, Ohio.

Van Dolah, R. F., Hyland, J. L., Holland, A. F., Rose, J. F. & Snoots, T. R. (1999). A Benthic Index Of Biological Intergrity For Assessing Habitat Quality In Estuaries Of The Southern Usa. *Marine Environment Research*, *48*, 269-283.

Venturini, N. & Tommasi, L. R. (2004). Polycyclic Aromatic Hydrocarbons And Changes In The Trophic Structure Of Polychaete Assemblages In Sediments Of Todos Os Santos Bay, Northeastern, Brazil. *Marine Pollution Bulletin*, *48*, 97-107.

Weber, R. R., Zanardi, E. & Bícego, M. C. (1998). Distribuição E Ocorrência Dos Hidrocarbonetos Biogênicos E De Petróleo, Na Água Do Mar Superficial E Nos Sedimentos De Superfície Da Região Da Plataforma Interna Do Canal De São Sebastião, Sp, Brasil. *Relatório Técnico Do Instituto Oceanográfico, 43*, 1-14.

Weisberg, S. B., Ranasinghe, J. A., Dauer, D. M., Schaffner, L. C., Diaz, R. J. & Frithsen, J. B. (1997). An Estuarine Benthic Index Of Biotic Integrity (Bibi) For Chesapeake Bay. *Estuaries, 20(1)*, 149-158.

Whitfield, M. (1974). The Hydrolysis Of Ammonia Ions In Sea Water – A Theoretical Study. *Journal Marine Biological Association, Uk 54*, 565-580.

Winger, P. V. & Lasier, P. J. (1991). A Vacuum-Operated Pore-Water Extractor For Estuarine And Freshwater Sediments. *Archives Of Environmental Contamination And Toxicology, 21*, 321-324.

Zamboni, A. J. & Abessa, D. M. S. (2002). Tríade Da Qualidade De Sedimentos. In: I., Nascimento, E. C. P. M. Sousa, M. G. Nipper, (Eds.). *Ecotoxicologia Marinha: Aplicações No Brasil*. Editora Artes Gráficas, Salvador. Cap.Xx, 233-243.

Zar, J. H. (1984). "Biostatistical Analysis". Englewood Cliffs, Prentice-Hall; *Inc, 718*.

In: Encyclopedia of Environmental Research
Editor: Alisa N. Souter

ISBN: 978-1-61761-927-4
© 2011 Nova Science Publishers, Inc.

Chapter 47

PRODUCTION FRESH WATER FISH WITH UNCONVENTIONAL INGREDIENTS IN EGYPT

Magdy M.A. Gaber[*]

National Institute of Oceanography and Fisheries, Cairo, Egypt

ABSTRACT

The evaluation of feed ingredients is crucial to nutritional research and feed development for aquaculture species. In evaluating unconventional ingredients for use in aquaculture feeds, there are several important knowledge components that should be understood to enable the judicious use of a particular ingredient in feed formulation. This includes information on (1) ingredient digestibility, (2) ingredient palatability and (3) nutrient utilization and interference.

Diet design, feeding strategy, fecal collection method and method of calculation all have important implications on the determination of the digestible value of nutrients from any unconventional ingredient. There are several ways in which palatability of unconventional ingredients can be assessed, usually based on variable inclusion levels of the unconventional ingredient in question in a reference diet and feeding of those diets under apparent satiate or self-regulating feeding regimes. The ability of fish to use nutrients from the test ingredient, or defining factors that interfere with that process, is perhaps the most complex and variable part of the ingredient evaluation process. It is crucial to discriminate effects on feed intake from effects on utilization of nutrients from unconventional ingredients (for growth and other metabolic processes). To allow an increased focus on nutrient utilization by the animals, there are several experimental carried out, which are based on variations in diet design and feeding regime used to determine the optimum level of substitutions of unconventional feed stuffs ingredients. Other issues such as ingredient functionality influence on immune status and effects on organoleptic qualities are also important consideration in determining the value of unconventional ingredients in aquaculture feed formulations.

---

[*] E-mail address: Gabermagdy@yahoo.com

INTRODUCTION

The rapid worldwide expansion of aquaculture and livestock strongly indicates that a crisis will be precipitated in the livestock and aquaculture feed industries in the near future. Food for humans is not included in this consideration because; livestock, fish and humans can all eat the same basic food commodities and feedstuffs are eaten first by humans.

The consumption of the world production of food fish in 2002 about, 101 million ton providing an annual person consumption 16.2 kg of food fish, while in Egypt total production 806400 tons of food fish providing an individual consumption 11.2 kg (FAO, 2004). From this comparison we found that Egypt faced to increase fish production.

Fresh water fish are the most efficient converters of feed to flesh requiring from 2-3 kg of basic feedstuffs to produce 1 kg of fish. In contrast to land animals, fish are fastidious eater in that require higher levels of dietary protein. In addition, the amino acid requirements to promote rapid growth of fish appear to be more rigid than for land animals. As an extremes example, the protein content in the diets of carnivorous fish such eels and yellow tail need to be almost of animals origin white the protein requirements of ruminants can be partially satisfied from is non protein nitrogen so sources sash as area

Fishmeal has traditionally been considered an important protein source for use in aquaculture diets for both carnivorous and omnivorous species, and many aquaculture formulations still have fishmeal included at levels in excess of 50%. However, being too reliant on any one ingredient presents considerable risk associated with supply, price and quality fluctuations. As a strategy to reduce risk, the identification, development and use of alternatives to fishmeal and oil in aquaculture diets remains a high priority. As a result of the volumes of fishmeal and oil used in aquaculture, especially for carnivorous species, aquaculture of these species is still perceived as a net fish consumer rather than producer, and this practice has raised concerns about the long-term sustainability of these industries (Naylor et al. 2000).

Substantial effort has been expended over the past decades in evaluating a wide range of potential alternatives to fishmeal and fish oils for use in aquaculture diets. Those ingredients can generally be classified into those being derived from either plant origin or terrestrial animal origin.

PROBLEMS ASSOCIATED WITH UNCONVENTIONAL FEEDSTUFFS FORMULATION PROBLEMS

Problem faced with the food supply for cultured fish, nutritionists have done more work evaluating alternate protein sources in aquaculture diets during the last 7 years than during the previous 50 years. A review of the literature shows that practically no potential feed material has been ignored. Table 1 showed categorizes these sources which divided into three groups; Feedstuffs faced with food supply problem for cultured fish, nutritionists have done more work evaluating alternate protein sources in aquaculture diets during the last 10 years than during the pervious 50 years. A review of the literature shows that practically no potential feed material has been ignored and unconventional ingredient are categorized these sources into two groups.

A. Commercialized Sources, which Divided into

1. Plant derived resources: Its include soybean meals, protein concentrates and oils (Kaushik et al. 1995; Refstie et al. 1998, 1999), canola meals, protein concentrates and oils (Higgs et al. 1982; Mwachireya et al. 1999; Burel et al. 2000; Forster et al. 2000; Glencross et al. 2003b, 2004a,b) and lupin meals and protein concentrates (Burel et al. 1998; Booth et al. 2001; Farhangi & Carter 2001; Glencross et al. 2003a, 2004c).
2. Animal sources: Its included resources such as rendered meat meals (Bureau et al. 1999, 2000; Stone et al. 2000; Sugiura et al. 2000; Williams et al. 2003 a,b), blood meals (Allan et al. 1999a,b; Bureau et al. 1999) and poultry meals (Bureau et al. 1999; Nengas et al. 1999).

B. Not Commercialized Sources as Single Cell, Protein Grasses, Leaf Protein, Yeast and Phytoplankton

As previously mentioned, the nutritive requirements of fish are such that they probably offer less flexibility in diet formulation than do those of most land animals. First of all, several species of fish selected for culture are almost pure carnivores, requiring a diet of high protein content. These fish have very poor utilization of carbohydrate as an energy source, and some evidence is developing that the inclusion of certain types of carbohydrates in the diet is, in fact, detrimental. Hence, it has been recently reported that soybean meal extracted with alcohol performed better in rations than unextracted meal. This improvement has been attributed to the removal of low molecular weight carbohydrates by the alcohol. The subject of carbohydrates in soybean meal will be discussed later.

Unconventional ingredients contain protein in amounts ranging from about 15 to 50 percent. This falls within the range of the protein requirements for optimum growth of several species of fish. However, proteins derived from vegetable sources are somewhat deficient in several key amino acids such as lysine, methionine, and tryptophan. This, of course, is not the only problem when unconventional sources of protein are used. A fish diet must provide a suitable energy source and be in proper balance with respect to:

- proteins,
- minerals,
- lipids,
- carbohydrates,
- Vitamins and growth factors.

Then, there are the factors that relate to the feeds tuffs comprising the diet. These include:

(a) Composition: The commodity must have a composition that allows it to be compounded into a balanced diet. For example, because of the high water content, it would be difficult to compound potential feeds tuffs such as sea plants, algae, etc., into diets.

(b) Physical form: Many feedstuffs must be modified for proper formulation into diets. A common example in salmon diets is the lint that remains in cottonseed meal. It readily plugs the dies when small diameter feeds are to be prepared.

Table 1. Alternate Sources of Protein that are Being Evaluated or have Potential as Partial or Whole Replacement for Fish Meal in Aquaculture Diets

Commercialized		Not commercialized
Vegetable	Animal	
Soy meal	Poultry byproducts	Insect larvae
Rapeseed meal	Feather meal	Single cell protein
Sunflower meal	Shrimp and crab meal	Grasses
Oat groats	Blood flour	Leaf protein
Cottonseed meal	Fish silage	Vegetable silage
Wheat middlings	Meat meal	Zooplankton (krill, etc.)
		Recycled wastes
		Yeast
		Phytoplankton
		Bacteria
		Algae
		Higher plants
Protein (range), %		
15-50	50-85	4-85

Table 2. Some Compounds Occurring in Feedstuffs that are Known and/or Suspected of Causing Physiological Abnormalities or Otherwise Impairing the Growth of Fish

Compounds	Found in
Glycosides	Grass and leaves
Phytates	All plant foods tuffs
Mycotoxins (aflatoxin)	Cereal-based meals not naturally occurring but produced by microorganisms
Cyclopropenoid fatty acids	Cottonseed oil and meal
Trypsin inhibitors	Soy and rapeseed meal
Mimosine	Leaves (Leucaena leucocephala)
Glucosinolates	Rapeseed meal
Haemoglutinins	Soyabean meal
Plant phenolics	
Gossypol	Cottonseed meal
Tannins	Rapeseed meal
Oxidized and polymerized lipids	Fish meal; poultry byproducts, krill meal

| Histamine and putrescine | Fish meal, primarily tuna |
| Nitrosamines | Fish meal |

(c) Palatability several potential feedstuffs contain compounds that are offensive to the olfactory receptors of the fish.
(d) Factors affecting bio-availability of nutrients
(e) Stability during storage: this primarily relates to the vitamin stability and also the stability of the lipid portion that may oxidize in either dry or frozen storage.
(f) Toxic factors (See Table 2)

WORLD PRODUCTION FROM AQUACULTURE

The world total aquaculture production in 2000 was liar million tons and comprise 3.6% of the world total production. FAO (2002) statistics indicate that china has remained the number one, producer both within Asia and within the world produced 629182 in 2000 which more than 6 times the 1990 production. Egypt also, showed impressive increase in tilapia production from 24916 tons in 1990 to 157421 tons in 2000. In 2000, of the 1.27 million tons of tilapia produced from aquaculture, 85% was grown in fresh water environment, while 14.1 % in brackish water (FAO 2002).as shown in figure (1).

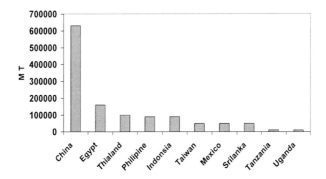

Figure 1. Tilapia aquaculture production by top ten countries (Data sources FAO, 2002).

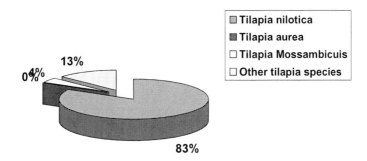

Figure 2. Percent shared world tilapia aquaculture production according to species (Data sources FAO, 2002).

Figure 3. Nile tilapia (Oreochromis niloticus).

I. UNCONVENTIONAL INGREDIENTS FOR TILAPIA FEED

1. Soybean Meal (SBM)

With regard to its composition soybean meal appears to be reasonably good component for aquaculture diets. It contains 40-50 percent protein, 5-6 percent ash, 1 percent lipid and about 40 percent carbohydrates.

However, it is limiting in sulfur containing amino acid, (methionine, lysine and cystine) and contains many endogenous anti nutrient including protease (trypsin) inhibitor phytohaemegylutimin and anti - vitamins. Several investigators have demonstrate that heating soybean and rattler severely not only increases its acceptability to fish, but also improves the availability of nutrients. This is first accomplished by deactivating trypsin inhibitors, and second by denaturing the protein and thus making it more digestible, and third by detoxifying natural toxicants. Several studies, however, indicate that when soybean meal is heated, its performance is still below the expected considering its amino acid composition. Recent studies with Nile tilapia, for example, show that even when soybean meal was fortified with lysine, and methionine, growth performance was comparable when menhaden meal was used as the source of protein (El-Saidy and Gaber, 1997 & 2002). While the studies with catfish, for example, show that even when soybean meal was fortified with lysine, cystine, and methionine, growth was significantly lower than when menhaden meal was used as the source of protein. It was concluded, among other things, that soybean meal must contain some anti-growth factors or that catfish do not utilize free amino acids. Other workers, however, have shown that the latter conclusion might not be true, since the carp and salmon can utilize free amino acids.

A. Production of Fingerlings from Nile Tilapia

In an experiment carried out by El-Saidy and Gaber, (1997). They used soybean meal as sole source of protein in the diet for production of Nile tilapia fingerlings by total replacement of fish meal protein (Table 3), in the diets with soybean meal protein and with addition of methionine and lysine (1.0 and 0.5% respectively).

Table 3. Composition and proximate composition for production of Nile tilapia fry reared in concrete tanks

Ingredients (%)	Diets	
	FM	SBM
Fish meal (66% C.P.)	20.0	--
Soybean meal (44 % C.P.)	30.0	55.0
Wheat bran (14% C.P.)	17.2	16.7
Yellow corn meal	24.0	18.0
Soybean oil	5.0	5.0
Mineral and vitamin premix[1]	0.3	0.3
L-Methionine	--	1.0
L-Lysine	--	0.5
Dicalcium phosphate	1.0	1.0
Molasses (as bender)	2.0	2.0
Proximate composition (%)		
Dry matter	94.6	94.6
Crude protein	33.1	33.2
Crude fat	13.8	12.5
Crude fiber	7.4	6.6
Ash	3.9	3.8
NFE[2]	36.4	38.5
Gross Energy (kcal/g^{-1} diet)	4.8	4.8

[1]-Vitamin and mineral premix (according to Xie et al. (1997).
[2]-N.F.E., Nitrogen free extract calculated as 100- %(Moisture+ Crude protein +Crude fat +Ash +Crude fiber)

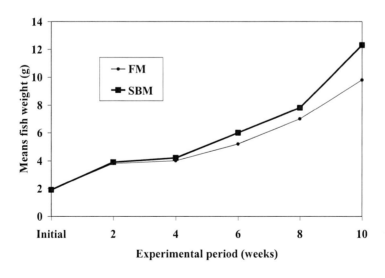

Figure 4. Effects of total replacement of fish meal with soybean meal supplemented with methionine and lysine on growth Nile tilapia fingerlings.

Table 4. Calculated level of amino acids in the two experimental diets and amino acid requirement for Nile tilapia fry

Diets		Requirement[a]	Amino acid
SBM	FM		
33.2	33.1		Crude protein (%)
			Essential amino acid
1.84	2.3		Arginine
0.69	0.89		Histidine
1.23	1.64	32	Isoleucine
2.15	2.77		Leucine
2.05	1.58	1.6	Lysine
1.34	0.76	0.65	Methionine[b]
1.31	1.58	1.18	Phenylalanine[b]
1.02	1.41	1.29	Threonine
0.39	0.45	1.95	tryptophan
1.28	1.81	1.02	Valine
		1.42	Non-essential amino acid
1.36	1.67	1.43	Alanine
2.70	2.91	0.32	Asparatic acid
0.48	0.39	1.06	Cystine
1.13	1.38		Glycine
4.64	4.47		Glutamic acid
1.29	1.28		Praline
1.25	1.28		serine
0.87	0.94		Tyrosine

a- According to Santiago and Lovell (1988)
b-The values of methionine and phenylalanine are the requirement in the presence of 2% cystine and 1% tyrosine of the diet, respectively

Two experimental diets were formulated. Fish meal (FM) diet, with 30% hexane-extracted soybean meal (SBM) and 20% fish meal, was formulated to be high quality commercial tilapia fish diet. The other diet contained 55% SBM and 0.5 % L-lysine supplementation (Table 4). Amino acid composition of the two diets (Table 4 and figure 4) was calculated from tabular provided for diet ingredients (NRC 1993).

Table 5. Economic information for Nile tilapia fry reared in concrete tanks

Item	Diets	
	FM	SBM
No. fry stocked/m^3	180	180
Cost fry (LE/m^3)	9	9
No fish harvested	175	177
Harvested (kg/m^3)	1,24	1,24
Food used(kg/m^3)	2,86	2,99
Fingerling cost (LE)[2]	36	36
Food cost[3]	9.6	5.3
Total cost (lE.)	18.6	14.3
Net profit (LE)	17.4	21.7

The economic calculation from the study is presented in table (5). It showed that the feed cost and the total cost (Lever Egyptian) increased with fish fed fish meal protein diet. From the economic information, it can be concluded that the highest net profit (Lever Egyptian) was achieved with fish fed soybean meal protein. The profit of using soybean in production of Nile tilapia fingerlings about three Egyptian pounds per cubic meter of cultured fish.

B. Production of Marketing Fish from Nile Tilapia

In an experiment carried out by Gaber and Hanafy (2008). They used soybean meal as soul source of protein in the diet for production marketing fish of Nile tilapia by total replacement of fish meal protein (Table 6), with soybean protein and supplementation of methionine and lysine (1% and 0.5% respectively).

Two isocaloric diets (17.0 kJ gross energy g^{-1} diet) were formulated to contain either fish meal protein or soybean meal protein (Table 6). Diet of soybean meal protein contained 45% hexane-extracted soybean meal (SBM) and was formulated according to El-Saidy & Gaber (2002). All ingredients were first ground to a small particle size (approximately 250 µm) in a Wiley mill. Dry ingredients were thoroughly mixed prior to adding water to 40% moisture. Diets were passed through a mincer with die into 3-mm diameter spaghetti-like strands, sun dried and stored in airtight containers, proximate composition of the experimental diets was determined according to AOAC methods (1995), while crude fiber in fish diets was determined according to methods of Bredon & Juko (1961). Total carbohydrate content (NFE) of diets was calculated by difference. Gross energy (GE) was calculated using the gross energy values for the macronutrients (23.4 kJ g'1 protein, 39.8 kJ g'1 fat and 17.2 kJ g^{-1} carbohydrate, fiber was not included in calculation).

The economic calculation from the study is presented in table (7). It showed that the feed cost and the total cost (Lever Egyptian) increased with fish fed fish meal protein daily. From the economic information, it can be concluded that the highest net profit (Lever Egyptian) was achieved with fish fed soybean meal protein.

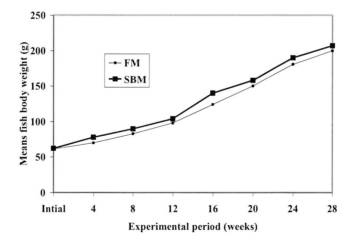

Figure 5. Effects of total replacement of fish meal with soybean meal supplemented with methionine and lysine on growth Nile tilapia for marking production.

Table 6. Composition and proximate composition for production of Nile tilapia fry reared in concrete tanks

Ingredients (%)	Diets	
	FM	SBM)
Fish meal (66% C.P.)	30.0	0.0
Soybean meal (44 % C.P.)	0.0	45.0
Wheat bran (14% C.P.)	23.0	20.0
Yellow corn meal	37.5	24.0
Soybean oil	5.0	5.0
Mineral and vitamin premix	1.0	1.0
L-Methionine	0.0	1.0
L-Lysine	0.0	0.5
Dicalcium phosphate	1.4	1.4
Vitamin C	0.1	0.1
Molasses (as bender)	2.0	2.0
Proximate composition (%)		
Moisture	10.7	11.0
Crude protein	26.0	26.0
Crude fat	9.8	9.7
Crude fiber	6.2	7.0
Ash	7.3	7.7
NFE[2]	40.0	38.6
Gross Energy (kcal/g-[1] diet	4.3	4.3

[1]-Vitamin and mineral premix (according to Xie et al. (1997).
[2]-N.F.E., Nitrogen free extract calculated as 100- %(Moisture+ Crude protein +Crude fat +Ash +Crude fiber)

In the present study, there are some factors leading to intensive fish culture in recirculation systems. Increasing the land cost and decreasing the freshwater supply is, therefore, the main reason for intensification of fish farming in Egypt, though additional advantages include savings in manpower and easier stock management. Increased fish yields in conventional, static ponds or reservoirs was accomplished by a combination of management procedures, the most important among them being the use of supplementary feed, polyculture and auxiliary aeration during the night (Sarig 1989). Higher yields were obtained in specially designed smaller units, 50-1000 m^3 (Zohar, et. al 1985; Van Rijn, et.al 1986), which differ from conventional ponds in design. These are made of concrete or are plastic-lined, and their configuration allows periodical removal of organic matter from the bottom. Most of these units are operated in a semi-closed mode, allowing optimal use of water and hence, minimal water discharge. Due to their reduced environmental impact, their development is supported by national and regional authorities. Pollution control is, therefore, another factor underlying the development of intensive systems. Finally, culture of fingerlings (mainly tilapia) during off-season in heated, indoor systems is rapidly expanding, and so, heat conservation can be counted as an additional factor promoting the use of intensive recirculation systems.

Several investigators have determined the effect of dietary soybean meal protein for tilapia and the results are not consistent. Among the plant protein sources, soybean meal has been shown to posses an acceptable amino acid profile for growth of many fish species (Murai, et. al. 1986; El-Saidy & Gaber 1997) and may be used as the major protein sources in many fish diets (Lovell 1989 and El-Saidy & Gaber 2002 & 2003). In the present study, the growth performance had no significant differences between tilapia groups fed fish meal protein and soybean protein diets, but showed significant increase with increasing feeding frequency four times daily. These results may attributed to fish at low feeding frequency might cause metabolic changes as depressing protein assimilation and accumulation of phospholipids in the muscle. According to Kayano, et al. (1993), the young fish fed several times a day could effectively assimilate dietary protein to muscle. In previous studies, it was found that, four feeding daily represent the optimum frequencies for fish (Holm, et. al 1990; Kayano, et. al. 1990 & 1993).

Feeding frequencies had significant effect on growth, feed consumption, net production, total production and specific growth rate (SGR) in Nile tilapia. By the end of the experiment, fish fed at higher feeding frequency had gained significantly more weight and more length than fish fed the lower feeding frequency. Fish fed at higher frequency consumed larger quantities of food than those fed less often, but individual meal size was smaller. This is consistent with studies conducted on other fish species. (Ishiwata 1969), where fish fed fewer meals per day tend to eat more per meal. Fish accomplished this by increasing stomach volume and became hyperphagic (Ruohonen and Grove 1998). However, although fish fed at higher frequency consumed larger quantities of food, when the interval between meals is short, the food passes through the digestive tract more quickly, resulting in less effective digestion (Liu & Liao 1999). The determining the optimal feeding frequency is important.

There was a strong trend for total production and net production (kg/m3) increase with increasing feeding frequency. These results agree with those of Essa, et al. (1989) and Kayano, et al. (1993) reported that production in fish culture are generally depended on daily feed consumption rate and feeding frequency. Production estimated, which are based on biomass estimates adjusted for mortality and corrected for growth rate (Chapman 1968). It is the basis for estimating economic yield for fish. Because net production and harvest value were not dependent on protein source but dependent on feeding frequency. In addition, final harvest and production values were directly related to feeding frequency and there are some feeding frequencies at which growth rate is reduced and when it occurs, production will be reduced. The critical level in our experiment was a feeding frequency four times daily for maximum growth in adult Nile tilapia

From the above results and the economic evaluation, we can conclude that, a diet containing soybean meal protein at four times daily feeding frequency recommended for adult Nile tilapia. These fish showed no significant effect to different dietary protein sources, but had significant increase with increasing feeding frequency Thus a soybean protein diet with four times feeding frequency would be cost effective and maintaining adequate growth and production of adult Nile tilapia in concrete tanks under the experimental condition.

2. Cotton Seed Meal (CSM)

Cottonseed meal (CSM),which ranks second to soybean in Egypt in terms of tonnage produced, is less expensive than fish meal and soybean on a per unit protein basis. Numerous studies have been conducted to determine the level of CSM that can be incorporated in Nile tilapia diets without affecting their growth performance (El-Saidy, 1999; El-Sayed, 1999; Mbahinzireki, Dabrowski, El-Saidy &Wisner, 2001). Results have shown that the amount of CSM that can be included in Nile tilapia diets depends mainly on the levels of free gossypol and available lysine. El-Saidy (1999) reported that prepressed solvent-extracted CSM could replace up to 50% of fish meal in juvenile Nile tilapia diets without requiring lysine supplementation. Free gossypol, when present in large quantity in the diet, has been shown to be toxic to monogastric animals including fish. Growth depression occurred in channel cat fish fed diets containing more than 900 mg free gossypol kg^{-1} diet (Dorsa, Robinette, Robinson & Poe 1982), whereas a diets containing as low as 290 mg free gossypol kg $diet^{-1}$ reduced the growth of rainbow trout (Herman 1970). Iron, as ferrous sulphate, has been successfully used to counteract the toxicity of free gossypol in diets of monogastric, terrestrial animals (Jones 1987; Martin 1990). High levels of supplemental iron used to counteract the toxicity of gossypol may be harmful to fish because it has been suggested that a delicate balance exists between the need of iron for host defense mechanisms and the need of iron to sustain microbial growth. Sealey, Lim and Klesius (1997) reported that high levels of dietary iron may lead to increased susceptibility of channel cat fish to Edwardsiella ictaluri infection. Moreover, no studies have been conducted to evaluate the effect of this compound to detoxify gossypol in Nile tilapia feeds containing cottonseed products.

Table 7. Economic information for Nile tilapia reared in concrete tanks for 24 weeks fed two protein sources

Item	Diets	
	FM	SBM
No. fish stocked/tank	100	100
No fish harvested	100	100
Harvested (kg/m^3)	4.78±0.27	4.86±0.19
Harvested kg/tank (4m^3)	19.11±1.1	19.45±0.73
Food used(kg/tank)	37.6	38.3
Fingerling cost $(LE)^2$	40.0	40.0
Food cost [3]	105.28	86.18
Total cost (LE)	145.28	126.18
Value of harvest (8.6 LE. kg^{-1})	162.44	165.33
Net profit (LE)	17.16	39.15

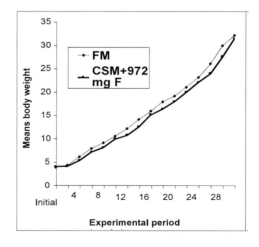

Figure 6. The of total replacement of fish meal with cotton seed meal supplemented with 972 mg iron on body weight of Nile tilapia fry during 30 weeks of feeding.

In an experiment carried out by El-Saidy and Gaber, (2004). They used cotton seed meal as soul source of protein in the diet for production of Nile tilapia fingerlings by total replacement of fish meal protein (Table 7 and Figure 6). They reported that; adding 972 mg Fe kg diet^{-1} from ferrous sulphate to the CSM-based diets that contained 972 mg free gossypol (1:1 iron to free gossypol ratio) for Nile tilapia reduce the negative effects of gossypol and improved growth performance, feed utilization and blood parameters and can totally replace fish meal in tilapia diets.

Results of these studies indicate that total replacement of FM with CSM (0.145% free gossypol) reduced the nutritional value of the diets. For CSM containing diets, supplemented with iron, as ferrous sulphate at 1:1 ratio of iron to free gossypol, had no effects on the nutritional value of the diets. In addition iron presents in practical diets at a level of 972 mg Fe kg diet^{-1} appears to sufficient to maintain normal function of growth performance, feed utilization and biological and hematological parameters of Nile tilapia.

Table 8. Ingredients and proximate composition of fish meal (FM) and cottonseed meal (CSM) diet supplemented with 972 mg Fe kg diet^{-1} for Nile tilapia fingerlings

Ingredients (%)	Diets	
	FM	CSM
Fish meal (60% C.P.)	50.0	0.0
Cotton seed meal (40 % C.P.)	0.0	45.0
Yellow corn meal	36.7	20.0
Wheat bran (14% C.P.)	5.0	24.0
Soybean oil	5.0	5.0
Mineral and vitamin premix	1.3	1.0
L-Methionine	0.0	1.0
L-Lysine	0.0	0.5
Molasses (as bender)	2.0	1.4
Supplemented iron (mg kg diet^{-1})	0.0	2.0
Proximate composition (%)		
Moisture	10.7	11.0
Crude protein	26.0	26.0
Crude fat	9.8	9.7
Crude fiber	6.2	7.0
Ash	7.3	7.7
NFE^2	40.0	38.6
Gross Energy (kcal/g-^{1} diet)	4.3	4.3

^{1}-Vitamin and mineral premix (according to Xie et al. (1997).
^{2}-N.F.E., Nitrogen free extract calculated as 100- %(Moisture+ Crude protein +Crude fat +Ash +Crude fiber)

Early studies have indicated that the amount of CSM that can be used in Nile tilapia feed depends mainly on the level of free gossypol and available lysine content of the meal. Due to unfavorable physiological effects of gossypol and to a reduction in the biological availability of lysine because of the binding properties of gossypol. Ofojekwu and Ejike (1984) and Robinson, Rawles and Stickney (1984) found that O. aureus fed CSM-based diets yielded poor performance. The authors attributed the poor performance to the gossypol contained in glanded and glandless CSM respectively. On the contrary, repressed, solvent-extracted CSM was successfully used as a single dietary protein source for O. mossambicus (Jackson, Capper & Matty 1982) and Nile tilapia (El-Sayed 1990). In the present study, regardless of supplemental levels of iron, fish fed diets that contained 67% CSM (972 mg free gossypol) supplemented with lysine to a level equal to that of the FM diet and supplemented with 972 mg Fe kg diet 1 exhibited better FBW and SGR than those fed diet 2 (67% CSM without additional iron). This may be due to addition of iron sulphate at a weight ratio of 1:1 of iron to free gossypol was effective in reducing the toxicity of free gossypol and improving their performances (Table 8). Although, Robinson and Rawles (1983) showed that supplementation of lysine to diets containing 44.6% glandless cottonseed flour or 54.4% glandless CSM as total replacement of soybean meal (SBM) did not improved growth and feed conversion. However, when glanded CSM with 0.022% free gossypol was used to totally replace SBM, supplementation of lysine is needed to improve the nutritional value of channel catfish feed to a level comparable with that of control (Robinson1991). The response of fish, based on RBC, Ht and Hb to dietary CSM was influenced by supplemental levels of dietary iron. For diets containing no CSM (FM based diets), there was an increase in these parameters. When FM was totally replaced by CSM (67%) the diet (SBM) supplemented with 972 mg Fe kg diet^{-1} exhibited superior results of Ht, Hb and RBC to fish meal diet. Our results are in agreement with those of Rojas and Scolt (1969), Wedegaertner (1981), Jones (1987) and Martin (1990). They found that addition of iron sulphate at a weight ratio of 1:1 of iron to free gossypol in pigs and broilers diets was effective in reducing the toxicity of free gossypol and improving animal performance. They suggested that iron inactivates gossypol by forming a strong complex compound in the intestinal tract, thus preventing it from absorbed (Wedegaertner, 1981). On the contrary to our results of the hematological values with Nile tilapia in the present study, Barros, Limand and Klesius (2002) reported that channel catfish fed a diet containing 50% CSM and supplemented with 671 mg Fe kg diet 1 had no improvement in growth of fish or the hematological values compared with treatments without dietary iron added. They attributed that to diets high in SBM may contain compounds or factors, which reduce iron absorption or availability. The present study indicated that GSI of males of Nile tilapia were not influenced by CSM diets with or without iron supplementation. Our results are in agreement with those of Dabrowski, Rinchard, Lee, Blom, Ciereszko and Ottobre (2000) and Rinchard, Lee, Dabrowski, Ciereszko, Blom and Ottobre (2003). They reported that GSI of male rainbow trout not negatively affected by increasing levels of CSM in their diets. Results of the present study indicate that GSI of females of Nile tilapia significantly influenced with iron supplementation and the lowest values were recorded with groups of fish fed diet 2 (100% CSM without iron supplementation. The same results were reported by Dabrowski et al. (2000) and Blom, Lee, Rinchard, Dabrowski and Ottobre (2001) in their studies on rainbow trout fish. The present study showed that ADC values of nutrients in CSM were comparable with those in other oilseed meals. El-Saidy and Gaber (2002) reported that ADCs of crude protein in soybean meal were 78.7^88.9% for Nile tilapia. Cheng and Hardy (2002) found the ADCs of crude protein in CSM were 81.6-87.9% for rainbow trout. These values are in agreement of our results where digestibility of crude protein in CSM-based diets ranged from 76.3% to 88.2%. The results are also in agreement with Mbahinzireki et al. (2001) who reported that ADCs of crude protein decreased as dietary gossypol level increased in tilapia (Oreochromis sp.) feeds. It is well known that the rate of digestion and nutrients assimilation in fish may be influenced by various physiological and a biotic factors, including fish size, ration level and temperature (Windell, Foltz & Sarokon 1978; NRC 1993). In the present experiment with Nile tilapia, the particle size of the ingredients tested was uniform and the size class of fish, feeding regime and temperature employed was consistent in the trial. The lower digestibility observed for the diet 2 can be attributed to free gossypol presented in CSM. Results of these studies indicate that total replacement of FM with CSM (0.145% free gossypol) reduced the nutritional value of the diets. For CSM containing diets, supplemented with iron, as ferrous sulphate at 1:1 ratio of iron to free gossypol, had no effects on the nutritional value of the diets. In addition iron presents in practical diets at a level of 972 mg Fe kg diet^{-1} appears to sufficient to maintain normal function of growth performance, feed utilization and biological and hematological parameters of Nile tilapia.

From the above results and the economic evaluation, we can conclude that, a diet containing cotton seed meal protein recommended for adult Nile tilapia. These fish showed no significant effect to different dietary protein sources, but had significant increase with cotton seed meal supplemented with iron at a level of 972 mg Fe kg diet^{-1}. Thus a cotton seed meal supplemented with iron at a level of 972 mg Fe kg diet^{-1} would be cost effective and maintaining adequate growth and production of adult Nile tilapia in concrete tanks under the experimental condition.

3. Sunflower Meal

Sunflower meal may be a possible alternative protein sources. Which is cheap and rich in both protein and lipid (Jackson et al. 1980) Sunflower seed meal is known to be used effectively by trout (Sanz et al. 1994) al low levels and by tilapia (El-Saidy and Gaber, 2002) at higher levels, with addition of methionine and lysine.

In an experiment carried out by El-Saidy and Gaber (2002). They used sunflower meal as soul source of protein in the diet for production fingerlings of Nile tilapia by total replacement of fish meal protein (Table 10 and figure 7), sunflower meal protein and supplementation of methionine and lysine (1% and 0.5% respectively).

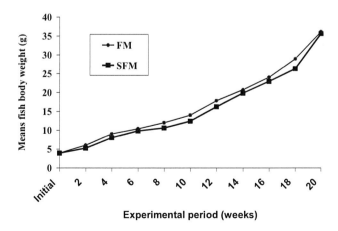

Figure 7. The effects of partial replacement of fish meal with 50% of sunflower meal on body weight of Nile Tilapia fry during 20 weeks.

Table 9. Economic information for Nile tilapia reared in concrete tanks for 24 weeks fed two protein sources. Values are means ±SD

Item	Diets	
	FM	SBM
No. fish stocked/tank	100	100
No fish harvested	100	100
Harvested (kg/m^3)	4.78±0.27	4.86±0.19
Harvested kg/tank (4m^3)	19.11±1.1	19.45±0.73
Food used(kg/tank)	37.6	38.3
Fingerling cost (LE)^2	40.0	40.0
Food cost ^{3}	105.28	86.18
Total cost (lE.)	145.28	126.18
Value of harvest (8.6 LE. kg^{-1})	162.44	165.33
Net profit (LE)	17.16	39.15

Table 10. Partial replacement of fish meal by sunflower meal for production Nile tilapia fingerlings

Ingredients	Diets	
	FM	DSM
Fish meal (66% C.P.)	50.0	25.0
Dehulled SFM (42 % C. P.)	--	35.72
Yellow corn meal	36.5	24.78
Wheat bran (14% C.P.)	5.0	5.0
Soybean oil	4.0	4.0
Molasses (as bender)	2.0	2.0
Mineral and vitamin premix^1	2.4	2.4
Vitamin C	0.1	0.1
L-Methionine	--	0.5
L-Lysine	--	0.5
Proximate composition (%)		
Moisture	7.61	7.31
Crude protein	32.5	32.6
Crude fat	14.36	13.98
Crude fiber	8.44	8.21
Ash	6.2	6.21
NFE^2	30.89	31.69
Gross Energy (kcal/g-^{1} diet)^4	4.5	4.5

^{1}-Vitamin and mineral premix (according to Xie et al. (1997).
^{2}-N.F.E., Nitrogen free extract calculated as 100- %(Moisture+ Crude protein +Crude fat +Ash +Crude fiber)

These study demonstrated that up to 50% dehulled sunflower meal (DSM) protein could be replace fish meal (FM) as protein source in the diet of Nile tilapia without affecting the overall growth performance of fish. El- Saidy et al. (1999) demonstrated that a 32% protein diet for Nile tilapia achieved the best growth and feed conversion which was suitable to examine alternative protein sources. The results of experiment carried out by El-Saidy and Gaber (2002) indicated that Nile tilapia can not be raised successfully by feeding diets formulated on dehulled sunflower meal alone as a sole source of protein. These finding are consistent with those reported by other investigator, where Jackson et al. (1982) reported that dehulled sunflower meal could not used as a sole source of protein. Also Sanz, et al. (1994) observed that dehulled sunflower meal had limited utilization in rainbow trout and resulted in poor growth and decreased feed conversion. Similar results were reported by Martinez (1984) for rainbow trout (Salmo gairdneri L.) fed sunflower meal. Tacon et al. (1984) found that sunflower meal was efficiently utilized as a partial replacement for fish meal protein in diets for rainbow trout. The palatability of the two diets (Table 9) is the same based in feed intake. No significant difference in weight gain among groups of fish fed two diets.

The calculated essential amino acid concentration (Table 10) in the diets met or exceeded those recommended by Santiago and Lovell; (1988). From this table indicated that all essential amino acid concentration is recovered by additional supplementations of l-methionine and L lysine as follow 1.0% an 0.5% respectively.

The above results and the economic evaluation, (Table 11) we can conclude that, a diet containing dehulled sunflower meal (DSM) protein can partially (50%) replace fish meal (FM) as a main source of protein in diets for Nile tilapia fingerlings. Thus a dehulled sunflower meal (DSM) would be cost effective and maintaining adequate growth and production of Nile tilapia fingerlings in concrete tanks under the experimental condition.

Table 11. Calculated amino acids in the two experimental diets and amino acid requirement for Nile tilapia fry (% of dry diet)

Diets		Requirement[a]	Amino acid
DSM	FM		
32.5	32.5	32	Crude protein (%)
			Essential amino acid
2.48	2.08	1.6	Arginine
0.83	0.84	0.65	Histidine
1.54	1.59	1.18	Isoleucine
2.67	2.73	1.29	Leucine
2.37	2.48	1.95	Lysine
1.35	0.95	1.02	Methionine [b]
1.49	1.42	1.42	Phenylalanine [b]
1.34	1.40	1.43	Threonine
0.38	0.37	0.32	tryptophan
1.75	1.81	1.06	Valine
			Non-essential amino acid
2.09	2.79		Alanine
5.57	4.9		Asparatic acid
0.44	0.38		Cystine
2.07	2.27		Glycine
9.15	7.8		Glutamic acid
2.34	1.98		Proline
1.4	1.38		Serine
0.99	1.08		Tyrosine

a- According to Santiago and Lovell (1988)
b-The values of methionine and phenylalanine are the requirement in the presence of 2% cystine and 1% tyrosine of the diet, respectively

Table 12. Economic information for Nile tilapia reared in concrete tanks for 24 weeks fed two protein sources

Item	Diets	
	FM	DSM
No. fish stocked/tank	100	100
No fish harvested	100	100
Harvested (kg/m^3)	4.78±0.27	4.86±0.19
Harvested kg/tank (4m^3)	19.11±1.1	19.45±0.73
Food used(kg/tank)	37.6	38.3
Fingerling cost (LE)^2	40.0	40.0
Food cost ^{3}	105.28	86.18
Total cost (lE.)	145.28	126.18
Value of harvest (8.6LE. kg^{-1})	162.44	165.33
Net profit (LE)	17.16	39.15

4. Linseed Meal

In Egypt, the linseed which known locally as "ketene" is widely grown and it is considered a plant oil, after oil extraction. The by-product which known as linseed meal are used for feeding young caws and calves particularly in the rural areas there is less information available on the use of linseed meal as a protein source in aquatic animals feeds. Therefore, a study was carried out to determine the feasibility of this linseed meal as partial replacement of fish meal protein in practical diet by El-Saidy and Gaber, (2001). As shown im table (13) and figure (8).

Table 13. Partial replacement of fish meal by sunflower meal for production Nile tilapia fingerlings

Ingredients	Diets	
	FM	LSM
Fish meal (66% C.P.)	50.0	12.5
Linseed meal (44 % C.P.)	--	69.3
Corn meal	36.5	--
Wheat bran (14% C.P.)	5.0	8.9
Soybean oil	4.0	4.0
Molasses (as bender)	2.0	2.0
Mineral and vitamin premix	2.4	2.4
Dicalcium phosphate	2.0	2.0
L-Methionine	-	0.5
L-Lysine	--	0.5
Proximate composition (%)		
Moisture	7.61	7.61
Crude protein	32.5	32.5
Crude fat	14.36	14.36
Crude fiber	8.44	8.44
Ash	6.2	6.2
NFE^2	30.89	30.89
Gross Energy (kcal/g-^{1} diet)	4.5	4.5

^{1}-Vitamin and mineral premix (according to Xie et al. (1997).
^{2}-N.F.E., Nitrogen free extract calculated as 100- %(Moisture+ Crude protein +Crude fat +Ash +Crude fiber)

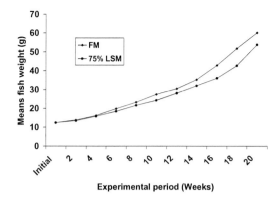

Figure 8. The effects of total replacement of fish meal with plant protein mixture meal (PPM) on growth rate of Nile tilapia fry during 16 weeks of feeding.

The present study exhibited that linseed meal protein can replace 75% in practical diets of Nile tilapia fingerlings. This is agreement with the results of Jakson et al. (1982), who fed Oreochromis mosambicus diets with varying level of linseed meal for 9 weeks. Nile tilapia showed very significant growth depression with linseed protein over 75%.

Table 14. Economic information for Nile tilapia reared in concrete tanks for 24 weeks fed two protein sources

Item	Diets	
	FM	LSM
No. fish stocked/tank	150	150
No fish harvested	150	150
Harvested (kg/m^3)	9.0	8.1
Food used(kg/tank)	23.9	21.3
Fingerling cost (LE)^2	22.5	22.95
Food cost ^{3}	59.95	21.95
Total cost (IE.)	82.45	43.76
Value of harvest (8.6 LE. kg^{-1})	150.0	150
Net profit (LE)	67.55	105.4

From the above results and the economic evaluation, (Table 14) we can conclude that, a diet containing Linseed meal (LSM) protein can partially (75%) replace fish meal (FM) as a main source of protein in diets for Nile tilapia fingerlings. Thus Linseed meal (LSM) would be cost effective and maintaining adequate growth and production of Nile tilapia fingerlings in concrete tanks under the commercial condition.

5. Krill

To achieve optimal production of fish in intensive aquaculture it is necessary to understand the factors affecting fish appetite. Extensive studies have led to improvements in nutritional quality of commercial feeds, but feeds will not be utilized if palatability is poor (Toften et al. 1995). A key factor in weaning fish fry to dry diets is the attractability of food.

Food attract ability and stimulation of ingestion involve stimuli, such as "smell" and "taste" of food particles (Kolkovski et al. 1997a, 2000). Mackie and Mitchell (1985) stated that chemical stimuli initiate search movement for food particle identification, and subsequent feeding. Tasting and feed intake involve taste buds that are chemically stimulated (Sorensen and Caprio 1997). These stimuli are extremely important in fish larvae and juveniles because the visual sense may not yet be fully developed and chemical sensors are the main ones used in food search (Dempsey 1978; Iwai 1980). Feed attractants have been characterized and isolated from different marine organisms such as squid, marine worms, mussel Mytilus edulis, clam, krill and brine shrimp Artemia sp. (Tandler et al. 1982; Mackie and Mitchell 1985; Mearns et al. 1987; Hara 1993; Kolkovski et al. 1997b). These food attractants can play an important role in acceptance of dry diets in fish during the weaning period as well as enhancing growth due to higher consumption (Kolkovski et al. 1997b). One possible application of feeding stimulants may be to mask different feeding deterrents that lower the palatability of diets. It would be of interest to test the influence of using different levels of krill meal supplemented to whole plant protein diets as attractant or stimulant of juvenile Nile tilapia.

Krill represent the largest potential available source. Various estimates made by fishery biologists show that over 300 million metric tons could be harvested annually. The problem here of course is that krill as widely dispersed primarily in the arctic and Antarctic regions and require a capital and energy for the harvest.

Table 15. Composition and proximate analysis of the fish meal and soybean meal (SBM) diets with krill meal (KM) supplementation fed to Nile tilapia, Oreochromis niloticus, juveniles

Ingredients	Diets	
	FM	SBM
Fish meal (60%c.p.)	20.0	0.0
Soybean meal	41.5	54.0
Wheat bran	14.0	14.0
Corn meal	15.0	15.0
Fish oil	5.0	5.0
Molasses	2.0	2.0
Vit.& Min. Premix[1]	2.0	2.0
L-Methionine	0.0	1.0
L-Lysine	0.0	0.5
Krill meal	0.0	6.0
Proximate analysis		
Dry matter	94.6	91.3
Crude protein	33.8	33.9
Crude fat	10.4	11.4
Ash	10.5	8.2
Crude fiber	3.9	7.1
NFE[2]	41.4	39.4
G.E. (Kcal/g)	4.6	4.6

[1]-Vitamin and mineral premix (according to Xie et al. (1997).
[2]-N.F.E., Nitrogen free extract calculated as 100- %(Moisture+ Crude protein +Crude fat +Ash +Crude fiber)

In feeding tilapia (Gaber 2005), krill provide an important source of caroteneaids and as attractant, while lead to increasing tilapia production. Also, showed that the supplementation of krill meal at 1.5 % in increasing rate in the diets of Nile tilapia as attractant or stimulant may lead to increasing feed intake, growth performances, feed utilization and soybean meal can totally replace fish meal based in diets for juvenile tilapia (Table 15):

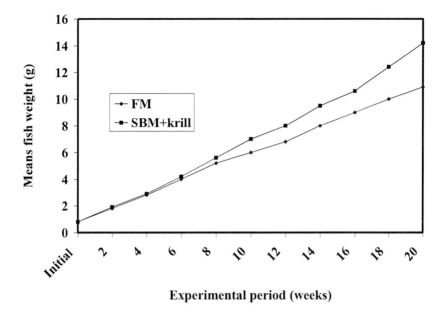

Figure 9. The effects of total replacement of fish meal with soybean meal supplemented with 10% krill meal on growth rate of Nile tilapia fry during 20 weeks of feeding.

Table 16. Economic information for Nile tilapia reared in concrete tanks for 24 weeks fed two protein sources

Item	Diets	
	FM	LSM
No. fish stocked/tank	150	150
No fish harvested	150	150
Harvested (kg/m^3)	9.0	8.1
Food used(kg/tank)	23.9	21.3
Fingerling cost (LE)[2]	22.5	22.95
Food cost [3]	59.95	21.95
Total cost (lE.)	82.45	43.76
Value of harvest (8.6 LE. kg^{-1})	150.0	150
Net profit (LE)	67.55	105.4

The FM (100%) fish meal protein and one diet with 100% soybean meal protein which was found the best diet for Nile tilapia fry in a previous work (El-Saidy and Gaber, 2002). was supplemented with krill meal (Specialty Marine Products, Alexandria, Egypt) dried under vacuum at 60 C. After the drying process, krill meal was added to the test diet.

In general, weight gain and feed utilization parameters were significantly higher for fish fed whole plant protein soybean meal diets supplemented with different levels of krill meal compared with the groups of fish fed the control diet. It also suggests that because food intake increased, fish grew more. This is in agreement with the results of Kolkovski et al. (2000) who used krill hydrolysis as a feed attractant for larvae and juvenile of three freshwater fish species. Based on feed intake data, the utilization of whole plant protein soybean meal diets increased with increasing krill meal levels. Tandler et al. (1982) found increased feeding when they used mussel Mytilus edulis extract in gilthead sea bream Sparus aurata diets. Feeding behavior in Atlantic salmon Salmo salar was elicited when gels flavored with shrimp extracts were offered to the fish (Mearns et al. 1987). These findings are in agreement with EL-Saidy and Gaber (2002) who reported that Nile tilapia fed diets containing 0% fish meal and 55% SBM supplemented with a 5% L-lysine had similar growth rate when compared with fish fed commercial diets containing fish meal. In the present study, the increased feed intake resulting from adding krill meal to tilapia plant protein diets can be due to "smell"-releasing chemical compounds leaching into the water and consequently increasing the food searching behavior in fish, and/or palatability palatability of food particles by direct interaction with the taste buds in the fish buccal cavity. In our study, when krill meal was added at different levels to plant protein diets, feed intake was increased by increasing krill meal levels in diets when compared to fish fed a control diet. This finding is in agreement with the results of Kolkovski et al. (1997b, 2000) who reported an increase in ingestion rate of gilthead sea bream when Artemia nauplii were added. These results can be explained by the enhanced food searching activity by juvenile fish. Sorensen and Caprio (1997), in a review of chemoreception in fish, concluded that traditional models of fish behavior have assumed that olfaction is responsible for recognition of food odor from a distance and initiation of food searching. Without exception, all studies have found normal feeding behavior to be mediated by mixtures of L-amino acids, nucleotides, betaine, and krill (Hara 1982; Jones 1992; Kolkovski et al. 1997a, 2000). In the present study, feeding rate of diets supplemented with krill meal were significantly different from control diet and diet 2 (unsupplemented with krill meal). This demonstrates the effect of krill meal on increasing the feeding rate of diet was correlated with higher growth rates and final weight of Nile tilapia fry. Supplementing the plant protein diet with 1.5% to 6% krill meal improved growth and survival rate of Nile tilapia juveniles.

Fish fed diet supplemented with krill meal had higher growth performance than control diet. Generally, the addition of krill meal to feeds would seem to have widespread application in the study of plant protein feeds and dietary preferences, as its use may aid in increasing the flavor of plant protein feeds. In addition, supplementation of plant protein soybean basal diets with krill meal can improve dry diet attractability, increase fry growth, and can potentially decrease the duration of the weaning period. More research is needed to identify the specific fractions in the krill meal that are responsible for improve growth comparison of Nile tilapia fry fed fish meal (FM) and soybean basal diet supplemented with krill meal.

From the above results and the economic evaluation, we can conclude that, a diet containing soybean meal (SBM) protein and supplemented with krill meal can completely replace fish meal (FM) as a main source of protein in the diets for Nile tilapia fingerlings. Thus soybean meal (SBM) protein and supplemented with krill meal would be cost effective and maintaining adequate growth and production of Nile tilapia fingerlings in concrete tanks under the commercial condition.

6. Poultry-Products and Feather Meal

Poultry by–products and feather meal have long been articles of commerce and their methods of production have been discussed elsewhere. Pool ton by-products are used by the pet food industries and feather meat is a dietary ingredient in poultry rations.

They appear to be excellent portion and lipid sources containing 69% crude protein, 10-21% lipid and about 10%- 25%. The so-called hydrolyzed feather meal is really anisomery it is at best slightly hydrolyzed product produced by cooking feathers the presence of calcium hydroxide to increase its digestibility.

Little if any free amino acids are found in this product. Feather meal has been evaluated in both fresh and salt-water species. It contains about 80-85% protein and is relatively good sauce of sapphire containing amino acids. Whether these amino acids are completely available has not been demonstrated good result, have been reported when it used in catfish diets at the 15% level.

In an experiment of Gaber (1996) for production of Nile tilapia fingerling by total Substitution of fish meal by mixture of poultry-by product and feather meal as show in table (17).

From the above results and the economic evaluation, we can conclude that, a diet containing poultry by product +feather meal protein can completely replace fish meal (FM) as a main source of protein in diets for Nile tilapia fingerlings. Thus Poultry by product +feather meal would be cost effective and maintaining adequate growth and production of Nile tilapia fingerlings in concrete tanks under the commercial condition.

Table 17. Composition and proximate analysis of the fish meal basal diet and Poultry by product +feather meal fed to Nile tilapia, Oreochromis niloticus, juveniles

Ingredients	Diets	
	FM	PBP+Fe.
Fish meal (60%c.p.)	40.0	--
Poultry by product +feather meal	--	40.0
Water plant meal	35.0	35.0
Wheat bran	10.0	10.0
Soybean oil	2.0	2.0
Vit.& Min. Premix[1]	2.0	2.0
Amino acid mixture	1.0	1.0
Proximate analysis		
Dry matter	97.79	95.39
Crude protein	32.8	32.7
Crude fat	13.59	13.12
Ash	15.18	12.66
Crude fiber	8.56	8.1
NFE[2]	27.66	28.81
G.E. (Kcal/g)	4.6	4.6

[1]-Vitamin and mineral premix (according to Xie et al. (1997).
[2]-N.F.E., Nitrogen free extract calculated as 100- %(Moisture+ Crude protein +Crude fat +Ash +Crude fiber)

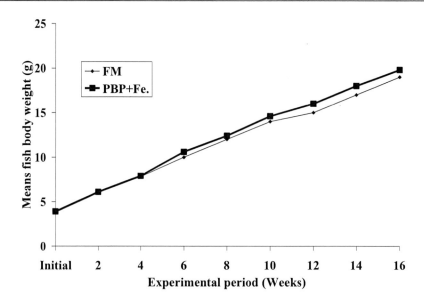

Figure 10. The effects of total replacement of fish meal with poultry by-product and feather meal on growth rate of Nile tilapia fry during 16 weeks of feeding.

Table 18. Economic information for Nile tilapia reared in concrete tanks for 24 weeks fed two protein sources

Item	Diets	
	FM	PBP +Fe
No. fish stocked/tank	180	180
No fish harvested	175	175
Harvested (kg/m^3)	1.76	1.96
Food used(kg/m^3)	2.77	2.96
Fingerling cost (LE)^2	36.0	36.0
Food cost ^{3}	9.7	4.4
Total cost (lE.)		
Net profit (LE)	17.3	22.0

7. Fish By-Product Protein Concentrates

It is wastes from processed fish and shrimp constitute a sizeable protein of commercial fishery produced. This discarded material is potentially valuable, for its contents of fish meal and other substances. Heu et al. (2003) reported that by-products of processing amount to 52% of shrimp processed. Also, wastes material from fish processing generally makes up 50% of the weight of the fish materials processed (Shih et al., 2003). These wastes material from fish processing may valuable sources of nutrients in fish diets.

Currently, the processing wastes material during the tilapia industry has considerable value as alternate animal protein sources due to 64% of the fish as waste lost during the various processing operation. (Maigualema and Gernat, 2003)

Many. authors reported that between 30% and 75% of dietary FM could be replaced by animal by-products (Tacon et at, 1983; Viola & Zohar 1984; Otubusin, 1987; El-Sayed 1988;

Fasakin et at) 2005). The differences among these results may have been related to protein source, quality and processing, fish species and size, experimental period and culture systems. However, plant proteins may partially replace the fish meal content of the diets, but total replacement of fish meal by plant protein sources usually depress growth and feed utilization (Gomes et al. 1995, El-Sayed 2006; Goda et al., 2007). Although a high proportion of phosphorus plant feedstuffs is present as phytic acid, which is not available to fish (Goda, 2007). Therefore, in formulating low- pollution diets, the protein and phosphorus levels, dietary protein /energy ratio and digestibility of feedstuffs must be carefully evaluated.

Fish meal is the main protein source in fish diets, and it is usually the individual feedstuff responsible for the major proportion of diet cost. Therefore, reduction in its inclusion level in practical fish diets is a priority (Pike, ci at 1990). The results of present study suggested that substitution of dietary FM with FPC at a level of 100% in Nile tilapia diet (Table 19 and Figure 11) has resulted in highest growth indices and feed utilization efficiency. Fish protein concentrates had been shown to improve the growth of Wile tilapia. These may be due that EPC has generally been reported to contain good quality protein compared to fish meal. The essential amino acid composition of the FPC source used in the present study (Table 20) was similar or higher than that of FM. Fish meals manufactured from fisheries by-catch and processing by-products from processing have been evaluated in diets for Coho salmon, OncorhynOhas kisutch (Rathbone et al.2001), rainbow trout, Oneoryhnchus mykiss (Hardy, et at 2005), and red drum, Scianops ocellatus (Li, et al. 2004; Whiteman & Gatlin III 2005). The results generally indicate that by-catch meals made from whole fish are suitable as total or partial replacements for commercial fish meal in aqua-feeds manufactured for carnivorous species. Also, co-dried fish silage prepared from fishery discards has received some, attentiQn as a tilapia feed ingredient (Fagbenroy et el 1994; Goddar et al.,2003; Goddard & Perret 2005). No previous reports were found on the evaluation of fish meals prepared from fishery by-catch or FPC in practical diets for tilapia, (Goddard et a1 2008), and the present study was planned to evaluate their potential use, in the diets of tilapia fingerlings.

Fish meal is recognized as the best single source of essential amino acids as a valuable attractant and is consequently the most widely used protein ingredient in aqua feeds (Hardy & Barrows 2002). The critical role of fish meal as a major ingredient in aqua-feeds and the issues associated with sustainability and future supplies are well recognized (New 1996; Nayloi *et a!* 2000) and there exists substantial research dedicated to identifying alternate protein ingredients. Although tilapia are able to utilize a wide range of ingredients froth both plant and animal sources (El-Sayed 1999), most commercial feeds used in semi intensive and intensive culture systems include fish meal and. soybean meal as the main, protein ingredients (Orachunwon et al 2001). Extending the range of raw materials from which a suitable fish meal is prepared may have the potential to reduce feeding costs. Fish meal content in commercial tilapia feeds is generally 15% or less of the total ingredients depending on the type of culture system used (New & Csavas 1995; Orachunwon *et* a!. 2001). Because global tilapia production currently exceeds 1.8 million tonnes each year and is predicted to increase (FAO 2006), it' may be assumed that tilapia grown .in semi-intensive and intensive systems consume a• significant proportion of the estimated 2:94 million tonnes of fish meal used. in aquaculture (Tacon, Hasan &. Subasinghe 2006). The conversion of fishery by-catch and fish waste processing into fish meal and fish protein concentrates faces a number of logistical and financial constraints. The raw materials are extremely variable and of a low value. In addition, regular supplies of sufficient quantity must be available to support the profitable operations of fish meal plants.

The values for weight gain and SGR in the present study were in agreement with or exceeded published data for tilapia fingerlings (Fontainhas-Fernandes, et al., 1999; Cho and Jo, 2002; Goda et al., 2007).

Table 19. Composition and proximate analysis of the fish meal diet and fish protein concentrate fed to juveniles Nile tilapia

Ingredients	Diets	
	FM	FPC
Fish meal (60% c. p.)	43.0	--
Fish protein conc. (80% c. p.)	--	31.0
Wheat bran	40.0	21.0
Corn meal	6.0	37.0
Soybean oil	7.0	7.0
Vit.& Min. Premix[1]	2.0	2.0
Dicalcium phosphate	2.0	2.0
Proximate analysis		
Dry matter	90.3	90.2
Crude protein	32.4	32.3
Crude fat	10.7	8.6
Ash	6.9	6.5
Crude fiber	6.1	5.7
NFE[2]		
G.E. (Kcal/g)	4.9	4.9

[1]-Vitamin and mineral premix (according to Xie et al. (1997).
[2]-N.F.E., Nitrogen free extract calculated as 100- %(Moisture+ Crude protein +Crude fat +Ash +Crude fiber)

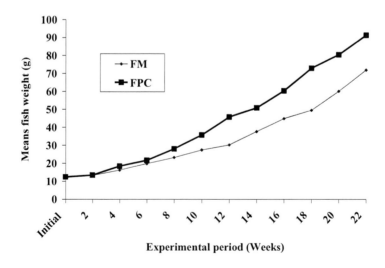

Figure 11. The effects of total replacement of fish meal with fish protein concentrate on growth rate of Nile tilapia fry during 20 weeks of feeding.

Table 20. Calculated amino acids in the two experimental diets and amino acid requirement for Nile tilapia fry (% of dry diet)

Diets		Requirement[a]	Amino acid
FPC	FM		
32.5	32.5		Crude protein (%)
		32	Essential amino acid
1.29	2.08		Arginine
1.17	0.84	1.6	Histidine
1.83	1.59	0.65	Isoleucine
2.86	2.73	1.18	Leucine
2.68	2.48	1.29	Lysine
1.05	0.95	1.95	Methionine [b]
1.81	1.42	1.02	Phenylalanine [b]
1.31	1.40	1.42	Threonine
0.49	0.37	1.43	tryptophan
1.96	1.81	0.32	Valine
0.78	0.38	1.06	Cystine
1.22	1.08		Tyrosine

a- According to Santiago and Lovell (1988)
b-The values of methionine and phenylalanine are the requirement in the presence of 2% cystine and 1% tyrosine of the diet, respectively

Table 21. Economic information for Nile tilapia reared in concrete tanks for 24 weeks fed two protein sources

Item	Diets	
	FM	FPC
No. fish stocked/tank	180	180
No fish harvested	175	175
Harvested (kg/m^3)	1.76	1.96
Food used(kg/m^3)	2.77	2.96
Fingerling cost (LE)^2	36.0	36.0
Food cost ^{3}	9.7	4.4
Total cost (lE.)		
Net profit (LE)	17.3	22.0

Preparations of Fish By-Product Protein Concentrate (FPC)

Fish by-product protein concentrate was prepared as method described by Reffai (1982). Fish by-product were minced by meat mincer then dried in an oven for 48 h (60°C). After drying, acetic acid (0.5%, concentration) was added to fish by product at a mixture 2:1, by weight respectively and dried in an oven for one hour at 60°C. then fish by-product dried again in an oven for 4 hours at 75°C. The dried fish by-product was soaked in ethanol-hexane azeotropic mixture (ethanol 21% and 79%) at ratio 1:2 respectively (w/v) for one hour. The extracted fish by-product was dried for 405 hours at 65 °C. Fish by product was packed in polyethylene bags and stored at -4 °C.

From the above results and the economic evaluation, we can conclude that, a diet containing fish by-product protein concentrate can completely replace fish meal (FM) as a main source of protein in diets for Nile tilapia fingerlings. Thus fish by-product protein concentrate would be cost effective and maintaining adequate growth and production of Nile tilapia fingerlings in concrete tanks under the commercial condition.

8. Brewers Dried Yeast (BDY)

Successful incorporation of yeast and bacterial SCP into fish diets allowed, replacement of 25-50% of fish meal component equivalent to an inclusion level of 15-30% by weight (Beck et al. 1979; Mahnken et al. 1980). Viola and Zohar (1984) also, showed that significant replacement of fish meal by single cell protein was achieved for hybrid tilapia (Oreochromis). Similarly, Windell et al. (1974) and Bergstrom (1979) reported best results of the inclusion of bacterial SCP in test diets for salmonids

Rumsey et al. (1991) found that the protein quality of Saccharomyces yeast (brewers or bakers yeast) in diets for rainbow trout is improved by a treatment to disrupt the cell walls, thereby making the protein more available. When yeast cell walls are disrupted, 50% of the protein in rainbow trout diets can be supplied by bakers yeast with equivalent growth and feed conversion ratio to a control diet with protein supplied by casein and gelatin. Although single cell proteins are potentially good protein sources, limited availability or high cost has so far limited their use in fish diets. This situation may change in the near future, however, as new processes are developed to recover single cell proteins from brewery waste, and upgrade its quality by air-classification to lower fiber content.

The investigation showed that up to 50% of soybean meal (SBM) in practical diets for Nile tilapia could be effectively replaced by Brewer dried yeast without a significant reduction in growth performance. This compares well with the work of Viola and Zohar (1984) in which 50% of fish meal protein in diets for hybrid tilapia (Oreochromis niloticus x Oreochromis aureus) was successfully replaced by bacterial SCP and in contrast, Kaushik and Luguet ('1980) and Spinelli et al. (1979) found that 80-100% replacement of fish meal with various SCP sources was possible in experimental diets for rainbow trout (Salmo gairdneri). Similar finding have also been reported for salmonids by Bergstrom (1979), Beck et al. (1979) and Matty and Smith (1978).

Gaber, 2009 (In press), reported that, the highest substitution of SBM amounting to 75-100% replacement with BDY resulted in substantial reduction in growth rates. Dabrowski (1982) reported similar effects with high SCP incorporation in juvenile rainbow trout and carp larvae (Cyprinus capio), respectively. Similarly poorer feed conversion and net protein utilization were both obtained for 75-100% replacement of SBM. These effects may have been due to several contribution factors, as dietary imbalance defined by deficiency of one or more limiting amino acids causing either a reduced efficiency in protein utilization (Atack, et al. 1979; Hilton, 1983), or alternatively leading to a depression in feeding intake. In the present study, fixed feeding rate of 6% during first 6 weeks then reduced to 4% till the end of the experiment lead to a noticeable reduction in gross consumption when tilapia fed diets with an increasing level of BDY.

It is suggested that the apparent differences would have resulted from an adlibitum feeding regime as suggested by Jobling (1983). The later author attributes this to compensatory effects elevating feed intake when ingredients of lower nutritional value are included.

Table 22. Composition and proximate analysis of the fish meal basal diet and 50% brewers dried yeast diet fed to Nile tilapia

Ingredients	Diets	
	SBM	BDY
Soybean meal (44% C.P.)	55	27.5
BDY (44% C.P.)	-	27.5
Wheat bran	16.7	16.7
Corn meal	18.0	18.0
Soybean oil	5.0	5.0
Vit.& Min. Premix[1]	1.8	1.8
Molasses	2.0	2.0
l-methionine	1.0	1.0
l-lysine	0.5	0.5
Proximate analysis		
Moisture	9.8	9.6
Crude protein	33.1	33.1
Crude fat	13.8	12.5
Ash	7.4	6.6
Crude fiber	3.9	3.8
NFE[2]	38.4	38.5
G.E. (Kcal/g)	4.8	4.8

[1]-Vitamin and mineral premix (according to Xie et al. (1997).
[2]-N.F.E., Nitrogen free extract calculated as 100- %(Moisture+ Crude protein +Crude fat +Ash +Crude fiber)

Table 23. Calculated level of amino acids in the two experimental diets and amino acid requirement for Nile tilapia fry

Diets		Requirement[a]	Amino acid
BDY	SBM		
33.2	33.1	32	Crude protein (%)
			Essential amino acid
1.65	1.85	1.6	Arginine
0.73	0.74	0.65	Histidine
1.31	1.28	1.18	Isoleucine
2.1	2.0	1.29	Leucine
2.22	2.0	1.95	Lysine
1.49	1.49	1.02	Methionine [b]
1.25	1.32	1.42	Phenylalanine [b]
1.19	1.08	1.43	Threonine
0.36	0.31	0.32	tryptophan
1.43	1.36	1.06	Valine

a- According to Santiago and Lovell (1988)
b-The values of methionine and phenylalanine are the requirement in the presence of 2% cystine and 1% tyrosine of the diet, respectively

The indication of dietary amino acid balance of the experimental diets (table 4) were confirmed by the amino acid requirements of tilapia (Santiago & Lovell 1988) with the amino acid profiles obtained for the test diets in this experiment. We found that phenylalanine was not considered to be limiting amino acid in the reference diet because several studies have shown that the presence of dietary tyrosine reduces the amount obtained for dietary phenylalanine levels necessary for maximum growth (Lovell, 1989). Arginine level showed a progressive reduction with graded BDY incorporation. For these reason must be established amino acid requirements levels for tilapia and most warm water fish are subject to variation due to many different techniques and approaches used. Supplemental L-methionine and L-lysine in fish diets have variable success. Viola and Lahava (1991) stated that addition L-lysine to a diet deficient in lysine for common carp, improved fish growth when compared with commercial diet. Also El-Saidy and Gaber (2002) reported that the additional of supplement L-methionine and L-lysine improved nutritional value of soybean in diets for Nile tilapia. In addition when BDY diets supplemented with cystine, arginine, lysine and tryptophan had beneficial effects on growth of fish (Grop et al. 1979; Beck et al. 1979; Murray and Marchant, 1986).

The composition of the experimental diets (Table 3) shows a relative increase in total nucleic acid content with increasing BDY protein up to maximum in diets 5. It is known that certain purin and pyrimidine bases do have growth depression properties, when fed at high levels to rats and chicks (Baker and Molitoris, 1974, Clifford and Story 1976). Similar effects were reported for rainbow trout by Tacon and Cooke (1980) and this might partly explain the reduction in the growth of tilapia fed high BDY diets. There is evidence that some dietary nucleotides affect the gut motility. Thus increasing the transit time of digestion thereby promoting an increased mineral uptake (Kim et al., 1968), and this may account for greater ash content observed in the present study.

The highest protein digestibility was observed in diet 1-2 and these relives were significantly higher than other diets (Table 5). Protein digestibility was generally ranging from 75% to 95% (Cho.and Kaushik 1990, NRC (1993). The ADC of protein from SBM was comparable with values found in other teleosts. The 25% BDY and 50% BDY diets has higher protein digestibility than the 75% BDY or 100% BDY diet 4 & 5) is likely due to the quality of raw material

Increasing levels of BDY in the diets has no effect on the over-all composition of the fish and apparent net protein utilization (NPU) was similar among all diet. Analysis of variance of NPU estimates showed no statistically significant differences. Increasing levels of BDY in the diet has no effect on the fish ability to assimilate protein.

The significantly better growth of fish fed diets 1 and 2 (25% BDY) might be due to the facts that the essential amino acid composition was well balanced and the levels of ANFs in diet 2 were below the level that induces toxic effects in Nile tilapia fingerlings. The results of the present study indicated that up to 25% of BDY in diet Nile tilapia can be included without affecting the growth and nutrient utilization. However, studies are underway to investigate the possibility of a higher dietary inclusion level of BDY after applying different processing techniques to detoxify or reduce the antinutrient contents.

From the above results and the economic evaluation, we can conclude that, a diet containing fish by-product protein concentrate can completely replace fish meal (FM) as a main source of protein in diets for Nile tilapia fingerlings. Thus fish by-product protein concentrate would be cost effective and maintaining adequate growth and production of Nile tilapia fingerlings in concrete tanks under the commercial condition.

Table 24. Economic information for Nile tilapia reared in concrete tanks for 24 weeks fed two protein sources

Item	Diets	
	FM	FPC
No. fish stocked/tank	180	180
No fish harvested	175	175
Harvested (kg/m^3)	1.76	1.96
Food used(kg/m^3)	2.77	2.96
Fingerling cost (LE)^2	36.0	36.0
Food cost ^{3}	9.7	4.4
Total cost (lE.)		
Net profit (LE)	17.3	22.0

9. Broad Bean Meal

In Egypt, the broad bean (Vicia faba) is commercially grown and tender pods are a popular vegetable particularly in new cultivated land. In recent years, intense research at the Agriculture Research Center, Giza, Egypt, has led to development of a high-yielding line. This study was carried out to determine the feasibility of using broad bean meal (BBM) protein as a replacement for FM protein in practical diets for Nile tilapia fry. These results suggest that BBM can replace 50% of the FM in diet for Nile tilapia fry, without adverse effects on fish performance.(Table 25 and figure).

Feed Intake, Growth Performance and Feed Utilization

Approximately 50% of fishmeal (FM) protein could be replaced by BBM, and result in growth rates of Nile tilapia comparable to a FM- based diet. These is the first time to our knowledge that broad bean meal has bean demonstrated to be effective in replacing FM in fish diets. There were no significant differences in weight gain and feed conversion (FCR) was significantly greater with control diet and diet contained 50% BBM which represented the highest level of substitution, which was not significantly different from the control.

This result is similar to previous reports utilizing, alfalfa leaf protein (Olvera-Novoa, Campos, Sabido & Martinez-Palacios, 1990), green gram (De Silva & Gunasekera 1989), Jack bean (Martinez-Palacious, Galvan Cruz, Overa Novoa, & Chavez, 1988) and winged bean (Hashim, Saat, & Wong, 1994) which showed no adverse effect on growth and feed utilization of tilapia fry when 30-37% of fishmeal protein was replaced.

This study clearly demonstrated that as much as 50% of the FM protein could be replaced by BBM protein in commercial production of Nile tilapia. Additional supplementation of methionine and lysine to these diets maintained a similar growth rate provided by the control diet. Jackson; Capper & Matty (1982) reported that, methionine and lysine are the most limiting amino acids in plant protein, frequently causing reduced growth. El-Saidy and Gaber (1997, 2002) reported that, soybean meal supplementation with 1% methionine or 1% methionine and 0.5% lysine can totally replace FM protein in Nile tilapia diets.

Table 25. Composition and proximate analysis of the fish meal diet and broad bean waste fed to Nile tilapia, juveniles

Ingredients	Diets	
	FM	BBM
Fish meal	50.0	25.0
Broad bean	--	37.0
Corn meal	35.5	17.8
Wheat bran	6.4	10.6
Fish oil	4.0	4.0
molasses	2.0	2.0
Vit.& Min. Premix[1]	2.0	2.0
vitamin C	0.1	0.1
L-methionine	--	1.0
L-lysine	--	0.5
Proximate analysis		
Moisture	9.4	9.6
Crude protein	31.2	31.2
Crude fat	11.8	11.7
Ash	9.1	9.8
Crude fiber	4.8	5.9
NFE[4]	33.7	31.8
G.E. (Kcal/g)	34.8	4.8
Cost (*Egyptian pound)[2]	3.0	2.3

[1]-Vitamin and mineral premix (according to Xie et al. (1997).
[2]-N.F.E., Nitrogen free extract calculated as 100- %(Moisture+ Crude protein +Crude fat +Ash +Crude fiber)

Table 26. Calculated level of amino acids in the two experimental diets (fish meal diet and broad bean waste) fed to Nile tilapia, and amino acid requirement for Nile tilapia fry (% of dry diet.)

Diets		Require-ment[a]	Amino acid
SBM	FM		
33.2	33.1		Crude protein (%)
			Essential amino acid
1.75	1.52	1.6	Arginine
0.89	0.97	0.65	Histidine
1.48	1.32	1.18	Isoleucine
1.9	1.81	1.29	Leucine
2.15	1.69	1.95	Lysine
1.62	0.88	1.02	Methionine [b]
1.4	1.22	1.42	Phenylalanine [b]
1.36	1.37	1.43	Threonine
0.36	0.31	0.32	tryptophan
1.49	1.36	1.06	Valine

a- According to Santiago and Lovell (1988)
b-The values of methionine and phenylalanine are the requirement in the presence of 2% cystine and 1% tyrosine of the diet, respectively

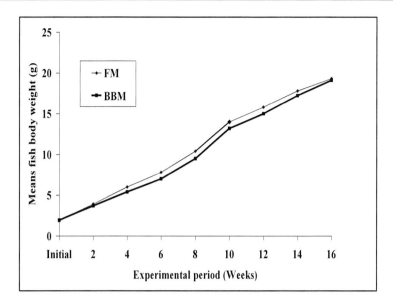

Figure 12. The effects of total replacement of fish meal with broad bean on growth rate of Nile tilapia fry during 16 weeks of feeding.

In the present study, total phenolic substances level in diet contained 50% BBM might be not sufficient enough for causing negative effect on growth and feed utilization in Nile tilapia. The adverse effect of tannic acid on tilapia has been studied by Al-Owafeir (1999) where found significant reduction in growth of tilapia fed diets containing as low as 0.27% tannic acid. On the other hand, Becker & Makkar (1999) reported that 2% inclusion of quebracho tannins (condensed tannins) were shown to be tolerated without any adverse effect on growth of common carp, whereas similar levels of hydrolysable tannis (tannic acid) reduced the feed acceptability after 4 weeks of feeding. A relatively higher concentration of total phenolics from mucuna beans (0.72%) in common carp diet has also been shown significantly reduce the growth performance and feed utilization in common carp (Hossain & Jauncey K. 1993 and Siddhuraju & Becker 2001).

The combination of aforementioned ANF with fiber could have not caused the significant decrease in PER and PPV in diet contained 50% BBM. Anderson, Jackson & Capper (1984) showed drastic reduction in growth, PER, FCR and whole body fat of Nile tilapia when more than 10% α-cellulose was included in the diets. Hilton, Atkinson & Slinger (1983) also reported a similar reduction in growth performance of rainbow trout when fed with a high fiber diet. Shiau, Yu, Hwa, Chen &Hsu (1988) observed a significant decrease in lipid content of Nile tilapia fed diets containing 6%, 10% and 14% of carboxymethyl cellulose (CMC) as compared to those fed 2% CMC. Hilton, Atkinson & Slinger (1983) mentioned that this phenomenon was associated with a decrease in gut passage time and diet digestibility. Furthermore, Shiau (1997) reported that importance of nutrient absorption dependence on the time for which nutrients are in contact with the absorptive epithelium. Dietary fiber apparently influences the movement of nutrients along the gastrointestinal tract and significantly affects nutrient absorption. Another exacerbating effect might be a change in enzyme activity, possibly through absorption or immobilization of enzymes by dietary fiber. It has also been shown that fiber can bind nutrients like fat, protein (Shah, Atallah, Mahoney

& Pellet 1982; Ward & Reichert, 1986) and minerals (Ward & Reichert, 1986), and reduce their bioavailability.

It is likely that the higher concentrations of crude lipid in the diets at increasing levels of BBM inclusion did not contribute towards dietary digestible energy, since Lipid in BBM will probably have contained a considerable amount of indigestible waxes and cutins. On the whole, higher level of broad bean meal inclusion would therefore have reduced the dietary energy available for protein synthesis and led to lower growth performance and nutrient utilization.

Apparent Digestibility Coefficients

The highest protein digestibility was observed in FM diet and diet contained 50% BBM (Gaber, 2006). Protein digestibility is generally high ranging from 75% to 95% (Cho & Kaushik 1990; NRC 1993). The ADC of protein from the fishmeal was comparable to values found in other teloest. The 50% BBM diet had a protein digestibility comparable FM diet. Grabner, & Hofer. (1985) and Siddhuraju & Becker (2001) they reported that the presence of higher concentration of total phenolic substance is known to reduce the protein digestibility and amino acid availability through phenolices-protein and/or phenolics-protein enzyme complex. Or the method of feces collection. However, Spyridakis, Metailer, Gabaudan & Raiza (1989), used the same method in study digestibility in European sea bass show that the nitrogen leaching was not significant. When the energy content of the feces and diets were determined, the results on energy digestibility support those for the different dietary components.

From the above discussion, it may be concluded that better growth of fish diet contained 50% BBM might be due to the facts that the essential amino acid composition was well balanced and the levels of ANFs in diet 3 were below the level that induces toxic effects in Nile tilapia. The results of the present study indicated that up to 50% of BBM in diet Nile tilapia can be included without affecting the growth and nutrient utilization. However, studies are underway to investigate the possibility of a higher dietary inclusion level of BBM after applying different processing techniques to detoxify or reduce the antinutrient contents.

Table 27. Economic information for Nile tilapia reared in concrete tanks for 24 weeks fed two protein sources

Item	Diets	
	FM	FPC
No. fish stocked/tank	180	180
No fish harvested	175	175
Harvested (kg/m^3)	1.76	1.96
Food used(kg/m^3)	2.77	2.96
Fingerling cost (LE)^2	36.0	36.0
Food cost ^{3}	9.7	4.4
Total cost (L E.)		
Net profit (LE)	17.3	22.0

From the above results and the economic evaluation, we can conclude that, a diet containing fish by-product protein concentrate can completely replace fish meal (FM) as a main source of protein in diets for Nile tilapia fingerlings. Thus fish by-product protein concentrate would be cost effective and maintaining adequate growth and production of Nile tilapia fingerlings in concrete tanks under the commercial condition.

10. Plant Protein Mixture (PPM)

The present study demonstrated the potential of PPM for inclusion in commercial Nile tilapia feeds, as well as being of immediate importance for feed production in Egypt. There is little information in the scientific literature concerning the use of PPM in Nile tilapia feeds, particularly the feeds produced under commercial conditions. Since tilapia production is in excess of 45% of the annual fish production in Egypt (CAPMS 1994), there is considerable benefit related to the replacement of part of the FM currently used in feeds.

A significant amount of research has been conducted on the replacement of FM with soybean meal as the protein source in feeds for Nile tilapia (El-Saidy & Gaber 1997, 2002b). They reported that soybean meal supplemented with 1% methionine only or 1% methionine plus 0.5% lysine can totally replace the FM in Nile tilapia diets.

In contrast to the soybean meal, there is less information available on the use of other oilseed by-product meals in feeds for Nile tilapia. Oilseed cakes and meals such as linseed meal (Hossain, Nahar & Kamal 1997; El-Saidy & Gaber 2001), sunflower meal (Sanz et aL 1994; El-Saidy & Gaber 2002a) and cottonseed meal (El-Sayed 1990; Middendorp & Huisman 1995; El-Saidy 1999; Mbahinzireki at al. 2001) have been used in Nile tilapia feeds.

The present study clearly demonstrated that as much as 100% of the FM protein could be replaced by PPM (El-Saidy and Gaber, 2003). Without, reducing the growth rates of Nile tilapia (Figure table 25). Many individual plant protein sources such as soybean meal, cottonseed meal, linseed meal and sunflower meal have been shown to be suitable replacements for more than 50% of FM in diets for tilapia. El-Saidy & Gaber (1997) found that the soybean meal plus methionine could completely replace FM for Nile tilapia without impairing the growth and feed efficiency. Linseed meal has been reported as a partial replacement for the PM in diets for Nile tilapia (El-Saidy & Gaber 2001) and common carp (Hasan, Macintosh & Jauncey 1997). The sunflower meal was also studied in rainbow trout (Sanz et. al. 1994). When the FM is replaced with a combination of ingredients, the interpretation of results is difficult because the interaction between the nutrients may be involved. In our study, 100% of PPM could replace the FM protein in Nile tilapia diets without any statistically significant effect on the growth rate; although in the group fed the 100% PPM diet a growth rate reduction at less than 15% of the control was recorded. The lower growth performance in the group fed 100% PPM is probably related to the lowest feed intake recorded in this group of fish.

The present study indicated that the feed intake values were similar to the values reported in El-Saidy & Gaber (1997). The feed intake values tend to be higher in fish fed diets 2, and this may be due to the relatively slow feeding habits of Nile tilapia (Shiau. Lin.Yu, Lin & Kwok 1990), Further, the colour of the diets (light brown) sometimes made them difficult to be seen in the aquarium bottoms and may have resulted in overfeeding. However, the feed supply must not be limiting in nutrition experiments, and overfeeding is more desirable than underfeeding (Tacon & Cowey 1985).

Table 28. Composition and proximate analysis of the fish meal basal diet and fish protein concentrate fed to Nile tilapia, juveniles

Ingredients	Diets	
	0% PPM	100% PPM
Fish meal (60% c. p.)	50.0	0.0
PPM. (40.6% c. p.)[1]	0.00	74.0
Corn meal	35.0	0.0
Wheat bran	6.5	16.2
Soybean oil	4.0	4.0
Molasses	2.0	2.0
Vit.& Min. Premix[2]	2.4	2.4
Vitamin C[3]	0.1	0.1
Amino acid		
Lysine	0.0	0.5
Methionine	0.0	0.5
Proximate analysis		
Moisture	7.84	8.28
Crude protein	33.64	33.2
Crude fat	14..61	14.77
Ash	7.94	9.56
Crude fiber	6.21	7.47
NFE[4]	29.73	26.72
G. E. (Kcal/g)	4.7	4.7
Cost kg^{-1}	2.65	1.4

1-PPM: Plant protein mixture, FM: Fish meal.
2-Premix supplied according to Xie, Cui and Liu (1997).
3-Phospitan C (Mg-L-ascorbyl-2-phosphate (Showa Den Ko.KK,, Tokyo, Japan.
4 N F E, Nitrogen free extract calculated as 100- %(moisture+ crude protein+ crude fat+ ash+ crude fiber)

Table 29. Calculated level of amino acids in the two experimental diets and amino acid requirement for Nile tilapia fry (% of dry diet)

Diets		Requirement[a]	Amino acid
(100% PPM)	(0% PPM)		
33.2	33.6	32	Crude protein (%)
			Essential amino acid
1.65	1.85	1.33	Arginine
0.73	0.74	0.54	Histidine
1.31	1.28	0.99	Isoleucine
2.1	2.0	1.09	Leucine
2.22	2.0	1.63	Lysine
1.49	1.49	1.02	Methionine [b]
1.25	1.32	1.82	Phenylalanine [b]
1.19	1.08	1.15	Threonine
0.36	0.31	0.32	Tryptophan
1.43	1.36	1.09	Valine

a- According to Santiago and Lovell (1988)
b-The values of methionine and phenylalanine are the requirement in the presence of 2% cystine and 1% tyrosine of the diet, respectively

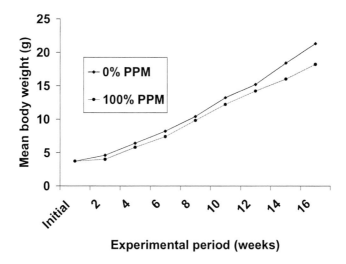

Figure 13. The effects of total replacement of fish meal with plant protein meal (PPM) growth rate of Nile tilapia fry during 16 weeks of feeding.

The apparent protein digestibility was not significantly different for up to 100% replacement of the FM protein diet; even though weight gain was lower in the 100% replacement. This result suggests that the apparent protein digestibility of PPM is slightly lower than that of FM for Nile tilapia. Although FM and PPM were not the only- protein sources in the diets, the other proteins (e.g. wheat bran and corn meal) were incorporated in the diets at different levels. Hossain et al. (1997) reported that the protein digestibility of linseed meal for Indian major carp, Labio rohita, was lower (81.55%) than the value (85.8%) for common carp (Hossain and Jauncy, 1989), but similar to the value (81.44%) reported for tilapia (Hossain, Nahar, Kamal & Islam 1992). In contrast, Smith, Peterson & Allred (1980) reported a lower ADO protein value of 67.7% for linseed meal in rainbow trout. The presence of mucilage in linseed meal may be a possible reason for the lower protein digestibility in PPM in our study (Mani. Nikolaiczuk & Maw 1949; El-Saidy & Caber 2001). Slominski. Kienzle, Jiang, Campbell, Pickard & Rakow (1999) reported the supplementation of yellow mustard (Sinapis alba) diets with mucilage (soluble fibre) depolymerizing enzymes and reduced digesta viscosity in broiler chickens.

The present study revealed that the nutritional utilization indices of the protein (PER, PPV and ABV) in fish led the test diets were not significantly different this means that the two diets tested have good protein quality and palatability.

Most of the works reviewed have evaluated PM replacements in tiiapia feeds from biological or nutritional viewpoints. Little attention has been paid to the economic analysis of these protein sources. Only a few studies have been conducted in this subject and these have indicated that those unconventional protein sources were more economical than the FM because of their local availability at low prices.

The cost benefit analyses of the present study clearly indicated that PPM diets are better protein sources for Nile tilapia than FM. Similar results were reported by other workers. The economic evaluation of cottonseed meal (El-Sayed 1990), corn gluten feed and meal (Wu, Rasati, Sessa & Brown 1995), and sunflower meal (El-Saidy & Caber 2002a) as single protein sources for Nile tilapia, and brewery waste (Oduro-Boateng & Bart-Plange 1988) for Tilapia indicated that these sources were economically superior to the FM, even at the total replacement levels.

In conclusion, our findings suggest that PPM could replace 100% of FM protein in the diets for juvenile Nile tilapia without adverse growth effects for up to 16 weeks. This observation is supported by the ADC protein and ADC energy values for diets containing mixtures of plant protein meals. In addition, these plant protein sources are locally available at much lower prices than imported FM. Further research will be required to determine the feasibility of using PPM composed of different combinations of ingredients and to examine the effects of PPM use in diets on large sizes of fish under the field conditions.

11. Yucca Schidigera

In an experiment carried by Gaber (2006) to study the effects of yucca on the replacement of fish meal (FM) protein with solvent-extracted plant by-products as soybean meal, cottonseed meal, sunflower meal, and linseed meal was tested in diets for juvenile Nile tilapia, Oreochromis niloticus.

Fish averaging 14.2 ± 2.9 g were divided into 18 groups and fed for 6 months on pelleted feed containing each of the plant protein meal supplemented with Yucca schidigera powder extract at 750 mg/kg. Methionine (1%) and lysine (0.5%) were added to each diet except the control diet (FM), while diet FM + Y was supplemented with yucca only.

Three groups of fish were fed each of six isonitrogenous (25% crude protein) and isocaloric (4.3 kcal/g) diets replacing 100% of FM protein (Table 22) and performance compared against a nutritionally balanced control and a commercial tilapia feed. After 6 mo of feeding, the fish fed plant protein diets supplemented with yucca exhibited growth performance not differing significantly from that of fish fed FMC + Y, while differing significantly from the control FMC and diet linseed meal (LSM). The highest apparent protein digestibility coefficient was observed for diets treated with yucca, which was significantly higher than that observed for the control diet FMC. No significant differences were found in whole body moisture of fish fed different experimental diets. An increase in the whole-body protein content was observed in fish fed diets supplemented with yucca, which was significantly different from that of the diet FMC.

As intensive aquaculture continues to expand, so does the requirement for high-quality protein sources (Hardy 1996). Fish meal (FM) is an expensive component of commercial tilapia feeds, and numerous studies have investigated the potential use of alternative protein sources in tilapia feed. Most published research on the use of plant proteins in tilapia feeds has focused on the inclusion of soybeanmeal (SBM; El-Saidy and Gaber 2002b; Wilson et al. 2004), cottonseed meal (CSM; Mbahinzireki et al. 2001; El-Saidy and Gaber 2004a), sunflower meal (SFM; El-Saidy and Gaber 2002a; Maina et al. 2003), and linseed meal (LSM; El-Saidy and Gaber 2001). Less attention has been paid to improving the metabolic efficiency and feed utilization of these plant-protein-based feed. Usually, plant proteins have some negative qualities such as amino acid imbalances and antinutritional factors. Natural and biologically active plant products could be used as synthetic growth stimulants and reduce the accumulation of nitrogenous wastes that limit production intensity in aquaculture (Francis et al. 2002). The use of Yucca schidigera extract has shown promise in the control of ammonia accumulation acting as a urease inhibitor, increasing bacterial use of ammonia (Jacques and Bastien 1989) and directly binding with ammonia (Headon and Dawson 1990). Also, yucca has shown improved performance and increased feed efficiency when Y. schidigera extract was incorporated into feeds for poultry (Johnston et al. 1982) and for Nile tilapia (El-Saidy and Gaber 2004b). Johnston et al. (1982) suggested that surface components of Y. schidigera extract could aid in nutrient absorption. Also, the use of Y. schidigera extract in poultry feed is a good alternative to maintaining metabolic and environmental ammonia levels within acceptable limits to improve productivity parameters (Al-Bar et al. 1993). The aim of this study was to assess the total replacement of FM with different sources of protein. This study demonstrated the potential of four different plant proteins for inclusion in commercial Nile tilapia feeds, an issue of immediate importance for feed production in Egypt. Because tilapia production is in excess of 45% of the annual fish production in Egypt (CAPMS 1994), there is considerable benefit related to the total replacement of FM, which is currently used in tilapia feeds. This study clearly demonstrated that as much as 100% of the FM protein could be replaced by SBM, CSM, SFM, and LSM supplemented with Y. schidigera in commercial production of tilapia. Additional supplementation of methionine and lysine to these diets maintained a similar growth rate as the rate provided by the FMC + Y diet. Jackson et al. (1982) reported that methionine and lysine are the most limiting amino acids in plant protein, frequently causing reduced growth. El-Saidy and Gaber (1997, 2002b) reported that SBM supplemented with 1% methionine or 1% methionine and 0.5% lysine can totally replace FM in Nile tilapia diets. Addition of Y. schidigera to formulated Nile tilapia diets has shown to reduce ammonia accumulation and increase fish growth. El-Saidy and Gaber (2004b) reported that adding yucca to Nile tilapia diets, especially in intensive culture systems, at 750 mg/kg diet could reduce ammonia and nitrite in the water and act as a growth stimulant to improve growth performance and feed utilization. In this study, addition of yucca to tilapia feeds caused a significant improvement in final average weight compared to the FMC diet. When SBM, CSM, and SFM were used with yucca in tilapia diets, an FBW increment of 68.3, 72.4, and 64.5% over the FMC diet was observed, respectively, while LSM produced a 51.5% increment over the FMC diet. The experiment showed not only an increase in body weight but also an improvement in FCR of fish fed diets supplemented with Y. schidigera extracts. Al-Bar et al. (1993) and Essa and Abd El-Hamied (2003) recorded similar results, and they reported an increase in body weight that was attributed to the natural components of Y. schidigera. Feed utilization parameters (Table 4) were significantly higher ($P \# 0.05$) for groups of fish fed diets supplemented with Y. schidigera. This is in agreement with results presented by Sarkar (1999), who used De-oderase (extracted from Y. schidigera) as feed supplement for Labeo rohita, Catla catla, and Cirrhinus mrigala in fishponds. In addition, the effectiveness of the feed to encourage growth, indicated by the FER, was greater in the diets supplemented with Y. schidigera than in the FMC diet. Diets supplemented with Y. schidigera (FMC + Y, SBM, CSM, and SFM) were consumed in high quantity, while the FMC diet had a significantly lower FI. This may be attributed to a lower passage of macromolecules of the FMC diet across the fish intestine compared to the diets supplemented with yucca. Onning et al. (1996) reported that saponins (present in yucca) significantly increased passage of macromolecules across rat intestine in vitro. Protein digestibility coefficients ranged from 70.5% (FMC) to 78.4% (FMC + Y). These results are consistent with the range of protein digestibility coefficients reported by other researchers (El-Saidy and Gaber 2002a, 2002b).

Table 30. Feed formulation and proximate composition of Yucca schidigera diets used in the study

Ingredients %	Diets					
	FMC	FMC+Y	SBM	CSM	LSM	SFM
Fish meal	30.0	30.0				
Soybean meal.			40.0			
Cottonseed meal				45.5		
Linseed meal					50.0	
Sunflower meal						50.0
Wheat bran	27.0	27.0	26.0	26.0	25.0	25.0
Corn meal	36.0	36.0	25.5	20.5	16.5	16.5
Fish oil	2.0	2.0	2.0	2.0	2.0	2.0
Vita. & min. premix[1]	2.5	2.5	2.5	2.5	2.5	2.5
Molasses	2.0	2.0	2.0	2.0	2.0	2.0
Yucca shidigera		0.075	0.075	0.075	0.075	0.075
Amino acid supplement						
L-methionine			1.0	1.0	1.0	1.0
L-lysine			0.5	0.5	0.5	0.5
Cr_2O_3	0.5	0.5	0.5	0.5	0.5	0.5
Total	100	100	100	100	100	100
Proximate analysis (%)						
Moisture	8.07	8.87	9.59	9.16	9.46	9.06
Crude protein	25.31	25.31	25.31	25.31	25.31	25.31
Crude fat	11.37	11.06	10.93	10.93	11.51	11.13
Crude ash	10.03	9.43	10.6	10.44	10.47	9.56
Crude fiber	6.1	6.2	6.32	7.52	7.34	7.45
NFE[2]	38.22	39.13	37.25	36.64	35.91	37.49
G.E.(kcal.g^{-1})	4.3	4.3	4.3	4.3	4.3	4.3
Cost (Kg^{-1})[3]	2.5	2.7	1.7	1.6	1.5	1.5

1. Premix supplied according to Xie et al. (1997)
2. NFE (Nitrogen free extract) =100-(%moisture+%C.protein+%C.fat+%C.fiber+ash)
3. Price in Egyptian pound: 1.00=US$0.14 based on 2005 exchange prices.

The source of the dietary protein appeared to be relatively unimportant when considering the high digestibility coefficients recorded for all the diets evaluated. El-Saidy and Gaber (2004) reported a similar outcome in a digestibility study involving yucca in diets for Nile tilapia and concluded that the choice of protein content, amino acid composition, and yucca supplement leads to increased protein digestibility coefficients. These may be because of surface properties of components of the Y. schidigera extract, which may aid nutrient absorption and lead to increased protein digestibility as previously reported by Johnston et al. (1982). Francis et al. (2002) have shown that the Quillaja saponin mixture, when presented in the diet of carp, stimulated some gut and liver enzymes. In conclusion, the use of different sources of plant protein together with Y. schidigera in Nile tilapia feeds appears to increase FI to similar levels recorded with high-quality FM diets. Supplementing Y. schidigera would seem to have a wide application in the study of plant protein feeds and their performance. This may aid in the identification of specific feeding stimulants for the fish farming industry. Also, supplementing plant-protein-based diets with Y. schidigera can improve fingerling growth and can identically decrease the duration of their grow-out period. More research is needed to identify the specific fraction in Y. schidigera that improved feeding utilization and growth of tilapia fingerlings under this culture conditions.

Table 31. Estimated amino acid composition of diets (g.100 g^{-1}diet)

Indispensable amino acid[1]	Required[2]	Diets					
		FMC	FMC+Y	SBM	CSM	LSM	SFM
Arginine	1.33	1.52	1.52	1.58	2.3	1.76	2.55
Histidine	0.54	0.64	0.64	0.53	0.65	0.55	0.77
Isoleucine	0.99	1.13	1.13	0.96	0.91	0.79	1.32
Leucine	1.09	2.03	2.03	1.86	1.59	1.43	2.38
Lysine	1.63	1.67	1.67	1.67	1.47	1.31	1.66
Methionine	1.02	0.64	0.64	1.27	1.34	1.25	1.42
Phenylalanine	1.82	1.05	1.05	1.11	1.25	.92	0.81
Threonine	1.15	1.0	1.0	0.87	0.81	.75	0.79
Tryptophane	0.32	0.27	0.27	0.35	0.38	0.35	0.4
Valine	1.09	0.81	0.81	0.11	1.16	0.99	1.57
Cystine		0.34	0.34	0.45	0.48	0.35	0.5
Tyrosine		0.51	0.51	0.74	0.64	0.57	0.86

1. Data obtained from National Research Council (1993) and Hasan et al (1997).
2. From Santiago and Lovell (1988).

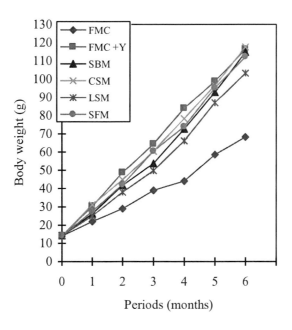

Figure 14. Changes in mean body weight of Nile tilapia fingerlings fed fish meal (FM) diets and plant protein basal diets supplemented with Yucca shidigera for six months. Values are means of triplicate groups.

From the above results and the economic evaluation, we can conclude that, a diet containing fish by-product protein concentrate can completely replace fish meal (FM) as a main source of protein in diets for Nile tilapia fingerlings. Thus fish by-product protein concentrate would be cost effective and maintaining adequate growth and production of Nile tilapia fingerlings in concrete tanks under the commercial condition.

COMMON CARP

Figure 15. Common carp (Syprinus carpio).

Physiology

Common carp can grow to a maximum length of 5 feet (1.5 meters), a maximum weight of over 80 lb (37.3 kg), and an oldest recorded age of at least 65 years There was one carp that was caught that weighed 88.6 pounds (40.1 kilograms) (Panek,. 1987) The wild, non-domesticated forms tend to be much less stocky at around 20% - 33% the maximum size.

Habitat

Although they are very tolerant of most conditions, common carp prefer large bodies of slow or standing water and soft, vegetative sediments. A schooling fish, they prefer to be in groups of 5 or more. They natively live in a temperate climate in fresh or brackish water with a 7.0 - 7.5 pH, a water hardness of 10.0 - 15.0 dGH, and an ideal temperature range of 37.4 - 75.2 °F (3 - 24 °C.

Diet

The common carp and its variants are omnivorous and will eat almost anything encountered. The common carp is happy to eat a vegetarian diet of water plants, but will also consume insects, crustaceans (including zooplankton), and even dead fish if the opportunity arises.

Reproduction

An egg-layer, a typical adult fish can lay 300,000 eggs in a single spawning. Research shows that carp can spawn multiple times in a season in some areas. The young are preyed upon by other predatorily fish such as the northern pike and largemouth bass.

Common carp have been introduced, often illegally, into many countries. Due to their habit of grubbing through bottom sediments for food and alteration of their environment, they may destroy, uproot and disturb submerged vegetation causing serious damage to native duck and fish populations.

Efforts to non-chemically eradicate a small colony from Tasmania's Lake Crescent have been successful, however the long-term, expensive and intensive undertaking is an example of the both the possibility and difficulty of safely removing the species once it is established. In Australia there is enormous anecdotal and mounting scientific evidence that introduced carp are the cause of permanent turbidity and loss of submergent vegetation in the Murray-Darling river system, with severe consequences for river ecosystems, water quality and native fish species In Victoria, Australia, Common carp has been declared as noxious fish species therefore there is no restriction on the quantity that a fisher can take.[5] In South Australia, it is an offence for this species to be released back to the wild. An Australian company churns common carp into plant fertilizer.

Common carp were brought to Egypt in 1952. In the late 1955 they were distributed widely throughout the country by the government as a food fish. However, common carp are no longer prized as a food fish in Egypt.

II. UNCONVENTIONAL INGREDIENTS FOR COMMON CARP

A. Cluster Bean Cyamposis Tetagonoloba

In Egypt, the cluster bean Cyamposis tetagonoloba, known locally as "Guar," is widely grown. The young leaves and tender pods make Guar a popular vegetable; particularly in newly cultivated land. In recent years, through intense research to propagate its growth at the Agriculture Research Center, Giza, Egypt, a high yielding line of cluster beans has been successfully produced.

This study was conducted to determine the feasibility of cluster bean seed meal protein as a possible replacement for fish meal protein in practical diets for common carp fingerlings. The cluster bean is a moderate and subtropical legume with seeds that are rich in protein. The protein composition and amino acid profile of cluster bean is similar to soybean, which is high in lysine but deficient in the sulphur amino acids (Table 32). Although antinutritional factors such as trypsin inhibitor are present in the cluster bean, these are easily inactivated by heat (boiling) treatment, without adverse effect on its protein quality as described by Liener (1980). In the current study approximately 50% of fishmeal (FM) protein could be replaced by CBM, and result in growth rates in common carp comparable to a FM-based diet. This is the first time to our knowledge that cluster bean meal has bean demonstrated to be effective in replacing FM in fish diets although other authors have demonstrated fishmeal replacement potential for a variety of seed meals. Hossain et al. (2001) indicated that up to 12% of untreated sesbania seed meal can be included in diet of common carp without adverse effect on growth performances.

Pongmaneerat et al. (1993) reported that soybean meal supplemented with methionine can totally replace fishmeal in the diet of common carp without adverse effect on carp growth. Chou et al. (1994) reported that up to 50% of winged bean seed meal protein can be successfully used as a partial replacement for fishmeal protein in practical diets of red tilapia fry without adverse effects on growth and feed utilization. El-Saidy and Gaber (2001) reported that up to 75% of fishmeal protein can be replaced by linseed meal protein in fingerling Nile tilapia diets. Also, El-Saidy and Gaber (2002b) demonstrated that dehulled sunflower meal can partially replace fishmeal as a main source of protein in Nile tilapia diets at not more than 50% without any adverse effects on its growth and feed utilization parameters. Based on feed intake, the palatability of control diet and diets 2 appeared to be comparable to each other. The highest level of substitution, which was not significantly different from the control in growth performance, was 50% CBM. This level of substitution is similar to previous reports utilizing alfalfa leaf protein (Olvera-Novoa et al. 1990), green gram (De Silva and Gunasekera 1989), Jack bean (Martines-Palacious et al. 1988), and winged bean (Hashim et al. 1994) which showed no adverse effect on growth and feed utilization by tilapia fry when 30–37% of fish meal protein was replaced. The ADC of protein from the fish meal was comparable to values found in other teleosts ranging from 75% to 95% (Cho and Kaushik 1990; NRC 1993). The 50% CBM diets had a higher protein digestibility than the control diet. However, Spyridakis et al. (1989) used the same fecal collection method in a digestibility study in European sea bass and showed that the nitrogen leaching was El-Saidy et al. insignificant. When comparing digestibility of two diets (present experiment) with digestibility of soybean meal (El-Saidy and Gaber 2002a), digestibility of soybean meal is higher, possibly due to the deficient methionine content of cluster bean meal compared to soybean meal (Table 33). Kaushik and Dabrowski (1983) reported that amino acid availability has been found to be highly reflective of digestibility of protein. The methionine content of diet 5 is 0.43 g/100 g diet or 1.3 g/100 g protein. This is lower than the requirement reported by Nose (1979) of 0.64 g/100 g diet for methionine. However, the fish used by Nose (1979) weighed 62 mg at the beginning of their experiment. Decreasing nutritional requirements with size are relatively common in vertebrates, thus differences in initial weight may explain the differences in values. Results of this experiment indicate that increasing CBM to levels more than 50% in common carp diet decreased growth rate. This may be related to methionine deficiency (Table 33) when compared to the EAA requirement for carp (Nose 1979). As in other animals, arginine, histidine, isoleucine, lysine, methionine, phenylalanine, threonine, tryptophan, and valine were considered essential amino acids (Tacon and Cowey 1985).

The nonessential amino acid cystine and tyrosine can only be synthesized by the fish from methionine, and phenylalanine requirement of the fish will partially depend on the cystine and tyrosine content of the diet (Lovell 1989). The PER obtained for diet 2 was not significantly different from the control diet (diet FM). Studies by Murai et al. (1982) showed that supplementation of deficient protein with adequate amount of EAA can indeed improve the use of protein sources, known to have recognized amino acid imbalance. However, Chance et al. (1966), Harper et al. (1970), Hughes and Rumsey (1983), and Robinson et al. (1984) have all pointed out the negative influence of diets with an imbalance of amino acids on the growth and nutrient utilization of the diet by fish. The poor metabolic utilization of CBM compared to fish meal could be due to a decrease in digestible protein (diet 5). The results of the present work demonstrate that common carp fed a diet containing 50% CBM protein had comparable growth performances to fish fed a fishmeal based diet. Good digestive utilization of protein and fat from cluster bean meal (CBM), when incorporated at 50% of total replacement of fishmeal provided good common carp growth. Furthermore, CBM could reduce the cost of diets, thus increasing commercial profitability.

Table 32. Composition and proximate analysis of the fish meal diet and 50% cluster bean fed for common carp

Ingredients	Diets	
	FM	50% CBM
Fish meal (59% c. p.)	55.0	27.5
Cluster bean (39% c.p.)	0.00	41.25
Soybean meal (44% c. p.)	00.0	0.0
Wheat bran	8.0	8.0
Soybean oil	6.0	6.0
Animal fat	2.0	2.0
Vit.& Min. Premix[1]	2.4	2.4
Vitamin C^2	0.1	0.1
Molasses	2.0	2.0
Corn starch	24.5	10.75
Proximate analysis		
Moisture	8.34	8.77
Crude protein	33.1	33.1
Crude fat	15..94	17.82
Ash	8.35	6.42
Crude fiber	5.23	6.05
NFE^3	29.04	27.84
G. E. (Kcal/g)	4.8	4.8
Cost LE kg^{-1}	2.65	1.4

1-Premix supplied according to Hossain et al.(2001).
2 Phospitan C (Mg-L-ascorbyl-2-phosphate (Showa Den Ko.KK,, Tokyo, Japan.
3 N F E, Nitrogen free extract calculated as 100- %(moisture+ crude protein+ crude fat+ ash+ crude fiber)

Table 33. Calculated level of amino acids in fish meal diet and 50% cluster bean and amino acid requirement for common carp (% of dry diet)

Diets		Requirement[a]	Amino acid
(100% PPM)	(0% PPM)		
33.2	33.6	32	Crude protein (%)
			Essential amino acid
1.98	2.08	1.52	Arginine
0.94	1.05	0.56	Histidine
1.53	1.49	0.92	Isoleucine
2.54	2.46	1.64	Leucine
2.37	2.31	2.12	Lysine
0.64	0.87	0.64	Methionine [b]
1.5	1.33	1.16	Phenylalanine [b]
1.32	1.4	1.32	Threonine
1.7	1.69	1.16	Valine

a- - according to Nose (1979).
b-The values of methionine and phenylalanine are the requirement in the presence of 2% cystine and 1% tyrosine of the diet, respectively

2. Evaluation of Linseed Meal as a Dietary Protein Source for Common Carp, Cyprinus Carpio L

A five-month feeding trial was conducted for common carp (Cyprinus carpio L.) with an average initial weight of 110.4 ± 3.45 g/fish. Fish were divided into 18 groups and fed on pelleted feed containing each of the plant protein meal supplemented with Yucca schidigera powder extract at 250 mgkg^{-1}. Methionine (1%) and lysine (0.5%) were added to each diet except the control diet (FMC) which was supplemented with yucca only to examine the effect of replacement soybean with linseed meal on growth, feed conversion ratio (FCR) protein efficiency ratio (PER) and body composition of common carp. Three groups of fish were fed each of six isonitrogenous (25.2% C.P) and isocaloric (4.3 kcal g^{-1}) and the soybean meal (SBM) was replaced with linseed meal (LSP) at a level 0%, 25%, 50%, 75% and 100%. The experiment was studied in three concrete ponds area of each was (40m^3) (4 x 10 x 1.0 m, width, long,, and height) and divided into six units (6m^2) by nets. The experimental fish was stocked at a rate 1 carp/m^2 (6 carp/unit). After five months of feeding, there were a significant differences in the final individual weights, weight gains, specific growth rates (SGR), feed conversion ratios (FCR), Feed efficiency ratio (FER), protein efficiency ratios (PER), and feed intake among fish groups ($P \leq 0.05$). Common carp fed the diet containing 50% (LSM) exhibited comparable growth to those fed a fishmeal-and soybean meal based diet. Digestibility of protein, energy and lipid decreased with increasing levels of (LSM) above 50% of total replacement soybean meal in the diet. Incorporation of (LSM) in diets not significantly affected the dry matter, protein, and ash of whole fish body. These results suggest that LSM can replace 50% of the soybean meal in diet for Common carp.

Dietary protein above 24% (Shiau et al., 1987, 1990) satisfied the growth requirement of common carp (Cowey & Cho, 1991; and Kaushik, 1995), the actual crude fat (6.34 %) and the actual energy level (4.14 kcalg^{-1}) were similar to the respective level of commercial carp feed in Egypt. The highest crude fiber content in diet (E) was because of its highest inclusion of LSM that contained 7.1%crude fiber (Table 2). In addition the fish were fed with feeding level at 2% of body weight according to Xie et al (1997) reported that the proportion of gross energy intake channeled into heat production showed the lowest value at an intermediate ration (2%) and was higher at lower or higher ration levels. Also, they concluded that, the decline in feeding efficiencies at higher ration could be caused by: (1) a decrease in apparent digestibility (2) an increase in the proportion of gross energy intake lost in excretory energy and (3) an increase in the proportion of gross energy intake spent in heat production.

As the protein in our diets was largely from plant, protein sources the methionine lysine and threonine content in both soybean and linseed was deficient. The supplementation diets with constant levels of methionine 0.5% lysine 0.5% and threonine 0.5% to covers the requirements recommended by Nose (1979) for common carp. Despite supplementation diets with methionine, lysine and threonine, the energy weight gain was 594.5% for carp and SGR 1.01 for carp. This can be attributed to those carp full utilized efficiently synthetic amino acids. Also the sparing effects cystine at low level can not be converted into methionine. However, at high level of cystine lowered the use of methionine for protein synthesis (Liou 1989).

In addition, other studies indicate that adequate treatments of synthetic amino acids incorporated into diets improves utilization in several fish including common carp (Murai *et al.*, 1981; 1982 & 1983). The studies by Murai et al., 1984) show that supplementation of deficient protein with adequate amounts of IAA can indeed improve the use of protein sources known to have recognize amino acid imbalances. Viola et al. (1992) observed that supplementation of a diet based on plant protein, with lysine improved efficiency of protein utilization and also resulted in a decrease in nitrogenous losses.

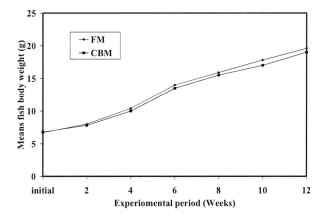

Figure 16. The effects of fish meal and cluster bean meal (CBM) on growth rate of Nile tilapia fry during 12 weeks of feeding.

Table 34. Composition and proximate analysis of soybean meal and linseed meal fed for common carp

Ingredients	Diets	
	SBM	50% LSM
Soybean (44%C.P.)[1]	40.0	30.0
Linseed (38%C.P.)	0.0	12.5
Wheat bran	26.0	28.0
Corn meal	25.5	21.0
Sunflower oil	5.0	5.0
Sodium diphosphate	2.0	2.0
Vit.. & minerals. premix [2]	1.0	1.0
L-methionine [3]	0.5	0.5
L-Lysine [3]	0.5	0.5
Molasses	2.0	2.0
Yucca schidigera(mg/kg)	0.750	0.750
Total	100	100
Chemical composition (%):		
Moisture	9.79	9.47
Crude protein	25.6	25.6
Crude fat	9.29	9.34
Crude fiber	6.31	6.62
Ash	7.25	6.71
NFE[4]	41.76	41.26
Gross Energy (kcal/g)	4.3	4.3

1-Soybean meal: heated at 110°C for sex h and solvent (hexane extracted)
2-Premix supplied according to Hossain et al.(2001).
3-L.methionine and L.lysine: dietary methionine and L.lysine (Commercial products)
4-Nitrogen free extract (NFE) = (moisture + crude protein +crude fiber +ash+ crude fat) – 100

Methionine, tryptophane, arginine and lysine are the amino acids likely to be adversely affected by the various processing techniques applied to oilseed meal and fish diets (Liener, 1980). The shortfall in methionin, lysine and threonine was treated by the inclusion of only 0.5 % of three amino acids as commercial synthetic amino acids. In Egypt, the feed do not incorporate FM in tilapia feed because of to its high cost and fierce price competition. Although our diets (2-6) had no inclusion of FM. our fish, did not grow slower than the fish fed on diet (1) with a high inclusion of FM .Total replacement of fish meal in a diet for carp with soybean meal as the major protein source necessitated adequate supplementation with L. methionine (Pongmaneerat et al.1993) in order to maintain optimal growth and feed efficiency (Table 34 and Figure 16).

Replacement of soybean meal with linseed meal has had variable success. In those study in which growth is reduced, several hypotheses have been suggested to explain the results: 1) suboptimal amino acid balance (NRC 1993); 2) inadequate levels of phosphorus in linseed meal (NRC 1993); 3) Presence of antinutritional factors (including trypsin inhibitors) (Liener 1980); and 4) inadequate levels of energy in LSM (El-Saidy & Gaber 2001). LSM has one of the best amino acid after SBM and the composition meet the requirements of amino acid for Nile tilapia and common carp. Calculated amino acid (Table 35) indicated that that all diets met the amino acid requirements of common carp (Nose, 1979).); however, the biological value of amino acid from linseed meal may be less than indicated. El-Saidy & Gaber (2001) stated that methionine availability may be reduced when linseed meal comprises large percentage (>50%) of diet. The use of plant-derived materials as oil seed cake and leaf meal is limited by the presence of a wide-variety of antinutritional substances. Especially cassava leaf and linseed meal, due to cyanogens containing feed materials, have generally shown reduced growth when compared to the respective control (Hossain & Jauncey, 1989). These results agree with our present experiment, we found that, there is growth depression in common carp.

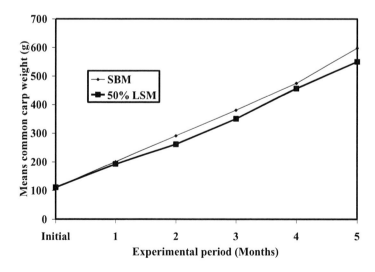

Figure 17. The effects soybean meal and linseed meal (LSM) on growth rate of Common carp during five month of feeding.

Table 35. Calculated level of amino acids in soybean meal and linseed meal for common carp (% of dry diet)

Diets		Requirement[a]	Amino acid
(100% PPM)	(0% PPM)		
			Essential amino acid
2.89	1.56	1.52	Arginine
0.83	0.60	0.56	Histidine
1.42	1.03	0.92	Isoleucine
2.50	1.87	1.64	Leucine
2.08	1.79	2.12	Lysine
0.91	0.80	0.64	Methionine[b]
1.60	1.10	1.16	Phenylalanine[b]
1.80	1.37	1.32	Threonine
1.04	1.10	1.16	Valine

a- according to Nose (1979).
b-The values of methionine and phenylalanine are the requirement in the presence of 2% cystine and 1% tyrosine of the diet, respectively

REFERENCES

Al-Bar, A., P. R. Ismail and H. S. Nakue. (1993). Effect of dietary *Yucca schidigera* extract of chicken and rabbits. *Journal of Animal Science* 71(1): 114-122.

Al-Hafedh Y.S. (1999) Effects of dietary protein on growth and body composition of Nile tilapia (Oreochromis niloticus L.). *Aquaculture Research* 30, 385-393.

Allan, G.L., Rowland, S.J., Parkinson, S., Stone, D.A.J.&Jantrarotai, W. (1999a) Nutrient digestibility for silver perch Bidyanus bidyanus: development of methods. *Aquaculture*, 170, 131–145.

Allan, G.L., Parkinson, S., Booth, M.A., Stone, D.A.J., Rowland, S.J., Frances, J.&Warner-Smith, R. (1999b) Replacement of fish meal in diets for Australian silver perch Bidyanus bidyanus: I. Digestibility of alternative ingredients. *Aquaculture*, 186, 293–310.

Al-Owafeir M. (1999) *The effect of dietary saponine and tannin on growth performance and digestion in Oreochromis niloticus and Clarrias gariepinus*. PhD thesis, Institute of Agriculture, University of Stirling, United Kingdom. p. 220.

Anderson J.; Jackson A.J. & Capper B.S. (1984) Effect of dietary carbohydrate and fiber on the tilapia Oreochromis niloticus (L.). *Aquaculture* 37, 303-314.

Association of Analytical Chemists (AOAC). 1995. *Official methods of analysis, 16th edition*. AOAC, Arlington, Virginia.

Atack, T. H., Jauncey, K. and Matty, A.J. (1979) *The evaluation of some single cell proteins in the diets of rainbow trout. II. The determination of net protein utilization, biological value and true digestibility*. In J. E. Halver and Tiwes (Editors), Finfish Nutrition and Fish Feed Technology. Vol. 1. Heenemann, Berlin, pp. 261-273.

Baker, D. H. and Molitoris, B. A. (1974) Utilization of nitrogen from selected purines and pyrimidines and from urea by the young chick. *Journal of Nutrition*, 104: 553-557.

Barros M.M., Lim C. & Klesius P.H. (2002) Effect of soybean meal replacement by cottonseed meal and iron supplementation on growth, immune response and resistance of channel catfish (Ictalurus punctatus) to Edwardsiella ictaluri challenge. *Aquaculture* 207, 263-279.

Beck, H., Gropp, J., Koops, H. and Tiews, K. (1979) *Single cell protein in trout diets*. In: J. E. Halver and Tiwes (Editors), Finfish Nutrition and Fish Feed Technology. Vol. 2. Heenemann, Berlin, pp. 105-116.

Becker K. & Makkar H. P.S. (1999) Effect of dietary tannic acid quebracho tannin on growth performance and metabolic rates of common carp (Cyprinus carpio L). *Aquaculture* 175, 327-335.

Bergstrom, E. (1979) Experiment on the use of single cell proteins in Atlantic salmon diets. In: J. E. Halver and Tiwes (Editors), *Finfish Nutrition and Fish Feed Technology*. Vol. 2. Heenemann, Berlin, pp. 105-116.

Blom J.H., Lee K.-J., Rinchard J., Dabrowski K. & Ottobre J. (2001) Reproduction efficiency and maternal-offspring transfer of gossypol in rainbow trout (Oncorhynchus mykiss) fed diets containing cottonseed meal. *Journal Animal Science* 79,1533-1542

Bredon R. M. &. Juko, C. D (1961) A semi-micro technique for crude fiber determination. *Journal Science Food and Agriculture* 12, 196-201.

Booth, M.A., Allan, G.L., Frances, J.&Parkinson, S. (2001) Replacement of fishmeal in diets of silver perch: VI. Effects of dehulling and protein concentration on the digestibility of four Australian grain legumes in diets for silver perch (Bidyanus bidyanus). *Aquaculture*, 196, 67–85.

Bureau, D.P., Harris, A.M.&Cho, C.Y. (1999) Apparent digestibility of rendered animal protein ingredients for rainbow trout (Oncorhynchus mykiss). *Aquaculture*, 180, 345–358.

Bureau, D.P., Harris, A.M.&Cho, C.Y. (2000) Feather meals and bone meals from different origins as protein sources in rainbow trout (Oncorhynchus mykiss). *Aquaculture*, 181, 281–291.

Burel, C., Boujard, T., Tulli, F.&Kaushik, S. (2000) Digestibility of extruded peas, extruded lupin, and rapeseed meal in rainbow trout (Oncorhynchus mykiss) and turbot (Psetta maxima). *Aquaculture*, 188, 285–298.

Burel, C., Boujard, T., Kaushik, S. J. (2001) Effects of rapeseed meal glucosinolates on thyroid metabolism and feed utilization in rainbow trout. *Gen. Comp. Endocrinol.*, 124, 343–358.

CAPMS (Central Agency for Public Mobilization and Statistics). (1994). *Statistics of fish production in Egypt,* Report No. 71-123413/92. Cairo, Egypt.

Carter, C. G. and Hauler, R. C. (2000). Fish meal replacement by plant meals in extruded feeds for Atlantic salmon, (Salmon Salar L.) *Aquaculture*, 71: 37-50.

Chance, R. E., I. T. Mertz, and J. E. Halver. (1966). Nutrition of salamonid fishes. Xii, Isoleucine, Leucine, Valine. Moreover, phenylalanine requirements of Chinook salmon and interrelations between isoleucine and leucine for growth. *Journal of Nutrition* 83:177–185.

Chapman D. W. (1968) Production. In: S. D. Gerking (Editor), Methods for Assessment of Fish Production in Fresh Water, *International Biology Programmer Handbook* No. 3, 313pp.

Cheng Z.J. & Hardy R.W. (2002) Apparent digestibility co-efficient and nutritional value of cotton seed meal for rainbow trout (Oncorhynchus mykiss). *Aquaculture* 212,361-372.

Cho C. Y. and Kaushik S. J. (1990) Nutritional energetic in fish: protein and energy utilization in rainbow trout. In: Bourne, G. H. (Editors) Aspects of food production, consumption and energy values. *World Review Animal Nutrition*, l. (61) 132-172.

Cho, S.H. and Jo, J-Y.(2002) Effect of dietary energy level and number of meals on growth and body compostion of Nile tilapia (Oreochromis niloticus L.) during summer and winter seasons. *Journal of the World Aquaculture Society* 33, 48-56.

Chou, L. M., A. D. Munro, T. J. Lam, T. W. Chen, L. K. K. Cheong, J. K. Ding, K. K. Hooi, H. W.Khoo, V. P. E. Phang, K. F. Shim, and C. H. Tan, editors. (1994). Winged bean seed meal: its successful use as a partial replacement for fish meal in practical diets for red tilapia fry. *The Third Asian Fisheries Forum.* Asian Fisheries Society, Manila, Philippines.

Clifford, A.J. and Story, D.L. (1976). Levels of purines in foods and their metabiolic effects on rats. *Journal of Nutrition*, 106: 435-442.

Cowey C.B. and C.Y. Cho (1991) *Nutritional Strategies & Aquaculture Waste*. Proceedings, Gulp, Guelph, Canada, 275 pp.

Dabrowski, K. (1982) Further study on dry diet formulation for common carp larvae. *Riv. Ital. Piscic. Ittiopatol.*, 27: 11-29.

Dabrowski K., Rinchard J., Lee K.-J., Blom J.H., CiereszkoA. & Ottobre J. (2000) Effects of diets containing gossypol on reproductive capacity of rainbow trout (Oncorhynchus mykiss). *Biology of Reproduction* 62, 227-234.

Dempsey, C. H. (1978). Chemical stimuli as a factor in feeding and intraspecific behavior of herring larvae. *Journal of Marine Biology Association U.K.* 58, 739-747.

De Silva S.S. & Gunasekera R.M. (1989) Effect of dietary protein level and amount of plant ingredient (Phaseolus sureus) incorporated into the diets on consumption, growth performance and carcass composition in Oreochromis niloticus (L.) fry. *Aquaculture*, 84: 315-370.

Dorsa, W. J., Robinette H.R., Robinson E. H. & Poe, W. E. (1982) Effects of dietary cotton seed meal and gossypol on growth of young channel catfish. *Transactions of the American Fisheries Society* 111,651-655.

El-Saidy D.M.S.D. (1999) Evaluationof cottonseed meal as partial andcomplete replacement of ¢shmeal in practical diets of Nile tilapia (Oreochromis niloticus) fingerlings. *Egyptian Journal of Aquatic Biology and Fisheries* 3,441^457.

El-Saidy D.M.S. & Gaber M.M (1997). Total replacement of fish meal by soybean meal, with various percentages of supplemental L-methionine, in diets for Nile tilapia (0reochromis niloticus L) *Annals of Agriculture Science of Moshtohor*, 35 (3):1223-1238.

El-Saidy D.M.S. & Gaber M.M (2001). *Linseed meal- Its successful use as a partial and complete replacement for fish meal in practical diets for Nile tilapia Oreochromis niloticus.* Pages 635-643 in Proceeding of the Second International Conference on Animal Production and Health in Semi-Arid Areas. Organized by Faculty of Environmental Agriculture Sciences, Suez Canal University, El-Arish-North Sinai, Egypt.

El-Saidy D.M.S. &. Gaber M.M (2002). Complete replacement of fishmeal by soybean with the dietary L-Lysine supplementation in Nile tilapia fingerlings. *Journal of the World Aquaculture Society* 33: 297-306.

El-Saidy D.M.S. & Gaber M.M. (2002). Evaluation of dehulled sunflower meal as a partial and complete replacement for fish meal in Nile tilapia, Oreochromis niloticus (L), diets.

In: *Proceeding of the First Annual Scientific Conference on Animal and Fish production* (Faculty of Agriculture, AL-Mansoura University, Egypt).September 24-25, 2002,pp.193-2050

El-Saidy D.M.S. & Gaber M.M (2002). Evaluation of hulled sunflower as a dietary protein source for Nile tilapia (Oreochromis niloticus L) *Annals of Agriculture Science of Moshtohor*, 40(2):831-841.

El-Saidy D.M.S. & Gaber M.M (2003). Replacement of fish meal with a mixture of different plant protein sources in juvenile Nile tilapia, Oreochromis niloticus (L), diets. *Aquaculture Research*, 34:1119-1127.

El-Saidy D.M.S. & Gaber M.M (2004b). Use of cottonseed meal supplemented with iron for detoxification of gossypol as total replacement of fish meal in Nile tilapia, Oreochromis niloticus (L) diets. *Aquaculture Research*, 35 859-865.

El-Saidy D.M.S. & Gaber M.M (2005). Effect of dietary protein levels and feeding rates on growth performance, production traits and body composition of Nile tilapia, Oreochromis niloticus (L) cultured in concrete tanks. *Aquaculture Research*, 36, 163-171.

El-Saidy D.M.S., Gaber M.M and Abd-Elshafy A.S. (2005). Evaluation of cluster bean meal, Cyamposis tetragonoloba, as a dietary protein source for common carp, Cyprinus carpio L. *Journal of the World Aquaculture Society*, 36 .(3).311-319.

El-Sayed A.-F.M. (1988) Total replacement of fish meal with animal protein sources in Nile tilapia (Oreochromis niloticus L.), feeds. *Aquaculture Research*, 29: 275-280.

El-Sayed A.-F.M. (1990) Long-term evaluation of cottonseed meal as a protein sources in Nile tilapia (Oreochromis niloticus L.).*Aquaculture*, 84: 315-320.

El-Sayed A.-F.M. (1999) Alternative dietary protein sources for farmed tilapia, Oreochromis spp. *Aquaculture* 179, 149^168

El-Sayed A.-F.M. (2006) *Tilapia culture*. CABI publishing, CAB International, Wallingford, UK., pp: 304.

Essa M. A., El-Sherif, Z. M ; Aboul-Ezz,. S. M &. Abdel-Moati, A.R (1989) Effect of water quality, food availability and crowding on rearing condition and growth of some economical fish species grown under polyculture system. Bulletin National Institute of Oceanography and Fisheries. *ARE* 15 (1), 125-134.

FAO (Food and Agriculture Organization of the United Nations), (2002) *The State of World Fisheries and Aquaculture* 2002. Rome, Box 2, p. 9.

FAO (Food and Agriculture Organization of the United Nations) (2006) *World Review of Fisheries and Aquaculture*. The State of World Review of Fisheries and Aquaculture.pp. 1-64. FAO Fisheries Department, Rome, Italy

Forster, I., Higgs, D.A., Donsanjh, B.S., Rowshandeli, M.&Parr, J. (2000) Potential for dietary phytase to improve the nutritive value of canola protein concentrate and decrease phosphorus output in rainbow trout (Oncorhynchus mykiss) held in 11C fresh water. *Aquaculture*, 179, 109–125.

Francis, G., Makkar, H.P.S. & Becker, K. (2001) Anti-nutritional factors present in plant-derived alternate fish feed ingredients and their effect in fish. *Aquaculture*, 199, 197–227.

Francis, G., H.P.S Makkar, and K. Becker. (2002). Dietary supplementation with a Quillaja saponin mixture improves growth performance and metabolic efficiency in common carp (Cyprinus carpio).*Aquaculture* 203: 311-320.

Focken U. and K. Becker (1993) *Body composition of carp (Cyprinus carpio L).* In: T. Braunbeck, W. hanke and H. Segner (Editors), Fish Ecotoxicology and Ecophysiology. VCH mbH, Weinheim, pp. 269-288.

Fagbenero, O.A.,

Farhangi, M.& Carter, C.G. (2001) *Growth, physiological and immunological responses of rainbow trout (Oncorhynchus mykiss) to different dietary inclusion levels of dehulled lupin (Lupinus angustifolius).* Aquaculture Research., 32, 329–340.

Gaber M. M. (1996).Partial and complete replacement of fish meal by poultry by-product and feather meal in diets of Nile tilapia (0reochromis niloticus L) *Annals of Agriculture Science of Moshtohor*, 34 (1):203-214.

Gaber M. M. A. (2005). The effects of different levels of krill meal on feed intake, digestibility and chemical composition of Nile tilapia (Oreockromis niloticus, L) fry *Journal of the World Aquaculture Society*, 36.(3).325-332.

Gaber M. M. A. (2006). The effects of plant protein-based diets supplemented with Yucca on growth, digestibility and chemical composition of Nile tilapia (Oreochromis niloticus, L) fingerlings. *Journal of the World Aquaculture Society*, 37 (1) 1-8.

Gaber M. M. A. (2006).Partial and complete replacement of fishmeal by broad bean in feeds for Nile tilapia. (Oreockromis niloticus, L) fry *Aquaculture Research*, 37, 986-993.

Gaber M. M. A. and Hanafy M. A. (2008). Relationship between dietary protein source and feeding frequency during feeding Nile tilapia, Oreochromis niloticus (L.) cultured in concrete tanks. *Journal of Applied Aquaculture*, 20 (03) 200-212.

Glencross, B.D.&Hawkins, W.E. (2004) A comparison of the digestibility of several lupin (Lupinus spp.) kernel meal varieties when fed to either rainbow trout (Oncorhynchus mykiss) or red seabream (Pagrus auratus). *Aquac. Nutr.*, 10, 65–73.

Glencross, B.D., Boujard, T.B.&Kaushik, S.J. (2003a) Evaluation of the influence of oligosaccharides on the nutritional value of lupin meals when fed to rainbow trout, Oncorhynchus mykiss. *Aquaculture*, 219, 703–713.

Glencross, B.D., Hawkins, W.E.&Curnow, J.G. (2003b) Evaluation of canola oils as alternative lipid resources in diets for juvenile red seabream, *Pagrus auratus. Aquac. Nutr.*, 9, 305–315.

Glencross, B.D., Curnow, J.G.&Hawkins, W.E. (2003c) Evaluation of the variability in chemical composition and digestibility of different lupin (Lupinus angustifolius) kernel meals when fed to rainbow trout (Oncorhynchus mykiss). *Anim. Feed Sci. Technol.*, 107, 117–128.

Glencross, B.D., Curnow, J.G., Hawkins, W.E., Kissil, G.W.&Petterson, D.S. (2003d) Evaluation of the feed value of a transgenic strain of the narrow-leaf lupin (Lupinus angustifolius) in the diet of the marine fish Pagrus auratus. *Aquac. Nutr.*, 9, 197–206.

Glencross, B.D., Hawkins, W.E.&Curnow, J.G. (2004a) Nutritional assessment of Australian canola meals. I. Evaluation of canola oil extraction method, enzyme supplementation and meal processing on the digestible value of canola meals fed to the red seabream (Pagrus auratus, Paulin). *Aquac. Res.*, 35, 15–24.

Glencross, B.D., Hawkins, W. E. and Curnow, J. G. (2004b) Nutritional assessment of Australian canola meals. II. Evaluation of the influence of canola oil extraction method on the protein value of canola meals fed to the red seabream (Pagrus auratus, Paulin). *Aquaculture Research*, 35, 25–34.

Glencross, B. D., Evans, D., Jones, J. B. & Hawkins, W. E. (2004c) Evaluation of the dietary inclusion of yellow lupin (Lupinus luteus) kernel meal on the growth, feed utilisation and tissue histology of rainbow trout (Oncorhynchus mykiss). *Aquaculture*, 235, 411–422.

Glencross, B.D., Carter, C.G., Duijster, N., Evans, D.E., Dods, K., McCafferty, P., Hawkins, W.E., Maas, R.&Sipsas, S. (2004d) A comparison of the digestive capacity of Atlantic salmon (Salmo salar) and rainbow trout (Oncorhynchus mykiss) when fed a range of plant protein products. *Aquaculture*, 237, 333–346.

Gomes, E. F., Rema, P. and Kaushik, S. J. (1995) Replacement of fish meal by plant proteins in the diet of rainbow trout (Oncorhynchus mykiss): digestibility and growth performance. *Aquaculture*, 130, 177–186.

Goda, A.M.A-S. (2007) Effect of Dietary Soybean Meal and Phytáse Levels on Growth, Feed Utilization and Phosphorus Discharge for. Nile tilapia Oreochrornis niloticus (L.). *Journal of Fisheries and Aquatic Science*, 2(4): 248-263.

Goda, A.M.A.S.; El-Husseiny, O.M.; Abdul-Aziz , G.M, Suloma, A. and Ogata, H. Y., (2007) Fatty acid and free amino acid composition of Muscles and Gonads from Wild and Captive Tilapia, Oreochromis niloticus (L.)(Teleostei: percjfonnes) : an Approach to development Bloodstock Diets. *Journal of Fisheries and Aquatic Science* 2 (2):86 —99.

Goddard J.S., McLean E. & Wille K. (2003) Co-dried sardine silage as an ingredient in tilapia, Oreochromis aureus, diets. *Journal of Aquaculture in the Tropics* 18, 257-264.

Goddard J.S. & Perret J.S.M. (2005) Co-drying fish silage for use in Aquafeeds. *Animal Feed Science and Technology* 118,337-342.

Goddard J.S., G. Al-Shagaa & A.Ali(2008) Fisheries by-catch and processing waste meals as ingredients in-diets for Nile.tilapia, Oreochromis niloticus. *Aquaculture Research*, (39) 518-525.

Gomes,E.F., P. Rema, A.Gouveia & A.Oliva-Teles (1995) Replacement of fish meal by plant proteins in diets for rainbow trout (Oncorhynchus mykiss) effect of the quality of the fish meal based control diets on digestibility and nutrient balances. *Water Science and technology*, 31: 2-05-211.

Grabner M. & Hofer R.(1985) The digestibility of the protein of broad bean (Vicia faba) and soybean (Glycin max) under vitro condition stimulating the alimentary tracts of rainbow trout (Salmo gairdneri) and carp (Cyprinus carpio). *Aquaculture* 48, 111-122.

Gropp, J., Koops, H. Tiews, K. and Beck, H. (1979). *Replacement of fishmeal in trout feeds by other feedstuffs.* In: T. V. R. Pillary and W. A. Dill (Editors), Advances in Aquaculture Fishing News Books Ltd., Farnham, Surrey, Great Britain, pp. 596-601.

Hara, T. J. (Editor) 1982. *Fish Chemoreception*, Elsevier Scientific Publication Company, New York, USA.

Hara, T. J. 1993. *Chemoreception.* Pages 191-218 in D.H. Evans editor. *The physiology of fish.* CRC Press, Inc., Boca Raton, Florida USA.

Hardy, R.W. (1996) Alternate protein sources for salmon and trout diets. *Animal Feed Science Technology,* 59: 71-80.

Hardy, R.W. and Barrows F. (2002) *Diet formulation and manufacure.* In Fish Nutrition (ed. By J. E. Halver & R. W. Hardy), 3rd ed. Pp. 505-600.

Hardy, R.W , Sealy W. M. and Gatlin D.M. (2005) Fisheries by catch and by product meals as protein sources for rainbow trout (Oncorhynchus mykiss). *Journal of the World Aquaculture Society*, 36: 393-400.

Harper, A. E., N. J. Benevenga, and R. M. Wohlhueter. (1970). Effects of ingestion of disproportionate amounts of amino acids. *Physiology Review* 50:428–558.

Hasan, M.R.; D.J. Macintosh and K. Jauncey. 1997. Evaluation of some plant ingredients as dietary as dietary protein sources for common carp (Cyprinus carpio L.) fry. *Aquaculture*, 151: 55-70.

Hashim R.; Saat N. A. M. & Wong C. H. (1994) *Winged bean seed meal: Its successful use as partial replacement for fishmeal in practical diets for Red tilapia fry.* Pages 660-662 in L.M. Chou and C.H. Tam editors. The Third Asian Fisheries Forum. Asian Fisheries Society Manila Philippines.

Headon, D.R. and K.A, Dawson. 1990. Yucca extract controls atmospheric ammonia levels. *Feed stuffs* 62(29): 2-4.

Heu, M., Kim, J. and Shadidi, F. (2003) Components and nutritional quality of shrimp processing by products. *Food Chemistry*, 82: 235-242.

Higgs, D.A., McBride, J. R., Markert, J. R., Dosanjh, B.S., Plotnikoff, M.D.& Clarke, W. C. (1982) Evaluation of tower and candle rapeseed protein concentrate as protein supplements in practical dry diets for juvenile Chinook salmon (Oncorhynchus tshawytscha). *Aquaculture*, 29, 1–31.

Hilton J. W. Atkinson J .I. & Slinger S.J. (1983) Effect of increased dietary fiber on the growth of rainbow trout (Salmo gairdneri) *Canadian Journal Fish Aquatic Science* 40, 81-85.

Holm J. C., Refstie T &. Bo, S (1990) The effect of fish density and feeding regimes on individual growth rate and mortality in rainbow trout, Oncorhynchus mykkiss. *Aquaculture* 89, 225-232.

Hossain, M.A. and K, Jauncey, 1989 Studies on the protein, Energy and Amino acid digestibility of common carp (Cyprinus carpio L).*Aquaculture* 83, 59-72.

Hossain A. J. & Jauncey K. (1993) *The effect of varying dietary phytic acid, calcium and magnesium levels on the nutrition of common carp, Cyprinus carpio* In: Kaushik, S. J. Luquent P. (Eds), *Fish Nutrition in Practice*, Proceeding of International Conference, Biarritz, France, June 24-27, 1991, pp. 705-715.

Hossain, M. A., U. Focken, and K. Becker. (2001). Evaluation of an unconventional legume seed, Sesbania aculata, as a dietary protein source for common carp, Cyprinus carpio (L.). *Aquaculture* 198:129–140.

Hughes, S. G. and Rumsey G. L. (1983) Dietary requirements of essential branched-chain amino acids by lake trout. *Transaction of the American Fisheries Society* 12:812–817.

Ishiwata N. (1969) Ecological studies on the feeding frequency on the food intake and satiation amount. *Bulletin of Japanese Society scientific and Fisheries*, 35: 979-984.

Iwai, T. (1980). *Sensory anatomy and feeding of fish larvae. Pages 124-145 in Fish behavior and its use in the capture and culture of fishes.* International Center for Living Aquatic Resources Management, Conference Proceedings.

Jackson, A.J., B.S. Capper and A.J. Matty. 1982. Evaluation of some plant protein in complete diets for the tilapia, Sartherodon mosambicus. *Aquaculture* 27: 97-109.

Jobling M. 1983 Effect of feeding frequency on feeding intake and growth of Arctic charr, Salvelinus alpinus L. *Journal Fish biology* 23, 177-185.

Jobling, M., Coves, D., Damsgard, B., Kristiansen, H.R., Koskela, J., Petursdottir, T.E., Kadri, S.&Gudmundsson, O. (2001) Techniques for measuring feed intake. In: *Food*

Intake in Fish (Houlihan, D.Boujard, T.&Jobling, M. eds), pp. 49–87. Blackwell Science, Oxford

Johnston, N. L., C.I Quarles. and D.J Fagerberg. 1982 Broiler performance with DSS40 Yucca saponin in combination with monensin. *Poultry Science* 61: 1052-1054.

Jones, K. A. 1992. Food search behavior in fish and the use of chemical lures in commercial and sports fishing. Pages 288-320 in T. J. Hara, editor. *Fish Chemoreception*. Chapman and Hall, London, England.

Jones L.A. (1987) Recent advances in using cottonseed products. *Proceedings of the Florida Nutrition Conference,* 12-13 March 1987, pp. 119-138. Daytona Beach, FL, USA.

Kaushik, S. J. and Luquet, P. (1980) Influence of bacterial protein incorporation and of sulphur amino acid supplementation to such diets on growth of rainbow trout, Salmo gairdneri Richardson. *Aquaculture*, 19: 163-175.

Kaushik, S. J. and K. Dabrowski. 1983. Nitrogen and energy utilization in juvenile carp fed casein, amino acids or a protein-free diet. *Reproduction Nutrition Development* 23:741–754.

Kaushik S.J. (1995) Nutrient requirements, supply and utilization in the context of carp culture. *Aquaculture* 129, 225-241.

Kaushik, S.J., Cravedi, J.P., Lalles, J.P., Sumpter, J., Fauconneau, B.&Laroche, M. (1995) Partial or total replacement of fish meal by soybean protein on growth, protein utilization, potential estrogenic or antigenic effects, cholesterolemia and flesh quality in rainbow trout, Oncorhynchus mykiss. *Aquaculture*, 133, 257–274.

Kayano Y; Jeong, T, D. S.; Oda, T & Nakagawa. H. (1990) Optimum feeding frequency on young red-spotted grouper, Epinephelns akara. *Suisanzoshku* 38: 319-326

KayanoY. T., Tao, S.; Yamamoto, S. &.Nakagawa, H (1993) Effect of feeding frequency on the growth and body constituents of young red- spotted grouper, Epinephelus akara. *Aquaculture* 110: 271-278.

Kim, T. S., Schulman, J. and Levino, R. A. (1968) Relaxant effect of cyclic adenosine 3, 5-monophosphate on the isolated rabbit ileum. *Journal of Pharmacology Experimental Thereby*, 163: 36.

Kolkovski, S., A. Arieli and A. Tandler. 1997a. Visual and chemical cues stimulate microdiet ingestion in sea bream larvae. *Aquaculture International* 5, 527-537.

Kolkovski, S., W.M. Koven and A. Tandler.(1997b). The mode of action of Artemia in enhancing utilization of microdiet by githerad seabream Sparus aurata larvae. *Aquaculture* 155, 193-205.

Kolkovski, S., A. S Czesny and K. Dabrowski. 2000. Use of krill hydrolysate as a feed attractant for fish larvae and juveniles. *Journal of the World Aquaculture Society* 31, 81-88.

Li, P., Wang, X., Hardy, R. W. and Gatlin, D. M.(2004).Nutritional; value of fisheries by catch and by product meal in the diet of red drum (Sciaenops ocellatus). *Aquaculture*, 236: 485-496.

Liener I. E. (1980) *Toxic constituents of plant food stuffs*, 2nd.edition. Academic Press, New York, USA.

Liou C.H. (1989) *Lysine and sulfur amino acid requirement of juvenile blue tilapia* (Oreochromis aureus, Walbaum.).Ph D Dissertation, Texas A & M University, College Station. TX, USA, 101 pp.

Liu F. G. & Liao, C. I.(1999) Effect of feeding frequency on the food consumption in hybrid striped bass, Moronc saxitiisi x M chrysops. *Fish Science* 64, 513-519.

Lovell R. T. (1989) *Nutrition and Feeding of Fish*. Van Nostrand Reinhold. New York, New York USA.

Mackie, A. M. and A. I. Mitchell, 1985. Identification of gustatory feeding stimulants for fish application in aquaculture. Pages 177-191 in C. B. Cowey, A. M. Mackie, and J. G. Bell, editors. *Nutrition and feeding in fish*. Academic Press, London, England.

Maigualema, M. A. and Gernat, A. G., (2003)The Effect of Feeding Elevated Levels of Tilapia .(Oreochrornus niloticus) By-prodUct Meal on Broiler Performance and Carcass Characteristics. *International Journal of Poultry Science* 2 (3): 195-199.

Maina, J. G., R. M. Beames, D. Higgs, P. N. Mbugua, G. Iwama, and S. M. Kisia. 2003. Partial replacement of fishmeal with sunflower cake and corn oil in diets for tilapia Oreochromis niloticus (Linn): effect on whole body fatty acids. *Aquaculture Research* 34: 601–609.

Martin S.D. (1990) Gossypol effects in animal feeding can be controlled. *Feedstuffs* 62, 14-17.

Martinez-Palacious C. A., Galvan Cruz R., Overa Novoa M. A. & Chavez C. (1988) The use of Jack bean (Canavalia ensiformis Leguminous) meal as partial substitute for fishmeal in diets for tilapia (Oreochromis mossambicus. C, chlidae). *Aquaculture* 68, 165-175.

Matty, A. J. and Smith, P. (1978) Evaluation of yeast, a bacterium and algae as a protein source for rainbow trout. I. Effect of protein level on growth, gross conversion efficiency and protein conversion efficiency. Aquaculture, 14: 235-246.

Mbahinzireki, G.B., K.J. Dabrowski, K.J. Lee, D. El-Saidy and E.R. Wisner. (2001). Growth, feed utilization and body composition of Tilapia (Oreochromis sp.) fed with cottonseed meal-based diets in a recirculating system. *Aquaculture Nutrition* 7: 188-200.

Mearns, K.J., O. F Ellingsen, K. J. B. Doving and S. Helmer. 1987. Feeding behavior in adult rainbow trout and Atlantic salmon elicited by chemical fractions and mixtures of compounds identified in shrimp extract. *Aquaculture* 64, 47-63.

Murray, A.P. and Marchant, R. (1986) Nitrogen utilization in rainbow trout fingerlings (Salmo gairdneri Richardson) *Aquaculture*, 54: 263-275

Murai T., Ogata, H ;Kosutarak P. &. Ari,. S (1986) Effect of amino acid supplementation and methanol treatment on utilization of soy flour by fingerling carp. *Aquaculture* 56: 197-206.

Murai T., T. Akiyama and T. Nose 1981 Use of crystalline amino acids coated with casein in diets for carp. *Bulletin of the Japanese Society of Scientific Fisheries* 47, 523-527.

Murai T., H. Ogata and T. Nose, 1982. Methionine coated with virous materials supplemented to soybean meal diet for fingerling Carp, Cyprinus carpio and channel catfish, Ictalurus punctatus. *Bulletin of the Japanese Society of Scientific Fisheries* 48, 83-88.

Murai T., T. Hirasawa, T. Akiyama. and T Nose. 1983 Effect of dietary ph and electrolyte concentration on utilization of crystal amino acids by fingerlings carp. *Bulletin of the Japanese Society of Scientific Fisheries* 49, 1377-1611.

Murai T., T .Akiyama and T. Nose 1984 Effect of amino acid balance on efficiency in utilization of diet by fingerlings carp. *Bulletin of the Japanese Society of Scientific Fisheries* 50, 893-897.

Mwachireya, S.A., Beames, R.M., Higgs, D.A.&Dosanjh, B.S. (1999) Digestibility of canola protein products derived from the physical, enzymatic and chemical processing of commercial canola meal in rainbow trout, Oncorhynchus mykiss (Walbaum) held in freshwater. *Aquac. Nutr.*, 5, 73–82.

Naylor, R.L., Goldburg, R.J., Primavera, J.H., Kautsky, N., Beveridge, M.C.M., Clay, J., Folke, C., Lubchenco, J., Mooney, H.&Troell, M. (2000) Effect of aquaculture on world fish supplies. *Nature*, 405, 1017–1024.

New M.B. & Csavas I. (1995) *The use of marine resources in aquafeeds. In: Sustainable Fish Farming (ed. by H. Reiner- sten & H. Haaland)*, pp. 43-78. AA Balkana, Rotterdam, the Netherlands.

New M.B. (1996) Responsible .uses of aquaculture feeds. *Aquaculture Asia*, 1: 3-15.

National Research Council (1993) *Nutrient requirement of fish*. National Academy Press, Washington, DC, USA. pp. 114.

Nengas, I., Alexis, M.N.&Davies, S.J. (1999) High inclusion levels of poultry meals and related byproducts in diets for gilthead seabream Sparus aurata L. *Aquaculture*, 179, 13–23.

Nose T., 1979. *Summary report on the requirements of essential amino acids for carp*. Pages 145-156, in: J.E. Halver and K. Tiewes, editors. *Finfish Nutrition and Fish Technology*, Vol. I. H. Heenemann GmbH, Berlin.

Ofojekwa P.C. & Ejike C (1984) Growth response and feed utilization in tropical cichlid (Oreochromis niloticus L.) fed on cottonseed-based artificial diets. *Aquaculture* 42, 27-36.

Olvera-Novoa, M. A., Campos, G. S., Sabido, G. M. and Martinez-Palacios, C. A., 1990. The use of alfalfa leaf protein concentrates as protein source in diets for tilapia (Oreochromis mosambicus). *Aquaculture* 90: 291-302.

Onning, G., Q. Wang, B.R Westrom, N.G. Asp, and S.W. Karlsson. 1996. Influence of oat saponin on intestinal permeability in vitro and vivo in the rate. British. *Journal of Nutrition* 76: 141-151.

Orachunwon C., Thammasarat S. & Lohawatanakul C. (2001) *Recent developments in tilapla feds*. In: Prbceedings of the Tilapia 2001 International Technical and Trade Conference on Tilapia, Kuala Lumpur, Malaysia (ed. by S. Subasinghe & I. Singh), pp. 113-122. INFOFISH. Kuala Lumpur, Malaysia.

Otubusin, S, O.(1987) Effects of different levels .of blood meal in pelleted feeds on tilapia (Oreochromis niloticus) production in floating bamboo net-cages. *Aquaculture*, 65: .263-266.

Panek, F. M. (1987). *Biology and ecology of carp, Pages 1-16 In: Cooper, E.L. (editor) Carp in North America*. American Fisheries Society, Bethesda, Maryland, USA.

Pike, I. H. Andorsdottir G. & Mundheim H. (1990) The Role of fish Meal in Diets for salmonids. *International Association of Fish Meal Manufacturers*, no.24. Hertfordshire, England.

Pongmaneerat J. Watanabe T. Takeuchi T. & Satoh S. 1993 Use of different protein meals as partial or total substitution for fish meal in carp diets. *Nippon Suisan Gakkaishi* 59, 157-163.

Refstie, S., Storebakken, T.&Roem, A.J. (1998) Feed consumption and conversion in Atlantic salmon (Salmo salar) fed diets with fish meal, extracted soybean meal or soybean meal with reduced content of oligosaccharides, trypsin inhibitors, lectins and soya antigens. *Aquaculture*, 162, 301–312.

Refstie, S., Svihus, B., Shearer, K. D. and Storebakken, T.(1999) Nutrient digestibility in Atlantic salmon and broiler chickens related to viscosity and non-starch polysaccharide content in different soybean products. *Animal Feed Science Technology*, 79: 331-345.

Rathbone C.K., Babbitt l.K., Dung F. M. & Hardy R. W. (2001) Performance of juvenile coho salmon Oncorhynehus Icisutch fed diets containing meals from fish wastes, de boned fish wastes, or skin-and-bone by-product as the protein ingredient. *Journal of the World Aquaculture Society* 32, 21-29.

Reffat, 0. C. A., (1982) *Utilization of fish flesh and viscera in production concentrate for human nutrition.* M. Sc. Thesis Fac. of Home Economic, Helwan Univ. Egypt. pp. 202.

Robinson, E. M., Poe, W. E. and Wilson, R. P.(1984) Effects of feeding diets containing an imbalance of branched chain amino acids on fingerling channel catfish. *Aquaculture* 37: 51-62.

Ruohonen K. &. Grove D. J (1998) Effect of feeding frequency on the growth and food utilization of rainbow trout (Salrno gairdneri} fed low-fat herring or dry pellets. *Aquaculture* 165, 111-121.

Rinchard J., Lee K-J., dabrowski K., CiereszkoA., Blom J.H. & Ottobre J.S. (2003) In£uence of gossypol from dietary cottonseed meal on hematology, reproductive steroids and tissue gossypol enantiomer concentrations in male rainbow trout (Oncorhynchus mykiss). *Aquaculture Nutrition* 9, 275-282.

Robinson E.H. & Rawles S.D. (1983*)* Use of defatted, glandless cottonseed £our and meal in channel catfish diets. In: *Proceedings of Annual Conference of Southeastern Association of Fish and Wild life Agencies* 37, 358-363.

Rumsey, G.L., Hughes, S.G., Smith, R.R., Kinsella, J.E., Shetty, K.J., (1991) Digestibility and energy values of intact, disrupted, and extracts from brewers dried yeast fed to rainbow trout (Oncorhynchus mykiss). *Animal Feed. Science Technology*, 33, 185-193.

Santiago, C.B. and Lovell, R.T (1988) Amino acid requirements for growth of Nile tilapia. *Journal of Nutrition*, 118(12) 1540-1546.

Sanz A., Morales M., Higuera, Dl. La and Cardenete G. (1994) Sunflower meal in rainbow trout (Oncorhynchus mykiss).diets*:* Protein and energy utilization. *Aquaculture*, 128: 287-300.

Sarkar, S.K. (1999). Role of plant glycocomponents (De-odorease) on water parameters in fish) ponds. *Journal of Environmental-Biology* 20 (2): 131-134.

Sarig S. (1989) *Introduction and state of art in aquaculture.* In: M. Shilo and S. Sarig (Editors), Fish culture in Warm Water Systems. Problems and Trends. CRC Press, Boca Raton, FL, pp. 2-19.

Schwarz F.J. & Kirchgessner M. (1988) Amino acid composition of carp (Cyprinus carpio L) with varying protein and energy supplies. *Aquaculture* 72, 307-317.

Sealey, W.M., Lim C. & Klesius P.H. (1997) Influence of the dietary level of iron from iron methionine and iron sulfate on immune response and resistance of channel catfish to Edwardsiella ictaluri. *Journal of the World Aquaculture Society* 28,142-149

Shah N.; Atallah M.T.; Mahoney P.R. & Pellet P. I. (1982) Effect of dietary fiber component on fecal nitrogen excretion and protein utilization in growing rats. *Journal of Nutrition* 112, 658-666.

Shiau, S.Y.; Chuang, J.I. & Sun, C.I. (1987) Inclusion of SBM in tilapia (Oreochromis niloticus x O. aureaus) diets at low protein level. *Aquaculture* 65, 251-261.

Shiau S.Y.; Yu H.L.; Hwa S.; Chen S.Y.; Hsu S. I., (1988) The influence of carboxymethyl cellulose on growth, digestion, gastric emptying time and composition of tilapia. *Aquaculture* 70, 345-354.

Shiau S.Y Lin S.F.; Yu S.L.; Lin A.L. & Kwork C.C. 1990 Defatted and full-fat SBM as partial replacements for fishmeal in tilapia (Oreochromis niloticus x O. aureaus) at low protein level. *Aquaculture* 86, 401-407.

Shiau S.Y. (1997) Utilization of carbohydrate in warmwater fish-with particular reference to tilapia, Oreochromis niloticus x O. aureus. *Aquaculture* 151, 79-96.

Slominski B.A., Kienzle H.D., Jiang P., Campbell L.D., Pickard M. & Rakow G.(1999) *Chemical composition and nutritive value of canola-quality Sinapis alba mustared*. In Proceedings of the 10th International Rapeseed Congress Canberra. Australia, pp. 416-421.

Shih, L., Chen, L., Vu, T., Chang, W., Wang, S., (2003) *Microbial reclamation of flsh processing wastes for the production of fish sauce. Enzyme & Microbial Technology*, 5j54-62.

Spyridakis, p.; Metailler, R.; Gabaudan, J. & Riaza, A., 1989. Studies on Nutrient digestibility in European sea bass (Dicentrachus Labrax): 1 Methodological aspect concerning faces collection. *Aquaculture* 77: 61 – 70.

Siddhuraju P. & Becker K. (2001) Preliminary nutritional evaluation of mucuna seed meal (Mucuna pruriens var. utilis) in common carp (Cyprinus carpio L.): an assessment by growth performance and feed utilization. *Aquaculture* 196, 105-123.

Sorensen, P.W. and J. Caprio. 1997. *Chemoreception.* Pages 375-405 in D. H. Evans, editor. *The Physiology of fishes*. CRC Press, Inc., Boca Raton, Florida, USA

Stone, D.A.J., Allan, G.L., Parkinson, S.&Rowland, S.J. (2000) Replacement of fish meal in diets for Australian silver perch, Bidyanus bidyanus. III. Digestibility and growth using meat meal products. *Aquaculture*, 186, 311–326.

Sugiura, S.H., Babbit, J.K., Dong, F.M. and Hardy, R.W. (2000) Utilization of fish and animal by-product meals in low-pollution feeds for rainbow trout Oncorhynchus mykiss (Walbaum). *Aquaculture Res.*, 31, 585–593.

Tacon, A.G.J. and Cooke, D.J. (1980) Nutritional value of dietary nucleic acids to trout. *Nutritional Reports International*, 22 (5): 631-640.

Tacon A. G. J., Haster., J.V; Featherstone, P. B. Ken;. K &. Jakson A. J (1983) Studies on the utilization of full-fat soybean and solvent extracted soybean meal in complete diet for rainbow trout. *Nippon Suisan Gakkaishi* 49, 1437-1443..

Tandler, A., B. A. Berg and G. Wm Kissil. 1982. Effect of food attractants on appetite and growth rate of gilthead seabream, Sparus aurats. *Journal of Fish Biology* 20, 673-681.

Tacon A.G. and .Cowey, .C.B. 1985. *Protein and amino acid requirements*. Pages 155-187 in P. Tytler and P. Calow, editors. Fish Energetic: New Perspectives. Crom Helm, Beckenham, Kent.

Tácon A.G.1,llasan M.R.& Subasinghe ltP.(2006) *Use of Fisheries Resources as Feed Inpots for Aquaculture Development: Trends and Policy*. FAO, Fisheries Circular.No1018.FAO, Rome, Italy, Taylor, J., R. Mahon. 1977. *Hybridization of Cyprinus carpio and Carassius auratus, the first two exotic species in the lower Laurentian Great Lakes*. Environmental Biology Of Fishes 1(2):205-208.

Toften, H., E. H. Jorgensen and M Jobling. 1995. *The study of feeding preferences using radiography: oxytetracycline as a feeding deterrent and squid extract as a stimulant in diets for Atlantic salmon.* Aquaculture Nutrition 1, 145-149.

Van Rijn J., Stutz, S. R;. Diab S &. Shilo M (1986) Chemical, physical and biological parameters of superintensive concrete fish ponds. *Bamidgeh* 38, 35-43.

Viola S., Lahav E. & Angeoni H. 1992 Reduction of feed protein levels and nitrogenous excretions by lysine supplementation in intensive carp culture. *Aquatic. Living Resources* 5, 277-285.

Viola. S. and G. Zohar., (1984)Nutrition studies with market sizç hybrids of rilapia (Oroehrotnis niloticus) in intensive culture, 3 protein levels and sources Bamid. 36:3.

Viola. S. and Lahava, E. (1991) The protein sparing effect of synthetic lysine in practical carp feeds. Fish Nutrition in practice. *IV International Symposium on fish nutrition and Feeding*, Biarritz, France. Abstract only.

Ward A. T. & Reichert R. D. (1986) Comparison of effect of cell wall and hull fiber from canola and soybean on the bioavailability for rats of minerals, protein and lipid. *Journal of Nutrition* 116, 233-241.

Wedegaertner T.C. (1981) Making the most cottonseed meal. *Feed Management Magazine* 32,13.

Whiteman K.WL .& Gatlin D.M. [II (2005) Evaluation of $htcis by-catch and by- product mals in diets for red drum Sêiaenops ocellatus L. *Aquaculture Research* 36,1572-1580.

Wilson, M.F., E.P. Luiz, M.B. Margarida, A C Pezzato, R. B. Valéria. (2004). Use of ideal protein concept for precision formulation of amino acid levels in fish meal free diets for juvenile Nile tilapia (Oreochromis niloticus L.) *Aquaculture Research* 35: 1110-1116.

Williams, K.C., Barlow, C.G., Rodgers, L.J.&Ruscoe, I. (2003a) Potential of meat meal to replace fish meal in extruded dry diets for barramundi, Lates calcarifer (Bloch). I. Growth performance. *Aquac. Res.*, 34, 23–32.

Williams, K.C., Patterson, B.D., Barlow, C.G., Ford, A.&Roberts, R. (2003b) Potential of meat meal to replace fish meal in extruded dry diets for barramundi, Lates calcarifer (Bloch). II. Organoleptic characteristics and fatty acid composition. *Aquac. Res.*, 34, 33–42.

Windell, J.I., Armstrong, R. and Cinebell, J. R. (1974) Substitution of brewer's single cell protein into pelleted fish feed. *Feedstuffs*, 46: 22-23.

Zeitler M. H., Kirchgessner M. & Schwarz F.J. 1983 Effects of different protein and energy supplies on carcass composition of carp (Cyprinus carpio L). *Aquaculture* 36, 37-48.

Zohar G.,. Rappaport, U & Sarig, S (1985) Intensive culture of tilapia in concrete tanks. *Bumidgeh* 37, 103-112.

Xie, S.; Cui, Y.; Yang, Yi. and Liu J. (1997) Energy budget of Nile tilapia, Oreochromis niloticus in relation to ration size. *Aquaculture*, 154, 57-68.

In: Encyclopedia of Environmental Research
Editor: Alisa N. Souter
ISBN: 978-1-61761-927-4
© 2011 Nova Science Publishers, Inc.

Chapter 48

THE POULTRY LITTER LAND APPLICATION RATE STUDY – ASSESSING THE IMPACTS OF BROILER LITTER APPLICATIONS ON SURFACE WATER QUALITY

Matthew W. McBroom[1] and J. Leon Young[2]

[1] Assistant Professor of Forest Hydrology, Stephen F. Austin State University Arthur Temple College of Forestry and Agriculture, Nacogdoches, Texas, USA

[2] Regents Professor of Agriculture and Director of the Soil, Plant, and Water Analysis Laboratory, Stephen F. Austin State University Arthur Temple College of Forestry and Agriculture, Nacogdoches, Texas, USA

ABSTRACT

Poultry production in the United States has grown dramatically in recent years which has resulted in the generation of large quantities of poultry litter. The high nutrient content of poultry litter makes it an excellent soil nutrient amendment. However, concern has arisen regarding potential negative impacts on stream water quality. Numerous studies have evaluated the edge of field effects of litter applications through plot studies while others have evaluated instream effects. However, an integration of both instream (watershed scale) as well as edge of field (plot studies) effects of litter applications on water quality was needed. Beginning in 1994, four related studies were conducted in East Texas by Stephen F. Austin State University, the Texas Commission on Environmental Quality, and the Angelina Neches River Authority to determine these potential effects. This project included: 1) a study to determine the effects on runoff water quality from 12 experimental plots with four different rates of surface litter application, 2) an upstream/downstream stream gaging study to evaluate water quality of two tributary streams of the Attoyac Bayou in areas of intense poultry production, 3) an assessment of water quality conditions within the Attoyac Bayou, 4) the use of a computer simulation model (AGNPS) to determine possible water quality effects of litter applications in the study areas. Storm water samples collected from the runoff plots and watershed gaging stations were analyzed for nutrients (total phosphorus, orthophosphorus, nitrate-nitrogen, Total Kjeldahl nitrogen, and potassium), total suspended sediment, pH, and conductivity.

At watershed gaging stations, weekly grab samples were also collected and analyzed for the above parameters along with dissolved oxygen, temperature, and bacteria. Surface plot data indicated that a vegetated filter strip of 4.5 m was effective in reducing nutrient losses at the edge of fields where litter is applied, though buffer effectiveness was seen to decline after multiple applications of litter at higher rates. For the stream sampling sites, significantly higher nitrogen concentrations were found from pastured sites receiving broiler litter than from upstream forested sites. However, these concentrations were still below levels that could result in adverse water quality impacts and likely result from multiple nonpoint pollution sources in the pastured watershed. Bacterial concentrations were higher in pastured watersheds, though significant contributions resulted from wildlife in the forested watersheds as well. Results from these studies indicate that with proper management, including the use of streamside buffers and appropriate application rates, poultry litter applications are not likely to result in water quality degradation in the Attoyac Bayou.

INTRODUCTION

Agricultural nonpoint source pollution is one of the major sources of water quality impairments in the United States. Over 45% of the nation's rivers and streams are reported as impaired and not supporting all of their designated uses, with 81% remaining unassessed (USEPA, 2007). Of this, sediment, nutrients, and pathogens are listed as the leading causes of impairments, and as a land use, agriculture is responsible for almost 40% of these listings (USEPA, 2007).

One component of this is the land application of animal manures as a nutrient amendment. Land applications of manure application not only provides nutrients and organic material for receiving crops, it also reduces the volume of waste generated by confined animal feeding operations (CAFOs). However, land applications of manures can cause water quality degradations when not properly managed (Daniel et al., 1994). In particular, rapid, concentrated growth of the poultry industry stimulated by increasing demands for low cholesterol meats has resulted in concerns about the potential water quality impacts of land applications of these large volumes of poultry litter. Overall, integrated broiler industry in the US produced 8.9 billion chickens (*Gallus gallus domesticus*) in 2007. The industry has shown steady growth since 1955 with about 1 billion birds to approximately 6 billion in 1990, with most of this growth being concentrated in southern states (USDA - National Agricultural Statistics Service, 2002; 2008).

The Poultry Litter Land Application Rate Study (PLLARS) was initiated in 1994 to determine the potential effects of land applied litter on water quality in the Attoyac Bayou watershed. PLLARS was a joint study between the Angelina Neches River Authority and Stephen F. Austin State University with administrative oversight provided by the Texas Commission on Environmental Quality. This study provided a comprehensive view of poultry litter applications by examining the nature of the litter, plot level, sub-watershed, and watershed levels of water quality, biological, and bacteriological impacts. The objectives of this study were to evaluate broiler litter production and characteristics, determine the effects of litter applications on soils and storm runoff from study plots, quantify the effects of applications on tributary streams, ascertain the status of the Attoyac Bayou, and to use a model (AGNEPS) to examine the overall effects of these activities.

BROILER LITTER PRODUCTION AND CHARACTERISTICS

In the broiler industry, the integrator company owns the breeding stock, a hatchery, feed mill and processing facility in a central location with contract growers located in a region approximately 60 miles in diameter. The integrator company provides baby chicks, feed and technical support to the contract grower who provides housing and care of the birds until the birds are taken for processing. This cycle takes 4 to 8 weeks depending on the size of the bird produced (1.6 to 3.2+ kg) and is repeated 5 to 7 times per year. The amount of manure produced per bird depends on the size of the bird and the formulation of the ration fed the bird and ranges from about 0.7 to almost 1.4 kg per bird (oven dry basis). Growers are paid according to a contracted rate based on the number of birds or pounds of bird produced. Traditionally, the manure has belonged to the contract grower. Texas produced about 616 million broilers in 2007 (USDA - National Agricultural Statistics Service, 2008). Assuming each broiler produced 1.1 kg of dry manure, Texas produced 677,600 metric tons of manure of broiler litter. This is enough to spread 2.7 metric tons per hectare on 250,963 hectares of agriculture land.

Broiler litter includes the manure plus some type of bedding material usually wood shavings or rice hulls. The material is primarily manure by the time the contract grower spreads it on agricultural land as a fertilizer. Broiler litter is a very nutrient dense material compared to other manures and as a result it is an excellent fertilizer material. The literature has numerous references to the nutrient content of broiler litter. Typically, litter will contain between 20 and 30% moisture, 3 to 4% N, 1.5 to 2.0% P and 2 to 3% K (based on the authors observation of several thousand litter samples over the last 20 years).

Litter has an N to P ratio of 2 to 1 while plants take up N and P in an 8 to 1 ratio (Edwards and Daniel, 1992a). Repeated application of litter on the same fields has resulted in litter being primarily an N source as the level of P in the soil has steadily increased. In Texas, litter is mostly applied to permanent sod grasses like bermudagrass (*Cynodon dactylon*) or bahiagrass (*Paspalum notatum*) which are used for cattle production, both hay and grazing. Anecdotal soil test results from farmer fields have shown soil test P values exceeding 1,000 mg kg^{-1}, with many fields testing higher than 100 mg kg^{-1}. Most P soil test procedures stop recommending P fertilizer at 40 to 60 mg kg^{-1}.

The P concentration in runoff water from freshly applied broiler litter has been shown to exceed environmental standards at the edge of field as indicated by results from runoff plots using simulated rainfall (Edwards and Daniel 1992b; 1993a; 1993b; 1993c; 1994). Phosphorus has become the application rate limiting factor because in watersheds where repeated applications of broiler litter have occurred it is possible for P to move into surface waters (Sharpley et al., 1999; Sharpley et al., 2003). Numerous strategies are available for managing the P inputs and outputs from a watershed (Sharpley et al., 2003) including vegetative filter strips or buffer zones (Chaubey et al., 1995).

STUDY AREA

The Attoyac Bayou watershed is located in deep East Texas, in portions of Nacogdoches, Rusk, San Augustine, and Shelby counties (Figure 1). The watershed covers 1685 km^{-2} (650 mi^{-2}). The Attoyac Bayou flows southward from its head near Mt Enterprise in Rusk County (32°04' N, 94°42' W) to its confluence with the Angelina River in Sam Rayburn Reservoir (31°22' N, 94°19' W). The dominant cover within the watershed is forest (64%), with pastureland (23%) and roads, crops, and urban (13%) occupying the rest (USSCS, 1980). The dominant agricultural activities in the watershed include poultry production and the production of beef cattle (hay production and pasture). The Attoyac Bayou remains on EPA's 303d list for bacteria with agriculture listed as a contributing cause.

The topography of the watershed is dominated by rolling hills with flat flood plains around larger order streams. Elevations range from 67 m (220 ft) above sea level at the confluence with the Angelina River to 213 m (700 ft) at the headwaters. Streams formed as a result of headward erosion in Eocene sedimentary deposits. Weathering of these formations have created numerous soils within the watershed. In floodplains, primarily Inceptisols and Entisols formed in sandy and loamy alluvial deposits dominate. Upland soils range from deep, well drained loamy sands to sandy clay loams, with Ultisols and Alfisols being predominant.

The climate of the region is humid and sub-tropical, with hot summers and cool winters. The 119 cm (47 in) of average annual rainfall is fairly evenly distributed through the year, with April and May receiving the most rainfall. On average, Nacogdoches receives 89 rain days per year. The mean annual temperature is 18.7°C (66°F) with an average summer temperature of 27.2°C (81°F) and an average winter temperature of 9.5°C (49°F) (Chang et al., 1996).

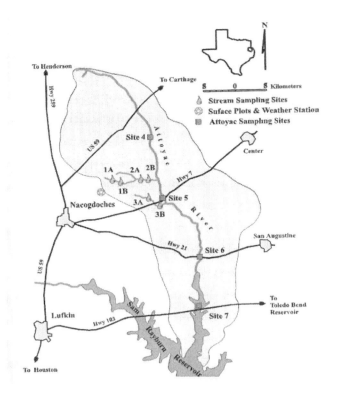

Figure 1. Map of sampling sites on the Waffelo and Terrapin Creeks in the Attoyac River Watershed in East Texas (From Cochran, 1996).

RUNOFF PLOT STUDY

Twelve runoff plots were established on private property, 8 km northeast of Nacogdoches in March, 1994. The soil classification at the site was a Nacogdoches clay loam, fine kaolinitic, thermic Rhodic Paleudalf. Slope at the site was approximately 5% and initial soil test P was in the very low class. Plot size was 1.8 m by 22.1 m with an area of 0.004 ha and the plots were established in Coastal Bermuda grass (Figure 2). The perimeter of each plot consisted of 15 cm metal flashing driven into the soil to prevent water from outside the plot from entering the plot and to direct water in the pot toward the RunOff Collection Apparatus (ROCA). The ROCA was constructed at the downhill end of the plot consisting of wooden collection apron at the soil surface, approach section, a five cm PVC pipe and a 1600 L metal storage tank (Figure 3). Broiler litter was applied to the plots at rates of 0, 5.6, 11.2 and 22.4 Mg ha^{-1} application^{-1} using a randomized block design with three replications. Blocks were determined using the results of water samples collected during four runoff events occurring before treatments were applied using a canonical correlation and multi-variate analysis of variance. A 4.6 m vegetative filter strip or buffer strip was maintained at the down hill end of the runoff plots. No broiler litter was applied to this section.

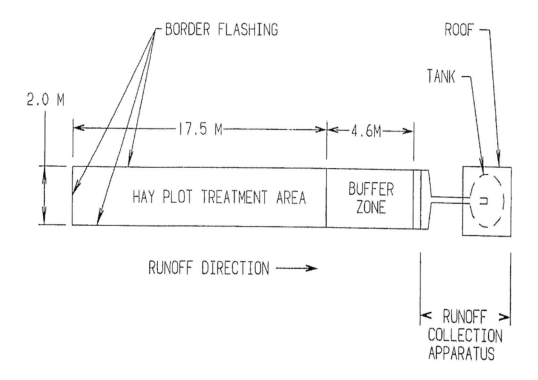

Figure 2. Aerial view of a runoff collection apparatus (ROCA) from the poultry litter land application rate study (From Stark, 1999).

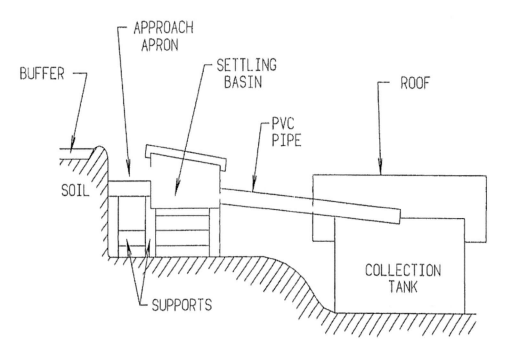

Figure 3. Cross sectional view of a runoff collection apparatus (ROCA) from the poultry litter land application rate study (From Stark, 1999).

Water volume after each rainfall event was measured in the collection tanks and samples or water were collected and preserved for laboratory analysis. Water preservation followed the guidelines of USEPA (1983) with laboratory analyses following the procedures of APHA (1992). Thirteen water quality parameters were monitored in samples collected from runoff events. Only the results for total P (TP) and phosphate (PO_4-P) will be reported in this chapter. Chemical elements (P) were measured using ICP spectroscopy and ions (PO_4) using ion chromatography. Litter N was measured using a Kjeldahl procedure while other mineral nutrients were measured on a nitric and perchloric acid digest using ICP. Soil analyses used the standard soil testing procedures of the SFASU Soil Testing Laboratory and consisted of extraction with the Texas macronutrient soil test extractant (1.4 M NH_4OAc + 1 M HCl + 0.025 M EDTA extraction solution), a 20:1 solution to soil ratio, and a shaking time of one hour.

The first litter application was made on June 2, 1995 followed by additional applications on January 17, 1996, February 3, 1997 and April 2, 1997. By the time of the 4th application of litter the plots had received 284, 570, 1139 kg ha^{-1} of P at the 5.6, 11.2 and 22.4 Mg ha^{-1} litter rates, respectively.

Only P (TP and PO_4-P) are discussed here since it is generally accepted as the long term broiler litter application rate limiting factor (Sharpley et al., 2003). Following the first application of broiler litter in June of 1995 there were 13 rain events but only 4 of those events produced enough runoff for complete water analysis on all 12 plots. The treatment TP and PO_4-P means from the two largest rainfall events are presented in Tables 1 and 2. There was no statistically significant broiler litter treatment effect on the two P parameters. When all 4 complete runoff events were combined using a split pot design with event in the whole plot and treatment in the sub-plot there was still no statistically significant treatment effect. These events occurred in September, October and November, 3 to 6 months after the litter was applied.

Table 1. Sequence of major runoff events and their effects on total P concentration in runoff water from the 12 runoff plots by litter application rate

Study	Date	Rainfall (cm)	Runoff (L)	No of Litter Applications	Rate per Application (Mg ha^{-1})			
					0	5.6	11.2	22.4
					Total P (mg L^{-1})			
Whiteside	10/03/1995	9.14	29	1	0.39	0.38	0.35	0.82
(1996)	12/18/1995	7.95	48	1	0.39	0.32	0.32	0.26
Blackerby	4/15/1996	2.16	17	2	2.82	3.27	3.62	3.63
(1996)	4/21/1996	1.78	11	2	2.51	3.84	1.37	2.49
Stark	2/13/1997	7.37	896	3	0.07 c	2.71 b	13.86 a	17.51 a
(1999)	2/21/1997	4.32	671	3	0.84 c	3.21 c	10.11 b	13.26 a
	3/03/1997	8.13	1304	3	0.82 c	3.06 c	8.97 b	13.21 a
Stark, 1999	Mean of 5 events after 4th application			4	1.70 b	2.18 b	8.08 a	9.07 a

Means in a row followed by the same letter are not significantly different ($\alpha = 0.05$) using Duncan's multiple range test.

Table 2. Sequence of major runoff events and their effects on PO$_4$-P concentration in runoff water from the 12 runoff plots by litter application rate

Study	Date	Rainfall (cm)	Runoff (L)	No of Litter Applications	Rate per Application (Mg ha^{-1})			
					0	5.6	11.2	22.4
					PO$_4$-P (mg L^{-1})			
Whiteside	10/03/1995	9.14	29	1	0.39	0.26	0.36	0.54
(1996)	12/18/1995	7.95	48	1	0.18	0.43	0.58	0.90
Blackerby	4/15/1996	2.16	17	2	0.16	0.05	0.26	0.13
(1996)	4/21/1996	1.78	11	2	0.08	0.07	0.10	0.05
Stark	2/13/1997	7.37	896	3	0.09 b	0.51 b	3.92 a	4.97 a
(1999)	2/21/1997	4.32	671	3	0.23 a	0.14 a	0.49 c	0.23 b
	3/03/1997	8.13	1304	3	0.10 c	0.57 c	2.74 b	3.97 a
Stark, 1999	Mean of 5 events after 4th application			4	0.13 b	0.30 b	2.48 a	2.94 a

Means in a row followed by the same letter are not significantly different ($\alpha = 0.05$) using Duncan's multiple range test.

On January 17 a second application of litter was made. There were only 4 runoff events in the 4 month period following application. This low level of rainfall is unusual in the East Texas region. The TP and PO$_4$-P results of the two larger events are presented in Tables 1 and 2. Again, the litter application rate did not have a statistically significant effect on the concentration of the two parameters in runoff water in either of the two events nor in the average of the four events using the split-plot design discussed earlier. No data were collected from the plots from April 21, 1996 until November 25, 1996 when results from one event were collected. Again there was no statistically significant litter treatment effect on TP or PO$_4$-P concentration in the runoff water (data not presented).

Following the February 3, 1996 application of litter there were 5 runoff events. This resulted in statistically significant litter rate effects for all three events on TP and on two of the three events for PO$_4$-P (Tables 1 and 2). Soil samples collected from 0 to 2.5 cm depth on March 31, 1997 had 6, 43, 183, and 467 mg P dm^{-3} at the 0, 5.6, 11.2 and 22.4 Mg ha^{-1} application^{-1} rates of broiler litter, respectively. There was little or no treatment effect on soil test P for samples taken deeper than 5 cm. Though the data are not reported here, the surface 2.5 cm soil test P was had much higher correlation with TP or PO$_4$-P in runoff water than soil samples taken from 0 to 5 cm, 0 to 10 cm, or 0 to 15 cm depths.

Tables 1 and 2 also illustrate mean TP and PO$_4$-P for 5 runoff events following the fourth application of litter which occurred on April 2, 1997. These events occurred between April 5, 1997 and August 9, 1997. Three of the five events, including the last event, produced statistically significant treatment effects on both TP and PO$_4$-P concentration. Even though there are increases in P concentrations in runoff water as a result of continued broiler litter application, the amount of total P transported from the plots equates to less than 0.5 kg ha^{-1} per rainfall even and the amount of PO$_4$-P is less than 100 g ha^{-1} event^{-1}, even at the highest litter application rate.

This somewhat unique study of P in runoff water from repeated applications of broiler litter over a two year period leads to several conclusions. In the early part of the study after the first two litter applications, low P soil and the vegetated filter strip or buffer zone were probably acting as a sinks for P in the runoff water. As the accumulation of labile P (soil test P) in the surface 2.5 cm (1 inch) of soil increased as a result of repeated litter applications, the soil itself was no longer a sink for dissolved P and the beneficial effects of the buffer zone were overpowered by the large amounts of P in the runoff water. The concentration of labile P or soil test P in the surface 2.5 cm of soil is a better predictor of P runoff potential than the more traditional sampling depth of 0 to 10 cm or 0 to 15 cm used to access soil fertility status as a basis for recommending fertilizer. In fact the 11.2 Mg ha^{-1} litter rate gave as soil test P of 10 mg dm^{-3} (mg kg^{-1} or ppm) on the 0 to 15 cm soil sample while the 0 to 2.5 cm sample gave 187 mg dm^{-3}. The deeper soil sample would have recommended a significant P fertilization rate while the shallower sample indicates no P is needed and the probability of P in the runoff water is quite high. This conundrum indicates that soil test P and soil test P sampling practices, which were developed for determining fertilization rates, my not be the best way to access heavily manured soils were P runoff potential is high. Soil test summaries from the SFASU Soil Testing Laboratory (Isom, 2002) and the results of this plot study indicate that a significant number of pastures in the broiler producing counties of East Texas my have excessive P runoff. On the other hand, the quantity of P transported per ha of area is quite small.

TRIBUTARY STREAM STUDY

In addition to plot level studies, it is important to monitor the effects of poultry litter applications at the watershed scale as well. This is primarily due to spatial differences in land use patterns which actually occur at different scales within watersheds. In other words, edge-of-field nutrient losses fail to account for downstream effects (Gburek et al., 2000). The analysis of water quality impacts of poultry litter on stream ecosystems is necessary to develop a better understanding of overall potential environmental impacts. In particular, the use of bioassessment techniques in conjunction with chemical analyses in lotic waters provides for a more comprehensive evaluation of aquatic systems (Plafkin et al., 1989). One disadvantage to watershed studies at this scale is, as noted below, that it is impossible to control for all of the land use activities within the watershed and isolate water quality impacts to poultry production alone.

METHODS

Study Watersheds

Monitoring was conducted on two tributary streams, the Waffelo and Terrapin Creeks, in the Attoyac basin draining watersheds with a high number of broiler farms (Figure 1). On these two streams, sampling was conducted on an upstream segment draining a predominately forested area ("A" site) on which there was limited litter application and one downstream on which litter applications regularly occurred ("B" site). Two sets of sampling sites were established on the Waffelo Creek (1A,1B; 2A,2B), with one set on the Terrapin (3A, 3B). Streams in the B sites typically had between 10 and 150 m (15-500 ft) buffers consisting of forest or unfertilized pasture between the streams and areas on which poultry litter was applied.

Broiler houses have been in operation in the immediate vicinity of the pastured stream monitoring sites since the late 1940s, with many additional houses being built more recently. Litter from these farms has been routinely land-applied to hay meadows and pastures around these houses since they have been in operation. The typical rate of application on these fields from these farms was about 2.7-4.5 metric tons ha^{-1} (3-5 tons ac^{-1}) annually.

Stream Monitoring

Storm event and weekly baseflow monitoring occurred on these six sites on both streams from March 1995 to December 1996. Additional baseflow bacteriological monitoring occurred weekly from July to October 2001. Stream discharge was determined by establishing a rating curve by manually measuring discharge at each site using a Gurley current meter for each depth encountered. An Isco 3230 bubbler flow meter was used to monitor stream depth and depths were converted to discharge using the rating curve developed for each site. An Isco 3700 pumping sampler was used to automatically collect storm runoff event samples at each site. A clear-vinyl bubbler tube (0.64 cm diameter) and translucent-vinyl sample collection tube (1.27 cm diameter) ran from the flow meter and sampler, respectively through a 3.8 cm poly-vinyl-chloride (PVC) pipe. The PVC pipe was secured to the stream bottom with steel rods. A steel strainer was inserted into the sampler tube to prevent debris from being sucked into the sampler tube. The pipes were oriented downstream to provide less resistance to streamflow, to prevent damage, and to prevent sediment accumulations in the pipe. Stream depths were measured manually and compared to depths recorded by the bubbler and the bubbler was found to be accurate.

Storm runoff event samples were automatically collected following a precipitation event large enough to cause overland flow and raise the creek above a pre-determined sample initiation level. The sampler continued collecting samples over a pre-determined time interval until all 24 one L polyprophylene bottles in the sampler were full or until the stream fell below the sampler initiation level.

Weekly grab samples were collected in 1 L polyprophylene bottles. At the time of grab sample collection, dissolved oxygen and water temperature were measured in the field with an Orion Model 820 dissolved oxygen meter. The benthic macroinvertebrate community was sampled in April and October 1995 by collecting bottom sediment and coarse particulate organic matter samples at each stream site by methods consistent with Plafkin et al. (1989), TNRCC (1993), and Lind (1985).

For determination of fecal coliform (FC) and e-coli bacterial concentrations, grab samples were collected monthly between March and December 1996 and weekly between July and October 2001. Samples were brought to SFASU on ice and were analyzed within 24 hours. Bacterial enumeration followed the membrane filtration method as described in Standard Methods by APHA (1992).

Benthic macroinvertebrate fauna were sampled during two periods in April and October 1995 and analyses were consistent with protocols recommended by Plafkin et al. (1989), TNRCC (1993), and Lind (1985). Detailed discussions of methodology are given in Cochran (1996). Analyses included the Shannon and Weaver (1964) diversity index, EPT (orders Ephemeroptera, Plecoptera, and Tricoptera) index (Twidwell and Davis, 1989), evenness (Pielou, 1975), and richness (Margalef, 1957).

Weather data were monitored at the project weather station. Air temperature and humidity were measured with a Weathertronics Model 5020-A hygorthermograph. This instrument was housed in a standard U.S. National Weather Service Stephenson shelter. Rainfall was measured with a Belfort Model 5-780 universal recording raingage. Charts were changed weekly on these instruments.

Laboratory and Data Analyses

Stream water samples were collected within 24 hours after a storm event and delivered in an ice chest to the SFASU Soil, Plant, and Water Analysis Laboratory for analysis of nitrate-nitrogen (NO_3-N), total Kjeldahl nitrogen (TKN), orthophosphate phosphorus (PO_4-P), total phosphorus (TP), potassium, total suspended solids (TSS), and pH. Laboratory analysis was conducted following Standard Methods for the Analysis of Water and Wastewater (APHA, 1992) and Methods for Chemical Analysis of Water and Wastes (USEPA, 1983). The Student's t-test was employed to determine differences ($\alpha = 0.05$) between and among sites. Complete data sets and analysis for the study period are further reported in Cochran (1996) and McBroom (1997).

RESULTS AND DISCUSSION

Nutrients

Based on 82 weekly grab samples and 63 storm runoff event samples collected during the 21-month study period, there were no evidence was found that would indicate significant water quality problems on the Waffelo or Terrapin Creeks. The Texas Commission on Environmental Quality (TCEQ) has yet to finalize nutrient criteria for rivers and streams, so nutrient concentrations cannot be compared to standards. All NO_3-N samples were below the EPA standard of 10 mg L^{-1} for drinking water. However, Sawyer (1947) and Vollenwieder (1968) have suggested that when concentrations of NO_3-N and PO_4-P exceeds 0.3 mg L^{-1} and 0.1 mg L^{-1}, respectively, then eutrophication can begin to become a problem. About 56% of the NO_3-N samples and 15% of the PO_4-P samples exceeded these suggested limits.

Generally, concentrations of the dissolved forms of nitrogen and phosphorus (NO_3-N and PO_4-P) were significantly greater from the pastured sites than forested for storm event samples (Table 3). Only NO_3-N was significantly greater for grab baseflow samples (Table 4). TKN and TP were not found to be statistically significant for either baseflow or storm event samples. When all sample types (grab and stormflow) were combined, TP was found to be significantly greater from pastured sites. As expected, storm event concentrations were significantly greater than baseflow. This highlights the importance of storm-event sampling for nonpoint pollution monitoring. Furthermore, one or two storms often result in the majority of the annual load for nonpoint source pollutants (McBroom et al., 2008a).

While significantly greater concentrations of nutrients were measured from the pastured sites, the overall magnitude of this difference was relatively small, especially when compared with results from other studies where land applications of poultry litter occur. For example, in the Blackland Prairie region of central Texas, Harmel et al. (2004) reported mean NO_3-N ranging from 1.04 to 4.78 mg L^{-1} from watersheds after two years of litter applications while PO_4-P concentrations ranged from 0.22 to 0.39 mg L^{-1}.

Table 3. Nutrient concentrations (mg L^{-1}) for six sampling sites on the Waffelo and Terrapin Creeks in the Attoyac Watershed for automatic stormflow samples collected from both forested (A) and pastured (B) sites

		Waffelo Creek				Terrapin Creek			
		1A	1B	2A	2B	3A	3B	All Forested	All Pastured
PO_4-P	Mean	0.011	0.011	0.007	0.051	0.011	0.013	0.009	0.030
	Std Dev	0.011	0.019	0.012	0.124	0.019	0.021	0.016	0.087
	N	13	11	51	57	63	57	127	125
TP	Mean	0.080	0.328	0.181	0.233	0.172	0.196	0.147	0.186
	Std Dev	1.722	0.316	0.158	0.226	0.202	0.205	0.182	0.241
	N	13	11	46	55	59	52	118	118
NO_3-N	Mean	0.146	0.176	0.187	0.447	0.547	0.680	0.359	0.530
	Std Dev	0.153	0.117	0.120	0.334	0.302	0.271	0.294	0.331
	N	13	11	53	59	63	58	129	128
TKN	Mean	0.770	0.818	1.072	1.114	1.030	0.794	1.013	0.967

Table 3. (Continued)

		Waffelo Creek				Terrapin Creek			
		1A	1B	2A	2B	3A	3B	All Forested	All Pastured
	Std Dev	1.142	0.708	0.922	0.901	0.829	0.815	1.101	1.283
	N	12	11	53	59	59	56	124	126
K	Mean	2.723	3.991	3.349	3.155	2.332	2.039	2.417	2.698
	Std Dev	0.702	2.031	4.959	1.489	1.092	0.645	0.950	1.518
	N	13	11	47	53	47	42	107	96

Note: Bold underlined means within pairs are significantly greater ($\alpha = 0.05$) using the paired T-test than the corresponding site.

Table 4. Nutrient concentrations (mg L^{-1}) for six sampling sites on the Waffelo and Terrapin Creeks in the Attoyac Watershed for grab baseflow samples collected from both forested (A) and pastured (B) sites

		Waffelo Creek				Terrapin Creek			
		1A	1B	2A	2B	3A	3B	All Forested	All Pastured
PO$_4$-P	Mean	0.010	0.006	0.007	0.006	0.006	0.007	0.007	0.006
	Std Dev	0.037	0.013	0.019	0.010	0.009	0.015	0.024	0.013
	N	74	76	82	82	80	80	236	238
TP	Mean	0.035	0.035	0.070	0.071	0.026	0.051	0.058	0.076
	Std Dev	1.676	0.038	0.067	0.046	0.028	0.147	0.089	0.133
	N	66	70	74	75	75	74	215	219
NO$_3$-N	Mean	0.100	0.144	0.151	0.392	0.647	0.788	0.303	0.446
	Std Dev	0.089	0.188	0.090	0.139	0.185	0.226	0.279	0.324
	N	74	76	82	82	80	80	236	238
TKN	Mean	0.525	0.496	0.595	0.851	0.489	0.464	0.563	0.604
	Std Dev	1.012	0.565	0.608	1.663	1.055	0.819	0.772	0.907
	N	70	72	77	76	75	74	222	222
K	Mean	3.128	3.506	2.828	2.536	1.289	1.901	2.458	2.706
	Std Dev	1.236	1.566	1.479	1.082	0.629	2.336	1.406	1.875
	N	69	72	77	77	74	74	220	223

Note: Bold underlined means within pairs are significantly greater ($\alpha = 0.05$) using the paired T-test than the corresponding site.

Nutrient concentrations were converted into mass losses by multiplying concentration by discharge and dividing by watershed area (Table 5). This allows for the examination of land use effects normalized by discharge and watershed area. As expected, greater total flow was observed from the downstream pastured sites. Furthermore, all nutrient parameters and sediment were greater from the pastured sites. However, differences were most pronounced on the Terrapin Creek, with only NO$_3$-N being significantly greater on the pastured site on the Waffelo. Losses recorded from these streams were within the range of those found by Chang et al. (1983) on several other streams in the East Texas area. These values are also within the rage found by Harmel et al. (2004). Furthermore, Harmel et al. (2006) compiled nutrient load data from across the United States into the Measured Annual Nutrient loads from Agricultural Environments (MANAGE) database.

Table 5. Total streamflow, nutrient, and sediment mass losses per hectare for six sampling sites on the Waffelo and Terrapin Creeks in the Attoyac Watershed for the 1996 water year (10/1995-9-1996) collected from both forested (A) and pastured (B) sites

Site	Flow —cm—	PO_4-P	TP kg ha^{-1}	NO_3-N	TKN	Sediment
Waffelo Creek						
1A	14.70	0.02	0.05	0.11	0.59	49.51
1B	16.95	0.01	0.06	0.13	0.51	251.17
2A	4.64	0.00	0.03	0.04	0.20	14.94
2B	5.29	0.00	0.03	0.16	0.27	16.79
Terrapin Creek						
3A	9.30	0.00	0.02	0.51	0.64	34.05
3B	17.02	0.01	0.11	1.23	0.65	46.89
All Forested	9.54	0.01	0.03	0.22	0.48	32.84
All Pastured	13.09	0.01	0.07	0.51	0.48	104.95

Note: Bold underlined means within pairs are significantly greater ($\alpha = 0.05$) using the paired T-test than the corresponding site.

Nutrient losses reported from the Waffelo and Terrapin creeks tend to be on the lower end of levels reported in the MANAGE database from other agricultural areas in the United States (Harmel et al., 2006).

Sediment

Total suspended sediment concentrations were not found to be different between the pastured and forested sites for either grab or baseflow samples (Table 6, 7). This indicates that there is no evidence for concluding that current land uses in these two watersheds are resulting in increased erosion. When concentrations were converted to mass losses however, significantly greater sediment losses were observed from site 3B and from all pastured sites combined (Table 5). This in part can be explained by significantly higher discharge rates being observed at sites 1B, 3B, and the combination of pastured sites. This is to be expected since discharge increases with watershed area. In addition, sites 1B and 3B, were immediately downstream of highway crossings, and greater sediment losses could have resulted from the effects of the bridge structures on stream banks. Stream banks can often be the greatest source of sediments for streams in easily eroded marine geology (McBroom et al., 2008b). Sediment yields measured in the current study fall within the range reported by Chang et al. (1983) for similar sized streams in East Texas. These losses are much lower than reported by McBroom et al. (2008b), for forested watersheds, partly due to the large watershed area in the current study. In general, smaller headwater streams are steeper, and their watersheds tend to have greater sediment losses per unit area than larger streams (Chang, 2006).

Table 6. Water quality parameters, sediment, and discharge for six sampling sites on the Waffelo and Terrapin Creeks in the Attoyac Watershed for automatic stormflow samples collected from both forested (A) and pastured (B) sites

		Waffelo Creek				Terrapin Creek		All	All
		1A	1B	2A	2B	3A	3B	Forested	Pastured
TSS	Mean	280.46	709.55	168.79	129.44	205.49	179.30	156.77	160.03
(mg L^{-1})	Std Dev	408.31	966.30	216.02	124.24	237.16	212.52	231.97	348.65
	N	13	11	53	59	63	58	129	128
pH	Mean	6.01	6.12	6.21	6.25	6.03	6.35	6.10	6.29
	Std Dev	0.329	0.431	0.30	0.25	0.40	0.32	0.37	0.31
	N	13	11	53	59	63	59	129	129
EC	Mean	73.85	88.64	111.33	99.31	69.32	73.59	86.84	86.64
(μmhos)	Std Dev	21.59	39.46	26.81	26.20	32.48	25.13	35.54	29.53
	N	13	11	52	59	63	59	128	129
Discharge	Mean	12.50	197.56	389.92	410.72	389.98	499.83	358.41	431.04
(L sec^{-1})	Std Dev	8.18	212.30	538.79	486.15	293.86	432.22	410.47	448.44
	N	11	11	48	53	58	53	118	117

Note: Bold underlined means within pairs are significantly greater ($\alpha = 0.05$) using the paired T-test than the corresponding site.

OTHER WATER QUALITY PARAMETERS

Mean dissolved oxygen concentrations were well above the TCEQ standard of 5.0 mg L^{-1} (Table 7). During summer months, concentrations were typically lower when water temperatures were higher. Overall, 14% of individual samples were below the TCEQ standard. However, this standard is based on 24 hour averages, and these were individual measurements. For smaller order streams, even in pristine forested areas, summer dissolved oxygen measurements may naturally fall below 5.0 mg L^{-1} (Ice and Sugden, 2003). In addition, dissolved oxygen was not found to be significantly different between forested and pastured sites.

Mean stream pH fell within the TCEQ standard of 6.0-8.5 (Table 6 and 7). Overall, stormflow and baseflow samples from pastured sites tended to have a significantly greater pH than was recorded from forested sites. This difference could be due to leaching of humic acids from forest leaf litter and applications of lime to pastured areas. However, the magnitude of this difference was relatively small.

Electronic conductivity was not high enough to suggest any potential sources of ionizing pollutants in these watersheds. Overall, pastured sites were not significantly different from forested sites. Stormflow and baseflow conductivities were similar.

Table 7. Water quality parameters, sediment, and discharge for six sampling sites on the Waffelo and Terrapin Creeks in the Attoyac Watershed for grab baseflow samples collected from both forested (A) and pastured (B) sites

		Waffelo Creek				Terrapin Creek		All	All
		1A	1B	2A	2B	3A	3B	Forested	Pastured
TSS	Mean	10.82	16.64	18.52	18.65	16.49	17.06	41.16	38.01
(mg L^{-1})	Std Dev	26.10	28.36	22.28	20.77	21.96	16.61	120.67	101.04
	N	74	76	82	81	81	80	237	237
DO	Mean	8.31	8.61	6.22	6.33	8.00	8.03	7.46	7.60
(mg L^{-1})	Std Dev	1.70	1.63	2.05	2.23	1.19	1.19	1.90	1.99
	N	57	57	67	67	66	67	190	191
pH	Mean	6.09	6.28	6.36	6.38	6.29	6.39	6.25	6.35
	Std Dev	0.24	0.26	0.26	0.24	0.22	0.21	0.27	0.24
	N	74	76	82	82	81	80	237	238
EC	Mean	97.08	113.24	114.85	101.99	59.10	70.43	90.25	94.58
(mhos)	Std Dev	14.48	18.92	29.96	30.00	30.80	29.61	35.42	32.00
	N	74	75	82	82	81	80	237	237
Discharge	Mean	12.50	17.22	118.90	120.26	123.49	265.22	88.05	135.59
(L sec^{-1})	Std Dev	8.18	18.22	328.68	225.82	181.30	212.86	227.74	207.07
	N	71	74	82	80	80	79	233	233

Note: Bold underlined means within pairs are significantly greater ($\alpha = 0.05$) using the paired T-test than the corresponding site.

VEGETATED FILTER STRIPS

The use of vegetative buffer strips is a very effective way to reduce the quantity of nutrients (McBroom et al., 2008a) and sediment (McBroom et al., 2008b) from land use activities. In agricultural areas, buffer zone widths of 20 meters have been found to reduce by 90% the amount litter constituents that may enter streams (Westerman et al., 1983; Chaubey et al., 1994; Mikkelsen and Gilliam, 1995). For the Waffelo and Terrapin Creeks, natural vegetative buffer zones between 15 and 150 meters were maintained between the litter application areas and stream banks in the pastured watersheds. This was in part due the fact that the floodplains around the streams were wet during the time of year when runoff was most likely to occur. Broiler litter applications did not occur during these periods as a result, and grazing or forestry was the predominant land uses. Furthermore, forested zones along stream banks also help buffer against changes in water quality parameters like temperature and dissolved oxygen. As noted in the plot study above, vegetative strips do have a finite buffering capacity. However, with the combination of trees and other woody and herbaceous vegetation, these buffers appear to be helping to reduce the overall quantity of nutrients and sediment leaving these watersheds.

BACTERIA

Bacterial Concentrations

Based on the 19 fecal coliform (FC) and 10 E-coli samples collected during the study period, it was found that overall FC ranged from 5 to 4,300 cfu/100 ml with medians equal to or exceeding the standard 400 cfu/100 ml water for contact use for single grab samples at all six sites (Table 8). From 36 to 79% of FC concentrations exceeded the contact standard. The contact standard also requires that concentrations be below 400 cfu/100 ml in more than 10% of all samples (TNRCC, 1995). Using this standard, 37 to 79% of FC concentrations were greater than this level. The contact standard is base on a five-sample, 30-day geometric mean concentration less than 200 cfu/100 ml. Two 30-day periods were analyzed, from 7/31/01 to 8/28/01, and 9/11/01 to 10/9/01. No FC samples exceeded the TCEQ 30-day geometric mean standard of 200 cfu/100 ml during these periods on Terrapin Creek. However, on Waffelo Creek, which was on EPA's 303d list during the sampling period, site 1B exceeded this standard for both periods and site 2A, a forested site, was in violation from 9/11/01 to 10/9/01.

Relatively high concentrations have been reported by many other researchers in agricultural areas (Doran and Linn, 1979; Edwards et al., 1997; Howell et al., 1995; Lindsey, 1975; Richardson, 1975; Robbins et al., 1972). Sherer et al. (1992) found that animal traffic cause increased turbulence and resuspension of sediment bound bacteria which resulted in erratic FC concentrations. In the current study area, cattle had direct access to streams at all pasture sites, possibly resulting in such resuspension.

In Nebraska, Doran et al. (1981) found that wildlife may contribute FC bacteria in excess of published standards, indicating that standards developed for point source pollution might be inappropriate for nonpoint source pollution for FC. In that study, more than 90% of FC samples in both grazed and ungrazed pastures exceeded the 200 cfu/100 ml criterion. In eastern Oregon, Tiedemann et al. (1987) reported that natural forest watersheds could have FC concentrations exceeding 500 cfu/100ml. Forested watersheds may not necessarily have lower FC concentrations than other land uses at all times.

Table 8. Descriptive Statistics for fecal coliform (FC) and E-coli (E-C) concentrations (cfu/100 ml) for three pairs of upstream forested (A) and downstream pastured (B) sites in the Attoyac River Watershed in East Texas

Statistic	1A FC	1A E-C	1B FC	1B E-C	2A FC	2A E-C	2B FC	2B E-C	3A FC	3A E-C	3B FC	3B E-C
Mean	200	19	550	67	560	81	245	30	308	16	260	21
Median	660	3	733	15	755	26	569	14	560	6	595	12
Std. Dev.	1060	46	988	154	610	118	653	39	756	28	752	34
Minimum	10	1	59	2	15	4	17	6	35	0	5	1
Maximum	4300	150	3140	504	1950	368	2100	116	2870	94	2800	114
N	19	10	19	10	19	10	19	10	19	10	19	10
% Obs. > TCEQ Std*	42.1	0	52.6	10	78.9	0	47.4	0	36.8	0	52.6	0

*TCEQ primary contact standard for FC is 400 cfu and 394 cfu for E-coli for a single sample.

In the current study, significant wildlife contributions are likely, especially in areas where instream wild hog rooting, wallowing, and defecation was observed, in both pastured and forested sites. Current broiler litter land-application rates on pasturelands did not result in FC concentrations significantly higher than the wildlife activity from forested watersheds.

USEPA (1986) indicates that E-coli may be a more reliable indicator of bacterial contamination than FC. In this study, only one sample collected from 7/31/01 to 10/9/01 exceeded the TCEQ single sample criteria of 394 cfu/100 ml. This was at site 1B where rural residences were in close proximity to the stream. E-coli concentrations are typically evaluated based on a geometric mean of 5 samples over a 30-day period. Two 30-day periods were analyzed, from 7/31/01 to 8/28/01, and from 9/11/01 to 10/9/01. No E-coli samples exceeded the TCEQ 30-day standard of 126 cfu/100 ml during these periods. Furthermore, no significant differences in E-coli concentrations were observed among monitoring sites. In addition there was less variation observed with E-coli data (Table 8). As a water quality indicator, E-coli tended to better reflect actual land use conditions, indicating that the E-coli standard may be more appropriate.

SOURCES OF CONTAMINATION

It has been proposed that the ratio of fecal coliform to fecal streptococcus (FC/FS) may be used as an indicator for possible bacterial with human sources for FC/FS ratios > 4.0, domestic animal sources for ratios 0.1 - 0.7, and wildlife sources for ratios < 0.1 (Geldreich, 1976). Other researchers have proposed modifications of this, with Howell et al. (1995) reporting a FC/FS ratio of 0.1 to 4.0 to indicate domestic animal contamination and Gary et al. (1983) using 2.5.

For the current study, mean FC/FS ratios ranged from 0.93 at 1A to 3.1 at 1B, indicating domestic sources of contamination at all sites based on Geldreich (1976) and Howell et al. (1995). At site 1A, a ratio < 0.1 would be expected, since it was fully covered by mature forests with no potential sources of bacteria other than wildlife. However, the FC/FS ratio was never less than 0.1, indicating that the calculated source of contamination by FC/FS ratio was inconsistent with land use at this site. On site 1B, the FC/FS ratio greater than 4.0 three out of nine months or 30% of the observations, suggesting possible sources of human contamination. There were residences immediately upstream of the 1B sampling-site which could have contributed human sources of bacteria.

Variations in the consistency of FC/FS ratios as indicators for sources of bacteria contamination in the study areas were observed. Many other studies have also reported inconsistencies in using the FC/FS ratio (Doran and Lin, 1979; Howell et al., 1995; Edwards et al., 1997; Boyer and Pasquarell, 1999). Some of the reasons that these variations exist are: 1) different mortality rates of FC and FS, 2) distance of animal activity to stream channels, 3) rainfall and runoff, 4) watershed characteristics, and 5) regrowth and residence durations. In this study, data were sufficient to characterize bacteria water quality conditions, but not sufficient to make a detailed reassessments of the FC/FS ratio criteria. A more effective means of determining bacterial sources would be through targeted monitoring of outfall areas or the use of DNA source tracking.

Aquatic Environment and Bacteria

Correlations were analyzed between FC concentrations and other water parameters including discharge rate, water temperature, DO, salinity, EC, and pH. Correlation coefficients (r) were low, ranging from 0.027 to –0.587. Water pH was the only parameter with significant r-values (-0.509 for A sites and –0.587 for B sites) at the 95% probability level. The r-values were greater than 0.16 for the other parameters. Tiedman et al. (1987) reported on five streams in Oregon and found significant correlations between FC and pH and turbidity, while discharge, conductivity, and temperature were not correlated. In the Reynolds Creek watershed in Idaho, researchers found that except for water temperature and chloride, total and FC concentrations did not show a significant relationship with other physical and chemical parameters (Stephenson and Street. 1978). Edwards et al. (1997) did not find a significant relationship between runoff and FC, while Robbins et al. (1972) did.

The relationships between FC and water parameters often display marked variations within and between studies. This reflects the complexity of nonpoint source pollution monitoring and management, where factors such as bacterial production, transport, and life span in conjunction with livestock management, manure application rate and schedule, precipitation and overland flow, soil conditions, and aquatic environment often result in inconsistent results. In order to account for this variation, sampling schemes must be implemented that will take these variables into account. As Stephenson and Street (1978) observed, variations in livestock management along the streams often overshadowed the effects of aquatic parameters. This makes relationships between hydrologic parameters and bacterial pollution and land use difficult to define and predictive models difficult to develop (Edwards et al., 2000).

Rapid Bioassessment

During the sample dates, dissolved oxygen levels ranged from 6.1 to 8.5 mg L^{-1}, well above the TCEQ standard of 5.0 for streams in East Texas. For benthic macroinvertebrates collected during this period, mean diversity index (H) and evenness (E) values indicated moderate stress for aquatic life (H >1.0, E>0.5). However, the EPT index was greater than 7, for all sites except 3B. This indicates high to exceptional habitat, while the 6 measured at 3B indicates intermediate habitat. The most commonly collected organisms included members of Ephemeroptera (Mayflies) and Chironomidae (bloodworms), with large numbers of Anisoptera (dragonfly), Zygoptera (damselfly), and Trichoptera (caddisfly) (Cochran, 1996). While insufficient data were collected for detailed statistical analyses to be conducted between sites, these data to indicate that the two study streams are supporting the appropriate aquatic organisms. This lends support the conclusion drawn from the physiochemical water quality analyses that current land uses are not significantly degrading these streams.

ATTOYAC BAYOU

Four sites along the Attoyac Bayou were grab-sampled weekly from April 1995 to March 1996. Sampling sites were located down the river, with site 4 being the farthest upstream, and site 7 being farthest downstream, where the Attoyac empties into Sam Rayburn Reservoir. In general, greater effects from broiler applications were expected further downstream, since the dominant land use in the headwater site was forest and more potential effects of agriculture and small municipalities were likely further downstream. Dissolved oxygen was measured in the field while the other parameters were analyzed for in the lab (pH, EC, NO_3-N, PO_4-P, K, TP, TSS, TKN). Discharge data were available from the USGS gage on site 6 at the Highway 21 Bridge (Station 08038000). Due to the lack of rainfall during the sampling period, these samples generally represent baseflow conditions for the Attoyac.

Average grab sample concentrations from the Attoyac Bayou were similar to that measured from the tributary streams (Table 9). ANOVA using Duncan's Multiple Range Test was performed on the water quality parameters and sampling sites were compared. Only for PO_4-P and NO_3-N were sites on the bayou found to be significantly different. In general, concentrations of PO_4-P declined from the headwaters to the confluence with Sam Rayburn. The magnitude of this difference is quite small. Most of the samples collected during the study period were near the method detection limit and these differences were due to two or three higher concentration samples collected from site 4. In addition, dilution from greater stream discharge could have accounted for this. Regardless, soluble P concentrations in the Attoyac Bayou were found to be low and not likely to result in eutrophication problem.

Soluble nitrogen was also found to be significantly different between sites (Table 9). Unlike with P, concentrations were lowest at the headwater site and increased further downstream, only to decrease again at the confluence with Sam Rayburn. It is possible that dilution of the NO_3-N was occurring at the confluence. As mentioned above, forest was the dominant land use for the headwater site. Greater inputs of NO_3-N would be expected at sites 5 and 6 from agricultural, residential, and small municipal sources. While not excessively high, especially when compared with concentrations measured from other agricultural areas (Harmel et al., 2004), these trends in NO_3-N concentrations suggest that future monitoring is warranted.

Concentrations of the other parameters were generally low and were not found to be significantly different among sampling sites (Table 9).

Table 9. Mean concentrations of water quality parameters collected from 4 sites on the Attoyac Bayou between April 1995 and March 1996

Site	PO4-P	TP	NO3-N	TKN	K	pH	EC	DO	TSS
			——mg L^{-1}——						
Site 4	0.019 a	0.107 a	0.235 b	0.532 a	2.141 a	7.1 a	104 a	8.2 a	22.9 a
Site 5	0.010 ab	0.087 a	0.457 a	0.521 a	2.037 a	7.1 a	97 a	8.1 a	30.1 a
Site 6	0.009 ab	0.070 a	0.518 a	0.587 a	1.852 a	7.2 a	101 a	8.1 a	24.2 a
Site 7	0.006 b	0.070 a	0.225 b	0.526 a	2.032 a	7.3 a	105 a	8.0 a	17.8 a
Average	0.011	0.083	0.358	0.542	2.016	7.2	102	8.1	23.8

Means with the same letter by parameter are not significantly different from each other at $\alpha = 0.05$.

Mean dissolved oxygen was above the TCEQ standard of 5.0 mg L^{-1}, with no samples on site 6, 2 out of 48 samples at sites 4 and 5, and 3 out of 48 samples on site 7 falling below this standard. This was most likely to occur during the summer when temperatures were highest. Stream pH was within the TCEQ standard of 6.0-8.0 throughout the study period. In general, no evidence was found that current land uses are impairing the nutrient and dissolved oxygen status of the Attoyac Bayou. However, further monitoring, especially for bacteria, may be justified.

AGNPS

The Agricultural Non-Point Source (AGNPS) model (Young et al., 1989) was used to simulate litter applications in the Waffelo Creek watershed. The 5,036 ha watershed was divided into 311 cells of 16.2 ha each. Each cell was then parameterized. Runoff, sediment, nitrogen, and phosphorus were then calculated from each cell and routed into adjoining downstream cells. Young et al. (1996) provides additional discussion on the AGNPS model.

Assuming 6.35 cm rainfall evenly distributed on the watershed with varying rates of litter applied to the whole watershed, the model predicted 0.05 mg L^{-1} of PO$_4$-P and 0.17 mg L^{-1} of NO$_3$-N in runoff waters leaving the watershed. One of the limitations of this model is that as a single event model it could not measure cumulative effects of litter applications. To compensate for this, soil N and P levels were increased to represent the effects of multiple years of broiler litter applications. This did not change the runoff outputs of NO$_3$-N and PO$_4$-P.

Model outputs were compared with field data, and it was found that when litter application rates were set to 0 in AGNPS, predicted runoff concentrations were similar to values actually measured at site 2B on the Waffelo. This suggests that the model may be overestimating runoff and concentrations of nutrients. These results should be interpreted with caution since limited calibration data were available and due to the inherent limitations of the model. Other models like the Soil and Water Assessment Tool (SWAT) may be more appropriate for modeling the hydrologic and water quality effects of poultry litter and for parsing out specific treatment differences (Green et al., 2007). However, these results do suggest that these soils can serve as a sink for large amounts of N and P from broiler litter applications.

CONCLUSION

While broiler litter applications can potential result in water quality degradations and eutrophication, the Attoyac Bayou and two of its tributary streams were not significantly impacted by the poultry industry. On the tributary streams, significant differences were noted in some of the nutrient parameters between pastured areas receiving broiler litter and upstream forested areas, but these differences could not be directly attributed to litter applications due to the multiple land use activities in the watershed. Furthermore, the overall magnitude of these differences was relatively small. No concerns were generated from analyses of the other water quality parameters measured.

Fecal coliform concentrations in the tributary streams frequently exceeded the TCEQ standard, and the Waffelo was listed on EPA's 303d list. However, E-coli concentrations were later compared with fecal coliforms, and it was found that E-coli was below the water quality standards, displayed less variation between sites, and is a better indicator of bacterial contamination. Since this time, the Waffelo Creek has been removed from the 303d list. The Attoyac Bayou was later added to the 303d list due to bacteria, and additional targeted monitoring is needed to determine bacterial sources and to begin to develop strategies to reduce bacterial loadings.

Other than for bacteria, no evidence was found of water quality problems in the Attoyac Bayou, with nutrient concentrations being low and differences along the bayou were relatively small. Dissolved oxygen was found to be adequate to fully support aquatic life. Stream pH was within the expected range for East Texas waters.

During the first two years of the surface plot study, the vegetated filter strip of 4.5 m was effective in reducing nutrient losses at the edge of fields where litter is applied. The vegetated filter strip was likely acting a sink for P in the runoff waters. However, as soil test P increased in the soils, the beneficial effects of the buffer were reduced, and runoff concentrations increased. This suggested that pastures in the broiler producing counties of East Texas may produce excess P in runoff. However, the overall quantity of P transported remained relatively small.

Recommendations from this study included testing soils for P saturation and only applying nutrients to meet receiving crop demands, maintaining vegetated buffer strips around streams which receive no litter, and by avoiding applications immediately before large rain events. Starting in 2001, following the conclusion of this study, the Texas Legislature required that water quality management plans be developed for every poultry farm in Texas. These management plans, developed and implemented by the Texas State Soil and Water Conservation Board and the local soil and water conservation district, included guidelines such as the recommendations from this study. When these recommendations and agricultural best management practices detailed in the water quality management plan are followed, little evidence exists that broiler litter applications will significantly impact water quality in the Attoyac Bayou.

ACKNOWLEDGMENTS

The Texas Commission on Environmental Quality provided funding and oversight for this project under the authority of the Texas Clean Rivers Act. Landowners Mr. and Mrs. George Burn, Mr. James Bennett, Mr. Marshall Anderson, Mr. A. T. Mast, the Kenalli estate, and the Texas Highway Department provided access to their properties for this study. The contributions of Dr. Mingteh Chang as a Co-Principal Investigator to the study are greatly appreciated. In addition, as graduate students, Mark C. Cochran, Laura Whiteside, Scott Blackerby, and Clay Stark performed much of the field and laboratory work and made this project a success. Finally, the invaluable contributions from staff from Stephen F. Austin State University and the Angelina and Neches River Authority are greatly acknowledged.

REFERENCES

APHA (American Public Health Association), 1992. Standard Methods for the Examination of Water and Wastewater. 18th ed. *Am. Public health Assoc.*, Washington, D.C.

Boyer, D. G., and G. C. Pasquarell, 1999. Agricultural land use impacts on bacterial water quality in karst groundwater aquifer. *JAWRA* 35(2): 291-300.

Blackerby, S. D., 1996. Evaluation of Non-Point Source Pollution Concentrations Due to Runoff from Agricultural Land Applied with Broiler Litter. M.S. Thesis, Steen Library, Stephen F. Austin State University.

Chang, M., J.D. McCullough, and A.B. Granillo, 1983. Effects of land use and topography on some water quality variables in forested East Texas. *Water Res. Bull.* 19: 191-196.

Chang, M., L.D. Clendenon, and H.C. Reeves, 1996. Characteristics of a Humid Climate, Nacogdoches, Texas. College of Forestry, Stephen F. Austin State University, Nacogdoches, Texas. 211 pp.

Chang, M., 2006. Forest Hydrology, An Introduction to Water and Forests, 2nd ed. CRC Press, Taylor and Francis Group, Boca Raton, FL.

Chaubey, I., D.R. Edwards, T.C. Daniel, P.A. Moore Jr., and D.J. Nichols, 1995. Effectiveness of Vegetative Filter Strips in Controlling Losses of Surface-Applied Poultry Litter Constituents. *Trans. ASAE.* 38: 1687-1692.

Cochran, M.C., 1996. Water Quality of East Texas Streams: Forested Versus Pastured Watersheds Receiving Poultry Litter Applications. M.S. Thesis, Stephen F. Austin State University, May 1996, 211 p.

Daniel, T.C., A.N. Sharpley, D.R. Edwards, R. Wedepohl, and J.L. Lemunyon, 1994. Minimizing surface water eutrophication from agriculture by phosphorus management. *J. Soil Water Cons.* 49: 30-38.

Doran, J.W., and D.M. Linn, 1979. Bacteriological quality of runoff water from pastureland. *Appl. Env. Microbiology* 37: 985-991.

Doran, J.W., J.S. Schepers, and N.P. Swanson, 1981. Chemical and bacteriological quality of pasture runoff. *J. Soil Water Cons.* 36(3): 166-171.

Edwards, D.R. and T.C. Daniel, 1992a. Environmental impacts of on-farm poultry waste disposal- a review. *Bioresource Tech.* 41:9-33.

Edwards, D.R. and T.C. Daniel, 1992b. Potential runoff quality effects of poultry manure slurry applied to fescue plots. *Amer. Soc. Of Agri. Engineers.* 35: 1827-1832.

Edwards, D.R. and T.C. Daniel, 1993a. Abstractions and runoff from fescue plots receiving poultry litter and swine manure. *ASAE.* 36: 405-411.

Edwards, D.R. and T.C. Daniel, 1993b. Drying interval effects on runoff from fescue plots receiving swine manure. *ASAE.* 6:1673-1678.

Edwards, D.R. and T.C. Daniel, 1993c. Effects of poultry litter application rate and rainfall intensity on quality of runoff from fescuegrass plots. *J. Environ. Qual.* 22:361-365.

Edwards, D.R. and T.C. Daniel, 1994. Quality of runoff from fescuegrass plots treated with poultry litter and inorganic fertilizer. *J. Environ. Qual.* 23:579-584.

Edwards, D. R., M. S. Coyne, P. F. Vendrell, T. C. Daniel, P. A. Moore, Jr., and J. F. Murdoch, 1997. Fecal Coliform and Streptococcus Concentrations in Runoff from Grazed Pastures in Northwest Arkansas. *JAWRA* 33(2): 413-422.

Edwards, D.R., B.T. Larson, and T.T. Lim, 2000. Runoff Nutrient and Fecal Coliform

Content from Cattle Manure Application to Fescue Plots. *JAWRA* 36(4): 711-721.

Gary, H.L., S.R. Johnson, and S.L. Ponce, 1983. Cattle Grazing Impact on Surface Water Quality in a Colorado Front Range Stream. *J. Soil Water Cons.* 38(2): 124-128.

Gburek, W.J., A.N. Sharpley, L. Heathwaite, and G.J. Folmar, 2000. Phosphorus management at the watershed scale: A modification of the phosphorus index. *J. Environ. Qual.* 29:130-144.

Geldreich, E.E., 1976. Fecal Coliform and Fecal Streptococcus Density Relationships in Waste Discharges and Receiving Waters. CRC Critical Review Environmental Control 6: 349-369.

Green, C.H., J.G. Arnold, J.R. Williams, R. Haney, and R.D. Harmel, 2007. Soil and water assessment tool hydrologic and water quality evaluation of poultry litter applications to small-scale subwatersheds in Texas. *Trans. ASABE* 50:1199-1209.

Harmel, R.D., H.A. Torbort, B.E. Harggard, R. Haney, and M. Dozier, 2004. Water quality impacts of converting to a poultry litter fertilization strategy. *J. Environ. Qual.* 33:2229-2242.

Harmel, R.D., S. Potter, P. Casebolt, K. Reckhow, C. Green, and R. Haney, 2006. Compilation of measured nutrient load data for agricultural land uses in the United States. *JAWRA* 42(5):1163-1178.

Howell, J. M., M. S. Coyne, and P. Cornelius, 1995. Fecal coliform in agricultural waters of the Bluegrass Region of Kentucky. *J. Environ. Qual.* 24: 411-419.

Ice, G.G. and B. Sugden, 2003. Summer dissolved oxygen concentrations in forested streams of northern Louisiana. *S. J. Appl. Forestry,* 27:92-99.

Isom, K. A., 2002. Three aspects of the broiler litter phosphorus problem in East Texas. M.S. Thesis, Steen Library, Stephen F. Austin State University.

Lind, O.T., 1985. Handbook of Common Methods in Limnology, 2nd ed. Kendall Hunt Publishing Company, Dubuque, IA.

Lindsey, L.R., 1975. An Analysis of the Coliform and Streptococcus Bacteria as Indicators of Fecal Pollution in the Attoyac River. M.S. Thesis, Stephen F. Austin State Univ., Nacogdoches, TX.

Margalef, R., 1957. Information theory in ecology. In: *Gen. Syst.* 3:36-71.

McBroom, M.W., 1997. A Physiochemical Analysis of Streams Receiving Runoff form Land-Applied Poultry Litter in East Texas. M.S. Thesis, December 1997, Stephen F. Austin State University, Nacogdoches, Texas.

McBroom, M. W., M. Chang, and M. C. Cochran, 1999. Water quality conditions of streams receiving runoff From land-applied poultry litter in East Texas. p. 549-552. In: R. Sakrison and P. Sturtevant, ed., Watershed Management to Protect Declining Species, Proceedings of the American Water Resources Association 1999 Annual Water Resources Conference, AWRA Technical Publication TPS-99-4, 561pp.

McBroom, M. W., M. Chang, and C. W. Wells, 2003. Bacteriological water quality of forested and pastured streams receiving land-applied poultry litter. In: Kolpin, Dana and John D. Williams (Editors), 2003. Agricultural Hydrology and Water Quality. AWRA's 2003 Spring Specialty Conference Proceedings. American Water Resources Association, Middleburg, Virginia, TPS-03-1, CD-ROM.

McBroom, M.W., Beasley, R.S., Chang, M., and Ice, G.G, 2008a. Water quality effects of clearcut harvesting and forest fertilization with best management practices. *J. Env. Qual.*, 37(1):114-124.

McBroom, M.W., Beasley, R.S., Chang, M., and Ice, G.G, 2008b. Storm runoff and sediment losses from forest clearcutting and stand reestablishment with best management practices in the Southeastern United States. Hydrological Processes, 22(10):1509-1522.

Mikkelsen, R.J., and J.W. Gilliam, 1995. Animal waste management and edge of field losses. In: Animal Waste and the Land-Water Interface, pp. 57-68, Ed. By K. Steele, Lewis Publishers, Boca Raton, FL.

Pielou, E.C., 1975. Ecological Diversity. John Wiley and Sons, New York, NY.

Plafkin, J.L., M.T. Barbour, K.D. Porter, S.K. Gross, and R.M. Hughes, 1989. Rapid bioassessment protocols for use in streams and rivers: Benthic macroinvertebrates and fish. USEPA, EPA/440/4-89/001, Assessment and Watershed Protection Division, Washington, D.C.

Richardson, L.J., 1975. An Analysis of the Coliform and Streptococcus Bacteria as Indicators of Fecal Pollution in the Angelina River. M.S. Thesis, Stephen F. Austin State Univ. Nacogdoches, TX.

Robbins, J.W., D.H. Howells, and G.J. Kriz, 1972. Stream pollution from animal production units. *J. Water Poll. Control Fed.* 44: 1537-1544.

Sawyer, C.N., 1947. Fertilization of lakes by agricultural and urban drainage. *J. New England Water Works Assoc.* 61:109-127.

Shannon, C.E. and W. Weaver, 1964. The Mathematical Theory of Communication. University of Illinois Press, Urban, IL.

Sharpley, A.N., T. Daniel, T. Sims, J. Lemunyon, R. Stevens, and R. Parry, 1999. Agricultural phosphorus and eutrophication. *Agricultural Research Service*, ARS-149.

Sharpley, A.N., T. Daniel, T. Sims, J. Lemunyon, R. Stevens, and R. Parry, 2003. Agricultural phosphorus and eutrophication, second edition. *Agricultural Research Service*, ARS-149.

Sherer, B.M., J.R. Miner, A.J. Moore, and J.C. Buckhouse, 1992. Indicator bacteria survival in stream sediment. *J. Env. Qual.* 21: 591-595.

Stark, C., 1999. Effects of Continued Broiler Litter Application on Runoff Water Quality. M.S. Thesis, Steen Library, Stephen F. Austin State University.

Stephenson, G. R., and L.V. Street, 1978. Bacterial variations in streams from a southwest Idaho rangeland watershed. *J. Env. Qual.* 7:150-157.

Tiedemann, A.R., D.A. Higgins, T.M. Quigley, H.R. Sanderson, and D.B. Marx, 1987. Responses of fecal coliform in streamwater to four grazing strategies. *J. Rng. Mgt.* 40(4): 322-329.

TNRCC, 1993. A Guide to Freshwater Ecology. Texas Natural Resources Conservation Commission (now Texas Commission on Environmental Quality), Austin, TX.

TNRCC, 1995. Texas surface water quality standards. July 1995. §§307.1-307.10, Texas Natural Resource Conservation Commission (now Texas Commission on Environmental Quality), Austin, TX, 50 pp. + appendices A-E.

Twidwell, S.R. and J.R. Davis, 1989. Assessment of six least disturbed unclassified Texas streams. Texas Water Commission, Austin, TX.

USDA - National Agricultural Statistics Service, 2002. Annual Broiler Production, 1950-2000.http://www.nass.usda.gov/Publications/Statistical_Highlights/2002/graphics/broiler.htm

USDA - National Agricultural Statistics Service, 2008. Broilers: Inventory By State, US. http://www.nass.usda.gov/Charts_and_Maps/Poultry/brlmap.asp

USEPA, 1983. Methods for chemical analysis of water and wastes. Revised March 1983, EPA/600/4-79/020, Environmental Monitoring and Support Laboratory, Office of Research and Development, USEPA, Cincinnati, OH.

USEPA, 1986. Ambient Water Quality Criteria for Bacteria – 1986. EPA-440/5-84-002, U. S. Government Printing Office, Washington, D.C.

USEPA, 2007 National Water Quality Inventory: Report to Congress. Office of Water, Washington DC, EPA 841-R-07-001

USSCS, 1980. Soil Survey of Nacogdoches County, Texas. United States department of Agriculture, Soil Conservation Service and Forest Service in cooperation with Texas Agricultural Experiment Station.

Vollenweider, R.A., 1968. Scientific fundamentals of the eutrophication of lakes and flowing water, with particular reference to nitrogen and phosphorus as factors in eutrophication. Paris, Rep. Organization for Economic Cooperation and Development, DAS/CSI/68.27, 192 pp.; Annex, 21 pp.; Bibliography, 61 pp.

Westerman, P.W., M.R. Overcash, and S.C. Bingham, 1983. Reducing runoff pollution using vegetated borderland for manure application sites. Project summary, USEPA, EPA-600/S2-83-022, Robert S. Kerr Environmental Research Laboratory, Ada, OK.

Whiteside, L. L., 1996. Poultry Litter Land Application Rate Study for Nacogdoches County,Texas.M.S. Thesis, Steen Library, Stephen F. Austin State University.

Young, J.L., M. Chang, M.C. Cochran, and L.L. Whiteside, 1996. Poultry Litter Land Application Rate Study, Final Report, October 1996, Presented to Texas Natural Resources Conservation Commission by Stephen F. Austin State University and Angelina and Neches River Authority, 60 pp. + appendices.

Young, R.A., C.A. Onstad, D.D. Bosch, and W.P. Anderson, 1989. AGNPS: A nonpoint-source pollution model for evaluating agricultural watersheds. *J. Soil Water Conserv.* 44:168-173.

In: Encyclopedia of Environmental Research
Editor: Alisa N. Souter
ISBN: 978-1-61761-927-4
© 2011 Nova Science Publishers, Inc.

Chapter 49

AGRICULTURAL RUNOFF: NEW RESEARCH TRENDS

Víctor Hugo Durán Zuazo[*,1,2,]
Carmen Rocío Rodríguez Pleguezuelo[1,2], *Dennis C. Flanagan*[2],
José Ramón Francia Martínez[1] *and Armando Martínez Raya*[1]

[1] IFAPA Centro Camino de Purchil. Apdo. 2027, 18080-Granada, Spain
[2,†] USDA-National Soil Erosion Research Laboratory,
West Lafayette, Indiana 47907-2077, USA

ABSTRACT

There is public concern worldwide about the impact of agriculture on the environment and the migration of agrochemicals from their target to nearby terrestrial and aquatic ecosystems and sometimes to the atmosphere and other times to the groundwater. To achieve the highest yields, farmers use many agrochemicals and practices, which frequently have repercussions for the nearby natural or/and semi-natural adjacent areas. Research is needed to identify the correct amount of fertilisers and the appropriate management to be applied in order to minimise non-target effects while allowing the farmers profitable yield from their agro-ecosystems. Some agrochemicals can leave the agro-ecosystem by the runoff water, by the percolation of soil water or by evaporating into the atmosphere as gases. The recent awareness about the relationship between agricultural activities and non-point source pollution is also growing in many parts of the world. The challenge for farmers, land managers, and land users is to maintain the quality of surface waters and the health of biological communities by reducing and managing the amount of sediment, nutrients, and other pollutants in agricultural runoff. Making changes to land use or correcting past abuses is often expensive. A considerable amount of research is underway by organisations with a commitment to, or responsibility for, managing the effects of land-use practices on the environment. Studies have been made concerning land-use practices to reduce environmental impact, such as more efficient fertiliser use, sustainable grazing practices, fencing of riverbank access, and re-planting of vegetation along streams. Finally, research has supported a range of monitoring activities to determine the quantity and fate of agro-

---

[*] Corresponding autor e-mail: victorh.duran@juntadeandalucia.es.
[†] Address: 275 S. Russell Street, West Lafayette, IN 47907-2077, USA

materials in agricultural runoff; this information is used to assess the health and status of agro-ecosystems and form a basis for evaluating the effectiveness of land-management practices to reduce threats.

1. INTRODUCTION

Agricultural runoff is surface water leaving cultivated fields as a result of receiving water in excess of the infiltration rate of the soil. Excess water is due primarily to precipitation, but it can also come from irrigation and snowmelt on frozen soils. Also, there is considerable concern about erosion of cultivated fields due to the rainfall and runoff processes, primarily related to the loss of valuable topsoil from the fields and subsequent losses in productivity. The potential for pollution of surface waters such as rivers and lakes due to agricultural runoff has been recognized and the nature and extent of such pollution is systematically assessed. Agricultural runoff is grouped into the category of non-point source (NPS) pollution because the potential pollutants originate over large, diffuse areas and the exact point of entry into water bodies cannot be precisely identified. These sources of pollution are particularly problematic in that it is difficult to capture and treat the polluted water before it enters a stream. The point sources of pollution such as municipal sewer systems usually enter the water body via pipes and it is comparatively easy to collect such water and reroute it through a treatment system prior to releasing it into the environment. Because of the NPS nature of agricultural runoff, efforts to minimize or eliminate pollutants are focused on practices to be applied on or near cultivated fields themselves. Agricultural runoff is considered to be the primary source of pollutants to streams and lakes, as well as estuaries. Runoff from agricultural fields introduces soil, organic matter, manure, fertilizer and pesticides into small streams, increasing the volume of stream discharge and changing water quality. Researchers such as Cooper [1] have reviewed the acute toxic and sub-lethal chronic effects of such runoff, and have identified pesticides as one of the major stressors of aquatic communities.

One important environmental objective is to achieve and maintain a good chemical and ecological status of surface water bodies, and the first step towards this is to locate water bodies that are likely not to meet this criterion. For such evaluations, water-monitoring programs are indispensable on the river-basin scale and even on larger scales, especially in the vast tracts of lands with agricultural production. More effective action is needed to further reduce agricultural runoff of sediment, nutrients, and other pollutants, thereby alleviate the present threat to the agroecosystems.

2. CONTROLLING AGRICULTURAL RUNOFF

Water quality and agriculture are closely linked because of the potential NPS pollution of lakes, rivers, streams, etc. by agricultural runoff [1, 2]. Contaminants such as sediments, bacteria (*e.g.* dairy waste), nutrients (*e.g.* nitrogen and phosphorus), and pesticides may be transported from agricultural fields during rainstorm runoff events. As Cooper [1] stated, pesticides have played a key role against food shortages and vector-borne diseases, and humankind would be vastly different without their use. However, non-target impacts from pesticide-associated agricultural runoff persist, and thus scientists are currently studying methods to minimize these risks.

One possible solution to minimize agricultural runoff into streams and other water bodies involves the development of constructed wetlands to replace lost edge-of-field wetlands and serve as buffers. The utility of constructed wetlands for mitigating several different kinds of contaminants has been extensively studied. Their effectiveness in removing organics, nutrients, and metals has been previously reported [3, 4, 5, 6, 7]. Few studies, however, have focused on the potential of constructed wetlands as buffers to mitigate pesticide losses [8, 9].

One of the most direct methods of controlling pollution from agricultural runoff is to minimize the potential for runoff. Other methods such as best management practices (BMPs) can be used to reduce the amounts of sediments and dissolved chemicals in surface runoff (Figure 1). BMPs are individual practices or combinations of management, cultural, and structural approaches which have been identified as the most effective and economical way of reducing damage to the environment. Often practices aimed at controlling one aspect of agricultural runoff can also be successful in reducing other components because of the interrelationships between runoff volume, erosion, transport, dissolution, and delivery. Maintaining good soil tilth and healthy vegetation can reduce runoff by promoting increased infiltration into the soil matrix. Terracing, contour ploughing, and use of vegetated waterways to safely convey surface runoff can result in decreased runoff by slowing the water leaving a field and allowing more time for infiltration. Construction of farm ponds to receive runoff can result in less total runoff from a farm, lowered peak runoff rates, and water storage for use in irrigation or livestock watering [10].

Control of water pollution in agricultural runoff is often effectively achieved by reducing soil erosion from the field, and the primary method of doing this is by maintaining some type of plant cover on the soil surface or reducing the area of bare soil [11, 12, 13]. Techniques include conservation tillage, strip tillage, plant strips, and the use of cover crops. Additional measures that can be employed at the edge of the field, or off-site, include farm ponds and vegetative filter strips, especially during critical runoff/erosion periods.

The loss of nitrogen and other plant nutrients can be reduced if fertilizer is applied in appropriate quantities and when the crop needs it. This often requires multiple applications, which can be difficult, time-consuming and costly. Because nitrogen fertilizers have been relatively inexpensive, growers tend to overapply rather than underapply. Efforts have been made to make the N less soluble by changing the form applied to the field so that it becomes available to the plants (and, thus, available for loss by runoff) more slowly [14].

Another method of controlling losses of agricultural chemicals is to minimize their use through such programs as integrated pest management (IPM), in which fields are regularly monitored and some crop damage is allowed until it becomes economically justified to apply pesticides. Development of chemicals that are more easily degraded is also desirable, so that they are less likely to persist long enough to be transported by agricultural runoff.

On other hand, agricultural activities associated with livestock management can have large impacts on the quality of pasture runoff and adjacent surface waters. Field spreading of livestock manure can contaminate streams and estuaries with fecal coliform bacteria (FCB), signalling the possible presence of feces-associated pathogens and thus outweighing beneficial uses. The installation of plant buffers between manure-application areas and surface waters is a common BMP [15].

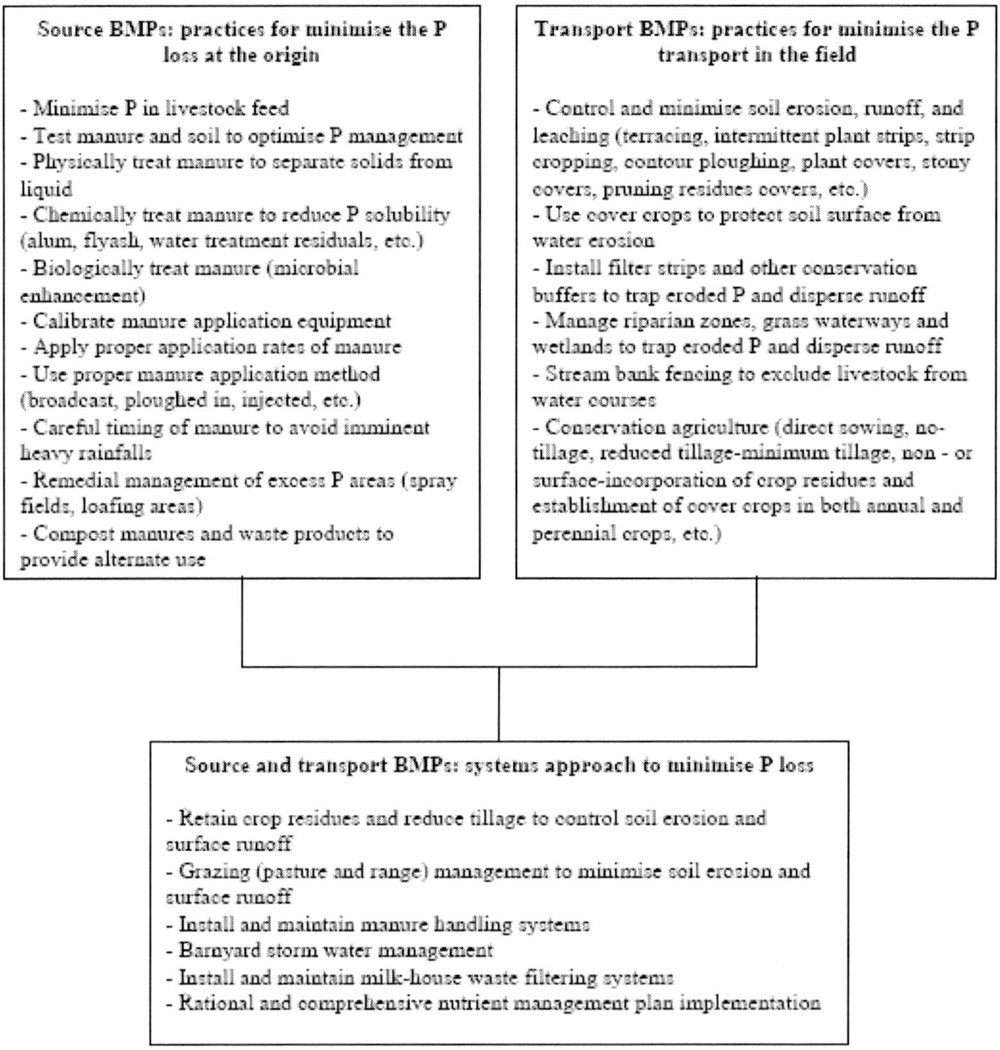

Source BMPs: practices for minimise the P loss at the origin	Transport BMPs: practices for minimise the P transport in the field
- Minimise P in livestock feed - Test manure and soil to optimise P management - Physically treat manure to separate solids from liquid - Chemically treat manure to reduce P solubility (alum, flyash, water treatment residuals, etc.) - Biologically treat manure (microbial enhancement) - Calibrate manure application equipment - Apply proper application rates of manure - Use proper manure application method (broadcast, ploughed in, injected, etc.) - Careful timing of manure to avoid imminent heavy rainfalls - Remedial management of excess P areas (spray fields, loafing areas) - Compost manures and waste products to provide alternate use	- Control and minimise soil erosion, runoff, and leaching (terracing, intermittent plant strips, strip cropping, contour ploughing, plant covers, stony covers, pruning residues covers, etc.) - Use cover crops to protect soil surface from water erosion - Install filter strips and other conservation buffers to trap eroded P and disperse runoff - Manage riparian zones, grass waterways and wetlands to trap eroded P and disperse runoff - Stream bank fencing to exclude livestock from water courses - Conservation agriculture (direct sowing, no-tillage, reduced tillage-minimum tillage, non - or surface-incorporation of crop residues and establishment of cover crops in both annual and perennial crops, etc.)

Source and transport BMPs: systems approach to minimise P loss

- Retain crop residues and reduce tillage to control soil erosion and surface runoff
- Grazing (pasture and range) management to minimise soil erosion and surface runoff
- Install and maintain manure handling systems
- Barnyard storm water management
- Install and maintain milk-house waste filtering systems
- Rational and comprehensive nutrient management plan implementation

Figure 1. Best management practices (BMPs) for phosphorus to reduce the impacts of land applied manures on P loss to surface water bodies.

Numerous studies have evaluated the influence of agricultural practices on microbiological quality of runoff water [16, 17, 18, 19]. To quantify this impact is difficult because the extent of bacterial pollution is related to climatological factors such as rainfall amount and intensity, as well as microbial populations and die-off. Bacterial transport varies with the initial population, soil conditions, temperature, sunlight, and organic matter [20]. For these and perhaps other reasons, empirical data correlating manure application with the quality of runoff water frequently contain contradictions [21].

In addition, fecal coliform bacteria (FCB) concentration is a common indicator of bacterial contamination, implying the potential presence of microorganisms that are pathogenic to humans [22]. Homeothermic animals shed large amounts of these bacteria in their feces, and pathogen presence in water implies fecal pollution. The US Environmental Protection Agency requires that FCB concentrations do not exceed 200 colony-forming units (cfu)/100 mL for contact recreation and 14 cfu/100 mL for shellfish harvesting [23]. Most pathogenic bacteria are removed by physical and chemical adsorption within the soil profile [20], and FCB concentrations therefore typically decline substantially when transported through soil, suggesting that transport to surface water occurs mainly by surface flow [24, 25, 26].

Management practices that are currently used to diminish the input of pollutants from animal waste to surface and groundwater include control of animal numbers [27, 28], control of animal diet [29], constructed wetlands, and riparian filterstrips [30, 31, 32]. However, there are some problems with vegetative systems: (1) vegetation in wetlands or riparian areas can take months to years to become completely established; (2) the systems are not effective when the vegetation is not growing, and can become nutrient sources rather than nutrient sinks [33, 34, 35]; (3) riparian filter strips or constructed wetlands are effective for only small quantities of runoff (relatively infrequent or low-intensity runoff events) because continuous application can quickly overload the system's ability to withdraw nutrients [36, 37, 33]; and (4) vegetative systems cannot be transported to the site of a waste spill or runoff area. Therefore, even when BMPs are used, livestock production can sometimes contribute large amounts of nutrients and enteric microorganisms to watercourses and water bodies. Pollution control is achieved through natural processes that promote leaching and that prevent or retard overland transport. Although these functions are well established, the degree to which they can be enhanced by the installation of larger buffers has not been quantified [38, 39]. In this context, according to Sullivan et al. [40], substantial FCB contamination of runoff occurs from manure-treated pasturelands, and it might be disproportionately associated with specific field or management conditions, such as the absence of a vegetated buffer or the presence of soils that allow low water infiltration and therefore generate larger volumes of runoff.

Mulching the soil surface in farmlands with a layer of plant residue is an effective method of conserving water and soil because it reduces agricultural runoff, increases infiltration of water into the soil, and retards soil erosion. Mulching the soil with elephant grass (*Pennisetum purpureum*) debris may benefit late cropping (second cropping) by increasing stored soil water for use during dry weather and reducing erosion and agricultural runoff on sloping land [41].

Sedimentation occurs when surface runoff carries soil particles from one area, such as a farm field, and deposits them in a water body, such as a stream or lake. Excessive sedimentation clouds the water, reducing the amount of sunlight reaching aquatic plants, and covers fish spawning areas and food supplies. In addition, other pollutants such as phosphorus, pathogens, or heavy metals are often attached to the sediment particles. Farmers could reduce erosion and sedimentation by 20 to 90% by applying management measures to control the volume and flow rate of runoff water, and reduce sediment transport [11, 12].

The application of anionic polyacrylamide (PAM) to soils and/or vegetative treatments may also provide a cost-effective way to dramatically reduce bacteria and nutrient loads in animal waste and thereby reduce pollution in waters receiving these effluents. PAM application can be used alone or in conjunction with vegetative strategies, which may then operate more effectively due to reduced contaminant loads in waste entering the system.

Wastes from a variety of agricultural runoff sources are major sources of nutrients, pesticides, and enteric microorganisms entering surface and ground waters. Water-soluble anionic polyacrylamide was found to be a highly effective erosion-preventing and infiltration-enhancing polymer, when applied at rates of 1-10 g m^{-3} in furrow irrigation water. Water flowing from PAM-treated irrigation furrows substantially reduced sediment, nutrients, and pesticides. Recently PAM and PAM mixtures [PAM + CaO and PAM + Al $(SO_4)_3$] have been shown to filter bacteria, fungi and nutrients from animal wastewater. Low concentrations of PAM [175-350 g PAM ha^{-1} as PAM or as PAM + CaO and PAM + $Al(SO_4)_3$ mixture] applied to the soil surface, resulted in dramatic decreases (10 fold) of total, coliform and faecal streptococci bacteria in cattle, fish, and swine wastewater leachate and surface runoff. PAM treatment also filtered significant amounts of NH_4, PO_4 and total P in cattle and swine wastewater.

Potential benefits of PAM treatment of animal-facility waste streams include: (1) low cost, (2) easy and quick application, (3) suitability for use with other pollution-reduction techniques. Research on the efficacy of PAM for removal of protozoan parasites and viruses and more thorough assessment of PAM degradation in different soils is still needed for a complete evaluation of PAM treatment as an effective waste-water treatment. Entry et al. [42] demonstrated the potential efficacy of PAM and its mixtures in reducing sediment, nutrients, and microorganisms from livestock and poultry effluents.

Alum [$Al_2 (SO_4)_3 nH_2O$] treatment of runoff has been used as a storm-water retrofit option and its effect on water quality and ecological impacts of treatment systems had been studied [43, 44]. Alum treatment of stormwater consistently provides removal efficiency of 85-95% for total phosphorus, >95% for total suspended solids (TSS), 35-70% for total nitrogen, 60-90% for metals, and 90-99% for total FCB. Although only positive chemical and ecological impacts have been reported in water bodies receiving alum treatment, current state policies require collection and disposal of the generated floc, an issue that has received considerable attention in recent years. On the other hand, according to Smith et al. [45], the addition of alum to various poultry litters reduced P runoff by 52 to 69%, the greatest reduction occurring when alum was used in conjunction with HAP (high available phosphorus) corn and phytase. This study demonstrated the potential added benefits of using dietary modification in conjunction with manure amendments in poultry operations. Thus, integrators and producers should consider the use of phytase, HAP corn, and alum to reduce potential P losses associated with poultry litter application to pastures.

Controlled drainage systems (CDS) are an innovative management practice to reduce water losses from agricultural fields with subsurface drainage tiles, as well as losses of associated agricultural chemicals such as nitrates and phosphates. Originally developed in North Carolina (USA) [46, 47, 48] as a means of reducing high levels of agricultural nutrients leaving tile drains that were adversely affecting off-site water bodies, these systems also can boost crop productivity by increasing soil moisture available for plant growth during droughty summer periods. A CDS has some type of control structure or housing, that allows placement of a series of vertical blocks or gates that prevent water from a tile system outlet from exiting a field, effectively raising the soil water table. The water table is maintained at a high level during the winter and early spring, then the blocks are removed a month or so prior to field operations and planting, so that the soil can dry enough to support tillage equipment. Following planting and other cultivation, blocks are added to an intermediate height, reducing excessive drainage and helping to maintain moisture in the lower soil profile for plant growth. Prior to harvest the blocks may all be removed to allow the soil to fully dry again so that it can support heavy harvesting equipment. After harvest and any fall tillage, the blocks are all inserted again so that the water table is at its highest level (perhaps a foot below the soil surface) during the winter. Several reseachers have shown large benefits of CDS on reducing nutrient losses from agricultural sites to off-site waters, particularly losses of nitrate-N (losses reduced up to 95%) [49, 50, 51]. Use of CDS in new tile systems and retro-fitting of existing tile systems in the Midwestern United States may be one part of a solution to reducing hypoxia problems in the Gulf of Mexico which result from high soluble nutrient losses from agricultural lands.

3. Plant Nutrients in Agricultural Runoff

For crop and livestock agriculture, P is an essential nutrient, which can migrate from farmland to off-site bodies of water, accelerating eutrophication [52, 53]. Recent water-quality appraisals have identified eutrophication as one of the most common water-quality hazards in America, Europe, Australia, and Asia [54, 55, 56]. Eutrophication restricts water use for fish farms, industry, and recreation, by spurring growth of undesirable algae and aquatic weeds, as well as depleting oxygen levels through these organisms' death and decomposition [57]. Intensive livestock farming has caused a vast transfer of P from grain- to animal-producing areas, where localized surpluses of P-manure can accumulate. The land application of manure can increase the potential loss of P to surface waters [58, 59]. More attention has focused on livestock as part of agriculture's role in water-quality damage and the need for BMPs for manures, which are an integral part of remedial strategies [60, 61, 62, 63].

Table 1. Best management practices (BMPs) at farm level to control the excess of nutrients

BMPs in crop lands	
Nitrogen leaching	Phosphorus loss
- Apply only recommended rates of nitrogen fertilizer, manure or sludge	- Use soil erosion control practices
- Know the accumulated levels of nitrogen collected from rotated legume crops, and past fertilizer, manure and bio-solid applications	- Test soils on a regular basis and avoid applications of manure or fertilizer that would result in soil levels above an optimum range
- Proper application timings including spit-applications when appropriate	- Obtain nutrient analysis of manure sources and use soil test recommendations and yield goals when calculating application amounts
- Realistic yield goals	- Adopt phosphorus based management plants for areas vulnerable to phosphorus runoff or leaching
- Correctly calibrate application equipment	- Do not spread manure on frozen ground
- Use of cover crops to utilize the residual nitrate in the soil at the end of the growing season	- Inject or incorporate applications of manure or fertilizer whenever possible
- In irrigation system farms periodic control of nitrate concentrations in irrigation waters	- Using P-rich organic sources like manure and sludge on soils with lower soil test rates for phosphorus

Controlling the fate of P applied in manure requires efforts at many scales, which range from field-specific BMPs to watershed-wide planning (Table 1). These practices are generally designed to use agricultural chemicals efficiently, encourage ground cover, impede surface runoff, and improve livestock waste management. On the other hand, the control of soil erosion is essential to avoid nutrient NPS pollution of surface waters, since eroded soil particles can carry nutrients, particularly phosphorus, into water bodies [11, 64]. Also, seeking agricultural nutrient non-point source pollution solutions, farmers may adopt physical approaches (*e.g.* waste-containment tanks or pits, sediment basins, terracing, fences, tree planting) and/or managerial ones (*e.g.* nutrient budgeting, rotational grazing, conservation tillage). Either way, sound management is indispensable to decrease agricultural pollution.

To check the effects of long-term localized P accumulations, system-level changes in farm and regional P balances are required. In any case, even when nutrient balances are maintained, P can migrate from land to water. Because manure contains high concentrations of nutrients in ratios beyond crop needs, appropriate application of manure requires a holistic approach to P management.

Modern farming practices are often wasteful in terms of using inorganic fertilizers. Therefore, much of the N can be lost to the environment in several ways and its control is urgent (Table 1). For example, large amounts of nitrate (NO_3^-) can be lost in both surface runoff as well as in water leaching down the soil profile below the plant root zone, and from there lost either in tile drainage (that reaches surface ditches and streams) or deep seepage (that can contaminate aquifers). Nitrogen entering the soil disturbs the ammonium ions (NH_4^+), which are normally in equilibrium with NH_4^+ in the soil. This disrupts the soil pH, thereby affecting plant growth. Whenever the soil pH is high enough, the equilibrium is tipped towards production of volatile ammonia gas (NH_3). This nitrogen release into the atmosphere eventually ends in NH_4^+ returning to the soil with rainfall. An additional problem can be denitrification of NO_3^- back to volatile nitrogen gas (N_2) by soil bacteria. Although not a direct cause of pollution, it is wasteful, since artificial fixation of nitrogen for fertilizers requires large energy inputs.

The growing use of phosphate fertilizers has resulted in the accumulation of P in soils. Problems arise because the means by which P is immobilized cannot accommodate the additional P that fertilizers add to soils. Consequently, greater soluble and sediment-bound phosphorus loads can be transported off-site with agricultural runoff, negatively affecting lakes, streams and other water sources because, as mentioned above, P leads to eutrophication. Both N and P upset aquatic systems by augmenting the growth of algae and aquatic weeds. After death, microbial decomposition of the detritus depletes the oxygen supply available to other organisms that depend on dissolved oxygen levels in the water.

Government programs are available to land owners and land managers to help design and pay for systems and practices to prevent and control NPS pollution. Several U.S. Department of Agriculture and state-funded programs provide cost-sharing, technical assistance, and economic incentives to implement NPS pollution-management practices. Many farmers use their own resources to adopt technologies and practices to minimize losses of increasingly expensive fertilizer inputs and limit water-quality impacts caused by their agricultural activities.

4. PESTICIDES IN AGRICULTURAL RUNOFF

Insecticides, herbicides, and fungicides, though used to kill pests and control the growth of weeds and fungi, can enter water through direct application, runoff, wind transport, and atmospheric deposition, killing fish and wildlife, poisoning food sources, and destroying animal habitats. Such NPS contamination from pesticides can be reduced by applying IPM techniques based on the specific soils, climate, pest history, and crop selection for a particular field. IPM helps limit pesticide use and manages necessary applications to minimize pesticide movement from the field. Although relatively small herbicide loads are carried by surface runoff water in relation to the amount applied to a cultivated field (from less than 0.5% up to 5%) [65], their residues can pose serious environmental risks. The main pathway for herbicide losses is surface runoff, and a rainstorm shortly after application can carry high chemical concentrations in the runoff [66, 67], wreaking serious consequences for water quality and wildlife habitats. Vegetative filter strips (VFS) have been proposed as a means to reduce surface water contamination caused by agricultural NPS [68, 69, 70]. A VFS acts as a natural dam or terrace and, by reducing runoff, the water has more time to penetrate and incorporate the pollutants in the soil and thus prevent off-site movement [71]. Also, VFS alters flow hydraulics, reducing runoff speed and increasing water infiltration [72]. The filter thus enhances sediment deposition and filtration by vegetation, pollutant adsorption into the soil and dead and living plant materials, and uptake of soluble pollutants by plants [73].

Infiltration was found to be the most important herbicide removal mechanism associated with VFS, especially for soluble or weakly adsorbed pesticides [74]. Watanabe and Grismer [75], investigating diazinon transport within a VFS, found that the pesticide was trapped on its surface and in the root-zone, where further adsorption, attenuation and presumably degradation may occur. Nevertheless, enhanced infiltration could cause more leaching, allowing the herbicide to reach the water table, changing the ecotoxicological impact from surface to subsurface water. Delphin and Chapot [76] evaluated this leaching and investigated the fate of atrazine and de-ethylatrazine transported in runoff effluents and trapped by a grass filter strip.

The plants in the VFS conferred higher organic-matter content to the filter zone than in the adjacent cultivated fields. This organic matter accumulation boosted the adsorption capacity and microbial activity for herbicide degradation, thereby reducing the amount of herbicide in surface runoff and leaching water [77]. Higher herbicide dissipation in the VFS soil is due both to enhanced degradation and the formation of non-extractable (bound) residues, which can become a long-term sink inside the filter [78].

5. AGRICULTURAL RUNOFF CONTROL IN MEDITERRANEAN MOUNTAINOUS CROPLANDS

Torrential rains are the major cause of runoff and soil erosion in agricultural mountainous areas in Mediterranean south-eastern Spain. Strong efforts have been made to curb cropland erosion, primarily by reducing agricultural runoff that risks pollution of water bodies [11, 12, 64]. In southern Spain (Andalusia), many orchards (*i.e.* vines, almonds, and olives) are under rainfed conditions and confined to slopes or rugged land, occupying large parts of mountains and hills of the Mediterranean agroecosystem. In such an environment, rainfall needs to be fully harvested and unnecessary runoff losses should be minimized, given the vital importance of water in this type of climate. Also, extensive rainfed crops are cultivated mainly in shallow soils under traditional agriculture techniques with a high risk of transporting pollutants. In this sense, according to Francia et al. [11] and Martínez et al. [12], agricultural runoff can be significantly reduced by the use of different plant strips running across the hill slope, especially aromatic and medicinal plants (AMP) in comparison with the runoff of other traditional soil-management systems used in the zone (Table 2).

Inappropriate wild harvesting of AMP by uprooting in mountainous areas endangers the soil resource. Durán et al. [13] on the southern flank of the Sierra Nevada Mountains (SE, Spain) studied the runoff response to cultivated AMP covers by comparing four harvest intensities (HI) of the biomass (0%, 25%, 50%, and 75%) of four aromatic shrubs (*Lavandula lantana* L., *Santolina rosmarinifolia* L. *Origanum bastetanum,* and *Salvia lavandulifolia* V.). The average runoff for HI-0, HI-25, HI-50, and HI-75 during the study period was 2.6, 3.2, 3.4, and 4.7 mm, respectively, differing significantly from that found by Durán et al. [79] for hilly areas and bare soil (182 mm) in the same study zone. This study demonstrated that the cultivation of AMP (even when removing 50% of the above ground biomass), rather than the harvest of wild AMP, protected the soil from rain erosivity and produced potentially profitable essential-oil yields [18]. Consequently, rational harvesting on agricultural semiarid slopes not only protects the soil against erosion and enhances soil quality but also minimizes the pollution risk from agricultural runoff in mountainous regions.

The coast of the provinces of Granada and Malaga (SE Spain) are economically important areas for intense subtropical fruit cultivation [80, 81]. The farmers in this zone construct terraces primarily to use the steeply sloping lands for agriculture.

Table 2. Agricultural runoff and soil loss prevention by plant strips in semiarid slopes in olive and almond orchards

Soil-management system	Olive orchards with 30% slope		Almond orchards with 35% slope	
	Agricultural runoff (mm yr^{-1})	Soil loss (Mg ha^{-1} yr^{-1})	Agricultural runoff (mm yr^{-1})	Soil loss (Mg ha^{-1} yr^{-1})
NT	39.0	25.6	n.a.	n.a.
CT	10.9	5.70	58.1	10.5
BS	19.8	2.10	23.8	1.66
NVS	8.6	7.1	n.a.	n.a.
LeS	n.a.	n.a.	47.8	5.18
ThS	n.a.	n.a.	26.1	0.50
SaS	n.a.	n.a.	31.5	2.10
RoS	n.a.	n.a.	23.2	0.60

NT, non-tillage without plant strips; CT, Conventional tillage; BS, non-tillage with barley strips; NVS, non-tillage with native vegetation strips; LeS, lentil strips; ThS, thyme strips; SaS, salvia strips; RoS, rosemary strips; n.a., not available.

According to Durán et al. [82], severe soil erosion (9.1 Mg ha^{-1} yr^{-1}) occurs frequently on the bare taluses of orchard terraces, especially those with sunny southern orientations (some having runoff up to 100 mm yr^{-1}). Use of AMP such as thyme (*Thymus serpylloides*) and sage (*Salvia officinalis*) compared to bare soil reduced runoff in the taluses of orchard terraces by 54 and 40%, respectively (Durán et al. [83]). Also, plant nutrient losses (NPK) from taluses of orchard terraces was controlled by plant covers [64]. Thus, terrace pollution and erosion (even destruction) were prevented by planting the taluses with plant covers having multiple uses (aromatic, medicinal, culinary, and mellipherous, etc.). Moreover, with plant establishment, an ecological balance can be at least partially restored, reducing pollution risk from agricultural runoff that is injurious to the environment as well as to humans.

6. IMPACT OF AGRICULTURAL RUNOFF ON RIVER SYSTEMS

In river systems, nutrient concentrations govern biological productivity as well as the status of aquatic systems. Shifts in nutrient concentrations in river systems are often determined more by anthropogenic activities that include discharge of waste inputs, sewage, runoff from the heavily fertilized fields and silting in the surrounding areas than by natural processes of weathering, and *in situ* eutrophication [84, 85, 86]. These factors can directly and indirectly affect benthic fauna assemblages. During the winter months, dissolved oxygen concentration in the river water can be depleted due to metabolic requirements for oxygen to degrade organic matter. Nutrient loadings to river systems can have far-reaching effects on the biological status of aquatic systems and water quality, and therefore there is a need to conduct environmental monitoring of agriculture in a critical rainy period. Water is known to be an effective medium for the transport of pollutants from source to sink, carrying all waste and nutrients generated on landscapes in catchment systems.

On the other hand, in most oceans worldwide, biological productivity is controlled by the supply of nutrients reaching surface waters. The relative balance between supply and removal of nutrients, including N, iron (Fe), and P, determines which nutrient limits phytoplankton growth. Although N limits productivity in much of the ocean, large areas of the tropics and subtropics suffer extreme nitrogen depletion. In such regions, microbial denitrification removes biologically available forms of N from the water column, leading to substantial deficits with regard to other nutrients. Despite naturally high nutrient concentrations and productivity, N-rich agricultural runoff fosters large phytoplankton blooms in vulnerable areas of the ocean [87]. Thus, runoff strongly and consistently influences biological processes, and in 80% of the cases stimulating blooms within days of fertilization and irrigation of agricultural fields. This underlines the present and future vulnerability to agricultural runoff in these ecosystems, and reflects the urgency to adopt proper environmental control measures.

6.1. Grassed Waterways

Great potential to reduce agricultural runoff, sediments and pollutants coming from watersheds is offered by grassed waterways (GWWs) that have large hydraulic roughnesses. For conservation planning the knowledge of overall effectiveness and its seasonal variation is highly relevant.

Another management tool widely considered important to decrease agricultural NPS pollution are grass or vegetative filter strips (VFS) located at field edges or around surface water bodies [88, 89]. Grassed waterways often attract less interest in this effort, even though they might have a greater impact at the catchment scale. The sediment yield of a catchment, according to Verstraeten et al. [90], could be reduced by 20% if ditches were replaced by GWWs, while an installation of VFS at the downslope end of fields with high soil loss consumed more agricultural area and resulted in only a 7% sediment reduction.

Even with the great potential of GWWs, most research studies have dealt with VFS and assessed their sediment-trapping efficiency, runoff reduction, and trapping of pollutants in plot experiments [91, 92, 76, 17], with a wide range of input parameters (inflow, rain on the plot), vegetation characteristics (length and density of grasses, mostly single or a few grass species), soil characteristics (soil type, soil moisture) and morphological parameters (slope and length of the plot). The runoff reduction varied from 6% [85] to 89% [86], and the sediment trapping from 15% [91] to 99% [92]. Only a few studies have examined the long-term trapping efficiency of VFS under natural conditions. In this regard, Schauder and Auerswald [93] found that a VFS located downslope from a hop garden trapped (17-year average) 55% of the sediments entering the filter. Besides the experimental studies, there are a few mathematical models of runoff reduction and sediment trapping in VFS [94, 95, 96, 97].

The effectiveness of GWWs in reducing runoff and sediment loads has been investigated [98, 99, 100, 101, 102]. In a landscape experiment, Chow et al. [98] found that a terrace/GWW system reduced the average runoff by 86% and the average sediment delivery by 95% in an area where potatoes were grown on different slopes. Another study measured the effects of two field GWWs between 1994 and 2000 [101]. One of the GWWs reduced runoff by 10% and trapped 77% of sediment, while the other reduced runoff by 90% and trapped 97% of sediment because of a low flow velocity due to its flat-bottomed cross section and rough vegetation.

Neither the VFS nor the GWW studies have considered the seasonal variation in effectiveness; even if the wide range of experimental set-ups give some clues on this issue. Knowledge of seasonal variation in effectiveness is highly relevant for conservation planning to ensure that a VFS or a GWW is effectively applied. For instance, for herbicides to be kept from entering surface-water bodies, it is necessary to ascertain the practice effect at the time the herbicide is applied and shortly thereafter. It was in this context that Fiener and Auerswald [103] pointed out the high potential of GWWs for reducing runoff and sediment delivery, particularly when combined with an intensive soil- and water-conservation system in the draining fields, and for conservation planning.

Therefore, vegetative biofilters including filter strips, GWWs, and natural grasslands may shield water resources from non-point source agricultural pollution, with minimal impact on farming profits [104, 105, 106, 107]. Such vegetative buffers function through infiltration reducing the volume of runoff, and through adsorption/sedimentation, lowering the concentration of pollutants in runoff [108]. After infiltration, dissolved chemicals may become sorbed to soil or organic matter [107] and may later degrade. Biofilters lower sediment concentrations [109, 89] and thus may also diminish the concentration of pesticides strongly sorbed to sediment particles. However, the effectiveness of biofilters for pesticides that are weakly to moderately sorbed (e.g. atrazine and metolachlor) is unclear [110, 111, 112]. Most studies appear to show that, regardless of the degree of sorption, infiltration is the main mechanism for reducing pesticide loads [108, 113] although adsorption/sedimentation has not been thoroughly investigated [113]. In practice, where runoff-generating rain falls about equally on the watershed area as on the receiving biofilter, infiltration will tend to be low unless evapotranspiration is very much higher in the receiving area, creating a 'sink' for the runoff. Infiltration thus seems unlikely to be effective for retaining herbicides where the ratio of watershed area to biofilter area is very high. Under these conditions, adsorption and sedimentation may assume greater relative importance than infiltration.

In short, different mechanisms determine the effects of VFS on agricultural-surface runoff: (i) dilution, which reduces herbicide concentration in runoff water passing through the filter, and (ii) the "sponge-like" effect, with chemicals temporarily trapped inside the filter, then released after being degraded. A relevant factor influencing VFS effectiveness in reducing agricultural runoff depth is grass cover: the denser the grass, the greater the hydraulic resistance and the lower the runoff volume compared to cropped soil. Finally, physical and chemical properties determine herbicide behavior in the environment and hence their interaction with the VFS. For these reasons, different climatic conditions and cropping systems could require different types of VFS to assure the best performance in reducing herbicide agricultural runoff.

6.2. Riparian Forest Buffer Systems

Riparian forest buffer systems (RFBSs) are streamside ecosystems that can be managed to reduce NPS pollution after it leaves fields but before it reaches streams and lakes in many types of watersheds. Many authors [114, 115, 116, 117, 118, 119] have pointed out that RFBSs are excellent nutrient and herbicide sinks that reduce the pollutant discharge from surrounding agroecosystems. According to Comerford et al. [120], the use of RFBSs is relatively well established as a BMP for water-quality improvement in forestry practices but has been far less widely applied as a BMP in agricultural areas or in urban or suburban scenarios. The RFBSs are especially important for small streams, where interaction between terrestrial and aquatic ecosystems is intense. Riparian vegetation has well-known beneficial effects on bank stability, biological diversity, and water temperatures of streams [121].

Compared to other NPS pollution control measures, RFBSs can lead to longer-term changes in the structure and function of agricultural landscapes. To provide long-term improvements in water quality, RFBSs must be designed with an understanding of the processes that remove or sequester pollutants entering the riparian buffer system, the effects of riparian management practices on pollutant retention, the effects of such forest buffers on aquatic ecosystems, the recovery time after harvest of trees or reestablishment of riparian buffer systems, and the effects of underlying soil and geologic materials on the chemical, hydrological, and biological processes.

The RFBSs provide four important functions:

- The first is the control of sediment and sediment-borne pollutants carried in surface runoff. A properly managed RFBS should provide a high level of such control regardless of the physiographic region. Research studies have indicated that forests are particularly effective in filtering fine sediments and promoting deposition of sediment as water infiltrates. The slope of the RFBS is the main factor limiting the

- effectiveness of the sediment-removal function. It is important to convert concentrated flow to sheet flow in order to optimize RFBS function. Conversion to sheet flow and deposition of coarse sediment, which could damage young vegetation, are the primary functions of the grass filter strip.
- The second function of an RFBS is to control nitrate (NO_3^-) influx into shallow groundwater moving toward streams. When groundwater moves in short, shallow paths through an RFBS, 90% of the NO_3^- input may be removed. In contrast, NO_3^- removal might be minimal in areas where water moves to regional groundwater, and high-NO_3^- groundwater then emerges in stream channels as base flow and bypasses most of the RFBS. The degree to which NO_3^- (or other groundwater pollutants) will be removed depends on the proportion of groundwater moving in or near the biologically active root zone and on the residence time of the groundwater in these biologically active areas.
- The third function of an RFBS is control of dissolved P in surface runoff or shallow groundwater. Control of sediment-borne P is generally effective. In certain situations, dissolved P can contribute a substantial amount of the total P load. Because most soluble P is bioavailable, the potential impact of a unit of dissolved P on aquatic ecosystems is greater than a equivalent unit of sediment-borne P. It appears that natural riparian forests have very low net dissolved P retention. In managing for increased P retention, effective fine-sediment control should be coupled with the use of vegetation which can increase P uptake into plant tissue.

The final function of an RFBS is to control the stream environment: adjusting stream temperature and controlling light quantity and quality, fostering habitat diversity, altering channel morphology, and enhancing food webs as well as species richness. These factors are important to the ecological health of a stream and are provided best by an RFBS.

In addition, constructed wetlands have become a widespread technology to treat contaminated surface and wastewaters [122, 123]. Wetlands are particularly appropriate for treating NPS of pollution, such as urban and agricultural runoff, because they function under a wide range of hydraulic loads, are capable of internal water storage, and can remove or transform a number of contaminants, including oxygen-demanding substances, suspended solids, and P and nitrogenous compounds [124]. Nevertheless, achieving cost-effective P removal to very low levels (<20 µg L^{-1}) in wetlands has proved challenging because the surface area required to reach low P levels can be unmanageably large [123]. In this regard, Dierberg et al. [125] indicates that the incorporation of submerged aquatic vegetation (SAV) (*Najas guadalupensis, Ceratophyllum demersum, Chara* spp., and *Potamogeton illinoensis*) communities into the stormwater-treatment areas may benefit Everglades restoration by removing P from agricultural runoff.

7. CONCLUSION

Water-quality protection and the degree of enforcement within watersheds requires more thorough scrutiny. Future research should include total-nutrient assays, especially total phosphate concentrations, in order to assess recent trends from historical phosphate data, and to achieve a better appraisal of the nutrient dynamics in aquatic systems. More data are required to evaluate changes in water quality over time, and determine the impact of other time-sensitive influences on water quality. Also, more research should concentrate on the relative impact of the various NPS pollutants, thus dictating which sources might require the most legislative efforts to improve water quality for the future.

Also, this review provides other basic reflections:

- Future research may be directed toward efficacy of inert new compounds such as polyacrylamide to remove specific bacterial species from wastewater, since PAM can already remove larger organisms like fungi and algae.
- Fomenting organic farming avoids the use of high amounts of pesticides and fertilizers.
- Avoiding large concentrations of animals can reduce nutrient pollution, and their waste can be used as fertilizer for crops. Regulations are needed on how much manure can be held in a storage area without environmental risk.
- Sediment loads needs to be decreased, especially those resulting from freshly-tilled fields.
- Conservation tillage is a good practice where ground cover after harvest remains on the fields throughout the year (conservation agriculture).
- Strip cropping, or growing plants in strips across the slope to reduce water erosion, is profitable and combats erosion as well as contaminant migration.
- Conservation tillage, which also includes the no-till method, can be very effective against sheet and rill erosion and the spread of pollution.
- Applying IPM techniques based upon soil type, climate, scouting history of the specific pest, and the type of crop is also useful.
- Surveys of rational pesticide use together with annual summaries should be made available for policy purposes, and manufacturers of pesticides should be included in these studies.
- Further analyses of pesticide-use patterns need to be carried out at the national, regional, and individual-user level in order to identify uses and practices which constitute significant ongoing risks (applicator, sustainability of the production system, environmental, food chain, etc.).
- Watershed management decision-support systems (predictive models) should be developed for evaluating the environmental impact of land-management activities designed to reduce agricultural NPS.

ACKNOWLEDGMENTS

Part of research leading to this publication was sponsored by following research project "Environmental Impact of Farming Subtropical Species on Steeply Sloping Lands. Integrated Measures for the Sustainable Agriculture" (RTA05-00008-00-00), granted by INIA, Spain.

REFERENCES

[1] Cooper, CM. (1993). Biological effects of agriculturally derived surface-water pollutants on aquatic systems: review. *J. Environ. Qual.* 22, 402-408.

[2] Maul, JD; Cooper, CM. (2000). Water quality of seasonal flooded agricultural fields in Mississippi, USA. *Agric. Ecosyst. Environ.* 81, 171-178.

[3] Wolverton, BC; McDonald, RC. (1981). Natural processes for treatment of organic chemical wastes. *Environ. Prof.* 3, 99-104.

[4] Gersberg, RM; Lyon, SR; Elkins, BV; Goldman, CR. (1984). The removal of heavy metals by artificial wetlands. In: Future of Water Reuse, Vol. 2. Proceedings of the Water Reuse Symposium, AWWA Research Foundation, Denver, CO, pp. 639-648.

[5] Wieder, RK; Lang, GE. (1984). Influence of wetlands and coal mining on stream water chemistry. *Water Air Soil Pollut.* 23, 381-396.

[6] Cooper, CM; Testa, IS; Knight, SS. (1994). Preliminary effectiveness of constructed wetlands for dairy waste treatment. In: Campbell, et al. (Eds.), Proceedings of the Second Conference on Environmentally Sound Agriculture. ASAE Publication 04-94, pp. 439-446.

[7] Hawkins, WB; Rodgers, Jr JH; Gillespie, JrWB; Dunn, AW; Dorn, PB; Cano, ML. (1997). Design and construction of wetlands for aqueous transfers and transformations of selected metals. *Ecotoxicol. Environ. Safety* 36, 238-248.

[8] Wolverton, BC; Harrison, DD. (1973). Aquatic plants for removal of mevinphos for the aquatic environment. *J. Miss. Acad. Sci.* 19, 84-88.

[9] Gilliam, JW. (1994). Riparian wetlands and water quality. *J. Environ. Qual.* 23, 896-900.

[10] Stewart, BA; Woolhiser, DA; Wischmeier,WH; Caro, JH; Frere, MH. (1976). Control of Water Pollution from Cropland, Volume II – An Overview. EPA- 600/2-75-026b. USEPA, Washington, DC.

[11] Francia, JR; Durán, ZVH; Martínez, RA. (2006). Environmental impact from mountainous olive orchards under different soil-management systems (SE Spain). *Sci. Tot. Environ.* 358, 46-60.

[12] Martínez, RA; Durán, ZVH; Francia, FR. (2006). Soil erosion and runoff response to plant cover strips on semiarid slopes (SE Spain). *Land Degr. Develop.* 17, 1-11.

[13] Durán, ZVH; Rodríguez, PCR; Francia; MJR; Cárceles, RB; Martínez, RA; Pérez, GP. (2008). Harvest intensity of aromatic shrubs vs. soil-erosion: an equilibrium for sustainable agriculture (SE Spain). *Catena* 73, 107-116.

[14] Owens, LB. (1994). Impacts of Soil N Management on the Quality of Surface and Subsurface Water in Soil Process and Water Quality. R. Lal and B.A. Stewards, Eds. Lewis Publishers Inc., Boca Raton, FL.

[15] US EPA (US Environmental Protection Agency). (2003). National Pollutant Discharge Elimination System Permit Regulation and Effluent Limitation Guidelines and Standards for Concentrated Animal Feeding Operations (CAFOs). Final Rule. Federal Register, Available from http://cfpub.epa.gov/npdes/afo/compliance. [21 March 2008].

[16] Castelle, AJ; Johnson, AW; Conolly, C. (1994). Wetland and stream buffer size requirements: A review. *J. Environ. Qual.* 23, 878-882.

[17] Fajardo, JJ; Bauder, JW; Cash, SD. (2001). Managing nitrate and bacteria in runoff from livestock confinement areas with vegetated filter strips. *J. Soil Water Conserv.* 56, 185-191.

[18] Johnson, JYM; Thomas, JE; Graham, TA; Townshend, I; Byrne, J; Selinger, LB; Gannon, VPJ. (2003). Prevalence of Escherichia coli O157:H7 and Salmonella spp. in surface waters of southern Alberta and its relation to manure sources. *Can. J. Microbiol.* 49, 326-335.

[19] Stoddard, CS; Coyne, MS; Grove, JH. (1998). Fecal bacterial survival and infiltration through a shallow agricultural soil: timing and tillage effects. *J. Environ. Qual.* 27, 1516-1523.

[20] Gerba, CP; Wallis, C; Melnick, JL. (1975). Fate of wastewater bacteria and viruses in soil. Journal of Irrigation and Drainage Division, Amer. Soc. Civil Eng. 3, 157-174

[21] Edwards, DR; Coyne, MS; Vendrell, PF; Daniel, TC; Moore, PA; Jr, Murdoch JF. (1997) Fecal coliform and streptococcus concentrations in runoff from grazed pastures in northwest Arkansas. *J. Amer. Water Res. Assoc.* 33, 413-422

[22] Entry, JA; Hubbard, RK; Thies, JE; Fuhrmann, JJ. (2000a). The influence of vegetation in riparian filterstrips on coliform bacteria: II. Survival in soils. *J. Environ. Qual.* 29, 1215-1224

[23] US EPA. US Environmental Protection Agency. (1976). Fecal coliform bacteria. In: Quality criteria for water. US Environmental Protection Agency, Washington, DC, pp 42-50

[24] Abu-Ashour, J; Joy, DM; Lee, H; Whitely, HR; Zelin, S. (1994). Transport of microorganisms through soil. *Water Air and Soil Pollut.* 75, 141-157

[25] Howell, JM; Coyne, MS; Cornelius, PL. (1996). Effect of particle size and temperature on fecal bacterial mortality rates and the fecal coliform/fecal streptococci ratio. *J. Environ. Qual.* 25, 1216-1220

[26] Huysman, F; Verstraete, W. (1993). Water facilitated transport of bacteria in unsaturated soil columns: Influence of cell surface hydrophobicity and soil properties. *Soil Biol. Biochem.* 25, 83-90

[27] Gary, HL; Johnson, SR; Ponce, SL. (1985). Cattle grazing impact on surface water quality in a Colorado front range stream. *J. Soil Water Conserv.* 40, 124-128.

[28] Jawson, MD; Elliott, LF; Saxton, KE; Fortier, DH. (1982). The effect of cattle grazing on indicator bacteria in runoff from a Pacific Northwest watershed. *J. Environ. Qual.* 11, 621-627.

[29] Diez, GF; Callaway, TR; Kizoulis, MG; Russell, JB. (1998). Grain feeding and the dissemination of acid resistant Escherichia coli from cattle. *Sci.* 281, 1666-1668.

[30] Coyne, MS; Gilifillen, RA; Villalba, A; Zhang, Z; Rhodes, RW; Dunn, L; Blevins, RL. (1998). Fecal coliform trapping by grass filter strips during simulated rain. *J. Soil Water Conserv.* 52, 140-145.

[31] Walker, SE; Mostaghimi, S; Dillaha, TA; Woest, FE. (1990). Modeling animal waste management practices: impacts on bacteria levels in runoff from agricultural land. *Trans. ASAE* 33, 807-817.

[32] Young, RA; Hundtrods, T; Anderson, W. (1980). Effectiveness of vegetated buffer strips in controlling pollution from feedlot runoff. *J. Environ. Qual.* 9, 483-487.

[33] Hubbard, RK; Newton, GL; Davis, JG; Lowrance, R; Vellidis, G; Dove, CR. (1998). Nitrogen assimilation by riparian buffer systems receiving swine lagoon wastewater. *Trans. ASAE* 41, 1295-1304.

[34] Snyder, NJ; Mostaghimi, S; Berry, DF; Reneau, RB; Hong, S; McClellan, PW; Smith, EP. (1998). Impact of riparian buffers on agricultural nonpoint source pollution. *J. Amer. Water Res. Assoc.* 34, 385-395.

[35] Jordan, TE; Correll, DT; Weller, DE. (1993). Nutrient interception by a riparian forest receiving inputs from adjacent cropland. *J. Environ. Qual.* 14, 467-472.

[36] Entry, JA; Hubbard, RK; Thies, JE; Furhmann, JJ. (2000b). The influence of vegetation in riparian filterstrips on coliform bacteria I. Movement and survival in surface flow and groundwater. *J. Environ. Qual.* 29, 1206-1214.

[37] Entry, JA; Hubbard, RK; Thies, JE; Furhmann, JJ. (2000c). The influence of vegetation in riparian filterstrips on coliform bacteria II. Survival in soil. *J. Environ. Qual.* 29, 1215-1224.

[38] Dosskey, MG. (2000). How much can USDA riparian buffers reduce agricultural nonpoint source pollution? In: Wigington PJ Jr, Beschta RL (eds), Riparian ecology and management in multiland use watersheds. American Water Resources Association, Middleburg, VA, pp 427-432.

[39] Roodsari, RM; Shelton, DR; Shirmohammadi, A; Pachepsky, YA; Sadeghi, AM; Starr, JL. (2005). Fecal coliform transport as affected by surface condition. *Trans. ASAE* 48, 1055-1061.

[40] Sullivan, TJ; Moore, JA; Thomas, DR; Mallery, E; Snyder, KU; Wustenberg, M; Wustenberg, J; Mackey, SD; Moore, DL. (2007). Efficacy of vegetated buffers in preventing transport of fecal coliform bacteria from pasturelands. *Environ. Manage.* 40, 958-965.

[41] Adekalu, KO; Olorunfeni, IA; Osunbitan, JA. (2007). Grass mulching effect on infiltration, surface runoff and soil loss of three agricultural soils in Nigeria. *Bioresour. Technol.* 98, 912-917.

[42] Entry, JA; Sojka, RE; Watwood, M; Ross, C. (2002). Polyacrylamide preparations for protection of water quality threatened by agricultural runoff contaminants. *Environ. Pollut.* 120, 191-200.

[43] Moore, PA Jr; Daniel, TC; Edwards DR. (1999). Reducing phosphorus runoff and improving poultry production with alum. *Poultry Sci.* 78, 692-698.

[44] Moore, PA Jr; Daniel, TC; Edwards, DR. (2000). Reducing phosphorus runoff and inhibiting ammonia loss from poultry manure with aluminum sulfate. *J. Environ. Qual.* 29, 37-49.

[45] Smith, DR; Moore, PA Jr; Miles, DM; Haggard, BE; Daniel, TC. (2004). Decreasing phosphorus runoff losses from land-applied poultry litter with dietary modifications and alum addition. *J. Environ. Qual.* 6, 2210-2216.

[46] Gilliam, JW; Skaggs, RW. (1986). Controlled agricultural drainage to maintain water quality. *J. Irrig. Drain. Eng.* 112, 254-263.

[47] Evans, RO; Parsons, JE; Stone, K; Wells, WB. (1992). Water table management on a watershed scale. *J. Soil Water Conserv.* 47, 58-64.

[48] Evans, RO; Skaggs, WR; Gilliam, WJ. (1995). Controlled versus conventional drainage effects on water quality. *J. Irrig. Drain. Eng.* 121, 271-276.

[49] Wesstrom, I; Messing, I; Linner, H; Lindstrom, J. (2001). Controlled drainage – effects on drain outflow and water quality. *Agric. Water Manage.* 47, 35-100.

[50] Wesstrom, I; Messing, I. (2007). Effects of controlled drainage on N and P losses and N dynamics in a loamy sand with spring crops. *Agric. Water Manage.* 87, 229-240.

[51] Evans, R; Gilliam, JW; Skaggs, RW. (1996). Controlled drainage management guidelines for improving water quality. AG 443, North Carolina Cooperative State Extension Service. http://www.bae.ncsu.edu/programs/extension/evans/ag443.html [22 April 2008]

[52] Carpenter, SR; Caraco, NF; Correll, DL; Howarth, RW; Sharpley, AN; Smith, VH. (1998). Non-point pollution of surface waters with phosphorus and nitrogen. *Ecol. Applic.* 8, 559-568.

[53] Sharpley, AN. (2000). Agriculture and phosphorus management: the Chesapeake Bay. Boca Raton, FL, CRC Press. 229 p.

[54] Heaney, SI; Foy, RH; Kennedy, GJA; Crozier, WW; O'Connor, WCK. (2001). Impacts of agriculture on aquatic systems: lessons learnt and new unknowns in Northern Ireland. *Mar. Freshwater Res.* 52, 152-163.

[55] MfE (1997). *In*: The State of New Zealand's Environment. Wellington, New Zealand, Ministry for the Environment and GP Publications. pp. 7.1-7.100.

[56] USGS (1999). The quality of our nation's waters: nutrients and pesticides. US Geological Survey Circular 1225. Denver, CO, US Geological Survey Information Services. http://www.usgs.gov [4 March 2008]

[57] NRC (2000). Watershed management for potable water supply: assessing the New York City strategy. Washington, DC, National Resource Council and National Academy Press.

[58] Kellogg, RL; Lander, C.H. (1999). Trends in the potential for nutrient loadings from confined livestock operations. *In*: The state of North America's private land. US Department of Agriculture—Natural Resources Conservation Service, Washington, DC, US Government Printing Office. 5 p.

[59] Withers, PJA; Lord, EI. (2002). Agricultural nutrient inputs to rivers and groundwaters in the UK: policy, environmental management and research needs. *Sci. Total Environ.* 282, 9-24.

[60] USDA-USEPA (1999). Unified national strategy for animal feeding operations, 9 March 1999, Washington, DC, US Department of Agriculture and US Environmental Protection Agency. http://cfpub.epa.gov/npdes/afo/ustrategy.cfm?program_id=7 [4 March 2008].

[61] USEPA (2000). The total maximum daily load (TMDL) program. EPA 841-F-00-0009. Office of Water (4503F). Washington, DC, US Environmental Protection Agency. http://www.epa.gov/owow/tmdl/overviewfs.html [4 March 2008].

[62] Withers, PJA; Davidson, IA; Foy, RH. (2000). Prospects for controlling diffuse phosphorus loss to water. *J. Environ. Qual.* 29, 167-175.

[63] Gillingham, AG; Thorrold, BS. (2000). A review of New Zealand research measuring phosphorus in runoff from pastures. *J. Environ. Qual.* 29, 88-96.

[64] Durán, ZVH; Martínez, RA; Aguilar, RJ. (2004). Nutrient losses by runoff and sediment from the taluses of orchard terraces. *Water Air Soil Pollut.* 153, 355-373.

[65] Wauchope, RD. (1978). The pesticide content of surface water draining from agricultural fields-a review. *J. Environ. Qual.* 7, 459-472.

[66] Brown, CD; Hodgkinson, RA; Rose, DA; Syers, JK; Wilcockson, SJ. (1995). Movement of pesticides to surface waters from a heavy clay soil. *Pestic. Sci.* 43, 131-140.

[67] Ng, HYF; Clegg, SB. (1997). Atrazine and metolachlor losses in runoff events from an agricultural watershed: the importance of runoff components. *Sci. Total Environ.* 193, 215-228.

[68] Daniels, RB; Gilliam, JW. (1996). Sediment and chemical load reduction by grass and riparian filters. *Soil Sci. Soc. Amer. J.* 60, 246-251.

[69] Lee, KH; Isenhart, TM; Schultz, RC; Mickelson, SK. (2000). Multispecies riparian buffers trap sediment and nutrients during rainfall simulations. *J. Environ. Qual.* 29, 1200-1205.

[70] Rankins, JA; Shaw, DR. (2001). Perennial grass filter strips for reducing herbicides losses. *Weed Sci.* 49, 647-651.

[71] Webster, EP; Shaw, DR. (1996). Impact of vegetative filter strips on herbicide loss in runoff from soybean (*Glycine max*). *Weed Sci.* 44, 662-671.

[72] Misra, AK; Baker, JL; Mickelson, SK; Shang, H. (1996). Contributing area and concentration effects on herbicide removal by vegetative buffer strips. *Trans. ASAE* 39, 2105-2111.

[73] Blanche, SB; Shaw, DR; Massey, JH; Boyette, M; Cade Smith, M. (2003). Fluometuron adsorption to vegetative filter strip components. *Weed Sci.* 51, 125-129.

[74] Klöppel, H; Kördel, W; Stein, B. (1997). Herbicide transport by surface runoff and herbicide retention in a filter strip—rainfall and runoff simulation studies. *Chemosphere* 35, 129-141.

[75] Watanabe, H; Grismer, ME. (2001). Diazinon transport through inter-row vegetative filter strips: micro-ecosystem modelling. *J. Hydrol.* 247, 183-199.

[76] Delphin, JE; Chapot, JY. (2001). Leaching of atrazine and desethylatrazine under a vegetative filter strip. *Agronomie* 21, 461-470.

[77] Staddon, WJ; Locke, MA; Zablotowicz, RM. (2001). Microbiological characteristics of a vegetative buffer strip soil and degradation and sorption of metolachlor. *Soil Sci. Soc. Amer. J.* 65, 1136-1142.

[78] Benoit, P; Barriuso, E; Vidon, P; Real, B. (2001). Isoproturon movement and dissipation in undisturbed soil cores from a grassed buffer strip. *Agronomie* 20, 297-307.

[79] Durán, ZVH; Francia, MJR; Rodriguez, PCR; Martínez, RA; Cárceles, RB. (2006a). Soil erosion and runoff prevention by plant covers in a mountainous area (SE Spain): implications for sustainable agriculture. *The Environmentalist* 26, 309-319.

[80] Durán, ZVH; Martínez, RA; Aguilar, RJ; Franco TD. (2003). El cultivo del mango (*Mangifera indica* L.) en la costa granadina, Granada, Spain.

[81] Durán, ZVH; Rodríguez, PCR; Franco, TD; Martín, PFJ. (2006b). El cultivo del chirimoyo (*Annona cherimola* Mill.), Granada, Spain.

[82] Durán, ZVH; Aguilar, RJ; Martínez, RA; Franco, DT. (2005). Impact of erosion in the taluses of subtropical orchard terraces. *Agric. Ecosys. Environ.* 107, 199-210.

[83] Durán, ZVH; Martínez, RA; Aguilar, RJ. (2002). Control de la erosión en los taludes de bancales en terrenos con fuertes pendientes. *Edafología* 9, 1-10.

[84] Carmago, JA; Alonso, A; de la Puente, M. (2005). Eutrophication downstream from small reservoirs in mountain rivers of Central Spain. *Water Res.* 39, 3376-3384

[85] Novotny, V; Bartosova, A; O'Reilly, N; Ehlinger, T. (2005). Unlocking the relationship of biotic integrity of impaired waters to anthropogenic stresses. *Water Res.* 39, 184-198

[86] Wetzel, RG. (2001). Limnology: lake and river ecosystems. Academic, London.

[87] Beman, JM; Arrigo, KR; Matson, PA. (2005). Agricultural runoff fuels large phytoplankton blooms in vulnerable areas of the ocean. *Nature* 434, 211-214.

[88] Norris, V. (1993). The use of buffer zones to protect water quality: A review. *Water Resour. Manage.* 7, 257-272.

[89] Dosskey, M. (2002). Setting priorities for research on pollution reduction functions of agricultural buffers. *Environ. Manage.* 30, 641-650.

[90] Verstraeten, G; Van Oost, K; Van Rompaey, A; Poesen, J; Govers, G. (2002). Evaluating an integrated approach to catchment management to reduce soil loss and sediment pollution through modelling. *Soil Use Manage.* 19, 386-394.

[91] Chaubey, I; Edwards, DR; Daniel, TC; Moore, PA; Nichols, DJ. (1995). Effectiveness of vegetative filter strips in controlling losses of surface-applied poultry litter constituents. *Trans. ASAE* 38, 1687-1692.

[92] Schmitt, TJ; Dosskey, MG; Hoagland, KD. (1999). Filter strip performance and processes for different vegetation, widths, and contaminants. *J. Environ. Qual.* 28, 1479-1489.

[93] Schauder, H; Auerswald, K. (1992). Long-term trapping efficiency of a vegetated filter strip under agricultural use. *Zeitschrift für Pflanzenernährung und Bodenkunde* 155, 489-492.

[94] Tollner, EW; Barfield, BJ; Vachirakornwatana, C; Haan, CT. (1977). Sediment deposition patterns in simulated grass filters. *Trans. ASAE* 20, 940-944.

[95] Muñoz, CR; Parson, JE; Gilliam, JW. (1993). Numerical approach to the overland flow process in vegetative filter strips. *Trans. ASAE* 36, 761-770.

[96] Muñoz, CR; Parsons, JE; Gilliam, JW. (1999). Modeling hydrology and sediment transport in vegetative filter strips. *J. Hydrol.* 214, 111-129.

[97] Deletic, A. (2001). Modelling of water and sediment transport over grassed areas. *J. Hydrol.* 248, 168-182.

[98] Chow, TL; Rees, HW; Daigle, JL. (1999). Effectiveness of terraces/grassed waterway systems for soil and water conservation: A field evaluation. *J. Soil Water Conserv.* 3, 577-583.

[99] Briggs, JA; Whitwell, T; Riley, MB. (1999). Remediation of herbicides in runoff water from container plant nurseries utilizing grassed waterways. *Weed Technol.* 12, 157-164.

[100] Fiener, P; Auerswald, K. (2003a). Concept and effects of a multipurpose grassed waterway. *Soil Use Manage.* 19, 65-72.

[101] Fiener, P; Auerswald, K. (2003b). Effectiveness of grassed waterways in reducing runoff and sediment delivery from agricultural watersheds. *J. Environ. Qual.* 32, 927-936.

[102] Fiener, P; Auerswald, K. (2005). Measurement and modelling of concentrated runoff in a grassed waterway. *J. Hydrol.* 301, 198-215.

[103] Fiener, P; Auerswald, K. (2006). Seasonal variation of grassed waterway effectiveness in reducing runoff and sediment delivery from agricultural watersheds in temperate Europe. *Soil Till. Res.* 87, 48-58.

[104] Mickelson, SK; Baker, JL. (1993). Buffer strips for controlling herbicide runoff losses. Paper No. 932084. American Society of Agricultural Engineers, St. Joseph, MI, USA.

[105] Arora, K; Mickelson, SK; Baker, JL; Tierney, DP; Peters, CJ. (1996). Herbicide retention by vegetative buffer strips from runoff under natural rainfall. *Trans. ASAE* 39, 2155-2162.

[106] Patty, L; Real, B; Gril, J. (1997). The use of grassed buffer strips to remove pesticides, nitrate and soluble phosphorus compounds from runoff water. *Pestic. Sci.* 49, 243-251.

[107] Barfield, BJ; Blevins, RL; Fogle, AW; Madison, CE; Inamdar, S; Carey, DI; Evangelou, VP. (1998). Water quality impact of natural filter strips on karst areas. *Trans. ASAE* 41, 371-381.

[108] Dosskey, M. (2001). Toward quantifying water pollution abatement in response to installing buffers on crop-land. *Environ. Manage.* 28, 577-598.

[109] Karssies, LE; Prosser, IP. (1999). Guidelines for riparian filter strips for Queensland irrigators. CSIRO Land andWater Technical Report, 32/99, Australia.

[110] Kookana, RS; Baskaran, S; Naidu, R. (1998). Pesticide fate and behaviour in Australian soils in relation to contamination and management of soil and water: a review. *Aust. J. Soil Res.* 36, 715-764.

[111] USDA (United States Department of Agriculture). (2000). Conservation buffers to reduce pesticide losses. National Resource Conservation Service (NCRS), USA. Available on-line: http://www.nrcs.usda.gov [4 March 2008].

[112] Connolly, RD; Silburn, DM; Freebairn, DM. (2002). Transport of sediment and pesticides in surface waters. In: Kookana, S., et al. (Eds.), Environmental Protection and risk assessment of organic contaminants. Science Publishers Inc., Enfield (NH), USA.

[113] Baker, JL; Mickelson, SK; Arora, K; Misra, AK. (2000). The potential of vegetated filter strips to reduce pesticide transport. American Chemical Society Symposium Series.

[114] Vellidis, GR; Lowrance, P; Gay, RW; Hubbard, RK. (2003). Nutrient transport in a restored riparian wetland. *J. Environ Qual.* 32, 711-726.

[115] Lowrance, RR; Todd, RL; Asmussen, LE. (1983). Waterborne nutrient budgets for the riparian zone of an agricultural watershed. *Agric. Ecosys. Environ.* 10, 371-384.

[116] Lowrance, R; Leonard, R; Sheridan, J. (1985). Managing riparian ecosystems to control nonpoint pollution. *J. Soil and Water Conserv.* 40, 87-91.

[117] Lowrance, R; Sharpe, JK; Sheridan, JM. (1986). Long-term sediment deposition in the riparian zone of a coastal plain watershed. *J. Soil and Water Conserv.* 41, 266-271.

[118] Lowrance, RG; Vellidis, RD; Wauchope, PG; Bosh, DD. (1997). Herbicide transport in a riparian forest buffer system in the coastal plain of Georgia. *Trans. ASAE* 40, 1047-1057.

[119] Peterjohn, WT; Dorrel, DL. (1984). Nutrient dynamics in an agricultural watershed: Observations on the role of a riparian forest. *Ecol.* 65, 1466-1475.

[120] Comerford, NB; Neary, DG; Mansell, RS. (1992). The effectiveness of buffer strips for ameliorating offsite transport of sediment, nutrients, and pesticides from silvicultural operations. Nat. Coun. Paper Ind. Air & Stream Improvement Tech. Bull. 631. New York, NY, USA.

[121] Karr, JR; Schlosser, IJ. (1978). Water resources and the land-water interface. *Sci.* 1, 229-234.

[122] Reed, SC; Crites, RW; Middlebrooks, EJ. (1995). Natural systems for water management and treatment. New York, NY. McGraw-Hill.

[123] Kadlec, R; Knight, R. (1996). Treatment wetlands. Boca Raton, FL: Lewis Publishers, 1996.

[124] Chescheir, GM; Skaggs, RW; Gilliam, JW. (1992) Evaluation of wetland buffer areas for treatment of pumped agricultural drainage water. Trans ASAE 35, 175-182.

[125] Dierberg, FE; DeBuska, TA; Jacksona, SD; Chimney, MJ; Pietro, K. (2002). Submerged aquatic vegetation-based treatment wetlands for removing phosphorus from agricultural runoff: response to hydraulic and nutrient loading. *Water Res.* 36, 1409-1422.

In: Encyclopedia of Environmental Research
Editor: Alisa N. Souter
ISBN: 978-1-61761-927-4
© 2011 Nova Science Publishers, Inc.

Chapter 50

EFFECTS OF AGRICULTURAL RUNOFF VERSUS POINT SOURCES ON THE BIOGEOCHEMICAL PROCESSES OF RECEIVING STREAM ECOSYSTEMS

Gora Merseburger[1,*], Eugènia Martí[2], Francesc Sabater[1] and Jesús D. Ortiz[2,*]

[1]Department of Ecology, University of Barcelona
Av. Diagonal, 645. E-08028 Barcelona, Catalonia, Spain
[2]Center of Advanced Studies of Blanes-Spanish Council of Research (CEAB-CSIC)
Camí d'accés a la Cala St. Francesc, 14, E-17300 Blanes, Girona (Catalonia, Spain)

ABSTRACT

A full understanding of the N cycling in lotic ecosystems is crucial given the increasing influence of human activities on the eutrophication of streams. Knowledge of stream processes involved in nutrient dynamics and metabolism in human-altered ecosystems is currently limited. On the other hand, N uptake and metabolism are likely to be linked in pristine streams, fact that remains unclear in human-altered streams. We aimed to examine N dynamics and the degree of linkage to ecosystem metabolism in two streams draining catchments with contrasting land uses, dominated by forest (reference) and agricultural activity (affected by agricultural runoff). In the two streams, we selected two reaches located upstream and downstream of a WWTP effluent input (i.e., point source) to compare the effect of point versus diffuse sources on the chemical and functional attributes of the two study scenarios. To achieve these objectives, we estimated rates of different biogeochemical processes involving N (i.e., uptake, nitrification and denitrification) and examined relationships between measured and estimated (based on *in situ* metabolism measurements) N demand upstream and downstream of the point source in the two streams. All measurements were done on 8 and 9 samplings through 2001-2003. Our results showed that land uses modulate the effect of point sources on chemical and functional attributes of the receiving streams. In the agricultural stream, diffuse sources from adjacent agricultural fields were likely to overwhelm the local effect of the

---

[*] current adress: Rheos ecology, Camí de Valls 81-87, E-43204 Reus, Tarragona (Catalonia, Spain) www.rheosecology.com

point source on nutrient concentrations. In contrast, nutrient concentrations significantly increased below the point source. These different effects on stream water chemistry were reflected on the effects of the point source on the functional attributes of the study streams. Rates of studied biogeochemical processes were higher downstream than upstream of the point source in the forested stream, whereas they were similar between the two reaches in the agricultural stream. On the other hand, ambient nutrient concentrations upstream of the point source were significantly higher in the agricultural than in the forested stream, except for NH_4^+-N. This trend was reflected on higher nitrate uptake and denitrification rates in the agricultural than in the forested stream. Finally, coupling between measured and estimated N demand becomes weaker with increasing nutrient inputs from human activities to streams. Overall, our results allow us to introduce the concept of the agricultural stream syndrome, by providing insights of stream ecosystem function in scenarios affected by agricultural runoff.

INTRODUCTION

During the last decades, the amount of land occupied by urban and agricultural activities has increased at the global scale (Crouzet *et al.* 1999, Walsh 2000). This occupation process is expected to continue given that population growth and related activities are predicted to increase during the 21st century (Palmer *et al.* 2005). Urban and agricultural activities generate point and diffuse sources of nutrients, respectively, decreasing the water quality of the receiving freshwater ecosystems, such as streams (Vitousek *et al.* 1997, Crouzet *et al.* 1999). The current excess of fertilization and the depletion of riparian vegetation make agriculture the major cause of freshwater degradation (EEA 2003, Paul and Meyer 2001). Field crops represent a diffuse, continuous and cumulative supply of nutrients that reach adjacent freshwater ecosystems draining agricultural landscapes through surface and subsurface runoff. This additional supply of fertilizers often lead to problems of eutrophication that have severe consequences on aquatic organisms and, therefore, on the structure and function of the receiving aquatic ecosystems. In developed countries, water quality issues are related to chemical contamination that limits some water uses or implies increasing water treatment costs. In developing countries, the consequences are more dramatic because populations do not have other option than use untreated waters for personal uses (Meybeck 2003), with implications for human health. In addition to agriculture, point source inputs, such as those from wastewater treatment plants (WWTPs), also cause the impairment of receiving streams (Walsh 2000, Walsh *et al.* 2005).

Diffuse and point source inputs not only increase ambient levels of nutrients and change the hydrological regime of stream ecosystems (House *and* Denison 1997, Kim *et al.* 2002, Albek 2003, Martí *et al.* 2004), but also affect their ecological integrity (Paul *and* Meyer 2001). This is especially relevant in regions where freshwater is scarce, such as the Mediterranean region, because human-derived inputs of nutrients can account for most of the stream nutrient loads (Gasith *and* Resh 1999). While point sources may cause an abrupt hydrological and chemical discontinuity along the stream, changes due to diffuse sources may not be as discrete at local scale. These contrasting patterns between human-derived sources of nutrients should have differentiated effects on the ecology of stream ecosystems.

Pristine streams play an important role in transforming and retaining nutrients from their catchments (e.g., Mulholland *et al.* 1985, Triska *et al.* 1989, Munn *and* Meyer 1990, Martí and Sabater 1996, Peterson *et al.* 2001). Nevertheless, ecosystem functioning in streams receiving nutrients from point and diffuse sources is still poorly known (Gibson 2004, Inwood *et al.* 2005, Meyer *et al.* 2005, Walsh *et al.* 2005). Improving existing understanding about the biogeochemistry of human-altered streams (i.e., receiving anthropogenic nutrient inputs) would provide insights on their capacity to retain and transform nutrients. This self-purification capacity is likely to contribute to improve water quality in receiving streams and downstream ecosystems, and thus, provide ecosystem services (Constanza *et al.* 1997, Meyer *et al.* 2005). Humans depend on water availability for municipal, industrial, agricultural, and recreational uses. Therefore, improving such understanding is needed to combine increasing human population and related activities with maintenance of the natural services of streams upon which humans depend (Palmer *et al.* 2005).

We aimed to study N dynamics in altered streams because human activities are increasing fluxes of N and N cycling has not yet been quantified in these ecosystems (Inwood *et al.* 2005). A full understanding of the N cycling in lotic ecosystems is crucial given the increasing influence of human activities on the eutrophication of streams and rivers (Inwood *et al.* 2005). On the other hand, N uptake and whole-stream metabolism are likely to be linked in pristine streams (Webster *et al.* 2003), but this relationship remains unclear for human-altered streams. Therefore, we examined N dynamics and whether it is linked or not to ecosystem metabolism in two streams draining catchments with contrasted land uses, dominated by forest (i.e., reference stream) and agricultural activity (i.e., stream affected by agricultural runoff). In the two streams, we selected two reaches located upstream and downstream of a WWTP effluent input (i.e., point source) in order to compare the effect of point versus diffuse sources on the functioning attributes of the two contrasting study scenarios. To achieve these objectives, we firstly estimated rates of different biogeochemical processes involving N (i.e., uptake, nitrificacion and denitrification) and, secondly, we examined relationships between N uptake and metabolism rates upstream and downstream of the point source in the two study streams. Increasing the current scientific understanding of this human-caused global change on N dynamics is important to lessen the impacts of human activities on stream ecosystems and contribute to the development of successful management plans.

STUDY SITES

The study was conducted in Tordera and Gurri streams, located in Catalonia (NE of Spain, Figure 1). The two streams are relatively close (ca. 40 km apart) and their hydrological regime is influenced by the Mediterranean climate of the region. In the two streams, discharge can vary orders of magnitude within a hydrological year and between years. Low flow occurs mostly in summer, and these streams become usually intermittent for some period during this season. Peak flows occur during spring and fall as a result of heavy rainfalls. Major differences between the two streams are related to the geology and land use composition of their catchments (Figure 1). Above the sampling sites, Tordera drains a siliceous catchment (80.2 km^2) and Gurri drains a calcareous catchment (37.7 km^2). In Tordera stream, land use within the sub-catchment upstream of the WWTP input is dominated by forests (87 %) of evergreen oak (*Quercus ilex*), pine (*Pinus sylvestris*), and oak (*Quercus humilis*) with some open land. Agricultural practices account for a small proportion of the catchment area (11 %) and urbanization (2 %) is concentrated in the lower part of the basin, surrounding the study area.

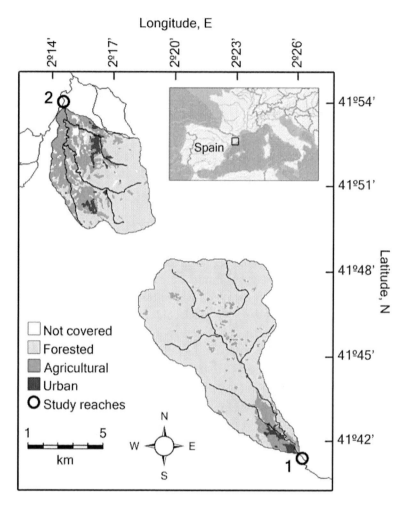

Figure 1. Location of the two study sites and catchment land use composition for the two study streams: 1, Tordera (forest-dominated); 2, Gurri (agricultural-dominated).

Hence, hereafter we refer to Tordera as the forested stream. In contrast, land use in Gurri catchment is dominated by agricultural practices (61 %) and to a lesser extent by a sparse forest (35 %) of pine (*Pinus sylvestris*) and oak (*Quercus humilis*). Urbanization is scarce (4 %) and dispersed throughout the catchment. Hence, hereafter we refer to Gurri as the agricultural stream.

In each stream we selected two reaches located upstream (100-m long in The forested stream and 51-m long in The agricultural stream) and downstream (104-m long in The forested stream and 97-m long in The agricultural stream) of a WWTP. We defined six sampling points at each reach. Downstream reaches were located few hundred meters from the point source inputs to avoid the confounding effects of strong gradients in ambient nutrient concentrations measured below the point source (Merseburger *et al.*, 2005). The reaches were located in the midterm of the streams at 200 m a.s.l. in the forested stream and at 500 m a.s.l. in the agricultural stream. No tributaries joined the streams along any of the selected reaches. The 4 reaches have a low sinuosity channel, a run-riffle sequence with few shallow pools, and a slope close to 1 %. In the forest stream, substrata type consists of cobbles (34%), pebbles (22%) and boulders (22%); and riparian vegetation is well developed along the reaches. In the agricultural stream, bedrock (69%) is the dominant substrata type with patches of very fine sediment (14%). In this stream, canopy cover in the downstream reach was sparser than in the upstream reach.

The WWTP effluents that discharge into the forested and the agricultural streams treat 5808 and 11666 inhabitant-equivalent, respectively; where 1 inhabitant-equivalent is the biodegradable organic matter load equivalent to a DBO_5 of 60 g O_2 day^{-1}. The two WWTPs have activated sludge biological treatment, but do not have technology to actively remove nitrogen or phosphorus. None of the WWTPs chlorinate the effluent. Hence, the effluents had relatively high nutrient concentrations, which were comparable between the two sites (Merseburger 2006). The major difference between the two effluents was the proportion of DIN as ammonium (NH_4^+-N) and nitrate (NO_3^--N). This indicates that the WWTP in the agricultural site had a high capacity to nitrify (mean ± SE % of DIN as NO_3^--N was 97.9 ± 0.9), whereas the WWTP in the forested stream did not (mean ± SE % of DIN as NH_4^+-N was 87.2 ± 5.5).

The point source affected the hydrology of the two streams, decreasing water residence time (i.e., increasing water velocity or discharge below the input; Table 1). These effects varied over time due to the irregular hydrologic regime characteristic of the Mediterranean climate. The contribution of the point source to downstream discharge was higher at lower discharges in the two streams. This effect was exacerbated in summer, when the point source accounted for 100 % of the downstream flow. Therefore, point source inputs not only altered the amount of stream flow, but also shifted the hydrologic regime of the study streams from being intermittent to permanent. These results support previous findings (Martí et al. 2004) and evidence the vulnerability of streams, especially those located in arid or semi-arid regions, to point source inputs (Gasith and Resh 1999). Ambient nutrient concentrations upstream of the WWTP were higher in the agricultural than in the forested stream (independent samples T-test, $df = 14$: NO_3^--N, $t = 8.122$, $P < 0.001$; DIN, $t = 8.199$, $P < 0.001$; soluble reactive phosphorus (SRP), $t = 6.157$, $P < 0.001$; dissolved organic carbon (DOC), $t = 7.112$, $P < 0.001$) except for NH_4^+-N concentration (independent samples T-test: $t = 0.495$, $df = 14$, $P = 0.629$; Table 1). In the forested stream, the WWTP input significantly increased concentrations of all studied nutrients and decreased the NO_3^-:NH_4^+ and DIN:SRP molar ratios (Table 1). The change in the NO_3^-:NH_4^+ ratio represented an increase in the average (±SE) percentage of DIN as NH_4^+-N from 5 ± 1 % in the upstream reach to 20 ± 4 % in the downstream reach. The point source did not affect the DOC:DIN molar ratio (Table 1). In contrast, in the agricultural stream, the point source only increased the concentration of PO_4^{3-}-P. In this stream, the nutrient molar ratios did not significantly vary between the two reaches (Table 1). In the agricultural stream, DIN was dominated by NO_3^--N (≥ 98 %) in the two reaches.

To achieve the objectives of this study, we measured a number of parameters described below in the two reaches of the forested and the agricultural streams on 8 and 9 sampling dates through 2001-2003, respectively.

Table 1. Range and mean ± standard error (SE) of hydrological and chemical parameters measured at the upstream and downstream reaches in the forested ($n = 8$) and the agricultural (upstream, $n = 8$; downstream, $n = 9$) streams

Parameter	Reach	Forested stream Range	Mean ± SE	P	Agricultural stream Range	Mean ± SE	P
Velocity (m/s)	up	0.03-0.47	0.27 ± 0.05	**	0.04-0.59	0.27 ± 0.07	*
	down	0.09-0.71	0.39 ± 0.07		0.17-0.85	0.39 ± 0.08	
Discharge (L/s)	up	1.5-405.9	171.1 ± 47.3	ns	5.8-117.0	48.6 ± 16.1	**
	down	15.0-504.7	227.9 ± 61.4		8.7-395.0	101.2 ± 40.5	
NH_4^+-N (mg N/L)	up	0.006-0.072	0.037 ± 0.009	***	0.011-0.150	0.050 ± 0.017	ns
	down	0.146-2.210	0.780 ± 0.309		0.014-0.210	0.078 ± 0.022	
NO_3^--N (mg N/L)	up	0.66-2.20	1.11 ± 0.18	**	2.69-13.38	7.43 ± 1.18	ns
	down	1.59-6.40	2.99 ± 0.58		4.47-10.87	7.33 ± 0.78	
DIN (mg N/L)	up	0.72-2.21	1.15 ± 0.18	**	2.76-13.68	7.53 ± 1.20	ns
	down	1.80-9.33	3.90 ± 0.89		4.54-11.02	7.48 ± 0.79	
SRP (mg/L)	up	0.006-0.040	0.014 ± 0.004	***	0.02-0.38	0.21 ± 0.05	*
	down	0.05-3.78	0.68 ± 0.45		0.12-0.69	0.38 ± 0.07	
DOC (mg/L)	up	0.55-2.39	1.16 ± 0.25	**	3.22-7.75	5.33 ± 0.62	ns
	down	1.01-2.71	1.90 ± 0.18		3.86-7.86	5.30 ± 0.40	
NO_3^-:NH_4^+	up	10-244	71 ± 30	**	41-598	287 ± 81	ns
	down	2-28	8 ± 3		32-502	175 ± 51	
DIN:SRP	up	40-493	243 ± 50	**	19-901	188 ± 104	ns
	down	5-197	64 ± 25		22-156	59 ± 14	
DOC:DIN	up	0.5-2.8	1.3 ± 0.3	ns	0.4-3.3	1.2 ± 0.4	ns
	down	0.3-1.3	0.8 ± 0.1		0.5-2.0	1.0 ± 0.2	

The table shows the degree of significance from paired *T*-test on these parameters between the two reaches within each study stream. Ratios between nutrients are molar

METHODOLOGY

Chemical parameters

At each sampling point, we collected 3 replicate water samples to analyze concentrations of NH_4^+-N, nitrite-nitrogen (NO_2^--N), NO_3^--N, SRP and DOC. Water samples were filtered *in situ* through pre-ashed fiberglass filters (Whatman® GF/F) and cold-stored for subsequent analysis.

Water samples were transported to the laboratory and placed in the refrigerator until chemical analyses were done. Chemical analyses were all completed within a week after the field experiment. Nutrient concentrations were analyzed using colorimetric methods. Specifically, concentration of NH_4^+-N was analyzed on a Bran-Luebbe® Technicon Autoanalyzer II (method 98-70W). Nitrite (NO_2^--N), NO_3^--N and SRP concentrations were analyzed on a Bran-Luebbe® TRAACS 2000 Autoanalyzer (methods J-003-88E, J-002-88E, and G-033-92C, respectively), using the cadmium-copper reduction method for NO_3^--N determination, and the molybdenum-blue colorimetric method for SRP determination (Murphy and Riley, 1962). DOC concentration was analyzed using a high-temperature catalytic oxidation Shimadzu TOC 5000 analyzer. Finally, chloride concentration was measured by capillary electrophoresis (Waters®, CIA-Quanta 5000, Romano and Krol, 1993).

Uptake Rates

We quantified N retention, in terms of uptake rates of NH_4^+-N and NO_3^--N in the two study streams. We calculated total uptake rates of DIN by summing those of NH_4^+-N and NO_3^--N. Uptake rate is the mass of nutrient retained per unit area of the stream bottom and per unit time (U, in mg N m^{-2} min^{-1}), and indicates the stream nutrient retention capacity (Stream Solute Workshop 1990). To measure U, we conducted short-term nutrient additions on several dates during 2002 and 2003 in the two reaches of the two streams. The short-term nutrient addition at constant rate is the most commonly used methodology to estimate stream nutrient retention metrics (Webster and Valett, 2006). However, the high ambient nutrient concentrations of the two streams constrained the application of this methodology, especially in the reaches located downstream of the point source input. Instead, we conducted slug additions (McColl, 1974, Meals et al., 1999, Wilcock et al., 2002), which were more suitable for this particular study and allowed comparison of results between the two reaches.

On each sampling date and at each reach, we conducted two nutrient additions — $NH_4^+ + PO_4^{3-}$ and $NO_3^- + PO_4^{3-}$ — on two consecutive days. NH_4^+, NO_3^- and PO_4^{3-} were added as NH_4Cl, $NaNO_3$ and $NaH_2PO_4 \cdot H_2O$, respectively. The added solution was set to result in low increases of ambient nutrient concentrations (c.a., 2-4 times ambient levels) and to vary N and P concentrations proportionally to stream ambient DIN:SRP ratios. The solution also contained chloride (as NaCl) used as conservative tracer to account for solute dilution and dispersion (Bencala, McKnight and Zellweger, 1987) and to signal the passage of the added solution at the bottom of the reach. All experiments were conducted at noon on unclouded days and under baseflow conditions.

For each slug addition, reagents were mixed in 10-50 L (depending on the stream discharge of each sampling date) of stream water in a carboy. The solution was added as a single pulse to the stream at a point of concentrated flow (i.e. high turbulence) about 10 m above the head of the reach to ensure a fast mixture between the added solution and the stream water along the selected reach. Once the solution was poured into the stream, we recorded conductivity and water temperature (WTW® LF 340 conductivity meter) at 5-second intervals with a data logger (Campbell CR 510) at the bottom of the reach. Data was recorded until conductivity returned again to ambient levels after the passage of the solution pulse. The time-curve conductivity data was used to calculate discharge (Q, L s^{-1}) and average water velocity (v, m s^{-1}) following Gordon, McMahon and Finlayson (1992). During this period, we simultaneously collected water samples with 30 ml plastic bottles at the bottom of the reach every 10-60 seconds following the frequency of conductivity changes over time. Water samples were filtered in situ through pre-ashed fiberglass filters (Whatman ® GF/F) and placed in a cooler for transportation to the laboratory where they were kept in the refrigerator until analysis, described in the previous section.

Nutrient retention metrics were calculated based on the comparison between the time-through curve of measured nutrient concentrations (NH_4^+-N and NO_3^--N) and that of the predicted nutrient concentrations estimated from chloride concentrations. The predicted concentrations assumed that nutrients were conservative elements; and thus, their variation over time was solely due to hydrological factors (i.e., advection, dispersion and dilution). Whereas, variation over time of measured nutrient concentrations was additionally influenced by retention/release processes. We estimated U, as described in Merseburger 2006, following the approach of the Stream Solute Workshop (1990).

Potential Nitrification Rates

We estimated potential nitrification rates based on longitudinal net decreases in NH_4^+-N concentration along a reach located below the point source in the two study streams (Merseburger et al. 2005). At the upstream reach of the forested stream, nitrification rates were estimated by using a multiple regression that included $\log_{10}$ of nitrification rates from below the input as dependent variable and $\log_{10}$ of NH_4^+-N concentration and stream discharge as independent variables ($R^2 = 0.43$, $P = 0.034$). In the agricultural stream, we were able to estimate just one nitrification rate because net longitudinal decline in NH_4^+-N concentration downstream of the point source occurred only on one sampling date (Merseburger et al. 2005). These nitrification rates were calculated as NH_4^+-N uptake rates (Stream Solute Workshop 1990), by dividing the product of NH_4^+-N concentration at the top of the reach (mg/L) times discharge (L/s) by the product of NH_4^+-N processing length (m) times wetted perimeter (m). These nitrification rates may overestimate actual rates because we assume that the decrease in NH_4^+-N concentration along the reach (i.e., total net removal) is all due to nitrification and part of it could also be due to NH_4^+-N assimilatory uptake. In streams with low NH_4^+-N concentrations (≤ 3 μg N/L), nitrification rates ranged from less than 3% to 20% of the total NH_4^+-N uptake rates (Dodds et al. 2000, Mulholland et al. 2000, Tank et al. 2000). In other pristine streams, nitrification rates accounted for 40-57% of the total NH_4^+-N uptake rates (Hamilton et al. 2001, Merriam et al. 2002, Ashkenas et al. 2004). Nitrification rates may represent a greater part of NH_4^+-N uptake rates in streams with higher NH_4^+-N concentrations, such as the agricultural stream.

Denitrification Rates

Denitrification is a net sink for NO_3^--N from stream waters, and thus, may minimize the effects of nutrient inputs from human activities on stream water chemistry (Ostrom et al. 2002). To estimate denitrification potential rates, we collected six substrata samples along each study reach. In the two streams, we collected sediment samples. In the agricultural stream, where streambed is mostly dominated by bedrock, we also collected biofilm samples.

All sediment samples were collected using a plastic cylinder (Ø = 26 cm) that was driven into the streambed to a depth of 5 cm. Biofilm samples were collected by scraping 100 cm^2 of colonized bedrock surface, as reported for measurement of denitrification potential of biofilm communities (Ventullo and Rowe 1982). All samples were transferred to plastic bags, transported in an icebox to the laboratory and stored in the refrigerator at ~4°C until the denitrification incubations were initiated (within 24 h following samples collection). On each sampling date, we also collected samples of stream water from each reach. Water samples were used for the laboratory incubations.

To estimate denitrification rates we used the acetylene (C_2H_2) inhibition technique (Tiedje et al. 1989, Holmes et al. 1996). Denitrification incubations were prepared by transferring a known weight of wet mass of the sample (approximately 350 g for sediment and 12 g for biofilm) into bottles of 300-mL (sediment) or 60-mL (biofilm), and by adding unfiltered stream water to bring the total volume to 250-mL (sediment incubations) or 30-mL (biofilm incubations). Incubation bottles were made anoxic by purging N_2 in the water for 10 min, time in which we had checked that O_2 concentration in water was < 1 mg/L. We then capped and sealed the bottles with septa-fitted screw-top lids. We added C_2H_2 (10 ml) with a syringe to each incubation bottle. Bottles were shacked to ensure C_2H_2 mixture within the substrata, water and headspace. Then, pressure was equalized by piercing for 10 s the bottle septa with a syringe needle. Bottles were incubated in the dark at ambient laboratory temperature. In preliminary experiments, gas samples were taken every 4 h during 24 h to determine appropriate sampling time. Subsequent incubations required only an initial and final (at 24 h) gas sampling for steady linear accumulation of nitrous oxide (N_2O). Gas samples were collected using a double needle with sealed, preevacuated vials. After collection of gas samples, an equal volume of N_2 (90 %) and C_2H_2 (10 %) was returned to each incubation bottle to avoid pressure changes. Vials containing gas samples were stored at ~4°C until N_2O analysis.

For N_2O analysis, we sampled 1 mL from the gas vials with a gastight syringe. We used a N_2O standard of 100 μg/mL. We determined N_2O using a gas chromatograph (Hewlett Packard 5890 A) equipped with a thermal conductivity detector (250 °C). The components were separated on a Porapak Q column, increasing from 60 °C (5 min) to 200 °C (2 min) at a rate of 10 °C/min, with helium as the carrier gas (60 ml/min).

Denitrification rates were calculated from the difference between final and initial headspace N_2O concentration. Total mass of N_2O in the headspace was calculated using the headspace N_2O concentrations and total microcosm volumes following correction for N_2O solubility in the liquid phase with an appropriate temperature-dependent Bunsen coefficient (Knowles 1979, Martin et al. 2001). In the case of the agricultural stream, in which we measured denitrification rates from two different substrata (i.e. biofilm and sediment), we calculated habitat weighted denitrification for the reach based on the percentage of coverage of the two substrata in the reach. All denitrification rates were expressed as mg N m^{-2} d^{-1}. Denitrification rates in incubations with stream water could underestimate actual denitrification rates because C_2H_2 blocks nitrification, a potentially important source of NO_3^--N to denitrifiers (Tiedje et al. 1989). However, linear accumulation of N_2O during the 24 h of our denitrification assays indicated no NO_3^--N limitation. In addition, we measured similar NO_3^--N concentrations in stream water before and after denitrification incubations, fact that supported the assumption of no NO_3^--N limitation during the incubation period. Finally, denitrification rates measured using the described method are potential rates due to optimum conditions of anoxia in the sediments and interstitial NO_3^--N concentrations equal to those of overlying stream water.

Metabolism

Whole-ecosystem metabolism was measured using the upstream-downstream dissolved oxygen change technique, introduced by Odum (1956) and refined by Marzolf *et al.* (1994). Measurements using this open system method include an entire section of a stream, and thus, results reflect metabolism rates at the ecosystem level. The basis of this method is that dissolved oxygen changes (Q_{DO}) between the two stations depend on GPP, R and the oxygen exchange (E) with the atmosphere (i.e., Q_{DO} = GPP − R ± E). Measurements of dissolved oxygen (DO) concentration and water temperature were recorded at 10-min intervals over a 24-h period using two oxygen meters (WTW® Oxi 340-A) placed at the top and bottom of the reach. The distance between the two stations depended on stream water velocity; it was commonly about 100 m, but was reduced to 30 m in the downstream reach of the agricultural stream in July 2003, when the upstream reach was dry and discharge at the downstream reach was very low (9 L/s).

Oxygen exchange with the atmosphere was estimated based on reaeration rates, which were measured 1 day after the oxygen measurements using *in situ* gas (butane) addition coupled with a hydrological tracer (Marzolf *et al.* 1994). Reaeration rates were calculated from the decline in dissolved butane concentration during steady-state corrected by hydrological dilution. Reaeration rates of oxygen have been commonly calculated from those of propane using a conversion factor of 1.39 (Rathbun *et al.* 1978). This factor is the quotient between propane and oxygen molecular mass, and thus, we calculated oxygen reaeration rates from those of butane using a correction factor of 1.82. Because butane additions were successful only on those sampling dates with lowest discharge, reaeration rates for the rest of dates were estimated using a number of indirect methods. In particular, we used the empirical equation of Owens (1974) and those compiled from literature by Genereux *and* Hemond (1992), based on physical attributes (i.e., channel slope, water velocity and depth) of streams. We also used the approach of Young *and* Huryn (1998) to calculate reaeration rates based on the night-time variation of mean DO deficits at two stations (i.e., upstream-downstream) and net DO changes. Results from indirect methods were compared with those available from *in situ* butane additions. Reaeration rates estimated with the equation of Owens (1974) were more similar to those from butane additions and the approach of Young *and* Huryn (1998) than those estimated with the rest of indirect methods for the two study streams, and thus, this equation was selected to estimate unavailable reaeration rates.

We calculated instantaneous rates of net oxygen change (mg O_2 m^{-2} min^{-1}) along the reach as the change at 10-min intervals in DO between the two stations, and corrected them for air-water oxygen exchange to estimate metabolism rates. The daily rate of ecosystem R (g O_2 m^{-2} d^{-1}) was estimated by extrapolating average night-time respiration through the daylight hours (Edwards *and* Owens 1962, Bott 1996). The daily rate of GPP (g O_2 m^{-2} d^{-1}) was determined by summing the differences between measured corrected net oxygen change rate and the extrapolated value of R during the daylight period. We calculated the GPP:R ratio. The simple balance between GPP and R does not necessarily indicate which source of carbon (autochthonous or allochthonous) is most important to ecosystem respiration (Rosenfeld *and*

Mackay 1987, Meyer 1989). However, as originally proposed by Odum (1956), it is generally accepted that the GPP:R ratio indicates whether an ecosystem is dominated by autotrophic (GPP:R > 1) or heterotrophic (GPP:R < 1) metabolism (e.g., Vannote *et al.* 1980, Mulholland *et al.* 2001, McTammany *et al.* 2003).

Coupling between N Uptake and Metabolism Rates

To examine relationships between N uptake and metabolism rates, we compared measured DIN uptake rates (i.e., sum of NH_4^+-N and NO_3^--N uptake rates) with N demand estimated from whole-stream metabolism measurements. To estimate N demand based on metabolism rates, we followed the approach of Webster *et al.* (2003). Total N demand was estimated by summing autotrophic and heterotrophic N demand. Autotrophic N demand was estimated by using a photosynthetic quotient (PQ, the molar ratio of O_2 evolved to CO_2 fixed) of 1.2 (Wetzel *and* Likens 2000) to convert O_2 production during gross primary production (GPP) to C fixation. Then we estimated autotrophic N demand as 70 % of GPP (Webster *et al.* 2003). Heterotrophic N demand was estimated from R measurements. We calculated heterotrophic R as whole-stream R minus autotrophic R (30 % GPP) and minus oxygen use by nitrification (2 moles O_2 per mole N oxidized). Respiration was converted from O_2 to C using a respiratory quotient (RQ, moles of CO_2 evolved per moles O_2 consumed) of 0.85 (Bott, 1996). We then calculated heterotrophic production as 0.28 times R (Cole *and* Pace, 1995). The N demand of this production was calculated using a C:N molar ratio of 5 (Fenchel *et al.* 1998). We examined Pearson's correlations between measured and estimated N demand separately for the two study streams using the SPSS® statistical package (for Windows, version 13.0, SPSS Inc., Chicago, Illinois).

Statistical Analyses

We used paired *T*-test analysis on log-transformed data from the two reaches to examine the effect of the WWTP input on hydrological, chemical and functional parameters in the two streams. Independent samples *T*-test on log-transformed data from the upstream reaches of the two streams was used to compare chemical and functional parameters between the two streams. All data was log-transformed to stabilize variances and normalize the data sets. The significance level used for all statistical tests was $P \leq 0.05$. All statistical analyses were performed with the SPSS® statistical package (for Windows, version 12.0, SPSS Inc., Chicago, Illinois).

RESULTS AND DISCUSSION

N Dynamics in the Forested Stream

In the forested stream, N demand as NO_3^--N tended to be higher than NH_4^+-N demand upstream and downstream of the point source (Figure 2).

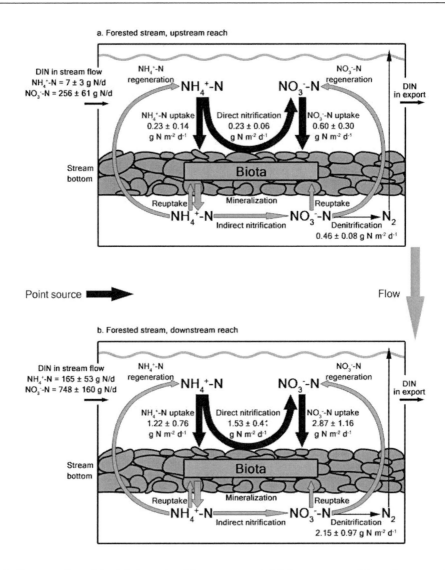

Figure 2. Conceptual model of N dynamics (a) upstream and (b) downstream of the point source in the forested stream. The figure shows mean ± standard error of the studied pathways (black arrows) of N cycling (all in g N m-2 d-1). Expanded from Peterson et al. (2001) for headwater stream ecosystems (grey arrows represent pathways represented by these authors that we have not directly quantified).

Percentage of demand for NH_4^+-N (upstream, 26 ± 13 %; downstream 29 ± 10 %) and NO_3^--N (upstream, 74 ± 13 %; downstream 71 ± 10 %) relative to total N demand was similar between the two reaches. We expected higher demand for NH_4^+-N than for NO_3^--N given that most benthic organisms (bacteria, fungi and algae) prefer to take up NH_4^+-N instead of NO_3^--N because uptake of the latter requires higher energetic cost. Uptake of NO_3^--N was greater than NH_4^+-N uptake early in a ^{15}N tracer addition conducted in a forested stream (Mulholland et al. 2000). Other studies have shown that increasing NO_3^--N concentrations may reduce heterotrophic demand for NH_4^+-N, reducing competition for this reduced N form between heterotrophs and nitrifiers (Bernhardt et al. 2002, Hall et al. 2002, Merseburger et al. 2005). Nitrifiers are poor competitors for NH_4^+-N (Verhagen and Laanbroek 1991, Verhagen et al.

1992), and thus, reducing this competition is likely to stimulate nitrification rates (Bernhardt et al. 2002, Hall et al. 2002, Merseburger et al. 2005). We estimated nitrification to be a 14 ± 4 % of total ecosystem respiration upstream of the point source, and a 41 ± 3 % downstream. These results support that NH_4^+-N inputs from the point source favor stream nitrifying activity (Figure 2). Mean potential nitrification rates were similar to those of NH_4^+-N uptake upstream and downstream of the point source. We assumed that NH_4^+-N was entirely taken up by nitrifying bacteria, in which case, other primary uptake compartments (i.e., benthic algae and other microbes) would take up DIN from NO_3^--N. Potential nitrification rates estimated for the forested stream were three orders of magnitude higher than those reported for other prairie (Dodds et al. 2000) and forested (e.g., Tank et al. 2000, Merriam et al. 2002, Mulholland et al. 2000) streams. Higher nitrification rates in the forested study stream than in these pristine streams may be due not only to our calculation method, but also to higher NH_4^+-N concentrations. High demand for NO_3^--N may be due in part to denitrifying activity upstream and downstream of the point source, given that mean potential denitrification rates were almost as high as mean NO_3^--N uptake (Figure 2). Quantification of denitrification rates should be viewed with caution because estimated denitrification rates were potential. Overall, thus, N as NH_4^+-N from the WWTP input was rapidly transformed to NO_3^--N via nitrification, and subsequently lost from the water column via denitrification. Our results strongly suggest that in-stream processes (i.e., NH_4^+-N and NO_3^--N retention, nitrification and denitrification) controlled DIN export in the forested stream, despite point source inputs. In-stream capacity to control nitrogen export has been reported for headwater streams, which in general export downstream less than half of the input of DIN from their watersheds (Peterson et al. 2001). Our results expand this potential capacity into human-altered streams.

N Dynamics in the Agricultural Stream

Land uses modulate the effect of point sources on chemical and functional attributes of the receiving streams (Merseburger 2006). In the case of the agricultural stream, diffuse sources from adjacent agricultural fields were likely to overwhelm the local effect of the point source on these stream attributes. Under this scenario, more than 95 % of N retention was in the form of NO_3^--N upstream and downstream of the point source (Figure 3). We estimated a nitrification rate of 0.05 g NH_4^+-N m^{-2} d^{-1} from the unique net longitudinal decline in NH_4^+-N concentration observed below the point source (July 2002). This rate is within the same range of values reported for other agricultural streams where NO_3^--N account for most of DIN (0.11 g NH_4^+-N m^{-2} d^{-1}; Kemp and Dodds 2002). We assume that nitrification rates may be similar between the two reaches in the agricultural stream (Figure 3), as were for the other studied chemical and functional attributes. Nitrification was likely to account for a low percentage (< 3 %) of whole-stream R in the agricultural stream. On the other hand, denitrification as a pathway to remove N from the water column was less important in the agricultural than in the forested stream (Figure 3). Low vertical water exchange in the agricultural stream due to the dominance of bedrock may contribute to counterbalance denitrification rates between the two streams in spite of water chemistry differences. Higher NO_3^--N fluxes in the agricultural than in the forested stream are likely to result in lower efficiency of denitrifiers to remove N from the water column. Hence, much of the NO_3^--N reaching the channel via diffuse sources in the agricultural stream was probably lost downstream.

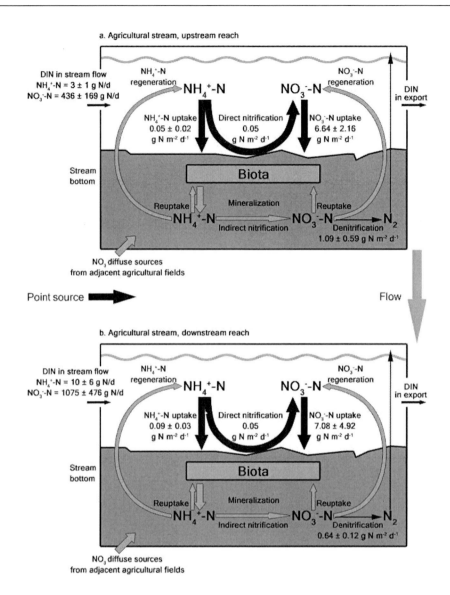

Figure 3. Conceptual model of N dynamics (a) upstream and (b) downstream of the point source in the agricultural stream. In addition to the point source, diffuse sources are also represented. The figure shows mean ± standard error of the studied pathways (black arrows) of N cycling (all in g N m-2 d-1). Expanded from Peterson et al. (2001) for headwater stream ecosystems (grey arrows represent pathways represented by these authors that we have not directly quantified).

This overview of results from the two study streams suggests that in-stream processes can buffer to some extent local inputs of nitrogen from point sources in the forested stream. However, in the agricultural stream, this capacity is exceeded by additional nitrogen reaching the stream from agricultural runoff. Our results show that downstream export of DIN is higher in the agricultural than in the forested stream regardless of point source inputs common to both.

Coupling between N Uptake and Metabolism

Whole-stream metabolism (i.e., primary production and respiration) is likely to drive nutrient cycling in pristine streams (Mulholland *et al.* 2001). Hence, N uptake and metabolism rates should be linked in these stream ecosystems. Webster *et al.* (2003) estimated N demand based on *in situ* metabolism measurements, in an attempt to address relationships between carbon metabolism and N uptake. These authors showed a good correspondence for near-pristine streams between measured N uptake and estimated N demand, supporting the importance of metabolism rates driving N cycling. Webster *et al.* (2003) emphasized misunderstanding concerning linkage between N demand and metabolism rates in streams influenced by nutrient inputs from human activities.

Human influences from point and diffuse sources in the forested and the agricultural streams were reflected in higher concentrations of DIN (in the order of mg N/L) than those (in the order of μg N/L) reported by Webster *et al.* (2003) for a variety of pristine streams. Measured and estimated N demand were not correlated in any of the study reaches of the two study streams (Pearson's correlations, $P > 0.05$). Nevertheless, the coupling between measured and estimated N demand was greater for streams studied by Webster *et al.* (2003) and the upstream reach of the forested stream (i.e., forested site) than for downstream of the point source and the two reaches of the agricultural stream (i.e., sites receiving nutrients from human activities; Figure 4a). We illustrate (Figure 4b) measured and estimated N demand using scales (*x*- and *y*-axes) one order of magnitude higher than those reported by Webster *et al.* (2003; Figure 4c). At the upstream reach of the forested stream, 100 % of the data fitted within these scales. In contrast, downstream of the point source and in the two reaches of the agricultural stream, only about a 50 % of the data fitted within these ranges. Our results suggest that coupling between measured and estimated N demand becomes weaker with increasing nutrient inputs from human activities to streams. Estimated total demand of N was correlated with heterotrophic N demand (Pearson's correlations: forested stream—upstream, $R^2 = 0.856$, $P = 0.014$, and downstream, $R^2 = 0.930$, $P = 0.002$; agricultural stream—upstream, $R^2 = 0.935$, $P = 0.001$, and downstream, $R^2 = 0.927$, $P = 0.001$), but not with autotrophic N demand (Pearson's correlations, $P > 0.05$ in all reaches). These latter results indicate dominance of heterotrophic activity in the two reaches of two study streams. A previous study focused on ecosystem metabolism showed that the two study streams are predominantly heterotrophic (Merseburger 2006). Heterotrophic bacteria can obtain N from organic substrates, and thus, their influence on water column N may be lower than N demand by primary producers (Webster *et al.* 2003). We expected dominance of heterotrophic N demand in the study streams to result in a decoupling between DIN retention and metabolism rates, with estimated N demand greater than measured N demand. Contrary to our expectations, estimated N demand was lower than measured N demand in the study streams (Figure 4). These results indicate the importance of processes contributing to N uptake that are not accounted with the methodology used to estimate metabolism rates, such as denitrification or uptake by roots of riparian vegetation submerged in the stream water.

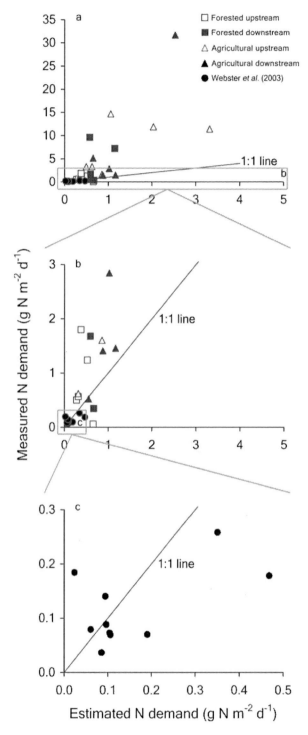

Figure 4. (a) Relationship between measured and estimated N demand for the upstream and the downstream reaches of the forested and the agricultural streams, and for data published by Webster et al. (2003). (b) Data from figure a by using a smaller scale for the y-axis (note that scales of x and y-axes are 10 times those shown in c). (c) Reported by Webster et al. (2003). The solid black lines represent 1:1 relationships.

Implications for Management Strategies

Understanding of the complex processes controlling nutrient cycling in stream ecosystems needs to be improved (Mulholland *et al.* 2000). Still nowadays, there are many gaps regarding knowledge of ecosystem processes in streams affected by human activities (Paul *and* Meyer 2001, Inwood *et al.* 2005, Meyer *et al.* 2005, Walsh *et al.* 2005). Improving this understanding will contribute to achieve the aims of the Water Framework Directive (WFD 2000/60/EEC) of the European Commission. The WFD aims to establish adequate policy practices to prevent further deterioration of streams within human landscapes, and to protect and enhance a good ecological status—defined as that status showing low levels of distortion resulting from human activity, but deviating only slightly from those normally associated with stream under undisturbed conditions—that should be achieved by 2015 (WFD 2000/60/EEC). To describe the ecological degradation of streams draining urban areas, Walsh *et al.* (2005) identified the urban stream syndrome based on several symptoms basically related to physical, chemical and biological parameters (Table 2). These authors highlighted the lack of functional symptoms to characterize the urban stream syndrome due to limited research on stream nutrient retention or ecosystem metabolism in urban streams. Hence, this study contributes to clarify existing understanding about the processes involved in the functioning of human-altered streams, and thus, to the scientific understanding needed to achieve the aims of the WFD and other policies in general. Moreover, our results allow expanding the concept of ecosystem syndrome due to human-alterations beyond urban scenarios with the introduction of the agricultural stream syndrome, by providing insights of stream ecosystem function in scenarios affected by agricultural runoff.

Table 2. Symptoms generally associated with the urban stream syndrome

Feature	Consistent response	Inconsistent response	Limited research
Hydrology	↑ Frequency of overland flow	Baseflow magnitude	
	↑ Frequency of erosive flow		
	↑ Magnitude of high flow		
	↓ Lag time to peak flow		
	↑ Rise and fall of storm hydrograph		
Water chemistry	↑ Nutrients (N, P)	Suspended sediments	
	↑ Toxicants		
	↑ Temperature		
Channel morphology	↑ Channel width	Sedimentation	
	↑ Pool depth		
	↑ Scour		

Consistent responses are those observed in multiple studies, whereas inconsistent responses are those that have been observed to increase (↑), decrease (↓), and/or remain unchanged with increased urbanization. Limited research implies the need for more studies before concluding whether responses are consistent or inconsistent

Table 2. (Continued)

Feature	Consistent response	Inconsistent response	Limited research
	↓ Channel complexity		
Organic matter	↓ Retention	Standing stock/inputs	
Fishes	↓ Sensitive fishes	Tolerant fishes	
		Fish abundance/biomass	
Invertebrates	↑ Tolerant invertebrates		Secondary production
	↓ Sensitive invertebrates		
Algae	↑ Eutrophic diatoms	Algal biomass	
	↓ Oligotrophic diatoms		
Ecosystem processes	↓ Nutrient uptake	Leaf breakdown	Net ecosystem metabolism
			Nutrient retention
			P:R ratio

From Walsh et al. (2005).

ACKNOWLEDGMENTS

We are indebted to Dr. N. Ubero-Pascal for his assistance in the field and to S. Pla for chemical analyses. Jesús D. Ortiz benefited from a fellowship of the Department of Universities, Research and the Information Society of the Generalitat, Government of Catalonia (Spain). This study was supported by fundings of the STREAMES European project (EVK1-CT-2000-00081). The Serveis Cientificotècnics of the University of Barcelona provided their facilities and technical help in N_2O analyses.

REFERENCES

Albek, E. (2003). Estimation of point and diffuse contaminant loads to streams by non-parametric regressions analysis of monitoring data. *Water, Air, and Soil Pollution, 147*, 229-243.

Ashkenas, L. R., Johnson, S. L., Gregory, S. V., Tank, J. L., and Wollheim, W. M. (2004). A stable isotope tracer study of nitrogen uptake and transformation in an old-growth forest stream. *Ecology, 85*, 1725-1739.

Bencala, K. E., McKnight, D. M., and Zellweger, G. W. (1987). Evaluation of natural tracers in an acidic and metal-rich stream. *Water Resources Research, 23*, 827-836.

Bernhardt, E. S., Hall, R. O., and Likens, G. E. (2002). Whole-system estimates of nitrification and nitrate uptake in streams of the Hubbard Brook experimental forest. *Ecosystems, 5*, 419-430.

Bott, T. L. (1996). Primary productivity and community respiration. In F. R. Hauer and G. A. Lamberti (Eds.), *Methods in Stream Ecology* (pp. 533-556). San Diego, California, USA: Academic Press.

Cole, J. J., and Pace, M. L. (1995). Bacterial secondary production in oxic and anoxic freshwaters. *Limnology and Oceanography*, 40, 1019-1027.

Constanza, R., d'Arge, R., de Groot, R., Farber, S., Grasso, M., Hannon, B., Limburg, K., Naeem, S., O'Neill, R.V., Paruelo, J., Raskin, R.G., Sutton, P., andVan den Belt, M. (1997). The value of the world's ecosystem services and natural capital. *Nature, 387,* 253-260.

Crouzet, P., Leonard, J., Nixon, S., Rees, Y., Parr, W., Laffon, L., Bøgestrand, J., Kristensen, P., Lallana, C., Izzo, G., Bokn, T., Bak, J., Lack, T. J. (1999). Nutrients in European ecosystems. *Report number 4*, 1-155.

Dodds, W. K., Evans-White, M. A., Gerlanc, N. M., Gray, L., Gudder, D. A., Kemp, M. J., López, A. L., Stagliano, D., Strauss, E. A., Tank, J. L., Whiles, M. R., and Wolheim, W. M. (2000). Quantification of the nitrogen cycle in a prairie stream. *Ecosystems*, 3, 574-589.

Edwards, R. W. and Owens, M. (1962). The effects of plants on river conditions IV. The oxygen balance of a chalk stream. *The Journal of Ecology*, 50, 207-220.

EEA (2003). Europe's environment: the third assessment. *10,* 1-343.

Gasith, A. and Resh, V. H. (1999). Streams in Mediterranean climate regions: abiotic influences and biotic responses to predictable seasonal events. *Annual Review of Ecology and Systematics*, 30, 51-81.

Genereux, D. P. and Hemond, H. F. (1992). Determination of gas exchange rate constants for a small stream on Walker Branch Watershed, Tennessee. *Water Resources Research*, 28, 2365-2374.

Gibson, C. A. (2004). Alterations in ecosystem processes as a result of anthropogenic modifications to streams and their catchments. *Ph. D. thesis*. University of Georgia, Athens, Georgia, USA.

Hall, R. O., Bernhardt, E. S., and Likens, G. E. (2002). Relating nutrient uptake with transient storage in forested mountain streams. *Limnology and Oceanography*, 47, 255-265.

Hamilton, S. K., Tank, J. L., Raikow, D. F., Wolheim, W. M., Peterson, B. J., and Webster, J. R. (2001). Nitrogen uptake and transformation in a midwestern U.S. stream: A stable isotope enrichment study. *Biogeochemistry*, 54, 297-340.

Holmes, R. M., Jones Jr, J. B., Fisher, S. G., and Grimm, N. B. (1996). Denitrification in a nitrogen-limited stream ecosystem. *Biogeochemistry*, 33, 125-146.

House, W. A. and Denison, F. H. (1997). Nutrient dynamics in a lowland stream impacted by sewage effluent: Great Ouse, England. *The Science of the Total Environment*, 205, 25-49.

Inwood, S. E., Tank, J. L., and Bernot, M. J. (2005). Patterns of denitrification associated with land use in 9 midwestern headwater streams. *Journal of the North American Benthological Society*, 24, 227-245.

Kemp, M. J. and Dodds, W. K. (2002). Comparisons of nitrification and denitrification in prairie and agriculturally influenced streams. *Ecological Applications*, 12, 998-1009.

Kim, K., Lee, J. S., Oh, C. W., Hwang, G. S., Kim, J., Yeo, S., Kim, Y., and Park, S. (2002). Inorganic chemicals in an effluent-dominated stream as indicators for chemical reactions and streamflows. *Journal of Hydrology*, 264, 147-156.

Knowles, R. (1979). Denitrification, acetylene reduction, and methane metabolism in lake sediments exposed to acetylene. *Applied and Environmental Microbiology, 38*, 486-493.

Lowrance, R. R. (1998). Riparian forest ecosystems as filters for nonpoint-source pollution. In M. L. Pace and P. M. Groffman (Eds.), *Successes, limitations and frontiers in ecosystem science* (pp. 113-141). New York, USA: Springer-Verlag.

Martí, E., Armengol, J., and Sabater, F. (1994). Day and night nutrient uptake differences in a calcareous stream. *Verhadlungen Internationale Vereinigung für Limnologie, 25*, 1756-1760.

Martí, E. and Sabater, F. (1996). High variability in temporal and spatial nutrient retention in Mediterranean streams. *Ecology, 77*, 854-869.

Martin, L. A., Mulholland, P. J., Webster, J. R., and Valett, H. M. (2001). Denitrification potential in sediments of headwater streams in the southern Appalachian Mountains, USA. *Journal of the North American Benthological Society, 20*, 505-519.

Marzolf, E. R., Mulholland, P. J., and Steinman, A. D. (1994). Improvements to the diurnal upstream-downstream oxygen change technique for determining whole-stream metabolism in small streams. *Canadian Journal of Fisheries and Aquatic Sciences, 51*, 1591-1599.

McColl, R. H. S. (1974). Self-purification of small freshwater streams: phosphate, nitrate, and ammonia removal. *New Zealand Journal of Marine and Freshwater Research, 8*, 375-388.

McMahon and Finlayson 1992

McTammany, M. E., Webster, J. R., Benfield, E. F., and Neatrour, M. A. (2003). Longitudinal patterns of metabolism in a southern Appalachian river. *Journal of the North American Benthological Society, 22*, 359-370.

Meals, D. W., Levine, S. N., Wang, D., Hoffmann, J. P., Cassell, E. A., Drake, J. C., Pelton, D. K., Galarneau, H. M., and Brown, A. B. (1999). Retention of spike additions of soluble phosphorus in a northern eutrophic stream. *Journal of the North American Benthological Society, 18*, 185-198.

Merriam, J. L., McDowell, W. H., Tank, J. L., Wolheim, W. M., Crenshaw, C. L., and Johnson, S. L. (2002). Characterizing nitrogen dynamics, retention and transport in a tropical rainforest stream using an *in situ* ^{15}N addition. *Freshwater Biology, 47*, 143-160.

Merseburger, G. C., Martí, E., and Sabater, F. (2005). Net changes in nutrient concentrations below a point source input in two streams draining catchments with contrasting land uses. *Science of the Total Environment, 347*, 217-229.

Merseburger, G. (2006). Nutrient dynamics and metabolism in Mediterranean streams affected by nutrient inputs from human activities. *Ph. D. thesis*. University of Barcelona, Barcelona, Spain.

Meybeck, M. (2003). Global analysis of river systems: from Earth system controls to Anthropocene syndromes. *Philosophical Transactions of the Royal Society of London B, 358*, 1935-1955.

Meyer, J. L. (1989). Can P/R ratio be used to assess the food base of stream ecosystems? A comment on Rosenfeld and Mackay (1987). *OIKOS, 54*, 119-121.

Meyer, J. L., Paul, M. J., and Taulbee, W. K. (2005). Stream ecosystem function in urbanizing landscapes. *Journal of the North American Benthological Society, 24*, 602-612.

Mulholland, P. J., Fellows, C. S., Tank, J. L., Grimm, N. B., Webster, J. R., Hamilton, S. K., Martí, E., Ashkenas, L., Bouden, W. B., Dodds, W. K., McDowell, W. H., Paul, M. J., and Peterson, B. J. (2001). Inter-biome comparison of factors controlling stream metabolism. *Freshwater Biology*, *46*, 1503-1517.

Mulholland, P. J., Tank, J. L., Sanzone, D. M., Wollheim, W. M., Peterson, B. J., Webster, J. R., and Meyer, J. L. (2000). Nitrogen cycling in a forest stream determined by a ^{15}N tracer addition. *Ecological Monographs*, *70*, 471-493.

Munn, N. L. and Meyer, J. L. (1990). Habitat-specific solute retention in two small streams: an intersite comparison. *Ecology*, *71*, 2069-2082.

Murphy, J., and J. P. Riley. (1962). A modified single solution method for determination of phosphate in natural waters. *Analytica Chimica Acta, 27,* 31-36.

Odum, H. T. (1956). Primary production in flowing waters. *Limnology and Oceanography*, *1*, 102-117.

Ostrom, N. E., L. O. Hedin, J. C. von Fischer, and G. P. Robertson. (2002). Nitrogen transformations and NO_{3-} removal at a soil-interface: a stable isotope approach. *Ecological Applications, 12,* 1027-1043.

Owens, M. (1974). Measurements on non-isolated natural communities in running waters. In R. A. Vollenweider (Ed.), *A manual on methods for measuring primary production in aquatic environments* (pp. 111-119). Oxford, United Kingdom: Blackwell Scientific Publications.

Palmer, M., Bernhardt, E., Chornesky, E., Collins, S., Dobson, A., Duke, C., Gold, B., Jacobson, R., kingsland, S., Kranz, R., Mappin, M., Martinez, M. L., Micheli, F., Morse, J., Pace, M., Pascual, M., Palumbi, S., Reichman, O. J., Simons, A., Townsend, A., and Turner, M. (2005). Ecology for a crowded planet. *Science*, *304*, 1251-1252.

Paul, M. J. and Meyer, J. L. (2001). Streams in the urban landscape. *Annual Review of Ecology and Systematics*, *32*, 333-365.

Peterson, B. J., Wolheim, W. M., Mulholland, P. J., Webster, J. R., Meyer, J. L., Tank, J. L., Martí, E., Bowden, W. B., Valett, H. M., Hershey, A. E., McDowell, W. H., Dodds, W. K., Hamilton, S. K., Gregory, S., and Morrall, D. D. (2001). Control of nitrogen export from watersheds by headwater streams. *Science*, *292*, 86-90.

Rathbun, R. E., D. W. Stephens, D. J. Schultz, and D. Y. Tai. (1978). Laboratory studies of gas tracers for reaeration. *Proceedings of the American Society of Civil Engineering, 104*, 215-229.

Romano, J. and Krol, J. (1993). Capillary ion electrophoresis, an environmental method for the determination of anions in waters. *Journal of Chromatography A*, *640*, 403-412.

Stream Solute Workshop (1990). Concepts and methods for assessing solute dynamics in stream ecosystems. *Journal of the North American Benthological Society*, *9*, 95-119.

Tank, J. L., Meyer, J. L., Sanzone, D. M., Mulholland, P. J., Webster, J. R., Peterson, B. J., Wolheim, W. M., and Leonard, N. E. (2000). Analysis of nitrogen cycling in a forest stream during autumn using a ^{15}N-tracer addition. *Limnology and Oceanography*, *45*, 1013-1029.

Tiedje, J. M., Simkins, S., and Groffman, P. M. (1989). Perspectives on measurement of denitrification in the field including recommended protocols for acetylene based methods. *Plant and Soil*, *115*, 261-284.

Triska, F. J., Kennedy, V. C., Avanzino, R. J., Zellweger, G. W., and Bencala, K. E. (1989). Retention and transport of nutrients in a third-order stream in northwestern California: hyporheic processes. *Ecology, 70*, 1893-1905.

Vannote, R., Minshall, G., Cummins, K., Sedell, J., and Cushing, C. (1980). The river continuum concept. *Canadian Journal of Fisheries and Aquatic Sciences, 37*, 130-137.

Ventullo, R. M. and Rowe, J. J. (1982). Denitrification potential of epilithic communities in a lotic environment. *Current Microbiology, 7*, 29-33.

Verhagen, F. J. M., Duyts, H., and Laanbroek, H. J. (1992). Competition for ammonium between nitrifying and heterotrophic bacteria in continuously percolated soil columns. *Applied and Environmental Microbiology, 58*, 3303-3311.

Verhagen, F. J. M. and Laanbroek, H. J. (1991). Competition for ammonium between nitrifying and heterotrophic bacteria in dual energy-limited chemostats. *Applied and Environmental Microbiology, 57*, 3255-3263.

Vitousek, P. M., Aber, J. D., Howarth, R. W., Likens, G. E., Matson, P. A., Schindler, D. W., Schlesinger, W. H., and Tilman, D. G. (1997). Human alteration of the global nitrogen cycle: sources and consequences. *Ecological Applications, 7*, 737-750.

Walsh, C. J. (2000). Urban impacts on the ecology of receiving waters: a framework for assessment, conservation and restoration. *Hydrobiologia, 431*, 107-114.

Walsh, C. J., A. H. Roy, J. W. Feminella, P. D. Cottingham, P. M. Groffman, and R. P. Morgan II. (2005). The urban stream syndrome: current knowledge and the search for a cure. *Journal of the North American Benthological Society, 24*, 706-723.

Webster, J. R., Mulholland, P. J., Tank, J. L., Valett, H. M., Dodds, W. K., Peterson, B. J., Bowden, W. B., Dahm, C. N., Findlay, S., Gregory, S. V., Grimm, N. B., Hamilton, S. K., Johnson, S. L., Martí, E., McDowell, W. H., Meyer, J. L., Morrall, D. D., Thomas, S. A., and Wollheim, W. M. (2003). Factors affecting ammonium uptake in streams - an inter-biome perspective. *Freshwater Biology, 48*, 1329-1352.

Webster, J.R. and Valett, H.M. (2006). Solute Dynamics. In F.R. Hauer and G.A. Lamberti (Eds.), *Methods in Stream Ecology* (pp. 169-186). Burlington, Massachusetts, USA: Academic Press.

Wetzel, R. G., and G. E. Likens. (2000). *Limnological analyses*. New York, USA: Springer-Verlag.

WFD (Water Framework Directive). (2000). Official Publication of the European Community, L327. Brussels.

Wilcock, R. J., Scarsbrook, M. R., Costley, K. J., and Nagels, J. W. (2002). Controlled release experiments to determine the effects of shade and plants on nutrient retention in a lowland stream. *Hydrobiologia, 485*, 153-169.

Young, R. G. and Huryn, A. D. (1998). Comment: improvements to the diurnal upstream-downstream dissolved oxygen change technique for determining whole-stream metabolism in small streams. *Canadian Journal of Fisheries and Aquatic Sciences, 55*, 1784-1785.

In: Encyclopedia of Environmental Research
Editor: Alisa N. Souter
ISBN: 978-1-61761-927-4
© 2011 Nova Science Publishers, Inc.

Chapter 51

PROCESSES FOR THE TREATMENT OF DIRTY DAIRY WATER: A COMPARISON OF INTENSIVE AERATION, REED BEDS AND SOIL-BASED TREATMENT TECHNOLOGIES

Joseph Wood[*,1] *and Trevor Cumby*[2]

[1]Centre for Formulation Engineering, Department of Chemical Engineering,
University of Birmingham, Edgbaston, Birmingham, B15 2TT, UK
[2]Fleggdale Nurseries, 8 Hatley Road, Potton
Sandy, Bedfordshire, SG19 2DX

ABSTRACT

In agricultural environments such as dairy farms, large volumes of dirty water are produced, which must be managed and disposed of. Dirty water is typically spread on to fields, but excess run off can lead to depleted oxygen levels, pollution and damage to aquatic life within surrounding water courses. In this Chapter, the available techniques for treatment of agricultural run off will be reviewed and compared.

A case study of treatment of dairy water at a UK dairy farm is presented in detail. At the farm, water with biochemical oxygen demand of typically 2500 mg/l and total solids of typically 6000 mg/l was produced from yard run-off, requiring treatment before disposal by land spreading. The performance of aeration, reed beds, overland flow and soil percolation plots, are compared based on extensive performance data from the working dairy farm. Average removals of BOD5 were over 95 % for the intensive aeration plant, 55 – 65 % for the reed bed, 55 % for the percolating soil plot and 80 – 90 % for the overland flow plot. However, it was shown that the reed bed could be highly effective as a polishing step following prior treatment by aeration. For total solids removal, the overland flow plot achieved the highest removals (42 – 57 %), followed by the intensive aeration plant (33 – 40 %). High loadings of solids applied to the reed beds caused operational problems such as blockages. Removals of total nitrogen, total ammoniacal nitrogen, phosphorus and COD in the treatment systems are also reported

---
[*] Corresponding author: j.wood +44 (0) 121 414 5295.

and compared. The highest average removals of these pollution indicators of total nitrogen 87.3 %, total ammoniacal nitrogen 98.1 %, phosphorus 63.7 % and COD 82.6 % occurred in the intensive aeration plant, with the other systems displaying lower removals. The costs of building and operating the different treatment systems are estimated, in order to evaluate the economics of different treatment options in addressing the problem of agricultural discharge. Finally an outlook on future treatment technologies is presented, including considerations of costs of different treatment options.

1. INTRODUCTION

Dairy Water

Dirty water is the general term used to describe contaminated water produced in agricultural environments, containing less than 3% dry matter and contaminants such as bedding, faeces, urine, silage effluent, milk, wash water and cleaning products [MAFF, 1998]. Large volumes of such dirty water are produced at most dairy farms, including those in the UK, which can present a serious pollution risk if discharged directly to the watercourses.

Currently, where possible, dirty water is stored on the farm and spread on to fields during the dry season. However, a large storage capacity is required to avoid spreading in wet conditions, when the risk is greatest of run off to water courses and possible pollution. The capacity of soil to absorb dirty water without surface run off is influenced by various factors including its moisture content, type and slope of land, so varies depending upon the particular farm site. If discharged untreated, and if direct run-off occurs, the dirty water can lead to depleted oxygen levels in water courses.

Figure 1. Lagoon at the Case Study Farm showing dirty water collected from the farm yard.

Figure 1 shows the lagoon utilized at The Case Study farm to store the wastewater during the cold months, thus graphically illustrating the large quantity of slurry and dirty water to be dealt with. However, even with this amount of storage capacity, some spreading of dirty water was necessary during the wetter winter period.

The dirty water produced contains organic matter, solids, nitrogen and phosphorous compounds and may include various pathogens, such as thermotolerant coliforms. Such components can cause detrimental effects on the environment if they exceed certain limits, as a result of their capacity to provoke eutrophication [Hickey et al., 1989]. With 5-day Biochemical Oxygen Demand (BOD_5) concentrations of up to 5000 mg/L and ammoniacal-nitrogen (NH_4-N) values often exceeding the 500 mg/L, the potential pollution threat is a permanent characteristic of dairy waste waters [Sun et al., 1998]. Therefore, successful treatment of dirty water is necessary to reduce the pollution loading so that year-round land application can remain as a viable treatment option for dairy farmers on sensitive sites [Cumby et al., 1999]. Without such measures, the alternatives are to provide even larger storage facilities, so that all dirty water can be retained during the winter and then spread in dry weather, or to seek permission for discharge to a watercourse following a full-scale treatment process. Whilst the latter approach is common in the food and dairy processing industries, they are very rare on farms. The extent to which pollutants must be removed from dirty water prior to permitted discharge to a water course depends upon the Discharge Consent Limits, which are maximum pollutant concentration values set by regulatory bodies such as the Environment Agency (UK) or Environmental Protection Agency (USA). The discharge consent limit partly depends upon the quantity of waste to be discharged and other environmental factors. Typical values of discharge consent limits in the dairy industry are BOD_5 12 – 20 mg/l, Total Kjeldahl Nitrogen (TKN) 15 mg/l, Total Suspended Solids (TSS) 30 mg/l [Copybook Solutions, 2008]. The purpose of this chapter is to present a case-study of an experimental trial of four different treatment processes to treat dirty water at the Case Study farm: intensive aeration plant (IAP), reed beds (RB), percolation soil plot (PSP) and overland flow plot (OFP). These processes were used to assess the effectiveness and costs of treating dirty water before land spreading, so that the environmental impact of any subsequent run-off would be minimised, and also to evaluate comprehensive treatment options that might enable consented discharge of treated effluent to water courses.

In the following sections descriptions of the four types of process used in this study are presented together with review of previous work for each system. The four processes (IAP, RB, PSP and OFP) are reviewed in below, the latter three processes sharing some similar features in using a planted media through which water is filtered.

Aeration

Aeration is the use of oxygen in the form of air to reduce pollutant concentration within the wastewater. The aim of aeration is to remove the bulk BOD_5 of the wastewater by oxidation processes. This can be achieved by: **(1)** facilitating the mass transfer of oxygen from air to water, such as a vigorously stirred reactor and **(2)** producing an environment with a large surface area where aerobic micro-biological organisms can colonize, for example a packed tower [Gray, 1999]. At the same time, ammonia may be removed from the water by nitrification processes under the action of nitrobacter microbes [Burton and Farrent, 1998], and under certain conditions denitrification also occurs [Vaoliene and Matuzevicius, 2007]. Less desirably, stripping of ammonia by air can also occur in aeration processes leading to emissions to the environment, although with appropriate control this removal route can be reduced to negligible levels [Burton and Farrent, 1998].

Previous studies have reported the application of a dual stage system for the treatment of dirty waters [Burton and Farrent 1998], in which two continuous stirred tank reactors (CSTRs) were used. However the aeration system used in the trials at The Case Study farm was based on a CSTR followed by a High Rate Trickling Filter (HRTF). The former oxygenates the wastewater, facilitating aerobic decomposition of the organic pollutants in the dirty water. Being a bed of coarse media (often stones or plastic), the latter provides a large surface where the micro-biological organisms can colonize. Microbes within the HRTF attach themselves to the inert packing allowing aerobic activity to take place in the surrounding areas. Trickling filters drain at the bottom where the wastewater is collected for sedimentation. Although aeration processes themselves do not generally lead to the removal of a large fraction of solids from the wastewater, they are usually combined with gravity settling stages, which facilitate the removal of suspended solids. Aeration systems offer a high level of reliability, but are also expensive to operate, owing to the costs of running pumps and blowers.

Much of the research concerning aeration has focused on improving its efficiency by reducing its intensity or duration. Burton and Farrent (1998) studied the aeration of pig slurry for both short intensive (2 days) and long (20 days) durations, and with both single and dual stage treatment tanks. They found the best combination of treatments to be single stage aeration for a long period of 20 days, as the dual stage system did not offer any significant advantage given the additional costs. Ndegwa et al (2007) studied the aeration of dairy wastewaters in batch reactors, and found that during the first 3-4 days of treatment, dissolved oxygen concentration was close to the detection limit, indicating that all dissolved oxygen was rapidly used up to oxidize the waste. Reduction of aeration rate by 50 % led to only a 14 % decrease in biological decay rate, therefore it was economically prudent to aerate at the lower rate. A maximum removal of 70 % of the COD and total volatile solids occurred after 8 days of aeration treatment. Zhang et al (2006) showed that aeration treatment could be used to effectively reduce odour of swine manure.

Reed Beds

RBs are engineered biological treatment systems, which have found application in the treatment of domestic, industrial and agricultural waters [Kern and Idler, 1999; Kadlec and Knight, 1996; Cooper and Green, 1995]. They are designed to optimise the microbiological, chemical and physical processes naturally occurring in the wetland.

RB systems have proved to be an appropriate technology for cleaning domestic wastewater [Cooper and Green, 1995; Ansola *et al*, 1995] and acid mine drainage [Karathanasis and Thompson, 1995; Watson *et al.*, 1989]. To a lesser extent they have also been employed for the treatment of stormwater [Green and Martin, 1996], industrial wastewater [Vrhovsek *et al*, 1996], landfill leachate [Bernard and Lauve, 1995]. In general, relatively little information exists about the treatment of agricultural wastewaters by RB treatment systems [Kern and Idler, 1999].

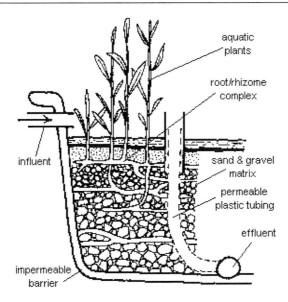

Figure 2. Cross section of a vertical flow RB.

Previous studies carried out at the University of Birmingham have reported the treatment of pig slurry by down flow RB treatment systems [Sun *et al.*, 1998 and 1999]. The operational problems usually produced by the high pollutant content typically found in agricultural effluents may have constituted a barrier to the application RBs in this area [Zhao *et al.*, 2004]. Consequently RB treatment systems are sometimes considered as a polishing step after an aerobic or anaerobic pre-treatment [Van Oostrom and Cooper, 1990].

Generally, treatment within RBs occurs in the root zone. Polluted water passing around the roots is mainly purified by the immense population of microorganisms flourishing at the *rhizosphere* of the plant [Gray, 1999]. This is the area around the *rhizome*, which is known as a persistent underground stem providing a means of vegetative propagation and support. This constitutes a large potential surface area for the growth of heterotrophic microorganisms. Therefore, it is usually in the *rhizosphere*, the root/rhizome complex, where interactions among the plant, microorganisms and soil take place, as shown in Figure 2.

The key advantage of using common reeds is that they transfer atmospheric oxygen to the *rhizome*, which passes through the roots, finally raising the oxygen levels in both the gravels and *rhizosphere*. As a result, pollutants are not simply stored in the RB; they are actually degraded into harmless components as result of the break down of a wide range of organic chemical products.

The decomposition of the organic matter in RB treatment systems could either be aerobic, anaerobic or a combination of both [Vesilind, 1975]. The removal of the solid content of dirty water is mainly based on two processes: sedimentation and filtration [Sun et al, 1999]. Such processes take place as the wastewater passes through vegetation matter or gravel media. The removal mechanisms for nitrogen include uptake by plants and other living organisms, ammonification, nitrification, denitrification, ammonia volatilization and cation exchange in case of the ammonium [Brix, 1993]. The main mechanism to remove the phosphorous content in RB treatment systems is plant uptake and subsequent harvesting [Lantzke *et al.*, 1998]. Alternatively, the phosphorous content of the effluent is also reduced through oxidation to phosphate (PO_3^-), which precipitates and adsorbs to soil particles and hydrolysis to orthophosphate.

The effectiveness of RB systems has been improved through the development of various "Constructed RBs". These have been used for more than twenty years to treat wastewaters such as local domestic sewage, airport runway run-off and agricultural effluents [Newman *et al.*, 2000]. Therefore, as a result of years of investigation, and depending on their operational characteristics, constructed wetlands are classified in two main categories [Price and Probert, 1997]: horizontal and vertical systems. Horizontal RBs work particularly well for low strength effluents and for effluents that have undergone some form of pre-treatment process. Vertical RBs are more appropriate for the treatment of strong effluents as the oxygen for the reeds is supplemented by oxygen diffused from the atmosphere. Previous studies have reported oxygen transfer rates between 29.7 – 57.1 $g/m^2/day$ [Sun et al, 2003]. The effect of loading rate and planting on the treatment of dairy farm wastewaters was studied by Tanner *et al.* (1995a) being mainly focused on the removal of nitrogen and phosphorus. In a serial publication [Tanner *et al.*, 1995b], they considered the removal of BOD_5, suspended solids and faecal coliforms. Total BOD_5 removal was in the range 50 – 80 % for 300 g/m^3 influent, and was sensitive to the wetland loading and hydraulic residence time. The study showed that the RBs have considerable potential for removal of nitrogen, phosphorus, BOD_5, suspended solids and faecal coliforms from dairy waters, following pre-treatment in oxidation ponds.

Percolation Soil and Overland Flow Systems

An alternative option to RBs is soil-based treatment. In this type of system an artificial barrier is provided to control the flow of wastewater through a hydrologically isolated plot of soil, usually planted with grass. Effluent and rainwater passing through the plot are prevented from percolating into the underlying sub-soil by an impermeable barrier such as plastic membrane, and all run-off from the plot is collected as a discharge stream of treated effluent. Such systems offer the advantage of being relatively inexpensive to set up and operate. Two configurations of soil based system are commonly employed, as described below:

Percolating Soil Plot: The wastewater is applied to the top of the soil plot and allowed to trickle down through the vegetation root system and soil. In these systems, the soil pores and soil surface provide a support matrix for microbial activity. Previous studies have demonstrated their application for the treatment of dirty water, being capable of reducing BOD_5, NH_4-N and phosphorous contents [Chadwick *et al.*, 2000]. Usually any increase in the soil depth results in an increase of purification efficiencies. This system is similar in concept to that of downflow RBs.

Overland Flow Plot: This is recognized as one of the principal categories of soil based treatment systems. The wastewater flows over land that is carefully graded to encourage sheet flow. Hence, the permeability of the soils employed in these systems is usually low. The treatment is encouraged by the slope, the lack of vertical percolation and the high rate of wastewater application. The common gradient is of around 2 % onto which the wastewater is evenly trickled from the top and allowed to run down the slope to a collection basin at the bottom [Metcalf and Eddy, 1991]. While filtration occurs, the wastewater is treated due to interaction with the uppermost layers of the soil, vegetation, microorganisms and the atmosphere. The collected treated wastewater is usually recycled back and filtered as many times as the necessary amount of treatment has been reached [Tyrrel and Leeds-Harrison, 2005]. The system has some similarities to a horizontal flow RB, except that the filtration medium is soil rather than gravel. Liu et al (1998) studied the use of a loamy sand irrigation plot for the treatment of swine manure. It was found that levels of nitrates (NO_3-N) and phosphorus (P) build up in the soil, showing that overland flow may have a limited capacity for treating agricultural effluents. Tyrell et al (2002) have demonstrated that overland flow is effective for removal of ammoniacal nitrogen from disposal of wastes such as landfill leachate. For the treatment of dirty water from dairy farms, Tyrell et al (2005) found that removal of BOD_5 and ammonia was sustainable, and not significantly influenced by variations of weather pattern. However, the data for the removal of solids was more variable, and the removal of phosphorus decayed over time, suggesting that this type of system may not be sustainable for the removal of phosphorus.

2. EXPERIMENTAL METHODS: THE CASE STUDY FARM

The focus of this chapter concerns a case study of dirty water treatment at The Case Study farm, in the UK. A daily volume of approximately 21 - 25 m^3 dirty water was produced from cattle in the farm yard, due to the run off of rainwater, milk, faeces, urine, bedding, uneaten forage and dairy parlour wash water containing cleaning chemicals. In the study reported in this Chapter, a set of experimental trials of the four treatment systems were carried out to treat nominally $1/50^{th}$ of the daily dirty water produced on the farm. The dirty water was obtained from the exiting dirty water collection and handling system, and pumped to two 10 m^3 storage tanks, one in a nearby field and one in the farm yard. Dirty water was treated in each of the four trial treatment systems, each having a nominal design flowrate of 500 l/day. However, the actual volume treated was adjusted in some cases, for example to prevent waterlogging and ponding of the soil based systems. The objectives of the study were to test each treatment system operating independently (Trial 1) and then to test the combination of treatment systems to achieve an optimal configuration in terms of treatment efficiency and cost (Trial 2). Each treatment system used in the trial is described in detail below.

Description of Aeration Plant

The aeration plant was located in the farm yard and consisted of a series of tanks, stirrers and equipment mounted onto two steel frames, which were located at the farm yard. The main treatment stages of the plant comprised (I) a CSTR of working volume 2.2 m^3, followed by (II) a HRTF. The air supplied to the CSTR was provided at a rate of 3 – 5 l/s at 30 kPa from a blower via three diffuser disks located at the bottom of the tank in order to provide an effective oxygen supply. Besides aerating the liquid, the air bubbles from the diffuser disks also agitated the vessel contents. The HRTF was packed with plastic packing rings on which biological film can grow. The dimensions of the HRTF were approximately 3.5 m height × 1.5 m diameter. The dirty water was recirculated from the holding tank below the trickling filter over the packing of the filter at a rate of typically 100 – 150 l/min, so that the entire contents of the tank were recirculated through the HRTF every few minutes. The dirty water trickled down through the packed bed by means of gravity while air was blown upwards from the bottom of the filter. The main plant items of the IAP are shown in Figure 3 and a photograph is displayed in Figure 4.

As observed in Figure 3, the six main pieces of equipment comprising the IAP were: (1) primary screening, (2) primary sedimentation, (3) aerobic treatment by CSTR, (4) secondary sedimentation, (5) aerobic treatment by HRTF and (6) tertiary sedimentation. Depending on the mode of operation, the effluent was pumped to the RBs either from the CSTR or from the HRTF.

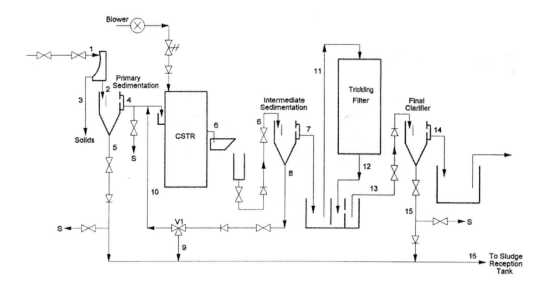

Figure 3. Flow diagram of the IAP.

Figure 4. Photograph of the IAP.

The aeration plant was controlled via a series of float switches for regulating the level in each vessel and timers incorporated in to the computer control sequence.

Description of the RBs

The system was erected near the IAP and consisted of 5 portable experimental RB units, and 5 storage tanks, located at the farm yard. Three individual RBs planted in metal tanks (RBs 1, 2 and 3) constituted the first stage of the treatment system, whilst the second stage was based on two interconnected RBs (4 and 5) planted in a single PVC tank. While RBs 1, 3, 4 and 5 were planted with common reed (*Pharagmites Australis*), RB 2 was planted with mace reed (*Typha Latifolia*).

During the treatment process, the wastewater was kept in tanks (T_1, T_2, T_3 and T_4) which provided storage for wastewater awaiting treatment and temporary storage for treated effluent. Although theoretically the oxidation of pollutants mainly took place in the RBs, some oxidation may have also occurred in the storage tanks. The dimensions of these tanks are listed in Table 1.

Wastewater entered the system via T_1 from either the farmyard storage tank containing raw dirty water or from the IAP. The feed for the system was pumped across T_2 before being brought into the RBs. The effluent coming from the first stage of the treatment plant was collected in T_3. This was the holding tank between the primary and the secondary stages. The treated effluent was stored in T_4, from where part of it was pumped back to tank T_2 constituting the recycle input. The rest was transferred from tank T_4 to T_5. Subsequently this was pumped to the 3 m^3 pyramid-bottom tank (T_6), where treated effluents from both RBs and IAP were collected. The layout of the pilot plant is shown in Figure 5, together with a photograph of the system in Figure 6.

The surface area of each RB was $1m^2$. All beds were filled with gravel to a depth of 60 cm and operated in a vertical mode with a tidal operation cycle. While RB_2 contained a homogeneous gravel structure, the rest of the sub-beds contained a multi-layered matrix. The second sub-bed (RB_2) was filled with pea gravel of a fixed diameter (~ 5mm). The layered structure of the bed media for RBs 1, 3, 4 and 5 is listed in Table 2.

Table 1. Dimensions of the tanks supporting the RB treatment system

Feature	Tank 1	Tank 2	Tank 3	Tank 4
Height (cm)	62	180	81	66
Length (cm)	240	-	124	154
Width (cm)	61	$\emptyset = 100$	92	89

Table 2. Description of the gravel layers of RBs 1, 3 4 and 5

Layer Thickness	Gravel particle size (diameter)
13 cm	$\emptyset = 0.6 - 0.8$ cm
29 cm	$\emptyset = 1.5 - 1.7$ cm
18 cm	$\emptyset = 3 - 3.5$ cm

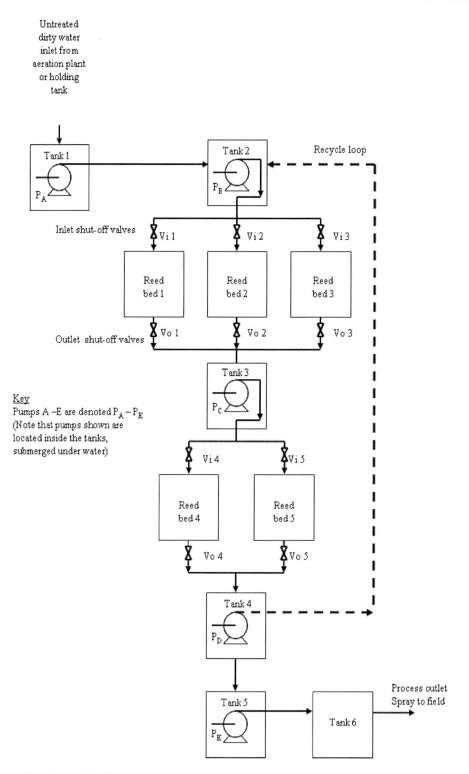

Figure 5. Plan view of the RB plant.

Figure 6. Photograph of the RB plant.

Aeration pipes were built into the matrix for natural ventilation. With a diameter of 10 cm and a length of 1 m, these enhanced the rate of the oxygen transfer into the rhizome area of the reeds. The pumps were submersible centrifugal pumps, located within the tanks. These were switched on and off by the timer sequence in the control cabin. The tanks and the RBs were connected together by reinforced plastic hose of 50 mm diameter.

The RBs were operated in tidal mode such that 200 litres of mixed raw dirty water and recycled treated water were applied to the beds within duration of 1 minute 40 seconds, followed by a period of 165 minutes drainage time. The filling of tanks with dirty water and application of dirty water pulses to the RB were controlled by an RS-328-134 programmable electronic sequence controller.

Description of the Percolation Soil Plot

The PSP was built at the field site and consisted of a bed of soil 1 m deep and surface dimensions of 7 m × 7 m, located in a field at the farm. Soil at the base of the plot was graded to a slope of approximately 2 % in order to facilitate flow of water to the bottom corner. Earth bunds were pushed into place and a geo-textile sheet was used to protect a plastic liner from puncture by roots and stones. This ensured hydrological isolation of the plot from the surrounding site. A collection drain was placed along the base of the two lower sides (in gravel) to facilitate movement of effluent to the outlet pipe. Figure 7 shows the plan of the PSP, and Figure 8 a photograph of the installed system.

The plot was filled with soil supplied by a local top-soil distributor, after grading to remove roots and stones. Care was taken to minimise the degree of compaction by the excavator. After the plot had been established, it was sown with *Lolium perenne*.

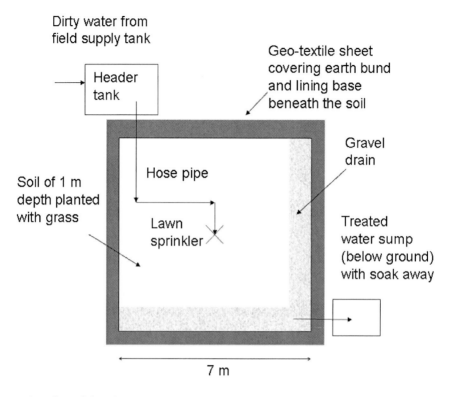

Figure 7. Plan view of the PSP.

Figure 8. Photograph of the PSP.

Dirty water was initially supplied to the system via a gutter system, but this was replaced after 3 months of operation by a lawn sprinkler. After percolating through the soil plot, dirty water was collected in a sump located alongside the system in the corner where the drain was located. Samples for analysis were removed from this sump. A tipping bucket system was used to monitor the flow of leachate draining from the plot in to the sump.

Description of the Overland Flow Plot

The overland flow system was constructed near the PSP in the form of a gently inclined plot of vegetated clay soil with a slope of approximately 1% and measuring $10 \times 7.0 \times 0.3$ m, which was hydrologically isolated with an 800 gauge polythene membrane of the sort usually used for horticultural poly tunnels. The overland flow system was installed in a field at the farm, close to the PSP. The soil for the plot was sourced from the Case Study Farm. A 1 % slope was used because experience had shown that steeper slopes lead to channelling of flow and uneven distribution of water across the slope. Six recirculation/distribution tanks each of 1 m^3 capacity were installed at the upper end of the slope to contain the dirty water. These tanks were interconnected to provide a large capacity reservoir, and were filled with dirty water. In front of these tanks was a distribution trough across the width of the plot. At the bottom of the slope was a gravel drain, which flowed into a sump in which a float operated pump (pump 2) re-circulated the water to the tanks. Figure 9 shows a plan of the OFP, whilst Figure 10 shows a photograph of the installed system.

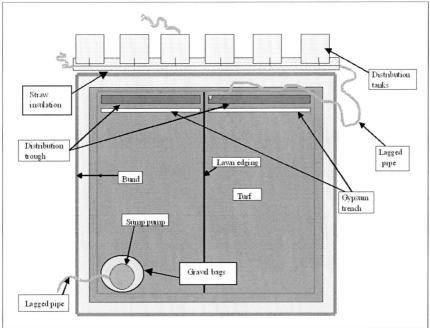

Fig. 2.5.1 Plan layout of the overland flow plot (OFP) highlighting the revisions to the plot design.

Figure 9. Plan view layout of the OFP.

Figure 10. Photograph of the OFP shortly after construction.

Once a day, at 6 pm, the dirty water in the tanks was pumped by pump 1 into the distribution trough and flowed across the plot to the sump, via the drain, for a period of two hours. The plot then drained for 22 hours allowing air to enter the soil. Pump 2 returned water that collected in the sump to the distribution tanks, so that the cycle could be repeated 24 hours later. In the initial phase of operation, which included all of Trial 1 plus approximately the first four months of Trial 2, the treated dirty water was sampled and removed after seven days and then replaced with more untreated dirty water. In the latter part of Trial 2, treatment of dirty water with a two-week cycle of operation was practiced.

Operation and Sampling Regime

It was necessary to ensure that both the systems installed at the farm yard and at the field received consistent supplies of dirty water. Untreated dirty water was obtained from a strainer box attached to The Case Study farm's lagoon and run off from the concrete yard areas near the lagoon. An 18 m^3 tank was installed at the farm yard in order to supply dirty water to both the farm yard site and the field site using a pipeline. Likewise, treated dirty water could be pumped from the farm yard site to the field site via a separate pipe. At the farm yard, the dirty water storage tank was equipped with a stirrer to ensure that the contents of the tank were well mixed. On the field site, the contents of the dirty water storage tank were recirculated using a small submersible pump, since a larger mixer could not be used due to limited power supply at the field.

The systems were operated as two trial phases. During Trial 1 each system was evaluated independently, with the exception of the RB that sometimes received partially treated dirty water. During Trial 2 some modifications were made to some of the treatment processes as well as combining certain treatment processes as described in further detail later on. Each system was designed to nominally treat 500 l/day dirty water. Table 3 shows the operating flows of each treatment process, which in some cases were adjusted from the design flowrate of 500 l/day due to operational factors.

Table 3. Operating flows of each treatment process

	Trial 1 (6/5/03 – 27/4/04)			Trial 2 (13/7/04 – 31/5/05)
IAP	550 l/day raw dirty water			750 l/day raw dirty water
RB	Mode A: 6/5/03 - 26/8/03	Mode B: 2/9/03 - 10/2/04	Mode C: 17/2/04 - 27/4/04	380 l/day from IAP CSTR
	500 l/day from IAP HRTF	50 l/day raw dirty water	380 l/day from IAP CSTR	
OFP	229 – 343 l/day raw dirty water			343 l/day raw dirty water
PSP	257 l/day raw dirty water			Not operational as a treatment process

During Trial 1, the IAP received untreated dirty water from the farm yard supply, which was treated in both the CSTR and HRTF stages of the process. The RBs were operated in three different modes during Trial 1. Initially (Mode A) the RB received treated dirty water from the IAP, as it was thought that raw dirty water would be too concentrated in pollutants for application to such a system. Subsequently (Mode B), the RB was fed with raw dirty water, but at a lower rate of only 50 l/day, in order to test its capability as an independent treatment system. Finally (Mode C), the RB was converted back to accepting treated dirty water from the IAP, as problems such as blockages of the bed surface were encountered when treating raw dirty water. In each case, the RB was operated with a recycle flow of treated dirty water, which was mixed with raw dirty water prior to application, in order to avoid shock loads of pollutants being applied to the beds. The recycle flows used in the three modes of Trial 1 were (Mode A) 1100 l/day, (Mode B) 1100 l/day and (Mode C) 420 l/day. During Trial 1, the OFP was operated by recirculating between 1600 – 2400 litres of water across the plot each day during a weekly period, before changing to a new batch of dirty water.

The aim of Trial 2 was to operate some of the treatment processes in combination, in order to examine whether they were more effective than individual treatment systems. The throughput of the CSTR of IAP was increased to 750 l/day, and this flow was then divided between the HRTF stage of the aeration plant (250 l/day) and the RB (380 l/day). Thus these treatments operated as polishing steps of water aerated by the CSTR, allowing for comparison between the performance of the HRTF and the RB. The possibility of using the OFP as a polishing treatment was also considered, but this was rejected as it was operating well with dirty water feed and it would have been logistically difficult to transfer dirty water from the aeration plant in the farmyard to the OFP in the field. Consequently the OFP continued to treat 343 l/day dirty water from the field dirty water tank. During the latter part of Trial 2 (14 December 2004 – 31 May 2005) the OFP was converted to operating on a fortnightly cycle, such that a batch of fresh dirty water was added each week for recirculation, but water was only discharged from the system after the two weeks of treatment. During Trial 2 the percolation soil plot was closed down due to operational problems caused by the collapse of the soil structure and therefore the system was not used to treat dirty water.

Sampling of each treatment system was usually carried out on a weekly basis, with samples of dirty water taken from the outlet storage tank of each treatment system, packed in to a cool container and sent for analysis. These samples were analysed for 5-day biochemical oxygen demand and total solids (analysis by weighing after evaporation at 105±2 °C) using standard methods.

Three 'intensive monitoring' periods were held per year, in which additional measurements (including ammoniacal-N, total-N, phosphorus and Chemical Oxygen Demand (COD)) were made on a daily basis for a period of 1 week each. Total nitrogen was measured using Kjeldahl analysis and ammoniacal-N by Aqua 800 Advanced Quantitative Analyser.

3. RESULTS

Dirty Water Characteristics

Seasonal Variations in the Dirty Water at the Case Study Farm and Between Sites 1 and 2

Figures 11 - 13 compare the BOD_5, TS and TSS concentrations of the untreated dirty water samples at Sites 1 (farm) and 2 (field), throughout Trials 1 and 2. These data show that considerable variations occurred from season to season, to the extent that the differences between the highest and lowest concentrations in both parameters readily exceeded an order of magnitude. This implies that the design of any dirty water treatment system should be able to handle a high turn down ratio, such that periods of highest demand can be accommodated, whilst allowing the extent (and hence cost) of the process to be scaled-down during less demanding periods.

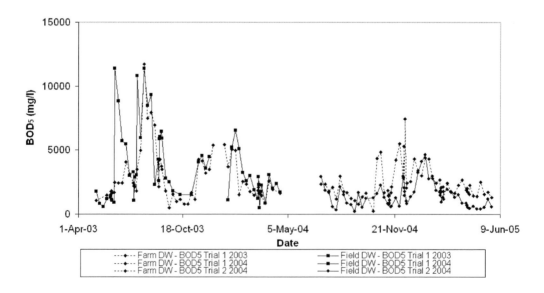

Figure 11. Comparison of the variations in the 5-day Biochemical Oxygen Demand (BOD5) concentration in the untreated dirty water in the two supply tanks during Trial 1 and Trial 2.

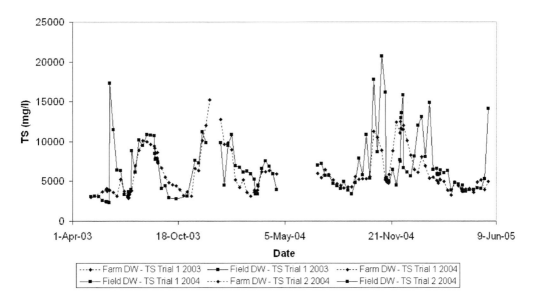

Figure 12. Comparison of the variations in the Total Solids (TS) concentration in the untreated dirty water in the two supply tanks during Trial 1 and Trial 2.

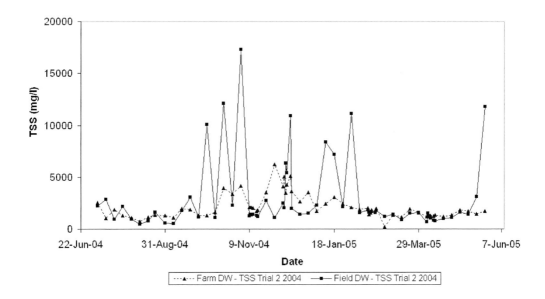

Figure 13. Comparisons of the variations of total suspended solids (TSS) concentrations in the untreated dirty water in the two supply tanks during Trial 1 and Trial 2.

The results in Figures 11 – 13 show that the composition of the dirty water from the farm and field sites followed similar trends and therefore the mixing and pumping systems installed to minimise differences in the composition of the dirty water between the two experimental sites were largely effective. Occasional differences occurred however, especially during the winter periods. The largest differences were observed for TS and TSS

(Figures 12 and 13) which show some high peak concentrations. In particular, some high TSS and TS values were observed occasionally in the samples from Site 2 (field). The TSS data suggested that these might have reflected sampling errors that derived from the limited mixing capacity available at this site. Accordingly, a revised sampling procedure introduced early in 2005, which increased the duration of mixing before sampling, helped to reduce the incidence of sampling errors during the latter part of Trial 2. The peak values observed in BOD_5 may have resulted from events at the farm, such as heavy rainfall washing additional manure from the yard or spillages of milk. The peak TS and TSS values may have resulted from the accidental disturbance of settled material from the bottom of the tanks from which the sample was taken.

Figure 11 illustrates that the BOD_5 of the raw dirty water regularly reached a value of 5000 mg/l, and occasionally exceeded 10,000 mg/l. These values are much higher than the usual BOD_5 applied to RBs, for example in the study of Sun et al (1998), the influent BOD_5 was 778 mg/l. Similarly, in studies reported in the literature TSS values applied to constructed wetlands [Dunne et al , 2005] were in the range 941 – 1078 mg/l, compared with values sometimes reaching over 15,000 mg/l in this study. Although pre-treatment or dilution were used at The Case Study farm to reduce the concentration of the pollutants actually being applied to the RBs, it should be noted that the strength of the wastewater produced at working dairy farms presents a particular challenge to the use of "natural" treatment systems such as RBs or soil-based systems.

BOD5 removal

The average removal performance of each system is first compared in Figure 14. In view of the ways in which the systems periodically worked in combinations, separate bars are shown for the CSTR, HRTF working independently and another bar for the IAP with these two stages connected in series. Data from the OFP system include an indication of its performance whilst operating with a two-week treatment cycle. All data were calculated using the untreated dirty water values obtained one week before the treated dirty water values, to reflect the weekly recharge mode of operation.

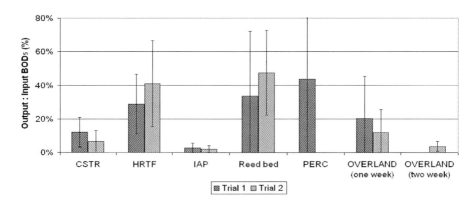

Figure 14. Comparison of system performance expressed as mean values of the ratios of output to input BOD5 concentrations. Error bars represent +/- one std.

These results show that the IAP (CSTR + HRTF) was the most effective system for reducing the BOD_5 concentration, and that the great majority of the BOD_5 removal occurred in the CSTR. However, the RB performance almost matched that of the HRTF, and so the CSTR + RB combination was almost as good, with an overall BOD_5 removal of 97 %, compared with 97.9 % for the IAP. Although the sample size was small, the overall performance of the OFP system operating with a two-week cycle was also effective. However, it must be noted that the treated effluent from this system was subject to considerable dilution with rainfall during the reported period of two-week operation.

In order to assess the reduction of BOD_5 by the treatment systems, it is noted that the BOD_5 concentration of the raw dirty water at the farmyard was mainly between 1000 and 5000 mg/l, with the exception of some unusually high values during the Autumn of 2003, exceeding 10,000 mg/l. Part of the treatment occurred in the CSTR, from which the treated effluent varied quite widely from 50 to 500 mg/l or higher, though the typical values were in the range 100 – 200 mg/l. Further analysis of the performance of each system can be made by examining the traces of BOD_5 output as a function of sampling date during each Trial. As observed from Figure 15, from March – September 2003, and August – November 2004, the final outlet BOD_5 from the HRTF was typically in the range 10 – 50 mg/l, with only occasional excursions above 100 mg/l. During other periods, such as October 2003 – February 2004, a lower level of removal occurred such that the final effluent had a BOD_5 of up to 500 mg/l. It was observed that the CSTR was generally very effective for removing large quantities of BOD_5, such that when loaded with relatively dilute wastes there was little requirement for further treatment. When loaded with more concentrated wastes, the HRTF acted as a secondary stage to remove additional loads of BOD_5. However, on some occasions the outlet BOD_5 from the HRTF was still higher than the 20 mg/l emphasising the difficulties of achieving the standards that might be required for dirty water disposal via consented discharge to water courses.

The performance of the RB varied according to the different modes of operation. During Mode A of Trial 1, the RB was receiving pre-treated effluent from the IAP, removal of an average 75.1 % BOD_5 occurred in the RB. Owing to the combined action of the aeration plants and RBs, 99.6 % of the BOD_5 was removed from the water. During this period the average BOD_5 of water at the outlet from the IAU was 47.4 mg/l whilst at the outlet from the RB it was 11.7 mg/l. This reduction in BOD_5 achieved by the IAP and RB systems operating in tandem is significant, since it indicates the extent to which dirty water might need to be treated for consented discharge to water courses. During Mode B of operation, the RB treated raw dirty water mixed with recycled treated water. The raw water had a BOD_5 of 2650 mg/l, the value at the inlet after mixing with recycle was 393 mg/l, whilst the outlet BOD_5 from the RB was 291 mg/l. Clearly the BOD_5 at discharge was much higher than observed in Mode A, when pretreatment of the water by aeration was taking place. The BOD_5 of the discharge during Mode B also greatly exceeded the consent discharge limit of 20 mg/l. During Mode B the average total solids of the raw water was also high, having a value of 6780 mg/l, which led to a number of operational problems such as blockages of pipework and blinding of the gravel surface of the RB. These problems compounded the poor removals of BOD_5 observed during Mode B. In Trial 1, Mode C, the performance of the RB was similar to Mode A. During Trial 2, when the RB was accepting partially treated water from the CSTR of the aeration plant, the value of the BOD_5 at the CSTR outlet was 135 mg/l, whilst the discharge from the RB was 64.3 mg/l. This value was intermediate between the performance of the bed in Trial 1 Modes A and B, however the value was above the typical discharge consent limit of 20 mg/l. Thus it could be concluded that pretreatment by aeration in the CSTR led to a significantly improved final effluent of the RB compared with no pre-treatment. However in Trial 2, as in Trial 1, Mode A, it was clear that a third stage of treatment would be necessary to meet typical requirements for consented discharge to water courses.

Figure 16 shows the data for the BOD_5 removal in the PSP and OFPs. Initially the PSP was able to reduce the BOD_5 concentrations significantly, even when very high values of BOD_5 occurred in the dirty water applied. However, the discharge from the plot was sometimes as high as 7500 mg/l. Over the summer of 2003 there was a gradual decline in the treatment efficiency of the system up to October 2003, during which ponding of water was observed on the surface of the plot. This was due to a collapse in the soil structure resulting from high loading of sodium. Consequently a soil with better drainage characteristics was sought, and the soil replaced in November 2003. However, due to problems of local supply, the fresh soil was found to be similar to the original soil and the drainage problems recurred during the winter of 2003-4. Consequently the percolation soil plot was closed as a treatment process at the end of Trial 1 and not included in Trial 2, although it was used to study the extent to which further treatment of partially treated dirty water takes place following application to land.

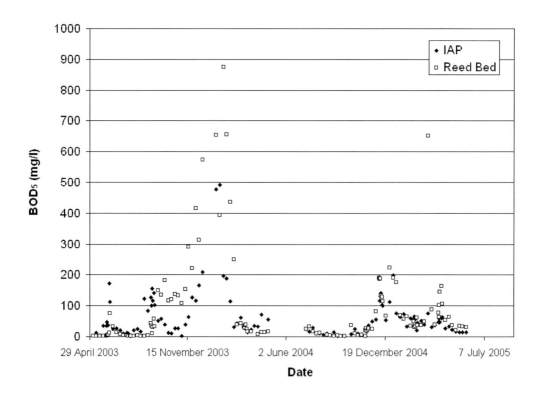

Figure 15. BOD5 at the outlet of IAP and RBs.

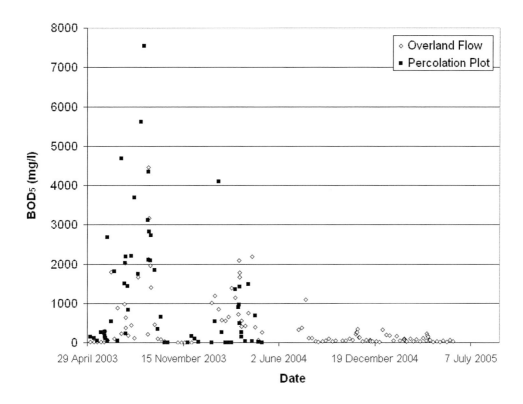

Figure 16. BOD5 at outlets of OFP and PSP.

During Trial 1, the OFP initially operated on a weekly basis, whereby 2400 l dirty water was recirculated over the plot and returned to the collection/feed tanks for a period of one week. The BOD_5 concentration was reduced by an average of 84 % when operating on a one-week basis. The performance was also quite insensitive to the BOD_5 of the influent water, which ranged from 585 – 11400 mg/l. It should be noted however, that because of the relatively large surface area, the OFP was more susceptible to variations due to precipitation and evapotranspiration than the other systems. During the wet periods the pollutants were diluted by rainfall, whilst during the dry period part of the water evaporated from the plot. The ratio of input volume to output volume averaged 0.89 but ranged from 0.46 in wet weather conditions to 2.0 in a dry summer week.

The OFP was found to be susceptible to drainage problems associated with soil structure similar to those experienced by the percolation soil system. This was found to be due mainly to the high sodium content of the parlour washings leading to clay dispersion, whereby sodium displaces calcium on the clay complex in soil, forcing charged soil particles apart and leading to collapse of the soil structure. This problem was remediated by rotavating 200 kg agricultural gypsum through the soil in October 2003 in order to replace the calcium in the soil. Additionally, the plot was divided in to two halves, such that dirty water was applied to each half whilst the other half was resting and draining. From December 2004 onwards, the plot was operated on a two week cycle whereby the same batch of dirty water was recirculated over the plot for a two week duration. This mode of operation was found to be more consistent, with 95.7 % removal occurring in the two weeks of operation, although much of this removal, of over 90 %, occurred in the first week of treatment with only a further 45 % removal of the remaining BOD_5 occurring during the second week.

The overall trends in removal of BOD_5 in each system follow the pattern that might be expected. Previous studies have shown intensive aeration [Zhang et al., 2006], RBs [Sun et al, 1998] and soil based systems [Tyrell and Leeds-Harrison, 2005] are effective for the removal of BOD_5. The IAP used a significant input of electrical power to continuously aerate the water and therefore may be expected to achieve a higher removal of material susceptible to oxidation. The similar performance of the HRTF to the RB, may be reflected by their similar modes of operation by trickling the water to be treated over a packing material, relying upon mass transfer of air to a biological film upon the packing surface. The common features of the RB, OFP and PSP are that they were gravity fed, vegetated, and subject to precipitation and evapotranspiration. The water trickles over the gravel surface of the RB and through the soil structure of the PSP and OFP, such that a combination of filtration and biological treatment occur in the process. Although such systems do not require constant pumping of air and water, and are thus cheaper to operate, they are subject to more variability in removal rates than the IAP. There are several reasons why this is the case. Firstly, they are reliant on maintenance of the soil or bed structure, such that collapse of soil structure or blockage of gravel media can lead to impaired performance. Secondly, during periods of high precipitation, the dirty water may be diluted, whilst in dry periods evaporation may lead to concentration of pollutants in the water. Thirdly, accumulated material from within the bed or soil may occasionally be released in to the outlet water, leading to unusually high BOD_5 readings. Additionally, all systems are influenced by temperature, whereupon a 10 °C temperature rise can lead to an increase of typically 25 % in the first order rate constant describing the BOD_5 removal process [Wood et al, 2007].

Total Solids Removal

The solids content of dirty water may be expressed in terms of TS, which encompasses both total suspended solids (TSS) and dissolved salts. TS was measured weekly for the whole duration of the trials, whilst TSS was measured only from August 2004 – May 2005. TSS tend to be removed by clarification and settling processes, whilst TS includes dissolved salts which may not be removed by settling or microbial activity associated with BOD_5 reduction.

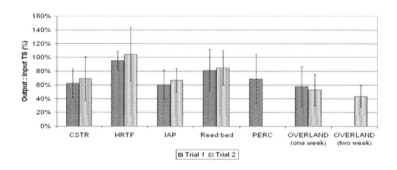

Figure 17. Comparison of system performance expressed as mean values of the ratios of output to input TS concentrations. Error bars represent +/- one standard deviation.

Figure 17 compares system performance based on the ratio of the output TS concentrations to the input TS concentrations from Trials 1 and 2. Since the PSP was not used as a dirty water treatment during Trial 2, its performance data relate to only Trial 1.

The results showed that the OFP system achieved the largest reduction in TS concentration, especially during the two-week treatment cycles, followed by the IAP. However, since the RB showed a lower output: input TS ratio compared with the HRTF, the combined CSTR-RB system would be more effective (combined TS removal of 41 %) compared with the CSTR-HRTF combination (combined TS removal of 33 %). The good average performance of the OFP relative to other systems showed that the soil is generally effective as a media for deep bed filtration, whereupon particles of solid are first transported towards the soil particles by sedimentation, interception, diffusion, inertia and hydrodynamic effects. Subsequently the particles must be held together by various forces including capillary effects, van der Waals forces and surface charges. Soil based systems should be effective for removal of both suspended particles as well as dissolved salts, which may become adsorbed upon the soil particles. By contrast, in the IAP the main mechanism of removal is sedimentation, which is predominantly controlled by particle size and density of the suspended material as well as fluid velocity and residence time within the sedimentation vessel. In general, all systems showed a lower percentage removal of TS compared with BOD_5. This could be because TS encompasses dissolved salts which are not removed effectively by settling, biological activity or filtration.

Further observations regarding the performance of the different treatment systems can be made from plots of TS at the plant outlet as a function of sample date. Figure 18 shows TS at the outlet of the aeration plant. Typically the TS of the raw dirty water were highly variable between the values 4000 – 10000 mg/l, and after treatment in the IAP ranged from 300 – 7000 mg/l. The separation mainly occurred by gravity settling, giving an underflow of sludge with TS concentration of between 10,000 – 20,000 mg/l. Comparison with the TS of the raw dirty water of Figure 12 showed that incidences of peak outlet TS broadly corresponded to periods of peak inlet concentrations, demonstrating that the IAP could not effectively remove solid matter when shock loaded with pollutants. Overall, the performance of the aeration plant in removing solids was less effective than removal of BOD_5, such that further treatment would be required for discharge to water courses.

As shown in Figure 18, the TS of the outlet of the RB ranged from 300 – 5000 mg/l. The outlet TS value was quite dependent upon the mode of operation, that is whether the RB was accepting partially treated or raw dirty water. In general the peaks and troughs of the recorded values for the RBs follow similar trends to those of the IAP, illustrating the sensitivity of both systems to high TS values of the raw dirty water. During Trial 1, Mode A, the RB accepted treated water from the aeration plant with an average inlet TS of 2670 mg/l and the average value for the final effluent after treatment in the RB was 1850 mg/l. During Mode B of the RB operation, the average removal of solids was as low, with an average discharge of 4130 mg/l TS. As previously mentioned, blockages and blinding of the bed by high solids loading impaired the performance of the RB. On some occasions negative removals were observed, whereby the outlet water contained more solids than the influent. Such occurrences could have resulted from accumulated solids being periodically released from the bed matrix of gravel.

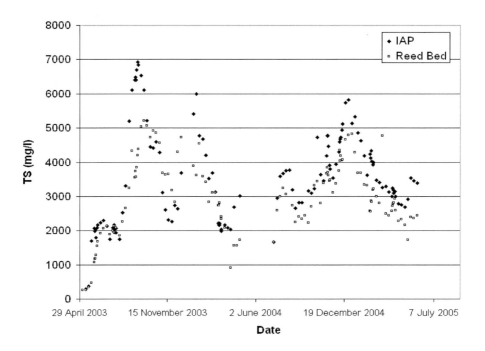

Figure 18. TS at outlets of IAP and RBs.

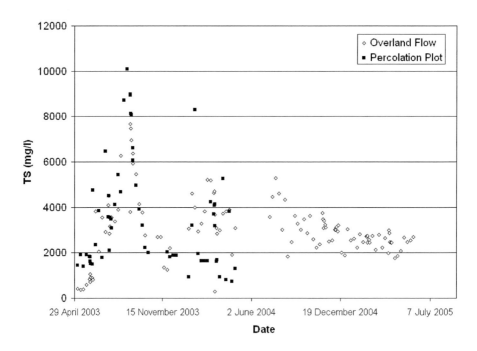

Figure 19. TS at outlets of OFP and PSP.

During Trial 1, Mode C and Trial 2, the performance of the RB in removing TS was between those observed during Modes A and B of Trial 1, with an average TS value in the final effluent of 3090 mg/l. In each case, the effluent would require further treatment to remove solids before discharge to water courses.

For the PSP, the removal of TS was quite variable (Figure 19), the outlet TS ranging over an order of magnitude from 1000 to 10000 mg/l. During some weeks substantial removal occurred whilst during others the outlet TS was similar to the inlet. Again, the removal of TS was affected by the collapse of soil structure, with a noticeable decrease in removal during the first phase of operation during 2003. This problem was considerably improved when the soil was replaced in October 2003, with higher removal rates observed from November 2003 – March 2004 corresponding to outlet TS of around 2000 mg/l. However, eventually in April 2004, the TS removal decreased again, indicating recurrence of soil structure problems.

In the OFP, the removal of TS was also subject to some variability (Figure 19), the outlet TS concentration ranging from 300 - 8000 mg/l. On some occasions, negative removals occurred, which could have been due to settled material from the storage tanks being mixed with fresh dirty water, sloughing of biological matter from soil during periods of heavy rainfall or washing of particles from soil when dispersed due to high sodium contents. Tyrell and Leeds-Harrison (2005) also previously showed that removal of TS in overland flow type systems is more variable than removal of other pollution indicators such as BOD_5.

In summary, the removal of TS in all systems was relatively poor. Firstly, the outlet TS of the systems followed a similar trend to the raw dirty water, showing that the systems were generally unable to deal with shock loadings of pollutants. Secondly, the absolute values of TS in the discharge were usually too high for discharge to water courses. For the IAP, the main removal mechanism of suspended solids is sedimentation and therefore increasing the residence time in the settling tanks could improve the fraction of solids removed. Alternatively a flocculant could be added to the dirty water to aid the formation of clusters which settle more easily. The RB and soil based systems mainly remove solids by mechanisms similar to those prevalent in deep bed filtration. Although such removal mechanisms are sometimes effective for relatively low TS loadings, in cases of higher TS values, the filtration media can become clogged or the particles washed through the bed in to the outlet. For all systems, considerable over sizing would be required in order to handle the levels of TS observed in dairy farm effluents, with associated increases of cost that would also be incurred.

Total Suspended Solids Removal

For the removal of Total Suspended Solids (TSS), system performance levels during Trial 2 are compared in Figure 20, showing that the OFP system, recorded the biggest percentage reductions in TSS. The IAP (i.e. CSTR + HRTF) was almost as good, although the specific values for both the CSTR and the HRTF were subject to large error bars which resulted from spurious values recorded on isolated occasions.

The results also show that the RB performed better than the HRTF, so that the combined ratio of the CSTR + RB was marginally better than that of the IAP. Generally the removal of TSS was higher than the removal of TS, indicating that the systems investigated were more effective at removing suspended materials by settling or filtration compared with dissolved salts. Suspended solids may be composed of phytoplankton biomass and detritus [Hickey et al, 1989; Wrigley and Toerien, 1990].

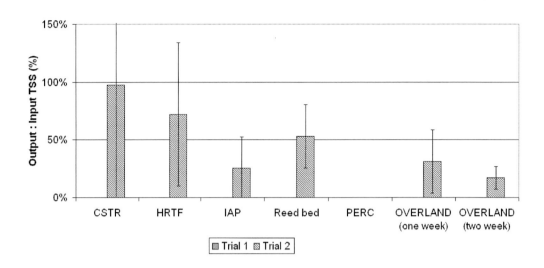

Figure 20. Comparison of system performance expressed as mean values of the ratios of output to input TSS concentration. Error bars represent +/- one standard deviation.

Ammonia, Nitrogen, Phosphorus and COD Removal

The eight intensive monitoring periods detailed in Table 4 enabled several properties to be measured that were not possible during the regular weekly monitoring operations. These included Total (Kjeldahl) Nitrogen (TN); Total Ammoniacal Nitrogen (TAN); Total Phosphorus (TP). Combined results for each of these, for all of the treatments are illustrated in Figures 21 – 23.

The removal mechanisms for nitrogen include uptake by plants and other living organisms, ammonification, nitrification, denitrification, ammonia volatilization and ammonium ion cation exchange [Brix, 1993]. Nitrogen containing pollutants may be degraded by ammonification to produce inorganic nitrogen according to the process:

Ammonification: $R\text{-}NH_2 \rightarrow NH_4^+$ Afterwards, the ammonia (NH_4-N) is mainly removed by nitrification-denitrification processes taking place within the RB [Gray, 1990]. Nitrification is the microbial oxidation of ammonium ions, which takes place in two stages as follows:

Nitrosomas: $NH_4^+ + 1.5O_2 \rightarrow NO_2^- + 2H^+ + H_2O$
Nitrobacter: $NO_2^- + 0.5O_2 \rightarrow NO_3^-$

Consequently, nitification is an oxygen demanding process.
Denitrification constitutes the second step of nitrogen removal [Gray, 1990]:

$NO_3^- \rightarrow NO_2^- \rightarrow NO \rightarrow N_2O \rightarrow N_2\uparrow$

The denitrification stage is influenced by process conditions such as temperature, pH, dissolved oxygen concentration (which should be less than 1 mg/L), nature and concentration of carbon organic matter, nitrate concentration, presence of facultative bacteria, cell retention time, and presence of toxic material [Do Canto et al, 2008].

Comparing all four treatment systems, Figures 21 and 22 show that the IAP (i.e. CSTR +HRTF) achieved the biggest reductions in both TAN and TN. However, it is important to note that the values for the IAP presented in Figures 21 and 22 exclude data from the first intensive monitoring period because this took place before the nitrification process had become established in the IAP.

The average TAN removal in the CSTR of the intensive aeration unit was 68.4 %, whilst further removal of 91 % of the remaining TAN occurred in the HRTF. Previously Burton and Farrent (1998) have shown aeration to be effective for the removal of TAN (up to 93 % removal), together with the production of nitrous oxide (around 12 % of total nitrogen (TN) in the raw feed). TAN can be removed from dirty water by either stripping as ammonia gas, or by nitrification process in which it is converted to nitrites or nitrates, as described above. In aerobically controlled processes, nitrification does occur, as evidenced by the recorded outlet concentrations of nitrates and nitrites from the IAP, shown in Table 4. Some of the highest recorded NO_3^- levels were recorded for the outlet from the HRTF. It is also possible that due to the flow of air through the water, some stripping of ammonia may have taken place. Significantly at 5 out of 8 intensive monitoring periods, the outlet concentration of TAN was below 10 mg/l, thus achieving a typical discharge consent standard. In cases where TAN was higher than this value, additional nitrification has been shown to take place during subsequent storage of the dirty water. The average removal of TN in the CSTR was 55.4 % and for the HRTF the value was 64.4 %. Proportionately less TN was removed than TAN, since TN includes organic matter that is relatively unaffected by short term biological treatment. However, some of the TN may be present as partially insoluble material which can be removed by settlement.

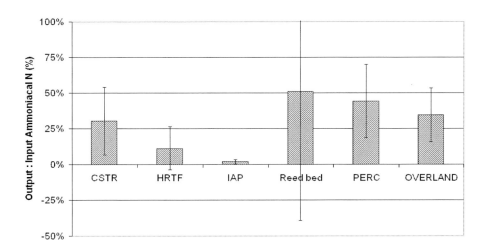

Figure 21. Total Ammoniacal Nitrogen (TAN) recorded during the eight intensive monitoring periods, including Trials 1 and 2, expressed as mean values of the ratios of output to input ammoniacal nitrogen concentrations. IAP data from the first intensive monitoring period are omitted since the nitrification process had not begun at that time. Error bars represent +/- one std deviation.

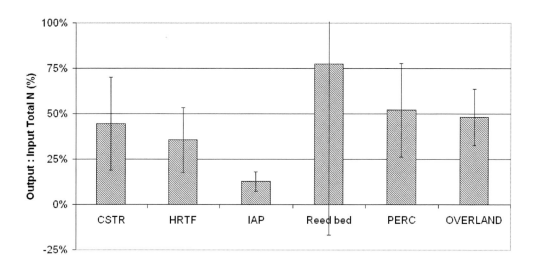

Figure 22. Total Nitrogen (TN) observed during the eight intensive monitoring periods, including Trials 1 and 2, expressed as mean values of the ratios of output to input TN concentrations. IAP data from the first intensive monitoring period are omitted since the nitrification process had not begun at that time. Error bars represent +/- one std deviation.

Table 4. Dirty water component removals recorded during intensive monitoring periods

	TN g/l	P mg/l	NO_3^- mg/l	NO_2^- mg/l	TAN mg/l	COD mg/l
Trial 1						
June 2003						
Raw farm	0.34	65			239	4890
Raw field	0.18	47.1			140	3150
Treated HRTF	0.18	31.1			135	745
Treated CSTR						
Treated RB	0.02	4.3			2.8	206
Treated percolation Treated	0.08	12.3			45.2	1650
overland	0.04	4.6			9.8	736
July 2003						
Raw farm	0.28	92.5	0.23	0.07	186	5350
Raw field	0.32	86.6	0.20	0.09	217	6810
Treated HRTF	0.02	8.3	19.7	0.10	2.2	611
Treated CSTR						
Treated RB	0.01	5.9	10.4	0.04	2.0	173
Treated percolation Treated	0.17	47.5	0.48	0.09	107	3220
overland	0.17	27.3	0.17	0.24	102	2180
Sept 2003						
Raw farm	0.74	107	0.34	0.28	600	8550
Raw field	0.71	116	0.30	0.22	583	11500
Treated HRTF	0.10	46.5	180	2.90	12.0	1610
Treated CSTR						
Treated RB	0.04	10.3	57.4	0.57	10.6	461
Treated percolation Treated	0.37	37.9	0.35	0.23	249	7090
overland	0.36	33.7	0.36	0.17	235	6250
Mar 2004						
Raw farm	0.37	70.9	0.26	0.22	222	5290
Raw field	0.41	70.6	0.24	0.15	271	6060
Treated HRTF	0.05	37.2	150	0.46	5.8	1060

	TN g/l	P mg/l	NO$_3^-$ mg/l	NO$_2^-$ mg/l	TAN mg/l	COD mg/l
Treated CSTR	0.18	42.2	0.18	0.14	122	1300
Treated RB	0.13	34.2	148	1.72	13.1	840
Treated percolation Treated	0.23	30.3	0.37	0.13	142	2960
overland	0.35	42.5	0.26	0.11	230	4330
	TN g/l	P mg/l	NO$_3^-$ mg/l	NO$_2^-$ mg/l	TAN mg/l	COD mg/l
Trial 2						
Nov 2004						
Raw farm	0.47	82.6	0.88	0.32	281	6250
Raw field	0.68	94.9	0.28	0.21	406	7470
Treated HRTF	0.03	12.9	294	1.66	1.4	760
Treated CSTR	0.23	73.4	136	10.0	24.7	3430
Treated RB	0.05	106	200	2.13	4.6	717
Treated percolation Treated overland	0.25	27.6	0.40	0.12	169	1918
Dec 2004						
Raw farm	1.19	173	1.14	0.37	874	13300
Raw field	0.92	149	0.54	0.26	589	10100
Treated HRTF	0.15	29.2	316	38.0	32.7	1860
Treated CSTR	0.59	36.7	26.8	0.35	376	3730
Treated RB	0.33	25.1	14.2	0.40	228	2120
Treated percolation Treated overland	0.15	13.9	18.6	0.34	109	1000
Feb 2005						
Raw farm	0.55	68.3	1.30	0.31	371	5610
Raw field	0.77	48.2	1.24	0.50	564	6150
Treated HRTF	0.10	28.9	246	3.28	3.7	1520
Treated CSTR	0.22	43.4	124	2.17	76.8	2270
Treated RB	0.18	21.7	37.2	1.82	129	1020
Treated percolation Treated overland	0.13	11.6	12.0	1.44	81.8	1080
April 2005						
Raw farm	0.39	72.2	0.88	0.63	228	4870
Raw field	0.55	55.3	1.40	0.67	424	3080
Treated HRTF	0.07	42.8	246	1.01	4.00	843
Treated CSTR	0.15	51.2	172	0.38	41.3	1330
Treated RB	0.15	43.2	116	0.83	47.6	1100
Treated percolation Treated overland	0.18	15.3	6.66	0.50	112	1080

As shown in Figure 21, the average removal of TAN from pretreated water within the RBs was 49 %, but with a very large error bar. As shown in Table 4, the RB displayed significant variability in removal of TAN, varying from 98 % removal (June 2003) to a 68.4 % increase of TAN (February 2005). TAN removal in RBs occurs by nitrification, adsorption and biomass assimilation. For TAN removed by adsorption, further nitrification would be expected to take place, however events such as heavy rainfall could cause adsorbed TAN to be desorbed from the gravel in to the flow of water, thus possibly explaining the occasional negative removals observed. The removal of TN in the RB was on average 22.4 %, which shows a relatively poor improvement compared with the raw water. This was surprising as RBs are normally found to be effective for removal of TN and TAN [Morris and Herbert, 1997; Cooper et al, 1997].

In the PSP, an average removal of 55.8 % TAN and 48 % TN occurred in the system, although again the performance declined with collapse of soil structure. For example, 85.4 % TAN and 72.7 % TN were removed from the system during the first week of operation in May 2003, but the performance gradually declined throughout Trial 1, until in March 2004, 68.7 % TAN and 47.5 % TN were removed.

In the OFP, the average outlet TN removal was 51.7 %. The final concentrations were quite dependent upon inlet concentrations, with the highest discharge concentration being 0.25 g/l. The average removal of TAN was 65.4 %, but the outlet concentrations recorded were still too high for discharge to a water course. In most, but not all cases, there was a rise in nitrate concentration, indicating that some nitrification of TAN to nitrates occurred. However the amount of nitrate was insufficient to account for all of the loss of TAN, suggesting that some of it could have been released as gaseous nitrogen to air via denitrification processes or ammonification could have occurred in the dirty water storage tanks.

Figure 23 compares the removal of phosphorus in each system, showing a close similarly in the performance of the OFP system and the IAP. It is also interesting to note that the RB performed slightly better than the HRTF.

Phosphorus removal can occur through several different mechanisms. In planted systems, such as RBs or soil plots, one mechanism is through plant uptake and harvesting [Lantzke et al, 1998]. The phosphorus content of the effluent can also be reduced through oxidation to phosphate (PO_3^-), which precipitates and adsorbs to soil particles. Inorganic phosphorus transformations to orthophosphate forms such as $H_2PO_4^-$, HPO_4^{2-} and PO_4^{3-} which can be retained by soils, depending upon the pH, Fe, Al and Ca mineral content and amount of native soil phosphorus present [Watson et al, 1989; Adler et al, 1996]. Chemical reactions with Fe, Al and Ca lead to the formation of settleable phosphorus compounds.

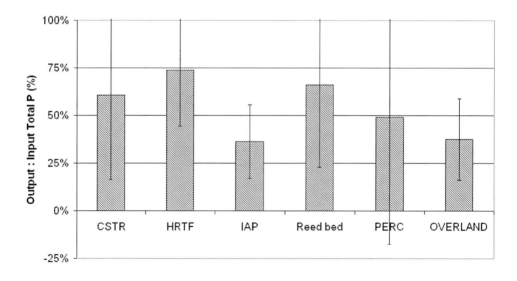

Figure 23. Comparison of system performance observed during eight intensive monitoring periods, including Trials 1 and 2, expressed as mean values of the ratios of output to input phosphorus concentrations. Error bars represent +/- one standard deviation.

For the IAP, settling and clarification are the main mechanisms by which phosphorus is removed, since much of the phosphorus is contained in insoluble matter. However the removal of P varied widely over the range 41 – 91 %. It was shown that the second stage (HRTF) of the aeration process was responsible for some of the removal. It was also known that reductions in pH tend to increase the solubility of phosphorus compounds in water, the raw dirty water had a pH of around 7.5, whilst the pH of the treated water typically was about 8.0. However, these factors alone would not account for the wide variability of the results for phosphorus removal.

The average removal of phosphorus in the RBs was 34.1 %. However, due to the relatively large error bar on this result, the actual phosphorus concentrations varied from 4.3 – 106 mg/l at the RB outlet. In the RB, removal of phosphorus may occur by adsorption on to the gravel matrix surrounding the roots of the reeds. Sun et al (1998) have reported removals of 34.7 % PO_4-P, so very similar to the average removal observed in this work. Tanner et al (1995) observed a higher removal of phosphorus of 48 – 75 % in planted constructed wetlands, although of a different, horizontal, type to those used in the study at the Case Study farm. It should also be noted that the RBs used at the Case Study farm had previously been installed at another site, and therefore the capacity of the gravel to adsorb phosphorus may have been decreased before the trial began. Gradually phosphorus may build up in the bed matrix, leading to breakthrough of phosphorus in the treated effluent. Depending on operating feed and conditions may take typically between 5 and 20 years after which regeneration of the beds must be performed by removal and cleaning of the gravel.

The average phosphorus removal in the PSP was 51 %, the actual outlet readings varying in the range 12.3 – 47.5 mg/l. From Table 4, the first reading in June 2003 showed a considerably lower phosphorus value at the outlet compared with later values. The collapse of soil structure in the autumn of 2003 may have led to a reduction in the ability of the soil particles to effectively adsorb phosphorus and hence the higher readings.

For the OFP the average removal of phosphorus was 62.4 %, and in this case the error bars were smaller than those of the RB. The actual discharge values of phosphorus varied in the range 4.6 – 42.5 mg/l. Tyrell and Leeds-Harrison (2005) have reported a gradual decrease in the capability of another overland flow system to remove phosphorus from dirty water, which led them to question the longer term sustainability of this type of technology for phosphorus treatment. This could be due to the finite capacity of the soil to adsorb phosphorus before breakthrough occurs. However for the system at the Case Study farm, there did not seem to be a noticeable decrease in treatment efficiency of the OFP over time. The final recorded phosphorus value of 15.3 mg/l in April 2005 was of comparable magnitude with earlier readings and considerably lower than the highest reading of 42.5 mg/l in March 2004. Therefore it did not appear that the OFP had become completely saturated with phosphorus.

Table 4 displays the COD readings recorded during the intensive monitoring periods, the raw dirty water having COD values in the range 4000 – 14000 mg/l. As indicated by the values in Table 4, the treatment systems showed variable levels of COD removal and average removals in each treatment systems were: IAP 82.6 %, RBs 41.5 %, PSP 43.7 %, OFP 49.3 %. The removal reported for the IAP was the highest of all treatment systems, in keeping with the trend for removal of BOD_5. The other systems showed similar average levels of COD removal. For the IAP the majority of COD removal occurred in the CSTR (64.6%), with the HRTF contributing a lower level of removal (33.3 %).

Costs and Future Outlook

From the experimental trials it was shown that the IAP operated reliably and was effective for pre-treatment of dirty water. The OFP and RBs demonstrated some significant pollutant removals, but when operating with raw dirty water the outlet pollutant concentrations were too high for discharge to water courses. Consequently, a combination of treatment systems was devised for the case study of treatment of the full flowrate of ~21 m^3/d wastewater produced at the Case Study farm. This is based on a CSTR (132 m^3 capacity), an OFP (2 treatment planes with a combined area of 1.09 ha) and a downflow RB (total area of 350 m^2) operating in series. The installed costs and operating costs were estimated as shown in Table 5. The costs of include the equipment required, auxiliary equipment such as pipes and pumps, their transport to site and installation, but do not include land costs. The operating costs are based on calculations of electrical power requirements for running water pumps and air blowers on the aeration plant. The design life was considered to be 15 years for the IAP and RB, and 5 years for the OFP, owing to the limited capacity of the soil to adsorb phosphorus. The capital cost was highest for the IAP, owing to the physical hardware requirement of tanks, blowers and control equipment. A significant proportion of the capital cost of the OFP was the price of the plastic liner required to be placed beneath the soil. A possible strategy for reducing this cost would be for sites where compacted clay could be used to line the site. The highest operating cost occurs for the aeration plant, owing to the need to run air blowers and water pumps for long durations. The simple annual operating costs are highest for the OFP, owing to its limited expected lifetime of only 5 years. The overall costs of the combined treatment system per cow are £161 per cow per year, which is relatively expensive. Additionally even the combined system would only reduce the BOD$_5$ to 50 mg/l and total solids to 1658 mg/l, thus the water would require further storage or settling to allow for further solids removal before discharge to a water course. Therefore the treatment and management of agricultural run off is likely to be expensive for dairy farmers.

Table 5. Estimated capital and operating costs for a treatment system including IAP (CSTR), OFP and RB in series. (Number of cows in the milking herd, 420; annual milk production 2.7 million litres; annual dirty water applied to land, 7600 m3)

	IAP (CSTR only)	OFP	RB
Average input BOD5, mg/l	3300	307	124
Average input TS, mg/l	7000	4599	2000
% reduction in BOD5 before spreading	90.7 %	96.3 %	59.5 %
% reduction in TS before spreading	34.3 %	56.5 %	17.1 %
Capital cost, £	£296,000	£190,000	£66,000
Design life, years	15	5	15
Running costs, £/yr	£6,759	£360	£134
Simple total annual cost, £/yr	£26,492	£38,360	£4534
Simple total annual cost per cow, £/yr	£66	£96	£11

4. CONCLUSION

In a substantial Case Study, four different treatment systems were trialled at a working dairy farm for the treatment of dirty water (a major example of agricultural run off). The results showed that the untreated effluent constituted a considerable pollution risk with BOD_5 and TS concentrations ranging from 600 to 1200 mg/l and from 300 to 18000mg/l respectively, underlining the need for careful management. Due to limitations on storage capacity at the Case Study farm, year-round disposal of dirty water by spreading on land was unavoidable. Therefore, this study compared the effectiveness and expected costs of treating dirty water before land spreading, thus reducing the potential environmental impact of any subsequent run-off. Comprehensive treatment options were also reviewed to assess the potential for industrial-style wastewater management involving consented discharge of treated effluent to water courses. The IAP (CSTR) was effective for the removal of BOD_5 (>90 %) and other pollutants such as TN, TAN and phosphorus, although less effective for the removal of TS (~34 %). However, the treated outlet from the IAP CSTR would require further treatment to remove TS. The RBs were effective for the polishing treatment of water which had been pre-treated in another system, with up to 75 % BOD_5 removal, but not for the removal of TS or the treatment of raw dirty water. Although initially showing some capability for pollutant removal, the PSP was found to be unreliable due to the difficulty of not being able to locally source a supply of coarse textured, free draining soil to avoid waterlogging. After modification, the OFP at The Case Study farm demonstrated effective removal of several pollutants including BOD_5 (>90 %), TSS (> 80 %) and TAN (~ 80 %), but when treating raw dirty water did not reduce pollutant concentrations sufficiently to enable possible consideration of consented discharge to a water course. Indeed, it was shown that given the extreme variability in the composition of the dirty water produced, such an industrial-style approach would require at least three treatment processes operated in sequence to ensure sufficient treatment capacity and reliability.

The IAP was also the most reliable system of those tested, since it did not accumulate any matter, which would lead to deterioration in performance. The RB and soil-based systems were subject to possible problems of clogging or freezing, which reduced their effectiveness on certain occasions. However, the RB and soil based systems are relatively cheap to operate and are therefore more sustainable in terms of their low power consumption. A possible combination of treatment systems including an intensive aeration CSTR, followed by OFP and/or a RB could be used to effect a more thorough pollutant removal.

ACKNOWLEDGMENTS

The authors are grateful to Department of the Environment, Farming and Rural Affairs (DEFRA), the Milk Development Council (MDC), BOC Foundation and ARM Reed Beds for their contributions to the sponsorship and management of the reported project. The project partners are thanked for their contribution during the experimental trials: Colin Burton, Trevor Cumby (Project Manager) and Elia Negro of Silsoe Research Institute; Marc Dresser, Peter Leeds-Harrison and Sorche O'Keefe of Cranfield University; David Chadwick of the Institute of Environmental and Grassland Research; John Gregory, Ken Smith and Ian Muir of ADAS UK Ltd; Andrew Barker and Garikoitz Fernandez of the University of Birmingham. Messrs Jim, Jonathan, Tim and Peter Harrison, plus their colleagues are acknowledged for supporting and hosting the trials, as are Direct Laboratories who performed the majority of the reported biochemical analyses.

REFERENCES

Adler, P.R.; Summerfelt, S.T.; Glenn, D.M.;Takeda, F. *Wat. Environ. Res.* 1996, *68*, 836 – 840.
Ansola, G.; Fernandez, C.; De Luis, E. *Ecol. Eng.* 1995, *5*, 13 – 19.
Bernard, J.M.; Lauve, T.E. *Wetlands* 1995, *15*, 176 – 182.
Brix, H. "Wastewater treatment in constructed wetlands: system design, removal processes and treatment performance" In: Moshiri, G.A. (Ed.) Constructed Wetlands for Water Quality Improvement. Lewis Publishers, Boca Raton, FL, 1993, 9 – 22.
Burton, C.H.; Farrent, J.W. *J. Agr. Eng. Res.*, 1998, *69*, 159 – 167.
Chadwick, D.R.; Brookman, S.; Leeds-Harrison, P; Tyrrel, S.; Fleming, S. and Pain, B. "Soil based treatment systems for dirty water". In Moore, J.A. (ed.) "Proceedings of the international symposium on animal, agricultural and food processing wastes" ASAE Iowa MI, 2000, pp. 728 – 737.
Cooper, P.; Green, B. *Water Sci. Technol.* 1995, *32*, 317 – 327
Cooper, P.F., Hobson, J.A. and Findlater, C. *Water Sci. Technol.* 1990, *22*, 57 – 64.
Cooper P.; Smith M.; Maynard H. *Water Sci. Technol.* 1997, *25*, 215 – 221.
Copybook Solutions, 2008. MBR at Dairygold - Effective Treatment of Dairy Processing Waste for over Five Years. Waste and Waste Water International. (http://www.water)
Cumby, T.R.; Brewer, A.J.; Dimmock, S.J. *Bioresource Technol.*1999, *67*, 155 – 160
Do Canto, C.S.A.; Rodrigues, J.A.D.; Ratusznei, S.M.; Zaiat, M.; Foresti, E. *Bioresource Technol.* 2008, *99*, 544 – 654.
Dunne, E.J.; Culleton, N.; O'Donovan, G.; Harrington, R.; Olsen, A.E. *Ecol. Eng.* 2005, *24*, 221 – 234.
Gray N.F. "Activated sludge. Theory and practice" Oxford University Press, New York, 1990.
Gray, N.F. Water Technology: An Introduction for Environmental Scientists and Engineers; Arnold, London, 1999.
Green, M.B.; Martin, J.R. *Water Environ. Res.* 1996, *68*, 1054 -1060.
Hickey, C.W.; Quinn, J.M.; Davies-Colley, R.J. *New Zeal. J. Mar. Fresh* 1989, *23*, 569-584.
Kadlec, R.H.; Knight, R.L. "Treatment wetlands" Lewis Publishers, Boca Raton, FL, 1996.
Karathanasis, A.D.; Thompson, Y.L. *Soil Sci. Soc. Am. J.* 1995, *59*, 1773 – 1779.
Kern, J.; Idler, C. *Ecol.Eng.* 1999, *12*, 13 – 25.
Lantzke I.R.; Heritage A.D.; Pistillo G. and Mitchell D.S. *Water Res.* 1998, *32*, 1280 – 1286.
Liu, F.; Mitchell, C.C.; Odom, J.W.; Hill, D.T.; Rochester, E.W. *Bioresource Technol.* 1998, *63*, 65- 73.
MAFF (1998) "Code of Good Agricultural Practice for the Protection of Water" Ministry of Agriculture, Fisheries and Food, UK.

Metcalf and Eddy, revised by Tchobanoglous, G. and Burton, F.L. "Wastewater Engineering – Treatment, disposal and reuse" 3rd edition. McGraw-Hill, New York, USA, 1991.

Morris M.; Herbert R. *Water Sci. Technol.* 1997, *35*, 197 – 204.

Ndegwa, P.M.; Wang, L.; Vadella, V.K. *Biosystems Eng.* 2007, *97*, 379 – 385.

Newman, J.M.; Clausen, J.C. and Neafsey, J.A. *Ecol. Eng.* 2000, *14*, 181 – 198.

Price T.; Probert D. *Appl. Energ.* 1997, *57*, 129 – 174.

Sun G.; Gray K.R.; Biddlestone A.J. *J. Agric. Engng. Res.* 1998, *69*, 63 – 71.

Sun G.; Gray K.R.; Biddlestone A.J. *Environ. Technol.* 1999, *20*, 233-237.

Sun, G.; Gray, K.R.; Biddlestone, A.J.; Allen, S.J.; Cooper, D.J. *Process Biochem.* 2003, *39*, 351 – 357.

Tanner, C.C.; Clayton, J.S.; Upsdell, M.P. *Water Res.* 1995a, *29*, 17 – 26.

Tanner, C.C.; Clayton, J.S.; Upsdell, M.P. *Water Res.* 1995b, *29*, 27 – 34.

Tyrrel, S.F.; Leeds-Harrison, P.B.; Harrison, K.S. *Water Res.* 2002, *36*, 291 – 299.

Tyrrel, S.F.; Leeds-Harrison, P.B. *Water Sci. Technol.* 2005, *51*, 73–79.

Vaboliene, G.; Matuzevicius, A.B. *J Environ. Eng. Landsc. Manag.*, 2007, 77 – 84.

Van Oostrom, A.J.; Cooper, R.N. "Meat processing effluent treatment in surface flow and gravel bed constructed wastewater wetlands" Cooper, P.F.; Findlater, B.C.; Eds.; Constructed Wetlands in Water Pollution Control. Pergamon Press, Oxford, 1990, pp. 321 – 332.

Vesilind, P.A. "Environmental pollution and control" Ann Arbor Science Publishers, UK, 1975, p. 169.

Vrhovsek, D.; Kukanja, V. and Bulc, T. *Water Res.* 1996, *30*, pp. 2287 – 2292.

Watson J.T.; Reed S.C.; Kadlec R.H.; Knight R.L. and Whitehouse A.E. "Performance expectations and loading rates for constructed wetlands" In: Hammer, D.A. (Ed.) "Constructed wetlands for wastewater treatment" Lewis Publishers, Chelsea, MI, 1989, pp. 319-347.

Wood, J., Fernandez, G., Barker, A., Gregory, J. and Cumby, T. *Biosystems Engng.* 2007, *98*, 455 – 469.

Wrigley, T.J.; Toerien, D.F. *Water Res.* 1990, *24*, 83 – 90.

Zhang, Z.; Zhu, J.; Park, K.J. *Water Res.* 2006, *40*, 162 – 174.

Zhao Y.Q.; Sun G..; Allen S.J. *Water Res.* 2004, *38*, 2907 – 2917.

Reviewed by Professor Jan Baeyens, Catholic University of Leuven, University of Antwerp Belgium. 2/6/08.

Chapter 52

EVALUATION OF INTERTWINED RELATIONS BETWEEN WATER STRESS AND CROP PRODUCTIVITY IN GRAIN-CROPPING PLAIN AREA BY USING PROCESS-BASED MODEL

Tadanobu Nakayama[*]
National Institute for Environmental Studies (NIES), 16-2 Onogawa,
Tsukuba, Ibaraki 305-8506, Japan

ABSTRACT

Water stress in Northern China has intensified water use conflicts between upstream and downstream areas and also between agriculture and municipal/industrial sectors. North China Plain (NCP) is one of the most important grain cropping areas in China. It is a giant alluvial plain formed by deposition by the Yellow, Hai, and Luan Rivers and their tributaries. Furthermore, there are some megalopolis, such as Beijing and Tianjin. This region has changed from water-rich in the 1950's to water-poor area at present, which indicates various ecosystem degradations such as dry-out Yellow River, nearly closed Hai River, groundwater degradation and seawater intrusion in the NCP. The author has so far developed the process-based model, called NIES Integrated Catchment-based Eco-hydrology (NICE) model (Nakayama, 2008a, 2008b, 2008c, 2009; Nakayama and Watanabe, 2004, 2006a, 2006b, 2008a, 2008b, 2008c; Nakayama et al., 2006, 2007), which includes surface-unsaturated-saturated water processes and assimilates land-surface processes describing the variation in phenology with satellite data. In this research, the author simulated the effects of irrigation on groundwater flow dynamics in the North China Plain by coupling the NICE with DSSAT-wheat and DSSAT-maize, two agricultural models. This combined model (NICE-AGR) was applied to the Hai River catchment and the lower reach of the Yellow River (530 km wide by 840 km long) at a resolution of 5 km. It reproduced excellently the soil moisture, evapotranspiration and crop production of summer maize and winter wheat, correctly estimating crop water use. So, the spatial distribution of crop water use was reasonably estimated at daily steps in the simulation area. In particular, the NICE-AGR reproduced groundwater levels better

---

[*] Tel.: +81-29-850-2564; fax: +81-29-850-2584.E-mail address: nakat@nies.go.jp (T. Nakayama).

than the use of statistical water use data. This indicates that the NICE-AGR does not need detailed statistical data on water use, making it very powerful for evaluating and estimating the water dynamics of catchments with little statistical data on seasonal water use. Furthermore, the simulation reproduced the spatial distribution of groundwater level in 1987 and 1988 in the Hebei Plain, showing a major reduction of groundwater level due mainly to over-pumping for irrigation. These results show that this model is very powerful to simulate the future crop productivity in relation to the water withdrawal in this area. This study is very important for evaluation of ecosystems and environments on the moving edge of freshwater and marine.

1. INTRODUCTION

The North China Plain is located in the eastern part of China between 35°00′ and 40°30′N and between 113°00′ and 119°30′E; it includes all the plains of Hebei Province, Beijing and Tianjin, and the northern parts of plains in Shandong and Henan provinces (Figure 1). The climate is semi-arid. The area is bounded by the Taihang Mountains to the west, the Yanshan Mountains to the north, the Bohai Sea to the east, and the Yellow River to the south. It is a giant alluvial plain formed by deposition by the Yellow, Hai and Luan rivers and their tributaries. Many palaeochannels of different stages form different geomorphologic features (Chen, 1996).

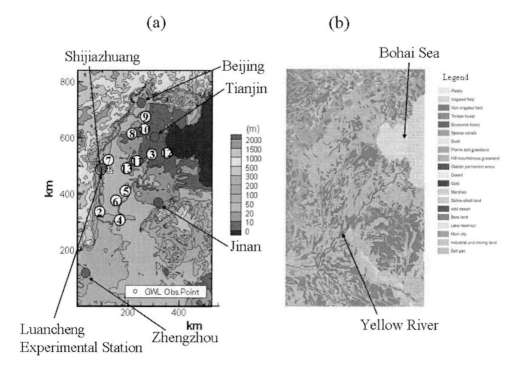

Figure 1. Study area in the North China Plain (NCP) about (a) mean elevation and (b) land cover. In Fig. 1a, the red line is the border of Hebei Plain. The number shows the groundwater observation stations of China Institute for Geo-Environmental Monitoring (2003) (G-1 – G-13), and the triangle shows the groundwater observation stations we measured (G-14 – G-31) in Table 4.

This is one of the most important grain cropping areas in China because of its large area (about 1.36×10^5 km^2) and huge population (about 112 million). Water resources are the key to agricultural development, and the demand for groundwater will increase in this area. With the development of the regional economy, the water use environment has changed. The groundwater has lowered dramatically over the previous half century owing to the over-pumping of groundwater and to drought, accompanied by the expansion of saline-alkaline land area (Brown and Halweil, 1998; Shimada, 2000; Chen *et al.*, 2003). Several studies have measured the water and heat balances in winter wheat and maize (corn) fields, focusing on the relationship between soil moisture of irrigated fields and water use efficiency of irrigated crops (Wang *et al.*, 2001), evaporation and evapotranspiration (Liu *et al.*, 2002; Shen *et al.*, 2002; Zhang *et al.*, 2002), energy fluxes (Shen *et al.*, 2004; Zhang *et al.*, 2004), soil water balance and related recharge (Kendy *et al.*, 2003) and solutes (Chen *et al.*, 2004).

Crop models such as CERES (Crop-Environment REsources Synthesis model) (Jones and Kiniry, 1986), EPIC (Erosion Productivity Impact Calculator crop model) (Thomson *et al.*, 2002), WOFOST (World Food Studies model) (Vandiepen *et al.*, 1989) and so on can estimate crop production, water use and management from simplified climate, site conditions and initial soil moisture. For instance, Doll and Siebert (2002) simulated the effect of climate change on world irrigation water use under designed climate change scenarios. Using DSSAT-wheat and DSSAT-maize (DSSAT: Decision Support Systems for Agro-technology Transfer), Adams *et al.* (1990) simulated the effects of climate change on US agriculture. Winter wheat (*Triticum aestivum* L.) and summer maize (*Zea mays* L.) are the two major crops in the northern region of the North China Plain. McVicar *et al.* (2002) investigated the distribution of winter wheat and maize in the Hebei Plain, where most of the study region in the study is located, in 1986 and 1988. Of crops harvested in early June, 90% of the yield came from winter wheat. Over 30% of the area in nearly two-thirds of the counties in the Hebei Plain grew winter wheat. Of crops harvested in September (so-called autumn crops), maize was the most important. According to statistics, grain crops in 1988 covered nearly 76% of the total cultivated land in Hebei (Governmental Office of Hebei Province, 1989), of which over 80% grew winter wheat and maize. Yang *et al.* (2002) showed that the seasonal fluctuation of groundwater in the piedmont region of the Taihang Mountains is closely linked with the water use of the two crops. On the other hand, few studies have simulated the effects of irrigation on the groundwater flow dynamics, although some studies have evaluated them by measuring isotopes such as oxygen-18 (^{18}O), carbon-14 (^{14}C), and tritium (^{3}H) (Chen *et al.*, 2002; Shimada *et al.*, 2002; Chen *et al.*, 2003). Therefore, it is urgently necessary to investigate the secular changes of land and water use, and land subsidence, and to forecast the future distribution of groundwater levels, by using numerical models.

The objectives of this study were to simulate water use in relation to the reduction of groundwater due to its over-drawing for irrigation and industrial uses, to help the evaluation of water use plans designed to maintain the groundwater level balance, and to help decision-making on the amount of water to be transferred from other catchments. Because the approach does not need detailed statistical data about water use in agriculture, it is applicable to other catchments where little water use information is available, and is very powerful at evaluating and forecasting the water dynamics of a catchment.

2. MODEL DESCRIPTION OF NICE-AGR

2.1. NICE Model

Previously, the author developed the National Institute for Environmental Studies (NIES) Integrated Catchment-based Ecohydrology (NICE) model, which includes surface-unsaturated–saturated water processes and assimilates land-surface processes describing the variations of *LAI* (leaf area index) and *FPAR* (fraction of photosynthetically active radiation) from MODIS (Moderate Resolution Imaging Spectroradiometer) satellite data (Nakayama, 2008a, 2008b, 2008c, 2009; Nakayama and Watanabe, 2004, 2006a, 2006b, 2008a, 2008b, 2008c; Nakayama et al., 2006, 2007). The author expanded NICE to include the effects of local topography on snow cover, the freezing/thawing soil layer and spring snowmelt runoff in the same catchment (Nakayama and Watanabe, 2006a), and included the interaction between the lake water and groundwater (Nakayama and Watanabe, 2008a). NICE includes the biophysical/soil moisture model and the groundwater flow model. The unsaturated layer divides the canopy into two layers (canopy layer and ground surface), and the soil into three layers (upper layer, intermediate layer, and lower layer) in the vertical dimension. The governing equations for the prognostic variables consists of temperatures, interception stores, soil moisture stores, and canopy conductance to water vapor.

Canopy, ground surface, and deep soil temperatures

$$C_c \frac{\partial T_c}{\partial t} = Rn_c - H_c - \lambda E_c - \xi_{cs} \qquad (1)$$

$$C_g \frac{\partial T_g}{\partial t} = Rn_g - H_g - \lambda E_g - \frac{2\pi C_d}{\tau_d}(T_g - T_d) - \xi_{gs} \qquad (2)$$

$$C_d \frac{\partial T_d}{\partial t} = \frac{1}{2\sqrt{365\pi}}(Rn_g - H_g - \lambda E_g) \qquad (3)$$

The subscript c refers to the canopy, g to the soil surface, and d to the deep soil. T_c, T_g, and T_d (K) are canopy, ground surface, and deep soil temperatures; Rn_c and Rn_g (W/m^2) are absorbed net radiation of canopy and ground; H_c and H_g (W/m^2) are sensible heat flux; E_c and E_g (kg/m^2/s) are evapotranspiration rates; C_c, C_g, and C_d (J/m^2/K) are effective heat capacities; λ (J/kg) is latent heat of vaporization; τ_d (s) is daylength; and ξ_{cs} and ξ_{gs} (W/m^2) are energy transfer due to phase changes in M_c and M_g (described next), respectively.

Interception stores

$$\frac{\partial M_c}{\partial t} = P - D_d - D_c - E_{ci}/\rho_w \qquad (4)$$

$$\frac{\partial M_g}{\partial t} = D_d + D_c - E_{gi}/\rho_w \qquad (5)$$

M_c and M_g (m) are water or snow/ice stored on the canopy and on the ground; P (m/s) is precipitation rate; D_d (m/s) is canopy throughfall rate; D_c (m/s) is canopy drainage rate; E_{ci} and E_{gi} (kg/m²/s) are interception loss of canopy and ground; and ρ_w (kg/m³) is density of water.

Soil moisture stores

$$\frac{\partial W_1}{\partial t} = \frac{1}{\theta_s D_1}\left[P_{w1} - Q_{1,2} - \frac{1}{\rho_w}E_{gs}\right] \qquad (6)$$

$$\frac{\partial W_2}{\partial t} = \frac{1}{\theta_s D_2}\left[Q_{1,2} - Q_{2,3} - \frac{1}{\rho_w}E_{ct}\right] \qquad (7)$$

$$\frac{\partial W_3}{\partial t} = \frac{1}{\theta_s D_3}\left[Q_{2,3} - Q_3\right] \qquad (8)$$

W_i is the soil moisture fraction of the i-th layer ($=\theta_i/\theta_s$); θ_i (m³/m³) is volumetric soil moisture in the i-th layer; θ_s (m³/m³) is the value of θ at saturation; D_i (m) is the thickness of the soil layer; $Q_{i,j}$ (m/s) is the flow between layers i and j; Q_3 (m/s) is gravitational drainage from recharge soil moisture store; E_{ct} (kg/m²/s) is canopy transpiration; E_{gs} (kg/m²/s) is ground evaporation; and P_{w1} (m/s) is infiltration of precipitation into the upper soil moisture store.

About the saturated layer, the NICE solves a partial differential equation describing three-dimensional groundwater flow for both unconfined and confined aquifers:

$$\frac{\partial}{\partial x}\left(K_{xx}\frac{\partial h_g}{\partial x}\right) + \frac{\partial}{\partial y}\left(K_{yy}\frac{\partial h_g}{\partial y}\right) + \frac{\partial}{\partial z}\left(K_{zz}\frac{\partial h_g}{\partial z}\right) + W = S_s\frac{\partial h_g}{\partial t} \qquad (9)$$

K_{xx}, K_{yy}, and K_{zz} (m/s) are values of hydraulic conductivity along the x, y, and z coordinate axes; h_g (m) is the potentiometric head; W (1/s) is the volumetric flux per unit volume representing sources and/or sinks of water; and S_s (1/m) is the specific storage. Equation (9) is discretized by using the finite-difference method on the assumption that the variables between two cells change linearly. Details are written in Nakayama and Watanabe (2004).

Although NICE could reproduce the observed groundwater fluctuations on the whole, it could not reproduce some fluctuations that were due to local water withdrawal (Figure 4 in Nakayama and Watanabe (2008a)). The discrepancy occurred because the annual time-series of actual groundwater use in agriculture was different from the area-averaged fluctuations used in the simulation, and because of differences of the actual and model inputted values about land cover and crops. Although more accurate input data on groundwater withdrawal are necessary for better spatial and temporal reproduction of the observed values, there is often little statistical data on the seasonal trends of water use, in particular in China. Therefore, it would be very useful to couple agricultural models such as DSSAT or WAVES

(WAter Vegetation Energy and Solute modelling; Yang *et al.* (2003)) with NICE in order to estimate the seasonal variation of water use by different agricultural crops.

2.2. DSSAT model

DSSAT is a computer program relying on crop soil and weather data bases to simulate multi-year outcomes of crop management strategies (Ritchie, 1998). The models chosen for the study include DSSAT-wheat and DSSAT-maize, two widely used models of the DSSAT 3.5 family developed by the International Consortium for Agricultural System Application with the intention of understanding the effects of agricultural water use on groundwater depletion in the North China Plain. DSSAT-wheat has been used to assess the irrigation pattern in winter wheat (Yang *et al.*, 2004a), and both models have been used to assess the effects of agricultural water use on groundwater level change in Gaocheng County in the piedmont region of the Taihang Mountains (Yang *et al.*, 2004b).

DSSAT-wheat and DSSAT-maize simulate, in daily steps, the phenology of wheat and maize from pre-sowing to harvest; photosynthesis and plant growth; biomass allocation to root, stem, leaf, and grain; and soil water and nutrient movement and deficit and their influence on the growth of the two crops. Since the intensive management of the farmland in the North China Plain leaves very little chance of nutrient deficit and there is great difficulty in assessing the fertilizer input over the whole region, no nutrient cycle was simulated in the study, and the nitrogen and phosphorus cycles were turned off. Figure 2 illustrates the simultaneous simulation of three processes: phenological development, crop growth represented by biomass accumulation and allocation, and water deficit and supply and their interaction.

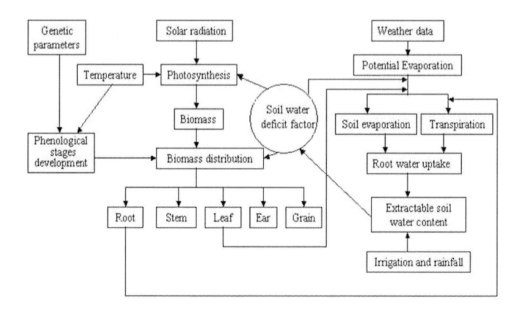

Figure 2. Structure of DSSAT-wheat and DSSAT-maize in simulating crop growth and crop water use.

For water assessment, the two models share a module that evaluates the soil water balance in daily steps as a function of precipitation, infiltration, transpiration, soil evaporation and drainage from the soil profile. The equation for calculating potential evaporation rate EEQ (mm/day) is derived from the study of Priestley and Taylor (1972):

$$EEQ = SOLARD \times (4.88 \times 10^{-3} - 4.37 \times 10^{-3} \times albedo) \times (T_D + 29), \qquad (10)$$

where $SOLARD$ (MJ/m^2/day) is solar radiation; $albedo$ is combined crop and soil albedo; and T_D (°C) is daytime temperature. The actual rates of soil evaporation and plant transpiration are calculated from the potential evaporation and the parameter calculated from actual and potential extractable soil water in each soil layer. The water supply from the roots in different soil layers is also considered in terms of root density and soil-extractable water content. Further equations and processes involved in water cycle simulation in the model can be found in Ritchie (1998) and Ritchie *et al.* (1998).

Plant growth is based on biomass formation, which depends on photosynthetically active radiation (PAR), leaf area index (LAI), temperature and other factors, for instance the effects of soil water shortage and nutrient deficit, although nutrient deficit is not considered in this study. The basic equation for the accumulation of biomass in wheat and maize can be expressed as:

$$PCARB = RUE \times IPAR, \qquad (11)$$

where $PCARB$ (g/m^2) is the potential biomass production, RUE (g/MJ) is radiation use efficiency (5.0 g/MJ for maize and 2.6–4.0 g/MJ for wheat), and $IPAR$ (MJ/m^2) is the fraction of PAR incepted by the crop, which relates to the development of LAI.

Genetic parameters are important for the control of crop phenological development, and influence the growth of LAI and crop yield. Only with proper parameters can crop growth be simulated correctly. Allocation of biomass to root, stem, leaf and grain as yield depends on the development of phenological stages. For instance, before flowering, much biomass is allocated to leaf for the growth of LAI. After flowering, much biomass is allocated to the growth of stem and ear. Then, during grain-filling, biomass is distributed mainly to grain growth. Differences in plant growth such as in LAI can influence the simulation of crop water use. On the other hand, the same soil moisture condition can have different effects on biomass formation and allocation. Therefore, changes of genetic parameters are often necessary to achieve a good simulation of crop growth, yield and water use. Phenological stages in DSSAT-maize are based primarily on degree-days. Ritchie (1998) and Ritchie *et al.* (1998) provide more detail on the growth of the two crops.

In the previous studies (Yang *et al.*, 2004a; Yang et al., 2004b), the authors separately calibrated DSSAT-wheat and DSSAT-maize and validated them against experimental data from 2000–2002. Through calibration, specific leaf area ("SLA" in the model) has been changed from its original value of 115 cm^2/g (which means 115 cm^2 of leaf area grows from 1 g of biomass allocated to leaf growth) to 180 cm^2/g to account for the quicker LAI growth of cultivars in the North China Plain (Yang *et al.*, 2004b). In the present study, the author used the revised DSSAT-wheat model but the original DSSAT-maize model. The author simulated the water balance for wheat and maize production in each county in the North China Plain.

The author validated the simulated values against statistical data, and then input them into NICE.

2.3. INTEGRATION OF MODELS

The groundwater withdrawal and the crop productivity are closely related with each other in the study area, and their relation varies in a growing stage and a kind of crop. Therefore, it is necessary and powerful to combine NICE with DSSAT. The drainage of water out of the bottom of the unsaturated soil layer to create base flow Q_3 (m/s) is given as follows (Sellers, et al., 1996):

$$Q_3 = f_{ice}\left[\sin\Theta_S K_S W_3^{2B+3} + 0.001\frac{\theta_S D_3 W_3}{\tau_d}\right], \quad (12)$$

where Θ_s is local slope angle, K_S (m/s) is hydraulic conductivity at saturation, B is an empirical constant, W_3 is the soil moisture fraction of the lowest layer of unsaturated flow ($=\theta_3/\theta_S$), θ_3 (m³/m³) is volumetric soil moisture in the lowest layer, θ_S (m³/m³) is the value of θ at saturation, D_3 (m) is the thickness of the lowest layer of unsaturated flow, τ_d (s) is the daylength (= 86 400) and f_{ice} is a progressive reduction in soil hydraulic conductivity as the soil freezes. The first term on the right-hand side of equation (12) covers gravitational drainage, and the second term is the contribution to base flow made by heterogeneities in the soil moisture fields of large river basins.

NICE uses the water flux q_f to combine unsaturated flow and saturated flow in natural regions (Nakayama and Watanabe, 2004), expressed by using the gradient of hydraulic potential between the deepest layer of unsaturated flow and the groundwater level in the expansion of equation (12) in the following equation:

$$q_f = -\overline{K}\nabla\Psi = -\overline{K}\frac{\Delta\Psi}{\Delta z} = -\overline{K}\frac{\Psi_g - \Psi_3}{D_3/2 + (D_g - h_g)} = \overline{K}\left(\frac{\psi_3}{D_3/2 + (D_g - h_g)} + 1\right), \quad (13)$$

where $\overline{K}$ (m/s) is the estimated effective hydraulic conductivity between unsaturated and saturated layers; Ψ_g ($= h_g$) and Ψ_3 ($= \psi_3 + D_g + D_3/2$) (m) are hydraulic potentials at the groundwater level and the lowest layer of unsaturated flow; z (m) is the vertical distance; D_g (m) is the distance between the top of the second layer and the bottom of the 20th layer (the standard level) in the groundwater model; and h_g (m) is the hydraulic head simulated by the groundwater model. When the groundwater level rises and enters the soil layer, the partial pressure of water vapour in the soil layer is set at the bottom of the unsaturated layer ($\Psi_3 = \psi_p$) to simulate soil moisture. After the water flux q_f is calculated at each time step, the flows between each unsaturated and saturated soil layer are simulated by using an improved backward-implicit scheme in order to simulate soil moisture in each unsaturated soil layer

(Nakayama and Watanabe, 2004). Furthermore, this flux is input into the groundwater flow model as recharge rate at the highest active cell as the upper boundary condition, and the groundwater flow model is simulated.

In an agricultural field, the additional water volume AW (m/s) is estimated from simulated values by NICE and DSSAT in the expansion of Doll and Siebert (2002) as follows:

$$AW = \min\left(Q_3, P_{w1} - \frac{1}{\rho_w} ET_{act}\right), \qquad (14)$$

where Q_3 (m/s) is drainage water in natural regions simulated by the equation (12), ET_{act} (kg/m^2/s) is crop-specific actual evapotranspiration rate simulated by DSSAT, P_{w1} (m/s) is the effective precipitation rate infiltrating into the upper soil moisture store, and ρ_w (kg/m^3) is the density of water. The second term of equation (14) means the net irrigation requirement proposed by Doll and Siebert (2002) but is very different from their value. NICE-AGR does not need the crop coefficient (depending on a growing stage and a kind of crop) for the calculation of ET_{act} and simulates directly ET_{act} without detailed site-specific information or empirical relation to calculate P_{w1} from the total precipitation (Doll and Siebert, 2002). The additional water infiltrates into the groundwater layer as recharge rate at $AW > 0$ in non-agricultural periods, whereas the deficit water is pumped up from the groundwater layer at $AW < 0$ in agricultural periods (as crop water requirement). The additional water at $AW < 0$ is theoretically pumped up from the groundwater (Doll and Siebert, 2002). NICE-AGR uses a minimum value of two terms in equation (14) in non-agricultural periods, because the author assumes that some of the drainage water stays in the soil layer and is held back from recharging the aquifer by the artificial bottom structure of the farmland. This additional water simulated in the agricultural fields in each county of the North China Plain was interpolated to each discretized grid cell (5 km-mesh) and then input into NICE as recharge rate at each time step.

3. DATA AND BOUNDARY CONDITIONS FOR SIMULATION

3.1. Meteorological Forcing Data

Six-hour reanalysed data of downward short- and long-wave radiation, precipitation, atmospheric pressure, air temperature, air humidity, wind speed at a reference level, *FPAR* and *LAI*, interpolated from the ISLSCP (International Satellite Land Surface Climatology Project) data with a resolution of 1° × 1° (Sellers, *et al.*, 1996), were input into each grid cell by interpolating these parameters in inverse proportion to the distance back-calculated in each grid. Daily weather data for 1987–1988 from the Luancheng Experimental Station of the Chinese Academy of Sciences, a station of the Chinese Ecological Research Network, and monthly average weather data from several meteorological stations were used to calibrate the data. The ISLSCP precipitation data had the least reliability, but maximum and minimum temperature and radiation were relative reliable. Therefore, at most sites where monthly total precipitation was available throughout 1987–1988, daily precipitation was corrected in accordance with the ratio of the monthly total precipitation of the ISLSCP data to the measured total precipitation in the corresponding month (see Figure 3). Those corrected sites cover the whole study area where groundwater fluctuation was influenced by agricultural water use. The correction is important because slight changes in precipitation will result in large variations in crop water requirement and, therefore, are directly linked to the fluctuation of groundwater level. Because recharge rate data in natural regions are scarce (except for research using the chloride mass balance method (Lee, 1996), these values were simulated by the SiB2 submodel of NICE at each time step and were input into the groundwater submodel. Details are given in Nakayama and Watanabe, 2004).

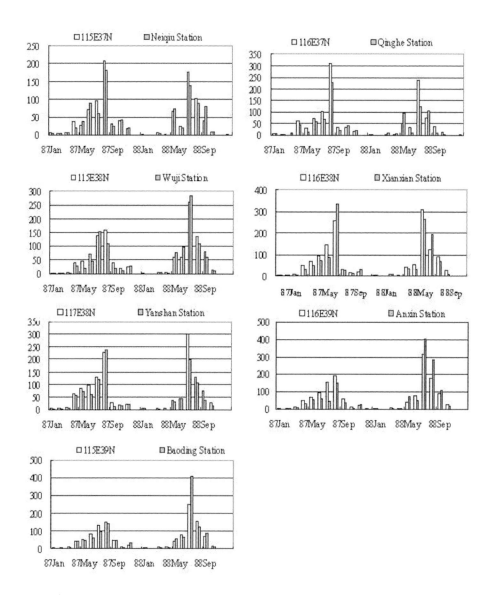

Figure 3. Comparison between ISLSCP precipitation data and measured month precipitation data.

3.2. Vegetation, Soil, and Geological Properties

The mean elevation of each 5-km grid cell was calculated by using the spatial average of a global digital elevation model (DEM; GTOPO30) with a horizontal grid spacing of 30 arc-seconds (approximately 1-km mesh) (U.S. Geological Survey, 1996) throughout the North China Plain, including the Hai River and the lower reach of the Yellow River (Figure 1). About 50 vegetation and soil parameters were calculated in each cell on the basis of vegetation class and soil texture obtained from the Vegetation and Soil Maps of China (1:400 000 000) (Chinese Academy of Sciences, 1988). The major parameters include vegetation cover, green fraction, albedo, surface roughness length and zero displacement height, soil conductivity and soil water potential at saturation, and some parameters of stomatal resistance that relate to environmental factors. Nine soil profiles representing soil texture changes across the study region were chosen and used in the simulation (Table 1). This information comes originally from Zhu *et al.* (1995) and includes soil texture, soil organic matter, pH and soil nutrient content in different soil layers down to 150 cm from the surface. Hydrologic parameters of the soil layers, such as saturation water content, lowest water content for crop growth, upper water content after free drainage and saturated water conductivity, were calculated from the input soil texture and soil organic matter by the model itself.

Table 1. Location of soil profiles used for DSSAT simulation

Sites and Soil type	Longitude	Latitude	Local name	Representing area.	Basic texture description
Kaifeng	114.5067	34.8085	Chao rang tu	Allusive plain of the Yellow river in Henan	Sandy loam
Dezhou	116.3900	35.9818	Meng yu qing qai tu	Allusive plain of the Yellow river in Shandong	Sandy loam with 30 cm of clay layer.
Huimin	117.465	37.5636	Qing bai tu	Allusive plain of the Yellow river near the coast	Sandy loam with 40 cm of clay layer.
Jiaohe	116.5909	38.1081	Er he tu	Riverside soil type in Hai river catchment	Clay soil with sandy layers.
Longyao	114.7841	37.3513	Dao meng jin tu	Piedmont region in the south part of Hai River catchment	Sandy clay loam
Guangzong	115.2273	37.0900	Gangmian Sha tu	Allusive plain of the old Yellow River	Sandy loam

Table 1. (Continued)

Sites and Soil type	Longitude	Latitude	Local name	Representing area.	Basic texture description
Wuyi	115.8636	37.8108	Qing Ru Rang Tu	Low plain region of the Hebei Plain	Sandy loam with slight saline
Mengcun	117.1136	38.0631	Zhong Ru Er He Tu.	Coast region of the Hebei Plain	Sandy saline loam
Luancheng	114.6667	36.8830	Chao Tu.	Piedmont of the Taihang Mountains.	Loam with a clay layer.

The geology around these catchments consists mainly of diluvium in the mountain areas (Taihang Mountains, Yanshan Mountains and piedmont plains) and alluvium in the lowland and coast areas (fan, flood and coastal plains). After calibration for fitting simulated hydraulic heads to observed heads, by three-dimensionally interpolating the scanned and digitized geological material in the horizontal plane and boring data in the vertical direction (Geological Atlas of China, 2002), the author could divide geological structure into four types with a resolution of 5-km in the horizontal direction and with 20 layers in the vertical direction on the basis of hydraulic conductivity (K_h and K_v), the specific storage of porous material (S_s) and specific yield (S_y) (Table 2). These parameters are in the range of previous researches at the study area such as $K_h = 7.75 \times 10^{-2} \sim 6.22 \times 10^{-1}$ (m/hour) (Chen et al., 2002), $S_y = 0.15$ (Chen et al., 2003), and others (Shimada, 2000). It can be seen that the Quaternary Formation in the Hebei Plain is deep, generally reaching 400–600 m in depth, and is divided into four aquifers with unconfined groundwater in the first layer and confined groundwater in the second, third and fourth layers at 38 °N (Figure 4a). The thickness of the respective aquifers is 20–40, 60–130, 80–220 and 50–350 m, and all aquifers consist of sand, gravel, fine sand and silt in the previous researches (Zhu et al., 1992, 1995; Shimada, 2000; Chen et al., 2002) (Figure 4b). The geological structures discretized in this study are very similar to those reported previously studies, showing that the geological structures the author categorized follow a real lithology and are sufficient for this simulation.

Table 2. Geological parameters used in numerical simulation

Type	Soil Type	Horizontal Hydraulic Conductivity Kh (m/hour)	Vertical Hydraulic Conductivity Kv (m/hour)	Specific Storage Ss (1/m)	Specific Yield Sy
1	gravel	5.0E+01	1.0E+01	1.0E-05	1.5E-01
2	fine to medium sand	5.0E+00	1.0E+00	1.0E-04	1.5E-01
3	finer than silt	5.0E-01	1.0E-01	1.0E-03	8.0E-02
4	rigid base	5.0E-01	1.0E-01	1.0E-03	4.0E-02

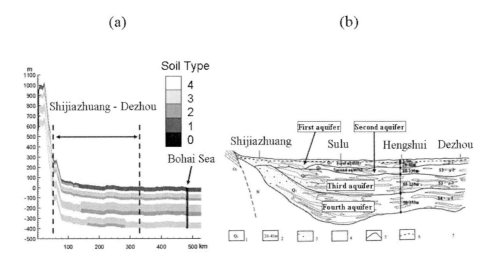

Figure 4. Vertical geological structure at the latitude of 38°N in the Hebei Plain, (a) in the simulation, and (b) in the previous research (Zhu et al., 1995). In the Fig. 4a, the soil type shows the same value at the Table 2. Soil type=0 shows the interface between the sea and land. Fig. 4b shows the cross sectional profile of the Hebei Plain between Shijiazhuang (38.07°N, 114.55°E) and Dezhou (37.50°N, 116.30°E). 1; stratum age, 2; stratum thickness, 3; gravel and boulder, 4; clay and sandy clay, 5; aquifer boundary, 6; water table.

3.3. Water Use in Agriculture

In the North China Plain, in particular, the Hebei Plain, almost all irrigation water is pumped up from underground; some industrial and domestic use water comes from other deep aquifers. Most of the agricultural land (Figure 1) in this area is covered by cultivated fields (wheat and maize; effective irrigation area is about 46.4 %) (Governmental Office of Hebei Province, 1988; Governmental Office of Hebei Province, 1989), where groundwater is commonly used. The groundwater withdrawals for agricultural use were input into the groundwater submodel in the following three cases (Figure 5).

Case 1 (Figure 5a): Water use given by statistical data and previous research. The statistical data of annual groundwater-use data for agriculture at each county of Hebei Province (Governmental Office of Hebei Province, 1988; Governmental Office of Hebei Province, 1989; Hebei Department of Water Conservancy, 1987; Hebei Department of Water Conservancy, 1988) was expanded spatially to all the simulation area by multiplying the weighting factor between this statistical data and annual groundwater-use data for agriculture in the North China Plain (1:4 500 000) (Institute of Hydrogeology and Engineering Geology of the National Geology Bureau, 1980) at the Hebei Province, and then digitized to a 5-km mesh of the simulation area (Figure 6a); the annual-average water use for agriculture is about 92.1 % of total water use in this area (Hebei Department of Water Conservancy, 1987; Hebei Department of Water Conservancy, 1988). Annual time-series data of groundwater use by agriculture per unit area were estimated from previous studies (Wang *et al.*, 2001; Liu *et al.*, 2002; Kendy *et al.*, 2003) (Figure 6b). Maize is grown in the summer (from June to September) and wheat in the winter (from October to June).

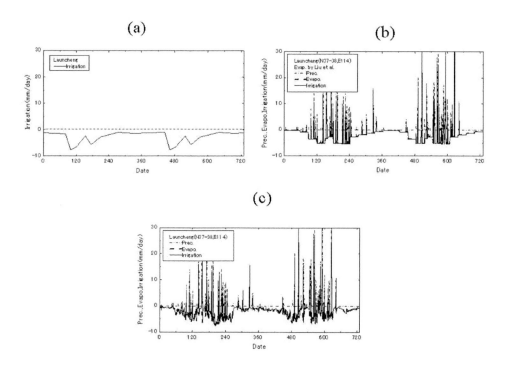

Figure 5. Comparison of additional water volume at agricultural field in Launcheng County; (a) annual-averaged irrigation estimated from the previous observation (Kendy et al., 2003), (b) estimated value by using the monthly evapotranspiration (Liu et al., 2002), and (c) estimated value by using the simulated result of the DSSAT. Dash-dotted line; precipitation, dashed line; evaporation, solid line; water use in irrigation.

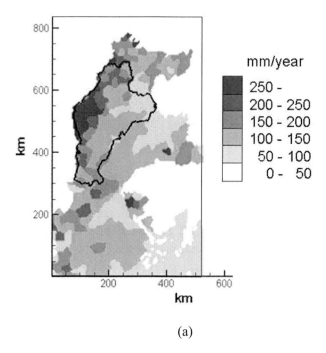

(a)

Figure 6. Continued on next page

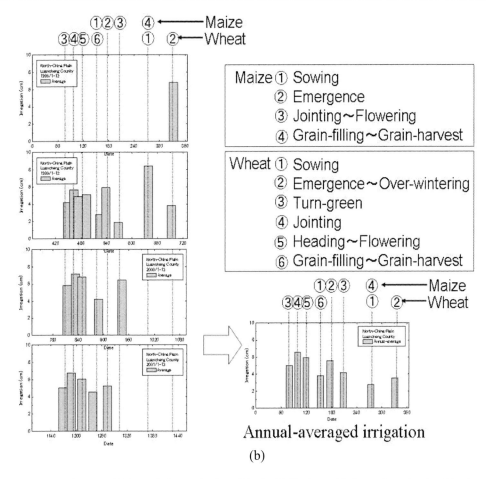

Figure 6. Groundwater withdrawals for agricultural in the NCP about (a) Annual total value (Institute of Hydrogeology and Engineering Geology of the National Geology Bureau, 1980) digitized to 5 km mesh, and (b) Annual time-series of groundwater use for agriculture per unit area estimated from the previous study (Wang et al., 2001; Liu et al., 2002; Kendy et al., 2003). In Fig. 6a, the black line is the border of Hebei Plain.

The irrigation rates for four consecutive years in Luancheng County, Hebei Province, at 16 sites (Kendy et al., 2003) were scanned and averaged to estimate the time-series of annual-averaged irrigation rate. There are generally three or four peaks of water withdrawal, corresponding to the growth stages the two crops. The estimated additional water in case 1 is useful on the assumption that the annual trend of water use for agriculture is similar both throughout the North China Plain and in different years.

Case 2 (Figure 5b): Water use estimated from monthly evapotranspiration. Additional water use was calculated in each cell from previous observation data (Liu et al., 2002) of monthly evapotranspiration (Table 3); the precipitation data in each cell were used in equation (14) on the assumption that the monthly evapotranspiration is similar in different years and that the daily evapotranspiration is similar within a month. The estimated additional water in case 2 is more accurate than that in case 1 because case 2 used the precipitation data during the full simulation from 1987 to 1988, and the monthly evapotranspiration is more constant than the water irrigation volume among different years.

Table 3. Previous observation data about the monthly evapotranspiration during the winter wheat and summer maize periods (Liu *et al.*, 2002)

Crop Type	Month	Cropping Practice	Evapotranspiration (mm)
winter wheat	October	Sowing~Emergence	59.2
winter wheat	November	Over-wintering	36.9
winter wheat	December	–	19.5
winter wheat	January	–	8.2
winter wheat	February	–	8.1
winter wheat	March	Turn green	28.3
winter wheat	April	Jointing	104.8
winter wheat	May	Heading~Flowering~Grain-filling	163.5
winter wheat	June	Grain-harvest	33.3
summer maize	June (11-30)	Sowing~Emergence	56.2
summer maize	July	Jointing	159.7
summer maize	August	Flowering~Grain-filling	165.0
summer maize	September (1-20)	Grain-harvest	53.3

Case 3 (Figure 5c): Water use estimated by DSSAT. Automatic irrigation was chosen to simulate the theoretical water requirement by the two crops models in the North China Plain to the north of the Yellow River. The additional water volume used in agricultural fields was estimated by using equation (14) from the DSSAT results in each county in the Hebei Plain. The interpolated value was applied to all cells in the simulation area on the assumption that the theoretical deficit water is pumped up from the groundwater layer in agricultural periods. The estimated additional water use has a higher temporal resolution than that in case 2, because DSSAT can simulate daily evapotranspiration. Before the application of the DSSAT results to NICE-AGR, experimental data were used to calibrate the model in order to confirm the correctness of the DSSAT simulation.

3.4. Boundary conditions

There were less available data about the observed groundwater level at the top of the mountains (upstream boundaries) in the study area because most of the areas are natural regions such as forest and bare land and non-irrigated field (Figure 1b), and because the groundwater level is very deep from the ground surface. Therefore, at the upstream boundaries where there are no observed data, the reflecting condition on the hydraulic head was used for the groundwater flow submodel on the supposition that the groundwater flow forms a watershed at the ridge in the same way as surface hydrology (Nakayama, 2008a, 2008b, 2008c, 2009; Nakayama and Watanabe, 2004, 2006a, 2006b, 2008a, 2008b, 2008c; Nakayama et al., 2006, 2007). At the eastern sea boundary (Bohai Sea), constant head was set at 0 m. Though there are some cone depressions formed in deep groundwater by the industrial water use and irrigation on the coast of the sea as shown later in Figure 8, the groundwater level theoretically rises up to the sea level near the sea. The hydraulic head values parallel to the ground level were input as the initial conditions for the groundwater flow submodel. In river cells (1747 cells) including the Yellow, Dugai (Tuhai), Majia, Fuyang, Hutuo, Ziya, Daqing, Yongding and Hai (Haihe) rivers, inflows or outflows from the riverbeds were considered depending on the difference of the hydraulic heads of groundwater and river. The time-series of observed river discharge at Zhengzhou, Henan Province (34°49'12"N, 113°40'00"E, mean elevation 122 m), beside the lower reach of the Yellow River were also input into the model (China Institute for Geo-Environmental Monitoring, 2003).

3.5. Running the Simulation and Validation

The simulation area is 530 km wide by 840 km long in the Albers (WGS 1984) co-ordinates, covering almost all of the Hai River catchment and the lower reach of the Yellow River. This area is discretized into a grid of 106 × 168 blocks with a grid spacing of 5 km in the horizontal direction and into 20 layers with a weighting factor of 1.1 layers (finer at the upper layers) in the vertical direction. The upper layer was set at 2 m depth, and the 20th layer was defined as an elevation of –400 m from the sea surface. The NICE-AGR simulation was conducted on an NEC SX-6 supercomputer. Simulations were performed for 1 January 1987 to 31 December 1988. In the agricultural fields, DSSAT simulated the winter wheat and maize rotations automatically following the sequence of 1987 winter wheat, 1987 maize, 1988 winter wheat and 1988 maize. Automatic irrigation was chosen to simulate the theoretical water requirement by the two crops. This method calculates the theoretical water requirements at different sites on the basis of the differences in local weather conditions, especially precipitation, and soil conditions. To achieve realistic crop water use, the author considered that the ratios of agricultural area to total area and the ratios of irrigated area to total agricultural area from statistics books (Governmental Office of Hebei Province, 1988; Governmental Office of Hebei Province, 1989; Hebei Department of Water Conservancy, 1987; Hebei Department of Water Conservancy, 1988) would give a practical spatial water use and included the irrigated ratio at each grid in order to input the more correct water use into the NICE-AGR.

The first 6 months were used as a warm-up period until equilibrium water levels were reached, and parameters were estimated by comparison of simulated steady-state values in a steady-state condition with observed values published in the literature (Clapp and Hornberger, 1978; Rawls et al., 1982). A time step of $\Delta t = 6$ h was used. Previous observed data (5-day interval, 13 points of groundwater level, G-1 – G-13; China Institute for Geo-Environmental Monitoring (2003) and the observed data (seasonal data, 18 points of groundwater level, G-14 – G-31) were used for validation of the simulations (Table 4). Three cases (section about water use in agriculture) were simulated in order to evaluate the effect of irrigation on the groundwater dynamics.

Table 4. Lists of groundwater observation stations of China Institute for Geo-Environmental Monitoring (2003) (G-1 – G-13) and the stations we measured (G-14 – G-31)

No.	Point Name	Lat.	Lon.	Elev. (m)
G-1	Luancheng	37.90	114.56	56.44
G-2	Handan City	36.60	114.41	56.34
G-3	Cangzhou	38.30	116.90	8.28
G-4	Daming	36.30	115.30	43.38
G-5	Nangong	37.20	115.60	30.34
G-6	Guangzong	36.90	115.20	34.84
G-7	Wuji	38.20	114.90	47.50
G-8	Rongcheng	39.00	116.00	10.96
G-9	Langfang	39.50	116.70	13.19
G-10	Bazhou	39.10	116.60	4.53
G-11	Xianxian	38.10	116.20	13.61
G-12	Huanghua	38.30	117.60	5.68
G-13	Shenzhou	37.90	115.70	20.79
G-14	Biaoling	38.02	114.78	53.71
G-15	Liuhaizhuang	37.91	114.90	42.04
G-16	Xiaochang'an	37.94	114.94	45.11
G-17	Liujiazhuang	37.91	114.90	45.76
G-18	Jiashizhuang	37.87	114.95	40.51
G-19	Shunzhong	37.98	114.82	48.8
G-20	Meihua	37.88	114.82	44.8
G-21	Datong	38.01	114.72	55.82
G-22	Gangshang	38.05	114.69	60.12
G-23	Liyang	37.97	114.72	53.03
G-24	Nandong	38.10	114.76	57.05
G-25	Zhidu	38.14	114.71	64
G-26	Hanjiawa	38.15	114.77	58.48
G-27	Nanwa	38.09	114.80	56.8
G-28	Nanmeng	38.19	114.78	58.43
G-29	Zengcun	38.24	114.74	62.41
G-30	Qianxiguan	38.27	114.80	58.76
G-31	Xiaoguozhuang	38.29	114.74	65.25

3.6. Calibration of DSSAT Models

Since there has been rapid development of crop cultivars in China, for the study of water use in 1987–1988, the author used experimental data from Luancheng Experimental Station, a long-term experimental base of the Chinese Ecological Research Network in Luancheng County, Hebei Province (Figure 1), for winter wheat in 1991–1992 and for maize in 1990 to develop genetic parameters in order to get good simulation of agricultural water use in 1987–1988.

In the field wheat experiment, winter wheat was planted on 10 October 1991 and harvested on 8 June 1992. The cultivar was Jimai 28, which was once widely planted. Detailed phenology and growth data such as *LAI* and stem and leaf biomass were recorded. Data entered into the model included meteorological data, experimental information, site soil conditions and initial soil moisture on 1 October 1991, which was treated as the starting point of the simulation. Daily meteorological data such as maximum and minimum daily air temperature, rainfall and solar radiation were input. Soil information including texture, organic matter content, soil fertility and pH was collected from long-term historical records. Wheat growth and phenological data were used to adjust genetic parameters responsible for phenological stages, growth rate and yield.

In the field maize experiment, maize was planted on 13 June 1990, matured on 12 September, and was harvested on 15 September. The cultivar was Luyu 13. Detailed phenology and growth data such as *LAI* and stem and leaf biomass and seasonal changes of soil moisture were recorded. Initial soil moisture on 13 June 1990 was treated as the starting point of the simulation. Maize growth and phenological data were used to adjust genetic parameters responsible for phenological stages, growth rate and yield.

4. RESULT

4.1. Calibration of Phenology Simulated by DSSAT Models

Table 5 shows the adjusted parameters from the field experiments from 1990 to 1992. Table 6 compares measured and simulated phenological date, yield and maximum *LAI* of the two crops after adjustment of genetic parameters. The results suggest that DSSAT-wheat and DSSAT-maize perfectly simulated phenological development, while keeping the maximum growth of *LAI* and yield very similar to the field measurements.

Figure 7 compares measured and simulated *LAI* and measured and simulated soil moisture content. *LAI* is a main factor controlling evapotranspiration and is important in energy partitioning in maize than wheat (Shen *et al.*, 2004). Winter wheat in the North China Plain has a much longer growing season than maize. The wheat *LAI* stays almost constant during the winter then starts to increase in spring (Wang *et al.*, 2001; Zhang *et al.*, 2002). The simulation reproduces this characteristic well, showing good consistency between the simulated growth pattern of *LAI* and the growth of the field crops ($r^2 = 0.95$ and $P_r < 0.01$ for winter wheat; $r^2 = 0.99$ and $P_r < 0.01$ for maize, where r^2 is the correlation between the simulated *LAI* and the observed value, and P_r is the probability of error), and soil moisture, a factor indicating crop water use, is very similar to measured soil moisture (no significant difference in paired test).

Table 5. Genetic parameters obtained from fitting DSSAT-wheat and DSSAT-maize processes in simulating the growth and yield of winter wheat in 1991-1992 growing season and maize in 1990

Parameters for winter wheat							
Parameters	P1V	P1D	P5	G1	G2	G3	PHINT
Value	1.0	5.0	-5.9	5.5	5.5	2.9	60.0
Parameters for maize							
Parameters	P1	P2	P5	G2	G3	PHINT	
Value	320	0.30	620	720	11.0	45.0	

Table 6. Comparison between simulated phenological and growth data by DSSAT-wheat and DSSAT-maize models and measured phenological and growth data from field experiment of winter wheat in 1991-1992 and maize in 1990

	Wheat		Maize	
	Measured	Simulated	Measured	Simulated
Flowering date (dap)	206	207	56	59
Physiological maturity date (dap)	240	240	93	94
Yield (kg/ha; dry)	5707	5869	7371	7388
Maximum LAI (m2/m2)	4.92	4.14	5.21	4.64

"dap" = days after planting

Furthermore, the simulated soil moisture in an agricultural field compared well to observed values from the Global Soil Moisture Data Bank (Entin, *et al.*, 2000; Robock *et al.*, 2000) (data not shown). Simulated soil moisture reproduces the observed values and fluctuates in response to changes in precipitation, evaporation and in particular, irrigation; and the redistribution increases and the influence of meteorological forcing is smoothed in deeper layers, which are characteristic of irrigation fields in a semi-arid region.

4.2. Evaluation of crop water use simulated by DSSAT-wheat and DSSAT-maize

The crop coefficient (K_c) is often used for estimating spatial crop water use (Chen *et al.*, 1995). To determine the reliability of the crop water use simulated by DSSAT, actual evaporation calculated by DSSAT was divided by potential evaporation calculated using the method of Priestly and Taylor (1972) to give K_c. Table 7 shows how K_c changes spatially in both crops in 1987–1988.

The crop coefficient of winter wheat is generally higher in the southern part of the region and lower in northern part. This is consistent with the study of Chen *et al.* (1995), which showed variation in K_c from 1.06 in the southern part of Henan Province to 0.86 in Hebei Province, though the value of K_c from DSSAT is slightly lower at the southern end of the study region.

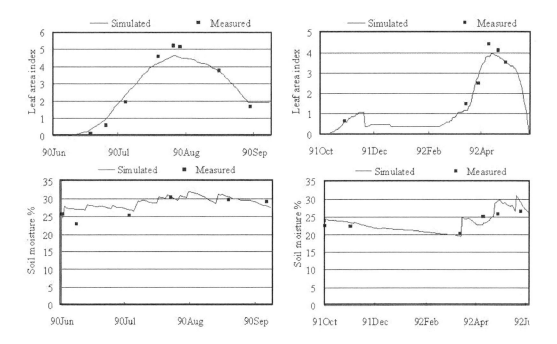

Figure 7. Comparison between simulated soil water content and field measured soil water storage in 0-160 cm soil profile and comparison between measured leaf area index growth and simulated leaf area index growth in winter wheat in 1991-1992 growing season and maize in 1990 growing season. The left two figures are for maize in 1990 and the right two figures are for winter wheat in 1991-1992 seasons.

Table 7. Crop coefficient (K_c) for winter wheat and maize in 1987-1988 in some sites from DSSAT simulation

Locations	87_wheat	87_maize	88_wheat	88_maize
114 E – 35 N	0.89	0.97	0.92	0.99
115 E – 35 N	0.87	0.94	0.92	0.97
116 E – 35 N	0.88	0.94	0.93	0.98
115 E – 36 N	0.89	1.02	0.93	1.01
116 E – 36 N	0.89	0.99	0.93	1.00
115 E – 37 N	0.88	1.04	0.88	1.02
116 E – 37 N	0.88	0.99	0.9	1.00
117 E – 37 N	0.81	0.98	0.85	0.99
118 E – 37 N	0.83	0.89	0.85	0.95
115 E – 38 N	0.84	1.03	0.83	1.13
116 E – 38 N	0.84	1.03	0.87	1.07
117 E – 38 N	0.83	0.98	0.89	0.98
116 E – 39 N	0.81	1.03	0.83	1.09
117 E – 39 N	0.87	1.06	0.89	1.04
Average	0.86	0.99	0.89	1.02

For maize, Chen *et al.* (1995) showed a general trend of higher K_c in the southern part of the region, with a highest value of 1.14 near the site, 114°E and 35°N, and a lower value in the northern part, with a lowest value of 0.84 near 116°E and 39°N. K_c calculated by DSSAT fell within the range reported by Chen *et al.* (1995), although the author did not find a clear trend. By means of a large weighing lysimeter, Liu *et al.* (2002) studied the evapotranspiration of the two crops continuously from 1995 to 2000 at Luancheng Station, the station used for the calibration of DSSAT in the study. Their study concluded that the average value of K_c was 0.93 for winter wheat and 1.02 for maize, both of which are near to the simulated values of 0.88 for winter wheat and 1.01 for maize. Those studies suggested the relative reliability of the water use simulated by the two DSSAT models.

4.3. Steady Simulation of Groundwater Dynamics

Figure 8 shows simulated and observed annual-averaged groundwater levels in the Hebei Plain (Shimada, 2000). The groundwater level has decreased year by year in the Hebei Plain (decreased by 20 to 50 m during the last 30 years), and most rapidly around cultivated fields. The groundwater level was only 3 m under the surface and the groundwater welled naturally at some places in 1959. The groundwater was basically balanced apart from a few cone depressions around the bigger cities. So the author assumed zero pumping in 1959 in the simulation. Because statistical data on water use in 1987 to 1988 are available (Governmental Office of Hebei Province, 1988; Governmental Office of Hebei Province, 1989; Hebei Department of Water Conservancy, 1987; Hebei Department of Water Conservancy, 1988), the second-order curve of year was estimated from these two periods between 1959 and 1987-1988 as though it was a curve of economic conditions (for example, GDP), and then the interpolated values were applied to 1975 and 1992 in this study. The meteorological data of averaged value from 1987 to 1988 was used at the simulation of each year because the effect of meteorological data on the groundwater dynamics is smaller than that of water use change.

The steady simulation reproduces the groundwater level of the observed values excellently, because the simulation includes the effect of irrigation in the Hebei Plain. Hydraulic gradients are steeper around the Taihang and Yanshan Mountains and are relatively flat in the plain area. Because the groundwater level is higher in the mountainous areas, groundwater constantly flows into the plain, and the mountainous area plays an important role as a recharge region. Furthermore, cone depressions occur around the bigger cities (Cangzhou, Hengshui, Baoding, *etc.*) owing to excessive groundwater use in 1975 and 1992. The simulation cannot reproduce these cone depressions because the numerical model does not explicitly include the processes of industry and domestic water use.

4.4. Seasonal Variations of Groundwater Dynamics Affected by Irrigation

Temporal variations in groundwater fluctuations in the Hebei Plain from 1 January 1987 to 31 December 1988 are plotted against the observed values (China Institute for Geo-Environmental Monitoring, 2003) in Figure 9. Generally, the groundwater level is drawn down from March to June–July mainly by winter wheat production, and then starts to recover as the rainy season starts.

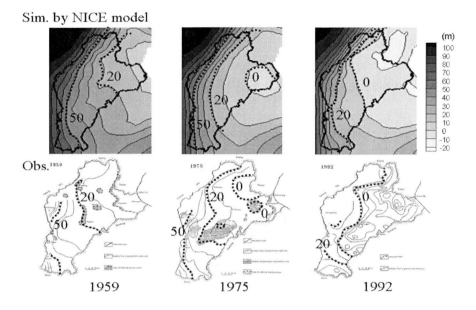

Figure 8. Simulated annually-averaged groundwater levels in the Hebei Plain from 1959 to 1992 together with the observed values (Shimada, 2000). Red color corresponds with higher value and blue color with lower value. Bold black line shows the bolder of the Hebei Plain. Bold red line, pink line, and blue line show the groundwater levels 0 m, 20 m, and 50 m above sea level, respectively.

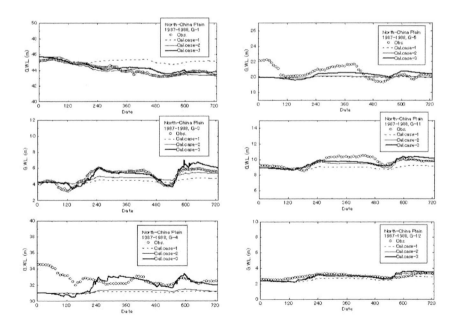

Figure 9. Temporal variations of groundwater fluctuations in the Hebei Plain from 1 January 1987 to 31 December 1988. Circle; observed value by the Chinese Academy of Sciences (1987–1988), dashed line; simulated value of case-1, solid line; simulated value of case-2, and bold line; simulated value of case-3.

The level varies less in the coastal area (G-12 in Figure 1a and Table IV), because the salty groundwater cannot be used for irrigation (Figure 6a). There were two- types of variation in groundwater level during the two years: (I) the groundwater level declines continually (G-1, 4, 5), and (II) the level declines only during the agricultural periods (G-3, 11, 12). Type I is predominant around the piedmont and mountainous areas, where only groundwater is used for irrigation. Type II occurs on the alluvial and coastal plain areas, because some of the groundwater is recharged from the rivers and canals. However, these characteristics are applicable only from 1 January 1987 to 31 December 1988, because the groundwater level has been decreasing continually since 1978 at an average rate of about 0.7 m/year, with the exception of two heavy rainfall years, 1988 and 1996, in the North China Plain (Chen *et al.*, 2003). Case 3 reproduces observations much better than cases 1 and 2, because case 3 can simulate both infiltration in non-agricultural periods and pumping in agricultural periods. The simulated groundwater levels reproduce actual levels very accurately in both the mountainous and plain areas, because the author used a three-dimensional groundwater submodel by the equation (9) and because SiB2 (Simple Biosphere model 2) for natural regions (Nakayama and Watanabe, 2004) and DSSAT for agricultural regions can correctly simulate the recharge rates in the upper layer (equations (13) and (14)).

The simulated seasonally averaged groundwater levels from 1 January 1987 to 31 December 1988 around the plain area (Figure 10) were greatly affected by the local topography, and fluctuated more where more groundwater is used for irrigation (Figure 6a).

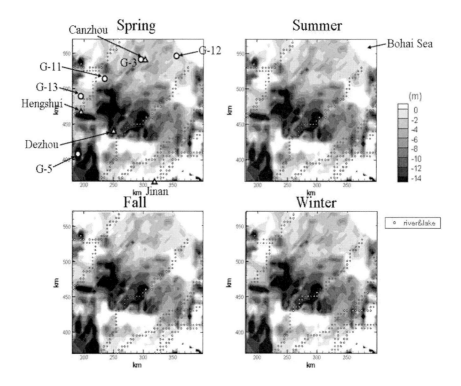

Figure 10. Simulated results of seasonally-averaged groundwater levels from the ground surface from 1 January 1987 to 31 December 1988 around the plain area of the Hebei Plain (X=180-400km, Y=370-570km in Fig. 1a). White circle is the groundwater observation points (Fig. 1a and Table 4), and the dot is the river and lake cells in the simulation.

The annually averaged groundwater levels in 1987 and 1988 are almost similar, but a modest increase in groundwater levels was due to heavy rainfall in 1988 (data not shown). The groundwater levels in 1987 and 1988 lie between those of 1975 and 1992 in Figure 8, which indicates that the unsteady simulation can reproduce the degradation of groundwater level over the previous half-century due to over-pumping of groundwater for irrigation in the Hebei Plain. The figure shows that the groundwater is deepest in spring (Chen *et al.*, 2002) and that this effect is predominant near the plain area because most of the groundwater withdrawal is used for winter wheat production (Figure 10). The groundwater levels start to rise as the rainy season starts in July. Therefore, it is very effective to reduce the area planed with winter wheat rather than summer maize or to improve the method of water withdrawal in order to stop the decline of groundwater levels in this region.

5. DISCUSSIONS

NICE was coupled to DSSAT to include the effect of water use in agricultural fields in the North China Plain, and to include mechanisms of vegetation–water relations and surface-unsaturated–saturated water processes (NICE-AGR). The seasonal variations in water use for the two crops were implicitly simulated as an input to the groundwater flow submodel. This method is very powerful for evaluating the water dynamics of catchments with little statistical data on seasonal water use.

Although the overall simulated values reproduced the observed values excellently, some discrepancies existed. There are some differences between the observed and simulated data in the maximum *LAI* and the crop yield (Figure 7 and Table 7). Although this is mainly caused by the inaccuracy of the observed values, the model has some imperfections. Although DSSAT ignored the nutrient deficit, there are in fact large nutrient deficits in the Hebei Plain. This is obvious, considering that the groundwater is highly contaminated by nutrients that are not absorbed by the crops and thus infiltrate into the aquifer. Therefore, it is necessary to include the nutrient cycle in the simulation for better reproduction. This also depends on whether the author can obtain more accurate statistical data on fertilizer inputs.

The unsaturated-layer simulation could not reproduce the rapid change of soil moisture in the vertical direction at higher precipitation (Figure 7). This occurred because the porosity and the hydraulic conductivity were treated as constants in the vertical direction in the unsaturated layer in this model, even though the soil texture changes with depth (Yu *et al.*, 2001). The rapid change in soil moisture can be reproduced better when the vegetation structure, including the root depth and density, is better parameterized. In the groundwater flow model, the rapid change of groundwater level is difficult to reproduce in the same way as soil moisture, because correct recharge rates from the unsaturated layer are not attainable by observation (Nakayama and Watanabe, 2004). For better recharge rate simulation, the unsaturated layers must be more correctly simulated and the horizontal resolution of grid size must be finer than the 5km-mesh. More groundwater level data are necessary near the mountainous areas because boundary conditions for groundwater levels are difficult to input correctly, except with constant head values at lake or sea level.

Furthermore, the simulation could not reproduce the groundwater level near the mountainous regions and some irrigated fields even in case 3 (Figure 9). This is because much more groundwater is used for irrigation than the theoretical water requirement of equation (14) in these regions. In this simulation, the ratio of agricultural area to total area and the ratio of irrigated area to total agricultural area are included at each grid cell (Governmental Office of Hebei Province, 1988; Governmental Office of Hebei Province, 1989; Hebei Department of Water Conservancy, 1987; Hebei Department of Water Conservancy, 1988). So, the finer mesh is necessary in order to reproduce the groundwater dynamics more correctly. In addition to the case that more water is pumped up from the groundwater, a part of water is in reality irrigated from the river or dam in different conditions such as crops and seasons, *et al*. Furthermore, it is necessary to input detailed data on domestic and industrial groundwater use in the bigger cities into NICE-AGR or to couple a water use model or inventory model in the cities in NICE-AGR in order to reproduce the cone depressions around the bigger cities (Shimada, 2000) (Figure 8). It is further necessary to include the water transfer to the bigger cities such as Beijing and Tianjin supplied by some dams of cutting up the rivers in the upper region. The more excellent reproduction of the simulation will need to get more data in the near future because the water dynamics are greatly affected by the artificial constructions, for example, the groundwater and the river discharge declines at the lower area of big dam.

In the study area, the groundwater has lowered dramatically over the previous half century owing to over-pumping of groundwater. So seawater intrusion into the aquifers has a greater impact on agriculture near the coastal regions, because salty water cannot be used for agriculture or domestic use. The tidal range in the Bohai Sea (the tidal mode up to the 4th degree: $M_2+S_2+K_1+O_1$) is about 1 m at the mouth of Yellow River and 4 m outside of Bohai Sea (Ogura, 1933), which affects greatly this seawater intrusion, in particular, at the spring tide (Shimada *et al.*, 2002). So if the area of seawater intrusion can be evaluated both qualitatively and quantitatively, the people can use the shallow and deep aquifers more efficiently to prevent the groundwater degradation in this area. To estimate the area of seawater intrusion in order to take measures against it, it will be necessary to include the density current and solute transport process of groundwater near the coastal regions in the model.

6. CONCLUSIONS

The author simulated the effects of irrigation on groundwater flow dynamics in the North China Plain by coupling the NIES Integrated Catchment-based Eco-hydrology (NICE) model with DSSAT agricultural models. This combined model (NICE-AGR) was applied to the Hai River catchment and the lower reach of the Yellow River at a resolution of 5 km. The simulation reproduced excellently the soil moisture, *LAI*, evapotranspiration and crop production of summer maize and winter wheat, and therefore, the spatial distribution of crop water use was reasonably estimated at daily steps in the simulation area. In particular, coupling NICE with the agricultural models (case 3) reproduced groundwater levels better than the use of statistical water use data (case 1 and 2). This indicates that the combined model does not need detailed statistical data on water use, making it very powerful for evaluating and estimating the water dynamics of catchments with little statistical data on seasonal water use. Furthermore, the simulation reproduced the spatial distribution of groundwater level in 1987 and 1988 in the Hebei Plain, showing the remarkable reduction of groundwater level in spring due mainly to over-pumping for irrigation.

ACKNOWLEDGMENTS

The author thanks Dr. M. Watanabe, Keio University, Japan, and Dr. Q. Wang, National Institute for Environmental Studies (NIES), Japan, for valuable comments about the study area. Some of the simulations were run on an NEC SX-6 supercomputer at the Center for Global Environment Research (CGER), NIES. The support of the Asia Pacific Environmental Innovation Strategy (APEIS) Project from the Japanese Ministry of Environment is also acknowledged.

REFERENCES

Adams, R.M., Rosenzweig, C., Peart, R.M., Ritchie, J.T., McCarl, B.A., Glyer, D., Curry, R.B., Jones, J.W., Boote, K.J., and Allen, L.H. (1990). Global climate change and US agriculture. *Nature* 345: 219-224.

Brown, L.R., and Halweil, B. (1998). China's water shortage could shake world food security. *World Watch* July/August:10-18.

Chen, J,Y,, Tang, C.Y., Sakura, Y., Kondoh, A., and Shen, Y.J. (2002). Groundwater flow and geochemistry in the lower reaches of the Yellow River: a case study in Shandang Province, China. *Hydrogeology Journal* 10: 587-599.

Chen, J.Y., Tang, C.Y., Sakura, Y., Kondoh, A., Shen, Y.J., and Song, X.F. (2004). Measurement and analysis of the redistribution of soil moisture and solutes in a corn field in the lower reaches of the Yellow River. *Hydrological Process*es 18: 2263-2273.

Chen, J.Y., Tang, C.Y., Shen, Y.J., Sakura, Y., Kondoh, A., and Shimada, J. (2003). Use of water balance calculation and tritium to examine the dropdown of groundwater table in the piedmont of the North China Plain (NCP). Environmental Geology 44: 564-571.

Chen, W. (1996). Preface. *Geomorphology* 18: 1-4.

Chen, Y.M., Guo, G.S., Wang, G.X., Kang, S.Z., Luo, H.B., and Zhang, D.Z. (1995). *Main crop water requirement and irrigation of China*. Beijing: Hydrologic and Electronic Press: Beijing; 73-102.

China Institute for Geo-Environmental Monitoring (CIGEM). (2003). China Geological Environment Infonet, *Database of groundwater observation in the People's Republic of China*, http://www.cigem.gov.cn.

Chinese Academy of Sciences. (1998). *Administrative division coding system of the People's Republic of China*. Chinese Academy of Sciences, Beijing.

Clapp, R.B., and Hornberger, G.M. (1978). Empirical equations for some soil hydraulic properties. *Water Resources Research* 14: 601-604.

Doll, P., and Siebert, S. (2002). Global modeling of irrigation water requirements. *Water Resources Research* 38: 8-1—8-10.

Entin, J.K., Robock, A., Vinnikov, K.Y., Hollinger, S.E., Liu, S., and Namkhai, A. (2000). Temporal and spatial scales of observed soil moisture variations in the extratropics. *Journal Geophysical Research* 105(D9): 11865-11877.

Geological Atlas of China. (2002). Geological Publisher: Beijing. (in Chinese).

Governmental Office of Hebei Province. (1988). *Hebei Agriculture Year Book in 1987.* Economy Science Publishing House: Beijing. (in Chinese).

Governmental Office of Hebei Province. (1989). *Hebei Agriculture Year Book in 1988.* Economy Science Publishing House: Beijing. (in Chinese).

Hebei Department of Water Conservancy. (1987). *Hebei year book of water conservancy for* Department of Hebei Water Conservancy. Hebei, China. (in Chinese).

Hebei Department of Water Conservancy. (1988). Hebei year book of water conservancy for 1988, Department of Hebei Water Conservancy. Hebei, China.

Institute of Hydrogeology and Engineering Geology of the National Geology Bureau. (1980). *Hydrogeological map for the People's Republic of China*. China Map Publishing House: Beijing. (in Chinese).

Jones, C.A., and Kiniry, J.R. (1986). *CERES-maize: A simulation model of maize growth and development*, Texas A and M University Press, Temple, TX. 193 pp.

Kendy, E., Pierre, G.-M., Walter, M.T., Zhang, Y., Liu, C., and Steenhuis, T.S. (2003). A soil-water-balance approach to quantify groundwater recharge from irrigated cropland in the North China Plain. *Hydrological Processes* 17: 2011-2031.

Lee, T.M. (1996). Hydrogeologic controls on the groundwater interactions with an acidic lake in karst terrain, Lake Barco, Florida. *Water Resources Research* 32: 831-844.

Liu, C., Zhang, X., and Zhang, Y. (2002). Determination of daily evapotranspiration of winter wheat and corn by large-scale weighting lysimeter and micro-lysimeter. *Agricultural and Forest Meteorology* 111: 109-120.

McVicar, T.R., Zhang, G.L., Bradford, A.S., Wang, H.X., Dawes, W.R., Zhang, L., and Li, L. (2002). Monitoring regional agricultural water use efficiency for Hebei Province on the North China Plain. *Australian Journal Agricultural Research* 53: 55-76.

Nakayama, T. (2009). Simulation of hydrologic and geologic changes affecting a shrinking mire. *River Research and Applications*, doi: 10.1002/rra.1253.

Nakayama, T. (2008a). Factors controlling vegetation succession in Kushiro Mire. *Ecological Modeling*, doi: 10.1016/j.ecolmodel.2008.02.017.

Nakayama, T. (2008b). Simulation of Ecosystem Degradation and its Application for Effective Policy-Making in Regional Scale, In River Pollution Research Progress, (Eds) Mattia N. Gallo and Marco H. Ferrari, Nova Science Publishers, Inc., Chapter 1, 1-89.

Nakayama, T. (2008c). Recovery of mire ecosystem by re-meandering of channelized rivers inflowing to Kushiro Mire. *Forest Ecology and Management*, 256, 1927-1938, doi: 10.1016/j.foreco.2008.07.017.

Nakayama, T., and Watanabe, M. (2004). Simulation of drying phenomena associated with vegetation change caused by invasion of alder (*Alnus japonica*) in Kushiro Mire. *Water Resources Research* 40(8): W08402.

Nakayama, T., and Watanabe, M. (2006a). Simulation of spring snowmelt runoff by considering micro-topography and phase changes in soil layer. *Hydrology and Earth System Sciences Discussions* 3:2101-2144.

Nakayama, T., and Watanabe, M. (2006b). Development of process-based NICE model and simulation of ecosystem dynamics in the catchment of East Asia (Part I), CGER's

Supercomputer Monograph Report, 11, NIES, 100p., http://www-cger.nies.go.jp/publication/I063/I063.html.

Nakayama, T., and Watanabe, M. (2008a). Missing role of groundwater in water and nutrient cycles in the shallow eutrophic Lake Kasumigaura, Japan. *Hydrological Processes* 22: 1150-1172, doi: 10.1002/hyp.6684.

Nakayama, T., and Watanabe, M. (2008b). Role of flood storage ability of lakes in the Changjiang River catchment. *Global and Planetary Change*, doi: 10.1016 j.gloplacha.2008.04.002.

Nakayama, T., and Watanabe, M. (2008c). Modelling the hydrologic cycle in a shallow eutrophic lake. *Verh. Internat. Verein. Limnol.* 30.

Nakayama, T., Yang, Y., Watanabe, M., and Zhang, X. (2006). Simulation of groundwater dynamics in North China Plain by coupled hydrology and agricultural models. *Hydrological Processes* 20(16): 3441-3466, doi: 10.1002/hyp.6142.

Nakayama, T., Watanabe, M., Tanji, K., and Morioka, T. (2007). Effect of underground urban structures on eutrophic coastal environment. *Science of the Total Environment* 373(1): 270-288, doi: 10.1016/j.scitotenv.2006.11.033.

Ogura, N. (1933). *A tide inshore Japan*, Japan Coast Guard's Report, 7, Japan Coast Guard, Japan (in Japanese).

Priestley, C.H.B., and Taylor, R.J. (1972). On the assessment of surface heat flux and evaporation using large-scale parameters. *Monthly Weather Review* 100: 81-92.

Rawls, W.J., Brakensiek, D.L., and Saxton, K.E. (1982). Estimation of soil water properties. *Transactions of the ASAE* 25: 1316-1320.

Ritchie, J.T. (1998). Soil Water Balance and Plant Water Stress. In *Understanding Options for Agricultural Production*, Tsuji GY, Hoogenboom G, Thornton PK (Eds), Great Bratain: Kluwer; 41-54.

Ritchie, J.T., Singh, U., Godwin, D.C., and Bowen, W.T. (1998). Cereal Growth, development and yield. In *Understanding Options for Agricultural Production*, Tsuji GY, Hoogenboom G, Thornton PK (Eds), Great Bratain: Kluwer; 79-98.

Robock, A., Konstantin, Y.V., Govindarajalu, S., Jared, K.E., Steven, E.H., Nina, A.S., Suxia, L., and Namkhai, A. (2000). *The global soil moisture data bank*. Bull American Meteorology Society 81: 1281-1299. http://climate.envsci.rutgers.edu/soil_moisture/.

Sellers, P.J., Randall, D.A., Collatz, G.J., Berry, J.A., Field, C.B., Dazlich, D.A., Zhang, C., Collelo, G.D., and Bounoua, L. (1996). A revised land surface prameterization (SiB2) for atmospheric GCMs. Part I : Model formulation, *Journal of Climate* 9: 676-705.

Sellers, P.J., et al. (1992). International Satellite Land Surface Climatology Project (ISLSCP), Initiative I, NASA,http://daac.gsfc.nasa.gov/CAMPAIGN_DOCS/ISLSCP/islscp_i1.html, CD-ROM.

Shen, Y., Kondoh, A., Tang, C., Zhang, Y., Chen, J., Li, W., Sakura, Y., Liu, C., Tanaka, T., and Shimada, J. (2002). Measurement and analysis of evapotranspiration and surface conductance of a wheat canopy. *Hydrological Processes* 16: 2173-2187.

Shen, Y., Zhang, Y., Kondoh, A., Tang, C., Chen, J., Xiao, J., Sakura, Y., Liu, C., and Sun, H. (2004). Seasonal variation of energy partitioning in irrigated lands. *Hydrological Processes* 18: 2223-2234.

Shimada, J. (2000). Proposals for the groundwater preservation toward 21st century through the view point of hydrological cycle. *Journal of Japan Association of Hydrological Sciences* 30: 63-72. (in Japanese).

Shimada, J., Tang, C., Iwatsuki, T., Xu, S., Tanaka, T., Sakura, Y., Song, X., and Yang, Y. (2002). Groundwater flow system of Hebei Plain in China and their recent groundwater environment change, *Water Resources Research Center (WRRC) Report* 22: 117-121, Kyoto University, Japan. (in Japanese).

Thomson, A.M., Brown, R.A., Ghan, S.J., Izaurralde, R.C., Rosenberg, N.J., and Leung, L.R. (2002). Elevation dependence of winter wheat production in Eastern Washington State with climate change: A methodological study. *Climatic Change* 54: 141-164.

US Geological Survey. (1996). GTOPO30 Global 30 Arc Second Elevation Data Set, USGS, http://www1.gsi.go.jp/geowww/globalmap-gsi/gtopo30/gtopo30.html.

Vandiepen, C.A., Wolf, J., Vankeulen, H., and Rappoldt, C. (1989). WOFOST - A simulation model of crop production. *Soil Use and Management* 5:16-24.

Wang H, Zhang L, Dawes WR, Liu C. 2001. Improving water use efficiency of irrigated crops in the North China Plain – measurements and modeling. *Agricultural and Forest Meteorology* 48: 151-167.

Yang, Y., Watanabe, M., Zhang, X.Y., and Hu, H.Z. (2004a). Estimation of groundwater use by crop production simulated by DSSAT-wheat and DSSAT-corn models in the piedmont region of North China Plain. *Hydrological Processes*. (accepted).

Yang, Y., Watanabe, M., Zhang, X.Y., Zhang, J.Q., Wang, Q.X., and Hayashi, S. (2004b). Using a wheat simulation model to optimize irrigation management to slow down groundwater depletion in northern China. *Agricultural Water Management* (revised).

Yang, Y., Watanabe, M., Wang, Z., Sakura, Y., and Tang, C. (2003). Prediction of changes in soil moisture associated with climate changes and their implications for vegetation changes: WAVES model simulation on Taihang Mountain, China. *Climatic Change* 57:163-183.

Yang, Y.H., Watanabe, M., Tang, C.Y., Sakura, Y., and Hayashi, S. (2002). Groundwater table and recharge changes in the piedmont region of Taihang Mountain in Gaocheng City and its relation to agricultural water use. *Water SA* 28: 171-178.

Yu, Z., Carlson, T.N., and Barron, E.J. (2001). On evaluating the spatial-temporal variation of soil moisture in the Susquehanna River Basin. *Water Resources Research* 37: 1313-1326.

Zhang, Y., Liu, C., Yu, Q., Shen, Y., Kendy, E., Kondoh, A., Tang, C., and Sun, H. (2004). Energy fluxes and the Priestley-Taylor parameter over winter wheat and corn in the North China Plain. *Hydrological Processes* 18: 2235-2246.

Zhang, Y., Liu, C., Shen, Y., Kondoh, A., Tang, C., Tanaka, T., and Shimada, J. 2002. Measurement of evapotranspiration in a winter wheat field. *Hydrological Processes* 16: 2805-2817.

Zhu, K.G., Du, G.H., Zhang, S.Y., Ma, T.S., Wei, X.F., et al. (1995). *Chinese soil cyclopedia, 4*. Chinese Agricultural Press: Beijing. (in Chinese).

Zhu, Y., et al. (1992). Comprehensive hydro-geological evaluation of the Huang-Huai-Hai Plain. Geological Publishing House of China: Beijing. 277p. (in Chinese).

In: Encyclopedia of Environmental Research
Editor: Alisa N. Souter

ISBN: 978-1-61761-927-4
© 2011 Nova Science Publishers, Inc.

Chapter 53

RUNOFF AND GROUND MOISTURE IN ALTERNATIVE VINEYARD CULTIVATION METHODS IN THE CENTER OF SPAIN

M.J. Marques[*,1,] *M. Ruiz-Colmenero and R. Bienes*[2]

[1]Agroenvironmental Dept. IMIDRA, El Encín. Ctra. A-2 km 38.2
28800 Alcalá de Henares, Madrid, Spain
[2]Geology Dept. University of Alcalá, 28871 Alcalá de Henares, Madrid, Spain

ABSTRACT

In the Mediterranean the best lands have gradually been dedicated to the growing of cereals, as this is the foundation of the public's diet, leaving the worst, most degraded and generally sloping areas for vineyards and olives. The traditional management of these two agricultural products also tends to keep the ground bare, thereby intensifying runoff, erosion and soil degradation.

The sector's economic importance is undeniable, given that in producing areas it is the foundation of the economy and rural employment. The use of alternative bare soil management methods for this type of crop grown on slopes is a priority. The vegetable coverings have already been tried out in more humid countries, but it´s not very common in semiarid regions. Thanks to the biomass generated by the coverings, it is an effective method for erosion control and to increase the organic matter content in soil. But one must keep in mind that the coverings necessarily establish a relationship with the crop. Among others, one must point out the competition for nutrients and, above all, for water at certain times of year. In semi-arid climates, this aspect is crucial.

This chapter aims to explore how these vegetable coverings affect water dynamics, measuring the soil moisture and the runoff generated over the course of one year, and if a clear competition effect exists for water, which could damage not irrigated vineyards.

The vineyard in this study is dryland, located in the center of Spain, near Madrid. Its surface (2 ha) was divided following three treatments: Traditional tillage, Grass sowing using Brachypodium distachyon, and using rye (Secale cereale).

---

[*] tel. +34918879459; E-mail: mjose.marques@madrid.org; fax: +34918879494

The average soil moisture data for the different periods of the vine's growth cycle show that the live cover of Brachypodium treatment presented less moisture than the other two treatments at 35 cm depth in spring and during the autum buds due to water consumption. The grape production of this treatment is less than the other two.

The runoff annual totals showed that the Bare soil treatment is similar to that of the Brachypodium treatment, equivalent to 8% of the total rainfall; the runoff of the Secale treatment was 9%. Regarding the Bare soil, the consequences of a certain storm depend on the time elapsed since the soil was last tilled.

INTRODUCTION

Across the European Union, agricultural terrain covers a substantial proportion of the land area (46%) with 24 % of arable lands, including 5.7 % for orchards and vineyards, which are traditionally worked using bare soil techniques (Van Camp et al., 2004). In Spain there are some 2 million hectares of olive groves and 1 million hectares of vineyards (data from the National Institute for Statistics in 2005). This is the largest vineyard area, by percentage, in the world. Viticulture is a mainstay of the economies of many Mediterranean climates around the world, and Spain is no exception, with its wine sector representing some 5% of the national food industry (Medrano et al., 2007).

Traditional vineyard management methods using bare soil allow for greater production. It may even be said that maintaining the soil bare is an essential part of exploiting the soil, since it has been determined that of the four most important factors in the success of such a crop, namely pruning, irrigation, fertilization and weed control, the latter exerts the most influence on yield during the first few years (Zabadal et al., 1991).

Unfortunately, the absence of vegetation cover on gradients results in high rates of water runoff and soil loss due to water erosion. It is not uncommon to find olive groves and vineyards either on or near sloping terrain. It is necessary to find alternatives to managing such crops (Montanarella, 2006). In traditional bare soil management, water erosion leads to a reduction in the permeability of the soil to rainwater (Green et al., 2003; Mahmood and Latif, 2003), mainly due to crusting (Augeard et al., 2005; Battany and Grismer, 2000). It is necessary to note as well the chronic lack of organic material present in these marginal farming soils. Even the Thematic Strategy for Soil Protection in Europe has established a link between poor organic material in the soil and water erosion in certain Mediterranean areas of France, Italy and Spain (Van Camp et al., 2004), which tends to aggravate the conditions needed to maintain productivity.

Published water runoff rates for vineyards are quite variable, as they depend on several factors, such as weather conditions, the gradient, the type of soil and how it is worked, etc. Published values range from 3% (Ramos and Martínez Casasnovas, 2006a), to 15% (Mekki et al., 2006), and even up to the 59% cited in Bini et al. (2006). Its consequences are diverse and, in Europe, include sedimentation and obstruction in rivers, in the silting of downstream reservoirs, in the loss of soil productivity and loss of nutrients (Ramos and Martinez-Casasnovas 2004, 2006b), since runoff takes with it organic material and nutrients, along with undesirable elements or molecules (Bini et al., 2006; van der Perk and Jetten, 2006) such as metals (Ribolzi et al., 2002; Boy and Ramos, 2005; Fernandez-Calvino, 2008), pesticides and herbicides (Spahr et al., 2000). Also swept away with the runoff are particles suspended in the water, for which figures as high as 56 mg L^{-1} have been published for rivers as a consequence of runoff from the drainage basins of vineyards (Leib et al., 2005).

There are several options for preventing soil degradation on gradients, such as planting following the contour line, or the creation of traditional terraces. These age-old practices are sometimes ignored or abandoned due to increased labor costs (Dunjó et al., 2003; Zalidis et al., 2002).

The classification used by the Ministry of Agriculture in Spain to agrologically characterize soils notes that difficulties can be encountered when trying to work soil on gradients in excess of 10%, and establishes an upper limit on gradients of 20% for working soil. A 50% value is set for those soils which do not allow for any kind of exploitation outside of a natural reserve.

The use of vegetation cover between the rows of vineyards not only halts erosion, but also has other positive consequences on the soil, such as a reduction in bulk density, an increase in its ability to retain water, an increase in biological activity (Sisa et al., 2000), and others. The European Thematic Strategy for Soil Protection (Van Camp et al., 2004) not only encourages the use of temporary or permanent vegetation cover to protect the soil, but also notes that the growth of green manures/catch crops may be limited to soil with a sufficient supply of water. Where low precipitation limits soil moisture, winter cover crops will compete with the main crops for available water.

In addition to preventing erosion, a decrease in runoff can affect the moisture content of soil. Infiltration rates are often calculated as the difference between rainfall and runoff. The water that escapes through runoff does not infiltrate the soil, thus leading to a lower moisture content.

From a viticulture standpoint, less than optimal water conditions, typical of Mediterranean climates, maintain a certain water deficit that maximizes the quality of the wine. Along with the use of Regulated Deficit Irrigation in watering the vines, the water condition of the vines should be considered if the use of vegetation cover is to be employed to control erosion in semiarid areas. This is because many Mediterranean vineyards are in areas which receive less than 400 mm of rainfall annually. The coexistence of the requirement for maximum water with the period of maximum water stress could result in the use of vegetation cover being harmful due to the crop's greater need for water. Generic potential evapotranspiration (PET) data can be used to assess the limit at which the water deficit could harm the plant; the soil moisture can be measured directly, but the most expedient approach is to determine the water status of the plants *in situ* (Linares et al., 2007). One of the most commonly used indicators is the measurement of leaf water potential with a Scholander chamber. This potential is the work required to be supplied per unit mass of water in the soil or in the plant tissues to obtain free water values at the same temperature and at atmospheric pressure. The measurement of the leaf water potential can be used as an indicator of the suitability of the use of vegetation cover in semiarid environments.

Doubts regarding the impact of vegetation cover in vineyard production, along with difficulties in changing the traditional habits of farmers (Alexandratos, 1995) keep these practices from being fully implemented. This could be addressed by making the results of specific applications in different geographical settings known.

In this chapter we provide information for determining whether the use of vegetation cover in the semiarid environment of the mid-Iberian peninsula could contribute to protecting the soil of vineyards on gradients without depleting yields.

MATERIALS AND METHODS

The area studied is in the "Vinos de Madrid" Designation of Origin area in the southeast of Madrid province, where the vineyards are situated on gypsum and limestone marl, usually on marginal soils. As for the climate, according to data from the nearest meteorological station (Figure 1), the average annual temperature is 14° C, with an average annual rainfall of 378 mm. Summers tend to be very hot, and give rise to convective storms that result in erosive episodes at a time when the soil is dry and unprotected by vegetation. During the year in the study, total rainfall amounted to 348 mm, similar to the annual average, but with a different distribution, with a very wet spring followed by a normal, dry and very dry summer, autumn and winter, respectively, as shown in Figure 1.

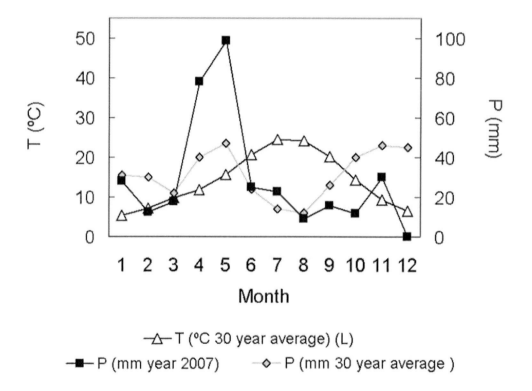

Figure 1. Climatic characteristics of the study area. Data from the Meteorological State Agency (AEMET, 1971-2000 Series. (Torrejón. Madrid. Spain. 40° 29' N – 3° 27' W).

The vineyard in question is located near the village of Campo Real (40° 21' N – 3° 22' W), at an elevation of 800 m, in a ridged area with soil classified as Calcic Haploxeralf (USDA, 2003), with a clay loam texture and stony (60%). The average gradient is 12 ± 2%. Table 1 lists some of the parameters for the soil in the study.

Since its planting in 1992, the vineyard has been worked using the traditional tillage in the region. The rows of vertically trellised vines face in the direction of the gradient. They are tilled using a chisel plow that digs some 10-12 cm into the soil. Two or three passes are made in the spring depending on the rainfall, with a deeper pass made in autumn.

This chapter provides the measurements for runoff and soil moisture for three types of techniques: a) traditional tillage; b) sowing of rye *Secale cereale* (70 kg ha^{-1}) cut in spring; and c) sowing of *Brachypodium distachyon* (40 kg ha^{-1}) that is allowed to self-sow.

Table 1. Mean and Standard Deviation (n=9) of different soil characteristics at 0 – 10 cm depth

Soil parameters	Mean	S.D.
Slope (%)	14	2
Texture	Sandy Loam	
Sand (%)	58.6	10.5
Silt (%)	17.8	5.8
Clay (%)	23.6	5.3
Field capacity	24.1	3.7
Wilting point (%)	10.4	1.9
Organic Matter (%)	1.34	0.10
CO_3^{2-} (%)	26.8	8.5
Active Limestone (%)	12.5	7.6
Bulk density (g cm^{-3})	1.2	0.07
Soil Density (g cm^{-3})	2.40	0.04
pH	8.7	0.10
EC 1:2.5 (mmhos cm^{-1})	0.25	0.09
Ca (meq/100g)	19.8	2.0
K (meq/100g)	1.08	0.03
Ca (meq/100g)	17.35	2.90
Mg (meq/100g)	1.06	0.25
Na (meq/100g)	0.04	0.01
CiC (meq/100g)	20.00	4.45

The *Secale cereale* was sown with a seeder with an 8-cm distance between rows. Smaller seeding rates, around 15% have result in a vegetative cover of 32% (Olmstead et al., 2001), that we consider insufficient for soil protection. The *Secale* is cut in spring, although the resulting chaff was left on the ground. This is because given this species' need for water, it is assumed *a priori* that it will compete more with the vines than *Brachypodium distachyon* which, though also a grass, is short (25 cm at maturity). The *Brachypodium* was sown manually since the elongated shape and low weight of the seeds was causing them to jam in the seeder. This grass has low water requirements, and has proven to be suitable in olive groves in the south of Spain (Saavedra and Alcántara, 2005), which is why the *Brachypodium* was allowed to complete its annual cycle and self-sow.

Each treatment was repeated successively every three of the vineyard's 66 rows, which covered a total of 2 ha (Figure 2). A total of nine 2-m² (4 x 0.5 m) erosion plots were set up in the center of the vineyard rows, three in each treatment.

Figure 2. Test set-up. The treatments were repeated every 3 of the vineyard's 66 rows. The test row was in the center, with the two edge rows situated on either side. The picture shows the center of the three rows with the *Brachypodium* treatment. To the left are three rows with bare soil, and to the right three rows with recently cut *Secale cereale*. The erosion plots are in the center of the test rows. There are a total of 9 plots, three in each treatment.

A Gerlach-type trough was placed at the base of each plot and connected using a tipping gauge that tipped every 80 ml of runoff and generated an electrical signal recorded by a Decagon Em50 data logger at 10-minute intervals. Also connected to this data logger were two 35-cm long ECH2O moisture sensors, one at a depth of 10 cm and the other at 35 cm, half a meter away horizontally from the first. There was also a rain gauge next to the plots in order to measure rain, with a resolution of 0.2 mm seg^{-1}.

The bulk density of soil crusts was measured by immersing the samples in petroleum (Busoni, 1997).

Evotranspiration can be used to determine the water requirements of the treatments, as it has been shown that over 99% of the water absorbed by plants is lost by evaporation and transpiration at the plant's surface. Crop evapotranspiration is mainly determined by climatic factors and hence water consumption can be estimated with reasonable accuracy using meteorological data (Wample and Smithyman, 2002)

The leaf water potential was also measured and combines in one reading the soil's water potential with a set of related values such as precipitation, the planting structure, the distribution of the roots, the depth, texture and moisture of the soil, the PET rate, etc. (Linares et al., 2007).

RESULTS AND DISCUSSION

Runoff and Soil Moisture

Data on monthly precipitation and the monthly runoff and average monthly soil moisture for 2007 are provided. The Secale sprouts faster than the Brachypodium, at around 20-30 days. Starting in March, the differences in the three treatments became noticeable, since until then the soil in all three could be considered bare, as in all three cases, the amount of cover was less than 15%. The top part of Figure 3 shows the monthly rainfall throughout 2007, as well as the runoff generated each month. In addition to precipitation, the most important factors in determining the soil moisture content are the tilling process and the vegetation cover. According to some studies (Hebrard et al., 2006), there is no correlation between the spatial variability in the soil moisture and local effects such as insolation, the gradient or the texture of the soil.

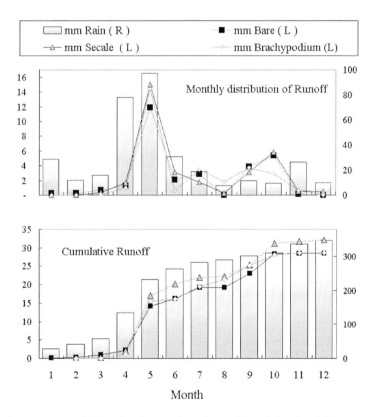

Figure 3. Plot of Means of rainfall, in the Right Axis and runoff (mm) for the different treatments, in the Left Axis.

In general, we can note from the figure that the runoff was proportional to the rainfall in all the treatments, although a considerable amount of runoff was occasionally recorded with moderate amounts of rain. Note in Figure 3, for example, the 6-mm runoff recorded in October with barely 10 mm of rainfall. Contrast this to the November values which show rainfall in excess of 27 mm and a runoff below 2 mm. In order to explain these monthly differences, we must keep in mind that runoff is not affected solely by precipitation, but rather by the intensity and duration of the most significant storms present during the period in question. For example, on October 1st, 8.2 mm of rain fell in 5 hours, with an average intensity of 36 mm h^{-1}. The runoff generated was 0.63, 0.57 and 0.29 mm in Secale, bare soil and Brachypodium, respectively, or, put another way, runoff rates of 7.7, 6.9 and 3.5 %. In contrast, the rainfall on November 22nd totaled 27 mm but it lasted 37 hours, with an average intensity of 6 mm h^{-1}, resulting in hardly any runoff, less than 0.1 mm in the three treatments, equivalent to runoff rates of around 1%. Something similar occurred with rainfall in April, which totaled 78 mm but over a period of 12 days of light or moderate rain, which only generated small amounts of runoff.

It seems that the differences in infiltration during normal periods of rain are slight, and only become apparent during the storms that are so frequent in this type of climate, as has been noted by other authors (Ferrero et al., 2002). The intensity and consequences of the storm that took place on May 20th, in which 43 mm fell in just 4.5 h, with intensity peaks of over 200 mm h^{-1} lasting a few minutes, is of particular interest. This storm generated considerable runoff (Table 2), which resulted in an average runoff rate of 25%.

Table 2. Mean and Standard Deviation of Runoff generated in the main storm that happens the 20th May 2007, 43 mm in 4.5 h

	Runoff volume (mm)		
Runoff registered for 30 minutes	Bare soil	*Secale*	*Brachyp.*
Maximum in 10 minutes	9.8	11.2	9.6
Mean ± Standard dev.	5.7 ± 3.5	8.5 ± 3.9	5.6 ± 3.8
	Soil moisture (%) (Mean ± SD)		
At 10 cm depth	Bare soil	*Secale*	*Brachyp.*
The week before	14.4 ± 0.4	12.8 ± 0.5	14.6 ± 0.4
The week after	18.4 ± 1.0	16.2 ± 1.2	19.8 ± 1.5
Increase	28%	26%	36%
At 35 cm depth			
The week before	16.5 ± 0.6	11.9 ± 0.3	15.6 ± 0.6
The week after	20.6 ± 1.7	14.5 ± 1.7	23.4 ± 1.9
Increase	25%	22%	50%

Soil moisture difference between one week before and after the storm in the different treatments. Volumetric soil moisture m3 m-3 is shown as %

The control treatment had been tilled and the rows of the Secale treatment had been harvested just ten days before this storm. The loss of soil was high as well, with the tilled soil in two of the three experimental plots silting up, which disrupted the automatic runoff measurement system. Before then, however, maximum 10-minute runoff rates of between 19 and 22 liters per plot had been recorded, as shown in the table. The maximum runoff rates were 0.96, 0.98 and 1.12 mm per minute in Brachypodium, bare soil and Secale, respectively.

Also noteworthy was the fact that the sediment yield was considerably lower in the Secale and Brachypodium treatments, with 19 and 26 g m^{-2} event^{-1} being recorded, respectively, compared with the at least 786 g m^{-2} event^{-1} for bare soil (Marques et al., 2008a).

Note the accumulated runoff and rainfall amounts in Figure 3. Until about April, the runoff data were similar for all the treatments. After the May storms, however, a greater amount of runoff in the rye crop is noticeable, becoming more apparent starting in autumn. This crop finished the year with an accumulated runoff equivalent to approximately 9% of the annual rainfall. The reason for the lower infiltration was the sealing of the soil, given that during severe storms, the chaff is swept up by the coursing water, leaving the rye stalks practically perpendicular to the soil surface and some 50% of the soil bare and with a layer a few millimeters thick in which the bulk density is far above that of the subsoil, on the order of 1.2 g cm^{-3} (Table 1). For example, the May storms left a 0.8-1.8 mm thick crust on the soil, with a bulk density of 1.71 ± 0.15; 1.65 ± 0.23 and 1.65 ± 0.15 g cm^{-3} in the Secale, Brachypodium and bare soil treatments, respectively.

The sealing of the soil gave rise to runoff even after rainfall amounts as low as 10 mm (Pla and Nacci, 2001). The differences found, though slight, reveal a higher sealing tendency in the Secale treatment than in the other two, as the bare soil is tilled and the Brachypodium treatment presented more soil protection.

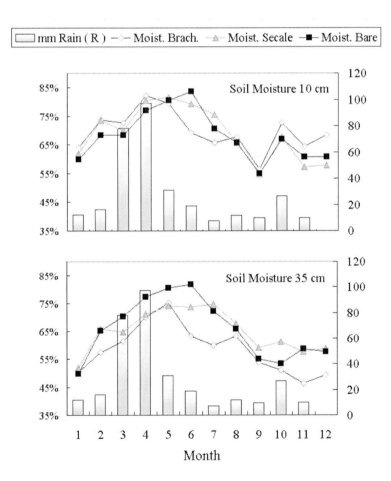

Figure 4. Plot of Means of Rainfall (mm) and Soil Moisture (%) along the year 2007.

Regarding the soil moisture at 35 cm depth, the Figure 4 shows how during the spring the bare soil treatment allows the increase of soil moisture, the Secale treatment maintains the soil moisture almost constant until the summer, and under the Brachypodium treatment it decreases, being progressively drier, even after the rains of November. The water consumption of this live cover, also produced a decrease in the soil moisture at 35 cm during the first spring buds and from October on, when it buds again, given that, as mentioned, the crop cover was not cut in this treatment and was allowed to self-sow.

The treatment of the tilled soil also exhibited irregular behavior, depending on the time elapsed since the soil was last tilled. In the 43-mm storm mentioned earlier, the soil had been recently tilled, which resulted in lower runoff in the bare soil.

This fact has been noted in research involving other vineyards with a gradient similar to ours, 11%, which, having been tilled recently, exhibited runoffs which were approximately half that of untilled vineyards (Dunjo et al., 2003). In other papers pending publication involving simulated rainfalls and with the control bare soil recently tilled, we noted higher infiltration in the tilled control soil than in soil with vegetation cover. The recorded runoff rate was 1.4% for bare soil, versus 12.7% in Brachypodium and 16.3% in Secale, meaning a 98.6% infiltration rate in bare soil.

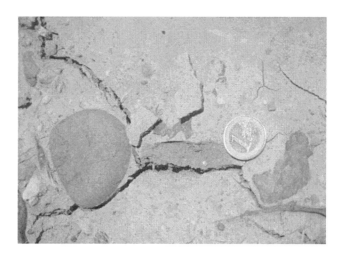

Figure 5. Soil sealing after the storms in May 2007.

Such behavior could be misleading when trying to decide on the most appropriate treatment for planting on a gradient, since although the infiltration is greater at first, as the soil receives more rainfall, a surface crust will form which, if the soil is not re-tilled, will gradually seal it and impede rain water from penetrating into the soil, resulting in a greater volume of runoff (Marques et al., 2008b). This was noted during the rains after June, when the bare soil treatment was also sealed, producing more runoff than in the other two, where the presence of vegetation, whether Brachypodium or Secale, hindered the production of runoff. The soil surface in the bare soil treatment, with a 1.5 mm crust (Figure 5).

It was also noted that the soil tilled had a moisture content equal to or greater than that of the other treatments at the time of budding, blooming and veraison, it was drier at ripening due to the increased temperature brought on by the absence of shade (Table 3).

The storm of 20th May also gave us the chance to determine how the different treatments make use of rainfall. Table 3 shows the volumetric moisture content in the soil the week before and following said event. Note how the Brachypodium treatment experienced the largest increase in moisture, followed by the bare soil and lastly by the Secale, which was also the one to undergo the greatest runoff.

Table 3. Results of Rainfall, Water Consumption, Soil Moisture and Leaf Water Potential during the growth season of vines and in the three different treatments

	Growth season periods		
	Budding (Mar- ½ Apr)	Blooming to veraison (½ Apr-Jul)	Ripening (Aug-Oct)
	Rainfall in		
Total: 246 mm	55 mm	169 mm	22 mm
	Water Consumption		
Total: 200-270 mm ([a])	5 % of the total (10-14 mm)	75 % of the total (150-203 mm)	20 % of the total (40-54 mm)
Mean ± SD Soil Moisture 10 cm (% of FC)			
Bare soil	73 ± 9 a	82 ± 6 a	65 ± 7 a
Secale	75 ± 7 a	79 ± 5 a	67 ± 16 a
Brachypodium	77 ± 6 a	74 ± 6 a	66 ± 9 a
Mean ± SD Soil Moisture 35 cm (% of FC)			
Bare soil	74 ± 7 a	81 ± 4 a	62 ± 10 a
Secale	68 ± 16 a	74 ± 13 ab	66 ± 19 ab *
Brachypodium	66 ± 6 a	69 ± 11 b **	58 ± 8 c *
Mean ± SD Leaf Water Potential (MPa)			
		(n=16)	(n=12)
Bare soil	n.m.	-0,75 ± 0,06 a	-0,95 ± 0,07 a
Secale	n.m.	-0,73 ± 0,07 a	-0,92 ± 0,05 ab
Brachypodium	n.m.	-0,72 ± 0,08 a	-0,87 ± 0,12 b **

The Water Consumption was estimated using the figures proposed by Wample and Smithyman (2002). The soil moisture is given as % of the Field Capacity (FC) of soil.
([a] : Annual Water consumption in rainfed Tempranillo cv (Cuevas et al., 1999; Centeno and Lissarrague, 2007)). Means with the same letter are not significantly different.
(* P<0.10; ** P< 0.05; n.m.= not measured).

If we examine the annual totals, we see that the Brachypodium showed a runoff similar to that of the bare soil. It seems that the near 100% cover of the Brachypodium plots favor infiltration as much as the recently tilled bare soil. In both cases, the accumulated runoff was 28 mm, equivalent to 8% of the total rainfall.

As concerns the average annual soil moisture content (Table 4), the average for all the treatments was 70 and 65 % of the field capacity of soil at 10 and 35 cm, that is 0.17 and 0.16 m^3 m^{-3}, respectively. The Brachypodium cover exhibited an above-average moisture content at 10 cm, but for the moisture at depth (35 cm), it was below average, ranging 60% of field capacity, that is 0.14 m^3 m^{-3}.

Table 4. Mean annual Soil Moisture (n=12 months) for different treatments, give as % of the Field Capacity (FC)

	Volumetric Soil moisture (% FC)	
	10 cm ± DT	35 cm ± DT
Mean	69% ± 11% a	65% ± 13% a
Bare soil	69% ± 10% a	68% ± 12% a
Brachypodium	70% ± 10% a	60% ± 10% b
Secale	69% ± 14% a	66% ± 17% a

Different letters indicate significant difference from the Mean (P<0.05)

Table 3 shows the recorded rainfall, the soil moisture and the leaf water potential of the vines for the three treatments. The data are distributed among the three main growing periods of the vine until its harvest, which took place in the first week of October 2007.

The rainfall between April and June tends to be highly variable, and the vine's water requirement is medium. The plant is in the budding stage and is growing at the maximum rate. In general the soil does not exhibit moisture problems since that is when the spring rains occur in a Mediterranean climate. Rainfall between July and September is very scarce, however, the vines require high amounts of water. The vine is in the veraison and ripening stages. It is during this time that the decrease in the soil moisture content is most manifest. From November to March, the water requirement of the vines is low, yet the rainfall is high. This is when the soil accumulates water in reserve for the following summer.

Throughout the vine's annual growing period, rainfall provided 246 mm of water, most of that during the period between blooming and veraison. It can be deduced from the data that a little hydric deficit was experienced; it started in August, as the vines were ripening, and even then it could be classified as moderate, as the Leaf Water Potential was around -0.9 MPa, this was probably because this spring was very wet. The midday water potentials in excess of -1.0 MPa do not represent a water stress for the vines, while levels of -1 and -1.2 MPa place the vines in a moderate hydric stress condition (Linares et al., 2007). Growth can stop at values of around -1.4 MPa due to a sudden drop in photosynthetic activity (Kriedemann and Smart, 1971).

In spite of this slight effect on Leaf Water Potential some differences in the grape production have been noticed, as in the bare soil treatment it was nearly the double of the other two. These results must be checked in future years because the storms in May almost destroyed the first bunches.

CONCLUSION

A total of 348 mm of rain fell in 2007, 32 mm of which were lost as runoff in the *Secale cereale*, versus the 28 mm in the other two treatments: traditional tilling and *Brachypodium distachyon*.

The traditional bare soil treatment makes the best use of the rainfall when freshly tilled, and exhibits an infiltration rate above 98%. This treatment, however, is prone to the formation of surface crusts which act to gradually seal the soil and increase runoff as time elapses since the last tilling. This can only be avoided through repeated tilling, though this practice accelerates the oxidation of organic matter in the soil and facilitates the shifting of soil during severe storms. This treatment also results in a higher grape yield (data pending publication), though two other factors must also be considered. First, in viticulture the yield is not always the prime consideration, but rather the quality of the product. A second consideration is the soil degradation that occurs as a medium- and long-term consequence of tilling the soil. The average vineyard lasts 50 years, and a high soil loss rate would not allow adequate output to be maintained until the end of the plant's production potential. The influence of these treatments on the quality of the wine is another objective of vine research. Currently the wines are in the malo-lactic fermentation stage.

The *Secale cereale* treatment has higher water consumption but after the cut in early spring, the soil moisture is statistically similar to that of the Bare soil treatment in the period from blooming to veraison, 74 % and 81% respectively in Table 3; that is 0.18 and 0.20 $m^3 m^{-3}$ respectively, where the water requirement is maximum. This treatment presents the highest soil moisture content during the vine's growing season, though in the annual tally, there is no difference between it and the traditional tilling treatment.

The *Secale* treatment generated more runoff (an 9% annual runoff rate) due to the crust formation in the first milimeters of soil. This could be prevented if the sowing covered more terrain because the mulch, which in principle should be able to protect it, tends to be washed away during severe storms on this sloping terrain, leaving it unprotected. This treatment requires annual sowing, harvest and the redistribution of the cut vegetative material if it is to be effective in protecting the soil.

The *Brachypodium* cover is the treatment that requires the least labor, since it is only sown the first year and allowed to self-sow after that. This grass increases the infiltration rate, but it is not enough to offset the evident water consumption rate, as it shows the less moisture in the soil. In the period from blooming to veraison, the treatment of *Brachypodium* has significant less soil moisture at 35 cm than the other two treatments, around 69% of the field capacity (Table 3), that is 0.17 $m^3 m^{-3}$.

Moreover, as a live cover, it can hinder water storage capacity in autumn and winter. It is then that the vine is dormant, but grasses can bud starting in October or November and consume water until the following spring, when the vine starts to bud. If this treatment is chosen for soil protection, the density of plants must be monitored. An increase in this parameter was noted in the second year of the treatment. It would be advisable, then, to sow at a rate below 40 kg ha^{-1} and to cut it once grown to reduce its ability to self sow the following year.

There are additional inconveniences to consider, one of them being financial cost. The *Secale* seeds used in this research cost 0.7 €/kg, and the *Brachypodium* seeds cost 6 €/kg. While this might be affordable at an experimental level, it is not from an agricultural standpoint. Low-cost seeds of these small grasses must be made available if they are to be more profitable to farmers.

We must also consider the greater difficulty in keeping the vineyard rows free from vegetation, especially if the appropriate machinery is not available and if the farm is managed ecologically, without herbicides to control the appearance of unwanted vegetation. In this sense, *Secale* competes more efficiently than *Brachypodium* with unwanted vegetation.

But there are significant medium- and especially long-term advantages. We must bear in mind that planting a vegetation cover is not just effective in preventing erosion, but also improves the properties of the soil, such as its organic matter content or its microporosity, increases the soil's biodiversity, and so on. If the ground is flat and has no serious shortages of organic material, then perhaps covering the ground in a semi-arid Mediterranean climate would not be advisable if it jeopardized the water supply. The above-mentioned properties will be analyzed at the end of the 4th year of the study to determine the medium-term impact of the treatments.

Unquestionably, the choice of species or even of species variety sown is very important. The way the land is worked must also be specific to each location and situation. We were able to note that the higher degree of moisture at depth in the bare soil treatment was due to the absence of weeds that increase evotranspiration, while the moisture in the soil with live vegetation cover was due to higher infiltration.

ACKNOWLEDGMENTS

Projects FP 06-DR3-VID IMIDRA and RTA2007-86 INIA, Bodegas y Viñedos Gosálbez-Ortí for allowing us the use of their vineyard and for the selfless cooperation.

REFERENCES

Alexandratos, N. In *World Agriculture : Towards 2010*, John Wiley and Sons Ed. FAO Roma, IT and Chichester, GB, 1995.
Augeard, B.; Cyril Kao, C.; Chaumont C. ; Vauclin, M. *Physics and Chemistry of the Earth, Parts A/B/C,* 2005 *30 (8-10),* 598-610.
Battany, M.C.; Grismer, M.E. *Hydrol. Proc.* 2000, *14(7),* 1289-1304.
Bini, C.; Gemignani, S.; Zilocchi, L. *Sci. Tot. Environ.* 2006, *369(1-3),* 433-440.
Boy, S.; Ramos, M. C. *Advances in Geoecology*, 2005, *36,* 511-517
Busoni, E. In *Metodi di analisi fisica del suolo*; Ministero per le Politiche Agricole (Ed). Milano, IT, 1997; 10-12.
Centeno, A and Lissarrague, J.R. In *Actas de Horticultura 2007, 48, 162-165*
Cuevas, E., Baeza, P. and Lissarrague, J.R. *Acta Horticulturae 1999, 493*, 253-259
Dunjo, G.; Pardini, G.; Gispert, M. *Catena* 2003, *52 (1),* 23-37.
Fernandez-Calvino, D.; Pateiro-Moure, M.; Lopez-Periago, E.; Arias-Estevez, M.; Novoa-Munoz, J. C. *Eur. J.l Soil Sci.* 2008, *59(2),* 315-326.
Ferrero, A.; Lisa, L.; Parena L.S.; Sudiro, L. In Proceedings of *ERB and Northern European FRIEND Project 5 Conference*; Demänovská dolina, SK, 2002
Green, T.R.; Ahuja, L.R.; Benjamin, J.G. *Geoderma* 2003, *116,* 3-27.
Hebrard, O.; Voltz, M.; Andrieux, P.; Moussa, R. *J.Hydrol.* 2006, *329 (1-2),* 110-121
Kriedemann, P.E.; Smart, R.E. *Photosynthetica* 1971, *5,* 6-15.

Leib, B. G.; Redulla, C. A.; Stevens, R. G.; Matthews, G. R.; Strausz, D. A. *Appl. Engin.Agric.* 2005, *21(4)*, 595-603.

Linares, R.; Baeza P.; Lissarrague, J.R. In *Fundamentos, Aplicación y Consecuencias del Riego en la Vid* P. Baeza, J. R. Lissarrague and P. Sánchez de Miguel Editorial Agrícola Española SA. (Eds.).. Madrid, ES, 2007; 37-45.

Mahmood, S.; Latif, M. *Irrigation and Drainage Systems* 2003, *17*, 367-279.

Marques M. J.; Bienes, R.; Garcia-Muñoz, S.; Muñoz-Organero, G. In *Proceedings of Geophysical Research Abstracts,* EGU General Assembly.Vienna, AT. 2008a; Vol. 10, SRef-ID: 1607-7962/gra/EGU2008-A-12188.

Marques M.J.; Bienes R.; Pérez-Rodríguez R.; Jiménez L. *Earth Surf. Proc. Land.* 2008b, *33,* 414-423.

Medrano, H.; Escalona, J.M.; Flexas, J. In *Fundamentos, Aplicación y Consecuencias del Riego en la Vid*. P. Baeza, J. R. Lissarrague and P. Sánchez de Miguel (Eds.) Editorial Agrícola Española SA; Madrid, ES., 2007; 15-34.

Mekki, I.; Albergel, J.; Ben Mechlia, N.; Voltz, M. *Physics and Chemistry of the Earth* 2006, *31 (17),* 1048-1061.

Montanarella, L. *Inflammatory Bowel Disease: Genetics, Barrier Function, Immunologic Mechanisms, and Microbial Pathways* 2006, *1072,* 3-11

Olmstead, M.A.; Wample, R.L.; Greene, S.L.; Tarara, J.M. *Am. J. Enol.Vitic.* 2001, *52(4),* 292-303.

Pla, I.; Nacci, S. (2001). In *Sustaining the Global Farm*; D.E. Stott, R.H. Mohtar and G.C. Steinhardt (Eds.); Catalonia, ES, 2001; 812-816.

Ramos, M. C.; Martinez-Casasnovas, J. A. *Catena* 2004, *55(1),* 79-90.

Ramos, M. C.; Martinez-Casasnovas, J. A. *Catena* 2006a, *68(2-3),* 177-185

Ramos, M. C.; Martinez-Casasnovas, J. A. *Agric. Ecosyst. Environ.* 2006b, *vol 113 (1-4),* 356-636

Ribolzi, O.;Valles, V.; Gomez, L.; Voltz, M. *Environ. Pollut* 2002, *112 (2),* 261-271

Saavedra, M.; Alcántara, C. In *Proceedings of Congreso Internacional sobre Agricultura de Conservación*. Córdoba, ES, 2005; 75-87..

Sisa, R.; Sixta, J., Ruzek, L.; Storkanova, G. *Rostlinna Vyroba* 2000, *46(2),* 55-61

Spahr, N.E.; Apodaca, L.E.; Deacon, J.R.; Bails, J.B.; Bauch, N.J.; Smith, C.M.; Driver, N.E. (2000). Water Quality in the Upper Colorado River Basin, Colorado, 1996–98: U.S. Geological Survey Circular 1214, 33 p., on-line at http://pubs.water

USDA *Key to Soil Taxonomy*. 9th Ed. Soil Survey Staff. NRCS Ed. Washington, US, 2003; Handbook 436. 332 p.

van der Perk, M.; Jetten, V. G. *Geomorphology* 2006, *79(1-2),* 3-12

Van-Camp. L.; Bujarrabal, B.; Gentile, A-R.; Jones, R.J.A.; Montanarella, L.; Olazabal, C.; Selvaradjou, S-K. *Reports of the Technical Working Groups Established under the Thematic Strategy for Soil Protection.*. Office for Official Publications of the European Communities Ed. Luxembourg, LU, 2004; EUR 21319 EN/1, 872 pp

Wample, R.L. and Smithyman R. 2002. In *FAO Deficit Irrigation Practices*. Water reports 22. p.89-100. FAO, Rome.

Zabadal, T.; Howell, G.S.; Dittmer T.W.. *HortScience* 1991, *26,* 761.

Zalidis G., Stamatiadis S., Takavakoglou V., Eskridge K. and Misopolinos N. *Agric., Ecosys. . Environ.* 2002, *88,* 137–146.

In: Encyclopedia of Environmental Research
Editor: Alisa N. Souter
ISBN: 978-1-61761-927-4
© 2011 Nova Science Publishers, Inc.

Chapter 54

MULTI-FUNCTIONAL ARTIFICIAL REEFS FOR COASTAL PROTECTION

Mechteld Ten Voorde[*,1], *José S. Antunes Do Carmo*[2] *and Maria Da Graça Neves*[3]

[1]Ph..D student, University of Coimbra, IMAR, LNEC,
Av. do Brasil, 101, 1700-066 Lisboa, Portugal
[2]Associate Professor, University of Coimbra, IMAR, Department of Civil Engineering,
3030-788 Coimbra, Portugal
[3]Research officer, LNEC, Av. do Brasil, 101, 1700-066 Lisboa, Portugal

LIST OF SYMBOLS

B = length of structure,
β = wave angle,
$\vec{c}$ = wave celerity vector,
c_g = wave group celerity,
g = gravity acceleration,
h = water depth at structure,
h_b = breaking depth,
h_c = water depth at crest of the structure,
H = wave height,
H_b = wave height at breakpoint,
H_0 = deep water wave height,
H/L = wave steepness,

$K_s = \dfrac{c_{g1}}{c_{g2}}$ is the shoaling factor,

---

[*] mvoorde@lnec.pt

$K_r = \dfrac{\cos\theta_1}{\cos\theta_2}$ is the refraction factor,

L = wavelength,
L_0 = deep water wave length,
T = wave period,
V_p = 'peel rate' of the wave,
$\vec{V}_p$ = velocity vector of the peel rate,
$\vec{V}_S$ = down-line velocity vector,
s = bottom slope,
S = distance from undisturbed shoreline to structure,
S_a = distance from the apex of the structure to the undisturbed shoreline,
SZW = natural surf zone width,
W = crest width,
$\tan\beta$ = bed slope in the vicinity of the structure,
α = peel angle,
α_1 = angle θ minus the decrease due to refraction on the reef slope,
α_2 = difference between the angles for the cases with and without wave focusing,
α_3 = angle due to deviation of the breaker line from the parallel to the bathymetry of the reef side,
θ = angle,
ξ_b = inshore Iribarren number,
ξ_∞ = deep water breaker parameter,
γ_b = breaker parameter.

1. INTRODUCTION

1.1. Definition of Coastal Zones

The nomenclature is not standardized, and various authors describe the same features using different names. This ambiguity is especially evident in the terminology used for the subzones of the shore and littoral areas. In the absence of a widely accepted standard nomenclature, coastal researchers would do well to accompany their reports and publications with diagrams and definitions to ensure that readers will fully understand their use of terms.

As shown in Figure 1, the coastal zone is divided into four main subzones: Coast, Shore, Nearshore or Shoreface and Continent shelf - Offshore.

Each of the following definitions applies (Army Corps of Engineers, 1992):

> The coast is a strip of land of indefinite width that extends from the coastline inland as far as the first major change in topography. Cliffs, frontal dunes, or a line of permanent vegetation usually mark this inland boundary. On barrier coasts, the distinctive back barrier lagoon/marsh/tidal creek complex is considered part of the coast. The area

experiencing regular tidal exchange can serve as a practical landward limit of the coast. The seaward boundary of the coast, the coastline, is the maximum reach of storm waves.

The shore extends from the low-water line to the normal landward line of storm effects, i.e., the coastline. Where beaches occur, the shore can be divided into two zones: backshore (or berm), above the high-tide shoreline, which is covered with water only during storms, and the foreshore (or beach face). The foreshore extends from the low-water line to the limit of wave uprush at high tide, so it is the portion exposed at low tide and submerged at high tide. The backshore is nearly horizontal while the foreshore slopes seaward. This distinctive change in slope, which marks the junction of the foreshore and backshore, is called the beach or berm crest.

The nearshore, also shoreface, is the seaward-dipping zone that extends from the low-water line offshore to a gradual change to a flatter slope, denoting the beginning is the continental shelf. The continental shelf transition is the toe of the shoreface. Its location can only be marked approximately, because of the gradual slope change. The nearshore is never exposed to the atmosphere, but is affected by waves that touch bottom. It is the zone of most frequent and vigorous sediment transport, especially the upper part.

The continental shelf is the shallow seafloor that borders most continents. The shelf floor extends from the toe of the shoreface to the shelf break where the steeply inclined continental slope begins. It has been common practice to subdivide the shelf into inner-, mid-, and outer zones, although there are no regularly occurring geomorphic features on most shelves that suggest a basis for these subdivisions.

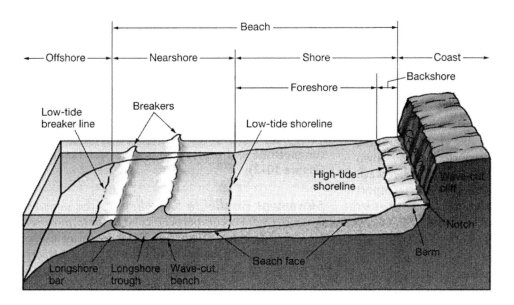

Figure 1. Landforms and terminology of coastal zones (Thurman and Trujillo, 1999).

Beaches include both the shore and nearshore subzones. They are composed of the material – sediment – primarily provided by the erosion of beach cliffs and/or by rivers that drain lowland areas. Other sources of materials, like shell fragments and the remains of microscopic organisms that live in the coastal waters, are less common. Beaches on volcanic islands are frequently composed of dark fragments of the basaltic lava that make up islands, or of coarse debris from coral reefs that develop around islands in low latitudes. Thus, around the world, there are beaches composed of crushed coral, beaches made of quartz sand, beaches made of rock fragments, beaches made of black (or even green) volcanic material, beaches composed of shell fragments, and even artificial beaches composed of scrap metal dumped at the beach (Thurman and Trujillo, 1999).

In general sediments consisting of sand and gravel occur on the upper and middle shoreface, muds of fluvial origin occur on the shelf and mixed sands and muds are found in the lower shoreface zone (transition zone to shelf). Bed sediments generally fine in seaward direction, supplied by offshore-directed bottom currents (rip currents and storm-induced currents). Sometimes relatively coarse-grained relict sediments are found on the shoreface.

1.2. Coastal Uses and Vulnerabilities

The coastlines of many countries have not only been shaped by natural forces but, over the last few centuries especially, they have also been strongly influenced by people.

Early communities adopted a subsistence way of life which, compared with today's lifestyles, had a low impact on the natural heritage of the coast; these communities lived in relative harmony with their environment.

This is easily explained because in past centuries the littoral was always an area with little attraction, to be avoided at all costs. Apart from fishing communities and ports, there was little to attract people to the coast - quite the contrary, in fact. Climatically it is an area of contrast, hot during the day and cold at night, windy, and without protection against the sun. In terms of resources there was little to exploit except for fisheries and harbours. On the other hand it was a dangerous area. People living there (mostly fishermen) were rude; pirate raids were frequent, killing and looting the local populace. Consequently, the littoral was for a long time a very sparsely populated area (Dias *et al*., 2002).

The awakening interest in the sea and the beaches appears all over Europe from the end of the 19th century. This was when the first bathing beaches started to spring up everywhere, particularly in France and Britain.

From the end of the 19th century to the first half of the 20th century the occupation of the littoral can be said to have been of a therapeutic character, as the objective of bathers when they left their houses carrying their bags and baggage to the beach resorts was to "go to the baths" This consisted of exposing oneself to the waves at the beach. The practice was prescribed medically, being indicated as a treatment for several physical illnesses or states of mind, and applied to adults and children. Baths in the sea were taken "as remedy, not for pleasure" (Colaço and Archer, 1943, in Dias *et al*., 2002).

The 20th century saw changes in the way of looking at the littoral. It became a place for pleasure, where the beach played an important role in leisure, and became a popular holiday destination.

It is a matter of fact that major social and economic changes over the last few hundred years in many countries, including Portugal, have resulted in dramatic changes to the character and natural heritage of the coastline. More intensive farming, coastal development and better access to the coast have all affected the natural coastal processes and led to a loss of habitat and wild coastal land. The bigger cities have grown and spread along the coastline and numerous towns have developed, mostly as fishing ports, on mainland and island coasts alike.

Some indicators are easily stated:

- 20% and 40% of people live within 30 km and 100 km of the coast, respectively;
- coastal populations are growing more rapidly than global populations, mainly in urban settings;
- global mean sea levels rose 10 to 20 cm in the 20th century;
- in the 21st century, human-induced climate change will contribute to a global rise of 20 to 100 cm, with a mid estimate of 50 cm;
- other climate factors relevant to the coast will also change, although details are unclear.

All around the world, mega-cities (cities with more than 10 million people) are increasing in number as the global human population continues to become more urban. By the end of 2030, three-fifths of the world population will be living in urban areas. Much of this urban growth occurs outside defined city boundaries and the resulting expansion of urban or "built-up" areas can be clearly seen from orbiting satellites. As a consequence, the degradation of coastal ecosystems is to be expected, as well as increased pollution of coastal waters.

Global warming will lead to dramatic coastal changes in the near future. In fact, the greenhouse effect and resulting warming of the earth's temperature may accelerate the mean sea level. A rise in the mean sea level would cause erosion, flooding, and saltwater intrusion in bays and estuaries.

An extensive study of coastal problems has been conducted under the EUROSION project (Eurosion, 2004). This project entailed 60 case studies, considered to be representative of European coastal diversity. The case studies reveal many erosion problems along the European Atlantic coast. The Atlantic Ocean borders Western Europe around the following EU countries: the United Kingdom, Ireland, France, Spain and Portugal. Generally speaking, the coastline around the Atlantic Ocean is made up of hard and soft cliffs interspersed with sandy and shingle beaches and dunes (Eurosion, 2004). The high relief, hard cliffs and rocky coastlines are mostly found in northern Spain, northern Portugal and parts of northern France. The softer coasts may be found in western Ireland (e.g. Donegal and Rosslare) and the southern United Kingdom (Sussex), where soft cliffs with shingle and sand beaches and smaller dunes alternate with small bays and estuaries. Larger, extensive dunes are found on the southwest coast of France (Aquitaine).

Erosion of the Atlantic coastline, as has occurred in Estela, on the coast of Portugal (Figure 2), is a consequence of natural and human-induced factors. The high-energy, storm generated waves from the Northern Atlantic and the macro-tidal regime (medium range 2-4 m, maximum up to 15 m in the Bay of Mont Saint-Michel, France), are the dominant erosive forces on the continental European Atlantic coastline. Together they create extreme conditions with strong alongshore tides and/or wave driven currents and cross-shore wave driven currents that can easily erode beaches and undermine cliffs. Climate change is expected to induce accelerated sea level rise (at present 2-4 mm/yr) at some point in the future, as well as a potential increase in storminess. Both will worsen erosion along the

Atlantic coast. Human interference, such as the construction of seawalls or groynes, damming of rivers and sand mining, has intensified erosion locally.

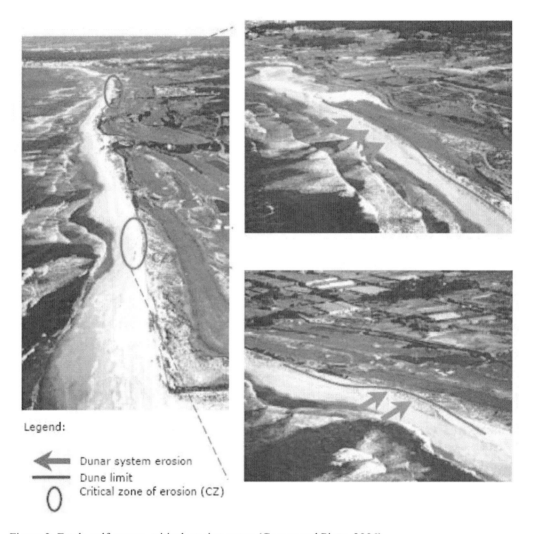

Figure 2. Estela golf course, critical erosion zones (Gomes and Pinto, 2006).

In both northern and southern Europe erosion threatens urbanization (the safety of human lives and investments), tourism and nature. Furthermore, in Spain, Portugal and France fishing and aquaculture are of great importance in the coastal zone. In the United Kingdom and Ireland, a lot of agricultural land is found in the coastal zone. The explosive growth of the population in the littoral zone, partly due to tourism, has increased the pressure on the coast, especially in France, Spain and Portugal and the south coast of England. It appears that most of the European Atlantic coastal areas are at high risk due to the low-lying coastal plains that are at risk of flooding.

The 'hold the line' policy option is often applied when seaside resorts or other recreational facilities are at risk. This is especially true in France, Spain and Portugal but it is often relevant in the south of England and Ireland, too, where tourism plays a leading role in the protected sites. Furthermore, high population densities and economic investments are protected by applying the 'hold the line' policy, as in the United Kingdom, Ireland and Portugal.

'Managed realignment' is possible at some of the seaside resorts and recreational facilities if the amount of capital at risk is relatively small and the recreation facilities or houses can be moved inland without too many problems. In a flooding area, a new defense line is usually defined (under the principle of 'Managed realignment'). 'Do nothing' is usually applied to cliff coasts where there are no flooding risks and therefore the amount of capital at risk is relatively low.

At many sites along the Atlantic coast, a mixture of hard and soft engineering solutions is adopted to deal with erosion. Various types of hard solutions had been applied in the cases considered. Although applied in nearly all cases, beach nourishments are executed on a much smaller scale (in terms of m^3) than in the North Sea and the Baltic regions. Whereas in the North Sea regions soft measures are often implemented to combat erosion, along the Atlantic coasts the soft solutions are often combined with hard me asures. The high energy conditions of the coast and the steep foreshore mean that nourished sediment is quickly transported in an offshore direction, and it might not return by means of the equilibrium movement during the year.

1.3. General Protection Measures

In order to combat erosion, several technical measures can be taken. Below the various types will be described:

- Hard measures

Hard measures are structures like seawalls, revetment/slope protections, groynes and detached breakwaters. The case studies show that hard measures can work really well if the consequences are accepted. If erosion is predicted for and accepted elsewhere, the overall performance can be good. On the Atlantic coast, structures need to be built that are strong to resist extreme events that bring very large waves.

- Soft measures

Soft measures are those like beach and dune nourishment, submerged nourishment, vegetation techniques and cliff stabilization. Soft measures can be a short term solution. Erosion can continue at the same rate and soon the action (measure) needs to be repeated. This is especially true in the Atlantic where wave heights and tidal amplitudes can be quite large.

- Combined measures

At many sites along the Atlantic coast, a combination of hard and soft engineering solutions is often adopted for dealing with the erosion issues, probably due to the high energy conditions of the coast.

- Innovative measures

These include measures like beach drainage systems. The 'Écoplage' system (applied in Sables-d'Olonne, France) consists of a gravity drain that lowers the water table beneath the beach. As a result, the beach is not saturated with water when waves break on the shore and the infiltration of the water into the sand is improved. The purpose of the system is to reduce swash velocities, sediment transport, and therefore erosion. The water flows by gravity from the drain to a pumping system. The water is then pumped into the sea or is used as (filtered) water in swimming pools or aquariums. The beach drainage system seems successful. It does not block the littoral drift like a groyne does. The treated beach is stabilized and the untreated beach is continuing to be eroded.

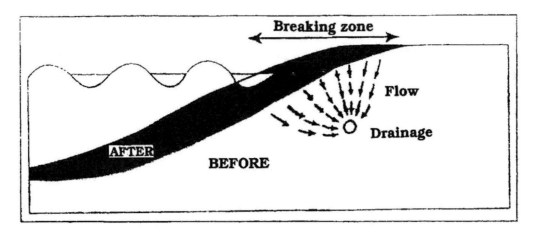

Figure 3. Scheme of the 'Écoplage' system installed in Sables-d'Olonne, France.

A completely different type of measure was tried at El Médano (beach with dunes) on the Canary Islands. In 1995 an experiment started which aimed at recovering eroded sediments. In this case, it was decided to place small obstacles in a completely bare exposed area. This involved scattering amounts of about 2 m^3 of volcanic gravel around the eastern slope of Bocinegro mountain. The experiment is based on the volcanic gravel's capacity to retain sand, on which plants germinate, beginning the process of soil regeneration and dune formation. The experiment was effective: the retention and accumulation of sand increased, above all on the smaller obstacles.

Another innovative solution was implemented to protect a sand dune system in Leirosa, south of Figueira da Foz, in the centre of the Portuguese West Atlantic coast, based on the use of geotextiles. In mid '90's the construction of an underwater effluent emissary was responsible for the disruption of this sand dune system. The use of hard machinery in a fragile area, aggravated by the erosion caused by the breakwater about 1 km north Leirosa, also contributed to the destruction of this system. The restoration of the Leirosa sand dunes started in 2000. The dunes were reconstructed mechanically according to the AIDS (Artificially Inseminated Dune Systems) method (Reis and Freitas, 2002). This process allowed the damaged dune system to be rebuilt the desired height and slope in a short time. The second part of the rehabilitation process consisted of increasing the stability that had been achieved by revegetation with Ammophila arenaria. This beach grass is widely used all over the European coast in the rehabilitation of degraded systems due to its unique capacity for sand stabilization and dune formation.

In February 2001, during a storm that affected the whole of the central region of Portugal, especially on the coast, the ocean front of the Leirosa dune system was destroyed in just one night.

Figure 4. Installation of the sand containers and first layers of geotextiles in the Leirosa sand dune system (Ten Voorde et al., 2008).

After this event it was decided to try the solution based on the use of geotextiles. This material can be as effective as any so called "hard engineering protection", with the advantage that it is adaptable to the morphology of the dune system and uses the local available sand (Bleck *et al.*, 2003).

In the Leirosa system geotextiles were applied to a dune extension of 120 m along the coast. A defense was created on the front bottom of the dune, at the +2.0 level (hydrographic zero) with sand containers about 6.40 m long, 3.20 m wide and 0.825 m high, in a pyramid arrangement parallel to the coastline (Figure 4).

The protective barrier was constructed to a height of 8 m by several layers of sand enveloped in geotextiles, in a so called "wrap around technique", which encapsulates the sand and can be installed quickly.

This technique allows the absorbsorption of the energy of upcoming wave attacks and prevents erosion of the fill material through the tensile forces which are activated by this stress. Furthermore, pore water pressures within the encapsulated sand fill are released thanks to the good drainage capacity of the geotextile.

The upper layer used a geotextile revetment throughout the area, about 8.60 m long, 4.30 m wide and 1.10 m high, followed by a 1.0 m layer of sand where dune vegetation was planted, turning this area into an attractive and safe coastal dune system (Reis *et al.*, 2005).

It is expected that the technique developed to protect the Leirosa sand dunes will become an important model to be usd for other dune systems with similar erosion problems.

Experience has shown that as yet there is no miracle solution to counteract the adverse effects of coastal erosion (EUROSION, 2004). The best results have been achieved by combining different types of coastal defense including hard and soft solutions, taking advantage of their respective benefits while mitigating their respective drawbacks. Since it can be seen that coastal erosion results from a combination of various natural and human-induced factors it is not surprising that miracle solutions to counteract the adverse effects don't exist. Nevertheless, the general principle of "working with nature" was proposed as a starting point in the search for a cost-effective measure. However, this observation also undeniably takes on board the idea that soft engineering solutions are preferable to hard ones. This is backed by a number of considerations derived from experience:

- Even tried and tested soft solutions - such as beach nourishment, which aroused tremendous enthusiasm in the 1990's - have suffered serious setbacks. Such setbacks have been caused by inappropriate nourishment scheme design induced by poor understanding of sediment processes (technical setback), difficult access to sand reserves which leads to higher costs (financial setback), or unexpected adverse effects on the natural system - principally the benthic fauna - (environmental setback). These are well covered by, respectively, the case of Vale do Lobo (Portugal) where 700 000 cubic metres and 3.2 millions euros of investment have been washed away by longshore drift within a few weeks only, the case of Ebro where the amount of sediment needed to recharge the beach sediments had been imported from another region, and the case of Sitges (Spain) where the dredging of sand to be supplied has caused irreversible damage to sea grass communities (Posidonia).
- Soft solutions, due to their particularity of working with nature, are found to be effective only in a medium to long-term perspective, i.e. when coastal erosion does not constitute a risk in the short-term (5 to 10 years). Their impacts do indeed slow down coastline retreat, but they don't stop it. The long term positive effect of soft solutions may be optimized by hard structures which make it possible to tackle an erosion problem efficiently but have a limited lifetime (in general no more than 10 years). This has been particularly well-documented for example in the case of Petite Camargue (France), where the presence of hard structures - condemned anyway – also turned out to provide sufficient visibility for soft defense such as dune restoration wind-screens to operate.

1.4. New Concepts for Coastal Protection

Various coastal structures can be used to solve, or at least, to reduce coastal erosion problems. Some of them can provide direct protection, like breakwaters, seawalls and dikes, and others, such as detached breakwaters and artificial reefs, provide indirect protection, reducing the hydraulic load on the coast to the level required to maintain the dynamic equilibrium of the shoreline. To achieve this objective, these structures are designed to allow the transmission of a certain amount of wave energy over the structure by overtopping (and also some transmission through the porous structure, in some types of breakwaters), and/or wave breaking and energy dissipation on a shallow crest (submerged structures) (Pilarczyk, 2003).

Rock walls, breakwaters or groynes usually serve the purpose of protecting land from erosion and/or enabling safe navigation into harbours and marinas, but other commercial value and multi-purpose recreational and amenity enhancement objectives can also be incorporated into coastal protection and coastal development projects. Offshore breakwaters/reefs can be permanently submerged, permanently exposed or inter-tidal. In each case, the depth of the structure, its size and its position relative to the shoreline determine the coastal protection level provided by the structure. Submerged breakwaters could be an interesting and efficient strategy, not only to protect a coastal system, but also to improve the bathing conditions of some coastal zones. Therefore, these so-called multifunctional artificial reefs (MFARs), are one of the new innovative concepts for coastal protection. The actual understanding of the functional design of these structures may still be insufficient for optimum design but it may be just good enough for these structures to be considered as serious alternatives for coastal protection (Pilarczyk, 2003).

The two main purposes of a MFAR (coastal protection and increasing the surfing possibilities in a certain area) are explained in greater detail below.

The construction of a MFAR can play a part in different kinds of coastal protection, like:

- Preventing coastal erosion;
- Increasing, in combination or not with sand nourishment, the stability of beaches.

These sorts of coastal protection are possible because:

- A MFAR can reduce the wave loads on the coast through a series of wave transformation processes occurring on the structure, viz., reflection and energy dissipation due to waves breaking on the structure and to flow circulation inside the porous media;
- A MFAR can create current circulation cells behind the reef which can cause sedimentation at the shoreline;
- A MFAR can be used to regulate wave action by refraction and diffraction.

With one kind of design, waves can break over a MFAR in such a way that surfers can enjoy great sport riding them. Surfing and bodyboarding are growing in popularity, and are practised especially by young people. Surfing means taking a wave board into the sea and waiting for a breaking wave to 'ride' on (Figure 5).

Figure 5. Waiting for and surfing a wave (source: www.surfline.come).

Multi-functional reefs are in fact a hard measure for tackling coastal problems, and they have several advantages over soft measures. The most suitable construction material for multi-functional artificial reefs is geotextile, used as a sand container. Sand-filled geotextile containers are becoming increasingly recognized as a tool for coastal defense. Geotextiles are a family of synthetic materials including polyethylene, polyester and polypropylene. In their common form they are flexible, permeable and durable sheet fabrics, resistant to tension and tear. They can be sewn or ultra-sonically welded to produce containers designed to retain sand or mortar for use as a construction material. Geotextile containers can be filled and placed using many different methods depending on the site location, fill material, container size, available plant, and type of geotextile fabric chosen. The advantages are several:

- Environmental impact

Construction with geotextile containers allows the use of local materials that would otherwise be unsuitable for coastal construction. Permission could be obtained to fill the containers with sandy materials taken from the seabed in the region of the project site itself. This means that the amount of 'foreign' materials introduced to an area is minimized. Unless exposed to high temperatures or pH levels, geotextile materials have been shown to be inert in the marine environment.

- Durability

Modern geotextiles are designed to withstand environmental degradation from abrasion, UV, chemical and biological influences, and as such a life span of the order of 25 years can be expected, notwithstanding vandalism or mechanical damage. Under accelerated testing, life spans of up to 100 years even in challenging marine conditions have been postulated. During the construction of the Narrowneck reef very effective underwater patching techniques were developed to repair damage that had arisen during the construction (Restall *et al.*, 2002). The holes are sealed with a silicone based adhesive and a patch is screwed down over the hole, using nylon wall screws, to provide added protection. Various coatings were trialed for the crest bags, with mixed success. But towards the end of the construction a durable composite (hybrid) material was developed and tested with great success. Initial trials utilized a spray-on polyurethane coating of varying thickness, however this product became rigid once exposed to water and actually made the products more susceptible to impact and wave damage. The composite material, consisting of two layers of non-woven geotextile, used towards the end of the project allows approximately $4kg/m^2$ of sand and shell particles to be retained within it. Once the geotextile is impregnated with these particles its puncture resistance shows significant improvement, while marine growth can protect it from UV degradation.

- Structure is removable

In the unlikely event that the structure has a previously unforeseen negative impact on the surrounding area, geotextile systems allow the reef to be quickly and easily removed. By filling the containers with locally obtained sand, the environmental impact of removal would be insignificant as the material released from the reef would be the same as or very similar to the natural beach sand.

- Provision of marine habitat

Marine ecosystem enhancement is a fourth advantage of the use of geotextile sand containers. These containers have appeared to provide an excellent substrate for marine flora and the development of a diverse ecosystem (Borrero and Nelson, 2003, Jackson *et al.*, 2004). Figure 6 shows, as an example, short algae and sea grasses on shallower containers in the Narrowneck reef off the Gold Coast of Australia.

Figure 6. Short algae and sea grasses on shallower containers of the Narrowneck reef (Jackson *et al*, 2004).

When the ecosystem on this reef was well developed, a clear zoning between areas of sea grass and kelp could be observed. Visually, the macroalgal communities dominated the reef, covering over 70% of the reef surface. It was also populated by a variety of other benthos, including coralline algae, sub-massive sponges, ascidians, octocorals (soft corals), hydroids and crinoids (feather starfish), echinoids (sea urchin) and abalone. Observations by the National Marine Science Centre indicated that "the biological communities associated with Narrowneck Artificial Reef appear to enhance biodiversity and productivity at a local scale and may also contribute to overall regional productivity" (Edwards, 2003).

- Soft reef surface

The use of sand-filled fabric containers maximizes safety on the reef by providing a relatively soft structure without sharp edges, reducing the risk of injury should a surfer come into contact with the reef.

- Cost

As with any construction method, the cost of a reef project is site dependant, as it relies on the availability of plant, materials, labour and suitable construction conditions. However in many situations, especially with large projects, it has been found that geotextile solutions can halve the cost of the equivalent rock structure (ASRLtd, 2008).

However, there are also some limitations:

- Experience of contractors

Coastal engineering contractors around the world are highly experienced in the use of rock and concrete for coastal construction. Experience in the use of sand-filled geotextile systems is less common, although the recent significant growth in the use of geotextiles for coastal projects worldwide is rapidly improving the skill base.

- Susceptibility to mechanical damage and vandalism

Testing has shown modern geotextiles to have good puncture and abrasion resistance; but the resistance of the fabric to mechanical damage and vandalism is clearly lower than that of rock or concrete. As mentioned before, very effective underwater patching techniques were developed to repair damage during the construction of the Narrowneck reef. Notwithstanding this fact, extra care must be taken while handling the units during construction, and in some locations it may be necessary to introduce measures to safeguard against vandalism.

- Lack of design guidance

At present there are relatively few guidelines available for the design of coastal structures using geotextiles. Despite the lack of official design guidance, there has been considerable research in the area of geotextiles and the conclusions of this research along with experience of a wide range of previous projects can be drawn on to allow successful design.

Several reasons may account for the growing interest of MFARs over conventional ways of protecting a local coastline:

- These structures have, by definition, a minimal visual intrusion that especially enhances their value in zones with strong aesthetic constraints;
- The structure becomes known better as a way of boosting surfing possibilities as well as a way to protect the coast;
- The water renewal induced by the high level of transmission is desirable to avoid stagnation and ensure satisfactory quality for recreational waters, especially in tideless seas, and water oxygenation for animals and plants living in the protected area leeward of the breakwater;
- The structure is interesting in economic terms. First, geotextile sand containers tend to be cheaper per unit volume than rubble-mound structures, and second, the surfing aspect can attract the tourism, which is good for the local economy;
- The expansion of the environmental value is a great benefit in these times, in which nature is being destroyed more and more, by the behavior of humans.

The following sections of this chapter focus on multi-functional reefs for coastal protection and the generation of surfing waves.

2. MULTI-FUNCTIONAL ARTIFICIAL REEFS

2.1. Multi-Functional Artificial Reefs Built So Far

Until half 2008 three artificial reefs have been built either with the purpose of enhancing of surfing possibilities (Artificial Surfing Reef), or jointly with other purposes, like coastal protection (Multi Functional Artificial Reef). Below, the reefs are described in chronological order of building. The description of each reef focuses on the results of environmental and monitoring studies.

Cable station (or Cables), Perth, Australia

The first ASR was built close to Perth in Australia and named Cable Station or Cables (Ranasinghe et al., 2001). Construction was completed in December 1999. The purpose of the surf reef was to produce surfable waves regularly. It was not intended as a shore protection structure, as the shoreline at Cable Station consists mainly of rocky outcrops and platforms, and so is naturally stable.

The overall dimensions are 80 m cross-shore by 90 m longshore. The reef (Figure 7) was constructed of granite.

Two experimental design studies (one in a wave flume and one in a wave basin) and three numerical surfability studies were undertaken for this reef. In addition, there was a beach response study in the design phase, and an environmental study was also undertaken in which the marine habitats were observed, before and after construction.

The researchers of this last study, Bowman Bishaw Gorham (2000) predicted the recovery of the diversity of marine life that previously inhabited the area. Finally, a post-construction performance study was carried out.

When 90 percent of the reef had been built its performance was studied. During the 6-month study period, 77 surfable days were identified. The researchers concluded that the performance of this reef exceeded expectations (Pattiaratchi, 2000),although this has been disputed by some members of the local surfing population.

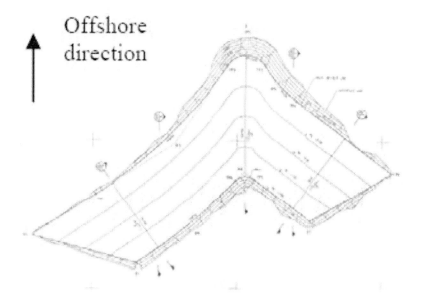

Figure 7. Plan view of design reef Cable station (Ranasinghe, 2001).

Narrowneck reef, Gold Coast, Australia

The Narrowneck reef is situated off the Gold Coast, Australia. The construction was finished in December 2000 (Ranasinghe et al., 2001). The primary objective of the reef was to widen the beach and mitigate storm erosion, by retaining and protecting the nourished beach (over 1 million cubic metres of nourishment). The secondary objective was to improve surfing.

The overall dimensions are 400 m cross-shore by 200 m longshore. The submergence is 1.5 m below lowest tide (Jackson *et al.*, 2005; Corbett *et al.*, 2005). The reef (Figure 8) was constructed of GSCs that weighed 160-300 tonnes, being typically 20 m long and up to 5 m diameter (Black, 2000). In all, 332 bags were placed. The total volume of the reef at the end of construction was110 000 cubic metres. The bags were filled with natural sand in a split-hull hopper dredge. Once filled, the bags were dropped on the seabed using bow and stern satellite positioning to align the dredger.

For this reef one physical study was carried out (Turner *et al.*, 2001). Besides the experimental study, four numerical studies were carried out for the reef design (Black and Mead, 2001). The models implemented were a refraction model WBEND, a multi-purpose model 3DD, a sediment transport model POL3DD, and a beach circulation and sediment transport model 2DBeach.

A pre-construction study (Ranasinghe *et al.*, 2001) had been undertaken to investigate the likely environmental impacts on the area. During and after construction an ARGUS video imaging system was installed to monitor the shoreline response to the reef. This system is capable of providing very accurate quantitative information. Also, a number of hydrographic surveys have been carried out. It appeared from a beach monitoring study (Jackson *et al.*, 2005) in the period 2000-2004 that the beach updrift of the reef was of the order of 40 m wider than at the start of monitoring. In the lee of the reef, an additional 30 m had been obtained. However, it has to be mentioned that before construction a nourishment program of over one million cubic meters of sand was carried out, so it is difficult to analyze the effect of nourishment and the effect of the reef. Another result of four years of monitoring the reef is that, according to the researchers, the size of the reef could have been smaller. Environmental research and analysis have provided a comprehensive list of the diverse marine species found on the reef. It has become evident that since construction the marine habitats created by the reef are of significant value (and much bigger then expected), both environmentally and for recreation in the form of diving and fishing.

The GSCs were predicted to be stable in the 8-10 m amplitude waves that occur during cyclones and up to now they have proved to be so.

Concerning surfing, there use of the Narrowneck area has significantly increased for all types of surfing and a number of competitions are held at Narrowneck because of its more reliable surfing conditions. However, Narrowneck's surfing attraction suffers from its proximity to numerous world class waves and the reef doesn't produce always 'perfect' waves - regardless of wind and wave conditions - as the general surfing public hoped (Jackson *et al.*, 2005).

Pratte's reef, El Segundo, America

Pratte's reef near Los Angeles in America was constructed between 2000 and 2001 (Borrero and Nelson, 2003). The purpose of the reef was just to enhance recreational surfing in order to offset the loss of surfing areas due to the construction of a groyne.

The overall dimensions are 30 m cross-shore with 70 m longshore. The submergence of the crest was 0.9 m below lowest tide. The reef (Figure 9) was constructed with about 200 GSCs.

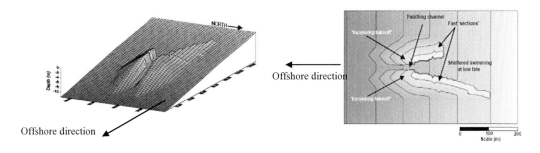

Figure 8. Narrowneck reef, Gold Coast, Australia - Plan view (up) and side view (down) of reef (Black, 2000).

The bags had a maximum volume of 7.9 cubic meters (the largest bags of the Narrowneck reef were about 400 cubic meters). The total volume was about 1600 cubic meters. The bags were filled in the port of Los Angeles and then loaded onto a barge, to be taken to the site. The bags were placed in the surf zone by a barge-mounted crane.

There is no available information about any design studies that have been carried out for this reef. After construction, however, shoreline monitoring, bathymetric surveys, diving surveys and surf quality surveys were undertaken. As a conclusion of the shoreline and bathymetry surveys it can be said that the reef had no effect on the shoreline and bathymetry. Any noticeable changes are due to natural seasonal changes. One remarkable finding of the diving survey was the rapid biological growth on the bags. Within weeks of the initial bag installation, algae had begun to grow on the bags and schools of small fish were attracted to the site. The diving survey also revealed that several reef bag units were ripped or shredded and were losing their fill material.

The conclusion of the surf quality of the reef is that there is an almost complete absence of surfers. The basic problem is that Pratte's reef is too small to significantly alter wave breaking and near shore coastal processes. The Narrowneck reef is 70 times larger (in terms of dimensions).

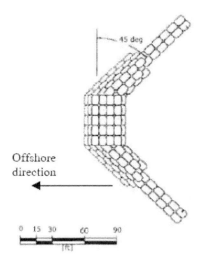

Figure 9. Plan view of design of Pratte reef (Henriquez, 2004).

2.2. Multi-Functional Artificial Reef Under Construction

Half 2008 just one MFAR was under construction, although several studies on the viability of MFAR are underway around the world. As before, the description of this reef concentrates on the results of environmental and monitoring studies.

Mount reef, Mount Maunganui, New Zealand

The reef is located on the east coast of New Zealand's north island. The ASR is designed to have a primary purpose of creating high quality surfing waves. Besides that, the Mount reef will be a research site for sustainable coastal protection and marine ecology. The sand banks, used by surfers, are constantly changing which means there are seldom consistent high-quality surfing waves at any particular spot.

The dimensions of the reef are 70 m cross-shore and 90 m longshore. The reef (Figure 10) will be constructed of 24 GSCs. The bags vary from 30 m long with a diameter of 1 m to 50 m long with a diameter of 3.5 m, This is a volume varying from 27 to 660 cubic metres. A smaller geotextile tube ("scour tube") will also be installed behind the reef to prevent sand being scoured from under the two big bags.

The total volume of the reef is 6 000 cubic meters, which is 1/20 of the Narrowneck reef. The reef will be constructed as follows: the empty bags will be tied onto a webbing lattice on dry land before being folded up and towed out to the reef site on a barge. The empty reef will then be offloaded and pulled down into position on the seabed using Ancor Locs. Once secured in place the sand is pumped into the bags to fill them all up and create the shape of the reef.

For this reef one physical study has been carried out and published (Moores *et al.*, 2006). No numerical studies have been published.

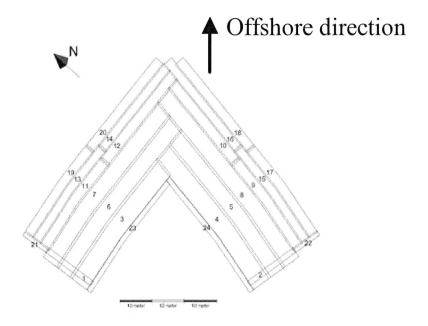

Figure 10. Plan view design Mountreef (source: www.mountreef.co.nz).

3. FUNCTIONALITY OF A MFAR

The viability of a MFAR is related to its functionality regarding the two main goals of the structure: surfing capability and coastal protection. In terms of coastal protection, it is very important to analyze both the hydrodynamics and morphodynamics. Concerning surf characteristics, there are several parameters to be checked, as described below.

3.1. Coastal Protection

3.1.1. Hydrodynamics in coastal zones

Most waves generated in the sea area by storm winds move across the ocean as swell. Generally, they release their energy along the margins of the continents in the surf zone, which is the zone of breaking waves.

According to the physical characteristics of the coastal platform responsible for the wave transformations as they travel cross-shore, wind-generated waves have been described by several different theoretical developments. In deep waters the small amplitude wave theory, which constitutes the first order of approximation of the Strokes theory, performs well. In transitional waters, as the waves become larger, higher orders of approximation to the Stokes theory can be can be used. Other nonlinear theoretical approaches should be used in shallow water conditions, like the cnoidal wave theory derived by Korteweg and de Vries (1895), and others. For very shallow water waves, a solitary wave theory developed by Boussinesq (1872), Serre (1953) and others should be used.

Figure 11, adapted from Kamphuis (2000), shows the applicability of the various wave theories.

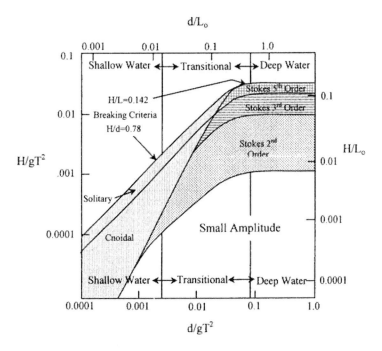

Figure 11. Applicability of various wave theories (Kamphuis, 2000).

The waves are transformed as they travel into shallower water. Once in shallow waters, a wave undergoes many physical changes before it breaks. The shoaling depths interfere with water particle movement at the base of the wave, so the wave velocity decreases. As one wave slows, the next waveform, which is still moving at unrestricted velocity, comes closer to the wave that is being slowed, thus reducing the wavelength. The energy in the wave, which remains the same, must go somewhere, so wave height increases. The crests become narrow and pointed, and the troughs become wide curves. This increase is called shoaling.

The combination of increasing wave height and decreasing wavelength causes an increase in the steepness (H/L) of the waves. In addition, energy is dissipated due to bottom friction.

If sections of a single wave crest are travelling in different water depths, the sections in deeper water will travel further per unit time, and therefore the wave will change direction or refract. Therefore, wave refraction is the gradual reorientation of waves propagating at an angle to the bottom slope or against a current. Figure 12 shows a refraction plane for the S. Lourenço fortification – Tagus estuary (Portugal) - corresponding to a wave 3.0 m high (offshore), a 12 sec period, and direction 225° (Antunes do Carmo and Seabra-Santos, 2002). As observed in this figure, refraction patterns are often interpreted by means of orthogonals, or wave rays. The first are lines drawn at right angles to the wave crests, and the second are lines indicating the direction of energy transmission.

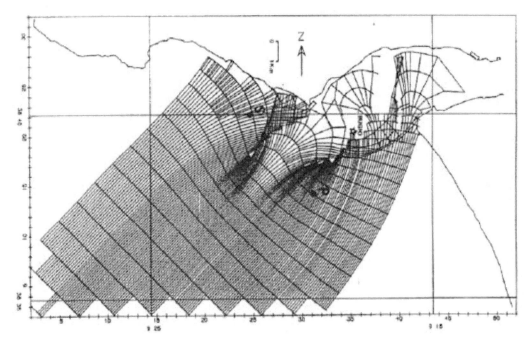

Figure 12. Refraction plane for the S. Lourenço fortification – Tagus estuary (Portugal) - corresponding to a wave 3.0 m high (offshore), 12 sec period, and direction 225° (Antunes do Carmo and Seabra-Santos, 2002).

In selected situations, like that presented in Figure 13, such as when waves pass a small cape, island or even piers and jetties, diffraction occurs; this is a lateral transfer of energy along wave crests.

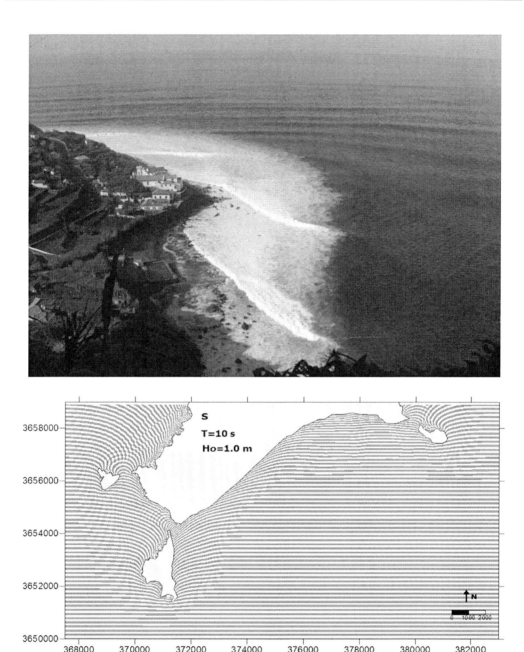

Figure 13. Madeira Island and wave crests at Porto Santo island computed by REFDIF (Fortes et al., 2006).

Incident waves may be reflected from beaches, cliffs, submarine shoals, bars and ridges, jetties, seawalls, etc. The reflected waves may be of the same dimensions as the incident waves, and if the wave travels in exactly the opposite direction then a standing wave can develop. Standing waves are the product of two waves of the same wavelength moving in opposite directions, resulting in no net movement. In confined basins standing waves are known as seiches.

3.1.2. Morphodynamics in Coastal Zones

Morphological changes in beaches depend on the nature of beach material, hydrodynamic processes (waves, tides and currents), the original profile and local boundary conditions. This last is related to the existence of headlands and bays, which control the energy acting on the beach and produce spatial variations in beach slope, grain size and sediment transport rates (Rey *et al.*, 2002).

Most shorefaces and shelves are underlain by relatively thick marine sedimentary sequences. In general, sediments consisting of sand and gravel occur on the upper and middle shoreface, muds of fluvial origin occur on the shelf, and mixed sands, and muds are found in the lower shoreface zone (transition zone to shelf) (Van Rijn, 2001). Bed sediments are generally fining in a seaward direction, supplied by offshore-directed bottom currents (rip currents and storm-induced currents).

The fluid in the shoreface zone may be homogeneous (well-mixed) or stratified with a surface layer consisting of relatively low fluid density (fresh warmer water) and a bottom layer of relatively high density (saline colder water). Strong horizontal density-related pressure gradients may occur in regions close to river mouths.

The upper shoreface (surf zone) is dominated by wave-driven processes. The surf zone can be seen as a subsystem of the shoreface zone. The shoreface zone influences the surf zone by providing boundary conditions, but the surf zone also affects the shoreface zone by generating strong rip currents that bring sediments to the shoreface. Furthermore, the surf zone is a source of free low-frequency energy propagating into the shoreface zone. The middle and lower shorefaces are affected by tide- and wind-driven currents and by Coriolis effects (Van Rijn, 2001).

Sand can be transported by wind-, wave-, tide- and density-driven currents (current-related transport), or by the oscillatory water motion itself (wave-related transport). The waves generally act as a sediment stirring agent, whereas the sediments are transported by the mean current. Wave-related transport may be caused by the deformation of short waves under the influence of decreasing water depth (wave asymmetry). Low-frequency waves interacting with short waves may also contribute to the sediment transport process (wave-related transport).

Tide- and wind-generated currents with near-bed velocities greater than 0.3 m/s are able, even at great depths, to move the sediments forming the bed surface. Although often rather weak, cross-shore currents combined with the stirring action of the waves are important for the long-term evolution of the shoreface morphology. Big storms are able to move sediments along the bed surface in water at depths of up to 100 m. Thus, the shoreface is an active morphodynamic zone, although the bed evolution processes may proceed rather slowly (Van Rijn, 2001).

While alongshore transport is the primary mechanism for changes in beach plan shape, cross-shore transport is the means by which the beach profile changes. The response time of beaches to variation in cross-shore transport can be as short as one tidal cycle (during storms) or as long as six months (seasonal variations). Predictions of beach response are very important for coastal designers and managers (CIRIA, 1996).

The variation in the processes across the beach results in characteristic beach profiles. The form of the beach profile will then have a feedback role in modifying the subsequent shoaling waves.

Dunes are created by the accumulation of wind-blown sand transported landward from the backshore and the higher portion of the inter-tidal foreshore. To successfully trap and retain this sand, dunes rely on vegetation, especially certain species of grass, which both reduce the wind velocity close to the dune face allowing deposition, and retain moisture, which increases the threshold of motion of sand grains. Figure 14 shows a healthy dune system on the central part of Portugal's west coast.

Figure 14. Healthy dune system on the west central Portuguese coast.

Dunes located on the backshore of a sandy beach are important in the development of the profile of that beach. They act as a reservoir of material which is available during storms and, if necessary, enables the beach profile to adjust to a flatter profile, and absorb the incoming wave energy (CIRIA, 1996).

3.1.3. Shoreline Response on a MFAR

Knowledge of the shoreline response on a reef is important for coastal protection purposes. Not only is the published information available on shoreline response to MFARs insufficient, but also relatively little is known about shoreline response to submerged structures in general. Ranasinghe *et al.* (2006a) recently made a compilation and review of reported field, laboratory and numerical modeling investigations and concluded that the key environmental and structural parameters governing shoreline response to submerged structures have not yet been adequately elucidated. The published reports of field experiences with submerged prototype structures mentioned in Ranasinghe *et al.* (2006a) are summarized in Table 1.

In brief, it was found that 70% of submerged structures constructed for beach protection to date have resulted in net erosion of the shoreline in their lee. It was also concluded that structure length, structure crest level, crest width, nearshore slope, littoral drift rates, and the presence or absence of concurrent sand nourishment, do not appear to govern the principal mode of shoreline response.

Table 1. Features of the sites and the submerged coastal structures reported in the published literature (Ranasinghe et al., 2006a)

Location	Reference	Structure type	Shoreline response	Nourishment	Longshore transport rate (m³/year)	B (m)	S (m)	W (m)	h (m)	h_c (m)	tan β
Delaware Bay, USA	Douglass and Weggel (1987)	Single breakwater +2 end groins	Erosion	Y	Negligible	300	75	Not reported	1	At MLW	Not reported
Keino-Matsubara Beach, Japan	Deguchi and Sawaragi (1986)	Single breakwater	Erosion	Y	Not reported	80	85	20	4	2 m below MLW	0.1 nearshore and 0.03 offshore
Niigata, Japan	Funakoshi et al. (1994)	Single breakwater +2 groins	Erosion	N	Exists, but not quantified	540	400	20	8.5	1.5 m below MWL	0.02
Lido di Ostia, Italy (#1)	Tomassicchio (1996)	Single breakwater	Erosion	Y	50,000	3000	100	15	4	1.5 m below MSL	0.05
Lido di Ostia, Italy (#2)	Tomassicchio (1996)	Single breakwater	Accretion	N	50,000	700	50	15	3–4	0.5 m below MSL	0.1
Lido di Dante, Italy	Lamberti and Mancinelli (1996)	Single breakwater	Accretion	Y	Negligible	770	150	12	3	0.5 m below MSL	0.02
Marche, Italy	Lamberti and Mancinelli (1996)	Multiple segmented breakwaters	Erosion	N	Negligible	Not reported	100–200	10–12	3	0.5 m below MSL	Not reported
Palm Beach, FL, USA	Dean et al. (1997)	Single breakwater	Erosion	N	100,000	1260	70	4.6	3	0.7 m below MLLW	0.04
Vero Beach, FL, USA	Stauble et al. (2000)	Segmented breakwater	Erosion	N	30,000	915	85	4.6	2.1–2.7	0.25 m–0.35 m below MLLW	0.03
Gold Coast, Australia	Jackson et al. (2002)	Multi-function surf reef	Accretion	Y	500,000	350	100–600	2	2–10	1 m below MLW	0.02

(Sh. re.=shore response, Er.=erosion, Ac.=accretion, No.=nourishment, B=length of structure, S = distance from undisturbed shoreline to structure, W = crest width, h = water depth at structure, hc = water depth at crest of the structure, tan β = bed slope in the vicinity of the structure)

In an innovative study, Black and Andrews (2001) quantified the shape and dimensions of salients and tombolos formed in the lee of natural reefs by visually inspecting aerial photographs of the coastlines of south eastern Australia and New Zealand. By analyzing natural shoreline adjustment, all physical inputs that act to shape salients and tombolos over long time scales are brought together. Results suggested that natural salients are larger than salients created in the lee of breakwaters and in laboratory studies. Analyses produced non-dimensional ratios, enabling the prediction of limiting parameters for salient and tombolo formations, determination of salient apex position with respect to the type of offshore obstacle (islands or reefs), and the length of shoreline affected (salient basal width). One of the suggestions emerging from the analysis was that, if all other parameters (length of reef/structure, B, distance from shoreline to reef/structure, S, wave climate, etc.) are equal, a larger salient would develop in the lee of a submerged reef/structure than in the lee of an emergent structure. However, as mentioned by Ranasinghe *et al.* (2006b), this conclusion is counter-intuitive because wave sheltering will be greater in the lee of emergent structures, leading to more favorable conditions for salient growth in their lee. A subsequent review of the methodology adopted by Black and Andrews (2001), indicated that the approach used to assess the critical length scales of natural reefs from aerial photographs incorporated several shortcomings (Ranasinghe *et al.*, 2006b). The most obvious shortcoming of the predictive empirical relationship proposed by Black and Andrews (2001) is the fact that erosion is not predicted to occur for any combination of B and S. This is highly questionable in view of the fact that a clear majority of submerged structures installed to date have resulted in (unintended) shoreline erosion (see Table 1).

In order to gain more insight into the environmental and structural conditions under which shoreline erosion and accretion occur in the lee of submerged structures, and to gain insight into the nearshore processes governing shoreline erosion and accretion in the lee of submerged structures, Ranasinghe *et al.* (2006b) completed a second study on this subject. The shoreline response to submerged structures, such as MFARs, was investigated. The processes governing this response were found to be different from those associated with emergent offshore breakwaters. This was indicated by the results of a series of 2DH numerical and 3D scaled physical modeling tests. Unlike the case of emergent offshore breakwaters, where shoreline accretion (salient development) is expected under all structural/environmental conditions, the principal mode of shoreline response to submerged structures can vary between erosive and accretive, depending on the offshore distance to the structure. In the case of the left picture of Figure 15 erosion was seen, and in the right one accretion.

The predominant wave incidence angle and structure crest level also have important implications for the magnitude of shoreline response, but not for the mode of shoreline response (i.e. erosion vs. accretion). With the geometry and the wave conditions tested by Ranasinghe and Turner (2006b) the most significant feature in the structure-induced nearshore circulation patterns under shore normal wave incidence is the 'switch' from a symmetric 2-cell circulation system to a symmetric 4-cell circulation system, as the structure is moved offshore (Figure 15).

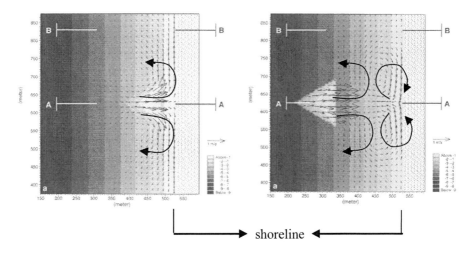

Figure 15. Distance apex structure-coast 100 m. (left) and distance apex structure coast 250 m. (right) (Ranasinghe et al., 2006).

Based on the results obtained in their study, a predictive empirical relationship was proposed as a preliminary engineering tool to assess shoreline response to submerged structures. This relationship is $S_a/SZW > 1.5$, where S_a is the distance between the apex of the structure and the undisturbed shoreline and SZW is the natural surf zone width.

Broad-crested nearshore structures are a relatively new field of research, and a 'standardized' method to describe their geometry has not yet been established in the literature. For more conventional shore-parallel rubble-mound structures it is probably most usual to define the characteristic length of 'S' as the distance from the shoreline to the centre-line of the crest. For the broad-crested structures Ranasinghe *et al.* (2006b) modeled, the distance to the apex of the crest (S_a) was chosen as the characteristic length (Figure 16).

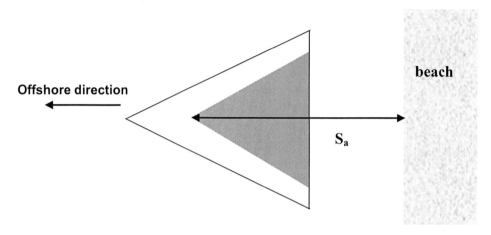

Figure 16. Characteristic length broad-crested breakwaters.

Much research has been done on currents and wave breaking on and around natural reefs. However, the topography of natural reefs is highly variable, some fringing reefs (reefs which front a continental land mass or island) may have quite flat seaward slopes, for example 1:15, whereas platform reefs (reefs located in open ocean) on the outersections may have very steep seaward slopes, for example 1:2 or steeper (Gourlay, 1993). No publications have been found that deal with the relation between the topography of the reef and the influence on hydrodynamics in general regarding the coastal protection behavior of such natural reefs.

In 2005 the DELOS project finished. The overall objective of this project was to promote the effective and environmentally compatible design of low crested structures (LCSs) to protect European shores against erosion and to preserve the littoral environment and the economic development of the coast. Much experimental and numerical research has been undertaken in this project. In all the experimental studies the LCS were small-crested and designed with a relatively steep slope. This means that the results are not useful for research into the coastal protection aspect of MFARs. Furthermore, no tests were conducted to investigate current velocity in the lee of an LCS in a basin. The focus of the basin tests was stability, hydrodynamics near the structure and wave transmission.

In order to get an idea of the shoreline response on a MFAR, and taking into account all the research described above, the mean current field can be used (Ranasinghe and Turner, 2006b). The numerical model results (current analysis) and the prototype field results (morphodynamic analysis) reported by these researchers showed that:

   a) under shore normal wave incidence, the mode of shoreline response to submerged structures is governed by the structure-induced nearshore circulation; and

b) under oblique wave incidence, the mode of shoreline response is governed by the interaction between the ambient longshore current and the structure-induced nearshore circulation pattern.

This means that the currents (the ambient longshore and the structure-induced nearshore circulation pattern) can be used as an indication of the shoreline response. Essentially, erosion occurs when the resultant current field contains divergent alongshore currents at the shoreline in the lee of the structure. Conversely, shoreline accretion occurs when convergent alongshore currents are generated at the shoreline in the lee of the structure.

The hydrodynamic processes that govern the development of nearshore circulation patterns around relatively simple delta-form MFARs are explained in Ranasinghe and Turner (2006) and are partially presented in this section. For reasons of simplicity, just shore normal waves are considered (i.e. no ambient longshore current due to oblique wave incidence at the shoreline). References to 'longshore flow/transport' in this section refer only to structure-induced alongshore flows/currents.

In the two-dimensional case of a submerged structure, such as the case of a submerged reef, longshore flow driven by an alongshore pressure gradient occurs behind the structure to satisfy continuity constraints. The total transport capacity of the longshore flow is determined by the longshore pressure gradient and the cross-sectional area of the longshore flow.

Figure 17 shows the cross-shore profile of surface elevation, wave height, and initial bed level at the two cross-sections presented in Figure 15 (left) and here when the structure is close to shore (apex at 100 m from the shore).

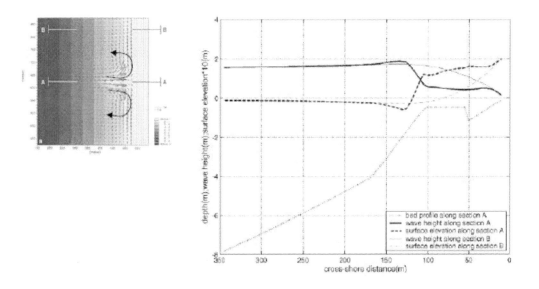

Figure 17. Wave height and surface elevation along different cross-sections for apex structure-coast distance 100 m (left, Ranasinghe et al., 2006).

One section is across the apex of the structure (Section A) while the other is along the plane bed away from the structure (Section B). Along Section A, the wave height decreases through the surf zone and over the top of the structure producing a higher surface elevation than at Section B. This alongshore gradient in surface elevation drives longshore flows away from the structure in both longshore directions in the lee of the structure. However, in this case, the small gap between the structure and the shoreline and the resulting relatively shallow trough region between the structure and the shoreline constrains the capacity for onshore flow over the structure. Therefore, the longshore flow in the lee of the structure is largely due to the alongshore surface elevation gradients resulting from alongshore gradients in the wave setup in the lee of the structure. However, because of the complex plane shape of the artificial surfing reef structure investigated here, currents are also generated along the two sides of the structure due to waves breaking obliquely on the sides of the structure. These 'along-structure' currents also contribute to the onshore flow over the structure and the alongshore flows in the lee of the structure. The resulting nearshore circulation pattern consists of two cells, symmetric about Section A, with divergent flow at the shoreline in the lee of the structure.

Figure 18 shows the cross-shore profile of surface elevation, wave height, and initial bed level at the two cross-sections when the structure is farther away from the shore (apex at 250 m from the shore).

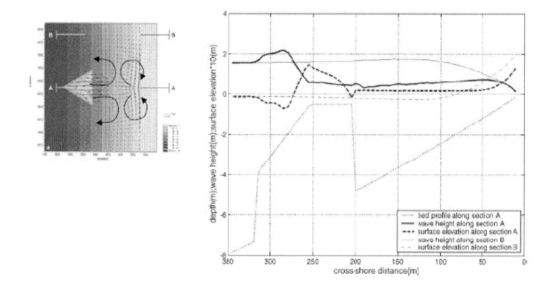

Figure 18. Wave height and surface elevation along different cross-sections for the case of distance apex structure-coast 250 m (left, Ranasinghe et al., 2006).

Again, one section is across the apex of the structure (Section A) while the other is along the plane bed away from the structure (Section B). In this case, the longshore flow is less constrained by the topography, thereby allowing larger cross-shore flow over the structure. In contrast to the previous case, here wave setup occurs through the surf zone on the structure, reaching a local maximum at the top of the structure slope, and decreases towards mean sea level over the flat top of the structure. Through the surf zone, part of the radiation stress gradient drives the onshore flow over the structure and the setup is reduced relative to the zero-onshore flow case (i.e. adjacent plane bed case). A small amount of wave dissipation occurs across the top of the structure which, together with the hydrostatic pressure gradient due to setup, maintains the onshore flow over the structure. A small positive set-up remains at the back of the structure providing the longshore pressure gradient to drive the longshore flow (directed away from the structure), with the maximum longshore transport occurring in the deepest water immediately behind the structure. As in the previous case, the 'along-structure' currents due to oblique wave breaking on the structure contribute to the onshore flow over the structure and to the alongshore flows in the lee of the structure. However, in contrast to the previous case the substantial distance between the back wall of the structure and the shoreline (i.e. 200 m) allows the waves to re-form to some extent in the lee of the structure. Therefore, wave setup occurs again at the shoreline in the lee of the structure when the re-formed waves break at the shoreline. However, the reformed wave height in the lee of the structure (section A) is lower than the height of the previously unbroken waves at the plane bed away from the structure (section B). Therefore, the wave setup at the shoreline in the lee of the structure is lower than that along the shoreline adjacent to the structure. This longshore gradient in the setup at the shoreline results in a longshore gradient of surface elevation which drives alongshore currents towards the lee area, where they converge before turning offshore, owing to mass conservation requirements. Therefore, in this case, the resulting nearshore circulation pattern consists of four cells, symmetric about Section A, with convergent flow at the shoreline in the lee of the structure.

More complex geometries (for example the presence of a platform) are however more difficult to explain with the theories set forth in the above paragraphs.

3.2. Surfing

Besides coastal protection, the creation of surfing conditions is an important aspect of a MFAR. This section gives a systematic explanation of the key surfing parameters used in the design of a MFAR, viz., the peel angle, the type of breaker, the wave height and the currents. However, due to the importance of waves for surfing, the phenomena of free waves are elucidated first.

3.2.1. Free Surface Waves

Waves form when the water surface is disturbed, by wind, earthquakes or planetary gravitational forces, for instance. Thus, we can identify at sea waves that have very short wave periods (order of 0.10 seconds, known as capillary waves) to tides, tsunamis and seiches (basin oscillations), where the wave periods are expressed in minutes or hours. Wind waves, which account for most of the total available wave energy, have periods from 1 to 30 seconds and wave heights that are seldom greater than 10 m and mostly of the order of 1 m, particularly in deep water conditions. As can be seen in Figure 19, most of the energy possessed by ocean waves is in wind-generated waves (Thurman and Trujillo, 1999).

During such disturbances energy and momentum are transferred to the water mass and transmitted in the direction of the impelling force. A proportion of the wave energy is dispersed by radial, inertial and convective means, but a large amount is not lost until waves encounter shallow coastal waters.

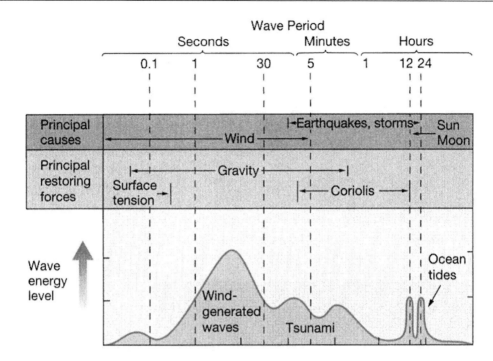

Figure 19. Distribution of energy in ocean waves (Thurman and Trujillo, 1999).

Wind waves can originate near the shore or offshore. In case of offshore generated waves, storms and depressions cause strong winds to blow over a stretch of ocean surface (fetch) for a certain amount of time. Wind energy is transformed into wave energy by the creation of high frequency waves known as choppy waves. The energy from the higher-frequency waves (lower periods) is in turn transferred to lower-frequency waves (longer periods): the so called **swell waves**. The swell waves travel faster than higher-frequency waves (speed of waves in deep water is $gT/(2\pi)$) and consequently the swell waves will separate from the higher frequency waves as the wave field propagates away from the region of generation. The higher-frequency waves of distant storms have dissipated most of their energy before they reach the coast, and so the wave field gets cleaned up. The swell waves with the longest wave periods can travel thousands of kilometers. The swell energy is distributed over a relatively narrow range of frequencies resulting in a slowly modulated wave field. Therefore, swell waves always arrive in sets. The periods of swell waves vary in general between 8 and 16 seconds. **Local waves** are formed by local winds, which cause the wave climate to be irregular and the waves to have short wave periods (in general between 4 and 7 seconds).

As mentioned, waves with lower frequency (longer wavelengths) travel faster, and thus leave the sea area first. The same is true for wave groups. Waves travel in wave groups. Figure 20 shows such a wave group in which L is the wavelength, L_g is the wave-grouplength, c is the wave celerity and c_g is the wave group celerity.

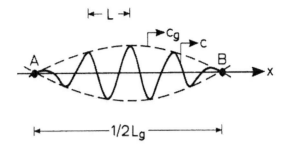

Figure 20. Wave group (Fredsøe and Deigaard, 1995).

The difference in speed of the single waves is the cause for the principle of wave dispersion, which is a sorting of waves by their wavelength. As a group of waves leaves a sea area and becomes a swell wave train, the group moves across the ocean surface at only half the velocity of an individual wave in the group. Progressively, the leading wave disappears. However, there is always the same number of waves in the group. As the leading wave disappears, a new wave replaces it at the back of the group. Most waves generated in the sea area by storm winds move across the ocean as swell. Generally, they release their energy along the margins of the continents in the surf zone, which is the zone of breaking waves.

3.2.2. Peel angle

Surfable waves never break all at once along the wave crest. If this does happen, the waves are closing-out and not suitable for surfing purposes. In order for a wave to be surfable, the wave has to break gradually (read peel) along the wave crest. The velocity with which this happens is called the 'peel rate' V_p of the wave (Figure 21).

The peel angle is one of the most important surfability parameters (α in Figure 21). The peel angle is the angle enclosed by the wave crest and the breaker line (Walker, 1974). Two other vectors that are shown in Figure 21 are $\vec{c}$ and $\vec{V}_S$. $\vec{c}$ is the wave celerity and $\vec{V}_S$ (called down-line velocity) is the absolute value of the vector sum of the velocities $\vec{c}$ and $\vec{V}_p$, it is the actual velocity experienced by the surfer. From Figure 21 it can be seen that another way to see the peel angle is as the angle enclosed by the velocity vectors of the peel rate $\vec{V}_p$ and the down-line velocity $\vec{V}_S$ (Henriquez, 2004).

In Figure 22 it can also be seen that the peel angle in fact is equal to the wave angle β between the normal on the bathymetry and the wave ray (the conventionally called 'wave angle') if the breaker line is parallel to the bathymetry. The angle θ is the so-called reef angle.

Whether a wave is surfable depends mainly on the value of the peel angle α, related to the down-line velocity as: $\vec{V}_S = \dfrac{\vec{c}}{\sin \alpha}$.

When the peel angle becomes too small, the down-line velocity will become very high and too fast for the surfer. The value of the peel angle, α, needs to be sufficiently large for a wave to be surfable. The velocity that a surfer can reach depends mainly on the wave height H_b at the breaking point and the skill of the surfer. Hutt *et al.* (2001) investigated what the peel angle α has to be for a given wave height H_b and surfer skill (Figure 23). The definition of these surfer skills is shown in Table 2.

Figure 21. Illustration of wave celerity vector $\vec{c}$, peel rate $\vec{V}_p$, down line velocity $\vec{V}_s$ and peel angle α (Source:surfermag.com).

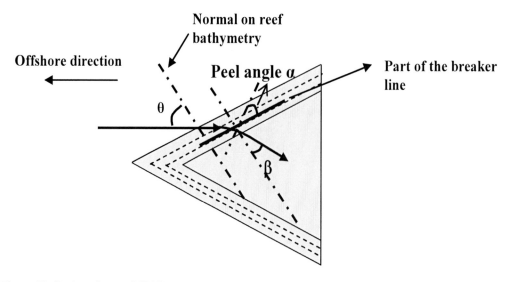

Figure 22. Peel angle on a MFAR.

Table 2. Rating of surfer skill level (Hutt et al., 2001)

Rating	Description of Rating	α (deg)	Hb (m)
1	Beginner surfers not yet able to ride the face of a wave and simply moves forward as the wave advances.	0	0.70 – 1.00
2	Learner surfers able to successfully ride laterally along the crest of a wave.	70	0.65 – 1.50
3	Surfers that have developed the skill to generate speed by 'pumping' on the face of the wave.	60	0.60 – 2.50
4	Surfers beginning to initiate and execute standard surfing manoeuvres on occasion.	55	0.55 – 4.00
5	Surfers able to execute standard manoeuvres consecutively on a single wave.	50	0.50 +
6	Surfers able to execute standard manoeuvres consecutively. Execute advanced manoeuvres on occasion.	40	0.45 +
7	Top amateur surfers able to consecutively execute advanced manoeuvres.	29	0.40 +
8	Professional surfers able to consecutively execute advanced manoeuvres.	27	0.35 +

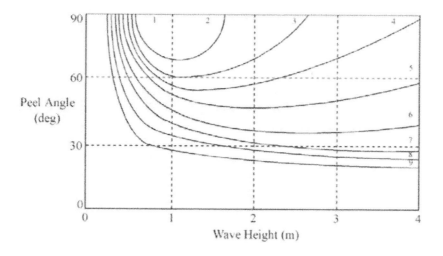

Figure 23. Peel angle as a function of wave height and surfer skill. The peel angle is on the y-axis, the wave height on the x-axis and the surfer skill is indicated by numbers in the graph (Hutt et al., 2001).

Henriquez *et al.* (2006) proved algebraically that, for an arbitrary peel angle, wave height and offshore depth (in the context of a geometrical optics approximation of linear wave theory), the maximum peel angle occurs at an angle on deep water of $arctan\sqrt{5} \approx 66°$. The algebraic proof means that regardless of wave conditions and offshore depth it will always show the maximum of the peel angle at an angle on deep water of 66°.

The relation between the peel angle, α, and the angle θ (Figure 22) is represented by the graph in Figure 24. The results shown in this figure are for a 10 s period and a wave height in deep water of 1.5 m. The calculations related to waves travelling over a shelf before reaching the reef (Figure 25). However, because linear shoaling is assumed, the results are the same for a wave travelling over a sloping bottom before reaching the reef (Figure 26).

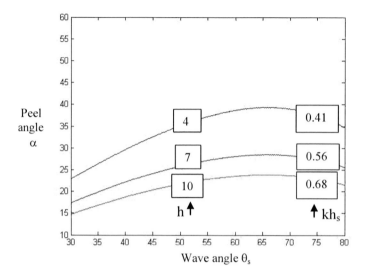

Figure 24. Peel angle α as a function of the reef angle θs and the shelf depth khs or hs for T = 10 s and H0 = 1.5 m.

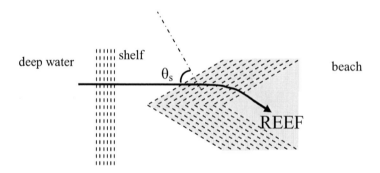

Figure 25. Peel angle with shelf.

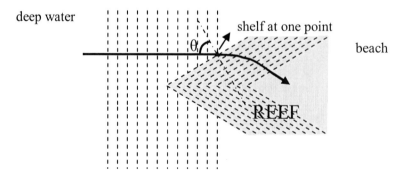

Figure 26. Peel angle without shelf.

Not only does the peel angle have a maximum, but it is clear from Figure 24 that it has an exponential growth for the same wave angle θ_s when the reef starts in shallower water. The physical explanations of these two phenomena are described below.

How Can the Peel Angle Have a Maximum?

When calculating the parameters like wave height, water depth and peel angle at the breaking point, with linear theory, it appears that the larger the wave angle in deep water, the less the water depth at the breaking point. Figure 27 shows this phenomenon for different wave angles in deep water. In this Figure, the angle of incidence in deep water, θ_0, can be seen as the angle θ in Figure 22 and the wave angle in the third column in this Figure, at the breaker line position, can be seen as the peel angle α in Figure 22. The second 'column' in Figure 27 are the values of the wave angles at the line of the breaker line for the angle on deep water of 45 degrees (wave conditions are a period of 10 seconds and a wave height of 1.5 m at deep water).

From this, it can be seen that the wave angle grows at the same water depth when the wave angle on deep water grows (law of Snellius).

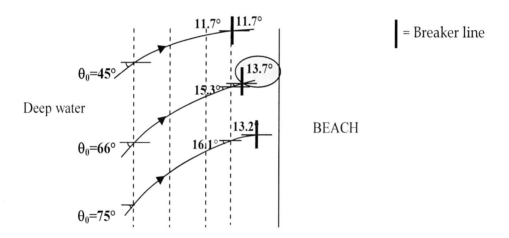

Figure 27. Wave breaking more shorewards for larger wave angle at deep water.

The fact that the wave travels more shoreward, for a larger wave angle on deep water, has to be caused by a stronger decline in the wave height so that the wave can travel on longer before it breaks. For that reason the development of the wave height along the wave ray is investigated. Two processes influence the wave height when a wave travels towards the coast, and these are shoaling and refraction. The variation in the wave height for different water depths is given by Eq. 2.

$$H_2 = H_1 * K_s * K_r \tag{2}$$

where $K_s = \dfrac{c_{g1}}{c_{g2}}$ is the shoaling factor and $K_r = \dfrac{\cos \theta_1}{\cos \theta_2}$ is the refraction factor. The wave travels from the location of wave height H_1 towards the location of wave height H_2. Shoaling is independent of the wave angle on deep water, refraction, however, is dependent on that angle. Figure 28 shows the factor K_s*K_r when the wave travels from deep to shallow water conditions for angles on deep water of 45 degrees and 75 degrees, considering the same wave conditions shown in Figure 24. Figure 29 shows the consequence for the wave height of what is demonstrated in Figure 26.

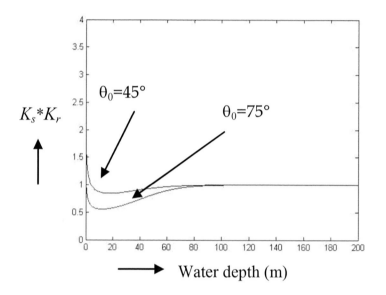

Figure 28. Multiplied factor of the shoaling and refraction factor.

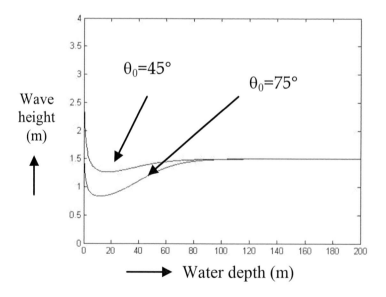

Figure 29. Wave height as a function of the shoaling and refraction factor.

Figure 29 shows that, for the same water depth, the wave height decreases when the wave angle on deep water increases. Because of this decrease, the wave can travel further towards the coastline before the breaking condition $H_b = 0.78*h$ is met. As a consequence of the law of Snellius, it can be continued longer and the wave angle can continue to diminish. The maximum peel angle occurs with an angle on deep water of 66 degrees no matter what the wave conditions. So, at an angle of 66 degrees the decrease of the peel angle by the effect of a longer continuation of refraction (law of Snellius) is larger than the increase of the peel angle through the effect of a larger wave angle on deep water.

Why is there Exponential Growth of the Peel Angle for the Same Wave Angle Θs when the Reef Starts in Shallower Water?

Another aspect of the peel angle is that it experiences exponential growth with decreasing depth of the start of the reef, for the same wave angle θ_s, as can be seen in Figure 24. The only process by which the wave angle changes along the wave ray when a wave travels from deep water towards the shoreline is refraction. The influence of refraction on the wave angle is given by Snellius' law (Eq. 3).

$$\frac{d}{dx}\left(\frac{\sin\theta}{c}\right) = 0 \tag{3}$$

where dx is in the direction of the wave ray, θ is the wave angle between the wave ray and the normal to the bathymetry and c is the wave velocity. Eq. 3 can be re-written as:

$$\sin\theta_2 = A\sin\theta_1 \; ; \; A = \frac{\tanh(kh)_2}{\tanh(kh)_1} \tag{4}$$

where k is the wave number, and the water depth h_1 is greater than the water depth h_2.

The difference in the water depth from point 1 to point 2, $\Delta h = h1 - h2$, is assumed to be the same for all water depths. In fact, the value of A in Eq. 4 gives information about the magnitude of refraction. The value of A in Eq. 4 is plotted in Figure 30 for different water depths. The period in these calculations is 10 s. It can be seen in Figure 30 that refraction for the same Δh is relatively greater in shallower water.

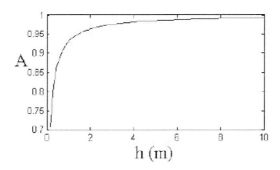

Figure 30. Relative refraction.

In order to show this effect on the exponential growth of the peel angle with decreasing depth, two cases are assumed (Figure 31):

- Case 1: two reefs in relatively deep water, starting with a difference of Δh in water depth;
- Case 2: two reefs in relatively shallow water, starting with a difference of Δh in water depth.

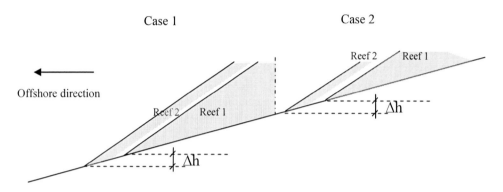

Figure 31. Two cases of reefs starting at different depths.

In Case 2 the loss in total refraction of the wave from θ to the wave angle β between the two reefs is larger than in Case 1. Figure 32 shows β for the deeper start of the reef in Cases 1 and 2, where Δx is the horizontal distance for the vertical distance Δh. This larger value of loss of refraction in case 2 means that the wave at the breaking point has refracted relatively less in Case 2 than in Case 1 (even though the initial refraction is higher (Figure 30)) leading to a stronger growth of the peel angle. This effect appears at every water depth and it is the reason that the peel angle experiences an exponential growth with decreasing depth, for the same wave angle θ. So it can be said that the peel angle grows exponentially for a lower start of the reef depth with the same angle θ because refraction is exponentially greater in shallow water.

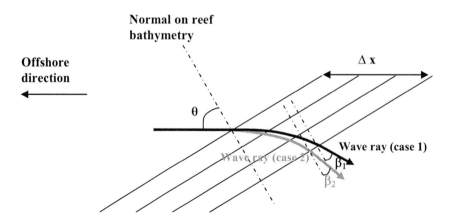

Figure 32. Different loss in refraction for the two cases of Figure 31.

3.2.3. Breaker type

The shape of a breaking wave is of great importance for surfing. Battjes (1974) developed a formula (Eq. 1) with which the breaker type can be predicted:

$$\xi_b = \frac{s}{\sqrt{\dfrac{H_b}{L_0}}} \tag{5}$$

where ξ_b is the inshore Iribarren number, s is the bottom slope, H_b the wave height at breakpoint and L_0 the deep water wave length. The value of the Iribarren number corresponds with every regime as in Table 3.

Table 3. Breaker type transition values

Regime	Range
Surging/collapsing	$\xi_b > 2.0$
Plunging	$0.4 > \xi_b > 2.0$
Spilling	$\xi_b < 0.4$

It should be noted that these results are based on experiments on plane slopes where the angle of incidence of the waves was zero. The slope is the slope that the wave experiences, so is along the path of the wave and not along the normal on the reef contours (Figure 33).

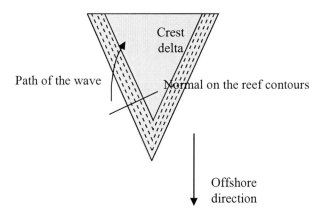

Figure 33. Path of the wave compared to the normal on the depth contours.

The main breaking types are described as follows (the technical part of terminology is from Galvin (1968), the relation between the breaker types and the Irribaren number from Battjes (1974), and the surfer interpretation from Henriquez (2004)).

- Spilling breakers

These breaking waves occur if the wave crest becomes unstable and flows down the front face of the wave producing a foamy water surface – surfers would say a 'soft' or 'weak' wave. This regime is considered surfable. Spilling waves are shown in Figure 34.

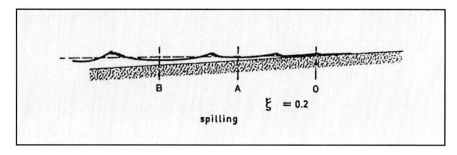

Figure 34. Spilling waves, $\xi_b = 0.2$ (Battjes, 1974).

- Plunging breakers

These breaking waves occur if the crest curls over the front face and falls into the base of the wave, resulting in a high splash - surfers call this a 'tubing' wave. This regime is preferred by most surfers, and a more spilling plunging wave is preferred over a more collapsing plunging wave. Plunging waves are shown in Figure 35 and Figure 36.

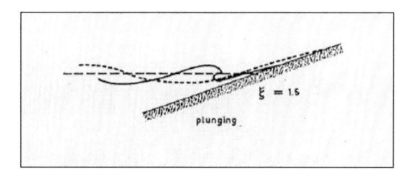

Figure 35. Plunging waves $\xi_b = 1.5$ (Battjes, 1974).

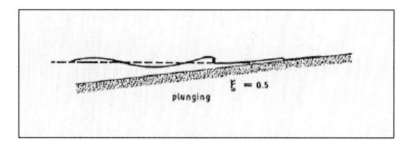

Figure 36. Plunging waves, $\xi_b = 0.5$ (Battjes, 1974).

- Collapsing breakers

These breaking waves occur if the crest remains unbroken and the front face of the wave steepens and then falls, producing an irregular turbulent water surface - surfers often encounter this regime at reef breaks when the tide is too low and the reef is not submerged enough to produce surfable waves, and so it is an unsurfable regime. Collapsing breaking waves are shown in Figure 37.

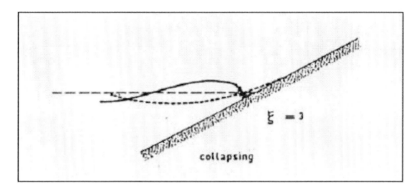

Figure 37. Collapsing waves, $\xi_b = 3$ (Battjes, 1974).

- Surging breakers

These breaking waves occur if the crest remains unbroken and the front face of the wave advances up the beach with minor breaking. This regime is also unsurfable. Surging breaking waves are shown in Figure 38.

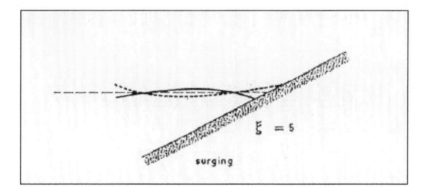

Figure 38. Surging waves, $\xi_b = 5$ (Battjes, 1974).

It should be noted that the breaker types mentioned above will only develop as the Iribarren number describes (according to the bottom slope and wave steepness) when the wave experiences the relevant conditions for enough length of the slope. So a wave that should in principle be plunging when it breaks on a MFAR, according to equation 1, will be more spilling when it breaks if the slope of the reef is relatively too short to reach the Iribarren number for the relevant bottom slope and wave steepness. In other terms, the wave breaks to some extent according to the bottom slope and not just according to the reef slope (Figure 39).

Figure 39. Breaker type.

3.2.4. Wave heights

By now every possible wave has been surfed. Surfers that surf on a long board are still surfing when waves are 0.15 m high, whereas those that are towed into waves ride the biggest waves they can find, up to 20 m. On the whole, waves between 0.5 m and 10 m are considered surfable (Henriquez, 2004). However, in order to study the viability of a MFAR, wave heights between 1m and 3 m are the most common values, and are adjusted according to the target level of surfers for which the MFAR are being designed.

3.2.5. Currents

Currents around a surf break are of vital importance when considering the surfability of the break. There could be waves in perfect surfing condition but yet unreachable due to strong currents. Usually these cases are rare but it is not ideal to have to be constantly paddling to keep positioning.

Rip currents can destroy good surfable waves. Rip currents are narrow strong currents that move seaward through the surf zone (Bowen, 1969). When the rip-current flows through the breaker zone the wave seems to get a rough surface and breaks in a hesitating manner making the waves unsuitable for surfing. Rip-currents can also be advantageous; the surfer can use the rip-current to get outside the breaker zone more easily (Henriquez, 2004).

3.2.6. Wave Focusing

Wave focusing, like refraction, is another physical process that has a large influence on surfing as it has a large influence on the peel angle. If wave focusing occurs, the wave rays converge, leading to an increase in wave height and consequently to its earlier breaking. This causes the breaker line to be nearer the intersection of the reef and the sea bottom than if there was no wave focusing, leading to less refraction and thus a higher value of the peel angle. The breaker line in Figure 40 and Figure 42 and the graph in Figure 41 can be divided into three. Part A, where wave focusing occurs; part B, where wave defocusing occurs, and part C, where there is neither wave focusing nor defocusing (α_2 and α_3 are zero, Figure 41).

When wave focusing occurs, the value of the peel angle can be divided into several contributions:

- α_1 is the angle θ minus the decrease due to refraction on the reef slope (Figure 41 and Figure 43);
- α_2 is the difference between the angles for the cases with and without wave focusing, due to less refraction by earlier breaking (see α_2 in Figure 41, $\alpha_2 = \alpha_4 - \alpha_1$ in Figure 43);
- α_3 is the angle due to deviation of the breaker line from the parallel to the bathymetry of the reef side (Figure 41 and Figure 44).

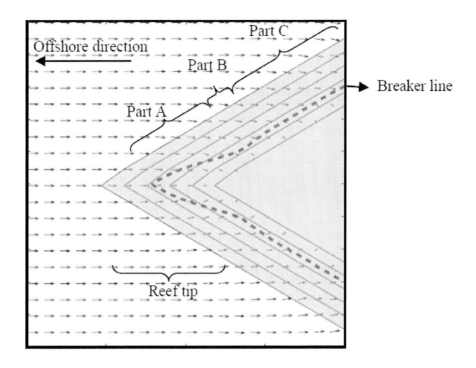

Figure 40. Breaker line (Henriquez, 2004).

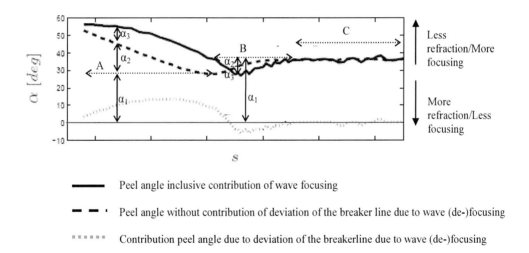

——— Peel angle inclusive contribution of wave focusing

— — · Peel angle without contribution of deviation of the breaker line due to wave (de-)focusing

· · · · · · Contribution peel angle due to deviation of the breakerline due to wave (de-)focusing

Figure 41. Contributions to the peel angle along the breaker line (basic Figure: Henriquez, 2004)

The transition from wave focusing to wave defocusing occurs at the minimum of angle $\alpha_1+\alpha_2$ on the right side of part A. This can be concluded because at that point the peel angle is equal to α_1, which is the value for the peel angle without no wave focusing.

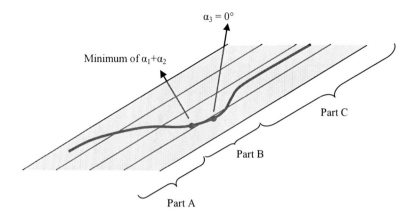

Figure 42. Enlargement of part of the breakerline; schematic positions of certain values of the peel angle.

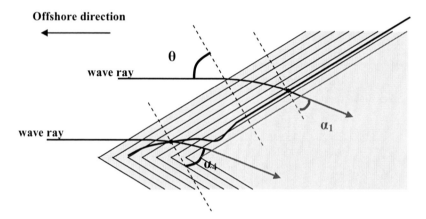

Figure 43. Angle α1 and α4.

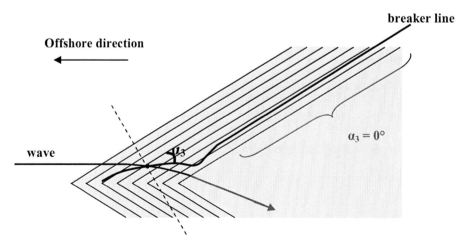

Figure 44. Angle α3.

The peel angle α_1 is different for the areas A and B. In area A, α_1 is the minimum value of $\alpha_1+\alpha_2$, corresponding to a peel angle value when no wave defocusing (after the wave focusing) would occur. For large values of s, the peel angle tends to a certain value. This is the value of α_1 in area B, which would occur if there were no wave focusing and no wave defocusing. α_1 is smaller in area A than in area B. This means that the position of the minimum of $\alpha_1+\alpha_2$ at the right end of area A is nearer the crest level of the reef than the location of the breakerline for large values of s in area C (Figure 42).

The angle α_2 is positive in area A and negative in area B (Figure 41). It is positive in area A because the breaking occurs further away from the crest level of the reef than it would if there were no wave focusing. The value in area B is negative because the breaking occurs nearer the crest level of the reef than it would if no wave focusing and no wave defocusing occur.

4. DESIGNING AN OPTIMAL GEOMETRY

It would be unrealistic to present a single optimal design of a MFAR. The optimal design depends on several aspects:

- Bathymetry;
- Local wave climate;
- Local currents and sediment transport;
- Environmental conditions;
- Budget.

However, for the design of any artificial reef that will be built to both protect the local coastline and to create surfing conditions, several design parameters have to be considered. What should be taken into account for these parameters in terms of hydrodynamics is described below. This chapter concentrates on hydrodynamics, but once an optimal geometry in relation to hydrodynamics has been determined, a thorough morphodynamic study (physical and/or numerical) is necessary to investigate the capacity of a MFAR to protect a local coastline.

4.1. MFAR-Angle

The choice of the MFAR-angle, θ, (Figure 45) is mainly related to the peel angle. As described in section 3, the peel angle has its maximum for $\theta = 66°$ and consequently this is the first option for this value. However, this maximum is found with linear refraction and without wave focusing. In order to get the proper form for the design peel angle, numerical simulations that take irregular waves and wave focusing into account, ought to be performed.

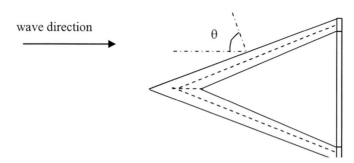

Figure 45. Reef angle (reef in plane view).

4.2. The Height of the Reef

The height of the reef depends on its horizontal dimensions, its distance from the shore, its submergence and its slopes. However, the height needs a minimum value in relation to the breaker type. For a certain reef slope, the height determines the length of the reef side. This should be long enough for the wave to feel the reef, otherwise the wave does not have enough space and will also break according to the slope of the bottom (see Figure 46).

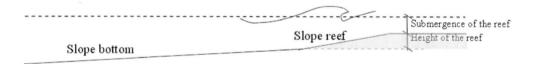

Figure 46. Schematic cross-section of the MFAR.

It is not yet known how long a reef needs to be to get the desired breaker type under certain wave conditions and slope of the reef (according to Eq. 1). The research of Smith and Kraus (1991) is closely related to the influence of the length of the reef slope on the breaker type. They conducted a laboratory study of a wave breaking over bars and artificial reefs and used the results to categorize the offshore breaker type differently from Battjes (1974) when the bottom has a barred profile. However, this categorization is general, for varying reef slopes and wave conditions, and the length of the reef slope is not taken into account, so it is not appropriate for this study.

To determine the exact length of the reef side required for the proper breaker type, as a function of the wave length, and thus the value of the reef height, numerical simulations should be performed. In order to choose the initial geometry for the numerical simulations, a fraction of the local wave length can be taken as the minimum length of the reef; take, for example, 1/4 times the local wave length. Thus, for a relatively long wave period of 14 seconds and a reef start at a depth of 4 m, the minimum length is 22 m. Assuming 1:10 slope, the minimum height is then 2.2 m for waves breaking on the crest of the reef. For the design, an extra value can be added, so waves can also break a little before reaching the crest, for example, 0.3 m. In this example, the minimum height that should be taken is 2.5 m. For shorter wave periods (or shorter wave lengths) a lower reef height should achieve the design breaker type.

4.3. The Geometry of the Reef

The choice of the form of the geometry is, like the choice of the MFAR-angle, mainly related to the peel angle. An initial design choice for the initial geometry shape for the numerical simulations can be a delta form composed of two rides (a left and a right ride) with a constant MFAR-angle θ = 66°, as described previously.

Such a delta structure, which will be designed to create surfable waves, could be placed on the sea bottom or on a platform. A platform is, in theory, positive for the coastal protection aspect of the reef, because it makes the (large) waves break over it across its whole width (Smit and Mocke, 2005). But the effect of the presence or absence of a platform on the current cells that will be formed needs to be studied with numerical simulations. Regarding the surfer parameters of peel angle, α, breaker type, here related to the corresponding inshore Iribarren number, ξ_b, and wave height at the breaking point, H_b, it has to be determined if the platform has a positive influence.

The peel angle should be taken between 40 and 60 degrees for the waves to be surfable for most surfers (see Table 2 and Figure 23), and the peel angle has a maximum for a MFAR angle of 66 degrees.

Some calculations have been made to gain more insight into peel angle values. To estimate the peel angle we have to know what the design wave height is. The larger the design wave height the smaller the peel angle will be (the wave breaks sooner, and so refracts less).

Calculations were performed for a reef consisting of a platform with a delta structure on top, having a MFAR-angle of 66 degrees and where the submergence of the reef is 1.5 m, for safety reasons (see 4.4, in this section).

The wave conditions were wave periods of 6, 10 and 14 s and wave heights of 1, 2, 3 and 4 m and the breaker criterion was chosen to be $H_b = 1.1 h_b$ (Table 1). Over gentle slopes, waves are expected to break when $H_b = \gamma^* h_b$, where H_b is the breaker height and h_b is the breaking depth, and γ is equal to 0.78 (Sverdrup and Munk, 1946). As the beach slope increases, as though over a reef, the value of γ increases for the same wave steepness in deep water. The chosen value of γ of 1.1 is based on the work of Kaminsky and Kraus (1993). These researchers derived an empirical formula in a review of seventeen data sets obtained by various investigators in laboratory experiments. The formula is:

$$\gamma_b = 1.20\, \xi_\infty^{0.27} \tag{5}$$

where γ_b is the breaker condition and ξ_∞ is the deep water breaker parameter. ξ_∞ is equal to ξ_b with the exception that the breaker wave height has to be replaced by the wave height in deep water, H_0. Assuming a reef slope of 1:10, the mean value of γ_b is 1.1 for the tested wave conditions, and the maximum is 1.3.

The peel angle was computed for two depths for the platform: 4.0 m, which is the minimum depth of the platform in the example cited in the part about the height of the reef

(1.5 m submergence + 2.5 m minimum height of the delta structure), and 5.0 m (Table 4). Waves that break before they have traveled a length equal to 1.0 m height of the delta are not taken into account, because they are not expected to result in plunging waves. Although the calculations use linear theory and the wave focusing is not taken into account, they give an indication about the values of the peel angle.

Table 4. Peel angles for a reef with a submergence of 1.5 m and reef start depths of 4.0 m (left) and 5.0 m (right)

a) Reef starts at 4.0 m depth.

T\H	1 m	2 m	3 m	4 m
6 s	*	35.7	44.6	-
10 s	*	38.6	49.5	X
14 s	-	41.8	X	X

b) Reef starts at 5.0 m depth.

T\H	1 m	2 m	3 m	4 m
6 s	*	31.9	39.1	-
10 s	*	33.8	41.8	X
14 s	-	36.0	45.3	X

Bold values = wave breaks on the reef
* = waves go over the reef
X = wave has already broken or breaks at the beginning of the reef
- = rarely existing wave conditions at west coast of Portugal

Results from Table 4 show that for wave heights in the range of 1-2 m the case with a platform at the minimum depth of 4.0 m offers more peel angle conditions within the range of amateurs, i.e. between 40 and 60 degrees. For cases with the platform lower than 5.0 m under the still water level the peel angles will be even lower than the values shown in Table 4b.

As the peel angle values are better for a start depth of the delta structure as small as possible, taking the example minimum height of the reef (Table 4), it can be concluded that a platform at the seaward end of the structure definitely has a positive influence on the peel angle.

However, if the platform is under the whole delta structure, it probably has a relatively negative influence on the peel angle at the shoreward end compared with the case without a platform, because the delta structure can be higher at this point.

Moreover, the influence of a platform on the other two surfer parameters, i.e., the breaker type and the wave height at the breaking point, has not yet been investigated. In order to analyze this, the qualitative development of the peel angle, α, the inshore Iribarren number, ξ_b, and the wave height at the breaking point, H_b, along the breaker line are presented for both cases presented before: with and without a platform under the delta structure. The development of α, $1/\xi_b$ and H_b is presented in Figure 47 and shows that, for the case without a platform, α, $1/\xi_b$ and H_b are smaller at the beginning of the reef and larger at the end. These differences are caused by more refraction at the beginning of the reef (by longer slope) and less refraction at its end (by shorter slope) in the case without a platform. When wave focusing no longer plays a role, α, ξ_b and H_b increase towards the end of the reef without a platform and stay constant in the case with a platform.

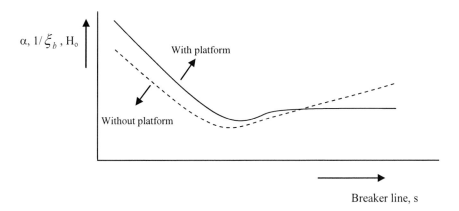

Figure 47. Relative development of the peel angle, wave height and Iribarren number along the breaker line.

From this analysis, it can be concluded that the choice of a platform does have an influence on the three surfer parameters: peel angle, breaker type and wave height at the breaking point. Wave defocusing is not taken into account in this analysis, for two reasons. First, because the area over which the wave focusing occurs is relatively small and second because the contributions to the peel angle of wave defocusing are relatively small compared with the area where wave focusing occurs, as described in the theoretical background.

Regarding the peel angle, the choice of a platform is not straightforward. A small variation in the peel angle along the ride does not represent a problem since surfers actually like some variation. However, the peel angle should be between 40 and 60 degrees, as mentioned before. Calculations have shown that these values of the peel angle are reached with the platform at the minimum depth for most wave conditions. So in terms of the peel angle, a platform in the first part of the reef is a better option. However, if the height at the last part of the reef is smaller in the case without a platform, the option of no platform would be better because refraction will be less.

Regarding the breaker type, the use of a platform is again a better option, because the ξ_b is smaller at the beginning and stays constant along the ride.

Regarding to the wave height at the breaker point, the preferred wave height depends on the skill of the surfer. However, it is easier to start surfing when the wave is somewhat higher, at the beginning, especially when the peel angle is small. So, it can be concluded that, for the wave height at the breaker point, a platform is a good option, since for this case the wave is higher at the beginning of the reef.

In conclusion, the use of a platform has a positive influence on the most important surfer parameters: the peel angle, the breaker type and the wave height at the breaker point.

However, not using a platform is better for the peel angle if the reef is less high, even though it should be higher than the minimum, as stated in the part about the height of the reef.

Moreover, constructing a platform only at the seaward part of the delta structure would give a maximum peel angle along the entire ride. In some cases this is actually the only possibility, because the platform intersects with the bottom (Figure 48). Positioning the reef more seawards in order to construct a platform under the whole delta is often not an option since, as will be explained in the part of about the horizontal dimensions, the distance between the shoreline and the structure is established with the objective of preventing coastline erosion.

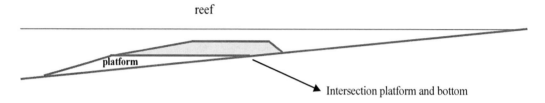

Figure 48. Intersection of platform and bottom.

Even though, in theory, a platform has a positive influence on the main surfer parameters, it still has to be taken into account that the exact influence can vary considerably with different geometries, and especially the reef angle is expected to be of much influence. One way to investigate this influence is to perform numerical simulations.

Another important factor that should be taken into account is that the form of the platform in the horizontal plane should be chosen so that its volume is as small as possible.

4.4. The Submergence of the Reef

The submergence of the reef is determined by two factors. Firstly it should be shallow enough for the design waves to break on the reef. Secondly it should be deep enough to ensure the safety of surfers.

With regard to the first factor, the submergence is dependent on the design breaking wave height and the breaking condition. Assuming a critical breaking condition of $H_b = 1.3\ h$, which gives the smallest breaking depth, the submergence of the reef with, for example, a 2.0 m design breaking wave height is 1.5 m.

Concerning the second factor, it is a known fact that the water depth during backflow under the wave trough can be very shallow and the reef may even become emerged ('suck dry'). However, surfing is a sport that involves risks. Surfers know that and most of them develop their own way to see if they can surf a certain section safely. Nonetheless, safety should be considered in the design. For diving in pools, FINA regulations suggest 1.8 m as an acceptable depth for submergence (Corbett *et al.*, 2005). But surfers tend to fall off their boards rather than dive vertically, reducing both the depth of the dive and the risk of serious injury (e.g. damage to the neck and spine). This fact and the physical experiments conducted by Corbett and Tomlinson (2002), in which the water depth above a reef with certain submergence was investigated for different wave heights, led to the design choice that submergence should be deeper than the design wave height in deep water. As an example, if the design wave height is 1.5 m, the minimum submergence for safety would be 1.5 m. The sucking dry phenomenon is probably hard to prevent completely, especially with high waves at low tide. Experience with the Narrowneck reef has indicated that, with a submergence of 1.5 m, the crest containers 'sucked dry' during larger wave conditions (>~2 m) at low tide (Jackson et al., 2005). In specific design studies numerical simulations will be needed to verify if and for what wave heights this phenomenon occurs.

Table 5. Peel angles for a reef with a submergence of 3.0 m and reef start at depths of 5.5 m (a) and 6.5 m (b)

a) Reef starts at 5.5 m depth.

T\H	1 m	2 m	3 m	4 m
6 s	*	*	*	-
10 s	*	*	*	**46.5**
14 s	-	*	**42.3**	**51.1**

b) Reef starts at 6.5 m depth.

T\H	1 m	2 m	3 m	4 m
6 s	*	*	*	-
10 s	*	*	*	**41.4**
14 s	-	*	*	**44.5**

Bold values = wave breaks on the reef
* = waves go over the reef
X = wave has already broken or breaks on the beginning of the platform
- = rarely existing wave conditions at west coast of Portugal

It should be noted that the design wave height and the corresponding design submergence mentioned above are for low tide, because this tidal level gives the critical submergence.

But many coastal zones have tidal conditions, so the influence of the tide on the peel angle should not be neglected. To investigate this influence on the peel angle, calculations assume a tidal level 1.5 m higher than that in Table 4, so the submergence is now 3.0 m and the depths of the platform are 5.5 and 6.5 m. Table 5 show the values of the peel angle obtained.

From Table 5 it can be seen that in both cases the peel angles are high enough for both amateur and professional surfers, but the waves that break over the delta have an H_0 of 3-4 m. However, by refraction and shoaling, the H_b on the reef will be even higher than 4 m. These wave heights cannot be surfed by surfers with skill levels 3 and 4. In conclusion, it can be said that higher tidal levels make surfing impossible for most amateurs. The tide also influences the breaker type by affecting the breaker wave height over the delta. Consequently, a guideline would be to pay attention to the fact that just one tidal level can be the 'design' level if the reef is designed for a certain category of surfers.

The minimum submergence and the minimum needed height of the reef for surfing, together with the distance offshore of the reef, and the slope of the sea bottom have a large influence on the effectiveness of a reef in terms of creating surfable waves. The slope of the sea bottom, the minimum submergence and the distance from the base of the delta to the shoreline will determine the height at the shore side of the reef. In order to verify achievement of the minimum height needed for surfing, some calculations have been performed. Table 6 shows the values of the reef heights at the shore side for a reef with a submergence of 1.5 m, a distance from the base of the delta to the shoreline of 175 m, a distance from the apex of the delta to the shoreline of 250 m (see 4.5) and different slopes of the sea bottom. In this case the cross-shore length of the delta is 75 m.

A height of 2.0 m at the base of the reef is here accepted as the minimum value for surfing. This is less than the value of the example presented in 4.2, where the minimum height was 2.5 m, because the reef will be higher further seawards where the wave rays that reach the end of the reef start. Table 6 shows that the minimum slope value for building a multifunctional artificial surfing reef that functions properly is 1:50. If a platfor m is chosen in the design, the intersection of the platform with the bottom, as a consequence of the chosen minimum height, gives a figure of 50 m under the seaward part of the delta. It is preferable if surfers can make all the ride from the seaward crest of the delta up to the base of the delta. Some simple refraction calculations can give an indication as to whether the wave rays at the shoreward intersection of the platform and the reef still reach the crest of the delta (Figure 49). Numerical simulations have to be performed to confirm these simple calculations.

Table 6. Reef heights for different bottom slopes

Distance from shoreline (m)	1:25	1:50	1:75	1:100
175 (base delta)	5.5 m	2.0 m	0.8 m	0.3 m
250 (apex delta)	8.5 m	3.5 m	1.8 m	1.0 m

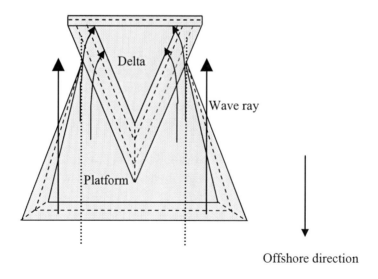

Figure 49. Wave rays reaching the crest delta.

4.5. The Horizontal Dimensions

The horizontal alongshore dimension depends on the length of the local coastline to be protected. The cross-shore dimension of the reef depends on the design length of the ride, which is the length of the breaker line at one of the sides of the reef (Figure 40). This dimension is limited, however. As mentioned in the state of art, Ranasinge and Turner (2006) found that the principal mode of shoreline response to submerged structures can vary from erosive to accretive, depending on the offshore distance to the structure (Figure 15). Based on these results, a predictive empirical relationship is proposed as a preliminary engineering tool to assess the shoreline response to submerged structures. This relationship is Sa/SZW > 1.5, where Sa is the distance from the apex of the structure to the undisturbed shoreline and SZW is the natural surf zone width. The distance Sa should clearly not be too large, because the effect of the structure on the morphodynamic processes adjacent to the shoreline will start to decline with increasing values. Even though this is a good preliminary engineering tool, it has to be ascertained whether Sa is a better characteristic cross-shore length for the mode of erosion or accretion for broad-crested structures than, for example, the distance from the base to the shoreline.

4.6. The Slope of the Reef Structure

A surfable wave for amateurs should be plunging, almost spilling. Based on experimental results with a 1 m wave height, Henriquez (2004) found that the inshore Iribarren number, ξ_b, should be between 0.6 and 0.9 to get surfable waves. These are plunging waves with relatively small inshore Iribarren numbers (see Table 3). Some calculations using linear theory have been performed for different wave heights to analyze which slopes will cause waves to break as a plunging breaker type. Figure 50 shows the inshore Iribarren numbers for a slope (that the wave meets) varying from 1:6 to 1:18, a wave height at breakpoint varying from 1 to 4 m and a wave period of 10 s. As can be seen, only a few combinations of slope and wave height give a plunging breaking wave in the surf range.

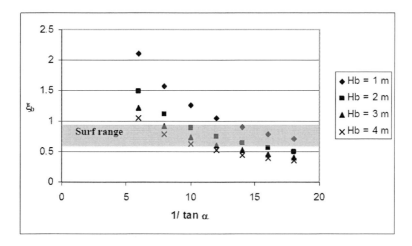

Figure 50. Surf range.

The values of ξ_b that lie in the surf range (given by Henriquez, 2004) are for slopes between 1:8 and 1:18, as can be seen in Table 7. It should be pointed out that the theory used is linear and that the surf range of 0.6 m to 0.9 m was found in experiments for a wave height of 1 m (prototype) at the wave maker, and here it is applied to a wave height range of 1 to 4 m at the breakpoint. However, it does give an indication as to what combination of reef slopes and breaking wave heights give surfable waves. Based on this indication an initial slope of the reef can be chosen to start with in numerical simulations.

Table 7. Inshore Iribarren number for different slopes and wave heights in the surf range

Tan α	H_b (m)	ξ_b
1:8	4	0.79
1:10	2	0.89
	3	0.73
	4	0.63
1:12	2	0.74
	3	0.61
1:14	2	0.64
1:16	1	0.79
1:18	1	0.70

If for example the breaking wave height is 2 m, the design choice of the side slope of the delta structure could be 1:10 because with this slope it can be expected that the design breaking wave height, H_b, of 2 m experiences slopes between 1:10 and 1:14 and that, consequently, the breaker type lies in the surf range. This will have to be confirmed with numerical simulations for each specific set of geometry and wave conditions.

The design choice for the slope at the shoreward end of the delta structure and for the slope of the platform at all sides should be as low as possible, in order to keep its volume smaller.

ACKNOWLEDGMENTS

This work is a contribution to the project "New protection concepts for the Portuguese coast" (PTDC/ECM/66516/2006), supported by the FCT – Portuguese Foundation for Science and Technology. The authors also acknowledge the financial sponsorship of M.Sc. ten Voorde's Ph.D. research by the Instituto de Investigação Interdisciplinar, University of Coimbra, Coimbra, Portugal.

REFERENCES

Antunes do Carmo, J.S. and Seabra-Santos, F.J., 2002. "Near-shore sediment dynamics computation under the combined effects of waves and currents." *Journal of Advances in Engineering Software*, Elsevier Science, Vol. 33, N° 1, p. 37-48.

Army Corps of Engineers, 1992. Egineering and Design – Coastal Geology, Department of the U.S. Army, publication number EM 1110-2-1810, CECW-EG.

ASRLtd, Reef construction: geotextile materials, 2008. http://www.asrltd.co.nz/ downloads_reef, accessed 25 March 2008.

Battjes, J.A., 1974. Surf similarity. Proc. 14th International Conference on Coastal Engineering, p. 466-479.

Black, K., 2000. Artificial surfing reefs for erosion control and amenity: theory and application. Proc. of the International Coastal Symposium 2000 (www.asrltd.co.nz).

Black, K. and Andrews, C., 2001. Sandy shoreline response to offshore obstacles: Part1. Salient and tombolo geometry and shape. *Journal of Coastal Research*, Special Issue 29, p. 82-93.

Black, K. and Mead, S., 2001. Design of the Gold Coast Reef for Surfing, Public Amenity and Coastal Protection: Surfing Aspects. *Journal of Coastal Research*, Special Issue No. 29, p. 115-130.

Bleck, M., Kübler, S. and Oumeraci, H., 2003. Sand-filled geotextiles containers for shore protection. In Coastal Structures 2004, Portland, Oregon.

Bowman Bishow Gorham, 2000. Cables Artificial Surf Reef: 12 Month Post-Construction Monitoring Survey. Report No. R99132:3, Department of Transport, Western Australia.

Borrero, J.C. and Nelson, C., 2003. Results of a comprehensive monitoring program at Pratte's reef. www.surfrider.org.

Boussinesq J., 1872. Théorie des ondes et des ramous qui se propagent le long d'un canal rectangulaire horizontal. *J. Math. Pure et Appl.*, 2, 17, p. 55-108.

Bowen, A.J. 1969. Rip currents: 1.Theoretical investigations. *Journal of Geophysical Research*, 74(54-67).

CIRIA (Simm J. D., Brampton, A. H., Beech, NW and Brooke J. S.), 1996. Beach management manual, Report 153, J. D. Simm (editor), ISBN 0-86017 4387.

Corbett, B. and Tomlinson, R., 2002, Noosa Main Beach Physical Modelling, Research Report No. 17, Griffith Centre for Coastal Management.

Corbett, B.B., Tomlinson, R.B. and Jackson, L.A., 2005. Reef breakwaters for coastal protection safety aspects and tolerances. Proc. of the 17th Australiasian Coastal and Ocean Engineering Conference (www.coastalmanagement.com.au).

Dias, J.A., Bernardo, P. and Bastos, R., 2002. "The occupation of the Portuguese littoral in 19th and 29th centuries", Littoral 2002, The changing coast. EUROCOAST/EUCC, Porto – Portugal. Ed. EUROCOAST – Portugal, ISBN 972-8558-09-0.

Edwards, R., 2003. A brief description of the Biological Assemblages Associated with Narrowneck Artificial Reef and Non-Woven Geotextile Substratum. Prepared for Soil Filters Australia.

Eurosion, 2004. Living with coastal erosion in Europe: Sediment and space for sustainability. A guide to coastal erosion management practices in Europe.

Fortes, C.J.E.M.; Coli, A.B.; Neves, M.G., CAPITÃO, R. (2006) – "Porto Santo Island. Offshore wave characterization and propagation." *Journal Coastal Research*, SI 39, 1600-1605

Fredsøe J. and Deigaard, R., 1995. "Mechanics of coastal sediment transport". Advances Series on Ocean Engineering – Volume 3. World Scientific Publishing Co., ISBN 9810208405 – ISBN 9810208413 (pbk).

Galvin, C.J., 1968. Breaker type classification on three laboratory beaches. Journal of geophysical research, 73(12), p. 3651-3659.

Gomes, F.V. and Pinto, F.T., 2006. EUROSIAN Case Study, Vagueira Mira (Portugal), Instituto de Hidráulica e Recursos Hídricos, FEUP, 18 p.

Gourlay, M.R., 1993. Wave set-up and wave-generated currents on coral reefs. Coasts 1993: Preprints of Papers: 11th Australasian Conference on Coastal and Ocean Engineering, 23-27 August 1993, Townsville, Qld. Instn Engrs, Aust., Barton, ACT, Nat. Conf. Publ. 93/4, Vol.2, p. 479 – 484.

Henriquez, M., 2004. Artificial surf reefs. M.Sc. thesis. Delft University of Technology (www.waterbouw.tudelft.nl).

Henriquez, M., Janssen, T.T., Van Ettinger, E.H.D. and Reniers, A.J.H.M, 2006. Refraction-controlled surfability. Proc. of the 5th International Surfing Symposium.

Hutt, J.A., Black, K.P. and Mead, S.T., 2001. Classification of surf Breaks in Relation to Surfing Skill. Journal of Coastal Research, special issue no. 29, p. 66-81.

Jackson, L.A., Reichelt, R.E., Restall, S., Corbett, B., Tomlinson, R. and McGrath, J., 2004. Marine ecosystem enhancement on a geotextile coastal protection reef – Narrowneck reef case study. Proceedings of the 29th International Conference on Coastal Engineering (Lisbon, Portugal), p. 3940-3952.

Jackson, L.A., Tomlinson, R., Turner, I., Corbett, B., D'Agata, M. and McGrath, J., 2005. Narrowneck artificial reef; results of 4 yrs monitoring and modifications. Proceedings of the 4th International Surfing Reef Symposium. (www.coastalmanagement. com.au).

Kaminski, G., and Kraus, N.C., 1993. Evaluation of depth-limited wave breaking criteria. Waves '93, Amer. Soc. Civil Engrs, p. 180-193.

Kamphuis, J.W., 2000. Introduction to Coastal Engineering and Management", World Scientific Press, 437 p.

Korteweg D. J. and De Vries G., 1895. On the change of form of long waves advancing in a rectangular canal, and on a new type of stationary waves. Phil. Mag. 39/5, p. 422-443.

Moores, A.E., Black, K.P., and Mead, S.M., 2006. Physical modeling of the Mount Maunganui Artificial Reef. *Proceedings of the First International Conference on the Application of Physical Modelling to Port and Coastal Protection.* (Porto, Portugal), p. 309-321.

Pattiaratchi, C., 2000. Design studies for an Artificial Surfing Reef: Cable Station, Western Australia. Proc. Coasts and Ports '99, p. 485-489.

Pilarczyk, K.W., 2003. Design of low-crested (submerged) structures – an overview. 6th International Conference on Coastal and Port Engineering in Developing Countries, Colombo, Sri Lanka.

Ranasinghe, R., Hacking, N. and Evans, P., 2001. Multi-functional artificial surf breaks: A review. NSW Department of Land and Water Conservation.(www.asrltd.co.nz).

Ranasinge, R. and Turner, I.L., 2006a. Shoreline response to multi-functional artificial surfing reefs: A numerical and physical modeling study. *Journal of Coastal Engineering* 53, p. 589-611.

Ranasinge, R. and Turner, I.L., 2006b. Shoreline response to submerged structures: A review. *Journal of Coastal Engineering* 53, p. 65-79.

Restall, S.J., Jackson, L.A., Heerten, G and, Hornsey, W.P., 2002. Case Studies showing the growth and development of geotextile sand containers. An Australian Perspective. Geotextiles and Geomembranes, vol 20, no 5.

Reis, C.S. and Freitas, H., 2002. Rehabilitation of the Leirosa sand dunes. In EuroCoast-Portugal Association (ed.) Littoral 2002, Porto, 22-26 September 2002. Porto, Portugal. III, p. 381-384.

Reis, C.S., Freitas, H. and Antunes do Carmo, J.S., 2005. Leirosa Sand Dunes: A Case Study on Coastal Protection. Proc. IMAM - Maritime Transportation and Exploitation of Ocean and Coastal Resources, Lisboa, 26-30 de September, 1469-1474. Ed. Taylor and Francis / BALKEMA. ISBN Vol. 2: 0 415 39374 4, cd-rom: 0 415 39433 3.

Rey S., I. Alejo J., Alcántara-Carrió and Vilas F., 2002. Influence of Boundary Conditions on Morphodynamics and Sedimentology of Patos Beach (Ría de Vigo, Nw of Spain), Littoral 2002, The Changing Coast. EUROCOAST / EUCC, Porto, Ed. EUROCOAST – Portugal, ISBN 972-8558-09-0.

Serre F., 1953. Contribution à l'étude des ecoulements permanents et variables dans les canaux. La Houille Blanche, p. 374-388.

Smith, E,R. and Kraus, N.C., 1991. Laboratory study of wave breaking over bars and artificial reefs. Journal of Waterway, Port, and Coastal Engineering, 117(4), p. 307-325.

Sverdrup, H.U. and Munk, W.H., 1946. Theoretical and Empirical Relations in Forecasting Breakers and Surf. Transactions of American Geophysical Union (27), p. 828-836.

Smit, F. and Mocke, G., 2005. Physical and numerical modelling of morphological and surf parameter response to an artificial reef. Proceedings 2005 Artificial Surfing Reef Conference.

Ten Voorde M., Neves, M.G.. and Antunes do Carmo, J.S., 2008. Preliminary study on the geometry of an artificial reef for coastal protection and surfing along the west coast of Portugal (in Portuguese), Revista de Gestão Costeira Integrada 8(1), p. 65-79.

Thurman, H.V, and Trujillo, A.P., 1999. "Essentials of Oceanography", 6th edition, Prentice-Hall, Inc., ISBN 0-13-727348-7.

Turner, I.L., Leyden, V.M., Cox, R.J., Jackson, L.A., McGrath, J.E., 2001. Three-dimensional scale physical model investigations of the gold coast artificial reef. *Journal of Coastal research*, special issue, no. 29, p. 131-146.

Van Rijn L.C., 2001. Sand transport and morphology of offshore sand mining pits/areas (EVK3-2001-00053 SAND PIT EU project).

Walker, J.R., 1974. Recreational Surf Parameters. Tech. rept. 30. University of Hawaii, James K.K. Look Laboratory of Oceanographic Engineering.

In: Encyclopedia of Environmental Research
Editor: Alisa N. Souter

ISBN: 978-1-61761-927-4
© 2011 Nova Science Publishers, Inc.

Chapter 55

MASSIVE SEDIMENTATIONS AT COASTAL AND ESTUARINE HARBORS: CAUSES AND MITIGATING MEASURES

Yu-Hai Wang[*]

Department of Sediment Research, China Institute of Water Resources and Hydropower Research, Chegongzhuangxilu 20, Beijing 100048, China

ABSTRACT

The entrances and/or access channels of harbors and other facilities constructed on coasts and estuaries where sediment supply is abundant might suffer massive sedimentations during extreme conditions, such as storm surges. Such massive sedimentation events often occur in a short time period, causing serious navigation problems and considerable economic losses. China's coast is stricken each year by a number of typhoons and temperate cyclones, and in recent years several harbors did experience massive sedimentations. Nonetheless, due to the complex interactions between coastal forces including wind, tidal current, wave, sediment transport, human activities and topographic configuration, etc., the underlying causes for such rapid, massive sedimentations have not been well understood and universally-applicable mitigation engineering measures are still lacking. Sea waves that approach the coast are becoming increasingly asymmetric and finally breaking due to the topographic effects in surf zones. This imposes shear stress upon the sea bed and initiates sediment transport. When the shear stress becomes high enough, ripples and dunes could be smoothed out and a thin layer of high concentration sediments is intensely transported on the sea bed. Such a sheet-flow transport process has been considered one of the major processes responsible for the massive sedimentation events. Nevertheless, good formulas that are capable of predicting longshore sediment transport rate in sheet-flow regime by stronger waves/currents are still lacking, though a good many efforts have been made to understand the sheet-flow mechanics, and formulas of various complexity have been proposed in recent years. This chapter introduces a formula that has a sound theoretical ground. It is able to predict longshore sand transport rate in sheet-flow regimes with

---

[*] wangyuhai-2166@126.com

higher precision, as verified with a recently-conducted physical model test as well as comparisons with other existing formulas. To mitigate the adverse massive sedimentations it is necessary to construct breakwaters or other infrastructure to impound the longshore sediment transport. However, an improper breakwater layout might even worsen the scenario. This chapter explores the optimum breakwater layout based upon the physical model test and presents a general guidance on breakwater layout design for coastal facilities. A new research perspective in coming years is also suggested.

1. INTRODUCTION

Coastal and estuarine harbors and other infrastructure, such as water intakes of nuclear/thermal power plants might be subject to massive sedimentations, which mean that a considerable amount of sediments are deposited at the harbors in a relatively short period of time, normally a few hours to days. Such massive sedimentations often occurs in extreme weather conditions, i.e. storm surge or typhoon approaching when wind-driven waves are violent and sediment supply is abundant. For example, massive amounts of sedimentation (0.5×10^9 kg mud and fine sand) could occur at the mouth of the Rotterdam Waterway during periods of less than one week in the period from late autumn until early spring during storms (Verlaan and Spanhoff 2000).

Massive sedimentations could significantly affect, or even stop the normal operation of harbors, causing dredging difficulties and considerable economic losses. For instance, on New Year's eve of 1974, the world's largest port at the time, Rotterdam port, saw a sudden incursion of 600,000 m³ of fluid mud following a storm in the North Sea, which caused the port to be closed for 3 weeks to deep-drafted vessels (Kirby et al. 2008). In the Port of Rotterdam about 10 million m³ of sediments has to be dredged annually (Verlaan and Spanhoff 2000), in the Port of IJmuiden it's about 3.2 million m³ (Van Rijn 2004), and in the Port of Hamburg the annually-dredged sediments is about 2.6 million m³ (Christiansen 1987).

In China the Tianjin new port was artificially dredged on the estuary of Hai river in 1954 (Figure 1).

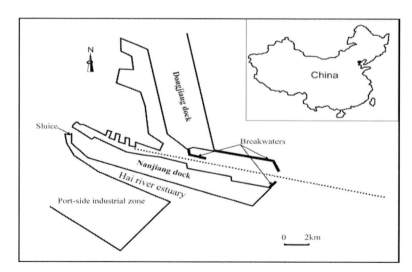

Figure 1. The layout of the Tianjin new port.

Shortly after its operation, the port experienced a series of massive sedimentations. During the period of 1954–1958, the total sedimentation volume is 28.03 million m^3, annual dredged mud is 5.31 million m^3, and the mean sedimentation rate was 4.54m/year (Sun 2003). The massive sedimentation events occurred in the form of fluid mud (0.005~0.007mm with clay fraction ≥40%). The sediment sources are predominantly supplied by the nearby Hai River supplemented with those agitated on the nearby shallow beaches by waves. If wind speed is larger than 7~8m/s and last for over 8 hours, fluid mud appeared in well-protected water areas or slowly-flowing water areas (Xu et al. 2003). In order to solve the sedimentation problem, the Ministry of Communications of P.R.C. called together experts and researchers from universities, research institutes and harbor services across China to conduct prototype observations and measurements, indoor tests and theoretical analysis. This had accumulated a great deal of first-hand data in the mid-1960s with preliminary solutions and measures put forward (Liu and Zhang 1993). Meanwhile, with the construction of anti-tide sluice and upstream reservoirs at the Hai river, the amount of mud input into the coast was considerably reduced. This together with a series of mitigation measures has led the magnitude of massive sedimentations due to fluid mud incursion to be reduced significantly though strong sedimentation still occurs sporadically in association with northerly (ENE-EN) strong winds (Sun 2003).

In the 1970s~1980s harbor construction in China had seen a rapid development. Nonetheless, harbors that lie at muddy coasts and estuaries, such as the Lianyun harbor and the Yangtze estuary port all experienced fluid-mud massive sedimentations. For example, when the No. 8310 typhoon was invading the Yangtze estuary in 1983, over 4 million m^3 mud was deposited in the Tongsha navigation course in the southern branch, the mean deposition thickness was 0.5m and the maximum was 1m and the sediments were hard to be dredged. This led the Tongsha navigation course to be closed and a new navigation course had to be explored in the northern branch of the estuary (Xu et al. 2003). By learning from lessons, Chinese engineers and researchers have basically understood the sedimentation mechanics at harbors constructed at muddy coasts and estuaries in the early 1990s and reported a series of research results (see Liu and Zhang 1993, Liu and Yu 1995).

In the 1990s, the rapid development of the Chinese economy has put high demand for more and more port construction often in less desirable sites, as far as the sediment regime is concerned. During this period, several harbors were constructed on silty open coasts where silt sediments are abundant. These harbors are all semi-enclosed with breakwaters extended to the shallow waters to protect the artificially-dredged harbor basins and access channels.

However, these harbors were subjected to several serious massive sedimentations. For example, on November 11~12, 2003 big wind-driven waves deposited over 1 million m^3 of silt at the entrance and access channel of the Jingtang harbor (Figure 2), the mean deposition thickness at the access channel was 1.9m and the maximum thickness was about 4m (Luo et al. 2007); at the same time, 9.7 million m^3 of silt was deposited into the outer access channel of the Huanghua harbor (Figure 3) in less than 10 hours, the maximum deposition thickness was more than 3m (Zhao and Han 2007). These deposited silts are not only huge in volume but also hard to be dredged, causing navigation difficulties and even stoppage of harbor operations. Table 1 briefly summarizes massive sedimentations that have occurred at various Chinese coastal and estuarine harbors in the past several decades.

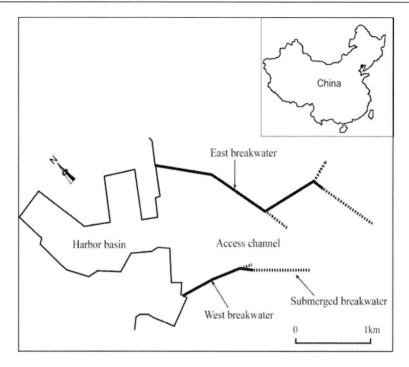

Figure 2. The layout of the Jingtang Harbor.

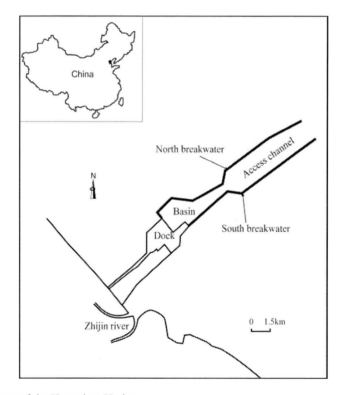

Figure 3. The layout of the Huanghua Harbor.

In order to promote the rapidly-growing shipping industry as the size of vessels rise inexorably, engineers have to come up with effective means to deal with the often severe sedimentation problems. Chinese engineers and researchers have conducted field observations and measurements, laboratory tests, theoretical analysis and numerical modeling to understand the underlying causes and explore effective mitigation engineering measures. Due to the complexity of coastal hydrodynamics and the special mobility of silt sediment, no universally accepted solution has been found so far.

As to the sedimentation problems with harbors at muddy estuaries the PIANC working group 43 has recently completed a report entitled 'Minimizing Harbor Siltation' (MHS) to summarize old, new, proven and promising methods for MHS that have been mostly exercised at NW European harbors (Kirby et al. 2008).

Nonetheless, due to the complex interactions between coastal forces including wind, tidal current, wave, sediment transport, human activities and topographic configuration, etc., the underlying causes for these massive sedimentations occurred on silty open coasts as well as muddy estuaries have not been well understood and universally-applicable mitigation engineering measures are still lacking, if feasible. This chapter is going to explore the casues, predicting means and mitigation engineering measures.

Table 1. Massive sedimentations at coastal and estuarine harbors in China

Harbor	Coast	D_{50} (mm) of sediments	Hydrodynamics		Notes
			Mean tidal range(m)	Mean wave height (m)	
Fangcheng Harbor	Sand	Bed sediment: 0.32~0.46	2.39	0.5	0.28m-thick deposition by one typhoon
Jingtang Harbor	Silt	Bedload: 0.075	0.88	1.0	1.9m-thick deposition at access channel, maximum 4m by one temperate cyclone
Daqinghekou Port	Silt	Bed sediment: 0.1	0.95	0.6	2.5m-thick silts were deposited in 70 days and the navigation channel was completely filled
Shantou Harbor	Silt	Bed sediment: 0.063~0.125	1.10	0.5	0.38m-thick silts were deposited in 60 days
Maojia Port	Silt	Bed sediment: 0.093	3.68	0.3	0.8m-thick silts were deposited at the access channel, which was completely filled

Table 1. (Continued)

Harbor	Coast	D_{50} (mm) of sediments	Hydrodynamics		Notes
			Mean tidal range(m)	Mean wave height (m)	
Huanghua Harbor	Silt	Bed sediment: 0.036	2.26	0.6	Access channel was completely filled by one temperate cyclone
Jinshanshihua test trench	Muddy silt	Bed sediment: 0.031; suspended sediment: 0.013	3.92	0.2	138cm-thick sediments were deposited in 90 days
Hangzhou Bay test trench	Silty mud	Bed sediment: 0.009~0.0103	3.35	0.4	128cm-thick sediments were deposited in 50 days
Yangtze Estuary Port	Silty mud	Bed sediment: 0.027; suspended sediment: 0.022	2.80	0.35	Fluid mud
Tianjin New Port	Mud	Bed sediment: 0.005; suspended sediment: 0.004	2.50	0.4	Fluid mud
Lianyun Harbor	Mud	Bed sediment: 0.004; suspended sediment: 0.004	3.40	0.52	Fluid mud

Note: some data are from Xu et al. (2003).

2. CAUSES

Those harbors that experienced massive sedimentations, i.e. the Jiangtang harbor (Figure 2) and the Huanghua harbor (Figure 3), are all located on silty open coasts. The harbor basins and access channels are all dredged at shallow silty beaches, using breakwaters to prevent from wave invasion and sediment incursion. The silt sediments are generally accepted in China's coastal engineering community as sediments with median diameter of 0.025-0.125mm (e.g., Cao et al. 2004, Zhao et al. 2006).

The common features of these sedimentation events are: 1) the more violent the wind-driven waves are, the severer the sedimentations; 2) sedimentation is less severe at a harbor basin and an inner access channel but becomes serious at an outer access channel. For example, a total of 9 million m³ silts were deposited at the Huanghua harbor in 2002, only 2 million m³ silts were deposited at the basin and inner access channel but 7 million m³ silts were deposited at the outer access channel; 3) annual sedimentation volume varies significantly, usually up to 2 times, whereas at muddy harbors it varies little—for example, the annual sedimentation quantity at the Tianjin new port is about 6 million m³, the maximum is about 7 million m³ and the minimum is about 5 million m³; 4) the deposited silt sediments are consolidated rapidly and very hard to be dredged (Cao et al. 2004).

"Wave mobilizes sediment but tidal current transports it" is a widely-accepted notion by the coastal engineering community. So the physical-model experiments and calculations using empirical formulas to predict the sedimentation rates and distributions at the above-mentioned silty-coast harbors were all conducted on the principle that the silts were transported by tidal flows in the mode of suspended sediments and deposited in a quiescent water or less-flowing water before the 1990s (Gao 2005). Nevertheless, the actual sedimentation volumes have far exceeded those predicted by these physical-model tests and empirical formulas. Engineering lessons have prompted engineers and researchers to realize that under the forces of big wind-driven waves there appears to be a bottom high-concentration sediment layer moving at a higher speed on the silty beaches; and it is these bottom high-concentration sediment layers (Figure 4), often called sheet flows, that are one of the major processes responsible for the massive sedimentations at harbor entrances and access channels (for example, Wang 2000, Cao et al. 2004, Luo 2004, Luo et al. 2007, Zhao and Han 2007).

Sheet flows means that when bed shear stress exerted by waves becomes high enough, ripples and dunes are smoothed out and the stationary bed is capped by a shear layer of particles in intense motion, this leads to large transport rates far in excess of those found at lower shear stress (Pugh and Wilson 1999). Sheet flows widely occur in river flooding (Dinehart 1992) in tidal estuaries, and in coastal waters in storm conditions; even in moderate wave conditions, sheet flows could occur in a shoaling area just before the breaking point (Hassan and Ribberink 2005) and in swash zones (Huges 1997), they even prevail at the peak time of a storm farther offshore and in the continental shelf region (Li and Amos 1999a).

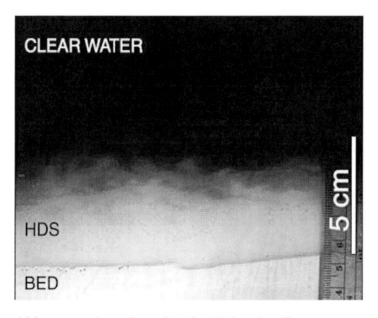

Figure 4. Bottom high-concentration sediment layer in a U-shaped oscillatory water tunnel (referred as 'HDS' on the figure, D_{50}=0.026mm, from Lamb et al. 2004, reproduced by permission of American Geophysical Union).

Nonetheless, arguments have existed regarding the role played by sheet flows in massive sedimentations at the entrances and access channels of harbors on silty open coasts. For example, as to the massive sedimentation at the outer access channel of the Huanghua harbor (Figure 3), one idea is that the deposits at the outer access channel were settled down by suspended sediments carried by tidal flows or longshore currents, while another idea argues that the deposits were from those mobilized by waves on the nearby shallow beaches and transported into the outer access channel in the form of sheet flows (Zhao and Han 2007); still other researchers consider that it is wakes and vortices produced by the head of breakwaters that are responsible for the access channel sedimentation (Gao 2005) or the eddy at the head of breakwaters as produced by the orthogonal alignment of the axis direction of access channel with the direction of prevailing waves and winds (Xu et al. 2003, Luo et al. 2007).

In fact, the occurrences of the massive sedimentation events at those silty coastal harbors are all related to the big waves from an approaching typhoon or temperate cyclone. So big waves are the dominant force to mobilize silt sediments on nearby beaches and transport them into the harbor entrance and/or access channels. As the entrances and access channels are protected by longer breakwaters, longshore transported sediments by obliquely incident waves and those mobilized by waves on nearby beaches or shoals are transported by the along-breakwater currents in the form of sheet flows to the entrance and/or access channels in extreme weather conditions, leading to serious sedimentation problems.

The site where massive sedimentations occur varies with the beach slope. For relatively steep beaches, massive sedimentation often occurs at the access channel near the entrance. As at the Jiangtang harbor (Figure 2), the beach slope within 1km from the shoreline is 1:60 while outside is 1:300~1:500, so major sedimentation always concentrates at the access channel just outside the entrance no matter how big the wave is (Zhang et al. 2003, Luo et al. 2007). This is because the water outside the entrance is deep enough so waves could not mobilize much sediment, and sediments entering into the access channel are predominantly transported by along-breakwater currents (Zhang et al. 2003, Wang 2000, Sun et al. 2005). On the other hand, on a very gentle shallow silty coast, i.e. the Huanghua harbor (slope 1:3500~1:4000), the breaker zone is much wider than the length of outer access channel in temperate cyclone weather. This enables waves to mobilize a large amount of sediments from the beach/shoal and transport them directly into the outer access channel; the sedimentation at the inner access channel by along-breakwater currents becomes relatively minor (Zhang et al. 2003). The wakes and vortices at the breakwater heads are only responsible for carrying suspended sediments arising from bed erosion around the breakwater head into the access channel, the deposit amount is limited. Generally speaking, the transiting suspended sediment from far-field source does not cause any significant sedimentation at the entrances/access channels in extreme weather conditions.

As for harbors constructed at muddy estuaries, sedimentation is largely governed by the position of a turbidity maximum and the river run-off in the estuarine area, but waves in stormy weather are often responsible for entraining the fluid mud (generally-defined as particles less than 0.625mm on the Wentworth sediment size scale, but this chapter refers to mud as the particles less than 0.025mm). Verlaan and Spanhoff (2000) reported that the massive sedimentations that occurred at the Waterway of Rotterdam was due to the presence of a wave-induced high-concentration mud suspension layer near the bottom with a thickness of a few decimeters, in which the 100 μm and the mud fraction contribute most whereas the 200 μm fraction contribute significantly only within the wave boundary layer; the residual flow created by the river discharge and tidal flow is responsible for importing the high-concentration suspension layer into the mouth of the Rotterdam Waterway and settling there.

At the Atchafalaya River Bar Channel connected to the Gulf of Mexico for offshore oil industry shipping, fluid mud began returning or reforming in the bar channel approximately two weeks after dredging, and an average of 6–8 weeks for well-defined fluid mud to build to a thickness of 2.5–3 m; annual dredged volumes of 7 to 8 million m^3 have historically failed to keep the channel free of mud for more than a few weeks (McAnally et al. 2007). In addition, under storm conditions harbors that were subject to considerable mud sedimentation also include Oostend and Zeebrugge in Belgium (Wens et al. 1990), the access channel to the Esbjerg port (SW Denmark, Clausen et al. 1981), the Charleston Harbor at the Cooper river (CTH 1971), the Emden, Bremerhaven and Brunsbüttel on the German North-Sea coast (Nasner 2006) and many other ports around the world.

The sedimentation volumes at muddy estuarine habors is more related to the exchange volume of waters entering and leaving the basins, which are governed by five major processes: 1) exchange flow by horizontal entrainment (mixing layer); 2) exchange flow by tidal filling; 3) exchange flow by fresh/salt driven density currents; 4) exchange flow by warm/cold driven density currents; 5) exchange flow by sediment-induced density currents (see Winterwerp 2005, Kirby et al. 2008).

3. PREDICTION OF MASSIVE SEDIMENTATION

3.1. Formulas

There has not existed a universally accepted, reliable formula to predict the massive sedimentation volume at the harbor basin and/or access channel. Based on the idea that suspended sediments settle at the basin and access channel Chinese researchers modified conventional empirical formulas by augmenting the sediment concentration under the co-actions of wind, tide and wave to upgrade the sedimentation rate in extreme weather conditions (for example, Liu 2004) or explore new numerical method (see Lu et al. 2005, Zhang et al. 2006). Alternatively, researchers reach this purpose by modifying conventional bedload transport formula (Luo 2004, Luo et al. 2007). Next different representative empirical formulas proposed by Chinese researchers are briefly introduced.

3.1.1. Empirical Formulas

1) *Liu's formula*

Liu and Zhang (1993) suggested empirical formulas that could be used to predict the sedimentation rates at navigation channels and basins of harbors on muddy/silty/sandy coasts:

a) navigation channel sedimentation rate (oblique flow-channel angle)

$$P=\frac{S\omega t}{\gamma_0}\left\{K_1[1-(\frac{h_1}{h_2})^3]\right\}\sin\beta+K_2[1-\frac{h_1}{2h_2}(1+\frac{h_1}{h_2})]\cos\beta \qquad (1)$$

b) navigation channel sedimentation rate (parallel flow-channel angle)

$$P = \frac{K_2 S \omega t}{\gamma_0}[1 - \frac{V_1'}{2V_1}(1 + \frac{h_1}{h_2})] \tag{2}$$

here P is the sedimentation rate, S is the sediment concentration in nearby shallow waters, h_1 and h_2 is the mean water depth of shallow beach/shoal and navigation channel, respectively; ω is the flocculating settling velocity of particles, generally set as 0.0004~0.0005m/s, but for 0.03mm<D_{50}<0.11mm silt sediments, ω should be taken as the value of the settling velocity of single particle, for D_{50}>0.11mm sands, ω should be taken as the settling velocity of single particle times $(0.11/D_{50})^2$; V_1 and V_1' is the mean flow velocity at the shallow beach/shoal and navigation channel, respectively; γ_0 is the dry bulk density of sediment; β is the flow-channel angle; K_1 and K_2 is the sedimentation coefficient for transverse flow and longitudinal flowing, set as 0.35 and 0.13, respectively; t is the sedimentation time.

c) sedimentation rate at a protected harbor basin

$$P = \frac{K_0 S \omega t}{\gamma_0}\left[1 - \left(\frac{h_1}{h_2}\right)^3\right] \exp\left[\frac{1}{2}\left(\frac{A}{A_0}\right)^{1/3}\right] \tag{3}$$

here A is the shoal area within the semi-enclosed basin, A_0 is the total water area of the semi-enclosed basin, K_0 is an empirical coefficient, taken as 0.14~0.17, other symbols are defined as previously explained.

The depth-averaged sediment concentration S (kg/m^3) in nearby shallow waters can be calculated as:

$$S = 0.0273 \gamma_s \frac{(|V_1| + |V_2|)^2}{gh} \tag{4}$$

in which,

$V_1 = |\overline{V}_b + \overline{V}_t|$ ——the combined flow velocity of wind-driven current and tidal current;

V_2 ——mean horizontal velocity of water particle under waves;

γ_s —— bulk density of particles (kg/m^3);

h —— the mean water depth at nearby beach/shoal;

g ——the gravity acceleration

The time-averaged flow velocity of wind-driven currents V_b and mean horizontal velocity of water particle under waves V_2 can be calculated as:

$\bar{V}_b = 0.02\bar{W}$ ($\bar{W}$ is the time-averaged wind speed)

$V_2 = 0.2\dfrac{H}{h}C$ (H, C is wave height and wave celerity, respectively)

While in breaking-wave conditions much more sediments are mobilized by waves and put into suspension. So the depth-averaged sediment concentration in shallow waters is calculated as (Liu 2004):

$$S_b = 0.0273\gamma_s \dfrac{(|V_1|+|V_{2b}|)^2}{gh_b} \qquad (5)$$

in which, $V_{2b} = \dfrac{1}{2}(g\gamma_b H_b)^{1/2}$, breaking-wave index $\gamma_b = H_b/h_b$ (H_b is the breaking-wave height, h_b is the water depth where wave is breaking).

2) Luo's formula

Luo (2004) and Luo et al. (2007) proposed a bottom-sediment transport rate formula under the actions of wind, wave and tide based on the idea that wave's orbital velocity mobilize sediment while wave's 'mass-transporting' velocity, tidal current velocity and wind-driven current speed all together transport the sediment. It is read as:

$$q_{sb} = \dfrac{k_b}{C_0^2}\dfrac{\gamma_s\gamma}{\gamma_s-\gamma}(U_{b\max}-U_c)\dfrac{U_b^2 V_m}{g\omega} \qquad (6)$$

here q_{sb} (kg/m. s) is the transport rate of bottom sediment per unit width; k_b is a coefficient, $k_b = 0.18 D_{50}^{0.365}$, $C_0 = C/\sqrt{g}$, C is the Chezy coefficient, γ_s, γ is the bulk density of sand particles and water, respectively; U_b is the depth-averaged, time-averaged orbital velocity of wave water particle in the half wave period; $U_{b\max}$ is the maximum orbital velocity of water particle motion just above the wave boundary layer, V_m is the combined speed of wind, wave and tide; ω is the settling velocity of particles and U_c is the critical flow velocity for particle initiation.

U_b, $U_{b\max}$, U_c, ω, V_m could be calculated respectively:

$$U_b = \dfrac{2h}{T}[1+4.263(\dfrac{H}{L})^{1.692}]\dfrac{1}{\sinh(2\pi h/L)}$$ (H is wave height, L is wave length and h is mean water depth);

$$U_{b\max} = 1.57 U_b$$

$$U_c = (\frac{H}{D_{50}})^{0.14}[17.6\frac{\gamma_s - \gamma}{\gamma}D_{50} + 6.5\times 10^{-7}\frac{10+H}{D_{50}^{0.75}}]^{0.5}$$

$$\omega = [(13.95\frac{\nu}{D_{50}})^2 + 1.09\frac{\gamma_s-\gamma}{\gamma}gD_{50}]^{0.5} - 13.95\frac{\nu}{D_{50}}$$

(ν is the kinematic viscosity of water);

$$V_m = |U_d| + |U_t| + |U_w|$$

(U_d is mean velocity of tidal current, U_w is the flow speed of wind-driven current);

$$U_t = 0.5\pi\frac{h^2}{TL}[1+57.4(\frac{H}{L})^{2.21}]\frac{1}{\sinh(2\pi h/L)}$$

(U_t is the 'mass-transporting' speed of wave);

$$U_w = 0.03W$$ (W is the time-averaged wind speed).

3) *Xu's formula*

Based upon the 'longshore wave energy flux' concept Xu (1996) proposed an empirical formula to predict the longshore sediment transport rate directly using in situ measured wave parameters at wave buoy. It reads:

$$q = k_1 H^{2.4}\sin\xi(\cos\xi)^{1.2}C_g^{0.2}(C_g/C) \qquad (7)$$

here C is wave speed, C_g is wave group speed, H is wave height, ξ is wave incidence angle and k_1 is a coefficient.

These parameters can be calculated as:

$$C = \{gh[y+(1+0.666y+-.445y^2+0.105y^3+0.272y^4)^{-1}]^{-1}\}^{-1/2}$$

, in which h is water depth, g is gravity acceleration and $y = w^2 h/g, w = 2\pi/T$, T is wave period (Hunt 1979);

$$C_g = n\cdot C;$$

$$n = \frac{1}{2}[1+\frac{2kh}{\sin h(2kh)}];$$

$k = 2\pi/L$, L is wave length.

3.1.2. Theoretical Formula

During the intense motion of bottom high-concentration sediment layer the motion of the water-sediment mixture could be treated as a Newtonian flow by simultaneously considering the applied external force on the surface and bottom of the mixture. This enables a mathematical model describing the mixture motion to be constructed as (Wang 2007 a,b,c):

$$\begin{cases} \dfrac{\partial h}{\partial t} + \dfrac{\partial (hu)}{\partial x} = 0 \\ \dfrac{\partial u}{\partial t} + u\dfrac{\partial u}{\partial x} + g\cos\alpha \dfrac{\partial h}{\partial x} = g\sin\alpha + \dfrac{\tau}{h\rho_m} - gC\dfrac{\rho_s - \rho_f}{\rho_m}\cos\alpha \tan\varphi \end{cases} \quad (8)$$

here h is the thickness of the water-sediment mixture, u is the depth-averaged velocity of particle movement within the mixture, C is the depth-averaged particle volumetric concentration of the mixture, τ is the shear stress exerted by water flow upon the top surface of the mixture, α is the slope angle (going-down-slope is positive value while going-up-slope is negative value), g is the gravity acceleration, φ is the dynamic internal frictional angle of particles, ρ_s is the density of particles, ρ_f is the density of water flow, and ρ_m is the density of the mixture, it can be defined as:

$$\rho_m = \rho_s C + \rho_f (1-C) \quad (9)$$

in which ρ_f could be further defined according to the water is clean water or washload-laden flow as:

$$\rho_f = \rho_s c + \rho_{water}(1-c) \quad (10)$$

in which ρ_{water} is the density of clean water and c is the volumetric concentration of washload in the water.

Eq. (8.1) is the mass continuity equation (for equilibrium bedload transport only). Eq. (8.2) is the momentum equation, the third item on the left of Eq. (8.2) is the pressure gradient, whereas the first item on the right is the gravitational force along the slope, the second item is the shear stress exerted by the overlying flow, and the third item is the frictional force, i.e. the shear stress exerted by the stationary bed. The frictional force is evaluated according to Bagnold's (1954) suggestion that the normal dispersive stress arising from grain/grain collisions is equal to the immersed weight of the sediment in an equilibrium situation and it together obeys the Coulomb stress relationship with the frictional force, i.e. $\tau = \sigma \tan\varphi$, τ is the shear stress, σ is the normal dispersive stress and φ is the dynamic frictional angle.

A special solution could be sought from the partial differential equations (8), it could be further derived into a formula to predict the sand transport rate under the action of asymmetrical waves, i.e.

$$Q_b = \frac{\frac{2}{3}\frac{1}{\kappa}\ln(\frac{A_{rms}}{r_s})\theta^{3/2}}{[\tan\varphi\cos\alpha - \sin\alpha - \frac{\rho_f}{(\rho_s-\rho_f)C}\sin\alpha]} \quad (11)$$

where Q_b is the non-dimensional bedload transport rate, κ is the von Karman's constant and taken as 0.407, C is set as 0.475, A_{rms} is the root-mean-square value of semi-excursion of orbital motion above the wave boundary layer, r_s is the roughness height of the stationary bed and is generally taken as 2D or $2D_{50}$ of particles in the stationary bed, θ is the Shields parameter, other symbols are defined as previously explained.

The detailed derivation could be referred to Wang (2007b). As distinctive from empirical formulas that often have fixed-value coefficients and limited application scopes, the coefficient of Eq. (11) appears to be a dynamic function that is able to automatically adjust according to the wave dynamics, sediment properties and boundary conditions (e.g. bed slope and roughness height).

The computed results using Eq. (11) compare well with the experimental measurements of Dohmen-Janssen and Hanes (2002) in a large-scale wave tank of Grober WellenKanal, GWK of the ForschungsZentrum Küste, Hannover and those at the large-scale oscillatory water tunnel of Delft Hydraulics (Hassan and Ribberink 2005) and field observations of Masselink and Hughes (1998) at the Australian Myalup beach.

Equation (11) could be further modified to predict longshore sand transport rate (m³/s) as:

$$q_b = \frac{\frac{2}{3}\frac{1}{\kappa}\ln(\frac{A}{r_s})\theta^{3/2}\cos\xi}{[\tan\varphi\cos\beta - \sin\beta - \frac{\rho_f}{(\rho_s-\rho_f)C}\sin\beta]} \bullet l \bullet \sqrt{g(s-1)D_{50}^3} \quad (12)$$

in which, q_b is the longshore sand transport rate, A is the semi-excursion of orbital motion above the wave boundary layer, ξ is the wave incidence angle, l is the beach width where significant longshore sediment transport occurs or the scope of concern, s is the relative density (ρ_s/ρ), the dynamic internal frictional angle $\varphi = 35.3D_{50}^{0.04}$ (φ's unit is degree, and D_{50}'s unit is millimeter, Zhang and Wang 1989).

It should be pointed out that the wave parameters in Eq. (12), i.e. A and ξ could be directly taken as the measured values at the wave buoy, avoiding troublesome calculations of breaking-wave parameters. Next the computation results using Eqs. (1), (6), (7) (12) are compared with the measurement results of a physical-model experiment recently conducted to investigate the sedimentation problem with the layout of the water-intake forbay of the Liaoning nuclear power plant (LNPP hereinafter for short).

3.2. Physical-Model Experiment

3.2.1. Background of LNPP

The LNPP is being constructed on the southeastern coast of the Liaodong Bay in China (Figure 5). Its first stage production capacity is 4×1000MW and the second stage production capacity is 2×1000MW and will be put into operation in 2014 with a total investment of 5 million RMB. LNPP will use sea water as cooling water and the summer intake capacity for a single unit is 50.2m^3/s and the winter intake capacity for a single unit is 37.6m^3/s. The sea water will self-flowing into the plant through a forbay (-11m deep below lower low water level) and four underground tunnels to the plant and eventually discharges out through underground tunnels and water-outlet building.

The LNPP coast consists of rocky heads and smaller bays with deeper water relatively closer to the shoreline, the minimum distance of the -10m contour from the shoreline is only 70m and the maximum distance is about 700m; the beach slope is steeper, about 1:20~1:30, the bottom sediments are mainly medium-coarse sand (D_{50}=0.24mm). By comparing the 1984 and 2004 bathymetric surveys it is found that the underwater topography has been generally stable, showing minor degree of erosion, for example, the retreat speed of -10m contour is about 1.5m/year and the erosion rate is about 0.02m/year.

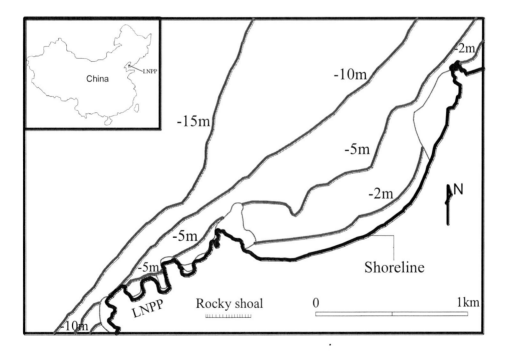

Figure 5. The LNPP coast configuration.

Tidal flows are the semi-diurnal mode, the flood-tide period is six hours plus five minutes while the ebb-tide period is six hours plus 21 minutes. The mean spring tidal range is 1.53m and the mean neap tidal range is 1.08m. The tidal flow runs parallel to the coast in the NE-SW direction and the ebb current is somewhat stronger than the flood current. The average flood/ebb tidal flow velocities are 0.5~0.6m/s and the measured maximum current velocity is 1.14m/s. Residual flow is weak and irregularly distributed and the measured maximum residual flow speed is less than 0.2m/s[1].

Waves are predominantly wind waves. They prevail in the N~NNE directions with an occurrence frequency of 21.4%, <0.5m height $H_{1/10}$ waves occur at the frequency of 72.4%, 0.6~1m height $H_{1/10}$ waves occur 15.5% and >2.0m height $H_{1/10}$ waves occur only 0.4% (Figure 6). The strongest waves in terms of magnitude and occurrence-frequency is the NNE waves. The maximum NNE $H_{1/10}$ is 4.3m with a mean period of 7.4s, as observed in 1992[2].

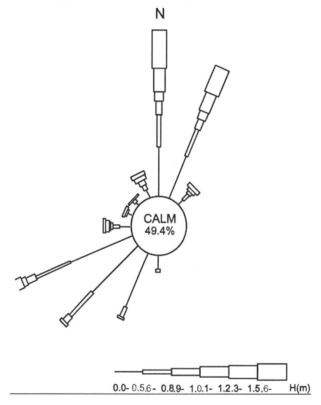

Figure 6. Local wave rose diagram.

The suspended sediment concentration is lower in normal weather conditions. It is on average 0.042kg/m³ in the spring tidal flow with a maximum value of 0.183 kg/m³, whereas the mean concentration is 0.009kg/m³ and the maximum concentration is 0.014 kg/m³ in the neap tidal flow[2]. The general character of the suspended sediment distribution is that the concentration decreases from the nearshore to offshore and from the bottom to the surface and it is larger in flood flow than in ebb flow. The suspended sediments are mainly from far-field sources and there is no sediment source input by nearby rivers. Only in gale weather, waves could mobilize and suspend a great deal of sands from the nearby shallow sea bed.

---

[1] Tianjin Coastal *Engineering* Company, LNPP Sea Hydrographic Survey Report, 2005 (in Chinese).

It is found that longshore sand transport occurs at the LNPP coast, as driven by the northerly-coming waves plus tidal currents, particularly, the ebb tidal current. In order to minimize sedimentation as well as preventing floating sea ice and drifting material from entering into the water-intake forbay of LNPP, two different layouts of the breakwaters are suggested (Figure 7). Physical-model experiments were conducted to investigate the sedimentations within the forbay.

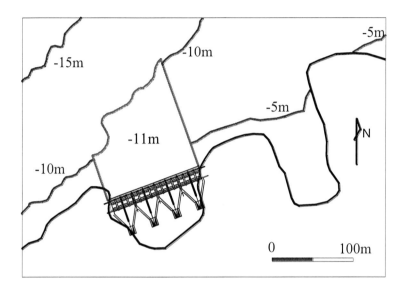

Figure 7a. The water-intake forbay layout without breakwater.

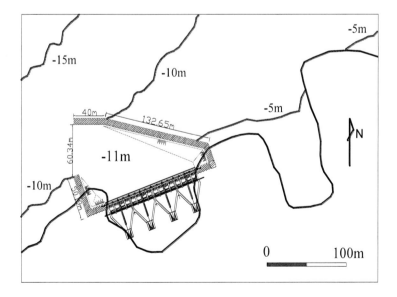

Figure 7b. The water-intake forbay layout with two breakwater.

3.2.2. Experiment Setup

The physical model experiment was conducted in a 40m long, 24m wide hall. The horizontal scale is 1:300 while the vertical scale is 1:100. The model satisfies the geometric similarity, flow similarity and wave similarity and other similarities (Table 2). Crashed coal (D_{50} =0.155mm, density=1.40g/l) was chosen to model the natural sand. Wave maker, wave gouge, tide generating facility and other instrument for measuring water level, velocity and sediment concentration were equipped (Figure 8). The model firstly calibrated the tidal flow filed with respects to water level, current velocity and direction, suspended load concentration at five sites surveyed in summer 2005 and beach evolution spanning the period of 1984-2004. The calibration results demonstrate that the physical model is capable of replaying the tidal, wave field and resultant sediment transport, and is able to forecast the hydrodynamic variations and sedimentation distribution due to the LNPP construction.

Table 2. Scale indexes used for the LNPP physical-model experiment

Scale		Symbol	Computation value	Experimental value
Geometry scale	Horizontal scale	λ_l	/	300
	Vertical scale	λ_h	/	100
Wave parameter scales	Wave-length scale	λ_L	/	100
	Wave-height scale	λ_H	/	100
	Wave-celerity scale	λ_C	10	10
	Wave-period scale	λ_T	10	10
	Orbital-velocity scale	λ_{v_m}	10	10
Tidal parameter scales	Current-velocity scale	λ_v	10	10
	Roughness scale	λ_n	1.24	1.24
	Flow-time scale	λ_t	30	30
	Flow-discharge scale	λ_Q	3.0×10^5	3.0×10^5
Sediment scales	Suspended-sediment diameter sacle	λ_d	0.898	0.898
	Volume-weight scale	λ_{ρ_s}	1.89	1.89
	Dry-volume-weight scale	λ_{γ_0}	1.91	1.91
	Settling-velocity scale	λ_w	3.33	3.33
	Sediment-concentration scale	λ_s	0.458	0.458
	Erosion/deposition timescale	λ_{t2}	125.1	Determined by calibration tests

To investigate the massive sedimentation within and outside the LNPP water-intake forbay a big wind wave with a return period of 50 years (NNE, significant wave height Hs=3.94m, Ts=8.51s)+ spring tide was used as the representative strong wave climate in this sea area. Such a wave climate is assumed to prevails two days.

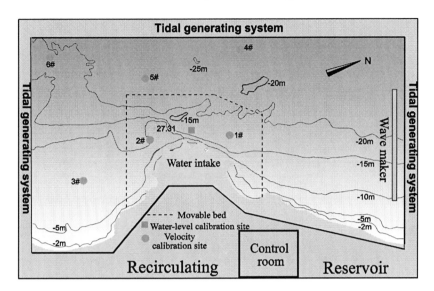

Figure 8. Setup of the physical model test.

3.2.3. Experimental Results

The flow fields within and outside the forbay layout without breakwater is shown on figure 9 while the sedimentation distribution is shown on figure 10.

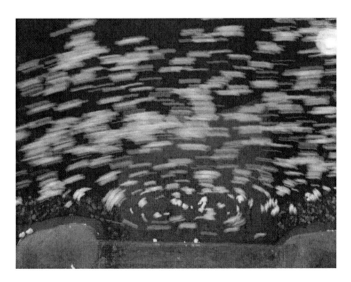

Figure 9a. The spring-flood flow at peak time, flowing from left to right. Note weak clockwise eddy appears in front of the water-intake building.

Figure 9b. The spring-ebb flow field at peak time, flowing from right to left.

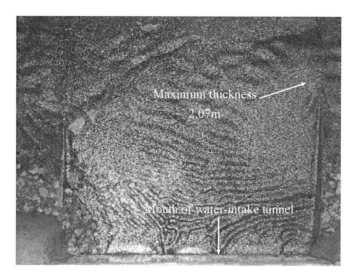

Figure 10. Sedimentation within the forbay by the big wind wave (Hs=3.94m, Ts=8.51s)+ spring tide.

It could be seen on figure 10 that deposits distribute fully within the forebay and a lot of sands have entered into the mouth of water-intake tunnels. It is also observed that sea water becomes turbid as a result that sands are transported near the bottom with a significant portion of finer sands put into suspension by the big wave's shearing actions. As the wave wanes, suspended sand immediately dropped to the bed and water became clear. By collecting, drying and weighing the deposits from the forbay and transforming into in situ deposition volume the massive deposition volume is 13,076.44m^3.

At the same time, the computed result using Eqs. (1), (6), (7) and (12) is 8376.27m^3, 46,069.49m^3, 605.04m^3 and 12,564.41m^3, respectively. This demonstrates that the theoretically-based formula (12) performs much better in predicting large sediment transport rate by big wind-driven waves than empirical formulas (1), (6) and (7). Though formula (1) attempted to augment its predicting capability of larger sediment transport rate by modifying the depth-averaged suspended sediment concentration S to take into account the breaking-wave effect it treats the sediment transport solely as the mode of suspension; meanwhile, formula (6) augmented its predicting capability by modifying its coefficients to take into account the co-actions of wind wave and tidal current under storm circumstances based upon in situ measurements, but it does not capture the physical scenery of sheet-flow transport that takes place in such circumstances; on the other hand, formula (12) has not only been derived from a theoretical model describing the bottom high-concentration sediment-water-mixture movement but also truly reflected the situation in which bedforms disappear and the near-bed high-concentration sediment layer becomes essentially plane, i.e. sheet flows. Next the sheet-flow mechanics is further explained in detail.

3.3. Sheet-Flow Transport

The transport and deposition of sheet flows could lead to significant morphodynamic evolution and massive sedimentations at the entrance and/or access channels of coastal and estuarine harbors and other facilities due to larger transport rate. It has become a hot research topic in international coastal engineering community in recent years. A large number of experimental observations, analytical and numerical studies have been conducted and many good results have been reported, e.g. the thickness of the sheet-flow layer, time-dependent particle concentration and velocity profile inside the sheet-flow layer, and the resulting sediment transport rates, etc.

For unidirectional flow, sheet flow occurs when the Shields parameter θ exceeds about 0.8 (Wilson 1966); nonetheless, there has no a universally-accepted criteria to judge the inception of sheet flow in oscillatory flow (see Table 3).

Manohar (1955) carried out one of the first experimental studies to investigate the inception of sheet flow in oscillatory flows using oscillatory trays and put forward a mobility number M.

$$M = 440 r^{-0.4}$$
$$r = [\frac{(\rho_s - \rho)gD^3}{\rho \upsilon^2}]^{0.5} \tag{13}$$

where ρ_s is the sediment density, ρ is the flow density, ν is the flow kinematic viscosity, g is the acceleration of gravity and D is the grain size.

Chan et al. (1972) used a horizontal tube (HT) to investigate the effect of the kinematic viscosity and the relative particle density on the inception of the sheet flows. They observed that the wave period has larger effects on the inception of sheet flow and suggested the following criteria:

$$\left\{ \psi(\frac{D_{50}}{\delta_w})^{0.8} \right\}_{cr} = 43.6 \text{ with } \psi = U_w^2/(s-1)gD_{50} \text{ and } \delta_w = \sqrt{\upsilon T_w / \pi} \tag{14}$$

in which, U_w is the wave orbital velocity, δ_w is the Stokes' boundary layer thickness, T_w is the wave period and D_{50} is the median particle diameter.

Dingler and Inman (1976) set the critical mobility number $M = \rho u_b^2 /(\rho_s - \rho)gD = 240$, in which u_b is the velocity amplitude of water motion above the wave boundary layer.

Dibajnia (1991) developed a new formula based on the Chan et al. (1972) data and proposed a new parameter ω_{pl}:

$$\omega_{pl} = \frac{\frac{1}{2}U_w^2}{(s-1)gW_s T_w} \tag{15}$$

here W_s is the settling velocity of particles, and s is the relative density.

In a similar way as that of Chan et al. (1972), Camenen and Larson (2006) proposed a criteria as:

$$\left\{\psi(\frac{D_{50}}{\delta_w})^{0.8}\right\}_{cr} = 70 \tag{16}$$

Bagnold (1956, 1966) also suggested the critical Shields parameter for the inception of sheet-flow regime is:

$$\theta_{cr} = C_0 \tan\varphi, \quad C_0 = 0.6\text{-}0.7, \quad \tan\varphi = 0.375\text{-}0.75 \tag{17}$$

Wilson (1989) suggested $A_b > 0.65D^{0.417}[g(s-1)T_w^2]^{0.583}$, where A_b is the semi-excursion of water movement above the wave boundary layer,.

Li and Amos (1999a) suggested $M = 12.13D^{-0.707}$ or $\theta_{cr} = 0.172D^{-0.376}$.

Though the thickness of a sheet-flow layer is normally of centimeter order it could be generally stratified into two sub-layers (Figure11): the upper suspended-sediment-dominated sublayer and the lower bedload-dominated sublayer (Ribberink and Al-Salem 1995, Kaczmarek and Ostrowski 2002, Malarkey et al. 2003, O'Donoghue and Wright 2004a,b). While the former is the transitional zone into outer suspension column where turbulence convection and diffusion processes dominate, over 90% of sediments in a sheet-flow layer is transported within the lower thin bedload sublayer, in which interparticle collisions are the predominant mechanism for momentum transfer (Ribberink and Al-Salem 1994, Ribberink 1998). This is especially true for medium-coarse sand ($D_{50} \geq 0.2$ mm). The detailed description of particle concentration and velocity profiles inside the sheet-flow layer could be referred to Wang and Yu (2007c).

Table 3 Critical parameters for judging the inception of sheet flows

Author(s)	D_{50}(cm)	ρ_s (g/cm^3)	T(s)	A_b (cm)	u_b (cm/s)	f_w	θ_w	M	Notes
(1) Manohar (1955)									
	0.024	2.49	3.12	36.83	74.10	0.00886	0.709	160	Oscillatory plates
	0.028	2.65	3.05	36.83	75.96	0.00925	0.590	127	
	0.061	2.54	2.67	36.83	86.56	0.01143	0.465	81	
	0.079	2.63	2.26	36.83	102.41	0.01233	0.515	84	
	0.101	2.6	2.31	36.83	99.97	0.01334	0.421	63	
	0.183	2.6	2.27	36.83	101.80	0.01634	0.295	36	
	0.198	2.63	2.21	36.83	104.5	0.01681	0.290	35	
(2) Carstens et al. (1969)									Oscillatory water tunnel
	0.30	2.47	3.52	45.03	80.4	0.00895	0.60	133	
(3) Horikawa et al. (1982)									Wave flume
	0.02	2.65			69.1		0.52	148	
	0.05	2.65			95.4		0.57	112	
	0.07	2.65			111.4		0.54	109	
(4) Ribberink and Al-Salem (1994)									
	0.021	2.65	6.5	67	72.5	0.00754	0.64	170	Oscillatory water tunnel
	0.021	2.65	9.1	101	74.5	0.00694	0.61	177	
	0.021	2.65	5	54	76.0	0.00790	0.73	184	
	0.021	2.65	12	142	83.0	0.00651	0.70	215	
(5) Dohmen-Janssen and Hanes (2002)									
	0.24	2.65	6.5	61	59	0.0064841	0.2908		Wave flume
	0.24	2.65	6.5	64.1	62	0.006395	0.31672		
	0.24	2.65	9.1	95.6	66	0.0057409	0.32219		
	0.24	2.65	9.1	107.2	68	0.005696	0.33934		
(6) Inman (1957)									Field observations
	0.0103	2.65	16.0	155	73.0	0.00568	0.95	320	
(7) Dingler and Inman (1976)									
	0.0157	2.65	8.5	112	83	0.00657	0.91	270	Field observations
	0.0162	2.65	7.5	101	84	0.00648	0.85	277	
	0.013	2.65	7	79	71	0.00664	0.80	244	

Notes: f_w is the wave friction index and θ_w is the Shields parameter under waves.

As far as the silt movement is concerned, Chinese researchers have also observed that there are silt transport in both bedload mode and suspension mode, both of which contribute significantly and can not be biased in formulating the sheet-flow transport of silt sediments (e.g. Zhao 1994, Han 1996). Cao and Jiao (2002) discussed how to determine the ratio of bedload silt to suspended silt in the transport process and navigation-channel sedimentation; Cao et al. (2004) further explored the silt transport formula based upon the concept of the ratio of bedload silt to suspended silt. This explains why the formulas (1) and (6) could not predict the massive sediment transport rate satisfactorily because they either solely treated the sediment transport as suspended sediment or solely as bedload transport; on the other hand, formula (12) is specially derived for sheet-flow transport regime and could satisfactorily predict the medium-coarse sand transport rate in sheet-flow regime as the majority of medium-coarse sand are transported as bedload.

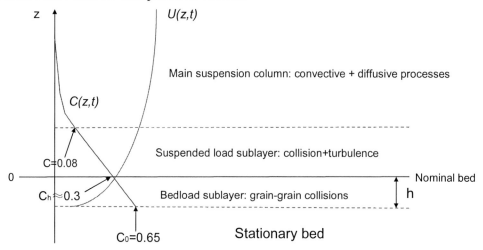

Figure 11. A conceptual illustration of the sheet-flow layer stratification.

Nevertheless, formula (12) could not satisfactorily predict the magnitude and direction of silt (D_{50}=0.025~0.125mm) transport due to that it does not take into account the phase-lag effect of silt sediments. The so-called phase-lag effect means that silt that is suspended in the first half wave period does not completely settles down to the bed and the remaining fraction of suspended silt is transported offshore in the second half wave period, vice versa, this reduces the onshore silt transport magnitude; so much so, the onshore silt transport might become zero or even transports oppositely in the offshore direction (Dohmen-Janssen et al. 2002). Such phase-lag effects become prominent under the asymmetrical waves, that is, the onshore orbital velocity is larger but last short while the offshore orbital velocity is smaller but lasts longer. The finer the sediment is, the shorter the wave period is and the bigger the orbital velocity is, the larger the phase-lag effect is (Dohmen-Janssen et al. 2002).

The earliest experimental observations on phase-lag effects were done by Watanabe and Isobe (1990), Dibajnia (1991) (D_{50}=0.2mm), Ribberink and Chen (1993) (D_{50}=0.13mm) and others. Dibajnia and Watanabe (1992) found that the 'quasi-steady' (i.e. transport rate responds simultaneously with the variations of bottom flow velocity or bottom shear stress) transport formulas could not predict silt transport rate with enough precision based upon silt-transport experiments conducted in a U-shaped oscillatory water tunnel at Tokyo University with shorter waves, as these types of formulas could only predict that silt moves in the direction of wave propagation. Therefore, in order to reflect the phase-lag effects, the authors tried to build a model to compute the suspended silt amount, the actually forward transported silt and those keeping suspension until the next half period, respectively. Dibajnia and Watanabe (1996, 1998), Dibajnia et al. (2001), Watanabe and Sato (2004) further modified this model. But due to that the model treats the silt transport solely as suspended mode to describe the phase-lag effects, the calculated results could not compare well with measured results in terms of the magnitude and direction of the net silt transport rate in a wave period.

In a similar way, Guizien et al. (2003) and many other researchers used the vertical 1D oscillatory boundary layer turbulent model + sediment diffusion equation to compute the phase-lag effects and silt transport rate; Liu and Sato (2006) modified the eddy coefficient and sediment vertical diffusion coefficient in respective wave half period to compute the phase-lag effects based on the two-phase flow theory.

On the other hand, Dohmen-Janssen et al. (2005) modified the 'quasi-steady' bedload transport rate formula of Ribberink (1998) by introducing a phase-lag index p describing the ratio of particle settling velocity and wave period, i.e. $p = \delta_s \omega / w_s$ (ω is the angular frequency, w_s is particle settling velocity and δ_s is the thickness of sheet-flow layer). This modification could reflect the reduction of silt transport magnitude due to phase-lag effects but failed to compute the offshore silt transport that is opposite to the incidence wave propagation. Different from Dohmen-Janssen et al.'s (2005) method, Camenen and Larson (2006) firstly introduced phase-lag index into the non-dimensional Shield parameter and calculated the Shields parameter for the respective wave half period, then combined the two half-period Shields parameter to obtain a representative Shields parameter to calculate the net silt transport rate in a whole wave period.

To sum up, researchers attempted to compute the sediment transport rate in sheet flow regime by either building empirical models or numerical method using two-phase flow theory or complex oscillatory boundary turbulent model. As Davies et al.'s (1997) comparisons show that numerical computations of sheet-flow sediment transport rate is not better than empirical, semi-empirical formulas do, because of the empirical treatments of the bottom boundary conditions of suspended-sediment concentration profile, i.e. 'reference height' and 'reference concentration', eddy coefficient of flow, and sediment diffusion coefficient, in particular, because of ineffective computation of phase-lag effects. On the other hand, Most of previous work attempted to introduce phase-lag index to formulas in low-medium transport regime to calculate sheet-flow large sediment transport rate. The results are resultantly unsatisfactory in comparison with experimental observations. This is because that the intrinsic mechanism for phase-lag effects has not been well understood and objective choice of hydrodynamic and sediment parameters, such as flow depth, particle settling velocity, wave orbital velocity and wave period, etc. is exercised to describe the phase-lag effect. Such modified formulas become very sensitive to one or several parameters.

Therefore, it is of great importance to study thoroughly the phase-lag effect to understand its influences on the magnitude and direction of net transport rate in the sheet-flow transport regime, especially, for silt sediments. This helps to work out good transport rate formulas to predict large transport rate under strong hydrodynamic conditions.

4. MITIGATION ENGINEERING MEASURES

4.1. Harbors on Silty and Sandy Coasts

Massive sedimentations at the basin, entrance and access channel of coastal and estuarine harbors or other facilities could cause serious even disastrous results. It is thus necessary to seek mitigating means to reduce or minimize the massive sedimentation. This chapter only focuses on the engineering measures.

At the very gentle, shallow silty coasts, harbors are often dredged with the basin and access channel protected by breakwaters and/or submerged sills. The wave breaker zone is often much wider than the length of access channel in extreme weather conditions. So the breakwaters have to extend outside the breaker zone to effectively prevent the bottom high-concentration sediment layer (sheet-flow layer) from entering the outer access channel and settling there. But this might cause a heavy financial burden and construction difficulties. One alternative option is to build a flow-through bridge to connect the semi-enclosed harbor basin that lies in the deep water outside the breaker zone. This could, on one hand, let the wave-driven sheet-flow sediments not to enter the access channel but to pass through the bridge, and on the other hand, makes effective use of the deep water resource (Zhang et al. 2005). One example is the Surabaya Port in Indonesia. Two container berths are constructed at the water depth of 10m where wave and current could not agitate sediment movement, the two berths are connected to the land by two 1540m-long lane-access bridges (Irie et al. 2002).

At the relative steep silty coasts, the semi-enclosed harbors more often than not experience massive sedimentations at the access channel near the entrance as a result of sheet-flow sediments settling there. Therefore, any mitigating measure must be able to effectively impound the sheet-flow sediment as driven by the wave-induced longshore currents. Researchers once suggested various solutions: to extend the length of breakwaters (Luo and Yang 2003), to add current-deflecting groins along the breakwater (Wang 2000, Sun et al. 2005, Ji 2006), whereas Luo et al. (2007) suggested that the axis direction of navigation channel should align parallel to the prevailing wave propagation and/or dominant flow direction, or if possible, the angle between them should be as smaller as possible.

The New Mangalore port is a semi-enclosed lagoon-type container port on the southwestern Indian coast (Figure 12). The bottom sediments are generally fine sand and silt: 0.11~0.17mm fine sand within the -5m contour and 0.03~0.09mm silt within the -5m~-15m contours; the beach slope is gentle: the slope from shoreline to -5m contour is about 1:125 with a width of 500m, while the slope becomes to be about 1:278 from -5m contour to -20m contour, the width is about 7km; the coast runs NNW15°, contours are parallel to the coastline and the breakwaters extend to the -5m contour (Ghosh et al. 2001). The access channel is almost perpendicular to the coast.

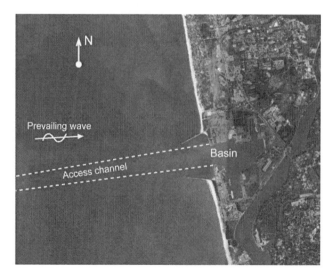

Figure12. Indian New Mangalore port.

Nevertheless, the port experienced massive sedimentations at each of its three development phases and had to be dredged considerably: at the first phase (1974-1975), the annual deposition quantum is 1.19 million m^3; at the second phase (1980-1992), the annual deposition quantum is 1.79~2.52 million m^3; at the third phase (1995-), the annual deposition quantum is 3.6~4.88 million m^3. The massive sedimentations usually occur at the southwestern monsoon season from June through September. The predominant wave direction during monsoon season is 266.5°, its significant height is 1.5~2m and period is 9.5s. It is noted that the wave propagation direction (266.5°) makes only a small angle of 7.5° with the extension of access channel (259°). This argues that a smaller angle could not prevent the massive sedimentations at the inner access channel and basin of the New Mangalore port. Moreover, it is observed that deposits at the inner access channel and harbor basin distribute evenly (Ghosh et al. 2001).

The sedimentations could be attributed to that the layout of breakwaters did not effectively prevent bottom high-concentration sediments from entering the inner access channel and harbor basin and settling there; at the same time, the width of the harbor entrance is 800m or so, that is about ten times the incidence wave length. As a result, the protection effects of the breakwaters are poor, leading sediments to directly enter the harbor basin under the actions of incidence waves.

Therefore, besides reasonable entrance width, one important measure is to layout the breakwaters properly to effectively impound the longshore sediment transport, more specifically, to prevent the bottom high-concentration sediment layer from entering the access channel. Generally speaking, the breakwaters should extend to the depth where big waves could not mobilize the bottom sediment, and the extension of breakwaters should have a less than 90° angle with the opposite-direction of the prevailing wave, the optimum angle ranges 45~60°. But the specific angle should be determined according to the coastal characteristics including geometry, lithology, hydrodynamics, sediment transport mechanics and the type of breakwater (i.e. tower-type, slope-type or compound-type).

As explained in section 3.2 the breakwater layout for the water-intake forbay of the LNPP is designed to protect the forebay from silting. The flow fields affected by the breakwaters are shown on Figs. 13. But it is observed that the designed breakwater layout could not effectively impound the longshore sand transport driven by the northerly-coming waves (Figure 14) and the longshore-transported sediments are deflected by the longer breakwater and transported to the entrance area of the forbay and settle there. Moreover, about 462m^3 sands would be deposited within the forbay by a 50-year-return-period wave + spring tidal flow in a two-day-long temperate cyclone.

The extension of the designed longer breakwater has an angle of about 100.5° with the predominant NNE wave's opposite-propagation direction. So the designed breakwater layout is optimized by shifting the longer breakwater northward by about 12° and extending the longer breakwater by 15m and extending the shorter breakwater by 21m with the axis direction of the entrance remained unchanged (Figure 15).

The physical-model test result indicates that in the same hydrodynamic conditions the optimized breakwater layout could effectively impound the longshore sand transport, which are mostly deposited near the shore and sediment amount within the forebay is about 307.6m^3 and sedimentation at the entrance area is insignificant (Figure 16).

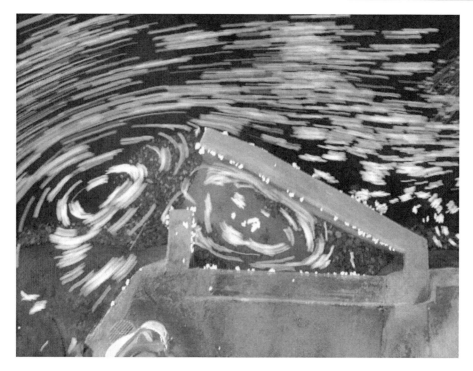

Figure 13a. The spring-ebb flow field at peak time. The eddy inside the forbay is clockwise and the eddy outside the forbay is anticlockwise.

Figure 13b. The spring-flood flow field at peak time. The eddy inside the forbay is clockwise.

Figure 14. Sedimentation within and outside the forbay of LNPP with the designed breakwater layout.

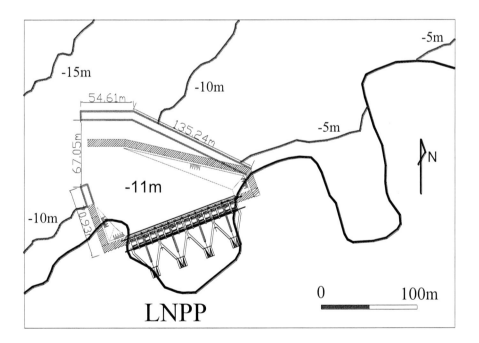

Figure 15. Optimized breakwater layout for the forebay of LNPP.

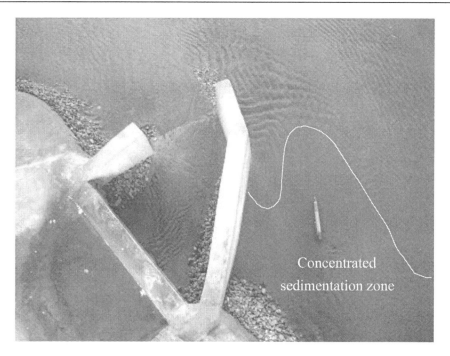

Figure16 The sedimentation distribution after the breakwater layout is optimized.

4.2. Muddy Harbors

As discussed in section 2, muddy sedimentations more often than not occur at estuarine harbor basins and sedimentation amount is closely related to the exchange volume of waters between the basin and ambient water system. Therefore, engineering measures have to be selected to reduce the exchange flow rates and the sediment quantum carried by the exchange flow in order to minimize the basin sedimentation. One of mature technologies is to build a current-deflecting-wall (CDW) to modify the flow field at the harbor entrance to reduce the exchange rates, and a submerged sill might be built together with the CDW to block bottom high-concentration muddy layer from entering the harbor basin (see Winterwerp 2005, Kirby et al. 2008). The CDW + sill system was originally built at Köhlfleet harbor and has been proved successful in significantly reducing the annual dredging deposits. Generally speaking, the CDW + sill system could reduce the sedimentation quantum in the harbor basin by up to 40% (Kuijper et al. 2005).

If very fine, organic matter are abundant in the ambient water system an open harbor basin might be appropriate as the dead water zone within the harbor basin could be avoided, but the openness has to be compromised with protection against wave attacks that significantly affect ships' anchoring (Winterwerp 2005).

Nonetheless, as pointed out by researchers (e.g. Winterwerp 2005, Kirby et al. 2008), a thorough understanding of the physics (i.e. hydrodynamic and sedimentary regime) which dictates sedimentation at a given harbour site must be firstly completed before a CDW + sill system is applied. This is because that the efficiency of a CDW +sill system critically depends on the local hydraulic conditions as well as on the shape and dimensions of the structure in relation to the geometry of the harbor entrance and the total harbor water volume.

As a result, its specific configuration must be determined by physical-model tests, otherwise, adverse effects might arise and harbour sedimentation becomes worsen. This technology might be further applied at the semi-enclosed harbors on open muddy coasts, so long as the water exchange rates could be reduced to a proper level.

In addition to mud source control, e.g. watershed management, bank and shore protection, etc. to effectively keep the mud out of the harbor area techniques might be applied to control the formation of fluid mud layer; these techniques aim to reduce slack water condition and eddy formation or reduce flocculation to keep mud sediment suspended, etc. (McAnally et al. 2007)

A example is the Semen Tuban port located at the north coast of Java Island in East Java province, Indonesia (Irie et al. 2002). Bottom sediments are less than 0.05mm in diameter. Here significant wave height is 3.9m and wave period is 10s, average tidal range is 1.46m. So the wave plus tidal current is the predominant force for sediment entrainment and transport. The port basin is located offshore at the depth of -6~-8m below lower water level (LWL) and connected to the land by a bridge, while the access channel is dredged at the depth of -8~-12m below LWL. The port basin is protected with underwater sills but the access channel is not. Topographic survey found that the underwater sills are very effective to reduce muddy sedimentation volume at the basin.

5. Research Prospects

5.1. Predicting Formulas

It is critically important to know the longshore and/or cross-shore sediment transport rate at the designing, siting stage of a coastal and estuarine engineering project as far as sedimentation problem is concerned. Researchers have suggested innumerous empirical, semi-empirical formulas with various complexities to predict the sand or mud transport rates. But reliable formula for predicting silt transport rate, especially in sheet-flow regime is still lacking.

As explained in section 3.3, under strong wave forces, silts are transported in the plane-bed sheet flow regime, in which both bedload movement and suspended silt transport occur. The bedload-type silts in the thin sheet-flow layer act like medium-coarse sands' movement whereas the suspended silts in the sheet-flow layer are much influenced by the phase-lag effects. It is thus incomplete or inaccurate to build transport model based upon 'sole suspended silt transport' or 'sole bedload silt transport'. Furthermore, as most of the proposed formulas are based upon transport mechanics of uniform sediments, it is of great importance conduct more and more experimental studies to understand non-uniform sediments transport driven by waves or wave-currents in terms of vertical sorting and/or longitudinal sorting as well as net transport rate, etc.

When numerical modelling is exercised, the bottom boundary condition for suspended silt concentration profile, such as the so-called 'reference concentration' and 'reference height', flow eddy coefficient and sediment vertical diffusion coefficient, in particular, the phase-lag unsteadiness must be determined as precisely as possible. In addition, the numerical models should be able to accurately compute the secondary concentration peak at flow reversals (see Li and Davies 1996, Guizien et al. 2003, Vittori 2003).

Due to the immature numerical techniques empirical, semi-empirical formulas would be used for a longer period of time to predict sediment transport rate. But empirical, semi-empirical formulas often have limited applications because the coefficients in the formulas are often derived from a smaller experimental and/or observational dataset. So formulas that have sound theoretical grounds are of critical importance. In this regard, more physical or numerical experiments and analytical studies have to be conducted in coming years to understand the phase-lag unsteadiness in silt sheet-flow transport regime and build good models to describe this effect and eventually come up with formulas to reliably predict the magnitude and direction of net transport rate of silt sediments.

5.2. Counter-Sedimentation Measures

No matter what type of sediment predominates in a coastal segment or estuary designers have to take into account the sedimentation problem at the design, siting, planning stage and find an optimized engineering layout. This might avoid the costly remedy measures to reduce sedimentation after the project has been put into operation.

The benefit of CDW + sill system in reducing sedimentation at muddy harbour basins is enormous. But further research is required to establish the effect of the entrance aspect ratio on the CDW-induced helical flow in the harbor entrance and on the exchange flow rates (Winterwerp 2005). In this regard, physical model tests are an indispensable tool to check the actual sedimentation-reducing measures. This calls for the development of more sophisticated flow and particle measurement instrument.

And the breakwater layout for the semi-enclosed harbors on silty and sandy open coasts plays a critical role in impounding longshore transported sand or bottom high-concentration silt layer from entering the access channel of a harbour. Site-specific physical-model experiments are also needed to optimize the layout.

In addition to physical-model experiments, sophisticated numerical techniques are also cost-effective to study the flow field and sedimentation patterns with a particular harbour layout. For this purpose the numerical models should be able to simulate truthfully the 3D flow effects (e.g. flow separation, flow stagnation, eddy formation, density and temperature stratifications, etc,) and fine, cohesive sediment transport and deposition. More often, these two types of techniques can be used together to obtain a reliable result to choose a cost-effective sedimentation-minimizing measure.

Though no counter-sedimentation method is universally applicable innovative and new method is still critically needed in the coming years. Moreover, new reliable, accurate in situ measurement instrument is needed to investigate the concentration, density, viscosity, shear stress and other properties of fluid mud layer and sheet-flow layer. This is helpful to understand better the mechanism of the formation, transport, accumulation and consolidation of fluid mud and sheet flows and build better process-based models to simulate the movements of these sediments.

6. CONCLUDING REMARKS

Massive sedimentations either occur on open silty coasts in association with the transport and deposition of near-bed high-concentration sediment layers (often termed as sheet flows) or occur at muddy estuarine harbors due to the formation, movement, accumulation and consolidation of fluid mud layers. Harbors and other coastal facilities subjected to such massive sedimentations could be considered as bad engineering projects from the point of view of sediment transport and deposition, as costly remedy measures have to be sought to reduce the dredging volumes after the harbors are put into operation.

Though considerable efforts have been made in the past several decades to understand the underlying causes and processes for such massive sedimentation events with various mitigation engineering measures suggested, there has still no a widely-accepted formula to predict the sediment transport rate with high precision, especially for silty sediments, nor has a universally-applicable mitigation engineering measre. This could be partially attributed to the complex interactions between coastal forces including wind, tidal current, wave, sediment transport, human activities and topographic configuration, etc.

As more and more environment-friendly policies have been implemented around the world the dreging and disposal of contaminated sediments becomes more restricted. As a result, it is critically important to put forward an optimum project layout in terms of minimizing sedimentation at the planning, siting and design stages. This requires a good understanding of the coastal and estuarine processes including meteorology, hydrodynamics, sediment transport mechanics, topography, etc. on a local and regional scale. Physical-model tests and good predicting formulas would be indispensable tools to achieve this purpose in coming years .

ACKNOWLEDGMENTS

The author wishes to thank Profs. Yu Guo-Hua, Zhou Yi-Ren and Tang Li-Qun for their constructive discussions in preparing this chapter. Special thanks goes to Prof. Robert Kirby for kind providing the PIANC Working Group 43 report "Minimizing Harbour Siltation". This study is a portion of Hydrological Simulation & Regulation of Watersheds funded by National Natural Science Foundation of China (No. 50721006), is also a portion of DALCOAST System supported by the 948 Project Office of the China Ministry of Water Resources (No. 200603).

REFERENCES

Bagnold, R.A. Experiments on a gravity-free dispersion of large solid spheres in a Newtonian fluid under shear. *Proc. R. Soc. Lond.* 1954, A255, 49-63.

Bagnold, R.A. The flow of cohesionless grains in fluids. Philosophical transactions of the Royal Society of London, Series A, 1956, 249, 235-297.

Bagnold, R.A. An approach to the sediment transport problem from general physics. U.S. Geological Survey Professional Paper, 1966, No. 422-I.

Camenen, B.; Larson, M. Phase-lag effects in sheet flow transport. *Coast. Eng.* 2006, 53, 531-542.

Cao, Z.; Jiao, G. Determination of the ratio of bedload silt to suspended silt in silt transport processes on silty coasts. *J. Waterway and Harbor,* 2002, 23 (1), 12-15 (in Chinese).

Cao, Z.; Kong, L.; Li, Y. Engineering sediment problems of silty coasts. *J. Waterway and Harbor*, 2004, 25 (suppl.), 26-30 (in Chinese).

Cao, Z.; Jiao, G.; Zhao, C. Computation of sediment transport and sedimentation on silty coasts. *Ocean Eng.* 2004, 22 (1), 59-65.

Carstens, M.R.; Neilson, F.M.; Altinbilek, H.D. Bed forms generated in the laboratory under an oscillatory flow: analytical and experimental study. U.S. Army Coastal Engineering Research Center, Technique Memory, 1969, No. 28, 39p.

Chan, K.; Baird, M.; Round, G. Behaviour of beds of dense particles in a horizontally oscillating liquid. *Proc. R. Soc. Lond.* 1972, A 330, 537–559.

Christiansen, H. New insights on mud formation and sedimentation processes in tidal harbors. *Proc. Int. Conf. on Coast. and Port Eng. in Developing Countries,* 1987, 2, 1332-1340.

Clausen, E.; Olsen, H.A.; Brink-Kjaer, O.; Mikkelsen, L.; Halse Nielsen, A. Sedimentation and dredging in the navigation channel to the port of Esbjerg. 25th Int. Navigation Congress, 1981, 231-239.

Committee on Tidal Hydraulics (CTH). Estuarine navigation projects. Technical bulletin 17, Corps of Engineers Committee on Tidal Hydraulics, Vicksburg, Miss. 1971.

Davies, A.G.; Ribberink, J.S.; Temperville, A.; Zyzerman, J.A. Comparisons between sediment transport models and observations made in wave and current flows above plane beds. *Coast. Eng.* 1997, 31, 163-198.

Dibajnia, M. Study on nonlinear effects in beach processes. PhD thesis, University of Tokyo, Japan, 1991.

Dibajnia, M.; Watanabe, A. Sheet flow under nonlinear waves and currents. Proc. 23rd Int. Conf. on Coast. Eng., 1992, 2015– 2028.

Dibajnia, M.; Watanabe, A. A transport rate formula for mixed-size sands. *Proc. 25th Int. Conf. on Coast. Eng.* 1996, 3791-3804.

Dibajnia, M.; Watanabe, A. Transport rate under irregular sheet flow conditions. *Coast. Eng.* 1998, 35, 167-183.

Dibajnia, M.; Moriya, T.; Watanabe, A. A representative wave model for estimation of nearshore local transport rate. *Coast. Eng. in Japan,* 2001, 43 (1), 1-38.

Dinehart, R.L. Evolution of coarse gravel bed forms: field measurements at flood stage. *Water Resour. Res.* 1992, 28(10), 2667-89.

Dingler, J.R.; Inman, D.L. Wave-formed ripples in nearshore sands. In: Proc. 15th Conf. Coast. Eng. ASCE, New York, 1976, 2109–2126.

Dohmen-Janssen, C.M.; Hanes, D.M. Sheet flow dynamics under monochromatic non-breaking waves. *J. Geophys. Res.* 2002, 107 (C10), 1301-1321.

Dohmen-Janssen, C.M.; Kroekenstoel, D.F.; Hassan, W.N.M.; Ribberink, J.S. Phase-lags in oscillatory sheet flow: experiments and bed load modeling. *Coast. Eng.* 2002, 46, 61-87.

Gao, J. Mechanism of massive sedimentaiton at the outer access channel of the Huanghua harbor in gale weather and harnessment. *Science and Technology Review* 2005, 1, 29-31 (in Chinese).

Ghosh, L.K.; Prasad, N.; Joshi, V.B.; Kunte, S.S.A study on sedimentation in access channel to a port, *Coast. Eng.* 2001, 43, 59-74.

Guizien, K.; Dohmen-Janssen, C.M.; Vittori, G. 1DV bottom boundary layer modeling under combined wave and current: Turbulent separation and phase lag effects. *J. Geophys. Res.* 2003, 108 (C1), 3016, doi:10.1029/2001JC001292.

Han,X. Sedimentation research of silty coasts. J. Waterway and Harbor, 1996, 3, 22-28 (in Chinese).

Hassan, W.N.; Ribberink, J.S. Transport processes of uniform and mixed sands in oscillatory sheet flow. *Coast. Eng.* 2005, 52, 745-70.

Horikawa, K.; Watanabe, A.; Katori, S. A laboratory study on suspended sediment due to wave action. Proc. 18th Int. Conf. Coast. Eng. ASCE, Cape Town, 1982, 1335-1352.

Hughes MG, Masselink G, Brander RW. Flow velocity and sediment transport in the swash zone of a steep beach. *Mar. Geol.* 1997, 138, 91-103.

Hunt J.N. Direct solution of wave dispersion equation. J. Waterw., Port, Coastal Ocean Div., Am. Soc. Civ. Eng. 105 4 (1979), pp. 457–459.

Inman, D.L. Wave-generated ripples in near-shore sands. U.S. Army Beach Erosion Board, Technique Memory, 1957, No. 100, 66p.

Irie, S.; Hidayat, R.; Morimoto, K.; Ono, N. Study of siltation protection in Asian ports. Proc. *12th Int. Offshore and Polar Eng. Conf. Kitakyushu*, 2002, 539-544.

Ji, Z.Z. Characteristics of layout of port water area on silty coasts. Ocean Eng. 2006, 24 (4), 81-85 (in Chinese).

Kaczmarek L.M.; Ostrowski, R. Modeling intensive near-bed sand transport under wave-current flow versus laboratory and field data. *Coast. Eng.* 2002, 45, 1-18.

Kirby, R.; Enriquez, J. ; Headland, J.; Lamers, K. ; Nasner, H.; Nakagawa, Y.; Paul, J.; Sas, M.; Vested, H.J.; Westermeier, F.; Winterwerp, J.C.; McAnally, W.H.; Nella, P. ; Norman, C.; Woolhouse, L. Minimizing Harbour Siltation. Rep. International Navigation Association/PIANC Working Group 43, 2008. The PIANC Headquarters, Brussels, Belgium.

Kuijper, C.; Christiansen, H.; Cornelisse, J.M.; Winterwerp, J.C. Reducing harbor siltation. II: Case Study of Parkhafen in Hamburg. *J. Waterway, Port, Coast., and Ocean Eng.* 2005, 131 (6), 267-276.

Lamb, M.P.; D'Asaro, E.; Parsons, J.D. Turbulent structure of high-density suspensions formed under waves. J. Geophys. Res. 2004, 109, C12026, doi:10.1029/2004JC002355, Copyright [2004] American Geophysical Union.

Li, M.Z.; Amos, C.L. Field observations of bedforms and sediment transport thresholds of fine sand under combined waves and currents. *Mar. Geol.* 1999a, 158, 147-60.

Li, M. Z; Amos, C.L. Sheet flow and large wave ripples under combined waves and currents: field observations, model predictions and effects on boundary layer dynamics. *Continental Shelf Res.* 1999b, 19, 637-663.

Li, Z.; Davies, A.G. Towards predicting sediment transport in combined wave-current flows. *J. Waterway, Port, Coast. Ocean Eng.* 1996, 122 (4), 157-164.

Liu, H.; Sato, S. A two-phase flow model for asymmetric sheetflow conditions. *Coast. Eng.* 2006, 53, 825-843.

Liu, J.; Zhang, J. Method to calculate harbor basin and navigation channel sedimentation and its application plus discussion on the sediment problem with the construction of new west breakwater for Lianyun Port. *Hydro-Science and Eng.* 1993, 4, 301-320 (in Chinese).

Liu, J.; Yu, G. Research and applications of coastal engineering sediments. *Hydro-Science and Eng.* 1995, 3, 221-233 (in Chinese).

Liu, J.; Navigation channel sedimentation on silty muddy coasts. *Hydro-Science and Eng.* 2004, 1, 6-11 (in Chinese).

Lu, Y.; Zuo, L.; Wang, H.; Li, H. Simulation of massive sedimentation at estuarine navigation courses under the action of wind-driven waves. Proc. 12th symp. China's coast. eng. Kunming, 2005, 33-39 (in Chinese).

Luo, G.; Yang, X.H. Choice of port entrance location on silty coasts. *China Port Eng.* 2003, 124 (3), 10-13 (in Chinese).

Luo, Z. Prediction of massive sedimentations at the outer access channel of the Huanghua harbor and discussions on anti-sedimentation measures. *Port and Waterway Eng.* 369 (10), 69-73 (in Chinese).

Luo, Z.; Ma, J.; Zhang, X. Preliminary discussions on the port layout and choice of navigation channel on silty coasts with reference to the rapid sedimentation at the Jingtang Port in stormy season. *China Harbor Eng.* 2007, 147 (1), 35-41 (in Chinese).

Malarkey, J.; Davies, A.G.; Li, Z. A simple model of unsteady sheet-flow sediment transport. *Coast. Eng.* 2003, 48, 171-88.

Manohar, M. Mechanics of bottom sediment movement due to wave action. U.S. Army Beach Erosion Board. *Technique Memory*, 1955, 75, 121p.

McAnally, W.H.; Teeter, A.; Schoellhamer, D.; Friedrichs, C.; Hamilton, D.; Hayter, E.; Shrestha, P.; Rodriguez, H.; Sheremet, A.; Kirby, R. Management of Fluid Mud in Estuaries, Bays, and Lakes. II: Measurement, Modeling, and Management. *J. Hydraul. Eng.* 2007, 133, 23-38.

Nasner, H.; Pieper, R.; Torn, P.; Kuhlenkamp, H. Prevention of sedimentation in brackish water harbors. Proc. *30th Int. Conf. Coast. Eng. San Diego,* 2006, 2994-3006.

O'Donoghue, T.; Wright, S. Concentrations in oscillatory sheet flow for well sorted and graded sands. *Coast. Eng.* 2004a, 50, 117-38.

O'Donoghue, T.; Wright, S. Flow tunnel measurements of velocities and sand flux in oscillatory sheet flow for well-sorted and graded sands. *Coast. Eng.* 2004b, 51, 1163–84.

Pugh, F.J.; Wilson, K.C. Velocity and concentration distributions in sheet flow above plane beds. *J. Hydraul. Eng.* 1999, 125, 117–25.

Ribberink, J.S.; Chen, Z. Sediment transport of fine sand under asymmetric oscillatory flow. Delft Hydraulics, Report H840, part VII, The Netherlands, 1993.

Ribberink, J.S.; Al-Salem, A.A. Sediment transport in oscillatory boundary layers in cases of rippled beds and sheet flow. *J. Geophys. Res.* 1994, 99(C6), 12707-27.

Ribberink J.S.; Al-Salem, A.A. Sheet flow and suspension in oscillatory boundary layers. *Coast. Eng.* 1995, 25, 205–25.

Ribberink, J.S. Bed-load transport for steady flows and unsteady oscillatory flows. *Coast. Eng.* 1998, 34, 59-82.

Sun, L. Review on the hydrology and sediment researchs of Tianjin Port. *Ocean Eng.* 2003, 21 (1), 78-86 (in Chinese).

Sun, L.; Sun, J.; sun, B.; Han, X. Research on sedimentation and mitigating engineering measures for the Jiangtang harbor. Proc. 12th symp. *China's coast. eng. Kunming,* 2005, 494-500 (in Chinese).

Van Rijn LC 2004 Principles of Sedimentation and Erosion Engineering in Rivers, Estuaries and Coastal Seas, AQUA Publications, Amsterdam.

Verlaan, P.A.J.; Spanhoff, R. Massive sedimentation events at the mouth of the Totterdam Waterway. *J. Coast. Res.* 2000, 16 (2), 458-469.

Vittori, G. Sediment suspension due to waves. *J. Geophys. Res.* 2003, 108 (C6), 3173.

Wang, C. Transport mechanics and harnessment measures for the silt sediments in the waters closer to the Jiangtang harbor. *Port Eng. Technology,* 2000, 1, 5-10 (in Chinese).

Wang, Y.-H. Intense transport of non-cohesive bedload sediment by steady currents or asymmetric waves, Ph.D dissertation, Nanjing Hydraulic Research Instiutute, 2007a (in Chinese).

Wang, Y.-H. Formula for predicting bedload transport rate in oscillatory sheet flows. *Coast. Eng.* 2007b, 54 (8), 594-601.

Wang, Y.-H.; Yu, G.-H. Velocity and concentration profiles of particle movement in sheet flows. *Advances in Water Resources,* 2007c, 30 (5), 1355-1359.

Watanabe, A.; Isobe, M. Sand transport rate under wave-current action. Proc. 22nd Int. Conf. on Coastal Eng. Delft, 1990, pp. 2495–2507.

Watanabe, A.; Sato, S. A sheet-flow transport rate formula for asymmetric, forward-leaning waves and currents. Proc. *29th Int. Conf. on Coast. Eng. Lisbon,* 2004, 1703-1714.

Wens, F.; De wolf, P.; Vantore, M., De Meyer, Chr. A hydrometeo system for monitoring shipping tranfic in narrow channels in relation with the problem of the nautical bottom in muddy areas. *27th Int. Navigation Congress, Osaka,* 1990, 5-14.

Wilson, K.C. Bed-load transport at high shear stress. *J. Hydraulic Eng.* 1966, 92, 49–59.

Wilson, K.C. Friction of wave-induced sheet flow. Coast. Eng. 1989, 12, 371-379.

Winterwerp, J.C. Reducing harbor siltation. I: methodology. *J. Waterway, Port, Coast., and Ocean Eng.* 2005, 131 (6), 258-266.

Xu, X. Direct calculation of longshore sediment transport rate using in situ measured wave parameters. *Ocean Eng.* 1996, 14 (2), 90-96 (in Chinese).

Xu, X.; She, X.; Cui, Z. Preliminary discussions on harbor and navigation channel massive sedimentation problems. Proc. 4th scientific symp. of China Society for Hydropower Eng. Beihai, 2003, 296-300 (in Chinese).

Zhang, Q.; Zhang, J.; Yang, H.; Li, S. Discussions on harbor layout on silty coasts. *China Harbor Eng.* 2005, 134 (1), 6-9 (in Chinese).

Zhang, Q.; Hou, F.; Xia, B.; Zhang, J.; Yang, H. 2D numerical modeling of the sedimentation at the outer access channel of the Huanghua harbor. *China Harbor Eng.* 2006, 145 (5), 6-9 (in Chinese).

Zhang, H.; Wang, J. Experimental study of repose angles of sand, gravel andmodel sand. *J. Sedimentary Res.* 1989, 3, 90–96 (in Chinese).

Zhao, C. Characteristics of bottom silt movement under the action of waves. *J. Waterway and Harbor,* 1994, 1, 34-39 (in Chinese).

Zhao, C.; Liu, F.; Cao,Z. Experimental studies of sediment transport at silty coasts. *J. Waterway and Harbor,* 2006, 23 (4), 259-261 (in Chinese).

Zhao, Q.; Han, H. The mechanism of rapid sedimentation at the Huanghua Port and nearshore 3D suspended sediment concentration field under the wave actions. *J. Waterway and Harbor*, 2007, 28 (2), 77-80 (in Chinese).

In: Encyclopedia of Environmental Research
Editor: Alisa N. Souter
ISBN: 978-1-61761-927-4
© 2011 Nova Science Publishers, Inc.

Chapter 56

RELATIVE SEA LEVEL CHANGES IN THE LAGOON OF VENICE, ITALY. PAST AND PRESENT EVIDENCE

Rossana Serandrei-Barbero[1], Laura Carbognin and Sandra Donnici[1]
Consiglio Nazionale delle Ricerche, Istituto di Scienze Marine,
Castello 1364A, 30122 Venezia, Italy
[1]EuroMediterranean Centre for Climate Change (CMCC), Venice, Italy

ABSTRACT

If no conservation work is carried out, subsiding coastal areas are prone to progressive submersion by the sea and to changes in their physiographical features.

In Italy, the Lagoon of Venice is a peculiar example on this situation, where the Relative Sea Level Rise (RSLR) is the result of the sea level rise and the land subsidence both natural and man-induced. In non-urbanised areas in the Lagoon, the mean rate of RSLR over the last 5000 years was about 1 mm/year, assuming the lowest figures in the last 1000 years. In built-up areas, where compaction of superficial marshy ground due to huge load of buildings must be added, the rate is between 1 and 2 mm/year. Over the last century, the RSLR has totalled about 23 cm, equal to a mean rate of nearly to 2 mm/year.

In the Lagoon of Venice, both natural and anthropogenic evidence of RSLR is revealed in sediments. Natural evidence goes back to the period prior to human settlement, and that due to man's activities dates back to historical times. The works in question were mainly carried out in order to avoid having to abandon settlements due to exceptional flooding.

In the archaeological areas of the Lagoon, human settlements act as proxies clarifying evolution. In some places, man has abandoned previous settlements: for example, the island of S. Lorenzo was urbanised in the first few centuries AD, remained inhabited until the XIV century, and was then abandoned, due to a new increase in sea level. At other times, man succeeded in preventing this process by infill works, or by raising paved levels above mean sea level, as on the island of S. Francesco del Deserto, which has been inhabited since the I century AD. Instead, other non-urbanised areas in the Lagoon, such as salt-marshes or beach cordons, have disappeared with the passing of time, and now lie well below mean sea level. This natural situation is clearly testified by foraminifera - useful proxies of past situations of environmental evolution and its local chronology.

In the urbanised areas, and chiefly in the historical centre of Venice, buildings have acted as proxies of the RSLR in historical times. In the last 100 years, technical improvements have allowed ground elevation and sea level to be regularly monitored. At present, inside the Lagoon, some areas lying at approximately sea level, such as high marshes, are undergoing a significant increase in flooding, mostly attributable to the RSLR, and are becoming perennially submerged areas due to the lack of intervention. In view of the considerable importance of the city of Venice, much conservation and restoration work is being undertaken, in a everlasting confrontation with the RSLR.

1. INTRODUCTION

The Lagoon of Venice is a low coastal area covering about 550 km^2, with an average depth of 0.6 m, criss-crossed by a complex network of channels. Three inlets allow the continual ebb and flow of the northern Adriatic tides in and out of the Lagoon. Tidal exchange is about 111 x 10^9 m^3/year (Gacic et al., 2005) and water dynamics inside the Lagoon are very active, because of human interventions carried out both in the past and in more recent times.

The creation of the Lagoon and its evolutionary phases, both natural and man-induced, was controlled by the Relative Sea Level (RSL), i.e., by interaction between eustacy and land subsidence, closely connected with the rate and nature of sediment accumulation. The Lagoon originated nearly 6000 years ago, when the rising Adriatic Sea flooded the upper Quaternary palaeo-plain (Gatto and Carbognin, 1981; Serandrei-Barbero et al., 2005). This event is easily identifiable in sediments, due to the ubiquitous presence of benthic foraminifera, a class of shelled protists (Sen Gupta, 1999) living in marine environments, generally ranging in size from less than 0.1 to 1 mm. Transitional environments like the Lagoon of Venice contain many species of these unicellular animals, with relatively high numbers of individuals and with assemblages showing the influence of the dominant physico-chemical parameters. Changes in environmental conditions are revealed by parallel significant changes in benthic foraminiferal assemblages, i.e., *ecological transitions* (Debenay and Guillou, 2002).

Due to the permanent flooding caused by the RSLR, *ecological transitions* in cores collected throughout the Lagoon show the existence of ancient intertidal morphologies under subtidal deposits. Evolution is therefore closely controlled by both land subsidence and sea level rise, that may be quantified by the mean sedimentation rate. In the transgressive phase (6840±40 to 5060±40 years BP), the mean sedimentation rate in the eastern Lagoon was 2.7 mm/year (Canali et al., 2007). This rate matches the results of Bortolami et al. (1977), who computed the average sedimentation rate as 3 mm/year more or less throughout the basin, between 7000÷5000/4000 years BP. In the last 5/4000 years, the sedimentation rate has gradually fallen to an average of 1.1 mm/year, as computed by Bortolami et al. (1984) and fitting the rate of 1.3 mm/year measured in the northern Lagoon (Serandrei-Barbero et al., 1997). Generally speaking, the fall in the Holocene sedimentation rate from 2.7 to 1.3 mm/year may be ascribed (Peltier, 2002) to the decreased velocity of the Sea Level Rise (SLR).

For the historical period, from about 1200 AD, many hydraulic works were carried out, the paramount of which was the diversion of the major lagoon tributaries into the sea, that caused drastic reductions in sedimentation rates, inducing, at the end, a decrease in the natural subsidence rate (Gatto and Carbognin, 1981; Serandrei-Barbero et al., 2006).

In non-anthropised lagoon areas, in conditions of sediment supply which were not sufficient to balance the increase in RSL, due to the combined effects of subsidence and eustacy, exceptional flood events developed towards a situation in which increasingly frequent flooding took place, and the whole area was definitively submerged and progressively buried by lagoonal sediments (Alberotanza et al., 1977; Favero and Serandrei-Barbero, 1978).

Instead, in anthropised areas, evolution took place in two directions:

- progressive abandon, in areas where practicality dictated that large-scale defence works were too costly, as at S. Leonardo in Fossa Mala (Lezziero et al., 2005), at Scanello (Serandrei-Barbero et al., 2004), and at S. Lorenzo di Ammiana (Favero et al., 1995);
- defence and raising works, in cases where, for historical or urbanising reasons, restoration of buildings or creation of access were considered necessary, as on S. Francesco del Deserto (Serandrei-Barbero et al., 1997; Bonardi et al., 1999). There, several interventions aiming at preserving inhabited areas above sea level were undertaken at different periods of time, in order to avoid high tide flooding.

In this context, the city of Venice, which has been urbanised for more than 1500 years, is an emblematic example of the everlasting fight against the RSLR.

Today, new types of analyses have enlarged our knowledge of lagoonal palaeo-environments, and new techniques can quantify present-day trends. The evolution of many sites has been recognised on a century-long scale by study of foraminifera in sediments deposited naturally, or by findings of artefacts revealing the safety leeway in anthropised areas. Both – foraminifera and artefacts – play the role of proxies. Throughout the last century, biological or anthropogenic indicators of flooding have been replaced by data on tides and ground levels, available from regular instrumental records.

2. MATERIALS AND METHODS

Sandy mud, muddy sand, mud, clay with calcareous bioclasts, and occasional peat horizons represent sedimentary sequences in the Lagoon, in which buried morphologies, sometimes with remains of ancient human settlements, are revealed by particular assemblages of benthic foraminifera. They are normally used in coastal sedimentary sequences to reconstruct and date palaeo-environmental changes (Reinhardt et al., 1994; Cearreta and Murray, 1996; Horton et al., 1999).

The conventional BP age determinations of mollusc shells cored in the lagoon sediments, were performed at the laboratories of the Australian Nuclear Science and Technology Organization (ANSTO) by Accelerator Mass Spectrometry (AMS), which requires only a few milligrams of carbon (Tuniz et al., 1998).

A new calibration of ages supplied from the buried palaeomorphologies was obtained by means of a regional correction factor. CALIB 5.0.1, a software programme from the

Quaternary Isotope Laboratory (Stuiver and Reimer, 1993, Stuiver et al., 2005), was used to calculate calibrations which uses the 04.14c database for marine samples (Hughen et al., 2004) and the intcal04.14c database for non-marine samples (Reimer et al., 2004). In the case of marine organisms, the different distribution of carbon isotopes in oceanic water with respect to that of the atmosphere is due both to the delay in surface exchange between atmospheric CO_2 and bicarbonate in the water, and the dilution of surface and deep waters. This global ocean reservoir correction was estimated at about 400 years. Regional causes may introduce a difference (Δr) in estimating the reservoir effect with respect to the global ocean model. For the Lagoon of Venice, the most reliable regional correction value of the reservoir effect, used for dating the mollusc shells, was $\Delta r=316\pm35$, obtained at the Lido di Venezia by Siani et al. (2000) and later confirmed by P. Reimer (personal communication).

Sedimentation rates in subtidal and intertidal environments for the last 3000 years were obtained from the calibrated ages, adopting the mean value between the lowest and the highest calibrated ages and assuming the age of the present-day seabed as year -50, that is the difference between 1950 (the reference year) and 2001 (the sampling year).

Regression curves were obtained by plotting the depth of the samples against their age; angular coefficient α of the regression line corresponds to the sedimentation rate. The closer the line to the experimental points, the higher the value of the correlation coefficient R^2; values of R^2 greater than 0.6 are significant.

The microfaunas and fragments of mollusc shells used as proxies in non-urbanised areas were obtained by normal washing techniques. Samples were dried, weighed, washed on a 0.063 mm sieve to disperse silt-clay fractions, dried at 50 °C, and then weighed again to determine the amounts of silt-clay present.

As regards archaeological remains, the height of floors and paving above mean sea level gives information about ancient levels. The ground floors of buildings were normally set at a safety leeway from high tide flooding. The safety height ranged from 20 cm for boat landing-stages or squares, to 130-150 cm for buildings, according to the type of construction (Canal, 1998).

Old Venetians needed to have a reference to define RSLR. As a proxy of mean sea level they adopted the height of the brown-green front left by algae and microorganisms on buildings facing the canals or on canal sides. This front corresponds to the average high-tide level and is called *Comune Marino*. Toward the end of XVIII century, the *Serenissima* Republic engraved a line and a letter "C" on the sides of some canals and buildings façades, in order to better highlight the algae front, i.e. the average level of high tides. Over centuries flooding events and their elevation were referred to this line. The "C" mark is indeed a very useful biological indicator, but it is only "rough marker" of RSLR since it supplies a non-uniform reference, showing considerable spatial variability. Again, it is very difficult to quantify each contribution, i.e. sea level rise and natural /anthropogenic subsidence at a regional /local scale. For this reason, starting nearly the beginning of the XX century, when technical improvement allowed, both the ground elevation and the sea level have been surveyed with proper instrumentation and the "C" reference was abandoned. From 1872 the tide-gauge had been adopted to monitor sea level height, and the mean sea level, m.s.l., (as an easy elaboration from recorded data) became the common reference; as from 1908 spirit levelling and (present-day) remote monitoring (GPS and SRI) have been adopted to survey ground level.

3. FLOODS OF PREHISTORIC AGE

3.1. Foraminifera as Proxies of RSLR

Proper palaeo-ecological interpretation of buried foraminiferal assemblages of Quaternary age depends on knowledge of processes operating in the present, allowing interpretation of past assemblages through comparisons between past and recent faunas.

In the Lagoon of Venice, subject daily to a wide range of variability, 80 taxa of benthic foraminifera are present (Serandrei-Barbero et al., 2008), which have been used as bioindicators, revealing their capacity to highlight the evolutionary trends of transition environments (Albani et al., 2007).

Sampling carried out in 1983, based on quantitative analysis of 559 bottom samples, defined the extent of many different foraminiferal assemblages in the entire Lagoon of Venice (Albani et al., 1991; Donnici et al., 1997; Serandrei-Barbero et al., 1989; 1999) and indicated the parameters controlling distribution of the various biofacies: exchange time of the lagoon water with the sea water, pollution, freshwater inputs, and the presence of intertidal morphologies. The last, in particular, lead to important considerations regarding the RSLR, as subtidal faunas above intertidal assemblages represent the transition from exceptional to permanent flood events.

Although vertical zoning within marshes exists (Scott and Medioli, 1980; Hayward and Hollis, 1994), in the present study, unless otherwise specified, the term generally refers to lagoon morphologies between the mean sea level and the mean high water level. *Trochammina inflata* (Montagu) is found in the lower parts of these morphologies, but *Ammonia beccarii* (Linnaeus) is the dominant taxon; in the higher parts, *Trochammina inflata* prevails (Scott and Medioli, 1980), its distribution and abundance depending on the duration and frequency of intertidal exposure (Horton et al., 1999). In view of the low tidal range of the Lagoon, the finding of a buried salt marsh environment allows us to determine the sea level position at that time within a few centimeters (Petrucci et al., 1983; Albani et al., 1984).

In cores 39 and 50 (Figure 1), the abundance of *Trochammina inflata* indicates the presence of salt marshes at depths of 30 and 68 cm and ages of 787-707 and 1883-1810 cal yrs BP, now buried by subtidal deposits due to the RSLR.

In core 29, littoral indicators at a depth of 93 cm among benthic foraminifera dated at 2715-2539 cal yrs BP (Table 1) reveal a buried coastal barrier, extending from the inlet of Malamocco to the coastal plain south of the Lagoon, between the rivers Brenta and Adige (Favero and Serandrei-Barbero, 1978). Compared with the coastline of the Po Delta, this barrier was estimated to be aged between 5000 and 2800 years BP and is currently buried by lagoon deposits, due to coastal progradation (Tosi, 1994), probably caused by reduced velocity of the SLR, as observed in other lagoons, where sedimentation rates exceed the local SLR (Andersen et al., 2006).

3.2. Flooding Preceding Urbanisation

In cores taken near the continental margin of the Lagoon (cores 1, 17, 27, 43, 50 and 53; Figure 1), the calibrated age of the pre-lagoonal substrate ranges between 905-853 cal yrs BP (peat at the base of core 1) and 2040-2147 cal yrs BP (freshwater gastropods at the base of core 43). In this time-span, the lagoon basin gradually extended to the margins of the mainland, as the RSLR was not sufficiently compensated by the sediment supply.

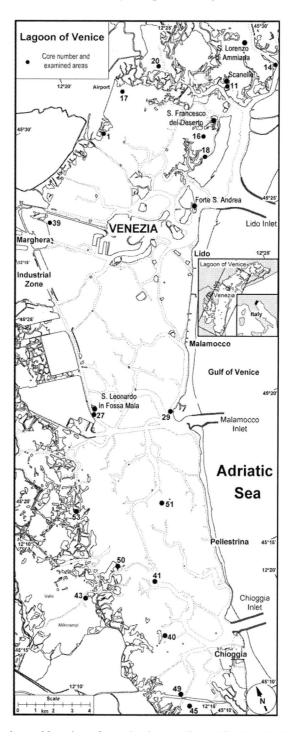

Figure 1. Lagoon of Venice and location of examined areas (from Albani et al. 2007, modified).

The sedimentation rate and the compaction, i.e. the natural subsidence, are closely correlated. The sedimentation rates in the subtidal and intertidal environments of the Lagoon are known from a number of conventional ages of mollusc shell fragments from various cores. Applying these ages, the long-term sedimentation rate for the entire Lagoon, given by subsidence and eustacy, is inferred to be about 1.1 mm/year between 2500 and 1500 years BP and about 0.5 -1 mm/year from 1500 y BP to the present (Serandrei-Barbero et al., 2006). The values referred to the former period, are consistent with the average Holocene sedimentation rate of about 1.3 mm/year calculated by both Bortolami et al. (1984) and Carminati et al. (2005), and the latter match the progressive decrease in sedimentation rate occurred over the last few centuries as reported by Brambati et al. (2003).

These results have been recalculated with calibrated ages (Tab. 1), updating the mean sedimentation rate.

In the northern part of the Lagoon, calibrated ages, all obtained from carbonatic shells of lagoonal molluscs, plot in the age vs. depth graph in two clearly defined clusters. Regression lines give accumulation rates of 0.8 mm/year at 2000-1000 BP and 0.5 mm/year between 1000 BP and the present (Figure 2a).

In the southern Lagoon, calibrated ages were calculated from both lagoon mollusc shells, mainly sampled in the intermediate portions of cores, and peat samples mainly at the core bottom. The sedimentation rates calculated from peat samples yield values of 0.6 mm/year at 3000-1000 cal yrs BP and 0.4 mm/year in the last 1000 years (Figure 2b) The calibrated ages calculated from shell samples with the corrected value of the reservoir effect, $\Delta r = 316\pm35$ (Siani et al., 2000), appear to be younger than those of peat samples. Conventional ages were calibrated, as an experiment, with values lower than Δr, i.e. 10 and -200 (with respect to the value of 400 used as correction factor for oceanic waters), showing better agreement with values obtained from the vegetal samples. In the last 1000 years, the sedimentation rates turn out to be respectively 0.7, 0.5 or 0.4 mm/year, according to the Δr value applied. The lowest correction value ($\Delta r = -200$) indicates a rate matching that obtained from peat samples (Figure 3).

This difference in Δr values in the southern rather than the northern part of the Lagoon may depend on the lower content of fragments of carbonatic rocks in southern river supplies, with respect to the rest of the Lagoon where, according to Siani et al. (2000), the value of the reservoir effect is higher because of the effect of old carbon carried into the Lagoon by tributaries in the past.

The reduced sedimentation rate over approximately the last 1000 years (Figure 2A, B) is closely related to the diverting of rivers into the sea (Favero et al., 1988). Since then, sediments entering the Lagoon come chiefly from the sea. They have been sufficient to form intertidal morphologies at the sides of the channels, where water loses speed and deposits the suspended sediments. There are many examples of now buried intertidal morphologies, both in areas which were later urbanised and in ones where the lack of sufficient interest in conserving them dictated their definitive burial.

The lowest mean sedimentation rates obtained from calibrated ages for the northern and southern lagoon areas, ranging between 0.4 and 0.8 mm/year, appear to be due to the fact that these lagoon areas were not subjected to anthropogenic influence as the central basin.

Table 1. Conventional ages (from Serandrei-Barbero et al., 2006) and calibrated ages (obtained by assessment of local reservoir effect) of samples used to calculate sedimentation rates in northern basin of Lagoon of Venice

ANSTO code	material	core	depth (cm)	d(13) per mil	percent Modern Carbon		Conventional 14C age		1s cal BP ranges		depositional environment
					pM	1s error	years BP	1s error	lower	upper	
OZG410*	Peat	1C	70	-25,0	89,34	0,35	905	32	853	905	alluvial
OZG792	Shell	11C	50	-2,9	78,91	0,39	1900	40	1080	1214	subtidal
OZG314	Shell	11C	96	0,0	72,72	0,35	2558	39	1782	1917	subtidal
OZG315	Shell	14C	132	0,0	74,76	0,31	2337	34	1523	1653	subtidal
OZG794	Shell	16C	45	-2,2	83,15	0,38	1480	40	659	755	subtidal
OZG316	Shell	16C	130	0,0	73,38	0,36	2486	40	1695	1829	subtidal
OZG411*	Peat	17C	120	-25,0	79,25	0,36	1869	37	1806	1867	alluvial
OZG796	Shell	18C	50	-2,8	81,67	0,36	1630	40	793	912	subtidal
OZG318	Shell	18C	122	0,0	77,19	0,35	2080	37	1265	1357	subtidal
OZG797	Shell	20C	48	0,0*	79,31	0,36	1860	40	1036	1171	subtidal
OZG319	Shell	20C	115	0,0	73,87	0,33	2433	36	1627	1773	subtidal
OZG321*	Peat	27C	115	-25,0	81,74	0,40	1620	40	1508	1555	alluvial
OZG322	Foraminifera	29C	93	0,0	67,30	0,40	3180	50	2539	2715	subtidal
OZG334	Peat	39C	53	-25,0	80,79	0,42	1714	42	1564	1630	subtidal
OZG335	Peat	39C	68	-25,0	79,07	0,39	1887	40	1810	1883	intertidal
OZG336	Peat	40C	42	-25,0	89,73	0,40	871	36	732	796	subtidal
OZG302	Shells	41C	55	0,0	82,29	0,43	1566	42	731	861	subtidal
OZG414	Peat	41C	100	-25,0	79,59	0,34	1834	34	1730	1817	subtidal
OZG303	Shells	43C	140	0,0	76,84	0,38	2117	40	2040	2147	alluvial
OZG304	Shells	45C	50	0,0	86,56	0,45	1160	42	430	514	subtidal
OZG305	Shells	45C	93	0,0	81,83	0,46	1611	45	778	902	subtidal
OZG306	Peat	45C	112	-25,0	78,62	0,43	1932	44	1857	1925	alluvial

Table 1. (Continued)

ANSTO code	material	core	depth (cm)	d(13) per mil	percent Modern Carbon		Conventional 14C age		1s cal BP ranges		depositional environment
					pM	1s error	years BP	1s error	lower	upper	
OZG803	Shell	49C	30	-5,2	85,22	0,39	1290	40	525	613	subtidal
OZG415	Peat	50C	30	-25,0	90,04	0,47	842	42	707	787	intertidal
OZG308	Peat	50C	108	-25,0	77,70	0,37	2026	38	1925	2007	alluvial
OZG804	Shell	51C	52	-3,1	84,34	0,41	1370	40	590	661	subtidal
OZG309	Peat	51C	134	-25,0	72,82	0,43	2548	48	2697	2747	subtidal
OZG805	Shell	53C	20	-4,0	90,28	0,42	820	40	61	151	subtidal
OZG310	Peat	53C	124	-25,0	79,11	0,37	1883	38	1810	1879	alluvial

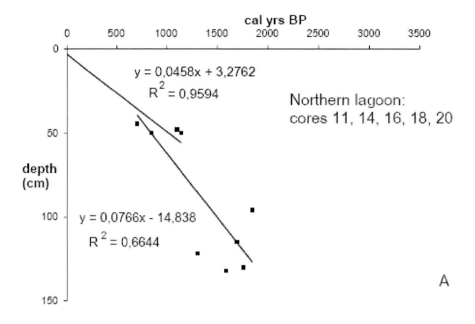

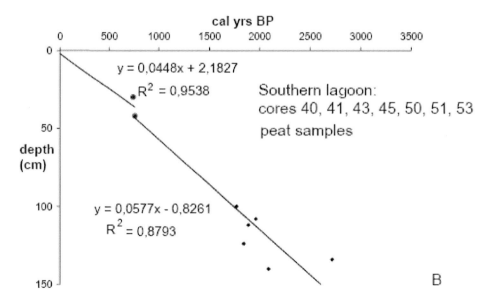

Figure 2, part A - 14C calibrated ages versus depth with respect to lagoon bottom in northern Lagoon. Angular coefficients of regression lines correspond to average sedimentation rates. In time interval 2000-1000 (1849-707) years BP, angular coefficient α indicates average deposition rate of 0.8 mm/year; in interval 1000 (1147) years BP to present, average rate is 0.5 mm/year. In both cases, R2 value is greater than 0.6. Part B - 14C calibrated ages versus depth with respect to lagoon bottom in southern Lagoon. Angular coefficients of regression lines correspond to average sedimentation rates. In time interval 2500-1000 (2722-764) years BP, angular coefficient α indicates average deposition rate of 0.6 mm/year; in interval 1000 (764) years BP to present, average rate is 0.4 mm/year. In both cases, R2 value is greater than 0.6.

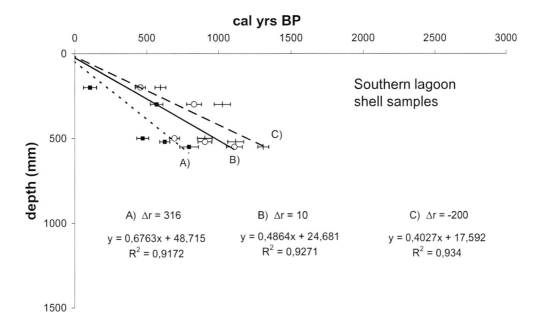

Figure 3. ^{14}C calibrated ages versus depth with respect to lagoon bottom of southern Lagoon in time interval 1000 (764) years BP to present. Angular coefficient of regression lines correspond to average sedimentation rates. Three regression lines are obtained when three different reservoir effect values are used to calibrate BP age of samples.

4. FLOODS OF HISTORIC AGE

4.1. Archaeological Remains and Buildings as Proxies of RSLR

Until the beginning of urbanisation, the impact of human settlement in the Lagoon was simply controlled by the evolution of towns and villages and the relative sea level rise. Problems arose with the true building of the territory and the different uses to which certain spaces were later put.

The Venetian chronicles of all ages report the occurrences of *acqua alta* (a local idiom meaning flooding, the first mention of which goes back to 589 AD).

Nevertheless, the old Venetians underestimated the phenomena of land subsidence and sea level rise (RSLR) which reduced the *franco altimetrico*, the safety margin between sea level and ground level, so that raising works, where possible, became necessary in order to make the city inhabitable.

It has been assessed that boat landing-grounds, where various types of vessels were moored, built and repaired, required a particularly low *franco altimetrico*, never exceeding 20 cm above sea level. Large, open, uninhabited spaces did not require reclamation and could thus be utilised in spite of their low *franchi altimetrici* (Canal, 1998). But in the public squares, beneath which the the Republic of Venice built cisterns to collect rainwater for drinking purposes, the access points for rainwater were set at least 50-60 cm above pavement level, as can still be seen today in many squares in Venice, where the cisterns still exist and were used until the XIX century to supply drinking-water.

In the Campo di S. Maria Formosa, the well-head, which in the XVI century was surrounded by two marble steps, now sunk into the raised pavement of the square, and the nearby façade of the church, built according to plans by Codussi in 1492, has lost about 50 cm of its basement.

Similar raising works were required in several buildings to maintain their floors above the level reached by exceptional high tides. Ground floors were normally built sufficiently height and, according to their type of construction, this safety leeway has been calculated at 130-150 cm (Canal, 1998).

For the Venice area, the XIII century was characterized by a major urban redevelopment, partly accomplished with the infill of most of the stretches of water still existing. The floors of Gothic buildings are now about 1 m under the present-day one (Dorigo, 1983), as later floors have had to be constantly raised over time in order to preserve them from high tides. The loss of about 1 m from Gothic times until today matches an RSLR value of 1.2 mm/year. At the basin scale, subsidence is mainly caused by the natural consolidation of recent lagoon deposits, but at local scale a superficial compaction of marshy ground by building overload worsened this process in past centuries. An exemplary case to demonstrate the RSLR as the sum of all natural and induced components - about 70 ± 12 cm from the XVIII century - is provided by the St. Andrew's Fort, a very large building on the island of the same name near the Lido inlet. The possibility of a visible control exists and can be verified by examining three images belonging to three different historical periods. The first is a "photographic painting" by Canaletto (c. 1730); the second is a daguerreotype of 1866; the third is a contemporary picture. The most notable aspect of each image is the position of the cannon openings and the seaweed line with respect to the water level, demonstrating unequivocally how the RSLR has changed with the passing of time (Carbognin et al., 1984). Recently, Camuffo and Sturaro (2003) analysed 11 Canaletto's and Bellotto's paintings with clearly visible seaweed line in buildings, and established an average sixty-cm submersion of Venice, i.e. RSLR, over the last three centuries.

However, the costly, continual interventions necessary to cope with increasing flooding were only carried out in places where economic, political or social interests required them. This was done all over the city of Venice but outside it, little work was done.

4.2. Flooding in Still Inhabited Areas: The Case of S. Francesco Del Deserto

Archaeological data, combined with information on foraminiferal faunas, clarify the palaeo-environmental evolution of the island of S. Francesco del Deserto in the northern Lagoon of Venice. In 1993, archaeological excavations were carried out north of the Basilica by the *Soprintendenza per i Beni Ambientali e Architettonici di Venezia* (Ammerman et al., 1995).

The excavation site, consisting of a rectangular trench 3 m long, 2.5 m wide, and 2.55 m deep from the present mean sea level, was wholly within a mud sequence of a lagoonal environment from the Sub-Boreal and Sub-Atlantic Chronozones, according to Orombelli and Ravazzi (1996).

The lagoonal palaeomorphologies identified by foraminiferal faunas are very similar to those of the present Lagoon. The sequence below 2.10 m from m.s.l., was deposited in an environment with strong sea-lagoon water exchange and in presence of an active channel north of the island, perhaps overlapping an older river course. A well-developed salt marsh, containing remains of a human settlement, including wooden poles placed for consolidation and dating back to I-V centuries AD, has been identified between 2.10 and 1.90 m above m.s.l. (Serandrei-Barbero et al., 1997). The remains of a wooden boat, found between the poles and the pavement, have been dated to 420 AD (Ammerman et al., 1995).

With the eustatic rise, the salt marsh began to be flooded and, despite the masonry pavement set 17 cm above the poles, the site was probably abandoned around the V-VI centuries.

This excavation was extended in 1995 with trench 8, which sampled a lagoonal sequence located north-east of the preceding work. The new samples amplify the palaeo-environmental reconstruction of the area.

The sequence of trench 8 lies about 3.70 m under the present-day mean sea level, and crosses a series of stratigraphic units (Figure 4). At -3.70 m present m.s.l. is a sterile clay rich in calcareous concretions, deposited in a continental environment.

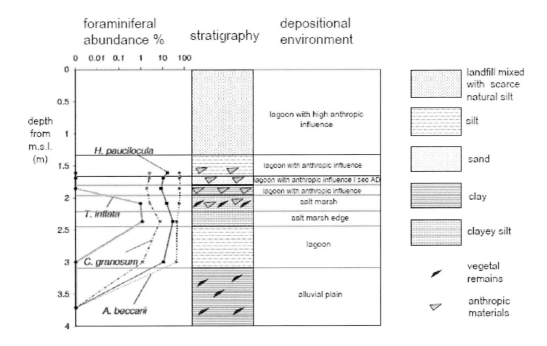

Figure 4. Stratigraphic record analysed in excavation 8 on island of S. Francesco del Deserto. In the samples at -2.37 m and at -2.09 m m.s.l., *Trochammina inflata* (Montagu) is present and *Ammonia beccarii* (Linnaeus) and *Haynesina pauciocula* (Cushman) are the dominant taxa, indicating a level very close to the sea level at that time.

Above and up to – 1.50 m m.s.l., alternating clay and silt are found, from which seven samples were taken and their micropaleontological contents analysed. The foraminifera indicate subtidal or intertidal environments. In the samples at -2.37 m and at -2.09 m, *Trochammina inflata* is present, but *Ammonia beccarii* and *Haynesina pauciocula* are the dominant taxa as in the low marshes worldwide (Scott and Medioli, 1980); the depth of this salt march is thus very close to the sea level of that time and at a level slightly deeper than the anthropised salt marsh identified in the 1993 excavation (Serandrei-Barbero et al., 1997) and compatible with the gradient of intertidal morphologies.

This level contains ceramic fragments dating back to the V century AD, confirming that the location was settled, favoured by the widespread occurrence of intertidal morpologies. Above the level containing the ceramic fragments, lagoonal conditions again prevailed but, at 1.69 m above sea level, the typical markers of lagoonal beds are found together with anthropogenic remains, in a sandy fraction. Being rich in beach indicators, this fraction reveals temporary abandon of the settlement before its gradual resettlement, as proven by the many fragments of ceramics, timber, bricks, and other human-related remains found in the silt above -1.40 m m.s.l.. These archaeological remains were found in a horizontal position, suggesting that the site was again below water level, matching the occurrence of a foraminiferal subtidal assemblage. But the further eustatic rise of the X and XI centuries (Bonardi et al., 1999) was hindered by a final landfill about 1.5 m thick, which allowed the human settlements on the island of S. Francesco to continue from historical times until today, providing an example of human impact on the landscape.

4.3. Flooding of Abandoned Areas: The Case of S. Lorenzo Di Ammiana

On the island of S. Lorenzo di Ammiana, in the northern part of the Lagoon, now abandoned and half-submerged, low tide exposes the remains of the foundations of two Hight Medieval buildings. These remains fit a palaeogeographic situation in which S. Lorenzo di Ammiana occupied a more central position between the the coastal barrier, further back than now, and the mainland, much nearer and already extensively anthropised, as shown by the remains of settlements which range from the prehistoric age to Roman and Medieval times. Archaeological studies reveal continual habitation from the I century AD, abandon of the island between the XI and XII centuries, the existence of a monastery until the XV century, and final abandon during the XVIII century (Favero et al., 1995; Canal, 1998).

Some boreholes drilled in the past reaching the base of the lagoonal sediments, describe the evolution of the island. Sandy mud, muddy sand, mud, and clay with occasional peat horizons represent the sedimentary sequences from boreholes dated from the Atlantic age to the present (Figure 5).

The foraminifera found in borehole drilled on the northern margin of the island, indicate a constantly submerged environment. The upper part of the core, from -3 m to the lagoon bottom, shows that sediment accumulation was influenced by human settlements, with ceramic fragments, charcoal, bricks, and yellowish replenishment sand. These sandy deposits made up of two distinct populations (lagoon silt, and well-classified fine-grained sand, i.e. bimodal), may represent a anthropogenic component in the sedimentary sequence, as the beach sands is scattered in the silt matrix which represents the sediment in situ.

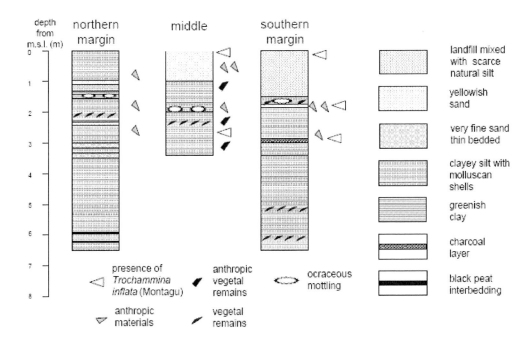

Figure 5. Sediments crossed by three boreholes on island of S. Lorenzo di Ammiana. The evidence of past and present levels close to the sea level is given by the presence of *Trochammina inflata* (Montagu).

Boreholes drilled in the middle of the island and at its southern margin, show the presence of *Trochammina inflata* (Montagu), the foraminiferal taxon characterising the salt marshes, associated with anthropogenic materials, including the yellowish sand noted above.

The *Trochammina inflata* specimens at the top of the two above boreholes, are of modern age, since the ground surface of the island is currently between the mean sea level and the high tide level. In the underlying sediments, the sequences of subtidal environment identify a greater exchange with the sea in the deeper intervals, due to the nearby coastal barrier, and evidence of salt marshes around -2.90 m m.s.l. and -1.80 m m.s.l., where *Trochammina inflata* specimens are present among ceramic fragments and other anthropic remains. These anthropised layers indicate sea levels similar to those of some archaeological areas of Venice (Tuzzato, 1991) between the end of the Roman Age and the VI-VII centuries AD.

The many findings of yellowish sand in the S. Lorenzo boreholes demonstrate reiterated but unsuccessful attempts to protect the island from the RSLR leading to the definitive abandon of the island in the XVII century. The few remains found today are submerged at high tide, and the northern margin of the island is permanently under water.

Environmental evolution very similar to that of S. Francesco del Deserto, has given opposite results at S. Lorenzo di Ammiana, with the total abandon of the area, its permanent flooding, and natural restoration of subtidal conditions above the anthropised levels -an example of the impact of human landforms on the landscape.

5. Floods in the Last Century

Documentation on Tide and Ground Level

Geological subsidence has always occurred in the Venetian area, at rates varying a on the geological events. Natural subsidence during the Holocene period likely reflects the consolidation of recent deposits and has played the major role since the birth of the lagoon. For a number of reasons (e.g. diverting of main lagoon tributaries into the sea) the average rate of 1.3 mm/year fell over recent centuries reaching the current figure 0-0.5 mm/year (Gatto and Carbognin, 1981, Bortolami et al., 1984, Carbognin et al., 1995). These rates of natural consolidation, are consistent with the average sedimentation rates, obtained from conventional ages in the southern and northern lagoon basins (Serandrei-Barbero et al., 1997, 2006). As previously noted, in the urbanized areas, a similar process, i.e. superficial compaction that locally enhanced the natural one, was induced during past centuries by the buildings' overload. Venice is a *town-setting made*; in the VI century, its inhabitants began to built-up the city in the middle of the lagoon, using material from the channels, until the final form of the present-day historical centre of Venice came into being (Figure 6a) at he most recent portions of the city are the ones which are sinking at the highest rate, and vice versa (Figure 6b) (Carbognin et al., 1984; Tosi et. al., 2002).

The process of RSLR includes land subsidence, both natural and anthropogenic, and sea level rise. A precise computation of RSLR can be supplied only for the last century, over which data are available from instrumental records.

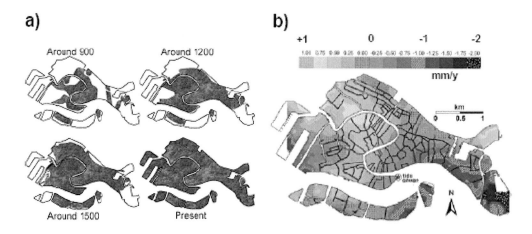

Figure 6. a) Sketch-map of growth of Venice area from 900 AD to present, showing city development (after Carbognin et al., 1984); b) Map of 1992-1996 vertical ground displacement of city (mm/year) obtained by satellite radar interferometry (SRI) at Venice with a pixel resolution of 25 m, overlapping an aerial photograph of city (after Tosi et al., 2002). The most stable city sectors generally corresponds to first urban settlement, and vice versa.

From the beginning of the 1900s, an overall relative land elevation loss of about 23 cm has occurred, consisting of about 12 cm of land subsidence, 3 cm of which natural and 9 cm anthropogenic, and 11 cm of sea level rise (Gatto and Carbognin, 1981). The relative 23 cm rise in the sea level has created a great concern definitely altering the relationship between land and water. Because of the peculiar environment, this RSLR has contributed to the increasing of the flooding phenomenon, both in the frequency (a more than seven-fold) (Figure 7) and degree, with immediate and indirect damage to the inhabitants and monuments in the city, and of hydrodynamics inside the lagoon, leading to erosion of its bed, silting-up of canals, and changes in the internal eco-morphology.

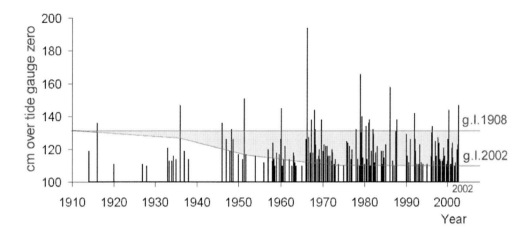

Figure 7. Increase of flooding events during the XX century due to the total RSLR. At present acque alte levels include levels which, in early 1900s, did not flood the city of Venice (updated after Gatto and Carbognin, 1981).

Comments have already been made about natural component of subsidence.

The anthropogenic component refers to groundwater exploitation, mainly performed for industrial purposes. Exploitation of the six artesian aquifers, located in the upper 350 m of the 1000-m-thick unconsolidated Venetian Quaternary formation, which progressively led to a considerable drawdown of aquifer pressure, began in the 1930s, grew with the post-war industrial development, and peaked in the 1950-1970 period (Serandrei-Barbero, 1972), together with the subsidence it caused ($\cong$ 5 mm/year); a peak rate of 17 mm/year was recorded between 1968–69 over the industrial zone. Drastic measures to curtail artesian overexploitation were taken soon after 1969: a general improvement occurred quite quickly because of the closure of artesian wells and the diversification of water supply. The 1973 levelling clearly showed a reversed trend of subsidence and mostly in the historical city where a slight but significant rebound was measured in 1975 (Carbognin et al., 1977). The 1993 and 2000 regional surveys confirmed the arrest of subsidence as a widespread phenomenon due to groundwater pumping (e.g. Carbognin et al., 1995; Carbognin and Tosi, 2003), and the stability of Venice and the adjacent mainland has also been verified in 2000 by land, space and sea observations (Tosi et al., 2002).

At present the city of Venice can be assumed to be stable, but recent investigations by means of an integrated monitoring system (Carbognin et al., 2004; Teatini et al., 2005; 2007), while they show the general stability of the central part of the lagoon area including the city of Venice, do reveal the sinking trend of the northern and southern extremities of the Venetian district, ranging between 5 and 15 mm/year outside the lagoon, and subordinately (2-3 mm/year) in the central and northern littoral strips.

The geodynamic, geological, geomorphological characteristics of the region, ongoing human activities and knowledge of the cause-and-effect relationships clearly support the above displacement rates. In particular another type of land subsidence (*geochemical subsidence*) is also continuing, chiefly in the southern zones, induced by both peat soil oxidation and salinisation of clayey sediments (e.g. Carbognin et al., 2006).

The relative land subsidence of the city is associated with sea level rise, as previously mentioned.

Evidence of sea level variation and geoclimatic fluctuations in the Venice area during the Pleistocene and Holocene, lie in the interpretation of quaternary sediment analyses (Kent et al., 2002), and in the context of periods corresponding to greater sedimentation alternating with periods without deposition or erosion (Bortolami et al., 1984; Tosi, 1994; Serandrei Barbero et al., 2005, Canali et al., 2007)

A complete analysis of eustacy over the XX century in the Upper Adriatic Sea was first performed by Carbognin and Taroni (1996) using accurate statistical analysis of the historical data for both Venice and Trieste, a coastal city located 120 km north-east of Venice, which is known to be stable. The authors provided a very reliable estimate of the sea level rise by a study based on a century-long records (1896-1993), separating out the effects of subsidence in the Venice area from the general sea levels rise. Actually, tide gauge measurements clearly show an up-and-down behaviour in the mean sea level depending upon short-term climatic fluctuations cycles and other phenomena influencing the behaviour of tides (Carbognin and Tosi, 2002). An analysis of coastal sea level records from seven tide gauges in the western Mediterranean and Adriatic (Tsimples and Baker, 2000), indicate a steadiness or a drop in the sea level trends since 1960s up until about 1995 in agreement with the result by Carbognin and Taroni (1996).

In spite of alternating figures, a linear growth rate for eustacy was computed at 1.13 mm/year, a value matching data provided by others tide gauges in the Mediterranean. In particular, the existence of a non-unique secular trend for Venice due to the influence of anthropogenic subsidence was statistically verified. The availability of mean sea-level data for many successive years has enabled results to be updated; the differences between tide gauge values recorded at Venice and Trieste since 1896 up to 2002 are shown in figure 8.

Although with yearly oscillations, gauge measurements in Venice and Trieste showed a steadily decreasing 20-year period between 1971 and 1993 (rate = -0.8 mm/year), followed by a serious increment in sea level (3.3 cm from 1994 to 2000); again, the more recent measurements indicate a new decreasing phase (-8 mm between 2001 and 2005) at both Venice and Trieste (Carbognin et al., 2007).

Basing on the time series analysis, it is clear that, in defining a "sea level trend " it is absolutely necessary to use homogeneous and prolonged time records ($\geq$ 60 years) from stable sites, as also suggested by IPCC, TAR Scientific Basis that believe 50 years the minimum period suitable to obtain a rather reliable trends (IPCC 2001; 2007).

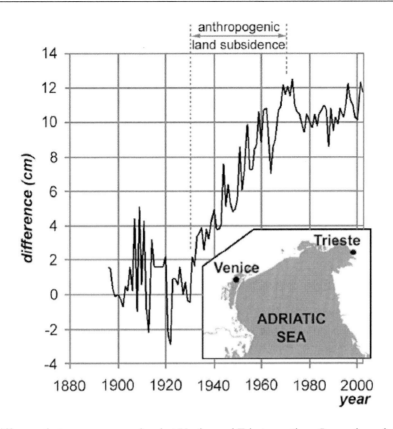

Figure 8. Difference between mean sea level at Venice and Trieste vs. time. Increasing relative difference in period 1931-1970, during which anthropogenic subsidence occurred, indicates an anomalous rise of sea level at Venice with respect to Trieste; no such anomaly is recorded before or after. Insert: location of Trieste with respect to Venice (after Carbognin et al., 2005).

5.2. Recent Submersion of Non-Urbanised Lagoonal Areas

Today, outside the city centre, where raising of ground level is under way in order to combat the relative rise in sea level, the lagoonal areas are being subjected to increasingly frequent flood events. They have led to rapid changes in the zones which lie close to the sea level and which are not of priority interest, like the high marshes.

In 2001, in the central and southern basins of the Lagoon, two samples of subtidal sediment were collected for micropalaeontological analysis. The species of foraminifera here present, were compared with the taxa in samples collected from the same sites in 1983 (Albani et al., 2007). In the central Lagoon, foraminiferal associations indicative of high marsh, are gradually being replaced by subtidal associations; this modification is due to erosion with loss of depth worsened by poor sediment supply and the man-made deepening and straightening of the main shipping channels. In the southern basin, the sample taken from the lagoon bottom in 2001, still shows foraminifera typical of high salt marsh. The association is also characterised by more than 60% of *Trochammia inflata,* the taxon typical of the highest sectors of intertidal morphologies (Petrucci et al., 1983). The Lagoon of Venice, characterised by shallow elevation and limited tidal range, is highly sensitive to variation in depth of even only 10-20 cm, which are sufficient to differentiate intertidal morphologies from subtidal environments (Petrucci et al., 1983; Albani et al., 1984): the features of high salt marsh in one sample, taken below average sea level, reveals rapid lowering of the bed, in any case after 1983.

These different types of behaviour are confirmed by recent surveys of land elevation (Carbognin et al., 2004; Teatini et al., 2007), which indicates conditions of relative stability in the last few years in the central basin, where erosion prevails, and greater lowering in the southern basin.

In the non-urbanised areas of the Lagoon, where no countermeasures against subsidence or reclamation works are being carried out, the trend which has been ongoing for thousands of years continues to change the morphology. In the southern basin, this process seems to be taking place at a dramatic speed.

5.3. Future Scenarios

It is well known that the international scientific community, while in agreement about the ongoing sea level rise, does not agree about future tendencies linked to climatic change. The understandable uncertainties surrounding climatic models lead scientists to propose scenarios, which predict trends rather than univocal estimates forecasting the rise in sea level. Among the authoritative contributions on the subject, we refer to studies carried by the Intergovernmental Panel on Climatic Change (IPCC).

With due consideration for the uncertainties regarding prospective scenarios, and having assessed the history of the region, based on regional historical data, i.e., phenomena that have been measured and analyzed and which form the basis of the results illustrated above, in 2000 it seemed plausible (Carbognin et al., 2000) to propose three scenarios of rising sea levels over the next century at Venice (Figure 9). These scenarios made in 2000, in relation to the frequency increase of *acqua alta*, considered tides higher than 80 cm datum - referred to the tide gauge zero- while actually the value above which the gates would close the lagoon to the sea is 110 cm datum, i.e. exceptional floodings. This implies a remarkable fall of the figures and the percentages reported below, when seen in term of the number of gate closure.

Scenario A (optimistic): this computation, made simply as an exercise, takes into account the fact, noted above, that no significant trend towards rising sea levels has been recorded between 1970 and 1993, and assumes that in future the only factor to cause a "rise" in sea level will be natural subsidence at a rate of 4 cm/century. Under this scenario, in 2100 approximately 70 flooding events over 80 cm datum could be expected each year, against the present 39.

Scenario B (realistic): taking into account the fact that, on average, sea level in the Adriatic has risen by 11 cm over the last century, it can be assumed that this trend will continue throughout the next century in addition to subsidence as described in scenario A (4 cm/century + 11 cm/century=15 cm/century). According to this scenario, in 2100 approximately 180 *acqua alta* events over 80 cm datum could be expected each year.

Scenarios C (pessimistic): taking into account the possibility of a further rise in sea level due to climatic variations resulting from increased greenhouse gas emissions into the atmosphere, but considering also that the estimates given in IPCC - SAR '95 (IPCC, 1995) are somewhat uncertain and may vary according to the model used, two median scenarios were examined. In the first case (scenario C1), a 27 cm rise in sea level in addition to a subsidence rate of 4 cm/century, would give an increase of flooding over 80 cm datum 240% greater than in the past. In the second case (scenario C2), a 49 cm rise in sea level in addition to subsidence rates of 4 cm subsidence, the percentage of flooding over 80 cm datum would get to 430%.

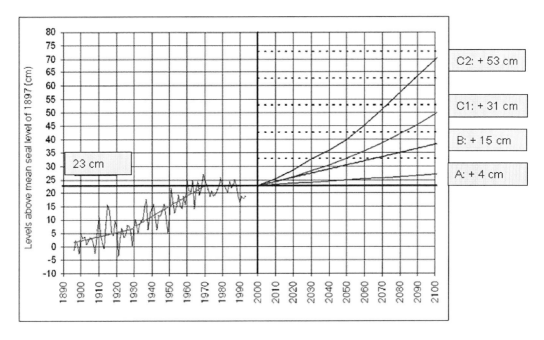

Figure 9. Scenarios of rising sea levels in Venice, compared with levels recorded over the last century in Venice (after Carbognin et al., 2000).

Taking into account the updated IPCC estimates of effects due to climatic changes associated with increased emission of greenhouse gases, APAT (2006) adopting, with due caution, statistic regressions applied to the historical series of mean sea levels for the period 1990-2100, indicates sea level increases between 25.3 cm and 31.3 cm, with a mean value of 27.7 cm.

6. CONCLUSION

Ever since the early years of the Christian era, in the Lagoon of Venice, due to the lack of emerging land, urbanisation involved intertidal flats which, being so close to sea level, did not require much landfill material and thus less labour than large-scale reclamation works.

The existence of ancient intertidal structures at the base of human settlements has been documented in the subsoil both of S. Francesco del Deserto, still inhabited, and the island of S. Lorenzo di Ammiana, now abandoned and semi-submerged.

In the city centre of Venice, buried palaeo-environments are more difficult to analyse, as they are sealed beneath artificial landfills of historical age. But there is much evidence that, over the centuries, the city underwent raising concurrently with the RSLR, and techniques to face exceptional flood events were very efficient, ranging from rain-water cisterns built above the ground level of the squares, to reclamation work on the same squares by raising their level.

Actually, RSLR has caused about 23 cm loss of elevation over the last 100 years, inducing a more than seven-fold increase in the frequency of flooding events in excess of 80 cm, and requiring extensive works to preserve the lagoon environment and the city. In order to prevent severe Adriatic Sea storms from flooding the city, a irreversible largest-scale engineering work of enormous impact under construction (the Modulo Sperimentale Elettromagnetico, called Mo.S.E. project) involves building mobile barriers at the three Venice Lagoon inlets. Barriers are projected to work up to a + 60 cm datum scenarious of RSLR.

Reasonable forecasts of possible RSLR during the next century must be provided in order to assess the efficiency of safeguarding measures. The understandable uncertainties surrounding climatic models have led scientists to propose scenarios and predicting trends that give intervals of variation, rather than univocal estimates, in forecasting the global increase in sea level. The proposed scenarios of relative sea level rise over the next century at Venice range from a lower (optimistic) limit of 4 cm to an upper (pessimistic) limit of 53 cm.

Its repercussions in terms of flooding events associated to each scenario, provide a large spread of possible conditions concerning the survival of Venice.

Besides the Mo.S.E. project, large-scale local defence works against more frequent medium-high tide flooding on a less invasive scale are now being undertaken in the old lagoon towns and the city of Venice. They include permanently "raising" the lower urban areas up to an altimetric level of 100 cm above m.s.l. in the city and 150 cm above m.s.l. in the minor populated centers of the lagoon, compatibly with architectural conditions and habitability, raising waterfronts and foundations and, more generally, the paving in those areas most seriously affected by flooding.

These raising works, that could also be further updated, allow pedestrians to move around the city continuously throughout the year but they are not sufficient to protect urban areas from increasing exceptional flood events (the two highest storm surges recorded over the last half-century, i.e. the ones of November 1966 and December 2008, reached 194 cm and 156 cm datum respectively).

The temporary closure of inlets alone could not be suitable to efficiently protect Venice from flooding due to the probably significant increase of high tides in the years to come. Again, although the Italian Government and the local Administrations have given their final approval, Mo.S.E. has still several opponents who believe it will cause severe threats to the lagoon ecosystem and to the city of Venice.

A new radical complementary solution recently proposed (Comerlati et al., 2004) consists of an anthropogenic raising of Venice. Based on hydrogeological and geomechanical data, preliminary simulations show that injections of seawater into a deep brackish sandy aquifer lying at 600-800 m, can uniformly raise Venice up to 30 cm over 10 years having no measurable environmental impact. This project, besides cutting off all the flooding within the 110-140 cm range, would turn extreme events into moderate or minor floods. Although the available data are not detailed enough for the definition of an operative project, they suffice for a prefeasibility study, and the results suggest that this solution might represent a complementary approach, with none of the environmental consequences charged to Mo.S.E..

Improvement of knowledge on the geological and geomechanical characteristics of the subsoil, are required for a feasibility study, but this project should not be rejected *a priori*.

The cases described in this chapter, in which similar environmental evolutions have had opposite results, show that the survival of Venice and the other islands in the Lagoon can only be guaranteed as long as economic and political pressures allow this urban landscape, of such historical significance, to be preserved. The RSLR must be combated with suitable and on differing time-scales means and an ad hoc management and planning of the Venice defence must be performed in advance.

REFERENCES

Albani, A.D.; Favero, V.; Serandrei-Barbero, R. *Geo-Marine Letters* 1984, 4, 43-47.

Albani, A.D.; Favero, V.; Serandrei-Barbero, R. *Rev. Esp. Micropaleontologia* 1991, 23, 29-45.

Albani, A.D.; Serandrei-Barbero, R.; Donnici, S. *Ecological Indicators* 2007, 7, 239-253.

Alberotanza, L.; Serandrei-Barbero, R.; Favero, V. *Boll. Soc. Geol. It.* 1977, 96, 243-269.

Ammerman, A.J.; De Min, M.; Housley, R.; McClennen, C.E. *Antiquity* 1995, 69, 501–510.

Andersen, T.J.; Pejrup, M.; Nielsen, A.A. *Marine Geology* 2006, 226, 115-125.

APAT - Agenzia per la Protezione dell'Ambiente e per il Servizi Tecnici, *Report* 2006, 69, 1-37.

8-448-0184-1, Roma.

Bonardi, M.; Canal, E.; Cavazzoni, S.; Serandrei-Barbero, R.; Tosi, L.; Enzi, S. *World Resource Review* 1999, 11 (2), 247-257.

Bortolami, G.; Fontes, J.Ch.; Markgraf, V.; Saliege, J.F. *Palaeogeography, Palaeoclimatology, Palaeoecology* 1977, 21, 139-156.

Bortolami, G.; Carbognin, L.; Gatto, P., *Land Subsidence*; IAHS 151; Johnson, A.I et Al. Eds: Wallingford, UK, 1984; pp 777-784.

Brambati, A.; Carbognin, L.; Quaia, T.; Teatini, P.; Tosi, L., *Episodes* 2003, 26, 264-268.

Camuffo D, Sturaro G. ; *Climatic Change* 2003, 58: 333–343.

Canal, E.. *Testimonianze archeologiche nella Laguna di Venezia. L'età antica*. Ed. del Vento, Cavallino di Venezia, Venezia, IT, 1998; pp 1-91.

Canali, G.; Capraro, L.; Donnici, S.; Rizzetto, F.; Serandrei-Barbero, R.; Tosi, L. *Palaeogeography, Palaeoclimatology, Palaeoecology* 2007, 253, 300–316.

Carbognin, L.; Taroni, G. *Atti IVSLA, Classe Scienze Fis., Mat. Nat.* 1996, 154, 281-298, Venezia, IT

Carbognin, L.; Tosi, L. *Marine Ecology* 2002, 23 Suppl.1, 38-50.

Carbognin, L.; Tosi, L., *Progetto ISES* ; Grafiche Erredici: Padova, IT, 2003; pp 1-95.

Carbognin, L.; Gatto, P.; Mozzi, G.; Gambolati, G.; Ricceri, G. *Land Subsidence*; IAHS 121; Rodda, J. C. Ed.: Washington, US, 1977; pp 65-81.

Carbognin, L.; Gatto, P.; Marabini, F. *The city and the Lagoon of Venice. A guidebook on the environment and land subsidence.*Booklet printed on the occasion of the Third International Symposium on Land Subsidence held in Venice, Italy; Publ. CNR: Venezia, IT, 1984; pp1-36.

Carbognin, L.; Tosi, L.; Teatini, P. In *Land Subsidence*; Eds, Barends F.B.J. et al.; Balkema: Rotterdam, ND, 1995; pp129-137.

Carbognin, L.; Cecconi, G.; Ardone, V. In *Land Subsidence*; Eds, Carbognin L.; Gambolati, G.; Johnson, A.I.; La Garangola (printer), Padova, IT, 2000; Vol. I, pp 309–324.

Carbognin, L.; Teatini, P.; Tosi, L. *Journal Marine Systems* 2004, 51, 345-353 doi:10.1016/j.jmarsys.2004.05.021, *51* (1-4), 345-353.

Carbognin, L.; Teatini, P.; Tosi, L. *Italian Journal Applied Geology* 2005, 1, 5–11. doi: 10.1474/GGA.2005-01.0-01.0001.

Carbognin, L.; Gambolati, G.; Putti, M.; Rizzetto, F.; Teatini, P.; Tosi, L. In *Management of Natural Resources, Sustainable Development and Ecological Hazards*; Brebbia, C.A. et Al. Eds.; WIT Press: Southampton, UK, 2006; pp 691–700.

Carbognin L.; Strozzi, T.; Teatini, P.; Tomasin, A.; Tosi, L. III Intern.TOPO-EUROPE Workshop, 2 th - 5 th May, Accademia Naz. Lincei, Roma, IT,. European Sciences Foundation Eds, Proceedings in press, Strasbourg Cedex, Fr, 2007, http://www.esf.org/topo-europe.

Carminati, E.; Doglioni, C.; Scrocca, D. In *Floodings and Environmental Challenges for Venice and Its Lagoon: State of Knowlede;* Fletcher, C.A.; Spencer, T. Eds.; Cambridge University Press: Cambridge, UK, 2005; pp 21-28.

Cearreta, A.; Murray, J.W. *J. Foraminiferal Res.* 1996, 26, 289–299.

Comerlati A.; Ferronato, M.; Gambolati, G.; Putti, M.; Teatini, P. *Journal Geophysical Research* 2004, 109, F03006, doi:10.1029/2004JF000119

Debenay, J.-P.; Guillou, J.-J. *Estuaries* 2002, 25, 1107-1120.

Di Maio, C.; Santoli, L.; Schiavone, P. *Mechanics of Materials* 2004, 36, 435–451.

Donnici, S.; Serandrei-Barbero, R.; Taroni, G. *Micropaleontology* 1997, 43, 440-454.

Dorigo, W. *Venezia Origini. Fondamenti, Ipotesi, Metodi*. Electa: Milano, IT, 1983; Vol. 1-2, pp 1-775.

Favero, V.; Serandrei-Barbero, R. *Mem. Soc. Geol. It.* 1978, 19, 337-343.

Favero, V.; Parolini, R.; Scattolin, M. *Morfologia storica della Laguna di Venezia*; Collana Ambiente 3; Venice Municipality Ed.:Venice IT, 1988; pp 1-79.

Favero, V.; Heyvaert, F.; Serandrei-Barbero, R. *IVSLA Rapporti e Studi* 1995, 12, 183–218, Venezia.

Gacic, M.; Kovacevic, V.; Mancero Mosquera, I.; Mazzoldi, A.; Cosoli, S. *Flooding and Environmental Problems of Venice and Venice Lagoon: State of Knowledge*; Cambridge University Press: London, UK, 2005; pp 431–444.

Gatto, P.; Carbognin, L. *Hydrological Sciences Bulletin* 1981, 26, 379-391.

Hayward, B.W.; Hollis, C.J. *Micropaleontology* 1994, 40, 185-222.

Horton, B. P.; Edwards, R. J.; Lloyd, J. M. *Marine Micropaleonotology* 1999, 36, 205-223.

Hughen, Ka.; Baillie, Mgl.; Bard, E.; Bayliss, A.; Beck, Jw.; Bertrand, Cjh.; Blackwell, Pg.; Buck, Ce.; Burr, Gs.; Cutler, Kb.; Damon, Pe.; Edwards, Rl.; Fairbanks, Rg.; Friedrich, M.; Guilderson, Tp.; Kromer, B.; McCormac, Fg.; Manning, Sw.; Bronk Ramsey, C.; Reimer, Pj., Reimer, Rw.; Remmele, S.; Southon, Jr.; Stuiver, M.; Talamo, S.; Taylor, Fw.; Van Der Plicht, J.; Weyhenmeyer, Ce. *Radiocarbon* 2004 , 26, 1059-1086.

IPCC (1995). http://www.ipcc.ch, Cambridge University Press.

IPCC (2001). Synthesis Report. http://www.ipcc.ch, Cambridge University Press, 396 pgs.

IPCC (2007). Synthesis Report. http://www.ipcc.ch, 73 pgs.

Kent, V.D.; Rio, D.; Massari, F.; Kukla, G.; Lanci, L. *Quaternary Science Review* 2002, 21, 1719-1727.

Lezziero, A.; Donnici, S.; Serandrei-Barbero, R. *Geografia Fisica Dinamica Quaternaria* 2005, Suppl. 7, 201–210.

Orombelli, G.; Ravazzi, C. *Quaternario - Italian J. Quat. Sciences* 1996, 9(2), 439-444.

Peltier, W.R. *Quaternary Science Reviews* 2002, 21, 377-396.

Petrucci, F.; Medioli, F.S.; Scott, D.B.; Pianetti, F.A.; Cavazzini, R. *Acta Naturalia Ateneo Parmense* 1983, 19, 63-77.

Reimer, P. J.; Baillie, M. G. L.; Bard, E.; Bayliss, A.; Beck, J. W.; Bertrand, C. J. H.; Blackwell, P. G.; Buck, C. E.; Burr, G. S.; Cutler, K. B.; Damon, P. E.; Edwards, R. L.; Fairbanks, R. G.; Friedrich, M.; Guilderson, T. P.; Hogg, A. G.; Hughen, K. A.; Kromer, B.; McCormac, F. G.; Manning, S. W.; Ramsey, C. B.; Reimer, R. W.; Remmele, S.; Southon, J. R.,;Stuiver, M.;Talamo, S.;Taylor, F. W.;van der Plicht, J.; Weyhenmeyer, C. E. *Radiocarbon* 2004, *46*, 1029-1058.

Reinhardt, E.G.; Petterson, R.T.; Schroeder-Adams, C.J. *Journal Foraminiferal Res* 1994, 24, 37–48.

Scott, D.B.; Medioli, F.S. *Cushman Foundation Foraminiferal Research* 1980, Special Pub. 17, 1-58.

Sen Gupta, B.K. In *Modern Foraminifera*; Sen Gupta, B.K. (Ed.); Kluwer Academic Publishers: Dordrecht, ND, 1999; pp. 7–36.

Serandrei-Barbero, R. *Technical Report* 31, Publ. National Research Council:Venezia, IT, 1972; pp1-97.

Serandrei-Barbero, R.; Albani, A.D.; Favero, V. *Boll. Soc. Geol. It.*1989, 108, 279-288.

Serandrei-Barbero, R.; Albani, A.D.; Zecchetto, S. *Palaeogeography, Palaeoclimatology, Palaeoecology* 1997, 136, 41-52.

Serandrei-Barbero, R.; Carbognin, L.; Taroni, G.; Cova, E. *Micropaleontology* 1999, 45, 1-13.

Serandrei-Barbero, R.; Albani, A.; Bonardi, M. *Palaeogeography, Palaeoclimatology, Palaeoecology* 2004, 202, 229-244.

Serandrei-Barbero, R.; Bertoldi, R.; Canali, G.; Donnici, S.; Lezziero A. *Quaternary International* 2005, 140/ 141, 37–52.

Serandrei-Barbero, R.; Albani, A.; Donnici, S.; Rizzetto, F. *Estuarine, Coastal Shelf Science* 2006, 69, 255– 269.

Serandrei-Barbero, R.; Albani, A.; Donnici, S. *Atlante dei Foraminiferi della Laguna di Venezia.*Istituto Veneto di Scienze, Lettere, Arti Printer: Venice, IT, in press.

Siani, G.; Paterne, M.; Arnold, M.; Bard, E.; Métivier, B.; Tisnerat, N.; Bassinot, F. *Radiocarbon* 2000, 42, 271-280.

Stuiver, M.; Reimer, P.G. *Radiocarbon* 1993, 35, 215–230.

Stuiver, M.; Reimer, P. J.; Reimer, R. W. 2005. CALIB 5.0. [www program and documentation].

Teatini, P.; Tosi, L.; Strozzi, T.; Carbognin, L.; Wegmüller, U.; Rizzetto, F. *Remote Sensing Environment* 2005, 98, 403-413.

Teatini, P.; Strozzi, T.; Tosi, L.; Wegmuller, U.; Carbognin, L. *Journal. Geophysical Research* 2007, 112, F01012, doi: 10.1029/2006JF000656,2007.

Tosi, L. *Quaternario - Italian J. Quat. Sciences* 1994, 7 (2), 589-596.

Tosi, L.; Carbognin, L.; Teatini, P.; Strozzi,T.; Wegmüller, U. *Geophysical Research Letters* 2002,. 29 (12), 10.1029/2001 GL 013211.

Tsimplis, M. N.; Baker, T. F. *Geophysical Research Letters* 2000, 27 (12), 1731-1734.

Tuniz, C.; Bird J.R.; Fink, D.; Herzog, G.F. *Accelerator Mass Spectrometry: Ultrasensitive Analysis for Global Science.* CRC Press: Boca Raton, US, 1998; pp 1- 371.

Tuzzato, S., *Quaderni di Archeologia del Veneto* 1991, 7, 92–103.

In: Encyclopedia of Environmental Research
Editor: Alisa N. Souter

ISBN: 978-1-61761-927-4
© 2011 Nova Science Publishers, Inc.

Chapter 57

SEDIMENT DYNAMICS AND COASTAL MORPHOLOGY EVOLUTION

François Marin[*]
Laboratoire Ondes et Milieux Complexes, FRE CNRS 3102
UFR des Sciences et Techniques, 76058 Le Havre Cedex, France

ABSTRACT

The action of waves and currents often generates bedforms in the coastal zone. These bedforms significantly affect the sediment transport. Bars on sandy beaches contribute to the natural protection of the shoreline. Feedback processes occur in the coastal area which forms a complex system. The coastal erosion is an important problem. Depending on the considered site, it may be more appropriate to let the erosion occur, to use "soft" methods such as beach nourishment to prevent the shoreline from erosion, or to protect the coastal zone with hard structures. The greatest care must be taken when hard structures, which may lead to significant erosion at their foot, are considered. A multidiscipline approach is essential to improve the forecast of the shoreline evolution, in particular within a context of climate change.

INTRODUCTION

Coastal erosion is a major problem. Great efforts are nowadays made to develop a coherent approach to this problem within the framework of the integrated coastal zone management. This chapter focuses on sandy coasts and underlines the feedback processes which take place in the coastal area. The estimation of sediment transport depends on the bed shear stress which is connected to the bed morphology. This sediment transport affects the beach profiles when these profiles actively take part in the natural protection of the shoreline.

---

[*] Address: 25 rue Philippe Lebon, B.P. 540, 76058 Le Havre Cedex, France; E-Mail: francois.marin@univ-lehavre.fr.

THE COASTAL AREA: A COMPLEX SYSTEM

Waves and currents in the coastal zone may induce sediment movements which often lead to the formation of bedforms (Sleath, 1984). Small scale bedforms called ripples with a typical wavelength of about 10 cm have a significant effect on the sediment transport, the dissipation of wave energy by enhancement of the bottom friction, and the dispersion of pollutants. Larger scale bedforms such as the offshore sandwaves have wavelengths of several hundreds of meters, and their heights can rise up to 10 m. Large parts of shallow seas are covered with such bed features (Németh et al., 2003). In the nearshore zone, bars consisting of ridges of sediment running roughly parallel to the shore are common features on sandy beaches. These structures provide a possible mechanism for the natural protection of beaches from the energy of incident waves, inducing waves reflection. Numerous studies have been carried out about small and big scale bedforms; however, many questions have still no answer. For example, different wave ripple types (orbital, suborbital, anorbital) have been suggested (Wiberg and Harris, 1994) depending on the values of the hydrodynamic and sedimentary parameters, but the physical processes governing the formation dynamics and the precise morphology of these structures are still unclear. The effect of the size heterogeneity on the bedforms morphology needs further research (Ezersky and Marin, 2008). The number and the position of bars on a sandy beach cannot be accurately predicted. The influence of the tide on the beach profiles is poorly understood. The beaches morphology and the sediment transport in the offshore zone are connected. The coastal erosion is an important problem; it may result from long-term processes or from episodic events (storms).

The sediments on the beaches come from different sources; these ones may be the zone of action of waves and currents on the bed, rivers, aeolian supplies, biogenic production, cliff recession. When a sandy coast is subjected to erosion, it is possible to let this erosion occur, to use "soft" methods as beach nourishment to prevent shorelines from erosion, or to protect the coastal zone with hard structures. The best choice among these three possibilities depends on the considered site. One method may be very well suitable for one site, but may lead to disastrous results for a neighbouring site. As far as the hard structures are concerned, several types are currently used: groins, offshore breakwaters, artificial headlands, revetments and seawalls (Dean and Dalrymple, 2001). The groins are well adapted for areas with significant alongshore transport. The sediments accumulate on the updrift side; however, erosion takes place on the downdrift side. A series of groins, instead of one isolated structure, is then in practice generally used. The principle of offshore breakwaters is to reduce the wave energy that reaches the coastline; multiple offshore breakwaters are often found in the field. Artificial headlands are breakwaters built along the shoreline. Revetments are shore parallel structures placed along the beach face or at the foot of bluffs. They induce wave breaking and loss of energy during the run-up process. Finally, seawalls are generally vertical walls fronting a bluff. These types of hard structures, which have to be the object of impact studies, may be very useful to stabilize a coastline. However, the greatest care must be taken when such structures are planned to be built. Recent works have underlined that seawalls or revetments may lead to significant erosion at the foot of the structures. Furthermore, the denaturation of the coastline has to be taken into account. Hard structures installed at the upper part of a beach interrupt the sediment exchanges between the beach and the bordering dune. Many problems arise because of the construction of tourism installations too close from the sea when the coastline naturally moves, in particular within a context of increasing influence of the seaside tourism on the economy and the society.

The local monitoring is important to increase the reliability of the prediction of the coastline evolution. The field studies, which are absolutely necessary and particularly adapted for the long-term evolution, are very expensive. It is then generally suitable to combine field and laboratory studies. The laboratory works are less expensive and allow to consider the effects of each physical parameter on the involved processes in the coastal zone. These studies may be based on physical models or on equations (including analytical and numerical models). They complete field works; scaling constraints and hypothesises have to be treated with great care for the physical and equations based models, respectively.

The forecasts associated with the climate change, such as an increase of the global sea level and of the frequency of storms, suggest an increase of the erosion problems in the future. Innovative coastal protection techniques are being tested, but much work has to be done in this area. The coastal authorities have to make predictions concerning the behaviour of the coastline over a timescale of order 50 years in order to satisfy the integrated coastal zone management planning requirements. At this aim, a multidiscipline approach (including social and human science, geography, geology, mechanics, physics, mathematics) is nowadays essential. The coastal area is a complex system with numerous interactions between its entities and feedback processes.

CONCLUSION

The coastal area is a complex system. A multidiscipline approach appears necessary to improve our knowledge of this system, and to better forecast the shoreline evolution, in particular within a context of climate change. The combination of field and laboratory studies is important to consider the involved processes.

REFERENCES

Dean, R.G.; Dalrymple, R.A. *Coastal processes with engineering applications;* Cambridge university press, 2001.

Ezersky, A.B.; Marin F. Segregation of sedimenting grains of different densities in an oscillating velocity field of strongly nonlinear surface waves. *Physical Rev. E,* 78 (2), doi: 10.1103/PhysRevE.78.022301.

Németh, A.A.; Hulscher, S.J.M.H.; De Vriend H.J. Offshore sand wave dynamics: Engineering practice and future solutions. *Pipeline Gas J.* 2003, 230 (4), 67-69.

Sleath, J.F.A. *Sea Bed Mechanics;* John Wiley, Hoboken, N.J., 1984.

Wiberg, P.L.; Harris, C.K. Ripple geometry in wave dominated environments. *J. Geophys. Res.* 1994, 99 (C1), 775-789.

In: Encyclopedia of Environmental Research
Editor: Alisa N. Souter

ISBN: 978-1-61761-927-4
© 2011 Nova Science Publishers, Inc.

Chapter 58

BIODIVERSITY OF SOURDOUGH LACTIC ACID BACTERIA

Luca Settanni and Giancarlo Moschetti*
University of Palermo, Palermo, Italy

ABSTRACT

Modern technology of baked goods largely uses sourdoughs because of the many advantages offered over baker's yeast. Sourdough is characterized by a complex microbial ecosystem, mainly represented by lactic acid bacteria (LAB) and yeasts, whose fermentation confers to the resulting product its characteristic features such as palatability and high sensory quality. Raw materials used in baking are not heat-treated; thus, they bring their wild microorganisms to the production process. Investigation of the composition and evolution of microbial communities of sourdough and raw materials is relevant in order to determine the potential activities of sourdough microorganisms. LAB are the main factor responsible for flavor development, improvement of nutritional quality as well as stability over consecutive refreshments of sourdough. LAB also establish some durable microbial associations, and the cell-cell communication process is crucial in determining sourdough performance during fermentation. The central aim of this chapter is to report on the knowledge of LAB hosted in raw materials and sourdoughs, their biodiversity and evolution before and during fermentation, their useful properties, their quorum sensing mechanisms and the methods routinely applied to their detection and monitoring in sourdough ecosystems.

INTRODUCTION

The discovery of fossil kernels revealed that the utilization of cereals by mankind commenced in the Neolithic era [223]. Cereal-based foods have been staples for humans for millennia. Nowadays, cereal grains constitute a major source of dietary nutrients throughout

---

* SENFIMIZO Department, Section of Phytopathology and Agricultural Microbiology, University of Palermo, Viale delle Scienze 4, 90128 Palermo, Italy. Tel: +39 091 7028843; fax: +39 091 7028855; e-mail: luca.settanni@unipa.it

the world [11]. They contain the macronutrients (proteins, lipids and carbohydrates) required by humans for growth and maintenance. Cereal grains also supply important minerals, vitamins and other micronutrients essential for optimal health, although they are deficient in some basic components (e.g., essential aminoacids such as lysine). It is becoming apparent that cereals in general have the potential for health enhancement beyond the simple provision of micronutrients and that their consumption can lower the risk of significant diet-related diseases quite substantially [225].

Dry cereals can only be eaten after grinding and mixing with water [187]. Such a mixture results in the formation of a product characterized by sour aroma when left on its own for a while [94], a phenomenon known as fermentation. The biological agents involved in the fermentation process include bacteria, yeasts and filamentous fungi, which determine the saccharification of starch in the raw materials and affect microbial protein supply [104]. In general, the preservation and microbial safety of fermented foods is ensured by lactic acid bacteria (LAB) [20], which also determine the microbial stability of the final product of fermentation [147]. In developing countries, particularly in areas where refrigeration, canning and freezing facilities are either inaccessible or unavailable, fermentation of starting substrates is still widely utilized as a means of food preservation.

The fermentation of cereals determines the nutritional value of grains [104] and represents the most simple and economical way to reduce the anti-nutritional factors (e.g., phytic acid), to convert toxic compounds (polypeptides responsible for the celiac sprue) and to enhance sensory properties and functional qualities of cereals [11,30]. Fermented cereals may have been the first example of fermented food employed by the mankind [94]. The first evidence of employment of fermented cereals in food preparation are dated at around 1500 BC as revealed by Egyptian mural paintings representing the baking of leavened dough [238], even though sourdough bread was already part of the European diet 5000 years ago [241]. Nowadays, cereal-based fermented foods and beverages are produced all over the world. Some products, such as sourdough and malt beer, are common to many societies and are well characterized, whereas the majority of fermented cereal products are regional and little is known about their chemical composition and microbial diversity.

Sourdough was used as a leavening agent in bread production until it was replaced by baker's yeast in the nineteenth century; from then on its use was reduced to artisan and rye bread. However, sourdough is still employed in the manufacture of a variety of products such as breads, cakes and crackers, with its application still on the increase [30] while being applied to a large variety of cereal flours throughout the world. Wheat sourdough is used in more than 30% of Italian bakery products [167], including more than 200 different types of sourdough breads [108]. In Italy, bread is traditionally produced from *Triticum aestivum* flour, but, in some regions of the South, sourdough breads are mostly manufactured with wheat flour species *Triticum durum* alone or in combination [29]. Rye as cereal for bread making is widely used in Germany, Poland, Russia and in Scandinavian countries [16].

The study of sourdough from a microbiological point of view barely started a hundred years ago [187]. Sourdough is reported to be a dough whose typical characteristics are mainly due to its microflora, basically represented by LAB and yeasts. Thanks to its microbial community, such a dough is metabolically active and can be reactivated. Sourdough microorganisms ensure acid production and leavening upon addition of flour and water. Since flour cannot be subjected to heat sterilization, the occurrence and number of certain types of microorganisms will strictly depend on a combination of available substrates and specific technological parameters [187]. Sourdoughs, on the basis of the technology applied, have been grouped into three types [13]: type I, type II and type III. Type I sourdoughs are produced with traditional techniques and are characterized by continuous, daily refreshments to keep the microorganisms in an active state. Type II sourdoughs, often used as dough-souring supplements during bread preparation, are semi-fluid silo preparations characterized by long fermentation periods (from 2 up to 5 days) and fermentation temperature sometimes > 30°C to speed up the process [13,94]. Type III sourdoughs are dried preparations containing LAB resistant to the drying process [94]. Unlike type I sourdoughs, doughs of types II and III require the addition of baker's yeast (*Saccharomyces cerevisiae*) as a leavening agent.

The modern biotechnology of baked goods largely uses sourdough as a natural leavening agent because of the many advantages it offers over baker's yeast, e.g., in the development of the characteristic flavour of bread, resulting in a final product with high sensory quality [96]. Many inherent properties of sourdough rely on the metabolic activities of its resident LAB: lactic fermentation, proteolysis, synthesis of volatile compounds, anti-mould and antiropiness are among the most important activities during sourdough fermentation [81,96]. Moreover, endogenous factors in cereal products (carbohydrates, nitrogen sources, minerals, lipids, free fatty acids and enzyme activities) and process parameters (temperature, dough yield, oxygen, fermentation time and number of sourdough propagation steps) markedly influence the microflora of sourdough and the features of leavened baked goods [94].

MICROORGANISMS OF BAKERY ENVIRONMENTS

Microorganisms responsible for sourdough fermentation are exclusively yeasts and LAB whose ratio is generally 1:100 [86,167].

Microbiota of Doughs

Yeasts found in sourdoughs belong to more than 20 species [90,185,211]. The frequently dominant *Saccharomyces cerevisiae* is often introduced through the addition of baker's yeast [29]. Typical yeasts associated with LAB in sourdoughs are *Saccharomyces exiguus*, *Candida humilis* (formerly described as *Candida milleri*) and *Issatchenkia orientalis* (*Candida krusei*) [30,75,106]. Other yeast species detected in sourdough ecosystems are *Pichia anomala* (as *Hansenula anomala*), *Saturnispora saitoi* (as *Pichia saitoi*), *Torulaspora delbrueckii*, *Debaryomyces hansenii*, *Pichia membranifaciens* [30] and *Candida famata* [156]. The extensive variability in the number and type of species found depends on several factors, including degree of dough hydration (Dough Yield [D.Y.] = weight of dough/weight of flour × 100), type of cereal used, leavening temperature, and sourdough maintenance temperature [86]. Yeasts are primarily responsible for the leavening of dough, while LAB determine the process of acidification, even though heterofermentative LAB partly contribute to the mass blowing [87,208].

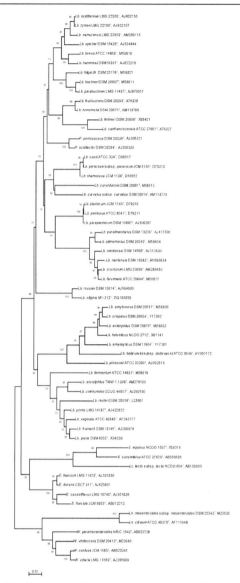

Figure 1. Phylogenetic tree of LAB commonly associated to or found in sourdough products based on 16S rRNA gene sequences. Sequence alignment was performed with CLUSTALX [220]. Sequence and alignment manipulations and calculation of similarity values and nucleotide compositions of sequences were performed with GeneDoc program version 2.5.000 (K. B. Nicholas and H. B. Nicholas, unpublished data). Positions available for analysis were ca. 1150 bp. Phylogenetic and molecular evolutionary analysis were conducted using MEGA version 3.1 [125]. Bar, 0.01 nucleotide substitution per site.

In contrast to the use of mostly homofermentative species of LAB in the majority of (fermented) food applications, heterofermentative species play a major role in sourdough fermentation [187], especially when sourdoughs are prepared in a traditional manner [26,29]. So far, a few less than 60 different LAB species have been reported to be isolated from sourdough [197], some of which, e.g., *Lactobacillus reuteri* and *Lactobacillus acidophilus*, may be of intestinal origin and, due to cross-contamination, are found in sourdoughs (Gänzle, oral communication). Although species belonging to *Leuconostoc, Weissella, Pediococcus*,

Lactococcus, Enterococcus and *Streptococcus* genera have been isolated from sourdough, *Lactobacillus* strains are the most frequently observed bacteria in this ecosystems and, due to their massive presence, lactobacilli represent the typical LAB associated to this complex food environment (Fig. 1).

Sourdough LAB, consisting of obligately and facultatively heterofermentative, and obligately homofermentative species are associated with type I, type II, type III sourdough and with type 0 dough. Type 0 dough, for which baker's yeast is the main fermenting agent, is not made with sourdough technology. However, yeast preparations often contain LAB, especially lactobacilli rather than *Pediococcus*, *Lactococcus* and *Leuconostoc* spp. [109], which in this case contributes only to a small degree to the acidification and aroma development of dough because of the short processing time.

Starting from glucose, homofermentative LAB mainly produce lactic acid through glycolysis (homolactic fermentation) (Fig. 2A) while heterofermentative LAB produce, besides lactic acid, CO_2, ethanol and/or acetic acid, depending on the presence of additional substrates acting as electron acceptors (mainly fructose, oxygen, malate and citrate) through 6-phosphogluconate/phosphoketolase (6-PG/PK) pathway (heterolactic fermentation) (Fig. 2B). Hexoses other than glucose enter these major pathways at the level of glucose-6-phosphate or fructose-6-phosphate after isomerization and/or phosphorylation [3]. Disaccharides are split by specific hydrolyses and/or phosphohydrolases to monosaccharides which then enter the major pathways. Pentoses are phosphorylated and converted to ribulose-5-phosphate or xylulose-5-phosphate by epimerases or isomerases and subsequently metabolized through the lower half of the 6-PG/PK pathway [3]. Utilization of pentoses is not restricted to LAB species which possess a constitutive phosphoketolase, the key enzyme of the 6-PG/PK pathway (obligately heterofermentative); facultative heterofermentative LAB, which ferment hexose through glycolysis because they possess a constitutive fructose-1,6-diphosphate aldolase (key enzyme of glycolysis), ferment pentoses in the same way as obligately heterofermentative species.

Under such circumstances, the phosphoketolase of facultative heterofermentative LAB is induced by the available pentose sugar [95]. Fermentation of pentoses results in the production of equimolar amounts of lactic and acetic acid; no CO_2 is formed and since no dehydrogenation steps are necessary to reach the intermediate xylulose-5-phosphate, acetyl phosphate is used by acetate kinase in a substrate level phosphorylation step yielding acetate and ATP. Obligately homofermentative LAB do not ferment pentoses [3].

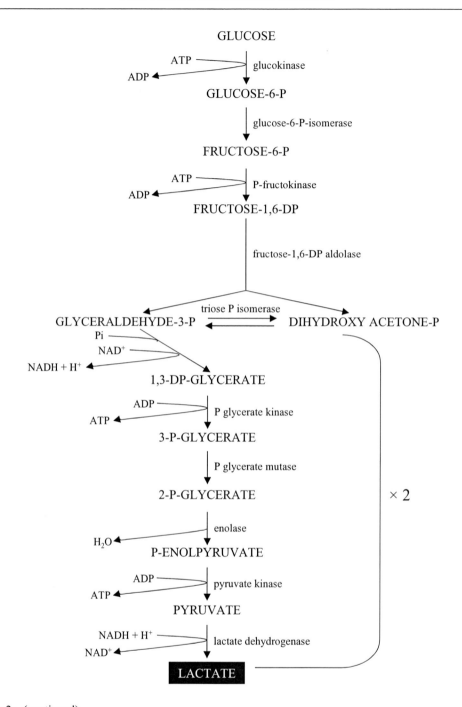

Figure 2a. (continued)

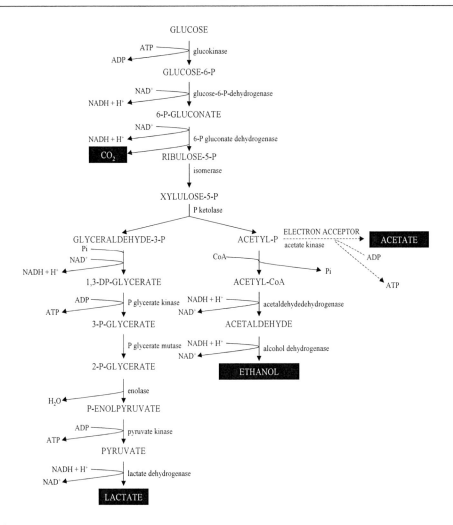

Figure 2b.

Figure 2. Lactic fermentation: A, homolactic fermentation (glycolysis, Embden-Meyerhof-Parnas pathway); B, heterolactic fermentation (pentose shunt, 6-phosphogluconate/phosphoketolase pathway).

Lactic Acid Bacteria Associated with Raw Materials

Several studies have been focused on the identification and characterization of LAB and yeast microbiota of mature sourdoughs, but only few works have been directed to the microbial ecology of raw materials used in bread-making. The first document is dated back to 1987 [69], when the authors analyzed samples of Italian wheat flours for the presence of different microbial groups. Subsequently, Corsetti et al. [28] isolated and identified LAB and yeasts from common wheat and organic flours. Both works have been performed with a phenotypic/biochemical approach that, in the case of the work of Corsetti and co-workers, revealed the following LAB species: *Lactobacillus alimentarius*, *Lactobacillus plantarum*, *Lactobacillus rhamnosus*, *Lactobacillus confusus* e *Lactobacillus viridescens*. Some of the species described at the time of writing have been re-classified and are now allotted into other genera such as *Weissella confusa* and *Weissella viridescens*.

A recently published paper has been carried out with the aim to characterize the cultivable LAB populations associated with durum wheat kernels cultivated in several Italian regions, bran and non-conventional (amaranth, chickpea, corn, rice, quinoa and potato) flours used to produce baked goods for celiac sprue affected people [32]. To this purpose, a genetic polyphasic strategy consisting of randomly amplified polymorphic DNA-polymerase chain reaction (RAPD-PCR) analysis, partial 16S rRNA gene sequencing, species-specific and multiplex PCRs has been employed, allowing the detection of 11 LAB species belonging to five genera: *Aerococcus* (*A. viridans*), *Lactococcus* (*Lc. garvieae*), *Enterococcus* (*E. durans*, *E. casseliflavus*, *E. faecalis*, *E. faecium* e *E. mundtii*), *Lactobacillus* (*Lb. coryniformis* subsp. *coryniformis*, *Lb. curvatus* subsp. *curvatus* e *Lb. graminis*) and *Pediococcus* (*P. pentosaceus*). Those findings demonstrated that, except *Lb. curvatus* which, together with *Lactobacillus brevis* was reported to be one of the more dominant lactobacilli of Portuguese maize sourdough [180], LAB species isolated from unprocessed raw materials are different from those commonly associated with sourdough fermentation. That work led to the confirmation that the matrix composition and the fermentation parameters play a defining role in the selection of LAB [35]. Furthermore, the lack of correspondence between grain/flour and sourdough LAB communities was explained by the fact that typical sourdough LAB may be present in raw materials as dormant (non-cultivable) flora and, for this reason, only detectable by means of culture-independent tools.

Lactobacilli in Sourdough

LAB characteristics of mature sourdoughs are mostly ascribable to Lactobacillus genus. Some *Lactobacillus* species of recent description (*Lb. spicheri*, *Lb. rossiae*, *Lb. acidifarinae*, *Lb. zymae*, *Lb. hammesii*, *Lb. nantensis*, *Lb. namurensis*, *Lb. crustorum*, *Lb. secaliphilus* and *Lb. siliginis*) have been first isolated from sourdough matrices. A selection occurs during propagation that leads to the establishment of, usually, one or two *Lactobacillus* species at numbers that are three or four orders of magnitude above those of the adventitious microbial flora [94].

The phylogram reported in Fig. 1 includes 44 species belonging to Lactobacillus genus. Probably due to their misidentification, some of them have been rarely or just once reported to be found in sourdough; as an example could be cited *Lb. rhamnosus* [209] detected before molecular techniques being available. Thirty-three species (Table 1) are effectively considered to be typical of sourdough environments [197].

Lactobacillus sanfranciscensis, *Lb. brevis* and *Lb. plantarum* are the most frequently lactobacilli isolated from sourdoughs [30,75,106,191]. However, some strains, initially classified as *Lb. plantarum* may actually belong to *Lactobacillus paraplantarum* or *Lactobacillus pentosus* species, since both phenotypic determination tools, such as carbohydrate fermentation patterns, and genetic methods, such as 16S rRNA gene sequence analysis, are not able to distinguish among *Lb. plantarum* group species. For a reliable identification of these species, a multiplex PCR with recA gene-derived primers was developed [227]. Traditional production by various stages of continuous propagation and production by commercial starter cultures allow the predominance of *Lb. sanfranciscensis* which is considered to be the key sourdough LAB [82]. *Lactobacillus sanfranciscensis* dominates type I sourdough fermentations; this observed predominance of *Lb. sanfranciscensis* strains is possibly the result of the selective pressures that arise through the environmental conditions pertinent to the applied sourdough fermentation technology [29]. In general, the competitiveness of heterofermentative lactobacilli in sourdoughs mainly relies on their combined use of maltose and electron acceptors [82,236]. The majority of *Lb. sanfranciscensis* strains only ferment glucose and maltose [29]. However, sucrose, raffinose, galactose, melibiose, ribose and fructose can be fermented by some *Lb. sanfranciscensis* strains [94]; in this case, fructose is used, under certain conditions, as external electron acceptor [82] since virtually all strains display mannitol dehydrogenase activity and reduce fructose to mannitol [72].

Table 1. *Lactobacillus* species associated with sourdough fermentation

Obligately heterofermentative	Facultatively heterofermentative	Obligately homofermentative
Lb. acidifarinae	Lb. alimentarius	Lb. acidophilus
Lb. brevis	Lb. casei	Lb. amylolyticus
Lb. buchneri	Lb. paralimentarius	Lb. amylovorus
Lb. fermentum	Lb. pentosus	Lb. crispatus
Lb. fructivorans	Lb. plantarum	Lb. crustorum
Lb. frumenti		Lb. delbrueckii subsp. delbrueckii
Lb. hammesii		Lb. farciminis
Lb. hilgardii		Lb. johnsonii
Lb. namurensis		Lb. mindensis
Lb. panis		Lb. nantensis
Lb. pontis		
Lb. reuteri		
Lb. rossiae		
Lb. sanfranciscensis		
Lb. secaliphilus		
Lb. siliginis		
Lb. spicheri		
Lb. zymae		

Gobbetti [81] reported on the *Lb. sanfranciscensis*/*Lb. plantarum* association in Italian wheat sourdough. *Lb. plantarum* may be superseded by another facultative heterofermentative species: *Lb. alimentarius* in its association with *Lb. sanfranciscensis* in sourdough made from durum wheat [29]. *Lactobacillus alimentarius* is capable of fermenting all four flour soluble carbohydrates (maltose, sucrose, glucose and fructose) and it is possible that this reduces direct metabolic competition with *Lb. sanfranciscensis*. Most of the *Lb. alimentarius* strains, due to a phenotypical misidentification, probably belong to *Lactobacillus paralimentarius*, a facultatively heterofermentative species first isolated from Japanese sourdough [18].

Scheirlinck et al. [191,192] found that the species *Lb. paralimentarius*, together with *Lb. plantarum, Lb. sanfranciscensis, Lactobacillus pontis* and *Lb. spicheri*, was dominant in type I sourdoughs produced in a Belgian bakery.

Lactobacillus brevis and *Lb. plantarum* have generally been found associated with *Lactobacillus fermentum* in Russian sourdoughs [111] and *Lb. fermentum* dominates Swedish sourdoughs [209] and German type II sourdoughs, although, in the latter case, it was demonstrated that it originated from the baker's yeast used [150]. Furthermore, Gobbetti et al. [86] reported that *Lb. acidophilus* is common in Umbrian (Italian region) sourdoughs, even though it is rarely isolated from sourdoughs of different origin [107,188]. Corsetti and coworkers [36], described a new sourdough associated species, *Lb. rossiae*, that seems to be wide diffused in sourdoughs of Southern and Central Italy [202]. From preliminary observations, *Lb. rossiae* is often associated with the key-sourdough *Lb. sanfranciscensis* (data not published). Moreover, *Lb. rossiae* has been found in environments other than sourdough (pig feaces) [48], while no other habitat is known for *Lb. sanfranciscensis* [93].

So far, some studies have been aimed to the sequencing of complete genome of lactobacilli, including *Lb. brevis* [141], *Lb. plantarum* [118] and *Lb. reuteri* (unpublished results), that are also found in sourdough. Furthermore, some papers dealing with investigation on their genome size, performed by pulsed field gel electrophoresis, [60,247] showed that genomes of *Lb. sanfranciscensis*, *Lactobacillus pontis* and *Lb. reuteri* are about 1.5 Mbp, thus indicating that the metabolic potential and adaptation to several niches of these species may be limited.

Biodiversity of LAB

The variety of microorganisms of a given sourdough environment forms a complex structure which contributes to the balance of the organoleptic characteristics as well as to its stability. In general, different methods are available to evaluate microbial biodiversity based both on phenotypic and genotypic characteristics using pure cultures.

The phenotype is the observable expression of the genotype and, basically, the investigation of biodiversity within a group of microorganisms is based on morphology, physiology and biochemical features. Phenotypic methods comprise all methods that are not directed toward DNA or RNA. One of the disadvantages of analyzing the phenotype is that the whole information potential of a prokaryotic genome is never expressed. Gene expression is directly related to the environmental conditions. Prokaryotic phenotype is unreliable to differentiate microorganisms at strain level since prokaryotes lack complex morphological features and do not generally show life cycles with different morphological stages, thus, the analysis of the prokaryotic phenotype mostly counts on the development of experimental techniques that test directly or indirectly different phenotypic properties, such as enzyme activities and substrate utilization. This approach does not easily discriminate among strains belonging to closely related species. Furthermore, phenotypic analysis represents the most tedious task during the process of microorganism grouping, as it requires time, skills, and, in order to avoid subjective observations, technical standardization [184]. However, phenotypic methods also include chemotaxonomic techniques which refer to the application of analytical methods for collecting information on various chemical constituents of the cell and the interpretation of results does not depend on subjective observations. In the latter case, such tools are used for species identification rather than strain differentiation.

Genotypic methods are tools of genomic information retrieval that are mostly directed toward DNA or RNA molecules. Genotypic typing methods basically target DNA. They may represent a complement or may be used as an alternative to the phenotypic methods. All these techniques are grouped in two classes: typing methods based on whole genome restriction fragment analysis; and typing methods based on the amplification of genome fragments by PCR [184]. DNA fragments are then subjected to the electrophoretic separation and visualized.

The different strategies routinely applied for strain differentiation in order to evaluate sourdough LAB biodiversity mainly rely on genotypic methods. In past years, these methods were also used for identification purposes by comparison of nucleic acid band patterns among unknown and type strains.

Restriction fragment length polymorphism (RFLP) is one of the first genotypic tools applied to the microbial recognition. This technique resulted useful to reveal the presence of an antibiotic producer *Lb. reuteri* strain in sourdoughs propagated in different years [73]. However, one of the limitations of the RFLP is due to the high number of bands generated during enzyme cutting that makes difficult the interpretation of band profiles. From this point of view PFGE overcomes this limit diminuishing the number of DNA bands to be compared. Different papers show the successful application of PFGE to the differentiation of LAB strains isolated from industrial and artisanal sourdough, e.g. *Lb. sanfranciscensis* [247] and *Lb. plantarum* [174]. In general, both techniques are being applied in combination with second generation methods, particularly when they are directed towards rRNA genes which need an amplification before restriction.

Genomic DNA polymorphisms are routinely studied by the fingerprinting method RAPD-PCR. One of the big advantages of this method, with regards to the first generation techniques, is due to its rapidity of execution, it can be easily performed starting from colonies, hence a high number of isolates may be analyzed at the same time. However, the low rigor of amplification may determine a certain variability of the resulting profiles. RAPD-PCR has been found a valuable method to clusterize sourdough LAB isolated throughout Italy [26]. Due to its versatility, this technique my be used to clarify different aspects of microbial biodiversity. Ehrmann and Vogel [59] reported on the use of RAPD-PCR not only as a survey of but also in monitoring the change of composition of lactobacilli in a type II model fermentation, demonstrating that the dominance of *Lactobacillus amylovorus* could be overcome by *Lb. pontis* and *Lactobacillus frumenti* over addition of an enzyme preparation. Settanni et al. [199] used this technique to follow the fate of starter cultures used to produce sourdoughs. The presence and identity of inoculated strains (*Lb. sanfranciscensis* CB1, *Lb. plantarum* 20 and *Lactococcus lactis* M30 and Q13) was monitored following isolation from dough, by microscopic inspection and comparing their RAPD profiles, generated with two oligonucleotide primers (P4 and M13), to those obtained from the original cultures. The same approach was used by Reale et al. [179] to confirm the presence of lactobacilli (*Lb. plantarum*, *Lb. brevis* and *Lb. curvatus*) inoculated as sourdough starter cultures at the end of fermentation.

Repetitive element sequence-based PCR (Rep-PCR) and the intergenic spacer region-based PCR (ITS-PCR) target specific sequences and have been both recently used to evaluate the diversity of a collection of *Lb. sanfranciscensis* strains from sourdough of Central and

Southern Italy [44]. Rep-PCR was also useful to examine the persistence of *Lb. amylovorus* DCE 471 during rye sourdough fermentation [133].

Second-generation typing methods are, however, characterized by a low interlaboratory reproducibility. Adaptor fragment length polymorphism (AFLP) as well as SAU-PCR are miscellaneous methods (genome restriction and fragment amplification) that avoid this problem. AFLP has been demonstrated to be useful for the discrimination of species closely related, both phenotypically and phylogenetically, such as *Lb. plantarum, Lb. paraplantarum* e *Lb. pentosus* that, as reported above, are often associated to sourdough environments [226]. Furthermore, this technique has also been applied to differentiate *Lactobacillus* spp. of different sources [70] and those isolated from Chinese [134] and Belgian [191] sourdoughs.

MICROBIAL INTERACTIONS

Knowledge to exploit and improve the stability of associated sourdough LAB and yeasts are necessary in order to prevent the loss of variety of regional specialities while simultaneously trying to meet consumer and industry demands [81].

Cell-Cell Communication

The term "sociomicrobiology" has been recently coined for the many social activities exhibited by microorganisms [173]. Bacteria were the first microbes to be studied in order to decipher their code of communication. They were for long time believed to exist as individual organisms in a given environment searching for nutrients to multiply. Nevertheless, bacteria are active in performing a census of their population, as well as to investigate the environment for development and to feel the presence of competitors [68]. Such actions are the result of an efficient intercellular communication that is based on the production, release, detection and reply to small signal molecules, which accumulate and trigger cascade events when a "quorum" concentration is reached. Hence, the term "quorum sensing" is used to describe the cell-cell communication. As a population of quorum-sensing bacteria grows, the individual organisms produce and secrete the signal molecules (autoinducers) into the extracellular environment. Thus, the concentration of external autoinducer is correlated with cell-population density. In a species-specific cell-cell communication, by monitoring the extracellular autoinducer concentration, the bacteria can "count" one another [67], acting as a group. When a signal accumulates to a sufficiently high concentration, the cognate response regulator is activated within the local population of cells, leading to coordinated gene expression [154]. In general, the coordination of population behaviour mediated by quorum sensing determines access to nutrients or specific environmental niches, collective defence against other competitor organisms or community escape where survival of the population is threatened [243]. Although several quorum sensing systems are known, the two most thoroughly described systems are the acyl-homoserine lactone (AHL) systems of many Gram-negative species and the peptide-based, also called auto-inducing peptide (AIP), signaling systems of many Gram-positive species [5,212].

While species-specific quorum sensing apparently allows recognition of self in a mixed population, it seems likely that in these contexts, bacteria also need a mechanism or mechanisms to detect the presence of other species. Additionally, it is conceivable that it is useful for bacteria to have the ability to calculate the ratio of self to other in mixed populations, and in turn, to specifically modulate behavior based on fluctuations in this ratio [63]. For example, the expression of quorum sensing systems may be manipulated by the activities of other bacteria within complex microbial consortia [243]. Recently, it become apparent that fungi, like bacteria, also use quorum regulation to affect population-level behaviors. Furthermore, considering the extent to which quorum-sensing regulation controls important processes in many distantly related bacterial genera, it is not surprising that cell density-dependent regulation also appears to be prevalent in diverse fungal species. Also for fungi quorum sensing is mediated by small diffusible signalling molecules that accumulate in the extracellular environment [105].

The studies regarding cell-cell communication in sourdough ecosystems are, so far, limited to the group of LAB. In particular, Di Cagno et al. [56] followed, by a proteomic approach, the growth of *Lb. sanfranciscensis* CB1 in mono-culture and then it was compared with that in co-cultures with *Lb. plantarum* DC400, *Lb. brevis* CR13 or *Lb. rossiae* A7. The authors found that, compared to mono-colture, *Lb. sanfranciscensis* CB1 at late stationary phase over-expressed 48, 42 and 14 proteins when co-cultured with strains DC400, CR13 and A7, respectively. Induced polypeptides, only in part common to all co-cultures, were identified as stress proteins, energy metabolism related enzymes and proline dehydrogenase, GTP-binding protein, *S*-adenosyl-methyltransferase, and Hpr phosphocarrier protein. By using primers designed from consensus amino acid sequences of phylogenetically related bacteria, two quorum sensing involved genes, *luxS* and *metF* were shown to be expressed in *Lb. sanfranciscensis* CB1.

This topic is under study and also the mechanisms of inter-kingdom communication between LAB and yeasts in sourdough environment needs investigations.

Positive Relations

Stable co-metabolism between LAB and yeasts is common in many foods, enabling the utilization of substrates that are otherwise non-fermentable (for example starch) by individual microorganisms and, thus, increasing the microbial adaptability to complex food ecosystems [52]. San Francisco French bread and Panettone are characterised by a unique and strict association (proto-cooperation) between *Lb. sanfranciscensis* and *S. exiguus* [65,86,213]. This association, optimal in continuously operating sourdough fermentation systems [237], has been found in sourdoughs from different countries [210,244] as well as in gluten-free baked goods [245]. Soluble carbohydrates are present in low amounts in rye and especially wheat flours and their concentration varies with the type of flour [189] and with the level of amylase activity on the damaged starch granules [168]. Wheat flour contains 1.55 to 1.85% soluble carbohydrates depending on the balance between starch hydrolysis by the flour and microbial enzymes and microbial consumption [144]. *Lactobacillus sanfranciscensis* hydrolyzes maltose by a maltose phosphorylase which produces a molecule of glucose-1-phosphate that is metabolized further, and a molecule of glucose, which is excreted outside the cell to avoid intracellular accumulation (Fig. 3) [59]. Maltose phosphorylase acts without the expenditure of ATP. Inside the *Lb. sanfranciscensis* cell cytoplasm, glucose-1-P is converted into glucose-6-P by the action of phosphoglucomutase and then further metabolized through the 6-PG/PK pathway. As long as there is plenty of maltose in sourdough, *Lb. sanfranciscensis* uses such an energetically more favourable pathway in which it is releasing glucose [187]. Maltose-negative yeasts (e.g., *S. exiguus*) may in turn utilize such excreted glucose. The lack of competition for maltose [213] possibly explains the observed stimulation of *S. exiguus* by *Lb. sanfranciscensis*. *Saccharomyces exiguus* preferentially uses glucose or sucrose and shows a high tolerance to the acetic acid produced by heterofermentative microorganisms [214]. Such an association presents ecological advantages to both *Lb. sanfranciscensis* and *S. exiguus*. However, maltose utilization by maltose-positive yeasts only begins when available glucose and fructose are depleted [4]. Amino acid production by yeasts stimulates *Lb. sanfranciscensis* growth even though synthetic medium is deficient of essential amino acids such as valine and isoleucine [84]. Sour Dough Bacteria (SDB) broth [119] is an optimal medium for the growth of *Lb. sanfranciscensis*. It is made with freshly-prepared yeast extract which contains a small peptide (Asp-Cys-Glu-Gly-Lys) identified as a growth stimulant factor for *Lb. sanfranciscensis* [8]; fresh yeast extract cannot be substituted by commercial yeast extracts, liver or protein hydrolysates.

Gobbetti et al. [83] observed a decrease in *Lb. sanfranciscensis* metabolism when associated with *S. cerevisiae*, a phenomenon which apparently is due to a faster consumption of maltose and, in particular, of glucose by the yeast. In contrast, Meignen et al. [146] reported a reduction of *S. cerevisiae* growth when it was mixed with *Lb. brevis*. The disappearance of *S. cerevisiae* from the microbial population of sourdough during consecutive fermentations is related to the repression of the genes involved in maltose fermentation, as a consequence of which maltose cannot be utilized [165]. This causes a rapid depletion of sucrose and a decline of particularly sensitive *S. cerevisiae* strains as a result of acetic acid produced by heterofermentative LAB that, at sourdough pH (4.0–4.5), is in the undissociated lipophilic and membrane-diffusable form, thus exhibiting its highest inhibitory activity [12].

In agreement with Hansen et al. [97], who demonstrated a greater adaptability of yeasts to grow in association with homofermentative rather than heterofermentative LAB, *S. cerevisiae* fermentation was positively influenced by *Lb. plantarum*, especially regarding the production of CO_2 that resulted increased compared to the CO_2 produced by the association *S. cerevisiae*/*Lb. sanfranciscensis* [87].

Recently, subdominant LAB populations of Central Italy sourdough ecosystems has been investigated [33]. The study evidenced that this community is mainly represented by non-*Lactobacillus* species. In particular, *E. faecium* and *P. pentosaceus* have been found in the range 10^4-10^6 UFC/g, while lactobacilli were about 2–3 orders of magnitude higher [230]. With the idea that subdominant LAB are originating from raw materials, the authors inoculated different mixtures of *E. faecium* and *P. pentosaceus* strains isolated from wheat kernels [32], in dual combination with *Lb. sanfranciscensis* of sourdough origin, into experimental doughs. The doughs were prepared using flour previously subjected to γ-ray treatment, in order to kill viable microbial cells while avoiding modifications of its chemical composition due to the thermal sterilization, so that raw material was chemically identical to that employed in the industrial processes. Single inoculums showed *E. faecium* and *P. pentosaceus* as stronger dough acidifier than the *Lb. sanfranciscensis* strain used. Furthermore both co-fermentation (*E. faecium*/*Lb. sanfranciscensis* and *P. pentosaceus*/*Lb.*

sanfranciscensis) confirmed this observation. Thus, the potential of non-*Lactobacillus* strains of raw materials is of paramount importance during the first steps of sourdough preparation, since these species may inhibit indigenous non-LAB microorganisms by lowering the pH and prepare the environment for the establishment of the typical species (e.g., *Lb. sanfranciscensis*) of mature sourdoughs.

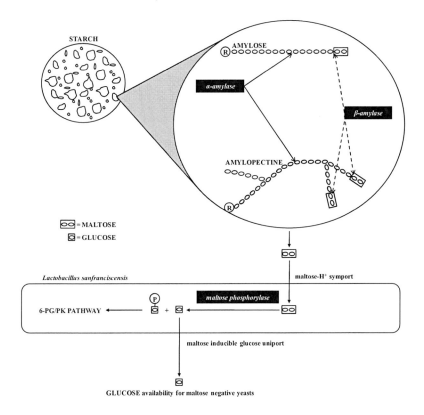

Figure 3. Utilization of maltose by *Lb. sanfranciscensis*.

Antagonistic Relations

In general, LAB play a crucial role in the preservation and microbial safety of fermented foods [20], thus promoting the microbial stability of the final products of fermentation [147]. Since LAB naturally occur in various food products, they have traditionally been used as natural food biopreservatives; protection of foods is due to the production of organic acids, carbon dioxide, ethanol, hydrogen peroxide and diacetyl, antifungal compounds such as fatty acids or phenyllactic acid, bacteriocins and antibiotics such as reutericyclin [198].

Fundamental features of an antimicrobial substance to be active under food conditions is that it is produced at active concentrations and that the effect is not masked by food components.

Basic Antagonism

Basic competitiveness of microorganisms is due to the production of primary metabolites that originate from the energetic metabolism, in case of sourdough LAB mainly organic acids.

During bread-making, environmental contamination of the products may occur with bacteria such as bacilli and clostridia, whose growth can be inhibited by acidity [240]. *Bacillus* spp., especially *B. subtilis* and *B. licheniformis* are spoilage agents of wheat bread due to rope formation [175], while representative strains of both species may also cause food-borne illness if present at levels over 10^5 CFU g^{-1} [123]. Although food-borne illness related to consumption of ropy bread is considered unlikely to happen due to the slimy appearance of the crumb, loaves with high counts of *B. subtilis* and *B. licheniformis*, showing no rope symptoms, may cause diarrhoea and vomiting [182]. Furthermore, *Bacillus* spores can survive the baking process where the temperature in the centre of the crumb remains at maximum of 97–101°C. However, rope formation occurs principally in wheat breads that have not been acidified, or in breads with high concentrations of sugars, fat or fruits [9]. The most natural way to prevent rope formation is represented by sourdough technology due to the activity of LAB. Pepe et al. [175] reported the prevention of growth of approximately 10^4 rope-producing *B. subtilis* G1 spores per cm^2 on bread slices for more than 15 days by adding heat-treated cultures of *Lb. plantarum* E5 and *Leuconostoc mesenteroides* A27. Menteş et al. [149] reported the effect of sourdoughs produced with *Lb. plantarum* LMO25 and *Lb. alimentarius* LMO7, previously characterized for their inhibitory activity against rope-forming *Bacillus* strains, in wheat bread [148]. The addition of 15% or 20% low pH (pH 3.5–4.0) sourdough to bread dough prevented the generation of visual rope caused by *B. subtilis* and *B. licheniformis* while sourdoughs with a higher pH (>4) were not effective in preventing rope caused by both *B. subtilis* and *B. licheniformis* at concentrations lower than 20%.

The acidity level of sourdough is the main determining factor of the inhibitory activity of LAB [239] also against moulds. The most frequent cause of spoilage in baked goods is represented by fungal growth, which may cause public health concerns related to mycotoxin production [131]. Spoiling fungi commonly associated with bakery losses belong to the genera of *Aspergillus, Cladosporium, Endomyces, Fusarium, Monilia, Mucor, Penicillium*, and *Rhizopus* [112,131]. Besides the common physical methods of food preservation (heat treatments, cold storage, modified atmosphere storage, drying, freeze-drying) [62], protection of baked goods from fungal spoilage is mainly achieved through the inactivation of contaminating spores by the use of (1) infrared and microwave radiation, (2) fungal inhibitors such as ethanol and propionic, sorbic, benzoic and acetic acid and some of their salts, (3) suitable packaging techniques such as modified atmospheres and (4) sourdough [15,17,91,131].

Sourdough addition seems to be the best preservation procedure of bread from spoilage, meeting consumer's demands for natural and additive-free food products [183]. Extended shelf-life and improved microbial safety of sourdough bread is attributed to LAB, whose fungistatic effects are mainly due to acetic acid rather than lactic acid production [181]. In this regard, heterofermentative lactobacilli produce most of the antifungal activity. Within this group *Lb. sanfranciscensis* CB1 displays the widest spectrum of antifungal activity due to the production of a mixture of acetic, caproic, formic, propionic, butyric and *n*-valeric acid, among which caproic acid was the organic acid with the higest antimould activity [27]. A combination of organic acid production and low pH generation also determined the anti-*Fusarium* (*F. proliferatum* and *F. graminearum*) activity of *Lactobacillus paracasei* subsp. *tolerans*, isolated from a traditional sourdough bread [102]. However, other factors may significantly contribute to their antagonistic activity of LAB.

Bacteriocins

Interactions among sourdough LAB depend on several mechanisms in which some competitive advantages are provided by bacteriocins, ribosomally synthesized, extracellularly released low molecular-mass peptides or proteins (usually 30–60 amino acids) which have a bactericidal or bacteriostatic effect on other bacteria [116], either in the same species (narrow spectrum) or across genera (broad spectrum) [38]. Bacteriocins from LAB have been classified in three classes on the basis of common, mainly structural, characteristics [163]. Bacteriocins target the cell envelope, and with the exception of the larger proteins (> 20 kDa) that degrade the murein layer (e.g. lysins and muramidases), use non-enzymatic mechanisms to disrupt the integrity of the target cell membrane and/or inhibit cell wall synthesis [229].

Microbial growth inhibition is due to a series of phenomena, first of all dissipation of the transmembrane electrical potential $\Delta\psi$. Strong cytotoxic effects presumably result from proton influx that leads to a drop of the intracellular pH and consequently to the inhibition of many enzymatic processes [157]. Also influx of sodium ions has a cytotoxic effect [23]. Pores in the cytoplasmic membrane (Fig. 4) clearly affect the energetic status of the cell, i.e. dissipation of proton motive force causes an arrest of ΔpH and $\Delta\psi$ dependent (e.g. transport) processes while certain bacteriocins cause ATP efflux [157]. Bacteriocin producer strains are protected against their own bacteriocins by a protection system referred to as immunity. Each bacteriocin has its own dedicated protein conferring immunity, which is expressed concomitantly with the bacteriocin [163].

Bacteriocins may contribute to the competitiveness of the producing strain in a fermented food ecosystem [20]. Bacteriocins produced by sourdough LAB have been purified and well characterized with regards to their *in vitro* activity. As examples can be cited plantaricin ST31 produced by *Lb. plantarum* ST31 [224] and bavaricin A produced by *Lactobacillus bavaricus* MI401 (previous designation of *Lactobacillus sakei*) [126]. Corsetti and co-workers [28,35] reported on the discovery of antimicrobial molecules produced by sourdough lactobacilli. Although these activities were not purified to homogeneity, such antimicrobial compounds displayed characteristics similar to bacteriocins, and were hence designated as bacteriocin-like inhibitory substances (BLIS) [215]. *Lb. pentosus* 2MF8 isolated from sourdough [35], was the first example of a sourdough-derived *Lactobacillus* strain that produced an antimicrobial compound with bacteriocin characteristics and reported to be active under sourdough conditions. The *in situ* activity of BLIS M30 produced by *Lc. lactis* subsp. *lactis* M30 and isolated from unmalted barley [101] was also tested in the sourdough ecosystem, where it was shown to possess a more potent inhibitory spectrum and activity than sourdough LAB, while it did not inhibit certain strains of the key sourdough bacterium *Lb. sanfranciscensis* [35]. The activity of the above BLIS M30, later identified as lacticin 3147-like bacteriocin [199], was also followed over a period of 20 days (corresponding to 20 refreshments) showing a strong influence on the microbial consortium of sourdough LAB and its aptitude to support the dominance of insensitive strains during sourdough fermentation (Fig. 5). De Vuyst et al. [49] reported on the application of a bacteriocin-producing *Lb. amylovorus* strain isolated from fresh corn steep liquor [50] in type II cereal fermentation. *Lb. amylovorus* DCE 471, which produces amylovorin L471, a class IIa bacteriocin [117] represented the only culture used as starter. Wheat and rye sourdough were acidified within 15 h to a final pH below 3.7 which would be in agreement with an industrial type II rye sourdough containing *Lb. amylovorus* at high concentration [160]. *Lb. amylovorus* DCE 471 persistence was tested over a period of 24 h and it showed the capability of producing amylovorin L471 under sourdough conditions [49] making it a suitable strain for use in production of type II sourdough. The same strain has also been demonstrated to be a competitive starter culture for this sourdough type, both on laboratory and pilot scale fermentations [133].

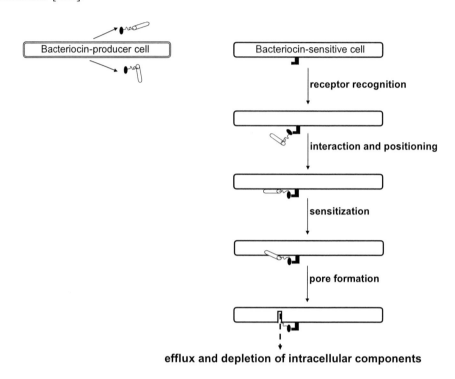

Figure 4. Model of action of bacteriocins.

LAB isolated from sourdough raw materials [31] show a higher ability, in terms of number of positive strains and inhibition activity, to produce bacteriocins than LAB associated with fermented doughs [35] where the adaptation to the particular environment (e.g. to nutritional sources) seems to play a major role for their persistence. Severeal *E. faecium* [34] and *E. mundtii* [200] strains of *T. durum* and flour origin have been characterized for their activity under different growth conditions in view of their use in sourdough biopreservation.

Although Rosenquist and Hansen [183] reported that a variety of bacteriocins from LAB were ineffective to control the growth of *Bacillus* spp. in bread, some sourdough LAB bacteriocins have shown a capability of inhibiting food-borne pathogens and/or food spoilage bacteria, including *Listeria monocytogenes*, *B. subtilis* and *Staphylococcus aureus*, thus their use as food additives or the application of the producer strains as starter or protective culture, might contribute to the manufacturing of safer products. Bacteriocins may lead to a reduction of chemical preservative used by food industry, thus facing the increasing consumer's demand for higher quality and naturalness of foods [198].

In general, the *in situ* efficacy of bacteriocin-producing starter strains in food fermentation is efficacy which can be negatively influenced by various factors, such as binding of the bacteriocins to food components (fat or protein particles) and food additives (e.g. triglyceride oils), inactivation by proteases or other inhibitors, changes in solubility and charge, changes in the cell envelope of the target bacteria [1,74,110,132]. However, some baking enzymes (hemicellulase, lipase, amyloglucosidase and amylase) have been recently found to have little or no effect on the antimicrobial effect of bacteriocin-like inhibitory substances produced by *Lb. sakei*, *P. pentosaceus* and *P. acidilactici* isolated from Lithuanian sourdoughs [162].

Antibiotics

Antibiotics are historically natural substances generally produced by microorganisms with a strong inhibitory effect against bacteria. Nowadays, many antibiotics are of plant origin or are chemically synthesized (in this case referred to as chemotherapeutic drugs) and they are active against yeasts, fungi and parasites. Antibiotics show several different chemical structures which determine their classification. Unlike bacteriocins, antibiotics, besides acting on citoplasmic membrane and cell-wall, also interfer with protein and nucleic acid synthesis and with primary metabolism [124].

The natural microflora associated with SER sourdough, an in house rye sourdough prepared for producing a commercially available baking aid [13], was followed over a period of 10 years. Considerable shifts in microflora composition were observed, apart for what concerned *Lb. reuteri* LTH2584, a reutericyclin producing strain first isolated in 1988 from SER. Reutericyclin is a tetrameric acid derivative with a molecular weight of 349 Da, and active against a broad range of Gram-positive bacteria including spoilage organisms and pathogens as *St. aureus*, *E. faecalis*, *L. monocytogenes* and *Bacillus cereus* as well as vegetative cells and spores of rope forming bacilli [71]. It is produced in active concentrations in dough and its production is relevant to dough ecology since it was shown to contribute to the stable persistence of *Lb. reuteri* in sourdough over about 50,000 generations of microbial growth [73]. Some *Lb. reuteri* strains were also found to produce reuterin (3-hydroxy-propanal), an antimicrobial compound representing the main product of glycerol metabolism that, like reutericyclin, is not classified as a bacteriocin. Reuterin was shown to be active against a wide range of Gram-positive as well as Gram-negative bacteria [236], yeasts and fungi [195] such as *Aspergillus* and *Fusarium* [24].

Antifungal Compounds

Antimicrobial substances resistant to baking conditions and active at the physical parameters pertinent to bread, can control the growth of spoilage organisms [182].

Lavermicocca et al. [127] purified and characterized two antifungal compounds produced by *Lb. plantarum* ITM21B, identified as phenyllactic and 4-hydroxy-phenyllactic acids, which still retained their fungicidal activities after baking. Lactobacilli able to produce the above two antifungal compounds, among which phenyllactic acid is the most potent one, were shown to delay *Aspergillus niger* and *Penicillium roqueforti* growth for up to 7 days and significantly prolong the shelf-life of bread [127,128]. Besides phenyllactic acid, *Lb. coryniformis* subsp. *coryniformis* Si3 and *Lb. plantarum* MiLAB14 isolated from plant material stored under anaerobic conditions were found to produce a broad-spectrum 3 kDa protein, as well as the cyclic dipeptides cyclo(L-Phe-L-Pro) and cyclo(L-Phe-trans-4-OH-L-Pro), all showing antifungal activity at concentrations in the order of mg ml^{-1} [143,206]. Similar results, in terms of antifungal compounds identified (phenyllactic acid and the two cyclic dipeptides) were reported by Dal Bello et al. [41] who followed the inhibition of *Fusarium culmorum* and *F. graminearum* by *Lb. plantarum* FST 1.7. *Lb. plantarum* was also found to produce antifungal hydroxylated fatty acids active at 10 μg ml^{-1} [206]. Schnürer and Magnusson [195] published a summary of the main antifungal compounds produced by different LAB: lactic and acetic acid, carbon dioxide, diacetyl, hydrogen peroxide, caproic acid, 3-hydroxy fatty acids, phenyllactic acid, cyclic dipeptides, reuterin and fungicins. The last are reported to be compounds of proteinaceous nature, since they loose their antifungal activity after treatment with proteolyitc enzymes. It has been found that fungicins are produced by strains of *Lb. casei*, *Lb. pentosus*, *Lb. paracasei* subsp. *paracasei* and *Lb. coryniformis* subsp. *coryniformis* [88,142,166].

In situ applications demonstrated the efficacy of sourdough fermented by antifungal-producing *Lb. plantarum* to inhibit the growth of common bread spoilage fungi (*As. niger*, *F. culmorum* and *Penicillium expansum*) and reduce the amounts of chemical additives in bread such as calcium propionate [186].

Technological Potential of Lactic Acid Bacteria

Contribution to Flavour Development

The flavour of wheat bread is assumed to be influenced by the nature of raw materials [98], fermenting microorganisms [30] as well as the conditions of the fermentation and baking process [94]. Czerny and Schieberle [40] found high odour activities in wholemeal and white wheat flours. In general, the baking process influences the typical aroma of bread crust, while dough fermentation is fundamental for the development of crumb flavour. Furthermore, the recipe and geographical origin make the difference in the aroma of breads consumed in different geographical areas [99].

Aromatic molecules are mostly formed during baking, e.g. the roasty aroma of wheat bread is due to the formation of compounds such as 2-acetyl-1-pyrrolin, giving the typical scent of wheat bread in the crust [194]. Furthermore, aromatic compounds are more numerous in crust than crumb. Bianchi et al. [10] identified a total of 89 compounds belonging to different chemical classes in the crust of Altamura bread, while 74 volatile compounds were detected in the crumb; some chemicals, such as ethanol and 3-methyl-1-butanol, were abundant in both bread parts.

Because of its lack in gluten proteins, rye flour has to be prefermented before bread can be manufactured from such rye-based dough and this procedure is one of the main reasons for the differences in the overall aromas between wheat and rye breads [114]. Nevertheless, sourdough fermentation is essential to achieve an acceptable flavour, since a comparison between chemically acidified bread and sourdough bread showed that the latter possessed a superior sensory quality [115]. The ratio between lactic and acetic acid, defined as the fermentation quotient (FQ), is an important factor that might affect the aroma profile, although it is also relevant for the structure of final products. Acetic acid, produced by heterofermentative LAB, is responsible for a shorter and harder gluten while lactic acid can gradually account for a more elastic gluten structure [138]. Regarding whole rye sourdough bread, optimal FQ drops into the range of 1,5–4,0 [208]. The attention towards the increasing of acetic acid content is also due to its antimicrobial effect as above reported.

Organic acids, alcohols, esters and carbonyls represent flavour compounds that strongly affect bread aroma [40,115,176,177,235]. In order to generate sufficient amounts of volatile compounds, the generation process needs multiple steps of about 12-24 h; when baker's yeast is used the fermentation is completed within a few hours [99]. Bacterial proteolysis during sourdough fermentation was shown to contribute much more to the development of typical sourdough flavours of baked breads when compared to breads produced from chemically acidified or yeasted doughs [98]. Several works have reported on the importance of amino acid conversion to flavour volatile compounds [30]. Furthermore, addition of amino acids (e.g. ornithine, leucine and phenylalanine) to doughs resulted in an enhanced conversion to flavour compounds [76,193]. At this proposal, proteolytic strains of sourdough LAB may affect amino acids level in doughs, but there is substantial evidence that the activation of cereal proteases is a major driving force for protein degradation in sourdoughs [137,217,218,219]. The key degradation reaction of amino acids during fermentation is the Ehrlich pathway leading to aldehydes or the corresponding alcohols, while during baking takes place the Strecker reaction which also leads to aldehydes, but also to the corresponding acids [40].

Dominating *Lactobacillus* spp. determine specific differences in the overall aroma profile of final bread. Ethyl acetate content was higher in firm wheat [43] and rye sourdoughs [140] fermented with heterofermentative LAB compared to sourdoughs fermented with homofermentative LAB. The same trend was observed for hexyl acetate in firm rye sourdough, while the content of aldehydes was higher in rye sourdoughs fermented with homofermentative cultures [140]. Regarding diacetyl, it has been found that its content was higher in sourdough manufactured with homofermentative compared to heterofermentative cultures [43,98,140]. However, due to evaporation during baking, the amounts of alcohols, esters and diacetyl in sourdough bread are much lower than in the corresponding sourdough [96,140].

Contribution of metabolites produced by yeasts in imparting taste and flavour to foods and, in particular to sourdough, is well documented [14,39,92,176,177]. Sourdoughs made with both LAB and yeasts resulted in more aroma compounds as compared to sourdoughs made from single starter based either on LAB (e.g. *Lb. brevis*) or yeast (e.g. *S. cerevisiae*) [146]. Guerzoni et al. [89] followed the exposure of *S. cerevisiae* and *Lb. sanfranciscensis* cells to oxidative, acid or osmotic sub-lethal stress observing a common or specific responses: γ-decalactone, 2(5H)-furanones and aldehydes were overproduced by LAB following oxidative stress; the acid stress induced both in yeasts and LAB, as well as in their co-cultures, a relevant accumulation of isovaleric and acetic acids and higher alcohols. Sourdoughs with a higher relative percentage of yeast-derived fermentation products are produced if a combination of *S. cerevisiae* with *Lb. sanfranciscensis* and *Lb. plantarum* is used during the sourdough fermentation process [85]. This observed increased production of aroma compounds in a mixed-starter process appears to be related to the proteolytic activity of LAB.

The arginine metabolism by *Lb. pontis* [217] and *Lb. sanfranciscensis* [46] has been demonstrated to have a defined impact on bread flavour. Gerez et al. [79] reported that gluten breakdown of lactobacilli and pediococci, besides reducing gluten-allergen compounds, also increased the basic amino acid concentration in broth cultures, mainly due to an increase in the amount of ornithine, which is considered to be the key flavour precursor in wheat bread [76], generating 2-acetyl-1-pyrrolin [217]. Phenylalanine metabolism, as studied in *Lb.*

plantarum and *Lb. sanfranciscensis*, besides producing phenyllactic acid (PLA) and its 4-hydroxy derivative (OH-PLA), whose antifungal activities are reported below (see paragraph 3.3.2), also generate flavour volatiles [234]. The growth of *Lb. sanfranciscensis* under osmotic stress generates a relevant accumulation of isovaleric acid, whose synthesis is associated with the branched chain amino acids metabolism [235].

Lipid metabolism by sourdough LAB may also partly contribute to the bread flavour formation. During growth in sourdough, *Lb. sanfranciscensis* and *Lb. reuteri* but not the facultative heterofermentative *Lb. sakei* convert the flavour-active compounds (E)-2-nonenal and (E,E)-2,4-decadienal to the corresponding alcohols with a much higher odour threshold [233]. Czerny and Schieberle [40] reported that fermentation of suspensions of wholemeal and white wheat flours LAB did not generate new odorants than those present in raw materials, but they demonstrated that many compounds, such as acetic acid or 3-methylbutanal, were increased, whereas aldehydes (formed from the degradation of unsaturated fatty acids) were decreased. The authors concluded that the odorant concentrations present before and after fermentation gave evidence that the main influence of the microorganisms on sourdough aroma is to either enhance or decrease specific volatiles already present in the flour. *Lb. sanfranciscensis* reduce the aldehydes to the corresponding unsaturated alcohols, because this reaction is coupled to the oxidation of NADH to NAD+, which enables this heterofermentative strain to produce additional ATP from glucose [233]. Therefore, careful selection of starter cultures is pivotal to modify the aroma of flour-based products [236].

Contribution to Dough Structure

The structure of bread, in particular crumb more than crust, is an important factor in consumer acceptance of bakery products. Cereals contain starch and non-starch polysaccharides, the latter composed of glucose (β-glucan), fructose (polyfructan), xylose and arabinose (arabinoxylan) [6]. Starch is partially digestible while some of the other polysaccharides, such as arabinoxylan, would represent dietary fibres. Wheat and rye flours contain, in addition to polyfructan, also nystose, kestose and other fructooligosaccharides of the inulin type [19] whose prebiotic effect is well documented [231]. Bacterial polysaccharides that are secreted into the environment are termed exopolysaccharides (EPS). Two classes of EPS from LAB can be distinguished, extracellularly synthesised homopolysaccharides (HoPS), composed of only one type of monosaccharide and are synthesised by extracellular glucan- and fructosyltransferases (glycosyltransferases) using sucrose as the glycosyl donor, and heteropolysaccharides (HePS) with (ir)regular repeating units. The repeating units of HePS are composed of 3–8 carbohydrate moieties that are synthesised intracellularly from sugar nucleotide precursors [51]. HePS application is currently limited to 'ropy' dairy starter cultures employed to improve the texture of yoghurt and other fermented milk products [129], while HoPS are generally applied to improve structural characteristics of baked goods. Sourdough lactobacilli have not been found to produce HePS while the secretion of alternan, dextran, fructan, inulin, levan, mutan and reutaran is known [221].

Lactobacillus sanfranciscensis has been well characterized for its contribution to the enhancement of polysaccharide content in sourdough due to the production of EPS

[120,121,122,221,222]. Formation of EPS is a well accepted characteristic of sourdough LAB, since this feature influences the viscosity of sourdough [236]. Furthermore, fructan produced by two strains of *Lb. sanfranciscensis* has been shown to stimulate bifidobacterial growth [42], thus acting as a prebiotic or in this case as a bifidogenic factor. Besides *Lb. sanfranciscensis*, cereal LAB isolates frequently produce oligo- and homopolysaccharides from sucrose, which can also improve the technological properties of gluten-free breads acting as hydrocolloids. Schwab et al. [196] employed *Lb. reuteri* LTH5448, which synthesize fructooligosaccharides and levan, and *Weissella cibaria* 10M, which synthesize isomaltooligosaccharides and dextran to ferment sorghum sourdough. The addition of sourdoughs fermented with oligo- and polysaccharide forming starter cultures can increase the content of prebiotic oligosaccharides in gluten-free breads.

LAB improve the quality of gluten-free products also because they delay their staling. *Lb. plantarum* and *Lb. sanfranciscensis* strains showed that the corresponding gluten-free sourdoughs became significantly softer during 24 h of fermentation compared to the chemically acidified control, although the positive effects were smaller than those found in wheat bread [158].

NUTRITIONAL AND HEALTHFUL PROPERTIES OF LACTIC ACID BACTERIA

Although optimal sensory characteristics represent the basis for any successful fermented food, consumers are particularly sensitive to its nutritional value and healthful aspects. Cereal-based products are utilized in human food throughout the world, thus their role in the nutritional pyramid is high.

Reduction of Anti-Nutritional Factors

Iron, potassium, magnesium and zinc are minerals provided by cereals which, however, also contain an anti-nutritional factors for humans [136,155] and animals, phytic acid or *myo*-inositol hexakiphosphate (IP_6) [172,228]. This characteristic is due to the central hexaphosphate ring, known to be the storage form of phosphorus in seeds, which being highly charged (six anionic groups) acts as a chelator of dietary minerals [164] preventing their absorption and thus reducing their bioavailability [135]. Phytic acid complexes basic amino acid group of proteins, resulting in a their subsequent reduced bioavailability [58]. Lower inositol phosphate derivatives showed health benefits in the protection of coronary heart, arteriosclerosis and neural tissue diseases [64].

Phytic acid may be hydrolysed by phytases, enzymes that play a pivotal role in upgrading the nutritional quality of phytate-rich foods and feeds [145]. Moreover, insoluble protein-phytate complexes are formed at low pH, as found in the stomach of monogastric animals, and protein digestibility is reduced. Dietary phytase supplementation has been shown to prevent the formation of such complexes or to aid in dissolving them faster, thus phytases may improve protein digestibility [113].

Phytases are widely present in plant materials such as wheat and rye flours, whose level depends on variety and crop year, but generally reported to be insufficient to significantly decrease the amount of phytic acid [37]. However, it has been recently published that a moderate decrease of pH by sourdough fermentation is sufficient to reduce phytate content of whole wheat flour through endogenous phytase activity [130]. Phytases are also produced by a multitude of microorganisms among which yeasts [228] and sourdough LAB [45,136,179].

Phytases show optimal activities at pH around 4.5 [66], hydrolyzing IP_6 into IP_5 and then into lower *myo*-inositol phosphate esters (IP_4-IP_1) which are less likely to bind minerals and form weaker mineral complexes [190]. Reale et al. [179] reported on the reduction of IP_6 of about 80-90 % in a dough prepared with a mixed starter culture (*Lb. plantarum*, *Lb. brevis*, *Lb. curvatus*) compared to the control dough, after 12 h fermentation; similar percentages of IP_6 level reduction were reported by De Angelis et al. [45] who showed that *Lb. sanfranciscensis* CB1, used to ferment a dough for 8 h, was able to decrease the Na-phytate concentration till 64-74 % compared to the unstarted sourdough. A study on the characterization of phytase activity of sourdough microorganisms showed that combining selected yeasts and LAB it is possible to reach high level of phytate biodegradation and the best combination was *S. cerevisiae*/*Lb. plantarum*/*Ln. mesenteroides* [22]. Moreover, some lactobacilli of intestinal origin (*Lb. reuteri* and *Lactobacillus salivarius*), showing high phytate degrading activity, were proven to have good properties for being used as starters in whole wheat breadmaking process [170]. These results emphasize microbial potential in improving the nutritional quality of cereal-based products.

Recently, also *Bifidobacterium* species (*B. catenulatum*, *B. longum* and *B. breve*) have been tested for phytate degrading activity to be employed as starters for whole wheat sourdough fermentation [169]. It resulted that the species *B. breve* and *B. longum* induced a high hydrolysis of phytic acid with simultaneous production of lower inositol phosphates. In addition, dough containing those strains had high pH and minor acidity than those containing a commercial starter (*Lb. plantarum*); thus, bifidobacteria could be used as breadmaking starters contributing to different acidification degrees [171].

Conversion of Toxic Compounds

Gluten content of flours commonly used in baking strongly limit the healthfulness of bread products for people affected by celiac sprue (CS), an autoimmune disease of the small intestinal mucosa. Ingestion of gluten causes self-perpetuating mucosal inflammation and subsequent loss of absorptive villi and hyperplasia of the crypts. Proteolytic enzymes of the endoluminal tract acting on prolamins of wheat (α-, β-, γ-, and ω-gliadin), rye (secalin) and barley (hordein), produce proline- and glycine-rich polypeptides that are responsible for the disease [204]. The list of proteins that liberate toxic peptides also includes high molecular weight glutenins [53]. Further proteolysis of such toxic peptides is made difficult by the position and abundance of proline residues [103]. For the above reasons, persons affected by CS cannot ingest gluten-containing products such as bread or pasta.

Some authors [203] found prolyl-endopeptidase produced by *Flavobacterium meningosepticum*, which is not interesting for breadmaking, which showed hydrolysing effect on a 33-mer peptide, one of the most potent peptides involved in triggering the disease, and the use of this endopeptidase has been proposed for an oral therapy for CS patients.

Lactobacilli have been shown to possess an outstanding potential in decreasing the CS-inducing effects of gluten. Di Cagno et al. [55] demonstrated active hydrolysis of various Pro-rich peptides, including the 33-mer peptide mentioned above, by some *Lactobacillus* species. This finding has been exploited to produce sourdoughs containing 30% of wheat flour (CS-inducing) and 70% of other (non-CS-inducing) [100] flours such as oat, buckwheat and millet, started with selected lactobacilli and fermented for 24 h. Following this, the mixed starter composed of *Lb. alimentarius, Lb. brevis, Lb. sanfranciscensis* and *Lb. hilgardii* was shown to almost completely hydrolize gliadin fractions and consequently the resulting bread was tolerated by CS patients [54]. With the same aim was also successfully tested VSL#3 probiotic preparation (VSL Pharmaceuticals, Gaithesburg, MD) (ca. 450 billion cells/sachet), containing *Streptococcus thermophilus, Lb. plantarum, Lb. acidophilus, Lb. casei, Lb. delbrueckii* spp. *bulgaricus, B. breve, B. longum* and *Bifidobacterium infantis* [47]. *Enterococcus faecalis* of Tunisian fermented wheat dough origin showed a high capacity to reduce the gliadin concentration from 45 g/kg to 18 g/kg [153]; similar conclusions were reported by Wieser et al. [242] during the selection of gluten degrading LAB. Gerez et al. [78] studied the functionality of LAB peptidase activities in the hydrolysis of gliadin-like fragments reporting that *Lb. plantarum* CRL 775 and *P. pentosaceus* CRL 792 displayed the highest tri- and di-peptidase activities. In that study, was also found that none of the LAB strains alone could hydrolyse 57–89 α-gliadin peptide, but the combination of the above two strains led to hydrolysis of 57% of that peptide in 8 h.

Sourdough based products show great potential in human diet. However, the complete suitability of sourdough for production of gluten free bread from wheat or rye flours is still under study while at present it has been proved that sourdough technology improves the sensory properties of gluten free bread [2]. Furthermore, selected *Lactobacillus* sourdough cultures may eliminate risks of contamination by gluten during processing of gluten free flour and enhance the nutritional properties of gluten free bread [57].

MONITORING OF SOURDOUGH MICROBIAL ECOLOGY

The outstanding importance of molecular analysis is reflected by its exploitation in monitoring (changes in) microbial populations in various ecosystems, such as complex food matrices, without any prior cultivation. The methods used at this proposal during food fermentation have recently been reviewed by Cocolin and Ercolini [25].

PCR-based methods, which target primarily a given gene (most commonly a ribosomal gene), are the most widely applied technologies, although a polyphasic approach, e.g., an approach employing both genotypic and phenotypic information [184], appears to be more helpful and realistic in describing microbial population dynamics [7]. Within culture-independent methods, developed to circumvent the limitations of conventional cultivation for analysis of microbial communities [232], PCR-denaturing gradient gel electrophoresis (PCR-DGGE) analysis is one of the most suitable and widely applied methods to study complex bacterial communities originating from various environments [161]. PCR-DGGE provides information about variation of PCR products of the same length but with different sequences upon differential mobility in an acrylamide gel matrix of increasing denaturant concentration; its application to food environments has been extensively documented [61,80,216]. PCR-

DGGE applied to template DNA directly extracted from a food matrix generates a specific profile of that product in that moment, given the conditions used. The fingerprint gives a picture of the microbiota of the product and can be taken into account as a specific trait of that food just like other biochemical, structural or sensorial properties [61].

With this in mind, PCR-DGGE is exploited to take a photograph of sourdough at any moment. PCR-DGGE has been successfully applied to the study of the LAB [30,106] and yeast [150] composition of fermented cereal-based products and to compare sourdough LAB communities subjected to different fermentation processes [151,152]. In particular, Meroth et al. [151] and Garofalo et al. [78] used the comparison of PCR-DGGE bands generated from unknown lactobacilli present in sourdough with an identification ladder constructed with 13 sourdough-associated *Lactobacillus* species. In this case, a second PCR-DGGE assay, using a different primer pair, was necessary in order to differentiate *Lactobacillus crispatus* from *Lb. acidophilus* and *Lb. sanfranciscensis* from *Lb. pontis*, as their migration distances were identical. Randazzo et al. [178] used this technique to study the microbial communities of traditional Sicilian sourdoughs. PCR-DGGE was useful for revealing the dominance of *Lb. sanfranciscensis* and *Lb. fermentum*, but did not discriminate between the closely-related species *Lactobacillus kimchii* and *Lb. alimentarius*. Similar results have been reported by Gatto and Torriani [77] who, in addition to the above species, also detected the presence of *Lactobacillus arizonensis-Lb. plantarum* as the prevailing species during fermentation. In addition, this technique was found useful to evaluate the taxonomic structure and stability of the bacterial community [192] and for tracking the dominant LAB [191] in Belgian sourdough ecosystems, as well as for assessing the dominant LAB during wheat sourdough propagation with nine strains of *Lb. sanfranciscensis* applied in single culture [205].

Another culture-independent, nucleic acid-based strategy representing a rapid and useful tool for monitoring bacteria during food fermentation is multiplex PCR, a methodology that permits, in the same reaction, simultaneous amplification of more than one locus [21]. Multiplex PCR assays targeting different genes have been successfully applied to identify lactobacilli from various environments [139,159,207,202,227,246]. With particular relevance to the sourdough ecosystem, Müller et al. [159] employed a multiplex PCR for simultaneously monitoring three *Lactobacillus* species, withouth prior cultivation, during laboratory-scale sourdough fermentation. Settanni et al. [202] developed a multiplex PCR assay utilizing primers that were based on sequences derived from the 16S rRNA-encoding DNA, the 16S-23S rRNA intergenic spacer region and its flanking 23S rRNA gene, allowing the *in situ* detection of a group of 16 sourdough-associated lactobacilli at group level. The same multiplex PCR has also been combined with a specific PCR-DGGE system to detect the above lactobacilli up to species level [201]. The systems were then used to identify common *Lactobacillus* species isolated from sourdough [230] to monitor changes in lactobacilli population composition (unpublished results).

CONCLUSION

Sourdough microflora are characterized by a large biodiversity in terms of yeast and LAB species and strains. Modern molecular techniques revealed a high number of new species, mainly belonging to the *Lactobacillus* genus, isolated from sourdough made with different types of flour both at artisanal and industrial level. So far, still very little is known about the interactions among sourdough microorganisms and between microorganisms and environmental factors; hence, further investigations of microbial dynamics, with particular regards to lactobacilli and their interactions with subdominant LAB species and other microorganism groups, in particular, dominant *Saccharomyces* are needed in order to better understand the role of the fermenting microbiota during sourdough fermentation. The actual trend of using defined starter cultures to initiate sourdough fermentation has lead to an intense study of sourdough microorganism properties, with a focus on the production of antimicrobial compounds to elongate the shelf-life of baked goods, the generation of flavour compounds to enhance consumers' attraction, and the ability to improve the nutritional quality and healthfulness of final products. Better-controlled sourdough-based processes, which are necessary to standardize the production of sourdough bread and to extend the use of this technology to products other than bread (e.g., healthful snacks, biscuits and several types of convenience foods) need a deeper investigation of microorganism properties. Culture-independent methods are potent and rapid tools that provide a complete microbial ecology imaging of sourdough ecosystems.

REFERENCES

[1] Aesen, I. M., Markussen, S., Møretrø, T., Katla, T., Axelsson, L., & Naterstad, K. (2003). Interactions of the bacteriocins sakacin P and nisin with food constituents. *International Journal of Food Microbiology*, 87, 35–43.

[2] Arendt, E. K., Ryan, L. A. M., & Dal Bello, F. (2007). Impact of sourdough on the texture of bread. *Food Microbiology*, 24, 165–174.

[3] Axelsson, L. (1998). Lactic Acid Bacteria: Classification and Phisiology. In S. Salminen, & A. von Wright (Eds.), *Lactic Acid Bacteria Microbiology and Functional Aspects* (pp. 1–72). New York, Marcel Dekker.

[4] Barber, S., Baguena, R., Benedito de Barber, C., & Martinez-Anaya, M. A. (1991). Evolution of biochemical and rheological characteristics and breadmaking quality during a multistage wheat sour dough process. *Lebensmittel Wissenschaft und Technologie*, 192, 46–52.

[5] Bassler, B. L. (2002). Small talk. Cell-to-cell communication in bacteria. *Cell*, 109, 421–424.

[6] Belitz, H. D., & Grosh, W. (1999). *Food Chemistry*. Berlin, Springer-Verlag.

[7] ben Omar, N., & Ampe, F. (2000). Microbial community dynamics during production of the Mexican fermented maize dough Pozol. *Applied and Environmental Microbiology*, 66, 3664–3673.

[8] Berg, R. W., Sandine, W. E., & Anderson, A. W. (1981). Identification of a growth stimulant for *Lactobacillus sanfrancisco*. *Applied and Environmental Microbiology*, 42, 786–788.

[9] Beuchat, L. R., & Ryu, J. H. (1997). Produce handling and processing practices. *Emerging Infectious Diseases*, 3, 459–465.

[10] Bianchi, F., Careri, M., Chiavaro, E., Musci, M., & Vittadini, E. (2008). Gas chromatographic–mass spectrometric characterisation of the Italian Protected Designation of Origin "Altamura" bread volatile profile. *Food Chemistry,* 110, 787–793.

[11] Blandino, A., Al-Aseeri, M. E., Pandiella, S. S., Cantero, & D., Webb, C. (2003). Cereal-based fermented foods and beverages. *Food Research International,* 36, 527–543.

[12] Blom, H., & Mörtvedt, C. (1991). Anti-microbial substances produced by food asociated micro-organisms. *Biochemical Society Transaction,* 19, 694–698.

[13] Böcker, G., Stolz, P. & Hammes, W. P. (1995). Neue Erkenntnisse zum Ökosystem Sauerteig und zur Physiologie des Sauerteig-Typischen Stämme *Lactobacillus sanfrancisco* und *Lactobacillus pontis. Getreide Mehl und Brot,* 49, 370–374.

[14] Brauman, A., Keleke, S., Malonga, M., Miambi, E., & Ampe, F. (1996). Microbiological and biochemical characterization of cassava retting, a traditional lactic acid fermentation for foofoo (cassava flour) production. *Applied and Environmental Microbiology,* 62, 2854–2858.

[15] Brock, M., & Buckel, W. (2004). On the mechanism of action of the antifungal agent propionate Propionyl-CoA inhibits glucose metabolism in *Aspergillus nidulans. European Journal of Biochemistry,* 271, 3227–3241.

[16] Bushuk, W. (2001). Rye production and uses worldwide. *Cereal Foods World,* 46, 70–73.

[17] Cabo, M. L., Braber, A. F., & Koenraad, P. M. (2002). Apparent antifungal activity of several lactic acid bacteria against *Penicillium disolor* is due to acetic acid in the medium. *Journal of Food Protection,* 65, 1309–1316.

[18] Cai, Y., Okada, H., Mori, H., Benno, Y., & Nakase, T. (1999). *Lactobacillus paralimentarius* sp. nov., isolated from sourdough. *International Journal of Systematic and Evolutionary Microbiology,* 49, 1451–1455.

[19] Campbell, J. M., Bauer, L. L., Fahey, G. C., Hogarth, A. J. C. L., Wolf, B. W., & Hunter, D. E. (1997). Selected fructooligosaccharide (1-kestose, nystose, and 1F-β-fructofuranosylnystose) composition of foods and feeds. *Journal of Agricultural and Food Chemistry,* 45, 3076–3082.

[20] Caplice, E., & Fitzgerald, G. F. (1999). Food fermentations: role of microorganisms in food production and preservetion. *International Journal of Food Microbiology,* 50, 131–149.

[21] Chamberlain, J. S., Gibbs, R. A., Ranier, J. E., Nguyen, P. N., & Caskey, C. T. (1988). Deletion screening of the Duchenne muscular dystrophy locus via multiplex DNA amplification. *Nucleic Acid Research,* 16, 11141–11156.

[22] Chaoui, A., Fais, M., & Belhcen, R. (2003). Effect of natural starters used for sourdough bread in Morocco on phytate biodegradation. *East Mediterranean Health Journal,* 9, 141–147.

[23] Cheng, J., Guffanti, A. A., & Krulwich, T. A. (1997). A two-gene ABC-type transport system that extrudes Na^+ in *Bacillus subtilis* is induced by ethanol or protonophore. *Molecular Microbiolgy,* 23, 1107–1120.

[24] Chung, T. C., Axelsson, L., Lindgren, S. E., & Dobrogosz, W. J. (1989). In vitro studies on reuterin synthesis by *Lactobacillus reuteri. Microbial Ecology in Health and Disease,* 2, 137–144.

[25] Cocolin, L., & Ercolini, D., (2008). *Molecular techniques in the microbial ecology of fermented foods*. New York, Springer.

[26] Corsetti, A., De Angelis, M., Dellaglio, F., Paparella, A., Fox, P. F., Settanni, L., & Gobbetti, M. (2003). Characterization of sourdough lactic acid bacteria based on genotypic and cell-wall protein analyses. *Journal of Applied Microbiology*, 94, 641–654.

[27] Corsetti, A., Gobbetti, M., Rossi, J., & Damiani, P. (1998). Antimould activity of sourdough lactic acid bacteria: identification of a mixture of organic acids produced by *Lactobacillus sanfrancisco* CB1. *Applied Microbiology and Biotechnology*, 50, 253–256.

[28] Corsetti, A., Gobbetti, M., & Smacchi, E. (1996). Antibacterial activity of sourdough lactic acid bacteria: isolation of a bacteriocin-like inhibitory substance from *Lactobacillus sanfranciscensis* C57. *Food Microbiology*, 13, 447–456.

[29] Corsetti, A., Lavermicocca, P., Morea, M., Baruzzi, F., Tosti, N., & Gobbetti, M. (2001). Phenotypic and molecular identification and clustering of lactic acid bacteria and yeasts from wheat (species *Triticum durum* and *Triticum aestivum*) sourdoughs of Southern Italy. *International Journal of Food Microbiology*, 64, 95–104.

[30] Corsetti, A., & Settanni, L. (2007). Lactobacilli in sourdough fermentation: a review. *Food Research International*, 40, 539–558.

[31] Corsetti, A., Settanni, L., Braga, T. M., de Fatima Silva Lopes, M., & Suzzi, G. (2008). An investigation on the bacteriocinogenic potential of lactic acid bacteria associated with wheat (*Triticum durum*) kernels and non-conventional flours. *LWT – Food Science and Technology*, 41, 1173–1182a.

[32] Corsetti, A., Settanni, L., Chaves-López, C., Felis, G. E., Mastrangelo, M., & Suzzi, G. (2007). A taxonomic survey of lactic acid bacteria isolated from wheat (*Triticum durum*) kernels and non-conventional flours. *Systematic and Applied Microbiology*, 30, 561–571a

[33] Corsetti, A., Settanni, L., Valmorri, S., Mastrangelo, M., & Suzzi, G. (2007). Identification of subdominant sourdough lactic acid bacteria and their evolution during laboratory-scale fermentations. *Food Microbiology*, 24, 592–600.b

[34] Corsetti, A., Settanni, L., Valmorri, S., & Suzzi, G. (2008). Effect of environmental and nutritional factors on the production of bacteriocin-like inhibitory substances (BLIS) by *Enterococcus faecium* strains. *Italian Journal of Food Science*, 20, 493–503b.

[35] Corsetti, A., Settanni, L., & Van Sinderen, D. (2004). Characterization of bacteriocin-like inhibitory substances (BLIS) from sourdough lactic acid bacteria and evaluation of their *in vitro* and *in situ* activity. *Journal of Applied Microbiology*, 96, 521–534.

[36] Corsetti, A., Settanni, L., Van Sinderen, D., Felis, G. E., Dellaglio, F., & Gobbetti, M. (2005). *Lactobacillus rossii* sp. nov. isolated from wheat sourdough. *International Journal of Systematic and Evolutionary Microbiology*, 55, 35–40.

[37] Cossa, J., Oloffs, K., Kluge, H., Drauschke, W., & Jeroch, H. (2000). Variabilities of total and phytate phosphorus contents as well as phytase activity in wheat. *Tropenlandwirt*, 101, 119–126.

[38] Cotter, P. D., Hill, C., & Ross, R. P. (2005). Bacteriocins: developing innate immunity for food. *Nature Reviews*, 3, 777–788.

[39] Czerny, M., Brandt, M. J., Hammes, W. P., & Schieberle, P. (2004). *Aroma potential of yeasts in wheat doughs*. Getreide, Mehl and Brot, 57, 340–344.

[40] Czerny, M., & Schieberle, P. (2002). Important aroma compounds in freshly ground wholemeal and white wheat flour-identification and quantitative changes during fermentation. *Journal of Agricultural and Food Chemistry*, 50, 6835–6840.

[41] Dal Bello, F., Clarke, C. I., Ryan, L. A. M., Ulmer, H., Schober, T. J., Ström, K., Sjögren, J., van Sinderen, D., Schnürer, J., Arendt, E. K., (2007). Improvement of the quality and shelf life of wheat bread by fermentation with the antifungal strain *Lactobacillus plantarum* FST 1.7. *Journal of Cereal Science*, 45, 309–318.

[42] Dal Bello, F., Walter, J., Hertel, C., & Hammes, W. P. (2001). *In vitro* study of prebiotic properties of levan-type exopolysaccharides from lactobacilli and non-digestible carbohydrates using denaturing gradient gel electrophoresis. *Systematic and Applied Microbiology*, 24, 232–237.

[43] Damiani, P., Gobbetti, M., Cossignani, L., Corsetti, A., Simonetti, M. S., & Rossi, J. (1996). The sourdough microflora. Characterization of hetero- and homofermentative lactic acid bacteria, yeasts and their interactions on the basis of the volatile compounds produced. *Lebensmittel-Wissenschaft und-Technologie*, 30, 63–70.

[44] De Angelis, M., Di Cagno, R., Gallo, G., Curci, M., Siragusa, S., Crecchio, C., Parente, E., & Gobbetti, M. (2007). Molecular and functional characterization of *Lactobacillus sanfranciscensis* strains isolated from sourdoughs. *International Journal of Food Microbiology*, 114, 69–82.

[45] De Angelis, M., Gallo, G., Corbo, M. R., McSweeney, P. L. H., Faccia, M., Giovine, M., & Gobbetti, M. (2003) Phytase activity in sourdough lactic acid bacteria: purification and characterization of a phytase from *Lactobacillus sanfranciscensis* CB1. *International Journal of Food Microbiology*, 87, 259–270.

[46] De Angelis, M., Mariotti, L., Rossi, J., Servili, M., Fox, P. F., Rollán, G. C., & Gobbetti, M. (2002). Arginine catabolism by sourdough lactic acid bacteria: purification and characterization of the arginine deiminase pathway enzymes from *Lactobacillus sanfranciscensis* CB1. *Applied and Environmental Microbiology*, 68, 6193–6201.

[47] De Angelis, M., Rizzello, C. G., Fasano, A., Clemente, M. G., De Simone, C., Silano, M., De Vincenzi, M., Losito, I., & Gobbetti, M. (2006b). VSL#3 proiotic preparation has the capacity to hydrolize gliadin polypeptides responsible for celiac sprue. *Biochimica and Biophysica Acta*, 1762, 80–93.

[48] De Angelis, M., Siragusa, S., Berloco, M. G., Caputo, L., Settanni, L., Alfonsi, G., Amerio, M., Grandi, A., Ragni, A., & Gobbetti, M. (2006a). Selection of potential probiotic lactobacilli from pig eaces to be used as additives in pelletted feedings. *Research in Microbiology*, 157, 792–801.

[49] De Vuyst, L., Avonts, L., Neysens, P., Hoste, B., Vancanneyt, M., Swings, J., & Callewaert, R. (2004). Applicability and performance of the bacteriocin producer *Lactobacillus amylovorus* DCE 471 in type II cereal fermentations. International *Journal of Food Microbiology*, 90, 93–106.

[50] De Vuyst, L., Callewaert, R., & Pot, B. (1996). Characterization of the antagonistic activity of *Lactobacillus amylovorus* DCE 471 and large scale isolation of its bacteriocin amylovorin L471. *Systematic and Applied Microbiology*, 19, 9–20.

[51] De Vuyst, L., De Vin, F., Vaningelgem, F., & Degeest, B. (2001). Recent developments in the biosynthesis and applications of heteropolysaccharides from lactic acid bacteria. *International Dairy Journal*, 11, 687–707.

[52] De Vuyst, L., & Neysen, P. (2005). The sourdough microflora: biodiversity and metabolic interactions. *Trends in Food Science and Technology*, 16, 43–56.

[53] Dewar, D. H., Amato., M., Ellis, H. J., Pollock, E. L., Gonzalez-Cinca, N., Wieser, H., & Ciclitira, P. J. (2006). The toxicity of high molecula weight glutenin subunits of wheat to patients with coeliac disease. *European Journal of Gastroenterology and Hepatology*, 18, 483–491.

[54] Di Cagno, R., De Angelis, M., Auricchio, S., Greco, L., Clarke, C., De Vincenzi, M., Giovannini, C., D'Archivio, M., Landolfo, F., Parrilli, G., Minervini, F., Arendt, E., & Gobbetti, M. (2004). Sourdough bread made from wheat and nontoxic flours and started with selected lactobacilli is tolerated in celiac sprue patients. *Applied and Environmental Microbiology*, 70, 1088–1096.

[55] Di Cagno, R., De Angelis, M., Lavermicocca, P., De Vincenzi, M., Giovannini, C., Faccia, M., & Gobbetti, M. (2002). Proteolysis by sourdough lactic acid bacteria: effects on wheat flour protein fractions and gliadin peptides involved in human cereal intolerance. *Applied and Environmental Microbiology*, 68, 623–633.

[56] Di Cagno, R., De Angelis, M., Limitone, A., Minervini, F., Simonetti, M. C., Buchin, S., Gobbetti, M. (2007). Cell-cell communication in sourdough lactic acid bacteria: *A proteomic study in Lactobacillus sanfranciscensis CB1*. Proteomics, 7, 2430–2446.

[57] Di Cagno, R., Rizzello, C. G., De Angelis, M., Cassone, A., Giuliani, G., Benedusi, A., Limitone, A., Surico, R. F., Gobbetti, M. (2008). Use of selected sourdough strains of *Lactobacillus* for removing gluten and enhancing the nutritional properties of gluten-free bread. *Journal of Food Protection*, 71, 1491–1495.

[58] Dvorakova, J. (1998). Phytase, sources, preparation and exploitation. *Folia Microbiologica*, 43, 323–338.

[59] Ehrmann, M. A., & Vogel, R. F. (1998). Maltose metabolism of *Lactobacillus sanfranciscensis*: cloning and heterologous expression of the key enzymes, maltose phosphorylase and phosphoglucomutase. *FEMS Microbiology Letters*, 169, 81–86.

[60] Ehrmann, M. A., & Vogel, R. F. (2005). Molecular taxonomy and genetics of sourdough lactic acid bacteria. *Trends in Food Science and Technology*, 16, 31–42.

[61] Ercolini, D. (2004). PCR-DGGE fingerprinting: novel strategies for detection of microbes in food. *Journal of Microbiological Methods*, 56, 297–314.

[62] Farkas, J. (2001). Physical methods for food preservation. In M. P. Doyle, L. R. Beuchat, & T. J. Mntville (Eds.), *Food microbiology: Fundamentals and frontiers* (pp. 67–592). Washington, ASM Press.

[63] Federle, M. J., & Bassler, B.L., 2003. Interspecies communication in bacteria. *Journal of Clinical Investigation*, 112, 1291–1299.

[64] Fisher, S. K., Novak, J. E., & Agranoff, B. W. (2002). Inositol and higher inositol phosphates in neural tissues: homeostasis, metabolism, and functional significance. *Journal of Neurochemistry*, 82, 736–754.

[65] Foschino, R., Terraneo, R., Mora, D., & Galli, A. (1999). Microbial characterization of sourdoughs for sweet baked products. *Italian Journal of Food Science*, 11, 19–28.

[66] Fretzdorff, B., & Brümmer, J. –M. (1992). Reduction of phytic acid during breadmaking of whole-meal breads. *Cereal Chemistry*, 69, 266–270.

[67] Fuqua, W. C., Winans, S. C., & Greenberg, E. P. (1994). Quorum sensing in bacteria: the *LuxR-LuxI* family of cell density-responsive transcriptional regulators. *Journal of Bacteriology*, 176, 269–275.

[68] Fuqua, C., Winans, S. C., & Greenberg, E. P. (1996). Census and consensus in bacterial ecosystems: the LuxR–LuxI family of quorum-sensing transcriptional regulators. *Annual Review of Microbiology*, 50, 727–751.

[69] Galli, A., & Franzetti, L. (1987). Ricerche sulla composizione microbiologica della farina di grano tenero". *Annals of Microbiology*, 37, 73–80.

[70] Gancheva, A., Pot, B., Vanhonacker, K., Hoste, B., Kersters, K. (1999). A polyphasic approach towards the identification of strains belonging to *Lactobacillus acidophilus* and related species. *Systematic and Applied Microbiology*, 22, 573–585.

[71] Gänzle, M. (1998). *Useful properties of lactobacilli for application as protective cultures in food*. PhD Thesis, Universität Hohenheim, Germany.

[72] Gänzle, M. G., Vermeulen, N, & Vogel, R. F. (2007). Carbohydrate, peptide and lipid metabolism of lactic acid bacteria in sourdough. *Food Microbiology*, 24, 128–138.

[73] Gänzle, M. G., & Vogel, R. F. (2003). Contribution of reutericyclin to the stable persistence of *Lactobacillus reuteri* in an industrial sourdough fermentation. *International Journal of Food Microbiology*, 80, 31–45.

[74] Gänzle, M., Weber, S., & Hammes, W. (1999). Effect of ecological factors on the inhibitory spectrum and activity of bacteriocins. *International Journal of Food Microbiology*, 46, 207–217.

[75] Garofalo, C., Silvestri, G., Aquilanti, L., & Clementi, F. (2008). PCR-DGGE analysis of lactic acid bacteria and yeast dynamics during the production processes of three varieties of Panettone. *Journal of Applied Microbiology*, 105, 243–254.

[76] Gassenmeier, K., & Schieberle, P. (1995). Potent aromatic compounds in the crumb of wheat bread (French-type)-influence of pre-ferments and studies on the formation of key odorants during dough processing. *Zeitschrift fur Lebensmittel Untersuchung und Forschung*, 201, 241–248.

[77] Gatto, V., & Torriani, S. (2004). Microbial population changes during sourdough fermentation monitored by DGGE analysis of 16S and 26S rRNA gene fragments. *Annals of Microbiology*, 54, 31–42.

[78] Gerez, C. L., De Valdez, G. F., & Rollán, G. C. (2008). Functionality of lactic acid bacteria peptidase activities in the hydrolysis of gliadin-like fragments. *Letters in Applied Microbiology*, 47, 427–432.

[79] Gerez, C. L., Rollán, G. C., & De Valdez, G. F. (2006) Gluten breakdown by lactobacilli and pediococci strains isolated from sourdough. *Letters in Applied Microbiology*, 42, 459–464.

[80] Giraffa, G. (2004). Studying the dynamics of microbial populations during food fermentation. *FEMS Microbiology Review*, 28, 251–260.

[81] Gobbetti, M. (1998). The sourdough microflora: interactions of lactic acid bacteria and yeasts. *Trends in Food Science and Technology*, 9, 267–274.

[82] Gobbetti, M., & Corsetti, A. (1997). *Lactobacillus sanfrancisco* a key sourdough lactic acid bacterium: a review. *Food Microbiology*, 14, 175–187.

[83] Gobbetti, M., Corsetti, A., & Rossi, J. (1994b). The sourdough microflora. Interactions between lactic acid bacteria and yeasts: metabolism of carbohydrates. *Applied Microbiology and Biotechnology*, 41, 456–460.

[84] Gobbetti, M., Corsetti, A., & Rossi, J. (1994c). The sourdough microflora. Interactions between lactic acid bacteria and yeasts: metabolism of amino acids. *World Journal of Microbiology and Biotechnology*, 10, 275–279.

[85] Gobbetti, M., Corsetti, A., & Rossi, J. (1995b). Interaction between lactic acid bacteria and yeasts in sour-dough using a rheofermentometer. *World Journal of Microbiology and Biotechnology*, 11, 625–630.

[86] Gobbetti, M., Corsetti, A., Rossi, J., La Rosa, F., & De Vincenzi, S. (1994a). Identification and clustering of lactic acid bacteria and yeasts from wheat sourdoughs of central Italy. *Italian Journal of Food Science*, 6, 85–94.

[87] Gobbetti, M., Simonetti, M. S., Corsetti, A., Santinelli, F., Rossi, J., & Damiani, P. (1995a) Volatile compound and organic acid production by mixed wheat sour dough starters: influence of fermentation parameters and dynamics during baking. *Food Microbiology*, 12, 497–507.

[88] Gourama, H. (1997). Inhibition of growth and mycotoxin production of *Penicillium* by *Lactobacillus* species. *Lebensmittel-Wissenschaft und-Technologie*, 30, 279–283.

[89] Guerzoni, M. E., Vernocchi, P., Ndagijimana, M., Gianotti, A., & Lanciotti, R. (2007). Generation of aroma compounds in sourdough: Effects of stress exposure and lactobacilli–yeasts interactions. *Food Microbiology*, 24, 139–148.

[90] Gullo, M., Romano, A. D., Pulvirenti, A., & Giudici, P. (2002). *Candida humilis*-dominant species in sourdoughs for the production of durum wheat bran flour bread. *International Journal of Food Microbiology*, 80, 55–59.

[91] Guynot, M. E., Marín, S., Sanchis, V., & Ramos, A. J. (2005). An attempt to optimize potassium sorbate use to preserve low pH (4.5–5.5) intermediate moisture bakery products by modelling *Eurotium* spp., *Aspergillus* spp. and *Penicillium corylophilum* growth. *International Journal of Food Microbiology*, 101, 169–177.

[92] Halm, M., Lillie, A., Sørensen, A. K., & Jakobsen, M. (1993). Microbiological and aromatic characteristics of fermented maize doughs for kenkey production in Ghana. *International Journal of Food Microbiology*, 19, 135–143.

[93] Hammes, W. P., Brandt, M. J., Francis, K. L., Rosenheim, M., Seitter, F. H., & and Vogelmann, S. (2005). Microbial ecology of cereal fermentations. *Trends in Food Science and Technology*, 16, 4–11.

[94] Hammes, W. P., & Gänzle, M. G. (1998). Sourdough breads and related products. In B. J. B. Wood (Ed.), *Microbiology of Fermented Foods* (Vol. 1, pp. 199–216). London, Blackie Academic and Professional.

[95] Hammes, W. P., & Vogel, R. F. (1995). The genus *Lactobacillus*. In B. J. B. Wood, & W. H. Holzapfel (Eds.), *The Genera of Lactic Acid Bacteria* (pp. 19–54). London, Blackie Academic and Professional.

[96] Hansen, A., & Hansen, B. (1996). Flavour of sourdough wheat bread crumb. Zeitschrift fur L*ebensmittel Untersuchung und Forschung*, 202, 244–249.

[97] Hansen, A., Lund, B., & Lewis, M. J. (1989a). Flavour production and acidification of sour doughs in relation to starter culture and fermentation temperature. *Lebensmittel Wissenschaft und Technologie*, 22, 145–149.

[98] Hansen, A., Lund, B., & Lewis, M. J. (1989b). Flavour of sourdough rye bread crumb. *Lebensmittel Wissenschaft und Technologie*, 22, 141–144.

[99] Hansen, A., & Schieberle, P. (2005). Generation of aroma compounds during sourdough fermentation: applied and fundamental aspects. *Trends in Food Science and Technology,* 16, 85–94.

[100] Hardman, C. M., & Fry, L. (1997). Absence of toxicity of oats in patients with dermatitis herpetiformis. *New England Journal of Medicine*, 337, 1884–1887.

[101] Hartnett, D. J., Vaughan, A., & van Sinderen, D. (2002). Antimicrobial-producing lactic acid bacteria isolated from raw barley and sorghum. *Journal of the Institute of Brewing*, 108, 169–177.

[102] Hassan, Y. I., & Bullerman, L. (2008). Antifungal activity of *Lactobacillus paracasei* subsp. *tolerans* against *Fusarium proliferatum* and *Fusarium graminearum* in a liquid culture setting. *Journal of Food Protection*, 71, 2213–2216.

[103] Hausch, F., Shan, L., Santiago, N. A., Gray, G. M., & Khosla, C. (2003). Intestinal digestive resistance of immunodominant gliadin peptides. *American Journal of Physiology*, 283, 996–1003.

[104] Herrera-Saldana, R. E., Huber, J. T., & Poore, M. H. (1990). Dry matter, crude protein, and starch degradability of five cereal grains. Journal of Dairy Science, 73, 2386–2393.

[105] Hogan, D. A. (2006). Talking to themselves: autoregulation and quorum sensing in fungi. *Eukaryotic Cell*, 5, 613–619.

[106] Iacumin, L., Cecchini, F., Manzano, M., Osualdini, M., Boscolo, D., Orlic, S., & Comi, G. (2009). Description of the microflora of sourdoughs by culture-dependent and culture-independent methods. *Food Microbiology*, 26, 128–135.

[107] Infantes, M., & Tourneur, C. (1991). Etude de la flore lactique de levains naturels de panification provenant de différentes régions françaises. *Sciences des Aliments*, 11, 527–545.

[108] INSOR (Istituto Nazionale di Sociologia Rurale) (2000). *Atlante dei prodotti tipici: il pane*. F. Angeli (Ed.). Roma, Agra RAI-ERI.

[109] Jenson, I. (1998). Bread and baker's yeast. In B. J. B. Wood (Ed.), *Microbiology of Fermented Foods* (pp. 172–198). London, Blackie Academic and Professional.

[110] Jung, D.-S., Bodyfelt, F., & Daeschel, M. (1992). Influence of fat and emulsifiers on the efficiency of nisin in inhibiting *Listeria monocytogenes* in fluid milk. *Journal of Dairy Science* 75, 387–393.

[111] Kazanskaya, L. N., Afanasyeva, O. V., & Patt, V. A. (1983). Microflora of rye sours and some specific features of its accumulation in bread baking plants of the USSR. In J. Holas, & F. Kratochvil (Eds.), *Developments in Food Science. Progress in Cereal Chemistry and Technology* (pp. 759–763). London, Elsevier.

[112] Keshri, G., Voysey, P., & Magan, N. (2002). Early detection of spoilage moulds in bread using volatile production patterns and quantitative enzyme assays. *Journal of Applied Microbiology*, 92, 165–172.

[113] Kies, A. K., De Jonge, L. H., Kemme, P. A., & Jongbloed, A. W. (2006). Interaction between protein, phytate, and microbial phytase. In vitro studies. *Journal of Agricultural and Food Chemistry*, 54, 1753–1758.

[114] Kirchhoff, E., & Schieberle, P. (2001). Determination of key aroma compounds in the crumb of a three-stage sourdough rye bread by stable isotope dilution assays and sensory studies. *Journal of Agricultural and Food Chemistry*, 49, 4304–4311.

[115] Kirchhoff, E., & Schieberle, P. (2002). Quantitation of odor-active compounds in rye flour and rye sourdough using a stable isotope dilution assay. *Journal of Agricultural and Food Chemistry*, 50, 5378–5311.

[116] Klaenhammer, T. R. (1988). Bacteriocins of lactic acid bacteria. *Biochimie*, 70, 337–349.

[117] Klaenhammer, T. R. (1993). Genetics of bacteriocins produced by lactic acid bacteria. *FEMS Microbiology Review*, 12, 39–85.

[118] Kleerebezem, M., Boekhorst, J., van Kranenburg, R., Molenaar, D., Kuipers, O. P., Leer, R., Tarchini, R., Peters, S. A., Sandbrink, H. M., Fiers, M. W., Stiekema, W., Lankhorst, R. M., Bron, P. A., Hoffer, S. M., Groot, M. N., Kerkhoven, R., de Vries, M., Ursing, B., de Vos, W. M., & Siezen, R. J. (2004). Complete genome sequence of *Lactobacillus plantarum WCFS1*. *Proceedings of the National Academy of Science of the United States of America*, 100, 1990–1995.

[119] Kline, L., & Sugihara, T. F. (1971). Microorganisms of the San Francisco sour dough bread process. II. Isolation and characterization of undescribed bacterial species responsible for the souring activity. *Applied Microbiology*, 21, 459–465.

[120] Korakli, M., Gänzle, M. G., & Vogel, R. F. (2002). Metabolism by bifidobacteria and lactic acid bacteria of polysaccharides from wheat and rye and exopolysaccharides produced by *Lactobacillus sanfranciscensis*. *Journal of Applied Microbiology*, 92, 958–965.

[121] Korakli, M., Pavlovic, M., Gänzle, M. G., & Vogel, R. F. (2003). Exopolysaccharide and kestose production by *Lactobacillus sanfranciscensis* LTH2590. *Applied and Environmental Microbiology*, 69, 2073–2079.

[122] Korakli, M., Rossman, A, Gänzle, M. G., & Vogel, R. F. (2001). Sucrose metabolism and exopolysaccharide production in wheat and rye sourdough by *Lactobacillus sanfranciscensis*. *Journal of Agricultural and Food Chemistry*, 49, 5194–5200.

[123] Kramer, J. M., & Gilbert, R. J. (1989). *Bacillus cereus* and other *Bacillus* sp. In M. P. Doyle (Ed.), *Foodborne bacterial pathogens* (pp. 22–70). New York, Marcel Dekker.

[124] Krasner, R. I. (2002). *The microbial challenge. Human-microbe interactions.* Washington, ASM Press.

[125] Kumar, S., Tamura, K., & Nei, M. (2004). MEGA3: Integrated software for molecular evolutionary genetics analysis and sequence alignment, *Briefings in Bioinformatic*, 5, 150–163.

[126] Larsen, A. G., Vogensen, F. K., & Josephsen, J. (1993). Antimicrobial activity of lactic acid bacteria isolated from sour doughs: purification and characterization of bavaricin A, a bacteriocin produced by *Lactobacillus bavaricus* MI401. *Journal of Applied Bacteriology*, 75, 113–122.

[127] Lavermicocca, P., Valerio, F., Evidente, A., Lazzaroni, S., Corsetti, A., & Gobbetti, M. (2000). Purification and characterization of novel antifungal compounds by sourdough *Lactobacillus plantarum* 21B. *Applied and Environmental Microbiology*, 66, 4084–4090.

[128] Lavermicocca, P., Valerio, F., & Visconti, A. (2003). Antifungal activity of phenyl-lactic acid against molds isolated from bakery products. *Applied and Environmental Microbiology*, 69, 634–640.

[129] Laws, A., & Marshall, V. M. (2001). The relevance of exopolysaccharides to the rheological properties in milk fermented with ropy strains of lactic acid bacteria. *International Dairy Journal*, 11, 709–721.

[130] Leenhardt, F., Levrat-Verny, M. A., Chanliaud, E., & Remesy, C. (2005). Moderate decrease of pH by sourdough fermentation is sufficient to reduce phytate content of whole wheat flour through endogenous phytase activity. *Journal of Agricultural and Food Chemistry*, 53, 98–102.

[131] Legan, J. D. (1993). Mould spoilage of bread: the problem and some solutions. *International Biodeterioration and Biodegradation*, 32, 33–53.

[132] Leroy, F., & De Vuyst, L. (1999). The presence of salt and a curing agent reduces bacteriocin production of *Lactobacillus sakei* CTC 494, a potential starter culture for sausage fermentation. *Applied and Environmental Microbiology*, 65, 5350–5356.

[133] Leroy, F., De Winter, T., Foulquié Moreno, M. R., & De Vuyst, L. (2007). The bacteriocin producer *Lactobacillus amylovorus* DCE 471 is a competitive starter culture for type II sourdough fermentations. *Journal of the Science of Food and Agriculture*, 87, 1726–1736.

[134] Li, H. X., Cao, Y. S., Liu, X. H., & Fu, L. L. (2007). Differentiation of *Lactobacillus* spp. isolated from Chinese sourdoughs by AFLP and classical methods. *Annals of Microbiology*, 57, 687–690.

[135] Lönnerdal, B. (2002). Phytic acid-trace element (Zn, Cu, Mn) interactions. *International Journal of Food Science and Technology*, 37, 749–758.

[136] Lopez, H. V., Ouvry, A., Bervas, E., Guy, C., Messager, A., Demigne, C., & Remesy, C. (2000). Strains of lactic acid bacteria isolated from sour doughs degrade phytic acid and improve calcium and magnesium solubility from whole wheat flour. *Journal of Agricultural and Food Chemistry*, 48, 2281–2285.

[137] Loponen, J., Mikola, M., Katina, K., Sontag-Strohm, T., & Salovaara, H. (2004). Degradation of HMW-glutenins during wheat sourdough fermentations. *Cereal Chemistry*, 81, 87–93.

[138] Lorenz, K. (1983). Sourdough processes. Methodology and biochemistry. *Baker's Digest*, 55, 85–91.

[139] Lucchini, F., Kmet, V., Cesena, C., Coppi, L., Bottazzi, V., & Morelli., L. (1998). Specific detection of a probiotic *Lactobacillus* strain in faecal samples by using multiplex PCR. *FEMS Microbiology Letters*, 158, 273–278.

[140] Lund, B., Hansen, A., & Lewis, M. J. (1989). *The influence of dough yield on acidification and production of volatiles in sour doughs*. Lebensmittel Wissenschaft und Technologie, 22, 150–153.

[141] Makarova, K., Slesarev, A., Wolf, Y., Sorokin, A., Mirkin, B., Koonin, E., Pavolv, A., Pavlova, N., Karamychev, V., Polouchine, N., Shakhova, V., Grigoriev, I., Lou, Y., Rohksar, D., Lucas, S., Huang, K., Goldstein, D. M., Hawkins, T., Plengvidhya, V., Welker, D., Hughes, J., Goh, Y., Benson, A., Baldwin, K., Lee, J. -H., Diaz-Muniz, I., Dosti, B., Smeianov, V., Wechter, W., Barabote, R., Lorca, G., Altermann, E., Barrangou, R., Ganesan, B., Xie, Y., Rawsthorne, H., Tamir, D., Parker, C., Breidt, F., Broadbent, J., Hutkins, R., O'Sulllivan, D., Steele, J., Unlu, G., Saier, M., Klaenhammer, T., Richardson, P., Kozyavkin, S., Weimer, B., & Mills, D. (2006). *Comparative genomics of the lactic acid bacteria. Proceedings of the National Academy of Science of the United States of America*, 103, 15611–15616.

[142] Magnusson, J., & Schnürer, J. (2001). *Lactobacillus coryniformis* subsp. *coryniformis* strain Si3 produces a broad-spectrum proteinaceous antifungal compound. *Applied and Environmental Microbiology*, 67, 1–5.

[143] Magnusson, J., Ström, K., Roos, S., Sjögren, J., & Schnürer, J. (2003). Broad and complex antifungal activity among environmental isolates of lactic acid bacteria. *FEMS Microbiology Letters*, 219, 129–135.

[144] Martinez-Anaya, M. A. (1996). Enzymes and bread flavour. Journal of Agricultural and *Food Chemistry*, 44, 2469–2480.

[145] Martinez, C., Ros, G., Periago, M. J., Lopez, G., Ortuno, J., & Rincon, F. (1996). Phytic acid in human nutrition. *Food Science and Technology International*, 2, 201–209.

[146] Meignen, B., Onno, B., Gélinas, P., Infantes, M., Guilois, S., & Cahagnier, B. (2001). Optimization of sourdough fermentation with *Lactobacillus brevis* and baker's yeast. *Food Microbiology*, 18, 239–245.

[147] Mensah, P., Tomkins, A. M., Drasar, B. S., & Harrison, T. J. (1991). Antimicrobial effect of fermented Ghanaian maize dough. *Journal of Applied Bacteriology*, 70, 203–210.

[148] Menteş, Ö., Ercan, R., & Akçelik, M. (2005). *Determination of the antibacterial activities of Lactobacillus strains isolated from sourdoughs produced in Turkey*. Gida 3, 155–164.

[149] Menteş, Ö., Ercan, R., & Akçelik, M. (2007). Inhibitor activities of two *Lactobacillus* strains, isolated from sourdough, against rope-forming *Bacillus* strains. *Food Control* 18, 359–363.

[150] Meroth, C. B., Hammes, W. P., & Hertel, C. (2003). Identification and population dynamics of yeasts in sourdough fermentation processes by PCR-denaturing gradient gel electrophoresis. *Applied and Environmental Microbiology*, 69, 7453–7461.b

[151] Meroth, C. B., Hammes, W. P., & and Hertel, C. (2004). Characterisation of the microbiota of rice sourdoughs and description of *Lactobacillus spicheri* sp. nov.. *Systematic and Applied Microbiology*, 27, 151–159.

[152] Meroth, C. B., Walter, J., Hertel, C., Brandt, M., & Hammes, W. P. (2003a). Monitoring the bacterial population dynamics in sourdough fermentation processes by using PCR-denaturing gradient gel electrophoresis. *Applied and Environmental Microbiology*, 69, 475–482.

[153] M'hir, S., Aldric, J. –M., El-Mejdoub, T., Destain, J., Mejri, M., Hamdi, M., & Thonart, P. (2008). Proteolytic breakdown of gliadin by *Enterococcus faecalis* isolated from Tunisian fermented dough. *World Journal of Microbiology and Biotechnology*, 24, 2775–2781.

[154] Miller, M. B., & Bassler, B. L. (2001). Quorum sensing in bacteria. *Annual Review in Microbiology*, 55, 165–199.

[155] Minihane, A. M., & Rimbach, G. (2002). Iron absorption and the iron binding and antioxidant properties of phytic acid. *International Journal of Food Science and Technology*, 37, 741–748.

[156] Mohamed, L., Zakaria, M., Ali, A., Senhaji, W., Mohamed, O., Mohamed, E., Hassan, B. E. L., & Mohamed, J. (2007). Optimization of growth and extracellular gluco-amylase production by Candida famata isolate. *African Journal of Biotechnology*, 22, 2590–2595.

[157] Moll, G. N., Konigs, W. N., & Driessen, A. J. M. (1999). Bacteriocins: mechanism of membrane insertion and pore formation. *Antonie van Leeuwenhoek*, 76, 185–198.

[158] Moore, M. M., Juga, B., Schober, T. J., & Arendt, E. K. (2007). Effect of lactic acid bacteria on properties of gluten-free sourdoughs, batters, and quality and ultrastructure of gluten-free bread. *Cereal Chemistry*, 84, 357–364.

[159] Müller, M. R. A., Ehrmann, M. A., & Vogel., R. F. (2000b). Multiplex PCR for the detection of *Lactobacillus pontis* and two related species in a sourdough fermentation. *Applied and Environmental Microbiology*, 66, 2113–2116.

[160] Müller, M. R. A., Wolfrum, G., Stolz, P., Ehrmann, M. A., & Vogel, R. F. (2001). Monitoring the growth of *Lactobacillus* species during a rye flour fermentation. *Food Microbiology*, 18, 217–227.

[161] Muyzer, G. (1999). DGGE/TGGE a method for identifying genes fromnatural ecosystems. *Current Opinions in Microbiology*, 2, 317–322.

[162] Narbutaite, V., Fernandez, A., Horn, N., Juodeikiene, G., & Narbad, A. (2008). Influence of baking enzymes on antimicrobial activity of five bacteriocin-like inhibitory substances produced by lactic acid bacteria isolated from Lithuanian sourdoughs. *Letters in Applied Microbiology*, 47, 555–560.

[163] Nes, I. F., Diep, D. B., Håvarstein, L. S., Brurberg, M. B., Eijsink, V., & Holo, H. (1996). Biosynthesis of bacteriocins in lactic acid bacteria. *Antonie van Leeuwenhoek*, 70, 113–128.

[164] Nolan, K.B., & Duffin, P.A. (1987). Effect of phytate on mineral bioavailability. In vitro studies on Mg^{++}, Cu^{++}, Ca^{++}, Fe^{++} and Zn^{++} solubilities in the presence of phytate. *Journal of the Science of Food and Agriculture*, 40, 79–85.

[165] Nout, M. J. R., & Creemers-Molenaar, T. (1987). Microbiological properties of some wheatmeal sourdough starters. *Chemie Mikrobiologie Technologie der Lebensmittel*, 10, 162–167.

[166] Okkers, D. J., Dicks, L. M. T., Silvester, M., Joubert, J. J., & Odendaal, H. J. (1999). Characterization of pentocin TV35b, a bacteriocin-like peptide isolated from *Lactobacillus pentosus* with a fungistatic effect on *Candida albicans*. *Journal of Applied Microbiology*, 87, 726–734.

[167] Ottogalli, G., Galli, A., & Foschino, R. (1996). Italian bakery products obtained with sour dough: characterization of the typical microflora. *Advances in Food Science*, 18, 131–144.

[168] Oura, E., Suomalainen, H., & Viskari, R. (1982). Breadmaking. In A. H. Rose (Ed.), *Economic microbiology* (pp. 512). London, Academic Press.

[169] Palacios, M. C., Haros, M., Rosell, C. M., & Sanz, Y. (2008). Selection of phytate-degrading human bifidobacteria and application in whole wheat dough fermentation. *Food Microbiology*, 25, 169–176. b

[170] Palacios, M. C., Haros, M., Sanz, Y., & Rosell, C. M. (2008). Selection of lactic acid bacteria with high phytate degrading activity for application in whole wheat breadmaking. LWT – *Food Science and Technology*, 41, 82–92. a

[171] Palacios, M. C., Haros, M., Sanz, Y., & Rosell, C. M. (2008). Phytate degradation by *Bifidobacterium* on whole wheat fermentation. *European Food Research Technology*, 226, 825–831.c

[172] Pandey, A., Szakacs, G., Soccol, C. R., Rodriguez-Leon, J. A., & Soccol, V. T. (2001). Production, purification and properties of microbial phytases. *Bioresearch and Technology*, 77, 203–214.

[173] Parsek, M. R., & Greenberg, E. P. (2005). Sociomicrobiology: the connections between quorum sensing and biofilms. *Trends in Microbiology*, 13, 27–33.

[174] Pepe, O., Blaiotta, G., Anastasio, M., Moschetti, G., Ercolini, D., & Villani F. (2004). Technological and molecular diversity of *Lactobacillus plantarum* strains isolated from naturally fermented sourdoughs. *Systematic and Applied Microbiology* 27, 443–453.

[175] Pepe, O., Blaiotta G., Moschetti G., Greco, T., & Villani F. (2003). Rope-producing strains of *Bacillus* spp. from wheat bread and strategy for their control by lactic acid bacteria. *Applied and Environmental Microbiology*, 69, 2321–2329.

[176] Plessas, S., Bekatorou, A., Gallanagh, J., Nigam, P., Koutinas, A. A., & Psarianos, C. (2008). Evolution of aroma volatiles during storage of sourdough breads made by mixed cultures of *Kluyveromyces marxianus* and *Lactobacillus delbrueckii* ssp. *bulgaricus* or *Lactobacillus helveticus*. *Food Chemistry*, 107, 883–889.

[177] Plessas, S., Fisher, A., Koureta, K., Psarianos, C., Nigam, P., & Koutinas, A. A. (2008). Application of *Kluyveromyces marxianus*, *Lactobacillus delbrueckii* ssp. *bulgaricus* and *L. helveticus* for sourdough bread making. *Food Chemistry*, 106, 985–990.

[178] Randazzo, C. L., Heilig, H., Restuccia, C., Giudici, P., & Caggia, C. (2005). Bacterial populations in traditional sourdough evaluated by molecular methods. *Journal of Applied Microbiology*, 99, 251–258.

[179] Reale, A., Mannina, L., Tremonte, P., Sobolev, A. P., Succi, M., Sorrentino, E., & Coppola, R. (2004). Phytate degradation by lactic acid bacteria and yeasts during the wholemeal dough fermentation: a ^{31}P NMR study. *Journal of agricultural and food chemistry*, 52, 6300–6305.

[180] Rocha, J. M., & Malcata, F. X. (1999). On the microbiological profile of traditional Portuguese sourdough, *Journal of Food Protection*, 62, 1416–1429.

[181] Röcken, W. (1996). Applied aspects of sourdough fermentation. *Advances in Food Science*, 18, 212–218.

[182] Rosenkvist, H., & Hansen, A. (1995). Contamination profiles and characterisation of *Bacillus* species in wheat bread and raw materials for bread production. *International Journal of Food Microbiology*, 26, 353–363.

[183] Rosenquist, H., & Hansen, A. (1998). The antimicrobial effect of organic acids, sour dough and nisin against *Bacillus subtilis* and *B. licheniformis* isolated from wheat bread. *Journal of Applied Microbiology*, 85, 621–631.

[184] Rosselló-Mora, R., & Amann, R. (2001). The species concept for prokaryotes. *FEMS Microbiology Review*, 25, 39–67.

[185] Rossi, J. (1996). The yeasts in sourdough. *Advances in Food Science*, 18, 201–211.

[186] Ryan, L. A. M., Dal Bello, F., & Arendt, E. K. (2008). The use of sourdough fermented by antifungal LAB to reduce the amount of calcium propionate in bread. *International Journal of Food Microbiology*, 125, 274–278.

[187] Salovaara, H. (1998). Lactic acid bacteria in cereal-based products. In S. Salminen, & A. von Wright (Eds.), *Lactic acid bacteria microbiology and functional aspects* (pp. 115–138). New York, Marcel Dekker.

[188] Salovaara, H., & Katunpää, H. (1984). An approach to the classification of lactobacilli isolated from Finnish sour rye dough ferments. *Acta Alimentaria Polonica*, 10, 231–239.

[189] Salovaara, H., & Valjakka, T. (1987). The effect of fermentation temperature, flour type, and starter on the properties of sour wheat bread. *International Journal of Food Science*, 22, 591–597.

[190] Sandberg, A. S., Brune, M., Carlsson, N. G., Hallberg, L., Skoglund, E., & Rossander-Hulthén, L. (1999). Inositol phosphates with different numbers of phosphate groups influence iron absorption in humans. *American Journal of Clinical Nutrition*, 70, 240–246.

[191] Scheirlinck, I., Van der Meulen, R., De Vuyst, L., Vandamme, P., & Huys, G. (2009). Molecular source tracking of predominant lactic acid bacteria in traditional Belgian sourdoughs and their production environments. *Journal of Applied Microbiology*, 106, 1081–1092.

[192] Scheirlinck, I., Van der Meulen, R., Van Schoor, A., Vancanneyt, M., De Vuyst, L. P. V., & Huys, G. (2008) Taxonomic structure and stability of the bacterial community in Belgian sourdough ecosystems assessed by culturing and population fingerprinting. *Applied and Environmental Microbiology*, 74, 2414-2423.

[193] Schieberle, P. (1990). The role of free amino acids present in yeast as precursors of the odorants 2-acetyl-L-pyrroline and 2-acetyltetrahydropyridine in wheat bread crust. *Zeitschrift fur Lebensmittel Untersuchung und Forschung*, 191, 206–209.

[194] Schieberle, P. (1996). Intense aroma compounds-Useful tools to monitor the influence of processing and storage on bread aroma. *Advances in Food Science,* 18, 237–244.

[195] Schnürer, J., & Magnusson, J. (2005). Antifungal lactic acid bacteria as biopreservatives. *Trends in Food Science and Technology,* 16, 70–78.

[196] Schwab, C., Mastrangelo, M., Corsetti, A., & Gänzle, M. (2008). Formation of oligosaccharides and polysaccharides by *Lactobacillus reuteri* LTH5448 and *Weissella cibaria* 10M in sorghum sourdoughs. *Cereal Chemistry*, 85, 679–684.

[197] Settanni, L., Conterno, L., & Cavazza, A. (2008). Il microbiota lattico degli impasti acidi e delle materie prime usate in panificazione. *Tecnica Molitoria,* 59, 631–642.

[198] Settanni, L., & Corsetti, A. (2008). Application of bacteriocins in vegetable food biopreservation. *International Journal of Food Microbiology,* 121, 123–138.

[199] Settanni, L., Massitti, O., Van Sinderen, D., & Corsetti, A. (2005b). *In situ* activity of a bacteriocin-producing *Lactococcus lactis* strain. Influence on the interactions between lactic acid bacteria during sourdough fermentation. *Journal of Applied Microbiology*, 99, 670–681.

[200] Settanni, L., Valmorri, S., Suzzi, G., & Corsetti, A. (2008). The role of environmental factors and medium composition on bacteriocin-like inhibitory substances (BLIS) production by *Enterococcus mundtii* strains. *Food Microbiology,* 25, 722–728.

[201] Settanni, L., Valmorri, S., Van Sinderen, D., Suzzi, G., Paparella, A., & Corsetti, A. (2006). Combination of multiplex PCR and PCR-Denaturing Gradient Gel Electrophoresis for monitoring common sourdough-associated *Lactobacillus* species. *Applied and Environmental Microbiology*, 72, 3739–3796.

[202] Settanni, L., Van Sinderen, D., Rossi, J., & Corsetti, A. (2005a). Rapid differentiation and *in situ* detection of 16 sourdough *Lactobacillus* species by multiplex PCR. *Applied and Environmental Microbiology*, 71, 3049–3059.

[203] Shan, L., Molberg, O., Parrot, I., Hausch, F., Filiz, F., Gray, G. M., Sollid, L. M., & Khosla, C. (2002). Structural basis for gluten intolerance in celiac sprue. *Science*, 297, 2275–2279.

[204] Silano, M., & De Vincenzi, M. (1999). Bioactive antinutritional peptides derived from cereal prolamins: a review. *Nahrung*, 43, 175–184.

[205] Siragusa, S., Di Cagno, R., Ercolini, D., Minervini, F., Gobbetti, M., & De Angelis, M. (2009). Taxonomic structure and monitoring of the dominant population of lactic acid bacteria during wheat flour sourdough type I propagation using *Lactobacillus sanfranciscensis* starters. *Applied and Environmental Microbiology,* 75, 1099–1109.

[206] Sjögren, J., Magnusson, J., Broberg, A., Schnürer, J., & Kenne, L. (2003). Antifungal 3-hydroxy fatty acids from *Lactobacillus plantarum* MiLAB 14. *Applied and Environmental Microbiology*, 69, 7554–7557.

[207] Song, Y. -L., Kato, N., Liu, C. -X., Matsumiya, Y., Kato, H., & Watanabe, K. (2000). Rapid identification of 11 human intestinal *Lactobacillus* species by multiplex PCR assays using group- and species-specific primers derived from the 16S-23S rRNA intergenic spacer region and its flanking 23S rRNA. *FEMS Microbiology Letters*, 187, 167–173.

[208] Spicher, G. (1983). Baked goods. In H. J. Rehm, & and G. Reed (Eds.), *Biotechnology* (pp. 1–80). Weinheim, Verlag Chemie.

[209] Spicher, G., & Lönner, C. (1985). Die mikroflora des sauerteiges. XXI. Mitteilung: die in sauerteigen schwedischer bäckereien vorkommenden lactobacillen. *Zeitschrift für Lebensmittel Untersuchung und Forschung*, 181, 9–13.

[210] Spicher, G., Rabe, E., Sommer, R., & Stephan, H. (1982). XV. Communication: on the behaviour of heterofermentative sourdough bacteria and yeasts in mixed culture. *Zeitschrift für Lebensmittel Untersuchung und Forschung*, 174, 222–227.

[211] Stolz, P. (1999). Mikrobiologie des Sauerteiges. In G. Spicher, & H. Stephan (Eds.), *Handbuch Sauerteig: Biologie, Biochemie, Technologie* (pp. 35–60). Hamburg, Behr's Verlag.

[212] Sturme, M. H., Kleerebezem, M., Nakayama, J., Akkermans, A. D., Vaughan, E. E., & de Vos, W. M. (2002). Cell to cell communication by autoinducing peptides in gram-positive bacteria. *Antonie Van Leeuwenhoek*, 81, 233–243.

[213] Sugihara, T. F., Kline, L., & McGgready, L. B. (1970). Nature of San Francisco sour dough in French bread process. II. *Microbiological aspects. Baker's Digest*, 44, 51–57.

[214] Suihko, M. L., & Mäkinen, V. (1984). Tolerance of acetate, propionate and sorbate by *Saccharomyces cerevisiae* and *Torulopsis holmii*. Food Microbiology, 1, 105–110.

[215] Tagg, J. R. (1991). *Bacterial BLIS*. ASM News, 57, 611.

[216] Temmerman, R., Huys, G., & Swings, J. (2004). Identification of lactic acid bacteria: culture-dependent and culture-independent methods. *Trends in Food Science and Technology*, 15, 348–359.

[217] Thiele, C., Gänzle, M. G., & Vogel, R. F. (2002). Contribution of sourdough lactobacilli, yeast, and cereal enzymes to the generation of amino acids in dough relevant for bread flavor. *Cereal Chemistry*, 79, 45–51.

[218] Thiele, C., Gänzle, M. G., & Vogel, R. F. (2003). Fluorescence labeling of wheat proteins for determination of gluten hydrolysis and depolymerization during dough processing and sourdough fermentation. *Journal of Agricultural and Food Chemistry*, 51, 2745–2752.

[219] Thiele, C., Grassl., S., & Gänzle, M. G. (2004). Gluten hydrolysis and depolymerization during sourdough fermentation. *Journal of Agricultural and Food Chemistry*, 52, 1307–1314.

[220] Thompson, J.D., Gibson, T.J., Plewniak, F., Jeanmougin, F., & Higgins, D.G. (1997). The CLUSTAL-X windows interface: flexible strategies for multiple sequence alignment aided by quality analysis tools. *Nucleic Acids Research*, 25, 4876–4882.

[221] Tieking, M., & Gänzle, M. G. (2005). Exopolysaccharides from cereal-associated lactobacilli. *Trends in Food Science and Technology*, 16, 79–84.

[222] Tieking, M., Korakli, M., Ehrmann, M. A., Gänzle, M. G., & Vogel, R. F. (2003). In situ production of exopolysaccharides during sourdough fermentation by cereal and intestinal isolates of lactic acid bacteria. *Applied and Environmental Microbiology*, 69, 945–952.

[223] Toderi, G. (1989). Frumento (*Triticum* spp.). In R. Baldoni, & L. Giardini (Eds.), *Coltivazioni erbacee*, (pp. 15–89). Bologna, Pàtron Editore.

[224] Todorov, S., Onno, B., Sorokine, O., Chobert, J. M., Ivanova, I., & Dousset, X. (1999). Detection and characterization of a novel antibacterial substance produced by *Lactobacillus plantarum* ST 31 isolated from sourdough. *International Journal of Food Microbiology*, 48, 167–177.

[225] Topping, D. (2007). Cereal complex carbohydrates and their contribution to human health. *Journal of Cereal Science*, 46, 220–229.

[226] Torriani, S., Clementi, F., Vancanneyt, M., Hoste, B., Dellaglio, F., & Kersters, K. (2001). Differentiation of *Lactobacillus plantarum*, *L. pentosus* and *L. paraplantarum* species by RAPD-PCR and AFLP. *Systematic and Applied Microbiology*, 24, 554–560.b

[227] Torriani, S., Felis, G. E., & Dellaglio, F. (2001). Differentiation of *Lactobacillus plantarum*, *L. pentosus*, and *L. paraplantarum* by *recA* gene sequence analysis and multiplex PCR assay with *recA* gene-derived primers. *Applied and Environmental Microbiology*, 67, 3450–3454.a

[228] Turk, M., Sandberg, A. S., Carlsson, N. G., & Andlid, T. (2000). Inositol hexaphosphate hydrolysis by baker's yeast. Capacity, kinetics, and degradation products. *Journal of Agricultural and Food Chemistry*, 48, 100–104.

[229] Twomey, D., Ross, R. P., Ryan, M., Meaney, B., & Hill, C. (2002). Lantibiotics produced by lactic acid bacteria: structure, function and applications. *Antonie van Leeuwenhoek*, 82, 165–185.

[230] Valmorri, S., Settanni, L., Suzzi, G., Gardini, F., Vernocchi, P., & Corsetti, A. (2006). Application of a novel polyphasic approach to study the lactobacilli composition of sourdoughs from the Abruzzo region (central Italy). *Letters in Applied Microbiology*, 43, 343–349.

[231] Van Loo, J., Cummings, J., Delzenne, N., Englyst, H., Franck, A., Hopkins, M., Kok, N., Macfarlane, G., Newton, D., Quigley, M., Roberfroid, M., van Vliet, T., & van den Heuvel, E. (1999). Functional food properties of Non-digestible oligosaccharides: a consensus report from the ENDO project (DGXII AIRII-CT-1095). *British Journal of Nutrition*, 81, 121–132.

[232] Vaughan, E. E., de Vries, M. C., Zoetendal, E. G., Ben-Amor, K., Akkermans, A. D. L., & de Vos, W. M. (2002). The intestinal LABs. *Antonie Van Leeuwenhoek*, 82, 341–352.

[233] Vermeulen, N., Czerny, M., Gänzle, M. G., Schieberle, P., & Vogel, R.F. (2007). Reduction of (E)-2-nonenal and (E;E)-2,4-decadienal during sourdough fermentation. *Journal of Cereal Science*, 45, 78–87.

[234] Vermeulen, N., Gänzle, M. G., & Vogel, R. F. (2006). Influence of peptide supply and cosubstrates on phenylalanine metabolism of *Lactobacillus sanfranciscensis* DSM20451^T and *Lactobacillus plantarum* TMW1.468. *Journal of Agricultural and Food Chemistry*, 54, 3832–3839.

[235] Vernocchi, P., Ndagijimana, M., Serrazanetti, D., Gianotti, A., Vallicelli, M., Guerzoni, M. E. (2008). Influence of starch addition and dough microstructure on fermentation aroma production by yeasts and lactobacilli. *Food Chemistry*, 108, 1217–1225.

[236] Vogel, R. F., Ehrmann, M. A., & Gänzle, M. G. (2002). Development and potential of starter lactobacilli resulting from exploration of the sourdough ecosystem. *Antonie van Leeuwenhoek*, 81, 631–638.

[237] Vollmar, A., & Meuser, F. (1992). Influence of starter cultures consisting of lactic acid bacteria and yeasts on the performance of a continuous sourdough fermenter. *Cereal Chemistry*, 69, 20–27.

[238] von Stokar, W. (1956). Der ursprung unseres hausbrotes. Brot und Gebäck, 10, 11–16.

[239] Voysey, P. A. (1990). Rope and pH of commercial bread. *FMBRA Bullettin,* 1, 13–20.

[240] Voysey, P. A., & Hammond, J. C. (1993). Reduced-additive breadmaking technology. In J. Smith (Ed.), *Technology of Reduced-Additive Foods* (pp. 80–94). London, Blackie Academic and Professional.

[241] Währen, M. (1985). Die entwicklungsstationen vom korn zum brot in 5. und 4. jahrtausend. Neueste untersuchungsergebnissen von ausgrabungsfunden. *Getreide Mehl und Brot*, 39, 373–379.

[242] Wieser, H., Vermeulen, N., Gaertner, F., Vogel, R. F. (2008). Effects of different *Lactobacillus* and *Enterococcus* strains and chemical acidification regarding degradation of gluten proteins during sourdough fermentation. *European Food Research Technology*, 226, 1495–1502.

[243] Williams, P. (2007). Quorum sensing, communication and cross-kingdom signalling in the bacterial world. *Microbiology*, 153, 3923–3938.

[244] Wlodarczyk, M. (1985). Associated cultures of lactic acid bacteria and yeasts in the industrial production of bread. *Acta Alimentaria Polonica*, 11, 345–359.

[245] Wlodarczyk, M., Jezynska, B., & Warzywoda, A. (1993). Associated cultures of lactic acid bacteria and yeast as starters in gluten-free baker's leavens. *Polish Journal of Food and Nutrition Science*, 43, 83–91.

[246] Yost, C. K., & Nattress, F. M. (2000). The use of multiplex PCR reactions to characterize populations of lactic acid bacteria associated with meat spoilage. *Letters in Applied Microbiology*, 31, 129–133.

[247] Zapparoli, G., Torriani, S., & Dellaglio, F. (1998). Differentiation of *Lactobacillus sanfranciscensis* strains by randomly amplified polymorphic DNA and pulsed-field gel electrophoresis. *FEMS Microbiology Letters*, 166, 324–332.

Chapter 59

AN OVERVIEW OF THE BIODIVERSITY AND BIOGEOGRAPHY OF TERRESTRIAL GREEN ALGAE

Fabio Rindi[1], Haj A. Allali[2], Daryl W. Lam[2]
and Juan M. López-Bautista[2]

[1]Martin Ryan Institute, National University of Ireland, Galway.
[2]Department of Biological Sciences, The University of Alabama, Tuscaloosa, AL, USA

ABSTRACT

Microscopic green algae are among the most widespread microorganisms occurring in terrestrial environments. For more than two centuries, generalizations on the diversity and biogeography of these organisms have been based entirely on morphological species concepts. However, ultrastructural and molecular data produced in the last 30 years have revealed a scenario in substantial contrast with morphological classifications. It has become clear that these organisms have been affected by an extreme morphological convergence, which has restricted their morphology to a narrow range, not indicative of their great genetic diversity. Their habit is very simple and uniform, usually referable to a few types (unicellular, uniseriate filamentous, sarcinoid colony) and offers very few characters useful for taxonomic and systematic purposes. These factors make the identification of terrestrial green algae and a correct interpretation of their biogeography very difficult. "Flagship" taxa with easily recognizable habit are the ones for which the best generalizations are possible. Examples of such taxa include the order Trentepohliales (for the highest diversity occurs in humid tropical regions of central-south America and south-eastern Asia) and members of the order Prasiolales (which are typically associated with polar and cold-temperate regions). In consideration of the recent developments, it is clear that many basic concepts about the biogeography of terrestrial green algae will have to be reconsidered critically. A deep understanding of this topic will require considerable work on many aspects of the biology of these organisms (systematics, distribution, dispersal, physiology), in which species circumscriptions based on molecular data will be a mandatory requirement.

INTRODUCTION

The bulk of organisms performing oxygenic photosynthesis in terrestrial environments represents a very heterogeneous and evolutionarily diverse assemblage. Although vascular plants are the most conspicuous and morphologically complex among these organisms, this group includes also a great number of microscopic algae. Species of microalgae occur virtually in every type of terrestrial habitat, including the most extreme, such as walls of urban buildings (Rindi, 2007a), biotic crusts in hot deserts (Lewis & Flechtner, 2002; Flechtner, 2007), Antarctic snow (Broady, 1996) and air at 2,000 m height (Sharma *et al.*, 2007). They are small in size (mostly 5-50 µm) and characterized by a simple morphology, often unicellular; for this reason, most species are not observable as individual specimens and become visually detectable only when producing large populations, typically in the form of black, green, red or brown patches (Figure 1).

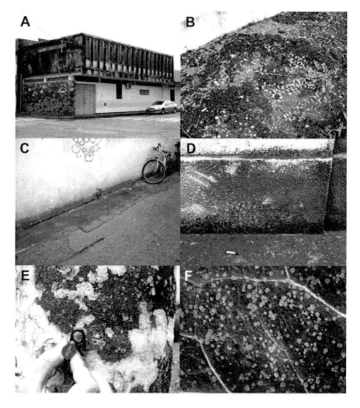

Figure 1. Examples of assemblages of terrestrial algae. Figure 1A: building covered by black stripes formed by cyanobacteria (Cayenne, French Guiana). Figure 1B: population of the cyanobacterium *Nostoc commune* growing on soil, mixed with plants. Figure 1C: the green alga *Klebsormidium flaccidum* (Klebsormidiophyceae) growing at the base of a urban wall (Pisa, Italy). Figure 1D: a population of the green alga *Rosenvingiella radicans* (Trebouxiophyceae) froming a green belt at the base of a wall (Galway, Ireland). Figure 1E: *Trentepohlia rigidula* (Ulvophyceae) forming red patches on the bark of a tree (Barro Colorado Island, Panama). Figure 1F: specimens of *Phycopeltis arundinacea* (Ulvophyceae) producing orange dishes on the surface of a leaf of ivy (Ashford Castle, Ireland).

Terrestrial microalgae belong primarily to three different evolutionary lineages: the blue-green algae (or Cyanobacteria), the green algae (Chlorophyta and Streptophyta) and the diatoms (Bacillariophyceae, Ochrophyta). From a numerical point of view, the green algae and the blue-green algae include the majority of the species described. It is generally reported that blue-green algae represent the main component of the terrestrial microalgal vegetation in tropical regions, whereas green algae represent the dominant element in temperate regions (Fritsch, 1907; John, 1988). In general, however, patterns of geographical distribution in terrestrial algae are poorly understood, mainly because the diversity of these organisms is itself poorly understood. Here, we discuss this topic for terrestrial green algae, reviewing the information available and pointing at directions for future work.

TERRESTRIAL GREEN ALGAE: DIVERSITY AND SYSTEMATICS

Green algae are photosynthetic eukaryotes bearing double membrane-bound plastids containing chlorophyll *a* and *b*, accessory pigments found in embryophytes (beta carotene and xanthophylls) and a unique stellate structure linking nine pairs of microtubules in the flagellar base (Lewis & McCourt, 2004). They are one of the most diverse groups of eukaryotes and include morphological forms ranging from flagellated unicells, coccoids, branched or unbranched filaments to multinucleated macrophytes and taxa with parenchymatic tissues (Pröschold & Leliaert, 2007). Dating based on molecular data suggests that the most recent common ancestor of all green algae may have existed 1100-1200 million years ago (Yoon *et al*., 2004). In general, however, reconstruction of the evolutionary history of these organisms is a very speculative matter, because of the limited fossil record. Phylogenetic calibrations have been possible only for a few groups in which the presence of a calcified cell wall has produced a good fossil record, such as the marine orders Bryopsidales and Dasycladales (Verbruggen *et al*., 2009).

From a systematic point of view, the green algae have been traditionally a difficult group. In the past the classification of these organisms has undergone several major rearrangements, due mostly to the fact that different criteria, based on different types of evidence (morphological, ultrastructural, molecular), have been adopted at different stages (Lewis & McCourt, 2004; Pröschold & Leliaert, 2007). Important advancements have been made in the last 30 years, in which new types of data have complemented the bulk of morphological information produced in the two previous centuries. The development of electron microscopy in the 70s of the last century revealed many important ultrastructural characters, which have since proved to be key features for the classification of the green algae. Mattox & Stewart (1984) proposed a new classification in which the ultrastructure of the basal body in flagellated cells and the cytokinesis during the mitosis were considered the two most important features. In the last 20 years, the advent of molecular systematics has represented a revolution for the classification of many algal groups, and green algae have been among the most affected. DNA sequence data have generally confirmed conclusions based on ultrastructure (Lewis *et al*., 1992; Friedl & Zeltner, 1994; Friedl, 1995) and have shown that in the green algae morphological characters are often not good indicators of phylogenetic relationships. It is now established that the green algae belong to a well-supported monophyletic group, the Viridiplantae, in which the land plants are also included (Figure 2).

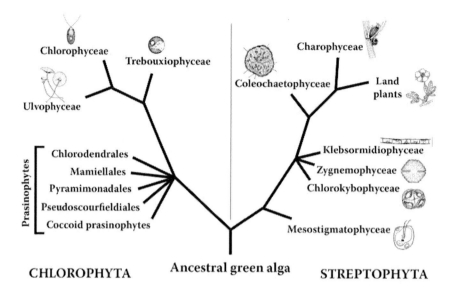

Figure 2. Schematic diagram showing the phylogenetic relationships among the major lineages of green algae based on DNA sequence data.

The Viridiplantae are subdivided in two major lineages, the Chlorophyta and the Streptophyta (Bremer, 1985). The Chlorophyta include green algae in which the cells normally divide with production of a phycoplast during the cytokinesis. This lineage consists of three groups that form well-supported clades and are recognized at the level of class (Chlorophyceae, Trebouxiophyceae and Ulvophyceae), and a non-monophyletic group called Prasinophytes.

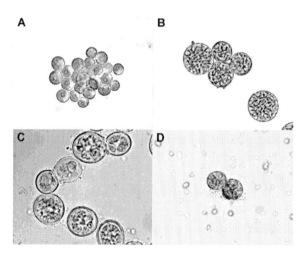

Figure 3. Examples of terrestrial green algae with unicellular morphology. Figure 3A: *Chlorella* cfr. *vulgaris* (cells 7-8 μm in diameter). Figure 3B: *Dictyochloropsis* sp. (cells 10-20 μm in diameter). Figure 3C: *Spongiochrysis hawaiiensis* (cells 10-15 μm in diameter). Figure 3D: *Trebouxia* sp. (cells 9-10 μm in diameter).

The Prasinophytes represent an unnatural agglomeration of unicellular organisms whose classification is still in need of rearrangement, viewed as the form of cell most similar to the ancestral green alga (believed to be a flagellate aquatic unicell, Lewis & McCourt, 2004). The Steptophyta include the land plants and groups of green algae that produce a phragmoplast in the cytokinesis. These consist of several groups (Charophyceae, Chlorokybophyceae, Coleochaetophyceae, Klebsormidiophyceae, Mesostigmatophyceae and Zygnemophyceae) which form well-supported clades, but whose relative relationships are not yet well-resolved.

About 800 species of green microalgae are known to occur in terrestrial environments. Soil is the type of habitat from which most of them have been reported (Metting, 1981); however, the range of habitats occupied by these organisms is extremely wide and includes natural rocks (Golubic, 1967), biotic crusts in deserts (Flechtner, 2007; Lewis, 2007), concrete walls (Rindi & Guiry, 2002), woodwork (John, 1988), iron rails (Schlichting, 1975), tree bark (Barkman, 1958), leaves and fruits (López-Bautista *et al.*, 2002) and hair of animals (Lewin & Robinson, 1979). Their morphology is very simple and typically referable to three different habits: 1) single cells (Figure 3; examples: *Chlorella*, *Chlorococcum*, *Stichococcus*, *Trebouxia*); 2) "sarcinoid" habit, i.e. packet-like colonies formed by a limited number of cells (Figure 4; examples: *Apatococcus*, *Desmococcus*, *Chlorosarcina*, *Chlorokybus*); 3) uniseriate filaments, either branched or not (Figure 5; examples: *Klebsormidium*, *Printzina*, *Rosenvingiella*, *Trentepohlia*). The study of terrestrial green algae has a long history, which dates back to the beginning of systematics.

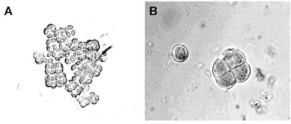

Figure 4. Examples of terrestrial green algae with sarcinoid morphology. Figure 4A: *Chlorosarcinopsis* sp. (cells 5-7 µm wide). Figure 4B: *Desmococcus olivaceus* (cells µm 4-7 wide).

Some species are among the earliest-known organisms, having been described by Linnaeus (1753, 1759) or other early authors (e.g., Agardh, 1824; Kützing, 1843, 1849; Nägeli, 1849). The systematics and taxonomy of these organisms have been investigated for two centuries and a half; nevertheless, the most important advancements have been made in the last 20 years, when molecular data have become available. It is now clear that the terrestrial green algae are a polyphyletic assemblage originated through many independent colonisations of the land by different lineages of aquatic algae, both marine and freshwater (Lewis & McCourt, 2004; Lewis, 2007; Cardon *et al.*, 2008). The capability to adapt to terrestrial conditions is taxonomically widespread, as indicated by the fact that terrestrial members occur in at least six different groups (Chlorophyceae, Trebouxiophyceae, Ulvophyceae, Chloroky-bophyceae, Klebsormidiophyceae and Zygnemophyceae; Figure 2); most of the species currently known belong to the class Trebouxiophyceae (Lewis & McCourt, 2004). The molecular data have also revealed that the genetic diversity of terrestrial green algae is much higher than suggested by their simple morphology. Different species, sometimes belonging to unrelated evolutionary lineages, have converged into an almost identical morphology, and several taxa described on morphological basis have been shown to be polyphyletic complexes of cryptic species. A typical example of this situation is the widespread genus *Chlorella*, which has been shown to be an artificial agglomeration of algae belonging to two different classes, the Trebouxiophyceae and the Chlorophyceae (Huss *et al.*, 1999); this genus is being gradually dismantled and subdivided into several separate genera (Krienitz *et al.*, 2004; Zhang *et al.*, 2008).

Other common genera which have been revealed as polyphyletic or paraphyletic include *Chlorococcum* (Buchheim *et al.*, 2002), *Klebsormidium*

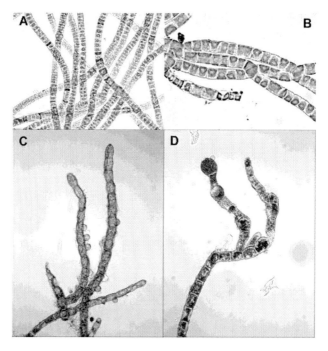

Figure 5. Examples of terrestrial green algae with filamentous morphology. Figure 5A: *Rosenvingiella radicans* (cells 8-12 µm wide). Figure 5B: *Klebsormidium flaccidum* (cells 6-8 µm wide). Figure 5C: *Trentepohlia* sp. (cells 10-14 µm wide). Figure 5D: *Printzina bosseae* (cells 12-15 µm wide).

(Mikhailyuk *et al.*, 2008), *Koliella* (Katana *et al.*, 2001), *Planophila* (Friedl & O'Kelly, 2002), *Printzina* (Rindi *et al.*, 2009) and *Trentepohlia* (Rindi *et al.*, 2009). It is now generally acknowledged that such morphological convergence is a typical trait of the evolution of terrestrial green algae, which represents a substantial complication for the systematics of these organisms. A direct consequence of this phenomenon is that the taxonomy and the phylogenetic relationships of many taxa will require a careful reassessment based on a polyphasic approach (Pröschold & Leliaert, 2007), in which molecular data will play a fundamental part.

Biogeography of Terrestrial Green Algae

The biogeography of microscopic organisms has traditionally been a complex matter, and in recent years it has become subject of great controversy. Two different schools of thought have been established. A view embraced primarily by researchers with ecological background assumes that free-living microrganisms have a cosmopolitan distribution and that eukaryotes smaller than 1 mm in size (such as most terrestrial microalgae) are distributed worldwide, wherever their habitat requirements are satisfied (Finlay, 2002; Fenchel & Finlay, 2003; Finlay & Fenchel, 2004). The rationale behind this view is that, due to their small size and large population sizes, these organisms will produce large amounts of new propagules (cysts, eggs, spores) capable of dispersal on very long distances, which will make the probability of local extinction low. Conversely, a view founded on taxonomic grounds believes that many microrganisms are not cosmopolitan; they have a more or less restricted distribution and the wrong impression of ubiquity is due to undersampling, misidentifications and reliance on morphological species concepts (Foissner, 2006, 2008). In recent years this theory has received the support of molecular evidence obtained from several groups of microscopic eukaryotes (Šlapeta *et al.*, 2005, 2006; Luo *et al.*, 2006; Evans *et al.*, 2008a, 2008b). For terrestrial green algae, the knowledge required to confirm or disprove either of these two scenarios is not yet available. The problems mentioned earlier in relation to systematics and taxonomy imply that many basic ideas about the biogeography of terrestrial microalgae will have to be reconsidered after a detailed reassessment of the taxonomy of many taxa. The information available in the literature about the geographical distribution of terrestrial green algae is generally vague and so far has been based entirely on morphological species concepts. Numerous floristic studies are available, especially for soil algae (Metting, 1981, and references therein) and algal assemblages growing on artificial substrata (Rindi, 2007a, and references therein). However, great caution should be used when trying to draw generalizations based on these studies, because taxonomic incongruities and differences in the nomenclature used make it difficult to compare the results of different investigations. In general, there is a certain number of species that, as defined on morphological basis, appear to be cosmopolitan or very widely distributed. Typical examples are *Apatococcus lobatus*, *Chlorella vulgaris*, *Desmococcus olivaceus*, *Klebsormidium flaccidum*, *Stichococcus bacillaris* and *Trentepohlia aurea*. These species have been reported from almost every region in which terrestrial green algae have been studied, and in taxonomic accounts they are defined as cosmopolitan, or widely distributed (Printz, 1939; Ettl & Gärtner, 1995; John, 1988, 2002); *Desmococcus olivaceus* has been defined the "commonest green alga in the world" (Laundon, 1985). Conversely, many other species are known in nature only from the type locality or from a restricted number of locations (*e.g.*, Thompson & Wujek, 1992; Broady & Ingerfield, 1993; Thompson & Wujek, 1997; Neustupa & Šejnohová, 2003; Neustupa, 2005; Novis, 2006; Rindi *et al.*, 2006).

Unfortunately, the small size and simple morphology of these organisms make their identification (and therefore the characterization of their distribution) very problematic. With the exception of relatively few species that producevisible populations, most terrestrial green algae are not observable with the unaided eye and their presence is only revealed by growth in culture (John, 1988; Broady, 1996). The case of species which are only known from the original culture and were never observed in nature is not infrequent. At the same time, the scarcity of features useful for identification and the homoplasious morphology of many lineages make the identification of these algae a difficult task, for which the expertise of specialized taxonomists is usually mandatory. In general, the most reliable conclusions can be drawn for "flagship" taxa with a characteristic morphology, which makes them easily recognizable in the field.

The order Trentepohliales (class Ulvophyceae) represents the best example of such a taxon. This order includes subaerial algae capable of growing on numerous substrata, such as natural rocks, concrete walls, plastic nets, tree bark, leaves, stems and fruits (Chapman, 1984; Thompson & Wujek, 1997; López-Bautista *et al.*, 2002). These algae produce large amounts of carotenoids (such as β-carotene and haematochrome) that give them a bright orange, yellow or red color, making them easily recognizable. Although identification at the species level may be very difficult, the recognition of members of this order and their attribution to one of the five genera currently accepted (*Cephaleuros*, *Phycopeltis*, *Printzina*, *Stomatochroon*, *Trentepohlia*) are generally straightforward. It is well documented that the Trentepohliales are most diverse and abundant in humid tropical regions, where they are the most common terrestrial green algae. In the tropics they are well known for their profuse development (Wee & Lee, 1980; John, 1988; Rindi & López-Bautista, 2008; Rindi *et al.*, 2008a) that may become cause of major practical nuisances, such as disfiguration of buildings (Wee & Lee, 1980) and infection of plants of commercial interest (Chapman & Waters, 2001). Because of such association with warm and humid climates, Aptroot & Van Herck (2007) suggested that the recent geographical expansion in western Europe of species of lichens with southern affinities may be due to the effect of global warming on their *Trentepohlia* phycobionts. Floristic studies based on morphology have reported the highest richness of Trentepohliales for Queensland, Australia (31 taxa), the area of Bogor in Indonesia (30 taxa), India (27 taxa), French Guiana (29 taxa) and Panama (24 taxa) (Rindi & López-Bautista, 2008, and references therein; Rindi *et al.*, 2008a). Molecular data have become available for the Trentepohliales only recently (López-Bautista & Chapman, 2003; López-Bautista *et al.*, 2006; Rindi *et al.*, 2009) and have revealed that the genetic diversity of this group is even higher than suggested by morphology. For example, *Cephaleuros virescens*, *Printzina lagenifera* and *Trentepohlia arborum* are considered widespread species with pantropical distribution (Printz, 1939; Cribb, 1970; Ettl & Gärtner, 1995; Thompson & Wujek, 1997). Morphologically they are well-defined and generally easy to identify. The phylogenies presented by López-Bautista *et al.* (2006) and Rindi *et al.* (2009), however, suggest that the morphologies typical of these species have evolved separately in several lineages. Therefore, these entities should be regarded as complexes of cryptic species rather than individual species of their own. Actual cryptic species are likely to have more restricted distributions, and further collections and molecular data will be necessary for a correct characterization of their systematics and biogeography. A similar situation was found for *Trentepohlia umbrina*, another trentepohlialean species reported as geographically widespread (Printz, 1939). Although showing an identical morphology, specimens from subtropical areas sequenced by Rindi *et al.* (2009) are separated with strong support and appear to represent a different species from European strains, which represent the genuine *Trentepohlia umbrina* (originally described by Kützing (1843), from tree bark in southern Germany). This suggests the possibility that the distribution of this species is relegated to temperate regions and that tropical strains will have to be accommodated into one or more separate species; even in this case, however, further molecular data are required. In recent studies, it has been shown that other species of *Trentepohlia* and *Printzina* have a wider geographical distribution than previously thought. For example, *Printzina bosseae* and *Trentepohlia dusenii*, which were previously known only from the old world (*P. bosseae* from tropical Asia and some regions of Oceania; *T. dusenii* from Cameroon, India and Queensland), were recently discovered in French Guiana by Rindi & López-Bautista (2008). Similarly *Trentepohlia minima* and *Trentepohlia treubiana*, previously known from several regions of Australasia, were recently recorded for Panama by Rindi *et al.* (2008a). It is likely that detailed surveys conducted in other tropical regions will reveal a wider geographical distribution for many other species of Trentepohliales.

The order Prasiolales belongs to the Trebouxiophyceae and represents one of the most versatile groups of green algae, including species distributed in marine, terrestrial and freshwater habitats (Rindi *et al.*, 2007). The distribution of these algae is associated with polar regions and cold-temperate regions with humid climate; the only members of this order recorded in geographical areas with warmer climates are freshwater species of *Prasiola* occurring in cold mountain streams (Naw & Hara, 2002; Rodríguez & Jiménez, 2005). *Prasiola crispa* is a particularly well-known case, because its characteristic thallus formed by dark-green, curled blades, is easily recognizable. This species occurs in polar and cold temperate regions of both hemispheres; it is the most common terrestrial alga in Antarctica, where it forms large populations at sites where penguin rookeries deposit large amounts of guano (Broady, 1996; Lud *et al.*, 2001). In the past Antarctic specimens have been placed into a separate subspecies (*Prasiola crispa* subsp. *antarctica* (Kützing) Knebel; Knebel, 1935), and molecular data have confirmed their distinctness from populations of the northern hemisphere; *rbc*L sequences suggest separation at the species level (Rindi, unpublished data). Marine species of *Prasiola* occur in the supralittoral and upper intertidal zone on rocky shores, producing large green patches in spots fertilized by seabird guano (Rindi, 2007b). Although also present in the southern hemisphere, marine *Prasiola* have their centre of diversity in the North Atlantic and North Pacific, where at least 6 species occur (*Prasiola borealis*, *P. delicata*, *P. furfuracea*, *P. linearis*, *P. meridionalis* and *P. stipitata*). In molecular phylogenies, these species form a well-supported monophyletic group (apart for *Prasiola borealis*, which occurs in a separate clade; Rindi *et al.*, 2007). The substitution rate of the *rbc*L gene in this group is unexpectedly low and does not match the considerable variation in vegetative morphology and life history. This suggests that marine *Prasiola* might have derived from a geologically recent trans-Arctic evolutionary radiation, a type of event that is well documented for several species of marine macroalgae (Lindstrom, 1987, 2001). Further molecular datasets based on different markers should help to clarify interspecific relationships and reveal geographical patterns in this group.

Klebsormidium is a genus of filamentous green algae occurring in many different terrestrial habitats, in particular soil, natural rocks and bases of concrete walls (Ettl & Gärtner, 1995; Lokhorst, 1996; John, 2002). Although it has been more frequently recorded in temperate and polar regions than in the tropics, its distribution is essentially cosmopolitan. *Klebsormidium flaccidum* is the type species of the genus and one of the most widely distributed terrestrial green algae (Printz, 1964; Ettl & Gärtner, 1995; John, 2002). Recent molecular data, however, have shown that taxonomic accounts of *Klebsormidium* based on morphology do not reflect its phylogeny, and the taxonomic identity of *K. flaccidum* requires clarification (Sluiman *et al.*, 2008; Mikhailyuk *et al.*, 2008; Rindi *et al.*, 2008b). Mikhailyuk *et al.* (2008) showed that strains of *Interfilum*, a freshwater genus with very different morphology, are nested within *Klebsormidium* and render it paraphyletic. Strains of

Klebsormidium flaccidum sequenced by Rindi *et al.* (2008) did not form a monophyletic group and were scattered among several clades in the phylogeny of the genus, revealing how complicated the delimitation of species in *Klebsormidium* is. Overall, the bulk of molecular information currently available for this genus points at the existence of a great genetic diversity hidden behind a simple and uniform morphology. Once the circumscription of species in *Klebsormidium* is reassessed on the basis of robust molecular phylogenies, we feel that the distribution of many species will turn to be more restricted than presently believed, both in terms of habitat and geography.

Conclusion

The cases illustrated above show how molecular data are reshaping species delimitations in terrestrial green algae and, indirectly, basic concepts about their biogeography. Until individual species are unambiguously defined and the taxonomy of the main groups is reassessed, it will be impossible to draw strong conclusions about the biogeography of terrestrial green algae. It is clear that for terrestrial and freshwater microalgae species concepts based only on morphological grounds cannot stand anymore. Species should be defined using a polyphasic approach combining as many different types of data as possible (morphological, molecular, ultrastructural, biochemical, physiological, ecological). The key importance of molecular data is now obvious, and the production of robust phylogenies based on multiple molecular markers will be essential to clarify systematics and taxonomy of many groups. The vast majority of the data available is currently represented by 18S rRNA gene sequences. The great popularity of this marker is due to its usefulness for inference at class and/or order level, which represents an important advantage for green microalgae. Since the morphology of these organisms is highly homoplasious and offers very few characters useful for identification, the 18S rRNA sequence will give an immediate indication of the phylogenetic placement of any green microalga for which an identification based on morphology is impossible or only tentative. However, the substitution rate of this gene is usually not suitable to clarify relationships at genus and species level; for this purpose, the use of other, more variable molecular markers is necessary. The most frequently used are ITS rRNA (Beck *et al.*, 1998; Müller *et al.*, 2005; Luo *et al.*, 2006; Yahr *et al.*, 2006; Sluiman *et al.*, 2008; Mikhailyuk *et al.*, 2008), the *rbc*L gene (Novis, 2006; Rindi *et al.*, 2007, 2008b, 2009), the 26S rRNA gene (Friedl & Rokitta, 1997) and the type 1 introns of the actin gene (Kroken & Taylor, 2000; Nelsen & Gargas, 2006, 2008). However, the taxon sampling and number of sequences available for these markers are still limited and mostly circumscribed to specific taxa. It is clear that to increase the amount of sequences (and expand substantially the taxon sampling) represents an absolute priority. Other types of molecular data for terrestrial green algae are scanty and restricted to Amplified Fragment Length Polymorphysms (AFLP); Müller *et al.* (2005) showed their usefulness in unravelling subspecific diversity in *Chlorella vulgaris*.

The availability of a large bulk of distributional data, based on accurate identifications, represents the other critical requirement for a clarification of the biogeography of terrestrial green algae. Due to the problems about species concepts and taxonomic identity outlined above, distributional records available in the literature need to be considered with great caution and re-evaluated case by case. Samples of terrestrial green algae are available in many herbaria or other public collections and represent a very valuable resource in this regard; but these records should also be subjected to a critical taxonomic re-examination. In general, the number of systematists working on terrestrial algae is much more limited than for other groups of algae and plants, and these organisms are comparatively undersampled. Therefore new surveys and investigations of natural history will be very important in the future and it is highly desirable that this type of work will receive high support from funding agencies. These surveys should ideally focus on geographical areas and habitats that have been underexplored, where it is expectable to find new species. This is particularly true for the tropics, which, for historical reasons, have received less coverage and less scrutiny by skilled algal taxonomists. Investigations in tropical regions continue to lead to the discovery of new species (Neustupa, 2003; Neustupa & Šejnohová, 2003; Neustupa, 2005; Neustupa et al., 2007; Rindi et al., 2006; Rindi & López-Bautista, 2007; Eliáš et al., 2008; Zhang et al., 2008; Neustupa et al., 2009) and the diversity of subaerial algae in these regions is probably immense, especially in rainforest environments. The high humidity and the great habitat diversity typical of rainforests are ideal to support a great diversity of terrestrial algae. The limited information available supports this idea (Neustupa, 2005; Rindi et al., 2006; Neustupa & Škaloud, 2008), and it is important that these environments are investigated in detail before human impacts cause irreparable damages. The records produced in future surveys should ideally be based on morphological and molecular data and made verifiable by conservation of voucher specimens. However, with the accumulation of sequence data and the improved characterization of algal biodiversity, new molecular data based on environmental sampling and cloning will become increasingly important in the future. This approach has shown a great potential in revealing the genetic diversity of several microalgal assemblages (Fawley et al., 2004; Kirkwood et al., 2008; Sherwood et al., 2008). In subaerial algal communities it has not yet been widely used, and it will certainly be an important tool for their study in the future; the possibility to unravel the presence of species that cannot be grown in culture represents its most important strength.

Investigations concerning other aspects of the biology of terrestrial algae can also provide valuable information to understand their biogeography. Range and mechanisms of dispersal are particularly important, as they contribute directly to determine the geographical distribution of species. Studies on this topic have been carried out in several regions, mainly by collection air samples, concentration of algal cells and examination in culture (Brown, 1971; Rosas et al., 1989; Roy-Ocotla & Carrera, 1993; Marshall & Chalmers, 1997; Tormo et al., 2001). The evidence available suggests that the diversity and abundance of airborne algal particles show climatic, topographical, geographical, diurnal and seasonal variation; tropical regions exhibit generally higher diversity and abundance than other climatic regions (Sharma et al., 2007). It is generally recognized that single cells 10-12 µm wide (such as species of Stichococcus) are ideal airborne algae, which can be carried in the atmosphere for thousands of km (Roy-Ocotla & Carrera, 1993). However, many studies on airborne algae do not provide identifications at species level; when provided, they are based on morphological concepts, with the same limitations already mentioned for distributional studies. Data on the physiology of species of terrestrial green algae (desiccation tolerance, survival ranges, optimal temperature, optimal range of irradiance for photosynthesis) are generally limited. Mechanisms of protection and adaptation to high UV radiation are the only aspect that has received considerable attention (Lud et al., 2001; Karsten, 2005; Hughes, 2006; Karsten,

2007a, 2007b and references therein), and detailed studies are available only for a few types of habitat, such as biotic crusts of deserts (Gray *et al.*, 2007) and algal communities growing on artificial surfaces (Ong *et al.*, 1992; Häubner *et al.*, 2006; Karsten *et al.*, 2007b). This is unfortunate, because physiological data (especially temperature ranges and optima) may provide very useful information to interpret distributional data and predict the geographical distribution of a species.

The information summarized here highlights how much work on all aspects of the biology of terrestrial green algae is necessary in order to achieve a deep understanding of their biogeography. Once the taxonomy of the most common species is sorted and reliable distributional information is available, the distribution records combined with physiological information will allow to design ecological niche models incorporating the effects of climatic parameters, which would be very useful to predict shifts in distribution due to climatic changes; at present, such models are inexistent for terrestrial algae. As a general conclusion, the biogeography of terrestrial green algae is a poorly-explored area, which holds a great potential for a bulk of exciting research and which deserves much greater attention than received so far.

ACKNOWLEDGMENTS

The ideas presented in this paper are largely based on work carried out in the course of a two-years project funded by the U.S. National Science Foundation (Systematics Program DEB-0542924 to J.M.L.-B.). We are grateful to Michael Guiry for the use of pictures.

REFERENCES

Agardh, C. A. (1824). *Systema algarum*. Lund, Sweden: Berling.

Aptroot, A. & van Herk, C. M. (2007). Further evidence of the effects of global warming on lichens, particularly those with *Trentepohlia* phycobionts. *Environmental Pollution, 146*, 293-298.

Barkman, J. J. (1958). *Phytosociology and ecology of cryptogamic epiphytes*. Assen, The Netherlands: Van Gorcum and Comp. N.V. - G.A. Hak and Dr. H.J. Prakke.

Beck, A., Friedl, T. & Rambold, G. (1998). Selectivity of photobiont choice in a defined lichen community: inferences from cultural and molecular studies. *New Phytologist, 139*, 709-720.

Bremer, K. (1985). Summary of green plant phylogeny and classification. Cladistics, *1*, 369-385.

Broady, P. A. (1996). Diversity, distribution and dispersal of Antarctic terrestrial algae. *Biodiversity and Conservation, 5*, 1307-1335.

Broady, P. A. & Ingerfield, M. (1993). Three new species and a new record of chaetophoracean (Chlorophyta) algae from terrestrial habitats in Antarctica. *European Journal of Phycology, 28*, 25-31.

Brown, R. M. (1971). Studies of Hawaiian freshwater and soil algae. I. The atmospheric dispersal of algae and fern spores across the island of Oahu, Hawaii. In B.C. Parker &

R.M. Brown (Eds.), *Contributions in phycology* (pp. 175-188). Lawrence, Kansas: Allen Press.

Buchheim, M. A., Buchheim, J. A., Carlson, T. & Kugrens, P. (2002). Phylogeny of *Lobocharacium* (Chlorophyceae) and allies: a study of 18S and 26S rDNA data. *Journal of Phycology*, *38*, 376-383.

Cardon, Z. G., Gray, D. W. & Lewis, L. A. (2008). The green algal underground: evolutionary secrets of desert cells. *BioScience*, *58*, 114-122.

Chapman, R. L. 1984. An assessment of the current state of our knowledge of the Trentepohliaceae. In D. E. G. Irvine, & D. M. John (Eds.), *Systematics of the green algae* (233-250). London, U.K.: Academic Press.

Chapman, R. L. & Waters, D. A. (2001). Lichenization of the Trentepohliales - complex algae and odd relationships. In J. Seckbach (Ed.), *Symbiosis*(361-371). The Netherlands: Kluwer Academic Publishers.

Cribb, A. B. (1970). A revision of some species of *Trentepohlia* especially from Queensland. *Proceedings of the Royal Society of Queensland*, *82*, 17-34.

Eliáš, M., Neustupa, J. & Škaloud, P. (2008). *Elliptochloris bilobata* var. *corticola* var. nov. (Trebouxiophyceae, Chlorophyta), a novel subaerial coccal green alga. *Biologia*, *63*, 791-798.

Ettl, H. & Gärtner, G. (1995). *Syllabus der Boden-, Luft- und Flechtenalgen*. Stuttgart, Jena and New York: Gustav Fischer Verlag.

Evans, K. M., Wortley, A. H. & Mann, D. G. (2008a). An assessment of potential diatom "barcode" genes (*cox*1, *rbc*L, 18S and ITS rDNA) and their effectiveness in determining relationships in *Sellaphora* (Bacillariophyta). *Protist*, *158*, 349-364.

Evans, K. M., Wortley, A. H., Simpson, G. E., Chepurnov, V. A. & Mann, D. G. (2008b). A molecular systematics approach to explore diversity within the *Sellaphora pupula* species complex (Bacillariophyta). *Journal of Phycology*, *44*, 215-231.

Fawley, M. W., Fawley, K. P. & Buchheim, M. A. (2004). Molecular diversity among communities of freshwater microchlorophytes. *Microbial Ecology*, *48*, 489-499.

Fenchel, T. & Finlay, B. J. (2003). Is microbial diversity fundamentally different from biodiversity of larger animals and plants? *European Journal of Protistology*, *39*, 486-490.

Finlay, B. J. (2002). Global dispersal of free-living microbial eukaryote species. *Science*, *296*, 1061-1063.

Finlay, B. J. & Fenchel T. (2004). Cosmopolitan metapopulations of free-living microbial eukaryotes. *Protist*, *155*, 237-244.

Flechtner, V. R. (2007). North American microbiotic soil crust communities: diversity despite challenge. In J. Seckbach (Ed.), *Algae and cyanobacteria in extreme environments* (539-551). Dordrecht, The Netherlands: Springer.

Foissner, W. (2006). Biogeography and dispersal of micro-organisms: a review emphasizing protests. *Acta Protozoologica*, *45*, 111-136.

Foissner, W. (2008). Protist diversity and distribution: some basic considerations. *Biodiversity and Conservation*, *17*, 235-142.

Friedl, T. (1995). Inferring taxonomic positions and testing genus level assignments in coccoid green lichen algae: a phylogenetic analysis of 18S ribosomal RNA sequences from *Dictyochloropsis reticulata* and from members of the genus *Myrmecia* (Chlorophyta, Trebouxiophyceae cl. nov.). *Journal of Phycology*, *31*, 632-639.

Friedl, T. & O'Kelly, C. J. (2002). Phylogenetic relationships of green algae assigned to the genus *Planophila* (Chlorophyta): evidence from 18S rDNA sequence data and ultrastructure. *European Journal of Phycology, 37*, 373-384.

Friedl, T & Rokitta, C. (1997). Species relationships in the lichen alga *Trebouxia* (Chlorophyta, Trebouxiophyceae): molecular phylogenetic analyses of nuclear-encoded large subunit rRNA gene sequences. *Symbiosis, 23*, 125-148.

Friedl, T. & Zeltner, C. (1994). Assessing the relationships of some coccoid green lichen algae and the Microthamniales (Chlorophyta) with 18S gene sequence comparisons. *Journal of Phycology, 30*, 500-506.

Fritsch, F. (1907). The subaerial and freshwater algal flora of the tropics. A phytogeographical and ecological study. *Annals of Botany, 21*, 235-275.

Golubic, S. (1967). Algenvegetation der Felsen. Eine ökologische Algenstudie im dinarischen Karstgebiet. In H. J. Elster, & W. Ohle (Eds.), *Die Binnengewässer 23* (pp. 1-183). Stuttgart, Germany: E. Schweizerbart'sche Verlagsbuchhandlung.

Gray, D. W., Lewis, L. A. & Cardon, Z. G. (2007). Photosynthetic recovery following desiccation of desert green algae (Chlorophyta) and their aquatic relatives. *Plant, Cell and Environment, 30*, 1240-1255.

Häubner, N., Schumann, R. & Karsten, U. (2006). Aeroterrestrial algae growing in biofilms on facades - response to temperature and water stress. *Microbial Ecology, 51*, 285-293.

Hughes, K. A. (2006). Solar UV-B radiation, associated with ozone depletion, inhibits the Antarctic terrestrial microalga *Stichococcus bacillaris*. *Polar Biology, 29*, 327-336.

Huss, V. A. R., Frank, C., Hartmann, E. C., Hirmer, M., Kloboucek, A., Seidel, B. M., Wenzeler, P. & Kessler, E. (1999). Biochemical taxonomy and molecular phylogeny of the genus *Chlorella* sensu lato (Chlorophyta). *Journal of Phycology, 35*, 587-598.

John, D. M. (1988). Algal growths on buildings: a general review and methods of treatment. *Biodeterioration Abstracts, 2*, 81-102.

John, D. M. (2002). Orders Chaetophorales, Klebsormidiales, Microsporales, Ulotrichales. In D. M. John, B. A. Whitton, & A. J. Brook (Eds.), *The freshwater algal flora of the British Isles* (433-468). Cambridge, U.K.: Cambridge University Press.

Karsten, U., Friedl, T., Schumann, R., Hoyer, K. & Lembcke, S. (2005). Mycosporine-like amino acids and phylogenies in green algae: *Prasiola* and its relatives from the Trebouxiophyceae (Chlorophyta). *Journal of Phycology, 41*, 557-566.

Karsten, U., Karsten, U., Lembcke, S. & Schumann, R. (2007a). The effects of ultraviolet radiation on photosynthetic performance, growth, and sunscreen compounds in aeroterrestrial biofilm algae isolated from building facades. *Planta, 225*, 991-1000.

Karsten, U., Schumann, R. & Mostaert, A. S. (2007b). Aeroterrestrial algae growing on man-made surfaces: what are the secrets of their ecological success? In J. Seckbach (Ed.), *Algae and cyanobacteria in extreme environments* (pp. 583-597). Dordrecht, The Netherlands: Springer.

Katana, A., Kwiatowski, J., Spalik, K., Zakryś, B., Szalacha, E. & Szymańska, H. (2001). Phylogenetic position of *Koliella* (Chlorophyta) as inferred from nuclear and chloroplast small subunit rDNA. *Journal of Phycology, 37*, 443-451.

Kirkwood, A. E., Buchheim, J. A., Buchheim, M. A. & Henley, W. J. (2008). Cyanobacterial diversity and halotolerance in a variable hypersaline environment. *Microbial Ecology, 55*, 453-465.

Knebel, G. (1935). Monographie der Algenreihe der Prasiolales, insbesondere von *Prasiola crispa. Hedwigia, 75*, 1-120.

Krienitz, L., Hegewald, E. H., Hepperle, D., Huss, V. A. R., Rohr, T. & Wolf, M. (2004). Phylogenetic relationship of *Chlorella* and *Parachlorella* gen. nov. (Chlorophyta, Trebouxiophyceae). *Phycologia, 43*, 529-542.

Kroken, S. & Taylor, J. W. (2000). Phylogenetic species, reproductive mode and specificity of the green alga *Trebouxia* forming lichens with the fungal genus *Letharia*. *Bryologist, 103*, 645-660.

Kützing, F. T. (1843). *Phycologia generalis*. Leipzig, Germany: F. A. Brockhaus.

Kützing, F. T. (1849). *Species algarum*. Leipzig, Germany: F.A. Brockhaus.

Laundon, J. R. (1985). *Desmococcus olivaceus* - the name of the common subaerial green alga. *Taxon, 34*, 671-672.

Lewin, R. A. & Robinson, P. T. (1979). The greening of polar bears in zoos. *Nature, 278*, 445-447.

Lewis, L. A. (2007). Chlorophyta on land: independent lineages of green eukaryotes from arid lands. In J. Seckbach (Ed.), *Algae and cyanobacteria in extreme environments* (571-582). Dordrecht, The Netherlands: Springer.

Lewis, L. A. & Flechtner, V. R. (2002). Green algae (Chlorophyta) of desert microbiotic crusts: diversity of North American taxa. *Taxon, 51*, 443-451.

Lewis, L. A. & McCourt, R. M. (2004). Green algae and the origin of land plants. *American Journal of Botany, 91*, 1535-1556.

Lewis, L. A., Wilcox, L. W., Fuerst, P. A. & Floyd, G. L. (1992). Concordance of molecular and ultrastructural data in the study of zoosporic green algae. *Journal of Phycology, 28*, 375-380.

Lindstrom, S. C. (1987). Possible sister groups and phylogenetic relationships among selected North Pacific and North Atlantic Rhodophyta. *Helgoländer Meeresuntersuchungen, 41*, 245-260.

Lindstrom, S. C. (2001). The Bering Strait connection: dispersal and speciation in boreal macroalgae. *Journal of Biogeography, 28*, 243-251.

Linnaeus, C. V. (1753). *Species plantarum. Vol. II*. Stockholm, Sweden.

Linnaeus, C. V. (1759). *Systema naturae per regna tria naturae. Vol. II*. Stockholm, Sweden.

Lokhorst, G. M. (1996). Comparative taxonomic studies on the genus *Klebsormidium* (Charophyceae) in Europe. *Cryptogamic Studies, Vol. 5*. Stuttgart, Germany: Gustav Fischer.

López-Bautista, J. M. & Chapman, R. L. (2003). Phylogenetic affinities of the Trentepohliales inferred from small-subunit rDNA. *International Journal of Systematic and Evolutionary Microbiology, 53*, 2099-2106.

López-Bautista, J. M., Rindi, F. & Guiry, M. D. (2006). Molecular systematics of the subaerial green algal order Trentepohliales: an assessment based on morphological and molecular data. *International Journal of Systematic and Evolutionary Microbiology, 56*, 1709-1715.

López-Bautista, J. M., Waters, D. A. & Chapman, R. L. (2002). The Trentepohliales revisited. *Constancea, 83*, http://ucjeps.berkeley.edu/constancea/83/lopez_etal/trentepohliales.html

Lud, D., Buma, A. G. J., van de Poll, W., Moerdijk, T. C. W. & Huiskes, H. L. (2001). DNA damage and photosynthetic performance in the Antarctic terrestrial alga *Prasiola crispa* ssp. *antarctica* (Chlorophyta) under manipulated UV-radiation. *Journal of Phycology*, *37*, 459-467.

Luo, W., Pflugmacher, S., Pröschold, T., Walz, N. & Krienitz, L. (2006). Genotype versus phenotype variability in *Chlorella* and *Micractinium* (Chlorophyta, Trebouxiophyceae). *Protist*, *157*, 315-323.

Marshall, W. A. & Chalmers, M. O. (1997). Airborne dispersal of Antarctic terrestrial algae and cyanobacteria. *Ecography*, *20*, 585-594.

Mattox, K. R. & Stewart, K. D. (1984). Classification of the green algae: a concept based on comparative cytology. In D. E. G. Irvine, & D. M. John, *The systematics of the green algae* (29-72). London, U.K.: Academic Press.

Metting, B. (1981). The systematics and ecology of soil algae. *Botanical Review*, *47*, 195-312.

Mikhailyuk, T. I., Sluiman, H. J., Massalski, A., Mudimu, O., Demchenko, E. M., Kondratyuk, S. Y. & Friedl, T. (2008). New streptophyte green algae from terrestrial habitats and an assessment of the genus *Interfilum* (Klebsormidiophyceae, Streptophyta). *Journal of Phycology*, *44*, 1586-1603.

Müller, J., Friedl, T., Hepperle, D. & Lorenz, M. (2005). Distinction between multiple isolates of *Chlorella vulgaris* (Chlorophyta, Trebouxiophyceae) and testing for conspecificity using amplified fragment length polymorphism and ITS rDNA sequences. *Journal of Phycology*, *41*, 1236-1247.

Nägeli, C. (1849). Gattungen einzelliger Algen, physiologisch und systematisch bearbeitet. *Neue Denkschriften der Allgemeine Schweizerischen Gesellschaft für die Gesammten Naturwissenschaften*, *10*, 1-139.

Naw, M. W. D. & Hara, Y. (2002). Morphology and molecular phylogeny of Prasiola sp. (Prasiolales, Chlorophyta) from Myanmar. *Phycological Research*, *50*, 175-82.

Nelsen, M. P. & Gargas, A. (2006). Actin type intron sequences increase phylogenetic resolution: an example from *Asterochloris* (Chlorophyta: Trebouxiophyceae). *Lichenologist*, 38, 435-440.

Nelsen, M. P. & Gargas, A. (2008). Dissociation and horizontal transmission of co-dispersed lichen symbionts in the genus *Lepraria* (Lecanorales: Stereocaulaceae). *New Phytologist*, *177*, 264-275.

Neustupa, J. (2003). The genus *Phycopeltis* (Trentepohliales, Chlorophyta) from tropical Southeast Asia. *Nova Hedwigia*, *76*, 487-505.

Neustupa, J. (2005). Investigations on the genus *Phycopeltis* (Trentepohliaceae, Chlorophyta) from South-East Asia, including the description of two new species. *Cryptogamie, Algologie*, *26*, 229-242.

Neustupa, J., Eliáš, M. & Šejnohová, L. (2007). A taxonomic study of two *Stichococcus* species (Trebouxiophyceae, Chlorophyta) with a starch-enveloped pyrenoid. *Nova Hedwigia*, *84*, 51-63.

Neustupa, J., Němcova, Y., Eliáš, M. & Škaloud, P. (2009). *Kalinella bambusicola* gen. et sp. nov. (Trebouxiophyceae, Chlorophyta), a novel coccoid *Chlorella*-like subaerial alga from Southeast Asia. *Phycological Research*, in press.

Neustupa, J. & Šejnohová, L. (2003). *Marvania aerophytica* sp. nov., a new subaerial tropical green alga. *Biologia*, *58*, 503-507.

Neustupa, J. & Škaloud, P. (2008). Diversity of subaerial algae and cyanobacteria on tree bark in tropical mountain habitats. *Biologia*, *63*, 806-812.

Novis, P. M. (2006). Taxonomy of *Klebsormidium* (Klebsormidiales, Charophyceae) in New Zealand streams and the significance of low-pH habitats. *Phycologia*, *45*, 293-301.

Ong, B. L., Lim, M. & Wee, Y. C. (1992). Effects of desiccation and illumination on photosynthesis and pigmentation of an edaphic population of *Trentepohlia odorata* (Chlorophyta). *Journal of Phycology*, *28*, 768-772.

Printz, H. 1939. Vorarbeiten zu einer Monographie der Trentepohliaceen. *Nytt Magasin for Naturvbidenskapene*, *80*, 137-210.

Printz, H. 1964. Die Chaetophoralen der Binnengewässer. Eine systematische übersicht. *Hydrobiologia*, *24*, 1-376.

Pröschold, T. & Leliaert, F. (2007). Systematics of the green algae: conflict of classic and modern approaches. In J. Brodie & J. Lewis (Eds.), *Unravelling the algae: the past, present and future of algal systematics*, (The Systematics Association Special Volume Series 75, 123-53). Boca Raton, London and New York: CRC Press.

Rindi, F. (2007a). Diversity, distribution and ecology of green algae and cyanobacteria in urban habitats. In J. Seckbach (Ed.), *Algae and cyanobacteria in extreme environments* (571-582). Dordrecht, The Netherlands: Springer.

Rindi, F. (2007b). Trebouxiophyceae - Prasiolales. In J. Brodie, C. A. Maggs & D. M. John (Eds.), *Green seaweeds of Britain and Ireland* (13-31). London, U.K.: British Phycological Society.

Rindi, F. & Guiry, M. D. (2002). Diversity, life history and ecology of *Trentepohlia* and *Printzina* (Trentepohliales, Chlorophyta) in urban habitats in western Ireland. *Journal of Phycology*, *38*, 39-54.

Rindi, F., Guiry, M. D. & López-Bautista, J. M. (2008b). Distribution, morphology and phylogeny of *Klebsormidium* (Klebsormidiales, Charophyceae) in urban environments in Europe. *Journal of Phycology*, *44*, 1529-1540.

Rindi, F., Lam, D. W. & López-Bautista, J. M. (2009). Phylogenetic relationships and species circumscription in *Trentepohlia* and *Printzina* (Trentepohliales, Chlorophyta). *Molecular Phylogenetics and Evolution*, *52*, 329-339.

Rindi, F., Lam, D. W. & López-Bautista, J. M. (2008a). Trentepohliales (Ulvophyceae, Chlorophyta) from Panama. *Nova Hedwigia*, *87*, 421-444.

Rindi, F. & López-Bautista, J. M. (2008). Diversity and ecology of Trentepohliales (Ulvophyceae, Chlorophyta) in French Guiana. *Cryptogamie, Algologie*, *29*, 13-43.

Rindi, F., López-Bautista, J. M., Sherwood, A. R. & Guiry, M. D. (2006). Morphology and phylogenetic position of *Spongiochrysis hawaiiensis* gen. et sp. nov., the first known terrestrial member of the order Cladophorales (Ulvophyceae, Chlorophyta). *International Journal of Systematic and Evolutionary Microbiology*, *56*, 913-922.

Rindi, F., McIvor, L., Sherwood, A. R., Friedl, T., Guiry, M. D. & Sheath, R. G. (2007). Molecular phylogeny of the green algal order Prasiolales (Trebouxiophyceae, Chlorophyta). *Journal of Phycology*, *43*, 811-822.

Rodríguez, R. R. & Jiménez, J. C. (2005). Taxonomy and distribution of freshwater *Prasiola* from central Mexico. *Cryptogamie, Algologie*, *26*, 177-188.

Rosas, I., Roy-Ocotla, G. & Carrera, J. (1989). Meteorological effects on variation of airborne algae in Mexico. *International Journal of Biometeorology*, *33*, 173-179.

Roy-Ocotla, G. & Carrera, J. (1993). Aeroalgae: response to some aerobiological questions. *Grana*, *32*, 48-56.

Schlichting, H. E., (1975). Some subaerial algae from Ireland. *British Phycological Journal*, *10*, 257-261.

Sharma, N. K., Rai, A. K., Singh, S., & Brown, R. M. (2007). Airborne algae: their present status and relevance. *Journal of Phycology*, *43*, 615-627.

Sherwood, A. R., Chan, Y. L. & Presting, G. G. (2008). Application of universally amplifying plastid primers to environmental sampling of a stream periphyton community. *Molecular Ecology Resources*, *8*, 1011-1014.

Šlapeta, J., Lopez-Garcia, P. & Moreira D. (2006). Global dispersal and ancient cryptic species. *Molecular Biology and Evolution*, *23*, 23-29.

Šlapeta, J., Moreira, D. & Lopez-Garcia, P. (2005). The extent of protist diversity. *Proceedings of the Royal Society - Biological Sciences*, *272*, 2073-2081.

Sluiman, H. J., Guihal, C. & Mudimu, O. (2008). Assessing phylogenetic affinities and species delimitations in Klebsormidiales (Streptophyta): nuclear-encoded rDNA phylogeny and ITS secondary structure models in *Klebsormidium, Hormidiella* and *Entransia. Journal of Phycology*, *44*, 183-195.

Thompson, R. H. & Wujek, D. E. (1992). *Printzina* gen. nov. (Trentepohliaceae), including a description of a new species. *Journal of Phycology*, *28*, 232-237.

Thompson, R. H. & Wujek, D. E. (1997). *Trentepohliales:* Cephaleuros, Phycopeltis *and* Stomatochroon. *Morphology, Taxonomy and Ecology*. Enfield, New Hampshire: Science Publishers.

Tormo, R., Recio, D., Silva, I. & Muñoz, A. F. (2001). A quantitative investigation of airborne algae and lichen soredia obtained from pollen trap in south-west Spain. *European Journal of Phycology*, *36*, 385-390.

Verbruggen, H., Ashworth, M., LoDuca, S. T., Vlaeminck, C., Cocquyt, E., Sauvage, T., Zechman, F. W., Littler, D. S., Littler, M. M., Leliaert, F. & De Clerck, O. (2009). A multi-locus time calibrated phylogeny of the siphonous green algae. *Molecular Phylogentics and Evolution*, *50*, 642-653.

Wee, Y. C. & Lee, K. B. (1980). Proliferation of algae on surfaces of buildings in Singapore. *International Biodeteration Bulletin*, *16*, 113-117.

Yahr, R., Vilgalys, R. & DePriest, P. T. (2006). Geographic variation in algal partners of *Cladonia subtenuis* (Cladoniaceae) highlights the dynamic nature of a lichen symbiosis. *New Phytologist*, *171*, 847-860.

Yoon, H. S., Hackett, J. D., Ciniglia, C., Pinto, G. & Bhattacharya, D. (2004). A molecular timeline for the origin of photosynthetic eukaryotes. *Molecular Biology and Evolution*, *21*, 809-818.

Zhang, J. M., Huss, V. A. R., Sun, X. P., Chang, K. J., Pang D. B. (2008). Morphology and phylogenetic position of a trebouxiophycean green alga (Chlorophyta) growing on the rubber tree, *Hevea brasiliensis*, with the description of a new genus and species. *European Journal of Phycology*, *43*, 185-193.

In: Encyclopedia of Environmental Research
Editor: Alisa N. Souter

ISBN: 978-1-61761-927-4
© 2011 Nova Science Publishers, Inc.

Chapter 60

CONSERVATION AND MANAGEMENT OF THE BIODIVERSITY IN A HOTSPOT CHARACTERIZED BY SHORT RANGE ENDEMISM AND RARITY: THE CHALLENGE OF NEW CALEDONIA

Roseli Pellens and Philippe Grandcolas*
UMR 7205 CNRS, Département Systématique et Evolution,
Muséum national d'Histoire naturelle, 45, rue Buffon, 75005 Paris, France.

ABSTRACT

New Caledonia is a peculiar hotspot, a small-sized island (ca. 17,000 km²), relatively isolated from any continent (ca. 1200 km from Australia), with moderately high mountains and complex orography. Its biota is very rich in endemic species and highly endangered. Our analysis of the number of references in systematics, ecology and conservation shows that its biota attracts attention of scientists since long ago. In a first and long period, references focused on the description of the biodiversity. More recently these descriptions were intensified and complemented by ecological and later by conservation studies. Recent researches on phylogenetics and biogeography indicated that the biota of New Caledonia is characterized by short range endemism and rarity, with three patterns of endemism: (I) species regionally endemic to New Caledonia distributed in the whole range of an ecosystem; (II) short range endemics with parapatric/allopatric distributions; (II) short range endemics with disjunct distributions. Researches on conservation showed that this biota is highly endangered due to three main threats: fire, mining and invasive species. In this chapter we detail these threats and elaborate a model based on their frequency and spatial distribution to understand how they could affect species with contrasting patterns of endemicity. Our analysis show that the conservation of the biodiversity in a context where species are dominantly short range endemics and rare is a main problem to be faced by New Caledonian authorities as well as by scientific researchers that must provide the basis for political decisions. Regardless the biogeographical pattern of endemism the chances of survival of rare species with short ranges in the case of large scale habitat destruction are quite low. In the case of threats

---

* Corresponding author: E-mail: pellens@mnhn.fr, E-mail: pg@mnhn.fr.

that are more restricted in area, the loss of a species disjunctly distributed is more problematic in terms of loss of phylogenetic diversity. In this case, speciation by niche conservatism can be hypothesized to be less frequent, thus each species can be implied to be more original and in stronger need of conservation by itself. In addition, due to the distances from one another, the number of closely related species in a small island can be much lower than in the case of species with parapatric/allopatric distribution. Fires, mining and introduced species need special control. The two firsts for the habitat destruction they promote over extensive areas and the later by the possibility of continuing to endanger even in areas officially protected.

INTRODUCTION

The biota of New Caledonia is very rich in endemic species and highly endangered. Therefore, the region has been classified as a hotspot of biodiversity (Myers et al. 2000). Hotspots are by definition these regions with conflicts of interest between conservation of amazingly rich and diverse biota and deleterious effects of man occupancy. We will review briefly the main characteristics of New Caledonia geography and biodiversity and then we will examine how they can be affected by the main local man-induced ecosystem disturbances.

New Caledonia is a peculiar hotspot indeed, a small-sized island with moderately high mountains (peaking at ca. 1600m) and complex orography, and relatively isolated from any continent (ca. 17,000 km^2; 1200 km from Australia). But it is also a very ancient piece of land and for this reason it has often been seen as a continental island and a Gondwanan refuge because of the occurrence of several relict taxa (e.g., *Amborella*, the endemic sister-group of all other flowering plants) and of a very old geological basement (80 My). In this context, high richness and endemism have been interpreted as the result of a long-term evolution in isolation. Recent biogeographical studies have revised this view, by considering both detailed geological studies and phylogenetic relationships of endemic taxa (Murienne et al., 2005; Grandcolas et al., 2008). The island could not have conserved in situ its relicts because this piece of Gondwana that collided with an island arc at the limit of two tectonic plates was submerged until 37 My ago. Accordingly, regionally endemic groups often resulted from several recent dispersal founding events since 37 My. Local short range endemism is also extremely important and is even necessarily more recent, at least partly dating back to quaternary climatic variations (e.g., Murienne et al., 2008a).

This new biogeographical paradigm came out with the recent reviving of evolutionary and ecological studies in New Caledonia. Until a few years ago, studies dealing with New Caledonian biota were mostly taxonomic studies or inventories produced at a rising pace (Figure 1a, b). This trend of an ever-increasing taxonomic production can also be detected with the examination of bibliographies (O'Reilly, 1955; Pisier, 1983; Chazeau, 1995) or by looking at the description series of the very rich local fauna (Tillier, 1988; Chazeau & Tillier, 1991; Matile et al., 1993; Najt &Matile , 1997; Najt & Grandcolas, 2002; Grandcolas, 2008a, 2009).

In lag with this trend, the number of evolutionary and ecological studies increased only recently (Figure 1a, b), showing that the background natural history knowledge brought back by taxonomic studies and inventories began to be fruitfully used further than for classification issues (Grandcolas, 2008b). This growing interest for understanding the origin and the functioning of New Caledonian biota took place in the context of higher threats to the local ecosystems that also determined a more recent increase in studies of conservation biology and ecosystem management (Figure 1a, b).

Looking at these three lagged trends (Figure 1a, b), one can identify three steps in the study of New Caledonian biodiversity: the description of the biota, the understanding of processes and then the management of the ecosystems and related conservation issues. As a matter of consequence, we are at a very critical moment for facing the threats to the ecosystem (Beauvais et al., 2006; Pascal et al., 2008), having now different cards in our hands to understand the situation and to propose some cautions or some remedies.

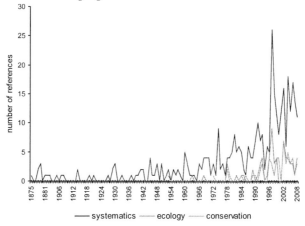

Figure 1a. Number of references dealing with systematics, ecology and conservation in New Caledonia, published from 1875 to 2008. Data from Zoological Records. The search for "systematics" was made with the word New Caledonia in the title "and" the combined results for the words new species, systematic*, taxonom*, phylogen*, evolut*, diversity and inventory in the topics. The search for "ecology" was made with the word New Caledonia in the title "and" the combined results for the words ecology, population, community, ecosystem and process in the topics. The search for "conservation" was made with the word New Caledonia in the title "and" the combined results for the words conservation, threat, invasive, introduction and extinction in the topics. In this search we considered references dealing with present terrestrial biota including fresh water, and excluding marine and paleontological data.

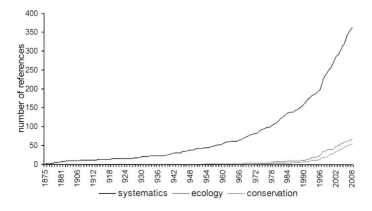

Figure 1b. Accumulation curves of the references dealing with systematics, ecology and conservation in New Caledonia, from 1875 to 2008. Data from Zoological Records (see details in the legend of Figure 1a).

In this perspective, the biodiversity of New Caledonia can be characterized with the following characteristics. First, patterns of short range endemism appear more and more frequent as distributional studies accumulate (Haase & Bouchet, 1998; Bradford & Jaffré, 2004; Pellens, 2004; Murienne et al., 2005, 2008a; Munzinger et al., 2008, Sharma & Giribet 2009). In this prevalent case, sister species apparently appeared by allopatric speciation and are often separated by very short distances. Conversely, some species look like relicts with disjunct distributions in so-called refuge areas (e.g., Pintaud et al., 2001). Such speciation has often taken place by niche conservatism without a strong adaptive differentiation and many short range endemics are therefore quite similar to each other. When such a differentiation occurred however, adaptation to the high local soil diversity including metalliferous soils is important, together with adaptation to climatic and orographic diversity. Short range endemism is not the only distinctive character of the local biota, since species rarity – population low density or high patchiness – is also commonplace. Both endemism and rarity are expectedly correlated as a result of several scale effects or specific mechanisms (Gaston, 1994). Consequently, many New Caledonian species have very small population size.

The recent degradation of the vegetation under the action of man adds one more biological constraint to the ecosystems, by fragmentation of the main forest types into large but dissected massifs of evergreen forest and into small and overdispersed pieces of sclerophyll forests, separated by cultivated areas, pastures or poor savannas.

We will successively detail the main threats to the ecosystems, and then we will infer the potential and specific impact of these threats, given the characteristics of New Caledonian biodiversity. Finally, we will recommend future directions for the study and conservation of local biota.

Main Threats to the New Caledonian Biodiversity

During the last two decades several studies/symposiums/meetings/reports were made in order to characterize the main dangers to the biodiversity in New Caledonia, and to elaborate recommendations to the conservation and management of natural resources. Although most of their recommendations take long to be put in practice, these studies are unanimous in calling attention to three major threats that work exclusively or in combination to put in danger the native biota: fire, mining, and introduced species. These threats along with insufficient number of protected areas, inadequate laws concerning introduced species, and developmental policies of economic help contrary of the maintenance of the local biodiversity contribute to destroy terrestrial, lotic and lagoon ecosystems, putting in danger a great number of species (Bouchet et al., 1995; Gabrié et al., 1995; Gargominy et al., 1996; Jaffré et al., 1998; Ekstrom et al., 2000; Gargominy, 2003; Beauvais et al., 2006). Here we will first make a brief review of these three threats for further exploring their impacts on endemic species.

Fires

The destruction of forests by fire is the most ancient and up-to-date problem in the New Caledonian mainland, being the major causes of habitat loss in the lowlands. Traditionally fires were used to clean pastures and cultivated fields before a new season of plantation, as a cultural practice. Nowadays it is often used as a cultural habit repeated every year in the same places without any immediate reason, i.e., even when there is not agriculture or pasture. It is also lit by hunters in grasslands and open savannas to enhance the growth of new grass, or to facilitate the access (Bouchet et al., 1995; Gabrié et al., 1995). The main outcome of these activities is that every year during the dry season uncontrolled fires destroy thousands of hectares of savannas, arboreous savannas, sclerophyll forests (also called dry forests) and shrublands (locally called *maquis* or *maquis minier* when talking about the vegetation from the areas disturbed by mining), which constitute the most sensitive ecosystems. To have an idea of its importance, 10,024 ha were destroyed in 1995, 3,340 in 1996, 21,678 in 1997, 2,558 in 1998, 5,075 in 1999 and 15,710 in 2000, i.e., a surface of 3.5% of the mainland burned in six years (Gargominy, 2003). The ecosystem most affected by fire is sclerophyll forests. Almost totally destroyed, it is now reduced to a few fragments that contain about than 2% of its original surface (Bouchet et al., 1995).

Mining

New Caledonia is exceptionally rich in rare metals, and presently the extraction of nickel (plus chromium and cobalt, although in minor quantities) constitute the base of its economic activity as well as its most striking environmental disturbance (Gabrié et al., 1995). Although the surface exploited is very small (0,9% of the mainland surface), the mining impacts spread over very large surfaces, totally changing the landscape of the mountains, and also impacting the rivers and the lagoon, as well as the areas near the roads used to nickel ore transportion. In the mountains, the opencast mineral exploitation creates a reddish landscape marked by the presence of openings, access roads, screes and enormous areas of soil erosion, that with time becomes recovered by a shurby secondary vegetation locally called *maquis minier*. In addition to that, another problem is that this kind of landscape degradation is produced in several mountains, very often promoted by small sites of prospection, hardly ever leading to the exploitation. It is useless to say that the first impact of the activities associated to mining is the elimination of the native vegetation, provoking immediate habitat loss, which certainly leads to local extinction of species whose ranges of distribution coincide with the areas impacted. Considering that the main vegetation of areas used for mineral exploitation is the shrublands "*le maquis*" with 92% of endemic species, the biodiversity loss caused by this activity can be enormous (Gabrié et al., 1995; Dupon, 1996; Gargominy, 2003; Pascal et al., 2008).

Introduced Species

As an insular ecosystem, New Caledonia attracts attention to the importance of the introduction of exotic species, especially because some of these species became invasive and provoked serious damages to native biota and ecosystems. This way, presently we have an extensive inventory of the introduced species in the archipelago (McKee, 1994; Gargominy et al., 1996; Beauvais et al., 2006) and some results permitting to evaluate their impact on the native biota (Gargominy et al., 1996; Le Breton et al., 2005; Jourdan, 1997; 2006; Jourdan &

Mille, 2006; Loope & Pascal, 2006; Meyer et al., 2006; Pascal et al., 2006). The number of introduced species in New Caledonia is rather high, having increased markedly during the last 50-60 years. Recent estimations indicate 1,412 species of plants (64 invasive, with nine of them in the IUCN list of "100 World's Worst Invasive Alien Species") (McKee, 1994; Gargominy et al., 1996; Meyer et al., 2006); 518 of invertebrates (six in IUCN the list of "100 World's Worst Invasive Alien Species") (Gargominy et al., 1996; Le Breton et al., 2005; Jourdan, 2006; Jourdan & Mille, 2006); and 42 of vertebrates (12 in the IUCN list of "100 World's Worst Invasive Alien Species") (Gargominy et al., 1996; Pascal et al., 2006). Although most of the introduced species remain in the surroundings of man, some invasive species spread their population through wild ecosystems, putting in danger wild populations of native species. One case well studied concerns the invasive fire ant *Wasmannia auropunctata*, considered "the most dangerous pest ever introduced in this archipelago" (Jourdan, 1997), being a main threat for human population as well as to the New Caledonian autochthon biodiversity. In the wild, they were shown to outcompete local ant species leading to their disappearance in the sites invaded (Jourdan, 1997; Le Breton et al., 2005); to cause the reduction of the diversity and density of lizards, especially geckos (Bauer & Sadlier 1993); and to endanger several hectares of Niaouli savannah due to the disruption of the ecological balance ultimately resulting in the obstruction of the photosynthesis (Cochereau & Potiaroa, 1995) (*W. auropunctata* leads to the reduction of predators and/or competitors of a mealybug that produces a honeydew used by a fungus with black spores that make a black film on the plant's leaves obstructing the photosynthesis). Among introduced vertebrates reported to have a severe effect in New Caledonia's native biota there are pigs, goats, cattle and deers. Individuals of these species graze the understory of humid and dry forests, trampling on young shoots preventing plants from regenerating, compacting the soil, reducing the litter mass, and changing microclimate at ground level (Pascal et al., 2006). Their activities strongly change the habitats, making difficult forest regeneration, and reducing the populations of understory organisms, as shown for species of the mollusc *Placostylus* (Brescia et al., 2008). Considering the rarity of many plant species in New Caledonian biota, the action of invasive vertebrates can be striking giving the final touch to the species extinction, as shown by Bouchet et al. (1995). In fact, species with very rare or very localized populations are extremely fragile, in a way that any action on their habitat lead to deterministic extinctions. Among plants there are several examples of species that are invading or becoming a danger. A recent evaluation of the subject made by Meyer et al. (2006) attracts attention to *Pinus caribea* largely used for reforestation in areas with acid soils and on ultramafic soils, now considered the only allochtone species able to develop in the ultramafic soils (Morat et al. 1999), becoming a potential danger to the species endemic of the *maquis* as well as to the vegetation of other canopy-opened ecosystems.

CHALLENGES FOR CONSERVATION OF SHORT RANGE ENDEMIC AND RARE SPECIES

Short range endemism and rarity are the key characteristics of New Caledonian biota (Grandcolas et al., 2008). Among endemic species one can find basically three main patterns of distribution. The first is the case of species regionally endemic to New Caledonia, distributed in the whole range of an ecosystem, as for example, *Araucaria montana,* one of the endemic species of *Araucaria* from New Caledonia that occurs in the *maquis* and in the dense and humid forests from middle and high altitude (Veillon, 1980) (Pattern I, Table 1). This kind of distribution is not necessarily the most frequent. The second is represented by short range endemic species with *parapatric* distributions, i.e., closely related species having contiguous distribution ranges, and *allopatric* distribution, i.e., with species separated by very short distances. In this last case, it is common to observe close related species separated by distances not greater than 5 km, with or without a clear geographical barrier, as is the case of species of the endemic cricket genus *Agnotecous* (Desutter-Grandcolas & Robillard, 2006), the endemic cockroach genus *Angustonicus* (Pellens, 2004; Murienne et al., 2005) and *Lauraesilpha* (Murienne et al., 2008a,b), or the endemic harvestman family Troglosironidae (Sharma & Giribet 2009). We assume that closely related species with these two patterns of geographic distribution are affected in a similar way by most threats, thus we consider them as a single case (Pattern II, Table 1). Short range endemics with parapatric/allopatric distribution are often inferred to have arisen by speciation with niche conservatism (Wiens, 2004; Murienne et al., 2008b, 2009). In this case, species maintain most of their characteristics and niche dimensions (like the specialization to particular habitats, or to particular forms of exploring habitat or food resources), in way that closely related species share most of their phylogenetic characteristics, and one species can represent better the diversity of its whole group. This aspect has important implications for conservation.

The third pattern of short range endemism often observed in New Caledonia is represented by related species with *disjunct distributions*, i.e., closely related species with short ranges of distribution but found in distant areas, often in relict or refuge habitats, as the case of the palms studied by Pintaud et al. (2001), for example (Pattern III, Table 1). In this case, speciation by niche conservatism can be hypothesized to be less frequent, thus each species can be implied to carry more autapomorphies. In addition, due to the distances from one another, the number of closely related species in a small island like New Caledonia can be much lower than in the case of species with parapatric/allopatric distributions. These aspects indicate that in a clade with species disjunctly distributed, the loss of a species is more problematical in terms of loss of phylogenetic diversity, than in the case of species with parapatric/allopatric distributions.

Independently of their pattern of distribution, species are most often very rare. Species distributed in wide areas (Pattern I) are often rare due to low population density or restricted microhabitats (even if this microhabitat is widely distributed). Short range endemics are rare due to the restricted areas of distribution, which makes them difficult to find regardless the population density. Nevertheless, the most common pattern observed in New Caledonia is even more extreme: rare species due to narrow ranges of distribution and low population densities, which makes very small population sizes.

In Table 1 we present an evaluation of the frequency and the effect of the spatial distribution of the main threats to the New Caledonian biota on the three types of endemic species. This evaluation is based on a model combining the three kinds of endemicity and the distribution and the intensity of the threats. Concentrated threats widespread all over the territory or all over a type of ecosystem (Type A) are the rarest ones. But their effects are the most dangerous being able to eliminate wide range species or several related short range species, independently of their pattern of endemism. In spite of the uncommonness of this kind of event, we have the concrete example of the of the sclerophyll forests, practically totally destroyed by fires that burned here and there every dry season during several decades, in which most of the species still existing are highly endangered (Bouchet et al., 1995). Conversely, impacts scarcely distributed, but widespread over the territory or over an ecosystem (Type B) are the most frequent (like fires at the edges of humid forests, the effects of several invasive species in different parts of the ecosystems, forest clearing for small plantations, hunting, selective logging, or the combination of these activities). The main effect of this kind of impact is the reduction of the population size of several species at the same time period. It can be of moderate consequences on species with Pattern I of endemism, but could lead to the extinction of at least one species with Patterns II and III, or highly endanger several related species. Nevertheless, due to the lower number of species and the higher phylogenetic information carried by each species, the impacts of this kind of threat is higher in species with disjunct distribution.

Table 1. Patterns of endemism and types of geographic distribution of the threats in New Caledonia. Pattern I: species regionally endemic to New Caledonia, distributed in the whole range of an ecosystem; Pattern II: related species with parapatric/allopatric distributions; Pattern III: related species with disjunct distributions. Type A: concentrated threat distributed in the whole range of an ecosystem; Type B: sparse threat distributed in the whole range of an ecosystem; Type C: concentrated threat limited to a small area/region; Type D: sparse threat limited to a small area/region. The frequency and the effect associated to the extension and distribution of the threat in related species of each pattern of endemism are evaluated as + low; ++ medium; +++ medium/high; ++++ high

Distribution of the Threat \ Pattern of Endemism	Pattern I	Pattern II	Pattern III
Type A	Frequency + Effect ++++	+ ++++	+ ++++
Type B	+++ ++	++++ +++	++++ ++++
Type C	+++ +	++++ +++	++++ ++++
Type D	++++ +	+++ +++	+++ ++++

Threats that are very concentrated and limited to a small area/region (Type C) are of moderate frequency. Mining is a good example of this kind of threat.Since this activity changes the landscape of a mountain by totally eliminating the vegetation and destroying the natural habitats, it is prone to lead one or several related species with parapatric/allopatric distribution to the extinction (Pattern II), due to the short scale of distribution of short range endemics and the short distances between different related species, and/or one with disjunct distribution. In opposition, its effects on species with Pattern I are moderate, i.e., it can lead to the reduction of population size, but individuals can survive elsewhere, due to the spatial restriction of the threat. Once more, due to lower number of species and higher phylogenetic information, species with disjunct distribution are the most affected.

Finally, threats that are sparse and limited to a small area/region (Type D) are of moderate to high frequency. This type of effect can be illustrated by invasive species as well as hunting, or selective logging. These activities usually occur in several places in a given area and are repeated several times. Their effects can be of minor importance on wide range species (Pattern I), due to their large scale of distribution, and moderate consequences on close related species with Pattern II, due to the fact that there will be some species that survive. Nevertheless, the impact on species with Pattern III of short range endemism can be strong, since they are usually very few related species in New Caledonia, therefore loosing a lot of genetic or phylogenetic information with the extinction of a single species.

CONCLUSION

In conclusion, the conservation of the biodiversity in a context where species are dominantly short range endemics and rare is a main problem to be faced by New Caledonian authorities as well as by scientific researchers that must provide the basis for political decisions in this respect.

By considering three types of endemic species facing different kinds of threats, we showed that the biodiversity of New Caledonia, which comprises a remarkably high number of short range endemics (Patterns II and III), is especially vulnerable to the most common threats. In most cases, small populations of short range endemics will be severely affected. In the long term, small and declining populations have few chances of lasting several years or decades, even if they can survive some time in very small untouched or less disturbed patches. This situation is even more problematical when one considers that population densities are often very low in New Caledonia, making population remnants especially small.

In this context, fires, mining and introduced species deserve special attention and control. The two firsts for the habitat destruction they promote over extensive areas and their repetition in space all over the territory that make them prone to lead several related species to be extinguished in a short time period. The later by the possibility of continuing to endanger even in areas officially protected.

Phylogenetic and genetic studies must also be developed in order to better understand the corollaries of the extinctions of some short range endemics, in terms of conservation of the biodiversity of New Caledonia whose evolutionary patrimonial value is invaluable.

ACKNOWLEDGMENTS

The present chapter has been made within the framework of the project ANR BIONEOCAL (grant ANR Biodiversité 2007, Philippe Grandcolas). We thank Dr. Tony Robillard and Dr. Eric Guilbert, both from DSE, MNHN, France, for kindly accepting reviewing this chapter.

REFERENCES

Bauer, A. M. & Sadlier, R. A. (1993). Systematics, biogeography and conservation of the lizards of New Caledonia. *Biodiversity Letters, 1*, 107-122.

Beauvais, M. L., Coléno, A. & Jourdan, H. (Eds.) (2006). *Les espèces envahissantes dans l'archipel néo-calédonien - Un risque environnemental et économique majeur*. IRD editions, Paris, 259 + CD.

Bouchet, P., Jaffré, T. & Veillon, J. M. (1995). Plant extinction in New Caledonia: protection of sclerophyll forest urgently needed. *Biodiversity and Conservation, 4*, 415-428.

Bradford, J. & Jaffré, T. (2004). Plant species microendemism and conservation of montane maquis in New Caledonia: two new species of *Pancheria* (Cunoniaceae) from the Roche Ouaïème. *Biodiversity and Conservation, 13*, 2253-2274.

Brescia, F., Pollabauer, C. M., Potter, M. A. & Robertson, A. W. (2008). A review of the ecology and conservation of Placostylus (Mollusca: Gastropoda: Bulimulidae) in New Caledonia. *Molluscan Research, 28*, 111-122.

Chazeau, J. (1995). *Bibliographie indexée de la faune terrestre de Nouvelle-Calédonie. Systématique, écologie et biogéographie*. ORSTOM Editions, Paris, 95.

Chazeau, J. & Tillier, S. (Eds.) (1991). Zoologia Neocaledonica 2. *Mémoires du Muséum national d'Histoire naturelle, 149*, 1-358.

Cochereau, P. & Potiaroa, T. (1995). Caféiculture et *Wasmannia auropunctata* (Hymenoptera, Formicidae, Myrmicinae) en Nouvelle-Calédonie, ORSTOM, Nouméa 20p.

Desutter-Grandcolas, L. & Robillard, T. (2006). Phylogenetic systematics and evolution of *Agnotecous* in New Caledonia (Orthoptera: Grylloidea, Eneopteridae). *Systematic Entomology, 31*, 65-92.

Dupon, J. E. (1986). *The effects of mining on the environment of high islands: a case study of nickel mining in New Caledonia*. South Pacific Regional Environment Programme, Nouméa, 6 p.

Ekstrom, J. M. M., Jones, J. P. G., Willis, J. & Isherwood, I. (2000). *The humid forests of New Caledonia: biological research and conservation recommendations for the vertebrate fauna of Grande Terre*. CSB Conservation Publications, 100.

Gabrié, C., Licari, M. L. & Mertens, D. (1995). *L'état de l'environnement dans les Territoires Français du Pacifique Sud : La Nouvelle Calédonie*, L'institut Français de l'environement, Paris, 115.

Gargominy, O. (2003). *Biodiversité et conservation dans les collectivités françaises d'autre mer*. Paris, Comité Français pour l'UICN, 246+X.

Gargominy, O., Bouchet, P., Pascal, M., Jaffré, T. & Tourneur, J. C. (1996). Conséquences des introductions d'espèces animales et végétales sur la biodiversité en Nouvelle-Calédonie. *Revue d'Ecologie (Terre Vie)*, *51*, 375-402.

Gaston, K. J. (1994). *Rarity*. Chapman & Hall, London, 205.

Grandcolas, P. (Ed.) (2008a). Zoologia Neocaledonica 6. Biodiversity studies in New Caledonia. Paris, *Mémoires du Muséum national d'Histoire naturelle*, *197*, 1-326.

Grandcolas, P. (2008b). Introduction. In: P. Grandcolas (Ed.), Zoologia Neocaledonica 6. Biodiversity studies in New Caledonia. *Mémoires du Muséum national d'Histoire naturelle, Paris*, *197*, 9-12.

Grandcolas, P. (Ed.) (2009). Zoologia Neocaledonica 7. Biodiversity studies in New Caledonia. *Mémoires du Muséum national d'Histoire naturelle*, *198*, in press.

Grandcolas, P., Murienne, J., Robillard, T., Desutter-Grandcolas, L., Jourdan, H., Guilbert, E. & Deharveng, L. (2008). New Caledonia: a very old Darwinian island? *Philosophical Transactions of the Royal Society of London*, *B*, *363*, 3309-3317.

Haase, M. & Bouchet, P. (1998). Radiation of crenobiontic gastropods on an ancient continental island: the *Hemistomia*-clade in New Caledonia (Gastropoda: Hydrobiidae). *Hydrobiologia*, *367*, 43-129.

Jaffré, T., Bouchet, P. & Veillon, J. M. (1998). Threatened plants of New Caledonia: is the system of protected areas adequate? *Biodiversity and Conservation*, *7*, 109-135.

Jourdan, H. (1997). Threats on Pacific islands: the spread of the Tramp Ant *Wasmannia auropunctata* (Hymenoptera: Formicidae). *Pacific Conservation Biology*, *3*, 61-64.

Jourdan, H. (2006). Les invertébrés menaçants pour l'archipel néo-calédonien: recommandations pour leur prévention. *In:* Beauvais, M.-L., Coléno, A. & Jourdan, H. (Eds.) *Les espèces envahissantes dans l'archipel néo-calédonien - Un risque environnemental et économique majeur*. IRD editions, Paris, CD 215-245.

Jourdan, H. & Mille, C. (2006). Les invertébrés introduits dans l'archipel néo-calédonien : espèces envahissantes et potentiellement envahissantes. Première évaluation et recommandations pour leur gestion. *In:* Beauvais, M.-L., Coléno, A. & Jourdan, H. (Eds.) *Les espèces envahissantes dans l'archipel néo-calédonien - Un risque environnemental et économique majeur*. IRD editions, Paris, CD 163-214.

Le Breton, J., Jourdan, H., Chazeau, J., Orivel, J. & Dejean, A. (2005). Niche opportunity and ant invasion: the case of Wasmannia auropunctata in a New Caledonian rain forest. *Journal of Tropical Ecology*, *21*, 93-98.

Loope, L. L. & Pascal, M. (2006). Quelques espèces animales envahissantes aux frontières de la Nouvelle-Calédonie et présentant un risque environnemental majeur. In: M. L. Beauvais, A. Coléno, & H. Jourdan (Eds.), *Les espèces envahissantes dans l'archipel néo-calédonien - Un risque environnemental et économique majeur*. IRD editions, Paris, CD 246-257.

Matile, L., Najt, J. & Tillier, S. (Eds.) (1993). Zoologia Neocaledonica 3. *Mémoires du Muséum national d'Histoire naturelle*, *157*, 1-218.

McKee, H. S. (1994). *Catalogue des plantes introduites et cultivées en Nouvelle-Calédonie*. Muséum national d'Histoire naturelle, Paris, 164.

Meyer, J. Y., Loope, L. L., Sheppard, A., Munzinger, J. & Jaffré, T. (2006). Les plantes envahissantes et potentiellement envahissantes dans l'archipel néo-calédonien : première évaluation et recommandations de gestion. In: M. L. Beauvais, A. Coléno, & H. Jourdan

(Eds.), *Les espèces envahissantes dans l'archipel néo-calédonien - Un risque environnemental et économique majeur.* IRD editions, Paris, CD 50-115.

Morat, P., Jaffré, T. & Veillon, J. M. (1999). Menaces sur les taxons rares et endémiques de la Nouvelle-Calédonie. *Bulletin de la Société Botanique du Centre-Ouest (SBCO), 19*, 129-144.

Morat, P., Jaffré, T. & Veillon, J. M. (2001). The flora of New Caledonia's calcareous substrates. *Adansonia, 23*, 109-207.

Munzinger, J., McPherson, G. & Lowry, P. P. (2008). A second species in the endemic New Caledonian genus *Gastrolepis* (Stemonuraceae) and its implications for the conservation status of high-altitude maquis vegetation: coherent application of the IUCN Red List criteria is urgently needed in New Caledonia. *Botanical Journal of the Linnean Society, 157*, 776-783.

Murienne, J., Grandcolas, P., Piulachs, M. D., Bellés, X., D'Haese, C., Legendre, F., Pellens, R. & Guilbert, E. (2005). Evolution on a shaky piece of Gondwana: is local endemism recent in New Caledonia? *Cladistics, 21*, 2-7.

Murienne, J., Pellens, R., Budinoff, R. B., Wheeler, W. & Grandcolas, P. (2008a) Phylogenetic analysis of the endemic New Caledonian cockroach *Lauraesilpha*. Testing competing hypothesis of diversification. *Cladistics, 24*, 802-812.

Murienne, J., Pellens, R. & Grandcolas, P. (2008b). Short range endemism in the cockroach Lauraesilpha (Blattidae, Tryonicinae) in New Caledonia: distribution and new species. In: P. Grandcolas (Ed.), Zoologia Neocaledonica 6, Systematics and Biodiversity in New Caledonia. *Mémoires du Muséum National d'Histoire Naturelle, 197*, 261-271.

Murienne, J., Guilbert, E. & Grandcolas, P. (2009). Species diversity in the New Caledonian endemic genera Cephalidiosus and Nobarnus (Insecta: Heteroptera: Tingidae), an approach using phylogeny and species distribution modeling. *Biological Journal of the Linnean Society, 95*, in press.

Myers, N., Mittermeier, R. A., Mittermeier, C. G., Fonseca, G. A. B. & Kent, J. (2000) Biodiversity hotspots for conservation priorities. *Nature, 403*, 853-858.

Najt, J. & Matile, L. (Eds.) (1997). Zoologia Neocaledonica 4. *Mémoires du Muséum national d'Histoire naturelle, 171*, 1-399.

Najt, J. & Grandcolas, P. (Eds.) (2002). Zoologia Neocaledonica 5. Systématique et endémisme en Nouvelle-Calédonie. *Mémoires du Muséum national d'Histoire naturelle, 187*, 1- 283.

O'Reilly, P. (1955). *Bibliographie méthodique, analytique et critique de la Nouvelle-Calédonie*. Societé des Océanistes, Paris, *361*.

Pascal, M., Barré, M., Garine-Wichatitsky, M., Lorvelec, O., Frétey, T. & Brescia, F. (2006). Les peuplements néo-calédoniens de vertébrés : invasions, disparitions. In: M. L. Beauvais, A. Coléno, A. & H. Jourdan, (Eds.), *Les espèces envahissantes dans l'archipel néo-calédonien - Un risque environnemental et économique majeur.* IRD editions, Paris, CD 111-162.

Pascal, M., Richer de Forges, B., Le Guyader, H. & Simberloff, D. (2008). Mining and other threats to the New Caledonia biodiversity hotspot. *Conservation Biology, 22*, 498-499.

Pellens, R. (2004). Nouvelles espèces d'*Angustonicus* (Insecta, Dictyoptera, Blattaria, Tryonicinae) et endémisme du genre en Nouvelle-Calédonie. *Zoosystema, 26*, 307-314.

Pintaud, J. C., Jaffré, T. & Puig, H. (2001). Chorology of New Caledonian palms and possible evidence of Pleistocene rain forest refugia. *Comptes rendus de l'Académie des Sciences de Paris, Sciences de la vie, 324*, 453-463.

Pisier, G. (1983). Bibliographie méthodique, analytique et critique de la Nouvelle-Calédonie. 1955-1982. *Publications de la Société d'Etudes Historiques de la Nouvelle-Calédonie, 34*, 1-350.

Sharma, P. & Giribet, G. (2009) A relict in New Caledonia: phylogenetic relationships of the family Troglosironidae (Opiliones: Cyphophthalmi). *Cladistics, 25*, 1-16.

Tillier, S. (1988). (Ed.) Zoologia Neocaledonica 1. *Mémoires du Muséum national d'Histoire naturelle, 142*, 1-158.

Veillon, J. M. (1980). Architecture des espèces néo-calédoniennes du genre Araucaria. *Candollea, 35*, 609-640.

Wiens, J. J. (2004). Speciation and ecology revisited: phylogenetic niche conservatism and the origin of species. *Evolution, 58*, 193-197.

Chapter 61

MACROINVERTEBRATE DISTRIBUTION ON EROSIONAL AND DEPOSITIONAL AREAS INCLUDING A FORMER GRAVEL-PIT. BIODIVERSITY AND ECOLOGICAL FUNCTIONING

A. Beauger, N. Lair** and J. L Peiry****

Laboratoire de Géographie Physique et Environnementale,
UMR 6042 CNRS. Maison des Sciences de l'Homme, 4 rue Ledru,
63057 Clermont-Ferrand cedex, France.

ABSTRACT

The current study was carried out on a longitudinal reach of an intermediate zone of the River Allier (France) that incorporates a flooded gravel pit, representing a significant depositional zone. The survey compared the macroinvertebrate communities of erosional and depositional forms along the study section. Fourty four taxa were recorded in the erosional forms and 46 in the depositional zones, 57 % being common. Globally, the biodiversity in the deepest areas including the former gravel pit, did not diverge significantly from the other parts of the studied section. Further analysis of 22 samples from each geoform type revealed that total and EPT richness per sample were higher in the erosional areas than in the others. In the two geoforms the feeding groups and locomotion/substratum relations differed also, both underlining ecological and functional differences. Body size of the four dominant taxa (Orthocladiinae, Chiromini, *Hydropsyche* and *Psychomyia pusilla*) present in both erosional and depositional zones was also analysed and only Orthocladiinae differed between geoforms and depth, the larger individuals being in depositional zones and deeper water. Overall these results suggest that in the context of a rapid but reliable water quality assessment for biomonitoring, surveys of shallow, and thus more accessible erosional areas of gravelbed rivers may provide adequate representation of whole system. The ability to omit sampling

---

* Corresponding authors: E-mail : *aude.beauger@univ-bpclermont.fr*
** E-mail: *nicole.lair@univ-bpclermont.fr*
*** E-mail: *j-luc.peiry@univ-bpclermont.fr*

of deeper depositional zones without adversely affecting the reliability of the outcome may prove both practicable and cost effective. The fact that erosional forms were seen to contain higher biodiversity than depositional areas and to shelter many of the taxa generally considered the most pollution-sensitive, gives further confidence.

INTRODUCTION

Erosional and depositional events induce natural, physical and hydrological disturbances which shape channel morphology and may thus influence the distribution of macroinvertebrates. Erosional forms are areas where the flow concentrates and scours the substratum. Depositional geoforms, on the other hand, are typically characterised by dispersed flow which allows sediment to settle and accumulate. During flood events these scenarios may be reversed, with sediment deposited in erosional forms and scoured from depositional ones (Petts & Amoros, 1996; Knighton, 1998). These forms correspond to riffles and pools, the two geomorphological units of alluvial channels common in both straight and meandering rivers (Richards, 1976). Erosional riffle and depositional pool forms have been described as "primary habitat classifiers of benthic invertebrates" (Merritt & Cummins, 1996) and have been the subject of a large literature (Minshall & Minshall, 1977; Dunne & Leopold, 1979; Brussock & Brown, 1991; Brown & Brussock, 1991; Carter & Fend, 2001; Pedersen, 2003; Brooks *et al.*, 2005). Previous authors have drawn attention to the differences between the faunas of the shallow fast-flowing erosional forms and the deeper, slow-flowing depositional areas of small alluvial channels. Benthic macroinvertebrate communities are clearly influenced by a suite of interrelated environmental variables, including temperature, oxygen, velocity, depth, substratum, all of which combine to determine distribution and density patterns (Ward, 1992). Other authors have also described seasonal effects at the scale of these geoforms, using taxonomic lists to compare species richness and density (Robson & Chester, 1999; Boyero & Bailey, 2001; Boyero, 2003; Schmera & Eros, 2004) or feeding habits (Baptista *et al.*, 2001; Huryn & Wallace, 1987). Despite this extensive work on small streams, the biodiversity and macroinvertebrate assemblages of larger alluvial systems is poorly documented (Gayraud *et al.*, 2003), with considerable scope for investigation.

Gravel pits incorporated into river systems function as depositional forms (Beauger, 2008), and we might expect this to be reflected in their invertebrate communities. However with the exception of the extraction phase, or the period when gravel-pits are being transformed in ponds or wetlands, the biodiversity and macroinverterbrate assemblages of these artificial water bodies have been little studied (Borcherding & Sturm, 2002; Borcherding *et al.*, 2002). In our recent studies, focused on a succession of riffles and pools located in the intermediate reaches of a temperate zone river incorporating a former gravel pit, previous results have shown that this section of river shelters a wide diversity of macroinvertebrates, including many pollution-sensitive taxa (Beauger *et al.*, 2006; Beauger & Lair, 2008). Our purpose in the current study was to analyse and compare communities in its erosional and depositional areas, including the gravel pit. Particular attention was paid to the distribution of biological and ecological traits at different depths of this part of the river, and also to the relative body size of taxa occurring across the different geoforms.

STUDIED AREA AND SAMPLING PROCESSES

The study site is situated on the alluvial plain of the River Allier, a tributary of the Loire, in the French Massif Central. It is designated a 6th order stream by the classification of Strahler (1957) (Figure 1; Table 1) (Beauger *et al.*, 2006). The study area was a 900m long stretch located at 160 km from the source and 400 m above sea level, characterised by a gravel bed and riffle-pool sequences. The catchment area is of mainly gnessic/granitic geology, with alluvial forest providing shade to areas of the floodplain and limiting bank erosion. The unit stream power (Table 1) indicates a strong potential for transport, and the potential for mobilization of gravels during floods.

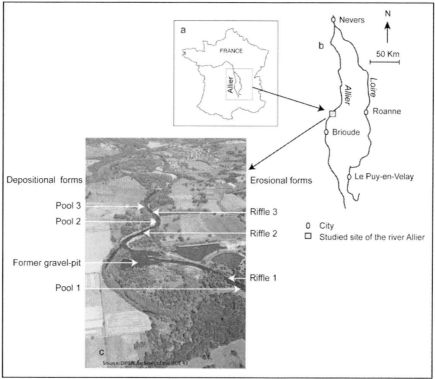

Figure 1. Map of the study site: a) location of the River Allier, b) position of the study reach, c) position of the three riffles at the study reach.

Table 1. Main characteristics of the studied area (* AELB data)

Characteristics	
*Watershed area km²	2750
*Mean annual discharge m³.s⁻¹	30
*10 years flood m³.s⁻¹	500
Slope m.m⁻¹	0.0016
Channel width m	40
Unit stream power at bankfull discharge W.m⁻²	90

A former gravel-pit has captured the river Allier in 1992 (Peiry, 2004), inducing a local break in the bedload transport with a deeper zone. The pit is filling with sediments, and an alluvial fan of particles can be distinguished in its upstream part. Overall, anthropogenic impact on the watershed is low, and limited to the immediate vicinity of the former gravel pit.

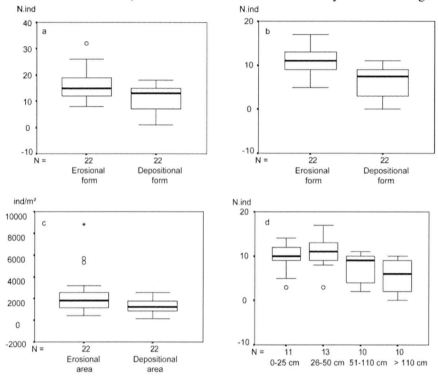

Figure 2. a) Box plot of richness in relation to geoforms. b) Box plot of EPT richness in relation with the forms. c) Box plot of density in relation to geoforms. d) Box plot of EPT richness in relation to depth.

Macroinvertebrate samples were captured from sites identified in the field as erosional and depositional geoforms. The extent of each form was mapped using a DGPS Trimble Pro XRS (submetric accuracy). The locations of the sampling points were also mapped and their hydraulic characteristics determined by measuring surface velocity (Sensa Z300 OTT probe), wide dominant grain size (cf. Malavoi and Souchon, 2002) and depth. The different classes of depths (0 – 25 cm, 26 – 50 cm, 51 – 110 cm and > 110 cm) were created from the distribution of depths at the sampling points. In the erosional forms {riffle, run and boil (Padmore et al., 1998)} animals were collected with a Surber net (sampling surface 0.05 m^2). In the depositional forms {main channel pool, marginal deadwater, damned pool, deep areas located at meander bends (Padmore et al., 1998) and former gravel pit} sampling was performed using a Petersen grab (sampling surface 0.04 m^2). Among the samples collected, 22 samples were taken at random from each of the two form types (including 8 from the former gravel pit) and retained for analysis. Organisms were sorted in situ, and preserved in 10% formaldehyde. Using the methodology of Tachet et al. (2000), insects were identified to genus (apart from the smallest unidentifiable larvae) or species (for monospecific genera), with the exception of the Diptera, which were identified to a level between tribe and family. Non-insect invertebrates, including representatives of the Mollusca, Acheta and Oligochaeta, were identified to the level of genus or family. Body lengths of each animal were measured to the nearest 0.5 mm.

The study took place from the 3rd to the 10th June 2003, following and during a period of hydrological stability, to avoid sediment transport and catastrophic drift. The water discharge was checked using a gauge station surveyed by the official Water Agency Loire-Bretagne, located at about 2 km from the studied area.

DATA ANALYSIS

Successive statistical analyses were performed in order to test for differences in biodiversity between erosional and depositional forms and between depth classes. The Kolmogorov-Smirnov statistic was used to test the hypothesis that data are normally distributed. Then, Kruskal-Wallis tests were used to determine significance of differences in depth, velocity and grain size between the erosional and depositional forms and of differences in community structure including total richness, Ephemeroptera + Plecoptera + Trichoptera (EPT) richness and total density {after ln (x+1) transformation of density records} between geoform types and depth classes. Using log-transformed density data, Correspondence Analyses (CA) were carried out between taxa contributing > 0.1% of relative abundance and samples, offering an overall view of macroinvertebrate distribution across all sampling points. Between-group analyses were then performed to test the differences between geoform types and depth classes. Further CAs with between-group analyses were carried out on data from depositional areas, testing for differences between the invertebrate communities (taxa > 0.1%) of the former gravel pit and the other deep areas. To study the influence of form type and depth on invertebrate body size, specimens of the four taxa present in both erosional and depositional geoforms and in every depth class (the tribes Orthocladiinae and Chiromini, the genus *Hydropsyche* and the species *Psychomyia pusilla)* were retained for measurement, and Kruskal-Wallis tests were performed on the ranks.

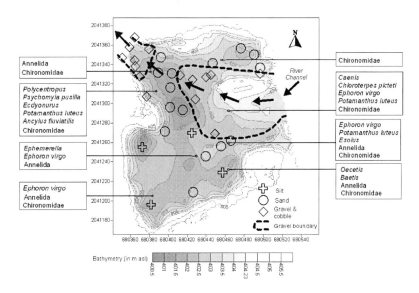

Figure 3. Distribution of macroinvertebrates in relation to depth in the former gravel pit.

Table 2. Principal characteristics (mean ± SD) of the depositional and erosional forms

	Erosional forms	Depositional forms	Depositional forms (without samples done in the former gravel-pit)	Depositional former gravel-pit
Depth (cm)	26 ± 14	102 ± 50	91 ± 36	137 ± 64
Velocity (cm.s^{-1})	67 ± 32	42 ± 14	54 ± 41	16 ± 20
Grain-size (mm)	78 ± 70	25 ± 17	28 ± 15	24 ± 12
Richness (N taxa)	16 ± 0	11 ± 5	14 ± 3	6 ± 2
EPT richness (N taxa)	11 ± 3	6 ± 3	8 ± 2	2 ± 1
Density (ind.m^{-2})	2332 ± 1963	1278 ± 687	1306 ± 599	846 ± 731

Finally, two ecological traits (saprobity and trophic status of the freshwater) and two biological traits (feeding habits and locomotion/substratum relations) were retained in order to qualify the communities of the two geoforms and classes of depth. These traits are the biological and ecological attributes of taxonomic entities, which describe their ecological niche and functional role within an ecosystem (*cf.* Tachet *et al.*, 2000). Trait categories were weighted using the log-transformed abundance of each taxon and analysed using Fuzzy Correspondance Analyses (FCA) (Chevenet *et al.*, 1994), with associated between-group analyses (between geoforms and between depth classes). Statistical analyses were carried out using SPSS (SPSS Inc. 1999) and ADE4 software (Thioulouse *et al.*, 1997).

RESULTS

Physical Characteristics of the Site

The spring 2003 was very dry. For the duration of the sampling period (and the preceding 10 days), river discharge remained low and practically unchanged (gauge: 10.4 ± 1.2 m^3.s^{-1}). The diurnal physico-chemical characteristics of the river were also stable, with pH = 7.8 (± 0.3), oxygen saturation = 81% (± 13) and conductivity = 95 (± 2.5) µS. cm^{-1}. The depth distribution of sampling sites differed significantly from normal ($p < 0.05$) and there were significant differences in depth, velocity and dominant grain size, between the erosional and depositional areas (Kruskal-Wallis test, $p < 0.05$) (Table. 2).

Distribution of the Macroinvertebrate Communities in Relation to Geoform Type and Depth

Overall, the dominant taxa (> 1% of total abundance) were the Plecoptera *Leuctra geniculata*, the Trichoptera *Brachycentrus*, *Hydropsyche* (6%), *Psychomyia pusilla* (7%),

Setodes and small larvae of the Glossosomatidae, the Ephemeroptera *Baetis*, *Ecdyonurus*, *Ephemerella*, *Ephoron virgo*, *Potamanthus luteus* and *Rhithrogena*, the Heteroptera *Micronecta*, the Coleoptera *Esolus* and the Diptera Chironomini (8%), Simuliini, Tanytarsini and Orthocladiinae (the most important taxon, representing 12% of the total sampled).

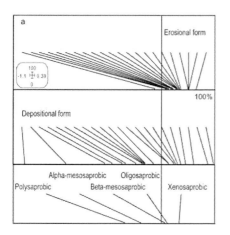
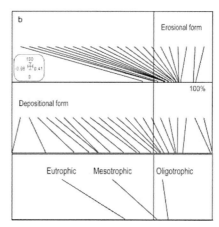

Figure 4. First factorial axis of samples taken from erosional and depositional forms of the River Allier, featuring the ecological traits "saprobity" (a) and "trophic status" (b) in the between-geoform analysis.

Overall, the survey recorded 44 taxa from the erosional forms and 46 taxa from the depositional areas. Of the total list, 57% taxa occurred in both erosional and depositional geoforms. The dominant organisms in the erosional forms were members of the Orthocladiinae (15%) and the genus *Ephemerella* (11%) whereas the depositional zones were dominated by Chironomini (17%) and the family Glossosomatidae (15%). The communities of the former gravel pit were dominated by two tribes of chironomid dipteran: the Chironomini (52%) and Tanytarsini (10%), and by the ephemeropteran *Ephoron virgo* (6%). The gastropod *Ancylus fluviatilis* represented 5% of the total community, but occurred only on the coarsest particles (size ranging from 32 to 64 mm) sampled from the former gravel pit. Rare items such as *Ephemera* and *Aphelocheirus aestivalis*, were recorded only in the erosional zones, while others, such as *Chloroterpes picteti* and *Thraulus bellus* appeared to be restricted to the depositional forms.

The CA with between-group analyses performed using whole community data identified no difference between the significant fauna (those contributing > 0.1%) of the two geoforms or the four depth classes. Similarly there was no difference when communities of the former gravel pit were compared with those of the other deep areas using taxonomic data (> 0.1%) from the depositional forms only.

While total richness was normally distributed across the sample set, the distributions recorded for EPT richness and total density differed significantly from normal ($p < 0.05$). Total richness, EPT richness and total density differed significantly between geoforms, with the highest values for each parameter being obtained in the erosional forms (Table 3, Figure 2 a, b, c). EPT richness was also influenced by depth, being greatest from 26 to 50 cm (Figure 2 d) and lowest at depths >110 cm. Richness was particularly low in the former gravel pit, which yielded just some Trichoptera (*Polycentropus*, *Psychomyia pusilla* and *Oecetis*) and some individuals of Ephemeroptera (*Baetis*, *Caenis*, *Chloroterpes picteti*, *Ecdyonurus*, *Ephemerella*, *Ephoron virgo* and *Potamanthus luteus*) (Figure 3).

Distribution of the Dominant Taxa in Relation to Body Size

Body size responses to abiotic variables differed between the four dominant taxa (Table 4). Of the 491 Orthocladiinae collected in the field, 428 were present in the erosional forms and 390 from the 26 to 50 cm depth class. The largest individuals however were recorded in the depositional forms (2.92 ± 0.94 mm) and at depths >110 cm (3.08 ± 1.22 mm) (Table 4).

Among the 284 Chironomini collected in the field, a majority, 195, were present in the depositional forms, and 125 were found in the > 110 cm depth class. There was no significant difference in the distribution of body size between geoforms and depth class, however specimens collected in the former gravel pit were dominant and larger than those recorded elsewhere (body size: 4.95 ± 2.10 mm).

Among the 240 *Hydropsyche* collected, 196 were occurred in erosional forms, and 141 within the 26 to 50 cm depth class. Only 12 animals were found deeper than 110 cm. Once again, however, there were no significant differences in body size distribution between geoforms and depth classes.

Table 3. Results of the Kruskal-Wallis test seeking effects of geoform type and depth class on richness, EPT richness and density
(NS = non significant)

	Geoforms	Depths
Richness	p = 0.015	NS
EPT richness	p = 0.000	p = 0.003
Density	p = 0.025	NS

Table 4. Results of Kruskal-Wallis tests performed on the body size of different taxa in relation to geoform and depth class
(NS = non significant)

	Geoforms	Depth
Orthocladiinae	p = 0.001	p = 0.014
Chiromini	NS	NS
Hydropsyche	NS	NS
Psychomyia pusilla	NS	p = 0.017

Table 5. Contribution to total variability of ecological and biological traits in different forms and depth classes (NS = non significant)

	Between-class analysis	Variability (%)	Significativity $p < 0.05$
Saprobity	Zone	12	0.007
	Depth	NS	NS
Trophic status	Zone	16	0.004
	Depth	NS	NS
Feeding habits	Zone	9	0.009
	Depth	16	0.021
Locomotion	Zone	13	0.001
	Depth	16	0.03

Finally, among the 271 *Psychomyia pusilla*, 214 were present in the erosional forms, with 204 occurring in the 26 to 50 cm stratum, while just 10 individuals were recorded in the shallowest depth class, < 26 cm. Geoform type had no significant observable effect on body size, but there were significant differences between depth classes with the largest animals sampled in the shallower waters ranging from 0 to 25 cm (4.50 ± 0.71 mm) and from 26 to 50 cm (3.99 ± 0.79 mm).

Distribution of Macroinvertebrate Communities in Relation to Biological and Ecological Trait

FCA with between-group analyses suggested that significant differences in saprobity between the two geoforms accounted for 12% of overall community variability (Table 5). The assemblages of erosional forms were typical of xeno-, oligosaprobic and ß-mesosaprobic waters whereas those of depositional areas were indicative of α-, β-mesosaprobic or polysaprobic water conditions (Figure 4 a). Among taxa collected from the gravel pit, 36% were indicators of β- mesosaprobic conditions, 26% corresponded to α- mesosaprobic waters and 10 % to polysaprobic waters. In the other depositional areas, the proportion of polysaprobic indicators remained low (5%).

Using faunal assemblages as indicators of trophic status, there was 16% of variability between the two geoforms. The erosional forms were characterized by taxa typical of oligo- and mesotrophic waters, whereas in the depositional forms, the fauna belonged to groups normally associated with mesotrophic and eutrophic conditions (Figure 4 b). In the gravel pit, the faunal present indicate that the water was partly meso- (40%) and partly eutrophic (41%) whereas in the other areas of the depositional forms, only 27% of taxa were characteristic of eutrophic conditions.

The FCA revealed significant differences in the feeding habits of macroinvertebrates living in the different geoforms and depth classes (accounting for 9% and 16% of variability respectively). In the erosional forms the organisms tended to be shredders, scrapers, grazers, predators and filterers, whereas in the depositional zones they were mainly deposit feeders, predators, filterers and parasites. Shredders were associated with the shallow areas (0-25 cm), while scrapers grazers and filterers dominated between 26 and 110 cm (Figure 5 a). The deeper areas (> 110 cm) were characterised by deposit feeders, predators and parasites. Deposit feeders dominated the gravel pit community (32%), while scrapers and grazers occupied other depositional forms (30%).

Further significant differences between sites were observed in terms of the locomotion modes of invertebrate faunas. These accounted for 13% of variability between geoforms and 16% of variability between depth class. In erosional forms the organisms were crawlers in shallow areas (< 50 cm) and swimmers in deeper water. In the depositional forms and at depths > 51 cm, the faunas were dominated by burrowers, fliers, interstitial and temporarily attached taxa (Figure 5 b). Crawlers (42%) and burrowers (23%) predominated at the bottom of the gravel pit, while in other depositional areas the benthic fauna comprised mainly crawlers (31%) and temporarily attached taxa (23%).

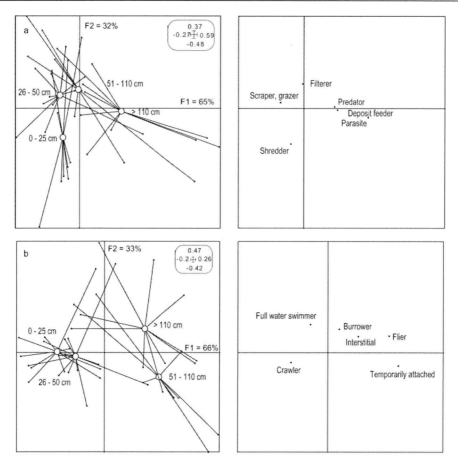

Figure 5. First factorial plan of the inertia centres of samples taken from different depths featuring the biological traits "feeding habit" (a) and "locomotion & substratum relation" (b) in the between-depth class analysis.

DISCUSSION

Considered as biological indicators, benthic macroinvertebrates are recognized to be used as an efficient tool to assess water quality, particularly in the upstream river sections. However, in most of bioassessment methods used in the world (Armitage et al., 1983; Kerans et al., 1992; Smith et al., 1999; Plotnikoff & Wiseman, 2001), they are collected in multiple habitats representing both erosional and depositional forms. Given that the distribution of such habitats is not consistent between systems, meaningful comparisons are often very difficult, while obtaining a good picture of the river biodiversity remains essential (Beauger, 2008; Beauger et al., 2009).

In this way, this study set out to analyse the main ecological characteristics of a river site and compare the biodiversity characterising its erosional and depositional geoforms. The section where sampling took place is a typical "transition zone" between upper and middle reaches of the Allier River (Beauger et al., 2006). Sampling took place during (and after) a week of hydrological stability. Overall, the fauna living in different habitats within the surveyed section remained statistically unchanged whatever the local conditions (depth, velocity and grain size) and the biodiversity recorded within the 22–sample series taken from each geoform was analogous, with a high proportion of taxa (57 %) occurring in samples from both geoforms types and in all depth classes.

Using both long-term survey and experimental data, previous authors have described significant biological differences between erosional and depositional forms within rivers. These differences have been attributed to physical conditions and grain size (the latter being dependant on the physical conditions) (Minshall & Minshall, 1977; Carter & Fend, 2001). Notably, these studies have recorded that most of the groups of organisms found in erosional forms were less abundant or absent in depositional areas. In the present study, there were significant differences in total richness and EPT richness between samples from different geoforms and the biodiversity, including that of pollution-sensitive EPT taxa, was indeed higher in the erosional forms. This difference can be attributed to characteristics of the erosional forms including: 1) frequent reworking of the bedload deposition, that prevent filling of the substratum with silt and organic matter (Keller and Melhorn, 1973); 2) enhanced oxygenation due to increased velocity and downstream hydrological exchanges by downwelling and upwelling processes (Vaux, 1968, Hendricks and White, 1991, Stanford and Ward, 1993, Mermillot-Blondin *et al.,* 2000; Lefebvre *et al.*, 2006) and 3) relatively large grain size (Beauger *et al.*, 2006). Similar differences were apparent in the observed faunal abundance, and were enhanced in the former gravel pit, which was characterized by the lowest total and EPT richness and lowest density. In this area, the accumulation of fine sediments at depth (Rempel *et al.*, 2000) certainly influences the biodiversity of the invertebrates, with the restricted potential for periphyton build-up a possible limiting factor for life (Quinn and Hickey, 1994).

While overall diversity was similar between sites, differences became more easily resolved when considering certain ecological and functional characteristics of the faunal assemblages. Feeding habits and locomotion/ substratum relations are known to be influenced by "local geomorphology and related physical parameters" (Wohl *et al.*, 1995) and differences between the two geoforms may be reflected in variations in saprobity and trophic status. In the Allier River, shredders typical of oligosaprobic and ß-mesosaprobic waters (*Leuctra*, *Ephemerella*) were associated mainly with the erosional forms and depths < 50 cm. Crawling scrapers and grazers such as *Ecdyonurus, Baetis,* and *Psychomyia pusilla*, adapted to withstand current (for example by dorso-ventral flattening) and to forage or avoid light in between coarse particles (Cummins, 1996), were found in the river at depths < 50 cm. In the depositional zones, higher levels of fermentable organic matter and organic elements derived from photosynthesis process may explain: 1) the preponderance of burrowing deposit feeders such as Chironomids, known to be well adapted (Minshall & Minshall, 1977; Vos, 2001), and 2) the presence of predators such as Tanypodinae, both of which are indicative of reduced primary production. A high concentration of organic matter (saprobity) would suggest that the depositional geoforms act as decantation areas and would explain the dominance of burrowing taxa.

Among the four dominant taxa recorded in this current study, depth effects were apparent in the Orthocladiinae and *Psychomyia pusilla*. The largest Orthocladiinae were found in the deepest areas (> 110 cm) of the depositional forms and particularly in the former gravel pit, probably because the low velocity and absence of coarse sediment transport "place low energetic demands" on benthic animals (Rempel *et al.*, 2000). The largest *Psychomyia pusilla* were sampled only at depths < 50 cm, characteristic of the erosional forms in which these taxa lived preferentially. Linked to the shear velocity, sometimes implicated in the body size distribution of aquatic fauna (Sagnes *et al.*, 2008), such preferences may also be due to the increased development of periphyton in these areas. However, *Psychomyia pusilla* were less numerous at depths < 25 cm (10 individuals), this is the zone of maximal irradiance, a factor known to exert direct effects on aquatic insects (Ward, 1992).

CONCLUSION

Knowledge of the relationships between environmental variables and aquatic communities is essential to understand the functioning of rivers. However, when considering their physical and biological descriptors as a whole, large alluvial streams remain relatively poorly documented (Gayraud *et al.*, 2003). The current study focused on the intermediate zone of a gravelbed river, with the additional aim of describing the influence of a former gravel pit now incorporated into the course of the river. The gravel workings aside, human impacts on this river are low, and with 57% of common taxa, including numerous pollution-sensitive items, retain a remarkable biodiversity whatever the geoform. Erosional and depositional areas presented two different systems of functioning, in which the biodiversity remains higher in the erosional forms. In the context of biomonitoring programs, this work suggests that sampling protocols focussing solely on the shallower, more accessible erosional geoforms, could yield sufficiently good representation of the biodiversity of the studied area, sampling of less accessible depositional areas proving to be useless. The fact that erosional forms were seen to contain higher biodiversity than depositional areas and to shelter many of the taxa generally considered the most pollution-sensitive, gives further confidence. A work in progress will seek to validate these results in other rivers.

REFERENCES

Armitage, P. D., Moss, D., Wright, J. F. & Furse, M. T. (1983). The Performance of a new Biological Water Quality Score System Based on Macroinvertebrates Over a Wide Range of Unpolluted Running-Water Sites. *Water Research*, *17*, 333-47.

Baptista, D. F., Dorvillé, L. F. M., Buss, D. F. & Nessiamian, J. L. (2001). Spatial and temporal organization of aquatic insects assemblages in the longitudinal gradient of a tropical river. *Revista Brasileira de Biologia*, *61*, 295-304.

Beauger, A., Lair, N., Reyes-Marchant, P. & Peiry, J. L. (2006). The distribution of macroinvertebrate assemblages in a reach of the River Allier (France), in relation to riverbed characteristics. *Hydrobiologia*, *571*, 63-76.

Beauger, A. & Lair, N. (2008). Keeping it simple: benefits of targeting riffle-pool macroinvertebrate communities over multi-substratum sampling protocols in the preparation of a new European biotic index. *Ecological Indicators*, *8*, 555-563.

Beauger, A., Lair, N., Peiry, J. L. & Reyes-Marchant, P. (2009). A Sampling Method to Assess Water Quality Based on Benthic Macroinvertebrates Living in Geomorphological Unit Riffles Typical of Gravel-bed Rivers. *7th International Symposium on Ecohydraulics. Restoration – Hydraulic or Water Quality or Biotic*; *9*.

Beauger, A. (2008). Impact de la capture d'un chenal fluviatile par une ancienne gravière, sur la distribution des macroinvertébrés benthiques dans trois seuils successifs. *Revue des Sciences de l'Eau, 21*, 87-98.

Borcherding, J. & Sturm, W. (2002). The seasonal succession of macroinvertebrates, in particular the zebra mussel (Dreissena polymorpha), in the River Rhine and two neighbouring gravel-pit lakes monitored using artificial substrates. *International Review of Hydrobiology, 87*, 165-181.

Borcherding, J., Bauerfeld, M., Hintzen, D. & Neumann D. (2002). Lateral migrations of fishes between floodplain lakes and their drainage channels at the Lower Rhine: diel and seasonal aspects. *Journal of Fish Biology, 61*, 1154-1170.

Brown, A. V. & Brussock, P. P. (1991). Comparisons of benthic invertebrates between riffles and pools. *Hydrobiologia, 220*, 99-108.

Brussock, P. P. & Brown, A. V. (1991). Riffle-pool geomorphology disrupts longitudinal patterns of stream benthos. *Hydrobiologia, 220*, 109-117.

Boyero, L. (2003). Multiscale patterns of spatial variation in stream macroinvertebrate communities. *Ecological Research, 18*, 365-379.

Boyero, L. & Bailey, R. C. (2001). Organization of macroinvertebrate communities at a hierarchy of spatial scales in a tropical stream. *Hydrobiologia, 464*, 219-225.

Carter, J. L. & Fend, S. V. (2001). Inter-annual changes in the benthic community structure of riffles and pools in reaches of contrasting gradient. *Hydrobiologia, 459*, 187-200.

Chevenet, F., Dolédec, S. & Chessel, D. (1994). A fuzzy coding approach for the analysis of long-term ecological data. *Freshwater biology, 31*, 295-309.

Cummins, K. W. (1996). Invertebrates. In Petts & Calow editors. *River Biota Diversity and Dynamics.* Oxford, UK. : Blackwell Sciences, 16.

Dunne, T. & Leopold, L. B. (1979). Water in Environmental Planning. Freeman. San Francisco.

Gayraud, S., Statzner, B., Bady, P., Haybachp, A., Schöll, F., Usseglio-Polatera, P & Bacchi, M. (2003). Invertebrate traits for the biomonitoring of large European rivers: an initial assessment of alternative metrics. *Freshwater Biology, 48*, 2045-2064.

Hendricks, S. P. & White, D. S. (1991). Physicochemical patterns within a hyporheic zone of a northern Michigan river, with comments on surface water patterns. *Canadian Journal of Fisheries and Aquatic Sciences, 48*, 1645-1654.

Huryn, A. D. & Wallace, J. B. (1987). Local geomorphology as a determinant of macrofaunal production in a moutain stream. *Ecology, 68*, 1932-1942.

Keller, E. A. & Melhorn, W. N. (1973). Bedforms and fluvial processes in alluvial stream channels: selected observations. In Binghamto Morisawa M. editor. *Fluvial geomorphology.* New York: New York State University Publications, 30.

Kerans, B. L., Karr, J. R. & Ahlstedt, S. A. (1992). Aquatic invertebrate assemblages: spatial and temporal differences among sampling protocols. *Journal of the North American Benthological Society, 11*, 377-390.

Knighton, D. (1998). Fluvial forms and processes. A new perspective. New York: Oxford University press.

Lefebvre, S., Marmonier, P. & Peiry, J. L. (2006). Nitrogen dynamics in rural streams : differences between geomorphologic units. *Annales de Limnologie. International Journal of Limnology, 42*, 43-52.

Malavoi, J. R. & Souchon, Y. (2002). Description standardisée des principaux faciès d'écoulement observables en rivière : clé de détermination qualitative et mesures physiques. *Bulletin Français de la Pêche et de la Pisciculture*, 365/366, 357-372.

Mermillod-Blondin, F., Creuze des Chatelliers, M., Marmonier, P. & Dole-Olivier, M. J. (2000). Distribution of solutes, microbes and invertebrates in river sediments along a riffle-pool sequence. *Freshwater Biology*, 44, 255-269.

Merritt, R. W. & Cummins, K. W. (1996). An introduction to the aquatic insects of North America. Kendall/Hunt Publishing Company, Dubuque, Iowa.

Minshall, G. W. & Minshall, J. N. (1977). Microdistribution of benthic invertebrates in a rocky mountain (U.S.A) stream. *Hydrobiologia*, 55, 231-249.

Padmore, C. L., Newson, M. D. & Charlton, M. E. (1998). Instream habitat in gravel-bed rivers: identification and characterization of biotopes. In Klingeman, P. C. et al. editors. *Gravel-bed rivers in the environment.* Englewood, Colorado: Water Resources Publications, 19.

Pedersen, M. L. (2003). Physical habitat structure in lowland streams and effects of disturbance. PhD Thesis. University of Copenhagen.

Peiry, J. L. (2004). Les kayaks de la recherche, où comment quantifier la recharge sédimentaire et les transports solides fluviatiles. Microscoop, Le Journal du CNRS en Délégattion Centre-Auvergne-limousin, Hors série n°13, 18-21.

Petts, G. E. & Amoros C. (1996). Fluvial hydrosystems. Petts G.E. & Amoros C. editors. London: Chapman & Hall,.

Plotnikoff, R. & Wiseman C. (2001). Benthic Macroinvertebrate Biological Monitoring Protocols for Rivers and Streams: 2001 Revision. Publ N°01-03-028. Dept. Ecology. Olympia, WA.; *34*.

Quinn, J. M. & Hickey, C. W. (1990). Characteristics and classification of benthic invertebrate communities in 88 New Zealand rivers in relation to environmental factors. *New-Zealand Journal of the Marine and Freshwater Research*, 24, 387-409.

Rempel, L. L., Richardson, J. S. & Healey, M. C. (2000). Macroinvertebrate community structure along gradients of hydraulic and sedimentary conditions in a large gravel-bed river. *Freshwater Biology*, 45, 57-73.

Richards, K. S. (1976). The morphology of riffle-pool sequences. *Earth Surface Processes*, 1, 77-88.

Robson, B. J. & Chester, E. T. (1999). Spatial patterns of invertebrates species richness in a river: the relationship between riffles and microhabitats. *Australian Journal of Ecology*, 24, 599-607.

Sagnes, P., Mérigoux, S. & Péru, N. (2008). Hydraulic habitat use with respect to body size of aquatic insect larvae : case of six species from a French Mediterranean type stream. *Limnologica, 38*, 23-33.

Schmera, D. & Eros, T. (2004). Effect of riverbed morphology, stream order and season on the structural and functional attributes of caddisfly assemblages (Insecta: Trichoptera). *Annales de Limnologie. International Journal of Limnology*, 40, 193-200.

Smith, M. J., Ky, W. R., Edward, H. D, Papas, P. J., Richardson, K. St, J., Simpson, J. C., Pinder, A. M., Dale, D. J., Horwitz, P. H. J., Davis, J. A., Yung, F. H., Norris, R. H. & Halse, S. A. (1999). AusRivAS: using macroinvertebrates to assess ecological condition of rivers in Western Australia. *Freshwater Biology*, 41, 269-282.

Stanford, J. A. & Ward, J. V. (1993). An ecosystem perspective of alluvial rivers: connectivity and the hyporheic corridor. *Journal of the North American Benthological Society*, *12*, 48-60.

Strahler, A. N. (1957). Quantitative analysis of watershed geomorphology. *Transactions of the American Geophysical Union*, *38*, 913-920.

Tachet, H., Richoux, P, Bournaud, M & Usseglio-Polatera, P. (2000). Invertébrés d'eau douce. Systématique, biologie, écologie. CNRS Editions. Paris: 2000.

Thioulouse, J, Chessel, D, Dolédec, S. & Olivier, J. M. (1997). A multivariate analysis and graphical display software. *Statistics and Computing*, *7*, 75-83.

SPSS Inc. (1999). SPSS version 10.0 for Windows. SPSS Inc., Chicago.

Vaux, W. G. (1968). Intergravel flow and interchange of water in a streambed. *Fisheries Bulletin*, *66*, 479-489.

Vos, J. H. (2001). Feeding of detritivores in freshwater sediments. PhD Thesis. University of Amsterdam.

Ward, J. V. (1992). Aquatic insect ecology. *Biology and habitat*. New York: John Wiley & sons, Inc;.

Wohl, D. L., Wallace, J. B. & Meyer, J. L. (1995). Benthic macroinvertebrate community structure, function and production with respect to habitat type, reach and drainage basin in the Southern Appalachians (U.S.A.). *Freshwater Biology*, *34*, 447-464.

In: Encyclopedia of Environmental Research
Editor: Alisa N. Souter
ISBN: 978-1-61761-927-4
© 2011 Nova Science Publishers, Inc.

Chapter 62

PLANT BIODIVERSITY HOTSPOTS AND BIOGEOGRAPHIC METHODS

Isolda Luna-Vega and Raúl Contreras-Medina

Departamento de Biología Evolutiva, Facultad de Ciencias,
Universidad Nacional Autónoma de México (UNAM),
Apartado Postal 70-399, 04510 México, D.F. Mexico.

ABSTRACT

Current loss of biodiversity places a premium on the task of recognizing and formulating proposals on potential areas for biological conservation based on scientific criteria; among these tasks, identification of hotspots has a relevant role on conservation of biodiversity. Among the approaches used in their recognition, biogeographic methods have a relevant role. In this study, we discuss the application of different biogeographic methods to identify plant biodiversity hotspots, based on the congruence among areas of endemism, panbiogeographic nodes, and Pleistocene refugia. Land plants are one of the best known biological groups that can be used as a test model for studies in biological diversity. Our study is based on previous biogeographic analyses of mosses, gymnosperms and angiosperms carried out in different places around the world, where panbiogeographic nodes and areas of endemism have been proposed for these plants, also including refugia proposed worldwide based on different biological groups. A remarkable congruence among these areas recognized by the application of different biogeographic methods for these plant groups is noted in western North America, Appalachian, Mesoamerica, southern Chile, southeastern Brazil, central Africa, Japan, southeastern China, Tasmania, New Caledonia, northeastern Australia, New Guinea, and New Zealand; this congruence indicates that these areas deserve special status in plant biodiversity and conservation. We propose that these areas identified by different biogeographic methods are important biodiversity areas and can be recognized as hotspots; these areas have a relevant role in plant biodiversity and are important in conservation due to their climatic conditions (refugia), the historical factors that have been involved in their evolution (nodes), and the restricted distribution of some plant taxa that inhabit them (areas of endemism).

INTRODUCTION

The current loss of biodiversity places a premium on the task of recognizing and formulating proposals on potential areas for biological conservation based on scientific criteria. Many human activities cause the loss of biodiversity, such as deforestation, animal husbandry, and plant cultivation (Challenger 1996). These similarly affect mosses and seed plants. Many bryophytes in temperate and tropical climates strongly depend on the presence of the seed plant vegetation that produces both the microclimatic conditions and the substrate for the bryophytes (Schofield 1985), and, in consequence, elimination of the forest which directly affects to seed plants, leads to the destruction of the resident bryophytes and animals associated with these plant communities. Due to this loss of biodiversity, a current main challenge is the recognition of important areas for conservation. Among this task, identification of hotspots has a relevant role in conservation of biodiversity (Myers et al. 2000).

Several methods have been proposed for the recognition of hotspots, the most frequent being the location of areas with high taxonomic richness; other methods include panbiogeography (Luna et al. 2000; Álvarez and Morrone 2004, Morrone 2009), complementary analysis (Ortega-Baes and Godínez-Álvarez 2006; Villaseñor et al. 2006), phylogenetic diversity (Forest et al. 2007), and endemism indices (Contreras-Medina and Luna 2007), among others.

In addition to criteria currently used, evolutionary biogeographic methods play a relevant role in the recognition of hotspots, as mentioned previously by various authors (Morrone and Espinosa 1998; Luna et al. 2000, 2009). The biogeographic regionalization of the Earth has been carried out since the nineteenth century (e.g. De Candolle 1820; Sclater 1858), and is continuously updated (e.g. Morrone 2001); it is based on the presence of endemic plant and animal taxa, to delimit areas of endemism. Endemism as a criterion has been used to characterize the floristic regions of the world (Delgadillo 1994). In fact, areas of endemism represent the basic units in cladistic biogeography and can help to determine priorities in conservation biology. These areas of endemism are also known as centers of endemism, distribution centers, centers of evolution, and core areas (Haffer 1985).

The concept of node is a contribution of the panbiogeographic method, originally developed by Croizat (1958, 1964). A node represents a geological and biological complex area (Craw 1982), recognized by the intersection of two or more generalized tracks obtained from the overlap of individual tracks of different animal or plant taxa (Morrone and Crisci 1995). Heads (1989) proposed that nodes have four main features: 1) presence of endemic taxa, 2) absence of widespread taxa, 3) phylogenetic and geographic relationships or affinities with several areas at once, and 4) phylogenetic and geographic boundary zones. Because of the biological component of these features, some authors have proposed the identification of nodes to identify priority areas for biodiversity conservation (Morrone and Crisci 1995; Grehan 1993; Morrone and Espinosa 1998; Luna et al. 2000). In fact, Craw (1989) recognized some relationship between areas of endemism and nodes because he defined a panbiogeographic node as an "area of endemism where two or more generalized tracks overlap".

The refuge model was proposed originally by Haffer (1969) to explain the high taxonomic diversity of birds in tropical South America, especially in the Amazonian area, which has been considered as a zone of taxonomic differentiation (Cracraft 1985). Refugia have been considered as areas with stable climatic conditions during dry climatic intervals following the reduction of once more extensive habitat due to glaciation; these areas allowed some organisms to survive during adverse climatic conditions. In the case of gymnosperms, the refuge model may be important in the spatial evolution of the genus *Pinus* (Millar 1998) and *Ceratozamia* (Pérez-Farrera et al. 2001), whereas for angiosperms and mosses these areas explain the presence of relict distributions (Toledo 1982; Schuster 1983; Hooghiemstra and Van der Hammen 1998). Delimitation of refugia is mainly based on geomorphological, geological, and palynological studies, and secondarily on distributional data of endemic taxa (Haffer 1982, 1985), although there exists some controversy on the criteria to recognize refugia (Amorim 1987; Morrone 2009). Refugia have been identified in both tropical and temperate areas of the world (Haffer 1982), and have been used in the design of biological reserves (Graham 1995), and in conservation programs for tropical South America (Brown 1987).

Because of the importance of these three mentioned approaches in biogeography and specifically in their application in biological conservation, we examine herein the use of different biogeographic methods to identify hotspots for mosses, gymnosperms, and angiosperms, based on the congruence among areas of endemism, panbiogeographic nodes, and Pleistocene refugia.

Biogeographic Studies and Distributional Patterns of Land Plants

Land plants are one of the best known biological groups and can be used as a test model for studies in biological diversity. In this study we included previous biogeographic analyses of mosses, gymnosperms and angiosperms carried out in different places of the world where panbiogeographic nodes and areas of endemism have been proposed for these plant groups, including those on refugia proposed worldwide.

In the case of mosses, studies by Schuster (1983), Delgadillo (1994), and Tan and Pócs (2000), presented the most important areas of endemism and diversity for bryophytes in the world (Figure 1). In relation to panbiogeographic analyses, we revised the studies of Tangney (1989, 2007) for mosses of the southern Hemisphere, where it is proposed four nodes located in New Guinea, New Caledonia, eastern Australia, and New Zealand (Figure 2); these analyses represent two of the few carried out with these non-vascular plants.

For gymnosperms we used the 23 areas of endemism proposed by Contreras-Medina and Luna (2002) based on the distribution of genera of Cycadales, Ginkgoales and Coniferales (Figure 3). Contreras-Medina et al. (1999) carried out a panbiogeographic study with gymnosperm genera and proposed seven nodes located in Japan, southeastern China, western North America, New Caledonia, New Zealand, northeastern Australia, and Tasmania (Figure 4).

In the case of flowering plants, we used as framework for areas of endemism the work of Takhtajan (1986), which presented a regionalization of the world mainly based on the distribution of genera and species of flowering plants (Figure 5). For angiosperms, there are many studies where panbiogeographic methods have been applied; studies include the genus *Cecropia* (Franco-Rosselli and Berg 1997), diverse genera and families of flowering plants distributed in the southern Hemisphere (Katinas et al. 1999), genera of Theaceae (Luna and

Contreras-Medina 2000), several genera of the cloud forests (Luna and Alcántara 2002), genera of Euphorbiaceae (Martínez and Morrone 2005), the genera *Rhus* (Andrés et al. 2006), *Piper* (Quijano-Abril et al. 2006), and *Bomarea* (Alzate et al. 2008), and several genera and species of New Caledonia (Heads 2008). In these studies, several nodes were proposed worldwide (Figure 6).

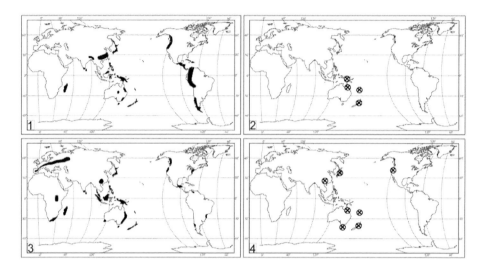

Figures 1-4. (1) Areas of endemism proposed for mosses. (2) Nodes proposed from panbiogeographic analysis of mosses. (3) Areas of endemism proposed for gymnosperms. (4) Nodes proposed from panbiogeographic analysis of gymnosperms.

From our results, a remarkable congruence among nodes, areas of endemism and refugia is noted (Figure 7); this fact is interesting, because these approaches have been seen as competing biogeographic methods. The congruence among areas of endemism and nodes is not general and complete, but some of the nodes are related with areas of endemism if we consider also other taxa. In the same way, those areas of endemism that do not coincide with nodes of these three groups of plants represent nodes for other taxa.

The remarkable congruence among these areas recognized from different biogeographic methods for these plant groups is noted in western North America, Appalachian, Mesoamerica, southern Chile, southeastern Brazil, central Africa, Japan, southeastern China, Tasmania, New Caledonia, northeastern Australia, New Guinea, and New Zealand (Figure 8); this congruence indicates that these areas deserve special status in plant biodiversity and conservation. We propose herein that such areas that are congruent using different biogeographic methods are important biodiversity areas and can be recognized as hotspots.

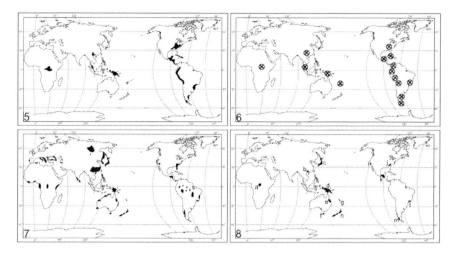

Figures 5-8. (5) Areas of endemism proposed for angiosperms. Only the provinces of Takhtajan (1986) that are congruent with nodes are drawn. (6) Nodes proposed from different panbiogeographic analyses of angiosperms. (7) Refugia proposed in the world. (8) Hotspots resulted from the overlap among areas of endemism, panbiogeographic nodes and Pleistocene refugia. (a) central Africa, (b) southeastern China, (c) Japan, (d) New Guinea, (e) northeastern Australia, (f) Tasmania, (g) New Caledonia, (h) New Zealand, (i) western North America, (j) Appalachian, (k) Mesoamerica, (l) southeastern Brazil, and (m) southern Chile.

Hotspots and Their Coincidence with Areas of Endemism, Nodes and Refugia

The term hotspot was originally used to refer to areas where high levels of species richness, endemism, and human threat coincide (Myers 1988). Currently, this concept has been used to refer to areas with extreme taxonomic richness (e.g. Prendergast et al. 1993; Gaston and Williams 1996; Contreras-Medina and Luna 2007). In this sense, Myers et al. (2000) proposed 25 hotspots distributed worldwide. A significant congruence among the areas herein proposed and the hotspots of Myers et al. (2000) is noted. We may add to the list six new areas, namely Appalachian, Japan, central Africa, Tasmania, northeastern Australia, and New Guinea, based on the overlap of areas recognized by the three different methods used in this study from land plant distribution.

Areas identified by the three different approaches (Figure 8) represent an independently valuable source to identify areas for conservation as suggested earlier by several authors (Brown 1987; Grehan 1993; Graham 1995; Morrone and Crisci 1995; Morrone and Espinosa 1998; Luna et al. 2000; Contreras-Medina et al. 2001); those areas detected by the three approaches have a relevant role in plant biodiversity and are important in conservation due to their climatic conditions (refugia), the historical factors that have been involved in their evolution (nodes), and the distributional congruence among plant taxa with restricted distribution inhabiting them (areas of endemism); their congruence reinforces the importance of those areas where the three approaches coincided in relation to conservation of plant biodiversity.

In previous works, the panbiogeographic method has been applied in conservation biology in New Zealand (Grehan 1989) and Mexico (Morrone and Espinosa 1998; Luna et al.

2000). Panbiogeography has been considered as an ideal tool for the advocates and designers of protected natural areas, because it emphasizes the importance of geographic distribution (Craw et al. 1999). Nodes have been considered as equivalent to hotspots (Grehan 1993; Luna et al. 2000; Contreras-Medina et al. 2001). The refugia have been consider as zones with high species richness of angiosperms (Toledo 1982), and for this reason must be consider in conservation plans (Brown 1987).

In the case of mosses, Tan and Iwatsuki (1996) established the following criteria in the recognition of hotspots in Asia: (1) zones with high diversity, (2) presence of several endemic taxa with narrow ranges, (3) a great variety of habitats and plant communities, (4) several complementary floristic elements, (5) areas included in natural official reserves, and (6) areas recently surveyed and confirmed as that by the authors. In this same study, Tan and Iwatsuki (1996) proposed 10 hotspots for mosses, based on the criteria above mentioned; from these hotspots, southeastern China and New Guinea are congruent with those proposed in this work; also noted is that moss diversity is highest in the Southern Hemisphere (Shaw et al. 2005) and this idea is congruent with those hotspots defined mainly by mosses.

For gymnosperms, Contreras-Medina et al. (2001) proposed five hotspots, based in the same methodology commented herein, where endemism and nodes play a relevant role in their recognition; these five hotspots are New Caledonia, Japan, southeastern China, Tasmania, and western North America. In relation to diversity, some areas as circum-Australian islands, Mesoamerica, western North America (including northeastern Mexico), and southeastern China have an important role in biodiversity of gymnosperms worldwide (Contreras-Medina et al. 1999). Australia, South Africa and Mexico (particularly Mesoamerica) contain the highest cycad diversity in the world, whereas eastern Australia, southeastern China, southern Mexico, New Caledonia and southeastern United States represent areas where diversity of conifers is also high (Mutke and Barthlott 2005).

In the case of flowering plants and from the revision of panbiogeographic studies, there are 15 nodes mainly located in the American continent. Due to the fact that more panbiogeographic research has been carried out in the Americas in relation to other regions of the world, more nodes are found in this continent. All of them coincide with the areas of endemism proposed by Takhtajan (1986), and some of them with the refugia proposed by Haffer (1985). Those areas detected by the three methods for angiosperms are central Africa, New Guinea, New Caledonia, southeastern China, Mesoamerica, Appalachian, southern Chile, and southeastern Brazil. In general, these areas are considered as important regions with high diversity and endemism of flowering plants (Takhtajan 1986; Myers et al. 2000; Mutke and Barthlott 2005). Currently, angiosperms are the most diverse and well distributed worldwide plant group, in relation to species diversity compared with other land plant divisions, so the detection of zones with special importance from a biogeographic viewpoint based on these plants emphasizes their importance in conservation biology.

From a reproductive biology perspective and compared to mosses, seed plants represent heterozygote systems with sexual reproduction and a longer life cycle; mosses distribute a single cell with haploid genome, usually in masses, in contrast, spermatophytes distribute a diploid embryo in a seed, particularly packed in a fruit in angiosperms (Frahm 2008); also is considered that seed plants are more geographically restricted, genetically stable and narrowly distributed than mosses (Delgadillo et al. 2003). These differences in reproductive biology suggest different distributional patterns, but it is interesting that in some areas of the world a distributional overlap among mosses and seed plants occur and frequently, at different taxonomic levels, both groups show similar patterns of distribution. At the species level, bryophytes show much wider distribution patterns than seed plants and numerous species also exhibit restricted distribution patterns shown by seed plants (Frahm 2008), and the same factors appear to have shaped these shared patterns (Schofield 1985).

Perspectives in the Study of Hotspots

Biodiversity in the real world is localized in different points in space and exhibit varying levels of biodiversity and representation. Biodiversity analyses require a research and management program with conceptual and methodological resources able to deal simultaneously in several places of the world, continents or countries (Craw et al. 1999). In this sense, we consider that the methodologies presented in this chapter are capable to locate and detect those relevant areas with conservation priorities. In this sense, biogeographic methods are cost effective because global biodiversity information can be developed from available data without losing time and money on redundant exploratory research and inventory compilation (Craw et al. 1999).

Once detected the areas considered as hotspots, regional studies are necessary in order to carry out conservation plans, integrating government, local habitants and scientists. Subsequent studies can be carried out in other areas such as phylogenetic diversity (Forest et al. 2007) and DNA barcoding (Lahaye et al. 2008), in order to emphasize the importance of selected areas.

Despite that some abiotic factors, such as evapotranspiration, humidity, habitat heterogeneity, and topography have been considered as core predictors of species richness (Kreft and Jetz 2007), and currently these factors are crucial to plant development and growth, we consider that historical elements are key factors in biodiversity that have played a key role in those areas recognized as hotspots. This historical perspective is fundamental to be included in the areas considered as hotspots, so in this chapter we analyze some historical biogeographic alternative methods.

Additional studies are necessary applying this methodological approach to animal taxa, such as vertebrates and insects, which represent two of the best known biological groups and can be used, such as land plants, as a test model for studies in biological diversity, in order to support or neglect the patterns herein observed and proposed; also it is necessary to incorporate additional information in relation to fern taxa, in order to include all groups of land plants.

ACKNOWLEDGMENTS

Frank Columbus invited us to contribute with this manuscript. Claudio Delgadillo made useful suggestions to a first draft of this manuscript. Figures were done by Othón Alcántara. RCM dedicates this chapter to his daughter Sandra Contreras Córdoba.

REFERENCES

Alzate, F., Quijano-Abril, M. A. & Morrone, J. J. (2008). Panbiogeographical analysis of the genus *Bomarea* (Alstroemeriaceae). *J. Biogeogr*, *35*, 1250-1257.

Álvarez-Mondragón, E. & Morrone, J. J. (2004). Propuesta de áreas para conservación de aves de México, empleando herramientas panbiogeográficas e índices de complementariedad. *Interciencia*, *29*, 112-120.

Amorim, D. S. (1987). Refugios cuaternarios e mares epicontinentais: uma analise de modelos, metodos e reconstrucoes biogeograficas para a regiao Neotropical, incluindo o estudo de grupos de Mycetophiliformia (Diptera: Bibionomorpha). PhD thesis, *Universidade de Sao Paulo, Sao Paulo*.

Andrés, R., Morrone, J. J. Terrazas, T. & López-Mata, L. (2006). Análisis de trazos de las especies mexicanas de *Rhus* subgénero *Lobadium* (Angiospermae, Anacardiaceae). *Interciencia*, *31*, 900-904.

Brown, K. S. (1987). Conclusions, synthesis, and alternative hypotheses. In:T. C. Whitmore, & G. T. Prance (Eds.), *Biogeography and quaternary history in tropical America*. Clarendon Press, Oxford, 175-196.

Challenger, A. (1998). *Utilización y conservación de los ecosistemas terrestres de México: pasado, presente y futuro*. 1st edition. CONABIO, Universidad Nacional Autónoma de México-Agrupación Sierra Madre. Mexico, D. F.

Cracraft, J. L. (1985). Historical biogeography and patterns of differentiation within the South American avifauna: Areas of endemism, In: P. A. Buckley, M. S. Foster, E. S. Morton, R. S. Ridgely, & F. G. Buckley (Eds.), *Neotropical Ornithology, Ornithological Monographs*, *36*, American Ornithologists Union, Washington, 49-84.

Craw, R. C. (1982). Phylogenetics, areas, geology and the biogeography of Croizat: a radical view. *Syst. Zool. 31(3)*, 304-316.

Craw, R. C. (1989). New Zealand biogeography: a panbiogeographic approach. *New Zealand J. Zool.*, *16*, 527-547.

Craw, R. C., Grehan, J. R. & Heads, M. J. (1999). *Panbiogeography: tracking the history of life*. Oxford University Press, New York. *229*.

Croizat, L. (1958). *Panbiogeography*. Published by the author, Caracas.

Croizat, L. (1964). *Space, time, and form: The biological synthesis*. Published by the author, Caracas.

Contreras-Medina, R., Luna, I. & Morrone, J. J. (1999). Biogeographic analysis of the genera of Cycadales and Coniferales (Gymnospermae): a panbiogeographic approach. *Biogeographica*, *75*, 163-176.

Contreras-Medina, R., Morrone, J. J. & Luna, I. (2001). Biogeographic methods identify gymnosperm biodiversity hotspots. *Naturwissenschaften*, *88*, 427-430.

Contreras-Medina, R. & Luna, I. (2002). On the distribution of gymnosperm genera, their areas of endemism and cladistic biogeography. *Australian Syst. Bot.*, *15(2)*, 193-203.

Contreras-Medina, R. & Luna, I. (2007). Species richness, endemism and conservation of Mexican gymnosperms. *Biodivers. Conserv.*, *16*, 1803-1821.

De Candolle, A. (1820). Geographie botanique. Dictionnaire des Sciences Naturelles. Paris, 359-422.

Delgadillo, C. (1994). Endemism in the Neotropical moss flora. *Biotropica*, *26*, 12-16.

Delgadillo, C., Villaseñor, J. L. & Dávila, P. (2003). Endemism in the Mexican flora: a comparative study in three plants groups. *Ann. Missouri Bot. Gard.*, *90*, 25-34.

Forest, F., Grenyer, R., Rouget, M., Davies, T. J., Cowling, R. M., Faith, D. P., Balmford, A., Manning, J. C., Proches, S., van der Bank, M., Reeves, G., Hedderson, T. & Savolainen, V. (2007). Preserving the evolutionary potential of floras in biodiversity hotspots. *Nature*, *445*, 757-760.

Frahm, J. (2008). Diversity, dispersal and biogeography of bryophytes. *Biodivers. Conserv.*, *17*, 277-284.

Franco-Rosselli, P. & Berg, C. C. (1997). Distributional patterns of *Cecropia* (Cecropiaceae): a panbiogeographic analysis. *Caldasia*, *19(1-2)*, 285-296.

Gaston, K. J. & Williams, P. H. (1996). Spatial patterns in taxonomic diversity. In K. J. Gaston (Ed.), *Biodiversity: a biology of numbers and difference*. Blackwell Science, Cambridge, 202-229.

Grehan, J. R. (1989). Panbiogeography and conservation science in New Zealand. *New Zealand J. Zool.*, *16*, 731-748.

Grehan, J. R. (1993). Conservation biogeography and the biodiversity crisis: A global problem in space/time. *Biod. Lett.*, *1*, 134-140.

Graham, R. W. (1995). The role of climatic change in the design of biological reserves: the paleoecological perspective for conservation biology.In: D. Ehrenfeld, (Ed.), *Readings from conservation biology: plant conservation*. Blackwell Science, Cambridge. 25-28.

Haffer, J. (1969). Speciation in Amazonian forest birds. *Science*, *165*, 131-137.

Haffer, J. (1982). General aspects of the refuge theory. In: G. T. Prance, (Ed.), *Biological diversification in the tropics*. Columbia University Press, New York, 6-24.

Haffer, J. (1985). Avian zoogeography of the Neotropical lowlands. In: P. A. Buckley, M. S. Foster, E. S. Morton, R. S. Ridgely, & F. G. Buckley (Eds.), *Neotropical Ornithology*. *Ornithological Monographs*, *36*, American Ornithologists Union, Washington, 113-146.

Heads, M. (1989). Integrating earth and life sciences in New Zealand natural history: the parallel arcs model. *New Zealand J. Zool.*, *16*, 549-585.

Heads, M. (2008). Panbiogeography of New Caledonia, South-west Pacific: basal angiosperms on basement terrenes, ultramafic endemics inherited from volcanic island arcs and old taxa endemic to young islands.*J. Biogeogr.*, *30*, 2153-2175.

Hooghiemstra, H. & Van der Hammen, T. (1998). Neogene and Quaternary development of the neotropical rain forest: the forest refugia hypothesis, and a literature overview. *Earth-Sci. Rev.*, *44*, 147-183.

Katinas, L., Morrone, J. J., Crisci, J. V., et al. (1999). Track analysis reveals the composite nature of the Andean biota. *Australian J. Bot.*, *47*, 111-130.

Kreft, H. & Jetz, W. (2007). Global patterns and determinants of vascular plant diversity. *PNAS*, *104*, 5925-5930.

Lahaye, R., van der Bank, M., Bogarin, D., Warner, J., Pupulin, F., Gigot, G. Maurin, O., Duthoit, S., Barraclough, T. G. & Savolainen, V. (2008). DNA barcoding the floras of biodiversity hotspots. *PNAS*, *105*, 2923-2928.

Luna, I. & Contreras-Medina, R. (2000). Distribution of the genera of Theaceae (Angiospermae: Theales): A panbiogeographic approach. *Biogeographica*, *76*, 79-88.

Luna, I. & Alcántara, O. (2002). Placing the Mexican cloud forests in a global context: a track analysis based on vascular plant genera. *Biogeographica*, *78*, 1-14.

Luna, I., Alcántara, O., Morrone, J. J. & Espinosa, D. (2000). Track analysis and conservation priorities in the cloud forests of Hidalgo, Mexico. *Divers. Distrib.*, *6*, 137-143.

Luna, I., Morrone, J. J. & Escalante, T. (2009). Conservation biogeography: a viewpoint from evolutionary biogeography. In: Columbus, F. (ed.). *Nature Conservation: Global, Environmental and Economic Issues*. Nova-Science Publishers, New York (in press).

Martínez, M. & Morrone, J. J. (2005). Patrones de endemismo y disyunción de los géneros de Euphorbiaceae sensu lato: un análisis panbiogeográfico. *Bol. Soc. Bot. México*, *77*, 21-33.

Millar, C. I. (1998). Early evolution of pines. In: D. M. Richardson, (Ed.). *Ecology and biogeography of Pinus*, Cambridge University Press, Cambridge, 69-91.

Morrone, J. J. (2001). *Biogeografía de América Latina y el Caribe*. SEA y M & T Tesis, Zaragoza, Spain.

Morrone, J. J. (2009). Evolutionary biogeography, an integrative approach with case studies. Columbia University Press, New York.

Morrone, J. J. & Crisci, J. V. (1995). Historical biogeography: Introduction to methods. *Annu. Rev. Ecol. Syst.*, *26*, 373-401.

Morrone, J. J. & Espinosa, D. (1998). La relevancia de los atlas biogeográficos para la conservación de la biodiversidad mexicana. *Ciencia* (México), *49*, 12-16.

Mutke, J. & Barthlott, W. (2005). Patterns of vascular plant diversity at continental to global scales. *Biol. Skr.*, *55*, 521-531.

Myers N. (1988). Threatened biotas: 'hot spots' in tropical forests. *Environmentalist*, *8*, 187-208.

Myers, N., Mittermeier, R. A. Mittermeier, C. G., da Fonseca, G. A. B. & J. Kent. (2000). Biodiversity hotspots for conservation priorities. *Nature*, *403*, 853-858.

Ortega-Baes, P. & Godínez-Álvarez, H. (2006). Global diversity and conservation priorities in the Cactaceae. *Biodivers. Conserv.*, *15*, 817-827.

Pérez-Farrera, M. A., Vovides, A. P. & Iglesias, C. (2001). A new species of *Ceratozamia* (Zamiaceae) from Chiapas, Mexico. *Bot. J. Linnean Soc.*, *137*, 77-80.

Prendergast, J. R., Quinn, R. M., Lawton, J. H., Eversham, B. C. & Gibbons, D. W. (1993). Rare species, the coincidence of diversity hotspots and conservation strategies. *Nature*, *365*, 335-337.

Quijano-Abril, M. A., Callejas-Posada, R. & Miranda-Esquivel. D. R. (2006). Areas of endemism and distribution patterns for Neotropical *Piper* species (Piperaceae). *J. Biogeogr.*, *33*, 1266-1278.

Schofield, W. B. (1985). *Introduction to Bryology*. Mcmillan Publishing Company, New York.

Schuster, R. M. (1983). Phytogeography of the Bryophyta. In: R. M. Schuster (Ed.), *New manual of bryology*, Vol. I. Hattori Botanical Laboratory, Nichinan, 462-626.

Sclater, P. L. (1858). On general geographical distribution of the members of class Aves. *J. Linn. Soc. Zool.*, *2*, 130-145.

Shaw, A. J., Cox, C. J. & Goffinet, B. (2005). Global patterns of moss diversity: taxonomic and molecular inferences. *Taxon*, *54*, 337-352.

Takhtajan, A. (1986). Floristic regions of the world. University of California Press. Berkeley.

Tan, B. C. & Iwatsuki, Z. (1996). Hot spots of mosses in east Asia. *An. Inst. Biol. Ser. Bot.*, *67*, 159-167.

Tan, B. C. & Pócs, T. (2000). Bryogeography and conservation of bryophytes. In: A. J. Shaw, & B. Goffinet (Eds.), *Bryophyte biology*. Cambridge University Press, Cambridge, 403-448.

Tangney, R. S. (1989). Moss biogeography in the Tasman Sea region. *New Zealand J. Zool., 16(4)*, 665-678.

Tangney, R. S. (2007). Biogeography of Austral pleurocarpous mosses: distribution patterns in the Australasian region. In: A. E. Newton, & R. S. Tangney (Eds.), *Pleurocarpous mosses, systematics and evolution*, CRC Press, New York, 393-407.

Toledo, V. M. (1982). Pleistocene changes of vegetation in tropical Mexico. In: G. T. Prance (Ed.), *Biological diversification in the tropics*. Columbia University Press, New York, 93-111.

Villaseñor, J. L., Delgadillo, C. & Ortiz, E. (2006). Biodiversity hotspots from a multigroup perspective: mosses and senecios in the Transmexican Volcanic Belt. *Biodivers. Conserv., 15*, 4045-4058.

In: Encyclopedia of Environmental Research
Editor: Alisa N. Souter

ISBN: 978-1-61761-927-4
© 2011 Nova Science Publishers, Inc.

Chapter 63

IDENTIFYING PRIORITY AREAS FOR BIODIVERSITY CONSERVATION IN NORTHERN THAILAND: LAND USE CHANGE AND SPECIES MODELING APPROACHES

Yongyut Trisurat[*1], *Naris Bhumpakphan*[1], *Utai Dachyosdee*[1], *Boosabong Kachanasakha*[2] *and Somying Tanhikorn*[2]

[1]Faculty of Forestry, Kasetsart University, Bangkok, Thailand.
[2]Department of National Park, Wildlife and Plant Conservation, Bangkok, Thailand.

ABSTRACT

Rapid deforestation occurred in northern Thailand over the last few decades and is expected to continue. The consequences of deforestation on biodiversity are substantial and widely recognized. The objectives of this chapter were to 1) predict future land-use change, 2) generate ecological niches of large mammals, and 3) assess wildlife concentrations and their hotspots as priorities for biodiversity conservation. Three land use change scenarios between 2002-2050 were evaluated. Scenario 1 is a continuation of land use trends and predicts that forest cover will decrease from the present 57% to only 45% in 2050. Scenario 2 is an integrated-management scenario and scenario 3 is a conservation-oriented scenario directed by government policy to maintain respectively 50% and 55% forest cover (natural forest and plantation) and strictly prevent encroachment in protected areas. Geographic Information Systems (GIS) and a spatially-explicit model (CLUE-s) were employed to explore land use changes. In addition, a machine learning algorithm based on maximum entropy theory (MAXENT) was used to generate ecological niches of 18 large mammal species as a proxy of biodiversity. The likely occurrences of selected wildlife species were aggregated and classified as wildlife concentrations. In addition, the predicted deforestation areas were overlaid on high

---

[*] Corresponding author: Tel: (662) 579-0176; Fax: (662) 942-8107; Email: fforyyt@ku.ac.th

wildlife concentration to determine threats to wildlife or wildlife hotspots both inside and outside protected areas.

The results reveal that forest cover in 2050 will mainly persist in the west and upper north of the region, which is rugged and not easily accessible. In contrast, the highest deforestation is expected to occur in the lower north and remnant habitats will disappear in this area. Current suitable habitats for most wildlife species are located in the west, north and east of the region. The predicted areas of high wildlife concentration or richness (≥ 7 species likely present) are found in the western forest complex and in the northeast of the region, covering approximately 16,000 km^2 or 9.3% of the region. In addition, wildlife hotspots derived from the trend scenario encompass approximately 3,100 km^2 or 1.8% and 74% is predicted in protected area coverage. These areas are identified as high priority for biodiversity conservation. In contrast, the hotspots are predicted of less than 0.5% for the scenarios 2 and 3. Based on the model outcomes, we recommend conservation measures to minimize the impacts of future deforestation on wildlife hotspots.

Keywords: Biodiversity; Deforestation; Land use modeling; Hotspots; GIS; Northern Thailand

INTRODUCTION

Biodiversity is the variation of *life* forms within a given *ecosystem*, *biome*, or for the entire *Earth*. It is a broad idea, so a variety of concepts have been created in order to measure and conserve biodiversity. However, the establishment of priorities for biodiversity conservation is a complex issue (Margales and Pressey, 2000). The first question is which areas given conservation dollars would contribute the most towards slowing the current rate of species extinction? Second, species are generally unevenly distributed, thus which means should be used to map species distribution (Gaston, 2000)? A further problem concerns which species we should evaluate because we cannot map all species and we have not even named most of them (Mittermeier *et al.*, 2004).

Biodiversity hotspots were created as a means of setting priorities for biodiversity conservation (Mittermeier *et al.*, 1998, 1999, 2000; Myers, 1988, 1990). The population definition is a *biogeographic* region with a significant reservoir of biodiversity that is threatened with destruction. The concept of biodiversity hotspots was originated by Myers *et al.* (1988, 1990) in two articles in "The Environmentalist". The hotspots idea was also promoted by Mittermeier *et al.* (1998) and was followed by an extensive global review (Mittermeier *et al.*, 1999, 2004) in the popular book "Hotspots: Earth's Biologically Richest and Most Endangered Terrestrial Ecoregions". Biodiversity hotspots are driven by two criteria: species endemism and degree of threat (Myers *et al.*, 2000). Among biodiversity threats, the major threat is deforestation and land-use change, particularly in many countries in Southeast Asia (Fox and Vogler, 2005). FAO (2005) estimated that tropical regions lost 15.2 million hectares of forests per year during the 1990s and Southeast Asia has experienced the highest rate of net cover change (0.71% per year) compared to other areas. In addition, forest loss in Thailand was ranked the highest of all countries in the Greater Mekong Sub-region and as fourth in the "Top 10" of tropical countries in terms of annual rate of loss in

1995 (CFAN, 2005). According to Charuphat (2000), forest cover in Thailand declined from 53% of the country area in 1961 to approximately 25% in 1998.

The impacts of land-use/land-cover change are well known and observed. Not only does deforestation cause habitat loss but also habitat fragmentation, diminishing patch size and core area, and isolation of suitable habitats (MacDonald, 2003). Fragmentation provides opportunities for pioneer species (light-demanding species) to invade natural habitat along the forest edge (Forman, 1995; McGarigal and Marks, 1995). In addition, recovery of degraded ecosystems to their original state is extremely difficult and time consuming. Fukushima *et al.* (2008) investigated the recovery of tree species composition in secondary forests in northern Thailand that were abandoned after swidden cultivation for more than 20 years. The results indicated that native species on recently abandoned poppy fields were mostly absent and that it will take more than 50 years to reach climax species composition.

Several models have been introduced to predict land-use change. The models range from simple system representations including a few driving forces to simulation systems based on profound understanding of situation-specific interactions among a large number of factors at different spatial and temporal scales, as well as environmental policies. Significant progress in the quantification and understanding of land-use/land-cover changes has been achieved over the last decade. Reviews of different land-use models were provided by Verburg *et al.* (2004), Matthews *et al.* (2007) and Priess and Schaldach (2008). The results of land-use change simulations are very useful to formulate effective conservation measures, especially identifying conservation-priority areas to cope with future deforestation (Menon *et al.*, 2001).

At the global level, Myers *et al.* (2000) identified at least 25 areas that qualify under this definition as areas which support nearly 60% of the world's plant, bird, mammal, reptile, and amphibian species, with a very high share of endemic species. Four of the world's 25 biodiversity 'hot spots' are in Southeast Asia. Yet to the best of our knowledge there has been no systematic study of biodiversity hotspots at national and regional levels in Thailand. The recent study on biodiversity important areas by the Office of Natural Resources and Environmental Policy and Planning or ONEP (2008) subjectively identifies 16 protected areas to represent biodiversity important areas or biodiversity hotspots of five main ecosystems, including forest, in-land water, montane forest, agriculture and semi-arid.

In this chapter, we focus on the northern region of Thailand, which contains the highest percentage of remaining forest cover compared to other regions. Even though approximately 30% of the region has been designated as national park and wildlife sanctuary (RFD, 2007), the effectiveness of protection to maintain forest cover and biodiversity is still an issue requiring attention. Due to ongoing human population growth and development projects for tourism in the region we may expect continuing deforestation, increasing habitat fragmentation and increasing threats from expansion of agriculture and infrastructure development. This chapter aims to 1) forecast land-use change and land-use patterns across the region based on three land demand scenarios, 2) generate ecological niches of large mammals as the proxy of biodiversity and 3) assess wildlife concentrations and their hotspots as priorities for biodiversity conservation.

MATERIALS AND METHODS

Study Area

Northern Thailand is situated between latitudes 14°56' 17" - 20°27' 5" North and longitudes 97° 20' 38" - 101° 47' 31" East. It covers 17 provinces and encompasses an area of 172,277 km^2 or one-third of the country's land area (Figure 1). The dominant topography is mountainous, oriented north-south. Average annual temperature ranges from 20 to 34 °C, the average annual rainfall ranges between 600 and 1,000 mm in low areas to more than 1,000 mm in mountainous areas. The rainy season is from May to October. According to the Local Administration Department (*http://www.dopa.go.th/* hpstat9/people2.htm), the total population is almost stable at 11 million over the last 10 years but the population in Tak, Mae Hong Son and Chiang Mai provinces has increased 20.6%, 19.2% and 7.6%, respectively. The growing population is leading to additional pressure on a limited land resource for the purpose of agricultural production and food self-sufficiency.

Northern Thailand was originally covered by dense forest. Dominant vegetation includes dry dipterocarp and mixed deciduous forests in low and moderate altitudes. Pine forest, hill evergreen forest and tropical montane cloud forest dominate high altitudes (Santisuk, 1988). Land Development Department or LDD (2003) indicated that forest cover in this region declined from 68% to 57% during 1961-1988. Except in protected areas (national parks and wildlife sanctuaries) covering approximately 53,200 km^2 or 30% of the region (RFD, 2007), the lowland forests have been removed due to extensive logging in the past and the expansion of agricultural land. These areas are now extensively managed for agriculture, with rice on irrigated land, and vegetables and fruit trees (e.g. longan, lychee) elsewhere. Meanwhile, secondary forest in mountainous northern Thailand is largely the result of swidden cultivation (Fukushima *et al.*, 2008). In addition, parts of swidden cultivation have had their cycles shortened or changed to monoculture cash crops over the last 50 years (Schmidt-Vogt, 1999; Fox and Vogler, 2005). According to the Office of Agricultural Economics or OAE (2007) approximately 50,000 ha of rubber was planted in this region during 2004-2006. The continuing rise of rubber price in the last decade has stimulated a huge land demand for rubber plantations.

Data on Land Use, Socio-Economic and Biophysical Factors

Land-use data were derived from the 1:50,000 land use map for 2002, produced by LDD (2003). The original land-use classes were aggregated into nine classes to facilitate land-use simulations indicating the disturbance of natural land cover by human use. The land use classes were (1) intact forest, (2) degraded forest, (3) forest plantation, (4) paddy, (5) upland crop, (6) tree crop, (7) miscellaneous area, (8) built-up area, and (9) water body. In addition, a set of biophysical and socio-economic factors that can describe the spatial variability of different land use types was generated. Soil types, topography and drainage were derived based on the land classes identified on the 1:100,000 soil map (LDD, 2001). Road networks were obtained from the Environmental GIS database of the Department of Environmental Quality Promotion. Altitude, aspect, slope, distance to main roads, and distance to streams and rivers were extracted and/or interpolated from topographic maps (Royal Thai Survey Department, 2002). A surface representing spatial variation in annual precipitation was interpolated from rainfall data recorded in the Meteorological Stations across the north of Thailand using Universal Kriging techniques. Current population data were obtained from the Local Administration Department. Protected area coverage was digitized from the National Gazette map. All spatial data were entered into ArcGIS software with a spatial resolution of 500x500 m as an appropriate size for regional assessment and supported by the resolution of the data available for the region.

Land-Use Modeling and Scenario Definition

For this research we chose one of the most commonly applied land-use models at the regional scale, the CLUE model (the Conversion of Land Use and its Effects; Verburg *et al.*, 2002). It is an effective tool that yields land use information important for environmental assessment and planning (Verburg and Veldkamp, 2004). In addition, this model deals with both the location and quantity issues in an integrated way. The dyna-CLUE version of the CLUE model was used to project land use transitions for different scenarios during the period 2002-2050. The dyna-CLUE model allows high-resolution modeling of land-use change for scenario conditions including regional changes in demand for land use types and specific, local trajectories of land use conversions as a result of succession, afforestation and land use policies. The model provides a spatially explicit simulation of land-use change based on a case-study specific parameterization in terms of location suitability, neighborhood interactions, conversion sequences and exogenous demand for land. The model can integrate different land-use change trajectories in their environmental and socio-economic context. The model requires four inputs that together create a set of conditions and possibilities for which the model calculates the best solution in an iterative procedure. These inputs are (1) land use requirements (demand), (2) location characteristics, (3) spatial policies and restrictions, and (4) land use type specific conversion settings. Land use requirements and spatial policies are scenario specific, whereas the location characteristics and conversion settings are equal for all scenarios.

Land use requirements or demand are calculated at the aggregate level as part of a specific scenario. The land use requirements constrain the simulation by defining the totally required change in land use. In this study, three land demand scenarios were developed. The estimated land requirements for northern Thailand in 2050 are shown in Figure 2.

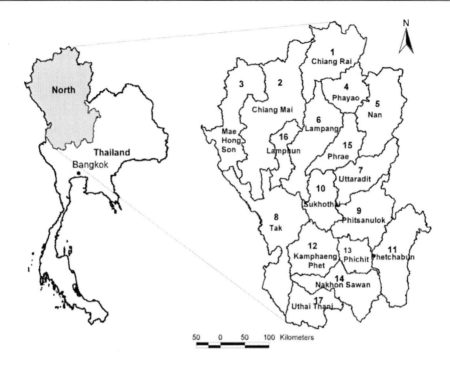

Figure 1. Location of provinces in northern Thailand

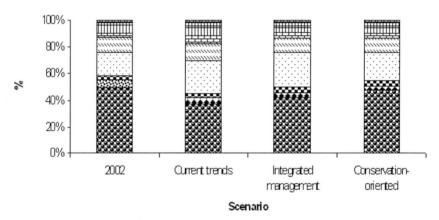

Figure 2. Land requirements for scenarios 1, 2 and 3 between 2002-2050

The trend scenario is based on a continuation of land use change of recent years (1998-2003). According to OAE (2007), the annual loss of forest cover in this period was 0.6%. On the other hand, the increment rates of plantation, paddy, upland crop, tree crop, miscellaneous area and built-up area were 0.2%, 0.8%, 0.2%, 2.5%, 0.7% and 1.1%, respectively. If the annual increments remain stable, natural forest cover will be 45% in 2050, while agricultural area will increase from 33% to 44%.

The second scenario is named integrated-management. The scenario is derived from the long-term National Environmental Policy (OEPP, 1997), which aims to maintain 50% forest cover at national level. This figure corresponds to the area of Class 1 Watersheds, conservation forest and headwater sources according to Tangtham (1992), and the estimate of vulnerable land for agriculture due to soil erosion according to the Land Development Department (2001) in the region. In spite of the high ambitions concerning the protection of forest it is assumed that current trends towards the establishment of rubber plantations continue in such a way that additional rubber plantation will cover an area of 480,000 ha (50% of the total suitable area) by 2050, which is targeted by the Department of Agriculture (2003). Half of this area will likely replace fruit tree plantations, and the remaining areas will be situated in marginal upland cropland, on idle land and partly in disturbed forest outside protected areas. The total agricultural area will cover 30% of the region in 2050. Urban area will increase according to population growth. In addition, water bodies are assumed to remain stable because large scale development of water resource is not feasible in this region and objected to by NGOs and civil society. The increment of plantation forest is 4,800 ha/year, which is somewhat higher than in previous years due to public concern on climate change and watershed degradation.

The third scenario is called conservation-oriented land use. This scenario aims to maintain 55% of the region as forest cover. Meanwhile, reforestation is preferable in the degraded Class 1 Watershed, which covers approximately 7,600 km^2 or 4.4% of the region. This is consistent with policy plans to rehabilitate degraded watersheds (NESDB, 2002). Additional rubber plantations will cover a smaller area than in the other scenarios (300,000 ha). The increment rate of urban area and the extent of water body will be the same as the integrated management scenario.

Location characteristics determine the relative suitability of a location for the different land use types. According to FAO (1995), land suitability classes are evaluated based on physical and economic conditions within any given area. Topographic factors (altitude, slope and aspect) represent limiting factors for agriculture. Other important land characteristics are annual precipitation, distance to available water, and soil texture. The socio-economic factors that influence deforestation include distance to village, distance to city, distance to main road, and population density. Distance to villages and population density are proxy indicators for local consumption, while distances to roads and distance to city are important proxies for the costs of transporting agricultural commodities to market.

In this study, we used logistic regression models to estimate the relation between occurrence of a land use type and the environmental and socio-economic conditions of a specific location (Table 1). Models were developed for seven land use classes (disturbed forest, plantation, paddy, upland crop, tree crop, miscellaneous area and built-up area). We did not estimate regression models for intact forest and water body because loss of forest is a result of other changes and the extent of water body is assumed stable. The goodness-of-fit of a logistic regression model is evaluated using the Receiver Operating Characteristic (ROC; Hosmer and Stanley, 2000). The value of area under curve (AUC) ranges between 0.5 (completely random) and 1.0 (perfect fit).

Spatial policies and restrictions indicate areas where land-use changes are restricted through strict protection measures such as protected areas. In scenario 1, there are no spatial policies implemented, thus forest encroachment may occur in protected areas if the location characteristics are favorable. In contrast, we impose land use policies for scenarios 2 and 3.

Under these scenarios, existing protected areas are designated as *restricted areas*, so that no further encroachment is allowed in these areas and natural succession is possible.

Land use type specific conversion settings influence the temporal dynamics of the simulations. Two sets of parameters are essential to characterize each land use type: conversion elasticities and land use transition settings. The conversion elasticities represent the advantage of keeping the same land use type at a specific location and are a proxy for the conversion costs. The elasticities are estimated based on capital investment, time and energy costs and expert judgment, ranging from 0 (easy conversion) to 1 (irreversible change) (Verburg et al., 2002). High values for this parameter were assigned to primary forest, paddy, built-up area and water body because these land use types are not likely to be displaced due to high investments or the time needed to establish primary forest. Medium values were given to disturbed forest and fruit trees. On the other hand, upland crop and miscellaneous area are highly dynamic land uses, thus low values were assigned for these land uses. In the land use transition settings we specified that minimally 30 years are needed for the natural succession of reforestation to primary forest and 20 years from disturbed forest succession to primary forest based on Obserhauser (1997).

When all inputs are provided, the dyna-CLUE model calculates the total probability for each grid cell of each land use type based on the suitability of location derived from the logit model and the conversion elasticity (Verburg et al., 2002). In case no constraints to a specific conversion are specified, the location will be allocated to the land use with the highest total suitability. The total allocated area of each land use equals the total land requirements specified in the scenario. In addition, we employed FRAGSTATS 3.0 software (McGarigal and Marks, 1995) to assess landscape structure and fragmentation indices of current and desired congregation areas of forest cover.

Species Distribution Modeling

The processes for mapping species niche distributions include five main steps: (1) selection of species; (2) collection of wildlife presence points; (3) generation of species distribution models; (4) determination of concentration of large mammals, and 5) determination of future deforestation and biodiversity hotspots.

1) Selection of species

The target species are those features or elements of wildlife that planners seek to conserve within a system of conservation areas (Groves, 2003). A set of four criteria were used to select the target species, namely, (1) wide-range distribution, (2) endangered status or ecosystem indicator (ONEP, 2007), (3) specialist availability and (4) public interest.

2) Collection of wildlife presence data

Department of National Park, Wildlife and Plant Conservation (DNP) kindly provided the mammalian occurrences in the protected area system. These data were recorded by wildlife scientists and trained personnel affiliated with the Wildlife Research Division of the DNP. Out of these databases, only those species with a minimum quantity of 30 records were chosen, so their distributions could be properly predicted using the statistical modeling in the next steps. Finally, 18 large mammals were selected for further analysis.

3) Generation of species distribution models

The species distribution maps were developed using a niche-based model or the maximum entropy method (MAXENT) (Peterson et al., 2001). The models operate by establishing a relationship between a species' known range and habitat variables within that region and then using this relationship to identify other regions the species may inhabit or to project potential range shifts under future land use types. The advantages of MAXENT include the following: 1) it requires only presence data and environmental information, 2) it can utilize both continuous and categorical variables, and 3) efficient deterministic algorithms have been developed that are guaranteed to converge to the optimal probability distribution (Philips et al., 2006).

We ran MAXENT using a convergence threshold of 10 with 1,000 iterations as an upper limit for each run. For each species, occurrence data was divided into two datasets. Seventy-five percent of the sample point data was used to generate species distribution models, while the remaining 25% was kept as independent data to test the accuracy of each model. Since our interest was to assess the predicted distributions of wildlife species, therefore we transformed the continuous probability of occurrence of each model output into a binary prediction. We applied a maximum training sensitivity plus specificity as a logistic threshold for each modeled species. We used AUC and ROC to assess the accuracy of each model similar to CLUE model (Hosmer and Stanley, 2000). In addition, a *descriptive statistic*, a boxplot (also known as a box-and-whisker diagram or plot) was used to graphically depict groups of relative contribution factors.

Table 1. Number of unique occurrence points for surrogate wildlife used for developing species distribution models

No.	Scientific name	Common name	IUCN status	Records
1	Bos javanicus	Banteng	EN	83
2	Bos gaurus	Gaur	VU	503
3	Canis aureus	Asiatic jackal (golden jakal)	LC	194
4	Capricornis sumatraensis	Serow	LC	174
5	Cercopithecidae spp.	Monkeys and langurs	LC	55
6	Cervus unicolor	Sambar	NT	673
7	Cuon alpinus	Dhole	EN	53
8	Elephas maximus	Asian elephant	EN	930
9	Hylobates spp.	Gibbons	EN	43
10	Muntiacus muntjak	Barking deer	LC	678
11	Naemorhedus griseus	Goral	VU	125
12	Panthera pardus	Leopard	NT	55
13	Panthera tigris	Tiger	EN	225
14	Prionailurus spp.	Small cats	EN/VU	283
15	Sus scrofa	Wild boar	LC	612
16	Tapirus indicus	Tapir	EN	174
17	Ursus spp.	Bears	VU	134
18	Viver spp.	Civets	LC	472
Total				5,466

Remarks: LC = Least concern; NT = Near threatened; VU = Vulnerable;
EN = Endangered

4) Determination of concentration of large mammals

The high concentrations of wildlife species were also evaluated by overlaying binary predicted distribution maps of all 18 species. The output maps included total extent of large mammals and species richness which was later classified into three classes: low concentration (occurrence of less than 4 species); moderate concentration (occurrence of 4-6 species); and high concentration (occurrence of equal or greater than 7 species).

Table 2. Significant location factors related to each land use location

Variables	Disturbed forest	Plantation	Paddy	Upland crop	Tree crop	Misc.	Built-up
Altitude	+ [1]	- [2]	-	-	n.s. [3]	+	+
Slope	n.s.	-	-	-	-	-	-
Soil texture [4]							
Loam	-	-	+	+	+	n.s.	n.s.
Sand	n.s.	-	-	n.s.	+	n.s.	n.s.
Laterite	+	-	n.s.	n.s.	+	+	-
Slope complex	n.s.	n.s.	-	-	n.s.	+	-
Clay	-	n.s.	+	n.s.	+		n.s.
Wetland	n.s.	-	+	-	n.s.	+	n.s.
Distance to stream	+	-	-	+	-	-	-
Distance to village	n.s.	n.s.	-	-	-	-	n.s.
Distance to main road	-	-	n.s.	-	-	n.s.	-
Distance to city	-	-	-	+	-	+	-
Population density	-	-	n.s.	-	n.s.	-	+
ROC	0.68	0.71	0.93	0.83	0.80	0.66	0.88

Remarks: 1/ positive correlated; 2/ negative correlated; 3/ n.s.: not significant at 0.05 level; 4/ category variable

5) Determination of future deforestation and biodiversity hotspots

We overlaid the current and predicted land-use maps in 2050 to determine the extent of projected deforestation and classifiy threat areas. Finally, we combined the threat areas of deforestation with the wildlife concentration map to define biodiversity hotspots in northern Thailand. In addition, we combined the hotspot areas with existing protected area coverage in order to identify whether they were situated inside or outside protected area networks.

RESULTS

Land-Use/Land-Cover Changes

Based on the empirical analysis eight variables were found to be important determinants of the location of disturbed forest in northern Thailand. Table 2 shows that disturbed forest is more likely to be found in high altitude and laterite soil. These areas contain many limiting factors for agriculture so further reclamation to agricultural use is limited. In contrast, areas that are close to a stream network, situated in dense population, accessible from main roads, in low altitude and have fertile soil (clay and loam) are at risk of encroachment because they are a prime target for agriculture. In addition, rugged terrain, poor soil and remoteness are limiting factors to future encroachment, particularly areas situated in the reserves. High values for the ROC statistic that measures the goodness-of-fit of the logistic regression models were found for paddy (0.93) and built-up area (0.88). Relatively high to moderate values were found for upland crops (0.83), tree crops (0.80) and forest plantation (0.71). Low values were observed for disturbed forest (0.68) and miscellaneous land use (0.66). These differences in goodness-of-fit are found because paddy requires specific land characteristics (e.g. poor drainage, clay texture), similar to built-up area that congregates in high population areas, close to roads and cities, and in low altitude. Sand and laterite soil are not systematic choices for upland crops. Disturbed forest and miscellaneous land use (including abandoned swidden cultivation) are found in all altitude zones, often on soils vulnerable to soil erosion (LDD, 2001; Thanapakpawin et al., 2006). Also because these classes represent a wide range of different activities lower ROC values are found.

Land-use/land-cover maps in 2000 and simulation results for 2050 for the three scenarios are shown in Figure 3. The results of the continuation of trends scenario without spatial policies and restrictions show that forest cover in the north decreases from 57% in 2002 to 45% in 2050 or a loss of approximately 12% of the region during the 48 years. The highest rate of deforestation is found in the four lower north provinces. Theses provinces still have annual loss of greater than 1% as in the past decade (Charuphat, 2000). The dominant topographies in these provinces are flat terrain to gentle slope with alluvial deposits, which are highly suitable for agriculture and development. Forest areas have been undergoing conversion for agriculture for several decades (Fukushima et al., 2008). The highest percentages of remaining forest cover in 2050 are found in the west such as Mae Hong Son Province (84%) followed by Chiang Mai Province (70%) which shows the lowest deforestation rate (6.2% or 0.1% per annum) followed by Nan (8.7% or 0.2% per annum). These provinces have mostly shallow, erosive soils of low fertility situated in slope complex terrains and accessibility to these areas is very difficult, so opportunities for agriculture are very restricted by natural barriers. In addition, the population in Mae Hong Son is less than 300,000 individuals and has the lowest density (20 persons/km^2) in Thailand (Department of Local Administration, 2007).

The integrated-management scenario, with protected areas and a slower rate of agricultural expansion, shows different land use patterns. This scenario assumes less demand for agriculture and rubber plantations, leading to higher remaining forest cover and tree crops. Forest cover is expected to increase from the west to the upper north and along the eastern national border. Mae Hong Son and Chiang Mai would have more than 75% forest cover in

2050, while Nan would have approximately 65%, which are relatively similar to the conditions in 2002. Substantial increases in forest cover are also predicted for Chiang Rai, Nan, Tak and Phitsanulok provinces under this scenario due to the restriction of further encroachment in the reserves and the regrowth of natural vegetation in abandoned agricultural areas (5,700 ha). Similarly to the trend scenario, the three provinces that show the highest forest loss are Phichit (4.0% per annum), Sukhothai (1.3% per annum) and Kamphaeng Phet (0.7% per annum). According to RFD (2007) these provinces have few protected areas compared to other provinces. In contrast, the remaining forest cover in Chiang Rai, Chiang Mai and Mae Hong Son is high and stable for both land use transition scenarios.

The results of the conservation-oriented scenario show that the extent and distribution pattern of remaining forest in 2050 are relatively similar to the conditions in 2002, except for the five provinces in the lower north. Mae Hong Son, Chiang Mai, Lampang, Tak and Phrae provinces still encompass more than 70% forest cover. In addition, the total extent of forest cover in Chaing Rai and Phayao is almost stable. The model predicted that Nan Province will gain approximately 5.6% forest cover which is mainly converted from secondary forest and abandoned swidden cultivation in high elevation through reforestation.

The results of the dyna-CLUE model reveal that the number of remaining forest patches increase for all scenarios over the 48 years simulated. The total number of patches increase from 1,250 patches in 2002 to 1,783 patches for the trend scenario, 1,515 patches for the integrated-management scenario, and 1,321 patches for the conservation-oriented scenario. This index corresponds to mean patch size index which reveals that the mean patch size of forests decreases from 7,930 ha in 2002 to 4,373 ha, 5,681 ha and 7,214 ha for the trend, integrated-management and conservation-oriented scenarios, respectively.

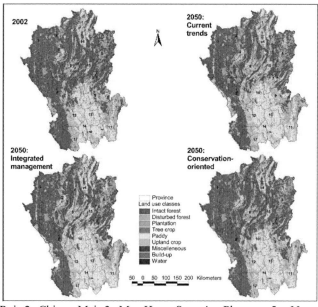

Indices: 1- Chiang Rai; 2- Chiang Mai; 3- Mae Hong Son; 4 – Phayao ; 5 – Nan ; 6 – Lampang ; 7 – Uttaradit ; 8 – Tak ; 9 – Phitsanoluk ; 10 – Sukhothai ; 11 – Phetchabun ; 12 – Kampaeng Phet ; 13 – Pichit ; 14 – Nakhon Sawan ; 15 – Phrae ; 16 – Lampun ; 17 – Uthai Thani

Figure 3. Land use patterns in 2050 simulated by the dyna-CLUE model for northern Thailand.

In addition, the largest patch of forest cover substantially declines from 54% of the remaining forest cover in 2002 to 39% for the trend scenario, 44% for the integrated-manahement scenario and 50% for the conservation-oriented scenario due to fragmentation. Small fragmented forest patches surrounded by agricultural land uses can be considered as disturbed forest or sink habitat (Forman, 1995) due to the fact that the whole patch corresponds to a border area.

Wildlife Distribution

Species occurrence points and model performance

We developed species distribution models for 18 large mammal species that had more than 30 sample points. Altogether, there were 5,466 records for wildlife species. The minimum number of records was 43 points for gibbon (*Hylobates* spp.) and the maximum number reached 930 records for Asian elephant. This is due to elephant signs and individual sightings being more easily observed than for other species. Elephant dung piles are stable and will last 85-145 days depending on the amount of rainfall. In addition, deterioration rate is less than approximately 0.0126 pile per day (Wanghongsa, 2004). Meanwhile, high occurrences (nearly 700 records) were also observed for barking deer and sambar deer. These two species are common species inhabiting a variety of habitats compared to other landscape species.

Figure 4 shows that 12 environmental factors are correlated with wildlife distribution. In general, human population density, distant to village and distant to city are prominent contribution factors to wildlife distribution in the northern landscape. The contribution of aspect, forest type and distant to stream are relatively low compared to other habitat factors. The box plot (Figure 4) indicates that the median of population density is the highest among 12 environmental factors followed by distant to city (district). However the variation of distant to city is larger. It contributes approximately 50% of banteng, elephant, gaur, sambar, tapir, small cat and tiger distributions because these species are very sensitive to human activities. However, its contribution with monkey and langur is less than 2%. The contribution of forest fragmentation to wild boar distribution is approximately 37%. The further from forest edge, the more likely the occurrence of wild boar. It is one of the game species for local villagers in rural areas.

The MAXENT models derived from the training and test dataset show good to excellent performance for predicting current and future distributions of all selected wildlife species. The AUC values for most wildlife species are equal to or greater than 0.9.

Spatial pattern distribution

The predicted distributions of wild boar cover an area of approximately 42,900 km^2 or 25% of the region area which is the highest compared to other species. It is followed by barking deer (23%) and bear (19%) (Table 3). These species are classified as common species and found in various habitats both inside and outside protected area networks. Figure 5 shows potential habitats of wild boar and barking deer. They are abundant in protected areas located in the west and in the north. Based on ground survey and species modeling, they are not likely to be found in the lower north due to lack of forest cover.

In contrast, potential habitats for banteng, elephant, gaur, goral, leopard, sambar, tapir and tiger cover less that 5% of the northern region. Most of them, except sambar are listed as rare and endangered species. Currently, they can be found in limited protected areas. Figure 6 shows that the distribution pattern of elephant is relatively similar to tiger. The largest suitable habitat patch is located in the core area of western Thailand. This area is named as the western forest complex which comprises of 11 national parks and 6 wildlife sanctuaries, covering approximately 19,000 km². It is the largest protected area complex in mainland Southeast Asia. Therefore, it facilitates the persistence of mega-fauna in the region. The results of our study are consistent with ground survey and previous study conducted by Trisurat and Pattanavibool (2008).

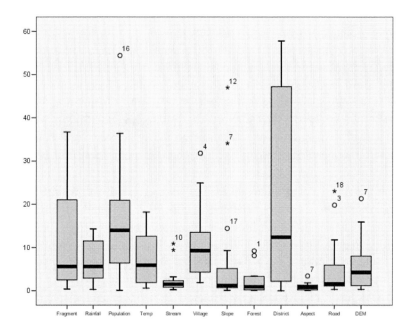

Figure 4. Box plots showing the contributions of environmental variables for mammal distribution models. The medians of range changes are shown as black horizon lines. The core boxes indicate the interquartile range whereas the lower and the upper whisker lines represent the 1st and 3rd quartiles. Single circles and asterisks represent mild and extreme outliers, respectively.

Location of congregation areas for population viability

We overlaid the current suitable habitats of all 18 species using GIS methods to derive current congregation areas of large mammals. The results suggested that the total area of congregation (≥ 7 spp.) covers approximately 16,000 km² or 9.3% of the region (Figure 7). Concentrations of large mammal species are clustered in three locations. The largest patch is situated in the Huai Kha Khaeng and Thung Yai Naresuan World Heritage Site in the west. The second area is located in Omkoi-Mae Tuan forest complex to the north of the first patch. The third patch can be found in the Chiang Dao-Lum Nam Pai complex, which is situated along the western border adjoining Myanmar. Besides these three clusters, concentration of large mammals are scattered in fragmented protected areas such as in the northeast of the region. The moderate concentration (4-6 species) covered 20,800 km² or 12.1% of the region while low concentration covered 33,400 km2 or 19.5% (Table 4). Figure 7 also indicates that it is difficult to find large mammals in the central and lower north due to remaining forest cover being minimal and accessibility to these areas being easy.

Table 3. Extent of suitable habitats for selected wildlife species in northern Thailand

Species	Suitable habitat	
	Km²	% of region*
Banteng	1,792	1.04
Gaur	3,997	2.32
Asiatic jackal (golden jakal)	20,777	12.06
Serow	16,452	9.55
Monkeys and langurs	30,338	17.61
Sambar	8,407	4.88
Dhole	19,588	11.37
Asian elephant	7,236	4.20
Gibbons	22,827	13.25
Barking deer	39,210	22.76
Goral	8,424	4.88
Leopard	4,514	2.62
Tiger	5,151	2.99
Small cats	13,076	7.59
Wild boar	42,897	24.90
Tapir	3,446	2.00
Bears	32,905	19.10
Civets	29,098	16.89

* regional area = 172,277 km²

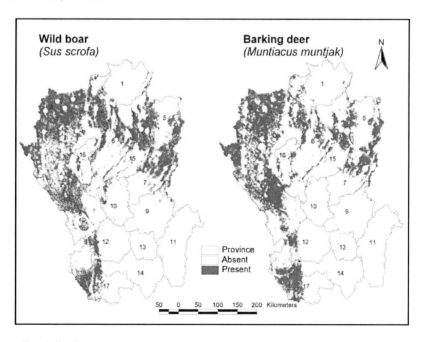

Figure 5. Predicted distributions of wild boar and barking deer in northern Thailand

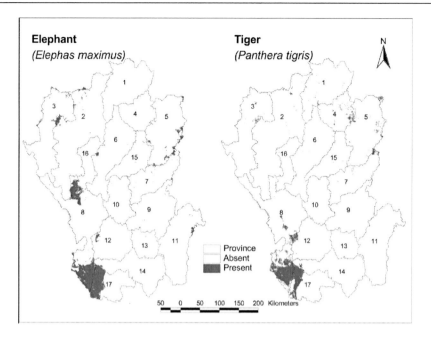

Figure 6. Predicted distributions of elephant and tiger in northern Thailand

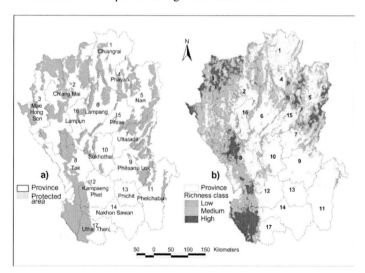

Figure 7. Distributions of concentration of large mammals in northern Thailand

Table 4. Estimated extent of each wildlife richness class and hot spots under different land-use scenarios

Richness	Region		Inside protected areas		
	km²	% region	km²	%region	%class
Low (1-3 spp.)	33,443	19.41	15,423	8.95	46.12
Medium (4-6 spp.)	20,800	12.07	13,888	8.06	66.77
High (≥ 7 spp.)	15,943	9.25	12,986	7.54	81.45
Absent	142,091	59.26	10,903	6.33	15.53
Total	172,277	100.00	53,200	-	-

It is observed that more than 80% of high wildlife concentration and 66% of moderate wildlife concentration are situated in the protected area system. In addition, 15% of existing protected areas are predicted as not suitable for selected large mammals due to habitat degradation. These results imply that persistence of large mammals in the region is highly dependent on effective management and protection of protected areas which cover approximately 30% of the region. Only a few species, especially least concerned or common species can be found outside protected areas.

Table 5. Extent of biodiversity hotspots in northern Thailand

Source of land use scenario	Region		Inside protected areas		
	km^2	%region	km^2	%region	%class
Trends	3,138	1.83	2,325	1.35	74.09
Integrated- management	764	0.44	0	0.00	0.00
Conservation-oriented	149	0.09	0	0.00	0.00

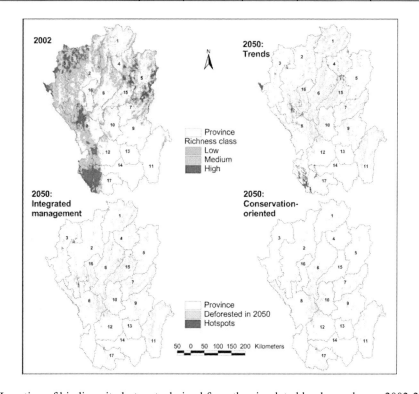

Figure 8. Location of biodiversity hotspots derived from the simulated land-use change 2002-2050

Biodiversity Threats

Based on the assumption of scenario 1, some of the currently protected areas are unlikely to be effective to maintain existing forest cover due to restrictions not being enforced. The results of CLUE-model indicate that the remaining forest cover inside the protected area network will decline from 86% of the reserve areas in 2002 to 76% in 2050. Future deforestation and its consequences will result in the extent of threats to wildlife or biodiversity hotspots of approximately 3,138 km^2 or 1.8% of the region under the trend scenario. Of this figure, approximately 2,350 km2 or 74% of biodiversity hotspots is predicted in protected area networks and they are widely spread in the northern region (Table 5 and Figure 8). The largest patch of biodiversity hotspots is located in Huai Kha Khaeng-Thung Yai Naresuan World Heritage Site. Forest patches are unlikely to stay preserved in the lower north of the region due to the low forest area remaining in this part of the region (Figure 3). In contrast to these results, the estimated area is 764 km^2 for the integrated-management scenario and 149 km^2 for the conservation-oriented scenario, which are less than 0.5% of the region. In addition, there are no threats to wildlife concentrations inside the protected area coverage.

Implications for Biodiversity Conservation

The northern region covers the highest percentage of protected areas in Thailand, which occupy approximately 30% of the region (RFD, 2007). They contribute significant positive impacts on maintaining forest cover and biodiversity. More than 85% of the reserves is under forest cover and contributes 49% of the total forest cover in the region. In addition, it will account for 56% and 52% of predicted forest cover in 2050 under the integrated-management and conservation-oriented scenarios, respectively. These results are relevant to the findings of Trisurat (2007), which indicated average forest cover in national protected area network was 84% due to effective law enforcement on the ground. In contrast, the assumption of poor law enforcement and protection (trend scenario) is likely to diminish large areas of intact forest cover and suitable habitats for large mammals both inside and outside protected area networks (Figures 3 and 8). This evidence confirms that protection status by law only is no guarantee of real protection on the ground.

Even though protected areas are effective means to protect forest cover and maintain biodiversity, establishment of more protected areas to cope with future deforestation is not appropriate. First, the existing protected areas in Thailand cover approximately 24.4% of the country's land area (Trisurat, 2007), nearly meeting the 25% target proposed by the National Forest Policy and the Ninth National Economic and Social Development Plan (2002–2006) (NESDB, 2002). In addition, declaration of more protected areas will probably create confrontation between conservation and future land demand for agriculture and human settlement (ICEM, 2003). Also, most remaining forest cover outside the existing protected area system is relatively small and fragmented and therefore not suitable to establish as new protected areas (IUCN, 1984).

Based on these findings we recommend specific conservation strategies for each land demand scenario. For land-use trend scenario, it is essential to put more efforts into maintaining biodiversity in the existing protected area network. If authorized institutes do nothing, the consequences of deforestation inside the protected area network will diminish substantial large mammal habitats. It is observed that significant loss is found in Huai Kha Khaeng and Thung Yai Naresuan wildlife sanctuaries. If this prediction is true, it will significantly affect the status of this natural world heritage site. Further deforestation and fragmentation would possibly disrupt landscape-species migration such as Asian elephant, tiger and guar (Simcharoen *et al.*, 2007). Intensive and regular patrolling to minimize deforestation is highly recommended for the trend scenario.

The integrated-management and conservation-oriented scenarios generate similar spatial patterns of biodiversity hotspots. They will slightly affect biodiversity in protected areas due to further deforestation in protected area networks are not allowed for these two land-use scenarios. Nevertheless, there are enclave communities inside protected areas living under poverty that heavily rely on natural resources. Raising conservation awareness of local people residing inside the protected areas and in the buffer zones is important to ensure that rising commodity and rubber prices will not stimulate further deforestation. Based on Fox and Vogler (2005), market price is one of most important driving factors on land-use/land-cover change in the tropics even with strict law enforcement. Besides, maintenance of ecological linkage and establishment of biodiversity corridors to link fragmented protected areas in the upper north and northeast of the study area are important. Many protected areas are too small to sustain native species and viable populations of large mammals, including tigers, leopards, and elephants (ICEM, 2003) which require contiguous forest blocks.

CONCLUSION

According to long-term forest monitoring, forest cover loss in northern Thailand will continue, especially outside protected areas unless strict protection measures are seriously undertaken. In this study, we used CLUE-s model to simulate land-use allocation in 2050 under different scenarios, generated the distributions of 18 large mammals and determined wildlife hotspots. The results of this study are summarized as follows:

1. The trend scenario is based on a continuation of the trends of land use conversion of recent years. It is expected that the existing forest cover of 57% of the region in 2002 will decrease to 45% by 2050. However, forest loss is likely to be strongly variable across the region. The remaining forest cover will be found mainly in the upper north and in the west where altitude is high and accessibility to these areas is difficult. The lowest loss and highest percentage of forest cover will be found in the northwest. The results of species modeling reveal that in addition, small patches are scattered in fragmented protected area networks.

    The consequences of deforestation will degrade the current concentrations of large mammal species to be clustered in three locations, namely the Huai Kha Khaeng and Thung Yai Naresuan World Heritage Site, Omkoi-Mae Tuan forest complex, and Chiang Dao-Lum Nam Pai complex. This scenario will create biodiversity hotspots of approximately 3,138 km^2, especially in the Huai Kha Khaeng and Thung Yai Naresuan World Heritage Site and Omkoi-Mae Tuan forest complex due to poor law enforcement on the ground. Intensive and regular patrolling to minimize deforestation in protected areas is highly recommended for this scenario because approximately 2,325 km^2 or 74% of the total hotspots are predicted in the reserve.

2. The integrated-management scenario is directed by the spatial policies that aim to maintain 50% forest cover. Under this scenario, much forested land in rugged terrain and protected areas will remain intact due to the land not being suitable for

agriculture and a restriction policy being undertaken in the existing protected area network, despite there being a high demand for rubber plantations. The estimated biodiversity hotspots derived from the simulated land-use map in 2050 cover 765 km^2 or 0.44% of the region area. In addition, none of the threat areas are expected inside protected areas due to effective law enforcement. Nevertheless, park officials must minimize future deforestation both in protected areas and in the buffer zones and maintain ecosystem connectivity of fragmented protected areas.

3. The conservation-oriented scenario, aims to maintain 55% of the region under forest cover. The results of model simulation show that the extent and pattern of remaining forest cover in 2050 are relatively similar to 2002, except in the lower north that will have less forest cover. A few provinces located along the national borders such as Nan Province will gain substantial forest cover which will be regenerated from secondary forest and abandoned swidden cultivation in high elevation. The estimated biodiversity hotspots derived from the conservation-oriented scenario cover 149 km^2 or 0.09% of the region area and no patch is located inside the existing protected areas. The proposed conservation measures to maintain biodiversity are more likely similar to the integrated- management scenario.

REFERENCES

CFAN (Forestry Advisers Network) (2005). Deforestation: tropical forests in decline. CIDA Forestry Advisers Network. Available at: *http://www.rcfa-cfan.org/english/ issues*

Charuphat, T. (2000). Remote sensing and GIS for tropical forest management. In: GIS Application Center (ed) *Proceedings of the Ninth Regional Seminar on Earth Observation for Tropical Ecosystem Management, Khao Yai*, Thailand, 20-24 November 2000. The National Space Development Agency of Japan, Remote Sensing Technology Center of Japan, RFD, and GIS Application Center/AIT, Khao Yai National Park Thailand, 42-49.

Department of Agriculture (2003). Map of potential area for para rubber plantation during 2004-2006 in the North and Northeast, Thailand. Department of Agriculture, Ministry of Natural Resources and Environment, Bangkok, Thailand.

Department of Local Administration (2007). Population census in Thailand from 1994 to 2007. Available at: *http://www.dopa.go.th/ hpstat9/people2.htm*.

FAO (Food and Agriculture Organization of the United Nations) (1995). Guidelines: land evaluation for irrigated agriculture. FAO soils bulletin 55. Food and Agriculture Organization of the United Nations, Rome, Italy.

FAO (Food and Agriculture Organization of the United Nations) (2005). Global forest assessment. Food and Agriculture Organization of the United Nations. Available at: *http://www.fao.org/forestry/site/fra/24690/ en*

Forman, R. T. T. (1995) Land Mosaics: the ecology of landscapes and regions. *Cambridge University Press, Cambridge*, UK.

Fox, J. & Vogler, J. B. (2005) *Land-use and landEnvironmental Management, 36(3)*, 394-403.

Fukushima, M., Kanzaki, M., Hara, M., Ohkubo, T. Preechapanya, P. & Chocharoen, C. (2008). Secondary forest succession after the cessation of swidden cultivation in the montane forest area in northern Thailand. *Forest Ecology and Management*, *255(5-6)*, 1994-2006.

Gaston, K. J. (2000). Global patterns in *biodiversity*. *Nature*, *405*, 220-227.

Groves, C.R. (2003). Drafting a conservation blueprint: a practitioner's guide to planning for biodiversity. The Nature Conservancy and Island Press, Washington.

Hosmer, D. W. & Stanley, L. (2000). Applied logistic regression, 2nd ed. Chichester, Wiley, New York, USA.

ICEM (International Center for Environmental Management) (2003). Thailand national report on protected areas and development; review of protected areas and development in the Lower Mekong River Region. Indooroopilly, Queensland, Australia.

IUCN (The World Conservation Union) (1984). Guidelines for protected area management categories. Gland, Switzerland

Land Development Department (2001). Soil erosion map. Land Development Department, Ministry of Agriculture and Co-operatives, Bangkok, Thailand.

Land Development Department (2003). Annual statistics report year 2003. Land Development Department, Ministry of Agriculture and Co-operatives, Bangkok, Thailand.

MacDonald, G. (2003). Biogeography: introduction to space, time and life. John Wiley & Sons, Inc., New York, USA.

Margales, C. R. & Pressey, R. L. (2000). Systematic conservation planning. *Nature*, *405*, 243-253.

Matthews, R., Gilbert, N., Roach, A., Polhill, J. G. & Gotts, N. M. (2007). Agent-based land-use models: a review of applications. *Landscape Ecology*, *22(10)*, 1447-1459.

McGarigal, K. & Marks, B. (1995). FRAGSTATS: Spatial pattern analysis program for quantifying landscape structure. Gen. Tech. Rep. PNW-GTR-351. Portland, USA.

Menon, S., Pontius, R. G., Rose, J. R. S., Khan, M. L. & Bawa, K. S. (2001). Identifying conservation-priority areas in the tropics: a land-use change modeling approach. *Conservation Biology*, *15(2)*, 501-512.

Mittermeier, R. A., Myers, N., Gil, P. R. & Mittermeier, C. G. (1999). Hotspots: Earth's Biologically richest and most endangered terrestrial ecoregions. Cemex, Conservation International and Agrupacion Sierra Madre, Monterry, Maxico.

Mittermeier, R. A., Myers, N., Mittermeier, C. G., Gustavo, A. B., da Fonseca GAB, K. & Jennifer, K. (2000). Biodiversity hotspots for conservation priorities. *Nature*, *403(24)*, 853-858.

Mittermeier, R. A., Myers, N., Thomsen, J. B., da Fonseca, G. A. B. & Oliveri, S. (1998). Biodiversity hotspots and major tropical wilderness areas: approaches to setting conservation priorities. *Conservation Biology*, *12*, 516-520.

Mittermeier, R. A., Robles, G., Hoffmann, M., Pilgrim, J. & contributors. (2004). Hotspots: earth's biologically richest and most endangered terrestrial ecoregions. CEMEX Mexico City, Mexico.

Myers, N. (1988). Threatened biotas: hotspots in tropical forests. *Environmentalist*, *8*, 187-208.

Myers, N. (1990). The biodiversity challenge: expanded hotspots analysis. *Environmentalist*, *10*, 243-256.

Myers, N., Mittermeier, R. A., Mittermeier, C. A., Fonseca, A. B. da & Kent, J. (2000). Biodiversity hotspots for conservation priorities. *Nature, 403*, 853-858.

NESDB (Office of National Economic and Social Development Board) (2002) The ninth national economic and social development plan (2002–2006). Prime Minister Office, Bangkok, Thailand.

OAE (Office of Agricultural Economics) (2007). Agricultural statistics of Thailand 2004. Ministry of Agriculture and Co-operatives, Bangkok, Thailand.

Obserhauser, U. (1997) Secondary forest regeneration beneath pine (*Pinus kesiya*) plantations in the northern Thai highlands: a chronosequence study. *Forest Ecology and Management, 99*,171-183.

OEPP (Office of Environmental Policy and Planning) (1997). Thailand policy and perspective plan for enhancement and conservation of national environmental quality, 1997-2016. Ministry of Science, Technology and Environment, Bangkok, Thailand.

ONEP (Office of Natural Resources and Environmental Policy and Planning) (2007). Thailand red data: vertebrates. Ministry of Natural Resources and Environment, Bangkok.

ONEP (Natural Resources and Environmental Policy and Planning) (2008). Biodiversity hotspots in Thailand. Ministry of Natural Resources and Environment, Bangkok.

Peterson, A. T. (2001). Predicting species' geographical distributions based on ecological niche modeling. *Condor, 103,* 599-605.

Philips, S. J, Anderson, R. P. & Schapire, R. E. (2006) Maximum entropy modeling of species geographical distributions. *Ecological Modelling, 190,* 231-259.

Priess, J. A. & Schaldach, R. (2008) Integrated models of the land system: a review of modelling approaches on the regional to global scale. *Living Reviews in Landscape Research 2*. Available at: *http://www.livingreviews. org/lrlr-2008-1*

RFD (Royal Forest Department) (2007). Forestry statistics year 2006. Ministry of Natural Resources and Environment, Bangkok, Thailand.

Royal Thai Survey Department (2002). Topographic map scale 1: 50,000. Ministry of Defense, Bangkok, Thailand.

Santisuk, T. (1988). An account of the vegetation of northern Thailand. *Geological Research 5*. Franz Steiner Verlag, Stuttgart.

Schmidt-Vogt, D. (1999). Swidden farming and fallow vegetation in northern Thailand. *Geological Research 8*. Franz Steiner Verlag, Stuttgart.

Simcharoen, S., Pattanavibool, A., Ullas, K., Nichols, J. D. & Kumar, N. S. (2007). How many tigers *Panthera tigris* are there in Huai Kha Khaeng Wildlife Sanctuary, Thailand? an estimate using photographic capture-recapture sampling. *Oryx, 41*, 447-453.

Tangtham, N. (1992). Watershed management. In: Royal Forest Department (ed) Thai Forestry Sector Master Plan, Vol 5: Subsectoral Plan for People and Forestry Environment. Ministry of Agriculture and Co-operatives, Bangkok, Thailand, 93-115.

Thanapakpawin, P., Richey, J., Thomas, D., Rodda, S., Campbell, B. & Logsdon, M. (2006). Effects of land-use change on the hydrologic regimes of the Mae Chaem river basin, NW Thailand. *Journal of Hydrology, 334(1-2)*, 215-230.

Trisurat, Y. (2007). Applying gap analysis and a comparison index to assess protected areas in Thailand. *Environmental Management, 39(2)*, 235-245

Trisurat, Y. & Pattanavibool, A. (2008). Assessing population viability for ecosystem management in the Western Forest Complex, Thailand. In M. S. Alonso and I. M. Rubio

(Eds). Ecological Management: New Research. Nova Science Publishers, New York. 143-164.

Verburg, P. H., Schot, P., Dijst, M. J. & Veldkamp, A. (2004). Land use change modelling: current practice and research priorities. *Geojournal, 61(4)*, 309-324.

Verburg, P. H., Soepboer, W., Limpiada, R., Espaldon, M. V. O., Sharifa, M. & Veldkamp, A. (2002). Land-use change modelling at the regional scale: the CLUE-S model. *Environmental Management, 30(3)*, 391-405.

Verburg, P. H. & Veldkamp, A. (2004). Projecting land use transitions at forest fringes in the Philippines at two spatial scales. *Landscape Ecology, 19(1)*, 77-98.

Wanghongsa, S. (2004). Monitoring of population dynamics of elephants in Khao Ang Rue Nai Wildlife Sanctuary (in Thai). Wildlife Research Division, Department of National Park, Wildlife and Plant Conservation, Bangkok.

In: Encyclopedia of Environmental Research
Editor: Alisa N. Souter
ISBN: 978-1-61761-927-4
© 2011 Nova Science Publishers, Inc.

Chapter 64

CHINA: A HOT SPOT OF RELICT PLANT TAXA

Jordi López-Pujol[1] and Ming-Xun Ren[2]

[1]Botanic Institute of Barcelona (CSIC-ICUB), Passeig del Migdia s/n, 08038 Barcelona, Spain.
[2]Key Laboratory of Aquatic Botany and Watershed Ecology, Wuhan Botanical Garden, Chinese Academy of Sciences, Wuhan 430074, China

ABSTRACT

Compared to other major regions of the North Temperate Zone, China harbours a large number of representatives of very ancient plant lineages. Relict taxa are generally survivors of formerly much more widespread lineages, having also suffered the extinction of their close relatives. Moreover, in some cases they have remained superficially unchanged for millions of years, thus being regarded as true 'living fossils'. Chinese relict lineages are mainly concentrated in the central and southern regions of the country (usually below latitudes of 35ºN), where the occurrence of numerous and extensive glacial refugia has been hypothesized. Some well known examples of relict plants still surviving in China are *Cathaya argyrophylla*, *Cyclocarya paliurus*, *Eucommia ulmoides*, *Ginkgo biloba*, and *Metasequoia glyptostroboides*. Most of the abovementioned species are now threatened because of their small distribution areas and increasing habitat destruction and fragmentation. The extirpation of such relict taxa would imply the loss of unique evolutionary history.

THE ABUNDANCE OF RELICT PLANT TAXA IN CHINA

China is one of the richest countries in the world in plant diversity, both in terms of taxonomical richness and endemism. It is widely acknowledged that more than 30,000 vascular plant taxa can be found in China, and endemics might account for half of these figures (López-Pujol et al., 2006; CSPCEC, 2008). Accordingly, China is recognized as one of the world's 'mega-diversity' countries (Mittermeier et al., 1997). In addition, large areas of

China are regarded as biodiversity hotspots of global significance (Davis et al., 1995; Olson & Dinerstein, 1998; Barthlott et al., 2005; Mittermeier et al., 2005).

There is also a wide consensus on the significant contribution of relict, ancient elements to this large biodiversity (e.g. Latham & Ricklefs, 1993; Axelrod et al., 1996; Qian & Ricklefs, 1999; Thorne, 1999; Qian, 2001; López-Pujol et al., 2006; Qian et al., 2006). Firstly, it should be noted that China has a disproportionately number of taxa belonging to the oldest plant taxonomic groups: around 2,200 bryophytes and approximately 2,600 pteridophytes (López-Pujol et al., 2006). These figures enable China to rank first in the world for these groups (Groombridge, 1994; Mutke & Barthlott, 2005). Of course, this not necessarily means that all the lower taxa (that is, genera, species and subspecies) included in these primitive groups are relict. China also harbours the highest number of gymnosperms of any country (250 taxa; that is, ca. 25% of the world's total), being Yunnan-Sichuan the main centre of gymnosperm diversity of the Earth (Farjon & Page, 1999; Mutke & Barthlott, 2005). Notably, the pteridophyte and the gymnosperm flora (at species level) of China are three times and almost two times more diverse than that of the United States, respectively (Qian & Ricklefs, 1999), two countries with nearly the same land area.

Compared to other Northern Hemisphere regions, the flora of China is also extremely rich for some of the main phylogenetically basal groups of angiosperms, such as the magnolids and ranunculids (Qian & Ricklefs, 1999; Qian, 2001). China has a significantly higher number of ancient endemic genera and families than the United States (Qian, 2001). In addition, most of the ca. 1,400 Asian endemic genera occurring in China are Tertiary relicts (that is, paleogenera) (Qian et al., 2006). Moreover, a more recent study (López-Pujol, 2008) revealed that the relict (pre-Quaternary) component of the modern endemic seed flora at the infrageneric (species and subspecies) level is considerable (around 40%; the remaining 60% are taxa of Pleistocene origin). For the Mediterranean endemic flora, the weight of the relict taxa is much lower (ca. 23%; Verlaque et al., 1997).

All these data draw a scenario in which the modern flora of China is still bearing a deep imprint of the Tertiary "boreotropical flora" and even older elements; in other words, China constitutes a centre of survival for relict lineages. Noteworthy is the occurrence, especially in the central and southern regions of China (see the next section), of numerous paleoendemic lineages whose distribution was much wider and today is restricted to a few, often disjunct refugia in East Asia. The fossil record shows that some of these lineages once existed in Europe or North America, but were extirpated from there during the Neogene as a consequence of climate deterioration (Latham & Ricklefs, 1993; Axelrod et al., 1996; Manchester, 1999; Manchester et al., 2009). Representative examples within the gymnosperms include the monotypic and/or oligotypic genera *Amentotaxus*, *Cathaya* (Box 1), *Cunninghamia*, *Ginkgo* (which is the unique representative of the monotypic family Ginkgoaceae but also of the entire order Ginkgoales), *Glyptostrobus*, *Keteleeria*, *Metasequoia* (Box 1), *Pseudolarix*, and *Taiwania*. From the angiosperms, several examples merit citation here: *Craigia*, *Cyclocarya*, *Davidia*, *Dipelta*, *Diplopanax*, *Emmenopterys*, *Eucommia* (the only representative of the Eucommiaceae), *Fortunearia*, *Pteroceltis*, *Sargentodoxa*, *Tapiscia*, *Tetracentron*, and *Trochodendron*.

Some of these paleoendemic lineages can be traced back to the Cretaceous or even the Jurassic, such as *Ginkgo*. This genus, which today is reduced to a single species (*G. biloba*; Figure 1) remaining in the wild in scattered populations in southern China (but widely planted around the world), appeared not later than the Early Jurassic (Royer et al., 2003). *Ginkgo biloba* is regarded as a genuine "living fossil" because it morphologically resembles to several *Ginkgo* species from the Jurassic and, in fact, it is almost indistinguishable from the Early Cretaceous *G. adiantioides* (Royer et al., 2003). The genus reached its greatest diversity during the Early Cretaceous, when it attained a nearly circumpolar distribution in the Northern Hemisphere. A decline of species and distribution began during the Late Cretaceous and continued along the Tertiary, disappearing from North America during the Miocene and from Europe at the end of Pliocene because of the Neogene global cooling (Del Tredici, 2000, Manchester et al., 2009). *Metasequoia glyptostroboides* and *Pseudolarix amabilis* are other suitable examples of morphological stasis during the last 100 My, and for which China also constitutes their last resting place (LePage & Basinger, 1995; LePage et al., 2005).

WHY THERE ARE SO MANY RELICT LINEAGES IN CHINA?

The strong relictual character of the modern Chinese flora is undoubtedly related to the tectonics and geologic history of China's landmass, with some of the events influencing the flora composition tracing back to the Mesozoic. Taking into account that the origin of vascular plants is often situated in Gondwana (Graham, 1993; Steemans et al., 2009), the collision of Gondwanan terranes (including southern China, Indochina and Sibumasu) with what is now northern China in the Triassic-Early Jurassic period, and the further impact of the Indian subcontinent during the Late Paleocene/Early Eocene period (Metcalfe, 1988; Şengör, 1997) would have extensively supplied China with pteridophytes and gymnosperms. This may help to explain the disproportionate number of lineages in China belonging to the two most ancient groups of vascular plants (Qian & Ricklefs, 1999; Qian, 2001).

Figure 1. Trees of *Ginkgo biloba* (and detail of the leaf) in autumn, planted in the Institute of Botany of Chinese Academy of Sciences (Xiangshan, Beijing) (photo by Jordi López-Pujol).

Box 1. *Cathaya* and *Metasequoia*: **two emblematic Chinese plant "living fossils"**

Perhaps the most paradigmatic examples of the relict plants still surviving in China are *Cathaya* and *Metasequoia*. Fossil evidences indicate that these two gymnosperm lineages appeared in the Cretaceous, and achieved a widespread distribution along the Northern Hemisphere in the Tertiary (Liu & Basinger, 2000; LePage et al., 2005). With the climate deterioration of the late Tertiary and the Quaternary, these lineages became progressively extinct, and today only one representative of each genus (*C. argyrophylla* and *M. glyptostroboides*) are surviving in the mountains of central and southern China. Surprisingly, these two species were not discovered until quite recent times (*C. argyrophylla* was described in 1958 and *M. glyptostroboides* in 1948) perhaps due to the rarity of these conifers. In fact, both species are endangered because of their limited ranges and population sizes: *C. argyrophylla* is discontinuously distributed in several subtropical mountain regions in Guizhou, Guangxi, Hunan provinces and Chongqing Municipality, with very small populations totaling less than 4,000 mature individuals (Ge et al., 1998). *Metasequoia glyptostroboides* has a more reduced occurrence (a small area in the juncture of Chongqing, Hubei and Hunan) but a similar population size (ca. 5,400 individuals; Leng et al., 2007). In addition, both species have low levels of genetic diversity and poor regeneration capability (Ge et al., 1998; Li et al., 2005; Wang & Ge, 2006; Leng et al., 2007), features which may compromise their long-term survival. In order to protect these relict and threatened taxa, several conservation measures have been implemented in recent decades, such as their inclusion in the national list of protected plant species (in the highest rank), the setting up of nature reserves for protecting them in-situ, and their cultivation in botanical gardens, both domestic and overseas (Chapman & Wang, 2002; Leng et al., 2007; Ma, 2007; Callaghan, 2009).

Left and top right: Type tree of *M. glyptostroboides* at Moudao Town (Hubei Province) (photo by Qin Leng), and artificial stand of the same gymnosperm species in the Institute of Botany of Chinese Academy of Sciences (Xiangshan, Beijing) (photo by Jordi López-Pujol). Bottom right: Leaves and cones of a wild individual of *C. argyrophylla*, in the Fanjingshan Mt. of Guizhou Province (photo by Ming-Xun Ren).

Box 2. China: also a hotspot of relict animals

Similarly to plants, China is still home of many ancient (Tertiary or older) animal lineages, both aquatic and terrestrial (Li, 2005). Mainly because of population declining and habitat fragmentation, a significant fraction of these have become severely threatened in recent decades, and today are included in red lists. Despite that some active protection measures have started to be implemented by the Chinese government in recent decades (Liu J. et al., 2003; Harris, 2007), some animal taxa are now in the brink of extinction or even might have disappeared from the wild, as in the case of Yangtze River dolphin (*Lipotes vexillifer*). This emblematic river dolphin, locally known as *baiji*, and the only representative of the family Lipotidae (a lineage dating from the Early Miocene; Ye et al., 2009), might now be extinct because no dolphins were observed in the late-2006 Yangtze River census (Turvey et al., 2007). This cetacean has declined severely in recent decades, mainly due to the construction of dams and other water development projects, water pollution, increased navigation, and incidental mortality through illegal fishing (Wang et al., 2006). If confirmed, this will represent the first global extinction of a cetacean due to human activities (Turvey et al., 2007). Other ancient aquatic animals which are also very endangered at present in the Yangtze River are the Chinese paddlefish (*Psephurus gladius*), the Chinese sturgeon (*Acipenser sinensis*), and the Yangtze sturgeon (*A. dabryanus*)

(Xie, 2003). Sturgeons (family Acipenseridae) and paddlefishes (Polyodontidae) are two lineages of living fossils which can be traced back to the Early Jurassic (Peng et al., 2007). The oldest of the three fish species is the Chinese paddlefish, which seems to have diverged from the only other extant paddlefish species (*Polyodon spathula*) ca. 70 Mya (in the Late Cretaceous), whereas *A. sinensis* and *A. dabryanus* diverged from each other about 10 Mya, in the late Miocene (Peng et al., 2007). It is also very ancient the Chinese alligator (*Alligator sinensis*), which separated from the American alligator (*A. mississippiensis*) ca. 50 Mya (Wu et al., 2003). This crocodilian is currently declining (the total population is fewer than 200 individuals) due to the destruction or modification of their habitats but also to hunting (Thorbjarnarson et al., 2002). The same threats are also menacing another relict animal, the Chinese giant salamander (*Andrias davidianus*) (Wang et al., 2004).

Left: 'Qi Qi', a Yangtze River dolphin (*Lipotes vexillifer*) in the Baiji Dolphinarium, in the Institute of Hydrobiology of Chinese Academy of Sciences (Wuhan) (photo by Ding Wang). Right: Wild individual of the Chinese alligator (*Alligator sinensis*), in Anhui Province (photo by Shao Mao).

In addition, China could have been close to the center of origin of flowering plants, which for some authors was located somewhere in SE Asia (Thorne, 1963 Takhtajan, 1969; 1987; Smith, 1973). The continuous land connection with southern Asian tropical areas (dating back at least 200 million years; Hallam, 1994) would have enabled a straightforward transferring of the most primitive angiosperm lineages to China. More recent studies suggest that angiosperms might have originated instead in Gondwana, but still in paleotropical latitudes (Morley, 2000; Barrett & Willis, 2001). With the finding of the early angiosperm *Archaefructus* fossils in Liaoning Province (NE China) in addition to other ancient angiosperms in neighboring areas, Sun et al. (2008) are postulating, however, for an "Eastern Asian origin of angiosperms". In any case, there is general agreement that China constituted an important area for the early diversification of angiosperms (Qian & Ricklefs, 1999; Barrett & Willis, 2001; Qian, 2001; Sun et al., 2008).

The most important factor to explain the current abundance of relict lineages in China is related, nevertheless, to more recent episodes in its geologic history. During the late Tertiary and the whole Quaternary, that is, from the Middle Miocene climatic optimum (ca. 15 Mya; Zachos et al., 2001) onwards, a progressive global cooling produced numerous plant extinctions in the Northern Hemisphere (Tiffney, 1985; Sauer, 1988; Axelrod et al., 1996;

Jackson & Weng, 1999), and these were centered on the relictual, termophilic elements of the "boreotropical flora" (Tiffney, 1985; Latham & Ricklefs, 1993). The overrepresentation of these elements in the modern flora of China, and the much larger overall taxonomical richness of the Asian country respect to other territories from the Northern Hemisphere (Europe and the United States, with comparable areas, harbour a much poorer flora: 11,500 and 18,000 species, respectively), is suggesting the existence of extensive refugial areas for plants during the Neogene (Axelrod et al., 1996; Qian & Ricklefs, 2000; López-Pujol, 2008). These would have prevented Tertiary (or older) lineages from extinction (i.e., acting as "centres of survival"), and, in most cases, also stimulated plant differentiation and speciation (i.e., "centres of evolution").

WHERE WERE THESE REFUGIA LOCATED?

The delimitation of areas of endemism is a widely used method to identify glacial refugia because of the role of these areas as centres of plant persistence and/or cladogenesis (Fjeldså and Lovett, 1997; Crisp et al., 2001; Jetz et al., 2004; Tribsch, 2004; Orme et al., 2005). This method, although being indirect, should be regarded as an extremely valuable tool when other more direct evidences (i.e. the paleontological data) are insufficient or poorly collected, which is the case of China (Liew et al., 1998; Liu H. et al., 2003; Manchester et al., 2009). In addition, phylogeographical studies, which constitute another indirect but suitable approach for locating the glacial refugia (Tribsch & Schönswetter, 2003; Gómez & Lunt, 2007), are still in their infancy in China.

Several studies carried out in the last 15 years and focused in the endemic plant taxa of China roughly indicated that the plant glacial refugia should be located in the central and southern parts of China (Ying et al., 1993; Wang & Zhang, 1994; Ying, 2001; Mittermeier et al., 2005; Tang et al., 2006). However, these approximations were not performed at the appropriate taxonomical level and/or geographical scale. These flaws, however, have been solved in the more recent study of López-Pujol (2008), in which the distribution patterns of a representative sample of the Chinese endemic flora (nearly 600 species or subspecies) were plotted on a 1° × 1° latitude/longitude grid. The most significant contribution was, nevertheless, that all the plant taxa included in that study were classified into paleo- and neoendemics and subsequently represented separately, which allowed to know if the areas of endemism (i.e. the glacial refugia) were mainly places for the survival of relict lineages or, alternatively, for their differentiation.

Up to 20 glacial refugia were delimited in the study of López-Pujol (2008), all located in central and southern China and, interestingly, coinciding with the main mountain ranges of these regions: Hengduan and Daxue Mountains, Yungui Plateau, central China Mountains (such as Qinling and Daba ranges), Nanling Mountains, eastern China Mountains (e.g. Tianmu), and also Hainan and Taiwan. The existence of such southern refugia (all below latitudes of 35°N, and most of them below 32°N) is also in broad agreement with the Quaternary vegetation reconstructions for China, which suggest the occurrence of extensive temperate and subtropical forests in these areas during the glacial periods (An et al., 1990; Winkler & Wang, 1993; Wang & Sun, 1994; Adams & Faure, 1997; Yu et al., 2000; Harrison et al., 2001). These vegetation types probably found suitable places for their persistence in the mountainous regions of southern China, which would have provided long-term stable habitats through varied topography and local buffering of the extreme late Tertiary and Pleistocene climatic conditions (Ying et al., 1993; Axelrod et al., 1996; Qian & Ricklefs, 1999). The presence of large forested areas in China during the Quaternary was possible due to the lack of major geographic barriers for the latitudinal plant migrations tracking the climatic oscillations. In eastern China, migrations were possible well beyond the Tropic of Cancer, and broad land connections with tropical areas of SE Asia allowed the warmest (tropical) elements to survive there in the cold periods (Guo et al., 1998). Southwards displacements in other parts of the Northern Hemisphere were, in contrast, much more limited: in Europe, for example, the large east-west oriented mountains and the Mediterranean Sea constituted formidable barriers, avoiding migrations beyond 40°N (Tiffney, 1985; Latham & Ricklefs, 1993; Hewitt, 1996).

Almost all the delimited refugial areas in China contained both paleo- and neoendemics. This is not surprising, because assemblages of relict and recently speciated taxa have been reported for other regions also characterized by a rugged topography and with a widely recognized role as refugia, e.g. California (Stebbins & Major, 1965), the Mediterranean Basin (Verlaque et al., 1997), or the tropical Andes (Fjeldså, 1994). However, considerable differences in the relative numbers of paleo- and neoendemics at each Chinese refugia were evidenced, with a clear trend consisting of the preponderance of neoendemics in the east fringe of the Tibetan Plateau (Hengduan Mountains *sensu lato*) and the dominance of paleoendemics towards the east. This distribution can be related to the divergent geological and tectonic history of the involved areas: the Tibetan Plateau is the most significant 'evolutionary front' of China probably due to its uninterrupted uplift from the late Tertiary (Li & Fang, 1999; Zheng et al., 2000; An et al., 2001), which enhanced differentiation. In contrast, the relative tectonic stability in central and southern China during most of the Tertiary (with a few exceptions such as Hainan and Taiwan) (Hsü, 1983; Wang, 1985; Ferguson, 1993; Sibuet & Hsu, 2004) may have maximized the persistence of relict plant lineages. The Three Gorges Region, the northeastern corner of Guangxi, and the area formed by SE Yunnan, SW Guangxi and SW Guizhou, were the most significant refugia for paleoendemics in China (Figure 2).

CONSERVATION CONSIDERATIONS

Many relict plant lineages in China are at risk, which is attributable to two factors which often act synergistically. Paleoendemics, as stated in former paragraphs, are often surviving in small and scattered populations, being thus "naturally" menaced. Any stochastic event, such a strong demographic fluctuation or a catastrophe (e.g. a flood or a severe drought), may put these species in the brink of extinction or even cause their extirpation from nature (Lande, 1988; Melbourne & Hastings, 2008). The delicate situation of these plant lineages, nevertheless, can be aggravated by a more modern source of threats: the human activities. China has suffered, especially in recent decades, a large modification of its natural environment, which has put a lot of pressure on plant biodiversity (e.g. Liu J. et al., 2003; López-Pujol et al., 2006; López-Pujol & Zhang, 2009). Massive habitat destruction, rampant pollution, over-exploitation of species for human use, and uncontrolled introduction of exotic species could have caused the extinction of at least 200 plant taxa since the 1950s (Zhang et al., 2000).

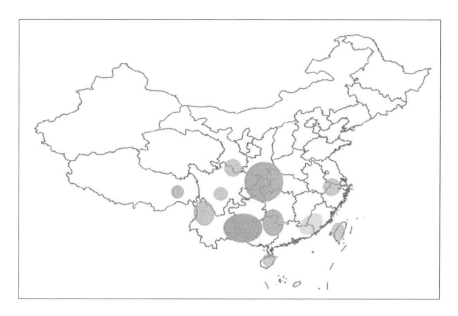

Figure 2. Approximate location of the most significant refugia for paleoendemics in China (redrawn from López-Pujol, 2008). Increasing darkness indicates higher concentrations of relict taxa.

The combined action of both natural and human threats is probably behind the fact that the majority of the most endangered species in China (those surviving in one or very few localities and total population sizes of just a few dozens of individuals) are probably paleoendemics (López-Pujol & Zhang, 2009). For example, habitat destruction has caused the extirpation of *Euryodendron excelsum* from Guangxi Province, and today this relict (which is the only representative of a genus endemic to China) is confined to a unique small population (with less than 30 mature individuals) in Yangchun County, in Guangdong (Shen et al., 2008). Deforestation is also the likely cause of the virtual extinction of one of the only two species of *Craigia*, *C. kwangsiensis* (it has not been seen since 1982; Tang et al., 2007). This genus, with first fossils dating from the Paleocene, was widespread in the Northern Hemisphere in mid-Tertiary time (Manchester et al., 2009) and today is restricted to SW China and N Vietnam.

The extirpation of such relict taxa would imply the loss of unique evolutionary history, or phylogenetic diversity (Faith, 1994; Mace et al., 2003). As previously discussed, China is still harbouring a very rich sample of ancient lineages, in some cases representing entire families or even groups with a higher taxonomical rank, e.g. Ginkgoales. The survival of *Ginkgo biloba* (Fig. 1) in southern China mountains during the Quaternary, and its further cultivation worldwide (for its horticultural value, timber, and medicinal properties) have meant the saving of an independent vascular plant lineage dating from the Early Permian, some 280 Mya (Willis & McElwain, 2002). Since the relict taxa are often taxonomically isolated (by former extinction of their close relatives and/or very long divergence), their loss may have devastating effects on the tree of life: their unique gene combinations and biological characters will be definitively lost. In contrast, the extirpation of phylogenetically young species is, a priori, less serious, because these usually have close relatives and thus their extinction from nature will not involve a significant loss of distinctive evolutionary history (Mace et al., 2003).

ACKNOWLEDGMENTS

The authors thank Song Ge, M.Carmen Martinell, Marta Ponseti, Hai-Qin Sun, Tsun-Shen Ying, Fu-Min Zhang, and Zhi-Yong Zhang for various helps. This study has been partially subsidized by a post-doctoral grant (EX2005-0922) from the Spanish Ministry of Education and Science (MEC) for the period 02/2006-01/2008 (Institute of Botany, Chinese Academy of Sciences) and a 'JAE-Doc' contract within the CSIC program 'Junta para la Ampliación de Estudios' to J.L.P., and a grant (O754101H01) from Wuhan Botanical Garden (Chinese Academy of Sciences) to M.X.R.

REFERENCES

Adams, J. M. & Faure, H. (Eds.). (1997). *Review and Atlas of Palaeovegetation: Preliminary land ecosystem maps of the world since the Last Glacial Maximum* [online]. Oak Ridge National Laboratory, Oak Ridge, USA. Available from: *http://www.esd.ornl.gov/projects/qen/* adams1.html [cited 2009 May 23].

An, Z., Kutzbach, J. E., Prell, W. L. & Porter, S. C. (2001). Evolution of Asian monsoons and phased uplift of the Himalaya-Tibetan plateau since Late Miocene times. *Nature, 411*, 62-66.

An, Z., Wu, X., Lu, Y., Zhang, D., Sun, X. & Dong, G. (1990). A preliminary study on the paleoenvironment change of China during the last 20,000 years. In: T. Liu (Ed.), *Loess, Quaternary geology and global change. Part II* (pp. 1-26). Beijing, China: Science Press (In Chinese).

Axelrod, D. I., Al-Shehbaz, I. & Raven, P. H. (1996). History of the modern flora of China. In: A. Zhang & S. Wu (Eds.), *Floristic characteristics and diversity of East Asian plants* (43-55). Beijing, China: China Higher Education Press.

Barrett, P. M. & Willis, K. J. (2001). Did dinosaurs invent flowers? Dinosaur-angiosperm coevolution revisited. *Biological Reviews, 76*, 411-447.

Barthlott, W., Mutke, J., Rafiqpoor, D., Kier, G. & Kreft, H. (2005). Global centers of vascular plant diversity. *Nova Acta Leopoldina, 92*, 61-83.

Callaghan, C. B. (2009). The Cathay silver fir: its discovery and journey out of China. *Arnoldia, 66*, 15-25.

Chapman, G. P. & Wang, Y. Z. (2002). *The plant life of China: Diversity and distribution*. Berlin and Heidelberg, Germany: Springer-Verlag.

Crisp, M. D., Laffan, S., Linder, H. P. & Monro, A. (2001). Endemism in the Australian flora. *Journal of Biogeography, 28*, 183-198.

CSPCEC (China's Strategy Plant Conservation Editorial Committee). (2008). *China's Strategy for Plant Conservation*. Guangzhou, China: Guangdong Press Group.

Davis, S. D., Heywood, V. H. & Hamilton, A. C. (1995). *Centres of plant diversity: A guide and strategy for their conservation. Volume II. Asia, Australasia and the Pacific.* Cambridge, UK: IUCN Publications.

Del Tredici, P. (2000). The evolution, ecology, and cultivation of *Ginkgo biloba*. In: T. A. van Beek (Ed.), *Ginkgo biloba* (7-23). Amsterdam, The Netherlands: Harwood Academic Publishers.

Faith, D. P. (1994). Phylogenetic pattern and the quantification of organismal biodiversity. *Philosophical Transactions of the Royal Society of London. Series B, Biological Sciences, 345*, 45-58.

Farjon, A. & Page, C. N. (Compilers) (1999). *Conifers: Status survey and conservation action plan.* Gland, Switzerland, and Cambridge, UK: IUCN Publications.

Ferguson, D. K. (1993). The impact of late Cenozoic environmental changes in East Asia on the distribution of terrestrial plants and animals. In: N. G. Jablonski (Ed.), *Evolving landscapes and evolving biotas of East Asia since the Mid-Tertiary. Proceedings of the Third Conference on the evolution of East Asian Environment* (145-196). Hong Kong, China: Centre of Asian Studies, University of Hong Kong.

Fjeldså, J. (1994). Geographical patterns for relict and young species of birds in Africa and South America and implications for conservation priorities. *Biodiversity and Conservation, 3*, 207-226.

Fjeldså, J. & Lovett, J. C. (1997). Geographical patterns of old and young species in African forest biota: the significance of specific montane areas as evolutionary centres. *Biodiversity and Conservation, 6*, 325-346.

Ge, S., Hong, D. Y., Wang, H. Q., Liu, Z. Y. & Zhang, C. M. (1998). Population genetic structure and conservation of an endangered conifer, *Cathaya argyrophylla* (Pinaceae). *International Journal of Plant Sciences, 159*, 351-357.

Gómez, A. & Lunt, D. H. (2007). Refugia within refugia: patterns of phylogeographic concordance in the Iberian Peninsula. In: S. Weiss & N. Ferrand (Eds.), *Phylogeography of Southern European refugia* (155-188). Dordrecht, The Netherlands: Springer.

Graham, L. E. (1993). *Origin of land plants*. New York, USA: John Wiley and Sons.

Groombridge, B. (Ed.) (1994). *Biodiversity data sourcebook. WCMC Biodiversity series No. 1.* Cambridge, UK: World Conservation Press.

Guo, Q., Ricklefs, R. E. & Cody, M. L. (1998). Vascular plant diversity in eastern Asia and North America: historical and ecological explanations. *Botanical Journal of the Linnean Society, 128*, 123-136.

Hallam, A. (1994). *An outline of Phanerozoic biogeography*. Oxford, UK: Oxford University Press.

Harris, R. B. (2007). *Wildlife conservation in China: Preserving the habitat of China's wild west*. Armonk, USA: M.E. Sharpe.

Harrison, S. P., Yu, G., Takahara, H. & Prentice, I. C. (2001). Diversity of temperate plants in east Asia. *Nature, 413*, 129-130.

Hewitt, G. M. (1996). Some genetic consequences of ice ages, and their role in divergence and speciation. *Biological Journal of the Linnean Society, 58*, 247-276.

Hsü, J. (1983). Late Cretaceous and Cenozoic vegetation in China, emphasizing their connections with North America. *Annals of the Missouri Botanical Garden, 70*, 490-508.

Jackson, S. T. & Weng, C. (1999). Late Quaternary extinction of a tree species in eastern North America. *Proceedings of the National Academy of Sciences of the United States of America, 96*, 13847-13852.

Jetz, W., Rahbek, C. & Colwell, R. K. (2004). The coincidence of rarity and richness and the potential signature of history in centres of endemism. *Ecology Letters, 7*, 1180-1191.

Lande, R. (1988). Genetics and demography in biological conservation. *Science, 248*, 1455-1460.

Latham, R. E. & Ricklefs, R. E. (1993). Continental comparisons of temperate-zone tree species diversity. In: R. E. Ricklefs & D. Schluter (Eds.), *Species diversity in ecological communities* (294-314). Chicago, USA: University of Chicago Press.

Leng, Q., Fan, S. H., Wang, L., Yang, H., Lai, X. L., Cheng, D. D., Ge, J. W. et al. (2007). Database of native *Metasequoia glyptostroboides* trees in China based on new census surveys and expeditions. *Bulletin of the Peabody Museum of Natural History, 48*, 185-233.

LePage, B. A. & Basinger, J. F. (1995). Evolutionary history of the genus *Pseudolarix* Gordon (Pinaceae). *International Journal of Plant Sciences, 156*, 910-950.

LePage, B. A., Yang, H. & Matsumoto, M. (2005). The evolution and biogeographic history of *Metasequoia*. In: B. A. LePage, C. J. Williams & H. Yang (Eds.), *The geobiology and ecology of Metasequoia* (3-114). Dordrecht, The Netherlands: Springer.

Li, X. T. (2005). *Rare animals of China*. Changsha, China: Hunan Education Press (In Chinese).

Li, J. & Fang, X. (1999). Uplift of the Tibetan Plateau and environmental changes. *Chinese Science Bulletin, 44*, 2117-2124.

Li, Y. Y., Chen, X. Y., Zhang, X., Wu, T. Y., Lu, H. P. & Cai, Y. W. (2005). Genetic differences between wild and artificial populations of *Metasequoia glyptostroboides*: Implications for species recovery. *Conservation Biology, 19*, 224-231.

Liew, P. M., Kuo, C. M., Huang, S. Y. & Tseng, M. H. (1998). Vegetation change and terrestrial carbon storage in eastern Asia during the Last Glacial Maximum as indicated by a new pollen record from central Taiwan. *Global and Planetary Change*, 16-17, 85-94.

Liu, Y. S. & Basinger, J. F. (2000). Fossil *Cathaya* (Pinaceae) pollen from the Canadian High Arctic. *International Journal of Plant Sciences, 161*, 829-847.

Liu, J., Ouyang, Z., Pimm, S. L., Raven, P. H., Wang, X., Miao, H. & Han, N. (2003). Protecting China's biodiversity. *Science, 300*, 1240-1241.

Liu, H., Xing, Q., Ji, Z., Xu, L. & Tian, Y. (2003). An outline of Quaternary development of *Fagus* forest in China: palynological and ecological perspectives. *Flora, 198*, 249-259.

López-Pujol, J. (2008). *Identification of glacial refugia in China through areas of plant endemism*. Post-doctoral Thesis. Institute of Botany, Chinese Academy of Sciences, Beijing, China (unpublished).

López-Pujol, J. & Zhang, Z. Y. (2009). An insight into the most threatened flora of China. *Collectanea Botanica*, 28: 95-110.

López-Pujol, J., Zhang, F. M. & Ge, S. (2006). Plant biodiversity in China: richly varied, endangered and in need of conservation. *Biodiversity and Conservation, 15*, 3983-4026.

Ma, J. (2007). A worldwide survey of cultivated *Metasequoia glyptostroboides* Hu & Cheng (Taxodiaceae: Cupressaceae) from 1947 to 2007. *Bulletin of the Peabody Museum of Natural History, 48*, 235-253.

Mace, G. M., Gittleman, J. L. & Purvis, A. (2003). Preserving the Tree of Life. *Science, 300*, 1707-1709.

Manchester, S. R. (1999). Biogeographical relationships of North American Tertiary floras. *Annals of the Missouri Botanical Garden, 86*, 472-522.

Manchester, S. R., Chen, Z. D., Lu, A. M. & Uemura, K. (2009). Eastern Asian endemic seed plant genera and their paleogeographic history throughout the Northern Hemisphere. *Journal of Systematics and Evolution, 47*, 1-42.

Melbourne, B. A. & Hastings, A. (2008). Extinction risk depends strongly on factors contributing to stochasticity. *Nature, 454*, 100-103.

Metcalfe, I. (1988). Origin and assembly of south-east Asian continental terranes. In: M. G. Audley-Charles, & A. Hallam (Eds.), *Gondwana and Tethys. Geological Society Special Publication No. 37* (101-118). Oxford, UK: Oxford University Press.

Mittermeier, R. A., Gil, P. R., Hoffman, M., Pilgrim, J., Brooks, T., Mittermeier, C. G., Lamoreux, J. & da Fonseca, G. A. B. (2005). *Hotspots revisited: Earth's biologically richest and most endangered ecoregions.* Washington DC, USA: Conservation International.

Mittermeier, R. A., Gil, P. R. & Mittermeier, C. G. (1997). *Megadiversity: Earth's biologically wealthiest nations.* Washington DC, USA: Conservation International.

Morley, R. J. (2000). *Origin and evolution of tropical rain forests.* Chichester, UK: Wiley.

Mutke, J. & Barthlott, W. (2005). Patterns of vascular plant diversity at continental to global scales. *Biologiske Skrifter, 55*, 521-537.

Olson, D. M. & Dinerstein, E. (1998). The Global 200, A representation approach to conserving the Earth's most biologically valuable ecoregions. *Conservation Biology, 12*, 502-515.

Orme, C. D. L., Davies, R. G., Burgess, M., Eigenbrod, F., Pickup, N., Olson, V. A., Webster, A. J. et al. (2005). Global hotspots of species richness are not congruent with endemism or threat. *Nature, 436*, 1016-1019.

Peng, Z., Ludwig, A., Wang, D., Diogo, R., Wei, Q. & He, S. (2007). Age and biogeography of major clades in sturgeons and paddlefishes (Pisces: Acipenseriformes). *Molecular Phylogenetics and Evolution, 42*, 854-862.

Qian, H. (2001). A comparison of generic endemism of vascular plants between East Asia and North America. *International Journal of Plant Sciences, 162*, 191-199.

Qian, H. & Ricklefs, R. E. (1999). A comparison of the taxonomic richness of vascular plants in China and the United States. *The American Naturalist, 154*, 160-181.

Qian, H. & Ricklefs, R. E. (2000). Large-scale processes and the Asian bias in species diversity of temperate plants. *Nature, 407*, 180-182.

Qian, H., Wang, S., He, J. S., Zhang, J., Wang, L., Wang, X. & Guo, K. (2006). Phytogeographical analysis of seed plant genera in China. *Annals of Botany, 98*, 1073-1084.

Royer, D. L., Hickey, L. J. & Wing, S. L. (2003). Ecological conservatism in the "living fossil" Ginkgo. *Paleobiology, 29*, 84-104.

Sauer, J. D. (1988). *Plant migration: The dynamics of geographic patterning in seed plant species.* Berkeley and Los Angeles, USA: University of California Press.

Şengör, A. M. C. (1997). Asia. In: E.M. Moores, & R.W. Fairbridge (Eds.), *Encyclopedia of European and Asian regional geology* (34-51). London, UK: Chapman & Hall.

Shen, S., Ma, H., Wang, Y., Wang, B. & Shen, G. (2008). Structure and dynamics of natural populations of the endangered plant *Euryodendron excelsum* H. T. Chang. *Frontiers of Forestry in China*, *4*, 14-20.

Sibuet, J. C. & Hsu, S. K. (2004). How was Taiwan created? *Tectonophysics*, *379*, 159-181.

Smith, A. C. (1973). Angiosperm evolution and the relationship of the floras of Africa and America. In: B. J. Meggers, E. S. Ayensu & W. D. Duckworth (Eds.), *Tropical forest ecosystems in Africa and South America: a comparative review* (49-61). Washington DC, USA: Smithsonian Institution Press.

Stebbins, G. L. & Major, J. (1965). Endemism and speciation in the California flora. *Ecological Monographs*, *35*, 1-35.

Steemans, P., Le Hérissé, A., Melvin, J., Miller, M. A., Paris, F., Verniers, J. & Wellman, C. H. (2009). Origin and radiation of the earliest vascular land plants. *Science*, *324*, 353.

Sun, G., Dilcher, D. L. & Zheng, S. L. (2008). A review of recent advances in the study of early angiosperms from northeastern China. *Paleoworld*, *17*, 166-171.

Takhtajan, A. L. (1969). *Flowering plants: Origin and dispersal*. Edinburgh, UK: Oliver & Boyd.

Takhtajan, A. (1987). Flowering plant origin and dispersal: the cradle of the angiosperms revisited. In: T. C. Whitmore (Ed.), *Biogeographical evolution of the Malay Archipelago* (26-31). Oxford, UK: Clarendon Press.

Tang, Y., Gilbert, M. G. & Dorr, L. J. (2007). Tiliaceae. In: Z. Y. Wu, P. H. Raven, & D. Y. Hong (Eds.), *Flora of China. Vol. 12 (Hippocastanaceae through Theaceae)* (240-263). Beijing, China, and St. Louis, USA: Science Press & Missouri Botanical Garden Press.

Tang, Z., Wang, Z., Zheng, C. & Fang, J. (2006). Biodiversity in China's mountains. *Frontiers in Ecology and the Environment*, *4*, 347-352.

Thorbjarnarson, J., Wang, X., Ming, S., He, L., Ding, Y., Wu, Y. & McMurry, S. T. (2002). Wild populations of the Chinese alligator approach extinction. *Biological Conservation*, *103*, 93-102.

Thorne, R. F. (1963). Biotic distribution patterns in the tropical Pacific. In: J. L. Gressitt (Ed.), *Pacific Basin biogeography* (311-354). Honolulu, USA: Bishop Museum Press.

Thorne, R. F. (1999). Eastern Asia as a living museum for archaic angiosperms and other seed plants. *Taiwania*, *44*, 413-422.

Tiffney, B. H. (1985). The Eocene North Atlantic land bridge: its importance in Tertiary and modern phytogeography of the Northern Hemisphere. *Journal of the Arnold Arboretum*, *66*, 243-273.

Tribsch, A. (2004). Areas of endemism of vascular plants in the Eastern Alps in relation to Pleistocene glaciation. *Journal of Biogeography*, *31*, 747-760.

Tribsch, A. & Schönswetter, P. (2003). Patterns of endemism and comparative phylogeography confirm palaeoenvironmental evidence for Pleistocene refugia in the Eastern Alps. *Taxon*, *52*, 477-497.

Turvey, S. T., Pitman, R. L., Taylor, B. L., Barlow, J., Akamatsu, T., Barrett, L. A., Zhao, X., et al. (2007). First human-caused extinction of a cetacean species? *Biology Letters*, *3*, 537-540.

Verlaque, R., Médail, F., Quézel, P. & Babinot, J. F. (1997). Endémisme végétal et paléogéographie dans le bassin Méditerranéen. *Geobios*, *21*, 159-166 (In French).

Wang, H. (Compiler) (1985). *Atlas of the palaeogeography of China*. Beijing, China: Cartographic Publishing House.

Wang, H. W. & Ge, S. (2006). Phylogeography of the endangered *Cathaya argyrophylla* (Pinaceae) inferred from sequence variation of mitochondrial and nuclear DNA. *Molecular Ecology, 15*, 4109-4123.

Wang, P. & Sun, X. (1994). Last glacial maximum in China: comparison between land and sea. *Catena, 23*, 341-353.

Wang, H. S. & Zhang, Y. L. (1994). The bio-diversity and characters of spermatophytic genera endemic to China. *Acta Botanica Yunnanica, 16*, 209-220 (In Chinese).

Wang, K., Wang, D., Zhang, X., Pfluger, A. & Barrett, L. (2006). Range-wide Yangtze freshwater dolphin expedition: The last chance to see baiji? *Environmental Science and Pollution Research, 13*, 418-424.

Wang, X. M., Zhang, K. J., Wang, Z. H., Ding Y. Z., Wu, W. & Huang, S. (2004). The decline of the Chinese giant salamander *Andrias davidianus* and implications for its conservation. *Oryx, 38*, 197-202.

Willis, K. J. & McElwain, J. C. (2002). *The evolution of plants.* Oxford, UK: Oxford University Press.

Winkler, M. G. & Wang, P. K. (1993). The late-Quaternary vegetation and climate of China. In: H. E. Wright Jr., J. E. Kutzbach, T. Webb III, W. F. Ruddiman, F. A. Street-Perrott & P. J. Bartlein (Eds.), *Global climates since the last glacial maximum* (221-261). Minneapolis, USA: University of Minnesota Press.

Wu, X., Wang, Y., Zhou, K., Zhu, W., Nie, J. & Wang, C. (2003). Complete mitochondrial DNA sequence of Chinese alligator, *Alligator sinensis*, and phylogeny of crocodiles. *Chinese Science Bulletin, 48*, 2050-2054.

Xie, P. (2003). Three-Gorges Dam: risk to ancient fish. *Science, 302*, 1149.

Xiong, Y., Brandley, M. C., Xu, S., Zhou, K. & Yang, G. (2009). Seven new dolphin mitochondrial genomes and a time-calibrated phylogeny of whales. *BMC Evolutionary Biology, 9*, 20.

Ying, T. S. (2001). Species diversity and distribution pattern of seed plants in China. *Biodiversity Science, 9*, 393-398 (In Chinese).

Ying, T. S., Zhang, Y. L. & Boufford, D. E. (1993). *The endemic genera of seed plants of China.* Beijing, China: Science Press.

Yu, G., Chen, X., Ni, J., Cheddadi, R., Guiot, J., Han, H., Harrison, S. P., et al. (2000). Palaeovegetation of China: a pollen data-based synthesis for the mid-Holocene and Last Glacial Maximum. *Journal of Biogeography, 27*, 635-664.

Zachos, J., Pagani, M., Sloan, L., Thomas, E. & Billups, K. (2001). Trends, rhythms, and aberrations in global climate 65 Ma to present. *Science, 292*, 686-693.

Zhang, P., Shao, G., Zhao, G., Le Master, D. C., Parker, G. R., Dunning, J. B. Jr. & Li, Q. (2000). China's forest policy for the 21st century. *Science, 288*, 2135-2136.

Zheng, H., Powell, C. M., An, Z., Zhou, J. & Dong, G. (2000). Pliocene uplift of the northern Tibetan Plateau. *Geology, 28*, 715-718.

In: Encyclopedia of Environmental Research
Editor: Alisa N. Souter

ISBN: 978-1-61761-927-4
© 2011 Nova Science Publishers, Inc.

Chapter 65

AN ASSESSMENT OF BIODIVERSITY STATUS IN SEMEN MOUNTAINS NATIONAL PARK, ETHIOPIA

*Behailu Tadesse and S.C. Rai**
Department of Geography, Delhi School of Economics,
University of Delhi, Delhi-110007, India.

ABSTRACT

The Semen Mountains, a U.N. World Heritage Site is located in the North Ethiopia, in the Gondar Administrative region. The park encompasses a wide altitudinal range: which extends from below 2000m to above 4000m asl. It includes extensive high plateau areas (roughly 3200 m to 4000m), steep escarpments and lower-lying parts (below 3000 to 2000 m) and flat areas. This results in a rich mosaic pattern of different habitats, which promotes species richness and high biodiversity. The Semen lowlands, primarily harbouring afro-montane vegetation, are richest in species, whereas the afro-alpine belt, although, the most spectacular part of the mountains, is the poorest. In general, species diversity decreases with increasing altitude. At present, approximately 550 taxa of angiosperms are recorded. The park is famous for their rare and endangered animals (including the endemic Walia ibex, Capra walie) also harbour endemic plant species, i.e. *Rosularia semiensis* and *Maytenus* cortii. The Semen Mountains is the historic plant collecting locality, and about 40% flowering plants originate from this region alone. But in the past few decades, the park is facing deterioration due to socio-political problems in the North of the country.

Keyword: Ethiopia, Semen, Afromontane, Afroalpine Ethiopia endemics, Gondar, World Heritage Site

---

* Corresponding author: E-mail: *raisc1958@rediffmail.com*

INTRODUCTION

Biological diversity or biodiversity, a term that first emerged some twenty year ago (Lovejoy, 1980 a & b; Wilson, 1984; Norse *et al.*, 1986; Wilson and Peters, 1988; Reid and Miller, 1989; McNeely *et al.*, 1990), describes the variety and variability of life on earth. It covers all forms of terrestrial and aquatic plants, animals and microorganisms, their genetic material and the ecosystem of which they are part. Biological diversity is a characteristic of an ecosystem. It can, in very simple terms, be defined by the number of species (species richness) found in the system. By this definition, species which differ widely from each other genetically or phenologically are more biologically diverse (Chopra, 1997). Global biodiversity is usually divided into three categories i.e., genetic, species and ecosystem diversity.

Biodiversity is important to human being for their sustenance, health, well-being and recreation. The benefits of biodiversity conservation can be grouped into three broad categories viz., 1st related to ecosystem services, 2nd related to biological resources and 3rd related to social benefits. Biodiversity data and information are necessary to support well-informed decision making at the global, national and local level. But biodiversity data are scattered, outdated and available in incompatible formats and resolutions. The continued loss of biodiversity along with the reporting requirement of international conventions such as the Convention on Biological Diversity (CBD), Ramsar Convention, World Heritage Convention etc. have called for extra efforts to generate better data and information.

There are currently about 30,350 protected areas in the world covering more than 13.23 million km^2 i.e., about 8.83% of the land on the earth (Green and Paine, 1997). As on December 1999, there are about 357 Man and Biosphere Reserves (MAB) worldwide. There are about 1011 Ramsar wetlands covering an area of over 71.80 million hectares. Currently, there are about 582 World Heritage Site of which 445 is cultural, 117 are natural and 20 are of mixed types (IUCN, 1998).

Moreover, baseline information on the status and distribution of biodiversity resources is necessary that can serve as a benchmark for monitoring. The United Nations Environment Programme (UNEP) outlines eight major categories of biodiversity data for country studies (UNEP, 1993). These datasets will serve three main purposes namely, the conservation of biodiversity, the sustainable use of biological resources and the equitable sharing of benefits from using those resources.

Increasing concern on biodiversity loss and reporting requirements of international biodiversity agreements have called for World's attention to inventory and monitor the wealth of biodiversity. But as yet, only a few biodiversity data and information are widely available. Therefore, a comprehensive review of the available data and information is necessary to see how well the available biodiversity data is reflected on the statistical and biological representations. Cox and Moore (1993) mapped the distribution of major terrestrial biomes of the world based on the physiognomy of the vegetation. An attempt in this direction of developing a data source of statistical and biological representation has ensured in this paper.

Ethiopia is mountainous tropical country with a large diversity of macro and micro climatic conditions that have contributed to the formation of diverse ecosystem containing a great diversity of life forms both fauna and flora. The Ethiopian highlands are separated into two plateaus by the Great Rift Valley, one in the north-west and the other in the south-east.

These outstanding biophysical features are a principal part of variety of flora and fauna for Ethiopia and Semen Mountains National Park. The Semen displays a spectacular landscape with high plateau, steep escarpments, deep gorges and sharp precipices. The present paper discusses plant diversity, phytogeographical affinities and enumerates selected endemic taxa and elaborates on their distribution ranges.

THE SEMEN MOUNTAINS NATIONAL PARK

The Semen (also: Simen, Simien, Semien or Semen) Mountains are located in the Noth of Ethiopia, in the Gonder (Also: Gondar) Administrative Region, 110 km NE of the town of Gondar. The Semen Mountain National park is located in between 13^0, 11' N to 38^0 04' E (Figure 1). The Semen Mountains National Park belongs to the world heritage sites and part of a high mountain massif in North-West Ethiopia. It consists of areas created by the major uplifting of the earth's surface, in contrast to its surrounding regions. The Semen displays a spectacular landscape with high plateaux, steep escarpments, deep gorges and sharp precipices (the result of heavy erosion of Trappean basalt layers which originated from lava outpourings of the Oligocene-Miocene volcanic systems (Puff and Nimomissa, 2001). It is a group of mountains which, however, are often separated from each other by deep valleys. The Semen lowlands (locally known as "Kolla"; below 3000 m) harbor patches of afromontane forest and the highlands are characterized by afroalpine vegetation.

METHODS

The present paper is based mainly of secondary data sources, collected from different organizations/departments i.e., Institute of Biological Diversity of Ethiopia, Wildlife Protection Authority, Ethiopia, Ministry of Agriculture and Rural Development, and Environmental Protection Authority, Ethiopia. Apart from secondary sources researcher visited the parks and collected some relevant information through primary survey.

BOTANICAL HISTORY

A large number of travelers and botanists visited the Semen Mountains National Park in the last century. Amongst them Wilhelm Georg Schimper, who collected sufficient information on flora and fauna of the park is central to the botanical history of Semen Mountains. He collected information from the Semen park on several occasions between 1838 and 1863 and, scattered over several years, spent a total of over 11 months in the mountains (Puff and Nemomissa, 2001). Many of his collection from the Semen are type specimens and quite a number of these are from Mt. Bwahit, with an altitude of 4430 m the highest point of the Semen proper and from Mt. Silki, 4420 m located 10-12 km NNE of the former.

According to Puff and Nemomissa (2001), type specimens of ca. 220 afromontane and alpine taxa of flowering plants (out of ca. 550 recorded; i.e., 40%) originate from the Semen

Mountains. It is likely that the number of types will increase, as data for numerous families not yet treated for FEE are still lacking. The vast majority of the type collections were made by Schimper. Unfortunately, none of Schimper's type specimens and those of other earlier collectors were housed in any Ethiopian institutions. The reason behind this may be that Systematic Botany as a science was not familiar to the natives during these periods in Ethiopia. The development of the Systematic Botany in Ethiopia is quite a recent phenomenon (Egziabher, 1986).

BIODIVERSITY STATUS

The Semen Mountains National Park (SMNP) represents one of the most outstanding natural areas and is a paramount specimen of world's natural heritage. Because of its rich biodiversity, its higher number of endemic species and its paramount bio-physical features, the SMNP is of international significance and has been declared as one of the first sites all over the world a "World Heritage Site" by the UNESCO World Heritage Committee in 1978. In 1996, the SMNP was inscribed on the *List of World Heritage Danger* due to the evidence of recent deterioration of the population of *Walia ibex*, due to agricultural encroachment, loss of biodiversity and the impacts of road construction.

The phytogeographical position, irregular and undulating topography with lofty hill ridges and deep valleys accompanied by wide variation in climate and soil have resulted in ecological diversity which has influenced the rich and fascinating vegetation in Semen Mountains National Park. Major portion of the area still covered with primary forests, though they are under threat of depletion due to various biotic and abiotic factors. Several forest types and subtypes with characteristic floristic compositions prevail in Semen Mountains National Park. The vegetation of Semen Mountains National Park is classified into four types. Distribution of these tree species is influenced by climate, topography and elevation (Table 1).

Table 1. Life zone, dominant vegetation and animal distribution in Semen Mountains National Park, Ethiopia

Climatic zone	Vegetation	Animals	
		Mammals	Birds
Above 3600m	Lobalia rhyncopetalum	Walia ibex ,Gelada baboon, Ethiopiamwolf, Africa wild cat, Klipspringer, Leopard, Meneliks bushbuck	Black-winged love bird, White colored pigeon, Golden black wood pecker, White winged cliff chat
2700-3600m	Erica arborea, Hypercium revolutum	Walia ibex, Gelada baboon, Klipspringer, Ethiopian Wolf, Narrow headed rat	Degolidark, Black headed oriol, Abyssinia cat bird, Ankober serine
2300-2700m	Evergreen broad leaved Montana forest, S. guineense, Junipur procera, Olea europaea	Colubus gureza, Anubis baboon, Bush pig, Menelik bush buck, Grass rat, Serval cat	Black tit, Rupples chatchough, Lamergier, Pallid harrier
Below 2000m	Accacia abyssinica, F. sur, F. sycomorus, phoenix reclinata	Grey duiker, Rock hyrax, Spotted hyena, Black faced monkey, Caracal, Common jackal	Spot breasted plover, Lander falcon

Table 2. Biodiversity sensitive national parks in Ethiopia

Name of the Park	Established (Year)	Area (km^2)	Elevation (m)	Major species
Abijatta Shalla	1974	514	1577	Flamingoes, Nyala, Ostriches, White-footed rat, Yellow-footed parrot, etc.
Awash	1966	756	1800	Accacia wood land, East African oryx, Soemmerrings gazeel, Dik-Dik, Cheeth, Leopard
Bale Mountains	1970	2200	>4000	Mountain Nyala, Semen fox, Meneliks bushbuck, Bohor Reedbuck, Dry evergreen forests
Gambella	1970	5061	1600	White-ears kob, Nile lechule, Elephant, African buffalo, Lion
Mago	2002	2162		Savanna acacia forest, Neri swamp
Nechisar	1974	514	1650	Plains zebra, Grant's gazeel, Dik-Dik, Swayness hartebeest, Anubis baboon
Omo	1980	4068	4000	Buffalo, Giraf, Elephant, Black Rhonocerros, Black-winged lovebird
Semen Mountains	1969	190	4620	Walia ibex, Ethiopian wolf, Meneliks bushbuck, Gellada baboon, Erica arboria, Lobellia rhinco pethalum, Rossa abyssinka, Hagenia
Yangudi Rassa	2002	4730	400-1459	African wild ass, Beisa oryx, Soemmerrings, Dorcas gazeel, Gerenuk, Grevy's zebra

Table 3. Families of flowering plants with highest species (taxa) diversity in Semen Mountains National Park (Puff and Nimomissa, 2001)

Family	Number of genera	Number of taxa	Percentage of the total (550 taxa)
Poaceae	31	63	11.5%
Asteraceae	22	63	11.5%
Fabaceae	20	38	6.9%
Cyperaceae	5	31	5.6%
Orchidaceae	10	21	3.8%
Rubiaceae	12	19	3.5%
Scrophulariaceae	10	18	3.3%

Taxa= in most cases species, in some infraspecific entities.

To protect the habitats and the biodiversity of the country, the Ethiopian Government has already declared eight national parks and some wildlife sanctuaries as protected areas, accounting for 1.40% of its total geographical land area (Table 2). The Semen Mountains National Park was established in 1969 by the Government of Ethiopia, and is considered to be one of the most important protected areas in the Ethiopia. The park was declared 'World Heritage Site" by UNESCO in 1978. The area is frequently visited by good number of tourists.

Floral Diversity

Semen Mountains National Park, accounts 0.0172% of the geographical areas of the Ethiopia, is custodian of more than 40% of the flowering plants of Ethiopia. Inaccessibility and remoteness marks the area as one of the richest botanical treasure house of the Ethiopia.

Several factors contribute to the Semen's high biodiversity: one is the mountain's geographical position; another is the presence of different altitudinal belts. Also microclimatic differences contribute to the biodiversity "wet" and "dry" types of afromontane forests, each with a characteristic set of core species, are documented (Neomissa and Puff, 2001); the Mesheha river valley and its tributaries lying in the rain shadow are the only place in the Semen where succulents like, for example, *Euphorbia abyssincia*, *Huernia* and *Mesembryanthemacear* are found. Puff and Nemomissa (2001) enumerates 550 taxa of flowering plants belonging to 95 families and 319 genera. Monocots constitute approximately a quarter of the flowering plants of Semen Mountains (137 out of the 550 taxa recorded). This high number is primarily due to the very species-diverse Poaceae, which has 31 genera and 63 taxa (11.50% of all recorded taxa). Also Cyperaceae has 5 genera and 31 taxa (5.6%) and Orchidaceae has 10 genera and 21 taxa (3.8%) are well represented. Amongst the dicots, the Asteraceae are the most diverse family; in terms of number of taxa (63) they are as richly represented as the Poaceae, but the number of genera present (22) is markedly lowers (Table 3). The second largest family is the Fabaceae. Also Rubiaceae and Scrophulariaceae are amongst the most diverse dicot families (Puff and Nemomissa, 2001).

Table 4. Families of flowering plants which are very poorly represented in the Semen Mountains (Puff and Nemomissa, 2001)

Family	Number of genera	Number of taxa	Percentage of the total (550 taxa)	Remarks
Primulaceae	2	2	<1%	Uncommon
Sterculiaceae	2	2	<1%	Rare
Sapotaceae	1	1	<1%	Rare
Hypoxidaceae	1	1	<1%	Rare
Discoreaceae	1	1	<1%	Rare
Juncaceae	1	1	<1%	Uncommon

On the other hand, there are also very poorly represented families, with only one or two genera and only a single or few species each (Table 4). Some of the taxa of these families are rare and /or restricted in their distribution.

The dominant families like *Poaceae, Asteraceae, Ericaceae, Cyperaceae, Orchidaceae, Rubiaceaeand Scrophulariaceae* are well represented and exhibit diversity and richness of the flora (Table 4). The mountains, primarily known to biologists for their rare and endangered animals (including the endemic *Walia Ibex, Capra walie*), also harbour endemic plant species (at least 12; e.g. Rosularia semiensis: afroalpine, or Maytenus cortii: afromontane). In addition to these, more than 30 taxa occurring in the Semen are Ethiopian endemics (e.g. the spectacular afroalpine Lobelia rhynchopetalum, or the afromontane Polyscias farinose). Numerous taxa, both from the afromontane and alpine zone, show a distribution stretching from the Semen southwards to the tropical E. African mountains (e.g. Psychotria orophilla, afromontane, or Trifolium cryptopodium, afroalpine), some times also with an extension to the W. African mountains (e.g. Galium simense). Others (primarily taxa occurring in the afromontane zone of the Semen) show wide distribution ranges extending to S. Africa (e.g. Halleria lucida, to the Drakensberg). Widely distributed Sudanian or Sudano-Zambezian species occurring in the "Semen lowlands" include Sterculia setigera and Stereospermum kunthianum, respectively.

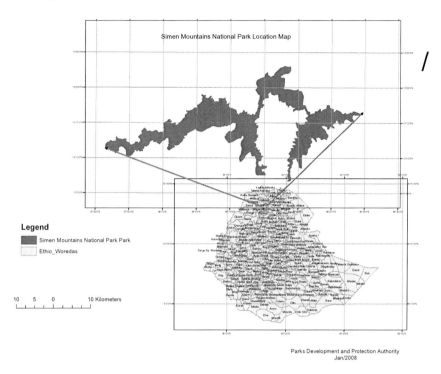

Figure 1. Location map of Semen Mountains National Park, Ethiopia.

As a consequence of the civil war between 1984 and 1991, during which the agricultural exploitation, deforestation and shooting of wildlife were intensified, and a general population growth of about 2-3% on average, the area has been suffering from a strong influx of settlers for the last decade. Villages are still growing, mostly by migration into the park and by absence of family planning. Human utilization of the area has expanded and increased very significantly since 1983, especially along the road. The pressure through settlement and cultivation is still rising.

In general, destruction of some habitats by man, either for agriculture, for grazing or for fuelwood collection pose a serious threats, especially in those areas of the Semen that are not included in the National Park. Semen Mountains have not been extensively explored in the recent decades due to the socio-political situation of the area. Afromontane forests of the park is much more species rich than the higher-lying afroalpine areas, has remained poor. At present, 19 taxa are newly recorded, but this number very possibly is incomplete. Most of the newly recorded taxa are afromontane (*Bulbophyllum josephii*, or *Galiniera saxifrage*), but others extend from the afromontane to the alpine region (*Crassula alba*). A Red Data List of endangered species and the conservation status of the flora of Semen Mountains are undoubtedly required.

Phytogeography

The majority of the flowering plants of the Semen Mountains have distribution which are neither confined to the Semen nor restricted to the Ethiopian mountains but also occur elsewhere. Sudanian and Sudano-Zambian species are few and exclusively found in dry parts of the Semen "lowlands", often along with afromontane elements. All of them have wide general distribution ranges. A good example is *Sterculia setigera,* listed as a Sudanian endemic by White (1983). Examples of even more widely distributed species, occurring both in the Sudanian and the Zambezian region, and sometimes also in regional transition zones (Puff and Nemomissa, 2001) include *Cussonia arborea* (Wickens, 1976).

High mountains of Africa are known for their characteristic endemic flora. Apart from several taxa confined to the Semen Mountains alone, quite a number of taxa recorded from the Semen are Ethiopian endemics. According to Puff and Nemomissa (2001), close to 10% of the Semen Mountains flowering plants are Ethiopian endemics.

Faunal Diversity

A zoological exploration of the Semen Mountains National Park is described by Dunbar (1975), who undertook an inventory of the most conspicuous mammals. The fauna of the area is exceptionally unique because 22 large mammals (*Walia ibex*, *Capra walie*) and 13 small mammals (e.g. *Arvicanthis abyssinicus*, *Cercopithecus aethiops*) have been recorded. The Semen Mountains are an afromontana island with the surrounding lowlands acting as an ecological barrier for many Semen and Ethiopian endemic species (e.g. Flagship species). These flagship mammals are the *Capra ibex walie*, *Canis simensis* and *Tragelaphus scriptus meniliks*. Besides this, 9 additional endemic small mammals recorded in the Semen Mountains National Park. These species are mainly found on the terraced escarpment of the ericaceous zone and around the afroalpine zone.

The Semen Mountains terrain is an important bird area. More than 185 bird species have been recorded (including 6 of the 16 species consider endemic to Ethiopia and 4 of them are globally threatened species) (Teshome, 1999). About thirty three highland biome species (69%) of the Ethiopian highland biome bird assemblage have been recorded from the Semen Mountains National Park. Other fauna such as reptiles, amphibians and fish are not well studied in the park; however, there are sightings of snakes, amphibians and fish in the source of the Adarmaz and Enisa river below the northern side of the escarpment in the lowlands (Teshome, 1999).

Because of the increasing population pressure and encroachment by human and livestock population resulted in widespread destruction of wildlife and their habitats. As a result most of the wildlife is in risk. The principal flagship large mammals of the Semen Mountains National Park are the Ethiopian wolf, Walia ibex, Menelik's bushbuck and Gelada baboon. Ethiopia wolf (*Canis simesis*) is poached for its skin which is considered to have medicinal value. Others were hunted or killed because the animals are locally believed to be a predator of sheep (Nowak, 1991).

The Walia ibex has become a national symbol. It is endemic to the Semen Mountains, one of the most highly endangered mammal species in the world and is threatened by extinction due to low numbers and the very restricted area of remaining habitat. Mainly domestic animals have occupied or diminished their habitats. In order to ensure large habitats for the *Walia ibex*, it is necessary to connect their habitats inside the SMNP by *Corridors* to potential and actual habitats outside the SMNP. Walia ibex are living on the steep slopes and the grassy ledges of the escarpment. They need to feed and sun themselves in mornings and evenings. As Walia ibex also feed on high plateaux when undisturbed, parts of the plateau near the escarpment need to be protected.

The Semen fox lives in the SMNP only in afro-alpine habitats in isolated mountain massifs above 3600 m asl. For comparison, in the Bale Mountain National Park, its habitat is above 3000 m. This variation is mainly due to population pressure. Being normally diurnal, this species has become nocturnal due to severe persecution. All its habitats up to 4400 m asl are occupied by grazing domestic animals accompanied by shepherds who chase the Simen fox or kill it in order to prevent losses of sheep.

MAJOR INITIATIVES TO CONSERVE THE BIODIVERSITY

The initiatives to conserve the biodiversity has primarily emendated from the pressure mounted by scientific community which can fall into two groups but are complimenting each other i.e., local and as well as international. The Convention on Biological Diversity (CBD), Convention on Migratory Species (CMS), Convention on International Trade on Endangered Species of Wild Fauna and Flora (CITES), Ramsar Convention and The World Heritage Convention etc. are the major international conventions regarding biodiversity conservation:

Convention on Biological Diversity (CBD)

The Convention on Biological Diversity's objectives is the first global comprehensive agreement to address all aspects of biological diversity: genetic resources, species, and ecosystems. It recognizes - for the first time - that the conservation of biological diversity is "a common concern of humankind" and an integral part of the development process. To achieve its objectives, the Convention-in accordance with the spirit of the Rio Declaration on Environment and Development - promotes a renewed partnership among countries. Its provisions on scientific and technical cooperation, access to genetic resources, and the transfer of environmentally sound technologies form the foundations of this partnership. As of January, 2000, 176 countries have ratified the convention. Ethiopia is a signatory to the CBD and adopted the Biodiversity Strategies and Action Plan in December 2006, under an agreement between the Government of Ethiopia and the United Nation Development Programme under the Global Environmental Facility Trust Fund. The overall goal of the Ethiopia's Strategy and Action Plan is to establishment of effective systems that ensure the conservation and sustainable use of country's biodiversity.

Convention on Migratory Species (CMS)

The Convention on Migratory Species (CMS) aims to protect those species of wild animals that migrate across or outside national boundaries. This includes conservation of terrestrial, marine and avian species over the whole of their migratory range. The convention was concluded in 1979 and came into force on 1 November 1983. As of December 1999, 68 states have ratified the convention. Ethiopia is not a signatory party to this programme.

The Convention on International Trade in Endangered Species of Wild Fauna and Flora (CITES)

CITES, is an international treaty drawn up in 1973 to protect wildlife against overexploitation and to prevent international trade from threatening species with extinction. The treaty entered into force on 1 July 1975 and now has a membership of 146 countries.

Ramsar Convention: Convention on Wetlands of International Importance

The Convention on Wetlands, signed in Ramsar, Iran, in 1971, is an intergovernmental treaty that provides the framework for national action and international cooperation for the conservation and wise use of wetlands and their resources. There are presently 117 Contracting Parties to the Convention.

The World Heritage Convention

The Convention Concerning the Protection of the World Cultural and Natural Heritage (the World Heritage Convention) was adopted by the General Conference of UNESCO in 1972. As of October 1999, more than 158 countries including Ethiopia have signed the convention. This is one of the most universal international legal instruments for the protection of the cultural and natural heritage. In Ethiopia, there are 8 World Heritage Sites, of which Semen Mountains National Park is the only Natural Heritage Site in Ethiopia. It is home to a number of extremely rare species including the Ethiopian wolf, Gelada baboon, Walia ibex. About 185 species of birds inhabited in the park.

CONCLUSIONS

The Semen Mountains National Park receives international attention as a World Heritage Site not only because of their outstanding physical features, scenic beauty and biodiversity, but also because they are refugia to rare and endemic animal and plant species. The mountains, being the type of locality of 220 afromontane and afroalpine flowering plants, furthermore, are central to the botanical history of Ethiopia. An attempt has been made to review the availability and accessibility of Semen Mountains National Park biodiversity data and information. Published reports and Internet resources were the major sources of information. More encouragingly, a number of INGO's such as WWF, IUCN, Conservation International, and Wetland International are actively contributing to the generation and maintenance of global biodiversity data and information. Users of global biodiversity data include national institutions, regional groupings, international institutions, international funding agencies, bilateral development agencies, international environmental and conservation groups and scientific communities.

An unwise utilization of the biological resources of the mountains could certainly bring about a disturbance of the ecological equilibrium which maintains the normal functioning of ecosystems of Semen and their structural components. It however, seems essential that any effective conservation strategies for the Semen Mountains should take the socio-economic aspects of the local inhabitants into consideration to bring about lasting and sustainable solutions to the degradation of the Semen at an alarming rate. A participatory approach, i.e., the involvement of local inhabitants in the conservation and sustainable uses of the biodiversity of the Semen is inevitable and could certainly prove successful.

REFERENCES

Cox, C. B. & Moore, P. D. (1993). *Biogeography: An ecological and evolutionary approach.* Blackwell Scientific Publications, London.

Chopra, Kanchan (1997). *Economic Valuation of Biodiversity*- Part I: Status Paper. Institute of Economic Growth, University of Delhi, Delhi.

Dunbar, R. J. M. (1975). Social dynamics of Gelada baboons. *Prinatol*, 6,1-157.

Egziabher, T. B. G. (1986). Issues in the development of Botany as a science in Ethiopia and the Ethiopian Flora Project. In: Hedberg, I. (ed.) Research on the Ethiopian Flora. *Proceeding of the first Ethiopian Flora symposium held in Uppsala*, May 22-26, 1984. Symb. Bot. Ups. 26, 1-8.

Green, M. J. B. & Paine, J. (1997). *State of the world's protected areas at the end of the twentieth century.* Paper presented at IUCN World Commission on Protected Areas Symposium on "Protected Areas in the 21st. Century: From Islands to Networks", Albany, Australia, 24-29 November, 1997.

IUCN (International Union for Conservation of Nature and Natural Resources) (1998). 1997 United Nations List of Protected Areas. Prepared by WCMC and WCPA. IUCN, Gland, Switzerland and Cambridge, UK. 1xii + 412.

Lovejoy, T. E. (1980a). *Conservation Biology*: *An evolutionary-ecological perspective*, v-x.Sinauer Associates, Sunderland, USA

Lovejoy, T. E. (1980b). *Changes in Biological Diversity*. The Global 2000 report to the president. Vol. 2 (the technical report), Penguin Books, USA.

McNeely, J. A., Miller, K. R., Reid, W. Mittermeier, R. & Werner, T. (1990). *Conserving the World's Biological Diversity*. IUCN, WRI, World Bank, WWF-US, CI, USA.

Nievergelt, B. (1996). Field Study on the Flora and Fauna of the Simen Mountains: A Summarized Report. University of Zurich, East Anglia, Vienna and Addis Ababa in association with EWCO and the EWNHS.

Nimomissa, S. & Puff, C. (2001). Flora and vegetation of the Simen Mountains National Park, Ethiopia. *Biol. Skr.*, *54*, 335-348.

Norse, E. A., Rosenbaum, K. L., Wilcove, D. S., Wilcox, B. A., Romme, W. H. & Johnston, Nowak, R. M. (1991). *Walker's Mammals of the World.* 5th Edition, Vol. 2 John Hopkins University Press, London.

Puff, C. & Nimomissa, S. (2001). The Simen Mountains (Ethiopia): comments on plant biodiversity, endemism, phytogeographical affinities and historical aspects. *Syst. Geogr.*, *71*, 975-991.

Reid, W. V. & Miller, M. R. (1989). *Keeping Options Alive.* World Resources Institute, Washington, USA.

Teshome, E. (1999). SMNP, Ethiopia, Conservation Values and Their Relevance to Ecotourism. A Dissertation Degree of M. Sc. in Environmental Forestry in the School of Agricultural and Forest Sciences, University of Wales, Bangor Gwynedd, UK.

UNEP, (1993). *Guidelines for Country Studies on Biological Diversity.* United Nations Environment Programme, Nairobi, Kenya.

White, F. (1983). The vegetation of Africa: a descriptive memoir to accompany the UNESCO/AETFAT/UNSO vegetation map of Africa. Paris.

Wilson, E. O. (1984). *Biophilia.* Harward University Press, Cambridge, USA

Wilson, E. O. & Peters, F. M. (Eds.), (1988). *Biodiversity.* National Academy Press, Washington DC. USA.

CHAPTER SOURCES

Chapters 1-12 - Versions of these chapters were also published in *Grassland Biodiversity: Habitat Types, Ecological Processes and Enviromental Impacts,* edited by Johan Runas and Thedor Gahlgren, published by Nova Science Publishers, Inc. They were submitted for appropriate modifications in an effort to encourage wider dissemination of research.

Chapters 13-24- Versions of these chapters were also published in *Conservation of Natural Resources*, edited by Nikolas J. Kudrow, published by Nova Science Publishers, Inc. They were submitted for appropriate modifications in an effort to encourage wider dissemination of research.

Chapters 25-37 and 47- Versions of these chapters were also published in *Freshwater Ecosystems and Aquaculture Research,* edited by Felice De Carlo and Alessio Bassano, published by Nova Science Publishers, Inc. They were submitted for appropriate modifications in an effort to encourage wider dissemination of research.

Chapters 38-46 - Versions of these chapters were also published in *Estuaries: Types, Movement Patterns and Climatical Impacts,* edited by Julian A Crane and Ashton E. Solomon, published by Nova Science Publishers, Inc. They were submitted for appropriate modifications in an effort to encourage wider dissemination of research.

Chapters 48-57 and Short Communications A and B- Versions of these chapters were also published in *Agrictural Runoff, Coastal Engineering and Flooding,* edited by Christopher A. Hudspeth and Timothy E. Reeve, published by Nova Science Publishers, Inc. They were submitted for appropriate modifications in an effort to encourage wider dissemination of research.

Chapters 58- 65 Versions of these chapters were also published in *Biodiversity Hotspots,* edited by Vittore Rescigno and Savario Maletta, published by Nova Science Publishers, Inc. They were submitted for appropriate modifications in an effort to encourage wider dissemination of research.

INDEX

A

AAS, 561
abalone, 1465
abatement, 722, 1229, 1346
abiotic, xxi, 413, 1223, 1367
Aboriginal, 402, 406, 407
Abraham, 508, 841, 852
absorption, xxix, 402, 431, 432, 441, 446, 573, 729, 779, 789, 820, 829, 832, 833, 836, 837, 842, 850, 1072, 1076, 1077, 1084, 1086, 1132, 1250, 1269, 1275, 1276, 1601, 1615, 1617
absorption coefficient, 446
abstraction, 709, 968
Abstraction, 968
Abundance, 214, 297, 312, 381, 382, 545, 578, 955, 982, 983, 1221
abundance data, 378, 381
AC, 355, 615, 1145
academics, 405
Acanthaceae, 225, 1160
acceleration, 931
acceptor, 1105
acceptors, 1103, 1105
accessibility, 40, 41, 46, 711, 878, 1693, 1697, 1701, 1732
accessions, 135
accidental, 584, 1101, 1388
accidents, 420, 568, 623
acclimatization, 889
accounting, 620, 669, 821, 876, 951, 973, 1081, 1194, 1663, 1727
accreditation, 584
accuracy, 18, 408, 1023, 1026, 1076, 1442, 1658, 1691
acetate, 54, 93, 149, 1105, 1109, 1110, 1112
acetic acid, 485, 495, 496, 1111, 1263, 1583, 1592, 1594, 1598, 1599, 1600, 1606
acetone, 1130
acetylation, 1108
acetylcholinesterase, xxxiii, 1071, 1073, 1094, 1095
acetylene, 1357, 1368, 1369
achievement, 1504
acidic, 55, 87, 431, 447, 686, 725, 827, 1366, 1434
acidification, xxi, 394, 413, 420, 421, 429, 432, 447, 585
acidity, 431, 432, 447, 802, 805, 827, 1030, 1594, 1602
ACL, 508
acrylate, 492
actinobacteria, 312
actinomycetes, 469
activated carbon, 1076
activation, 504, 1599
activation energy, 504
active compound, 1600, 1612
acute, 609, 1326
acylation, 129, 150
ad hoc, 1571
Adams, 600, 944, 947, 948, 949, 950, 976, 978, 980, 987, 988, 990, 1231, 1409, 1433, 1573
adaptability, xxiii, 47, 245, 263, 550, 995, 1591, 1592
adaptation, 46, 113, 140, 178, 211, 216, 217, 342, 397, 398, 399, 404, 554, 559, 565, 573, 780, 788, 797, 803, 877, 901, 1108, 1208, 1588, 1596, 1633, 1644
adaptations, 795, 808, 1124
ADC, 1250, 1266, 1270, 1274, 1280
additives, xxix, 486, 493, 494, 819, 835, 839, 1596, 1598, 1608
adductor, 1077
adenosine, 1292

adhesives, 494, 502, 511
adjustment, 202, 625, 685, 781, 1075, 1425, 1476
administration, 433
administrative, 196, 406, 407, 1300
administrators, 865
adsorption, 476, 1211, 1329, 1333, 1336, 1344, 1399, 1401
adult, xviii, 283, 284, 285, 289, 297, 300, 301, 303, 305, 321, 324, 370, 534, 537, 553, 559, 560, 632, 962, 964, 970, 986, 1207
adults, 284, 632, 759, 812, 963, 1456
advancement, 263
advancements, xiv, 13, 1625, 1627
advantages, xxxi, xlii, 395, 438, 487, 568, 684, 686, 911, 998, 1193, 1246, 1450, 1464, 1579, 1581, 1589, 1592, 1595, 1691
adverse effects, xxxii, 184, 450, 464, 924, 949, 969, 1072, 1267, 1280, 1461, 1462, 1541
advocacy, 736
AEA, 393, 394, 395, 398
aerobic, 395, 1102, 1104, 1105, 1373, 1374, 1375, 1378
aeronautical, 503
aesthetic, 432, 454, 457, 1194, 1466
aesthetics, 967, 1197
aflatoxin, 1240
agar, 1103
age, 28, 44, 187, 235, 332, 336, 485, 488, 494, 510, 756, 772, 773, 774, 783, 785, 787, 789, 793, 815, 919, 1158, 1163, 1278, 1419, 1439, 1551, 1552, 1553, 1555, 1556, 1557, 1559, 1562, 1563, 1569
agencies, 272, 276, 595, 596, 597, 723, 736, 754, 755, 866, 867, 1633, 1733
agent, 409, 487, 489, 495, 502, 504, 508, 1105, 1474, 1580, 1581, 1583, 1606, 1614
agents, 185, 188, 322, 337, 352, 464, 489, 492, 500, 502, 503, 527, 529, 599
agglomeration, 1627, 1628
aggregates, 384, 470, 476, 477
aggregation, xxviii, 155, 156, 476, 800, 805, 806, 808, 809, 810, 811, 812
aging, 485, 492, 493, 494, 505, 506, 507, 510, 623
agricultural crop, 467, 1412
agricultural residue, 476, 477, 478, 479
agricultural sciences, 459
agricultural sector, 439, 440, 701, 702, 710, 714, 716
agrochemicals, xxxvii, 430, 433, 450, 1325
agroforestry, xxiii, 371, 456, 550, 559
aid, 1038, 1040, 1395
AIDS, 1460
air, xvii, 29, 243, 246, 393, 394, 419, 422, 423, 426, 441, 448, 450, 454, 490, 1077, 1101, 1117, 1129, 1186, 1358, 1373, 1377, 1384, 1392, 1397, 1400, 1402, 1415, 1425
air pollutant, 393
air pollutants, 393
air quality, 454, 1186
air temperature, xvii, 243, 246, 891, 1129, 1415, 1425
alanine, 496, 919
Alaska, 140, 332, 814
Albanians, 616
Alberta, 1341
albumin, 490
Alcalase, 497, 500, 501
alcohol, 169, 226, 469, 1204, 1207
alcoholism, 610
alcohols, 1599, 1600
aldehydes, 487, 493, 501, 504, 505, 507, 1599, 1600
aldolase, 1583
alfalfa, 170, 1267, 1280, 1294
algae, xxv, xlii, 531, 571, 619, 692, 891, 892, 897, 899, 901, 902, 904, 908, 909, 910, 943, 950, 974, 1012, 1076, 1077, 1144, 1154, 1158, 1170, 1176, 1178, 1239, 1293, 1331, 1332, 1339, 1360, 1465, 1469, 1552, 1623, 1624, 1625, 1626, 1627, 1628, 1629, 1630, 1631, 1632, 1633, 1634, 1635, 1636, 1637, 1638, 1639, 1640
Algal, 1031, 1366
Algeria, 118, 121, 606, 709
algorithm, xliv, 18, 155, 156, 201, 648, 676, 1683
alien species, 757, 872, 873, 878, 882, 884, 975
aliphatic polymers, 475
alkali, 29, 473
alkaline, xxii, 322, 357, 483, 489, 492, 493, 495, 497, 499, 500, 506, 1409
alkaline hydrolysis, 499
alkalinity, 898, 1002
alkane, 1217, 1224
alkanes, 1121, 1205, 1208, 1216, 1220, 1226
alleles, xx, 375, 378, 379, 381, 382, 383
allopatric speciation, 1644
alluvial, xxxix, 434, 435, 437, 1022, 1302, 1407, 1408, 1430, 1556, 1557
almonds, 1333
alpha, 5, 1073
alpha-tocopherol, 851
Alps, 356, 377, 378, 386
alternative, xxxix, 39, 209, 372, 384, 395, 584, 1188, 1376, 1437, 1536, 1589, 1667, 1677, 1678
alternative energy, 721
alternatives, 396, 559, 1373, 1438, 1463
alters, 313, 331, 334, 339, 766, 767, 1333
aluminum, 395, 1130, 1342

Aluminum, 1213
Amazon, 543, 999
Amazonian, 546
ambient air, 1129
ambient air temperature, 1129
amendments, 478, 1108, 1110, 1113, 1330
AMF, xix, 310, 345, 346, 347, 348, 349, 350, 351, 352, 353, 354
amide, 493, 496, 500, 503, 507
amine, 474
amines, 503
amino groups, 485, 500, 502, 504, 505
ammonia, 4, 439, 490, 492, 503, 725, 832, 836, 837, 840, 942, 1002, 1030, 1171, 1172, 1206, 1219, 1220, 1228, 1275, 1291, 1332, 1342, 1368, 1373, 1375, 1377, 1396, 1397
ammonium, 2, 4, 314, 329, 332, 497, 793, 1332, 1353, 1370, 1375, 1396
ammonium salts, 497
amorphous, 469, 474, 476
amphibia, 236, 895, 1059, 1685
amphibians, xxiv, 593, 1730
amplitude, 748, 751, 752, 936, 999, 1127, 1136, 1468, 1471, 1532
amplitudes, xxvii, 742, 747, 749, 999, 1015, 1127, 1459
AMS, 1551
Amsterdam, 169, 214, 281, 333, 335, 617, 1039, 1097, 1121, 1234, 1547
amylase, 1591, 1597
anaemia, 610
anaerobic, 394, 395, 973, 1102, 1103, 1104, 1105, 1108, 1109, 1111, 1115, 1116, 1117, 1118, 1119, 1121, 1122, 1375
anaerobic digestion, 394, 395, 1108
anaerobic sludge, 973, 1118
analgesic, 609
analysis of variance, 290, 292, 1081, 1303
anatomy, 277, 771, 1291
angiogenesis, 850
angiosperm, 1162, 1712, 1716
Angiosperms, 165
animal behavior, 244, 261
animal feeding operations, 1300, 1343
animal health, 163
animal husbandry, 244, 252, 735, 1672
animal waste, 1329, 1330, 1341
animal welfare, 595
Animals, xxiv, 603, 613, 614, 616, 617, 1206
anion, 1206
anionic surfactant, 488, 489
anionic surfactants, 488, 489
annealing, 490

annual rate, 429, 701, 1684
anomalous, 1567
ANOVA, 5, 290, 291, 292, 628, 777, 845, 859, 1081, 1084, 1087, 1140, 1207, 1317
anoxia, 782, 783, 914, 1358
anoxic, 941, 944, 950, 970, 973, 974, 1006, 1120, 1357, 1367
antagonism, 692
antagonists, 317
Antarctic, 174, 1036, 1039
anthocyanin, 79, 80, 81, 82, 83, 86, 88, 89, 171
Anthocyanins, 169
anthracene, 1216
anthropic, 414, 420, 421, 422, 426, 427, 432, 443, 444, 455, 1211, 1563
anthropic factors, 455
anthropological, 188, 197
anthropology, 188, 217
antibacterial, 164, 492
antibiotic, 174, 1589
antibiotics, 450, 487
antigenecity, 486
antioxidant, 168, 842, 1073, 1091, 1094, 1095, 1096, 1097, 1615
APA, 822, 826, 868
apathy, 598
APC, 561
apex, 1454, 1476, 1477, 1478, 1479, 1480, 1504, 1505
Appalachian Mountains, 140, 1368
appetite, 843, 1255, 1296
appraisals, 1331
appropriate technology, 1374
aptitude, 436, 1595
aquaculture, xxviii, xxix, xxx, xxxi, xxxvi, 585, 819, 820, 821, 823, 825, 826, 834, 835, 838, 839, 840, 855, 856, 857, 865, 867, 868, 869, 871, 872, 884, 911, 912, 913, 919, 1091, 1125, 1137, 1142, 1237, 1238, 1241, 1242, 1255, 1261, 1274, 1285, 1293, 1294, 1295, 1458
aquaporin-3, 917
aquaria, 873
aquarium, 872, 873, 874, 875, 880, 882, 884, 886, 1271
aquatic habitat, 365, 1163
aquatic habitats, 365, 771, 812
aquatic life, xxxviii, 643, 1316, 1371
aquatic systems, xxix, 2, 572, 669, 694, 757, 839, 882, 907, 1307, 1332, 1335, 1339, 1340, 1343
aqueous solution, 497, 501, 1206
aqueous solutions, 501
Aquifer, 457, 1166

aquifers, 432, 437, 440, 575, 703, 1197, 1332, 1411, 1418, 1419, 1432, 1565
Arab world, 699, 712
Arabia, 100
Arabian Gulf, 1124
Arabian Peninsula, 888
Arabidopsis thaliana, 794
Arbacia punctulata, 1206
arbuscular mycorrhizal fungi, 310, 330, 331, 334, 335, 336, 337, 339, 342, 346, 354, 355, 356, 357, 358
arbuscular mycorrhizal fungus, 170, 334, 335, 336
archaea, 331, 1120
Archaea, 1105, 1107, 1109, 1112, 1113, 1116, 1121
archaeological sites, 457
architecture, 313, 315, 338, 346, 358, 414, 781, 798
Arctic, 1031
Arctic Ocean, 1031
Argentina, xxxii, 266, 280, 550, 993, 995, 996, 997, 999, 1001, 1003, 1006, 1010, 1021, 1022, 1023, 1029, 1030, 1031, 1032, 1033, 1034, 1035, 1036, 1037, 1038, 1039, 1040, 1041
arginine, 486, 1266, 1280, 1284, 1599, 1608
arid, 20, 113, 158, 182, 183, 184, 185, 186, 187, 191, 195, 196, 197, 202, 203, 204, 205, 206, 207, 208, 214, 215, 217, 218, 259, 341, 423, 576, 961, 1012, 1124, 1137, 1139, 1145, 1156, 1158, 1353
arithmetic, 1208
Arizona, 432, 1142, 1143, 1145
Arizona State University, 1142
Arkansas, 1320, 1341
Army, 1454, 1507, 1544, 1545, 1546
Army Corps of Engineers, 1454, 1507
aromatic compounds, 1101, 1598, 1610
aromatic hydrocarbons, 1093, 1101, 1102, 1117, 1205
aromatics, 1205
arrest, 914, 920, 1565, 1595
arrhythmia, 486
ARS, 1, 1322
arsenic, 356, 444, 489
arson, 1345
ART, 375
Artemia, 1256, 1258, 1292
arteriosclerosis, 1601
arthropods, 811
asbestos, 426
ash, 285, 311, 472, 490, 500, 1077, 1129, 1372
Asia, xvi, xxxi, xlii, 93, 243, 244, 377, 391, 459, 562, 600, 604, 610, 611, 613, 617, 771, 868, 872, 874, 879, 888, 889, 891, 893, 897, 899, 904, 906, 923, 1241, 1294, 1331, 1433, 1434, 1623, 1631, 1638, 1676, 1680, 1684, 1685, 1696, 1708, 1712, 1714, 1717, 1718, 1719, 1720
Asian, 175, 332, 428, 459, 609, 611, 614, 1167, 1545
aspartic acid, 496
aspect ratio, 1542
assessment tools, 688
assets, 719, 725
assignment, 546
assimilation, 259, 329, 406, 466, 1247, 1250, 1342, 1399
assumptions, 5, 442, 620
asthma, 610
Astragali, 163
asymmetry, 569, 935, 936, 1474
Athens, 1094, 1095, 1097, 1367
Atlantic, xxii, xxxiii, xxxiv, 336, 361, 370, 371, 373, 374, 513, 514, 515, 517, 519, 520, 521, 523, 526, 527, 529, 531, 533, 542, 543, 544, 545, 547, 554, 562, 585, 924, 938, 961, 977, 999, 1005, 1021, 1031, 1036, 1037, 1100, 1119, 1147, 1148, 1153, 1157, 1168, 1457, 1458, 1459, 1460, 1560, 1562
Atlantic Ocean, xxxiii, 361, 961, 977, 1005, 1021, 1036, 1100, 1457
Atlas, 548, 987, 1038, 1172, 1181, 1418, 1434
atmosphere, xxi, xxxvii, 348, 391, 398, 413, 429, 441, 462, 463, 464, 466, 477, 479, 482, 503, 582, 589, 720, 724, 727, 730, 771, 1105, 1106, 1109, 1325, 1332, 1358, 1376, 1455, 1552, 1568, 1594, 1633
atmospheric deposition, 34, 429, 431, 433, 443, 447, 1333
atmospheric pressure, 1415, 1439
atoms, 57, 402, 462
ATP, 820, 913, 1183, 1583, 1592, 1595, 1600
atrophy, 843
attachment, 1106
attacks, 136, 143, 1461, 1540
attitudes, 408, 595, 1190
attractant, 1256, 1257, 1258, 1261, 1292
attractiveness, 1194
attribution, 1630
audit, 689, 723
Australasia, 1631, 1717
Australia, xxxi, xliii, xliv, 322, 377, 391, 392, 396, 406, 410, 428, 459, 460, 923, 937, 959, 965, 975, 976, 980, 983, 986, 987, 1124, 1331, 1346, 1465, 1467, 1469, 1476, 1507, 1508, 1641, 1642, 1668, 1671, 1673, 1674, 1675, 1676, 1703, 1733
Austria, 107, 385, 753, 754, 756, 786, 909

authorities, xliii, 270, 276, 423, 711, 730, 866, 879, 881, 999, 1246, 1577, 1641, 1649
authority, 235, 865, 881, 1189, 1319
autoimmune disease, 1602
autonomy, 699
autotrophic, 466, 954, 1359, 1363
availability, xxxii, 194, 284, 310, 314, 315, 316, 317, 319, 321, 326, 327, 328, 331, 332, 334, 335, 349, 355, 369, 446, 447, 449, 452, 455, 544, 623, 624, 924, 926, 942, 947, 950, 954, 964, 967, 968, 969, 994, 1037, 1090, 1091, 1102, 1137, 1174, 1189, 1351, 1465, 1566, 1690, 1732
averaging, 928, 1039, 1041, 1125
avian, 966, 969, 1732
avoidance, 395, 913
awareness, xxv, xxxvii, 197, 209, 395, 405, 454, 603, 620, 722, 725, 1325, 1701
Azerbaijan, 614

B

BAC, 446
Bacillus, 496, 497
Bacillus subtilis, 496, 497, 1606, 1617
back, 184, 252, 253, 254, 957, 963, 970, 975, 1170, 1197, 1224
background, xxi, xxv, 413, 441, 443, 445, 456, 619, 732, 807, 925, 1032, 1125, 1501, 1629, 1642
background information, 807, 1125
backwaters, xxxiv, 751, 1148
bacterial, 312, 313, 318, 331, 333, 334, 335, 497, 932, 953, 1080, 1107, 1108, 1109, 1115, 1120, 1121, 1139, 1308, 1315, 1316, 1319, 1320, 1328, 1339, 1341
bacterial contamination, 1315, 1319, 1328
bacterial fermentation, 1109
bacterial pathogens, 1613
bacteriocins, 1593, 1595, 1596, 1597, 1605, 1610, 1612, 1616, 1618
bacteriostatic, 1595
bacterium, 841, 1117, 1119, 1121, 1293, 1595, 1610
baggage, 1456
baicalein, 113
Bali, 1035
Baluchistan, 217
ban, 203
Bangladesh, 428
banking, 701, 913
banks, xviii, xxxi, 283, 287, 357, 376, 434, 435, 440, 451, 452, 558, 737, 747, 782, 792, 857, 867, 911, 942, 950, 971, 987, 1012, 1101, 1164, 1218, 1311, 1313, 1470

barley, 57, 168, 169, 170, 332, 1334, 1595, 1602, 1612
barrier, 108, 446, 489, 524, 937, 978, 1009, 1125, 1145, 1375, 1376, 1454, 1461, 1553, 1562, 1563
barriers, 490, 557, 565, 840, 916, 933, 988, 996, 1021, 1570, 1693, 1714
BAS, 1224
basal layer, 845, 846
basic education, 701
basic needs, 729
basic research, 889
basic services, 698, 738
basidiomycetes, 340, 341
basophils, 843
baths, 488, 489, 490, 495, 497, 498, 500, 1456
bathymetric, 1469, 1525
beaches, xl, xlii, 930, 935, 1127, 1455, 1456, 1457, 1463, 1473, 1474, 1508, 1513, 1516, 1517, 1518, 1575, 1576
bedding, 1152, 1154, 1301, 1372, 1377
beef, 271, 360, 1302
beer, 1580
beetles, xv, 37, 39, 40, 42
behavior, 185, 194, 203, 244, 248, 261, 316, 370, 508, 511, 547, 1037, 1337, 1446, 1466, 1478, 1591
behaviors, 595, 622, 1591
Beijing, xxxix, 243, 246, 261, 262, 263, 615, 616, 1181, 1183, 1184, 1407, 1408, 1432, 1433, 1434, 1436, 1511, 1709, 1710, 1716, 1718, 1720, 1721
Belgium, 428, 431, 447, 615, 692, 798, 906, 907, 1032, 1109, 1405, 1519, 1545
beliefs, 188
benchmark, 202, 203
beneficial effect, 320, 325, 691, 785, 850, 1266, 1306, 1319, 1337
beneficiaries, 711
benefits, xxxi, 180, 208, 326, 327, 350, 395, 398, 426, 450, 455, 477, 527, 529, 669, 719, 722, 803, 863, 886, 911, 913, 1186, 1193, 1330, 1461, 1601, 1666, 1724
benthic diatoms, 946, 998, 1014, 1029, 1030
benthic invertebrates, 955, 1656, 1667, 1668
benzene, 51, 54, 57, 426, 1102
benzo(a)pyrene, 1216
beverages, 1580, 1606
bias, 201, 1719
binding, 396, 476, 503, 1093, 1143, 1591, 1596, 1615
bioaccumulation, 444, 573, 1072, 1089, 1092, 1097, 1102

bioassay, xxxiii, 1071, 1072, 1073, 1075, 1078, 1089, 1090, 1091, 1092, 1094, 1095, 1206, 1228
bioassays, xxxiii, 1071, 1072, 1204
bioavailability, 322, 339, 506, 507, 823, 826, 835, 838, 1072, 1202, 1270, 1297, 1601, 1616
biochemical processes, 482, 932
biochemistry, 108, 165, 1614
biocompatibility, 486
biocompatible, 505
bioconversion, 1102
biodegradable, xxii, 394, 395, 483, 486, 487, 492, 493, 494, 502, 504, 505, 506, 507, 511, 1353
biodegradable materials, 395, 505
biodegradable wastes, 394
biodegradation, 470, 473, 475, 502, 507, 1103, 1110, 1111, 1119, 1121, 1602, 1606
bioethanol, 468
biofilms, 986, 994, 1002, 1107, 1112, 1116, 1117
biofuel, 360
biofuels, 467, 468, 481
biogas, 720, 1111
Biogeochemical process, 1145
biogeography, xlii, xliii, 238, 313, 333, 335, 539, 924, 1028, 1623, 1629, 1630, 1632, 1633, 1634, 1641, 1650, 1672, 1673, 1678, 1679, 1680, 1681, 1717, 1719, 1720
bioindicators, 994, 1553
biological activity, 498, 1393, 1439
biological control, 1178
biological fluids, 1072, 1087, 1088, 1091
biological processes, 28, 310, 901, 1139, 1335, 1337
biological responses, 687, 852, 932, 1211
biological systems, xxviii, 799, 809
biomacromolecules, 465, 474
biomarker, 1072, 1073, 1075, 1089, 1091, 1092, 1094
biomarkers, xxxiii, 1031, 1071, 1073, 1075, 1091, 1092, 1093, 1094, 1095, 1096, 1097
biomonitoring, xxvi, xliii, 683, 684, 685, 687, 688, 689, 690, 691, 811, 814, 1072, 1092, 1097, 1655, 1666, 1667
biopolymer, 475, 486, 508
biopolymers, 475, 511
biopreservation, 921, 1596, 1618
bioreactors, 1103, 1120
bioremediation, 1103, 1104, 1118, 1120
biosphere, 414, 441, 462, 464, 466, 721, 1142
biosynthesis, 45, 48, 143, 156, 497, 1103, 1608
biosynthetic pathways, 55
biota, 312, 320, 321, 336, 464, 543, 969, 972, 975, 1012, 1089, 1101, 1117, 1138, 1202, 1216

biotechnological, xxiii, 489, 493, 494, 550, 555, 559, 560
biotechnology, 405, 560, 1581
biotic, xxi, xxviii, 28, 44, 184, 315, 326, 386, 413, 463, 465, 572, 577, 687, 692, 694, 799, 810, 963, 972, 986, 1090, 1092, 1250, 1345, 1367, 1624, 1627, 1634, 1666, 1726
biotic factor, xxviii, 799, 810, 1092, 1250
birds, xv, xxiv, xxxii, 37, 40, 41, 370, 371, 552, 593, 594, 599, 601, 715, 924, 966, 967, 972, 982, 988, 989, 1073, 1178, 1300, 1301, 1672, 1679, 1717, 1732
Birmingham, Alabama, 851
birth, 596, 623, 1564
Biscayne Bay, 1226, 1233
bison, 605, 607
bivalve, 951, 952, 957, 958, 959, 986, 1091, 1096, 1221
Black Sea, 1175, 1182
blades, 1631
blame, 582, 747
blastoderm, 914, 915
blastula, 914, 916
blends, 467, 511
blocks, xvi, 221, 222, 233, 632, 1330, 1357, 1423, 1701
blood, 171, 486, 487, 493, 510, 609, 836, 842, 850, 852, 913, 922, 1239, 1249, 1294
blood circulation, 487, 609
blood plasma, 493
blood vessels, 486, 842
body composition, 837, 838, 1282, 1285, 1288, 1293
body fat, 1269, 1293
body size, 687, 802, 803, 808, 811, 906, 1656, 1659, 1662, 1663, 1666, 1668
body weight, xxix, 486, 825, 839, 843, 1248, 1251, 1275, 1277, 1282
boiling, xxii, 483, 611
bonding, 120, 474, 699
bonds, 469, 485, 492, 494, 498, 499, 500, 502, 503, 506
bone, xxix, 484, 486, 610, 820, 822, 823, 850, 1286, 1295
bone marrow, 610
bones, 484, 486, 493, 842
BOP, 379
border crossing, 622, 624, 626, 627, 628, 629
boreal forest, 338, 794
boreholes, 716, 738, 1562, 1563
Borneo, 305, 562
borosilicate glass, 1106
Boston, 337, 601
Botanical Garden, 162, 224, 236, 279

Botswana, 210, 212, 215, 217, 218
bottleneck, 193, 557
bottom-up, 179, 418
boundary conditions, 813, 1431, 1474, 1524, 1535
boundary value problem, 643, 647, 674
bounds, 648
Boussinesq, 1471, 1507
Bradyrhizobium, 330
brain, 484
branched polymers, 469, 498, 499, 507
branching, 60, 354, 781
Brazilian, xvii, xviii, xxii, xxiii, 69, 92, 265, 266, 268, 269, 270, 271, 272, 276, 277, 278, 279, 280, 360, 370, 513, 514, 521, 523, 531, 533, 542, 544, 545, 547, 549, 550, 554, 557, 560, 562, 563, 586, 1102, 1109, 1118, 1119, 1120, 1230, 1231, 1234
breaches, 934, 962, 969
breakdown, 185, 503, 694, 842, 932, 1108, 1131, 1366, 1599, 1610, 1615
break-even, 711
breakwaters, xli, 1459, 1462, 1476, 1477, 1478, 1507, 1512, 1513, 1516, 1518, 1527, 1535, 1536, 1537, 1576
breeding, xx, 41, 371, 375, 376, 383, 385, 439, 562, 573, 604, 622, 635, 733, 888, 912, 963, 964, 982, 1074, 1094, 1301
brevis, 1045, 1586, 1587, 1588, 1589, 1591, 1592, 1599, 1602, 1603, 1615
bridges, xxxii, 476, 498, 971, 993, 996, 1009, 1536
Britain, 377, 428, 771, 906, 1290, 1456, 1639
British Columbia, 1102, 1117
broad spectrum, 527
broilers, 491, 1301
Brownian motion, 672, 1138
browsing, 178, 186, 203, 594
Brussels, 459, 460, 615, 1370, 1545
bubbles, 1377
budding, 1446, 1448
budgetary resources, 702
Buenos Aires, 280, 993, 1001, 1030, 1031, 1033, 1034, 1035, 1036, 1038
buffalo, 1727
buffer, xxxvii, 2, 41, 42, 183, 436, 447, 526, 964, 1078, 1300, 1301, 1303, 1306, 1313, 1319, 1329, 1337, 1340, 1341, 1342, 1344, 1345, 1346, 1347, 1362, 1701, 1702
buildings, xli, 414, 432, 454, 729, 735, 1549, 1550, 1551, 1552, 1560, 1562, 1564, 1624, 1630, 1636, 1640
Burkina Faso, 213, 216
burn, 271, 516
burning, xviii, 283, 456, 462, 464, 582

Burundi, 731
bushes, 524
businesses, 880, 1189
butane, 1358
butterfly, xviii, 284, 286, 289, 290, 296, 297, 301, 302, 303, 304, 305, 306, 307, 386
by-products, 433, 490, 510, 610, 820, 1259, 1260, 1261, 1274

C

Ca^{2+}, 464, 490, 493
cabinets, 378, 728
cacao, 371
cachexia, 843
cadmium, 1096, 1102, 1355
caffeic acid, 71, 72, 73, 74, 109, 160, 167
Cairo, 1237, 1286
calcium, 463, 489, 490, 826, 827, 834, 1259, 1291, 1391, 1598, 1614, 1617
calcium carbonate, 463
calculus, 673
calibration, 995, 1023, 1030, 1039, 1131, 1189, 1205, 1318, 1413, 1418, 1428, 1528, 1551
calving, 193
cambisols, 431
Cambodia, 612, 617
Cameroon, 709, 1157, 1631
campaigns, 1170, 1178
Canada, 132, 336, 391, 392, 460, 618, 975, 1109, 1183, 1209, 1231, 1234
canals, 439, 710, 711, 714, 715, 716, 780, 1143, 1430, 1552, 1565
CAP, 22
capacity, xxi, xxiii, 2, 10, 420, 421, 422, 430, 431, 433, 440, 441, 447, 457, 461, 464, 476, 487, 496, 499, 504, 527, 530, 550, 555, 572, 586, 597, 598, 599, 601, 602, 1313, 1333, 1351, 1353, 1355, 1361, 1362, 1372, 1373, 1377, 1383, 1388, 1401, 1402, 1403, 1441, 1447, 1449, 1460, 1461, 1479, 1480, 1497, 1525, 1553
Capacity, 420, 599, 1447, 1448
capacity building, 599, 723, 733
Cape Town, 236, 979, 980, 981, 982, 987, 988, 1545
capillary, 1076, 1355, 1393, 1481
capital cost, 1402
caprolactam, 503
captive populations, 912
carbohydrate, 346, 470, 772, 773, 776, 780, 788, 1239, 1245, 1285, 1296, 1586, 1600
carbohydrates, 324, 346, 462, 472, 772, 773, 774, 776, 917, 1239, 1242, 1580, 1581, 1587, 1591, 1608, 1610, 1620

Carbon, xxi, 28, 160, 340, 356, 357, 461, 462, 464, 465, 466, 467, 482, 1104, 1105, 1156, 1196, 1197, 1204, 1232, 1556, 1557
carbon atoms, 462
carbon dioxide, xxi, 398, 413, 462, 464, 475, 477, 478, 588, 589, 590, 591, 721, 1104, 1114, 1593, 1598
carbon neutral, 468
carbonates, 463, 465, 1204, 1213, 1223
carboxylates, 322
carboxylic, 51
carboxymethyl cellulose, 1269, 1296
carcinogenic, 498
cardiac arrhythmia, 486
Caribbean, 888
carnivores, xxiv, 362, 363, 364, 365, 366, 368, 369, 370, 372, 573, 593, 1239
carotene, 1625, 1630
carotenoids, 82, 174, 1630
carp, 750, 825, 827, 836, 837, 841, 852, 914, 916, 1242, 1264, 1266, 1269, 1271, 1273, 1276, 1278, 1279, 1280, 1281, 1282, 1283, 1284, 1285, 1286, 1287, 1288, 1289, 1290, 1291, 1292, 1293, 1294, 1295, 1296, 1297
carpets, 502
carrier, 493, 511, 1357
cartilage, 484, 485
case studies, xxvi, 683, 693, 719, 937, 1457, 1459, 1680
case study, xxxviii, 213, 217, 399, 439, 458, 614, 739, 934, 986, 1104, 1142, 1371, 1377, 1402, 1433, 1508, 1650
casein, 1264, 1292, 1293
cash, 698, 701, 702, 737, 1686
cash crops, 1686
cash flow, 698
cast, 509
casting, xxii, 483, 492, 506, 507
catabolism, 1608
catalase, 1091
catastrophes, 570
catchments, xxxviii, xxxix, 408, 705, 925, 926, 927, 928, 929, 967, 968, 975, 1170, 1178, 1349, 1350, 1351, 1367, 1368, 1408, 1409, 1418, 1431, 1432
categorization, 1498
category a, 967
catfish, 827, 843, 854, 878, 1242, 1250, 1259, 1286, 1287, 1293, 1295
Catholic, 213, 1405
cation, 2, 56, 476, 1206, 1375, 1396
cats, xxiv, 594, 599

cattle, xviii, xxiv, 196, 202, 252, 265, 266, 270, 271, 272, 276, 360, 484, 486, 487, 488, 491, 493, 494, 495, 496, 506, 510, 524, 594, 596, 601, 604, 609, 610, 699, 701, 784, 797, 1301, 1302, 1314, 1330, 1341, 1377, 1646
causal relationship, 638
causality, 800, 805, 806, 809, 811
causation, 809
cavities, 1207
CCA, 1002
CD-rom, 458
CE, 460, 1346
CEC, 443, 476
celiac sprue, 1580, 1586, 1602, 1608, 1609, 1618
cell, xlii, 18, 20, 250, 469, 470, 472, 473, 484, 486, 558, 622, 623, 625, 631, 632, 1031, 1076, 1107, 1124, 1318, 1341, 1397, 1415, 1417, 1421, 1432, 1477, 1579, 1588, 1590, 1591, 1595, 1597, 1605, 1607, 1609, 1619, 1676, 1690
cell culture, 558
cell differentiation, 484
cell membranes, 820
cell surface, 1341
cellulose, 4, 467, 469, 470, 472, 479, 509, 1154, 1158, 1269, 1296
Cellulose, 469, 470, 472
cellulosic, 1154
cement, 423, 1101
Census, 516, 517, 518, 597, 602, 1610
Census Bureau, 597, 602
centigrade, 749
Central African Republic, 699, 704, 709
Central America, 325, 566
Central Asia, 611, 891
Central Europe, 377, 686
centralized, 527
ceramic, 1562, 1563
ceramics, 1562
cereals, xiv, xxxix, 13, 22, 471, 1437
CERES, 1409, 1434
certification, 592, 724
cesium, 356
cetacean, 1711, 1720
C-glycosyl flavones, 57, 61, 65, 66
CH_4, 588, 590
Chad, 699, 704, 708, 709
Chaetoceros, 1010, 1011, 1045, 1046
challenges, xxxii, 41, 180, 187, 193, 209, 213, 398, 598, 719, 726, 727, 729, 814, 924, 1190
changing environment, 376, 995, 1198
channels, xiv, xxxv, xl, 14, 22, 1009, 1010, 1014, 1100, 1101, 1124, 1125, 1127, 1165, 1170, 1178, 1202, 1203, 1211, 1215, 1219, 1221,

1315, 1338, 1511, 1513, 1516, 1517, 1518, 1519, 1531, 1547, 1550, 1555, 1564, 1567, 1656, 1667
character, 485, 486, 489, 494, 498, 501, 505, 506, 507, 526, 726, 742, 743, 754, 1187, 1456, 1526, 1644, 1709
charcoal, xviii, 265, 277, 465, 473, 1562
charge density, 474
chemical characteristics, xxxii, 107, 423, 473, 961, 993, 994, 1011, 1124, 1660
chemical composition, 468, 470, 473
chemical content, 45, 46, 47, 113
chemical degradation, 423
chemical industry, 428
chemical interaction, 465
chemical properties, 205, 433, 447, 485, 1204, 1337
chemical reactions, 466, 1367
chemical stability, 729
chemical structures, 48, 51, 57, 71, 81, 87, 90, 94, 97, 98, 109, 117, 118, 124, 136, 1597
chemicals, 136, 426, 427, 485, 488, 492, 496, 503, 715, 737, 906, 908, 1078, 1198, 1327, 1330, 1332, 1336, 1337, 1367, 1377, 1598
chemotherapy, 840
chewing, 487
Chicago, 341, 371, 411, 546, 614, 616, 676, 1197, 1359, 1669, 1718
chicken, 491, 1285
children, 406, 411, 735, 860, 1456
Chile, xliv, 371, 1671, 1674, 1675, 1676
Chinese government, 1711
Chinese medicine, 609
chitin, 329
chitosan, 492, 509
Chl, 1132, 1138, 1139, 1140, 1175, 1176, 1178, 1179
chloride, 493, 1307, 1316, 1355, 1356, 1416
chlorine, 725, 1112
chloroform, 1102, 1107, 1206
chlorogenic acid, 71, 72, 73
chlorophenol, 1103, 1104, 1105, 1108, 1109, 1110, 1115, 1117, 1118
chlorophyll, xxxiii, xxxv, 251, 324, 692, 898, 901, 902, 905, 943, 980, 1073, 1123, 1129, 1130, 1138, 1139, 1169, 1171, 1172, 1175, 1177, 1183, 1184, 1625
chloroplast, 554, 562, 1636
cholecalciferol, 838
cholera, 733
cholesterol, 1300
cholinesterase, 1094
chorion, 914, 915, 916
CHP, 394

Christianity, 605
Christmas, 599
chromatography, 82, 165, 168, 916, 1076, 1108, 1205, 1304
chromium, 439, 445, 446, 458, 499, 510, 798, 1645
chromosome, 385
chromosomes, 69
chronic disease, 505
chronic diseases, 505
chronic stress, 1178
chronology, xli, 402, 408, 1550
Cincinnati, 1234, 1323
circulation, xxxi, 441, 487, 609, 727, 887, 889, 930, 936, 941, 1006, 1009, 1106, 1109, 1124, 1139, 1182, 1228, 1463, 1468, 1477, 1478, 1479, 1480, 1481
CITES, xxv, 235, 603, 605, 606, 607, 608, 609, 610, 613, 615
cities, xxvi, 423, 432, 514, 697, 726, 874, 1012, 1014, 1073, 1101, 1174, 1189, 1428, 1432, 1456, 1457, 1693
citizens, 449, 1189, 1190
citizenship, 697
citrus, 20, 353, 356, 1073
City, xxvii, 237, 597, 741, 742, 743, 744, 747, 750, 752, 753, 754, 860, 864, 874, 875, 876, 881, 1343, 1436, 1703
civil engineering, 451
civil society, 450, 1689
civil war, 716, 735, 736, 737, 1729
civilization, 406, 420, 514, 1186
cladocerans, 951, 952
clams, 1095, 1102
clarity, 966, 972, 1193
classes, xx, 51, 90, 187, 192, 200, 201, 206, 247, 361, 364, 367, 375, 434, 436, 437, 438, 440, 447, 520, 688, 773, 888, 944, 1026, 1175, 1589, 1595, 1598, 1600, 1628, 1658, 1659, 1660, 1661, 1662, 1663, 1665, 1686, 1689, 1692, 1693
classical, xv, 43
classical methods, 1614
classification, xv, 43, 47, 59, 60, 69, 75, 77, 89, 91, 108, 120, 160, 162, 170, 182, 200, 201, 217, 259, 263, 384, 418, 437, 459, 487, 533, 551, 593, 670, 685, 693, 805, 937, 995, 1024, 1204, 1209, 1264, 1303, 1439, 1508, 1597, 1617, 1625, 1627, 1634, 1642, 1657, 1668
clay, 2, 430, 435, 436, 443, 476, 932, 937, 943, 955, 1149, 1151, 1155, 1156, 1157, 1158, 1159, 1162, 1165, 1204, 1302, 1303, 1344, 1383, 1391, 1402, 1417, 1418, 1419, 1440, 1513, 1551, 1552, 1561, 1562

clay minerals, 476
clean air, 1186
clean energy, 716
clean technology, 509
cleaning, 452, 1372, 1374, 1377, 1401
cleavage, 54, 55, 914, 917
climate change, xiv, xvii, xxiii, xxxv, xlii, 27, 265, 357, 391, 393, 396, 398, 399, 402, 405, 411, 448, 464, 468, 478, 550, 620, 636, 678, 680, 687, 724, 725, 726, 816, 905, 974, 987, 994, 1117, 1185, 1187, 1191, 1192, 1197, 1409, 1433, 1436, 1457, 1575, 1577, 1689
climates, xxxix, 48, 269, 565, 567, 739, 777, 779, 897, 999, 1124, 1168, 1437, 1438, 1439, 1630, 1631, 1672, 1721
Climatic change, xiii, 13
climatic factors, 192, 928, 1442
clinical problems, 486
clone, 1109
cloning, 405, 1609, 1633
close relationships, 832
closure, xxxi, xxxv, 923, 925, 929, 931, 933, 935, 939, 941, 943, 945, 949, 950, 951, 956, 964, 965, 970, 972, 986, 990, 1169, 1174, 1565, 1568, 1570
clothing, 487, 489, 493, 507, 605
clouds, 402, 1329
cluster analysis, 45, 108, 155, 1015
clustering, 803, 1607, 1611
clusters, 59, 322, 474, 1395, 1555, 1696
CMC, 830, 1269
CNS, 1204
CO_2, xxi, 349, 356, 357, 391, 392, 393, 394, 395, 398, 461, 462, 463, 464, 465, 466, 468, 476, 478, 479, 588, 590, 721, 729, 730, 795, 840, 1105, 1109, 1110, 1112, 1134, 1191, 1359, 1552, 1583, 1592
coagulation, 490, 493, 1138
coal, 305, 462, 463, 467, 1340, 1528
Coast Guard, 1435
coastal areas, xli, 584, 586, 926, 973, 995, 1001, 1031, 1039, 1090, 1091, 1097, 1176, 1458, 1549
coastal communities, 820, 855
coastal ecosystems, 2, 815, 1457
coastal management, 869, 1167
coastal region, 942, 1089, 1148, 1432
coastal settings, xxxii, 993, 996
coastal zone, xlii, 217, 962, 971, 1140, 1175, 1454, 1455, 1458, 1463, 1471, 1503, 1575, 1576, 1577
coatings, 503, 509, 511, 1464
cobalt, 1645

Cocconeis, 895, 897, 1007, 1011, 1016, 1017, 1018, 1019, 1021, 1023, 1029, 1046
cockroach, 530, 531, 539, 546, 547
cockroaches, 531, 535, 543, 544, 547
coding, 1032, 1120, 1433, 1667
coefficient of variation, 700
co-existence, 315
cofactors, 1103
coffee, 360, 372, 515, 517, 519, 524
coliforms, 1319, 1373, 1376
collaboration, 179, 406, 446, 557, 1186
Collaboration, 405
collagen, xxii, xxix, 483, 484, 485, 486, 487, 488, 489, 490, 491, 492, 493, 494, 495, 496, 497, 498, 499, 500, 501, 502, 503, 504, 505, 506, 507, 508, 510, 511, 512, 840, 841, 845, 846, 848, 849, 850, 852, 853, 854
collagen materials, 485
collagen sponges, 486
collateral, 864
College Station, 1292
collisions, 1523, 1532
Colombia, 872, 878, 1234
colonisation, 326
colonization, xviii, 270, 309, 314, 316, 322, 323, 332, 334, 350, 351, 514, 516, 517, 539, 556, 563, 579, 632, 743, 907, 959, 1039, 1224
colonizers, 547, 568
Colorado, 1012, 1127, 1142, 1143, 1232, 1321, 1341, 1451
Columbia, 163, 373, 1102
Columbia University, 163, 373
combined effect, 803, 1507, 1551
combustion, xxi, 394, 461, 463, 464, 465, 477, 592, 1102
commerce, 874, 1187, 1259
commercial, xxix, xxxi, xxxiii, 378, 398, 432, 495, 496, 497, 498, 500, 501, 596, 638, 718, 725, 729, 820, 821, 823, 869, 872, 879, 881, 884, 911, 912, 913, 1010, 1099, 1140, 1186, 1244, 1255, 1258, 1259, 1260, 1261, 1264, 1266, 1267, 1271, 1274, 1278, 1281, 1282, 1284, 1292, 1294, 1462, 1586, 1592, 1602, 1621, 1630
commercialization, 560
commodities, 360
commodity, 1239, 1701
common findings, 758
commons, 544, 546
communication, xlii, 260, 287, 405, 512, 680, 730, 857, 880, 954, 1167, 1552, 1579, 1582, 1590, 1591, 1605, 1609, 1619, 1621
Community Based Organizations, 209

compaction, xli, 259, 418, 420, 421, 429, 430, 448, 449, 1381, 1549, 1555, 1560, 1564
comparative analysis, 979
comparative method, 814, 1227
compatibility, 323
compensation, 789, 1091, 1128, 1131
compensatory effect, 1264
competence, 408, 558
competition, xix, xxx, xxxix, xl, 159, 310, 315, 316, 319, 322, 326, 329, 332, 333, 334, 335, 337, 339, 341, 343, 345, 350, 352, 353, 357, 383, 385, 417, 418, 543, 558, 573, 575, 577, 601, 678, 732, 738, 766, 767, 768, 771, 788, 789, 796, 812, 816, 840, 858, 866, 867, 871, 880, 881, 898, 901, 907, 908, 953, 955, 975, 1284, 1360, 1437, 1587, 1592
competitive advantage, 351, 896, 1595
competitiveness, xxx, 722, 727, 729, 855, 866, 1587, 1593, 1595
competitor, 1590
competitors, xxiv, 316, 330, 570, 594, 802, 1360, 1590, 1646
compilation, 222, 584, 1029, 1475, 1677
complement, 179, 183, 349, 851, 1589
complex carbohydrates, 1620
complex interactions, xl, 206, 316, 620, 1140, 1511, 1515, 1543
complexity, xxviii, xli, 41, 174, 319, 401, 402, 403, 404, 407, 408, 411, 469, 530, 621, 630, 643, 799, 805, 809, 915, 1193, 1207, 1226, 1316, 1366, 1511, 1515
compliance, xxvi, 398, 683, 684, 719, 723, 1131, 1340
compost, 441, 452
composting, 394, 395
comprehension, xxiii, 550, 1211, 1226
compression, xxii, 483, 509, 725, 729
computation, 404, 410, 680, 1507, 1524, 1535, 1564, 1568
computer, xxxvii, 404, 437, 450, 622, 629, 1032, 1299, 1379, 1412
computer simulation, xxxvii, 622, 1299
computer technology, 437
computing, 403, 404
concentrates, 1470, 1497, 1518, 1656
concept map, 1192
conception, 351, 456
conceptual model, 348, 350, 482, 539, 743, 988, 1192, 1193
concordance, 1717
concrete, 60, 971, 1384, 1465, 1466, 1647
condensation, 54, 727
conditioning, 510, 725, 728, 729

conductance, 1410, 1435
conduction, 727
conductivity, xxxvii, 5, 200, 438, 440, 843, 890, 917, 1003, 1108, 1299, 1312, 1316, 1355, 1357, 1411, 1414, 1417, 1418, 1431, 1660
conference, 509, 736, 868, 1095
confidence, xliv, 19, 76, 926, 968, 1656, 1666
confidence interval, 19
configuration, xxxiv, xl, 253, 629, 801, 846, 1103, 1148, 1246, 1377, 1511, 1515, 1525, 1541, 1543
configurations, 624, 629, 1376
confinement, 1341
conflict, xx, xxii, 194, 375, 513, 574, 732, 733, 734, 736, 738, 972, 1639
conflict resolution, 732
confrontation, xlii, 1550, 1700
confusion, 237
Congo, 699, 731
congruence, xliv, 1671, 1673, 1674, 1675
conifer, 551, 553, 562, 747, 1717
connective tissue, 484, 485, 486, 497, 506, 573, 841, 843, 846
connectivity, xx, 2, 303, 305, 360, 369, 374, 454, 557, 636, 749, 750, 751, 753, 810, 818, 965, 1189, 1192, 1197, 1669, 1702
consciousness, 403, 426, 432
consensus, xxviii, 60, 396, 595, 596, 602, 640, 799, 865, 1208, 1591, 1610, 1620, 1708
consent, 1373, 1389, 1397
constant rate, 1355
constituents, 9, 61, 63, 64, 67, 69, 76, 95, 108, 160, 161, 165, 169, 173, 175, 472, 474, 610, 771, 1292, 1313, 1345, 1588, 1605
Constitution, 162, 523
constraints, 354, 454, 559, 620, 622, 1466, 1479, 1577
constructed wetlands, 771, 781, 782, 792, 794, 797, 1327, 1329, 1338, 1340, 1376, 1388, 1401, 1404, 1405
construction, xxxv, 178, 196, 209, 353, 355, 419, 432, 499, 516, 521, 526, 706, 708, 712, 725, 744, 746, 747, 750, 752, 905, 969, 971, 1006, 1125, 1169, 1173, 1179, 1192, 1340, 1384, 1458, 1460, 1463, 1464, 1465, 1466, 1467, 1468, 1469, 1470, 1507, 1513, 1528, 1536, 1546, 1552, 1560, 1570, 1576, 1711, 1726
consulting, 434
consumer markets, 586
consumerism, 390
consumers, xxxii, 326, 450, 584, 698, 728, 874, 924, 946, 954, 955, 1601, 1605
consumption patterns, 396

containers, 716, 1245, 1461, 1464, 1465, 1466, 1503, 1507, 1509
contaminant, xxxiii, xxxv, 426, 1071, 1072, 1073, 1081, 1084, 1089, 1090, 1091, 1092, 1096, 1097, 1101, 1202, 1215, 1216, 1220, 1329, 1339, 1366
contaminants, xxxiii, xxxv, xxxvi, 437, 441, 448, 452, 969, 1071, 1072, 1073, 1074, 1090, 1091, 1092, 1095, 1101, 1102, 1185, 1202, 1203, 1211, 1218, 1220, 1223, 1224, 1228, 1229, 1327, 1338, 1342, 1345, 1346, 1372
contaminated sites, 421, 426, 428, 442, 458, 1103, 1104
contaminated soil, xxi, 414, 446, 448, 453, 458
contaminated soils, xxi, 414, 446, 453, 458
contaminated water, 1218, 1372
Continental, 1031, 1037, 1181, 1233, 1545, 1718
continental shelf, 930, 1167, 1455, 1517
contingency, 810
continuity, 18, 434, 529, 571, 1479, 1523
contour, 18, 1327, 1439, 1525, 1536
control group, 845, 846, 847, 849, 850
control measures, 721, 723, 874, 903, 1101, 1335, 1337
controversial, xix, 359, 499, 501, 504
convection, 1532
convective, 1440, 1481
convention, 180, 210, 724, 1731, 1732
Convention on International Trade in Endangered Species, 610
convergence, xlii, 670, 814, 1623, 1628, 1691
conversion, xxxv, 9, 10, 462, 464, 468, 559, 975, 981, 1185, 1358, 1599, 1687, 1690, 1693, 1701
conversion efficiency, 1293
COOH, 489
cooking, 716, 718, 720, 735, 1259
cooling, 488, 506, 718, 913, 914, 915, 916, 918, 922, 1525, 1709, 1712
cooperation, 408, 450, 457, 544, 709, 723, 731, 732, 734, 1323, 1450, 1591, 1731, 1732
coordination, 709, 711, 867, 1590
Copenhagen, 238, 459, 592
copepods, 951, 952, 954, 955, 956, 957, 1175
copper, 673, 1097, 1102, 1205, 1355
coral, xl, 635, 1186, 1456, 1508
coral reefs, xl, 635, 737, 1186, 1456, 1508
Coriolis effect, 1474
corn, 60, 172, 174, 1073, 1330, 1409, 1433, 1434, 1436, 1586, 1595
corporations, 597
correlation, 90, 108, 128, 159, 206, 259, 293, 347, 364, 379, 380, 381, 382, 383, 435, 436, 443, 446, 447, 553, 554, 576, 577, 645, 670, 786, 809, 901, 1081, 1157, 1193, 1223, 1303, 1306, 1425, 1443, 1552
correlation analysis, 159
correlation coefficient, 259, 379, 381, 1081, 1223, 1552
correlations, 158, 159, 206, 247, 259, 379, 381, 443, 1084, 1090, 1223, 1224, 1225, 1316, 1359, 1363
corridors, 38, 369, 372, 526, 557, 622, 635, 1187
corruption, 727
cortex, 346
cortisol, 842, 852, 853
cosmetic, 487, 488, 493, 494, 496, 504, 507, 510
cosmetics, 489
cost of living, 597
Costa Rica, 360, 370, 371, 545, 867
cost-effective, 452, 1329, 1338, 1461, 1542
costs, xxxviii, 276, 326, 371, 394, 397, 398, 428, 430, 453, 486, 489, 490, 492, 502, 504, 526, 527, 599, 1130, 1350, 1372, 1373, 1374, 1402, 1403, 1439, 1456, 1462, 1689, 1690
costs of compliance, 398
costs of production, 866
cost-sharing, 1332
cotton, 3, 4, 10, 486, 701, 702, 709, 711, 712, 713, 1073, 1248, 1249, 1251, 1286, 1287
cotyledon, 60
cough, 610
Coulomb, 1523
coumarins, 52, 171
Council of Ministers, 709
counterbalance, 325, 336, 1361
countermeasures, 1568
coupling, xxxviii, xxxix, 985, 1350, 1363, 1407, 1432
court proceedings, 598
courtship, 573
coverage, 599, 1357
covering, xlv, 266, 378, 423, 514, 560, 685, 704, 714, 994, 1002, 1423, 1450, 1465, 1550, 1684, 1686, 1696, 1724
cows, 224, 1402
crab, 372, 957, 958, 982, 1127
crabs, 957, 1102, 1187
cracking, 440
cracks, 488, 743
Crassostrea gigas, 1141, 1143
CRC, 459, 479, 509, 984, 1097, 1142, 1320, 1321, 1343, 1574
creativity, 404, 600
credibility, 1190
Creeks, 1302, 1307, 1309, 1310, 1311, 1312, 1313
creosote, 339, 1127

Crete, 121
crew, 859
crises, 187
critical habitat, 598
critical state, 1100
critical value, 943
CRM, 844, 846, 847
Croatia, 1093
crop production, xxxix, 2, 454, 467, 468, 596, 702, 710, 713, 735, 1407, 1409, 1432, 1436
crop residue, xxi, 461, 467, 468, 469, 470, 471, 475, 477, 479
crop residues, xxi, 461, 467, 468, 469, 470, 471, 475, 477, 479
crop rotations, 456
croplands, 247, 252
crosslinking, xxii, 483, 485, 488, 493, 494, 499, 503, 504, 506, 507, 508, 510
cross-linking, 420
cross-linking, 511
crosslinking reactions, 504
cross-sectional, 1479
crown, 384, 553
CRP, xiii, 1, 2, 3, 4, 5, 7, 9, 10
crude oil, 673, 1096, 1121
crust, 422, 725, 1445, 1446, 1449, 1598, 1600, 1618, 1635
crustaceans, 958, 959, 1178, 1187, 1216, 1221
cryobanking, 912, 918
cryopreservation, xxxi, 558, 559, 911, 912, 913, 914, 916, 918, 919, 920, 922
cryopreserved, 558, 559
cryptography, 404, 410
crystal growth, 922
crystalline, xvi, 221, 447, 469, 476, 1293
crystals, 912, 917
cues, 963, 1292
cultivars, 78, 79, 81, 83, 88, 165, 173, 356, 377, 378, 384, 385, 386, 387, 1413, 1424
cultivation, xiv, 13, 20, 22, 178, 196, 223, 275, 360, 421, 429, 456, 465, 476, 477, 702, 1105, 1109, 1115, 1118, 1162, 1330, 1334, 1603, 1604, 1672, 1685, 1686, 1693, 1694, 1702, 1703, 1710, 1715, 1717, 1729
cultural heritage, 450, 458
cultural norms, 612, 865
cultural tradition, 356
cultural values, 454
cculture conditions, 1109, 1277
culture media, 558
culture medium, 1105, 1110, 1111
cure, 503, 511, 610, 611, 809, 1370
curing, 502, 503, 504, 1614

curing process, 503
curiosity, 340
currency, xxxv, 714, 1185
current limit, 685
curricula, 737
customers, 729
cutin, 471, 472, 473, 474, 475
cyanobacteria, 943, 946
Cyanobacteria, 973
Cyanophyta, 892, 894, 897, 898
cycles, 28, 312, 319, 368, 405, 462, 464, 526, 584, 620, 643, 750, 771, 772, 786, 787, 790, 935, 957, 980, 1100, 1124, 1132, 1140, 1197, 1393, 1412, 1435, 1566, 1588, 1686
cycling, xxvii, xxxvii, 312, 316, 319, 338, 339, 431, 455, 757, 769, 771, 790, 791, 792, 795, 797, 942, 962, 1031, 1138, 1349, 1351, 1360, 1362, 1363, 1365, 1369
cyclone, 1516, 1518, 1537
cyclones, xl, 929, 988, 1468, 1511
Cylindrotheca closterium, 1011, 1048
cysteine, 486
cystine, 1242, 1244, 1253, 1263, 1265, 1266, 1268, 1272, 1280, 1281, 1282, 1285
cytokines, 851
cytokinesis, 1625, 1626, 1627
cytology, 1638
cytoplasm, 1592
Czech Republic, 483, 685, 773, 779

D

Daidzein, 136
daily living, 252
dairy, xxxviii, 1074, 1326, 1340, 1371, 1372, 1373, 1374, 1376, 1377, 1388, 1395, 1402, 1403
dairy industry, 1373
danger, xxii, xxiii, 426, 513, 565, 574, 577, 649, 1644, 1646
Darfur, 702, 718, 734, 738
Darfur region, 738
data analysis, 259, 387, 408, 409, 692, 811, 1022
data collection, xxvi, 408, 423, 451, 586, 683, 731, 1131
data set, 30, 438, 813, 997, 1009, 1023, 1026, 1028, 1308, 1359, 1499
database, xiv, 14, 226, 250, 433, 441, 455, 481, 586, 591, 605, 1117, 1131, 1191, 1310, 1311, 1552, 1686
database management, 455
datasets, 451, 995, 1024, 1028, 1188, 1189, 1631, 1691, 1724

dating, xxi, xxxiv, 414, 556, 596, 1147, 1148, 1149, 1154, 1158, 1165, 1168, 1552, 1561, 1562, 1642, 1711, 1712, 1715
DCA, 291, 292, 293, 1020, 1022, 1023, 1026
death, 315, 327, 486, 622, 623, 1036, 1331, 1332
death rate, 315, 641
debates, 182
debt, 341
debts, 865
decay, 546, 727, 1079, 1134, 1374
deciduous, 286, 305, 306, 333, 1162, 1163, 1165
decision makers, 187, 414, 419, 451, 754, 1186
decision making, 180, 187, 564, 1189, 1190, 1198, 1229, 1724
Decision Support Systems, 1409
decisions, xxiii, xxxv, xliii, 183, 185, 186, 189, 191, 194, 202, 204, 207, 418, 550, 595, 598, 599, 620, 928, 1185, 1187, 1189, 1193, 1197, 1198, 1641, 1649
decomposition, xiv, 27, 28, 34, 35, 314, 327, 335, 463, 468, 470, 471, 473, 476, 510, 531, 686, 772, 781, 786, 787, 950, 1331, 1332, 1374, 1375
deconstruction, 473
decoupling, 1363
deep-sea, 464
Deer, 611, 615, 616, 617
defecation, 259, 1315
defence, xxix, 839, 1091, 1096, 1551, 1570, 1571, 1590
defense, 328, 332, 1091, 1459, 1461, 1462, 1464
defense mechanisms, 840, 1248
defenses, 328
deficiencies, 672
deficiency, 317, 574, 781, 820, 842, 843, 1091, 1179, 1264, 1280
deficit, 395, 431, 432, 455, 486, 706, 742, 753, 1412, 1413, 1415, 1422, 1431, 1439, 1448
deficits, 1335, 1358, 1431
definition, 181, 390, 403, 407, 418, 437, 544, 612, 935, 998, 1039, 1183, 1209, 1466, 1484, 1570, 1642, 1684, 1685, 1724
deforestation, xxi, xliv, 270, 276, 413, 419, 429, 464, 477, 514, 521, 523, 526, 530, 542, 716, 736, 742, 744, 747, 856, 905, 1672, 1683, 1684, 1685, 1689, 1690, 1692, 1693, 1700, 1701, 1702, 1729
deformation, 448, 1474
degradation process, 422, 449, 1102, 1158
degradation rate, 471
degraded area, 354, 426, 457, 1202, 1226
degrading, 457, 572, 622, 1103, 1104, 1105, 1108, 1109, 1110, 1111, 1118, 1120, 1121, 1316

dehalogenation, 1103, 1111, 1112, 1114, 1115
dehydration, 486, 558, 914, 917
dehydrochlorination, 1089
delirium, 609
delivery, xxxv, 508, 612, 1185, 1187, 1194, 1327, 1336, 1345
Delta, 29, 907, 1142, 1157, 1165, 1553
demand, xxv, xxxviii, 390, 394, 395, 420, 448, 449, 467, 468, 500, 582, 603, 604, 611, 612, 1349, 1359, 1360, 1363, 1364, 1371, 1386, 1409, 1513
Democratic Republic of Congo, 699, 731
demography, 306, 371, 435, 530, 783, 790, 1718
Demonstration Project, 1187
denaturation, 494, 496, 1576
denaturing gradient gel electrophoresis, 1107, 1120
denaturing gradient gel electrophoresis (DGGE), 1107
dendrogram, 60, 155
denitrification, xxxviii, 1119, 1139, 1332, 1335, 1349, 1351, 1356, 1357, 1361, 1363, 1367, 1369, 1374, 1375, 1396, 1400
denitrifying, 1361
Denmark, 88, 428, 489, 495, 496, 497, 498, 500, dentin, 484
Department of Agriculture, 11, 372, 595, 797, 1332, 1343, 1346, 1689, 1702
Department of the Interior, 883
Department of Transportation, 342
dependent variable, 364, 1356
depolymerization, 1619
deposition, xxxix, 15, 20, 22, 23, 34, 421, 429, 431, 433, 443, 447, 568, 726, 748, 782, 790, 997, 1012, 1029, 1141, 1149, 1151, 1153, 1155, 1164, 1168, 1194, 1213, 1333, 1337, 1345, 1346, 1407, 1408, 1475, 1513, 1515, 1528, 1530, 1531, 1537, 1542, 1543, 1558, 1566, 꽹1665
deposition rate, 20, 1558
deposits, xxxiv, 14, 464, 520, 568, 699, 743, 783, 937, 994, 1021, 1032, 1033, 1143, 1147, 1149, 1150, 1151, 1152, 1154, 1157, 1158, 1159, 1165, 1166, 1167, 1302, 1329, 1518, 1530, 1537, 1540, 1550, 1553, 1555, 1560, 1562, 1564, 1693
depression, 1248, 1255, 1264, 1266, 1284
deprivation, 970
derivatives, 44, 51, 52, 57, 59, 69, 70, 71, 76, 77, 79, 83, 84, 90, 92, 93, 95, 96, 99, 101, 107, 108, 111, 129, 136, 138, 146, 155, 156, 163, 172, 174, 569, 675, 1601
dermatitis, 1611
dermatitis herpetiformis, 1611

dermis, 484, 841, 842, 844, 845, 846, 850
desalination, 1142
desert, 108, 193, 214, 244, 247, 332, 334, 335, 339, 340, 373, 574, 1012, 1125, 1127, 1132, 1137
desiccation, 40, 805, 811, 947, 950, 1633, 1636, 1639
designers, 390, 1474, 1542, 1676
desire, 514
destination, 621, 1456
destruction, xiii, xxiii, xliii, xlv, 234, 259, 260, 261, 338, 341, 406, 435, 565, 569, 570, 574, 585, 620, 621, 634, 687, 738, 747, 807, 967, 971, 972, 1335, 1460, 1641, 1644, 1649, 1672, 1684, 1707, 1711, 1714, 1715, 1729, 1730
detectable, 260, 363, 367, 1219, 1586, 1624
detection, xlii, 5, 40, 260, 261, 371, 495, 687, 1076, 1105, 1114, 1205, 1317, 1374, 1579, 1586, 1590, 1604, 1609, 1612, 1614, 1615, 1618, 1676
detergents, xxxvi, 1202, 1208, 1218, 1220, 1228
detoxification, 1288
detritus, 317, 468, 937, 955, 964, 1332, 1395
devaluation, 714
developed countries, xxi, 391, 392, 396, 413, 479, 723, 728, 949, 1350
developing countries, xx, 211, 389, 391, 392, 393, 396, 423, 428, 479, 591, 604, 698, 726, 872, 907, 1350, 1580
developing nations, 398, 399
deviation, 73, 290, 292, 295, 534, 685, 824, 859, 1151, 1205, 1392, 1396, 1397, 1398, 1400, 1454, 1494
devolution, xxx, 855, 867
diagenesis, 1166, 1167
diamond, 989, 1165
Diamond, 346, 356
diatom assemblages, xxxii, 994, 995, 997, 998, 1002, 1003, 1005, 1007, 1009, 1011, 1012, 1014, 1019, 1021, 1022, 1023, 1026, 1027, 1028, 1029, 1035, 1036, 1038, 1039
diatom classification, 995
diatom communities, 1002
diatom frustule, 1001, 1003, 1029
diatoms, xxxi, xxxii, 693, 887, 895, 897, 902, 908, 909, 944, 946, 993, 994, 995, 996, 997, 998, 999, 1001, 1002, 1003, 1005, 1007, 1009, 1010, 1011, 1012, 1014, 1016, 1017, 1018, 1021, 1022, 1026, 1027, 1028, 1029, 1030, 1031, 1032, 1034, 1036, 1037, 1038, 1039, 뗌1040, 1175, 1366, 1625
diazotrophs, 317, 318
dichlorophenol (DCP), 1103, 1108
dicotyledon, 44

diet composition, 821
dietary, 40, 455, 1330, 1342
dietary fiber, 1269, 1291, 1295
dietary habits, 455
dietary intake, 853
dietary supplementation, 845, 850, 851, 853
diets, 486, 976, 981
differential equations, 673, 674, 1524
differentiation, xv, 43, 61, 67, 71, 95, 110, 315, 329, 383, 386, 414, 443, 484, 554, 556, 562, 563, 568, 1167, 1589, 1618, 1644, 1673, 1678, 1713, 1714
diffraction, 1463, 1472
diffusion, 322, 441, 487, 488, 489, 490, 494, 526, 527, 529, 635, 783, 917, 943, 1393, 1532, 1534, 1535, 1541
diffusion process, 1532
diffusion rates, 322
digestibility, xxix, xxxvi, 493, 819, 821, 822, 823, 826, 836, 838, 1237, 1250, 1259, 1261, 1266, 1269, 1270, 1273, 1274, 1275, 1276, 1280, 1282, 1285, 1286, 1289, 1290, 1291, 1295, 1296, 1601
digestion, 4, 394, 395, 509, 836, 916, 1076, 1108, 1205, 1247, 1250, 1266, 1285, 1296
digitalis, 958, 960
dignity, 737
dihydroquercetin, 100, 101
diluent, 1080
dimensionality, 645
dimeric, 1095
dinoflagellates, 944, 953
dipeptides, 1597
diploid, 69, 74, 165, 1676
directives, 423
dirt, 363
disability, 697
disadvantages, xxvi, 683, 727, 1588
disaster, 1182
discharges, xxxiii, 443, 747, 748, 928, 933, 938, 942, 968, 984, 999, 1018, 1071, 1073, 1101, 1174, 1175, 1181, 1211, 1213, 1215, 1220, 1229, 1353, 1525
discipline, 402, 403
discontinuity, 749, 1350
discrete assemblages, xxxii, 993, 996
discriminant analysis, 685
discrimination, 410, 1154, 1590
diseases, xxiv, xxix, xxx, 328, 342, 428, 456, 485, 505, 570, 594, 596, 609, 733, 839, 840, 842, 871, 880, 881, 1326, 1580, 1601
disequilibrium, 211
disinfection, 1102

dispersion, 73, 369, 527, 530, 543, 555, 771, 788, 1355, 1356, 1391, 1483, 1543, 1545, 1576
displacement, 316, 516, 622, 766, 791, 881, 1417, 1564, 1566
dissatisfaction, 859
disseminate, 1104
dissociation, 846
dissolved oxygen, xxxiii, xxxvii, 572, 820, 840, 843, 858, 941, 942, 980, 1002, 1123, 1128, 1139, 1180, 1204, 1206, 1211, 1219, 1223, 1300, 1308, 1312, 1313, 1316, 1318, 1321, 1332, 1335, 1358, 1370, 1374, 1396
distillation, 679
distilled water, 781, 1076, 1204, 1206
distinctness, 45, 60, 155, 1631
district heating, 726, 727, 729
disturbances, xxii, 255, 257, 513, 530, 531, 532, 539, 544, 572, 621, 630, 631, 632, 633, 634, 758, 759, 1481, 1642, 1656
disulfide, 489, 492, 498
disulfide bonds, 492, 498
diuron, 1074
divergence, 572, 1715, 1717
diversification, xxx, 56, 90, 171, 304, 319, 815, 840, 855, 863, 866, 867, 868, 1565, 1652, 1679, 1681, 1712
diversity, xv, xvii, xviii, xx, xxii, xxiii, xxvii, xxviii, xxxii, xxxvi, xlii, xliii, xliv, xlv, 39, 40, 41, 43, diving, xxxii, 924, 966, 1468, 1469, 1502
DNA, 312, 341, 342, 385, 554, 561, 562, 820, 913, 917, 1107, 1118, 1143, 1315, 1586, 1588, 1589, 1604, 1606, 1621, 1625, 1626, 1637, 1677, 1679, 1721
DNA damage, 1637
DOC, 1353, 1354, 1355
dogs, xxiv, 594
DOI, 370, 385, 387, 508, 510, 562, 788
domestic demand, 582, 611
domestication, 560, 604, 613, 617, 912
dominance, xix, xxxii, 77, 253, 254, 255, 258, 261, 271, 273, 300, 310, 311, 316, 319, 324, 328, 348, 577, 759, 764, 893, 897, 899, 902, 905, 906, 924, 931, 935, 944, 946, 951, 961, 963, 964, 974, 1001, 1014, 1022, 1155, 1162, 1165, 1208, 1221, 1222, 1223, 1224, 1361, 1363, 1589, 1595, 1604, 1665, 1714
donors, 736, 1105, 1109
DOP, 379
dosage, 842
double counting, 703
double helix, 496
double jeopardy, 577

dough, 493, 1580, 1581, 1583, 1589, 1592, 1594, 1597, 1598, 1602, 1603, 1605, 1610, 1611, 1613, 1614, 1615, 1616, 1617, 1619, 1621
draft, 434, 701, 1032, 1677
drainage, xiii, 1, 2, 3, 4, 5, 6, 7, 8, 9, 10, 14, 18, 41, 432, 440, 442, 443, 465, 478, 566, 567, 574, 611, 699, 713, 714, 746, 771, 905, 937, 999, 1072, 1073, 1168, 1170, 1179, 1211, 1322, 1330, 1332, 1342, 1343, 1347, 1374, 1381, 1390, 1391, 1411, 1413, 1414, 1415, 1417, 1439, 1459, 1461, 1667, 1669, 1686, 1693
drawing, 729, 1409
dream, 600
drinking, xvii, 244, 251, 252, 253, 254, 261, 437, 501, 610, 1197, 1309, 1559
drinking water, xvii, 244, 261, 437, 501, 701, 709, 712, 726, 734, 736, 905, 1197, 1309
Drosophila, 915, 921
drought, 24, 179, 196, 274, 355, 383, 526, 544, 702, 713, 716, 719, 733, 734, 735, 736, 744, 748, 756, 1197, 1409, 1714
droughts, 179, 184, 206, 266, 570, 940
drug delivery, 508
drug delivery systems, 508
drugs, xxv, 173, 493, 508, 610, 612, 615, 619, 1102, 1597
dry matter, 262, 472, 501, 785, 797, 1282, 1372
dry recyclables, 394, 395
drying, xxii, 483, 488, 490, 496, 498, 502, 506, 507, 524, 574, 611, 1076, 1117, 1163, 1257, 1263, 1290, 1434, 1530, 1581, 1594
DSC, 503, 504, 505, 506
DSM, 18
duration, 38, 303, 928, 929, 931, 933, 957, 962, 963, 968, 988, 990, 1003, 1131, 1143, 1374, 1381, 1388, 1391, 1392, 1444, 1553, 1660
dust, 441, 452, 531, 1137
duty free, 879
dyes, 496, 1102
dynamic environment, 1223
dynamic systems, 904

E

early warning, xxxiii, 204, 1071, 1072, 1092
earnings, 701, 702, 712, 716
ears, 599, 1727
earth, xiv, xxxiv, 13, 174, 260, 391, 411, 441, 451, 462, 465, 1148, 1679, 1724
Earth Science, 1030, 1032, 1038, 1120, 1167
earthquakes, 725, 1481
East Asia, 391, 611, 617, 1434, 1638, 1708, 1716, 1717, 1719
East Timor, 212

Easter, 860
Eastern Europe, 377, 396
eating, 531
echinoderms, 1229
ECM, 1506
E-coli, 1314, 1315, 1319
ecological data, 370, 998, 1032, 1208, 1667
Ecological Economics, 216, 399, 1198, 1199
ecological indicators, xvi, 177, 178, 183, 185, 187, 194, 195, 207
ecological information, xxxii, 993, 996, 997
ecological preferences, 1001, 1037
ecological requirements, xxxii, 40, 993, 996, 999, 1032
ecological restoration, 457
Ecological Society of America, 339
ecological systems, 187, 206, 212, 213, 218, 410, 686
ecologists, 180, 182, 183, 185, 193, 197, 199, 200, 202, 215, 275, 315, 349, 376, 595, 633, 1028
economic activity, 266, 391, 820, 1645
economic change, 866, 1456
economic crisis, 874
economic damage, 428
economic development, 702, 703, 712, 720, 722, 726, 731, 734, 736, 738, 856, 1478
economic evaluation, 1247, 1251, 1253, 1255, 1258, 1259, 1263, 1266, 1270, 1273, 1278
economic growth, 390, 396, 698, 720, 726
economic incentives, 718, 1332
economic losses, xl, 757, 1213, 1511, 1512
economic performance, 701
economic problem, 209
economic reform, 710
economic reforms, 710
economic values, 1197
economics, xiii, xxxix, 209, 216, 397, 399, 679, 680, 789, 855, 856, 1198, 1372
economies of scale, 866
economy, xxx, xxxix, 60, 360, 467, 499, 500, 596, 597, 604, 697, 701, 702, 710, 711, 716, 721, 725, 734, 737, 738, 794, 866, 871, 872, 911, 1095, 1140, 1187, 1409, 1437, 1466, 1513, 1577
Ecotourism, 1734
edema, 846
editors, 355, 358, 615, 617, 764, 767, 1287, 1291, 1293, 1294, 1296, 1667, 1668
education, 276, 406, 522, 523, 526, 527, 529, 697, 723, 725, 735, 737, 1186, 1189
Education, 1198, 1716, 1718
EEA, 418, 423, 428, 436, 449, 459, 692, 1350, 1367

efficiency, 315, 321, 351, 396, 398, 408, 489, 490, 493, 496, 497, 498, 526, 698, 713, 714, 720, 724, 727, 733, 772, 789, 795, 820, 821, 825, 835, 840, 866, 917, 940, 1076, 1077, 1084, 1086, 1095, 1113, 1261, 1264, 1271, 1274, 1282, 1283, 1284, 1286, 1288, 1293, 1330, 1336, 1345, 1361, 1374, 1377, 1390, 1401, 1409, 1413, 1434, 1436, 1540, 1570, 1612
effluent, xxix, xxxviii, 10, 438, 725, 819, 821, 823, 828, 829, 831, 832, 834, 835, 836, 928, 973, 981, 1102, 1111, 1174, 1349, 1351, 1353, 1367, 1372, 1373, 1375, 1376, 1378, 1379, 1381, 1389, 1393, 1395, 1400, 1401, 1403, 1405, 1460
effluents, xxix, 436, 437, 438, 440, 450, 458, 820, 825, 826, 835, 944, 973, 1012, 1101, 1119, 1142, 1202, 1329, 1330, 1333, 1353, 1375, 1376, 1377, 1379, 1395
egg, 511, 803, 805, 813, 973, 1279
eggs, 1207
Egypt, ix, 118, 121, 164, 217, 609, 615, 699, 700, 702, 705, 706, 707, 708, 709, 712, 731, 732, 736, 1237, 1238, 1241, 1246, 1247, 1254, 1257, 1267, 1271, 1275, 1279, 1282, 1284, 1286, 1287, 1288, 1295
eigenvalues, 291
elaboration, 236, 544, 882, 1552
Elam, 376, 385
elasticity, 503, 1690
elastin, 489, 490, 496, 497, 498, 510
elders, 188, 198, 201
election, 188, 370, 602
electrical conductivity, 843
electrical power, 1392, 1402
electricity, 394, 440, 709, 717, 718, 720, 738
electrolyte, 1293
electron, 402, 917, 1103, 1105, 1107, 1108, 1109, 1111, 1117, 1583, 1587, 1625
electron microscopy, 1107, 1117, 1625
elephants, 1701, 1705
elongation, 779, 782
e-mail, 549, 1325, 1579
embryo, 554, 573, 1676
embryogenesis, 558, 559, 563
embryonic development, 484, 1208, 1220
embryos, xxxvi, 558, 573, 1202, 1206, 1207, 1208
emergency, 472, 735, 736
emergency relief, 735, 736
emerging economies, 455
emerging issues, xix, 345
emigration, 962, 963

emission, xxi, xxiii, 393, 394, 396, 398, 402, 461, 464, 478, 503, 510, 581, 582, 588, 721, 729, 1101, 1109, 1569
emission source, 1101
emitters, 396, 398, 399
emotion, 595
employment, xxxix, 582, 717, 725, 856, 865, 872, 875, 1437, 1580
employment opportunities, 872
empowerment, 212
EMS, 723
encapsulated, 476, 934, 1461
encoding, 917, 1604
encouragement, 558, 730, 733
endangered species, 267, 599, 612, 754, 884, 885, 1073, 1696, 1715, 1729
Endangered Species Act, 598
endosperm, 554
endothelial cells, 841
endotoxins, 902
endowments, 724
end-users, 684
enemies, 312, 332
eenergy consumption, 488, 493, 585, 717, 720, 730
energy efficiency, 725
energy input, 1332
energy recovery, 395, 721
energy supply, 467
energy transfer, 1410
enforcement, 698, 719, 722, 725, 737, 880, 903, 975, 1338, 1700, 1701, 1702
engagement, 398, 544, 599
engineering, xl, 395, 414, 451, 457, 464, 486, 643, 680, 725, 771, 1115, 1459, 1461, 1462, 1465, 1477, 1505, 1511, 1515, 1516, 1517, 1531, 1535, 1540, 1541, 1542, 1543, 1546, 1547, 1570, 1577
England, 95, 104, 211, 213, 214, 337, 397, 429, 693, 755, 767, 788, 868, 884, 1036, 1121, 1292, 1293, 1294, 1322, 1367, 1458, 1611
enlargement, 19, 23, 635, 840
entanglement, 405, 409
entertainment, 419, 451
enthusiasm, 1462
Entomoneis alata, 1050
entropy, xliv, 644, 1683, 1691, 1704
entropy model, 1704
environment factors, 46
environmental aspects, 591
environmental awareness, 197
environmental change, xvi, 177, 179, 180, 181, 183, 184, 185, 187, 188, 189, 191, 192, 193, 196, 197, 198, 199, 201, 204, 206, 207, 209, 210, 212, 245, 407, 563, 570, 572, 620, 684, 732, 757, 995, 1038, 1039, 1166, 1181, 1551, 1717, 1718
environmental characteristics, xxxii, 310, 994, 1100
environmental conditions, xv, xxiii, xxviii, xxxii, xxxv, 44, 47, 178, 184, 293, 328, 383, 422, 436, 440, 468, 473, 555, 565, 568, 569, 598, 685, 689, 769, 772, 786, 799, 800, 801, 802, 803, 805, 806, 808, 809, 810, 811, 813, 816, 950, 961, 993, 995, 996, 997, 1004, 1075, 1090, 1124, 1139, 1140, 1164, 1185, 1198, 1220, 1477, 1550, 1587, 1588
environmental contaminants, 1097
environmental contamination, 426, 427, 1072, 1095, 1594
environmental control, 1035, 1335
environmental degradation, xxvii, 178, 195, 202, 209, 582, 584, 731, 741, 820, 1100, 1209, 1464
environmental effects, 404, 582
environmental factors, xiv, xxxvi, 27, 182, 314, 383, 384, 451, 728, 789, 812, 950, 1005, 1007, 1029, 1037, 1202, 1223, 1373, 1417, 1605, 1618, 1668, 1695
environmental impact, xxiii, xxviii, xxxvii, 218, 393, 395, 420, 436, 437, 443, 450, 457, 581, 582, 584, 585, 591, 592, 618, 723, 734, 736, 819, 820, 821, 881, 1142, 1246, 1307, 1325, 1339, 1373, 1403, 1464, 1468, 1570
environmental issues, xxi, 189, 408, 414, 584, 1186
environmental management, 180, 181, 201, 209, 212, 217, 397, 399, 404, 405, 407, 408, 584, 602, 680, 684, 719, 764, 978, 986, 1186, 1343
environmental protection, 208, 390, 408, 426, 698, 718
Environmental Protection Agency (EPA), 42, 444, 446, 690, 694, 1144, 1186, 1234, 1328, 1340, 1341, 1343, 1373
environmental quality, 418, 450, 1072, 1704
environmental regulations, 719, 725
environmental resources, 450, 454
environmental risks, 1333
environmental standards, 1301
environmental stress, 684, 687, 692, 1072
environmental sustainability, 185, 726
environmental variables, 183, 685, 989, 994, 995, 1002, 1011, 1013, 1026, 1031, 1039, 1223, 1228, 1656, 1666, 1696
enzymatic, xxii, 53, 54, 316, 317, 319, 469, 473, 483, 485, 489, 490, 494, 495, 497, 498, 500, 501, 502, 505, 507, 510, 1090
enzyme, 313, 561, 823, 836, 1073, 1091, 1094, 1269, 1270, 1289, 1581, 1583, 1588, 1589, 1612

enzymes, 314, 317, 318, 319, 469, 473, 476, 485, 489, 490, 493, 496, 498, 504, 507, 1091, 1093, 1094, 1095, 1096, 1103, 1269, 1273, 1276, 1591, 1597, 1598, 1601, 1602, 1608, 1609, 1616, 1619
EPC, 1261
Epi, 533
epidemiology, 397, 766
epidermis, 484, 487, 489, 490, 494, 497, 842, 844, 845, 846, 850
epiphytes, 520, 995, 1040, 1041
epistemological, 188
epithelial cell, 484
epithelial cells, 484, 841
epithelium, 573, 845, 850, 1269
epoxides, 503
epoxy, 470, 511
epoxy resins, 511
EPR, 403
EPS, 1600
equilibrium, 183, 184, 185, 186, 197, 203, 206, 212, 216, 218, 362, 404, 440, 454, 457, 529, 557, 644, 685, 727, 751, 778, 779, 937, 943, 1134, 1332, 1340, 1423, 1459, 1462, 1523, 1733
equipment, 253, 254, 582, 716, 724, 728, 733, 859, 861, 874, 1131, 1330, 1331, 1377, 1378, 1402
equity, 697, 867
ER, 200, 235, 247
Eritrea, 699, 731, 736
Erk, 1096
erythrocytes, 841, 843
Escherichia coli, 1341
ESR, 170
ESR spectroscopy, 170
essential oils, 110
ester, 470
esters, 160
estimating, xxxix, 306, 363, 373, 374, 403, 562, 928, 946, 1096, 1198, 1407, 1426, 1432, 1552
estimator, 540
Estonia, 41, 386
estuarine diatom distribution, xxxii, 994
estuarine diatom ecology, xxxii, 993, 995
estuarine diatoms, xxxii, 993, 995, 996, 999, 1026, 1029, 1040
estuarine environments, xxxii, 985, 993, 994, 995, 999, 1001, 1002, 1023, 1028, 1038, 1102
estuarine gradients, xxxii, 993, 994, 996
estuarine settings, 996, 1022
estuarine systems, 924, 958, 976, 990, 994, 1035, 1221
ethanol, 467, 479, 1263, 1583, 1593, 1594, 1598, 1606

Ethanol, 1117
ethers, 91, 92, 93, 100, 101, 114, 121, 132, 173, 504
Ethiopia, 192, 211, 213, 215, 1165
ethylene, 913, 922
ethylene glycol, 913, 922
etiology, 840
eucalyptus, 361
eukaryote, 1635
euphotic zone, 940
Eurasia, 266
Euro, 406, 407, 490
Europe, xvi, xxxi, 23, 69, 93, 104, 243, 244, 271, 303, 377, 396, 421, 422, 424, 428, 429, 430, 431, 432, 447, 456, 457, 460, 557, 584, 585, 596, 610, 615, 684, 685, 686, 687, 692, 693, 729, 797, 879, 888, 906, 923, 1039, 1331, 1345, 1367, 1438, 1456, 1457, 1458, 1507, 1630, 1637, 1639, 1708, 1709, 1713, 1714
European Commission, 395, 886, 1365
European Community, 459, 1370
European diatoms, xxxii, 993, 995
European market, 586
European Union, 391, 395, 414, 584, 906, 1438
Europeans, 406
eutrophication, xiii, xxxv, xxxvii, 1, 408, 420, 571, 585, 943, 944, 947, 973, 975, 1169, 1170, 1178, 1179, 1180, 1181, 1182, 1183, 1309, 1317, 1318, 1320, 1322, 1323, 1331, 1332, 1335, 1349, 1350, 1351, 1373
evacuation, 832
evaporation, xxxiii, 4, 705, 732, 858, 932, 940, 968, 1123, 1124, 1125, 1133, 1139, 1386, 1392, 1409, 1411, 1413, 1420, 1426, 1435, 1442, 1599
evapotranspiration, xxxix, 431, 432, 433, 701, 746, 1337, 1391, 1392, 1407, 1409, 1410, 1415, 1420, 1421, 1422, 1425, 1428, 1432, 1434, 1435, 1436, 1439, 1442, 1677
Everglades, 795, 1338
everyday life, 524
evolutionary process, 560
excavations, 1560
exchange rate, 1135, 1367, 1540, 1541
exchange rates, 1540, 1541
exclusion, 40, 196, 273, 276, 315, 332, 334, 402, 476, 812, 955, 958
excretion, 823, 825, 827, 828, 829, 830, 831, 832, 834, 835, 836, 837, 1072, 1077, 1295
execution, 709, 733, 737, 1589
exercise, 188, 642, 1568
exopolysaccharides, 1600, 1608, 1613, 1620
exoskeleton, 802
expenditures, 722

experimental condition, 772, 826, 904, 1247, 1251, 1253
experimental design, 759, 889, 890, 1467
expert, 459, 616
expertise, xxxi, 724, 755, 887, 889, 1630
experts, 191, 408, 734, 753, 1197, 1207, 1513
explicit knowledge, 404
exploitation, xxiii, 178, 196, 197, 202, 203, 204, 208, 316, 420, 523, 530, 543, 550, 551, 555, 560, 576, 604, 612, 679, 680, 701, 727, 856, 857, 902, 1439, 1565, 1603, 1609, 1645, 1714, 1729
exploration, xxiii, 349, 351, 404, 550, 856, 1167, 1208, 1621, 1730
exporter, 879
exports, 702, 710
exposure, 4, 325, 364, 747, 777, 849, 904, 914, 918, 947, 949, 950, 964, 1004, 1072, 1073, 1080, 1087, 1092, 1094, 1096, 1178, 1553, 1599, 1611
Exposure, 1088, 1144
extensor, 495
externalities, 867
extinction, xvi, xxii, xxiii, xxiv, xxv, xxx, xlv, 222, 341, 406, 513, 548, 565, 569, 570, 571, 572, 573, 574, 575, 576, 577, 578, 579, 594, 600, 619, 628, 629, 634, 635, 647, 651, 652, 656, 663, 665, 667, 671, 720, 812, 815, 871, 882, 885, 1078, 1164, 1629, 1643, 1645, 1646, 1648, 1649, 1650, 1684, 1707, 1711, 1713, 1714, 1715, 1718, 1720, 1730, 1732
extraction, 168, 196, 390, 394, 493, 494, 524, 582, 611, 721, 725, 727, 972, 1118, 1130, 1205, 1206, 1254, 1289, 1304, 1645, 1656
extracts, xxxiii, 87, 143, 163, 165, 1071, 1087, 1088, 1089, 1090, 1092, 1205, 1258, 1275, 1295, 1592
extrapolation, 207
extrusion, xxii, 483, 506, 507, 837
exudate, 113, 115, 132
eye, 330

F

fabric, 354, 1464, 1465, 1466
fabricate, 509
facies, xxiii, 514, 1167
facilitators, 527
factories, 426, 439, 747, 928, 1074, 1101
faecal, 1330, 1376
faecal coliforms, 1376
failure, 14, 305
false negative, 1226
family planning, 1729
family structure, 554

famine, 196
FAO, 216, 432, 467, 478, 479, 1097, 1450, 1451
Far East, 609
farm income, 713
farm land, 2
farm size, 356, 702
farmers, xiv, xvii, xxxiii, xxxvii, 13, 180, 243, 270, 271, 272, 274, 275, 419, 449, 450, 457, 552, 560, 702, 709, 710, 711, 713, 714, 719, 720, 874, 877, 881, 1123, 1325, 1332, 1334, 1373, 1402, 1439, 1449
farming, xiii, xv, xvii, 1, 23, 37, 41, 205, 209, 211, 243, 284, 306, 370, 386, 421, 435, 440, 478, 492, 499, 502, 517, 585, 1145, 1163, 1331, 1332, 1336, 1339, 1438, 1456
farmland, xv, 3, 37, 38, 40, 41, 370, 969, 1331, 1412, 1415
farmlands, 1329
farms, xiii, xxx, xxxi, xxxviii, 1, 2, 3, 4, 7, 9, 10, 38, 246, 247, 250, 251, 370, 435, 439, 440, 527, 735, 840, 867, 872, 873, 874, 875, 876, 877, 880, 881, 911, 912, 913, 1074, 1125, 1137, 1140, 1307, 1331, 1371, 1372, 1373, 1377, 1388
fat, 488, 489, 844, 1243, 1245, 1246, 1249, 1252, 1254, 1256, 1259, 1262, 1265, 1268, 1269, 1272, 1276, 1281, 1282, 1283, 1295, 1296, 1594, 1596, 1612
fats, 489
fatty acid, 470, 489, 490, 495, 497
fatty acids, 48, 50, 470, 489, 490, 495, 497, 1108, 1240, 1293, 1581, 1593, 1598, 1600, 1619
fax, 1407, 1437
fear, 685, 719
feather star, 1465
February, xxxiv, 2, 20, 200, 236, 364, 558, 592, 612, 613, 614, 615, 951, 966, 970, 974, 1092, 1123, 1134, 1139, 1190, 1304, 1306, 1389, 1399, 1460
feces, 259, 820, 821, 829, 1270, 1327, 1328
federal government, 517, 874, 879
federal law, 523
Federal Register, 1340
feed additives, xxix, 839
feedback, 201, 311, 319, 320, 321, 323, 327, 328, 330, 337, 342, 407, 794, 1474, 1575, 1577
feed-back, 454
feeding, xxxii, 313, 321, 334, 407, 491, 531, 573, 622, 924, 957, 959, 965, 981, 984, 986, 1072, 1076, 1097, 1106, 1108, 1178, 1341
feedstock, 467, 481
feedstuffs, 837, 1238, 1240, 1241, 1261, 1290
feelings, 404

feet, 414, 1125
females, 534, 537, 554, 611, 623
fencing, xxxvii, 1325
fermentation, xlii, 467, 1109, 1449, 1579, 1580, 1581, 1582, 1583, 1585, 1586, 1587, 1589, 1590, 1591, 1592, 1593, 1595, 1596, 1598, 1599, 1600, 1601, 1602, 1603, 1604, 1605, 1606, 1607, 1608, 1610, 1611, 1613, 1614, 1615, 1616, 1617, 1618, 1619, 1620, 1621
fermentation technology, 1587
fern, 163, 169
fertiliser, xxxvii, 421, 1325
fertility, 35, 205, 244, 328, 349, 357, 419, 423, 424, 430, 432, 433, 435, 447, 452, 455, 468, 766, 781, 796, 816, 1183, 1306, 1425, 1693
fertilization, xiii, 2, 9, 10, 170, 355, 356, 377, 449, 452, 554, 693, 913, 918, 1207, 1306, 1321, 1335, 1350, 1438
fertilizer, xiii, 1, 2, 4, 6, 9, 284, 354, 440, 494, 502, 942, 1010, 1101, 1174, 1179, 1301, 1306, 1320, 1326, 1327, 1331, 1332, 1339, 1412, 1431
fertilizers, 9, 10, 420, 426, 439, 441, 449, 450, 468, 492, 499, 968, 1229, 1327, 1332, 1339, 1350
ferulic acid, 71, 72, 73, 74, 108, 109, 156
Feynman, 404, 410
fiber, 254, 509, 846, 1192, 1243, 1245, 1246, 1249, 1252, 1254, 1256, 1259, 1262, 1264, 1265, 1268, 1269, 1272, 1276, 1281, 1282, 1283, 1285, 1286, 1291, 1295, 1297
fiber content, 1264, 1282
fibers, xxix, 484, 485, 486, 492, 505, 508, 840, 842, 843, 846, 848, 849, 850
fibrillar, 469, 484, 485, 486
fibrils, 484
fibroblasts, xxix, 840, 841, 845, 850
fidelity, 997, 998, 1036, 1191
field crops, 1425
filament, 1112
fill material, 1461, 1464, 1469
film, 487, 488, 489, 492, 505, 509, 1076, 1121, 1377, 1392
film formation, 509
film thickness, 1076
films, xxii, 363, 483, 486, 487, 492, 502, 503, 505, 506, 507, 509, 511, 512
filters, xix, 359, 1076, 1130, 1171, 1344, 1345, 1354, 1356, 1368, 1374
filtration, 497, 498, 500, 1186, 1308, 1333, 1375, 1376, 1392, 1393, 1395
financial, 376, 582, 597, 620, 669, 698, 710, 711, 713, 719, 720, 722, 727, 856, 859, 861, 864, 874, 877, 975, 1116, 1229, 1261, 1449, 1462, 1506, 1536

financial resources, 597, 698, 722, 861
financial support, 376, 720, 874, 877, 1116, 1229
fines, 1220
fingerprints, xv, 43, 47, 59, 138, 443
Finland, 399, 1166, 1181
fire, xviii, 203, 211, 216, 217, 223, 265, 269, 270, 271, 272, 273, 274, 276, 277, 278, 279, 280, 281, 319, 456, 523, 526, 563
fire resistance, 278
first generation, 467, 1589
FISH, 1107
Fish and Wildlife Service, 754, 767
fish oil, 1238
fish production, 582
fisheries, xiii, xxiii, xxvii, xxx, 575, 581, 582, 584, 591, 592, 672, 702, 741, 742, 751, 752, 855, 856, 857, 859, 860, 861, 865, 866, 867, 868, 869, 872, 881, 883, 884, 905, 907, 909, 913, 984, 986, 999, 1095, 1127, 1140, 1142, 1186, 1194, 1261, 1292, 1456
fishing, xxiii, xxx, 581, 582, 584, 585, 586, 588, 589, 590, 591, 614, 638, 646, 855, 856, 857, 858, 859, 860, 861, 862, 864, 865, 866, 867, 972, 1127, 1187, 1292, 1456, 1457, 1458, 1468, 1711
fitness, xix, 310, 312, 313, 326, 337, 376, 386, 420, 781
fixation, 57, 317, 327, 332, 466, 468, 570, 1107, 1332, 1359
FL, 479, 1097, 1185, 1320, 1322, 1340, 1343, 1347, 1404
flame, 1076, 1108, 1205
flame ionization detector, 1108, 1205
flammability, 729
flank, 1334
flavone, 59, 60, 61, 63, 64, 69, 76, 77, 91, 93, 95, 102, 110, 112, 113, 115, 116, 118, 119, 120, 121, 126, 127, 128, 132, 133, 135, 139, 148, 154, 166, 169, 172, 173, 174, 175
flavonol, 58, 59, 61, 69, 76, 77, 91, 93, 95, 100, 103, 106, 120, 121, 128, 135, 139, 140, 141, 142, 150, 154, 155, 156, 157, 159, 160, 161, 162, 166, 167, 171, 175
flavor, xlii, 1258, 1579, 1619
flavour, 1581, 1598, 1599, 1600, 1605, 1614
flaws, 690, 1713
flexibility, 47, 189, 484, 601, 864, 965, 1239
flexor, 495
flight, 289, 803
float, 1379, 1383
floating, 1527
flocculation, 932, 940, 1541

flood, xiv, 13, 14, 15, 17, 22, 391, 933, 934, 935, 936, 937, 938, 951, 952, 968, 971, 976, 978, 1003, 1124, 1127, 1149, 1152, 1191, 1197, 1302, 1418, 1435, 1525, 1526, 1529, 1538, 1544, 1551, 1553, 1565, 1567, 1569, 1570, 1656, 1657, 1714
flooding, xiv, xxxiv, xli, 14, 222, 454, 731, 738, 754, 758, 761, 784, 790, 792, 941, 950, 954, 959, 962, 976, 978, 1003, 1147, 1163, 1197, 1457, 1458, 1459, 1517, 1549, 1550, 1551, 1552, 1559, 1560, 1563, 1565, 1568, 1569, 1570
floods, xiii, xiv, xxv, 13, 14, 22, 23, 417, 570, 611, 637, 638, 661, 703, 705, 736, 738, 746, 926, 933, 968, 979, 1006, 1022, 1124, 1570, 1657
floodwalls, 750
flora, 204, 235, 237, 262, 266, 267, 268, 279, 280, 311, 422, 523, 545, 557, 638, 749, 750, 751, 754, 881, 1073, 1096, 1119, 1140, 1163, 1464, 1586, 1636, 1652, 1678, 1679, 1708, 1709, 1713, 1716, 1718, 1720, 1724, 1725, 1728, 1729, 1730
flora and fauna, 422, 523, 557, 638, 749, 751, 754, 881, 1073, 1140, 1725
flour, 486, 833, 1240, 1250, 1293, 1580, 1581, 1586, 1587, 1591, 1592, 1596, 1598, 1600, 1602, 1603, 1605, 1606, 1608, 1609, 1611, 1612, 1613, 1614, 1616, 1617, 1618
flow field, 1529, 1530, 1537, 1538, 1540, 1542
flow rate, 931, 934, 935, 937, 1076, 1077, 1329, 1540, 1542
flowering period, 296
flowers, xviii, 76, 78, 79, 80, 81, 84, 85, 86, 88, 89, 90, 93, 160, 164, 165, 166, 169, 170, 171, 172, 271, 284, 285, 286, 289, 290, 291, 293, 294, 295, 296, 297, 299, 300, 301, 302, 303, 304, 306, 1716
fluctuations, xv, xxvii, 44, 184, 207, 305, 654, 668, 673, 734, 741, 747, 749, 752, 787, 860, 899, 951, 959, 979, 1004, 1021, 1022, 1023, 1028, 1124, 1136, 1145, 1152, 1158, 1164, 1168, 1170, 1173, 1174, 1238, 1411, 1428, 1429, 1566, 1591
fluid, xxxiii, 727, 1012, 1071, 1080, 1087, 1089, 1393, 1474, 1512, 1513, 1518, 1519, 1541, 1542, 1543, 1581, 1612
fluid extract, xxxiii, 1071, 1087, 1089
fluorescence, 917, 1105, 1107
fluorescence in situ hybridization, 1107
flushing, 453, 935, 945, 964, 969, 974, 1150
fluvial, xl, 978, 1003, 1004, 1014, 1022, 1100, 1148, 1149, 1456, 1474
fluvial-lacustrine, 1022
FMC, 1274, 1275, 1276, 1277, 1282

focus group, 209
focusing, 182, 188, 210, 376, 384, 397, 999, 1409, 1454, 1494, 1495, 1497, 1500, 1501
foils, 492, 505, 506, 507
food additives, 486, 493, 494, 1596
food chain, xxi, xxvi, 404, 414, 420, 423, 426, 456, 577, 637, 638, 639, 640, 641, 642, 643, 647, 649, 651, 661, 663, 667, 669, 670, 676, 677, 680, 1175, 1339
food industry, 487, 492, 506, 730, 1438, 1596
food intake, 1258, 1291
food particles, 902, 1256, 1258
food production, xxii, 417, 419, 454, 483, 1287, 1606
food products, 487, 1593, 1594
food security, xx, xxi, 375, 413, 702, 711, 712, 886, 1433
food spoilage, 1596
food web, xxv, xxvi, 311, 637, 641, 642, 644, 645, 676, 683, 691, 694, 816, 946, 977, 985, 1139, 1145, 1338
foodstuffs, 486, 506, 1179
footwear, 489, 493, 507
forage crops, 385
forbs, 244, 256, 271, 273, 349, 350, 356
force, 312, 316, 396, 419, 724, 816, 930, 1481, 1518, 1523, 1541, 1595, 1599, 1732
Ford, 1297
forecasting, 668, 1191, 1192, 1409, 1568, 1570
foreign exchange, 702, 712
forensic, xxi, 414, 457
forest ecosystem, xvii, 265, 458, 531, 1368, 1720
forest fire, 421, 464, 643
forest fires, 421, 464
forest formations, 556
forest fragments, xxii, 513, 515, 519, 520, 523, 524, 530, 531, 532, 534, 535, 536, 537, 538, 540, 543, 544, 546, 547, 552, 557
forest habitats, 794
forest management, 424, 635, 1702
forest resources, 544, 699
forest restoration, 528, 529
Forest Service, 11, 279, 372, 595, 1323
forestry, xxi, 284, 361, 414, 435, 450, 523, 524, 543, 596, 1313, 1337
Forestry, 333, 523, 524, 561, 596, 601, 1299, 1320, 1321
formaldehyde, xxii, 483, 493, 502, 503, 505, 507, 509, 510, 511, 1204, 1207, 1658
formamide, 1107
formula, xli, 51, 590, 1491, 1499, 1511, 1519, 1521, 1522, 1524, 1530, 1532, 1533, 1534, 1535, 1541, 1543, 1544, 1547

fortification, 1472
fossil, xxi, 391, 394, 395, 461, 462, 463, 464, 467, 476, 568, 579, 585, 593, 635, 994, 996, 997, 1003, 1009, 1020, 1022, 1023, 1026, 1028, 1029, 1036, 1162, 1163, 1579, 1708, 1709, 1719
fossil fuel, xxi, 391, 394, 395, 461, 462, 463, 464, 467, 585
fossil fuels, 395, 462, 463, 464, 467, 585
fossils, xlv, 568, 994, 1149, 1707, 1710, 1711, 1712, 1715
foundations, 456, 677, 1562, 1570, 1731
founder effect, 570
Fox, 573, 578
fractal dimension, 621, 622, 629
fragility, 422, 459, 485, 492, 493, 543, 843
fragmentation, xix, xxii, xxiii, 42, 216, 223, 237, 359, 360, 361, 370, 371, 372, 373, 374, 376, 387, 486, 513, 514, 530, 531, 532, 539, 543, 548, 565, 570, 620, 621, 622, 624, 625, 626, 627, 628, 629, 630, 634, 635, 636
fragmented forests, 547
fragmented landscapes, 306, 366, 548, 622, 624, 629
fragments, xix, xxii, xl, 359, 370, 409, 500, 513, 514, 515, 519, 520, 523, 524, 525, 526, 529, 530, 531, 532, 534, 535, 536, 537, 538, 539, 540, 541, 542, 543, 544, 546, 547, 548, 552, 557, 572, 844, 1155, 1164, 1455, 1552, 1555, 1562, 1563, 1589, 1603, 1610, 1645
France, xli, xliii, 43, 95, 104, 170, 309, 392, 399, 406, 513, 544, 546, 548, 692, 790, 868, 994, 1029, 1032, 1033, 1037, 1038, 1157, 1168, 1291, 1297, 1438, 1456, 1457, 1458, 1459, 1460, 1462, 1549, 1551, 1560, 1561, 1562, 1563, 1569, 1575, 1641, 1650, 1655, 1666
FRE, 1575
free radicals, 842
free riders, 858, 859
freeze-dried, 1078, 1080
freezing, 728, 912, 917, 918, 921, 1124, 1403, 1410, 1580
freight, 1101
frequencies, 39, 365, 570, 634, 751, 995, 1000, 1007, 1020, 1028, 1197, 1247, 1482
frequency dependence, 330, 337
frequency distribution, 809
fresh water, 252, 953, 958, 975, 990, 1124, 1168
freshwater species, 872, 908, 951, 964, 1631
friction, 936, 1472, 1533, 1576
frost, 554, 777
fructose, 773, 1583, 1587, 1592, 1600
fruits, 87, 515, 560, 701, 1073, 1594, 1627, 1630
FSP, 532, 534, 535, 536, 538, 541, 542

fuel, xxi, 178, 270, 272, 287, 391, 394, 395, 413, 461, 462, 463, 464, 467, 585, 592, 972
functional approach, 1189
functional architecture, 338
funding, 454, 684, 689, 698, 731, 1319, 1633, 1733
funds, 180, 714, 734
fungal, 155, 313, 318, 320, 321, 322, 323, 324, 326, 328, 331, 333, 334, 336, 337, 342, 343, 346, 356, 357, 358, 1164
fungal infection, 1164
fungi, xix, 174, 310, 312, 313, 314, 317, 318, 319, 320, 321, 322, 323, 324, 325, 326, 328, 329, 330, 331, 333, 334, 335, 336, 337, 338, 339, 342, 343, 345, 346, 352, 354, 355, 356, 357, 358, 469, 531, 1330, 1333, 1339, 1360, 1580, 1591, 1594, 1597, 1598, 1612
fungicide, 347, 349, 357, 1102
fungicides, 1333
fungus, 136, 138, 154, 155, 170, 318, 323, 325, 326, 331, 332, 334, 335, 336, 337, 338, 1102, 1646
furniture, 502
fusion, 29
futures, 1193

G

Gallus gallus domestic, 1300
gamete, 911
gas, 395, 398, 399, 430, 449, 461, 463, 467, 482, 511, 588, 590, 976, 1076, 1103, 1106, 1108, 1109, 1119, 1139, 1205, 1332, 1357, 1358, 1367, 1369, 1397
gas chromatograph, 1076, 1108, 1119, 1205, 1357
gas exchange, 976, 1367
gas separation, 1103
gases, xxiii, xxxvii, 504, 581, 582, 586, 589, 590, 591, 1121, 1325
gasification, 395
gasoline, 467
gastrointestinal tract, 1269
gauge, 19, 974, 1383, 1442, 1552, 1566, 1568
GC, 50, 1107, 1142
GDP, 390, 393, 701, 731, 1428
GDP per capita, 731
GE, 1143, 1144, 1340
gel, 493, 494, 501, 502, 505, 506, 507, 1107, 1120, 1205, 1588, 1603, 1608, 1615, 1621
gel formation, 502
gelatin, 498, 507, 509, 510, 511, 512
Gelatine, 487
gels, xxii, 483, 487, 501, 502, 505, 506, 507, 508, 511
gemma, 1011, 1020, 1062
gender, 187, 623, 1207

gene, xx, 162, 369, 375, 376, 523, 530, 549, 554, 555, 556, 557, 561, 562, 570, 572, 1118, 1143, 1582, 1586, 1590, 1603, 1604, 1606, 1610, 1620, 1715
gene combinations, 1715
gene expression, 836, 1590
gene pool, xx, 375
gene regulation, 1143
generation, xxxvi, 406, 407, 516, 554, 630, 929, 1117, 1173, 1194, 1205, 1299, 1466, 1482, 1589, 1590, 1594, 1599, 1605, 1619, 1690, 1733
genes, 421, 530, 764, 1120, 1589, 1591, 1592, 1604, 1616, 1635
genetic control, 47
genetic diversity, xx, xxiii, xlii, 375, 376, 377, 378, 381, 382, 383, 384, 386, 387, 550, 553, 554, 555, 557, 560, 561, 562, 564, 570, 1623, 1627, 1630, 1632, 1633, 1710
genetic drift, 572
genetic factors, 579
genetic marker, 554
genetics, 182, 337, 376, 387, 562, 579, 880, 913, 1609, 1613
Geneva, 218, 387, 411, 460, 482
genistein, 140
genome, 918, 1588, 1589, 1590, 1613, 1676
genomic, 312, 554
genomics, 1614
genotype, 558, 1588
genotypes, 82, 337
genre, 1652, 1653
geochemical, xxxiv, 441, 445, 460, 462, 932, 1148, 1149, 1155, 1156, 1159, 1166, 1566
geochemistry, 35, 454, 1166, 1433
Geographic Information System, xvii, 243, 246
geographical origin, 1598
geography, 174, 397, 415, 565, 597, 613, 1577, 1632, 1642
geological history, xxiii, 565, 568
geology, 3, 35, 314, 414, 597, 686, 928, 937, 941, 994, 1166, 1311, 1351, 1418, 1577, 1657, 1678, 1716, 1719
Geomembranes, 1509
geometrical optics, 1485
geometry, 1158, 1477, 1478, 1497, 1498, 1499, 1506, 1507, 1509, 1537, 1540, 1577
geophysical, 1508
Georgia, 1034, 1346, 1367
germ cells, 918
Germany, 29, 107, 302, 342, 383, 386, 392, 428, 460, 592, 768, 795, 836, 907, 980, 1078, 1580, 1610, 1630, 1636, 1637, 1716

germination, 301, 310, 323, 759, 766, 781, 782, 794
gestation, 573
GHG, 391, 392, 393, 394, 395, 396, 398
GIS, xiv, xvii, 3, 14, 200, 217, 243, 245, 246, 250, 287, 304, 364, 437, 455, 1197
gland, 610, 1078, 1094, 1095, 1096
glass, 395, 449, 502, 505, 1130, 1204, 1207
glass transition, 505
glass transition temperature, 505
global climate change, 464, 468, 905
global consequences, 766
Global Positioning System, xvii, 243, 246, 248
global scale, 757, 1350, 1680, 1704, 1719
global warming, xxiii, 395, 581, 582, 585, 586, 588, 589, 590, 591, 728, 754, 980, 1630, 1634
Global Warming, 394, 581, 590, 1198
globulin, 492
Glomus intraradices, 170, 318, 325, 332
glucoamylase, 1615
glucose, 69, 76, 87, 129, 467, 469, 470, 773, 921, 1103, 1106, 1111, 1583, 1587, 1591, 1592, 1600, 1606
glucosidases, 469
glucoside, 64, 69, 78, 84, 86, 89, 101, 102, 119, 120, 127, 128, 135, 149, 163, 164, 169, 170, 171, 172
glucosinolates, 1286
glue, xxii, 483, 493, 494, 498, 501
glutamic acid, 496
glutamine, 493
glutaraldehyde, 493, 505, 508
glutathione, xxxiii, 1071, 1073, 1078, 1093, 1094, 1095
glutathione peroxidase, xxxiii, 1071, 1073, 1093
glycerol, 913, 914, 915, 920, 1597
glycine, 486, 496, 919, 1602
glycol, 913, 922
glycolysis, 1583, 1585
glycoproteins, 918
glycoside, 59, 64, 68, 69, 77, 82, 85, 93, 101, 109, 121, 122, 126, 127, 128, 129, 135, 139, 141, 149, 153, 157, 158, 160, 161, 163, 166, 170, 172, 173, 174
glycosides, 57, 58, 59, 60, 61, 63, 64, 68, 69, 76, 77, 78, 84, 85, 87, 93, 100, 101, 109, 110, 111, 112, 120, 121, 122, 127, 128, 129, 130, 135, 139, 140, 142, 143, 145, 146, 149, 150, 151, 153, 154, 155, 156, 157, 158, 159, 162, 163, 164, 165, 167, 168, 169, 171, 172, 174, 175
glycosyl, 49, 61, 86, 126, 135
glycosylated, 91, 120, 151, 159
glycosylation, 56, 57, 90, 120, 142, 150
GNP, 716

goals, xv, xvi, 37, 178, 179, 180, 181, 186, 188, 190, 196, 209, 210, 449, 456, 529, 595, 598, 599, 1189, 1193, 1331, 1471
gold, 1509
gonads, 573, 918
goods and services, 396, 710, 967, 975
governance, 866
government, xliv, 179, 181, 196, 275, 276, 395, 397, 414, 415, 423, 433, 517, 522, 527, 557, 596, 597, 1677, 1683, 1711
government intervention, 864
government policy, xliv, 1683
governments, 395, 397, 423, 433, 557, 698, 722, 726
GPP, 466, 1358, 1359
GPS, xvii, 18, 226, 243, 245, 246, 248, 249, 250, 253, 261, 1128, 1203, 1552
gracilis, 353, 1060, 1069
grades, 162, 223, 1002
grading, 440, 859, 862, 864, 1381
graduate students, 1319
grain, xxxix, 60, 169, 455, 467, 485, 489, 500, 555, 977, 1009, 1125, 1149, 1151, 1166, 1204, 1212, 1220, 1223, 1229, 1331, 1407, 1409, 1412, 1413, 1474, 1523, 1531
grain size, 977, 1009, 1149, 1151, 1166, 1220, 1223, 1229, 1474, 1531, 1658, 1659, 1660, 1665
grains, 108, 443, 1159, 1161, 1162, 1475, 1543, 1577, 1579, 1580, 1612
granite, 1467
granites, 1149
grants, 718, 1116, 1141
granules, 1119, 1153, 1154, 1591
graph, 31, 32, 257, 296, 299, 1485, 1494, 1555
graphite, 1076
grasses, xvii, xx, 57, 61, 69, 165, 265, 266, 268, 270, 271, 272, 273, 284, 285, 298, 300, 301, 318, 349, 350, 356, 362, 375, 377, 383, 385, 556, 594, 737, 764, 767, 770, 1163, 1301, 1336, 1449, 1465
gravitational force, 1481, 1523
gravity, 553, 556, 712, 714, 931, 1104, 1120, 1374, 1378, 1392, 1393, 1453, 1459, 1520, 1522, 1523, 1531, 1543
grazers, xviii, xxiv, 182, 185, 194, 203, 265, 271, 333, 593, 594, 771, 899, 1663, 1665
Great Britain, 377, 428, 906, 1290
Great Lakes, 886, 1296
Greece, viii, xxxiii, 118, 121, 347, 356, 1071, 1072, 1073, 1074, 1092, 1093, 1094, 1095, 1096, 1097
green alga, xlii, 897, 902, 904, 943, 1012, 1623, 1624, 1625, 1626, 1627, 1628, 1629, 1630, 1631, 1632, 1633, 1634, 1635, 1636, 1637, 1638, 1639, 1640
green belt, 1624
green front, 1552
greenhouse, xx, xxi, xxvii, 18, 312, 321, 325, 347, 348, 349, 353, 378, 389, 391, 396, 399, 461, 482, 590, 721, 722, 724, 725, 757, 758, 759, 762, 789, 1457, 1568, 1569
Greenhouse, 398, 399, 482, 590, 592
greenhouse gas, xx, xxi, 312, 389, 391, 396, 399, 461, 482, 590, 1568, 1569
greenhouse gas (GHG), 391
greenhouse gas emissions, xx, 389, 399, 721, 724, 725, 1568
greenhouse gases, xxi, 461, 482, 590, 721, 722, 724, 1569
greenhouses, 349
greening, 193, 207, 216, 1637
grid resolution, 19
gross domestic product, 360, 390
gross national product, 716
ground water, 458, 577, 1198, 1329
grounding, 188
groundwater, xxi, xxvii, xxxvii, xxxix, 414, 420, 423, 426, 432, 437, 438, 439, 440, 441, 446, 452, 698, 702, 703, 705, 706, 707, 708, 709, 713, 714, 716, 726, 738, 742, 743, 746, 754, 808, 943, 1101, 1320, 1325, 1329, 1338, 1342, 1407, 1408, 1409, 1410, 1411, 1412, 1414, 1415, 1416, 1418, 1419, 1421, 1422, 1423, 1424, 1428, 1429, 1430, 1431, 1432, 1433, 1434, 1435, 1436, 1565
grouping, 383, 808, 1588
growth dynamics, 790, 791
growth factor, 1239, 1242
growth rate, 313, 315, 320, 338, 423, 641, 699, 701, 759, 774, 779, 780, 781, 784, 794, 813, 820, 825, 829, 835, 875, 1029, 1178, 1247, 1255, 1257, 1258, 1260, 1262, 1264, 1267, 1269, 1271, 1273, 1275, 1279, 1282, 1283, 1284, 1291, 1296, 1425, 1566
GSCs, 1468, 1470
Guangdong, 354, 1715, 1716
Guangzhou, 345, 1716
guidance, xli, 744, 1466, 1512
guidelines, xxii, 189, 192, 210, 428, 456, 513, 515, 805, 806, 972, 976, 997, 1209, 1304, 1319, 1343, 1466
Guinea, xliv, 372, 1671, 1673, 1674, 1675, 1676
Gulf of Mexico, xxxv, 2, 856, 857, 868, 873, 1124, 1145, 1185, 1331, 1519
gymnosperm, 1673, 1678, 1708, 1710

H

H1, 1487, 1526
H_2, 1105, 1109, 1110, 1112, 1487
habitat quality, 807
habitation, 1562
hair, 490, 496, 1627
hair follicle, 490
halogenated, 394, 1102, 1103
halophyte, 1124
halophytes, 1127, 1143
Hamiltonian, 647, 648, 674, 675
handling, 1377, 1466
hands, 1643
hanging, 396
haploid, 1676
haplotypes, 385
harbors, xl, 444, 514, 1009, 1511, 1512, 1513, 1515, 1516, 1517, 1518, 1519, 1531, 1535, 1536, 1541, 1542, 1543, 1544, 1546
harbour, xxxiii, 313, 314, 514, 542, 1006, 1099, 1101, 1540, 1542
hardness, 254, 256, 257, 258, 1278
hardwood forest, 743, 744
hardwoods, 747
harm, 417, 488, 1439, 1456
harmful effects, 1074, 1179
harmony, 417, 1456
Harvard, 234, 636
harvest, 20, 596, 610, 612, 615, 1330, 1334, 1337, 1339, 1412, 1448, 1449
harvesting, xxviii, 203, 208, 419, 643, 646, 672, 713, 718, 735, 769, 772, 784, 785, 786, 788, 790, 792, 793, 1001, 1321, 1329, 1330, 1334, 1375, 1400
hatchery, 1301
Hawaii, vii, xxiv, 593, 594, 595, 596, 597, 598, 599, 600, 601, 602, 764, 767, 1509, 1634
hay meadows, 1307
hazardous substance, 419
hazardous substances, 419
hazardous waste, 725
hazardous wastes, 725
hazards, 426, 726, 1331
HDPE, 492
healing, xxix, 606, 613, 839, 841, 842, 843, 845, 846, 848, 849, 850, 851, 852, 853, 854
health, xxi, xxx, xxxvii, 37, 163, 182, 183, 191, 193, 195, 339, 391, 393, 404, 405, 413, 419, 426, 428, 445, 446, 450, 452, 455, 456, 457, 459, 501, 577, 604, 610, 612, 638, 686, 701, 715, 719, 726, 733, 735, 736, 854, 883, 884, 905, 933, 946, 978, 1072, 1091, 1101, 1140, 1186, 1190, 1202, 1226, 1320, 1325, 1338, 1350, 1580, 1594, 1601, 1620, 1724
health care, 604, 612, 701, 735, 736
health condition, 1091
health effects, 393
health problems, xxi, 413
health risks, 426, 905, 1140
health status, 1072
healthcare, 604
healthfulness, 1602, 1605
hearing, 598
heart, 484, 519, 609
heat, xlii, 363, 394, 426, 499, 504, 1132, 1143, 1409, 1410, 1435, 1579, 1580, 1594
heat shock factor 1, 1143
heat storage, 1132
heat treatment, 1594
heating, 54, 427, 490, 496, 497, 498, 503, 1132, 1139
heavy metal, xxxiii, 354, 419, 420, 423, 426, 428, 433, 441, 443, 446, 447, 452, 453, 458, 572, 1071, 1073, 1074, 1079, 1080, 1089, 1097, 1101, 1329, 1340
heavy metals, 420, 423, 426, 428, 433, 441, 443, 446, 447, 452, 453, 458, 572, 771, 868, 1073, 1074, 1080, 1089, 1102, 1329, 1340
Heisenberg, 403, 410
helium, 1357
helix, 484, 496
Helix, 485
hematocrit, 843
hematology, 851, 1295
hemicellulose, 469
hemisphere, xxxiii, 278, 552, 779, 924, 956, 1100, 1631
herbal, 44, 45, 610
herbicide, xxxi, 887, 889, 906, 1333, 1336, 1337, 1344, 1345
herbicides, 433, 492, 1074, 1102, 1333, 1336, 1337, 1344, 1345, 1438, 1450
herbivores, xviii, xxiv, 245, 248, 250, 251, 261, 265, 309, 310, 326, 328, 593, 602, 604
herbivory, 214, 215, 310, 321
herbs, 44, 87, 189, 331, 456, 604, 610, 1163
hermaphrodite, 554
heterogeneity, xv, xix, 37, 58, 74, 77, 302, 305, 306, 310, 311, 313, 314, 315, 316, 322, 331, 333, 334, 337, 340, 342, 372, 404, 566, 595, 691, 760, 764, 791, 800, 811, 812, 814, 817, 987, 1211, 1226, 1576, 1677
heterogeneous, xviii, 77, 186, 187, 194, 199, 284, 302, 313, 314, 322, 323, 374, 937, 1223
heterotrophic, 1359, 1360, 1363, 1370, 1375

heterotrophic microorganisms, 1375
heterozygosity, xx, 375, 378, 379
heterozygote, 1676
hexachlorobenzene, 1102
hexachlorocyclohexane, 1102
hexane, 1076, 1205, 1244, 1245, 1263, 1283
high pressure, 1205
high resolution, 245, 246, 260, 1035
high risk, 1334, 1458
high scores, 967
high strength, 486
high temperature, 486, 1464
high winds, 891
higher quality, 560, 1596
high-frequency, 942
highlands, xxiii, 223, 233, 270, 276, 278, 549, 550, 555, 556, 563, 567, 704, 705, 1704, 1724, 1725
Highlands, 737
high-performance liquid chromatography, 165, 168
highways, 432
hiring, 863
Hiroshima, 243
histidine, 1280
histochemistry, 843
histological, 496
histology, 852, 1290
historical data, 1566, 1568
historical reason, xxiv, 521, 593, 1633
holistic, 187, 390, 396, 403, 410, 454, 458, 967, 1332
holistic approach, 187, 396, 458, 967, 1332
Holland, 10, 11, 1109, 1234
Holocene, xvii, xviii, xxxiv, 265, 269, 556, 563, 604, 613, 1006, 1009, 1021, 1022, 1023, 1026, 1029, 1030, 1032, 1033, 1035, 1039, 1127, 1147, 1148, 1152, 1153, 1154, 1156, 1157, 1158, 1159, 1163, 1165, 1166, 1167, 1168, 1550, 1555, 1564, 1566, 1721
homeostasis, 1609
homes, 738
homogeneity, 60, 84, 378, 877, 1081, 1595
homogenized, 319, 434, 600, 1076, 1078, 1080, 1204, 1205, 1206
homogenous, 39, 78, 155, 187, 932, 946
Honda, 172
honey, 560
Hong Kong, 611, 617, 906, 1037, 1717
horizon, 22, 395, 414, 429, 443
horizontal transmission, 1638
hormones, 450, 563
horse, 253
horses, 256, 784
Horticulture, 345

host, xviii, xix, 284, 302, 312, 313, 317, 318, 320, 321, 322, 323, 325, 326, 327, 328, 330, 334, 335, 337, 345, 346, 351, 352, 408, 476, 597, 815, 882, 1248
hot spots, 312, 444, 1680, 1685, 1698
hotspots, xiii, xvi, xliv, xlv, 221, 222, 235, 237, 266, 280, 305, 547, 679, 1652, 1671, 1672, 1673, 1674, 1675, 1676, 1677, 1678, 1679, 1680, 1681, 1683, 1684, 1685, 1690, 1692, 1699, 1700, 1701, 1702, 1703, 1704, 1708, 1719
House, 578, 615, 1350, 1367, 1434, 1436, 1720
household, 198, 399
household income, 860, 865
households, 189
housing, 369, 552, 1301, 1330
Housing and Urban Development, 238
HPLC, 50, 155, 157, 161, 171
HRV, 245
human activity, 390, 402, 423, 441, 638, 759, 815, 882, 1009, 1365
human development, xxxv, 420, 448, 1185, 1197
Human Development Index, 701
human dimensions, 530, 601
human health, xxi, xxx, 391, 413, 419, 426, 445, 446, 452, 457, 871, 1186, 1202, 1350, 1620
human research, 730
humanity, xx, 402, 413, 414, 605, 635
humans, 196, 414, 426, 441, 443, 446, 448, 454, 569, 574, 577, 596, 604, 605, 1186, 1189, 1328, 1335, 1351, 1466
humic acid, 1312
humic substances, 474
humidity, xxxiii, 40, 158, 452, 555, 1099, 1308, 1415, 1633, 1677
humus, 447, 474, 475, 482, 1163
Hungary, 447
Hunter, 1606
hunting, xxiv, 40, 522, 524, 594, 596, 597, 604, 612, 1648, 1649, 1711
hurricanes, xxv, 637, 638, 661
husbandry, 244, 252, 735, 1672
hybrid, 29, 141, 211, 406, 1264, 1293, 1464
hybridization, xxx, 141, 758, 871, 881, 1107, 1116, 1120
hydrate, 490
hydro, 40, 101, 104, 485, 487, 492, 493, 505, 979, 1093, 1101, 1102, 1104, 1114, 1117, 1119, 1183, 1205, 1216, 1217, 1436
hydrocarbon, 1101, 1118, 1120, 1205
hydrocarbons, 725, 729, 868, 1093, 1101, 1102, 1104, 1114, 1117, 1119, 1205, 1216, 1217
hydrochloric acid, 490, 493
hydrocortisone, 854

hydrodynamic, 926, 928, 931, 932, 946, 1100, 1223, 1393, 1474, 1479, 1528, 1535, 1537, 1540, 1576
hydrodynamics, 926, 928, 931, 937, 989, 1031, 1138, 1144, 1229, 1471, 1478, 1497, 1515, 1537, 1543, 1565
hydrogels, 504, 505, 507, 511
hydrogen, 120, 469, 474, 489, 500, 505, 506, 507, 727, 973, 1213, 1593, 1598
hydrogen bonds, 469, 506
hydrogen peroxide, 500, 1593, 1598
hydrogen sulfide, 489, 973, 1213
hydrogeology, 438
hydrologic, 437, 438, 440, 1189, 1198, 1316, 1318, 1321, 1353, 1434, 1435
hydrological, xiii, 2, 4, 10, 13, 22, 422, 425, 435, 457, 460, 964, 980, 987, 994, 1003, 1012, 1153, 1337, 1350, 1351, 1354, 1356, 1358, 1359, 1381, 1435
hydrological conditions, 964
hydrological cycle, 1435
hydrology, xiii, xxxix, 1, 2, 435, 451, 454, 572, 577, 926, 977, 985, 988, 1124, 1140, 1197, 1345, 1353, 1407, 1422, 1432, 1435, 1546
hydrolysates, xxii, 483, 486, 487, 490, 492, 494, 496, 498, 499, 501, 502, 503, 504, 505, 507, 510, 511
hydrolysis, xxii, 469, 483, 485, 486, 488, 494, 497, 498, 499, 500, 501, 507, 1258, 1376, 1591, 1602, 1603, 1610, 1619, 1620
hydrolytic stability, 493
hydrolyzed, 446, 469, 496, 497, 498
hydrophilic, 40, 101, 104, 487, 492, 493, 505
hydrophility, xxii, 483
hydrophobic, 474
hydrophobic interactions, 474
hydrophobicity, 474, 1341
hydrosphere, 441, 462
hydrostatic pressure, 1480
hydroxide, 476, 489, 490, 1259
hydroxides, 443
hydroxyapatite, xxix, 820
hydroxyl, 49, 54, 57, 89, 108, 118, 155, 469, 485, 504
hydroxyl groups, 469, 504
hydroxylation, 54, 57, 69, 93, 111, 120, 485
hydroxyproline, 496, 497
hygiene, 725, 728, 737, 738, 973
hyperbolic, 227, 229, 231, 232
hyperplasia, 846, 1602
hypothesis, xiv, xxvi, xxxii, 27, 28, 35, 155, 156, 203, 269, 315, 317, 321, 322, 327, 335, 539, 540, 541, 542, 547, 639, 642, 644, 645, 683, 688, 825, 832, 986, 994, 1029, 1224, 1652, 1659, 1679
hypothesis test, xxvi, 683, 688
hypoxia, 783, 802, 1179, 1331
Hypoxia, 1184
hypoxic, xxxv, 941, 944, 973, 1169, 1179, 1180
hysteresis, 661, 928

I

IAP, 1373, 1378, 1379, 1385, 1388, 1389, 1390, 1392, 1393, 1394, 1395, 1397, 1398, 1400, 1401, 1402, 1403
ice, 4, 432, 487, 551, 556, 586, 725, 747, 749, 752, 758, 851, 912, 913, 914, 915, 917, 920, 921, 922, 1037, 1124, 1207, 1308, 1411, 1527, 1717
Iceland, 727
id, 592, 1343, 1559, 1569
Idaho, 1316, 1322
ideal, 568, 673, 784, 969, 1163, 1278, 1297, 1494, 1633, 1676
identity, xxvi, xxviii, 47, 60, 313, 316, 320, 323, 326, 343, 348, 349, 352, 354, 357, 358, 595, 630, 683, 799, 805, 1114, 1589, 1631, 1632
IEA, 468, 480
ileum, 1292
Illinois, 1322, 1359
illumination, 1639
image, xvii, 200, 201, 218, 244, 245, 248, 250, 251, 252, 259, 260, 261, 262, 364, 794, 1100, 1114, 1560
image analysis, 218, 245, 794
imagery, 190, 193, 195, 263, 925
images, xvi, 177, 200, 201, 247, 250, 260, 262, 263, 433, 916, 1029, 1560
imaging, 260, 1468
imbalances, xxi, 413, 1274, 1283
immersion, 950, 1129
immigration, 407, 571, 964
immobilization, 319, 1103, 1106, 1269
immune response, 842, 853, 854, 1285, 1295
immune system, 573, 840
immunity, 488, 1595, 1607
immunocompetence, 578
impact assessment, 450, 618, 723, 736, 881, 972
implementation, xv, 177, 178, 180, 181, 184, 187, 190, 191, 196, 197, 199, 201, 204, 207, 208, 209, 210, 214, 215, 276, 398, 441, 526, 527, 542, 564, 986, 1101, 1115
import substitution, 880
imports, 611, 878, 880
impotence, 610
improvements, xxvii, xxix, xli, 689, 697, 698, 734, 753, 819, 821, 1193, 1255, 1337, 1370, 1550

in situ, xxiii, xxxviii, 50, 199, 254, 387, 511, 550, 555, 559, 560, 561, 997, 1001, 1104, 1115, 1116, 1335, 1349, 1354, 1356, 1358, 1363, 1368, 1439, 1522, 1530, 1531, 1542, 1547, 1562
in situ hybridization, 1107, 1116
in transition, 726, 727
in vitro, 155, 508, 559, 563, 836, 1275, 1294, 1595, 1607
in vivo, 56, 836
INA, 1502
inactivation, 458, 496
inactive, 121
inbreeding, 554, 570, 912
incentive, 397
incentives, 395, 420
incidence, 339, 554, 733, 881, 1388, 1477, 1478, 1479, 1487, 1491, 1522, 1524, 1535, 1537
Incidents, 1176
incineration, 394
inclusion, xxxvi, 29, 364, 612, 686, 822, 835, 844, 955, 1192, 1237, 1239, 1261, 1264, 1266, 1269, 1270, 1271, 1274, 1282, 1284, 1289, 1290, 1294, 1710
income, xxx, 407, 420, 450, 552, 560, 698, 719, 720, 855, 857, 859, 860, 861, 862, 863, 864, 865, 869, 872
incomplete combustion, 1102
incubation, xiv, 27, 29, 32, 33, 34, 35, 473, 475, 1111, 1357
incubation period, xiv, xxxi, 27, 887, 897, 1111, 1357
incubation time, 1111
independence, 5, 733
independent variable, 364, 367, 1356
indeterminism, 409
India, ix, xxxi, xxxiv, 140, 167, 171, 391, 392, 432, 455, 609, 610, 613, 618, 739, 888, 893, 923, 924, 1147, 1148, 1153, 1157, 1158, 1163, 1165, 1166, 1167, 1168, 1630, 1723
Indian, xxiv, xxxiv, 171, 391, 585, 594, 924, 985, 1148, 1166, 1167, 1536
Indian Ocean, 924, 985
Indians, 515
indication, xxxii, 91, 202, 204, 207, 262, 439, 453, 924, 931, 944, 1073, 1090, 1091, 1388, 1479, 1500, 1504, 1506
indicators, xvi, xxxviii, 28, 177, 178, 179, 182, 183, 185, 187, 189, 193, 194, 195, 197, 199, 201, 205, 206, 207, 209, 210, 214, 216, 255, 257, 260, 261, 305, 397, 399, 407, 418, 433, 442, 524, 530, 577, 591, 976, 994, 997, 1002, 1030, 1031, 1032, 1034, 1038, 1165, 1209, 1315,

1367, 1372, 1395, 1439, 1457, 1551, 1553, 1562, 1663, 1664, 1689
indices, xvii, 17, 19, 244, 260, 261, 262, 364, 384, 1207, 1208, 1209, 1211, 1228, 1672, 1690
indigenous, xv, 177, 178, 179, 180, 181, 183, 184, 185, 186, 187, 189, 190, 191, 193, 197, 210, 212, 213, 215, 218, 312, 353, 371, 377, 432, 612, 1593
Indigenous, 177, 181, 190, 211, 213, 214, 215, 216, 217, 219, 613, 615
indigenous knowledge, xv, 177, 178, 179, 180, 181, 183, 184, 185, 186, 187, 189, 190, 191, 193, 197, 210, 212, 213, 612, 737
indirect effect, 326, 419, 730
Indochina, 617
Indonesia, 211, 305, 391, 392, 888, 909, 1536, 1541, 1630
induction, 69, 167, 558, 563, 843, 845, 846, 848, 1091
industrial application, 496, 497, 502, 508
industrial emissions, 411, 426, 443, 721
industrial processing, xxii, 483, 485, 486, 487, 488, 489, 494, 499, 500, 507
industrial production, 489
industrial sectors, xxxix, 1407
industrial wastes, 1202
industrialisation, 392, 396, 417, 418, 716, 720
industrialization, 582, 759, 1101
industrialized countries, 454, 604
industries, 426, 502, 503, 504, 514, 597, 610, 717, 718, 744, 826, 1073, 1090, 1214, 1229, 1238, 1259, 1373
industry, xxviii, xxx, 284, 405, 420, 428, 456, 486, 487, 492, 502, 504, 506, 510, 582, 584, 585, 596, 597, 610, 706, 709, 721, 722, 723, 724, 725, 727, 728, 730, 743, 751, 819, 820, 821, 856, 857, 865, 871, 872, 873, 874, 876, 877, 878, 880, 882, 884, 912, 913, 972, 1102, 1118, 1187, 1260, 1277, 1300, 1301, 1318, 1331, 1373, 1428, 1438, 1515, 1519, 1590, 1596
inert, 414, 454, 496, 1339, 1374, 1464
inertia, xxvi, 684, 690, 1393, 1664
infancy, 559, 1713
infection, 163, 168, 326, 343, 609, 852, 919, 1164, 1248, 1630
inferences, xxxii, 191, 277, 443, 993, 996, 999, 1026, 1634, 1680
infertile, 324, 327
infestations, 841
Infiltration, 436, 1333, 1337, 1439
infinite, 402, 404
inflammation, 841, 850, 853, 1602
inflammatory, 609

inflammatory cells, xxix, 840, 841, 844, 846, 849, 850
inflammatory responses, 850
inflation, 701, 864
information exchange, 731
information processing, 404
information retrieval, 1589
information sharing, 454
Information System, 200
information systems, 448
information technology, xvii, 23, 243
infrared, 251, 254, 261, 455
infrared spectroscopy, 455
infrastructure, xli, 417, 419, 432, 679, 717, 722, 723, 724, 729, 735, 880, 1202, 1512, 1685
ingest, 827, 902, 1602
ingestion, 426, 446, 456, 829, 956, 1256, 1258, 1291, 1292
ingredients, xxix, xxxvi, 610, 611, 612, 715, 819, 820, 821, 822, 823, 826, 829, 835, 836, 837, 838, 843, 1237, 1238, 1239, 1244, 1245, 1251, 1261, 1264, 1271, 1274, 1285, 1286, 1288, 1290, 1291
inhalation, 426, 446, 456
inheritance, 238, 554
inherited, 554
inhibition, 765, 782, 1090, 1091, 1108, 1357, 1595, 1596, 1597
inhibitor, 1242, 1275, 1279
inhibitors, 1072, 1078
inhospitable, 621
initiation, 779, 1258, 1308, 1521
injection, 464, 1207
injections, 1570
innate immunity, 1607
Innovation, 1433
inoculation, 138, 155, 318, 325, 347, 348, 353, 358, 1106
inoculum, 326, 1104, 1105, 1109, 1112, 1113
inorganic, xiii, xxxv, 1, 2, 4, 314, 326, 328, 329, 335, 341, 426, 452, 463, 466, 493, 571, 937, 942, 946, 950, 973, 1136, 1169, 1182, 1320, 1332, 1396
inorganic salts, 571
inositol, 836, 1601, 1602, 1609
insecticide, xxxi, 887, 889, 906, 1102
insecticides, 492, 1074, 1076
insects, xxii, xxiv, 284, 306, 334, 513, 515, 530, 594, 700, 813, 815, 1278, 1658, 1666, 1668, 1677
insertion, 1615
insight, 181, 441, 620, 685, 691, 810, 811, 813, 859, 931, 1477, 1499, 1718
instability, 424, 494, 499

institutional change, 733
institutions, 181, 209, 399, 404, 406, 523, 544, 584, 597, 722, 1726, 1733
instruments, 458, 526, 1308, 1732
insulation, 394, 499, 511
integration, xxx, xxxvi, 178, 180, 183, 185, 187, 190, 191, 199, 210, 360, 404, 409, 530, 777, 855, 866, 882, 1023, 1104, 1208, 1220, 1299
integrity, xxxii, xxxv, 450, 454, 572, 577, 690, 697, 743, 751, 754, 884, 904, 913, 917, 924, 967, 1185, 1188, 1345, 1350, 1595
intensity, xiii, 2, 5, 13, 14, 17, 19, 22, 23, 396, 407, 454, 478, 525, 540, 610, 630, 633, 1206, 1320, 1328, 1329, 1340, 1374, 1444
interaction, 181, 187, 189, 192, 245, 251, 313, 315, 316, 317, 322, 331, 333, 349, 484, 621, 632, 634, 930, 933, 1124, 1193, 1224, 1226, 1337, 1376, 1410, 1412, 1479, 1550
interaction effect, 849
interdisciplinary, 184, 186, 187, 197, 405, 409, 454, 526, 529
interest groups, 406
interest rates, 641, 672
interface, 316, 399, 414, 489, 565, 567, 940, 943, 1104, 1346, 1369, 1419, 1619
interference, xxxvi, 311, 557, 851, 902, 969, 1237, 1458
Intergovernmental Panel on Climate Change, 394, 399, 589, 592
Intergovernmental Panel on Climate Change (IPCC), 394
intermediaries, 877
intermolecular, 476, 485
intermolecular interactions, 476
International Energy Agency, 480
international relations, 732
International Rescue Committee, 736
International Rescue Committee (IRC), 736
international standards, 710
international trade, 872, 1732
internet, 1117
interpretation, 20, 408, 584, 1203, 1207, 1208, 1220, 1223, 1228, 1491, 1553, 1566
interrelations, xxviii, 799, 805, 806, 808, 1286
interrelationships, 1327
interrogations, 404
interstitial, 529, 1358
interval, 261, 287, 441, 503, 505, 506, 1021, 1022, 1308, 1320, 1423, 1558, 1559
intervention, xxv, xxx, xli, 276, 417, 426, 427, 435, 442, 453, 527, 625, 637, 647, 855, 864, 867, 1550

interviews, xvi, xxiv, 177, 188, 189, 198, 406, 581, 586
intestinal tract, 823, 1250
intestine, xxix, 820, 822, 829, 1275
intima, 485
intrinsic, xv, 43, 44, 45, 47, 51, 420, 436, 437, 439, 471, 1118, 1535
introns, 1632
intrusions, 932
invasions, 339, 342, 578, 620, 757, 759, 761, 762, 764, 765, 766, 767, 771, 788, 881, 885, 975, 1652
invasive, 182, 281, 313, 320, 321, 324, 331, 332, 339, 351, 352, 355, 358, 532, 547, 596, 959, 976, 1570
invasive species, 182, 281, 320, 351, 532, 547, 596
invertebrates, xv, 37, 38, 40, 514, 520, 543, 552, 573, 600, 691, 762, 813, 817, 897, 954, 955, 957, 959, 966, 970, 983, 989, 1030, 1094, 1142, 1366, 1646, 1656, 1658, 1665, 1667, 1668
Investigations, 458, 558, 563
investment, 222, 235, 276, 315, 489, 641, 668, 698, 709, 713, 728, 734, 861, 864, 1462, 1525, 1690
investment capital, 861
investments, 712, 803, 805, 808, 1458, 1690
investors, 668
IOC, 1097
ionic, 488
ionic strength, 488
ionization, 1108, 1205
ions, 354, 493, 1304, 1332, 1396, 1595
Iowa, 338, 619, 694, 1144, 1404, 1668
IP, 561, 1346
IPCC, 390, 391, 394, 397, 398, 399, 402, 411, 464, 480, 589, 590, 592, 1566, 1568, 1569, 1572
IPPC, 397
IR, 342, 357, 358
Iran, 392, 1732
IRC, 405, 736
Ireland, 792, 1035, 1343, 1457, 1458, 1623, 1624, 1639, 1640
iron, 699, 780, 1096, 1213, 1248, 1249, 1250, 1251, 1285, 1288, 1295, 1335, 1615, 1617, 1627
irradiation, 1211
irrigation, xxxix, 432, 439, 449, 456, 611, 702, 706, 708, 709, 710, 711, 712, 713, 714, 715, 731, 734, 735, 906, 928, 968, 1326, 1327, 1329, 1331, 1335, 1377, 1407, 1409, 1412, 1415, 1419, 1420, 1421, 1422, 1423, 1426, 1428, 1430, 1431, 1432, 1433, 1436, 1438
IS, 1340
Islam, 605, 1273
Islamabad, 43
Islamic, 609
island, xli, 339, 539, 570, 595, 596, 1145, 1457, 1470, 1472, 1473, 1478, 1549, 1560, 1561, 1562, 1563, 1569
island formation, 339
islands, xxiv, xxxv, xl, 328, 334, 442, 552, 569, 593, 594, 596, 597, 598, 611, 814, 857, 888, 987, 1009, 1202, 1455, 1476, 1571, 1650, 1651, 1676, 1679
ISO, 397, 592
isoelectric point, 493
isoenzymes, 554, 556
isoflavone, 146, 153, 163, 168, 170, 172
isoflavones, 136, 139, 141, 143, 149, 154, 168, 170, 171
isoflavonoid, 136, 138, 143, 144, 145, 149, 150, 165
isoflavonoids, 54, 56, 59, 136, 138, 143, 145, 149, 151, 154, 167, 174
isolated islands, 569
isolation, xxii, xxiv, xxvi, 68, 170, 376, 510, 514, 557, 565, 566, 568, 569, 571, 577, 593, 621, 622, 624, 625, 626, 627, 628, 629, 630, 634, 683, 881, 939, 967, 1381, 1589, 1607, 1608, 1642, 1685
isoleucine, 1280, 1286, 1592
isomerization, 54, 1583
isopods, 951, 957, 959, 1229
isorhamnetin, 76, 100, 101, 142, 150, 151, 152, 153, 156, 158, 159
Isorhamnetin, 168
isotope, xiv, 27, 29, 270, 327, 981, 1119, 1139, 1168, 1366, 1367, 1369, 1612
Isotope, 1552
isotopes, 28, 332, 355, 1145, 1409, 1552
Italy, x, xiv, xli, 13, 22, 23, 24, 95, 104, 164, 221, 392, 413, 415, 416, 417, 421, 424, 425, 428, 429, 430, 433, 434, 435, 439, 444, 445, 446, 447, 458, 460, 461, 481, 616, 617, 692, 1096, 1097, 1165, 1288, 1296, 1438, 1549, 1571, 1579, 1580, 1588, 1589, 1590, 1592, 1607, 1611, 1620, 1624, 1702
iteration, 622, 623, 626, 632
IUCN, xxv, 222, 226, 233, 236, 603, 604, 605, 606, 607, 608, 609, 610, 611, 613, 615, 617

J

January, 14, 446, 615, 1304, 1305, 1423, 1428, 1429, 1430
Japanese, 167, 262, 263, 284, 304, 305, 306, 307, 596, 1125, 1433, 1435, 1436
Java, 1035, 1541
joints, 486, 609
Jordan, 114, 129, 160, 615, 1342

JT, 399, 618
Judaism, 605
judge, 259, 1531
judgment, 1690
Jun, 1172
jurisdiction, 865
justification, 866
juveniles, 853, 963, 964, 969, 1256, 1258, 1259, 1262, 1268, 1272, 1292

K

kaempferol, 67, 69, 76, 77, 78, 84, 87, 88, 93, 100, 101, 102, 104, 121, 130, 141, 142, 146, 149, 150, 151, 152, 153, 156, 157, 158, 159, 162, 169
kaolinite, 476
Kazakhstan, 611
kelp, 1465
Kentucky, 1321
Kenya, xvi, 177, 192, 195, 196, 198, 201, 202, 203, 205, 206, 207, 208, 211, 212, 213, 214, 215, 216, 217, 219, 233, 236, 237, 479, 699, 704, 708, 731, 1734
keratin, xxii, 483, 487, 489, 490, 492, 498, 509
kerosene, 716
kidney, 827, 832, 842, 852
kidneys, 505, 821
kill, 615, 970, 974, 1333, 1592, 1731
killing, 276, 1333, 1456
kinetics, 492, 503, 506, 511, 795, 843, 851, 916, 917, 921, 1143, 1620
King, 36, 162, 165, 454, 459, 508, 621, 636
Kolmogorov, 1081
Korea, 392, 611, 617
Korean, 611
Kyoto protocol, 476
Kyoto Protocol, 395

L

labeling, 352, 509, 1619
labor, 213, 353, 596, 1439, 1449
laboratory method, 200
laboratory studies, 1476, 1577
laboratory tests, 1515
labour, 1465, 1569
lack of control, 866
lactate dehydrogenase, 921
lactic acid, xlii, 490, 1579, 1580, 1583, 1594, 1598, 1606, 1607, 1608, 1609, 1610, 1611, 1612, 1613, 1614, 1615, 1616, 1617, 1618, 1619, 1620, 1621
lactones, 164, 169, 351

Lafayette, 11, 1325
lakes, xxv, 408, 523, 568, 574, 576, 637, 670, 685, 692, 693, 738, 783, 789, 791, 796, 803, 817, 882, 895, 900, 901, 906, 907, 908, 909, 910, 966, 969, 980, 983, 990, 1168, 1322, 1323, 1326, 1332, 1337, 1435, 1667
lamina, 424
laminar, 424
laminated, 1166
land abandonment, 421
Land Use Policy, 768
landfill, 394, 395, 397, 453, 1374, 1377, 1562, 1569
landfill gas, 394
landfills, 435, 451, 504, 1215, 1229, 1569
landings, 856
landscapes, xix, 10, 41, 183, 186, 189, 190, 192, 194, 196, 199, 204, 206, 210, 216, 304, 306, 359, 366, 370, 371, 372, 374, 404, 454, 548, 595, 598, 613, 620, 622, 624, 625, 626, 627, 628, 629, 630, 636, 694, 725, 739, 760, 768, 788, 801, 812, 813, 815, 817, 1193, 1335, 1337, 1350, 1365, 1368, 1702, 1717
land-use, xiii, xiv, xxxvii, xliv, 13, 14, 217, 218, 223, 926, 928, 1325, 1683, 1684, 1685, 1686, 1687, 1689, 1692, 1698, 1699, 1700, 1701, 1702, 1703, 1704
language, 404
languages, 730, 731
Laos, 612
large-scale, 559, 1434, 1435, 1524, 1551, 1569, 1570
larvae, xv, 37, 40, 302, 307, 803, 946, 952, 955, 957, 963, 964, 983, 990, 1207, 1220, 1240, 1256, 1258, 1264, 1287, 1291, 1292, 1658, 1661, 1668
larval, 284, 951, 955, 957, 959, 961, 962, 963, 980, 985, 988, 1219
larval development, 1219
larval stages, 951
laser, 1116
Last Glacial Maximum, 556
Late Pleistocene, 563, 1009, 1157
Late Quaternary, 277, 280, 560, 563, 1036, 1165, 1166, 1167
lateral roots, 783
latex, 502
latexes, 502
Latin America, xxxiii, 1099, 1120
lattice, 1470
law, xxii, 276, 406, 513, 523, 526, 543, 1487, 1489
law enforcement, 698, 737, 1700, 1701, 1702
laws, 577, 598, 815, 816, 880, 1644
LC, 484, 485, 606, 607, 608, 609, 1547
LCA, xxiii, 581, 582, 584, 585, 586, 591, 592

LCS, 1478
LDCs, 613
leachate, 10, 1330, 1374, 1377, 1383
leaching, 2, 9, 416, 432, 446, 451, 454, 464, 968, 1258, 1270, 1280, 1312, 1329, 1331, 1332, 1333
leadership, 526, 597, 598, 599, 864
leakage, 465
learning, xliv, 188, 405, 409, 1513, 1683
leather, 40, 426, 446, 458, 485, 486, 487, 489, 493, 499, 500, 501, 504, 508, 509, 510, 511
leather processing, 509
Lebanon, 128
legal issues, 398
legal protection, 274, 371
legend, 296, 299, 367, 1643
legislation, xiii, 13, 180, 423, 687, 737, 906, 975
legislative, 1339
legume, 317, 327, 339, 349, 823, 1279, 1291, 1331
legumes, 56, 272, 317, 327, 330, 337, 349
Legumes, 327, 337
Leguminosae, 165, 171, 268, 294, 525, 1160
Lepidoptera, 305, 306
Leptocylindrus, 1055
lesions, 841, 845, 846, 850, 851
lettuce, 332
leucine, 496, 1286, 1599
Levant, 614
levees, 750
liberalisation, 711
lice, 851
lichen, 357, 1634, 1635, 1636, 1638, 1640
LIFE, 814
life cycle, 581, 582, 584, 585, 588, 591, 723, 749, 750, 772, 954, 957, 1588, 1676
Life Cycle Assessment, xxiii, 581, 592
life expectancy, 643
life forms, 182, 189, 995
life quality, 450
life sciences, 1679
life span, 1316, 1464
life style, 406
lifecycle, 390, 394, 395
life-cycle, 950
lifestyles, 180, 1456
lifetime, 181, 626, 1191, 1402, 1462
ligand, 476
light, xxxii, xxxiii, 95, 187, 247, 251, 256, 273, 300, 314, 346, 348, 356, 492, 543, 669, 716, 746, 754, 758, 759, 761, 770, 777, 783, 795, 844, 891, 899, 924, 940, 947, 949, 950, 985, 994, 1001, 1014, 1029, 1039, 1071, 1073, 1078, 1080, 1087, 1089, 1091, 1092, 1105, 1107, 1138, 1139, 1148, 1164, 1174, 1271, 1338, 1444, 1665, 1685
light conditions, 947
lignin, 469, 470, 471, 472, 473, 474, 475, 783, 1158
likelihood, 10, 201, 328, 391, 1107
limestone, 447, 463, 464, 703, 1440
limitation, 218, 408, 506, 985, 1026, 1178, 1182, 1189, 1357
limitations, xxxii, 190, 325, 427, 687, 993, 995, 996, 1002, 1027, 1159, 1208, 1318, 1368, 1403, 1465, 1589, 1603, 1633
lindane, 1074, 1076, 1084, 1094
linear, 14, 30, 184, 200, 206, 469, 1131, 1139, 1178, 1357, 1485, 1486, 1487, 1497, 1500, 1505, 1506, 1566
linear function, 676
linear model, 1131, 1139
linear regression, 30, 1131
linguistic, 515
linkage, xx, xxxviii, 121, 360, 378, 1349, 1363, 1701
links, xiv, xix, xx, 14, 171, 179, 187, 338, 343, 345, 349, 352, 354, 358, 389, 508, 1193
lipid metabolism, 1610
lipid peroxidation, 842, 1093, 1094
lipids, 469, 471, 495, 498, 1076, 1239, 1240, 1580, 1581
lipoid, 488, 495, 497, 498
lipophilic, 95, 175
lipoproteins, 842
liposomes, 508
liquid chromatography, 165, 168, 916
liquid fuels, 467
liquid phase, 1080, 1357
liquids, 489
liquor, 490, 491, 509
Listeria monocytogenes, 1596, 1612
literacy, 735
lithosphere, 441, 462, 464
liver, 484, 830, 842, 1276, 1592, 1720
liver enzymes, 1276
livestock, xv, xvi, xvii, xxiv, 37, 40, 177, 178, 181, 182, 183, 184, 186, 188, 189, 190, 191, 192, 193, 194, 195, 196, 197, 198, 200, 202, 203, 204, 205, 206, 207, 209, 210, 212, 218, 234, 243, 251, 252, 253, 263, 284, 433, 436, 437, 438, 439, 440, 441, 449, 450, 452, 458, 594, 596, 697, 699, 700, 701, 734, 737, 738, 840, 1238, 1316, 1327, 1329, 1330, 1331, 1332, 1341, 1343, 1730
Livestock, xvii, 196, 202, 205, 213, 217, 244, 246, 251, 252, 256, 261, 440
living conditions, 44
living environment, 997

LMW, 318
load model, 1544
loading, 29, 940, 954, 967, 973
local action, xvi, 178, 181, 209, 210
local community, xvi, 177, 178, 180, 181, 184, 188, 196, 199, 209, 320
local conditions, 971, 1188, 1197, 1664
local government, 276, 433
location, 4, 5, 38, 158, 159, 198, 203, 246, 251, 361, 378, 414, 459, 630, 632, 935, 1075, 1100, 1128, 1135, 1137, 1138, 1151, 1178, 1301, 1450, 1455, 1464, 1487, 1497, 1546, 1554, 1562, 1567
loci, 378, 555, 561
locus, 378, 527, 554, 1604, 1606, 1640
logging, 305, 522, 523, 524, 525, 747, 750, 782, 1162, 1648, 1649, 1686
logistics, 394
London, 161, 165, 166, 172, 174, 175, 211, 212, 213, 214, 216, 217, 218, 236, 277, 306, 330, 337, 338, 341, 355, 372, 373, 385, 399, 410, 411, 508, 509, 510, 548, 613, 614, 983, 988, 990, 1030, 1032, 1038, 1121, 1165, 1168, 1231, 1345, 1368, 1404, 1543, 1572
long distance, 555, 631
long period, 182, 184, 186, 456, 516, 925, 1374
long-term, 398, 448, 468, 473, 502, 514, 543, 598, 599, 612, 1332, 1333, 1336, 1337, 1425, 1449, 1450, 1462, 1474, 1555, 1576, 1577
Los Angeles, 1468, 1469
loss of appetite, 843
losses, xxi, xxxvii, 3, 202, 205, 206, 207, 356, 413, 416, 421, 424, 425, 428, 430, 433, 455, 460, 462, 477, 497, 555, 557, 572, 577, 932, 953, 968, 1178, 1213, 1300, 1307, 1310, 1311, 1319, 1322, 1326, 1327, 1330, 1332, 1333, 1334, 1335, 1342, 1343, 1344, 1345, 쨈1346
Louisiana, 919, 1321
love, 1726
low molecular weight, 474
low power, 1403
low temperatures, xiv, 27, 35, 287, 567, 725, 779, 922, 1136
low-density, 1124
lower prices, 1274
low-intensity, 1329
LTD, 254, 1192
luminescence, 1073, 1078, 1087, 1090
lungs, 484
Luo, 1433, 1513, 1517, 1518, 1519, 1521, 1536, 1546, 1629, 1632, 1638
luvisols, 431
Luxembourg, 460, 1451

LV, 563, 614
lying, xxxiv, xli, xlv, 1138, 1147, 1151, 1158, 1164, 1458, 1550, 1570, 1723, 1728, 1729
lymphocytes, 843
lysimeter, 1428, 1434
lysine, 485, 493, 1239, 1242, 1243, 1244, 1245, 1248, 1249, 1251, 1253, 1258, 1265, 1266, 1267, 1268, 1271, 1274, 1275, 1276, 1279, 1282, 1283, 1284, 1297, 1580
lysozyme, 850

M

M.O., 479
mace, 1379
machine learning, xliv, 1683
machinery, xiv, 13, 319, 430, 448, 452, 711, 718, 1449, 1460
machines, 440
macroalgae, 949, 950, 1014, 1016, 1017, 1631, 1637
macrobenthic, xxxvi, 955, 957, 958, 959, 969, 986, 988, 1039, 1178, 1179, 1202, 1203, 1228
macrobenthos, 959, 988, 1183, 1226
macromolecular chemistry, 510
macromolecules, 475, 1275
macronutrients, 790, 943, 1245, 1580
macrophages, 841, 842, 852
macropores, 2
mad cow disease, 486
Madison, 1346
magazines, 880
magnesium, 464, 500, 1291, 1601, 1614
magnetic, 1077, 1119
magnetic resonance, 916
magnitude, xiii, 13, 183, 192, 322, 398, 424, 464, 621, 625, 627, 631, 632, 633, 634, 668, 669, 748, 780, 801, 803, 930, 931, 939, 940, 947, 1027, 1309, 1312, 1317, 1318, 1351, 1361, 1363, 1365, 1386, 1395, 1401, 1477, 1489, 1513, 1526, 1534, 1535, 1542, 1586, 1592
maintenance, xix, 310, 320, 322, 323, 324, 329, 330, 335, 341, 369, 376, 384, 433, 456, 523, 526, 527, 558, 560, 563, 612, 636, 951, 1090, 1109, 1131, 1351, 1392
Maintenance, 270, 319, 337
maize, xxxix, 274, 275, 329, 332, 467, 471, 472, 473, 475, 479, 1407, 1409, 1412, 1413, 1419, 1422, 1423, 1425, 1426, 1427, 1428, 1431, 1432, 1434
major cities, 432
majority, xviii, xxix, xxxi, 2, 60, 108, 138, 207, 222, 266, 391, 423, 489, 492, 494, 523, 687, 702, 752, 819, 832, 833, 876, 923, 924, 926, 961, 966, 967, 1209, 1222, 1223, 1309, 1389, 1401,

1404, 1477, 1534, 1580, 1582, 1587, 1625, 1632, 1662, 1715, 1726, 1730
malaria, 701, 733
Malaysia, 617, 1294
males, 554, 573, 578, 610, 623
maltose, 1587, 1591, 1592, 1593, 1609
mammal, xxiv, xxv, xliv, 362, 367, 371, 372, 524, 593, 619, 1683, 1685, 1695, 1696, 1700, 1701, 1730
Mammalian, 373, 374
mammalian cells, 917
mammals, xxiv, xliv, 368, 370, 371, 372, 373, 374, 514, 520, 524, 545, 594, 604, 614, 841, 851, 1683, 1685, 1690, 1691, 1692, 1696, 1698, 1699, 1700, 1701, 1730
man, xix, xli, 196, 359, 360, 368, 369, 532, 637, 643, 644, 671, 720, 909, 930, 1006, 1549, 1550, 1567, 1636, 1642, 1644, 1646, 1729
management practices, xviii, xxxvii, 2, 10, 40, 179, 180, 201, 208, 272, 276, 283, 407, 448, 454, 460, 476, 477, 478, 479, 1189, 1319, 1321, 1322, 1326, 1327, 1328, 1331, 1332, 1337, 1341, 1507
mandates, 455
manganese, 780
mango, 1344
mangroves, xxxii, xxxiii, xxxv, 515, 924, 948, 977, 988, 989, 1100, 1165, 1202, 1213, 1216, 1218
manipulation, 620, 647, 742, 957
man-made, 196, 930, 1006, 1567
manpower, 1246
manufacture, 394, 485, 486, 487, 488, 489, 493, 498, 501, 502, 504, 507, 509, 586, 591, 1580
manufacturing, 398, 489, 508, 509, 584, 586, 721, 724, 725, 1596
manure, 455, 456, 725, 1300, 1301, 1316, 1320, 1323, 1326, 1327, 1328, 1329, 1330, 1331, 1332, 1339, 1341, 1342, 1374, 1377, 1388
mapping, 23, 213, 216, 217, 434, 435, 439, 451, 455, 458, 459, 1186, 1690
mares, 1678
marginal costs, 861
marine environment, xxxiv, 957, 958, 1002, 1074, 1090, 1093, 1147, 1148, 1182, 1183, 1464, 1550
marine fish, xxxii, 586, 825, 860, 861, 865, 924, 962, 963, 964, 969, 1289
marine species, 860, 861, 884, 885, 886, 911, 912, 951, 961, 963, 964, 965, 975, 1011, 1468
Marine Stewardship Council (MSC), 584
maritime, 1001, 1031
markers, xx, xxxiii, 48, 69, 90, 93, 112, 113, 115, 128, 164, 167, 168, 174, 375, 378, 380, 553, 554, 561, 562, 1071, 1073, 1092, 1097, 1562, 1631, 1632
market, xx, 389, 397, 398, 419, 449, 453, 457, 526, 586, 612, 1186, 1689, 1701
market economy, 710
market failure, 866
marketing, 584, 702, 710, 712, 737, 861, 865, 866, 1245
marketplace, 615
markets, 395, 467, 610, 611, 1186, 1187
marrow, 610
marsh, xxxii, 349, 355, 761, 766, 776, 779, 780, 791, 792, 793, 796, 924, 948, 949, 950, 957, 1003, 1005, 1022, 1031, 1033, 1034, 1038, 1124, 1127, 1128, 1134, 1140, 1142, 1454, 1553, 1561, 1562, 1567
marshes, xxxiii, xli, 949, 998, 1003, 1009, 1023, 1073, 1123, 1124, 1127, 1187, 1549, 1550, 1553, 1562, 1563, 1567
Marx, 1322
Maryland, 884, 1294
MAS, 472
masonry, 1561
mass, 18, 20, 29, 165, 505, 563, 569, 611, 718, 773, 774, 811, 891, 916, 949, 950, 953, 959, 977, 1077, 1205, 1310, 1311, 1355, 1357, 1358, 1373, 1392, 1416, 1439, 1478, 1481, 1521, 1522, 1523, 1581, 1595, 1606, 1646
mass loss, 1310, 1311
mass spectrometry, 165, 168
mass transfer, 1373, 1392
Massachusetts, 337, 1370
mass-transport, 1521, 1522
mast cells, 853
materials, xxii, xxxvii, xl, xlii, 23, 209, 254, 353, 390, 394, 395, 414, 417, 425, 431, 447, 476, 483, 485, 487, 489, 492, 493, 499, 501, 502, 503, 504, 505, 506, 507, 508, 511, 512, 584, 604, 610, 716, 717, 721, 773, 797, 890, 1138, 1191, 1260, 1261, 1284, 1293, 1326, 1333, 1337, 1395, 1455, 1464, 1465, 1507, 1563, 1579, 1580, 1585, 1586, 1592, 1596, 1598, 1600, 1602, 1617
mathematics, 677, 1577
matrix, xix, xx, 273, 306, 359, 360, 369, 370, 374, 418, 420, 434, 435, 446, 531, 533, 599, 621, 622, 623, 624, 625, 626, 627, 628, 629, 630, 634, 674, 675, 805, 841, 842, 850, 1157, 1211, 1327, 1376, 1379, 1381, 1393, 1401, 1562, 1586, 1603
maturation, 510, 558
Maya, 579, 961, 985
MB, 960, 961, 1345

meat, 181, 486, 487, 493, 506, 510, 524, 1239, 1259, 1263, 1296, 1297, 1621
mechanical properties, 492, 498, 504
mechanistic explanations, 811
MED, 1095, 1096
media, 69, 162, 558, 597, 599, 1373, 1374, 1375, 1379, 1392, 1393, 1395, 1463
median, 1516, 1532, 1568, 1695
mediation, 317, 325, 326, 329
medical, 485, 507, 508, 511, 604, 612, 613, 617, 735, 913
medical care, 604, 612
medicinal plants, 524, 560, 1334
medicine, xxv, 487, 488, 489, 504, 603, 604, 605, 606, 609, 610, 612, 613, 614, 615, 616, 617, 728, 735
medicines, xxv, 506, 603, 604, 609, 610, 611, 612, 613, 614, 615, 617
Mediterranean, xiii, xxxix, 13, 14, 20, 41, 113, 118, 127, 138, 217, 279, 421, 424, 431, 447, 456, 785, 984, 1074, 1090, 1093, 1094, 1096, 1097, 1124, 1145, 1183, 1333, 1350, 1351, 1353, 1367, 1368, 1437, 1438, 1439, 1448, 1450, 1566, 1606, 1668, 1708, 1714
Mediterranean climate, 447, 1351, 1353, 1367, 1438, 1439, 1448, 1450
Mediterranean countries, 118
medium composition, 1618
melt, 743, 746
melting, 501, 1012
melting temperature, 501
membership, 856, 859, 860, 865, 1732
membrane permeability, 913, 917
membranes, 484, 485, 489, 820, 842, 917, 921, 922
men, 596
mental activity, 403
mercaptans, 1094
mercury, 1101, 1118, 1129, 1194, 1215
Mercury, 511, 1230, 1232
mesophyll, 251
Mesopotamia, 432, 609, 616
Mesozoic, 1149
meta-analysis, 815, 1036
metabolic, 45, 46, 48, 50, 54, 90, 105, 150, 155, 159, 421, 1090, 1335
Metabolic, 51, 54, 57, 106, 1102
metabolic changes, 1247
metabolic pathways, 159
metabolism, xxxvii, 71, 90, 171, 174, 622, 773, 798, 836, 852, 853, 1091, 1093, 1103, 1109, 1110, 1111, 1112, 1140, 1286, 1349, 1351, 1358, 1359, 1363, 1365, 1366, 1368, 1369, 1370, 1591, 1592, 1593, 1597, 1599, 1600, 1606, 1609, 1610, 1613, 1620
metabolite, xv, 43, 45, 48, 49, 50, 56, 1089
metabolites, xv, 43, 44, 45, 47, 48, 49, 50, 51, 57, 59, 90, 172, 173, 313, 326, 1074, 1076, 1081, 1082, 1091, 1108, 1593, 1599
metabolized, 1108, 1583, 1591
metabolomics, 46, 48
metal content, 441
metal ion, 354
metal ions, 354
Metallothionein, 1078, 1094, 1097
metallothioneins, 1093, 1096
metallurgy, 426, 427
metals, xxxvi, 420, 423, 426, 427, 428, 431, 433, 441, 443, 446, 447, 452, 453, 458, 572, 720, 771, 868, 1073, 1074, 1076, 1080, 1084, 1089, 1095, 1102, 1202, 1205, 1213, 1214, 1215, 1216, 1220, 1327, 1329, 1330, 1340, 1438, 1645
metaphor, 411, 687
meteor, 1035
meteorological, 447, 1124, 1183, 1415, 1425, 1426, 1428, 1440, 1442
meter, 250, 253, 746, 857, 1077, 1106, 1131, 1204, 1245, 1307, 1308, 1355, 1442
methane, 394, 395, 588, 589, 591, 1104, 1105, 1109, 1111, 1115, 1120, 1368
methanogenesis, 1106, 1109, 1111, 1114, 1115
methanol, 913, 914, 920, 1105, 1109, 1110, 1293
methodology, 216, 365, 405, 433, 586, 588, 591, 592, 673, 685, 686, 690, 742, 995, 1076, 1159, 1181, 1308, 1355, 1363, 1476, 1547, 1604, 1658, 1676
methylation, 57, 60, 90, 95, 104, 105, 108, 113
methylene, 140, 503, 1206
methylene blue, 1206
metric, 385, 396, 611, 628, 1301, 1307
metropolitan area, 944, 973
Mexican, xxiii, 565, 566, 567, 568, 573, 574, 576, 577, 578, 1142, 1144
Mexico, viii, ix, xxiii, xxx, xxxiii, xxxv, 2, 213, 237, 360, 372, 373, 392, 565, 566, 567, 568, 573, 574, 575, 576, 578, 579, 608, 855, 856, 857, 860, 864, 868, 869, 871, 873, 874, 876, 877, 878, 879, 881, 882, 883, 884, 885, 888, 1123, 1124, 1125, 1126, 1132, 1142, 1143, 1144, 1145, 1185, 1331, 1519, 1639, 1671, 1675, 1676, 1678, 1680, 1681, 1703
Mexico City, 237
Mg^{2+}, 464, 1014
Miami, 1145, 1233
microaggregates, 476

microalgae, 943, 946, 947, 954, 967, 973, 984, 1012
microalgal assemblages, xxxii, 993, 1142, 1633
microbes, xix, 28, 310, 311, 312, 313, 314, 316, 317, 319, 320, 325, 327, 328, 329, 335, 346, 462, 1361, 1373
Microbes, 309, 311, 317, 1374
microbial, xix, xxii, 36, 310, 312, 313, 314, 316, 317, 319, 320, 321, 325, 326, 330, 332, 334, 336, 339, 341, 343, 355, 463, 468, 473, 475, 476, 483, 489, 490, 492, 493, 494, 496, 497, 498, 500, 507, 953, 954, 986, 1102, 1103, 1104, 1107, 1108, 1111, 1112, 1114, 1115, 1116, 1117, 1118, 1120, 1137, 1183, 1328, 1332, 1333, 1335, 1376, 1392, 1396
Microbial, 317, 319, 331, 333, 334, 336, 342, 1102, 1107, 1116, 1118, 1119, 1120, 1451
microbial cells, 1107, 1592
microbial communities, xix, xlii, 310, 312, 320, 325, 330, 336, 343, 355, 468, 766, 1104, 1108, 1579, 1603, 1604
microbial community, 316, 319, 320, 343, 1108, 1111, 1114, 1115, 1116, 1117, 1118, 1580
microbiota, 310, 319, 325, 329, 340, 1104, 1106, 1111, 1585, 1604, 1605, 1615, 1618
microclimate, 790, 1646
microcosm, 337, 340, 347, 348, 349, 352, 354, 357, 1357
microcosms, xxxi, 324, 334, 336, 347, 348, 355, 356, 887, 889, 897, 901, 904, 906, 1118
micro-environments, 435
microflora, 310, 320
microhabitats, 531, 690, 1647, 1668
micrometeorological, 262
micronutrients, 441, 780, 1580
microorganism, 317, 476, 1103, 1588, 1605
microorganisms, xlii, 312, 313, 317, 319, 321, 329, 331, 335, 339, 356, 414, 463, 469, 472, 477, 530, 841, 920, 971, 1094, 1105, 1114, 1115, 1119, 1240, 1328, 1329, 1330, 1341, 1375, 1376, 1552, 1579, 1580, 1588, 1590, 1591, 1593, 1597, 1598, 1600, 1602, 1605, 1606, 1724
micro-organisms, 136
micro-organisms, 338
micro-organisms, 964
micro-organisms, 984
microsatellites, 554, 556
microscope, 844, 890, 1105, 1107, 1207
microscopy, 853, 916, 921, 1031, 1107, 1114, 1116, 1117, 1625
microstructure, 1621
microwave, 1076
microwave radiation, 1594

microwaves, 1205
migrant, 371, 962, 963, 967
migrants, 554
migration, xxxvii, 437, 554, 556, 737, 841, 849, 850, 851, 946, 951, 957, 964, 968, 969, 982, 1325, 1339, 1604, 1700, 1719, 1729
military, 420, 595, 596, 597
milk, 181, 183, 205, 509, 609, 1372, 1377, 1388, 1402, 1600, 1612, 1613
milligrams, 1551
mimicking, 634
mineral resources, 725
mineralization, 317, 471, 473, 476, 787, 1115, 1119
mineralized, 426
mineralogy, 430
minerals, 318, 337, 463, 474, 476, 521, 1166
mines, 305, 426
mining, xliii, 322, 420, 423, 971, 972, 975, 989, 1166, 1174, 1340, 1458, 1509, 1641, 1644, 1645, 1649, 1650
Ministry of Education, 1716
Ministry of Environment, xiv, 13, 527, 529, 542, 1433
Minneapolis, 1168, 1721
Minnesota, 601, 1168
minors, 71
Miocene, 567, 568, 1709, 1711, 1712, 1716, 1725
miscarriage, 609
misleading, 1446
mission, 1186
missions, 391, 392, 393, 394, 395, 396, 398, 399, 433, 449, 507, 588, 589, 591, 595, 721, 722, 724
Mississippi, 1, 2, 3, 11, 1183, 1340
Mississippi River, 2, 1183
Missouri, 160, 162, 224, 236, 279, 1679, 1717, 1719, 1720
misunderstanding, 1363
misuse, 714, 718
mites, 174, 356, 955, 956
mitochondria, 913, 917
mitochondrial, 385, 562
mitochondrial DNA, 385, 1721
mitosis, 1625
mixing, 895, 899, 932, 936, 939, 940, 941, 942, 980, 998, 1139, 1387, 1389, 1519, 1580
mixture analysis, 1205
ML, 23, 480, 563, 943, 960, 961, 973, 1107, 1340
MMA, 274, 277, 279, 280, 514, 520, 527, 547, 1101, 1119
mobility, 186, 196, 202, 419, 443, 1515, 1531, 1532, 1603
model system, 334, 355, 639, 680, 913, 919

modeling, 20, 338, 339, 374, 405, 459, 620, 621, 629, 756, 777, 787, 791, 1186, 1188, 1192, 1193, 1318, 1433, 1436, 1475, 1477, 1508, 1509, 1515, 1544, 1545, 1547, 1652, 1684, 1687, 1691, 1695, 1701, 1703, 1704
modelling, 450, 685, 686, 688, 691, 742, 787, 796, 905, 928, 934, 976, 1031, 1344, 1345, 1412, 1509, 1541, 1611, 1704, 1705
modernisation, 737
modernization, 179
modification, xxx, 23, 155, 166, 417, 475, 493, 512, 547, 572, 721, 871, 882, 1009, 1076, 1148, 1163, 1165, 1174, 1321, 1330, 1403, 1535, 1567, 1711, 1714
modifications, xiv, xxvii, xxxiv, 13, 56, 435, 450, 454, 493, 572, 741, 746, 750, 753, 1006, 1009, 1148, 1164, 1201, 1315, 1342, 1367, 1384, 1508, 1592, 1735
MODIS, xvii, 244, 245, 247, 250, 261, 262, 1410
modulus, 503
moieties, 469
moisture, xxxix, xl, 28, 254, 263, 429, 488, 492, 499, 504, 505, 507, 556, 735, 770, 784, 1245, 1272, 1274, 1276, 1281, 1283, 1301, 1330, 1336, 1372, 1407, 1409, 1410, 1411, 1413, 1414, 1415, 1425, 1426, 1431, 1432, 1433, 1434, 1435, 1436, 1437, 1438, 1439, 1441, 1442, 1443, 1444, 1446, 1447, 1448, 1449, 1450, 1475, 1611
moisture content, 492, 1372, 1425, 1439, 1443, 1446, 1447, 1448, 1449
molar ratio, 503, 943, 1353, 1359
molar ratios, 943, 1353
molasses, 1268
molds, 844, 1613
mole, 1078, 1359
molecular biology, 1115
molecular markers, 553
molecular mass, 505, 1358
molecular structure, 475
molecular weight, xxii, 318, 473, 474, 483, 487, 494, 496, 497, 500, 501, 504, 505, 507, 1078, 1239, 1597, 1602
molecular weight distribution, 497
molecules, 48, 51, 54, 90, 270, 471, 475, 507, 511, 764, 917, 1091, 1438, 1589, 1590, 1591, 1595, 1598
mollusks, 997, 1207, 1216, 1221
molybdenum, 1355
momentum, 722, 930, 1481, 1523, 1532
money, 598, 599, 1677
Mongolia, xvii, 212, 243, 244, 245, 246, 251, 260, 261, 262, 263, 611, 615

monocotyledon, 44, 60
monomeric, 54, 1095
monomers, 470, 492
monosaccharide, 469, 1600
monoterpenoids, 121
monsoon, 223, 1157, 1163, 1166, 1167, 1537
Montana, 602, 1726
Montenegro, 1103, 1117, 1119
montmorillonite, 476
morale, 598
morning, 249
Morocco, 1606
morphological, xv, 43, 44, 45, 60, 108, 121, 145, 334, 336, 378, 407, 553, 937, 994, 1009, 1163, 1165, 1207, 1336, 1509
morphology, xxviii, xlii, 23, 164, 340, 415, 434, 563, 572, 752, 756, 769, 771, 772, 783, 789, 790, 796, 808, 815, 917, 918, 930, 934, 937, 1004, 1107, 1114, 1138, 1148, 1158, 1167, 1338, 1365, 1461, 1474, 1509, 1565, 1568, 1575, 1576, 1588, 1623, 1624, 1626, 1627, 1628, 1629, 1630, 1631, 1632, 1639, 1656, 1668
morphometric, xxxiv, 45, 792, 849, 1148
mortality, 321, 349, 601, 621, 622, 623, 624, 626, 627, 628, 629, 630, 634, 641, 646, 648, 653, 654, 656, 658, 660, 662, 663, 665, 669, 670, 673, 772, 773, 774, 777, 778, 841, 914, 940, 959, 977, 1125, 1178, 1208, 1224, 1247, 1291, 1315, 1341, 1711
mortality rate, 349, 621, 622, 627, 628, 629, 641, 646, 648, 653, 654, 663, 665, 669, 670, 673, 1315, 1341
mosaic, xix, xlv, 267, 271, 276, 302, 305, 314, 359, 361, 362, 370, 450, 570, 812, 1723
mosquitoes, 733
motion, 402, 1138, 1474, 1475, 1517, 1521, 1523, 1524, 1532
mountains, xiv, xliii, xlv, 27, 28, 32, 35, 129, 222, 516, 567, 568, 1333, 1422, 1641, 1642, 1645, 1710, 1714, 1715, 1720, 1723, 1725, 1728, 1730, 1732, 1733
movement, 2, 9, 38, 40, 324, 325, 352, 363, 369, 490, 544, 562, 622, 623, 925, 926, 935, 951, 1100, 1333, 1344, 1381, 1412, 1459, 1472, 1473, 1523, 1531, 1532, 1533, 1536, 1541, 1543, 1546, 1547
moving window, 364
Mozambique, 223, 233
mRNA, 318, 917
MS, 1, 4, 155, 560, 561, 613, 960, 961, 1341
MSC, 592
MSW, 394, 395, 605
MTs, 1091

mucosa, 1602
mucus, 850
multidimensional, 405, 455, 631, 632, 633, 790
multidimensional data, 455
multidimensional scaling, 631, 632, 633
multidisciplinary, 457, 564, 990
multiple regression, 859, 1356
multiple regression analysis, 859
multivariate, 29, 32, 45, 155, 277, 370, 387, 455, 1228
multivariate analysis, 32, 155, 277, 387, 811, 1669
municipal solid waste, 395
municipal solid waste (MSW), 395
muscle, 1077
muscles, 484
muscular dystrophy, 843, 1606
mussels, xxxiii, 1071, 1072, 1073, 1074, 1075, 1076, 1077, 1078, 1080, 1081, 1082, 1083, 1084, 1085, 1086, 1087, 1088, 1089, 1090, 1091, 1092, 1094, 1095, 1096, 1097, 1102
mutant, 318
mutations, 383, 570
Myanmar, 1638, 1696
mycelium, 318, 324, 333, 334, 343, 355, 356
mycorrhiza, 318, 329, 337, 338, 341, 343, 355
myricetin, 142, 150, 153
mythology, 605

N

NaCl, 488, 824, 830, 1080, 1109, 1355
NAD, 1600
NADH, 1600
NAFTA, 868, 879
Namibia, 217, 218
NAP, 385
naphthalene, 87
narratives, 184, 187, 198, 203
NASA, 263, 1435
Nash, 171
nation, 391, 597
national, 390, 398, 406, 410, 411, 419, 441, 458, 513, 514, 522, 523, 530, 546, 548, 597, 599, 611, 1339, 1343, 1438
National Academy of Sciences, 333, 341, 635
national action, 191, 204, 209
national borders, 1702
national character, 390
national culture, 304
national parks, 597, 699, 1686, 1696, 1727
National Pollutant Discharge Elimination System, 1340
national product, 716, 876
National Research Council, 1277, 1294, 1573

National Science Foundation, 1634
national strategy, 1343
National Weather Service, 1308
Native American, 605
Native Americans, 605
native plant, xxiv, 321, 325, 351, 594
native population, 750, 1091
native species, xx, xxiv, xxvii, xxx, 321, 351, 360, 526, 528, 568, 571, 572, 573, 594, 757, 759, 760, 761, 767, 871, 881, 882, 1646, 1685, 1701
natural capital, 635, 1198, 1367
natural disaster, xxv, 423, 637, 649, 661, 663, 670
natural disasters, xxv, 423, 637, 649, 661, 663, 670
natural disturbance, xxiv, 594, 634, 638
natural enemies, 332
natural environment, 339, 401, 402, 403, 406, 1075, 1118, 1221, 1223
natural food, 1593
natural gas, 783
natural habitats, xix, 314, 359, 361, 376, 1649
natural resource management, 209, 595, 596, 751
natural resources, xiii, xxi, xxv, 179, 407, 408, 413, 414, 436, 450, 454, 457, 514, 523, 526, 530, 575, 577, 582, 584, 597, 598, 599, 604, 612, 619, 701, 702, 703, 724, 726, 737, 738, 744, 1644, 1701
Natural Resources Conservation Service, 2, 1343
natural selection, xxviii, 161, 384, 572, 686, 799
nature conservation, 370, 377, 384
Navicula, 944, 946, 1002, 1004, 1011, 1016, 1017, 1019, 1020, 1030, 1057, 1058
NC, 356, 613, 1217
ND, 1157, 1572, 1573
Nebraska, 1314
NEC, 1423, 1433
neck, 1502
negative consequences, xxi, 413
negative effects, 272, 358, 432, 557, 781, 811, 1179, 1228, 1249
negative relation, 40, 347, 383, 859, 1139
neglect, 1677
nematode, 1038
nematodes, xxxii, 312, 319, 321, 338, 924, 955, 956, 957
neovascularization, 846
Nepal, 360, 370, 606, 616
nerve, 1077
nerves, 609
nervous system, 611
nesting, 1187
Netherlands, 332, 333, 338, 447, 509, 616, 617, 764, 779, 794, 797, 798, 817, 887, 907, 908, 910,

1039, 1144, 1182, 1234, 1294, 1546, 1634, 1635, 1636, 1637, 1639, 1717, 1718
network, xiv, 13, 54, 324, 325, 338, 355, 440, 459, 469, 511, 1037, 1073, 1150, 1550, 1693, 1699, 1700, 1702
network polymers, 511
neutral, 320, 321, 335, 347, 489, 497, 498, 689, 809
neutrophils, 841, 842
Nevada, 1334
New England, 755, 767, 1322, 1611
New Frontier, 336
New Jersey, 213, 341, 374, 579
New South Wales, 406, 969, 980, 986
New York, iv, 160, 163, 164, 213, 214, 216, 217, 329, 337, 341, 342, 373, 385, 411, 459, 481, 578, 579, 600, 613, 617, 924, 1032, 1093, 1142, 1144, 1166, 1168, 1230, 1232, 1322, 1343, 1346, 1368, 1370, 1404, 1405, 1544
New Zealand, xliv, 87, 377, 598, 693, 1343, 1368, 1470, 1476, 1639, 1668, 1671, 1673, 1674, 1675, 1678, 1679, 1681
Newton, 213, 954, 964, 985, 1342
Newtonian, 1523, 1543
NGO, 527, 529
NGOs, 208, 736, 1689
NH2, 505, 1396
NHS, 399
Ni, 426, 441, 447, 636, 1074, 1076, 1084, 1205, 1208, 1214, 1215, 1216, 1224, 1226
nickel, 1097, 1645, 1650
NIE, 379
Nielsen, 1107, 1120, 1544, 1571
nifedipine, 508
Nigeria, 614, 615, 616, 709, 1342
NIR, 250, 251, 254
NIST, 1205
nitrate, xxxvii, xxxviii, 2, 4, 329, 332, 334, 341, 356, 439, 716, 942, 1030, 1140, 1144, 1171, 1184, 1299, 1308, 1331, 1332, 1338, 1341, 1346, 1350, 1353, 1366, 1368, 1397, 1400
nitrates, 423, 449, 450, 1002, 1003, 1330, 1377, 1397, 1400
nitric acid, 1204
nitric oxide, 563
nitrification, xxxviii, 1349, 1356, 1357, 1359, 1361, 1366, 1367, 1373, 1375, 1396, 1397, 1398, 1399, 1400
nitrifying bacteria, 1361
nitrite, 4, 716, 1002, 1030, 1171, 1172, 1275, 1354
Nitrite, 9, 1355
nitrogen gas, 1332
nitrogen-fixing bacteria, 312, 332
nitrous oxide, 588, 589, 591, 1357, 1397

Nitzschia, 895, 944, 946, 1002, 1004, 1007, 1009, 1011, 1016, 1018, 1019, 1020, 1021, 1023, 1059, 1060, 1061
Nitzschia panduriformis, 1020
Nixon, 1170, 1174, 1183, 1367
NMR, 155, 160, 169, 472, 1617
NOAA, xvii, 211, 244, 262, 1096
nodes, xliv, 4, 287, 290, 781, 1671, 1672, 1673, 1674, 1675, 1676
nodulation, 325
nodules, 317, 327
noise, 393, 808, 972
non-biological, 466
non-crystalline, 476
non-destructive, 245, 251, 254, 612
nonionic, 488, 489
nonionic surfactants, 488, 489
nonlinear, 1471, 1544, 1577
non-linear, 455
non-native, xxiv, 594, 596
non-native species, xxiv, 594
nonparametric, 380
non-profit, 596, 597
non-renewable, 467
nontoxic, 1080
non-uniform, 1152, 1158, 1541, 1552
normal, 17, 19, 22, 54, 106, 266, 274, 930, 969, 971, 1081, 1092, 1158, 1175, 1440, 1444, 1455, 1477, 1478, 1479, 1483, 1489, 1491, 1512, 1523, 1526, 1552
normal distribution, 1081, 1175
norms, 181, 612
North Africa, 93, 100, 132, 888
North America, vii, xxvii, xxx, xxxi, xliv, 140, 266, 270, 321, 377, 579, 596, 598, 611, 685, 692, 693, 729, 757, 758, 762, 764, 765, 768, 788, 793, 871, 872, 879, 882, 888, 903, 923, 924, 1023, 1165, 1168, 1294, 1343, 1367, 1368, 1369, 1370, 1635, 1637, 1667, 1668, 1669, 1671, 1673, 1674, 1675, 1676, 1708, 1709, 1717, 1718, 1719
North American Free Trade Agreement, 879
North Carolina, 613, 1035, 1330, 1343
Northeast, 516, 613, 614
Northern China, xxxix, 1407
Northern Hemisphere, 995
Northern Ireland, 1343
Norway, 88, 177, 389, 852, 988
NPP, 466
NPS, 2, 9, 1326, 1332, 1333, 1336, 1337, 1338, 1339
NRC, 183, 215, 1201, 1343
NTU, 941, 1002, 1014
nuclear, 396, 526, 530, 554, 562, 1142, 1512, 1525

nuclear power, 396, 1525
nuclear power plant, 1525
nucleation, 273, 920, 921
nuclei, 108
nucleic acid, 48, 50, 314, 820, 1121, 1158, 1266, 1296, 1589, 1597, 1604
nucleic acid synthesis, 1597
nucleotide sequence, 162
nucleotides, 1258, 1266
nucleus, 57, 87, 403, 604
Nuevo León, 871, 877, 883
nuisance, 771, 1170, 1176
null, 689
numerical analysis, 171
numerical computations, 1535
nurse, 273, 278, 328
nutraceutical, 843, 849, 850
nutrient concentrations, xxxiv, xxxviii, 2, 4, 5, 10, 314, 942, 973, 994, 1003, 1136, 1140, 1169, 1170, 1175, 1176, 1179, 1309, 1319, 1335, 1350, 1352, 1353, 1355, 1356, 1368
nutrient cycling, 319, 338, 339, 431, 455, 1031, 1363, 1365
nutrient enrichment, 692, 781, 790, 1073
nutrient transfer, 324
nutrition, xix, 310, 317, 318, 320, 337, 339, 341, 346, 356, 789, 821, 835, 836, 840, 880, 1271, 1291, 1295, 1297, 1615
nylon, 503, 511, 1464

O

obligate, 328, 346
obligation, 523, 1189
obligations, 210, 391, 543
observations, 181, 189, 192, 193, 202, 208, 229, 231, 234, 235, 328, 384, 446, 515, 524, 531, 610, 931, 935, 995, 1015, 1031, 1163, 1315, 1393, 1430, 1513, 1515, 1524, 1531, 1533, 1534, 1535, 1544, 1545, 1565, 1588, 1667
obsolete, 409, 551
obstacles, 557, 747, 804, 1460, 1507
obstruction, 1009, 1438, 1646
occlusion, 487
ocean waves, 1481, 1482
Oceania, 600, 1631
oceans, 394, 463, 464, 465, 583, 585, 725, 1335
ODS, 729
OECD, 390, 399
officials, 208, 209, 1702
offshore, xl, 930, 987, 1011, 1125, 1455, 1456, 1459, 1472, 1474, 1476, 1477, 1481, 1482, 1485, 1498, 1504, 1505, 1507, 1509, 1517, 1519, 1526, 1534, 1535, 1541, 1576

offshore oil, 1519
Ohio, 1234
76, 1281, 1283, 1284, 1289, 1293, 1306, 1334, 1336, 1390, 1391, 1413, 1417, 1447, 1519, 1693
oil production, 719, 856
oil samples, 29
oils, 110, 164, 195, 322, 462, 609, 1596, 1646
oilseed, 1250, 1271, 1284
Oklahoma, 217, 616
oligomers, xxii, 54, 483, 503, 504
oligotrophic conditions, 1140
olive, xiv, 13, 18, 320, 430, 434, 436, 1334, 1340, 1438, 1441
olives, xxxix, 1073, 1333, 1437
Oman, 100
omission, 1028
one dimension, 405
online, 481, 601, 613, 614, 615, 617, 1038
on-line, 989
on-line, 1346
on-line, 1451
oocyte, 913
openness, 1540
operating costs, 859, 861, 865, 1402
operations, xxix, 44, 596, 698, 736, 747, 819, 820, 865, 913, 971, 972, 975, 979, 1215, 1229, 1262, 1300, 1330, 1343, 1346, 1396, 1513
operator, 226
opportunism, 865
opportunities, 193, 328, 329, 398, 557, 568, 699, 719, 736, 758, 760, 805, 867, 872, 968, 1186, 1193, 1685, 1693
opportunity costs, 371
opposition, 46, 48, 54, 72, 121, 155, 156, 404, 542, 598
optical, 254
optical fiber, 254
optics, 1485
optimism, 411, 447
optimization, 271, 639, 643, 647, 669, 676, 678, 854, 905
orchid, 340
ordinary differential equations, 640, 674, 675
Oregon, 989, 1030, 1036, 1314, 1316, 1507
ores, 1119, 1344
organ, xxviii, 273, 307, 769, 771, 772, 787
organelles, 912
organic C, 470, 475
organic compounds, 318, 453, 462, 475, 720, 1080, 1089, 1101, 1103, 1118
organic materials, 476
organic matter, xiv, xxxix, 27, 28, 36, 206, 312, 417, 418, 419, 420, 421, 429, 430, 435, 439, 443,

448, 452, 455, 462, 464, 465, 468, 474, 476, 482, 531, 771, 781, 788, 903, 946, 955, 1093, 1111, 1134, 1139, 1154, 1158, 1166, 1167, 1174, 1175, 1178, 1211, 1246, 1308, 1326, 1328, 1333, 1335, 1336, 1353, 1373, 1375, 1397, 1417, 1425, 1437, 1449, 1450, 1540, 1665
organic soils, 478, 781
organic solvent, 488, 497
organic solvents, 488, 497
organism, xxxiii, 466, 485, 802, 806, 873, 1071, 1072, 1090
organization, 407, 527, 529, 530, 562, 634
Organization for Economic Cooperation and Development, 1323
organizations, 406, 407, 423, 595, 596, 597, 598, 599
organize, 754, 866, 1104
organizing, 209, 404, 411, 754, 816
organochlorinated, 1091, 1101, 1104
organochlorine compounds, 1073, 1081, 1101
organophosphates, 1090
organs, xxiv, 75, 259, 273, 603, 604, 770, 771, 772, 783, 787, 790, 796, 842
orientation, 353, 936, 1015
ornamental plants, 78
ornithine, 1599
Oromo, 191
Orthophosphate, 1145
oryx, 607, 1727
oscillations, 899, 1148, 1481, 1566, 1714
osmosis, 915
osmotic, 1207
osmotic stress, 912, 1600
ossification, 691
outliers, 1696
outreach, 599, 1189
overexploitation, xxv, 574, 575, 576, 577, 603, 636, 972, 975, 1565
overgrazing, 196, 214, 244, 245, 259, 260, 452, 967
overharvesting, 620
overlap, 134, 277, 703, 706, 707, 1672, 1675, 1676
overlay, xiv, 14, 200
overload, 1329, 1560, 1564
oversight, 1300, 1319
ownership, 859, 861, 862
ox, 28, 54, 102, 167, 173, 177, 923, 941, 1684
oxidation, 51, 162, 414, 466, 485, 490, 498, 501, 842, 1078, 1355, 1373, 1375, 1376, 1379, 1392, 1396, 1400, 1449, 1566, 1600
oxidative, 51, 1091, 1093, 1095
oxidative damage, 1095
oxidative stress, 1091, 1599
oxide, 500, 563, 588, 589, 591, 1357, 1397

oxygen consumption, 1077
oxygen saturation, 941
oxygenation, 54, 91, 108, 1466
oyster, xxxiii, 856, 857, 858, 859, 860, 862, 864, 865, 867, 868, 1123, 1125, 1126, 1127, 1131, 1140, 1142
oysters, 860, 1102, 1125, 1141
ozone, 393, 585, 725, 727, 1636

P

P. sulcata, 1019
Pacific, xxiv, 459, 585, 593, 600, 617, 767, 868, 1038, 1124, 1133, 1143, 1167, 1341, 1433, 1631, 1637, 1650, 1651, 1679, 1717, 1720
packaging, xxii, 483, 487, 494, 502, 504, 506, 507, 586, 587, 589, 591, 1594
PAHs, xxxvi, 1101, 1102, 1202, 1205, 1208, 1216, 1217, 1220, 1224, 1225, 1226
pain, 609, 610
paints, 582, 585
pairing, 1119
Pakistan, 43, 428, 432, 1165
palaeo-environments, 1551, 1569
paleoecology, xxxii, 993, 996, 1029, 1033, 1036, 1038
paleoenvironmental changes, xxxii, 993, 994
Panama, 545, 1624, 1630, 1639
pancreatic, 490
paper, 22, 395, 460
Paper, 397, 1345, 1346, 1544
paradigm, 402, 743, 814, 1642
paradox, 403
Paralia sulcata, 1001, 1004, 1009, 1010, 1011, 1014, 1017, 1018, 1019, 1020, 1023, 1061
parallel, 60, 226, 387, 403, 404, 846, 850, 1423, 1454, 1461, 1478, 1483, 1494, 1520, 1526, 1536, 1550, 1576, 1679
parallelism, 849
paralysis, 609
parameter, 31, 32, 34, 45, 169, 258, 437, 438, 440, 586, 621, 629, 958, 1128, 1129, 1131, 1210, 1316, 1317, 1413, 1436, 1449, 1454, 1499, 1509, 1524, 1528, 1531, 1532, 1533, 1535, 1577, 1661, 1690
parasite, 155, 852, 1178
Parasite, 174
parasites, xviii, 309, 570, 572, 585, 880, 881, 1330, 1597, 1663
Parasites, 578
parenchyma, 274, 1164
Paris, 399, 410, 411, 457, 480, 510, 513, 546, 548, 617, 1323
parity, 216

Parkinson, 933, 937, 969, 970, 971, 985, 988
partial differential equations, 643, 674, 1524
Partial Least Squares, 1023
partial least squares regression, 1039
participants, 189, 201, 409, 1190
participatory research, 178, 189, 193
particle density, 1531
particles, 318, 402, 403, 424, 497, 932, 937, 943, 1012, 1077, 1138, 1212, 1213, 1329, 1332, 1336, 1376, 1391, 1393, 1395, 1400, 1401, 1438, 1464, 1517, 1518, 1520, 1521, 1523, 1524, 1532, 1544, 1596, 1658, 1661, 1665
particulate matter, 1144
partition, 316
partnership, 351, 527, 635, 1190, 1731
partnerships, 179, 398, 527, 1186, 1189
passive, 446
pasta, 1602
pastoral, 178, 181, 183, 185, 189, 195, 197, 200, 201, 202, 203, 205, 206, 209, 211, 212, 213, 604
pasture, 192, 244, 245, 262, 275, 424, 456, 478, 596, 693, 1302, 1307, 1314, 1320, 1327, 1645
pastures, 192, 266, 270, 271, 272, 274, 275, 377, 515, 517, 519, 701, 735, 1306, 1307, 1314, 1319, 1330, 1341, 1343, 1644, 1645
pathogenic, 136, 138, 174, 312, 313, 320, 321, 338, 1328
pathogens, xxiv, 108, 310, 313, 317, 320, 321, 326, 328, 330, 337, 339, 343, 572, 594, 596, 758, 840, 841, 842, 850, 880, 1300, 1327, 1329, 1373, 1596, 1597, 1613
pathology, 136, 851, 853, 880
pathways, 2, 9, 48, 55, 151, 159, 163, 644, 743, 783, 796, 873, 884, 901, 1103, 1191, 1192, 1193, 1194, 1360, 1362, 1583
patients, 486, 609
PCA, 29, 33, 35, 443, 1211, 1224, 1225, 1226
PCBs, 1076, 1081, 1090, 1102
PCP, 1102, 1103, 1104, 1105, 1106, 1108, 1109, 1110, 1111, 1112, 1113, 1114, 1115, 1116, 1117, 1118, 1119, 1120
PCR, 1107, 1340, 1344, 1586, 1589, 1590, 1603, 1604, 1609, 1610, 1614, 1615, 1618, 1619, 1620, 1621
peace, 736
peak concentration, 1388
peanuts, 51
peat, xxxiv, 395, 440, 781, 1147, 1167, 1551, 1554, 1555, 1562, 1566
pedestrians, 1570
peers, 1192

pelargonidin, 76, 79, 80, 81, 82, 83, 84, 87, 89, 90, 166
pendulum, 596
penis, 610
Pennsylvania, 337
peptidase, 1603, 1610
peptide, 485, 493, 498, 499, 500, 836, 1590, 1592, 1602, 1610, 1616, 1620
Peptide, 494
peptide bonds, 498, 499, 500
peptides, 510, 1595, 1602, 1609, 1612, 1618, 1619
PER, 1076
per capita, 390, 397
perceived outcome, 758, 760
percentile, 17, 18, 22
perception, 196, 207, 406, 407, 526, 527, 601
perceptions, 181, 183, 190, 191, 194, 205, 206, 211, 212, 364, 402, 405, 1190
percolation, xxxvii, xxxviii, 1325, 1371, 1373, 1376, 1385, 1390, 1391, 1398, 1399
performance indicator, 868
perfusion, 850
periodic, 202, 526, 948, 1100, 1173, 1331
peripheral blood, 842, 852
peritoneal cavity, 841
peri-urban, 942, 944, 973
permafrost, 355
permeability, xix, xx, 359, 360, 430, 435, 465, 841, 914, 915, 916, 1294, 1376, 1438
permeation, 504, 915, 921
permit, xix, 229, 359, 423, 434, 543, 544, 731, 856, 881, 970, 972
peroxidation, 842, 1093, 1094
peroxide, 500, 1593, 1598
personal, 404, 1350, 1552
personal communication, 260, 287, 1552
personal relations, 189
personal relationship, 189
personality, 403, 598
Perth, 1467
perturbation, 183
perturbations, xxiv, 404, 464, 539, 594, 997, 1170
Peru, 432, 479, 872, 878
pest control, xxiv, 594
pest management, 1327
pesticide, xxxiii, 421, 890, 1071, 1074, 1092, 1095, 1327, 1333, 1337, 1339, 1343, 1346
pesticides, 369, 407, 427, 450, 468, 572, 715, 888, 904, 906, 968, 1073, 1074, 1089, 1090, 1091, 1092, 1093, 1094, 1101, 1326, 1327, 1329, 1333, 1336, 1339, 1343, 1344, 1346, 1438
pests, 206, 312, 353, 1333
PET, 1439, 1442

petrochemical, xxxiii, 1010, 1099, 1101
petroleum, 467, 717, 718, 721, 738, 868, 1101, 1442
Petroleum, 1167, 1233, 1234
Petrology, 1166, 1232
pets, xxiv, 594, 599, 605
PGPR, 317
PGR, 376
pH values, 447, 1134, 1204
phagocytosis, 841
pharmaceutical, 488, 497, 504, 507, 611, 612
pharmaceutical industry, 504
pharmaceuticals, 610, 616
pharmaceutics, 487
pharmacological, 610
pharmacopoeia, 604
phase transitions, 921
phenol, 51, 1079, 1101, 1102, 1103, 1107
phenolic, xv, 44, 51, 54, 56, 57, 58, 70, 71, 72, 73, 74, 115, 146, 161, 170, 331, 1102
phenolic acid, 51, 57, 70, 71, 170, 331
phenolic acids, 51, 71
phenolic compounds, xv, 44, 51, 56, 57, 72, 73, 74, 115, 1102
phenotype, 1588, 1638
phenotypic, 342
phenylalanine, 1244, 1253, 1263, 1265, 1268, 1272, 1280, 1281, 1285, 1286, 1599, 1620
pheromone, 1102
Philadelphia, 411
Philippines, 867, 868, 872, 1287, 1291, 1705
philosophy, 457
phobia, xxvi, 684
phosphatases, 322
phosphate, xiii, xxix, xxxv, 1, 4, 314, 317, 318, 322, 332, 715, 779, 781, 789, 820, 821, 826, 829, 833, 834, 835, 837, 844, 903, 942, 943, 1136, 1169, 1171, 1172, 1174, 1175, 1178, 1243, 1246, 1254, 1262, 1272, 1281, 1304, 1332, 1338, 1368, 1369, 1376, 1400, 1583, 쳅1601, 1602, 1617
Phosphate, 317, 329, 943, 1139, 1182
phosphates, 1002, 1330, 1602, 1609, 1617
phospholipids, 820, 1247
phosphorous, xiv, 27, 29, 30, 32, 34, 35, 206, 420, 678, 895, 899, 942, 973, 1373, 1375, 1376
phosphorylation, 1583
photographs, 193, 363, 365, 433, 794, 935, 1476
photolysis, 1102
photomicrographs, 1164
photosynthesis, 262, 464, 468, 475, 779, 791, 1029, 1412, 1624, 1633, 1639, 1646, 1665
photosynthesize, 324
photosynthetic, 940, 967, 1039, 1359, 1448

photosynthetic performance, 1636, 1637
Phragmites australis, 169, 949, 950
phylogenesis, 416
phylogenetic, 165, 312, 416, 547
phylogeny, 174, 357, 546
phylum, 346, 357
physical characteristics, 182, 191, 194, 195, 452, 475, 507, 555, 928, 1037, 1471
physical environment, 181, 407, 762, 964
physical factors, 163
physical features, 1726, 1732
physical properties, 430
physical structure, 473
physicochemical, xxxiv, 509, 576, 1124, 1140
physico-chemical changes, 435, 969
physicochemical characteristics, 576, 890
physico-chemical characteristics, 961
physico-chemical parameters, 435, 849, 958, 1550
physicochemical properties, 509
physico-chemical properties, 433
physics, xx, 397, 401, 402, 404, 1540, 1544, 1577
physiological, 46, 317, 336, 338, 346, 354, 357, 426, 1072, 1075, 1084, 1093, 1097, 1223
Physiological, 1077, 1094, 1143, 1240, 1426
physiological factors, 905
physiological mechanisms, 789
physiology, xliii, 48, 171, 332, 355, 454, 769, 783, 912, 1290, 1588, 1623, 1633
phytochemicals, 138
phytoplankton, xxxi, xxxii, 642, 678, 680, 887, 890, 891, 892, 893, 895, 899, 900, 901, 903, 904, 905, 906, 908, 909, 910, 923, 942, 943, 944, 945, 946, 947, 950, 954, 955, 973, 977, 980, 985, 986, 987, 1001, 1003, 1010, 1011, 1014, 1031, 1034, 1037, 1073, 1077, 1139, 1144, 1145, 1154, 1158, 1175, 1176, 1178, 1182, 1183, 1335, 1345, 1395
phytoplanktonic, 998, 999, 1001, 1011, 1012, 1014, 1027, 1154
phytoremediation, 442, 452, 453
phytotoxicity, 331, 426
pig, 594, 1074, 1374, 1375
pigmentation, 82, 918, 1639
pigments, 79, 80, 83, 87, 88, 93, 136, 161, 165, 173, 1031
pigs, xxiv, 440, 593, 835, 837, 843, 1250, 1646
pilot study, 216
pinus, 275
pioneer species, 337, 1685
pipelines, 251, 252
planetary, 432, 1481

plankton, xiii, 681, 829, 888, 889, 890, 899, 906, 980, 981, 995, 998, 1001, 1003, 1014, 1028, 1029, 1040, 1041, 1170
planned economies, 729
planning, xxxv, 180, 397, 398, 404, 407, 423, 433, 435, 436, 447, 449, 451, 460, 524, 558, 709, 714, 725, 726, 733, 754, 867, 969, 971, 982, 1179, 1185, 1332, 1335, 1336, 1542, 1543, 1571, 1577, 1687, 1703, 1729
planning decisions, 433, 971
plant establishment, 332, 1335
plant growth, 20, 261, 314, 315, 317, 319, 320, 321, 330, 332, 333, 342, 349, 353, 501, 780, 784, 792, 1134, 1154, 1330, 1332, 1412, 1413
plant growth promoting rhizobacteria, 317, 330
plant type, 320
plasma, 493
plasma levels, 842
plasma membrane, 912, 915, 917
plastic, xxii, 464, 483, 503, 504, 586, 587, 1075, 1077, 1355, 1357, 1374, 1376, 1377, 1381, 1402
plasticity, 332, 334, 780, 1143
plasticizer, 493, 505
plastics, 394, 504
plastid, 162, 1640
platform, 567, 857, 1471, 1478, 1481, 1499, 1500, 1501, 1502, 1503, 1504, 1506
platforms, 1190, 1467
play, xix, xxi, 9, 10, 179, 181, 201, 318, 327, 328, 345, 347, 354, 413, 421, 451, 461, 479, 492, 530, 604, 606, 929, 935, 936, 937, 949, 963, 968, 975, 1102, 1132, 1224, 1350, 1463, 1551
playing, 466, 727, 962, 1115
pleasure, 1456
Pleistocene, 563, 566, 567, 568, 1006, 1009, 1015, 1022, 1030, 1154, 1157, 1158, 1166, 1167, 1168, 1566
Pliocene, 566, 567, 568, 1157, 1168, 1709, 1721
ploidy, 74, 106, 387
ploughing, 20, 450, 452, 1327
PLS, 1023, 1039
poison, 610
poisoning, 1333
Poland, 392, 1580
polar, xlii, 156, 157, 158, 495, 556, 1623, 1631, 1637
polarization, 848, 853
policy, xliv, 24, 180, 197, 210, 211, 370, 399, 418, 419, 423, 425, 449, 457, 670, 686, 710, 711, 712, 722, 723, 725, 727, 1186, 1193, 1339, 1343, 1365, 1458, 1683, 1689, 1702, 1704, 1721
policy instruments, 722

policy variables, 670
policymakers, 720, 729
political problems, xlv, 197, 1723
politicians, 451
politics, xxi, 413, 456, 732
pollen, 76, 108, 109, 277, 319, 338, 553, 554, 555, 562, 1159, 1161, 1162, 1163, 1166, 1168, 1640, 1718, 1721
pollination, xxv, 530, 553, 554, 619
pollinators, xxiv, 326, 593
pollutant, 573, 1097, 1101, 1333, 1337, 1373, 1375, 1402, 1403
pollutants, xiii, xxxvii, 1, 2, 312, 393, 433, 437, 443, 446, 452, 573, 720, 758, 840, 888, 970, 1097, 1101, 1174, 1202, 1309, 1312, 1325, 1326, 1329, 1333, 1334, 1335, 1336, 1337, 1338, 1339, 1340, 1373, 1374, 1375, 1379, 1385, 1388, 1391, 1392, 1393, 1395, 1396, 1403, 1576
polyacrylamide, 1107, 1329, 1339
polyamides, xxii, 483, 503, 504
polyamine, 563
polyaromatic hydrocarbons, 1205
polyaromatic hydrocarbons (PAHs), 1205
polychlorinated biphenyl, 1081, 1083, 1096, 1102, 1117
polychlorinated biphenyls (PCBs), 1081, 1102
polycyclic aromatic hydrocarbon, 1101, 1102
polydispersity, 501
polyester, 354, 1464
polyethylene, 492, 509, 586, 589, 591, 1206, 1464
polymer, 469, 485, 502, 503, 504, 505, 511, 1329
polymer chain, 485
polymer chains, 485
polymerase, 1120, 1586
polymerase chain reaction, 1120, 1586
polymers, xxii, 469, 470, 471, 472, 473, 474, 475, 483, 486, 494, 498, 499, 503, 504, 507, 511
polymorphism, xv, 43, 44, 45, 47, 48, 50, 73, 104, 107, 159, 1589, 1590, 1638
polymorphisms, xv, 44, 104, 1589
Polynesian, 596, 597
polynomial functions, 29
polypeptides, 492, 499, 1580, 1591, 1602, 1608
polyphenols, 160, 165, 470, 473
polyploid, 74, 106
polyploidization, 105
polyploidy, 105, 106
polypropylene, 1464
polysaccharide, 1295, 1600
polysaccharides, 469, 475, 505
polythene, 1383
polyurethane, 1103, 1106, 1111, 1464

polyurethane foam, 1103, 1106, 1111
ponds, xxv, 524, 637, 766, 814, 876, 884, 888, 906, 1246, 1282, 1295, 1297, 1327, 1376, 1656
pools, xx, xxi, xxviii, 28, 271, 287, 314, 316, 317, 319, 323, 375, 376, 461, 462, 463, 464, 465, 474, 748, 800, 808, 1127, 1163, 1353, 1460, 1502, 1656, 1667
poor, xxxii, xxxix, 110, 111, 196, 199, 234, 259, 315, 322, 423, 505, 531, 573, 610, 924, 939, 954, 967, 1011, 1091, 1360, 1389, 1395, 1399, 1407, 1438, 1462, 1537, 1567, 1644, 1693, 1700, 1701, 1710, 1729
poor performance, 803, 1249
poor relationships, 688
POPs, 1101
population control, 671
population density, 515, 622, 623, 636, 698, 699, 840, 1590, 1647, 1689, 1695
population growth, xxvi, xxxv, 397, 516, 697, 716, 731, 738, 1178, 1185, 1187, 1350, 1685, 1689, 1729
population size, xxv, 376, 383, 385, 386, 419, 545, 557, 569, 570, 573, 577, 579, 619, 620, 621, 622, 623, 624, 625, 626, 627, 628, 629, 630, 634, 649, 678, 812, 963, 979, 1629, 1644, 1647, 1648, 1649, 1710, 1715
population structure, 561
pore, 509, 1204, 1219, 1224, 1461
pores, 473, 1376
porosity, 430, 452, 509, 783, 1106, 1431
porous, 493, 494, 1418, 1462, 1463
ports, 929, 1106, 1187, 1456, 1457, 1519, 1545
Portugal, 447, 839, 887, 911, 1037, 1453, 1456, 1457, 1458, 1460, 1462, 1472, 1475, 1500, 1503, 1506, 1507, 1508, 1509
positive correlation, 159, 443, 577
positive feedback, 320, 321, 323, 327, 328
positive relation, 40, 322, 346, 383, 940, 1139
positive relationship, 40, 322, 346, 383, 760, 814, 859, 940, 1139
postmortem, 997, 998
potassium, xxxvii, 332, 826, 1299, 1308, 1601, 1611
potato, 1586
potatoes, 1336
potential benefits, 398, 862
potential energy, 467
poultry, xxxvi, 486, 491, 492, 498, 823, 1239, 1240, 1259, 1260, 1275, 1289, 1294, 1299, 1300, 1302, 1303, 1304, 1306, 1307, 1309, 1318, 1319, 1320, 1321, 1330, 1342, 1345
poverty, 423, 516, 517, 701, 736, 1701
powder, 511, 610

power, 187, 216, 252, 253, 366, 394, 396, 398, 404, 409, 420, 426, 427, 455, 1095, 1173, 1384, 1392, 1402, 1403, 1512, 1525, 1657
power generation, 724, 730, 744, 755, 1173
power plant, 426, 427, 1512
power plants, 426, 427, 1512
power relations, 187
power stations, 398, 420
PRC, 509
precipitation, xiii, xvii, 1, 5, 6, 9, 10, 14, 17, 18, 19, 22, 23, 24, 243, 246, 274, 414, 421, 425, 431, 432, 463, 551, 703, 737, 755, 858, 891, 926, 927, 937, 988, 1012, 1124, 1133, 1136, 1139, 1153, 1163, 1173, 1187, 1197, 1207, 1308, 1316, 1326, 1391, 1392, 1411, 1413, 1415, 1416, 1420, 1421, 1423, 1426, 1431, 1439, 1442, 1443, 1687, 1689
precursor cells, 841
predation, xxx, 321, 573, 641, 692, 814, 871, 881, 896, 901, 963, 967, 1178
predators, xxiv, 362, 371, 570, 572, 593, 594, 642, 648, 650, 802, 896, 946, 1187, 1646, 1663, 1665
predictability, 185, 645, 805
prediction, 348, 350, 351, 1128, 1182, 1476, 1577, 1691, 1700
predictive model, 188, 367, 435, 437, 1316, 1339
predictive models, 188, 367, 437, 1316, 1339
predictor variables, 30
predictors, 1677
pre-existing, 363
preference, 194, 195, 197, 200, 271
premium, 41
preparation, iv, 273, 488, 490, 493, 494, 495, 497, 498, 500, 501, 505, 510, 524, 604, 610, 611, 718, 808, 811, 817, 1078, 1580, 1581, 1589, 1593, 1603, 1608, 1609, 1666
prescription drug, xxv, 619
prescription drugs, xxv, 619
preservation, xxiii, xxxi, xxxii, 360, 371, 390, 474, 475, 476, 509, 522, 523, 524, 543, 550, 555, 558, 570, 571, 595, 676, 728, 911, 913, 922, 993, 998, 1154, 1158, 1304, 1435, 1580, 1593, 1594, 1609
preservative, 1596
president, 1733
President, 460
press, 371, 984, 1022, 1032, 1182, 1184, 1208
pressure, xvi, xxxv, 158, 178, 184, 185, 195, 196, 203, 204, 206, 243, 245, 246, 250, 251, 252, 254, 256, 257, 258, 259, 260, 262, 274, 419, 443, 455, 465, 486, 506, 598, 612, 933, 947, 971, 1072, 1090, 1169, 1205, 1357, 1414, 1415,

1439, 1458, 1474, 1479, 1480, 1523, 1565, 1686, 1714, 1729, 1730, 1731
pressure gradient, 1474, 1479, 1480, 1523
Pretoria, 972, 978, 982, 983
prevention, xx, 389, 423, 449, 451, 606, 701, 723, 725, 840, 842, 970, 1182, 1334, 1344, 1594
preventive, 423
price competition, 1284
prices, 419, 454, 490, 1701
primary function, 1338
primate, 238
principles, 186, 197, 342, 358, 369, 372, 464, 512, 676, 684, 690, 805, 814, 835, 934
prior knowledge, 186
priorities, 526, 547, 600, 601, 1345
pristine, xxxvii, 943, 944, 1312, 1349, 1351, 1356, 1361, 1363
private, 519, 523, 542, 595, 1303, 1343
private property, 523, 1303
probability, 323, 380, 402, 543, 620, 621, 622, 623, 624, 626, 627, 628, 629, 630, 631, 632, 634, 641, 644, 645, 685, 734, 783, 873, 882, 1179, 1306, 1316, 1425, 1629, 1690, 1691
probability density function, 644
probability distribution, 641, 645, 1691
probe, 1107, 1110, 1112, 1129, 1658
probiotic, 1603, 1608, 1614
procurement, 397, 710
producers, xxxi, xxxii, 397, 526, 527, 544, 728, 866, 874, 876, 877, 878, 880, 887, 889, 923, 946, 976, 1330, 1363
production capacity, 1525
production costs, 489, 492, 502, 724, 862
production function, 1192
productive capacity, 422
productivity growth, 725
productivity rates, 953
professionals, 598, 754
professions, 616
profit, 419, 596, 597, 668, 711, 754, 865, 1244, 1245, 1248, 1252, 1254, 1255, 1257, 1260, 1263, 1267, 1270
profitability, 859, 1281
profits, 596, 1336
progeny, 561
program, 407, 408, 433, 445, 529, 544, 586, 623, 1076, 1119, 1126, 1127, 1131, 1132, 1133, 1134, 1141, 1168, 1192, 1202, 1343, 1412, 1468, 1507, 1573, 1582, 1677, 1703, 1716
programming, 643, 644, 673, 674, 676
project, xvi, xxxvi, 177, 196, 197, 210, 404, 447, 526, 529, 543, 544, 668, 698, 719, 731, 883, 905, 941, 946, 989, 1104, 1117, 1125, 1131, 1141, 1142, 1186, 1188, 1189, 1190, 1191, 1192, 1299, 1308, 1319, 1339, 1366, 1403, 1457, 1464, 1465, 1478, 1506, 1509, 1541, 1542, 1543, 1570, 1620, 1634, 1650, 1687, 1691
prokaryotes, 1588, 1617
prokaryotic, 1117
proliferation, xix, xxxii, 310, 313, 315, 322, 335, 841, 850, 924, 949, 973
proline, 1591, 1602
promote, xxiii, 450, 532, 543, 550, 560, 565, 569, 573, 577, 1329, 1478, 1515
propagation, 560, 630, 782, 881, 1375, 1508, 1534, 1535, 1536, 1537, 1581, 1586, 1604, 1618
propane, 913, 1358
property, 405, 475, 502, 505, 523, 524, 527
prophylactic, 840
propylene, 913
prosperity, 719
proteases, 314, 318, 489, 490, 494, 498, 500, 1596, 1599
protected area, xvi, xxxv, 221, 369, 406, 1089, 1202, 1466
protected areas, xvi, xxxv, xliv, 221, 369, 406, 1202, 1644, 1651, 1683, 1685, 1686, 1689, 1693, 1695, 1696, 1698, 1699, 1700, 1701, 1702, 1703, 1704, 1724, 1727, 1733
protective coating, 503
protein, xvii, xxii, 244, 247, 248, 261, 270, 326, 329, 472, 474, 483, 484, 485, 486, 487, 490, 491, 493, 495, 496, 497, 498, 499, 500, 504, 505, 506, 508, 509, 510, 1078, 1087, 1088, 1093, 1117
protein components, 490, 495, 497, 498, 499
protein films, 505
protein hydrolysates, 490, 1592
protein synthesis, 1270, 1282
proteins, xxii, 270, 471, 475, 483, 484, 486, 487, 489, 490, 491, 492, 493, 495, 498, 505, 509, 510, 512, 826, 917, 921, 922, 1078, 1158, 1239, 1261, 1264, 1273, 1274, 1285, 1286, 1290, 1580, 1591, 1595, 1598, 1601, 1602, 1619, 1621
Proteins, 489, 490, 491, 495, 508
proteobacteria, 1119
proteoglycans, 484
proteolysis, 1581, 1599, 1602
proteolytic enzyme, 485, 496, 504, 507
protocol, 476, 971, 1107, 1204, 1205, 1206
protocols, 558, 1308, 1322, 1369
prototype, 1475, 1478, 1506, 1513
protozoa, 312
protozoan, 989

protozoan parasites, 1330
proxy, xxxiv, 210, 1039, 1148, 1149, 1159, 1166, 1552
prudence, 404
pruning, 1438
Pseudomonas, 469
Pseudo-nitzschia, 1064, 1065
PSI, 1117
PSP, 1373, 1381, 1382, 1383, 1385, 1390, 1391, 1392, 1393, 1394, 1395, 1400, 1401, 1403
psychology, 397
public, xxxvii, 37, 188, 303, 395, 397, 404, 406, 407, 420, 426, 427, 454, 457, 527, 529, 557, 595, 596, 599, 600, 604, 610, 612, 1101, 1190, 1325, 1468, 1559
public access, 527
public awareness, 725
public concern, xxxvii, 1325, 1689
public health, 37, 426, 427, 604, 610, 612, 715, 1101, 1594
public interest, 1690
public opinion, 457, 1190
publishing, 692, 1288
Puerto Rico, 763, 764
pulp, 489, 1102
pulse, 1022, 1355
pulses, 1381
pumping, xxxix, 1218, 1307, 1387, 1392, 1408, 1409, 1428, 1430, 1431, 1432, 1433, 1460, 1565
pumps, 707, 711, 713, 714, 733, 735, 738, 780, 1374, 1381, 1402
pure water, 402, 407, 687
purification, xxv, 488, 498, 619, 637, 638, 1351, 1368, 1376, 1608, 1613, 1616
purines, 1285, 1287
purity, 497, 500, 503
PVC, 1303, 1307, 1379
pyramidal, 553
pyrene, 1216
pyrimidine, 1266
pyrite, 1166
pyrolysis, 395, 453, 465
pyrophosphate, 476
pyruvate, 1105, 1108, 1109

Q

QA, 1205
quality control, 645, 912
quality improvement, 794, 1337
quality indicators, 433
quality of life, 433, 577, 697, 728, 1189
quality of service, 698
quality standards, 1319, 1322
quanta, 402, 410
quantification, 23, 189, 448, 449, 995, 1685, 1717
quantitative reconstruction, 1023
quantitative technique, 1022
quantum, xx, 401, 402, 403, 404, 405, 408, 409, 410, 411, 1537, 1540
quantum computers, 404
quantum dot, 917
quantum dots, 917
quantum mechanics, 411
quantum state, 405
quantum theory, 405
quartz, xl, 1149, 1456
Quaternary diatom assemblages, xxxii, 993, 996
Queensland, 372, 1346, 1630, 1635, 1703
quercetin, 67, 69, 76, 78, 87, 93, 100, 101, 102, 104, 106, 120, 121, 130, 135, 138, 142, 146, 150, 151, 156, 158, 159
Quercetin, 102, 106, 135
Quercus, 339, 1351, 1352
Quercus ilex, 339, 1351
questioning, 402, 409
questionnaire, 189
questionnaires, 188, 189
quinone, 87
quinones, 52
Quinones, 51, 163
quotas, 398, 859

R

RA, 355, 562, 1340, 1341, 1343, 1344
race, 385, 697, 840, 913, 1614
racing, 859, 1709
racism, 410
radar, 1564
radial distance, 202
radiation, xxxi, 146, 166, 253, 254, 256, 488, 568, 725, 727, 738, 747, 777, 778, 887, 891, 898, 930, 940, 967, 1102, 1410, 1413, 1415, 1425, 1480, 1594, 1631, 1633, 1636, 1638, 1720
Radiation, 1651
radical, xx, 22, 401, 1570
radicals, 842
radio, xxxiv, 718, 1147
radiography, 1297
radiometer, xvii, 244, 254, 259, 260, 263
radius, 200, 254, 364, 367, 523, 622, 623, 936
rain forest, 238, 548, 562, 601, 614, 635, 700, 1651, 1653, 1679, 1719
rainwater, 287, 1376, 1377, 1438, 1559
random, 187, 380, 539, 541, 542, 570, 611, 623, 624, 626, 628, 630, 634, 1128

rangeland, 179, 186, 202, 205, 211, 212, 215, 216, 218, 1322
RAPD, 554, 561, 562
ratification, 179
ratings, 202
rats, xxiv, 593
raw material, xxii, 390, 394, 395, 417, 479, 483, 487, 488, 489, 494, 499, 501, 506, 507, 510, 551, 584, 610
raw materials, xlii, 390, 394, 395, 417, 489, 501, 584, 610, 717, 1261, 1579, 1580, 1585, 1586, 1592, 1596, 1598, 1600, 1617
RB, 1340, 1342, 1344, 1373, 1374, 1375, 1376, 1377, 1379, 1380, 1381, 1384, 1385, 1389, 1392, 1393, 1394, 1395, 1396, 1398, 1399, 1400, 1401, 1402, 1403
RC, 617, 1142, 1145, 1340, 1344
RDA, 29, 33, 200
RDP, 785, 1107, 1117
reactants, 506
reaction mechanism, 503
reaction temperature, 504
reaction time, 498, 500
reactions, 402, 414, 466, 481, 504, 842, 1103, 1109, 1367, 1400, 1621
reactive nitrogen, 1186
reactive oxygen, 767, 1073, 1095
reactive oxygen species, 1073, 1095
reactivity, 502
reading, 521, 1401, 1442
reagent, 1078, 1080
reagents, 1355
realism, 716
reality, 188, 191, 212, 405, 409, 410, 686, 801, 807, 809, 1432
reclamation, 20, 22, 449, 452, 477, 1559, 1568, 1569, 1570
Reclamation, 452
recognition, xliv, 187, 319, 403, 408, 450, 457, 622, 686, 701, 722, 864, 997, 1023, 1258, 1589, 1591, 1630, 1671, 1672, 1676
recolonization, 302, 572
recommendations, iv, xvi, xxii, 178, 197, 513, 515, 1145, 1319, 1331, 1644, 1650
reconcile, 370, 420, 527
reconstruction, 735, 997, 998, 1006, 1022, 1023, 1030, 1033, 1154, 1159, 1167, 1561, 1625
recovery, xxv, 195, 201, 203, 207, 208, 278, 335, 395, 420, 426, 481, 557, 559, 563, 577, 620, 630, 647, 721, 744, 754, 773, 808, 813, 945, 976, 986, 1077, 1101, 1202, 1337, 1467, 1636, 1685, 1718

recreation, xxvii, 451, 638, 741, 1186, 1187, 1192, 1197, 1329, 1331, 1459, 1468, 1724
recreational, 751, 1194, 1351, 1458, 1459, 1462, 1466, 1468
recruiting, 964
recurrence, 1395
recycling, xxii, 272, 394, 395, 483, 490, 503, 504, 677, 718, 721, 949, 1137
Red Cross, 423
red light, 251
Red List, xxv, 222, 226, 236, 603, 606, 607, 608, 609, 610, 612, 613, 617, 1652
redevelopment, 1560
redistribution, 346, 352, 354, 773, 935, 964, 1426, 1433, 1449
redox, 443, 1198
reduction, xxi, xxiii, xxxix, 19, 395, 396, 411, 413, 414, 435, 477, 478, 498, 511, 517, 519, 526, 527, 550, 551, 555, 570, 571, 573, 574, 577, 584, 596, 621, 1208, 1219, 1330, 1336, 1344, 1345, 1355, 1368, 1389, 1392, 1393, 1401, 1402, 1408, 1409, 1414, 1433, 1438, 1439, 1535
redundancy, 818, 888
reed beds, xxxviii, 1371, 1373
reef, 935, 1038, 1454, 1463, 1464, 1465, 1466, 1467, 1468, 1469, 1470, 1475, 1476, 1478, 1479, 1480, 1483, 1486, 1487, 1489, 1490, 1491, 1492, 1493, 1494, 1497, 1498, 1499, 1500, 1501, 1502, 1503, 1504, 1505, 1506, 1507, 1508, 1509
reefs, xl, 635, 1186, 1207, 1456, 1462, 1464, 1466, 1476, 1478, 1490, 1498, 1507, 1508, 1509
reference system, 752
refining, xxii, 483
reflection, 402, 889, 1463, 1576
reflexes, 1101
reform, 419, 701, 710, 866
reforms, 710, 729
refrigeration industry, 727, 728
refuge, 1192
refugees, 735
regenerate, 184, 421, 624
regeneration, xxiii, 196, 272, 276, 306, 322, 335, 417, 550, 557, 558, 853, 1401, 1460, 1646, 1704, 1710
regional, xiv, xxi, xxiii, 13, 23, 180, 190, 279, 398, 413, 435, 456, 458, 514, 529, 539, 543, 544, 574, 938, 941, 975, 995, 1009, 1163, 1189, 1192, 1197, 1332, 1338, 1339, 1409, 1434, 1465, 1543, 1551, 1552, 1565, 1568
regional cooperation, 709
regionalization, 1672, 1673

regions of the world, 514, 975, 1157, 1672, 1676, 1680
regression, xxxiv, 30, 34, 35, 364, 807, 831, 836, 859, 861, 864, 1023, 1039, 1080, 1131, 1147, 1148, 1164, 1165, 1356, 1552, 1558, 1559, 1689, 1693, 1703
regression analysis, 30, 34, 35, 364, 831, 859
regression line, 831, 1552, 1558, 1559
regression model, 34, 859, 861, 864, 1689, 1693
regressions, 1366, 1569
regrowth, 793, 1315, 1694
regular, xxxi, 360, 923, 925, 944, 957, 972, 973, 1103, 1117, 1331, 1396, 1455, 1551
regulation, xv, xix, 43, 46, 50, 156, 157, 304, 320, 334, 345, 353, 354, 419, 423, 436, 440, 458, 636, 1102, 1143, 1186, 1192, 1197, 1591
regulations, xxx, 48, 72, 150, 159, 437, 441, 457, 585, 639, 719, 725, 872, 877, 881, 882, 1102, 1502
regulators, 1198, 1609, 1610
regulatory bodies, 1373
rehabilitate, 196
rehabilitation, 178, 201, 310, 329, 710, 712, 714, 716, 722, 754, 795, 961, 972, 989, 1460
rehabilitation program, 310, 329
relationships, xv, xviii, xx, xxiv, xxxviii, 44, 162, 165, 167, 184, 185, 187, 188, 189, 205, 238, 247, 251, 258, 260, 261, 309, 317, 322, 375, 383, 403, 406, 408, 452, 454, 561, 577, 578, 594, 599, 633, 933, 934, 946, 990, 1139, 1140, 1183, 1186, 1190, 1192, 1194, 1202, 1216, 1223, 1316, 1349, 1351, 1359, 1363, 1364, 1566, 1642, 1653, 1666, 1672, 1719
relatives, xlv, 554, 1636, 1707, 1715
relativity, 1157
relevance, xviii, 178, 189, 266, 326, 329, 339, 806, 934, 1178, 1604, 1613, 1640
reliability, xliv, 113, 408, 446, 889, 1374, 1403, 1415, 1426, 1428, 1577, 1656
relief, 196, 362, 378, 520, 610, 735, 736, 937, 1457
religion, 604, 605
religious traditions, 605
remediation, xxi, 414, 417, 421, 426, 428, 448, 452, 453, 457, 458, 1115, 1229
remodelling, 23
remote sensing, xvii, 190, 195, 217, 243, 245, 455, 1199
renewable energy, 450, 467, 716, 739
renewable fuel, 467
renewable resource, 417
rent, 859, 861
repair, 735, 841, 843, 849, 850, 851, 853, 854, 1464, 1466

reparation, 488, 497, 505, 1603
replacement, 138, 319, 327, 328, 508, 527, 534, 850, 1103, 1242, 1243, 1245, 1248, 1249, 1250, 1251, 1252, 1253, 1254, 1255, 1257, 1260, 1261, 1262, 1264, 1267, 1269, 1271, 1273, 1274, 1275, 1279, 1281, 1282, 1284, 1285, 1286, 1287, 1288, 1289, 1291, 1292, 1293
replication, 688, 690, 824, 1131
representative samples, 189
repression, 1592
reproduction, 336, 369, 602, 622, 623, 648, 771, 779, 788, 804, 806, 810, 812, 840, 880, 902, 922, 1072, 1077, 1092, 1178, 1411, 1431, 1432, 1676
reproductive organs, 770
reproductive state, 1090
reptile, 237, 1685
reptiles, xxiv, 514, 520, 593
requirements, xx, xxxii, 40, 182, 189, 190, 204, 209, 310, 317, 320, 369, 370, 389, 450, 486, 491, 494, 572, 709, 713, 723, 724, 725, 730, 738, 758, 761, 800, 802, 829, 835, 852, 912, 955, 976, 986, 993, 994, 996, 999, 1032, 1238, 1239, 1265, 1280, 1282, 1284, 1286, 1291, 1292, 1294, 1295, 1296, 1335, 1340, 1390, 1402, 1423, 1433, 1441, 1442, 1481, 1577, 1629, 1687, 1688, 1690, 1724
research, xiii, xxxvii, xxxix, xli, 1, 14, 397, 398, 405, 406, 408, 422, 427, 430, 433, 445, 448, 450, 451, 456, 485, 524, 526, 527, 528, 529, 531, 544, 564, 584, 585, 587, 597, 605, 610, 635, 1325, 1336, 1338, 1339, 1343, 1345, 1365, 1366, 1374, 1407, 1416, 1419, 뗌1446, 1449, 1466, 1468, 1470, 1478, 1498, 1506, 1508, 1509, 1512, 1513, 1531, 1542, 1545, 1576
research and development, 398
Research and Development, 219, 1142, 1143, 1144, 1185, 1323
research institutions, 597, 710, 867
researchers, xl, xliii, 187, 188, 189, 197, 203, 312, 316, 329, 760, 780, 781, 826, 996, 997, 1026, 1116, 1186, 1275, 1314, 1315, 1316, 1454, 1467, 1468, 1478, 1499, 1513, 1515, 1517, 1518, 1519, 1533, 1534, 1535, 1540, 1629, 1641, 1649
reservation, 475, 509, 595, 1435
reserves, xxii, xxiii, 237, 369, 387, 407, 452, 513, 519, 520, 521, 523, 524, 526, 531, 532, 534, 535, 536, 537, 538, 539, 540, 541, 542, 543, 550, 699, 773, 774, 779, 784, 791, 833, 950, 1090, 1142, 1462, 1673, 1676, 1679, 1693, 1694, 1700, 1710

reservoir, 464, 523, 572, 968, 1074, 1173, 1383, 1475, 1552, 1555, 1556, 1559
reservoirs, 462, 528, 1109, 1173, 1174, 1344, 1438, 1513
resettlement, 1562
residential, 284, 967, 971, 1125, 1138, 1317
residuals, 806
residues, xxi, 450, 452, 461, 466, 467, 468, 469, 470, 471, 472, 473, 474, 475, 476, 477, 478, 479, 485, 490, 500, 501, 510, 697, 738, 1093, 1095, 1101, 1103, 1333, 1602
resilience, 183, 186, 206, 213, 216, 218, 404, 454, 677, 678, 719
resin, 503
resins, xxii, 483, 503, 507, 511
resistance, xxix, 174, 271, 278, 384, 473, 485, 496, 498, 504, 507, 569, 684, 690, 698, 764, 766, 767, 839, 840, 842, 853, 914, 931, 994, 1285, 1295, 1307, 1337, 1417, 1464, 1466, 1612
resistence, 166
resolution, xxxix, 18, 19, 245, 246, 260, 277, 364, 409, 622, 689, 732, 808, 928, 1035, 1167, 1407, 1415, 1418, 1422, 1431, 1432, 1442, 1564, 1638, 1687
resource allocation, 785, 788, 866
resource availability, 369, 738
resource management, 186, 189, 197, 209, 422, 595, 596, 643, 751
resource utilization, 323, 753
respect, 2, 10, 74, 135, 180, 225, 226, 395, 398, 406, 407, 420, 426, 434, 436, 437, 440, 446, 447, 449, 454, 456, 477, 478, 490, 499, 598, 632, 699, 722, 779, 801, 808, 901, 904, 937, 1087, 1111, 1191, 1239, 1476, 1528, 1552, 1555, 1558, 1559, 1560, 1567, 1649, 1668, 1669, 1713
respiration, 456, 464, 466, 687, 772, 777, 778, 780, 782, 789, 803, 1072, 1076, 1077, 1084, 1086, 1183, 1358, 1361, 1363, 1367
respiratory, 356, 611, 1077, 1359
response time, 1474
responsibilities, 190
restaurant, 600
restaurants, 860
restoration, xxi, xxviii, xlii, 261, 304, 310, 325, 340, 395, 414, 420, 426, 442, 452, 457, 458, 465, 478, 526, 527, 528, 529, 577, 661, 678, 691, 693, 753, 754, 799, 800, 808, 811, 813, 814, 815, 816, 817, 932, 1039, 1115, 1190, 1338, 1370, 1460, 1462, 1550, 1551, 1563
restrictions, xvii, xxiii, 265, 396, 514, 517, 648, 874, 1687, 1689, 1693, 1699
resveratrol, 143, 168

retail, 876
retention, 429, 430, 477, 980, 1106, 1173, 1181, 1182, 1337, 1338, 1344, 1346, 1355, 1356, 1361, 1363, 1365, 1366, 1368, 1369, 1370, 1397, 1460
retinol, 851
returns, 950
revenue, 712
rewards, 668
Reynolds, 162, 173, 312, 313, 316, 317, 318, 319, 320, 323, 325, 327, 328, 329, 339, 343, 358, 1316
rhamnetin, 142, 150, 151, 153, 156, 159
Rhamnetin, 159
rheological properties, 493
rheumatic, 610
rheumatic pain, 610
rhizobia, 317, 327, 329
Rhizobium, 330, 332, 343
rhizome, 87, 770, 771, 772, 773, 774, 775, 777, 778, 779, 780, 781, 782, 783, 785, 787, 788, 789, 793, 1375, 1381
rhizosphere, xix, 309, 311, 312, 313, 314, 316, 317, 321, 326, 330, 331, 332, 334, 336, 338, 339, 341, 343, 351, 453, 1375
Rhodophyta, 1637
rhythms, 186, 991
ribonucleic acid, 318
ribose, 1587
ribosomal, 1117, 1118
ribosomal RNA, 1635
rice, 60, 261, 275, 343, 357, 611, 1118, 1301, 1586, 1615, 1686, 1701
rice field, 611
rigidity, 494, 501, 502, 506, 507
rings, 54, 56, 113, 259, 461, 500, 1377
Río de la Plata, 998, 999, 1000, 1001, 1002, 1021, 1027, 1028, 1029, 1030, 1031, 1032, 1034, 1035, 1036, 1037, 1041
rrisk assessment, 435, 448, 888, 906, 910, 1203, 1346
risks, 202, 423, 426, 459, 464, 477, 584, 610, 621, 634, 698, 864, 876, 886, 906, 1072, 1140, 1327, 1333, 1339, 1459, 1502, 1603
Rita, 361, 372
river basins, 732, 1158, 1182, 1414
river flows, xxvii, 706, 708, 741, 743, 750, 928
River Po, 1434
river systems, 981, 982, 1148, 1335, 1368, 1656
rocks, 192, 224, 225, 337, 414, 417, 426, 447, 454, 464, 531, 727, 735, 1074, 1555, 1627, 1630, 1631
rocky, 234, 930, 935, 989, 1207, 1457, 1467, 1525

rodent, 553
rodents, 715
rods, 861, 1108, 1112, 1114, 1307
rolling, 362, 520, 1302
Rome, 456, 1451
room temperature, 506, 914
root, xix, 2, 47, 48, 87, 136, 160, 163, 167, 171, 173, 259, 309, 313, 314, 315, 316, 317, 321, 322, 326, 327, 330, 332, 333, 334, 335, 338, 340, 343, 346, 351, 352, 357, 358, 385, 431, 432, 446, 453, 463, 468, 470, 471, 690, 735, 761, 769, 770, 772, 775, 779, 780, 781, 782, 783, 787, 792, 796, 797, 933, 950, 1332, 1333, 1338, 1375, 1376, 1412, 1413, 1431, 1524
root growth, 315, 317, 333, 334, 335, 761
root system, 259, 313, 314, 315, 332, 346, 358, 735, 769, 772, 796, 1376
root-mean-square, 1524
roots, xix, 103, 136, 146, 147, 150, 164, 170, 273, 310, 312, 313, 315, 316, 317, 320, 322, 324, 327, 331, 332, 336, 338, 340, 343, 350, 352, 353, 356, 407, 446, 468, 470, 604, 713, 777, 779, 780, 783, 787, 1363, 1375, 1381, 1401, 1413, 1442
ROS, 1091
rotations, 456, 709, 1423
rotifer, 896, 900, 903
rotifers, 888, 898, 900, 901, 902, 903, 906, 907, 908, 909, 952, 955
roughness, 257, 260, 1417, 1524
routes, 474
routing, 968
rowing, 113, 116, 1636
Royal Society, 277, 330, 337, 338, 548, 636, 677, 793, 814, 977, 979, 981, 982, 983, 985, 1368, 1543, 1635, 1640, 1651, 1717
rubber, 1640, 1686, 1689, 1693, 1701, 1702
rubidium, 356
rules, xxviii, 409, 754, 799, 808, 809, 880, 885, 1152, 1158
rumination, xvii, 244, 743,
rural, xxxix, 42, 179, 188, 212, 284, 306, 372, 435, 449, 460, 515, 517, 523, 526, 529, 542, 544, 595, 610, 1315, 1437
rural areas, xxvi, 284, 435, 449, 697, 716, 733, 739, 1254, 1695
rural development, 529
rural population, 179, 515
rural poverty, 701
Russia, 140, 218, 391, 392, 611, 888, 1580
Russian, 615
rust, 384
rutin, 69, 101, 120

Rwanda, 731
rye, xl, 467, 1437, 1441, 1445

S

SA, 329, 341, 481, 561, 592, 976, 981, 987, 988, 989, 1368, 1370, 1373, 1405, 1436, 1451
SAC, 37, 41
sacred, 612
safeguard, 427, 557, 1466
safeguards, 450
safety, 752, 833, 1458, 1465, 1499, 1502, 1507, 1551, 1552, 1559, 1560, 1580, 1593, 1594
saline, 432, 455, 932, 940, 949, 1021, 1114, 1211, 1212, 1409, 1418, 1474
saline water, 455, 940
salinity gradient, 932, 939, 940, 958, 963, 1003, 1007, 1009, 1019, 1124, 1125
salinity levels, 951
salinization, 432, 449, 452
salmon, 825, 827, 837, 850, 851, 852, 853, 854, 1240, 1242, 1258, 1261, 1286, 1290, 1291, 1293, 1294, 1295, 1297
Salmonella, 1341
salt, xxxii, xxxiii, xli, 349, 355, 432, 449, 452, 488, 490, 924, 931, 948, 949, 950, 959, 985, 998, 999, 1023, 1031, 1034, 1038, 1073, 1123, 1124, 1127, 1143, 1187, 1519, 1549, 1553, 1561, 1562, 1563, 1567, 1614
salt tolerance, 1143
salts, 432, 489, 490, 497, 499, 500, 571, 1105, 1392, 1393, 1395, 1594
saltwater, 770, 1197, 1457
sample, xliii, 4, 5, 29, 35, 76, 189, 195, 210, 226, 234, 363, 365, 471, 472, 531, 535, 537, 540, 959, 1001, 1003, 1004, 1014, 1018, 1019, 1026, 1027, 1028, 1040, 1041, 1076, 1080, 1116, 1127, 1134, 1141, 1151, 1157, 1162, 1204, 1205, 1206, 1208, 1209, 1212, 1228, 1306, 1307, 1308, 1309, 1314, 1315, 1316, 1317, 1357, 1388, 1389, 1393, 1567, 1655, 1661, 1665, 1691, 1695, 1713, 1715
sampling error, 1388
samplings, xxxviii, 1028, 1074, 1075, 1081, 1349
sanctuaries, 699, 1686, 1696, 1700, 1727
sandstones, 447
Santos Estuarine System, xxxv, 1202, 1203, 1209, 1210, 1214, 1217
Santos Estuarine System (SES), xxxv, 1202
SAP, 708
saponin, 1276, 1288, 1292, 1294
saponins, 170, 175
sapphire, 1259
SAR, 1568

SAS, 217, 1207
satellite, xvi, xvii, xxxix, 177, 193, 207, 244, 245, 246, 248, 250, 252, 259, 260, 261, 364, 927, 1029, 1119, 1407, 1410, 1468, 1564
satellite imagery, 193
satellite technology, 250
satellites, 246, 1457
saturation, 431, 432, 447, 641, 651, 655, 743, 834, 941, 1124, 1139, 1197, 1219, 1319, 1411, 1414, 1417, 1660
Saudi Arabia, 100
savannah, 699, 701, 1646
savings, 394, 701, 728, 1246
SBF, 514, 520, 547
SBR, 525
scaffold, 508
scaffolds, 509, 511
scaling, 631, 632, 633, 815, 816, 818, 928, 933, 1577
scaling law, 816
scaling relations, 933
scaling relationships, 933
Scandinavia, 377
Scanning electron, 1107
scanning electron microscopy, 1117
scarcity, 317, 318, 551, 636, 719, 732, 735, 803, 996, 1139, 1629
scatter, 35, 446, 504, 801, 810
scattering, 1460
scavenger, 1091
scent, 1598
Schiff, 485
Schiff base, 485
Schmid, 375, 376, 377, 378, 383, 384, 385, 386
scholarships, 1116
school, 210, 403, 404, 685, 737, 738, 1469, 1629
schooling, 1278
Schrödinger equation, 402
science, xxi, xxvi, 185, 218, 338, 402, 403, 408, 411, 413, 454, 455, 457, 458, 459, 599, 636, 677, 683, 684, 689, 690, 762, 809, 980, 990, 1031, 1095, 1368, 1577, 1679, 1726, 1733
scientific community, xviii, xxi, 266, 413, 454, 464, 1568
scientific knowledge, 179, 181, 216, 218, 522
scientific method, 178, 179, 180, 183, 187, 195, 199
scientific understanding, 809, 1351, 1365
scientists, xxi, 402, 406, 413, 432, 450, 455, 457, 468, 1327, 1568, 1570
scope, 585, 639, 670, 674, 1094, 1097, 1524, 1656
scores, 291, 292, 293, 378, 967, 1211
SCP, 1264
screening, 61, 165, 1378, 1606
SEA, 987, 1680

sea ice, 1124, 1527
sea level, xxxiv, xli, 246, 285, 514, 926, 931, 935, 936, 974, 984, 1035, 1037, 1127, 1147, 1148, 1149, 1152, 1155, 1158, 1164, 1166, 1187, 1197, 1302, 1423, 1429, 1431, 1457, 1480, 1549, 1550, 1551, 1552, 1553, 1559, 1560, 1561, 1562, 1563, 1564, 1566, 1567, 1568, 1569, 1570, 1577, 1657
sea urchin, xxxvi, 1202, 1206, 1228, 1465
seabed, 582, 1464, 1468, 1470, 1552
seafood, 582, 584, 1199
sea-level, 994, 1021, 1032, 1034, 1035, 1040, 1165, 1187, 1197, 1566
sea-level rise, 1187, 1197
seals, 584
search, 342, 358, 360, 364, 365, 516, 522, 526, 531, 613, 614, 1117, 1192, 1370, 1461, 1643
search terms, 758
searches, xvi, 221, 226, 1194
searching, 623, 1590
seasonal changes, 301, 303, 785, 899, 928, 962, 964, 1132, 1425, 1469
seasonal flu, xviii, 283, 285, 295, 296, 301, 899, 979, 1409
seasonal growth, 343, 782
seasonal pattern, xxxiv, 1011, 1123, 1131, 1133, 1135, 1138, 1139
seasonal variations, 186, 200, 961, 1036, 1431, 1474
seasonality, xxxi, 705, 866, 887, 909, 926, 963, 968, 999, 1028, 1092, 1096
Seattle, 614
seawater, xxxiv, xxxix, 925, 932, 939, 941, 942, 952, 955, 987, 1074, 1076, 1077, 1093, 1123, 1124, 1128, 1129, 1130, 1134, 1140, 1144, 1171, 1206, 1207, 1211, 1407, 1432, 1570
seaweed, 407, 1560
second generation, 467, 1589
Second World, 744
secondary data, 1725
secondary metabolism, 171
sectoral policies, 710
secular, 1409, 1566
secular trend, 1566
security, xx, xxi, 196, 375, 413, 672, 697, 1433
sedative, 610
sedentary, 196, 202, 957
Sediment Quality Triad, xxxv, 1202, 1230, 1231, 1233
seed, xxiv, 47, 48, 162, 207, 321, 330, 351, 378, 552, 554, 556, 558, 560, 593, 736, 760, 768, 782, 792, 794, 865, 950, 987, 1248, 1249, 1251, 1279, 1284, 1286, 1287, 1291, 1296, 1672, 1676, 1708, 1719, 1720, 1721

seeding, 353, 450, 451, 452, 891, 1441
seedlings, 321, 324, 353, 358, 523, 526, 527, 759, 761, 781
seeds, 138, 146, 147, 149, 153, 169, 170, 174, 273, 301, 354, 552, 553, 555, 558, 560, 561, 950, 1441, 1449
segregation, 315, 394, 570
seismic, 423
selecting, 189, 205, 384, 503, 555
selectivity, 342
selenium, 1093
Self, 173, 411, 481, 511, 1368
self-organizing, 404
self-regulation, 867
self-sufficiency, 1686
seller, 612, 728
SEM, 1114
semen, 163, 912, 919
semiarid, xxxix, 259, 341, 1012, 1334, 1340, 1437, 1439
semi-natural, xviii, xxxvii, 38, 283, 284, 285, 286, 287, 302, 303, 304, 305, 306, 338, 356, 376, 386, 456, 1325
semi-structured interviews, xvi, 177, 188, 189, 198, 406
Senate, 601
senescence, 9, 471, 472, 775
sensations, 403
senses, 690
sensing, xvii, xlii, 190, 195, 200, 217, 243, 245, 455, 1167, 1199, 1579, 1590, 1591, 1609, 1610, 1612, 1615, 1616, 1621, 1702
sensitivity, xiv, 27, 383, 422, 443, 449, 485, 505, 569, 620, 629, 633, 745, 865, 904, 913, 914, 916, 917, 967, 972, 994, 1193, 1393, 1691
sensors, 245, 246, 259, 1256, 1442
SEPA, 41, 1323
separation, 327, 372, 380, 411, 476, 489, 490, 491, 494, 499, 1103, 1393, 1542, 1545, 1589
sequencing, 1586, 1588
series, xvii, 23, 24, 95, 140, 162, 182, 217, 244, 247, 260, 403, 414, 437, 569, 578, 624, 632, 928, 995, 1001, 1009, 1014, 1015, 1021, 1097, 1108, 1148, 1158, 1330, 1377, 1379, 1388, 1402, 1411, 1419, 1421, 1423, 1463, 1477, 1513, 1561, 1566, 1569, 1576
serine, 1244, 1726
serum, 1105
SES, xxxv, 1202, 1203, 1208, 1211, 1212, 1213, 1214, 1215, 1216, 1218, 1219, 1220, 1222, 1223, 1224, 1226, 1227, 1228
seta, 1124, 1143

settlements, xvi, xli, 177, 178, 196, 197, 198, 199, 201, 202, 203, 204, 205, 206, 207, 208, 517, 736, 971, 1009, 1549, 1551, 1562, 1569
settlers, 596
severe stress, 1092, 1178
severity, 620
sewage, xxxvi, 422, 423, 426, 427, 433, 439, 440, 441, 450, 455, 687, 698, 753, 758, 781, 792, 797, 942, 944, 947, 968, 970, 973, 981, 1031, 1101, 1202, 1211, 1213, 1214, 1215, 1220, 1229, 1335, 1367, 1376
sex, 573, 919, 1283
sex ratio, 573
sexual reproduction, 1676
SH, 254, 256, 257, 258, 355, 1078
shade, 273, 371, 552, 761, 842, 1370, 1446, 1657
shallow lakes, 803
Shanghai, 615, 617, 1182
shape, 28, 50, 60, 181, 200, 226, 330, 335, 364, 367, 486, 493, 497, 519, 553, 725, 804, 845, 849, 928, 931, 1018, 1186, 1441, 1470, 1474, 1476, 1480, 1491, 1499, 1507, 1540, 1656
shaping, 184, 326, 356, 486, 487, 493, 495, 497, 562, 600, 1158
sharing, 60, 180, 183, 408, 454, 618, 1332, 1373, 1724
shear, xl, 1511, 1517, 1523, 1534, 1542, 1543, 1547, 1575, 1666
sheep, xvii, xxiv, 196, 202, 243, 246, 247, 248, 249, 250, 252, 261, 262, 492, 594, 596, 598, 604, 605, 609, 699, 701, 1730, 1731
Sheep, xvii, 243
shelf life, 1608
Shell, 1080, 1556, 1557
shellfish, 582, 858, 868, 1187, 1329
shelter, xxiii, xliv, 476, 514, 520, 523, 612, 735, 971, 1308, 1656, 1666
shelters, 196, 1656
shipping, 253, 1101, 1187, 1193, 1515, 1519, 1547, 1567
ships, 582, 585, 873, 1540
shock, 749, 919, 1143, 1385, 1393, 1395
shocks, 405
shoot, 332, 334, 349, 351, 357, 383, 385, 463, 470, 772, 773, 775, 776, 779, 781, 782, 783, 785, 786, 789, 793
shoots, 170, 771, 772, 774, 775, 777, 780, 781, 794, 1646
shorebirds, 983
shoreline, xlii, 14, 857, 929, 930, 1002, 1454, 1455, 1462, 1463, 1467, 1468, 1469, 1475, 1476, 1477, 1478, 1479, 1480, 1481, 1489, 1502,

1504, 1505, 1507, 1518, 1525, 1536, 1575, 1576, 1577
shores, 1142, 1478, 1631
short period, 22, 928, 936, 974, 1512
shortage, 736, 1413, 1433
shortfall, 1284
short-term, 183, 207, 356, 610, 954, 1355, 1462, 1566
Short-term, 1037
shoulder, xvii, 243
shrimp, 509, 856, 857, 859, 860, 861, 862, 864, 1137, 1140, 1145, 1187, 1256, 1258, 1260, 1291, 1293
shrinkage, 714
shrubland, 269, 280
shrublands, 218, 274
shrubs, 189, 192, 194, 201, 234, 273, 338, 339, 362, 699, 1127, 1334, 1340
side chain, 485, 493, 496, 504
sieve tube, 60
sign, 34, 194, 432
signalling, 1327, 1591, 1621
signals, xxxiii, 44, 748, 997, 1071, 1072
significance level, 1359
signs, xxxvi, 782, 843, 846, 944, 1101, 1202, 1215, 1228, 1695
silica, 895, 1014, 1175, 1205
silicate, xxxv, 942, 1169, 1171, 1172, 1174, 1175, 1179, 1181, 1182
silver, 835, 845, 1285, 1286, 1296, 1716
similarity, xv, 43, 60, 145, 152, 363, 365, 378, 568, 1022, 1109, 1507, 1528, 1582
simulation, xxxvii, xxxix, 158, 159, 227, 229, 231, 339, 532, 535, 540, 541, 622, 623, 624, 629, 631, 647, 648, 653, 663, 667, 668, 793, 988, 1119, 1299, 1344, 1407, 1411, 1412, 1413, 1417, 1418, 1419, 1421, 1422, 1423, 1425, 1427, 1428, 1430, 1431, 1432, 1434, 1436, 1685, 1687, 1693, 1702
simulations, xvi, 222, 226, 227, 232, 540, 622, 623, 625, 626, 631, 633, 636, 648, 652, 672, 745, 785, 1165, 1344, 1423, 1433, 1497, 1498, 1499, 1502, 1503, 1504, 1506, 1570, 1685, 1686, 1690
Sinai, 1287
Singapore, 617, 1032, 1640
sink habitat, 1695
sinusitis, 610
SiO2, 942
siphon, 706, 708
skeleton, 49, 54, 56, 84, 150, 440
skewness, 1151
skills, 408, 1484, 1588

skin, 484, 486, 487, 488, 490, 494, 496, 508, 609, 841, 842, 844, 845, 846, 847, 848, 849, 850, 851, 852, 853, 854, 1295, 1730
SLA, 1413
Slovenia, 447
sludge, 422, 423, 426, 427, 433, 439, 440, 441, 446, 450, 455, 499, 500, 725, 781, 792, 973, 1101, 1117, 1118, 1120, 1331, 1353, 1393, 1404
sludge disposal, 450
small mammals, 373, 374, 524
smelting, 1097
Smithsonian, 236, 370, 374
Smithsonian Institution, 236
smooth muscle, 484
snakes, 599, 1730
SOC, xxi, 270, 461, 463, 466, 468, 470, 473, 476, 477, 478
social activities, 186, 1590
social behavior, 547
social benefits, 1724
social capital, 399
social class, 187
social conflicts, xxx, 855, 856
social context, 457
social costs, 737
social development, 530, 716, 1704
social factors, 187, 456
social interests, 1560
social life, 716
social norms, 181
social organization, 370
social relations, 866
social responsibility, 527
social services, 701
social structure, 237
social systems, 405
social workers, 735
society, xx, 188, 192, 390, 397, 401, 404, 410, 418, 419, 421, 426, 450, 456, 459, 527, 644, 711, 726, 730, 737, 1186, 1197, 1577, 1689
socioeconomic, 597
sodium, 489, 490, 492, 493, 826, 844, 1105, 1390, 1391, 1395, 1595
software, xxiv, 200, 226, 230, 231, 247, 287, 364, 378, 379, 387, 532, 534, 581, 586, 588, 591, 859, 1107, 1128, 1157, 1192, 1193, 1207, 1208, 1551, 1613, 1660, 1669, 1687, 1690
Soil acidification, 431, 433, 458
soil analysis, 35
Soil and Water Assessment Tool (SWAT), 1318
soil erosion, 23, 24, 271, 407, 420, 425, 430, 435, 464, 468, 477, 526, 596, 744, 937, 967, 1327,

1329, 1331, 1332, 1333, 1334, 1645, 1689, 1693
soil particles, 1329, 1332, 1376, 1391, 1393, 1400, 1401
soil pollution, xxi, 413, 420, 428, 1101
soil seed bank, 782
soil type, xxiii, 4, 205, 246, 313, 330, 424, 431, 434,
solar, xxv, 253, 254, 256, 620, 942, 1132, 1139, 1413, 1425
solar energy, xxv, 620
solid phase, 498, 500
solid waste, xxix, 395, 399, 440, 587, 819, 1120
Solomon I, 219
Solomon Islands, 219
sols, 317
solubility, 495, 496, 498, 502, 503, 505, 506, 507, 827, 851, 1136, 1357, 1401, 1596, 1614
solution, xxx, 44, 276, 318, 322, 395, 409, 426, 440, 495, 496, 497, 500, 597, 643, 644, 647, 648, 667, 670, 674, 675, 698, 734, 796, 805, 844, 855, 867, 915, 917, 921, 1080, 1115, 1205, 1206, 1207, 1304, 1327, 1331, 1355, 1369, 1459, 1460, 1461, 1515, 1524, 1545, 펨1570, 1687
solutions, 486, 488, 490, 492, 493, 499, 502, 527, 598, 1332, 1459, 1461, 1462, 1465, 1513, 1536, 1577
solvent, 489, 503, 504, 1118
solvents, 488, 497, 725, 1102
Somalia, 223
somatic cell, 922
Sorghum, 69, 70, 169, 471, 478
sorption, 488, 729, 1336, 1344
sorting, 1483, 1541
South Africa, viii, xxxi, 212, 392, 616, 795, 867, 923, 924, 925, 926, 927, 928, 929, 930, 931, 932, 933, 934, 935, 936, 937, 938, 940, 942, 943, 944, 946, 947, 948, 949, 951, 953, 954, 955, 957, 959, 961, 965, 966, 967, 968, 969, 971, 972, 973, 974, 975, 976, 977, 978, 979, 980, 981, 982, 983, 984, 985, 986, 987, 988, 989, 990, 991, 1676
South America, xviii, 69, 88, 265, 266, 269, 274, 550, 566, 568, 878, 880, 888, 999, 1030, 1034, 1036, 1037, 1039, 1167, 1672, 1678, 1717, 1720
South Asia, 613
South Korea, 392, 611, 617
South Pacific, 1650
Southampton, 1572
Southeast Asia, 562, 600, 617, 888, 891, 899, 909, 1638, 1684, 1685, 1696

Southern Africa, 213, 616, 907, 976, 979, 982, 984, 987, 990, 1231
sowing, xl, 20, 718, 1412, 1437, 1441, 1449
soy, 165, 509
soy isoflavones, 165
soybean, 3, 10, 1344
soybeans, 274, 275
SP, 361, 371, 372, 561, 1120, 1144, 1201, 1203
SPA, 1073
space, xv, 44, 84, 158, 253, 273, 311, 312, 315, 316, 326, 403, 404, 432, 434, 451, 514, 516, 527, 531, 599, 632, 633, 729, 758, 801, 812, 874, 881, 916, 1023, 1028, 1191, 1498, 1507, 1565, 1649, 1677, 1679, 1703
Spain, x, xl, 23, 113, 129, 312, 337, 392, 548, 585, 765, 911, 1030, 1198, 1231, 1325, 1333, 1334, 1339, 1340, 1344, 1349, 1351, 1366, 1368, 1437, 1438, 1439, 1440, 1441, 1457, 1458, 1462, 1509, 1640, 1680, 1707
spatial, xix, xxxix, 18, 23, 182, 185, 186, 190, 193, 194, 202, 206, 245, 281, 310, 314, 315, 316, 322, 331, 333, 338, 339, 340, 364, 366, 367, 372, 397, 418, 447, 454, 455, 554, 555, 571, 599, 621, 622, 629, 947, 948, 957, 997, 1002, 1003, 1011, 1014, 1026, 1028, 1035, 1039, 1131, 1133, 1140, 1171, 1178, 1186, 1193, 1194, 1195, 1306, 1368, 1407, 1411, 1417, 1423, 1426, 1432, 1434, 1436, 1443, 1474, 1552
spatial heterogeneity, 322, 340
spatial information, 23
spawning, 957, 963, 964, 969, 1207, 1329
specialists, 328, 450, 572, 599, 634, 816
specialization, xxiii, 60, 61, 70, 328, 329, 565, 1647
speciation, xxiii, xliii, 565, 567, 568, 569, 578, 579, 1637, 1642, 1644, 1647, 1713, 1717, 1720
species richness, xvi, xx, xxii, xxvii, xlv, 38, 182, 195, 200, 202, 203, 218, 222, 226, 227, 229, 231, 232, 235, 305, 323, 335, 341, 347, 364, 365, 371, 373, 375, 376, 378, 379, 381, 382, 384, 513, 514, 532, 534, 535, 538, 541, 542, 543, 757, 758, 759, 760, 761, 767, 768, 815, 816, 818, 888, 900, 950, 951, 957, 969, 1208, 1221, 1222, 1224, 1338, 1656, 1668, 1675, 1676, 1677, 1692, 1719, 1723, 1724
specific surface, 476
specific tax, 955, 1632
specificity, 311, 313, 319, 323, 325, 328, 329, 336, 352, 357, 1691
spectral signatures, 201
Spectrophotometer, 1076
spectrophotometric, 1078, 1097, 1144
spectrophotometric method, 1097

spectrophotometry, 1130, 1144, 1206
spectroscopy, 170, 260, 455, 916, 921, 1304
spectrum, 266, 527, 630, 631, 634, 965, 1594, 1595, 1597, 1610, 1614
speech, 524
speed, 313, 403, 404, 941, 1003, 1333, 1415, 1482, 1483, 1485, 1513, 1517, 1521, 1522, 1525, 1526, 1555, 1568, 1581
spending, 248
sperm, xxxi, 911, 912, 913, 918, 919, 1207
spheres, 1543
spills, 1101
spin, 803
spine, 567, 1502
spin-offs, 803
spiritual, 605
spleen, 852, 854
spoil, 357
sponge, 509, 746, 1337
sponges, 486, 1465
sporadic, 1170
SPR, 18, 19
Spring, 964, 1034, 1074, 1075, 1080, 1084, 1086, 1088, 1142, 1321
springs, 435, 523, 524, 526, 544, 574, 575, 576
sprue, 1580, 1586, 1602, 1608, 1609, 1618
SPSS, 217, 1081, 1157, 1359, 1660, 1669
SPSS software, 1157
SR, 385, 1145, 1340, 1341, 1343
Sri Lanka, xxxi, 734, 872, 905, 906, 907, 909, 923, 924, 1508
St. Petersburg, 1187
stability, xx, xlii, 56, 185, 207, 213, 310, 322, 329, 339, 346, 357, 375, 376, 387, 403, 410, 429, 432, 450, 452, 454, 487, 488, 490, 493, 494, 498, 523, 577, 645, 649, 671, 678, 679, 729, 805, 867, 899, 942, 954, 958, 962, 964, 981, 983, 985, 1039, 1241, 1337, 1460, 1463, 1478, 1565, 1568, 1579, 1580, 1588, 1590, 1593, 1604, 1618, 1659, 1664, 1714
stabilization, xxi, 22, 461, 465, 466, 468, 474, 475, 476, 477, 750, 867, 922, 1039, 1459, 1460
stabilize, 464, 516, 1359, 1576
stable isotopes, 332, 355, 1145
stages, 72, 73, 191, 284, 302, 303, 463, 472, 488, 558, 584, 586, 587, 630, 951, 1104, 1374, 1377, 1379, 1385, 1388, 1396, 1408, 1413, 1421, 1425, 1448, 1543, 1552, 1586, 1588
stainless steel, 1204
stakeholder, 595, 597, 598, 602
stakeholder groups, 595, 597
stakeholders, 209, 210, 397, 398, 451, 595, 597, 598, 599, 600, 1189

standard deviation, 73, 290, 292, 295, 534, 641, 648, 671, 672, 824, 859, 1151, 1205, 1392, 1396, 1400
standard error, 5, 807, 859, 1081, 1354, 1360, 1362
standard of living, 467
standardization, 912, 913, 1588
standards, xxv, 398, 619, 1076, 1205, 1301, 1309, 1314, 1319, 1322, 1389
Standards, 1149, 1340
starch, xxii, 467, 483, 505, 506, 507, 508, 511, 512, 772, 773, 793, 833, 1281, 1295, 1580, 1591, 1600, 1612, 1621, 1638
starch granules, 1591
starch polysaccharides, 1600
starches, 511
starvation, 789
stasis, 610, 1709
State Department, 747, 749, 755, 756
state intervention, xxx, 855, 867
states, xix, xxxi, 46, 57, 92, 149, 178, 179, 180, 266, 359, 402, 409, 410, 551, 554, 645, 646, 647, 650, 675, 680, 690, 709, 718, 731, 732, 744, 762, 763, 856, 864, 865, 874, 875, 876, 923, 929, 934, 1124, 1193, 1300, 1456, 1732
statistical analysis, 24, 29, 1151, 1566
statistical inference, 995
statistics, 201, 205, 252, 374, 379, 677, 700, 868, 880, 1144, 1241, 1409, 1423, 1703, 1704
steel, 1101, 1204, 1214, 1215, 1229, 1307, 1377
steel mill, 1101
sterile, 318, 779, 1105, 1561
steroids, 1295
stiffness, 485, 494
stimulant, 610, 1256, 1257, 1275, 1297, 1592, 1605
stimulus, 605, 850
stochastic, 341
stochastic model, 647, 648, 652, 653, 657, 658, 661, 662, 663, 667, 668, 669, 671
stochastic processes, 641, 672
stock, 185, 186, 202, 946, 953, 954, 956, 1074, 1105, 1301, 1366
stock price, 641
stomach, 827, 1247, 1601
storms, 424, 933, 1138, 1309, 1440, 1444, 1445, 1446, 1448, 1449, 1455, 1474, 1475, 1482, 1512, 1570, 1576, 1577
stormwater, 1330, 1338, 1374
stoves, 720
strain, xxv, 619, 1588, 1589, 1592, 1595, 1597, 1600, 1608, 1614, 1618
strains, 1102, 1103, 1109, 1120
strategic, 398
strategic planning, 398, 738, 1179

stratification, xxxi, 187, 688, 899, 923, 932, 939, 941, 942, 985, 1039, 1534
stratified sampling, 690
stream water quality, xxxvi, 1286
strength, 315, 485, 486, 488, 502, 503, 504, 506, 584, 629, 1376, 1388
streptococci, 1330, 1341
Streptomyces, 493
stress, xxxiii, xxxv, xxxvi, xxxix, xl, 391, 420, 422, 448, 687, 690, 733, 734, 735, 782, 783, 800, 840, 842, 851, 852, 853, 888, 930, 942, 1030, 1071, 1072, 1075, 1089, 1090, 1091, 1092, 1093, 1097, 1140, 1143, 1178, 1185, 1187, 1202, 1223, 1316, 1407, 1439, 1448, 1461, 1480, 1511, 1517, 1523, 1534, 1542, 1547, 1575, 1591, 1599, 1600, 1611, 1636
stress factors, 1178
stress level, 1075, 1090, 1091
stressors, xxvi, xxxv, 683, 684, 687, 688, 692, 1072, 1090, 1185, 1187, 1192, 1326
stretching, 254, 514, 927, 1728
striatum, 145, 172, 1019
stroke, 609
strong interaction, 326
strontium, 356
structural characteristics, 1600
structural modifications, 435
structural protein, 472
structural variation, 56
structuring, 320, 331, 384, 555, 904, 959
students, 1319
style, 406, 686, 691, 1403
subcutaneous tissue, 498, 842
suberin, 470, 471, 473, 474, 475
subjective, 405
subjectivity, 806, 1208
sub-Saharan Africa, 178, 185, 212, 702, 734
subsidy, 698
subsistence, 360, 370, 737, 1456
subsistence farming, 370
substances, xxiv, 393, 426, 433, 441, 443, 448, 455, 456, 463, 474, 488, 490, 494, 495, 499, 500, 501, 503, 506, 507, 588, 589, 603, 604, 609, 610, 1090, 1101, 1338
substitutes, 508
substitution, 49, 60, 90, 102, 103, 111, 121, 1261, 1264, 1267, 1280, 1294, 1582, 1631, 1632
substitutions, xxxvi, 45, 49, 54, 56, 57, 60, 77, 90, 110, 721, 880, 1237
substrate, 318, 414, 476, 742, 747, 750, 751, 752, 772, 782, 786, 788, 789, 796, 797, 798, 897, 1078, 1103, 1108, 1109, 1114, 1178, 1464, 1553, 1583, 1588, 1672

substrates, 319, 572, 780, 895, 897, 1103, 1104, 1106, 1108, 1109, 1363, 1580, 1583, 1591, 1652, 1667
subsurface flow, 2, 4, 9, 743, 751, 1125
subtraction, 18, 20
subtropical forests, 1713
suburban, 449, 1337
succession, 212, 262, 300, 303, 306, 319, 323, 327, 328, 332, 334, 335, 337, 342, 355, 357, 414, 630, 752, 782, 897, 899, 900, 905, 1010, 1034, 1183, 1434, 1656, 1667, 1687, 1690, 1703
sucrose, 773, 913, 914, 917, 922, 1078, 1587, 1592, 1600, 1601
Sudan, vii, xxvi, 214, 613, 697, 698, 699, 700, 701, 702, 704, 705, 706, 707, 708, 709, 711, 712, 713, 714, 715, 716, 717, 718, 719, 721, 722, 724, 729, 731, 732, 733, 734, 735, 736, 737, 738, 739
suffering, 244, 259, 455
sugar, 57, 59, 63, 87, 120, 142, 160, 361, 362, 368, 369, 467, 469, 474, 596, 968, 971, 1583, 1600
sugar beet, 467
sugar cane, 361, 362, 968, 971
sugarcane, 360, 467, 596, 942
sugars, 48, 50, 69, 121
sulfate, 490, 1103, 1120, 1121, 1166, 1295, 1342
Sulfide, 1234
sulfur, 492, 1204, 1242, 1292
sulfuric acid, 490
sulphate, 57, 64, 164, 1117, 1205
sulphur, 1212, 1213, 1220, 1279, 1292
Sumatra, 1035
summaries, 1306, 1339
summer, xvii, xxxiii, xxxix, 3, 4, 14, 29, 192, 244, 247, 248, 253, 254, 273, 551, 926, 939, 944, 947, 951, 961, 962, 964, 1010, 1011, 1100, 1101, 1125, 1131, 1133, 1134, 1135, 1136, 1138, 1140, 1171, 1175, 1177, 1179, 1184, 1211, 1302, 1312, 1318, 1330, 1351, 1353, 1390, 1391, 1407, 1409, 1419, 1422, 1431, 1432, 1440, 1446, 1448, 1525, 1528
Sun, 1169, 1172, 1182, 1183, 1184, 1296, 1373, 1375, 1376, 1388, 1392, 1401, 1405, 1435, 1436, 1513, 1518, 1536, 1546, 1547, 1640, 1712, 1713, 1716, 1720, 1721
sunlight, 1211, 1328, 1329
supernatant, 1078, 1080, 1130, 1204
superposition, 403, 405, 409
supplemental, 1118
supplementation, xxix, 820, 823, 825, 826, 833, 834, 836, 837, 839, 842, 843, 845, 849, 850, 851, 853, 1244, 1245, 1248, 1250, 1251, 1256, 1257, 1258, 1267, 1273, 1275, 1280, 1282, 1284,

1285, 1287, 1288, 1289, 1292, 1293, 1297, 1601
supplier, 1080
supply, xvii, xl, 244, 252, 317, 332, 335, 349, 355, 419, 467, 488, 572, 928, 933, 942, 983, 994, 1092, 1175, 1181, 1191, 1197, 1332, 1335, 1343, 1350, 1377, 1384, 1385, 1386, 1387, 1390, 1403, 1412, 1413, 1439, 1450, 1511, 1512, 1551, 1554, 1560, 1565, 1567, 1580, 1620
support services, 1186
suppression, 301, 350, 523, 767, 842, 850
supramolecular, 474, 484
Supreme Court, 744, 754
surface area, 452, 473, 474, 476, 597, 705, 751, 779, 897, 901, 931, 959, 1003, 1338, 1373, 1375, 1379, 1391
surface component, 1275
surface layer, 432, 433, 1124, 1474
surface properties, 1276
surface roughness, 1417
surface water, xxxvii, 430, 1010, 1030, 1171, 1175, 1176, 1177, 1211, 1301, 1320, 1322, 1325, 1326, 1327, 1328, 1329, 1331, 1332, 1333, 1335, 1336, 1341, 1343, 1344, 1346
surface wave, 1577
surfactants, 488, 489, 496, 499, 507, 510, 1224
surfing, 1463, 1466, 1467, 1468, 1470, 1471, 1480, 1481, 1483, 1485, 1491, 1494, 1497, 1501, 1502, 1503, 1504, 1507, 1509
surgical, 487
surplus, 420, 431, 716, 737, 820, 827
surveillance, 865, 1093
survey, xv, xvii, xliii, 18, 19, 39, 44, 45, 60, 63, 73, 138, 160, 161, 162, 170, 171, 201, 222, 238, 244, 251, 252, 253, 254, 261, 410, 429, 431, 437, 442, 447, 449, 459, 485, 611, 615, 694, 793, 859, 868, 876, 904, 942, 953, 981, 982, 983, 989, 1032, 1094, 1128, 1135, 1137, 1138, 1171, 1183, 1184, 1469, 1541, 1552, 1589, 1607, 1655, 1661, 1665, 1695, 1696, 1717, 1718, 1725
survival, xxvii, xxviii, xliii, 284, 287, 319, 323, 371, 517, 530, 544, 552, 556, 570, 577, 604, 610, 611, 612, 635, 638, 736, 738, 749, 769, 775, 779, 780, 799, 805, 806, 866, 873, 914, 916, 918, 921, 922, 950, 957, 969, 984, 986, 1072, 1206, 1220, 1258, 1322, 1341, 1342, 1570, 1571, 1590, 1633, 1641, 1708, 1710, 1713, 1715
survival rate, 914, 1258
surviving, 530, 950, 958, 1206
survivors, xlv, 1707

susceptibility, 421, 423, 578, 999, 1091, 1119, 1248
suspensions, 489, 1545, 1600
sustainability, xxvi, xxvii, xxxv, 185, 208, 212, 341, 389, 390, 397, 405, 422, 454, 528, 529, 530, 581, 582, 584, 585, 637, 638, 639, 640, 643, 644, 645, 649, 650, 669, 670, 676, 677, 678, 679, 680, 698, 721, 726, 727, 742, 751, 754, 820, 859, 865, 867, 868, 880, 1185, 1188, 1191, 1238, 1261, 1339, 1401, 1507
sustainable development, 389, 395, 399, 404, 422, 455, 460, 544, 584, 612, 644, 697, 722, 730, 856, 866
sustainable growth, 730
swamps, 224, 225, 369, 477, 611, 1021
Sweden, 304, 447, 585, 685, 789, 791, 1181, 1634, 1637
swelling, 485, 492, 493
Switzerland, 302, 375, 377, 378, 384, 386, 387, 428, 592, 615, 617, 1703, 1717, 1733
symbiont, 337
symbioses, 322, 325, 328, 341, 349
symbiosis, xix, 310, 318, 324, 325, 332, 333, 334, 335, 336, 338, 341, 346, 347, 356, 357, 1640
symbiotic, 531
symbolic, 605
symbols, 252, 434, 1453, 1520, 1524
symptom, 574
symptoms, xxx, xxxv, 754, 782, 855, 866, 1169, 1176, 1179, 1365, 1594
synchronization, 555, 911
syndrome, xxxviii, 486, 787, 1097, 1350, 1365, 1370
synergistic, 469, 1192
synergistic effect, 1192
synthesis, 51, 69, 179, 197, 277, 330, 539, 543, 560, 635, 776, 788, 813, 816, 841, 851, 1104, 1148, 1270, 1282, 1581, 1595, 1597, 1600, 1606, 1678, 1721
synthetic polymers, xxii, 483, 486, 494
systematics, 60, 162, 164, 172, 546, 1121

T

tactics, 804, 808, 816, 817, 818
tags, 524
Taiwan, 611, 1182, 1713, 1714, 1718, 1720
Taiwan Strait, 1182
TAMU, 1201
tangible, 1186
tanks, 251, 252, 255, 257, 733, 827, 828, 829, 831, 832, 834, 843, 874, 889, 1243, 1244, 1246, 1247, 1248, 1251, 1252, 1253, 1254, 1255, 1257, 1258, 1259, 1260, 1263, 1264, 1266, 1267, 1270, 1271, 1278, 1288, 1289, 1297, 1304, 1332, 1374, 1377, 1379, 1381, 1383,

1384, 1386, 1387, 1388, 1391, 1395, 1400, 1402
Tanning, 493
tannins, 469, 470, 473, 1269
Tanzania, v, 213, 214, 215, 216, 221, 222, 223, 233, 234, 235, 236, 237, 238, 708, 731, 791, 796
taphonomic constrains, xxxii, 993
taphonomy, 1037, 1038
TAR, 1566
target, xx, xxxvii, 325, 375, 379, 381, 582, 687, 743, 744, 751, 752, 860, 913, 1078, 1325, 1326, 1494, 1589, 1595, 1597, 1603, 1690, 1693, 1700
targets, 386, 395, 396, 398
tariff, 728, 879
taxes, 395, 397, 718
taxonomic, xxxii, 44, 45, 46, 56, 61, 69, 74, 75, 76, 77, 78, 110, 115, 161, 163, 164, 165, 166, 168, 169, 170, 175, 238, 943, 946, 993, 996, 999, 1001, 1002, 1011, 1021, 1143, 1207, 1221, 1229
taxonomic descriptions, xxxii, 993, 996, 1001
taxonomy, 73, 136, 164, 238, 387, 415, 531, 889, 1002, 1117, 1609, 1627, 1628, 1629, 1632, 1634, 1636
taxons, 1652
tea, 610
teams, 864
technical assistance, 877, 1332
technical efficiency, 713
technical support, 725, 1301
technicians, 183, 1116
technological advances, 730
technological change, 398
technologies, xxxix, 449, 453, 459, 467, 490, 499, 526, 529, 687, 710, 722, 724, 729, 733, 821, 835, 1115, 1332, 1372, 1540, 1731
technology, xvii, xxii, xxix, xlii, 23, 180, 190, 243, 250, 398, 435, 437, 452, 453, 457, 483, 506, 507, 509, 527, 529, 563, 591, 644, 676, 684, 689, 716, 724, 725, 728, 819, 856, 881, 1095, 1290, 1338, 1353, 1374, 1401, 1409, 1541, 1579, 1581, 1583, 1587, 1594, 1603, 1605, 1621
technology transfer, 724, 725
Teflon, 1078, 1080, 1171
temperament, 372
temperate zone, 93, 374, 636
temperature dependence, 28
temperature gradient, xiv, 27, 28, 939
temporal, xiii, 2, 9, 23, 182, 185, 186, 190, 217, 245, 314, 328, 331, 386, 418, 420, 453, 455, 570, 599, 620, 621, 630, 631, 632, 633, 634, 928,
945, 946, 947, 968, 983, 985, 997, 1003, 1014, 1026, 1028, 1037, 1038, 1096, 1182, 1186, 1215, 1368, 1411, 1422, 1436
temporal distribution, 928
tendon, 498
tendons, 484, 485, 486, 494, 495, 496, 497, 498, 510
Tennessee, 1367
tensile, 1461
tension, 732, 734, 1077, 1464
tensions, 732, 734
terraces, 1334, 1343, 1344, 1345, 1439
terrestrial ecosystems, xix, 310, 314, 329, 345, 346, 465
territorial, 433, 435, 440, 456, 530
territory, xiv, 13, 14, 360, 415, 416, 435, 440, 442, 446, 447, 457, 514, 517, 519, 523, 529, 596, 810, 840, 999, 1559, 1647, 1649
test data, 1695
test procedure, 1080, 1301
test statistic, 19
testing, xxvi, 190, 408, 541, 683, 684, 688, 806, 997, 1078, 1301, 1304, 1319, 1464, 1635, 1638, 1659
testosterone, 573
Texas, xxxvi, 924, 1201, 1230, 1231, 1299, 1300, 1301, 1302, 1304, 1305, 1306, 1309, 1310, 1311, 1314, 1316, 1319, 1320, 1321, 1322, 1323, 1434
textile, 502, 1381
textiles, 394
textural character, 1150, 1166
texture, 194, 260, 430, 438, 440, 452, 1149, 1417, 1418, 1425, 1431, 1440, 1442, 1443, 1600, 1605, 1689, 1692, 1693
TGA, 503, 511
Thailand, viii, xi, xxxi, xliv, 611, 868, 887, 889, 891, 892, 899, 904, 905, 906, 907, 909, 910, 1095, 1683, 1684, 1685, 1686, 1687, 1688, 1692, 1693, 1694, 1696, 1697, 1698, 1699, 1700, 1701, 1702, 1703, 1704
Thalassionema, 1001, 1068
Thalassiosira, 1010, 1011, 1018, 1019, 1068, 1069
Thalassiosira eccentric, 1010, 1018, 1019, 1068
thawing, 1410
theoretical approaches, 1471
theoretical assumptions, 742
theory, 402, 405, 474, 539, 570, 598, 599, 1321, 1471, 1485, 1487, 1499, 1500, 1502, 1505, 1506, 1507, 1535
therapeutic use, 610
therapy, 1602
thermal analysis, 503, 507, 511
thermal treatment, 395

thermodynamic properties, 729
thermogravimetry, 511
thermometer, 1129
thermoplastics, 506, 507
Third World, 217, 479
Thomson, 211, 341, 987, 1142, 1143, 1409, 1436
thorns, 610
thoughts, 188
threat, xvi, xxiii, 222, 235, 421, 429, 432, 452, 565, 566, 570, 571, 572, 573, 574, 575, 610, 949, 950, 1326, 1373, 1643, 1646, 1648, 1649, 1675, 1684, 1692, 1702, 1719, 1726
threatened, xvi, xviii, 221, 222, 223, 235, 284, 286, 302, 376, 562, 577, 579, 599, 1197, 1342
threatening, 196, 274, 276
threats, xxx, xxxvii, xliii, xliv, 417, 418, 419, 420, 421, 447, 455, 611, 724, 768, 794, 814, 871, 881, 882, 904, 975, 1326, 1570, 1641, 1643, 1644, 1647, 1648, 1649, 1652, 1684, 1685, 1700, 1711, 1714, 1715, 1729
three-dimensional, 1194, 1411, 1418, 1430
threonine, 1280, 1282, 1284
threshold, 22, 444, 620, 623, 629, 630, 635, 1475, 1600, 1691
thresholds, 22, 203, 206, 370, 636, 1545
throat, 609, 610
thrombosis, 610
thyroid, 1286
Tibet, 355
tides, 758, 935, 937, 951, 952, 998, 1022, 1100, 1124, 1127, 1139, 1140, 1457, 1474, 1481, 1550, 1551, 1552, 1560, 1566, 1568, 1570
tiger, 305
timber, xvi, 221, 223, 234, 551, 560, 596, 601, 1186, 1562, 1715
timber production, 551
time frame, 722
time periods, 193
time series, xvii, 244, 247, 691, 1566
timing, 2, 270, 314, 333, 335, 963, 970, 1341
TIP, xiii, 1, 2, 4, 5, 6, 7, 9, 10
tissue, xxi, 44, 47, 48, 76, 138, 164, 274, 314, 461, 462, 471, 475, 484, 485, 486, 497, 558, 563, 573, 786, 841, 842, 843, 846, 849, 850, 851, 853, 1075, 1076, 1077, 1078, 1080, 1088, 1091, 1096, 1290, 1295, 1338, 1601
tissue engineering, 486
TOC, 1204, 1212, 1213, 1220, 1223, 1225, 1226, 1355
Tokyo, 171, 254, 261, 304, 305, 306, 1534, 1544
tolerance, 420, 994, 1002, 1023, 1028, 1029, 1041, 1143
tolls, 563

toluene, 1117, 1118
tomato, 353, 357
tones, xxix, 839
top-down, 179, 418, 598
topographic, xiv, xl, 14, 226, 263, 440, 523, 598, 1511, 1515, 1543
topsoil, 419, 421, 445, 447, 1326
torus, 243, 624
total energy, 402, 717, 1077
total organic carbon, 200, 206, 1204, 1212
total organic carbon (TOC), 1204
total product, 778, 857, 949, 1238, 1241, 1247
tourism, 420, 451, 523, 597, 865, 1101, 1187, 1202, 1458, 1466, 1576, 1685
tourist, 432
toxic, xxxvi, 87, 325, 426, 431, 441, 444, 446, 447, 456, 490, 973, 1089, 1090, 1108, 1114, 1115, 1170, 1176, 1202, 1206, 1219, 1220, 1228, 1326, 1397
toxic effect, 441, 446, 1266, 1270
toxic gases, 490
toxicities, 1224
toxicity, xxxv, 432, 446, 487, 493, 505, 511, 729, 906, 907, 914, 922, 1072, 1073, 1080, 1087, 1091, 1102, 1119, 1202, 1203, 1204, 1206, 1207, 1208, 1209, 1211, 1213, 1216, 1218, 1219, 1220, 1223, 1224, 1226, 1228, 1229, 1248, 1250, 1609, 1611
toxicological, 457, 1208
trace elements, 426, 441, 443, 444, 458
tracers, 1366, 1369
tracking, 186, 248, 249, 250, 929, 1315, 1604, 1618, 1678, 1714
tracks, 524, 972, 1672
traction, 599
tradable permits, 398
trade, xxx, 316, 317, 331, 370, 371, 384, 387, 582, 597, 598, 610, 611, 612, 614, 615, 616, 617, 732, 803, 806, 871, 872, 873, 874, 880, 882, 886, 1186, 1187, 1191, 1193, 1197, 1732
trade agreement, 370
trade-off, 316, 317, 331, 371, 384, 387, 597, 803, 806, 1191, 1193, 1197
traditional healers, 616
Traditional Medicine, 611, 617
traditional medicines, xxv, 603, 604, 609, 610, 611, 613
traditions, 356, 605
traffic, 426, 427, 1314
training, 197, 201, 210, 527, 691, 723, 735, 880, 1025, 1131, 1167, 1691, 1695
traits, xxviii, 47, 60, 287, 302, 328, 334, 336, 342, 376, 553, 555, 556, 560, 562, 569, 620, 687,

761, 765, 767, 793, 799, 800, 802, 803, 804, 805, 806, 807, 808, 810, 814, 815, 816, 817, 1288, 1656, 1660, 1661, 1662, 1664, 1667
trajectory, 403, 674, 675
trans, 143, 405, 493, 629, 630
transaction costs, 865
transactions, 865, 866, 1543
transcript, 318
transfer, 179, 318, 324, 327, 332, 334, 337, 338, 340, 343, 352, 355, 356, 357, 358, 426, 437, 451, 456, 465, 486, 493, 526, 527, 529, 585, 995, 997, 1009, 1023, 1026, 1035, 1036, 1038, 1040, 1132, 1175, 1194, 1197, 1224, 1331, 1375, 1376, 1381, 1385, 1410, 1432, 1472, 1532
transfer of technologies, 526
transference, 612
transformation, 14, 411, 420, 450, 494, 502, 510, 511, 517, 519, 542, 578, 833, 868, 980, 1102, 1103, 1119, 1197, 1366, 1367, 1463, 1659
transformation processes, 1463
transformations, 211, 314, 410, 517, 1340, 1369, 1400, 1471
transglutaminase, 509
transgression, xxxiv, 1006, 1147, 1148, 1163, 1164, 1165
transition, xl, 360, 503, 505, 506, 507, 547, 935, 997, 1163, 1455, 1456, 1474, 1491, 1495, 1553, 1664, 1690, 1694, 1730
transition economies, 726, 727
transition temperature, 505
transitions, 327, 1550, 1687, 1705
translation, 853, 1188
translocation, xxvii, xxx, 446, 576, 769, 771, 772, 773, 774, 775, 779, 787, 788, 872, 880
transmission, 448, 853, 1077, 1462, 1466, 1472, 1478, 1638
transparency, 572, 698
transparent, xxxv, 1077, 1185, 1190
transpiration, 108, 1411, 1413, 1442
transport processes, 1544
transportation, xxxi, 394, 419, 717, 743, 864, 911, 912, 1076, 1187, 1189, 1356
traps, xvi, 38, 221, 226, 227, 228, 363, 364, 365, 366, 368, 503, 524, 531
trauma, 609
travel, 586, 631, 1471, 1472, 1482, 1487, 1489
trawling, 1178
Treasury, 399
treaties, 734
tree cover, 203
trees, 189, 192, 194, 201, 207, 208, 234, 273, 274, 276, 286, 320, 325, 332, 353, 357, 362, 464, 520, 524, 525, 551, 552, 553, 554, 558, 560,

1107, 1162, 1163, 1313, 1337, 1686, 1690, 1718
trend, xxv, xxxviii, 18, 19, 390, 402, 429, 446, 536, 582, 619, 1213, 1226, 1350, 1395, 1401, 1421, 1428, 1565, 1566, 1568
trial, xxix, 251, 409, 814, 840, 1251, 1282, 1373, 1377, 1384, 1401
Triassic, 459
tribal, 69, 77
tribes, 44, 64, 77, 90, 91, 136, 164, 406, 1659, 1661
trickle down, 1376
trifolii, 343
triggers, 850
triglycerides, 489, 495, 497
trimmings, 486, 495, 497
Trinidad, 307
tritium, 1409, 1433
tropical forest, 237, 305, 321, 335, 530, 531, 545
tropical forests, 237, 321, 335, 530, 545, 797, 1680, 1702, 1703
tropical rain forests, 635, 700, 1719
tropical savannas, 266, 272, 274, 278
trucks, 253, 728, 873
Trust Fund, 1731
trypsin, 490, 1242, 1279, 1284, 1295
tryptophan, 1239, 1244, 1253, 1263, 1265, 1266, 1268, 1272, 1280
tsunamis, 1481
T-test, 1310, 1311, 1312, 1313, 1353, 1354, 1359
tubers, 75
tubular, 1103
tundra, 218
Tunisia, 43, 155, 158, 161, 606
turbulence, 940, 941, 1211, 1314, 1355, 1532
turbulent, 1492, 1534, 1535
Turbulent, 1545
Turkey, 128, 163, 175, 1615
turnover, 28, 36, 317, 320, 476, 685, 906, 995
turtle, 1096
turtles, 1073, 1096, 1187
Tuscany, 417, 424, 425, 430, 434
two-dimensional, 168, 1192, 1479
typhoon, 1512, 1513, 1515, 1518
typology, 401, 403, 441, 685, 686
tyrosine, 486, 1244, 1253, 1263, 1265, 1266, 1268, 1272, 1280, 1281, 1285
Tyrosine, 1244, 1253, 1263, 1277

U

U.S. Department of Agriculture, 11, 595, 1332
U.S. Department of the Interior, 883
U.S. Geological Survey, 595, 754, 1417, 1451, 1544
U.S. military, 595, 596

ubiquitous, 535, 536, 539, 543, 1550
Ukraine, 392
ulcer, 853
ulna, 1019, 1068, 1069
ultrastructure, 1615, 1625, 1636
ultraviolet, 146, 166
UNCED, 178, 179, 218
uncertainty, 28, 401, 404, 409, 492, 579, 584, 598, 601, 1189, 1192
unclassified, 1322
underlying mechanisms, 342, 358, 806
UNDP, 617
UNEP, 214, 216, 259, 263, 411, 1090, 1094, 1097, 1205
UNESCO, 196, 198, 211, 214, 215, 218, 1726, 1727, 1732, 1734
UNFCCC, 391, 395
uniform, xlii, 645, 937, 958, 1112, 1152, 1158, 1251, 1541, 1545, 1552, 1623, 1632
United Kingdom (UK), 989, 1037, 1039, 1285, 1369, 1457, 1458, 1459
United Nations (UN), 179, 218, 389, 395, 399, 402, 592, 611, 638, 681, 724, 730, 886, 1142, 1234, 1288, 1702, 1724, 1733, 1734
United Nations Environment Program, 1234
universe, 403, 600
universities, 181, 1513
updating, 1555
uranium, 426
urban, xxxv, 304, 414, 419, 420, 421, 422, 423, 426, 427, 432, 433, 434, 435, 443, 449, 452, 455, 460, 515, 517, 575, 595, 597, 599, 604, 614, 701, 737, 759, 764, 768, 884, 942, 944, 971, 973, 1009, 1010, 1014, 1072, 1100, 1185, 1187, 1191, 1192, 1195, 1197, 1199, 1302, 1322, 1337, 1338, 1350, 1365, 1369, 1370, 1435, 1457, 1560, 1564, 1570, 1571, 1624, 1639, 1689
urban areas, 435, 515, 517, 614, 701, 973, 1365, 1457, 1570
urban centers, 1010
urban population, 599
urban settlement, 1009, 1564
urbanisation, 417, 418, 422, 423, 432, 433, 434, 716, 726, 1559, 1569
urbanization, 575, 874, 949, 981, 1101, 1202, 1352, 1365, 1458
urbanized, xxxv, 360, 423, 1185, 1197, 1564
urea, xxii, 483, 502, 503, 505, 507, 511, 715, 837, 890, 1107, 1285
urine, 259, 610, 832, 1372, 1377
Uruguay, xxxi, 266, 268, 271, 867, 923, 924, 999, 1031, 1034, 1036, 1201, 1233

USA,
USDA, 4, 460, 758, 769, 771, 797, 1300, 1301, 1322, 1325, 1342, 1343, 1346, 1440, 1451
USEPA, 1206, 1300, 1304, 1308, 1315, 1322, 1323, 1340, 1343
USSR, 1612
Utah, 260
uterus, 484
UV, 95, 108, 146, 1105, 1107, 1464, 1633, 1636, 1638
UV light, 95, 1105, 1107
UV radiation, 1633
UV-radiation, 1638

V

vaccinations, 736
vacuum, 500, 1205, 1257
Valdez, 1144
validation, 411, 1423
validity, 384, 402, 407
valine, 496, 1280, 1592
valleys, 224, 235, 556, 709, 1725, 1726
valorization, xxiii, 550
valuation, xxxv, 459, 678, 733, 990, 1185, 1186, 1189, 1194, 1198, 1228, 1321, 1702
van der Waals, 1393
van der Waals forces, 1393
vandalism, 1464, 1466
vapor, 1410
variable, xiii, 1, 2, 9, 10, 404, 423, 440, 441, 453, 568, 633, 1208, 1212, 1213, 1229, 1356, 1377, 1393, 1395, 1401, 1438, 1448, 1478
variables, 29, 32, 33, 34, 35, 183, 184, 187, 190, 200, 223, 261, 313, 364, 367, 435, 436, 640, 641, 644, 646, 647, 648, 650, 663, 673, 674, 675, 686, 706, 758, 769, 859, 861, 862, 863, 989, 994, 995, 1002, 1011, 1013, 1026, 1031, 1039, 1131, 1139, 1140, 1179, 1206, 1210, 1223, 1224, 1226, 1228, 1316, 1320, 1356, 1410, 1411, 1509, 1656, 1662, 1666, 1691, 1693, 1696
variance, xx, 5, 33, 35, 262, 290, 292, 375, 378, 380, 383, 385, 1081, 1211, 1224, 1226, 1303
varieties, xv, xxx, 43, 45, 82, 90, 135, 165, 169, 552, 871, 873, 877, 878, 893, 1289, 1610
vascular system, 787
vascularization, 841, 844, 845
vector, 287, 456, 644, 645, 646, 673, 675, 884, 885, 1326, 1453, 1454, 1483, 1484
vegetable oil, 467
vegetables, 446, 701, 709, 713, 1686
vegetative cover, 1441
vehicles, 420, 1101

velocity, 646, 687, 748, 891, 926, 937, 1136, 1153, 1154, 1336, 1353, 1355, 1358, 1393, 1454, 1472, 1475, 1478, 1483, 1484, 1489, 1520, 1521, 1522, 1523, 1526, 1528, 1531, 1532, 1534, 1535, 1545, 1550, 1553, 1577, 1656, 1658, 1659, 1660, 1665
Veneto region, 458
Venezuela, 266, 370, 879
ventilation, 729, 770, 783, 1381
Venus, 1094
vermiculite, 476
Vermont, 1192
versatility, 1589
vertebrates, xxiv, 222, 514, 543, 552, 553, 593, 612, 616, 762, 841, 1280, 1646, 1677, 1704
vesicle, 914
vessels, xxiii, 486, 581, 582, 586, 588, 591, 842, 849, 1164, 1512, 1515, 1559
vested interests, 596
veterinary medicine, 614
victims, 215
Victoria, 954
video, 915, 1468
video microscopy, 915
Vietnam, 893, 1715
Viking, 617
village, xvi, 221, 223, 224, 225, 226, 227, 228, 233, 234, 245, 251, 252, 253, 254, 255, 256, 258, 259, 260, 261, 1440, 1689, 1692, 1695
vineyard, xl, 424, 425, 1437, 1438, 1439, 1440, 1442, 1449, 1450
violence, 735
violent, 1512, 1516
viruses, 312, 1330, 1341
viscera, 1295
viscometric measurements, 507
viscosity, 464, 499, 1273, 1295, 1522, 1531, 1542, 1601
visible, 254, 261, 594, 1144, 1560
vision, 402, 405, 408, 409, 474, 591, 1189
visions, 678
visualization, 1189, 1208, 1210
vitamin A, 842, 852
vitamin C, 842, 849, 850, 851, 854, 1268
Vitamin C, 1246, 1252, 1272, 1281
vitamin E, xxix, 839, 842, 843, 844, 845, 846, 847, 848, 849, 850, 851, 852, 853, 854
vitamin supplementation, 849
vitamins, 506, 840, 842, 850, 851, 852, 1242, 1580
vitrification, 558
VO, 385, 399
voiding, 912
volatile organic compounds, 720
volatility, 641, 673
volatilization, 1375, 1396
volcanic activity, 464, 567
vulnerability, 204, 399, 423, 433, 436, 437, 438, 439, 440, 446, 457, 458, 459, 572, 759, 859, 1335, 1353
vulnerable people, 719

W

wages, 672, 859
Wales, 306, 397, 406, 969, 980, 986, 1734
walking, 289, 524
war, 716, 735, 736, 737, 1565, 1729
warrants, 1214, 1215
Washington, 10, 11, 212, 215, 219, 236, 399, 410, 460, 481, 601, 602, 615, 616, 676, 677, 790, 1121, 1198, 1234, 1294, 1320, 1322, 1323, 1340, 1341, 1343, 1436, 1451, 1571, 1609, 1613, 1678, 1679, 1703, 1719, 1720, 1734
waste disposal, 420, 421, 423, 436, 443, 445, 451, 698, 718
waste management, xiii, xx, 389, 393, 394, 395, 397, 398, 399, 577, 795, 821, 1322, 1332, 1341
waste treatment, 1340
waste water, 455, 575, 973, 1073, 1121, 1373
wastes, 394, 504, 1073, 1202, 1377, 1389
wastewater, 10, 771, 785, 796, 797, 968, 984, 1074, 1101, 1118, 1121, 1330, 1339, 1341, 1342, 1350, 1373, 1374, 1375, 1376, 1379, 1388, 1402, 1403, 1405
wastewater treatment, 968, 1074, 1101, 1118, 1350, 1405
Wastewater treatment, 1404
wastewaters, 490, 501, 1338, 1374, 1376
water chemistry, xxxviii, 687, 788, 1124, 1340, 1350, 1356, 1361
water ecosystems, 751
water permeability, 915
water policy, 686
water quality, xxvi, xxxiii, xxxv, xxxvi, xliii, 37, 42, 407, 449, 454, 477, 571, 577, 643, 683, 685, 687, 698, 732, 738, 753, 794, 797, 840, 909, 910, 933, 946, 950, 967, 968, 976, 988, 1002, 1004, 1006, 1008, 1030, 1123, 1125, 1126, 1131, 1141, 1185, 1279, 1288, 1299, 1300, 1304, 1307, 1309, 1313, 1315, 1316, 1317, 1318, 1319, 1320, 1321, 1322, 1326, 1330, 1333, 1335, 1337, 1339, 1340, 1341, 1342, 1343, 1345, 1350, 1351, 1655, 1664
water quality standards, 1319, 1322
water resources, xiii, xxvii, 408, 409, 436, 455, 574, 596, 697, 701, 702, 703, 704, 705, 706, 707,

712, 715, 723, 726, 729, 732, 733, 734, 735, 737, 738, 856, 903, 909, 933, 987, 1336
water rights, 744
water shortages, 733
water supplies, 20, 577, 698, 731, 734, 909
water table, 3, 347, 358, 439, 1330, 1333, 1419, 1459
water vapor, 1410
water vapour, 1414
water-soluble, 490, 493, 494, 497, 500, 502
waterways, 420, 972, 1327, 1335, 1336, 1345
wave number, 1489
wave propagation, 1534, 1535, 1536, 1537
wavelengths, 251, 254, 1482, 1576
weak interaction, 476
weakness, 610, 629, 781
wealth, 179, 194, 197, 582, 725, 735, 1724
weapons, 331
weathering, 318, 414, 416, 417, 454, 463, 464, 1074, 1335
web, xxvi, 306, 311, 571, 642, 683, 691, 694, 816, 946, 977, 985, 1117, 1145, 1189, 1193
weedy, 255, 258, 325, 630, 631, 634, 1162, 1163
Weibull, 302, 306
weight gain, 272, 825, 849, 1253, 1258, 1262, 1267, 1273, 1282
weight ratio, 781, 1250
welfare, xxv, 595, 619, 735, 840
wellbeing, 450, 451, 459
well-being, xxxv, 208
well-being, 523
wells, 252, 435, 523, 707, 713, 735, 736, 946, 1565
West Africa, 211, 615, 771
Western Australia, 322, 1507, 1508, 1668
Western Cape Province, 962
Western Civilization, 406
Western Europe, 1457
wheat, xxxix, 20, 22, 57, 60, 275, 341, 424, 467, 1407, 1409, 1412, 1413, 1419, 1422, 1423, 1425, 1426, 1427, 1428, 1431, 1432, 1434, 1435, 1436, 1580, 1585, 1586, 1587, 1591, 1592, 1594, 1598, 1599, 1600, 1601, 1602, 1603, 1604, 1605, 1607, 1608, 1609, 1610, 1611, 1613, 1614, 1616, 1617, 1618, 1619
wholesale, xxiv, 594
wild animals, xxiii, 524, 550, 605, 610, 614, 1732
wild type, 481
wilderness, 1703
wildland, 679
wildlife, xv, xxiv, xxv, xxxvii, xliv, xlv, 37, 372, 421, 593, 594, 595, 596, 597, 598, 599, 603, 612, 615, 617, 699, 767, 771, 913, 918, 1187, 1197, 1198, 1300, 1314, 1315, 1333, 1683, 1684, 1685, 1686, 1690, 1691, 1692, 1695, 1696, 1697, 1698, 1699, 1700, 1701, 1727, 1729, 1730, 1732
wildlife conservation, xxv, 603, 913, 918
wind, xxi, xl, 259, 383, 413, 417, 418, 421, 524, 532, 553, 555, 562, 930, 932, 936, 937, 939, 940, 941, 942, 1018, 1125, 1127, 1137, 1138, 1170, 1333, 1415, 1462, 1468, 1471, 1474, 1475, 1481, 1511, 1512, 1513, 1515, 1516, 1517, 1519, 1520, 1521, 1522, 1526, 1529, 1530, 1543, 1546
wind speeds, 738, 1125
wine, 450, 610, 1438, 1439, 1449
Wisconsin, 356, 600, 617, 788, 790, 791, 793
withdrawal, xxxix, 706, 710, 732, 1408, 1411, 1414, 1421, 1431
wood, 178, 211, 275, 337, 462, 465, 502, 511, 517, 521, 522, 523, 524, 531, 543, 551, 716, 743, 1102, 1159, 1164, 1165, 1301, 1371, 1726, 1727
woodland, 18, 20, 22, 215, 302, 305, 306, 333, 338, 356, 362, 737
woods, 434, 1164
wool, 464, 486, 492
workers, 524, 735, 873, 1242, 1273, 1585, 1595
workforce, 527, 596
working groups, 1190
World Bank, 219, 423, 713, 1733
World Health Organization (WHO), 604, 1102, 1121
World Resources Institute, 399
worldwide, xxx, xxxii, xxxvii, xliv, 60, 235, 357, 390, 391, 392, 393, 452, 468, 479, 530, 574, 582, 606, 611, 684, 685, 687, 726, 732, 786, 871, 872, 912, 993, 1072, 1170, 1238, 1325, 1335, 1465, 1562, 1606, 1629, 1671, 1673, 1674, 1675, 1676, 1715, 1718, 1724
worm, 1031
worms, 1256
wound healing, xxix, 839, 851, 852, 853, 854
WRC, 975
WRI, 391, 392, 393, 396, 398, 399
writing, 276, 604, 1586

X

xenobiotic, 1095, 1096

Y

Yale University, 410
yeast, xlii, 318, 1105, 1109, 1111, 1113, 1264, 1265, 1293, 1295, 1579, 1580, 1581, 1583, 1585, 1588, 1592, 1599, 1604, 1610, 1612, 1615, 1618, 1619, 1620, 1621

yield, xxvii, xxxvii, 349, 377, 378, 384, 421, 425, 430, 432, 451, 455, 460, 487, 488, 494, 496, 497, 498, 507, 510, 678, 697, 702, 705, 711, 816, 972, 973, 1247, 1325, 1331, 1336, 1409, 1413, 1418, 1425, 1426, 1431, 1435, 1438, 1444, 1449, 1555, 1581, 1614, 1666
yolk, 913, 914, 915, 916
young people, 1463
Yugoslavia, 127

Z

Zea mays, 172, 481, 1409
Zimbabwe, 210, 234, 237
zinc, 836, 1080, 1089, 1097, 1101, 1601
Zinc, 1215
Zn, 426, 428, 441, 444, 1074, 1076, 1084, 1205, 1208, 1214, 1215, 1216, 1224
Zone 3, 200
zoning, 1465, 1553
zoogeography, 1679
zooplankton, xxxi, xxxii, 573, 887, 889, 890, 891, 897, 899, 900, 901, 904, 905, 906, 907, 908, 909, 924, 946, 947, 951, 952, 953, 954, 955, 964, 973, 979, 980, 981, 983, 984, 985, 987, 990, 1093, 1175, 1278